FBO000 CAS: 7782-63-0
FERROUS SULFATE HEPTAHYDRATE
mf: $O_4S \cdot Fe \cdot 7H_2O$ mw: 278.05

PROP: Pale blue green monoclinic crystals or granules; odorless with a salt taste. D: 2.99-3.08. Sol in water; insol in alc.

SYNS: COPPERAS ◇ FEOSOL ◇ FER-IN-SOL ◇ FERO-GRADUMET ◇ FERROUS SULFATE (FCC) ◇ FESOFOR ◇ FESOTYME ◇ GREEN VITROL ◇ HAEMOFORT ◇ IRONATE ◇ IRON(II) SULFATE (1:1), HEPTAHYDRATE ◇ IRON VITROL ◇ IROSUL ◇ MOL-IRON ◇ PRESFERSUL ◇ SULFERROUS

USE IN FOOD:

Purpose: Clarifying agent, dietary supplement, nutrient supplement, processing aid, stabilizer.

Where Used: Baking mixes, cereals, infant foods, pasta products, wine.

Regulations: FDA - 21CFR 182.5315, 182, 8315, 184.1315. GRAS when used in accordance with good manufacturing practice. BATF - 27CFR 240.1051. Limitation of 3 ounces/1000 gallons of wine.

ACGIH TLV: TWA 1 mg(Fe)/m^3

SAFETY PROFILE: Poison by intravenous, intraperitoneal, and subcutaneous routes. Moderately toxic by ingestion. Mutagenic data. When heated to decomposition it emits toxic fumes of SO_x.

TOXICITY DATA and CODEN

mmo-esc 30 μmol/L CIWYAO 49,144,50
orl-rat LDLo: 1389 mg/kg EQSSDX 1,1,75

CAS... ...ican Chemical So... ...racts Se... ...AS n... S...
See Intro... ... *xi*

PROP: — Physical properties including solubility and flammability data. May contain a definition of the entry.
See Introduction: paragraph 7, *p. xii*

SYNS: — Synonyms for the entry. A complete synonym cross-index is located in Section III.
See Introduction: paragraph 8, *p. xii*

Standards and Recommendations — Here are listed the OSHA PEL, ACGIH TLV, DFG MAK, and NIOSH REL workplace air levels. U.S. DOT Classification and labels are also listed.
See Introduction: paragraph 13, *p. xiii*

TOXICITY DATA: — Data for skin and eye irritation, mutation, teratogenic, reproductive, carcinogenic, human, and acute lethal effects.
See Introduction: paragraphs 15, 16, 17, 18, *and* 19, *pp. xv-xxx*

CODEN: — A code which represents the cited reference for the toxicity data. Complete bibliographic citations in CODEN order are listed in Section III.
See Introduction: paragraph 20, *p. xxx*

FOOD ADDITIVES HANDBOOK

FOOD ADDITIVES HANDBOOK

Richard J. Lewis, Sr.

VNR VAN NOSTRAND REINHOLD
New York

DISCLAIMER

Extreme care has been taken in preparation of this work. However, neither the publisher nor the author shall be held responsible or liable for any damages resulting in connection with or arising from the use of any of the information in this book.

Library of Congress Catalog Card Number 89-16516
ISBN 0–442–20508–2

Printed in the United States of America

Van Nostrand Reinhold
115 Fifth Avenue
New York, New York 10003

Van Nostrand Reinhold International Company Limited
11 New Fetter Lane
London EC4P 4EE, England

Van Nostrand Reinhold
480 La Trobe Street
Melbourne, Victoria 3000, Australia

Nelson Canada
1120 Birchmount Road
Scarborough, Ontario M1K 5G4, Canada

16 15 14 13 12 11 10 9 8 7 6 5 4 3 2 1

Library of Congress Cataloging-in-Publication Data

Lewis, Richard J., Sr.
Food additives handbook.

Includes index.
1. Food additives—Toxicology—Handbooks, manuals, etc. I. Title.
RA1270.F6L49 1989 363.19′2 89-16516
ISBN 0-442-20508-2

To Julie and Rich
and especially
Grace

Contents

Preface

The material in this handbook was assembled to serve the information needs of the food chemicals industry. Food additives are a vital part of the current food industry and serve many purposes. New additives are constantly being developed. Older additives fall under regulatory review as new toxicological studies are released. Workplaces must be designed to eliminate risk from bulk handling of additive chemicals which may possess toxic and reactive properties. The format of this handbook was designed to collect, in one place, fundamental data relating to these concerns.

Section 201 of the Food, Drug, and Cosmetic Act defines a food additive as "Any substance the intended use of which results in or may be reasonably expected to result in its becoming a component of the food." Following this broad definition, this handbook includes direct additives, indirect additives, packaging material components, pesticides added directly during food processing, pesticides remaining as residues on food, and selected animal drugs with residue limitations in finished foods. All of these substances may appear in the final food product either by direct design or as a residue.

Entries are listed in alphabetical order by substance name. Each entry is assigned a unique number in the form of three letters and three numbers, for example AAA123. The Purpose, Food, CAS, and Synonym cross-indexes use these numbers to facilitate rapid and easy location of an entry. It is much easier to remember a six-digit alpha-numeric number code than a complex and often lengthy entry name.

Each entry contains five classes of data: identifiers, properties, food specific information, occupational restrictions, and toxicological data. The introduction contains detailed discussions of the types, sources, and limitations of the data. The combined information provides the basis for evaluating an entry.

This volume ties together the many federal regulations which impinge on the use of the listed substances. Included are those of the Food and Drug Administration, the U.S. Department of Agriculture, the Bureau of Alcohol, Tobacco, and Firearms, the Occupational Safety and Health Administration, and the Department of Transportation. These regulations govern the shipping, workplace handling, and food addition of these substances.

Recommendations for workplace concentrations of bulk additives are included to aid in designing a safe workplace. Following the OSHA standards as revised in March 1, 1988 are the American Conference of Governmental Industrial Hygienists' (ACGIH) Threshold Limit Values, the National Institute for Occupational Safety and Health (NIOSH) Recommended Exposure Levels, the German Research Society's (MAK) values, and the U.S. Department of Transportation (DOT) classifications.

I hope the design of this handbook will promote rapid access to the assembled data and serve the information needs of the food industry and those interested in the use of food additives.

Richard J. Lewis, Sr.

Introduction

For each entry the following data are provided when available: the entry number, entry name, CAS number, DOT number, molecular formula, molecular weight, a description of the material and physical properties, and synonyms. The next section contains items relating to the use of the substance in food and food-related materials and includes the purpose served in foods, the food or food materials in which it is used, and the applicable regulations of the FDA, USDA, and the BATF. Following, where available, are IARC reviews, NTP Carcinogenesis Testing Program results, EPA Extremely Hazardous Substances List, the EPA Genetic Toxicology Program, the Community Right To Know List. The next grouping consists of the U.S. Occupational Safety and Health Administration's (OSHA) Permissible Exposure Levels, the American Conference of Governmental Industrial Hygienists' (ACGIH) Threshold Limit Values (TLV's), German Research Society's (MAK) values, National Institute for Occupational Safety and Health (NIOSH) Recommended Exposure Levels, and U.S. Department of Transportation (DOT) classifications. These are followed by a Safety Profile which discusses the toxic and other hazards of the entry. Each entry concludes with selected toxicity data with references for reports of primary skin and eye irritation, mutation, reproductive, carcinogenic, and acute toxic dose data. Each of the above information items is discussed in detail below.

1. *Entry Number* identifies each entry by a unique number consisting of three letters and three numbers, for example, AAA123. The first letter of the entry number indicates the alphabetical position of the entry. Numbers beginning with "A" are assigned to entries indexed with the A's. Entries in the four Cross-indexes are referenced to its appropriate entry by the entry number.

2. *Entry Name* The name of each material is selected, where possible, to be the most commonly used designation.

3. *Chemical Abstracts Service Registry Number* (*CAS*) is a numeric designation assigned by the American Chemical Society's Chemical Abstracts Service and uniquely identifies a specific chemical compound. This entry allows one to conclusively identify a material regardless of the name or naming system used.

4. *DOT:* indicates a four digit hazard code assigned by the U.S. Department of Transportation. This code is recognized internationally and is in agreement with the United Nations coding system. The code is used on transport documents, labels, and placards. It is also used to determine the regulations for shipping the material.

5. *Molecular Formula* (*mf:*) or atomic formula (*af:*) designates the elemental composition of the material and is structured according to the Hill System (see *Journal of the American Chemical Society*, 22(8): 478-494, 1900) in which carbon and hydrogen (if present) are listed first, followed by the other elemental symbols in alphabetical order. The formula for compounds that do not contain carbon are ordered strictly alphabetically by element symbol. Compounds such as salts or those containing waters of hydration have molecular formulas incorporating the CAS dot-

disconnect convention, in which the components are listed individually and separated by a period. The individual components of the formula are generally given in order of decreasing carbon atom count, and the component ratios given. A lower case ''x'' indicates that the ratio is unknown. A lower case ''n'' indicates a repeating, polymer-like structure. The formula is obtained from one of the cited references or a chemical reference text, or derived from the name of the material.

6. *Molecular Weight* (*mw:*) or atomic weight (*aw:*) is calculated from the molecular formula using standard elemental molecular weights (carbon = 12.01).

7. *Properties* (*PROP:*) are selected to be useful in evaluating the hazard of a material and designing its proper storage and use procedures. A definition of the material is included where necessary. The physical description of the material may include the form, color and odor to aid in positive identification. When available, the boiling point, melting point, density, vapor pressure, vapor density, and refractive index are given. The flash point, autoignition temperature, and lower and upper explosive limits are included to aid in fire protection and control. An indication is given of the solubility or miscibility of the material in water and common solvents.

8. *Synonyms* for the entry name are listed alphabetically. Synonyms include other chemical names, common or generic names, foreign names (with the language in parentheses), or codes. Some synonyms are followed by a notation in parentheses indicating it is the preferred name of one of the following organizations: (DOT) U. S Department of Transportation, (ACGIH) American Conference of Governmental Industrial Hygienists, (MAK) German Research Society, (FCC) National Research Council's *Food Chemicals Codex*, (USDA) U. S. Department of Agriculture. Some synonyms consist in whole or in part of registered trademarks. These trademarks are not identified as such. The reader is cautioned that some synonyms, particularly common names, may be ambiguous and refer to more than one material.

9. *Purpose:* The purposes for which the entry is used in foods or food products are listed here. Proposed uses reported in the research literature are not listed.

10. *Where Used:* This section contains the food and food products in which the entry substance is used. Also included are foods for which the entry may be an indirect additive or contaminant. The term ''various'' is used when the entry has general utility, but no specific foods have been identified. Many flavoring agents are in this category since they are blended to form final flavors.

11. *Regulations:* U.S. Federal regulations applying to the use of the entry in food or food products are listed here. Those of the Food and Drug Administration (FDA) are encoded in Volume 21 of the Code of Federal Regulations (21CFR). The regulations of U. S. Department of Agriculture (USDA) are encoded in Volume 9 (9CFR). The Bureau of Alcohol, Tobacco, and Firearms (BATF) regulates alcoholic beverage additives in Volume 27 of the Code of Federal Regulations (27CFR). References are given to the appropriate regulations and specific limitations or prohibitions are listed. The citations are given to facilitate review of the complete text of the regulations and the latest changes as reported in the *Federal Register*.

Also included in this section are the USDA Compound Evaluation System (CES). These are given in the format ''USDA CES Ranking: A-4 (1985).'' The USDA CES ranks pesticides, drugs, and biologics in a two part ranking. The first is a letter ranging from ''A'' for ''high health hazard'' to ''D'' indicating ''negligible health hazard.'' The second part of the rating is an exposure index ranging from 1 indicating ''high probability of exposure'' to 4 indicating ''negligible probability of exposure.'' A ''Z'' in either rating indicates insufficient information was available to the USDA panel to assign a rating.

12. *Reviews and Status* lines supply additional information to enable the reader to make knowledgeable evaluations of potential chemical hazards. Two types of reviews are listed: (a) International

Agency for Research on Cancer (IARC) monograph reviews, which are published by the United Nations World Health Organization (WHO), and (b) the National Toxicology Program (NTP) Carcinogenesis Testing Program reports.

a. Cancer Reviews. In the U.N. International Agency for Research on Cancer (IARC) monographs, information on suspected environmental carcinogens is examined, and summaries of available data with appropriate references are presented. Included in these reviews are synonyms, physical and chemical properties, uses and occurrence, and biological data relevant to the evaluation of carcinogenic risk to humans. The over 40 monographs in the series contain an evaluation of approximately 900 materials. Single copies of the individual monographs (specify volume number) can be ordered from WHO Publications Centre USA, 49 Sheridan Avenue, Albany, New York 12210, telephone (518) 436-9686.

An IARC entry indicates that some carcinogenicity data pertaining to a compound has been reviewed by the IARC committee. It indicates whether the data pertain to humans or to animals and whether the conclusion indicated the existence of "sufficient," "limited" "inadequate," or "no evidence" of carcinogenicity.

This cancer review reflects only the conclusion of the IARC Committee based on the data available at the time of the Committee's evaluation. Hence, for some materials there may be disagreement between the IARC determination and the tumorigenicity information in the toxicity data lines.

b. NTP Status. This entry indicates that the material has been tested by the National Toxicology Program (NTP) under its Carcinogenesis Testing Program. These entries are also identified as National Cancer Institute (NCI), which reported the studies before the NCI Carcinogenesis Testing Program was absorbed by NTP. To obtain additional information about NTP, the Carcinogenesis Testing Program, or the status of a particular material under test, contact the Toxicology Information and Scientific Evaluation Group, NTP/TRTP/NIEHS, Mail Drop 18-01, P.O. Box 12233, Research Triangle Park, NC 27709.

c. EPA Extremely Hazardous Substances List. This list was developed by the U.S. Environmental Protection Agency (EPA) as required by the Superfund Amendments and Reauthorization Act of 1986 (SARA). Title III Section 304 requires notification by facilities of a release of certain extremely hazardous substances. These 402 substances were listed by EPA in the *Federal Register* November 17, 1986.

d. Community-Right-To-Know List. This list was developed by the EPA as required by the Superfund Amendments and Reauthorization Act of 1986 (SARA). Title III, Sections 311-312 require manufacturing facilities to prepare Material Safety Data Sheets and notify local authorities of the presence of listed chemicals. Both specific chemicals and classes of chemicals are covered by these Sections.

e. EPA Genetic Toxicology Program. (GENE-TOX) This status line indicates that the material has had genetic effects reported in the literature during the period 1969-1979. The test protocol in the literature is evaluated by an EPA expert panel on mutations and the positive or negative genetic effect of the substance is reported. To obtain additional information about this program, contact GENE-TOX program, USEPA, 401 M Street, SW, TS796, Washington, DC 20460, Telephone (202) 382-3513.

13. *Standards and Recommendations* contains regulations by an agency of the United States Government, and recommendations by expert groups. "OSHA" refers to standards promulgated under Section 6 of the Occupational Safety and Health Act of 1970. "DOT" refers to materials regulated for shipment by the Department of Transportation. Because of frequent changes to and litigation of Federal regulations, it is recommended that the reader contact the applicable agency for information about the current standards for a particular material. Omission of a material or regulatory notation from this edition does not imply any relief from regulatory responsibility.

a. OSHA Air Contaminant Standards. The values given are for the revised standards which were published in January 13, 1989 and take effect on September 1, 1989 through December 31, 1992. These are noted with the entry ''OSHA PEL:'' followed by ''TWA'' or ''CL'' meaning either time-weighted average or ceiling value, respectively, to which workers can be exposed for a normal 8-hour day, 40-hour work week without ill effects. For some materials, TWA, CL, and Pk (peak) values are given in the standard. In those cases, all three are listed. Finally, some entries may be followed by the designation ''(skin).'' This designation indicates that the compound may be absorbed by the skin and, even though the air concentration may be below the standard, significant additional exposure through the skin may be possible.

b. ACGIH Threshold Limit Values. The American Conference of Governmental Industrial Hygienists (ACGIH) Threshold Limit Values are noted with the entry ''ACGIH TLV:'' followed by ''TWA'' or ''CL'' meaning either time-weighted average or ceiling value, respectively, to which workers can be exposed for a normal 8-hour day, 40-hour work week without ill effects. The notation ''CL'' indicates a ceiling limit which must not be exceeded. The notation ''skin'' indicates that the material penetrates intact skin, and skin contact should be avoided even though the TLV concentration is not exceeded. STEL indicates a short-term exposure limit, usually a 15-minute time-weighted average, which should not be exceeded. Biological Exposure Indices (*BEI:*) are, according to the ACGIH, set to provide a warning level ''...of biological response to the chemical, or warning levels of that chemical or its metabolic product(s) in tissues, fluids, or exhaled air of exposed workers...''

The latest annual TLV list is contained in the publication *Threshold Limit Values and Biological Exposure Indices.* This publication should be consulted for future trends in recommendations. The ACGIH TLV's are adopted in whole or in part by many countries and local administrative agencies throughout the world. As a result, these recommendations have a major impact on the control of workplace contaminant concentrations. The ACGIH may be contacted for additional information at 6500 Glenway Ave., Cincinnati, Ohio 45211, USA.

c. DFG MAK. These lines contain the German Research Society's Maximum Allowable Concentration values. Those materials which are classified as to workplace hazard potential by the German Research Society are noted on this line. The MAK values are also revised annually and discussions of materials under consideration for MAK assignment are included in the annual publication together with the current values. *BAT:* indicates Biological Tolerance Value for a Working Material which is defined as, ''...the maximum permissible quantity of a chemical compound, its metabolites, or any deviation from the norm of biological parameters induced by these substances in exposed humans.'' *TRK:* values are Technical Guiding Concentrations for workplace control of carcinogens. For additional information, write to Deutsche Forschungsgemeinschaft (German Research Society), Kennedyallee 40, D-5300 Bonn 2, Federal Republic of Germany. The publication *Maximum Concentrations at the Workplace and Biological Tolerance Values for Working Materials* can be obtained from Verlag Chemie GmbH, Buchauslieferung, P.O. Box 1260/1280, D-6940 Weinheim, Federal Republic of Germany, or Verlag Chemie, Deerfield Beach, Florida.

d. NIOSH REL. This line indicates that a NIOSH criteria document recommending a certain occupational exposure has been published for this compound or for a class of compounds to which this material belongs. These documents contain extensive data, analysis, and references. The more recent publications can be obtained from the National Institute for Occupational Safety and Health, U.S. Department of Health and Human Services, 4676 Columbia Pkwy., Cincinnati, Ohio 45226.

e. DOT Classification. This is the hazard classification according to the U.S. Department of Transportation (DOT) or the International Maritime Organization (IMO.) This classification gives an indication of the hazards expected in transportation, and serves as a guide to the development of proper labels, placards, and shipping instructions. The basic hazard classes include compressed gases, flammables, oxidizers, corrosives, explosives, radioactive materials, and poisons. Although a material may be designated by only one hazard class, additional hazards may be indicated by adding labels or by using other means as directed by DOT. Many materials are regulated under general headings such as "pesticides" or "combustible liquids" as defined in the regulations. These are not noted here as their specific concentration or properties must be known for proper classification. Special regulations may govern shipment by air. This information should serve *only as a guide* since the regulation of transported materials is carefully controlled in most countries by federal and local agencies. Because of frequent changes to regulations, it is recommended that the reader contact the applicable agency for information about the current standards for a particular material. United States transportation regulations are found in 40 CFR, Parts 100 to 189. Contact the U.S. Department of Transportation, Materials Transportation Bureau, Washington, D.C. 20590.

14. *Safety Profiles* are text summaries of the toxicity and a discussion of the symptomatology caused by exposure. The discussion of target organs and specific effects reported for human exposure was expanded to include additional details. Materials which are incompatible with the entry are listed here. Fire and explosion hazards are briefly summarized in terms of conditions of flammable or reactive hazard. Where feasible, fire-fighting materials and methods are discussed. A material with a flash point of 100°F or less is considered dangerous; if the flash point is from 100 to 200°F, the flammability is considered moderate; if it is above 200°F, the flammability is considered low (combustible).

Also included in the Safety Profiles are disaster hazard comments which serve to alert users of materials, safety professionals, researchers, supervisors, and fire fighters to the dangers that may be encountered on entering storage premises during a fire or other emergency. Although the presence of water, steam, acid fumes, or powerful vibrations can cause the decomposition of many materials into dangerous compounds, of special concern are high temperatures (such as those resulting from a fire) since these can cause many otherwise mild chemicals to emit highly toxic gases or vapors such as NO_x, SO_x, acids, and so forth, or evolve vapors of antimony, arsenic, mercury, and the like.

The final items in an entry are toxicity data lines and associated CODEN bibliographic references. The types and details of the contents of the data lines are discussed below.

15. *Skin and Eye Irritation Data* lines include, in sequence, the tissue tested (skin or eye); the species of animal tested; the total dose and, where applicable, the duration of exposure; for skin tests only, whether open or occlusive; an interpretation of the irritation response severity when noted by the author; and the reference from which the information was extracted. Only positive irritation test results are included.

Materials that are applied topically to the skin or to the mucous membranes can elicit either (a) systemic effects of an acute or chronic nature or (b) local effects, more properly termed "primary irritation." A primary irritant is a material that, if present in sufficient quantity for a sufficient period of time, will produce a nonallergic, inflammatory reaction of the skin or of the mucous membrane at the site of contact. Primary irritants are further limited to those materials that are not corrosive. Hence, concentrated sulfuric acid is not classified as a primary irritant.

a. Primary Skin Irritation. In experimental animals, a primary skin irritant is defined as a chemical that produces an irritant response on first exposure in a majority of the test subjects. However, in some instances compounds act more subtly and require either repeated contact or special environmental conditions (humidity, temperature, occlusion, etc.) to produce a response.

The most standard animal irritation test is the Draize procedure (*Journal of Pharmacology and Experimental Therapeutics*, 82: 377-419, 1944). This procedure has been modified and adopted as a regulatory test by the Consumer Product Safety Commission (CPSC) in 16 CFR 1500.41 (formerly 21 CFR 191.11). In this test a known amount (0.5 mL of a liquid or 0.5

grams of a solid or semisolid) of the test material is introduced under a one square inch gauze patch. The patch is applied to the skin (clipped free of hair) of twelve albino rabbits. Six rabbits are tested with intact skin and six with abraded skin. The abrasions are minor incisions made through the stratum corneum, but are not sufficiently deep to disturb the dermis or to produce bleeding. The patch is secured in place with adhesive tape, and the entire trunk of the animal is wrapped with an impervious material, such as rubberized cloth, for a 24-hour period. The animal is immobilized during exposure. After 24 hours the patches are removed and the resulting reaction evaluated for erythema, eschar, and edema formation. The reaction is again scored at the end of 72 hours (48 hours after the initial reading), and the two readings are averaged. A material producing any degree of positive reaction is cited as an irritant.

As the modified Draize procedure described above has become the standard test specified by the U.S. government, nearly all of the primary skin irritation data either strictly adheres to the test protocol or involves only simple modifications to it. When test procedures other than those described above are reported in the literature, appropriate codes are included in the data line to indicate those deviations.

The most common modification is the lack of occlusion of the test patch, so that the treated area is left open to the atmosphere. In such cases the notation ''open'' appears in the irritation data line. Another frequent modification involves immersion of the whole arm or whole body in the test material or, more commonly, in a dilute aqueous solution of the test material. This type of test is often conducted on soap and detergent solutions. Immersion data are identified by the abbreviation ''imm'' in the data line.

The dose reported is based first on the lowest dose producing an irritant effect and second on the latest study published. The dose is expressed as follows:

(1) Single application by the modified Draize procedure is indicated by only a dose amount. If no exposure time is given, then the data are for the standard 72-hour test. For test times other than 72 hours, the dose data are given in mg (or an appropriate unit)/duration of exposure, e.g., 10 mg/24H.

(2) Multiple applications involve administration of the dose in divided portions applied periodically. The total dose of test material is expressed in mg (or an appropriate unit)/duration of exposure, with the symbol ''I'' indicating intermittent exposure, e.g., 5 mg/6D-I.

The method of testing materials for primary skin irritation given in the Code of Federal Regulations does not include an interpretation of the response. However, some authors do include a subjective rating of the irritation observed. If such a severity rating is given, it is included in the data line as mild (''MLD''), moderate (''MOD''), or severe (''SEV''). The Draize procedure employs a rating scheme which is included here for informational purposes only, since other researchers may not categorize irritation response in this manner.

Category	Code	Skin Reaction (Draize)
Slight (Mild)	MLD	Well defined erythema and slight edema (edges of area well defined by definite raising)
Moderate	MOD	Moderate to severe erythema and moderate edema (area raised approximately 1 mm)
Severe	SEV	Severe erythema (beet redness) to slight eschar formation (injuries in depth) and severe edema (raised more than 1 mm and extending beyond area of exposure)

b. Primary Eye Irritation. In experimental animals, a primary eye irritant is defined as a chemical that produces an irritant response in the test subject on first exposure. Eye irritation study procedures developed by Draize have been modified and adopted as a regulatory test by CPSC in 16 CFR 1500.42. In this procedure, a known amount of the test material (0.1 mL of a liquid or 100 mg of a solid or paste) is placed in one eye of each of six albino rabbits; the other eye remains untreated, serving as a control. The eyes are not washed after instillation and are examined at 24, 48, and 72 hours for ocular reaction. After the recording of ocular reaction at 24 hours, the eyes may be further examined following the application of fluorescein. The eyes may also be washed with a sodium chloride solution (U.S.P. or equivalent) after the 24-hour reaction has been recorded.

A test is scored positive if any of the following effects are observed: (1) ulceration (besides fine stippling); (2) opacity of the cornea (other than slight dulling of normal luster); (3) inflammation of the iris (other than a slight deepening of the rugae or circumcorneal injection of the blood vessel); (4) swelling of the conjunctiva (excluding the cornea and iris) with eversion of the eyelid; or (5) a diffuse crimson-red color with individual vessels not clearly identifiable. A material is an eye irritant if four of six rabbits score positive. It is considered a nonirritant if none or only one of six animals exhibits irritation. If intermediate results are obtained, the test is performed again. Materials producing any degree of irritation in the eye are identified as irritants. When an author has designated a substance as either a mild, moderate, or severe eye irritant, this designation is also reported.

The dose reported is based first on the lowest dose producing an irritant effect and second on the latest study published. Single and multiple applications are indicated as described above under "Primary Skin Irritation." Test times other than 72 hours are noted in the dose. All eye irritant test exposures are assumed to be continuous, unless the reference states that the eyes were washed after instillation. In this case, the notation "rns" (rinsed) is included in the data line.

c. Species Exposed. Since Draize procedures for determining both skin and eye irritation specify rabbits as the test species, most of the animal irritation data are for rabbits, although any of the species listed in Table 2 may be used. Much of these data come from studies conducted on volunteers (for example for cosmetic or soap ingredients) or from persons accidentally exposed. When accidental exposure, such as a spill, is cited, the line includes the abbreviation "nse" (nonstandard exposure). In these cases it is often very difficult to determine the precise amount of the material to which the individual was exposed. Therefore, for accidental exposures an estimate of the concentration or strength of the material, rather than a total dose amount, is generally provided.

16. *Mutation Data* lines include, in sequence, the mutation test system utilized, the species of the tested organism (and where applicable, the route of administration or cell type), the exposure concentration or dose, and the reference from which the information was extracted.

A mutation is defined as any heritable change in genetic material. Unlike irritation, reproductive, tumorigenic, and toxic dose data which report the results of whole animal studies, mutation data also include studies on lower organisms such as bacteria, molds, yeasts, and insects, as well as in vitro mammalian cell cultures. Studies of plant mutagenesis are not included. No attempt is made to evaluate the significance of the data or to rate the relative potency of the compound as a mutagenic risk to man.

Each element of the mutation line is discussed below.

a. Mutation Test System. A number of test systems are used to detect genetic alterations caused by chemicals. Additional ones may be added as they are reported in the literature. Each test system is identified by the 3-letter code shown in parentheses. For additional information about mutation tests, the reader may wish to consult the *Handbook of Mutagenicity Test Procedures*, edited by B. J. Kilbey, M. Legator, W. Nichols, and C. Ramel (Amsterdam: Elsevier Scientific Publishing Company/North-Holland Biomedical Press, 1977).

(1) Mutation in Microorganisms (mmo) System utilizes the detection of heritable genetic alterations in microorganisms which have been exposed directly to the chemical.

(2) Microsomal Mutagenicity Assay (mma) System utilizes an in vitro technique which allows enzymatic activation of promutagens in the presence of an indicator organism in which induced mutation frequencies are determined.

(3) Micronucleus Test (mnt) System utilizes the fact that chromosomes or chromosome fragments may not be incorporated into one or the other of the daughter nuclei during cell division.

(4) Specific Locus Test (slt) System utilizes a method for detecting and measuring rates of mutation at any or all of several recessive loci.

(5) DNA Damage (dnd) System detects the damage to DNA strands including strand breaks, crosslinks, and other abnormalities.

(6) DNA Repair (dnr) System utilizes methods of monitoring DNA repair as a function of induced genetic damage.

(7) Unscheduled DNA Synthesis (dns) System detects the synthesis of DNA during usually nonsynthetic phases.

(8) DNA Inhibition (dni) System detects damage that inhibits the synthesis of DNA.

(9) Gene Conversion and Mitotic Recombination (mrc) System utilizes unequal recovery of genetic markers in the region of the exchange during genetic recombination.

(10) Cytogenetic Analysis (cyt) System utilizes cultured cells or cell lines to assay for chromosomal aberrations following the administration of the chemical.

(11) Sister Chromatid Exchange (sce) System detects the interchange of DNA in cytological preparations of metaphase chromosomes between replication products at apparently homologous loci.

(12) Sex Chromosome Loss and Nondisjunction (sln) System measures the nonseparation of homologous chromosomes at meiosis and mitosis.

(13) Dominant Lethal Test (dlt) A dominant lethal is a genetic change in a gamete that kills the zygote produced by that gamete. In mammals, the dominant lethal test measures the reduction of litter size by examining the uterus and noting the number of surviving and dead implants.

(14) Mutation in Mammalian Somatic Cells (msc) System utilizes the induction and isolation of mutants in cultured mammalian cells by identification of the gene change.

(15) Host-Mediated Assay (hma) System uses two separate species, generally mammalian and bacterial, to detect heritable genetic alteration caused by metabolic conversion of chemical substances administered to host mammalian species in the bacterial indicator species.

(16) Sperm Morphology (spm) System measures the departure from normal in the appearance of sperm.

(17) Heritable Translocation Test (trn) Test measures the transmissibility of induced translocations to subsequent generations. In mammals, the test uses sterility and reduced fertility in the progeny of the treated parent. In addition, cytological analysis of the F1 progeny or subsequent progeny of the treated parent is carried out to prove the existence of the induced translocation. In Drosophila, heritable translocations are detected genetically using easily distinguishable phenotypic markers, and these translocations can be verified with cytogenetic techniques.

(18) Oncogenic Transformation (otr) System utilizes morphological criteria to detect cytological differences between normal and transformed tumorigenic cells.

(19) Phage Inhibition Capacity (pic) System utilizes a lysogenic virus to detect a change in the genetic characteristics by the transformation of the virus from noninfectious to infectious.

(20) Body Fluid Assay (bfa) System uses two separate species, usually mammalian and bacterial. Test substance is first administered to host, from whom body fluid (e.g., urine, blood) is subsequently taken. This body fluid is then tested in vitro, and mutations are measured in the bacterial species.

b. Species. Those test species that are peculiar to mutation data are designated by the 3-letter codes shown below.

	Code	Species
Bacteria	bcs	*Bacillus subtilis*
	esc	*Escherichia coli*
	hmi	*Haemophilus influenzae*
	klp	*Klebsiella pneumoniae*
	sat	*Salmonella typhimurium*
	srm	*Serratia marcescens*
Molds	asn	*Aspergillus nidulans*
	nsc	*Neurospora crassa*
Yeasts	smc	*Saccharomyces cerevisiae*
	ssp	*Schizosaccharomyces pombe*
Protozoa	clr	*Chlamydomonas reinhardi*
	eug	*Euglena gracilis*
	omi	other microorganisms
Insects	dmg	Drosophila melanogaster
	dpo	Drosophila pseudo-obscura
	grh	grasshopper
	slw	silkworm
	oin	other insects
Fish	sal	salmon
	ofs	other fish

If the test organism is a cell type from a mammalian species, the parent mammalian species is reported, followed by a colon and the cell type designation. For example, human leukocytes are coded "hmn:leu." The various cell types currently cited in this edition are as follows:

Designation	Cell Type
ast	Ascites tumor
bmr	bone marrow
emb	embryo
fbr	fibroblast
hla	HeLa cell
kdy	kidney
leu	leukocyte
lng	lung
lvr	liver
lym	lymphocyte
mmr	mammary gland
ovr	ovary
spr	sperm
tes	testis
oth	other cell types not listed above

In the case of host-mediated and body fluid assays, both the host organism and the indicator organism are given as follows: host organism/indicator organism, e.g., "ham/sat" for a test in which hamsters were exposed to the test chemical and *S. typhimurium* was used as the indicator organism.

For in vivo mutagenic studies, the route of administration is specified following the species designation, e.g., "mus-orl" for oral administration to mice. See Table 1 for a complete list of routes cited. The route of administration is not specified for in vitro data.

c. Units of Exposure. The lowest dose producing a positive effect is cited. The author's calculations are used to determine the lowest dose at which a positive effect was observed. If the author fails to state the lowest effective dose, two times the control dose will be used. Ideally, the dose should be reported in universally accepted toxicological units such as milligrams of test chemical per kilogram of test animal body weight. While this is possible in cases where the actual intake of the chemical by an organism of known weight is reported, it is not possible in many systems using insect and bacterial species. In cases where a dose is reported or where the amount can be converted to a dose unit, it is normally listed as milligrams per kilogram (mg/kg). However, micrograms (μg), nanograms (ng), or picograms (pg) per kilogram may also be used for convenience of presentation. Concentrations of gaseous materials in air are listed as parts per hundred (pph), per million (ppm), per billion (ppb), or per trillion (ppt).

Test systems using microbial organisms traditionally report exposure data as an amount of

Table 1. Routes of Administration to, or Exposure of, Animal Species to Toxic Substances

Abbreviation	Route	Definition
Eyes	eye	Administration directly onto the surface of the eye. Used exclusively for primary irritation data. See Ocular
Intraaural	ial	Administration into the ear
Intraarterial	iat	Administration into the artery
Intracerebral	ice	Administration into the cerebrum
Intracervical	icv	Administration into the cervix
Intradermal	idr	Administration within the dermis by hypodermic needle
Intraduodenal	idu	Administration into the duodenum
Inhalation	ihl	Inhalation in chamber, by cannulation, or through mask
Implant	imp	Placed surgically within the body location described in reference
Intramuscular	ims	Administration into the muscle by hypodermic needle
Intraplacental	ipc	Administration into the placenta
Intrapleural	ipl	Administration into the pleural cavity by hypodermic needle
Intraperitoneal	ipr	Administration into the peritoneal cavity
Intrarenal	irn	Administration into the kidney
Intraspinal	isp	Administration into the spinal canal
Intratracheal	itr	Administration into the trachea
Intratesticular	itt	Administration into the testes
Intrauterine	iut	Administration into the uterus
Intravaginal	ivg	Administration into the vagina
Intravenous	ivn	Administration directly into the vein by hypodermic needle
Multiple	mul	Administration into a single animal by more than one route
Ocular	ocu	Administration directly onto the surface of the eye or into the conjunctival sac. Used exclusively for systemic toxicity data
Oral	orl	Per os, intragastric, feeding, or introduction with drinking water
Parenteral	par	Administration into the body through the skin. Reference cited is not specific concerning the route used. Could be ipr, scu, ivn, ipl, ims, irn, or ice
Rectal	rec	Administration into the rectum or colon in the form of enema or suppository
Subcutaneous	scu	Administration under the skin
Skin	skn	Application directly onto the skin, either intact or abraded. Used for both systemic toxicity and primary irritant effects
Unreported	unr	Dose, but not route, is specified in the reference

chemical per liter (L) or amount per plate, well, disk, or tube. The amount may be on a weight (g, mg, μg, ng, or pg) or molar (millimole (mmol), micromole (μmol), nanomole (nmol), or picomole (pmol)) basis. These units describe the exposure concentration rather than the dose actually taken up by the test species. Insufficient data currently exist to permit the development of dose amounts from this information. In such cases, therefore, the material concentration units used by the author are reported.

Since the exposure values reported in host-mediated and body fluid assays are doses delivered to the host organism, no attempt is made to estimate the exposure concentration to the indicator organism. The exposure values cited for host-mediated assay data are in units of milligrams (or other appropriate unit of weight) of material administered per kilogram of host body weight or in parts of vapor or gas per million (ppm) parts of air (or other appropriate concentrations) by volume.

17. *Toxicity Dose Data* lines include, in sequence, the route of exposure; the species of animal studied; the toxicity measure; the amount of material per body weight or concentration per unit of air volume and, where applicable, the duration of exposure; a descriptive notation of the type of effect reported; and the reference from which the information was extracted. Only positive toxicity test results are cited in this section.

All toxic dose data appearing in the book are derived from reports of the toxic effects produced by individual materials. For human data, a toxic effect is defined as any reversible or irreversible noxious effect on the body, any benign or malignant tumor, any teratogenic effect, or any death that has been reported to have resulted from exposure to a material via any route. For humans, a toxic effect is any effect that was reported in the source reference. There is no qualifying limitation on the duration of exposure or for the quantity or concentration of the material, nor is there a qualifying limitation on the circumstances that resulted from the exposure. Regardless of the absurdity of the circumstances that were involved in a toxic exposure, it is assumed that the same circumstances could recur. For animal data, toxic effects are limited to the production of tumors, benign (neoplastigenesis) or malignant (carcinogenesis); the production of changes in the offspring resulting from action on the fetus directly (teratogenesis); and death. There is no limitation on either the duration of exposure or on the quantity or concentration of the dose of the material reported to have caused these effects.

The report of the lowest total dose administered over the shortest time to produce the toxic effect was given preference, although some editorial liberty was taken so that additional references might be cited. No restrictions were placed on the amount of a material producing death in an experimental animal nor on the time period over which the dose was given.

Each element of the toxic dose line is discussed below:

a. Route of Exposure or Administration. Although many exposures to materials in the industrial community occur via the respiratory tract or skin, most studies in the published literature report exposures of experimental animals in which the test materials were introduced primarily through the mouth by pills, in food, in drinking water, or by intubation directly into the stomach. The abbreviations and definitions of the various routes of exposure reported are given in Table 1.

b. Species Exposed. Since the effects of exposure of humans are of primary concern, indications are given, when available, whether the results were observed in man, woman, child, or infant. If no such distinction was made in the reference, the abbreviation "hmn" (human) is used. However, the results of studies on rats or mice are the most frequently reported and hence provide the most useful data for comparative purposes. The species and abbreviations used in reporting toxic dose data are listed alphabetically in Table 2.

c. Description of Exposure. In order to better describe the administered dose reported in the literature, six abbreviations are used. These terms indicate whether the dose caused death (LD)

TABLE 2. Species (With Assumptions For Toxic Dose Calculation From Non-Specific Data*)

Abbreve.	Species	Age	Weight	Consumption Food g/day	(Approx.) Water mL/day	1 ppm in Food Equals, in mg/kg/day	Approximate Gestation Period days
Bird—type not specified	brd		1 kg				
Bird—wild bird species	bwd		40 gm				
Cat, adult	cat		2 kg	100	100	0.05	64 (59-68)
Child	chd	1-13 Y	2 kg				
Chicken, adult	ckn	8 W	800 gm	140	200	0.175	
Cattle	ctl		500 kg	10,000		0.02	284 (279-290)
Duck, adult (domestic)	dck	8 W	2.5 kg	250	500	0.1	
Dog, adult	dog	52 W	10 kg	250	500	0.025	62 (56-68)
Domestic animals (Goat, Sheep)	dom		60 kg	2,400		0.04	G: 152 (148-156) S: 146 (144-147)
Frog, adult	frg		33 gm				
Guinea Pig, adult	gpg		500 gm	30	85	0.06	68
Gerbil	grb		100 gm	5	5	0.05	25 (24-26)
Hamster	ham	14 W	125 gm	15	10	0.12	16 (16-17)
Human	hmn	Adult	70 kg				
Horse, Donkey	hor		500 kg	10,000		0.02	H: 339 (333-345) D: 365
Infant	inf	0-1 Y	5 kg				
Mammal (species unspecified in reference)	mam		200 gm				

Man	man	Adult	70 kg				
Monkey	mky	2.5 Y	5 kg	250	500	0.05	165
Mouse	mus	8 W	25 gm	3	5	0.12	21
Non-mammalian species	nml						
Pigeon	pgn	8 W	500 gm				
Pig	pig		60 kg	2,400		0.041	114 (112-115)
Quail (laboratory)	qal		100 gm				
Rat, adult female	rat	14 W	200 gm	10	20	0.05	22
Rat, adult male	rat	14 W	250 gm	15	25	0.06	
Rat, adult	rat	14 W	200 gm	15	25		
Rat, weanling	rat	3 W	50 gm	15	25	0.3	
Rabbit, adult	rbt	12 W	2 kg	60	330	0.03	31
Squirrel	sql		500 gm				44
Toad	tod		100 gm				
Turkey	trk	18 W	5 kg				
Woman	wmn	Adult	50 kg				270

Values given above in Table 2 are within reasonable limits usually found in the published literature and are selected to facilitate calculations for data from publications in which toxic dose information has not been presented for an individual animal of the study. See, for example, Association of Food and Drug Officials, Quarterly Bulletin, volume 18, page 66, 1954; Guyton, American Journal of Physiology, volume 150, page 75, 1947; The Merck Veterinary Manual, 5th Edition, Merck & Co., Inc., Rahway, N.J., 1979; and The UFAW Handbook on the Care and Management of Laboratory Animals, 4th Edition, Churchill Livingston, London, 1972. Data for lifetime exposure are calculated from the assumptions for adult animals for the entire period of exposure. For definitive dose data, the reader must review the referenced publication.

or other toxic effects (TD) and whether it was administered as a lethal concentration (LC) or toxic concentration (TC) in the inhaled air. In general, the term "Lo" is used where the number of subjects studied was not a significant number from the population or the calculated percentage of subjects showing an effect was listed as 100. The definition of terms is as follows:

TDLo - Toxic Dose Low - the lowest dose of a material introduced by any route, other than inhalation, over any given period of time and reported to produce any toxic effect in humans or to produce carcinogenic, neoplastigenic, or teratogenic effects in animals or humans.

TCLo - Toxic Concentration Low - the lowest concentration of a material in air to which humans or animals have been exposed for any given period of time that has produced any toxic effect in humans or produced a carcinogenic, neoplastigenic, or teratogenic effect in animals or humans.

LDLo - Lethal Dose Low - the lowest dose (other than LD50) of a material introduced by any route, other than inhalation, over any given period of time in one or more divided portions and reported to have caused death in humans or animals.

LD50 - Lethal Dose Fifty - a calculated dose of a material which is expected to cause the death of 50% of an entire defined experimental animal population. It is determined from the exposure to the material by any route other than inhalation of a significant number from that population. Other lethal dose percentages, such as LD1, LD10, LD30, and LD99, may be published in the scientific literature for the specific purposes of the author. Such data would be published if these figures, in the absence of a calculated lethal dose (LD50), were the lowest found in the literature.

LCLo - Lethal Concentration Low - the lowest concentration of a material in air, other than LC50, which has been reported to have caused death in humans or animals. The reported concentrations may be entered for periods of exposure which are less than 24 hours (acute) or greater than 24 hours (subacute and chronic).

LC50 - Lethal Concentration Fifty - a calculated concentration of a material in air, exposure to which for a specified length of time is expected to cause the death of 50% of an entire defined experimental animal population. It is determined from the exposure to the material of a significant number from that population.

The following table summarizes the above information:

Category	Exposure Time		Route of Exposure	TOXIC EFFECTS Human	Animal
TDLo	Acute or Chronic	NEO, ETA, REP	All except Inhalation	Any Non-Lethal	CAR, TER,
TCLo	Acute or Chronic	NEO, ETA, REP	Inhalation	Any Non-Lethal	CAR, TER,
LDLo	Acute or Chronic	Death	All except Inhalation	Death	
LD50	Acute	Death (Statistically Determined)	All except Inhalation	Not Applicable	
LCLo	Acute or Chronic	Death	Inhalation	Death	
LC50	Acute	Death (Statistically Determined)	Inhalation	Not Applicable	

d. Units of Dose Measurement. As in almost all experimental toxicology, the doses given are expressed in terms of the quantity administered per unit body weight, or quantity per skin surface area, or quantity per unit volume of the respired air. In addition, the duration of time over which the dose was administered is also listed, as needed. Dose amounts are generally expressed as milligrams (one thousandth of a gram) per kilogram (mg/kg). In some cases, because of dose size and its practical presentation in the file, grams per kilogram (g/kg), micrograms (one millionth of a gram) per kilogram (μg/kg), or nanograms (one billionth of a gram) per kilogram (ng/kg) are used. Volume measurements of dose were converted to weight units by appropriate calculations. Densities were obtained from standard reference texts. Where densities were not readily available, all liquids were assumed to have a density of one gram per milliliter. Twenty drops of liquid are assumed to be equal in volume to one milliliter.

All body weights have been converted to kilograms (kg) for uniformity. For those references in which the dose was reported to have been administered to an animal of unspecified weight or a given number of animals in a group (e.g., feeding studies) without weight data, the weights of the respective animal species were assumed to be those listed in Table 2 and the dose is listed on a per kilogram body weight basis. Assumptions for daily food and water intake are found in Table 2 to allow approximating doses for humans and species of experimental animals where the dose was originally reported as a concentration in food or water. The values presented are selections which are reasonable for the species and convenient for dose calculations.

Concentrations of a gaseous material in air are generally listed as parts of vapor or gas per million parts of air by volume (ppm). However, parts per hundred (pph or per cent), parts per billion (ppb), or parts per trillion (ppt) may be used for convenience of presentation. If the material is a solid or a liquid, the concentrations are listed preferably as milligrams per cubic meter (mg/m^3) but may, as applicable, be listed as micrograms per cubic meter ($\mu g/m^3$), nanograms per cubic meter (ng/m^3), or picograms (one trillionth of a gram) per cubic meter (pg/m^3) of air. For those cases in which other measurements of contaminants are used, such as the number of fibers or particles, the measurement is spelled out.

Where the duration of exposure is available, time is presented as minutes (M), hours (H), days (D), weeks (W), or years (Y). Additionally, continuous (C) indicates that the exposure was continuous over the time administered, such as ad libitum feeding studies or 24-hour, 7-day per week inhalation exposures. Intermittent (I) indicates that the dose was administered during discrete periods, such as daily, twice weekly, etc. In all cases, the total duration of exposure appears first after the kilogram body weight and slash, followed by descriptive data; e.g., 10 mg/kg/3W-I indicates ten milligrams per kilogram body weight administered over a period of three weeks, intermittently in a number of separate, discrete doses. This description is intended to provide the reader with enough information for an approximation of the experimental conditions, which can be further clarified by studying the reference cited.

e. Frequency of Exposure. Frequency of exposure to the test material varies depending on the nature of the experiment. Frequency of exposure is given in the case of an inhalation experiment, for human exposures (where applicable), or where CAR, NEO, ETA, REP, or TER is specified as the toxic effect.

f. Duration of Exposure. For assessment of tumorigenic effect, the testing period should be the life span of the animal, or until statistically valid calculations can be obtained regarding tumor incidence. In the toxic dose line, the total dose causing the tumorigenic effect is given. The duration of exposure is included to give an indication of the testing period during which the animal was exposed to this total dose. For multigenerational studies, the time during gestation when the material was administered to the mother is also provided.

g. Notations Descriptive of the Toxicology. The toxic dose line thus far has indicated the route of entry, the species involved, the description of the dose, and the amount of the dose. The next entry found on this line when a toxic exposure (TD or TC) has been listed is the toxic effect. Following a colon will be one of the notations found in Table 3. These notations indicate the organ system affected or special effects that the material produced, e.g., TER = teratogenic effect. No attempt was made to be definitive in reporting these effects because such definition

TABLE 3. Notations Descriptive of the Toxicology

Abbreviation	Definition (not limited to effects listed)
ALR	Allergic systemic reaction such as might be experienced by individuals sensitized to penicillin.
BCM	Blood clotting mechanism effects—any effect which increases or decreases clotting time.
BLD	Blood effects—effect on all blood elements, electrolytes, pH, protein, oxygen carrying or releasing capacity.
BPR	Blood pressure effects—any effect which increases or decreases any aspect of blood pressure.
CAR	Carcinogenic effects—see paragraph 9g in text.
CNS	Central nervous system effects—includes effects such as headaches, tremor, drowsiness, convulsions, hypnosis, anesthesia.
COR	Corrosive effects—burns, desquamation.
CUM	Cumulative effects—where material is retained by the body in greater quantities than is excreted, or the effect is increased in severity by repeated body insult.
CVS	Cardiovascular effects— such as an increase or decrease in the heart activity through effect on ventricle or auricle; fibrillation; constriction or dilation of the arterial or venous system.
DDP	Drug dependence effects—any indication of addiction or dependence.
ETA	Equivocal tumorigenic agent—see text.
EYE	Eye effects— irritation, diplopia, cataracts, eye ground, blindness by affecting the eye or the optic nerve.
GIT	Gastrointestinal tract effects—diarrhea, constipation, ulceration.
GLN	Glandular effects—any effect on the endocrine glandular system.
IRR	Irritant effects—any irritant effect on the skin, eye, or mucous membrane.
MLD	Mild irritation effects— used exclusively for primary irritation data.
MMI	Mucous membrane effects—irritation, hyperplasia, changes in ciliary activity.
MOD	Moderate irritation effects— used exclusively for primary irritation data.
MSK	Musculo-skeletal effects— such as osteoporosis, muscular degeneration.
NEO	Neoplastic effects—see text.
PNS	Peripheral nervous system effects.
PSY	Psychotropic effects—exerting an effect upon the mind.
PUL	Pulmonary system effects—effects on respiration and respiratory pathology.
RBC	Red blood cell effects—includes the several anemias.
REP	Reproductive effects—see text.
SEV	Severe irritation effects—used exclusively for primary irritation data.
SKN	Skin effects—such as erythema, rash, sensitization of skin, petechial hemorrhage.
SYS	Systemic effects—effects on the metabolic and excretory function of the liver or kidneys.
TER	Teratogenic effects— nontransmissible changes produced in the offspring.
UNS	Unspecified effects—the toxic effects were unspecific in the reference.
WBC	White blood cell effects—effects on any of the cellular units other than erythrocytes, including any change in number or form.

requires much detailed qualification and is beyond the scope of this book. The selection of the dose was based first on the lowest dose producing an effect and second on the latest study published.

18. *Reproductive Effects Data* lines include, in sequence, the route of exposure, the species of animal tested, the type of dose, the total dose amount administered, the time and duration of administration, and the reference from which the information was extracted. Only positive reproductive effects data for mammalian species are cited. Because of differences in the reproductive systems among species and the systems' varying responses to chemical exposures, no attempt is made to extrapolate animal data or to evaluate the significance of a substance as a reproductive risk to humans.

Each element of the reproductive effects data line is discussed below.

a. Route of Exposure or Administration. See Table 1 for a complete list of abbreviations and definitions of the various routes of exposure reported. For reproductive effects data, the specific route is listed either when the substance was administered to only one of the parents or when the substance was administered to both parents by the same route. However, if the substance was administered to each parent by a different route, the route is indicated as "mul" (multiple).

b. Species Exposed. Reproductive effects data are cited for mammalian species only. Species abbreviations are the same as those used for toxic dose data and are shown in Table 2. Also shown in Table 2 are approximate gestation periods.

c. Type of Exposure. Only two types of exposure, TDLo and TCLo, are used to describe the dose amounts reported for reproductive effects data. These two terms are also used to describe toxic dose data.

d. Dose Amounts and Units. The total dose amount that was administered to the exposed parent is given. If the substance was administered to both parents, the individual amounts to each parent have been added together and the total amount shown. Where necessary, appropriate conversion of dose units has been made. The dose amounts listed are those for which the reported effects are statistically significant. However, human case reports are cited even when no statistical tests can be performed. The statistical test is that used by the author. If no statistic is reported, a Fisher's Exact Test is applied with significance at the 0.05 level, unless the author makes a strong case for significance at some other level.

Dose units are usually given as an amount administered per unit body weight or as parts of vapor or gas per million parts of air by volume. There is no limitation on either the quantity or concentration of the dose or the duration of exposure reported to have caused the reproductive effect(s).

e. Time and Duration of Treatment. The time when a substance is administered to either or both parents may significantly affect the results of a reproductive study, because there are differing critical periods during the reproductive cycles of each species. Therefore, to provide some indication of when the substance was administered, which should facilitate selection of specific data for analysis by the user, a series of up to four terms follows the dose amount. These terms indicate to which parent(s) and at what time the substance was administered. The terms take the general form:

(uD male/vD pre/w-xD preg/yD post)

where u = total number of days of administration to male prior to mating

v = total number of days of administration to female prior to mating

w = first day of administration to pregnant female during gestation

x = last day of administration to pregnant female during gestation

y = total number of days of administration to lactating mother after birth of offspring

If administration is to the male only, then only the first of the above four terms is shown following the total dose to the male, e.g., 10 mg/kg (5D male). If administration is to the female only, then only the second, third, or fourth term, or any combination thereof, is shown following the total dose to the female, e.g.,

10 mg/kg (3D pre)

10 mg/kg (3D pre/4-7D preg)

10 mg/kg (3D pre/4-7D preg/5D post)

10 mg/kg (3D pre/5D post)

10 mg/kg (4-7D preg)

10 mg/kg (4-7D preg/5D post)

10 mg/kg (5D post) (NOTE: This example indicates administration to the lactating mother only after birth of the offspring.)

If administration is to both parents, then the first term and any combination of the last three terms are listed, e.g., 10 mg/kg (5D male/3D pre/4-7D preg). If administration is continuous through two or more of the above periods, the above format is abbreviated by replacing the slash (/) with a dash (-). For example, 10 mg/kg (3D pre-5D post) indicates a total of 10 mg/kg administered to the female for three days prior to mating, on each day during gestation, and for five days following birth. Approximate gestation periods for various species are shown in Table 2.

f. Multigeneration Studies. Some reproductive studies entail administration of a substance to several consecutive generations, with the reproductive effects measured in the final generation. The protocols for such studies vary widely. Therefore, because of the inherent complexity and variability of these studies, they are cited in a simplified format as follows. The specific route of administration is reported if it was the same for all parents of all generations; otherwise the abbreviation "mul" is used. The total dose amount shown is that administered to the F0 generation only; doses to the Fn (where n = 1, 2, 3, etc.) generations are not reported. The time and duration of treatment for multigeneration studies are not included in the data line. Instead, the dose amount is followed by the abbreviation "(MGN)", e.g., 10 mg/kg (MGN). This code indicates a multigeneration study, and the reader must consult the cited reference for complete details of the study protocol.

19. *Carcinogenic Effects Data* are given in more detail because of the importance attached to reports of the carcinogenic activity of materials. Tumorigenic citations are classified according to the reported results of the study only to aid the reader in selecting appropriate references for in-depth review and evaluation. The classification, ETA (equivocal tumorigenic agent), denotes those studies reporting uncertain, but seemingly positive, results. The criteria for the three classifications are listed below. These criteria are used to abstract the data in individual reports on a consistent basis and do not represent a comprehensive evaluation of a material's tumorigenic potential to humans.

The following nine technical criteria are used to abstract the toxicological literature and classify studies that report positive tumorigenic responses. No attempts are made either to evaluate the various test procedures or to correlate results from different experiments.

(1) A citation is coded "CAR" (carcinogenic) when review of an article reveals that all of the following criteria are satisfied:

(a) A statistically significant increase in the incidence of tumors in the test animals. The statistical test is that used by the author. If no statistic is reported, a Fisher's Exact Test is applied with significance at the 0.05 level, unless the author makes a strong case for significance at some other level.

(b) A control group of animals is used and the treated and control animals are maintained under identical conditions.

(c) The sole experimental variable between the groups is the administration or non-administration of the test material (see (9) below).

(d) The tumors consist of autonomous populations of cells of abnormal cytology capable of invading and destroying normal tissues, or the tumors metastasize as confirmed by histopathology.

(2) A citation is coded "NEO" (neoplastic) when review of an article reveals that all of the following criteria are satisfied:

(a) A statistically significant increase in the incidence of tumors in the test animals. The statistical test is that used by the author. If no statistic is reported, a Fisher's Exact Test is applied with significance at the 0.05 level, unless the author makes a strong case for significance at some other level.

(b) A control group of animals is used, and the treated and control animals are maintained under identical conditions.

(c) The sole experimental variable between the groups is the administration or non-administration of the test material (see (9) below).

(d) The tumors consist of cells that closely resemble the tissue of origin, that are not grossly abnormal cytologically, that may compress surrounding tissues, but that neither invade tissues nor metastasize; or

(e) The tumors produced cannot definitely be classified as either benign or malignant.

(3) A citation is coded "ETA" (equivocal tumorigenic agent) when some evidence of tumorigenic activity is presented, but one or more of the criteria listed in (1) or (2) above is lacking. Thus, a report with positive pathological findings, but with no mention of control animals, is coded "ETA."

(4) Since an author may make statements or draw conclusions based on a larger context than that of the particular data reported, papers in which the author's conclusions differ substantially from the evidence presented in the paper are subject to review.

(5) All doses except those for transplacental carcinogenesis are reported in one of the following formats.

(a) For all routes of administration other than inhalation: cumulative dose in mg (or appropriate unit)/kg/duration of administration.

Whenever the dose reported in the reference is not in the units discussed herein, conversion to this format is made. The total cumulative dose is derived from the lowest dose level that produces tumors in the test group.

(b) For inhalation experiments: concentration in ppm (or mg/m^3)/total duration of exposure. The concentration refers to the lowest concentration that produces tumors.

(6) Transplacental carcinogenic doses are reported in one of the following formats.

(a) For all routes of administration other than inhalation: cumulative dose in mg/kg/(time of administration during pregnancy).

The cumulative dose is derived from the lowest single dose that produces tumors in the offspring. The test chemical is administered to the mother.

(b) For inhalation experiments: concentration in ppm (or mg/m^3)/(time of exposure during pregnancy).

The concentration refers to the lowest concentration that produces tumors in the offspring. The mother is exposed to the test chemical.

(7) For the purposes of this listing, all test chemicals are reported as pure, unless otherwise stated by the author. This does not rule out the possibility that unknown impurities may have been present.

(8) A mixture of compounds whose test results satisfy the criteria in (1), (2), or (3) above is included if the composition of the mixture can be clearly defined.

(9) For tests involving promoters or initiators, a study is included if the following conditions are satisfied (in addition to the criteria in (1), (2), or (3) above):

(a) The test chemical is applied first, followed by an application of a standard promoter. A positive control group in which the test animals are subjected to the same standard promoter under identical conditions is maintained throughout the duration of the experiment. The data are not used if no mention of positive and negative control groups is made in the reference.

(b) A known carcinogen is first applied as an initiator, followed by application of the text chemical as a promoter. A positive control group in which the test animals are subjected to the same initiator under identical conditions is maintained throughout the duration of the experiment.

The data are not used if no mention of positive and negative control groups is made in the reference.

20. *Cited Reference* is the final entry of the irritation, mutation, reproductive, tumorigenic, and toxic dose data lines. This is the source from which the information was extracted. All references cited are publicly available. No governmental classified documents have been used for source information. All references have been given a unique six-letter CODEN character code (derived from the American Society for Testing and Materials *CODEN for Periodical Titles* and the CAS *Source Index*), which identifies periodicals, serial publications, and individual published works. For those references for which no CODEN was found, the corresponding six-letter code includes asterisks (*) in the last one or two positions following the first four or five letters of an acronym for the publication title. Following the CODEN designation (for most entries) is the number of the volume, followed by a comma; the page number of the first page of the article, followed by a comma; and a two-digit number, indicating the year of publication in this century. When the cited reference is a report, the report number is listed. Where contributors have provided information on their unpublished studies, the CODEN consists of the first three letters of the last name, the initials of the first and middle names, and a number sign (#). The date of the letter supplying the information is listed. All CODEN acronyms are listed in alphabetical order and defined in the CODEN Section.

Acknowledgments

I wish to thank my wife, Grace, for her constant help and advice in every aspect of writing this handbook. I extend heartfelt thanks to Susan Munger, Executive Editor, Chemistry and Industrial Safety for help in the conceptualization of this handbook and encouragement in its execution. My best to Alberta Gordon and Louise Kurtz for their expert professional advice and assistance in converting the manuscript to this volume.

Section I

Chapter 1

FOOD ADDITIVES

Their Regulatory Status
Their Use by the Food Industry

Robert W. Dean, Ph.D.*
Regional Nutritionist
Food and Nutrition Service
United States Department of Agriculture

HISTORY AND LEGISLATION

In the United States at the turn of the century, the food industry was at a turning point regarding production, processing, and distribution of food. Changes in population from agricultural to industrial had resulted in a need to establish distribution systems to supply the growing population of the cities with food. Conditions under which food was produced and processed were in many cases unsanitary and dangerous. There was little control over the use of chemical additives as preservatives or as coloring agents for foods. The United States had enacted a food inspection law in 1890 largely to establish confidence among foreign purchasers of U.S.-produced foodstuffs.

However, in 1906, Upton Sinclair published his book *The Jungle,* based on the meat-packing industry in Chicago. Although planned to be sociological in nature, focusing on poor working conditions, housing, exploitation of workers, etc., his description of how meat products themselves were produced led to the passage of the Meat Inspection Act of 1907 and the Federal Food and Drug Act of 1906.

At the time, it was relatively common for floor sweepings to be added to pepper, ash leaves to tea, copper and lead salts to cheese and candy, copper salts to pickles and peas, prussic acid to wine, alum to bread, acorns to coffee, and brick dust to cocoa. Fruit jams or jellies were made with water, dextrose, grass seed, and color. Some 80 dyes were in use in the United States for coloring food. The same batch of dye used for coloring textiles might be used for coloring food.[1]

One of the early champions of food laws was Dr. Harvey Washington Wiley, Indiana State Chemist from 1874 to 1883, and Chief of the Bureau of Chemistry of the U.S. Department of Agriculture from 1883 to 1912. In 1903 he established a "poison squad" of young men who consumed foods treated with known amounts of chemicals commonly used in foods. The objective was to establish if these compounds were injurious to health. The culmination of his efforts

*The opinions expressed in this chapter are soley those of the author. No opinions, statements of fact, or conclusions contained here can properly be attributed to the USDA, its staff, its members or its contracting agencies.

[1] The 1906 Act allowed only seven synthetic colors based on known composition and physiology. It is interesting to note that only two of those original seven dyes are presently permitted in food: Blue No. 2 and Red No. 3.

was the passage of the Federal Food and Drug Act of 1906 often referred to as "the Pure Food Law."

Dr. Wiley was much opposed to saccharin as a sweetener but was overruled by President Theodore Roosevelt.

The 1906 Act consisted of only six pages. It prohibited the manufacture and shipment in interstate commerce of "adulterated" or "misbranded" foods and drugs. The government could go to court against products which violated the law, but manufacturers were not given guidelines to assure their compliance. Weight or measure was not required on the label. It was only required that a contents statement, if used, be truthful. There was a confusing mix of state laws governing weights and measures.

Because of these factors, food manufacturers sought remedy through the Gould Amendment to the federal law. The Amendment, enacted in 1913, required net contents to be declared with tolerances established for variations.

The Bureau of Chemistry enforced the 1906 Act until 1927, when the research and enforcement functions were separated and the Food, Drug and Insecticide Administration was established. The name was changed to the Food and Drug Administration (FDA) in 1931, and in 1940 the FDA was transferred to what is now known as the Department of Health and Human Services.

The U.S. Department of Agriculture (USDA) retains authority for food products containing 2% meat or more, and enforcement of meat and poultry regulations is the responsibility of the USDAs Food Safety and Inspection Service (FSIS), previously known as the Meat Inspection Department. The lack of standards for composition of foods was a serious defect of the 1906 Act.

A growing consumer movement arose in the 1930s, marked by publication of the book *Your Money's Worth* by Stuart Chase and F. J. Schlink.

In 1937 a pharmacist compounded an "Elixir Sulfonilamide," containing diethylene glycol, which doctors prescribed for their patients, many of whom were children. As a result of consuming the product, 107 people died.

A result of all this was the passage into law of the Food, Drug and Cosmetic Act (FD&C Act) of 1938. This Act placed more emphasis on administrative procedures, gave authority specifically for on-site plant inspections, and provided for the establishment of standards of identity for individual food products. In addition, under the Act of 1906, a system had been set up for certification of individual lots or batches of synthetic colors to be certain that they conformed to chemical specifications. Certification in this case, however, was not mandatory. The new Act made certification mandatory. At this time there were 15 synthetic colors approved for use in food. The Act created three new categories of colors: colors approved for foods, drugs, and cosmetics (FD&C colors), colors approved for use in drugs and cosmetics only (D&C colors), and colors approved for external use only (external D&C colors). The Act of 1938 also allowed the government to obtain federal court injunctions against violators.

Meanwhile, due to economic factors and to protect growers and farmers from either fraud or damage to plants and animals, and in some cases damage to themselves, the federal government had begun to regulate the composition of insecticides beginning with the Insecticide Act of 1910 and culminating in the federal Insecticide, Fungicide and Rodenticide Act of 1947. This Act also regulated herbicides.

One of the important features of the Act was the requirement that these economic "poisons" be registered with the Secretary of Agriculture. (This function is now vested in the Environmental Protection Agency.) The Act also required that standard lead arsenate, basic lead arsenate, calcium arsenate, magnesium arsenate, zince arsenate, zinc arsenite, sodium fluoride, and sodium fluorosilicate, which are white powders, be colored to prevent their being confused with flour, sugar, salt, baking powder, or other edibles in the manufacture of foods.

Any economic poison which is highly toxic to man was required to have a skull and crossbones on the label, have the word "poison" prominently displayed in red on the label with a contrasting color background, and give an antidote on the label.

An interesting sidelight to this Act is a prohibition regarding a reference to registration with the Federal Government being mentioned on the label, presumably to prevent an implication that the product is endorsed by the government. The Act was amended in 1959 to include nematocides, plant growth regulators, defoliants, and dessicants.

In July, 1954, the federal FD&C Act was amended with respect to pesticides in or on raw agricultural commodities. Public Law 83–518, commonly known as the Miller Amendment, required the FDA to establish tolerances for the amounts of economic poisons allowed to remain in or on fruits, vegetables, and agricultural produce. This is now a function of the Environmental Protection Agency (EPA), established in 1970. The authority for the establishment of tolerances is given in Section 408, a new section added to the FD&C Act. The EPA is responsible for establishing tolerances for pesticides, while the FDA is responsible for monitoring and assuring compliance with the tolerances for agricultural commodities.

The EPA may set a zero tolerance for a pesticide for the following:

(a) "A safe level of the pesticide chemical in the diet of two different species of warm blooded animals has not been reliably determined."
(b) "The chemical is carcinogenic to or has other alarming physiological effects upon one or more of the species of the test animals used, when fed in the diet of such animals."
(c) "The pesticide chemical is toxic but is normally used at times when, or in such a manner that, fruit, vegetables or other raw agricultural commodities will not bear or contain it."
(d) "All residue of the pesticide chemical is normally removed through good agricultural practice, such as washing or brushing or through weathering or other changes in the chemical itself, prior to introduction of the raw agricultural commodity into interstate commerce."

Some community action groups have criticized pesticide regulation from the point of view that under Section 408 of the FD&C Act, a carcinogen may be used on produce because it is a pesticide, while under Section 409 of the same Act the same substance is a food additive and is banned from use under the Delaney Clause (discussed later in this chapter).

In 1988, the EPA transferred the tolerances established for pesticides from the part of the Code of Federal Regulations reserved for the FDA (Title 21) to that reserved for the EPA (Title 40). This was an administrative decision, and no new regulations were involved.

In establishing tolerances for Raw Agricultural Commodities (RACs), the EPA includes meat, fat, milk, and byproducts of cattle, hogs, horses, goats, sheep, and animal feeds, as well as fruits and vegetables intended for human consumption. The FD&C Act was amended in June, 1960 to exempt foods treated with pesticides from having their presence listed on the label as seen by the consumer. However, a shipping container containing an RAC which has a residue of a pesticide applied after harvest is required to have a label which gives the usual name of the pesticide and its function.

In September, 1958, the FD&C Act was amended "to prohibit the use in foods of additives which have not been adequately tested to establish their safety." The term food additive was defined as follows:

> any substance the intended use of which results or may reasonably be expected to result, directly or indirectly, in its becoming a component or otherwise affecting the characteristics of any food (including any substance intended for use in producing, manufacturing, packing, processing, preparing, treating, packaging, transporting or holding food; and including any source of radiation intended for any such use), if such substance is not generally recognized among experts qualified by scientific training and experience to evaluate its safety, as having been adequately shown through scientific procedures (or in the case of a substance used in food prior to January 1, 1958, through either scientific procedures, or experience based on common use in food) to be safe under the conditions of its intended use . . .

As a result of the Amendment, called the Food Additives Amendment of 1958, the term "Generally Recognized As Safe" (GRAS) came into use, and industry and government developed lists of substances which were considered to be "GRAS" for use in food, either in specific foods at specific levels or in accordance with "good manufacturing practice." Specific examples of the use of the term "GRAS" will be discussed later in this chapter.

Pesticide chemicals in or on RACs were excluded as well as substances given sanction prior to the enactment of the Amendment or given prior sanction with regard to the Poultry Products Act (Public Law 85–172) or the Meat Inspection Act of 1907.

A food was deemed to be adulterated if it was "intentionally subjected to radiation." Any food additive employed for a given use was deemed unsafe unless a regulation was issued under the FD&C Act describing the conditions under which the additive could be safely used. Section 409 of the FD&C Act, as amended, described the means by which a petition for the use of an additive in a given situation could be filed with the FDA. The petition had to include relevant data bearing on the effect that the food additive was intended to produce and the quantity required to produce the effect, as well as a description of methods to determine the quantity of the additive present as well as any other substances formed in the food as a result of the use of the additive. Full reports of investigations made with regard to the safety of the additive were required as well as disclosure of the methods and controls used in the investigations. The FDA could either grant or deny the petition within 90 to 180 days of filing.

The FDA could not grant the petition if "a fair evaluation of the data before the Secretary (of Health and Human Services) fails to establish that the proposed use of the food additive under the conditions of use to be specified in the regulation will be safe; provided that no additive shall be deemed to be safe if it is found to induce cancer when ingested by man or animal, or if it is found, after tests which are appropriate for the evaluation of the safety of food additives, to induce cancer in man or animal."

The petition could not be approved if it promoted deception of the consumer, or if the additive did not accomplish the intended effect. A tolerance level could not be set above the level required to accomplish the effect.

The prohibition of food additives causing cancer at any level or tolerance, previously referred to, is known as the Delaney Amendment. The Amendment is named after Representative James Delaney of New York, who introduced it into the legislation.

The Delaney Clause or Amendment has been an extremely controversial piece of food additive legislation and regulation. Its proponents state that it is a vital cog in protecting the health of the public, while its detractors state that it inhibits research and development and scientific progress. Its detractors point out that many foods contain carcinogens which occur naturally as a part of these foods.

In the early 1950s, three instances of diarrhea in children occurred as the result of excessive levels of FD&C Orange No. 1 and FD&C Red No. 32 in candy and on popcorn. Because of these incidents, and as a result of chronic toxicity animal feeding studies by FDA, FD&C Orange No. 2 and FD&C Red No. 32 were delisted.[2]

The efforts of the FDA to limit quantities of colors used in foods were opposed in the courts, and in late 1958 the Supreme Court held that under the original FD&C Act of 1938, the FDA did not have the authority to set limits or tolerances regarding use of colors in foods. Shortly afterward, FD&C Yellow Numbers 1 and 2, and 3 and 4 were delisted. This meant that there were no longer any oil soluble colors approved for use by the food industry.

This set the stage for the passage of the Color Additive Amendments of 1960 and also for the development of the FD&C lakes. The Code of Federal Regulations defines the term lakes

---

[2] Orange 1 and Orange 2 are now known as Orange 3 and Orange 4 and are not certified for use in food because of current lack of interest in their use by the food industry.

as a ''straight color extended on a substratum by adsorption, coprecipitation, or chemical combination that does not include any combination of ingredients made by a simple mixing process.'' A lake consists of a water soluble certified FD&C dye on a substratum of aluminum hydrate or aluminum hydroxide. The lake itself must be certified by the FDA. The lakes are useful in coloring oils or other food products where water is not present or is not desirable. They color by dispersion rather than solution. The lakes were added to the list of certified color additives in 1959.

The Color Additive Amendments of 1960 were divided into two sections. Title I deals mainly with procedures for dealing with all permitted colors both certified and uncertified. Colors are no longer classified as ''food additives'' but as ''color additives.'' The term ''color'' is to include black and white and intermediate grays. Provisions calling for listing and certification of harmless ''coal-tar colors'' were removed from the FD&C Act. All colors both ''synthetic'' and ''natural'' are to be treated alike, with the FDA having authority to decide which color additives will be exempted from certification. Pretesting requirements for all color additives are uniform. Pesticides or plant growth regulators, which enhance or promote the development of color in fruit or other produce, were excluded from the definition of color additives. The FDA is given the authority of proscribing conditions of use, specifications and tolerance limitations, as well as labeling and packaging requirements for the color itself and foods that contain it.

In determining whether a proposed use of a color additive is safe, the FDA must take into account the probable consumption of it and of any substance formed as a result of its use. It also evaluates the following:

> the cumulative effect, if any, of such additive in the diet of man or animals, taking into account the same or any chemically or pharmacologically related substance or substances in such diet; safety factors which, in the opinion of experts qualified by scientific training and experience to evaluate the safety of color additives for use or uses for which the additive is proposed to be listed, are generally recognized as appropriate for the use of animal experimentation data; and the availability of any needed practicable methods of analysis for determining the identity and quantity of (I) the pure dye and all intermediates and other impurities contained in such color additive, (II) such additive in or on any article of food, . . . and (III) any substance formed in or on such article because of the use of such additive.

Title I also includes a clause similar to the Delaney Clause which states the following:

> a color additive (i) shall be deemed unsafe and shall not be listed for any use which will or may result in ingestion of all or part of such additive, if the additive is found by the Secretary (of Health and Human Services) to induce cancer when ingested by man or animal, or if it is found by the Secretary, after tests which are appropriate for the evaluation of the safety of additives for use in food, to induce cancer in man or animal, and (ii) shall be deemed unsafe, and shall not be listed, for any use which will not result in ingestion of any part of such additive, if after tests which are appropriate for the evaluation of the safety of additives for such use, or often other relevant exposure of many or animal to such additive, it is found by the Secretary to induce cancer in man or animal.

In cases involving the cancer clause, an advisory committee may be set up, selected by the National Academy of Sciences, to give a report and recommendation on any matter arising under the clause.

A color additive cannot be listed if it promotes deception of the consumer by its use in a food or otherwise results in misbranding or adulteration.

Title II of the Color Amendments deals with provisional listings of color additives which were currently in use as of July 12, 1960, when the Amendments became law. Provisional listings of color additives were to extend for two and one-half years after passage to allow for scientific studies of the safety of these colors to be completed. However, the FDA was given the power to extend the closing dates, which has happened in many cases.

The Color Amendments shifted the burden of testing and proof of safety to industry. However, in 1957, prior to the passage of the Amendments, the FDA commenced feeding studies on all of the colors currently in use in foods. Some of the studies in dogs ran for as long as seven years. All of the colors in use at the time of the passage of the Color Amendments were provisionally listed, pending completion of these studies, following which industry petitioned for permanent listing.

In August, 1957, the Poultry Products Act (Public Law 85–172) was enacted. This law called for compulsory inspection of poultry and poultry products by the USDA. The Food Safety and Inspection Service (FSIS) of the USDA is responsible for carrying out the terms of the Act. The Act was designed to prevent the introduction of "unwholesome and adulterated poultry products in the channels of interstate or foreign commerce." The Act calls for inspected poultry products leaving a plant to bear, among other information on the label, a statement of ingredients, if two or more, including artificial flavors, colors, or preservatives.

The Fair Packaging and Labeling Act became law in 1966. The purpose of the Act was to inform consumers of the quantity and composition of the contents of packaged goods and promote practices that facilitate valid price comparisons through information on the label. The FDA was given authority for the mandatory parts involving food, drugs, devices and cosmetics, and the Federal Trade Commission (FTC) for other "consumer commodities." Standard weights and measures and standardized package sizes were the objectives of the Act. A target of the Act was nonfunctional slack fill consumer packages.

The Department of Commerce was given authority to work with industry in developing voluntary standards for packaging and labeling of commodities. The Department of Treasury was given authority to regulate imports in terms of the Act. One of the reasons for passage of the law by Congress was that insufficient information was provided in ingredient statements in terms of the percentage of either costly or inexpensive ingredients.

A major change took place in regulation of the meat packing industry with the passage of the Wholesome Meat Act of 1967. This Act amended the original Act of 1906–1907, renamed the Federal Meat Inspection Act, and designated it as Title I of the Wholesome Meat Act. Title II concerned the registration and regulation of establishments, persons or firms dealing with animal slaughter or carcasses intended for human consumption or not intended for human consumption. Title III of the Wholesome Meat Act deals with provisions allowing states to set up meat inspection programs equivalent to the federal system and substituting for it. Title IV contains a number of miscellaneous provisions including a statement that the provisions of the Wholesome Meat Act "shall not derogate from any authority conferred by the federal FD&C Act prior to enactment of the Wholesome Meat Act."

Under Title I of the Act, inspection and regulation was extended to include horses, mules, and other equines as well as cattle, sheep, swine, and goats. Two new sections were added to Title I. Both Section 1, consisting of definitions, and Section 2, dealing with changes in the wording of the 1907 Act, are concerned with adulteration and misbranding of meat and meat food products.

A carcass, part of a carcass, meat, or meat food product was defined as being adulterated; if it contains any added substance injurious to health, but if the substance was not added, it is not considered adulterated if the quantity present is not injurious to health; if it contains any substance administered to the live animal which is poisonous or deleterious and makes it unfit for human food. Pesticide chemicals, food additives, and color additives may either be permitted or not permitted, depending on whether they are considered safe or unsafe under the applicable sections of the FD&C Act (Section 408 for pesticides, 409 for food additives, and 706 for color additives). The FSIS may also prohibit a substance allowed under the FD&C Act. An example is the fact that sorbic acid, a food additive approved under the FD&C Act, is not permitted in meat salads because such use might mask spoilage caused by food poisoning organisms. It is permitted in oleomargarine, however, to retard mold growth.

Other definitions of adulterated include the following: being intentionally subjected to radiation[3]; substitution for, omission, or abstraction of a valuable ingredient; addition of a substance to increase bulk or weight, contamination with filth; storage in a container composed of a poisonous or deleterious substance; holding under unsanitary conditions; being the product of an animal which has not died by slaughter.

Among definitions of misbranding are the following: failure to specify on the label of a product for which there is a standard the common names of the optional ingredients (other than spices, flavoring, and coloring); failure to state on the label that it contains any artificial flavoring, artificial coloring, or chemical preservative; if it is represented as a product for special dietary uses and the label does not bear information, such as "vitamin, mineral, and other dietary properties" as required by the FDA.

The Wholesome Poultry Products Act of 1968 is very similar to the Wholesome Meat Act of 1967. Its almost identical provisions apply to poultry and poultry products, poultry being defined as "any domesticated bird, whether live or dead."

The FDA issued regulations in 1973 regarding the nutritional labeling of food products. Only food products for which the label makes a nutritional claim are required to give label information regarding nutrient content. Similar requirements for meat products have been developed by FSIS in the form of operational memoranda.

The Toxic Substances Control Act of 1976 gave the EPA the power to regulate intrastate and interstate commerce in chemical substances and mixtures and to require this testing for carcinogenesis, mutagenesis, tetrogenesis, behavioral disorders, and cumulative or synergistic effects. However, pesticides and foods, food additives, drugs, cosmetics, or devices as defined in Section 201 of the Federal FD&C Act are excluded. Foods are defined as meat and meat foods, poultry and poultry products, and eggs and egg products.

Developmental damage in a number of infants resulting from a lack of sufficient chloride in a commercial infant formula resulted in the enactment of the Infant Formula Act of 1980 which established nutrient requirements for infant formulas and gave the FDA authority to establish requirements for quality control, record keeping, reporting and recall procedures for these products. The original act exempted formulas for infants with inborn errors of metabolism or other medical or dietary problems.

CURRENT REGULATIONS

Regulatory agencies, such as the FDA, USDA-FSIS, and the EPA publish regulations, which they develop as a result of legislation requiring them to assume authority in an area, in the *Federal Register* on a daily or weekly basis. Once a year these regulations are published in total in the *Code of Federal Regulations* (CFR).

Different Title numbers are assigned to different agencies in the CFR. The FDA is assigned Title 21, the FSIS is Title 9, and the EPA is Title 40.

The various titles in the *Code of Federal Regulations* may be purchased at Government Book Stores maintained in Federal Buildings by the Government Printing Office in many major cities in the U.S.

[3] Pork is now permitted to be subjected to irradiation to kill trichina organisms.

Current Regulations of the Food Safety and Inspection Service Regarding Food Additives

Substances approved for use in meat products with appropriate label declarations are given in Title 9, Section 318.7 of the CFR in paragraph (c)(4), which includes a table giving class of substance, substance purpose, products in which allowed, and amount allowed. Identical information for substances approved for use in poultry products is included in Title 9, Section 381.147 paragraph (f)(1), Table 1.

Part 319 and Sections 381.155 to 381.171 of Title 9 give Standards of Identity and Composition for meat and poultry products respectively, and specify the substances which may be used in these products.

Other "harmless artificial flavorings" may be added to products with the approval of the Administrator of the FSIS in specific cases.

Coloring matter and dyes other than those listed may be applied to products, mixed with rendered fat, applied to artificial and natural casings and applied to casings enclosing products if approved by the Administrator in specific cases. When a dye or coloring matter is applied to casings, no penetration of coloring into the product is allowed.

The Administrator may grant approval to a new substance or for a new use or a new level of use for an approved substance if the following applies:

(a) the substance has been previously approved by the FDA for use in meat or meat food products as a food additive, color additive, or is generally recognized as safe (GRAS) and is listed in Title 21 (assigned to the FDA) of the CFR in Parts 73, 74, 81, 172, 173, 179, 182, or 184;
(b) its use complies with FDA requirements; or
(c) the use of the substance will not cause the product in which it is used to be adulterated or misbranded or in some other way not in compliance with the requirements of the Meat Inspection Act, and the use of the substance is functional and suitable for the product, and it is used at the lowest level at which the stated effect can be accomplished.

When approval is given to a new substance or new use or new level of use, the FSIS publishes a final rule in the Federal Register which amends Section 318.7 or 381.147 and gives the new information.

There are special requirements covering the use of nitrite and sodium ascorbate or iscascorbate (erythrobate) in bacon to avoid the production of cancer-causing nitrosoamines. Sodium nitrite at an ingoing level of 120 parts per million (ppm) or potassium nitrite at an ingoing level of 148 ppm and sodium ascorbate or erythrobate at a level of 550 ppm must be used. The ascorbate level should be based on a molecular weight of 198 requiring an adjustment calculation to reach 550 ppm when hydrated forms are used. Lower levels of potassium nitrite are allowed with the same level of ascorbate if a plant has a USDA-approved "partial quality control program."

No substance may be used in any product or on any product that "conceals damage or inferiority or makes the product appear to be better or of greater value than it is."

Paprika or its oleoresin is not permitted in conjunction with fresh meat except in chorizo sausage or in other products for which there is a standard of identity. Sorbic acid, calcium sorbate, and sodium sorbate may not be used in any product and sulfurous acid and salts of sulfurous acid may not be used as well. Niacin or nicotinamide may not be used in or on fresh meat.

Preservatives and other substances, including borax, may be used in the preparation of products for export from the United States. However, there must be specifications used from the foreign purchaser, the product must not conflict with the laws of the country to which it will be exported, and the container must be labeled to show that the product is intended for export only. Special precautions are required to prevent the introduction of borax into the domestic meat supply.

No nonmeat ingredients or meat, product, meat food product, meat byproduct used as an ingredient of a meat product is allowed to contain any residue of pesticides, food additives, color additives, or other substances, which is more than that permitted by the FD&C Act.

Animal drug residues are allowed in meat and meat food products if they are from drugs that have been approved by the FDA unless the Administrator of the FSIS determines otherwise.

For new food products, the FSIS requires premarket approval of labeling and formulation, while the FDA does not.

The National Residue Program of the Food Safety and Inspection Service

The National Residue Program was started in 1967. At that time the primary concern was with pesticide contamination from the environment of animals used for food. Since that time the program has been expanded to include analyses of approximately 100 drug and chemical residues.

The importance of a residue program was emphasized in 1979 when polychlorinated biphenyl (PCBs) leaking from a transformer contaminated animal feed and resulted in contaminated poultry, beef, and eggs in 19 states. The FSIS, in order to protect the public and avoid losses to the food industry, at that time, set up a contamination response system. This system triggers a coordinated response among FSIS, EPA, FDA, and state government agencies to determine the source of contamination and its extent.

The present residue program is set up to allow FSIS to detect potentially harmful residues of pesticides, animal drugs, and biologics in meat and poultry products. The USDAs legal authority governs animals as they are presented for slaughter only. On-farm uses of pesticides and drugs are regulated by other agencies. The FDA is responsible for testing the safety of animal drugs, and the EPA for testing pesticides and toxic substances. These two agencies can prohibit the use of a substance which is a danger to human health. They also set the conditions of use and the maximum levels permissible in meat and poultry.

Samples are taken in slaughtering and processing plants by more than 7,500 FSIS inspectors. To utilize its resources best, the FSIS has developed a compound evaluation system for individual compounds based on both the hazard of the compound to human health and the likelihood of human exposure as a contaminant in meat or poultry consumed at a concentration that affects human health.

The compound evaluation system ranks compounds on a scale that ranges from A-1 (high health hazard/high probability of exposure of humans to a toxic concentration from meat or poultry) to D-4 (negligible health hazard/negligible probability of exposure of humans to a toxic concentration from meat or poultry). For both health hazard and probability of exposure, there is a category Z which indicates that insufficient information is available for an evaluation. Other categories of health hazard used are B (moderate health hazard) and C (low health hazard). Other categories used for probability of exposure are 2 (moderate probability of exposure) and 3 (low probability of exposure).

Table 1-1 gives the rankings for 86 compounds evaluated under this system as of January, 1988.

Some drugs, such as penicillin would be a special case requiring no exposure since once an individual is sensitized to the drug, a second exposure can lead to a hyperallergenic response, ranging from a simple rash to a life-threatening experience.

Table 1-1. List of Compounds Ranked Under Compound Evaluation System (CES)

Compound	Original Rank	2-Value Ranking	Year of CES Ranking
Aflatoxin	D	A-4	1985
Alachlor	D	A-2	1985
Albendazole	D	A-2	1987
Aldicarb	D	A-4	1986
Aldrin	A	A-3	1986
Ampicillin	A	B-2	1985
Ampicillin trihydrate	A	B-2	1985
Arsanllic acid	A	C-1	1987
Atrazine	A	C-3	1985
Azaperone	D	B-4	1986
Benomyl	D	B-3	1986
BHC	D	B-2	1987
Cadmium	D	B-4	1985
Captan	D	B-3	1987
Carbadox	B	A-3	1987
Carbarsone	A	C-2	1987
Carbaryl	D	B-2	1988
Carbofuran	C	C-3	1986
Carboxin	D	C-4	1987
Chloramphenicol	A	A-2	1985
Chloramphenicol palmitate	A	A-2	1985
Chlordane (technical)	A	A-2	1987
Chlorpyrifos	B	B-4	1986
Cloprostenol	—	B-4	1988
Coumaphos and oxygen analog	A	B-2	1988
2, 4, D (technical)	B	B-2	1987
Dalapon	A	A-3	1985
Daminozide	D	B-3	1985
Decoquinate	D	Z-4	1986
Dibutyltin dilaurate	D	A-1	1988
Dichlorvos	C	B-4	1987
O,O-diethyl S-[2-(ethylthio)ethyl] phosphorodithioate	D	A-2	1988
Dimethoate	D	B-3	1986
Dinoprost tromethamine	—	B-4	1988
Diphenylamine	D	B-4	1985
Endrin	A	A-3	1986
Ethylene dibromide	D	A-4	1986
Fenbendazole	B	B-3	1987
Fenthion	B	C-3	1985
Furazolidone	B	A-1	1987
Gentamicin sulfate	A	B-2	1986
Halofuginone	D	B-1	1988
Heptachlor and heptachlor expoxide	A	A-1	1987
Hexazinone	D	D-4	1985
Hygromycin B	C	A-Z	1988
Ipronidazole	B	Z-4	1986
Ipronidazole hydrochloride	B	Z-4	1986
Ivermectin	D	B-1	1986
Lead	D	B-4	1985
Levamisole	A	C-2	1985
Levamisole hydrochloride	A	C-2	1985

Table 1-1. (*Continued*)

Compound	Original Rank	2-Value Ranking	Year of CES Rank
Lindane	A	A-2	1986
Mebendazole	B	B-4	1986
Methoxychlor	A	D-4	1987
Methyl bromide	D	B-4	1986
Methylene chloride	D	A-2	1986
Monensin	A	B-3	1985
Naled	B	B-4	1987
Neomycin sulfate	A	B-3	1986
Paraquat	A	A-4	1986
PCB's	D	A-4	1985
Pentachlorophenol (PCP)	D	B-1	1985
Permethrin	D	B-2	1987
Propazine	A	C-4	1988
Prometryne	C	C-3	1985
Roxarsone	A	C-1	1987
Silvex	D	A-3	1986
Simazine	A	C-3	1988
Streptomycin	A	A-3	1986
Sulfamethazine	A	B-1	1985
Sulfaquinoxaline	A	B-1	1987
Sulfathiazole	A	B-1	1987
Thiabendazole	A	B-2	1987
2,4,5-T	D	A-3	1985
Tetracycline hydrochloride	A	B-3	1986
Toxaphene	A	A-2	1985
Trichlorfon	C	B-3	1985
Trifluralin	C	C-4	1986
Triphenyltin hydroxide	D	B-4	1986
Tylosin	A	Z-3	1986
Xylazine	D	Z-4	1986
Zeranol	B	C-2	1986
Zinc	D	D-4	1985

Presence of Sulfanamide and Sulfamethazine Residue in Swine and the Milk Supply

Approximately 94% of the sulfanamides used in the U.S. as animal drugs is sulfamethazine. This product is available for use in animal feed and drinking water as oblets, boluses, and as an injectable solution. Sulfamethazine, when used in this manner, is defined as a new animal drug [Section 201 (w) of the FD&C Act] and requires an approved New Animal Drug Application (NADA) by the FDA or permission under an interim marketing provision.

Sulfamethazene, because of its antibacterial activity, has been widely used in the swine industry to promote weight gain and feed efficiency and to control dysentary, pneumonia, abscesses, and atrophic rhinitis. In 1968, the FDA established a tolerance of 0.1 ppm for "negligible residues of sulfamethazine in the uncooked edible tissues of chickens, turkeys, cattle, and swine." The drug is required to be withdrawn 15 days prior to slaughter. However, the FSIS has found through the National Residue program over the past 10 years that 4% of the hogs slaughtered have exceeded this tolerance. The FSIS feels that 1% is an acceptable level. Some states considerably exceed the 4% figure.

Sulfamethazine is not permitted for treating lactating dairy cattle, but it is available over the counter. In a check of 49 samples of whole milk collected in 10 large cities in the U.S., the FDA found 36 positive samples which showed a range from 0.8 parts per billion (ppb) to 40.3 ppb sulfamethazine.

In 1988 the FDAs National Center for Toxicological Research (NCTR) completed a study with male and female mice that indicated some increase in thyroid folicular cell adenomas for both males and females, and an increase in hepatocellular adenomas or carcinomas in females but not male mice when fed levels of 300 to 4,800 ppm sulfamethazine.

One result of the situation with sulfamethazine was the holding of a hearing by the FDA on May 25 and 26, 1988, to consider whether sulfamethazine represented an ''imminent hazard to the public health'' and should be immediately withdrawn from the market.

The FSIS also increased its testing program for sulfamethazine residues in pork and developed a Sulfa-On-Site (SOS) test, not requiring a laboratory analysis, which could be carried out by the USDA inspector in the packing plant. The FSIS found that levels in swine liver and muscle tissue existed in an approximately 3:1 ratio.

Meanwhile, the FDA released the results of an additional study from the NCTR indicating an increase in thyroid follicular adenomas and adenocarcinomas when male and female rats were fed a diet containing 10 to 2,400 ppm sulfamethazine.

The data from both the rat and mouse study is being reviewed by a pathology working group as part of the National Toxicology Program. The FDA will not make a final decision until this review is completed. However, the FDA is proposing to revoke Section 510.450 of Part 510 of Table 21 of the CFR (New Animal Drugs) which gives permission for interim marketing of drugs containing sulfamethazine and other sulfanomides pending new drug approval and will publish notices for an opportunity for a hearing on denial of approval to NADAs for sulfamethazine. A lower tolerance could result.

Current Regulations of the Food and Drug Administration (FDA) Regarding Food Additives

The FDA is part of the U.S. Public Health Service which is part of the U.S. Department of Health and Human Services.

SUBSTANCES PROHIBITED FROM USE IN HUMAN FOOD:

The following substances cannot be added directly to human food or used as human food.

Calamus (the dried rhizome of *Acorus Calamus* L.) and its derivatives. Prohibited in 1968. Has carminative properties. Has been used as a flavoring compound as oil or extract. Calamus oil (Jimma variety) has been reported to induce malignant tumors in rats.

Cinnamyl anthranilate ($C_{16}H_{15}NO_2$), the ester of cinnamyl alcohol and anthranilic acid. Prohibited in 1985 because of liver cancers in male and female mice and kidney and pancreatic cancers in male rats. Flavoring agent.

The cobaltous salts, $CoC_4H_6O_4$, $CoCl_2$, and $CoSO_4$. Prohibited in 1966 because of toxic effects on the heart. Used as foam stabilizers in malt beverages.

Coumarin, 1,2 benzopyrone, tonka beans, tonka extract. Prohibited in 1954 because of toxicity to the liver. Used as vanilla flavor.

Cyclamate and its derivatives. This includes calcium, sodium, magnesium, and potassium salts of cyclohexane sulfamic acid, $(C_6H_{12}NO_2S)_2$ Ca, $(C_6H_{12}NO_2S)$Na, and $(C_6H_{12}NO_2S)$K. Prohibited in 1970 because of evidence of bladder cancer, birth defects, mutations and testicular damage in animals. Artificial sweetener. A metabolite, cyclohexylamine, has been associated with testicular atrophy and other effects in test animals.

Diethylpyrocarbonate, the chemical pyrocarbonic acid diethyl ester, $C_6H_{10}O_5$. Prohibited in 1972 because ammonia and diethylpyrocarbonate react to form urethan, a carcinogen which has been detected in orange juice, wine, and beer to which diethylpyrocarbonate was added as an antimicrobial.

Dulcin, the chemical 4-ethoxyphenylurea $C_9H_{12}N_2O_2$. Prohibited in 1950 because of liver cancer. Artificial sweetener.

Monochloroacetic acid, the chemical chloroacetic acid, $C_2H_3ClO_2$. Highly toxic. Prohibited in 1941. Used as a preservative. The substance is allowed in food packaging adhesives, however, with a migration level of up to 10 ppb. The FDA recognized that analytical methods do not detect monochloroacetic acid at the 10 ppb level. A list of substances permitted for use in adhesives as "components of articles intended for use in packaging, transporting, or holding food" is given in Title 21, Section 175.105 of the Code of Federal Regulations (CFR).

Nordihydroquaiaretic acid (NDGA), 4,4′-(2,3-dimethyltetramethylene) dipyrocatechol, $C_{18}H_{22}O_4$. Prohibited by the FDA in 1968 and by the USDA in 1971. However, it is allowed to be used as an antioxidant in packaging materials in contact with food (limit of addition to food, 0.005% by migration) under prior sanction. (In use prior to 1958.)

P-4000, the chemical 5-nitro-2-n-propoxyaniline, $C_9H_{12}N_2O_3$.

Safrole, 4 allyl-1,2-methylenedioxy-benzene, $C_{10}H_{10}O_2$, a natural component of the sassafras plant. Its derivatives, isosafrole and dihydrosafrole, as well as oil of sassafras and sassafras bark are also prohibited. Sassafras leaves and (camphor) tree are permitted as natural flavorings provided they are safrole free. Prohibited in 1960. A carcinogen.

Thiourea, thiocarbamate, CH_4N_2S. Prohibited in 1948. Goitrogenic. Chronic administration causes hepatic tumors in rats. Bone marrow depression has also been reported. Has been proposed as an antimycotic (mold inhibitor) as a dip for citrus fruits.

Chlorofluorocarbon propellants. The use of these compounds as propellants in self-pressurized containers is prohibited.

This list is not all inclusive since any substance to be used in food must meet all the applicable requirements of the FD&C Act.

SUBSTANCES NOT ALLOWED IN HUMAN FOOD

By Indirect Addition from Food Contact Surfaces

Flectol H. 1,2-dihydro-2,2,4-trimethylquinoline.

Mercaptoimidazoline, or 2-mercaptoimidazoline, both with formula $C_3H_6N_2S$.

4,4-Methylenebis (2-chloroaniline), $C_{13}H_{12}Cl_2N_2$.

If any "added or detectable" levels of these substances, either added directly or indirectly, are found in food, there is a violation of the FD&C Act.

COLOR ADDITIVES WHICH CANNOT BE USED IN FOOD AT THE PRESENT TIME

Technically the FDA does not classify a "color additive" as a "food additive."

FD&C Red No. 1 (Ponceau 3R). Delisted in 1960 because of liver damage in rats, mice, and dogs at 0.5% of the diet.

FD&C Red No. 4 (Ponceau SX). Deleted in 1965 because of damage to the urinary bladder and the adrenal cortex in dogs. However, Red 4 was allowed to be used until 1976 to color maraschino cherries at a level of 150 ppm in cherries or syrup, and in short term (six weeks) ingested drugs at 5 mg per day.

FD&C No. 2 (Amaranth). Deleted in 1976 because suspected of being a carcinogen when ingested by rats. Russian studies had also indicated embryotoxicity.

FD&C Violet No. 1. Deleted in 1973 because of suspected carcinogenicity in rats.

FD&C Green No. 1 (Guinea Green) and FD&C Green No. 2 (Light Green) were delisted in 1966 because of little economic importance.

Carbon Black. Delisted in 1975 because of carcinogenic contaminants created during its production. Still allowed in Canada.

Graphite. Delisted in 1977 because of the presence of polynuclear aromatic hydrocarbons, some of which are carcinogenic.

COLOR ADDITIVES THAT MAY BE ADDED TO FOODS PROVIDED THEY ARE FROM CERTIFIED LOTS OF COLOR

Certification means that the color is from a batch of color, samples of which have been submitted to the FDA to determine that they meet the chemical criteria spelled out in the CFR for that color additive. In general, synthetically produced color additives are required to be certified. Except where noted, the only limitations on the use of these colors are that they be used with suitable diluents listed in Section 73.1 of Title 21 of the CFR and that they be used in amounts "consistent with good manufacturing practice, shall meet specifications given, and may not be used to color foods for which there is a standard of identity unless allowed by the standard of identity."

FD&C Blue No. 1 Dye (Brilliant Blue FCF).

FD&C Blue No. 2 Dye (Indigotine).

FD&C Green No. 3 Dye (Fast Green FCF).

Orange B Dye—may only be used for coloring the casings or surfaces of frankfurters and sausages to the extent that the quantity does not exceed 150 ppm based on the weight of the finished food.

FD&C Red 3 Dye (Erythrosine).

FD&C Red 40 Dye (Allura Red AC) and Red 40 Lake. A lake is a dye fixed on aluminum hydroxide to make it dispersible in fat-containing foods.

Citrus Red No. 2 Dye. May only be used for coloring the skins of oranges. The oranges treated can only contain 2 ppm based on the weight of the whole fruit. The dye is to be used only for oranges that are not "intended or used for processing (or if so used are designated in the trade as "packinghouse elimination").

FD&C Yellow No. 5 Dye (Tartrazine). Foods, including butter, cheese, and ice cream, which contain this color additive must have FD&C Yellow No. 5 listed specifically in the ingredient statement. The reason is that Yellow 5 is believed to be the cause of allergies in some people, particularly those who are also sensitive to aspirin. Normally the presence of any color additives at all is not required to be indicated on the label of butter, cheese, or ice cream, while other foods require that "artificial coloring" be indicated in the ingredient statement without identifying the identity of the color additive present.

FD&C Yellow No. 6 Dye (Sunset Yellow FCF). The FDA had required that food containing this color be labeled specifically, as is the case for Yellow No. 5 (including butter, cheese, and ice cream) by January 1, 1989, because of allergic reactions. However, the Certified Color

Manufacturers Association filed a petition with the Federal Court of Appeals in Washington D.C., and the FDA agreed to withdraw their rule and submit a proposed regulation in the Federal Register allowing comments regarding labeling requirements.

Lakes. All of the above colors are water-soluble with the exception of Citrus Red 2, Orange B, and Red 40 Lake, which are oil soluble or dispersible. Each of the previously listed colors can be prepared in lake form and thus be dispersible in oil-containing foods. The lakes, with the exception of Red 40 lake, are all "provisionally" listed by the FDA, which means in effect that they are under scrutiny and can be either permanently listed or can be delisted as animal studies of their toxicity are completed. In the meantime they can be used in foods. Red 40 lake has permanent rather than provisional status, since it was not in use prior to the passage of the Color Additive Amendments in 1960.

Diluents used in conjunction with certified color additives are exempt from certification if they are GRAS or are specifically listed in the CFR Title 21, Section 73.1.

COLOR ADDITIVES THAT MAY BE ADDED TO FOODS WITHOUT BEING CERTIFIED

Color additives that are from natural rather than synthetic sources may be added to foods without any requirement that they be certified.

Among these substances are the following: annato extract; dehydrated beets (beet powder); caramel (solid or liquid with antifoaming agents that meet FDA specifications); carmine and cochineal extract from the insect *Dactylopius coccus costa* pasteurized to destroy *Salmonella* bacteria; toasted partially defatted cooked cottonseed flour; grape color extract from Concord grapes (may not be added to beverages and may not contain pesticide residues in excess of those permitted on grapes by Section 408 of the FD&C Act); grape skin extract; enocianins (may be used to color still and carbonated beverages, ades, beverage bases, and alcoholic beverages); fruit or vegetable juices meeting applicable standards of identity if there are any; hexane extracted carrot oil; paprika oleoresin; saffron; and turmeric.

None of the above color additives can be used in a food product unless it is specified in the Federal Standard of Identity for the product, or there is no Standard of Identity for that product.

COLORING SUBSTANCES THAT MAY BE ADDED TO ANIMAL FEED OR ANIMAL FOOD WITHOUT REQUIRING CERTIFICATION

Among these are the following: ultramarine blue (may be added to salt for animal feed if it does not exceed 0.5% of the weight of the salt); synthetic iron oxide (may only be used in dog and cat food—not to exceed 0.25% by weight of the finished food); dried algae meal with no more than 0.3% of the antioxidant ethoxyquin; Tagetes meal and extract (dried ground petals of the Aztec marigold with no more than 0.3% ethoxyquin); and corn endosperm oil.

Dried algae meal, Tagetes meal or extract, and corn endosperm oil are added to chicken feed to enhance the yellow color of chicken skin and eggs. As additives they must be labeled with a statement indicating the concentration of xanthophyl and instructions on how to obtain a feed containing the proper level of xanthophyls and associated carotenoids. For algae meal and Tagetes meal or extract, the feed must have a label stating, "Ethoxyquin, a preservative, added to retard oxidation and destruction of carotine, xanthophyl, and vitamins A and E." The phrase "a preservative" may be omitted if desired. The feed itself must contain no more than 150 ppm ethoxyquin.

SYNTHETIC COLOR ADDITIVES THAT MAY BE ADDED TO FOOD AND THAT ARE NOT REQUIRED TO BE CERTIFIED

Three carotenoid pigments, ferrous gluconate, and titanium dioxide are in this category. These compound may not be used in foods for which there is a Standard of Identity, unless the standard permits their addition.

Canthaxanthin (β-carotene-4,4′-dione) is identical to the compound found in nature. It has been permitted in food since 1969. Its color is in the yellow to red range of the spectrum (''tomato red''). It is restricted to 30 mg per lb of solid or semisolid food and 30 mg per pint of liquid food. Canthaxanthin may also be used in broiler chicken feed at a level of 4.41 mg per kg (4.0 g per ton) as a supplement to other sources of xanthophyl and associated carotenoids to enhance the yellow color of broiler chicken skin.

β-apo-8′-Carotenal (Apocarotenal) has vitamin A activity. It has been allowed in foods since 1963. Its color is in the orange to orange red range of the spectrum. It is restricted to 15 mg per lb of solid or semisolid food or pint of liquid food.

β-Carotene may be obtained synthetically or from natural sources and has vitamin A activity. It provides a predominantly yellow color and is often substituted for FD&C Yellow No. 5 because of labeling considerations. It is not corrosive to cans, as is FD&C Yellow No. 5 and FD&C Yellow No. 6. The amount used in food must be consistent with good manufacturing practice (GMP).

Ferrous gluconate may be used to color ripe olives in amounts consistent with GMP. The FDA accepts the specifications found in the third edition of *Food Chemicals Codex* as defining this compound in terms of quality and purity.

For many food and color additives, the FDA accepts the specifications given in the *Codex*. The *Codex* is published by the Committee on Food Chemicals of the Food and Nutrition Board, Commission on Life Sciences, National Research Council, National Academy of Sciences, Washington D.C. The *Codex* has published two supplements since its original publication in 1981 (first supplement in 1983, second in 1986). In addition to specifications for purity and methods of identification, the *Codex* gives instructions on packaging and storage for many of the substances listed. All of the substances listed are permitted in foods or in food processing by the FDA or in other countries where the *Food Chemicals Codex* specifications are recognized.

Titanium dioxide (TiO_2) may not exceed 1% by weight of the food in which it is used. TiO_2 may contain silicon dioxide and/or aluminum hydroxide as dispersing aids but not more than a total of 2% either singly or combined.

FLAVORING AGENTS ACCEPTABLE FOR USE IN FOODS

There are a number of synthetic flavoring substances and adjuvants that the FDA regards as ''generally recognized as safe for their intended use'' (GRAS) by listing them specifically in Section 182.60 of Title 21 of the CFR. In addition, the FDA recognizes that they cannot list all substances that are GRAS for their intended use in food.

> General recognition of safety may be based only on the views of experts qualified by scientific training and experience to evaluate the safety of substances that are GRAS. The basis of such views may be either: (1) scientific procedures, or (2) in the case of a substance used in food prior to January 1, 1958, through experience based on common use in food. General recognition of safety requires common knowledge about the substance throughout the scientific

community knowledgeable about the safety of substances directly or indirectly added to food. (CFR 21 Part 170.30)

GRAS requires that, if based on scientific procedures, the information and evidence available must be the same as that required for a food additive petition, including published studies corroborated by unpublished studies and other data or information.

In 1965, the Flavor and Extract Manufacturers Association of the U.S. (FEMA) published a list of 1,130 GRAS flavoring substances based on a survey of its members, data from the National Association of Chewing Gum Manufacturers and data from several large candy producers. Average maximum use levels were given in ppm for approximately 10 categories of use. An expert panel of pharmacologists and toxicologists reviewed and evaluated the data received. Discussions always had to be unanimous. The list has been maintained, and as additions and deletions were made, as of 1985 the list contained 1,750 substances. In addition to "experience in common use in food," scientific procedures used as criteria for GRAS decisions include chemical structure and analogies, metabolic rate and, when necessary, toxicological tests. Both natural and synthetic substances are included on the FEMA GRAS list. The FEMA, in the 1965 list, placed an asterisk next to items on an FDA "white list." Currently, the FDA lists in CFR 21, Section 172.515 approximately 800 synthetic flavoring substances and adjuvants that may safely be used in foods if they are used in accordance with good manufacturing practice in the minimum quantity required to produce the intended effect. These substances are not GRAS but are food additives.

FDA LISTS OF NATURAL AND EXTRACTS OF NATURAL SUBSTANCES THAT ARE GRAS AS FLAVORING AGENTS FOR FOODS

The FDA lists the following:

- Spices and other natural seasonings and flavorings that are GRAS for their intended use within the meaning of Section 409 of the FD&C Act (CFR 21, Section 182.10). It is a list of 83 items with botanical names of plant sources, ranging from alfalfa herb and seed and all spice through vanilla and zedoary bark.
- Essential oils, solvent free oleoresins and natural extractives, including distillates that are GRAS for their intended use within the meaning of Section 409 of the FD&C Act (CFR 21, Section 182.20). It is a list of 160 items with botanical names of plant sources, including items such as citrus peel, coffee, caraway, kola nut, pepper, onion, paprika, saffron, spearmint, tea, wild cherry bark, and turmeric.
- Other solvent free natural extractives that are used in conjunction with spices, seasonings, and flavorings and that are GRAS are the following:
  - Apricot kernel (persic oil)—*Prunus armeniaca* L.
  - Peach kernel (persic oil)—*Prunus persica Sieb, et Zucc*
  - Peanut stearine—*Arachis hypogaea* L.
  - Persic oil (see above)
  - Quince seed—*Cydonia oblonga Miller*
- Other spices, seasonings, essential oils, oleoresins, and natural extracts that are GRAS are as follows:
  - Ambergis from *Physeter macrocephalus* L.
  - Castoreum from *Castor fiber* L. and *C. canadensis Kuhl*

- Civet (zibeth, zibet, zibetum) from Civet cats, *Viverra civetta* Schreber, and *Viverra zibetha* Schreber.
- Cognac oil, white and green—Ethyl oenanthate, so-called
- Musk (Tonquin musk) from Musk deer, *Moschus moschiferus* L.

OTHER SUBSTANCES THAT THE FDA CONSIDERS GRAS WHEN USED IN ACCORDANCE WITH GOOD MANUFACTURING PRACTICE (GMP)

The CFR Title 21 in Part 182 lists a number of substances that are GRAS for their intended use. In addition to spices, oleoresins, essential oils, etc. and some indirect additives, a number of substances classified as either "multiple purpose GRAS food substances, anticaking agents, chemical preservatives, dietary supplements, emulsifying agents, sequesterants, stabilizers, or nutrients," are listed. A number of compounds appear on both the dietary supplement and nutrient list. The fact that a compound is listed does not mean that the FDA has found that the compound is "useful as a supplement to the diet of humans."

Using these substances in accordance with GMP entails that the quantity of a substance added to food does not exceed the amount required to accomplish the physical, nutritional or other effect intended or that the quantity of the substance that "becomes a component . . . as a result of its use in the manufacturing, processing, or packaging . . . and which is not intended to accomplish any physical or other technical effect in the food itself, shall be reduced to the extent reasonably possible."

In addition, the substance is to be food grade and be prepared and handled as a food ingredient. In general, no numerical level of use is specified.

Substances in Part 182 are subject to re-evaluation by the FDA and can be affirmed as GRAS and placed under Part 184 (direct food substances) or Part 186 (indirect food substances) of Title 21. They can also be regulated as food additives in Parts 170 through 180 of Title 21 or be banned from use in food (Part 189). Part 184 lists specific substances affirmed as GRAS. The direct food ingredients listed in that part have been reviewed by the FDA and found to be GRAS for the purposes and conditions described. In most cases, the ingredient specifications are as described in the 3d edition of the *Food Chemical Codex* and, in many cases, the levels of use in specific classes of food are given. In some cases, no limitations other than GMP are given but classes of food in which the ingredient may be used are specified. In all cases the purpose for which the ingredient is to be used is specified. Ingredients appearing in Part 184 are also GRAS as indirect human food ingredients subject to regulations governing individual indirect food additives (Parts 174–179.45).

Uses of an ingredient significantly deviating from the uses or conditions spelled out in Part 184 may require a manufacturer to submit a GRAS petition or may call for a food additive regulation.

For each of the substances in Part 184, a Chemical Abstracts Service Registry Number is given.

SUBSTANCES THAT THE FDA AFFIRMS AS GRAS WHEN INDIRECTLY ADDED TO FOOD IN ACCORDANCE WITH GOOD MANUFACTURING PRACTICE

The FDA recognizes some substances as GRAS as indirect food ingredients. These substances are not authorized for direct addition to a food. They are present in food through "migration from their immediate wrapper, container or other food contact surface.'' The substances or ingredients listed must be used in accordance with GMP, which includes the requirement that they be of purity suitable for their intended use, and be used at a level no higher than reasonably required to achieve the intended effect in the food-contact article.

If the FDA limits the ingredient as GRAS to one or more current GMP conditions of use—which might include category of food contact surface, technical effect, or functional use—and the ingredient is used under conditions which are significantly different from those described, a manufacturer must independently establish that his/her use is GRAS, and submit a GRAS petition to the FDA. Otherwise, regulation will be as a food additive.

The substances are not authorized to be added to food through extraction from a food contact surface or for purposes of deceiving the consumer.

The list of these substances is found in the CFR Title 21, Part 186.

FLUORINE COMPOUNDS, GLYCINE, AND NITRITES/NITRATES

The FDA has a special concern regarding the addition of fluorine compounds to foods and also regarding reports that adverse effects occurred when experimental animals were fed high levels of glycine. Fluorine is limited to the amount present in public water supplies or to the amount permitted in bottled water (1.4 mg/L fluoride at air temperature of 79.3°F to 90.5°F, more at lower temperatures). Previously, glycine was considered GRAS when used with GMP. Now, however, glycine and its salts have been removed from most food products. A food additive petition is required, supported by toxicity data to show that any proposed level of glycine or its salts added to food for human consumption will be safe. Glycine continues to be GRAS for use in animal feed since it is considered an essential nutrient in certain of these feeds.

There are also special regulations requiring that nitrites and nitrates combined in curing premixes with spices and other flavoring and seasoning ingredients that contain or are a source of secondary or tertiary amines, including essential oils, disodium inosinate, disodium guanylate, hydrolysates of animal or plant origin (e.g. hydrolyzed vegetable protein), spice oleoresins, soy products, and spice extractives have an additive petition filed on their behalf. This petition must be supported by data indicating that nitrosoamines are not formed: nitrosoamines are carcinogenic. Separately packaged nitrites and/or nitrates may be used if a warning is placed on the labels of both nitrite/nitrate and spice as flavoring premix warning not to combine until just prior to actual use.

FOOD ADDITIVES PERMITTED BY THE FDA FOR DIRECT ADDITION TO FOOD FOR HUMAN CONSUMPTION—CFR TITLE 21, PART 172

These substances, which are not GRAS, are permitted with specified conditions and levels of use. The substance used is ''food grade'' and ''is prepared and handled as a food ingredient.'' Nutrient substances permitted are not necessarily useful or required to supplement the diet of humans: GMP is to be followed. Categories included are the following: preservatives, including antioxidants, coatings, films and related substances; special dietary and nutritional additives; anticaking agents; flavoring agents; gums; chewing gum bases and related substances; specific use additives; and multipurpose additives.

SECONDARY DIRECT FOOD ADDITIVES PERMITTED BY THE FDA IN FOOD FOR HUMAN CONSUMPTION—CFR TITLE 21, PART 173

Substances in this category include the following: defoaming agents; boiler water additives; ion exchange resins used for water or food treatment; enzymes derived from microorganisms; polymer chemicals used as flocculents in clarifying cane or beet sugar juice and liquor; microbial cultures used for producing citric acid by fermentation; ion exchange membranes used to adjust the ratio of citric acid to total solids in grapefruit juice; polymers used as tableting adjuvants or for clarifying beer, wine, or vinegar; various solvents used for preparation of oleoresins from spices; chemicals used for delinting cottonseed; and combustion gas used for displacing oxygen in packaging beverages and other foods, with the exception of fresh meat. Chemicals used to control growth of micro-organisms in cane and beet sugar mills are also permitted, as well as fixing agents for the immobilization of enzymes used in preparing certain food products, such as high fructose corn syrup. Solvents, such as ethyl acetate, methylene chloride, and trichloroethylene are allowed for use in removing caffeine from green coffee beans.

In some cases maximum allowable residues of the additive in food are specified. In other cases GMP is specified, while in other cases where liquid, water, or food contact polymers or resins, minimum flow rates and maximum permitted temperatures may be specified.

FOOD ADDITIVES THAT THE FDA PERMITS IN FOOD OR IN CONTACT WITH FOOD ON AN INTERIM BASIS PENDING ADDITIONAL STUDY

The following substances are not GRAS:

- acrylonitrite copolymers, used in plastic beverage bottles;
- mannitol, a sugar alcohol;
- brominated vegetable oil, used in producing "cloud" or opacity in fruit juice drinks and soft drinks; and
- saccharin and its ammonium, calcium, and sodium salts. (Saccharin was removed from the GRAS list in 1971.)

The FDA considered banning saccharin use in food in 1977 because of the development of cancerous tumors in the bladders of rats exposed to the substance. However, Public Law 95–203 passed in November of that year postponed any ban for eighteen months, and Congress subsequently extended the ban several times until the present. Currently, the CFR Title 21, Section 180.37 (c) states that "authority for such use should expire when the Commissioner (of the FDA) receives the final reports on the ongoing studies in Canada and publishes an order on the safety of saccharin and its salts based on those reports and other available data."

Saccharin must meet the specifications established in the third edition of the *Food Chemical Codex*. Use is limited in beverages to 12 mg per fluid ounce; as a kitchen or table sweetener to 20 mg for each teaspoonful of sugar sweetening equivalency; and in processed foods in amounts not to exceed 30 mg per designated serving. Labels of beverages, cooking or table use products, and processed foods are required to show quantity information regarding saccharin content.

Brominated vegetable oil was removed from the GRAS list in 1970 because of data developed by the Canadian Food and Drug Directorate indicating damage to rat tissues such as heart, liver, kidney, testicles, and thyroid. It is presently permitted as a stabilizer for flavor oils in fruit flavored beverages at 15 ppm or less in the finished beverage. Continued use depends on the results of 6-month reports on toxicological studies on the safety of the additive.

Mannitol is a hexahydric alcohol 1,2,3,4,5,6 hexanehexol ($C_6H_{14}O_6$). The label and labeling of food "whose reasonably foreseeable consumption may result in a daily ingestion of 20 grams of mannitol shall bear the statement, "Excess consumption may have a laxative effect." Mannitol used must meet the specifications of the third edition of the *Food Chemicals Codex*. It is poorly digested by the body. It is allowed at a 98% level in pressed mints and 5% in hard candy and cough drops; 31% is allowed in chewing gum, 40% in soft candy, 8% in confections and frostings, 15% in commercial nonstandardized jams and jellies, and 2.5% in other foods. Continued use of mannitol depends on "timely and adequate progress reports" of appropriate feeding studies which are being carried out and "no indication of increased risk to public health during that period."

USE OF SULFITING AGENTS IN FOOD

Sulfur dioxide (SO_2), sodium sulfite (Na_2SO_3), sodium metabisulfite ($Na_2S_2O_5$), sodium bisulfite $NaHSO_3$), potassium metabisulfite ($K_2S_2O_5$), and potassium bisulfite ($KHSO_3$) are sulfiting agents which, until July, 1986, were classified as GRAS for use in food by the FDA when used with GMP.

As a result of several deaths resulting from the consumption of fresh fruits or vegetables which had been treated with sulfiting agents, the FDA withdrew their GRAS status for this use in 1986.

At present, sulfiting agents are also prohibited from use in meat or other foods that contain thiamine since this use results in destruction of the vitamin.

The problem is that some asthmatics react moderately or severely when exposed to sulfites. For this reason, the FDA also proposed banning the use of these agents on "fresh" potatoes in 1987.

Sulfites are used by the industry because they inhibit microbial action and inhibit undesirable browning in fresh fruits and vegetables. Their use is still considered GRAS in other foods for which no prohibition exists. In 1986, however, the FDA required that when a sulfiting agent is present in any food above a 10 ppm sulfite level, its presence must be declared on the label. The method of analysis to be used is a modified Monier Williams Procedure based on *Official Methods of Analysis*, 14th edition, published by the Association of Official Agricultural Chemists and found in the Appendix of 21 CFR, Part 101.

The FDA presently allows the term "sulfiting agents" to be used on the label rather than naming specific sulfiting agents used when they are present in a food in a significant amount (10 ppm or more), and they have no technical or functional effect. The FDA plans to amend the regulations in this regard.

It is possible that in the future the FDA may revoke the GRAS status for all uses of sulfiting agents in food.

The FDA has not prohibited the use of sulfiting agents on grapes since the EPA regulates this use as a fungicide on a raw agricultural commodity under the Federal Insecticide, Fungicide and Rodenticide Act. The EPA is reviewing the safety of using sulfites on grapes and has developed regulations regarding this subject.

The FDA has concluded that at a 10 ppm sulfite level, a person would have to eat 300 grams of a food to consume 3 mg of sulfur dioxide. The Food Research Institute reported to the FDA that an ingested dose of 5 mg of potassium metabisulfite in a capsule, equivalent to 3 mg of sulfur dioxide, resulted in an adverse reaction. This is the lowest dose to date that has resulted in an adverse reaction.

USE OF ASPARTAME IN FOOD

Aspartame, at present, is not considered GRAS. Aspartame, 1 methyl N-L-a-aspartyl-L-phenylalanine ($C_{14}H_{18}N_2O_5$) may be used in food unless precluded from use by a standard of identity.

It may be used as a sweetener in dry free flowing sugar substitutes for table use, in units not to exceed the sweetening equivalent of two teaspoons of sugar, but not for cooking. It may also be used in tablets for sweetening coffee and tea, in cold breakfast cereals, in chewing gum, and in dry bases for beverages, instant coffee, gelatins, puddings and fillings, dairy analog toppings, carbonated beverages, and syrup bases, chewable multivitamin food supplements, and in noncarbonated refrigerated single-strength or frozen concentrates of fruit-juice based drinks, fruit flavored drinks and ades, or imitation fruit flavored drinks and ades. It may not be used as a sweetener in frozen stick type confections and novelties, breath mints, and ready to serve, liquid concentrate, or dry base tea. Aspartame may also be used as a flavor enhancer in chewing gum.

Any intermediate mix containing aspartame for manufacturing purposes has to have a statement on the principal display panel giving the concentration in the mix. The label of any food containing aspartame must state the following on the principal display panel or the information panel: "PHENYLKETONURICS: CONTAINS PNENYLALANINE."

When aspartame is used as a sugar substitute for table use, the label must include instructions not to use in cooking or baking.

IONIZING RADIATION FOR FOOD PROCESSING

The process of irradiation of food is considered a food additive. Previous to 1986 when the FDA allowed several major uses of irradiation for the treatment of food, food law had, in general, discouraged it. In fact, after the Chernobyl reactor explosion in Russia, the FSIS set up a program for screening groups of canned hams imported into this country for radioactivity.

At present, the main barrier to acceptance of irradiated foods appears to be consumer acceptance.

The FDA alllows gamma rays from sealed units of cobalt-60 or cesium-137, electrons generated by machine not to exceed 10 million electron volts, and X-rays not to exceed 5 million electron volts as sources of ionizing radiation for treating food. Permitted uses are as follows:

Use	Limitations
For control of *Trichinella spiralis* in pork carcasses or fresh, nonheat-processed cuts of pork carcasses.	Minimum dose 0.3 kGy (30 krad); maximum dose not to exceed 1 kGy (100 krad).
For growth and maturation inhibition of fresh foods.	Not to exceed 1 kGy (100 krad).
For disinfestation of arthropod pests in food.	Do.
For microbial disinfection of dry or dehydrated enzyme preparations (including immobilized enzymes).	Not to exceed 10 kGy (1 Mrad).
For microbial disinfection of the following dry or dehydrated aromatic vegetable substances: culinary herbs, seeds, spices, teas, vegetable seasonings, and blends of these aromatic vegetable substances. Turmeric and paprika may also be irradiated when they are to be used as color additives.	Not to exceed 30 kGy (3 Mrad).
The blends may contain sodium chloride and minor amounts of food ingredients ordinarily used in such blends.	

Foods that are irradiated are required to be marked with the following logo on retail packages along with the statement "Treated with radiation" or "Treated by irradiation":

The FSIS will also require that all meat and poultry products that contain irradiated pork bear this symbol and information while the FDA does not require foods that include irradiated ingredients to include that information.

If a food is irradiated and shipped to another manufacturer for further processing, the label and labeling of the food and the invoices or bills of lading used must contain the statement, "Treated with radiation—do not irradiate again," or the statement, "Treated by irradiation—do not irradiate again."

BIBLIOGRAPHY

Aurand, Leonard W., Edwin A. Wood and Marion R. Wells. *Food Composition and Analysis.* New York: Van Nostrand Reinhold, 1987.

Columbia University Press. "Harvey Washington Wiley." *Columbia Encyclopedia,* p. 1900. New York: Columbia University Press, 1944.

Engel, R. E., A. R. Post, and R. C. Post. "Implementation of Irradiation of Pork for Trichina Control." *Food Technology* (1988). Vol. 42, No. 7, p. 71.

Food and Nutrition Board, National Research Council. *Food Chemicals Codex.* 3d ed. Washington, D.C.: National Academy Press, 1981.

________. First Supplement to the 3d ed. *Food Chemicals Codex.* Washington, D.C.: National Academy Press, 1983.

________. Second Supplement to the 3d ed. *Food Chemicals Codex.* Washington, D.C.: National Academy Press, 1986.

Furia, Thomas E. *Hankbook of Food Additives.* 2d ed. Vol. I. Cleveland, OH: CRC Press, 1972.

________. *Handbook of Food Additives.* 2d. ed. Vol. II. Boca Raton, FL: CRC Press, 1980.

________. *Regulatory Status of Direct Food Additives.* Boca Raton, FL: CRC Press, 1980.

Hall, Richard L. and Bernard L. Oser. "GRAS Substances: Recent Progress in the Consideration of Flavoring Ingredients Under the Food Additives Amendment." No. 3. *Food Technology* (February 1965). Vol. 19, No. 2 (Part 2), p. 151.

________. "GRAS Substances." No. 4. *Food Technology* (1970). Vol. 24, No. 5, pp. 25–28, 30–32, and 34.

Hoffman-LaRoche Inc. "Carotenoids, Beta Carotene, Apocarotenal, Canthaxanthin." Nutley, NJ: Food Business Unit, Roche Chemical Division, Hoffmann-LaRoche Inc., 1986.

Indiana State Board of Health, Indiana Dept. of Education. *Nutrition Health Education Resource Guide.* Indianapolis, IN, 1986.

Jacobson, Michael F. *Eaters Digest and Nutrition Scoreboard.* Garden City, NY: Anchor Press/Doubleday, 1985

National Center for Toxicological Research. "Chronic Toxicity and Carcinogenicity Studies of Sulfamethazine in $B6C3F_1$ Mice." NCTR Technical Report for Experiment No. 418. Jefferson, AR: National Center for Toxicological Research, 1988.

Oser, Bernard L., and Richard A. Ford. "GRAS Substances: Recent Progress in the Consideration of Flavoring Ingredients Under the Food Additives Amendment." No. 6. *Food Technology* (1973). Vol. 27, No. 1, p. 64.

________. "GRAS Substances." No. 7. *Food Technology* (1973). Vol. 27, No. 11, p. 56.

________. "GRAS Substances." No. 8. *Food Technology* (1974). Vol. 28, No. 9, p. 76.

________. "GRAS Substances." No. 9. *Food Technology* (1975). Vol. 29, No. 8, p. 70.

________. "GRAS Substances." No. 10. *Food Technology* (1977). Vol. 31, No. 1, p. 65.

________. "GRAS Substances." No. 11. *Food Technology* (1978). Vol. 32, No. 2, p. 60.

________. "GRAS Substances." No. 12. *Food Technology* (1979). Vol. 33, No. 7, p. 65.

Oser, Bernard L., Richard A. Ford and Bruce K. Bernard. "GRAS Substances: Recent Progress in the Consideration of Flavoring Ingredients Under the Food Additives Amendment." No. 13. *Food Technology* (1984). Vol. 38, No. 10, p. 66.

Oser, Bernard L., and Richard L. Hall. "GRAS Substances: Recent Progress in the Consideration of Flavoring Ingredients under the Food Additives Amendment." No. 5. *Food Technology* (1972). Vol. 26, No. 5, p. 35.

Oser, Bernard L., et al. "GRAS Substances: Recent Progress in the Consideration of Flavoring Ingredients Under the Food Additives Amendment." No. 14. *Food Technology* (1985), Vol. 39, No. 11, p. 108.

Public Law 61–152. 36 Stat. "Insecticide Act of 1910." April 26, 1910.

Public Law 80–104. 61 Stat. "Federal Insecticide, Fungicide and Rodenticide Act of 1947." June 25, 1947.

Public Law 83–518. 68 Stat. The Miller Amendment." July 22, 1954.

Public Law 85–172. 71 Stat. "Poultry Products Inspection Act." August 28, 1957.

Public Law 85–929. 72 Stat. "Food Additives Amendment of 1958." September 6, 1958.

Public Law 86–618. 74 Stat. "Color Additive Amendments of 1960." July 12, 1960.

Public Law 89–755. 80 Stat. "Fair Packaging and Labeling Act." November 3, 1966.

Public Law 90–201. 81 Stat. "Wholesome Meat Act." December 15, 1967.

Public Law 90–492. 82 Stat. "Wholesome Poultry Products Act." August 18, 1968.

Public Law 94–469. 90 Stat. "Toxic Substances Control Act." October 11, 1976.

Public Law 96–359. 94 Stat. "The Infant Formula Act of 1980." September 26, 1980.

Roberts, Howard R. *Food Safety*. New York: John Wiley & Sons, 1981.

Sachs, Shearin. "Discussion of Irradiation Logo and Irradiated Pork." *USDA News* p. 7. (May 1986).

Taylor, R. J. *Food Additives*. New York: John Wiley & Sons, 1980.

U.S. Dept. of Health and Human Services. "Federal Food, Drug and Cosmetic Act, as Amended, and Related Laws." HHS Publication No. 86–1051, 1985.

U.S. Dept. of Health and Human Services. "Laws Enforced by the U.S. Food and Drug Administration." HHS Publication No. 80–1051, 1981.

U.S. Dept. of Health and Human Services. "Requirements of Laws and Regulations Enforced by the U.S. Food and Drug Administration." HHS Publication No. 85–1115, 1984.

U.S. Environmental Protection Agency. "Protection of Environment." *Code of Federal Regulations* (1 July 1988). Title 40.

U.S. Environmental Protection Agency. "Tolerances for Pesticides in Food and Animal Feeds; Transfer of Regulations." *Federal Register* (29 June 1988). Vol. 53, No. 125, p. 24666.

U.S. Food and Drug Administration. "Aspartame, Food Chemicals Codex." *Federal Register* (21 April 1988). Vol. 53, No. 77, p. 13134.

U.S. Food and Drug Administration. "Food and Drugs." *Code of Federal Regulations* (1 April 1988). Title 21.

U.S. Food and Drug Administration. "Food Labeling; Declaration of Sulfiting Agents." *Federal Register* (9 July 1986). Vol. 51, No. 131, p. 25012.

U.S. Food and Drug Administration. "Policies Concerning Uses of Sulfiting Agents." *Federal Register* (24 October 1988). Vol. 53, No. 205, p. 41922.

U.S. Food and Drug Administration. "Sulfamethazine; Availability of National Center for Toxicological Research's Technical Report." *Federal Register* (18 May 1988). Vol. 53, No. 96, p. 17850.

U.S. Food and Drug Administration. ''Sulfamethazine in Food-Producing Animals; Public Hearing Before the Commissioner (of the FDA).'' *Federal Register* (4 May 1988). Vol. 53, No. 86, p. 15886.

U.S. Food and Drug Administration. ''Sulfiting Agents; Revocation of GRAS Status for Use on Fruits and Vegetables Intended to be Served or Sold Raw to Consumers.'' *Federal Register* (9 July 1986). Vol. 51, No. 131, p. 25021.

U.S. Food Safety and Inspection Service. ''Animals and Animal Products.'' *Code of Federal Regulations* (1 January 1988). Title 9.

U.S. Food Safety and Inspection Service. ''Changes in Testing Procedures for Sulfamethazine.'' *Federal Register* (21 April 1988). Vol. 53, No. 77, p. 13137.

U.S. Food Safety and Inspection Service. ''Compound Evaluation System.'' Washington, D.C.: Residue Evaluation and Planning Division, Science Program, U.S. Food Safety and Inspection Service, 1988.

U.S. Food Safety and Inspection Service. ''Implementation of the Sulfa-on-Site Test for Sulfamethazine in Swine.'' *Federal Register* (21 April 1988). Vol. 53, No. 77, p. 13136.

U.S. Food Safety and Inspection Service. ''FSIS Facts: Food Additives.'' FSIS Publication No. FSIS-16, 1982.

U.S. Food Safety and Inspection Service. ''Meat and Poultry Inspection.'' FSIS Publication No. FSIS-33, 1987.

Warner-Jenkinson Co. ''Color Status Insert'' in *Red Seal Report* (53: Summer 1988). Warner-Jenkinson Co., St. Louis, Mo.

Warrick, Sheridan. ''The Drug-Blenders' Fatal Blunder.'' *Hippocrates* (September/October 1988).

Chapter 2

INDIRECT FOOD ADDITIVES

Seymour G. Gilbert, Ph.D.
Professor Emeritus of Food Science
Cook College
Rutgers University

HISTORY

The existence of toxicity from contamination of food supply is as old as life itself, but the importance of indirect food additives to food from accidental or deliberate adulteration in processing or distribution has become part of our increasing industrialized civilization. This importance has been exponentially increased with the accelerated reliance on stored and packaged foods needed to provide a constant supply of varied and plentiful food. In contrast the food supply of the nonindustrialized nations has been at the risk of famine from current crop failures of natural or man-made causes, such as war and pestilence in the absence of substantial amounts of stored packaged food.

Concern for the safety of packaged foods, however, has grown as the complexity of form and materials has increased. While other sources of contamination of biological origin are no less important, this review will not deal with them. The major developments in U.S. legislation are those of the last 50 years, beginning with the 1938 Food, Cosmetic Act and subsequent the 1958 Amendment which became operational in 1960. The latter legislation created two new principles: 1) the manufacturer of the food was responsible for proving safety prior to sale; and 2) that known or deliberate introduction of materials to food, and that the regulations also applied to indirect or incidental additives as well as to known or deliberate addition of materials to food. These are substances not normally expected to be present in the food as part of the ingredient formulation so that they exist as either contaminants of the as manufactured food or become so as a result of migration from the environment to the food. Since packaging is the mode of protection from such environmental contamination, the possible migration of substances of adverse nature from the package itself has been of considerable attention. Thus the 1958 Amendment resulted in a proliferation of regulations which in essence defined food-grade or acceptable packaging forms and materials for formulating and packaging foods.

While recorded incidents of actual harm from packaging are largely outside the U.S., the first and still major body of legislation regulating packaging is that of the U.S. Code of Federal Regulations (CFR) based on the previous stated laws along with some state generated ones, notably that of the Pennsylvania State Department of Agriculture. Two agencies in the U.S. dominate the scene—the Department of Agriculture, which was the original source and, its offshoot, the Food and Drug Administration, (FDA) which became the dominant regulatory agency.

The major concern of the pharmaceutical industry and pertinent Federal Regulations is the stability of initially effective products in prolonged storage up to 3 years. The two aspects of this concern are the reaction rate of deterioration from inherent and environmental factors and the migration of material into or out of the product from the polymeric containers.

The first aspect—reaction rate—is influenced by the effectiveness of the package as a barrier to environmental hazards, ranging from components of the gas phase (water vapor and oxygen) to liquids; and to biologic hazards, such as microbial invasion and recent human tampering.

The last factor—tampering—has caused extensive changes in package design requiring re-evaluation of even long established OTC pharmaceutical products.

Storage life in relation to package is determined by the following three main determinants:

(1) The mechanisms whereby environmental hazards affect the product's stability. These are primarily the reaction rates as affected by the concentration of those hazards as reactants. Equation 1 is a simplified Arrehenius statement of such rates (Gilbert 1985).

$$\frac{dQ}{dt} = k(C_c)^n A\, e^{-H/RT} \tag{1}$$

(2) The rate of entrance of these hazards into the packaged product is usually measured by the permeability constant of the package wall and its physical parameters determined by package design, such as wall thickness, material composition, surface area, amount of product, leakage rate including effects of normal expected abuse in shipping, handling and use in ultimate consumption on retention of barrier properties of the package.

(3) The actual environmental condition under expected use conditions.

Since the permeability factor and reaction rate have time parameters, the overall equation for shelf-life is

$$t_c = \frac{1}{P_p} \times \frac{1}{E_p} \times \frac{1}{R_p} \quad \text{or} \quad t_c = P_k \frac{1}{A} \tag{2}$$

This equation required two determinations—the intrinsic stability relative to potential hazard concentration and the barrier properties of selected package design.

The critical factors can be determined in "no package" tests using hermetic containers, such as sealed ampules (Johnson, 1988) or a new methodology based on Inverse Gas Chromatography (IGC) (Gilbert, 1984) for reaction rates.

When water uptake or loss is a concern, the sorption isotherms become important, particularly with solid dose forms. In such cases the modified IGC method offers speed (a few hours instead of weeks), and the ability to measure kinetic effects related to structure and composition of adjuvants, such as humectants and dispersents (Greenway and Gilbert 1989).

MIGRATION

The basic mechanism of indirect addition of possibly harmful ingredient addition to food from packaging involves the mass transfer from package wall to contained food. This process involves the thermodynamics and kinetics of mass transfer. The first defines the potential for the differential movement, with the second providing the time for a given amount of additive to transfer. The potential for migration is related to the solubility of the migrant in the two phases; food and packaging material. The time span is related to the diffusivity and the concentration difference in these two phases at any time. While the rate of migration is essentially parabolic over the period from first contact to equilibrium between the two phases (Hayakawa 1974), the practically important rate is the linear or steady state phase. Food law regulations, however, are usually based on the final partition concentration in the food for a toxicologically significant migrant as the maximum risk.

The partition coefficient *Kp*, the respective mass or volume of the two phases, and the concentration of the migrant at equilibrium or prolonged contact time are related by the Equation 3 (Gilbert 1979).

$$Ms = Mp\ Wp\ /\ (KpWp + Ws) \tag{3}$$

where

Ms = migrant concentration in food (w/w)
Mp = initial migrant concentration i polymer
Kp = partition coefficient of polymer/food
Wp = weight of polymer (p)
Ws = weight of food

This equation assumes time independence, but foods generally are sold and used prior to an expiratory date usually based on the anticipated useful shelf-life. For drugs in the U.S., this period is three years while foods may range from weeks to a year or two. This means that the diffusivity may be critical for any slow migration. The Unilever group (Figge and Rudolph, 1979) have classified migrants into three groups according to diffusivity. The influence of initial concentration and state of the potential migrant can also be critical for even carcinogens where polymeric bonding forces produce highly restricted nonlinear or concentration dependent desorption. In such cases, the normal Fickian diffusivity must be modified by a factor to recognize the immobilizing factor at very low initial concentrations of the indirect additive. An important case is that of vinyl chloride monomer in the polyvinyl chloride polymer matrix used in direct food contact. It has been shown that in the ppb range of residual monomer the risk of carcinogenesis from its migration is essentially zero (Gilbert, 1979). The modified diffusivity relation is given as Equation 4.

$$t^{1/2} = \frac{.05L}{D}\ (1 + KpR) \tag{4}$$

Where $t^{1/12}$ is the time for half of the initial migrant concentration to transfer, *L* is the polymer wall thickness, *Kp* the partition coefficient, and *R* is the ratio of mass of polymer to food. It is obvious that time is infinite as diffusivity *D* becomes zero. *R* has significance when the mass of food is larger than that of the package wall. For concentration dependent partition as in the Freundlich transport model, the term *Kp* is not constant but refers to the slope of the desorption isotherm of the conditions at $t^{1/2}$. Unlike the Henry or Langmuir isotherm, which have a constant *Kp* relation, polar polymers, such as PVC, PET, PAN, nylon have heterogenity of sorption sites so that *Kp* is constant only at infinite dilution. Thus the assumption of constancy is not of practical importance in migration involving finite concentrations. *Kp* is actually a thermodynamic parameter related to fugacity or free energy with both entropic and enthalpic terms. Thus both site binding energy and matrix structure can be important factors in the summation of energy terms.

These are some of the considerations used in negating the ''absolute zero'' interpretation of the Delaney Clause of the 1958 Amendment in the vinyl chloride and acrylonitrile monomer migration controversy over interpretation of FDA regulations in the U.S. Foreign food additive regulations, on the other hand, have always used a finite lower acceptable migrant level in the contacting food. (Gilbert 1979a).

COMPATABILITY

In product development of new pharmaceuticals, as well as new packages, the compatibility of the product with packaging material is an important evaluation that requires studies on the identification and extent of migration of components of the packaging material into the drug product capable of reacting with the drug formulation and, conversely, the loss of drug or other essential components of the pharmaceutical product into the package. (Varsano & Gilbert, 1973).

Transfer from packaging materials to drug contents generally does not involve major macromolecular components, such as the polymer itself, which is intrinsically odorless, tasteless, and nontoxic, but is concerned with minor constituents that can and do affect the quality of the contained product by sensory or toxicological hazards. The more complex the contents, the more difficult is the quantification of actual migration during the period of time from packaging to consumer use. This period is defined as the shelf-life and is finite for practically all products and rarely exceeds three years for drugs.

The formulations used in manufacturing plastic materials vary with the type of product for which they are used with most of those formulations not divulged by manufacturers. Various quantities of unreacted components and some of the ingredients may be converted to different chemical compounds during processing. Additives used by manufacturers may frequently contain minor or major amounts of other compounds apart from the main active component as contaminants.

Thus the prudent drug manufacturer will establish control procedures that supplement and verify the master files given to the FDA as part of the documentation of a New Drug Application (NDA). The following are methods and examples of their use for such quality assurance.

METHODOLOGY

The classical procedure for estimation of migration uses sorption data obtained at equilibrium with prolonged contact between product and package. This concept assumes that concentration dependance does not prevail even at very low initial migrant levels in the polymers in contact with product—a case of great importance with trace carcinogens. Thus the existence of such traces and their potential for migration becomes the dominant question in safety for packaged prodcuts.

These two facets—presence of migrants and their partition—are the key questions to be answered by experimental methods. The first has been addressed in a new and more rigid test procedure. Recent studies showed that product deterioration from migration can be efficiently assessed by using a prolonged Soxhlet extraction (Miskevich 1981, Kim and Gilbert 1989). The rationale is that partition or solvency can be defined by an exhaustive extraction with an array of solvents of differing polarity used with relatively large amounts of the contacting package structure. If a migrant is found at this stage at a significant level, the identification and quantitative analysis method can be applied to the actual conditions of use by analysis of a product itself.

Although migrants could be depleted to various extents from polymeric containers to the contacting phase during long term storage, depending on partition coefficients and phase ratios, Soxhlet extraction with organic solvents can be used to predict the maximum attainable migrants during long-term storage as a worst case or maximum risk from type and length of contact.

Test For Migration Potential

The Soxhlet extracts are made with about 200 grams of polymer to be in direct contact with the drug preparation. Six solvents ranging in polarity from water to heptane are used for at least a 48 hour period in 200 mL amounts. The extracts are then concentrated by distillation followed by nitrogen flushing to a final volume of 2 ML. One mL of concentrate is placed in a 5 mL screw cap vial and the solvent evaporated to the dryness in the water bath kept at 40° C using nitrogen flushing. Volatile derivates are then formed for analysis by gas chromatography and mass spectrometry.

EXAMPLES

Gaskets are one of the basic components of aerosol valves. In the presence of a high concentration of organic propellant, such as freons, the migration potential from contact polymeric matrix, such as rubber to the product can be significant in taste and odor changes or chemical interactions as well as in toxicology. Extraction of large samples with appropriate solvents was made to produce enough concentrated material for a comprehensive analysis of potential migrants in trace quantities.

Sample Preparation

Approximately 4 grams of aerosol gasket components were extracted with methylene chloride for 48 hours using a Soxhlet apparatus. The total extract was concentrated to 5 mL using a Snyder distilling column. The apparatus was covered with aluminum foil to protect it from the light. A control was prepared using an empty thimble without sample. Each gasket was also stored in a mixture of Freon 11 and Freon 12 at room temperature for 4 months, respectively.

The extracts were subjected to gas chromatography (GC) equipped with a flame ionization detector to obtain a chromatographic profile of each sample. The instrument used was HP-5710A (Hewlett Packard, CA) and the column used was 1.8 m × 3.2 mm stainless-steel column packed with 5% OV—101 on 80/100 mesh supelcoport (Supelco Inc., Bellefonte, PA). The flow rates of helium, hydrogen, and air were 30 mL/min, 30 mL/min, and 240 mL/min, respectively. The temperature of the injection port and detector were kept at 250° C and 350° C. The integrator used was HP-3390AC, and the sensitivity of $10^{-11} \times 32$ AMU (Atomic Mass Unit) was used.

Standard solutions were prepared covering the desired range for the potential migrants identified by GC/MS. The calibration curve of each compound is obtained by subjecting each standard solution to the Selected Ion Monitor (SIM) mode GC/MS at the same condition used for the analysis of the sample.

Extracts were also directly subjected to GC/MS with a GC/MS (Finnegan Mat, Model 8230, Germany) with the same column as above and helium used as a carrier gas at a flow rate of 1 mL/min. The injection port temperature was kept at 280° C. Column temperature was programmed from 50° C for 10 min to 280° C at the rate of 10° C/min for extract S and A and from 100° C for 4 min to 280° C at the rate of 10° C/min for extract A and D. Scan speed is 0.8 sec/ decade. Mass spectra of each peak obtained were compared to that of the published data for tentative identification, and the majority of them were confirmed by matching with those of authentic compounds. For confirmation and quantitative analysis the SIM mode of the GC/MS (HP 5990A) was used with a megabore column of fused silica 15 m × 0.53 mm with bonded phase DB-1, 1 μm thickness (J & W Scientific, Inc., Rancho Cordova, CA). Helium flow rate was 5 mL/min and column temperature program and the injection port temperature were the same as above.

The freon extract obtained from each gasket was separated by gas chromatograph (Varian 3700), equipped with Flame Ionization Detector. The same conditions as used for the GC/MS analyses were used. Diethyl terephthalate was used as the internal standard to quantify GC responses. By using GC/MS, a large number of compounds were tentatively identified from rubber gaskets of aerosol valves. Each type of gasket showed a different profile of potential migrants.

The characteristics of migrants of differently formulated sealing gaskets made by different manufacturers were compared. Different types of gaskets showed a considerably different profile of potential migrants. Major potential migrant in most of the gaskets were antioxidants, indicating that the choice of antioxidant should be carefully made by manufacturers of rubber articles. The identification of several antioxidants in one type of gasket indicated that the tested rubber articles are composed of several elastomeric polymers.

The presence of several compounds, such as triphenyl phosphine oxide and sulfur containing compounds showed that chemical additives for rubber article formulation can be converted to different chemical compounds during vulcanization so that direct identifications of fabricated articles are required in addition to formulation data. Similar considerations apply to other food and drug contacting polymeric materials. The second factor—partition coefficient—is usually tested in the conventional FDA method. This consists of a prolonged period of contact between product and package (three years for drugs). Analysis of the stored product provides unequivocal partition data but is too lengthy for development of new drugs or their packages. Changes in package design and/or materials can be disastrous for marketing with this test. The FDA has recognized the principle of equivalent or simulated solvent testing with an accelerated contact procedure to provide rapid assessment of migration potential. If the migrant in question is volatile, an inverse gas chromatographic method has been developed to provide rapid and highly sensitive partition data. (Varsano & Gilbert 1973).

SUMMARY

Numerous studies have been made on the migration of indirect additives from polymeric packaging materials and packaging articles to a contact phase. These studies have been reviewed in terms of polymer type, specific migrant, nature of the contact phase, level of migrant transferred, and analytical method. Only a few studies have been published on the comprehensive analysis of indirect additives migrated from polymeric packages. (Downes, 1987).

The most recent studies show that possible product contamination from packaging can be efficiently assessed. The rationale is that partition or solvency can be defined by an exhaustive extraction with an array of solvents of differing polarity used with relatively large amounts of the contacting package structure. A prolonged Soxhlet extraction with one-half ratio of resin to solvent will provide for so large an exaggeration of equilibrium partition that there can be no question that any potential migrant will be undetected in toxicologically significant amounts. The extract itself can be exhaustively examined by the variety of current analytical procedures useful for trace analysis including HPLC, GC/MS, MS, GC, UV, TLC, IR, NMR, AA, and radio-analytical techniques, depending on the specific migrants. If a migrant is found in the extract at a significant level in any solvent, the relation of this partition to the actual situation can be examined by the simulated solvent extraction methods of the FDA and ASTM to provide equilibrium data. If further questions arise as to the validity of the FDA test conditions, the identification and quantitative analysis method can be applied to the actual conditions of use by analysis of product itself.

Where toxicological studies do not provide valid data on the allowable threshold exposure of deliberate additives to a package wall, the use of this method will give data which will still allow for the estimation of the degree of risk involved. In a particular packaging system with an incidental presence of a compound of known or suspected carcinogenicity, these methods would provide a justifiable lower limit of detection for the presence of such compounds. Thus, a clear, unequivocal, estimation of the possible migrants and their quantitative transfer can be obtained for the packaging of pharmaceutical products. The author's hope that these studies lead to the development of a scientifically justified methodology.

REFERENCES

1. Gilbert, S. G., 1985. Food/package compatibility. *J. Food Technol.* 39:54–56.
2. Johnson, J. B., 1988. The Effect of the State of Water on the Kinetics of the Degradation of Retinoic Acid. Ph. D. Thesis, New Brunswick, NJ: Rutgers University.
3. Gilbert, S. G., 1984 *Inverse Gas Chromatography: Advances in Chromatography.* Marcel Dekker. Vol. 23, pp. 200–228.
4. Greenway, G. W. and Gilbert, S. G., 1989. A Rapid Method for Determining the Sorption Isotherms of Moisture Sensitive Foods. Submitted to *J. Food Sci.*
5. Hayakawa, K-I, 1974. Predicting an equilibrium state value from transient data. *J. Food Sci.* 39:272–275.
6. Gilbert, S. G., 1979a. Modeling the migration of indirect additives to food. *Food Tech.*, April, pp. 63–65.
7. Figge, K. and Rudolph, F., 1979. Diffusion in the plastics packaging/contents systems. *Angewewandte Makromolekulare Chemie* 78 1169:1570.
8. Gilbert, S. G., 1979b. "A scientific basis for regulation of indirect food additives from packaging materials." *J. Food Quality* 2:251–256.
9. Gilbert, S. G., 1980a. Scientific aspects of recent changes in packaging regulations in the United States." Third International Symposium on Migration, pp. 22–24 (October 1980). Hamburg, Germany.
10. Varsano, J. and Gilbert, S. G. 1969. "Pharmaceuticals in plastic packaging, drug and cosmetic industry." Vol. 104, (1) 148–149 (Part I), Jan.
11. Miskevich, R. D., 1981. Development of a system for the determination of migration in food packaging-application to components of a PET polymer used in beverage packaging. M.Sc. Thesis, New Brunswick, NJ: Rutgers University.
12. Kim, H. and Gilbert, S. G., 1987. Determination of potential migrants from commercial polyethylene terephthalate (PET) Bottle Wall. *Frontiers of Flavor,* Charolombous, G. (ed.) Amsterdam, Netherlands: Elsevier Science Publishers.
13. Varsano, J. L. and Gilbert S. G., 1973. Evaluation of interactions between polymers and low molecular weight compounds by gas chromatography. Part I. Methodology and Interaction Evaluation. Part II. Thermodynamics and elucidation. *Pharmaceutical Sciences,* Vol. 62, (1) January 96–98.
14. Giacin, J. R. and Bzozowska, A. *J. Plastic Film & Sheeting,* Vol. 1, 292, (1985) also 1987 in *Food Product-Package Compatability.* Proc. Michigan State Packaging Conf.: (eds.) Gray, J. I., B. R. Harte, and J. Miltz, Technomics Publishing Co. Lancaster, PA.
15. Downes, T. W., 1987. Practical and Theoretical Considerations in Migration, in: *Food Product-Package Compatability.* Proc. Michigan State Packaging Conf.: (eds.) Gray, J. I., B. R. Harte, and J. Miltz, Technomics Publishing Co. Lancaster, PA.

Section II
FOOD ADDITIVE CHEMICALS

Key to Abbreviations

abs – absolute
ACGIH – American Conference of Governmental Industrial Hygienists
alc – alcohol
alk – alkaline
amorph – amorphous
anhy – anhydrous
approx – approximately
aq – aqueous
atm – atmosphere
autoign – autoignition
aw – atomic weight
af – atomic formula
BATF – U.S. Bureau of Alcohol, Tobacco, and Firearms
bp – boiling point
b range – boiling range
CAS – Chemical Abstracts Service
cc – cubic centimeter
CC – closed cup
CL – ceiling concentration
COC – Cleveland open cup
conc – concentrated
compd(s) – compound(s)
conc – concentration, concentrated
contg – containing
cryst, crys – crystal(s), crystalline
d – density
D – day(s)
decomp, dec – decomposition
deliq – deliquescent
dil – dilute
DOT – U.S. Department of Transportation
EPA – U.S. Environmental Protection Agency
eth – ether
(F) – Fahrenheit
FCC – Food Chemical Codex
FDA – U.S. Food and Drug Administration
flash p – flash point
flam – flammable
fp – freezing point
g, gm – gram
glac – glacial
gran – granular, granules
GRAS – generally regarded as safe
hygr – hygroscopic
H, hr – hour(s)
htd – heated
htg – heating
IARC – International Agency for Research on Cancer
incomp – incompatible
insol – insoluble
IU – International Unit
kg – kilogram (one thousand grams)
L,l – liter
lel – lower explosive level
liq – liquid
M – minute(s)
m^3 – cubic meter
mg – milligram
misc – miscible
μ, u – micron
mL, ml – milliliter
mg – milligrams
mm – millimeter
mod – moderately
mp – melting point
mppcf – million particles per cubic foot
mw – molecular weight
mf – molecular formula
mumem – mucous membrane
mw – molecular weight
NIOSH – National Institute for Occupational Safety and Health
ng – nanogram
nonflam – nonflammable
NTP – National Toxicology Program
OC – open cup
org – organic
OSHA – Occupational Safety and Health Administration
PEL – permissible exposure level
petr – petroleum
pg – picogram (one trillionth of a gram)
Pk – peak concentration
pmole – picomole

powd – powder
ppb – parts per billion (v/v)
pph – parts per hundred (v/v)(percent)
ppm – parts per million (v/v)
ppt – parts per trillion (v/v)
prep – preparation
PROP – properties
refr – refractive
rhomb – rhombic
S,sec – second(s)
sl, slt, sltly – slightly
sol – soluble
soln – solution
solv(s) – solvent(s)
spont – spontaneous(ly)
subl – sublimes
TCC – Tag closed cup
tech – technical
temp – temperature
TLV – Threshold Limit Value
TOC – Tag open cup
TWA – time weighted average
U, unk – unknown, unreported
μ, u – micron
uel – upper explosive limits
μg, ug – microgram
ULC, ulc – Underwriters Laboratory Classification
USDA – U.S. Department of Agriculture
vac – vacuum
vap – vapor
vap d – vapor density
vap press – vapor pressure
vol – volume
visc – viscosity
vsol – very soluble
W – week(s)
Y – year(s)
% – percent(age)
> – greater than
< – less than
<= – equal to or less than
=> – equal to or greater than
° – degrees of temperature in Celsius (Centigrade)
°(F),°F – temperature in Fahrenheit

A

AAF900 CAS: 55589-62-3
ACESULFAME POTASSIUM
mf: $C_4H_4KNO_4S$ mw: 201.24

PROP: White crystalline solid; odorless with sweet taste. Mp: 250°. Very sol in water, DMF, DMSO; sol in alc. About 200 times sweeter than sucrose.

SYNS: ACESULFAME K ◇ POTASSIUM ACESULFAME ◇ POTASSIUM 6-METHYL-1,2,3-OXATHIAZINE-4(3H)-1,2,2-DIOXIDE ◇ SUNETTE

USE IN FOOD:

Purpose: Nonnutritive sweetener.

Where Used: Beverage (dry base), chewing gum, coffee (instant), dairy product analogs (dry base), gelatins (dry base), pudding desserts (dry base), puddings (dry base), tabletop sweetener, tea (instant).

Regulations: FDA - 21CFR 172.800. Use at a level not in excess of the amount reasonably required to accomplish the intended effect.

SAFETY PROFILE: When heated to decomposition emits toxic fumes of SO_x.

AAG250 CAS: 75-07-0
ACETALDEHYDE
DOT: 1089
mf: C_2H_4O mw: 44.06

PROP: Colorless, fuming liquid; pungent, fruity odor. Mp: −123.5°, bp: 20.8°, lel: 4.0%, uel: 57%, flash p: −36°F (CC), d: 0.804 @ 0°/20°, autoign temp: 347°F, vap d: 1.52. Misc in water, alc and ether.

SYNS: ACETALDEHYD (GERMAN) ◇ ACETIC ALDEHYDE ◇ ALDEHYDE ACETIQUE (FRENCH) ◇ ALDEIDE ACETICA (ITALIAN) ◇ ETHANAL ◇ ETHYL ALDEHYDE ◇ FEMA No. 2003 ◇ NCI-C56326 ◇ OCTOWY ALDEHYD (POLISH)

USE IN FOOD:

Purpose: Flavoring agent.

Where Used: Various.

Regulations: FDA - 21CFR 182.60. GRAS when used at a level not in excess of the amount reasonably required to accomplish the intended effect.

On Community Right-To-Know List. EPA Genetic Toxicology Program. IARC Cancer Review: Animal Sufficient Evidence IMEMDT 36,101,85; Human Inadequate Evidence IMEMDT 36,101,85

OSHA PEL: TWA 100; STEL 150 ppm
ACGIH TLV: TWA 100 ppm; STEL 150 ppm
DFG MAK: 50 ppm (90 mg/m^3) DOT Classification: Flammable Liquid

SAFETY PROFILE: Poison by intratracheal and intravenous routes. A human systemic irritant by inhalation. A narcotic. Human mutagenic data. An experimental teratogen. An experimental tumorigen by inhalation. Other experimental reproductive effects. A skin and severe eye irritant. A common air contaminant. Highly flammable liquid. Mixtures of 30-60 percent of the vapor in air ignite above 100°. It can react violently with acid anhydrides; alcohols; ketones; phenols; NH_3; HCN; H_2S; halogens; phosphorus; isocyanates; strong alkalies; and amines. Reactions with cobalt chloride; mercury(II) chlorate; or mercury(II) perchlorate form sensitive, explosive products. Polymerizes violently in the presence of traces of metals or acids. Reaction with oxygen may lead to detonation. When heated to decomposition it emits acrid smoke and fumes.

TOXICITY DATA and CODEN

eye-hmn 50 ppm/15M JIHTAB 28,262,46
skn-rbt 500 mg open MLD UCDS** 12/13/63
eye-rbt 40 mg SEV UCDS** 12/13/63
mma-sat 10 μL/plate EVHPAZ 21,79,77
sce-hmn:ipr 500 μg/kg MUREAV 88,389,81
ipr-rat TDLo:400 mg/kg (8-15D preg):TER SEIJBO 23,13,83
ipr-mus TDLo:640 μg/kg (10D preg):TER TJADAB 27,231,83
ihl-rat TCLo:1410 ppm/6H/65W-I:ETA TXCYAC 31,123,84
ihl-ham TCLo:2040 ppm/7H/52W-I:ETA EJCODS 18,13,82
ihl-hmn TCLo:134 ppm/30M:PUL JAMAAP 165,1908,57
orl-rat LD50:1930 mg/kg AMIHBC 4,119,51
ihl-rat LDLo:4000 ppm/4H AMIHBC 10,61,54

AAT250 CAS: 64-19-7
ACETIC ACID
DOT: 2789/2790
mf: $C_2H_4O_2$ mw: 60.06

PROP: Clear, colorless liquid; pungent odor. Mp: 16.7°, bp: 118.1°, flash p: 109°F (CC), lel: 5.4%, uel: 16.0% @ 212°F, d: 1.049 @ 20°/4°, autoign temp: 869°F, vap press: 11.4 mm @ 20°, vap d: 2.07. Misc water, alc, and ether.

SYNS: ACETIC ACID (AQUEOUS SOLUTION) (DOT) ◇ ACETIC ACID, GLACIAL (DOT) ◇ ACIDE ACETIQUE (FRENCH) ◇ ACIDO ACETICO (ITALIAN) ◇ AZIJNZUUR (DUTCH) ◇ ESSIGSAEURE (GERMAN) ◇ ETHANOIC ACID ◇ ETHYLIC ACID ◇ FEMA No. 2006 ◇ GLACIAL ACETIC ACID ◇ METHANECARBOXYLIC ACID ◇ OCTOWY KWAS (POLISH) ◇ VINEGAR ACID

USE IN FOOD:

Purpose: Acidifier, boiler water additive, flavor enhancer, flavoring agent, pH control agent, pickling agent, solvent.

Where Used: Baked goods, catsup, cheese, chewing gum, condiments, dairy products, fats, fats (rendered), gravies, mayonnaise, meat products, oils, pickles, relishes, salad dressings, sauces.

Regulations: FDA - 21CFR 182.1005. GRAS with a limitation of 0.26 percent in baked goods, 0.8 percent in cheese and dairy products; 0.5 percent in chewing gum; 9.0 percent in condiments and relishes; 0.5 percent in fats and oils; 0.3 percent in gravies and sauces; 0.6 percent in meat products, and 0.15 percent in all other food categories when used in accordance with good manufacturing practice. USDA - 9CFR 318.7. Sufficient for purpose.

OSHA PEL: TWA 10 ppm ACGIH TLV: TWA 10 ppm; STEL 15 ppm DFG MAK: 10 ppm (25 mg/m^3) DOT Classification: Label: Corrosive, Flammable Liquid

SAFETY PROFILE: A human poison by an unspecified route. Moderately toxic by various routes. A severe eye and skin irritant. Caustic; can cause burns, lachrymation, and conjunctivitis. Human systemic effects by ingestion: changes in the esophagus, ulceration or bleeding from the small and large intestines. Human systemic irritant effects and mucous membrane irritant. Experimental reproductive effects. Mutagenic data. A common air contaminant. A combustible liquid. Moderate fire and explosion hazard when exposed to heat or flame; can react vigorously with oxidizing materials. To fight fire, use CO_2, dry chemical, alcohol foam, foam and mist. When heated to decomposition it emits irritating fumes.

Potentially explosive reaction with 5-azidotetrazole; bromine pentafluoride; chromium trioxide; hydrogen peroxide; potassium permanganate; sodium peroxide; and phosphorus trichloride. Potentially violent reactions with acetaldehyde and acetic anhydride. Ignites on contact with potassium-tert-butoxide. Incompatible with chromic acid; nitric acid; 2-amino-ethanol; NH_4NO_3; ClF_3; chlorosulfonic acid; (O_3 + diallyl methyl carbinol); ethylenediamine; ethylene imine; (HNO_3 + acetone); oleum; $HClO_4$; permanganates; $P(OCN)_3$; KOH; NaOH; n-xylene.

TOXICITY DATA and CODEN

skn-hmn 50 mg/24H MLD TXAPA9 31,481,75
skn-rbt 50 mg/24H MLD TXAPA9 31,481,75
eye-rbt 50 μg open SEV AMIHBC 4,119,51
mmo-esc 300 ppm/3H AMNTA4 85,119,51
cyt-grl-par 40 μmol/L NULSAK 9,119,66
orl-hmn TDLo:1470 μg/kg:GIT AIHAAP 33,624,72
ihl-hmn TCLo:816 ppm/3M:NOSE,EYE,PUL AMIHAB 21,28,60
orl-rat TDLo:700 mg/kg (18D post):REP NTOTDY 4,105,82
itt-rat TDLo:400 mg/kg (1D male):REP FESTAS 24,884,73
orl-rat LD50:3310 mg/kg JIHTAB 23,78,41
ihl-rat LCLo:16000 ppm/4H AIHQA5 17,129,56
orl-mus LD50:4960 mg/kg JIHTAB 23,78,41

AAU000 CAS: 150-84-5
ACETIC ACID, CITRONELLYL ESTER
mf: $C_{12}H_{22}O_2$ mw: 198.34

PROP: Found in oils of Citronella Ceylon, Geranium, and about 20 other oils (FCTXAV 11,1011,73). Colorless liquid; fruity odor. D: 0.883-0.893, refr index: 1.440-1.450, flash p: +212°F. Sol in alc, fixed oils; ins in glycerin, propylene glycol, water @229°.

SYNS: ACETIC ACID-3,7-DIMETHYL-6-OCTEN-1-YL ESTER ◇ CITRONELLYL ACETATE (FCC) ◇ 2,6-DIMETHYL-2-OCTEN-8-OL ACETATE ◇ 3,7-DIMETHYL-6-OCTEN-1-YL ACETATE ◇ FEMA No. 2311

USE IN FOOD:

Purpose: Flavoring agent.

Where Used: Mayonnaise, salad dressings, sauces.

Regulations: FDA - 21CFR, 172.515, 182.60. GRAS when used at a level not in excess of the amount reasonably required to accomplish the intended effect.

SAFETY PROFILE: Mildly toxic by ingestion. A human skin irritant. Combustible liquid. When heated to decomposition it emits acrid smoke and irritating fumes.

TOXICITY DATA and CODEN

skn-hmn 20 mg/48H MLD FCTXAV 11,1011,73
skn-rbt 500 mg/24H FCTXAV 11,1011,73
orl-rat LD50:6800 mg/kg FCTXAV 11,1011,73

AAV000 CAS: 108-21-4
ACETIC ACID ISOPROPYL ESTER
DOT: 1220
mf: $C_5H_{10}O_2$ mw: 102.15

PROP: Colorless, aromatic liquid. Mp: −73°, bp: 88.4°, lel: 1.8%, uel: 7.8%, fp: −69.3°, flash p: 40°F, d: 0.874 @ 20°/20°, autoign temp: 860°F, vap press: 40 mm @ 17.0°, d: 3.52. Sltly sol in water; misc in alc, ether, fixed oils.

SYNS: ACETATE d'ISOPROPYLE (FRENCH) ◇ ACETIC ACID-1-METHYLETHYL ESTER (9CI) ◇ 2-ACETOXYPROPANE ◇ FEMA No. 2926 ◇ ISOPROPILE (ACETATO di) (ITALIAN) ◇ ISOPROPYLACETAAT (DUTCH) ◇ ISOPROPYLACETAT (GERMAN) ◇ ISOPROPYL ACETATE (ACGIH,DOT,FCC) ◇ ISOPROPYL (ACETATE d') (FRENCH) ◇ 2-PROPYL ACETATE

USE IN FOOD:

Purpose: Flavoring agent.

Where Used: Various.

Regulations: FDA - 21CFR 172.515. Use at a level not in excess of the amount reasonably required to accomplish the intended effect.

OSHA PEL: TWA 250 ppm; STEL 310 ppm
ACGIH TLV: TWA 250 ppm; STEL 310 ppm
DOT Classification: Flammable Liquid; Label: Flammable Liquid

SAFETY PROFILE: Moderately toxic by ingestion. Mildly toxic by inhalation. Human systemic irritant effects by inhalation and systemic eye effects by an unspecified route. Narcotic in high concentration. Chronic exposure can cause liver damage. Highly flammable liquid. Dangerous fire hazard when exposed to heat, flame or oxidizers. Moderately explosive when exposed to heat or flame. Dangerous; keep away from heat and open flame; can react vigorously with oxidizing materials. To fight fire, use foam, CO_2, dry chemical.

TOXICITY DATA and CODEN

eye-hmn 200 ppm/15M JIHTAB 28,262,46
ihl-hmn TCLo:200 ppm:IRR AMIHAB 21,28,60
unk-hmn TCLo:200 ppm:EYE JIHTAB 28,262,46
orl-rat LD50:3000 mg/kg 14CYAT 2,1879,63
ihl-rat LCLo:32000 ppm/4H AMIHBC 10,61,54

AAX250 CAS: 9003-20-7
ACETIC ACID VINYL ESTER POLYMERS
mf: $(C_4H_6O_2)_n$

PROP: Clear, water-white solid resin. Sol in benzene, acetone; ins in water.

SYNS: ACETIC ACID ETHENYL ESTER HOMOPOLYMER ◇ ASAHISOL 1527 ◇ ASB 516 ◇ AYAA ◇ AYAF ◇ BAKELITE AYAA ◇ BAKELITE LP 90 ◇ BOND CH 18 ◇ BOOKSAVER ◇ BORDEN 2123 ◇ BASCOREZ ◇ CEVIAN A 678 ◇ D 50 ◇ DANFIRM ◇ DARATAK ◇ DCA 70 ◇ DUVILAX BD 20 ◇ ELMER'S GLUE ALL ◇ EP 1463 ◇ FORMVAR 1285 ◇ GELVA CSV 16 ◇ GOHSENYL E 50 Y ◇ KURARE OM 100 ◇ LEMAC 1000 ◇ MERCKOGEN 6000 ◇ MOVINYL 114 ◇ NATIONAL 120-1207 ◇ POLYVINYL ACETATE (FCC) ◇ PROTEX (POLYMER) ◇ RHODOPAS M ◇ SOVIOL ◇ SP 60 ESTER ◇ TOABOND 40H ◇ UCAR 130 ◇ VA 0112 ◇ VINAC B 7 ◇ VINYL ACETATE HOMOPOLYMER ◇ VINYL ACETATE POLYMER ◇ VINYL ACETATE RESIN ◇ VINYL PRODUCTS R 10688 ◇ WINACET D

USE IN FOOD:

Purpose: Color diluent, manufacture of paper and paperboard, masticatory substance in chewing gum base.

Where Used: Chewing gum, confectionery, food supplements in tablet form, packaging materials.

Regulations: FDA - 21CFR 73.1, 181.30. Use in accordance with good manufacturing practice.

IARC Cancer Review: Animal Inadequate Evidence IMEMDT 19,341,79.

SAFETY PROFILE: When heated to decomposition it emits acrid smoke and irritating fumes. A commercial plastic resin used in adhesives, paints, paper coatings, textile treatment and nonwoven binders.

AAX750 CAS: 93-29-8
ACETISOEUGENOL
mf: $C_{12}H_{14}O_3$ mw: 206.26

PROP: White crystals; clove odor. Flash p: 153°F. Sol in alc, chloroform, ether; ins in water.

SYNS: 4-ACETOXY-3-METHOXY-1-PROPENYLBENZENE ◇ ACETYLISOEUGENOL ◇ FEMA No. 2470 ◇ ISOEUGENOL ACETATE ◇ ISOEUGENYL ACETATE (FCC) ◇ 2-METHOXY-4-PROPENYLPHENYL ACETATE

USE IN FOOD:

Purpose: Flavoring agent.

Where Used: Various.

Regulations: FDA - 21CFR 172.515. Use at a level not in excess of the amount reasonably required to accomplish the intended effect.

SAFETY PROFILE: Moderately toxic by ingestion. Combustible liquid. When heated to decomposition it emits acrid smoke and irritating fumes.

TOXICITY DATA and CODEN

orl-rat LD50:3450 mg/kg FCTXAV 13,681,75

ABB500 CAS: 513-86-0
ACETOIN
DOT: 2621
mf: $C_4H_8O_2$ mw: 88.12

PROP: Slightly yellow liquid or crystalline solid; buttery odor. D: 1.016, bp: 147-148°, refr index: 1.417, mp: 15°, flash p: 106°F. Misc with water, alc, propylene glycol; ins vegetable oil.

SYNS: ACETYL METHYL CARBINOL ◇ 2-BUTANOL-3-ONE ◇ DIMETHYLKETOL ◇ FEMA No. 2008 ◇ 3-HYDROXY-2-BUTANONE ◇ 1-HYDROXYETHYL METHYL KETONE ◇ γ-HYDROXY-β-OXOBUTANE

USE IN FOOD:

Purpose: Flavoring agent.

Where Used: Various.

Regulations: FDA - 21CFR 182.60. GRAS when used at a level not in excess of the amount reasonably required to accomplish the intended effect.

DOT Classification: IMO: Flammable Liquid; Label: Flammable Liquid

SAFETY PROFILE: Experimental reproductive effects. Mildly toxic by subcutaneous route. A moderate skin irritant. Flammable liquid. When heated to decomposition it emits acrid smoke and fumes.

TOXICITY DATA and CODEN

orl-rat TDLo:12600 mg/kg (42D male):REP FCTXAV 10,131,72
skn-rbt 500 mg/24H MOD CNREA8 33,3069,73
scu-rat LDLo:14 g/kg FCTXAV 17,509,79

ABC500 CAS: 93-08-3
2′-ACETONAPHTHONE
mf: $C_{12}H_{10}O$ mw: 170.22

PROP: White crystalline solid; orange blossom odor. Flash p: 264°F. Sol in fixed oils; sltly sol in propylene glycol; ins in glycerin.

SYNS: ORANGE CRYSTALS ◇ β-ACETONAPHTHALENE ◇ β-ACETYLNAPHTHALENE ◇ 2-ACETYLNAPHTHALENE ◇ ACETONAPHTHONE ◇ β-ACETONAPHTHONE ◇ 2-ACETONAPHTHONE ◇ FEMA No. 2723 ◇ METHYL-β-NAPHTHYL KETONE (FCC) ◇ METHYL-2-NAPHTHYL KETONE ◇ β-METHYL NAPHTHYL KETONE ◇ 1-(2-NAPHTHALENYL)ETHANONE ◇ β-NAPHTHYL METHYL KETONE ◇ 2-NAPHTHYL METHYL KETONE

USE IN FOOD:

Purpose: Flavoring agent.

Where Used: Various.

Regulations: FDA - 21CFR 172.515. Use at a level not in excess of the amount reasonably required to accomplish the intended effect.

SAFETY PROFILE: Moderately toxic by ingestion. A human skin irritant. Combustible liquid. When heated to decomposition it emits acrid smoke and fumes.

TOXICITY DATA and CODEN

skn-hmn 500 mg/24H FCTXAV 13,681,75
orl-mus LD50:599 mg/kg MDZEAK 8,244,67

ABC750 CAS: 67-64-1
ACETONE
DOT: 1090
mf: C_3H_6O mw: 58.09

PROP: Colorless liquid; fragrant mint-like odor. Mp: −94.6°, bp: 56.48°, refr index: 1.356, flash p: 0°F (CC), lel: 2.6%, uel: 12.8%, d: 0.7972 @ 15°, autoign temp: (color) 869°F, vap press: 400 mm @ 39.5°, vap d: 2.00. Misc in water, alc, and ether.

SYNS: ACETON (GERMAN, DUTCH, POLISH) ◇ DIMETHYLFORMALDEHYDE ◇ DIMETHYLKETAL ◇ DIMETHYL KETONE ◇ FEMA No. 3326 ◇ KETONE PROPANE ◇ β-KETOPROPANE ◇ METHYL KETONE ◇ PROPANONE ◇ 2-PROPANONE ◇ PYROACETIC ACID ◇ PYROACETIC ETHER ◇ RCRA WASTE NUMBER U002

USE IN FOOD:

Purpose: Color diluent, extraction solvent.

Where Used: Fruits, spice oleoresins, vegetables.

Regulations: FDA - 21CFR 73.1. No residue. 21CFR 173.210. Limitation of 30 ppm in spice oleoresins.

On Community Right-To-Know List.

OSHA PEL: TWA 250 ppm; STEL 1000 ppm ACGIH TLV: TWA 750 ppm; STEL 1000 ppm DFG MAK: 1000 ppm (2400 mg/m^3) DOT Classification: Flammable Liquid; Label: Flammable Liquid NIOSH REL: TWA 590 mg/m^3

SAFETY PROFILE: Moderately toxic by various routes. A skin and severe eye irritant. Human systemic effects by inhalation: changes in EEG, changes in carbohydrate metabolism, nasal effects, conjunctiva irritation, respiratory system effects, nausea and vomiting, and muscle weakness. Human systemic effects by ingestion: coma, kidney damage, and metabolic changes. Narcotic in high concentration. In industry, no injurious effects have been reported other than skin irritation resulting from its defatting action, or headache from prolonged inhalation. A common air contaminant. Highly flammable liquid. Dangerous disaster hazard due to fire and explosion hazard; can react vigorously with oxidizing materials.

Potentially explosive reaction with nitric acid + sulfuric acid; bromine trifluoride; nitrosyl chloride + platinum; nitrosyl perchlorate; chromyl chloride; thiotrithiazyl perchlorate; and 2,4,6-trichloro-1,3,5-triazine + water. Reacts to form explosive peroxide products with 2-methyl-1,3-butadiene; hydrogen peroxide; and peroxomonosulfuric acid. Ignites on contact with activated carbon; chromium trioxide; dioxygen difluoride + carbon dioxide; and potassium-tert-butoxide. Reacts violently with bromoform; chloroform + alkalies; bromine; and sulfur dichloride. Incompatible with CrO; (nitric + acetic acid); NOCl; nitryl perchlorate; permonosulfuric acid; NaOBr; (sulfuric acid + potassium dichromate); (thio-diglycol + hydrogen peroxide); trichloromelamine; air; HNO_3; chloroform; and H_2SO_4. To fight fire, use CO_2, dry chemical, alcohol foam.

TOXICITY DATA and CODEN

skn-rbt 500 mg/24H MLD 28ZPAK -,42,72
skn-rbt 395 mg open MLD UCDS** 5/7/70
eye-hmn 500 ppm JIHTAB 25,282,43
eye-rbt 3950 μg SEV AJOPAA 29,1363,46
sln-smc 47600 ppm ANYAA9 407,186,83
cyt-ham:fbr 40 g/L FCTOD7 22,623,84
ihl-mam TCLo:31500 μg/m^3/24H (1-13D preg):REP GTPZAB 26(6),24,82
orl-man TDLo:2857 mg/kg 34ZIAG -,64,69
ihl-man TDLo:440 μg/m^3/6M GISAAA 42(8)42,77
ihl-hmn TCLo:500 ppm:EYE JIHTAB 25,282,43
ihl-man TCLo:12000 ppm/4H:CNS AOHYA3 16,73,73
ihl-man TDLo:10 mg/m^3/6H GISAAA 42(8)42,77
orl-mus LD50:3000 mg/kg PCJOAU 14,162,80
ihl-rat LCLo:16000 ppm/4H AIHAAP 23,95,62

ABE000
ACETONE PEROXIDE

PROP: Liquid or absorbed on cornstarch. The trimeric form is crystalline. Mp: 97°.

USE IN FOOD:

Purpose: Bleaching agent, dough conditioner, maturing agent.

Where Used: Bread, flour, rolls.

Regulations: FDA - 21CFR 172.802. Use at a level not in excess of the amount reasonably required to accomplish the intended effect.

SAFETY PROFILE: Severe skin and eye irritant. Flammable by spontaneous chemical reaction; can react vigorously with reducing materials. The trimeric form is shock-sensitive and static-electricity-sensitive and may detonate.

ABH000 CAS: 98-86-2
ACETOPHENONE

mf: C_8H_8O mw: 120.16

PROP: Colorless liquid or plates; sweet, pungent odor. Mp: 19.7°, bp: 202.3°, flash p: 180°F (OC), d: 1.026 @ 20°/4°, vap d: 4.14, vap press: 1 mm @ 15°, autoign temp: 1060°F. Very sol in propylene glycol, fixed oils; sol in alc, chloroform, ether; sltly sol in water; ins in glycerin.

SYNS: ACETYLBENZENE ◇ BENZOYL METHIDE ◇ DYMEX ◇ FEMA No. 2009 ◇ HYPNONE ◇ KETONE

METHYL PHENYL ◇ METHYL PHENYL KETONE ◇ 1-PHENYLETHANONE ◇ PHENYL METHYL KETONE ◇ USAF EK-496

USE IN FOOD:

Purpose: Flavoring agent.

Where Used: Various.

Regulations: FDA - 21CFR 172.515. Use at a level not in excess of the amount reasonably required to accomplish the intended effect.

SAFETY PROFILE: Poison by intraperitoneal and subcutaneous routes. Moderately toxic by ingestion. A skin and severe eye irritant. Mutagenic data. Narcotic in high concentration. A hypnotic. Combustible liquid. To fight fire use foam, CO_2, dry chemical. When heated to decomposition it emits acrid smoke and fumes.

TOXICITY DATA and CODEN

skn-rbt 515 mg open MLD UCDS** 12/27/71
eye-rbt 771 μg SEV AJOPAA 29,1363,46
cyt-smc 10 mmol/tube HEREAY 33,457,47
orl-rat LD50:815 mg/kg GTPZAB 26(8),53,82
orl-mus LD50:740 mg/kg GTPZAB 26(8),53,82

ABX750 CAS: 123-54-6
ACETYL ACETONE
DOT: 2310
mf: $C_5H_8O_2$ mw: 100.13

PROP: Colorless to sltly yellow liquid; pleasant odor. Mp: −23.2°, bp: 139° @ 746 mm, flash p: 105°F (OC), d: 0.952-0.962, refr index: 1.402, vap d: 3.45, autoign temp: 644°F. Misc in alc, ether, chloroform, acetone, glacial acetic acid, propylene glycol; ins in glycerin, water.

SYNS: ACETOACETONE ◇ DIACETYLMETHANE ◇ FEMA No. 2841 ◇ PENTANEDIONE ◇ 2,4-PENTANEDIONE (FCC)

USE IN FOOD:

Purpose: Flavoring agent.

Where Used: Various.

Regulations: FDA - 21CFR 172.515. Use at a level not in excess of the amount reasonably required to accomplish the intended effect.

DOT Classification: IMO: Flammable or Combustible Liquid; Label: Flammable Liquid

SAFETY PROFILE: Moderately toxic via ingestion, intraperitoneal and inhalation routes. A skin and severe eye irritant. Flammable liquid when exposed to heat or flame. To fight fire use alcohol foam, CO_2, dry chemical. Incompatible with oxidizing materials.

TOXICITY DATA and CODEN

skn-rbt 488 mg open MLD UCDS** 7/8/71
eye-rbt 4760 μg SEV AJOPAA 29,1363,46
orl-rat LD50:1000 mg/kg JIHTAB 26,269,44
ihl-rat LCLo:1000 ppm/4H JIHTAB 31,343,49
skn-rbt LD50:5000 mg/kg UCDS** 7/8/71

ABY900 CAS: 140-40-9
2-ACETYLAMINO-5-NITROTHIAZOLE
mf: $C_5H_5N_3O_3S$ mw: 187.19

PROP: Needles from alc. Mp: 264-265°. Sol in aq solutions of NaOH and NH_3.

SYNS: ACETAMIDO-5-NITROTHIAZOLE ◇ ACINITRAZOLE ◇ AMINITROZOLE ◇ ENHEPTIN-A ◇ GYNOFON ◇ N-(5-NITRO-2-THIAZOLYL)ACETAMIDE ◇ PLEOCIDE ◇ TRICHORAD ◇ TRICHORAL ◇ TRITHEOM

USE IN FOOD:

Purpose: Animal drug.

Where Used: Turkeys.

Regulations: FDA - 21CFR 556.20. Limitation of 0.1 ppm in turkeys.

SAFETY PROFILE: When heated to decomposition emits toxic fumes of NO_x.

ACA900
ACETYLATED MONOGLYCERIDES

PROP: Esters of glycerin with acetic acid and edible fat-forming fatty acids. (FCC III) May be white to pale yellow liquids or solids; bland taste. Sol in alc, acetone; ins in water.

SYN: ACETYLATED MONO- and DIGLYCERIDES

USE IN FOOD:

Purpose: Coating agent, emulsifier, lubricant, solvent, texture modifying agent.

Where Used: Baked goods, cake shortening, desserts (frozen), fruits, ice cream, margarine, meat products, nuts, oleomargarine, peanut butter, puddings, shortening, whipped toppings, whipped toppings (dry-mix).

Regulations: FDA - 21CFR 172.828. Use at a level not in excess of the amount reasonably required to accomplish the intended effect. USDA - 9CFR 318.7. Sufficient for purpose, 0.5 percent in oleomargarine or margarine. Must conform to FDA specifications for fats or fatty acids derived from edible oils.

SAFETY PROFILE: When heated to decomposition it emits acrid smoke and irritating fumes.

ACI400 CAS: 10599-70-9
3-ACETYL-2,5-DIMETHYLFURAN
mf: $C_8H_{10}O_2$ mw: 138.16

PROP: Yellow liquid; strong roasted nutlike odor. D: 1.027-1.048, refr index: 1.475-1.496 (25°). Sol in alc, propylene glycol, fixed oils; sltly sol in water.

SYNS: 2,5-DIMETHYL-3-ACETYLFURAN ◇ FEMA No. 3391

USE IN FOOD:

Purpose: Flavoring agent.

Where Used: Various.

Regulations: GRAS when used at a level not in excess of the amount reasonably required to accomplish the intended effect.

SAFETY PROFILE: When heated to decomposition it emits acrid smoke and irritating fumes.

ACQ275 CAS: 65-82-7
N-ACETYL-l-METHIONINE
mf: $C_7H_{13}NO_3S$ mw:191.24

PROP: Colorless or white crystals or powder; odorless. Sol in water, alc, alkali, and mineral acids; ins in ether.

SYN: N-ACETYL-l-AMINO-4-(METHYLTHIO)BUTYRIC ACID

USE IN FOOD:

Purpose: Dietary supplement, nutrient.

Where Used: Various.

Regulations: FDA - 21CFR 172.372. Limitation of 3.1 percent l- and dl-methionine (expressed as the free amino acid) by weight of the total protein in the food.

SAFETY PROFILE: When heated to decomposition emits toxic fumes of NO_x.

ADA350 CAS: 22047-25-2
2-ACETYL PYRAZINE
mf: $C_6H_6N_2O$ mw: 122.13

PROP: Colorless to pale yellow crystals or liquid; sweet popcornlike odor. Mp: 75-78°, d: 1.100-1.115 @ 20°, refr index: 1.530-1.540 @ 25°. Sol in acids, alc, ether, water @ 230°.

SYN: FEMA No. 3126

USE IN FOOD:

Purpose: Flavoring agent.

Where Used: Various.

Regulations: GRAS when used at a level not in excess of the amount reasonably required to accomplish the intended effect.

SAFETY PROFILE: Skin and eye irritant. When heated to decomposition emits toxic fumes of NO_x.

ADA375 CAS: 1072-83-9
2-ACETYLPYRROLE
mf: C_6H_7NO mw: 109.12

PROP: Light beige to yellow crystals; breadlike odor. Sol in acids, alc, ether, water @ 230°

SYNS: FEMA No. 3202 ◇ METHYL 2-PYRROLYL KETONE

USE IN FOOD:

Purpose: Flavoring agent.

Where Used: Various.

Regulations: GRAS when used at a level not in excess of the amount reasonably required to accomplish the intended effect.

SAFETY PROFILE: When heated to decomposition emits toxic fumes of NO_x.

ADD400
ACETYL TRIBUTYL CITRATE

USE IN FOOD:

Purpose: Plasticizer.

Where Used: Packaging materials.

Regulations: FDA - 21CFR 181.27. Use in accordance with good manufacturing practice.

SAFETY PROFILE: When heated to decomposition it emits acrid smoke and irritating fumes.

ADD750 CAS: 77-89-4
ACETYL TRIETHYL CITRATE
mf: $C_{14}H_{22}O_8$ mw: 318.36

SYNS: CITRIC ACID, ACETYL TRIETHYL ESTER ◇ TRICARBALLYLIC ACID-β-ACETOXYTRIBUTYL ESTER ◇ TRIETHYL ACETYLCITRATE

USE IN FOOD:

Purpose: Plasticizer.

Where Used: Packaging materials.

Regulations: FDA - 21CFR 181.27. Use in accordance with good manufacturing practice.

SAFETY PROFILE: Moderately toxic by intraperitoneal route. Mildly toxic by ingestion. When heated to decomposition it emits acrid smoke and fumes.

TOXICITY DATA and CODEN

orl-rat LD50:8000 mg/kg NPIRI* 2,2,75
ipr-mus LD50:1150 mg/kg JPMSAE 53,774,64
orl-cat LDLo:7500 mg/kg TXAPA9 1,283,59

ADF600
ACID HYDROLYZED PROTEINS

PROP: Liquid, paste, or powder. Sol in water.

SYNS: HYDROLYZED MILK PROTEIN ◇ HYDROLYZED PLANT PROTEIN (HPP) ◇ HYDROLYZED VEGETABLE PROTEIN (HVP)

USE IN FOOD:

Purpose: Flavoring agent.

Where Used: Bologna, salami, sauces, stuffing.

Regulations: USDA - 9CFR 318.7. Use at a level not in excess of the amount reasonably required to accomplish the intended effect.

SAFETY PROFILE: When heated to decomposition it emits acrid smoke and irritating fumes.

ADG375 CAS: 59-51-8
ACIMETION

mf: $C_5H_{11}NO_2S$ mw: 149.23

PROP: White crystalline platelets; characteristic odor. Sol in water, dilute acids and alkalies; very sltly sol in alc; ins in ether.

SYNS: BANTHIONINE ◇ CYNARON ◇ DYPRIN ◇ LOBAMINE ◇ MEONINE ◇ MERTIONIN ◇ METHILANIN ◇ (±)-METHIONINE ◇ dl-METHIONINE (9CI, FCC) ◇ METIONE ◇ NESTON

USE IN FOOD:

Purpose: Dietary supplement, nutrient.

Where Used: Various.

Regulations: FDA - 21CFR 172.320. Not to be used in infant foods.

EPA Genetic Toxicology Program.

SAFETY PROFILE: Moderately toxic by ingestion and other routes. Experimental reproductive effects. When heated to decomposition it emits toxic fumes of SO_x and NO_x.

TOXICITY DATA and CODEN

orl-rat TDLo:40 g/kg (1-20D preg):REP JRPFA4 33,109,73
orl-rat TDLo:2250 mg/kg (90D pre):REP YACHDS 5,2041,77
orl-mus LDLo:4 g/kg YACHDS 5,2041,77

ADS400
ACRYLATE-ACRYLAMIDE RESINS

USE IN FOOD:

Purpose: Clarifying agent, production aid.

Where Used: Corn starch hydrolyzate, sugar liquor or juice.

Regulations: FDA - 21CFR 173.5. Limitation of 10 ppm of the cane or beet liquor for clarification. Limitation of 2.5 ppm in cane or beet juice or liquor for mineral scale control.

SAFETY PROFILE: When heated to decomposition it emits acrid smoke and irritating fumes.

ADX600
ACRYLONITRILE COPOLYMERS

USE IN FOOD:

Purpose: Component of packaging material.

Where Used: Packaging material.

Regulations: FDA - 21CFR 180.22. Limitation of 0.003 mg/sq in. for volume to surface ratio of 10 mm/sq in. or more for single-use articles. Limitation of 0.3 ppm for volume to surface ratio of less than 10 mm/sq in. for single-use articles. Limitation of 0.003 mg/sq in. for repeated-use articles. 21CFR181.32.

SAFETY PROFILE: When heated to decomposition it emits acrid smoke and irritating fumes.

ADY500 CAS: 9003-54-7
ACRYLONITRILE POLYMER WITH STYRENE

mf: $(C_8H_8 \cdot C_3H_3N)_x$

SYNS: ACRILAFIL ◇ ACRYLONITRILE-STYRENE COPOLYMER ◇ ACRYLONITRILE-STYRENE POLYMER ◇ ACRYLONITRILE-STYRENE RESIN ◇ ACS ◇ AS 61CL ◇ BAKELITE RMD 4511 ◇ CEVIAN HL ◇ DIALUX ◇ ESTYRENE AS ◇ KOSTIL ◇ LITAC ◇ LURAN ◇ LUSTRAN ◇ POLYSTYRENE-ACRYLONITRILE ◇ 2-PROPENENITRILE POLYMER WITH ETHENYLBENZENE ◇ REXENE 106 ◇ SANREX ◇ SN 20 ◇ STYRENE-ACRYLONITRILE COPOLYMER ◇ STYREN-ACRYLONITRILEPOLYMER ◇ TERULAN KP 2540 ◇ TYRIL

USE IN FOOD:

Purpose: Coatings, films.

Where Used: Packaging materials.

Regulations: FDA - 21CFR 181.32. No restrictions.

Cyanide and its compounds are on the Community Right-To-Know List.

SAFETY PROFILE: Moderately to highly toxic by ingestion. When heated to decomposition it emits toxic fumes of NO_x and CN^-.

TOXICITY DATA and CODEN

orl-rat LD50:1800 mg/kg CEHYAN 25,22,80
orl-mus LD50:1000 mg/kg CEHYAN 25,22,80

AEC500 CAS: 64365-11-3
ACTIVATED CARBON

af: C aw: 12.01
DOT: 1362

PROP: Black porous solid, coarse granules or powder. Ins in water, organic solvents.

SYNS: CARBON, ACTIVATED (DOT) ◇ CHARCOAL, ACTIVATED (DOT)

USE IN FOOD:

Purpose: Decolorizing agent, odor-removing agent, purification agent in food processing, taste-removing agent.

Where Used: Fats (refined animal), grape juice (black), grape juice (red), sherry (cocktail), sherry (pale dry), wine.

Regulations: FDA - 21CFR 240.361, 240.365, 240.401, 240.405, 240 527, 240.527a. GRAS when used in accordance with good manufacturing practice. BATF - 27CFR 240.1051. Limitation of 0.9 percent in wine. Limitation of 0.25 percent in pale dry sherry or cocktail sherry. Limitation of 0.4 percent in red and black grape juice. USDA - 9CFR 318.7. Sufficient for purpose.

DOT Classification: Flammable Solid; Label: Spontaneously Combustible

SAFETY PROFILE: It can cause a dust irritation, particularly to the eyes and mucous membranes. Combustible when exposed to heat. Dust is flammable and explosive when exposed to heat or flame or oxides.

AEN250 CAS: 124-04-9
ADIPIC ACID

DOT: 9077
mf: $C_6H_{10}O_4$ mw: 146.16

PROP: White monoclinic prisms. Mp: 152°, flash p: 385°F (CC), d: 1.360 @ 25°/4°, vap press: 1 mm @ 159.5°, vap d: 5.04, autoign temp: 788°F, bp: 337.5°. Very sol in alc. Sol in acetone, water = 1.4% @15°; 0.6% @ 15° in ether.

SYNS: ACIFLOCTIN ◇ ACINETTEN ◇ ADILACTETTEN ◇ ADIPINIC ACID ◇ FEMA No. 2011 ◇ 1,4-BUTANEDICARBOXYLIC ACID ◇ 1,6-HEXANEDIOIC ACID ◇ KYSELINA ADIPOVA (CZECH) ◇ MOLTEN ADIPIC ACID

USE IN FOOD:

Purpose: Flavoring agent, leavening agent, neutralizing agent, pH control agent.

Where Used: Baked goods, baking powder, beverages (nonalcoholic), condiments, dairy product analogs, desserts (frozen dairy), drinks (powdered), edible oils, fats, gelatin, gravies, margarine, meat products, oils, oleomargarine, puddings, relishes, snack foods, vegetables (canned).

Regulations: FDA - 21CFR 172.515, 184.1009. GRAS with a limitation of 0.05 percent in baked goods, 0.005 percent in nonalcoholic beverages, 5.0 percent in condiments and relishes, 0.45 percent in dairy product analogs, 0.3 percent in fats and oils, 0.0004 percent in frozen dairy desserts, 0.55 percent in gelatin and puddings, 0.1 percent in gravies, 0.3 percent in meat products, 1.3 percent in snack foods, 0.02 percent in other food categories when used in accordance with good manufacturing practice. USDA - 9CFR 318.7. Sufficient for purpose.

DOT Classification: ORM-E; Label: None

SAFETY PROFILE: Poison by intraperitoneal route. Moderately toxic by other routes. A severe eye irritant. Combustible when exposed to heat or flame; can react with oxidizing materials. When heated to decomposition it emits acrid smoke and fumes.

TOXICITY DATA and CODEN

eye-rbt 20 mg/24H SEV 28ZPAK -,51,72
orl-rat LDLo:3600 mg/kg 28ZPAK -,51,72
orl-mus LD50:1900 mg/kg JAFCAU 5,759,57

AET750 CAS: 1402-68-2
AFLATOXIN

USE IN FOOD:

Purpose: Contaminant.

Where Used: Corn, peanuts.

Regulations: USDA CES Ranking: A-4 (1985).

IARC Cancer Review: Human Limited Evidence IMEMDT 10,51,76

SAFETY PROFILE: Human poison by ingestion. Moderately toxic by other routes. An experimental tumorigen and teratogen. Experimental reproductive effects. Mutagenic data.

TOXICITY DATA and CODEN

dlt-mus-ipr 68 mg/kg NATUAS 219,385,68
par-ham TDLo:4 mg/kg (8D preg):TER DABBBA 34,5251,73
par-ham TDLo:6 mg/kg (8D preg):TER DABBBA 34,5251,73
orl-rat TDLo:2250 μg/kg (10-21D preg): REP,ETA CNREA8 33,262,73
orl-rat TDLo:7788 μg/kg/13W-C:ETA NATUAS 202,1016,64
orl-hmn LDLo:229 μg/kg/8W LANCAO 1,1061,75
orl-mky LD50:1750 μg/kg FCTXAV 14,227,76

AEX250 CAS: 9002-18-0
AGAR

PROP: Extracted from the red algae *Rhodopyceae*. Unground: in thin, translucent, membranous pieces; ground: pale buff powder. Sol in boiling water, insol in cold water and organic solvents.

SYNS: AGAR-AGAR ◇ AGAR AGAR FLAKE ◇ AGAR-AGAR GUM ◇ BENGAL GELATIN ◇ BENGAL ISINGLASS ◇ CEYLON ISINGLASS ◇ CHINESE ISINGLASS ◇ DIGENEA SIMPLEX MUCILAGE ◇ GELOSE ◇ JAPAN AGAR ◇ JAPAN ISINGLASS ◇ LAYOR CARANG ◇ NCI-C50475

USE IN FOOD:

Purpose: Emulsifier, stabilizer, thickener.

Where Used: Baked goods, baking mixes, candy (soft), confections, frostings, glazes, meat (thermally processed canned jellied), piping gels.

Regulations: FDA - 21CFR 184.1115. GRAS with a limitation of 0.8 percent in baked goods and mixes, 2.0 percent in confections and frostings, 1.2 percent in soft candy, 0.25 percent in all other foods when used in accordance with good manufacturing practice. USDA - 9CFR 318.7. Sufficient for purpose. Limitation of 0.25 percent in thermally processed canned jellied meat food products.

NTP Carcinogenesis Bioassay (feed); No Evidence: mouse, rat NTPTR* NTP-TR-230,82.

SAFETY PROFILE: Mildly toxic by ingestion. When heated to decomposition it emits acrid smoke and fumes.

TOXICITY DATA and CODEN

orl-rat LD50:11 g/kg FDRLI* 124,-,76
orl-mus LD50:16 g/kg FDRLI* 124,-,76

AFH400 CAS: 3011-89-0
AKLOMIDE
mf: $C_7H_5ClN_2O_3$ mw: 200.60

PROP: Gray scales from alcohol. Mp: 172°.

SYNS: AKLOMIX ◇ 2-CHLORO-4-NITROBENZAMIDE

USE IN FOOD:

Purpose: Animal drug.

Where Used: Chicken.

Regulations: FDA - 21CFR 556.30. Limitation of 4.5 ppm in chicken liver and muscle, 3 ppm in chicken skin and fat. 21CFR 558.35. Not to be fed to birds laying eggs for human consumption.

SAFETY PROFILE: When heated to decomposition emits toxic fumes of Cl^- and NO_x.

AFH600 CAS: 302-72-7
dl-ALANINE
mf: $C_3H_7NO_2$ mw: 89.09

PROP: White crystalline powder; odorless with a sweet taste. Mp: 198° (decomp). Sol in water, sltly sol in alc.

SYN: dl-2-AMINOPROPANOIC ACID

USE IN FOOD:

Purpose: Dietary supplement, nutrient.

Where Used: Various.

Regulations: FDA - 21CFR 172.320. Limitation 6.1 percent by weight.

SAFETY PROFILE: When heated to decomposition emits toxic fumes of NO_x.

AFH625 CAS: 56-41-7
l-ALANINE
mf: $C_3H_7NO_2$ mw: 89.09

PROP: White crystalline power; odorless with a sweet taste. Sol in water; sltly sol in alc; ins in ether.

SYN: l-2-AMINOPROPANOIC ACID

USE IN FOOD:

Purpose: Dietary supplement, nutrient.

Where Used: Various.

Regulations: FDA - 21CFR 172.320. Limitation 6.1 percent by weight.

SAFETY PROFILE: When heated to decomposition emits toxic fumes of NO_x.

AFI850 CAS: 70536-17-3
ALBUMIN MACRO AGGREGATES

SYNS: MAA ◇ ALBUMIN

USE IN FOOD:

Purpose: Binder, fining agent.

Where Used: Sausage (imitation), soups, stews, wine.

Regulations: BATF - 27CFR 240.1051. Limitation of 1.5 gallons of solution/1000 gallons of wine (solution containing 2 pounds albumin/gallon of brine solution).

SAFETY PROFILE: Poison by intravenous route. When heated to decomposition it emits acrid smoke and irritating fumes.

TOXICITY DATA and CODEN

ivn-rat LD50:17 mg/kg IJNMCI 5,51,78
ivn-mus LD50:18 mg/kg IJNMCI 5,51,78

AFK250 CAS: 309-00-2
ALDRIN

DOT: 2761
mf: $C_{12}H_8Cl_6$ mw: 364.90

PROP: Crystals. Mp: 104-105°. Insol in water; sol in aromatics, esters, ketones, paraffins, and halogenated solvents.

SYNS: ALDREX ◇ ALDREX 30 ◇ ALDRIN, cast solid (DOT) ◇ ALDRINE (FRENCH) ◇ ALDRITE ◇ ALDROSOL ◇ ALTOX ◇ COMPOUND 118 ◇ DRINOX ◇ ENT 15,949 ◇ HEXACHLOROHEXAHYDRO-endo-exo-DIMETHANONAPHTHALENE ◇ 1,2,3,4,10,10-HEXACHLORO-1,4,4a,5,8,8a-HEXAHYDRO-1,4,5,8-DIMETHANONAPHTHALENE ◇ 1,2,3,4,10,10-HEXACHLORO-1,4,4a,5,8,8a-HEXAHYDRO-exo-1,4,-endo-5,8-DIMETHANONAPHTHALENE ◇ 1,2,3,4,-10,10-HEXACHLORO-1,4,4a,5,8,8a-HEXAHYDRO-1,4-endo-exo-5, 8-DIMETHANONAPHTHALENE ◇ HHDN ◇ NCI-C00044 ◇ OCTALENE ◇ RCRA WASTE NUMBER P004 ◇ SEEDRIN

USE IN FOOD:

Purpose: Pesticide.

Where Used: Various.

Regulations: USDA CES Ranking: A-3 (1986).

EPA Genetic Toxicology Program. EPA Extremely Hazardous Substances List. Community Right-To-Know List. IARC Cancer Review: Human Inadequate Evidence IMEMDT 5,-25,74; Animal Inadequate Evidence IMEMDT 5,25,74; NCI Carcinogenesis Bioassay (feed); Clear Evidence: mouse NCITR* NCI-CG-TR-21,78; Inadequate Studies: rat NCITR* NCI-CG-TR-21,78

OSHA PEL: TWA 250 $\mu g/m^3$ (skin) ACGIH TLV: TWA 0.25 mg/m^3 DFG MAK: 0.25 mg/m^3 DOT Classification: Poison B; ORM-A NIOSH REL: TWA 0.15 mg/m^3

SAFETY PROFILE: Poison by ingestion, skin contact, intravenous, intraperitoneal and other routes. An experimental tumorigen, neoplastigen, carcinogen and teratogen. Human systemic effects by ingestion: excitement, tremors and nausea or vomiting. Experimental reproductive effects. Human mutagenic data. Continued acute exposure causes liver damage. When heated to decomposition it emits toxic fumes of Cl^-. An insecticide.

TOXICITY DATA and CODEN

cyt-hmn:leu 19125 µg/L PHTHDT 6,147,79
cyt-rat-ipr 9560 µg/kg PHTHDT 6,147,79
scu-rat TDLo:10 mg/kg (2D pre):REP ENVRAL 16,131,78
orl-mus TDLo:25 mg/kg (9D preg):TER TJADAB 9,11,74
orl-rat TDLo:200 mg/kg/2Y-C:NEO FCTXAV 2,551,64
orl-mus TDLo:270 mg/kg/80W-I:CAR NCITR* NCI-CG-TR-21,78
orl-mus TD :540 mg/kg/80W-I:CAR NCITR* NCI-CG-TR-21,78
orl-rat TD :188 mg/kg/2Y-C:ETA TXAPA9 11,88,67
orl-hmn TDLo:14 mg/kg:CNS 34ZIAG -,83,69
orl-chd LDLo:1250 µg/kg 34ZIAG -,83,69
orl-rat LD50:39 mg/kg SPEADM 74-1,-,74
skn-rat LD50:98 mg/kg TXAPA9 14,515,69

AFK920
ALGAE, BROWN

PROP: Various seaweeds harvested in coastal waters of the northern Atlantic and Pacific oceans.

SYN: BROWN ALGAE

USE IN FOOD:

Purpose: Flavor adjuvant, flavor enhancer.

Where Used: Flavorings, seasonings, spices.

Regulations: FDA - 21CFR 184.1120. Use at a level not in excess of the amount reasonably required to accomplish the intended effect.

SAFETY PROFILE: When heated to decomposition it emits acrid smoke and irritating fumes.

AFK925
ALGAE MEAL, DRIED

PROP: Mixture of algae cells from *Spongiococcum*, molasses, corn steep liquor, and a maximum of 0.3% ethoxyquin.

USE IN FOOD:

Purpose: Color additive.

Where Used: Chicken feed.

Regulations: FDA - 21CFR 73.275. Feed must be supplemented sufficiently with xanthophyll and associated carotenoids to accomplish the intended effect. Must meet the tolerance limitation for ethoxyquin 21CFR 573.380.

SAFETY PROFILE: When heated to decomposition it emits acrid smoke and irritating fumes.

AFK930
ALGAE, RED

PROP: Various seaweeds harvested in coastal waters of Pacific ocean.

SYN: RED ALGAE

USE IN FOOD:

Purpose: Flavor adjuvant, flavor enhancer.

Where Used: Flavorings, seasonings, spices.

Regulations: FDA - 21CFR 184.1121. Use at a level not in excess of the amount reasonably required to accomplish the intended effect.

SAFETY PROFILE: When heated to decomposition it emits acrid smoke and irritating fumes.

AFK940
ALGANET

USE IN FOOD:

Purpose: Coloring agent.

Where Used: Casings, fats (rendered).

Regulations: USDA - 9CFR 318.7. Sufficient for purpose.

SAFETY PROFILE: When heated to decomposition it emits acrid smoke and irritating fumes.

AFL000 CAS: 9005-32-7
ALGINIC ACID

PROP: Extracted from brown seaweeds. White to yellow white fibrous powder; odorless and tasteless. Sol in alkaline solutions; ins in organic solvents.

SYNS: KELACID ◇ LANDALGINE ◇ NORGINE ◇ PLOYMANNURONIC ACID ◇ SAZZIO

USE IN FOOD:

Purpose: Emulsifier, stabilizer, thickener.

Where Used: Soup mixes, soups.

Regulations: FDA - 21CFR 184.1011. GRAS when used in accordance with good manufacturing practice.

SAFETY PROFILE: Moderately toxic by intraperitoneal route. When heated to decomposition it emits acrid smoke and irritating fumes.

TOXICITY DATA and CODEN

ipr-rat LD50:1600 mg/kg AIPTAK 111,167,57
ipr-mus LDLo:1000 mg/kg TXAPA9 23,288,72

AFU500
ALLSPICE

PROP: Contains eugenol. Distilled from the fruit of *Pimenta officinalis* Lindley (Fam. *Myrtaceae*). Yellow to red-yellow liquid; odor and taste of allspice. D: 1.018-1.048, refr index: 1.527-1.540 @ 20°.

SYNS: PIMENTA OIL ◇ PIMENTA BERRIES OIL ◇ PIMENTO OIL

USE IN FOOD:

Purpose: Flavoring agent.

Where Used: Cakes, fruit pies, mincemeat, plum pudding, sauces soups.

Regulations: FDA - 21CFR 182.20. GRAS when used at a level not in excess of the amount reasonably required to accomplish the intended effect.

SAFETY PROFILE: A weak sensitizer which may cause dermatitis on local contact. Eugenol is moderately toxic. Combustible.

AFW750 CAS: 140-67-0
p-ALLYLANISOLE
mf: $C_{10}H_{12}O$ mw: 148.22

PROP: Isolated from rind of *Persea Gratissima Garth,* and from Oil of Estragon; found in oils of Russian Anise, Basil, Fennel, Turpentine, and others (FCTXAV 14,601,76).

SYNS: 4-ALLYL-1-METHOXYBENZENE ◇ CHAVICOL METHYL ETHER ◇ ESDRAGOL ◇ ISOANETHOLE ◇ p-METHOXYALLYLBENZENE ◇ 1-METHOXY-4-(2-PROPENYL)BENZENE ◇ METHYL CHAVICOL ◇ NCI-C60946 ◇ TARRAGON

USE IN FOOD:

Purpose: Flavoring agent.

Where Used: Bakery products, beverages (alcoholic), beverages (nonalcoholic), chewing gum, confections, fish, ice cream, salads, sauces, vinegar.

Regulations: FDA - 21CFR 182.10. GRAS when used at a level not in excess of the amount reasonably required to accomplish the intended effect.

SAFETY PROFILE: An experimental carcinogen and neoplastigen. Moderate acute toxicity by many routes. A skin irritant. Mutagenic data. When heated to decomposition it emits acrid smoke and irritating fumes. A spice used in foods, liqueurs and perfumes.

TOXICITY DATA and CODEN

skn-rbt 500 mg/24H MOD FCTXAV 14,601,76
mma-sat 1 μmol/plate MUREAV 60,143,79
dnd-mus-ipr 80 mg/kg CRNGDP 5,1613,84
orl-mus TDLo: 97 g/kg/1Y-C:NEO CNREA8 43,1124,83
scu-mus TDLo: 140 mg/kg/22D-I:CAR JNCIAM 57,1323,76
orl-rat LDLo: 1230 mg/kg FCTXAV 14(6),601,76

AGA500 CAS: 123-68-2

ALLYL CAPROATE

mf: $C_9H_{16}O_2$ mw: 156.25

PROP: Bp: 186-188°. Insol in water; sol in alc and ether.

SYNS: ALLYL HEXANOATE (FCC) ◇ FEMA No. 2032 ◇ 2-PROPENYL-N-HEXANOATE

USE IN FOOD:

Purpose: Flavoring agent.

Where Used: Candy, dessert gels, puddings.

Regulations: FDA - 21CFR 172.515. Use at a level not in excess of the amount reasonably required to accomplish the intended effect.

SAFETY PROFILE: Poison by ingestion and skin contact. Mutagenic data. An irritant to human skin. When heated to decomposition it emits acrid smoke and irritating fumes.

TOXICITY DATA and CODEN

skn-hmn 20 mg/48H MLD FCTXAV 11,1079,73
mrc-bcs 18 μg/disc OEKSDJ 9,177,78
orl-rat LD50: 218 mg/kg FCTXAV 2,327,64
skn-rbt LD50: 300 mg/kg FCTXAV 11,477,73

AGC000 CAS: 1866-31-5

ALLYL CINNAMATE

mf: $C_{12}H_{12}O_2$ mw: 188.24

PROP: Colorless to light yellow liquid; cherry odor. D: 1.052 @ 25°/25°; bp: 150-152° @ 15 mm. Insol in water; sol in alc; very sol in ether.

SYNS: ALLYL-3-PHENYLACRYLATE ◇ PROPENYL CINNAMATE ◇ VINYL CARBINYL CINNAMATE

USE IN FOOD:

Purpose: Flavoring agent.

Where Used: Baked goods, candy.

Regulations: FDA - 21CFR 172.515. Use at a level not in excess of the amount reasonably required to accomplish the intended effect.

SAFETY PROFILE: Moderately toxic by ingestion. Human skin irritant. When heated to decomposition it emits acrid smoke and irritating fumes.

TOXICITY DATA and CODEN

skn-hmn 20 mg/48H FCTXAV 15,611,77
orl-rat LD50: 1520 mg/kg FCTXAV 2,327,64

AGC500 CAS: 2705-87-5

ALLYL CYCLOHEXANEPROPIONATE

mf: $C_{12}H_{20}O_2$ mw: 196.32

PROP: Colorless liquid; pineapple odor. D: 0.945-0.950, refr index: 1.457-1.463, flash p: +212°F. Misc in alc, chloroform, ether; ins glycerin, water.

SYNS: 3-ALLYLCYCLOHEXYL PROPIONATE ◇ ALLYL HEXAHYDROPHENYLPROPIONATE ◇ FEMA No. 2026

USE IN FOOD:

Purpose: Flavoring agent.

Where Used: Various.

Regulations: FDA - 21CFR 172.515. Use at a level not in excess of the amount reasonably required to accomplish the intended effect.

SAFETY PROFILE: Poison by ingestion. When heated to decomposition it emits acrid smoke and irritating fumes. Combustible liquid.

TOXICITY DATA and CODEN

orl-rat LD50:585 mg/kg FCTXAV 2,327,64
orl-gpg LD50:380 mg/kg FCTXAV 2,327,64

AGE250 CAS: 93-15-2
4-ALLYL-1,2-DIMETHOXYBENZENE
mf: $C_{11}H_{14}O_2$ mw: 178.25

PROP: Colorless to pale yellow liquid; clove, carnation odor. D: 1.032-1.036, refr index: 1.532, flash p: 212°F. Sol in fixed oils; ins in glycerin, propylene glycol.

SYNS: 1-ALLYL-3,4-DIMETHOXYBENZENE ◇ 4-ALLYLVERATROLE ◇ 1,2-DIMETHOXY-4-ALLYLBENZENE ◇ 1-(3,4-DIMETHOXYPHENYL)-2-PROPENE ◇ ENT 21040 ◇ 1,3,4-EUGENOL METHYL ETHER ◇ EUGENYL METHYL ETHER ◇ FEMA No. 2475 ◇ METHYL EUGENOL (FCC) ◇ VERATROLE METHYL ETHER

USE IN FOOD:

Purpose: Flavoring agent.

Where Used: Various.

Regulations: FDA - 21CFR 172.515. Use at a level not in excess of the amount reasonably required to accomplish the intended effect.

SAFETY PROFILE: Poison by intravenous route. Moderately toxic by ingestion and intraperitoneal routes. A skin irritant. Mutagenic data. Combustible liquid. When heated to decomposition it emits acrid smoke and irritating fumes. Some other alkenylbenzenes have carcinogenic activity.

TOXICITY DATA and CODEN

skn-rbt 500 mg/24H FCTXAV 13,681,75
dnd-mus-ipr 80 mg/kg FCTXAV 13,857,75
orl-rat LD50:1179 mg/kg TXAPA9 31,421,75

AGH250 CAS: 142-19-8
ALLYL HEPTANOATE
mf: $C_{10}H_{18}O_2$ mw: 170.28

PROP: Colorless to pale yellow liquid; fruity, sweet, pineapple odor. D: 0.880, refr index: 1.426, flash p: 154°F.

SYNS: ALLYL ENANTHATE ◇ ALLYL HEPTOATE ◇ ALLYL HEPTYLATE ◇ FEMA No. 2031 ◇ 2-PROPENYL HEPTANOATE

USE IN FOOD:

Purpose: Flavoring agent.

Where Used: Various.

Regulations: GRAS when used at a level not in excess of the amount reasonably required to accomplish the intended effect.

SAFETY PROFILE: Moderately toxic by ingestion and skin contact. A human skin irritant. Combustible liquid. When heated to decomposition it emits acrid smoke and irritating fumes.

TOXICITY DATA and CODEN

skn-hmn 20 mg/48H MLD FCTXAV 15,611,77
skn-rbt 500 mg/24H MOD FCTXAV 15,611,77
orl-rat LD50:500 mg/kg TXAPA9 6,378,64
orl-mus LD50:630 mg/kg TXAPA9 7,18,65
skn-rbt LD50:810 mg/kg FCTXAV 15,611,77

AGI500 CAS: 79-78-7
ALLYL-α-IONONE
mf: $C_{16}H_{24}O$ mw: 232.40

PROP: Colorless to yellow liquid; fruity, woody odor. D: 0.928-0.935, refr index: 1.503-1.507, flash p: +212°F. Sol in alc; ins water @ 265°.

SYNS: CETONE V ◇ FEMA No. 2033 ◇ 1-(2,6,6-TRIMETHYL-2-CYCLOHEXEN-1-YL)-1,6-HEPTADIEN-3-ONE

USE IN FOOD:

Purpose: Flavoring agent.

Where Used: Various.

Regulations: FDA - 21CFR 172.515. Use at a level not in excess of the amount reasonably required to accomplish the intended effect.

SAFETY PROFILE: A skin irritant. Combustible liquid. When heated to decomposition it emits acrid smoke and irritating fumes.

TOXICITY DATA and CODEN

skn-rbt 500 mg MLD FCTXAV 11,1079,73

AGJ250 CAS: 57-06-7
ALLYL ISOTHIOCYANATE
DOT: 1545
mf: C_4H_5NS mw: 99.16

PROP: Colorless to pale yellow liquid; irr odor with mustard taste. Mp: −80°, bp: 150.7°, flash p: 115°F, d: 1.013-1.016 @ 25°/25°, vap press: 10 mm @ 38.3°, vap d: 3.41, refr index: 1.527-1.531. Misc with alc, carbon disulfide, ether.

SYNS: AITC ◇ ALLYL ISORHODANIDE ◇ ALLYL ISOSULFOCYANATE ◇ ALLYL ISOTHIOCYANATE, stabilized (DOT)

◇ ALLYL MUSTARD OIL ◇ ALLYLSENFOEL (GERMAN) ◇ ALLYL SEVENOLUM ◇ ALLYL THIOCARBONIMIDE ◇ ARTIFICIAL MUSTARD OIL ◇ CARBOSPOL ◇ FEMA No. 2034 ◇ ISOTHIOCYANATE d'ALLYLE (FRENCH) ◇ 3-ISOTHIOCYANATO-1-PROPENE ◇ MUSTARD OIL ◇ NCI-C50464 ◇ OIL OF MUSTARD, ARTIFICIAL ◇ OLEUM SINAPIS VOLATILE ◇ 2-PROPENYL ISOTHIOCYANATE ◇ REDSKIN ◇ SENF OEL (GERMAN) ◇ SYNTHETIC MUSTARD OIL ◇ VOLATILE OIL OF MUSTARD

USE IN FOOD:

Purpose: Flavoring agent.

Where Used: Baked goods, condiments, horseradish flavor (imitation), meats, mustard oil (artificial), pickles, salad dressings, sauces.

Regulations: FDA - 21CFR 172.515. Use at a level not in excess of the amount reasonably required to accomplish the intended effect.

IARC Cancer Review: Animal Limited Evidence IMEMDT 36,55,85; NTP Carcinogenesis Bioassay (gavage); No Evidence: mouse NTPTR* NTP-TR-234,82; Clear Evidence: rat NTPTR* NTP-TR-234,82.

SAFETY PROFILE: Poison by ingestion, skin contact, intravenous, subcutaneous, and intraperitoneal routes. An experimental neoplastigen, tumorigen and teratogen by skin contact and other routes. Other experimental reproductive effects. An allergen. May cause contact dermatitis. Mutagenic data. Combustible liquid. Highly reactive. When heated to decomposition (above 250°) or contact with acid or acid fumes it emits highly toxic fumes of CN^-, SO_x and NO_x. To fight fire, use foam, CO_2, dry chemical.

TOXICITY DATA and CODEN

eye-rbt 2 mg AEPPAE 219,119,53
mma-sat 1 μmol/plate CBINA8 38,303,82
mmo-esc 5 μg/plate KEKHB8 (9),11,79
scu-rat TDLo: 100 mg/kg (8-9D preg): TER FCTXAV 18,159,80
scu-rat TDLo: 200 mg/kg (8-9D preg): TER FCTXAV 18,159,80
orl-rat TDLo: 12875 mg/kg/2Y-I: NEO NTPTR* NTP-TR-234,82
skn-mus TDLo: 12 g/kg/12W-I: ETA AICCA6 11,699,55
orl-rat LD50: 14800 μg/kg PHARAT 14,435,59
orl-mus LD50: 310 mg/kg TXAPA9 42,417,77

AGJ500
ALLYL MERCAPTAN

mf: C_3H_6S mw: 74.15

PROP: Water-white liquid with a strong garlic odor, darkens on standing. D: 0.925 @ 23°/4°, bp: 68°, flash p: 14°F.

SYN: 2-PROPENE-1-THIOL

USE IN FOOD:

Purpose: Flavoring agent.

Where Used: Baked goods, condiments.

Regulations: FDA - 21CFR 172.515. Use at a level not in excess of the amount reasonably required to accomplish the intended effect.

SAFETY PROFILE: Poison by inhalation and ingestion. Strong irritant to skin and mucous membranes. When heated to decomposition it emits highly toxic fumes of SO_x. Very dangerous fire hazard. To fight fire, use water mist or spray, alcohol foam, CO_2, or dry chemical.

AGM500 CAS: 4230-97-1
ALLYL OCTANOATE

mf: $C_{11}H_{20}O_2$ mw: 184.31

PROP: Colorless liquid; fruity odor. D: 0.8550.861, refr index: 1.425, flash p: +151°F. Sol in alc, fixed oils; sltly sol in propylene glycol; ins in glycerin, water @ 260°.

SYNS: ALLYL CAPRYLATE ◇ FEMA No. 2037 ◇ OCTANOIC ACID ALLYL ESTER ◇ OCTANOIC ACID-2-PROPENYL ESTER

USE IN FOOD:

Purpose: Flavoring agent.

Where Used: Beverages, candy, dessert gels, puddings.

Regulations: FDA - 21CFR 172.515. Use at a level not in excess of the amount reasonably required to accomplish the intended effect.

SAFETY PROFILE: Moderately toxic by ingestion. A skin irritant. When heated to decomposition it emits acrid smoke and irritating fumes.

TOXICITY DATA and CODEN

skn-rbt 310 mg/24H MOD FCTXAV 16,637,78
orl-rat LD50: 570 mg/kg FCTXAV 16,637,78

AGQ750 CAS: 7493-74-5
ALLYL PHENOXYACETATE

mf: $C_{11}H_{12}O_3$ mw: 192.23

PROP: Colorless to light yellow liquid; heavy fruit odor.

SYN: ACETATE P.A.

USE IN FOOD:

Purpose: Flavoring agent.

Where Used: Beverages, candy.

Regulations: FDA - 21CFR 172.515. Use at a level not in excess of the amount reasonably required to accomplish the intended effect.

SAFETY PROFILE: Moderately toxic by ingestion and skin contact. When heated to decomposition it emits acrid smoke and irritating fumes.

TOXICITY DATA and CODEN

orl-rat LD50:475 mg/kg FCTXAV 13,681,75
skn-rbt LD50:820 mg/kg FCTXAV 13,681,75

AGX250
ALUMINUM AMMONIUM SULFATE

mf: $Al_2(SO_4)_3(NH_4)_2SO_4 \cdot 24H_2O$ mw: 906

PROP: Colorless crystals; odorless with sweet taste. D: 1.645, mp: 94.5°, bp: loses 20 waters 120°. Sol in water, glycerin; insol in alc.

USE IN FOOD:

Purpose: Buffer; neutralizing agent.

Where Used: Baking powder.

Regulations: FDA - 21CFR 182.1127. GRAS when used in accordance with good manufacturing practice.

SAFETY PROFILE: A mild astringent used as a general-purpose food additive. Irritating if inhaled or ingested. Upon decomposition it emits toxic fumes of NO_x and SO_x.

AGY100
ALUMINUM CALCIUM SILICATE

USE IN FOOD:

SYN: CALCIUM ALUMINUM SILICATE

Purpose: Anticaking agent.

Where Used: Table salt, vanilla powder.

Regulations: FDA - 21CFR 182.2122. GRAS when used in accordance with good manufacturing practice.

SAFETY PROFILE: A nuisance dust.

AHD600 CAS: 7047-84-9
ALUMINUM MONOSTEARATE

mf: $C_{18}H_{37}AlO_4$ mw: 344.48

PROP: Powder.

SYN: STEARIC ACID ALUMINUM DIHYDROXIDE SALT

USE IN FOOD:

Purpose: Anticaking agent; binder; emulsifier; stabilizer.

Where Used: Packaging materials, various foods.

Regulations: FDA - 21CFR 172.863. Must conform to FDA specifications for salts of fats or fatty acids derived from edible oils. 21CFR 181.29. Use in accordance with good manufacturing practice.

SAFETY PROFILE: When heated to decomposition it emits acrid smoke and irritating fumes.

AHD650 CAS: 1976-28-9
ALUMINUM NICOTINATE

mf: $C_{18}H_{12}AlN_3O_6$ mw: 393.30

PROP: Solid.

SYNS: ALUNITINE ◇ MICALEX ◇ NICOTINIC ACID, ALUMINUM SALT ◇ 3-PYRIDINECARBOXYLIC ACID, ALUMINUM SALT ◇ TRIS(NICTINATO)ALUMINUM

USE IN FOOD:

Purpose: Dietary supplement; nutrient.

Where Used: Special dietary foods.

Regulations: FDA - 21CFR 172.310. Use in accordance with good manufacturing practice.

SAFETY PROFILE: When heated to decomposition it emits toxic fumes of NO_x.

AHE750 CAS: 20859-73-8
ALUMINUM PHOSPHIDE

DOT: 1397

mf: AlP mw: 57.95

PROP: Dark gray or dark yellow crystals. D: 2.85 @ 25°/4°. Mp: >1000°.

SYNS: AlP ◇ AL-PHOS ◇ ALUMINUM FOSFIDE (DUTCH) ◇ ALUMINUM MONOPHOSPHIDE ◇ CELPHIDE ◇ CELPHOS ◇ DELICIA ◇ DETIA GAS EX-B ◇ FOSFURI di ALLUMINIO (ITALIAN) ◇ FUMITOXIN ◇ PHOSPHURES d'ALUMIUM (FRENCH) ◇ RCRA WASTE NUMBER P006

USE IN FOOD:

Purpose: Fumigant.

Where Used: Brewer's corn grits, Brewer's malt, Brewer's rice.

Regulations: FDA - 21CFR 193.20. Fumigant residue tolerance of 0.01 ppm on processed foods. 21CFR 561.40. Limitation of 0.1 ppm residue in animal feeds.

EPA Extremely Hazardous Substances List.

ACGIH TLV: TWA 2 mg(Al)/m^3 DOT Classification: Label: Flammable Solid and Dangerous When Wet; IMO: Flammable Solid; Label: Dangerous When Wet and Poison

SAFETY PROFILE: A human poison by inhalation and ingestion. Dangerous; in contact with water, steam or alkali it slowly yields PH_3, which is spontaneously flammable in air. Explosive reaction on contact with mineral acids produces phosphine. When heated to decomposition it yields toxic PO_x.

TOXICITY DATA and CODEN

orl-hmn LD50:20 mg/kg 85ARAE 3,38,76
ihl-mam LCLo:1 ppm PCOC** -,25,66

AHF100 CAS: 10043-67-1
ALUMINUM POTASSIUM SULFATE

mf: $AlK(SO_4)_2 \cdot 12H_2O$ mw: 474.38

PROP: Transparent crystals or white crystalline powder; odorless with sweet taste. Sol in water, glycerin; ins in alc.

SYN: POTASSIUM ALUM

USE IN FOOD:

Purpose: Buffer; firming agent; neutralizing agent.

Where Used: Various.

Regulations: FDA - 21CFR 182.1129. GRAS when used in accordance with good manufacturing practice.

SAFETY PROFILE: A nuisance dust.

AHG000 CAS: 11138-49-1
ALUMINUM SODIUM OXIDE

DOT: 1819/2812

mf: $NaAlO_2$ mw: 82.0

PROP: White, hygroscopic powder. Mp: 1650°.

SYNS: β-ALUMINA ◇ β''-ALUMINA ◇ NALCO 680 ◇ SODIUM ALUMINATE, SOLID (DOT) ◇ SODIUM ALUMINUM OXIDE ◇ SODIUM POLYALUMINATE

USE IN FOOD:

Purpose: Boiler water additive.

Where Used: Various.

Regulations: FDA - 21CFR 173.310. Use at a level not in excess of the amount reasonably required to accomplish the intended effect.

ACGIH TLV: TWA 2 mg(Al)/m^3 DOT Classification: ORM-B; Label: None, solid; Corrosive Material; Label: Corrosive, solution

SAFETY PROFILE: Moderate irritant to skin, eyes and mucous membranes. A corrosive substance. When heated to decomposition it emits toxic fumes of Na_2O.

AHG500
ALUMINUM SODIUM SULFATE

mf: $NaAl(SO_4)_2 \cdot 12 H_2O$ mw: 458.29

PROP: Colorless crystals. Mp: 61°; d: 1.675. Anhydrous: sol in alc; sltly sol in water. Dodecahydrate: sol in water, alc.

SYNS: SODA ALUM ◇ SODIUM ALUMINUM SULFATE

USE IN FOOD:

Purpose: Buffer; firming agent; neutralizing agent.

Where Used: Baked goods.

Regulations: FDA - 21CFR 182.1131. GRAS when used in accordance with good manufacturing practice.

SAFETY PROFILE: A weak sensitizer. A general-purpose food additive. Local contact may cause contact dermatitis. An irritant. When heated to decomposition it emits toxic fumes of SO_x and Na_2O.

AHG750 CAS: 10043-01-3
ALUMINUM SULFATE (2:3)

DOT: 1760/9078

mf: $O_{12}S_3 \cdot 2Al$ mw: 342.14

PROP: White powder; sweet taste. Mp: decomp @ 770°, d: 2.71. Solubility in water = 36.4% @ 20°.

SYNS: ALUM ◇ ALUMINUM TRISULFATE ◇ CAKE ALUM ◇ DIALUMINUM SULPHATE ◇ DIALUMINUM TRISULFATE ◇ SULFURIC ACID, ALUMINUM SALT (3:2)

USE IN FOOD:

Purpose: Animal glue adjuvant; firming agent.

Where Used: Packaging materials, pickle relish, pickles, potatoes, shrimp packs.

Regulations: FDA - 21CFR 173.3120, 182.1125. GRAS when used in accordance with good manufacturing practice.

ACGIH TLV: TWA 2 mg(Al)/m^3 DOT Classification: ORM-E; Label: None, solid; ORM-B; Label: None, solution

SAFETY PROFILE: Moderately toxic by ingestion and intraperitoneal routes. Experimental reproductive effects. Hydrolyzes to form sulfuric acid which irritates tissue, especially lungs. When heated to decomposition it emits toxic fumes of SO_x.

TOXICITY DATA and CODEN

scu-mus TDLo:27371 μg/kg (30D male):REP JRPFA4 7,21,64
orl-mus LD50:6207 mg/kg BJIMAG 23,305,66
ipr-mus LD50:1735 mg/kg COREAF 256,1043,63

AHJ100
AMBRETTE SEED OIL

PROP: Volatile oil from seeds of *Abelmoschus moschatus* Moench, syn. *Hibsiscus Abelmoschus L.* (Fam. *Malvaceae*). Amber liquid; odor of ambrettolide. Sol in fixed oils, ins in glycerin, propylene glycol.

SYN: AMBRETTE SEED LIQUID

USE IN FOOD:

Purpose: Flavoring agent.

Where Used: Various.

Regulations: FDA - 21CFR 182.20. GRAS when used at a level not in excess of the amount reasonably required to accomplish the intended effect.

SAFETY PROFILE: When heated to decomposition it emits acrid smoke and irritating fumes.

AHJ750 CAS: 52645-53-1
AMBUSH

mf: $C_{21}H_{20}Cl_2O_3$ mw: 391.31

SYNS: AI3-29158 ◇ BW-21-Z ◇ ECTIBAN ◇ EXMIN ◇ FMC 33297 ◇ FMC 41655 ◇ ICI-PP 557 ◇ KESTREL (Pesticide) ◇ NDRC-143 ◇ NIA 33297 ◇ OUTFLANK ◇ OUTFLANK-STOCKADE ◇ PERMETHRIN (USDA) ◇ PERMETRIN (HUNGARIAN) ◇ PERMETRINA (PORTUGUESE) ◇ 3-PHENOXYBENZYL (±)-3-(2,2-DICHLOROVINYL)-2,2-DIMETHYLCYCLOPROPANECARBOXYLATE ◇ (3-PHENOXYPHENYL)METHYL-3-(2,2-DICHLORETHENYL)-2,2-DIMETHYLCYCLOPROPANECARBOXYLATE ◇ POUNCE ◇ PP 557 ◇ S-3151 ◇ SBP-1513 ◇ TALCORD ◇ WL 43479

USE IN FOOD:

Purpose: Pesticide.

Where Used: Various.

Regulations: USDA CES Ranking: B-2 (1987).

SAFETY PROFILE: Poison by inhalation, intravenous and intracerebral routes. Moderately toxic by ingestion. Experimental reproductive effects. Mutagenic data. A skin irritant. An insecticide. When heated to decomposition it emits toxic fumes of Cl^-.

TOXICITY DATA and CODEN

skn-rbt 500 mg/24H MLD NTIS** AD-A047 284
cyt-mus-orl 150 mg/kg PHABDI 21,227,81
orl-rat TDLo:250 mg/kg (6-15D preg):REP BECTA6 29,84,82
orl-rat LD50:410 mg/kg NTIS** AD-A047-284
ihl-rat LC50:685 mg/m^3 YKYUA6 30,1635,79
orl-mus LD50:15 g/kg YKYUA6 30,1635,79

AIV500 CAS: 69-53-4
AMINOBENZYLPENICILLIN

mf: $C_{16}H_{19}N_3O_4S$ mw: 349.44

SYNS: ACILLIN ◇ ADOBACILLIN ◇ ALPEN ◇ AMBLOSIN ◇ AMCILL ◇ AMFIPEN ◇ d-(-)-α-AMINOBENZYLPENICILLIN ◇ d-(-)-α-AMINOPENICILLIN ◇ 6-(d(-)-α-AMINOPHENYLACETAMIDO)PENICILLANIC ACID ◇ (AMINOPHENYLMETHYL)-PENICILLIN ◇ AMIPENIX S ◇ AMPERIL ◇ AMPI-BOL ◇ AMPICILLIN (USDA) ◇ d-AMPICILLIN ◇ d-(-)-AMPICILLIN ◇ AMPICILLIN A ◇ AMPICILLIN ACID ◇ AMPICILLIN ANHYDRATE ◇ AMPICIN ◇ AMPIKEL ◇ AMPIMED ◇ AMPIPENIN ◇ AMPLISOM ◇ AMPLITAL ◇ AMPY-PENYL ◇ AUSTRAPEN ◇ AY-6108 ◇ BINOTAL ◇ BONAPICILLIN ◇ BRITACIL ◇ BRL ◇ BRL 1341 ◇ COPHARCILIN ◇ CYMBI ◇ DIVERCILLIN ◇ DOKTACILLIN ◇ GRAMPENIL ◇ GUICITRINA ◇ GUICITRINE ◇ LIFEAMPIL ◇ MARISILAN ◇ NSC-528986 ◇ NUVAPEN ◇ OMNIPEN ◇ P-50 ◇ PENBRISTOL ◇ PENBRITIN ◇ PENBRITIN PAEDIATRIC ◇ PENBRITIN SYRUP ◇ PENBROCK ◇ PENICLINE ◇ PENTREX ◇ PENTREXL ◇ PFIZERPEN A ◇ POLYCILLIN ◇ PONECIL ◇ PRINCIPEN ◇ QIDAMP ◇ RO-AMPEN ◇ SEMICILLIN ◇ SK-AMPICILLIN ◇ SYNPENIN ◇ TOKIOCILLIN ◇ TOLOMOL ◇ TOTACILLIN ◇ TOTALCICLINA ◇ TOTAPEN ◇ ULTRABION ◇ ULTRABRON ◇ VICCILLIN ◇ VICCILLIN S ◇ VICILLIN ◇ WY-5103

USE IN FOOD:

Purpose: Animal drug.

Where Used: Beef, milk, pork.

Regulations: FDA - 21CFR 556.40. Tolerance of 0.01 ppm in uncooked edible tissues of swine, cattle, and milk. USDA CES Ranking: B-2 (1985).

EPA Genetic Toxicology Program.

SAFETY PROFILE: Poison by intracerebral and other unspecified routes. Moderately toxic by intraperitoneal route. Human systemic effects by ingestion: fever, angranulocytosis, and other blood effects. Experimental reproductive effects. Mutagenic data. When heated to decomposition it emits very toxic fumes of NO_x and SO_x.

TOXICITY DATA and CODEN

dnr-esc 20 μL/plate MUREAV 97,1,82
pic-esc 10 ng/plate CNREA8 43,2819,83
ivn-mus LD50:4990 mg/kg NIIRDN 6,57,82
orl-man TDLo:400 mg/kg/4W-I:BLD,MET AIMEAS 69,91,68

AJB750 CAS: 21087-64-9

4-AMINO-6-tert-BUTYL-3-(METHYLTHIO)-1,2,4-TRIAZIN-5-ON

mf: $C_8H_{14}N_4OS$ mw: 214.32

SYNS: 4-AMINO-6-tert-BUTYL-3-(METHYLTHIO)-1,2,4-TRIAZIN-5-ONE ◇ 4-AMINO-6-tert-BUTYL-3-METHYLTHIO-as-TRIAZIN-5-ONE ◇ 4-AMINO-6-(1,1-DIMETHYLETHYL)-3(METHYLTHIO)-1,2,4-TRIAZIN-5(4H)-ONE ◇ BAY 61597 ◇ BAY DIC 1468 ◇ BAYER 94337 ◇ BAYER 6159H ◇ BAYER 6443H ◇ DIC 1468 ◇ LEXONE ◇ METRIBUZIN ◇ SENCOR ◇ SENCORAL ◇ SENCORER ◇ SENCOREX

USE IN FOOD:

Purpose: Herbicide.

Where Used: Barley (milled fractions, except flour), potato chips, potatoes (processed), sugarcane molasses, wheat (milled fractions, except flour).

Regulations: FDA - 21CFR 193.25. Herbicide residue tolerance of 3 ppm in barley, milled fractions (except flour), 3 ppm in processed potatoes including potato chips, 2 ppm in sugarcane molasses, 3 ppm in wheat milled fractions (except flour). 21CFR 561.41. Limitation of 3 ppm in sugarcane bagasse, 2 ppm in tomato pomace.

EPA Genetic Toxicology Program.

SAFETY PROFILE: Poison by ingestion. When heated to decomposition it emits very toxic fumes of NO_x and SO_x. A selective residue herbicide.

TOXICITY DATA and CODEN

orl-rat LD50:1100 mg/kg FMCHA2 -,C156,83
orl-mus LD50:698 mg/kg 28ZEAL 5,154,76

AMU250 CAS: 1918-02-1

4-AMINO-3,5,6-TRICHLOROPICOLINIC ACID

mf: $C_6H_3Cl_3N_2O_2$ mw: 241.46

PROP: Crystals. Mp: 218°.

SYNS: AMDON GRAZON ◇ 4-AMINO-3,5,6-TRICHLORO-2-PICOLINIC ACID ◇ 4-AMINO-3,5,6-TRICHLORPICOLINSAEURE (GERMAN) ◇ ATCP ◇ BOROLIN ◇ CHLORAMP (RUSSIAN) ◇ K-PIN ◇ NCI-C00237 ◇ PICLORAM ◇ PICLORAM (ACGIH) ◇ TORDON ◇ TORDON 10K ◇ TORDON 22K ◇ TORDON 101 MIXTURE ◇ 3,5,6-TRICHLORO-4-AMINOPICOLINIC ACID

USE IN FOOD:

Purpose: Pesticide.

Where Used: Barley (milled fractions, except flour), oats (milled fractions, except flour), wheat (milled fractions, except flour).

Regulations: FDA - 21CFR 193.350. Pesticide residue tolerance of 3 ppm in barley, milled fractions (except flour), 3 ppm in oats, milled fractions (except flour), 3 ppm in wheat, milled fractions (except flour).

NCI Carcinogenesis Bioassay (feed); No Evidence: mouse NCITR* NCI-CG-TR-23,78; Clear Evidence: rat NCITR* NCI-CG-TR-23,78

ACGIH TLV: TWA 10 mg/m^3

SAFETY PROFILE: An experimental carcinogen, neoplastigen, tumorigen and teratogen. Moderately toxic by ingestion. Mutagenic data. When heated to decomposition it emits very toxic fumes of Cl^- and NO_x. An herbicide and defoliant.

TOXICITY DATA and CODEN

mmo-smc 100 mg/L TGANAK 18,455,84
orl-rat TDLo:5 g/kg (6-15D preg):TER FCTXAV 10,797,72
orl-rat TDLo:7500 mg/kg (6-15D preg):TER FCTXAV 10,797,72
orl-rat TDLo:209 mg/kg/80W-C:CAR JTEHD6 7,207,81
orl-mus TDLo:340 g/kg/80W-C:NEO JTEHO6 7,207,81
orl-rat TD:417 g/kg/80W-C:CAR JTEHD6 7,207,81
orl-rat TD:208 g/kg/80W-C:ETA NCITR* NCI-CG-TR-23,78
orl-rat LD50:2898 mg/kg GNAMAP 15,38,76
orl-mus LD50:1061 mg/kg GNAMAP 15,38,76

AMY700 CAS: 1407-03-0
AMMONIATED GLYCYRRHIZIN

PROP: From roots of *Glycyrrhiza glabra.*

SYN: MONOAMMONIUM GLYCYRRHIZINATE

USE IN FOOD:

Purpose: Flavor enhancer, flavoring agent, surface-active agent.

Where Used: Baked goods, beverages (alcoholic), beverages (nonalcoholic), candy (hard), candy (soft), chewing gum, herbs, plant protein products, seasonings, vitamin or mineral dietary supplements.

Regulations: FDA - 21CFR 184.1408. GRAS with a limitation of (as glycyrrhizn) 0.05 percent in baked goods, 0.1 percent in alcoholic beverages, 0.15 percent in non-alcoholic beverages, 1.1 percent in chewing gum, 16.0 percent in hard candy, 0.15 percent in herbs and seasonings, 0.15 percent in plant protein products, 3.1 percent in soft candy, 0.5 percent in vitamin or mineral dietary supplements, 0.1 percent in all other foods except sugar substitutes when used in accordance with good manufacturing practice. Not permitted to be used as a nonnutritive sweetener in sugar substitutes.

SAFETY PROFILE: When heated to decomposition it emits acrid smoke and irritating fumes.

ANA300 CAS: 9005-34-9
AMMONIUM ALGINATE
mf: $(C_6H_7O_6NH_4)_n$ mw: 193.16 (calc.)

PROP: White to yellow powder. Sol in water; ins in alc, chloroform, ether.

SYN: ALGIN

USE IN FOOD:

Purpose: Boiler water additive, emulsifier, stabilizer, thickener.

Where Used: Frostings, fruits (fabricated), gelatins, gravies, ice cream, jams, jellies, oils, puddings, sauces, sweet sauces, toppings.

Regulations: FDA - 21CFR 173.310, 21CFR 184.1133. GRAS with a limitation of 0.4 percent in confections and frostings, 0.5 percent in fats and oils, gelatins and puddings, 0.4 percent in gravies and sauces, 0.4 percent in jams and jellies, 0.5 percent in sweet sauces, 0.1 percent in all other foods when used in accordance with good manufacturing practice. USDA - 9CFR 318.7. Sufficient for purpose.

SAFETY PROFILE: When heated to decomposition emits toxic fumes of NO_x.

ANB250 CAS: 1066-33-7
AMMONIUM BICARBONATE (1:1)
DOT: 9081
mf: $HCO_3 \cdot H_4N$ mw: 79.1

PROP: Hard, colorless to white crystals; faint ammonia odor, stable at room temp, volatile. Decomp @ 60°, mp: 107.5° (rapid heating). D: 1.586. Sol in water; insol alc.

SYNS: ACID AMMONIUM CARBONATE ◇ AMMONIUM CARBONATE ◇ AMMONIUM HYDROGEN CARBONATE ◇ CARBONIC ACID, MONOAMMONIUM SALT ◇ MONOAMMONIUM CARBONATE

USE IN FOOD:

Purpose: Alkali, dough strengthener, leavening agent, pH control agent, texturizer.

Where Used: Baking powder, cookies.

Regulations: FDA - 21CFR 184.1135. GRAS when used in accordance with good manufacturing practice.

DOT Classification: ORM-E; Label: none.

SAFETY PROFILE: Poison by intravenous route. When heated to decomposition it emits toxic fumes of NO_x and NH_3.

TOXICITY DATA and CODEN

ivn-mus LD50:245 mg/kg AJVRAH 29,897,68

ANE000 CAS: 506-87-6
AMMONIUM CARBONATE
mf: $(NH_4)_2CO_3$ mw: 96.09

PROP: Colorless crystals. Decomposes on standing to ammonium bicarbonate. Sltly sol in water.

SYNS: AMMONIUMCARBONAT (GERMAN) ◇ CARBONIC ACID, AMMONIUM SALT ◇ CARBONIC ACID, DIAMMONIUM SALT ◇ DIAMMONIUM CARBONATE

USE IN FOOD:

Purpose: Buffer, leavening agent, miscellaneous and general-purpose food chemical, neutralizing agent, pH control agent, yeast nutrient.

Where Used: Baked goods, baking powder, caramel, gelatins, puddings, wine.

Regulations: FDA - 21CFR 184.1137. GRAS when used in accordance with good manufacturing practice. BATF - 27CFR 240.1051. Limitation of 0.2 percent. The natural fixed acids shall not be reduced below 5 g/L.

SAFETY PROFILE: Poison by subcutaneous and intravenous routes. When heated to decomposition it emits toxic fumes of NO_x and NH_3.

TOXICITY DATA and CODEN

ivn-mus LD50:96 mg/kg AJVRAH 29,897,68
ivn-dog LDLo:200 mg/kg HBAMAK 4,1289,35
scu-frg LDLo:250 mg/kg HBAMAK 4,1289,35

ANE500 CAS: 12125-02-9
AMMONIUM CHLORIDE

DOT: 9085
mf: $H_4N \cdot Cl$ mw: 53.50

PROP: White crystals; salty taste. Bp: 520°, mp: 337.8°, d: 1.520, vap press: 1 mm @ 160.4° (subl). Sol in water, alc, glycerin.

SYNS: AMMONIUMCHLORID (GERMAN) ◇ AMMONIUM MURIATE ◇ CHLORID AMONNY (CZECH) ◇ SAL AMMONIA ◇ SAL AMMONIAC

USE IN FOOD:

Purpose: Dough conditioner, dough strengthener, flavor enhancer, leavening agent, processing aid, yeast food.

Where Used: Baked goods, condiments, relishes.

Regulations: FDA - 21CFR 184.1138. GRAS when used in accordance with good manufacturing practice.

ACGIH TLV: TWA 10 mg/m^3; STEL 20 mg/m^3 DOT Classification: ORM-E

SAFETY PROFILE: Poison by subcutaneous, intravenous and intramuscular routes. Moderately toxic by other routes. A severe eye irritant. Explosive reaction with potassium chlorate or bromine trifluoride. Violent reaction (ignition) with bromine pentafluoride; NH_4; NO_3; and IF_7. Reaction with hydrogen cyanide may give the explosive nitrogen trichloride. When heated to decomposition it emits very toxic fumes of NO_x, Cl^-, and NH_3.

TOXICITY DATA and CODEN

eye-rbt 500 mg/24H SEV 28ZPAK -,15,72
orl-rat LD50:1650 mg/kg 28ZPAK -,15,72
orl-dog LDLo:600 mg/kg HBAMAK 4,1289,35

ANF800 CAS: 3012-65-5
AMMONIUM CITRATE

DOT: 9087
mf: $C_6H_{14}N_2O_7$ mw: 226.19

PROP: Granules or crystals. D: 1.48. Sol in water; sltly sol in alc.

SYNS: AMMONIUM CITRATE, DIBASIC (DOT) ◇ DIAMMONIUM CITRATE

USE IN FOOD:

Purpose: Stabilizer.

Where Used: Packaging materials.

Regulations: FDA - 21CFR 181.29. Use in accordance with good manufacturing practice.

DOT Classification: ORM-E

SAFETY PROFILE: A skin and eye irritant. When heated to decomposition it emits acrid smoke and irritating fumes.

ANK250 CAS: 1336-21-6
AMMONIUM HYDROXIDE

DOT: 2672
mf: $H_4N \cdot HO$ mw: 35.06

PROP: Clear, colorless liquid solution of ammonia; very pungent odor. D: 0.90, mp: −77°. Sol in water. Soln contains not more than 44% ammonia.

SYNS: AMMONIA AQUEOUS ◇ AMMONIA SOLUTION (DOT) ◇ AQUA AMMONIA

USE IN FOOD:

Purpose: Alkali, boiler water additive, leavening agent, pH control agent, surface-finishing agent.

Where Used: Baked goods, caramel, cheese, fruits (processed), puddings.

Regulations: FDA - 21CFR 184.1139. GRAS when used in accordance with good manufacturing practice.

DOT Classification: Corrosive Material; Label: Corrosive NIOSH REL: CL 50 ppm

SAFETY PROFILE: A human poison by ingestion. An experimental poison by inhalation and ingestion. A severe eye irritant. Human systemic eye and other systemic irritant effects by inhalation. Mutagenic data. Incompatible with acrolein; nitromethane; acrylic acid; chlorosulfonic acid; dimethyl sulfate; halogens; (Au + aqua regia); HCl; HF; HNO_3; oleum; β-propiolactone; propylene oxide; $AgNO_3$; Ag_2O; (Ag_2O + C_2H_5OH); $AgMnO_4$; H_2SO_4. Dangerous; liquid can inflict burns. Use with adequate ventilation. When heated to decomposition it emits NH_3 and NO_x.

TOXICITY DATA and CODEN

eye-rbt 750 μg SEV AJOPAA 29,1363,46
eye-rbt 44 μg SEV AROPAW 25,839,41
mmo-sat 10 μL/plate ANYAA9 76,475,58
mmo-esc 10 μL/disc ANYAA9 76,475,58
orl-hmn LDLo:43 mg/kg 34ZIAG -,95,69
ihl-hmn LCLo:5000 ppm 34ZIAG -,95,69
ihl-hmn TCLo:700 ppm:EYE JISMAB 61,271,71
ihl-hmn TCLo:408 ppm:IRR JISMAB 61,271,71
orl-rat LD50:350 mg/kg JIHTAB 23,259,41

ANR500 CAS: 7783-28-0
AMMONIUM PHOSPHATE DIBASIC
mf: $H_6N_2 \cdot H_3O_4P$ mw: 132.08

PROP: White crystals or powder; salty taste. D: 1.619, mp: 155° (decomp). Sol in water; insol in alc.

SYNS: AMMONIUM PHOSPHATE ◇ DIAMMONIUM HYDROGEN PHOSPHATE ◇ DIBASIC AMMONIUM PHOSPHATE ◇ SECONDARY AMMONIUM PHOSPHATE

USE IN FOOD:

Purpose: Buffer, dough conditioner, firming agent, leavening agent, pH control agent, yeast food.

Where Used: Baked goods, beverages (alcoholic), condiments, puddings, wine.

Regulations: FDA - 21CFR 184.1141b. GRAS when used in accordance with good manufacturing practice. BATF - 27CFR 240.1051. Limitation of 0.17 percent as a yeast nutrient in wine production. Limitation of 0.8 percent in the production of sparkling wines.

SAFETY PROFILE: Low to moderate toxicity. When heated to decomposition it emits very toxic fumes of PO_x, NO_x, and NH_3.

ANR750 CAS: 7772-76-1
AMMONIUM PHOSPHATE, MONOBASIC
mf: $NH_4H_2PO_4$ mw: 115

PROP: Brilliant white crystals or powder. D: 1.803 @ 19°; mp: 190°. Sol in water.

USE IN FOOD:

Purpose: Buffer, dough conditioner, leavening agent, yeast food.

Where Used: Baked goods, baking powder, frozen desserts, margarine, whipped toppings, yeast food.

Regulations: FDA - 21CFR 184.1141a. GRAS when used in accordance with good manufacturing practice.

SAFETY PROFILE: Incompatible with NaOCl. A general-purpose food additive.

ANT100
AMMONIUM POTASSIUM HYDROGEN PHOSPHATE

USE IN FOOD:

Purpose: Stabilizer.

Where Used: Packaging materials.

Regulations: FDA - 21CFR 181.29. Use in accordance with good manufacturing practice.

SAFETY PROFILE: When heated to decomposition it emits acrid smoke and irritating fumes.

ANT500
AMMONIUM SACCHARIN
mf: $C_7H_8N_2O_3S$ mw: 200.21

PROP: White crystals or crystalline powder; intense sweet taste. Sol in water.

SYN: 1,2-BENZISOTHIAZOLIN-3-ONE 1,1-DIOXIDE AMMONIUM SALT

USE IN FOOD:

Purpose: Nonnutritive sweetener.

Where Used: Bacon, bakery products (nonstandardized), beverage mixes, beverages, chewing gum, desserts, fruit juice drinks, jams, vitamin tablets (chewable).

Regulations: FDA - 21CFR 180.37. Limitation of 12 mg per fluid ounce of beverages; 12 mg per fluid ounce of fruit juice drinks; 12 mg per fluid ounce of beverage mixes, 30 mg per serving in processed foods. USDA - 9CFR 318.7. Limitation of 0.01 percent.

SAFETY PROFILE: When heated to decomposition emits toxic fumes of NO_x.

ANU750 CAS: 7783-20-2
AMMONIUM SULFATE (2:1)
DOT: 2506
mf: $H_8N_2O_4S$ mw: 132.16

PROP: White crystals. Mp: > 280° (decomp); d: 1.77. Sol in water; ins in alc.

SYNS: AMMONIUM SULPHATE ◇ DIAMMONIUM SULFATE ◇ SULFURIC ACID, DIAMMONIUM SALT

USE IN FOOD:

Purpose: Dough conditioner, firming agent, miscellaneous and general-purpose food chemical, processing aid, yeast food.

Where Used: Baked goods, gelatins, puddings.

Regulations: FDA - 21CFR 184.1143. GRAS with a limitation of 0.15 percent in baked goods, 0.1 percent in gelatins and puddings when used in accordance with good manufacturing practice.

Community Right-To-Know List.

DOT Classification: ORM-B; Label: None

SAFETY PROFILE: Moderately toxic by several routes. Incandescent reaction on heating with potassium chlorate. Reaction with sodium hypochlorite gives the unstable explosive nitrogen trichloride. Incompatible with ($K + NH_4NO_3$); KNO_2; ($NaK + NH_4NO_3$). When heated to decomposition it emits very toxic fumes of NO_x, NH_3, and SO_x.

TOXICITY DATA and CODEN

orl-man TDLo:1500 mg/kg GISAAA 42(2),100,77
orl-rat LD50:3000 mg/kg CNJMAQ 12,216,48

AOA100 CAS: 61336-70-7
AMOXICILLIN TRIHYDDRATE
mf: $C_{16}H_{19}N_3O_5S \cdot 3H_2O$ mw: 419.50

SYNS: α-AMINO-p-HYDROXYBENZYLPENICILLIN TRIHYDRATE ◇ BRL 2333 TRIHYDRATE ◇ (2S-(2-α,5-α,6-β(S*)))-6-((AMINO(4-HYEROXYPHENYL)ACETYL)AMINO)-3,3-DIMETHYL-7-OXO-4-THIA-1-AZABICYCLO (3.2.0)HEPTANE-2-CARBOXYLIC ACID TRIHYDRATE

USE IN FOOD:

Purpose: Animal drug.

Where Used: Meat, milk.

Regulations: FDA - 21CFR 556.38. Tolerance of 0.01 ppm in milk and uncooked edible tissues of meat.

SAFETY PROFILE: Moderately toxic. An experimental teratogen. Experimental reproductive effects. When heated to decomposition it emits toxic fumes of SO_x and NO_x.

TOXICITY DATA and CODEN

orl-rat TDLo:162 g/kg (25W male):REP KSRNAM 7,3074,73
orl-rat TDLo:2800 mg/kg (7-13D preg):TER KSRNAM 7,3113,73
orl-mus TDLo:9100 mg/kg (7-13D preg):TER KSRNAM 7,3113,73
ipr-rat LD50:2870 mg/kg KSRNAM 7,3040,73

AOD125 CAS: 7177-48-2
AMPICILLIN TRIHYDRATE
mf: $C_{16}H_{19}N_3O_4S \cdot 3H_2O$ mw: 403.50

SYNS: AMCAP ◇ AMCILL ◇ AMINOBENZYLPENICILLIN TRIHYDRATE ◇ α-AMINOBENZYLPENICILLIN TRIHYDRATE ◇ AMPERIL ◇ AMPICHEL ◇ AMPIKEL ◇ AMPINOVA ◇ AMPLIN ◇ ANCILLIN ◇ CYMBI ◇ DIVERCILLIN ◇ LIFEAMPIL ◇ MOREPEN ◇ NCI-C56086 ◇ PEN A ◇ PENSYN ◇ POLYCILLIN ◇ PRINCILLIN ◇ RO-AMPEN ◇ TRAFARBIOT ◇ UKOPEN ◇ VIDOPEN

USE IN FOOD:

Purpose: Animal drug.

Where Used: Meat.

Regulations: USDA CES Ranking: B-2 (1985).

SAFETY PROFILE: An experimental teratogen. Experimental reproductive effects. When heated to decomposition it emits toxic fumes of SO_x and NO_x.

TOXICITY DATA and CODEN

orl-rat TDLo:1500 mg/kg (6-11D preg):REP ANTBAL 18,815,73
orl-rat TDLo:2800 mg/kg (7-13D preg):TER KSRNAM 7,3113,73
orl-mus TDLo:28 g/kg (7-13D pre):TER KSRNAM 7,3113,73

AOD175 CAS: 121-25-5
AMPROLIUM
mf: $C_{14}H_{19}ClN_4$ mw: 278.78

PROP: Crystals from methanol + ethanol. Decomp 248-249°. Sol in water, methanol, 95% eth; insol in isopropanol, butanol, dioxane, acetone, ethyl acetate, acetonitrile, isooctane.

SYNS: 1-[(4-AMINO-2-PROPYL-5-PYRIMIDINYL) METHYL]-2-METHYLPYRIDINIUM CHLORIDE ◇ 1-(4-AMINO-2-n-PROPYL-5-PYRIMIDINYLMETHYL)-2-PICOLINIUM CHLORIDE ◇ CORID

USE IN FOOD:

Purpose: Animal drug.

Where Used: Beef, chicken, eggs (chicken), eggs (turkey), pheasants, turkey.

Regulations: FDA - 21CFR 556.50. Limitation in chickens and turkeys of 1 ppm in liver and kidney, 0.5 ppm in muscle. Limitation of 8 ppm in egg yolks, 4 ppm in whole eggs. Limitation in calves of 2.0 ppm in uncooked fat, 0.5 ppm in uncooked muscle, liver, kidney. Limitation in pheasants of 1 ppm in uncooked liver, 0.5 ppm in uncooked, muscle. 21CFR 558.55.

SAFETY PROFILE: When heated to decomposition emits toxic fumes of Cl^-.

AOG500 CAS: 122-40-7
α-AMYL CINNAMALDEHYDE
mf: $C_{14}H_{18}O$ mw: 202.32

PROP: Yellow liquid; floral jasmine odor. D: 0.963, refr index: 1.554, bp: 174-175° @ 20 mm. Sol in fixed oils; ins in glycerin, propylene glycol

SYNS: α-AMYL CINNAMIC ALDEHYDE ◇ α-AMYL-β-PHENYLACROLEIN ◇ FEMA No. 2061 ◇ JASMINALDEHYDE ◇ α-PENTYLCINNAMALDEHYDE

USE IN FOOD:

Purpose: Flavoring agent.

Where Used: Various.

Regulations: FDA - 21CFR 172.515. Use at a level not in excess of the amount reasonably required to accomplish the intended effect.

SAFETY PROFILE: Moderately toxic by ingestion. A mild skin irritant. When heated to decomposition it emits acrid smoke and irritating fumes.

TOXICITY DATA and CODEN

skn-gpg 5%/2W MLD ADVEA4 58,121,78
orl-rat LD50:3730 mg/kg FCTXAV 2,327,64

AOG600
AMYL CINNAMATE
mf: $C_{14}H_{18}O_2$ mw: 218.28

PROP: Colorless to pale yellow liquid; slt cocoa odor. D: 0.992-0.997, refr index: 1.535, flash p: +100°. Sol in fixed oils; sltly sol in propylene glycol; ins in glycerin @ 310°.

SYNS: FEMA No. 2063 ◇ ISOAMYL CINNAMATE ◇ ISOAMYL 3-PENTYL PROPENATE

USE IN FOOD:

Purpose: Flavoring agent.

Where Used: Various.

Regulations: FDA - 21CFR 172.515. Use at a level not in excess of the amount reasonably required to accomplish the intended effect.

SAFETY PROFILE: Combustible liquid. When heated to decomposition it emits acrid smoke and irritating fumes.

AOJ900
AMYL HEPTANOATE
mf: $C_{12}H_{24}O_2$ mw: 200.32

PROP: Colorless to pale yellow liquid; fruity taste. D: 0.859, refr index: 1.422.

SYN: FEMA No. 2073

USE IN FOOD:

Purpose: Flavoring agent.

Where Used: Various.

Regulations: GRAS when used at a level not in excess of the amount reasonably required to accomplish the intended effect.

SAFETY PROFILE: When heated to decomposition it emits acrid smoke and irritating fumes.

AOM125 CAS: 9032-08-0
AMYLOGLUCOSIDASE

PROP: A powder derived from *Rhizopus niveus* with diatomaceous earth as a carrier.

USE IN FOOD:

Purpose: Degrading agent.

Where Used: Distilled spirits, vinegar.

Regulations: FDA - 21CFR 173.110. Limitation of 0.1 percent by weight of the gelatinized starch.

SAFETY PROFILE: When heated to decomposition it emits acrid smoke and irritating fumes.

AON350
AMYL PROPIONATE
mf: $C_8H_{16}O_2$ mw: 144.21

PROP: Colorless liquid; fruity, apricot-pineapple odor. D: 0.866, refr index: 1.405-1.409, flash p: 41°. Sol in alc, fixed oils; insol in glycerine, propylene glycol, water @ 160°.

SYNS: FEMA No. 2082 ◇ ISOAMYL PROPIONATE

USE IN FOOD:

Purpose: Flavoring agent.

Where Used: Various.

Regulations: FDA - 21CFR 172.515. Use at a level not in excess of the amount reasonably required to accomplish the intended effect.

SAFETY PROFILE: Combustible liquid. When heated to decomposition it emits acrid smoke and irritating fumes.

AON600
AMYRIS OIL, WEST INDIAN TYPE

PROP: Extracted from *Amryris balsamifera L.* (Fam. *Rutaceae*). Clear, pale yellow viscous liquid; odor of sandalwood. Sol in mineral oil, propylene glycol; ins in glycerin.

SYN: SANDALWOOD OIL, WEST INDIAN OIL

USE IN FOOD:

Purpose: Flavoring agent.

Where Used: Bakery products, beverages (nonalcoholic), chewing gum, confections, ice cream, puddings.

Regulations: FDA - 21CFR 172.510. Use at a level not in excess of the amount reasonably required to accomplish the intended effect.

SAFETY PROFILE: When heated to decomposition it emits acrid smoke and irritating fumes.

AOO780
ANGELICA ROOT OIL

PROP: Extracted from roots of *Angelica archangelica L.* A pale yellow to amber liquid; pungent odor with bitter-sweet taste. Sol in fixed oils; sltly sol in mineral oil; ins in glycerin, propylene glycol.

USE IN FOOD:

Purpose: Flavoring agent.

Where Used: Various.

Regulations: FDA - 21CFR 182.10. GRAS when used at a level not in excess of the amount reasonably required to accomplish the intended effect.

SAFETY PROFILE: When heated to decomposition it emits acrid smoke and irritating fumes.

AOO790
ANGELICA SEED OIL

PROP: Extracted from seeds of *Angelica archangelica L.* A light yellow liquid; sweet taste. Sol in fixed oils; sltly sol in mineral oil; ins in glycerin, propylene glycol.

USE IN FOOD:

Purpose: Flavoring agent.

Where Used: Various.

Regulations: FDA - 21CFR 182.10. GRAS when used at a level not in excess of the amount reasonably required to accomplish the intended effect.

SAFETY PROFILE: When heated to decomposition it emits acrid smoke and irritating fumes.

AOT500 CAS: 123-11-5
p-ANISALDEHYDE
mf: $C_8H_8O_2$ mw: 136.15

PROP: Colorless oil; hawthorn odor. D: 1.123 @ 20°/4°, refr index: 1.571-1.574, mp: 2.5°, bp: 247-248°, flash p: 121°. Misc in alc, ether, fixed oils; sol in propylene glycol; ins in glycerin, water.

SYNS: ANISIC ALDEHYDE ◇ FEMA No. 2670 ◇ 4-METHOXYBENZALDEHYDE ◇ p-METHOXYBENZALDEHYDE (FCC)

USE IN FOOD:

Purpose: Flavoring agent.

Where Used: Various.

Regulations: FDA - 21CFR 172.515. Use at a level not in excess of the amount reasonably required to accomplish the intended effect.

SAFETY PROFILE: Moderately toxic by ingestion. A skin irritant. Mutagenic data. Combustible liquid. When heated to decomposition it emits acrid smoke and irritating fumes.

TOXICITY DATA and CODEN

skn-rbt 500 mg/24H MOD FCTXAV 12,807,74
mmo-sat 400 μL/plate BECTA6 24,590,80
orl-rat LD50: 1510 mg/kg FCTXAV 2,327,64

AOU250 CAS: 8007-70-3
ANISE OIL

PROP: Consists of (80-90%) of Anethole. Small quantities of methyl chavicol, p-methoxyacetophenone and other materials also. Found in the dried ripe fruit of *Impinella anisum L.* (FCTXAV 11,855,73). D: 0.978-0.988 @ 25°/25°.

SYNS: ANISEED OIL ◇ ANIS OEL (GERMAN) ◇ OIL OF ANISE ◇ STAR ANISE OIL

USE IN FOOD:

Purpose: Flavoring agent.

Where Used: Bakery products, beverages (alcoholic), beverages (nonalcoholic), candy, chewing gum, confections, ice cream, liquors, meat, pastries (sweet), soups.

Regulations: FDA - 21CFR 182.10, 182.20. GRAS when used at a level not in excess of the amount reasonably required to accomplish the intended effect.

SAFETY PROFILE: Moderately toxic by ingestion. A weak sensitizer. May cause contact dermatitis. Combustible liquid. When heated to decomposition it emits acrid smoke and irritating fumes.

TOXICITY DATA and CODEN

orl-rat LD50:2250 mg/kg FCTXAV 11,855,73

AOV000 CAS: 94-30-4
p-ANISIC ACID, ETHYL ESTER
mf: $C_{10}H_{12}O_3$ mw: 180.21

PROP: Colorless liquid; fruity, anise odor. D: 1.103 @ 25/25, refr index: 1.522-1.526, mp: 7-8°, bp: 269-270°, flash p: +100°. Sol in alc, ether; sltly sol in water.

SYNS: ETHYL ANISATE ◇ ETHYL-p-ANISATE (FCC) ◇ ETHYL-4-METHOXYBENZOATE ◇ ETHYL-p-METHOXYBENZOATE ◇ FEMA No. 2420

USE IN FOOD:

Purpose: Flavoring agent.

Where Used: Various.

Regulations: FDA - 21CFR 172.515. Use at a level not in excess of the amount reasonably required to accomplish the intended effect.

SAFETY PROFILE: Moderately toxic by ingestion. Combustible liquid. When heated to decomposition it emits acrid smoke and irritating fumes.

TOXICITY DATA and CODEN

orl-rat LD50:2040 mg/kg FCTXAV 14,659,76

AOX750 CAS: 100-66-3
ANISOLE
mf: C_7H_8O mw: 108.15

PROP: Mobile liquid, clear straw color; phenol, anise odor. Vapor d: 3.72, mp: −37.3°, bp: 153.8°, flash p: 125°F (COC), d: 0.983-0.988, refr index: 1.513-1.518, vap press: 10 mm @ 42.2°, autoign temp: 887°F. Insol in water; sol in alc and ether.

SYNS: FEMA No. 2097 ◇ METHOXYBENZENE ◇ METHYL PHENYL ETHER ◇ PHENYL METHYL ETHER

USE IN FOOD:

Purpose: Flavoring agent.

Where Used: Various.

Regulations: FDA - 21CFR 172.515. Use at a level not in excess of the amount reasonably required to accomplish the intended effect.

SAFETY PROFILE: Moderately toxic by ingestion. A skin irritant. Combustible liquid. To fight fire, use foam, CO_2, dry chemical. When heated to decomposition it emits acrid fumes.

TOXICITY DATA and CODEN

skn-rbt 500 mg/24H MOD FCTXAV 17,241,79
orl-rat LD50:3700 mg/kg TXAPA9 6,378,64
orl-mus LD50:2800 mg/kg JPETAB 88,400,46

AOY400
ANISYL ACETATE
mf: $C_{10}H_{12}O_3$ mw: 180.20

PROP: Colorless to slt yellow liquid; fruity, balsamic odor. D: 1.104, refr index: 1.511-1.516, flash p: +99°. Sol alc, most oils; ins in glycerin, propylene glycol.

SYNS: FEMA No. 2098 ◇ p-METHOXYBENZYL ACETATE

USE IN FOOD:

Purpose: Flavoring agent.

Where Used: Various.

Regulations: FDA - 21CFR 172.515. Use at a level not in excess of the amount reasonably required to accomplish the intended effect.

SAFETY PROFILE: Combustible liquid. When heated to decomposition it emits acrid smoke and irritating fumes.

APE300 CAS: 60837-57-2
ANOXOMER

PROP: A polymer consisting of 1,4-benzenediol, 2-(1,1-dimethylethyl)-polymer with diethylbenzene, 4-(1,1-dimethylethyl)phenol, 4-methoxyphenol, 4,4′(1-methylethylidene)bis-(phenol) and 4-methylphenol.

USE IN FOOD:

Purpose: Antioxidant.

Where Used: Various.

Regulations: FDA - 21CFR 172.105. Not more than 5,000 ppm based on fat and oil content of the food.

SAFETY PROFILE: When heated to decomposition it emits acrid smoke and irritating fumes.

API750 CAS: 87-29-6
ANTHRANILIC ACID, CINNAMYL ESTER
mf: $C_{16}H_{15}NO_2$ mw: 253.32

PROP: Reddish yellow powder; balsamic odor. Mp: 60°, flash p: +100°. Sol in alc, chloroform, ether; ins in water.

SYNS: 2-AMINOBENZOIC ACID-3-PHENYL-2-PROPENYL ESTER ◇ CINNAMYL ALCOHOL ANTHRANILATE ◇ CINNAMYL-2-AMINOBENZOATE ◇ CINNAMYL-o-AMINOBENZOATE ◇ CINNAMYL ANTHRANILATE (FCC) ◇ FEMA No. 2295 ◇ NCI-C03510 ◇ 3-PHENYL-2-PROPENYLANTHRANILATE ◇ 3-PHENYL-2-PROPEN-1-YL ANTHRANILATE

USE IN FOOD:

Purpose: Flavoring agent.

Where Used: Baked goods, beverages, candy.

Regulations: FDA - 21CFR 172.515, 189.113. Prohibited from direct addition or use in human food.

IARC Cancer Review: Animal Limited Evidence IMEMDT 31,133,83; Animal Inadequate Evidence IMEMDT 16,287,78; NCI Carcinogenesis Bioassay (feed); Clear Evidence: mouse, rat NCITR* NCI-CG-TR-196,80.

SAFETY PROFILE: An experimental neoplastigen. Combustible liquid. When heated to decomposition it emits toxic fumes of NO_x.

TOXICITY DATA and CODEN

ipr-mus TDLo: 12 g/kg/8W-I: NEO CNREA8 33,3069,73
orl-rat LD50: 5000 mg/kg FCTXAV 13,681,75
skn-rbt LD50: 5000 mg/kg FCTXAV 13,681,75

APJ250 CAS: 134-20-3
ANTHRANILIC ACID, METHYL ESTER
mf: $C_8H_9NO_2$ mw: 151.18

PROP: Plates from alc or colorless liquid; grape odor. D: 1.161-1.169, mp: 23.8°, bp: 225-230° @ 15 mm, flash p: 104°. Very sol in water, propylene glycol, hot abs alc (23/100); insol in ether, chloroform, glycerin.

SYNS: 2-AMINOBENZOIC ACID METHYL ESTER ◇ o-AMINOBENZOIC ACID METHYL ESTER ◇ 2-CARBOMETHOXYANILINE ◇ o-CARBOMETHOXYANILINE ◇ FEMA No. 2682 ◇ 2-(METHOXYCARBONYL)ANILINE ◇ METHYL-2-AMINOBENZOATE ◇ METHYL-o-AMINOBENZOATE ◇ METHYL ANTHRANILATE (FCC)

USE IN FOOD:

Purpose: Flavoring agent.

Where Used: Various.

Regulations: FDA - 21CFR 182.60. GRAS when used at a level not in excess of the amount reasonably required to accomplish the intended effect.

SAFETY PROFILE: An experimental tumorigen. Moderately toxic by ingestion. A skin irritant. Combustible liquid. When heated to decomposition it emits toxic fumes of NO_x.

TOXICITY DATA and CODEN

ipr-mus TDLo: 2250 mg/kg/8W-I: ETA CNREA8 33,3069,73
skn-rbt 500 mg/24H MOD FCTXAV 12,807,74
orl-rat LD50: 2910 mg/kg FCTXAV 2,327,64

AQB000 CAS: 31282-04-9
ANTIHELMYCIN
mf: $C_{20}H_{37}N_3O_{13}$ mw: 527.60

SYNS: HYGROMIX-8 ◇ HYGROMYCIN B (USDA)

USE IN FOOD:

Purpose: Animal drug, animal feed drug.

Where Used: Animal feed, eggs, pork, poultry.

Regulations: FDA - 21CFR 556.330. Limitation of zero in eggs, swine, poultry. 21CFR 558.274. USDA CES Ranking: A-Z (1988).

SAFETY PROFILE: Poison by intraperitoneal route. When heated to decomposition it emits toxic fumes of NO_x.

TOXICITY DATA and CODEN

ipr-rat LD50: 63 mg/kg GISAAA 38,11,73
ipr-gpg LD50: 13 mg/kg GISAAA 38,11,73

AQO300 CAS: 1107-26-2
β-APO-8′-CAROTENAL
mf: $C_{30}H_{40}O$ mw: 416.65

PROP: Fine crystalline powder with dark metallic sheen. Sol in chloroform; sltly sol in acetone; ins in water.

SYNS: APO ◇ APOCAROTENAL

USE IN FOOD:

Purpose: Color additive.

Where Used: Beverages (orange), cheese, desserts, ice cream.

Regulations: FDA - 21CFR 73.90. Limitation of 15 mg/pound of solid or per pint of liquid food.

SAFETY PROFILE: When heated to decomposition it emits acrid smoke and irritating fumes.

AQP885 CAS: 37321-09-8
APRAMYCIN
mf: $C_{21}H_{41}N_5O_{11}$ mw: 539.60

PROP: Mp: 245-247°. Sol in water; sltly sol in lower alcs.

SYNS: AMBYLAN ◇ O-4-AMINO-4-DEOXY-A-D-GLUCOPYRANOSYL-(1-8)-O-(8R)-2-AMINO-2,3,7-TRIDEOXY-7-(METHYLAMINO)-D-GLYCERO-A-D-ALLO-OCTODIALDO-1,5:8,4-DIPYRANOSYL-(1-4)-2-DEOXY-D-STREPAMINE ◇ APRALAN

USE IN FOOD:

Purpose: Animal drug.

Where Used: Pork.

Regulations: FDA - 21CFR 556.52. Limitation of 0.1 ppm in kidney of swine. 21CFR 558.59.

SAFETY PROFILE: When heated to decomposition it emits acrid smoke and irritating fumes.

AQQ500 CAS: 9000-01-5
ARABIC GUM
mw: 240,000

PROP: A gum from the stems and branches of *Acacia senegal (L.)* Willd. or of *Acacia* (Fam. *Leguminosae*). Sol in water; ins in alc.

SYNS: ACACIA ◇ ACACIA DEALBATA GUM ◇ ACACIA GUM ◇ ACACIA SENEGAL ◇ ACACIA SYRUP ◇ AUSTRALIAN GUM ◇ GUM ARABIC ◇ GUM OVALINE ◇ GUM SENEGAL ◇ INDIAN GUM ◇ NCI-C50748 ◇ SENEGAL GUM ◇ STARSOL NO. 1 ◇ WATTLE GUM

USE IN FOOD:

Purpose: Emulsifier, flavoring agent, formulation aid, humectant, stabilizer, surface-finishing agent, thickener.

Where Used: Beverage bases, beverages, candy (hard), candy (soft), chewing gum, confections, cough drops, dairy product analogs, fats, fillings, frostings, gelatins, nut products, nuts, oils, puddings, quiescently frozen confection products, snack foods.

Regulations: FDA - 21CFR 184.1330. GRAS with a limitation of 2.0 percent in beverage and beverage bases, 5.6 percent in chewing gum, 12.4 percent in confections and frostings, 1.3 percent in dairy product analogs, 1.5 percent in fats and oils, 2.5 percent in gelatins, puddings, and fillings, 46.5 percent in hard candy and cough drops, 8.3 percent in nuts and nut products, 6.0 percent in quiescently frozen confection products, 4.0 percent in snack foods, 85.0 percent in soft candy, 1.0 percent in all other foods when used in accordance with good manufacturing practice.

NTP Carcinogenesis Bioassay (feed); No Evidence: mouse, rat NTPTR* NTP-TR-227,82.

SAFETY PROFILE: Inhalation or ingestion has produces hives, eczema, and angiodema. A weak allergen. Combustible. When heated to decomposition it emits acrid smoke.

TOXICITY DATA and CODEN

orl-rbt LD50:8000 mg/kg FDRLI* 124,-,76

AQR800 CAS: 9036-66-2
ARABINOGALACTAN

PROP: Derived from water extraction of Western larch wood having galactose units and arabinose units in the ratio of approx. 6:1. Mp: >200° (decomp). Sol in water.

SYNS: (+)-ARABINOGALACTAN ◇ LARCH GUM ◇ POLYARABINOGALACTAN

USE IN FOOD:

Purpose: Binder, bodying agent, emulsifier, stabilizer.

Where Used: Dressings (nonstandardized), essential oils, flavor bases, nonnutritive sweetener, pudding mixes.

Regulations: FDA - 21CFR 172.615. Use at a level not in excess of the amount reasonably required to accomplish the intended effect.

SAFETY PROFILE: When heated to decomposition it emits acrid smoke and irritating fumes.

AQV980 CAS: 74-79-3
l-ARGININE
mf: $C_6H_{14}N_4O_2$ mw: 174.20

PROP: White crystalline powder. Sol in water; sltly sol in alc; ins in ether.

SYN: l-1-AMINO-4-GUANIDOVALERIC ACID

USE IN FOOD:

Purpose: Dietary supplement, nutrient.

Where Used: Various.

Regulations: FDA - 21CFR 172.320. Limitation 6.6 percent by weight.

SAFETY PROFILE: When heated to decomposition emits toxic fumes of NO_x.

AQW000 CAS: 1119-34-2
l-ARGININE MONOHYDROCHLORIDE
mf: $C_6H_{14}N_4O_2 \cdot ClH$ mw: 210.70

PROP: White crystalline powder; odorless. Mp: 222-235° (decomp). Very sol in water; sltly sol in alc.

SYNS: ARGAMINE ◇ ARGININE HYDROCHLORIDE ◇ l-ARGININE HYDROCHLORIDE ◇ ARGININE MONOHYDROCHLORIDE ◇ ARGIVENE ◇ DETOXARGIN ◇ l-HYDROCHLORIDE ARGININE ◇ LEVARGIN ◇ MINOPHAGEN A ◇ R-GENE

USE IN FOOD:

Purpose: Dietary supplement, nutrient.

Where Used: Various.

Regulations: FDA - 21CFR 172.320. Limitation 6.6 percent by weight.

SAFETY PROFILE: Moderately toxic by intraperitoneal route. Mildly toxic by ingestion. An experimental teratogen. When heated to decomposition it emits very toxic fumes of NO_x and HCl.

TOXICITY DATA and CODEN

ipr-rat TDLo:90 mg/kg (1-6D preg):TER AJEBAK 51,553,73
orl-rat LD50:12 g/kg JPMSAE 62,49,73

ARA250 CAS: 98-50-0
ARSANILIC ACID

mf: $C_6H_8AsNO_3$ mw: 217.06

PROP: Needles from aq solns. Mp: 232°, bp: decomp, $-H_2O$ @ 15°. Very sol in hot water, alc; insol in ether and benzene.

SYNS: 4-AMINOBENZENEARSONIC ACID ◇ p-AMINOBENZENEARSONIC ACID ◇ AMINOPHENYLARSINE ACID ◇ p-AMINOPHENYLARSINE ACID ◇ p-AMINOPHENYLARSINIC ACID ◇ 4-AMINOPHENYLARSONIC ACID ◇ p-AMINOPHENYLARSONIC ACID ◇ p-ANILINEARSONIC ACID ◇ ANTOXYLIC ACID ◇ 4-ARSANILIC ACID ◇ p-ARSANILIC ACID ◇ ATOXYLIC ACID

TOXICITY DATA and CODEN

orl-rat LD50:216 mg/kg TXAPA9 18,185,71
ipr-rat LDLo:400 mg/kg JPETAB 80,393,44
ipr-mus LD50:291 mg/kg JMCMAR 9,221,66
ivn-mus LD50:100 mg/kg CSLNX* NX#06774

USE IN FOOD:

Purpose: Animal feed drug.

Where Used: Animal feed.

Regulations: FDA - 21CFR 558.6. Use at a level not in excess of the amount reasonably required to accomplish the intended effect. USDA CES Ranking: C-1 (1987).

IARC Cancer Review: Animal Inadequate Evidence IMEMDT 23,39,80. Arsenic and its compounds are on the Community Right-To-Know List.

OSHA PEL: TWA 10 $\mu g/m^3$ ACGIH TLV: TWA 0.2 $mg(As)/m^3$

SAFETY PROFILE: Poison by ingestion, intravenous, and intraperitoneal routes. A human carcinogen. Flammable, decomposes with heat to yield flammable vapors. When heated to decomposition or on contact with acid or acid fumes emits highly toxic fumes of As and NO_x.

ARA500 CAS: 127-85-5
ARSANILIC ACID, MONOSODIUM SALT

mf: $C_6H_7AsNO_3 \cdot Na$ mw: 239.05

PROP: Tetrahydrate: white, odorless, crystalline powder; faint salty taste. Sol in water; somewhat sol in alc.

SYNS: (4-AMINOPHENYL)ARSONIC ACID SODIUM SALT ◇ ANHYDROUS SODIUM ARSANILATE ◇ ARSANILIC ACID SODIUM SALT ◇ ATOXYL ◇ NCI-C61176 ◇ SODIUM AMINARSONATE ◇ SODIUM-p-AMINOBENZENEARSONATE ◇ SODIUM AMINOPHENOL ARSONATE ◇ SODIUM-p-AMINOPHENYLARSONATE ◇ SODIUM ANILARSONATE ◇ SODIUM-ANILINE ARSONATE ◇ SODIUM ARSANILATE ◇ SODIUM-p-ARSANILATE ◇ SODIUM ARSONILATE

USE IN FOOD:

Purpose: Animal feed drug.

Where Used: Animal feed.

Regulations: FDA - 21CFR 558.6. Use at a level not in excess of the amount reasonably required to accomplish the intended effect.

Arsenic and its compounds are on the Community Right-To-Know List.

OSHA PEL: TWA 10 $\mu g/m^3$ ACGIH TLV: TWA 0.2 $mg(As)/m^3$

SAFETY PROFILE: Poison by subcutaneous route. Can cause blindness. When heated to decomposition it emits very toxic fumes of As and NO_x.

TOXICITY DATA and CODEN

scu-rat LD50:75 mg/kg BIZEA2 184,360,27
scu-mus LD50:400 mg/kg MEIEOO 10,1230,83

ARA750 CAS: 7440-38-2
ARSENIC

DOT: 1558

af: As aw: 74.92

PROP: Silvery to black, brittle, crystalline and amorphous metalloid. Mp: 814° @ 36 atm, bp: subl @ 612°, d: black crystals 5.724 @ 14°; black amorphous 4.7, vap press: 1 mm @ 372° (subl). Insol in water; sol in HNO_3.

SYNS: ARSEN (GERMAN, POLISH) ◇ ARSENICALS ◇ ARSENIC-75 ◇ ARSENIC BLACK ◇ COLLOIDAL ARSENIC ◇ GREY ARSENIC ◇ METALLIC ARSENIC

USE IN FOOD:

Purpose: Animal drug.

Where Used: Chicken, eggs, pork, turkey.

Regulations: FDA - 21CFR 556.60. Tolerance of 0.5 ppm in muscle, 2 ppm in uncooked edible by-products, 0.5 ppm in eggs from chickens and turkeys. Limitation of 2 ppm in liver and kidney, 0.5 ppm in muscle and by-products of swine.

IARC Cancer Review: Human Sufficient Evidence IMEMDT 23,39,80; Human Inadequate Evidence IMEMDT 2,48,73. Arsenic and its compounds are on the Community Right-To-Know List.

OSHA PEL: TWA 10 $\mu g/m^3$ ACGIH TLV: TWA 0.2 $mg(As)/m^3$ DFG TRK: 0.2 mg/m^3 calculated as As in that portion of dust that can possibly be inhaled. DOT Classification: Poison B; Label: Poison NIOSH REL: CL 2 $\mu g(As)/m^3$

SAFETY PROFILE: A human carcinogen. Poison by subcutaneous, intramuscular, and intraperitoneal routes. Human systemic skin and gastrointestinal effects by ingestion. An experimental teratogen and tumorigen. Mutagenic data. Flammable in the form of dust when exposed to heat or flame or by chemical reaction with powerful oxidizers such as bromates; chlorates; iodates; peroxides; lithium; NCl_3; KNO_3; $KMnO_4$; Rb_2C_2; $AgNO_4$; NOCl; IF_5; CrO_3; ClF_3; ClO; BrF_3; BrF_5; BrN_3; RbC_3BCH; CsC_3BCH. Slightly explosive in the form of dust when exposed to flame. When heated or on contact with acid or acid fumes, emits highly toxic fumes; can react vigorously on contact with oxidizing materials. Incompatible with bromine azide; dirubidium acetylide; halogens; palladium; zinc; platinum; NCl_3; $AgNO_3$; CrO_3; Na_2O_2; hexafluoro isopropylideneamino lithium.

TOXICITY DATA and CODEN

cyt-mus-ipr 4 mg/kg/48H-I EXPEAM 37,129,81
orl-rat TDLo: 605 μg/kg (35 W preg): REP GISAAA (8)30,77
orl-mus TDLo: 120 mg/kg (preg): TER TJADAB 15,31A,77
ipr-mus TDLo: 40 mg/kg (preg): TER TJADAB 15,31A,77
imp-rbt TDLo: 75 mg/kg: ETA ZEKBAI 52,425,42
orl-man TDLo: 7857 mg/kg/55Y: SKN CMAJAX 120,168,79
orl-man TDLo: 7857 mg/kg/55Y: GIT CMAJAX 120,168,79
ims-rat LDLo: 20 mg/kg NCIUS* PH 43-64-886,SEPT,70
scu-rbt LDLo: 300 mg/kg ASBIAL 24,442,38

ARL250 CAS: 8022-37-5
ARTEMISIA OIL

PROP: Chief constituent is Thujone, and found in the plant *Artemisia absinthium L* (FCTXAV 13,681,75).

SYNS: ABSINTHIUM ◇ ARTEMISIA OIL (WORMWOOD) ◇ OIL, ARTEMISIA

USE IN FOOD:

Purpose: Flavoring agent.

Where Used: Beverages (alcoholic).

Regulations: FDA - 21CFR 172.510. Finished food must be thujone free. Use at a level not in excess of the amount reasonably required to accomplish the intended effect.

SAFETY PROFILE: Moderately toxic by ingestion. An allergen. Habitual users develop, "absinthism" with tremors, vertigo, vomiting and hallucinations. May cause a contact dermatitis. When heated to decomposition it emits acrid smoke and irritating fumes.

TOXICITY DATA and CODEN

orl-rat LD50: 960 mg/kg FCTXAV 13,681,75

ARN000 CAS: 50-81-7
l-ASCORBIC ACID

mf: $C_6H_8O_6$ mw: 176.14

PROP: White crystals. Mp: 192°, flash p: +99°. Sol in water; sltly sol in alc; insol in ether, chloroform, benzene, petroleum ether, fixed oils and fats.

SYNS: ASCORBIC ACID ◇ l(+)-ASCORBIC ACID ◇ ASCORBUTINA ◇ CEVITAMIC ACID ◇ CEVITAMIN

◇ FEMA No. 2109 ◇ 3-KETO-l-GULOFURANOLACTONE ◇ l-3-KETOTHREOHEXURONIC ACID LACTONE ◇ NATRASCORB INJECTABLE ◇ NCI-C54808 ◇ 3-OXO-l-GULOFURANOLACTONE ◇ VITACIN ◇ VITAMIN C ◇ VITAMISIN ◇ VITASCORBOL ◇ XITIX ◇ l-XYLOASCORBIC ACID.vs,1

USE IN FOOD:

Purpose: Antioxidant, dietary supplement, nutrient, preservative.

Where Used: Beef (cured), cured comminuted meat food product, pork (cured), pork (fresh), sausage, wine.

Regulations: FDA - 21CFR 182.3013, 182.3041, 182.5013, 182.8013. GRAS when used in accordance with good manufacturing practice. BATF - 27CFR 240.1051. USDA - 9CFR 318.7. Limitation of 75 ounces to 100 galions pickle at 10 percent pump level, 0.75 ounces to 100 pounds of meat. Not to exceed 500 ppm or 1.8 mg/sq in. of surface ascorbic acid, erythorbic acid or sodium ascorbate singly or in combination; and/or not to exceed either 250 ppm or 0.9 mg/sq in. of surface of citric acid or sodium citrate, singly, or in combination on fresh pork cuts.

NTP Carcinogenesis Bioassay (feed); No Evidence: mouse, rat NTPTR* NTP-TR-247,83; NTPTR* NTP-TR-214,82.

SAFETY PROFILE: Moderately toxic. Human blood systemic effects by intravenous route. Mutagenic data. Combustible liquid. When heated to decomposition it emits acrid smoke and irritating fumes.

TOXICITY DATA and CODEN

mmg-sat:500 μg/plate ABCHA6 45,327,81
cyt-ham:ovr 300 mg/L FCTXAV 18,497,80
ivn-man TDLo:2300 mg/kg/2D:BLD AIMEAS 82,810,75
ivn-mus LD50:518 mg/kg RPOBAR 2,269,70

ARN125 CAS: 134-03-2
ASCORBIC ACID SODIUM SALT

mf: $C_6H_8O_6 \cdot Na$ mw: 199.13

PROP: Minute white to yellow crystals; odorless. Decomp at 218°. Freely sol in water; very sltly sol in alc; ins in chloroform, ether.

SYNS: l-ASCORBIC ACID SODIUM SALT ◇ ASCORBICIN ◇ ASCORBIN ◇ CEBITATE ◇ CENOLATE ◇ ISKIA-C ◇ MONOSODIUM ASCORBATE ◇ NATRASCORB ◇ NATRI-C ◇ SODASCORBATE ◇ SODIUM ASCORBATE (FCC) ◇ SODIUM-l-ASCORBATE ◇ VITAMIN C ◇ VITAMIN C SODIUM

USE IN FOOD:

Purpose: Antioxidant, dietary supplement, nutrient.

Where Used: Beef (cured), cured comminuted meat food product, pork (cured), pork (fresh), sausage, wine.

Regulations: FDA - 21CFR 182.3731. GRAS when used in accordance with good manufacturing practice. USDA - 9CFR 318.7. Limitation of 87.5 ounces per 100 gallons of pickle at 10 percent pump level; 7/8 ounce per 100 pounds meat, 10 percent to surfaces of cured cuts prior to packaging. Not to exceed 500 ppm or 1.8 mg/sq in. of surface ascorbic acid, erythorbic acid or sodium ascorbate singly or in combination; and/or not to exceed either 250 ppm or 0.9 mg/sq in. of surface of citric acid or sodium citrate, singly, or in combination on fresh pork cuts.

SAFETY PROFILE: Human mutagenic data. Used in vitamin C preparations, antioxidant in chopped meat and other food, also in curing meat. When heated to decomposition it emits toxic fumes of Na_2O.

TOXICITY DATA and CODEN

sce-hmn:lym 100 μmol/L MUREAV 60,321,79
cyt-ham:ovr 20 mmol/L CNREA8 39,4145,79

ARN150
ASCORBYL PALMITATE

mf: $C_{22}H_{38}O_7$ mw: 414.54

PROP: White-to-yellowish powder; slt odor. Very sltly sol in water, vegetable oil.

SYN: PALMITOYL, l-ASCORBIC ACID

USE IN FOOD:

Purpose: Antioxidant.

Where Used: Beverages, bread, breakfast foods, lemon drinks, margarine, oleomargarine, potato flakes, rolls, shortening.

Regulations: FDA - 21CFR 182.3149. GRAS when used in accordance with good manufacturing practice. USDA - 9CFR 318.7. Limitation of 0.02 percent in margarine or oleomargarine.

SAFETY PROFILE: When heated to decomposition it emits acrid smoke and irritating fumes.

ARN180
ASCORBYL STEARATE

USE IN FOOD:

Purpose: Antioxidant.

Where Used: Margarine, oleomargarine.

Regulations: USDA - 9CFR 318.7. Limitation of 0.02 percent individually or in combination with other antioxidants approved for use in margarine.

SAFETY PROFILE: When heated to decomposition it emits acrid smoke and irritating fumes.

ARN810 CAS: 70-47-3
l-ASPARAGINE

mf: $C_4H_8N_2O_3 \cdot H_2O$ mw: 150.13

PROP: White crystalline powder; slt sweet taste. Mp: 234°. Sol in water; ins in alc, ether.

SYN: l-α-AMINOSUCCINAMIC ACID

USE IN FOOD:

Purpose: Dietary supplement, nutrient.

Where Used: Various.

Regulations: FDA - 21CFR 172.320. Limitation 7.0 percent by weight.

SAFETY PROFILE: When heated to decomposition emits toxic fumes of NO_x.

ARN825 CAS: 22839-47-0
ASPARTAME

mf: $C_{14}H_{18}N_2O_5$ mw: 294.34

PROP: White crystalline powder; odorless with a sweet taste. Sltly sol in water, alc.

SYNS: 3-AMINO-N-(α-CARBOXYPHENETHYL)SUCCINAMIC ACID N-METHYL ESTER, stereoisomer ◇ ASPARTYLPHENYLALANINE METHYL ESTER ◇ N-l-α-ASPARTYL-l-PHENYLALANINE 1-METHYL ESTER (9CI) ◇ CANDEREL ◇ DIPEPTIDE SWEETENER ◇ EQUAL ◇ METHYL ASPARTYLPHENYLALANATE ◇ 1-METHYL N-l-α-ASPARTYL-l-PHENYLALANINE ◇ NUTRASWEET ◇ SWEET DIPEPTIDE

USE IN FOOD:

Purpose: Flavor enhancer, sugar substitute, sweetener.

Where Used: Beverage syrup base (carbonated), beverages (carbonated), beverages (dry base), breath mints, cereals (cold breakfast), chewable multivitamin food supplements, chewing gum, coffee (dry base instant), confections (frozen stick-type), dairy product analog topping (dry base), fillings (dry base), fruit flavored drinks and aides (noncarbonated refrigerated single strength and frozen concentrate), fruit juice based drinks (noncarbonated refrigerated single strength and frozen concentrate), gelatins (dry base), imitation fruit flavored drinks and aides (noncarbonated refrigerated single strength and frozen concentrate), puddings (dry base), tea beverages.

Regulations: FDA - 21CFR 172.804. Use at a level not in excess of the amount reasonably required to accomplish the intended effect.

SAFETY PROFILE: Human systemic effects by ingestion: allergic dermatitis. Experimental reproductive effects. When heated to decomposition it emits toxic fumes of NO_x.

TOXICITY DATA and CODEN

orl-rat TDLo:275 g/kg (2W male/2W pre-16D post):REP NETOD7 1,79,78
orl-rat TDLo:449 g/kg (2W male/2W pre-21D post):REP NETOD7 1,79,78
orl-mus TDLo:4 g/kg (15-18D preg):REP RCPBDC 9,385,85
orl-wmn TDLo:3710 μg/kg:SKN AIMEAS 104,207,86

ARN830 CAS: 617-45-8
dl-ASPARTIC ACID

mf: $C_4H_7NO_4$ mw: 133.10

PROP: Colorless to white crystals; acid taste. Mp: 280° (decomp). Sltly sol in water; ins in alc, ether.

SYN: dl-AMINOSUCCINIC ACID

USE IN FOOD:

Purpose: Dietary supplement, nutrient.

Where Used: Various.

Regulations: FDA - 21CFR 172.320. Limitation 7.0 percent by weight.

SAFETY PROFILE: When heated to decomposition emits toxic fumes of NO_x.

ARN850 CAS: 56-84-8
l-ASPARTIC ACID

mf: $C_4H_7NO_4$ mw: 133.10

PROP: Colorless to white crystals; acid taste. Mp: 270°. Sltly sol in water; ins in alc, ether.

SYN: l-AMINOSUCCINIC ACID

USE IN FOOD:

Purpose: Dietary supplement, nutrient.

Where Used: Various.

Regulations: FDA - 21CFR 172.320. Limitation 7.0 percent by weight.

SAFETY PROFILE: When heated to decomposition emits toxic fumes of NO_x.

ARW150
AVERMECTIN B_{1a}

mf: $C_{48}H_{72}O_{14}$ mw: 873.09

SYNS: AVM ◇ C-076

USE IN FOOD:

Purpose: Insecticide, miticide.

Where Used: Animal feed, citrus oil.

Regulations: FDA - 21CFR 193.473. Limitation of 0.10 ppm in citrus oil. (Expired 5/8/1988) 21CFR 561.441. Limitation of 0.10 ppm in dried citrus pulp when used for animal feed. (Expires 5/8/1988)

SAFETY PROFILE: When heated to decomposition it emits acrid smoke and irritating fumes.

ASH500 CAS: 86-50-0
AZINPHOS METHYL

mf: $C_{10}H_{12}N_3O_3PS_2$ mw: 317.34

PROP: Crystals or brown, waxy solid. D: 1.44, mp: 74°. Sltly sol in water; sol in organic solvents.
DOT: 2783

SYNS: AZINFOS-METHYL (DUTCH) ◇ AZINPHOS-METILE (ITALIAN) ◇ AZINPHOS-METHYL (ACGIH, DOT) ◇ AZINPHOS METHYL, liquid (DOT) ◇ BAY 9027 ◇ BAY 17147 ◇ BAYER 9027 ◇ BAYER 17147 ◇ BENZOTRIAZINEDITHIOPHOSPHORIC ACID DIMETHOXY ESTER ◇ BENZOTRIAZINE derivative of a METHYL DITHIOPHOSPHATE ◇ CARFENE ◇ COTNION METHYL ◇ CRYSTHION 2L ◇ CRYSTHYON ◇ DBD ◇ S-(3,4-DIHYDRO-4-OXO-BENZO(α)(1,2,3)TRIAZIN-3-YLMETHYL)-O,O-DIMETHYL PHOSPHORODITHIOATE ◇ S-(3,4-DIHYDRO-4-OXO-1,2,3-BENZOTRIAZIN-3-YLMETHYL)- O,O-DIMETHYL PHOSPHORODITHIOATE ◇ O,O-DIMETHYL-S-(BENZAZIMINOMETHYL) DITHIOPHOSPHATE ◇ O,O-DIMETHYL-S-(1,2,3-BENZOTRIAZINYL-4-KETO)METHYL PHOSPHORODITHIOATE ◇ O,O-DIMETHYL-S-(3,4-DIHYDRO-4-KETO-1,2,3-BENZOTRIAZINYL-3-METHYL) DITHIOPHOSPHATE ◇ DIMETHYLDITHIOPHOSPHORIC-ACID N-METHYLBENZAZIMIDE ESTER ◇ O,O-DIMETHYL-S-(4-OXO-3H-1,2,3-BENZOTRIZIANE-3-METHYL)PHOSPHORODITHIOATE ◇ O,O-DIMETHYL-S-(4-OXOBENZOTRIAZINO-3-METHYL)-PHOSPHORODITHIOATE ◇ O,O-DIMETHYL-S-(4-OXO-1,2,3-BENZOTRIAZINO(3)-METHYL) THIOTHIONOPHOSPHATE ◇ O,O-DIMETHYL-S-((4-OXO-3H-1,2,3-BENZOTRIAZIN-3-YL)-METHYL)-DITHIOFOSFAAT (DUTCH) ◇ O,O-DIMETHYL-S-((4-OXO-3H-1,2,3-BENZOTRIAZIN-3-YL)-METHYL)-DITHIOPHOSPHAT (GERMAN) ◇ O,O-DIMETHYL-S-4-OXO-1,2,3-BENZOTRIAZIN-3(4H)-YLMETHYL PHOSPHORODITHIOATE ◇ O,O-DIMETIL-S-((4-OXO-3H-1,2,3-BENZOTRIAZIN-3-IL)-METIL)-DITIOFOSFATO (ITALIAN) ◇ ENT 23,233 ◇ GOTHNION ◇ GUSATHION ◇ GUTHION (DOT) ◇ GUTHION, liquid (DOT) ◇ 3-(MERCAPTOMETHYL)-1,2,3-BENZOTRIAZIN-4(3H)-ONE O,O-DIMETHYL PHOSPHORODITHIOATE ◇ 3-(MERCAPTOMETHYL)-1,2,3-BENZOTRIAZIN-4(3H)-ONE-O,O-DIMETHYL PHOSPHORODITHIOATE-S-ESTER ◇ METHYLAZINPHOS ◇ N-METHYLBENZAZIMIDE, DIMETHYLDITHIOPHOSPHORIC ACID ESTER ◇ METHYL GUTHION ◇ METILTRIAZOTION ◇ NA 2783 (DOT) ◇ NCI-C00066 ◇ R 1582

USE IN FOOD:

Purpose: Insecticide.

Where Used: Citrus pulp (dried), soybean oil, sugarcane bagasse.

Regulations: FDA - 21CFR 193.150. Insecticide residue tolerance of 1 ppm in soybean oil. 21CFR 561.180. Limitation of 5 ppm in dried citrus pulp, 1.5 ppm in sugarcane bagasse.

NCI Carcinogenesis Bioassay (feed); Inadequate Studies: rat NCITR* NCI-CG-TR-69,78; No Evidence: mouse NCITR* NCI-CG-TR-69,78. EPA Genetic Toxicology Program. EPA Extremely Hazardous Substances List.

OSHA PEL: TWA 0.2 mg/m^3 (skin) ACGIH TLV: TWA 0.2 mg/m^3 (skin) DFG MAK: 0.2 mg/m^3 DOT Classification: Poison B; Label: Poison, liquid mixture

SAFETY PROFILE: Poison by inhalation, ingestion, skin contact, intravenous, intraperitoneal, and possibly other routes. An experimental tumorigen and teratogen. Other experimental reproductive effects. Human mutagenic data. When heated to decomposition it emits very toxic fumes of PO_x, SO_x, and NO_x.

TOXICITY DATA and CODEN

mmo-ssp 25 mmol/L MUREAV 117,139,83
cyt-hmn:lng 120 mg/L CNJGA8 17,455,75
orl-rat TDLo:190 mg/kg (6-22D preg/21D post):REP ARTODN 43,177,80

orl-rat TDLo:12500 μg/kg (6-15D preg):TER ARTODN 43,177,80
orl-rat TDLo:85 mg/kg (6-22D preg):REP NTIS** PB288-457
orl-mus TDLo:16 mg/kg (8D preg):TER TCMUD8 5,3,85
orl-rat TDLo:5110 mg/kg/78W-C:ETA NCITR* NCI-CG-TR-69,78
orl-rat TD:121 g/kg/78W-C:ETA JEPTDQ 1(6),829,78
orl-rat LD50:7 mg/kg JPPMAB 13,435,61
ihl-rat LC50:69 mg/m^3/1H NTIS** PB277-077
skn-rat LD50:220 mg/kg SPEADM 74-1,-,74

ASM300 CAS: 123-77-3
AZODICARBONAMIDE

mf: $C_2H_4N_4O_2$ mw: 116.08

PROP: Yellow to orange-red crystalline powder. Mp: above 180° (decomp). Sltly sol in dimethyl sulfoxide; ins in water, organic solvents.

USE IN FOOD:

Purpose: Adjuvant in foamed plastic, maturing agent for flour.

Where Used: Bread dough, cereal flour.

Regulations: FDA - 21CFR 172.806. Limitation of 45 ppm in cereal flour, 45 ppm bread dough. 21CFR 178.3010. Limitation of 5 percent in finished foamed polyethylene.

SAFETY PROFILE: Flammable solid. When heated to decomposition emits toxic fumes of NO_x.

B

BAC250 CAS: 1405-87-4
BACITRACIN

PROP: An antibiotic. White to pale buff, hygroscopic powder; odorless or slt odor. Freely sol in water, alc, methanol, and glacial acetic acid; insol in acetone, chloroform, and ether.

SYNS: AYFIVIN ◇ BACIGUENT ◇ BACI-JEL ◇ BACILIQUIN ◇ BACITEK OINTMENT ◇ FORTRACIN ◇ PARENTRACIN ◇ PENITRACIN ◇ TOPITRACIN ◇ USAF CB-7 ◇ ZUTRACIN

USE IN FOOD:

Purpose: Animal drug.

Where Used: Beef, chicken, eggs, milk, pheasants, pork, turkey.

Regulations: FDA - 21CFR 556.70. Tolerance of 0.5 ppm in uncooked tissues of cattle, swine, chickens, turkeys, pheasants, quail and in milk and eggs.

SAFETY PROFILE: A poison by intraperitoneal and intravenous routes. Moderately toxic by ingestion and subcutaneous routes. Mutagenic data. When heated to decomposition it emits acrid smoke and irritating fumes.

TOXICITY DATA and CODEN

dnd-esc 5 μmol/L MUREAV 89,95,81
ipr-rat LD50:190 mg/kg PSEBAA 64,503,47
scu-mus LDLo:1300 mg/kg PSEBAA 64,503,47
ivn-mus LD50:360 mg/kg PSEBAA 64,503,47

BAC260 CAS: 55852-84-1
BACITRACIN METHYLENE DISALICYLATE

PROP: White to brownish-gray powder; disagreeable odor. Sol in water, pyridine, ethanol; less sol in acetone, ether, chloroform, pentane, benzene.

SYN: BACITRACIN METHYLENEDISALICYLATE

USE IN FOOD:

Purpose: Animal feed drug.

Where Used: Animal feed.

Regulations: FDA - 21CFR 558.76. Use at a level not in excess of the amount reasonably required to accomplish the intended effect.

SAFETY PROFILE: When heated to decomposition it emits acrid smoke and irritating fumes.

BAC265
BACITRACIN ZINC

USE IN FOOD:

Purpose: Animal feed drug.

Where Used: Animal feed.

Regulations: FDA - 21CFR 558.78. Use at a level not in excess of the amount reasonably required to accomplish the intended effect.

SAFETY PROFILE: When heated to decomposition it emits acrid smoke and irritating fumes.

BAD400
BAKERS YEAST EXTRACT

PROP: From ruptured cells of *Saccharomyces cerevisiae*. Liquid, paste or powder. Water sol.

SYN: AUTOLYZED YEAST EXTRACT ◇ BAKERS YEAST GLYCAN

USE IN FOOD:

Purpose: Emulsifier, flavoring agent, nutrient supplement, stabilizer, thickener, yeast food.

Where Used: Cheese spread analogs, dough, frozen dessert analogs, meat carcasses (freshly dressed), poultry, salad dressings, snack dips (cheese flavored), snack dips (sour cream flavored), soups, sour cream analogs, wine.

Regulations: FDA - 21CFR 172.325, 172.896, 172.898. Limitation of 5 percent in salad dressings. 21CFR 184.1983. GRAS when used in accordance with good manufacturing practice. BATF - 27CFR 240.1051. USDA - 9CFR 318.7. Limitation of 1.5 percent of hot carcass weight when applied. 9CFR 381.147. Sufficient for purpose.

SAFETY PROFILE: When heated to decomposition it emits acrid smoke and irritating fumes.

BAE750
BALSAM OF PERU

PROP: Dark brown, viscid liquid; vanilla odor. Sol in fixed oils; sltly sol in propylene glycol; insol in glycerin. Extracted from *Myroxylon pereirae Klotzsch*.

SYNS: BALSAM PERU OIL (FCC) ◇ PERUVIAN BALSAM

USE IN FOOD:

Purpose: Flavoring agent.

Where Used: Various.

Regulations: FDA - 21CFR 182.20. GRAS when used at a level not in excess of the amount reasonably required to accomplish the intended effect.

SAFETY PROFILE: A mild allergen. Combustible when heated. When heated to decomposition it emits acrid smoke and irritating fumes.

BAR250 CAS: 8015-73-4
BASIL OIL

PROP: Contains about 55% methyl chavicol and 35% of alcohols calculated as lenatoal and other compounds found in the leaves of *Ocimum resilium* L. (FCTXAR 11,855,73). a pale yellow liquid; floral, spicy odor. Sol in fixed oils, propylene glycol; ins in glycerin.

SYNS: BASIL OIL, EUROPEAN TYPE (FCC) ◇ OCIMUM BASILICUM OIL ◇ OIL of BASIL

USE IN FOOD:

Purpose: Flavoring agent.

Where Used: Various.

Regulations: FDA - 21CFR 182.10. GRAS when used at a level not in excess of the amount reasonably required to accomplish the intended effect.

SAFETY PROFILE: Moderately toxic by ingestion. A skin irritant. When heated to decomposition it emits acrid smoke and irritating fumes.

TOXICITY DATA and CODEN

skn-mus 100 % MLD FCTXAV 11,867,73
orl-rat LD50: 1400 mg/kg FCTXAV 11,855,73

BAR275
BASIL OIL, COMOROS TYPE

PROP: From steam distillation of *Ocimum basilicum* L. Light yellow liquid; spicy odor. Sol in fixed oils, mineral oil; sltly sol in propylene glycol; ins in glycerin.

SYNS: BASIL OIL EXOTIC ◇ BASIL OIL, REUNION TYPE

USE IN FOOD:

Purpose: Flavoring agent.

Where Used: Tomato sauces, vegetables.

Regulations: FDA - 21CFR 182.10, 182.20. GRAS when used at a level not in excess of the amount reasonably required to accomplish the intended effect.

SAFETY PROFILE: When heated to decomposition it emits acrid smoke and irritating fumes.

BAT500
BAY OIL

PROP: Consists mainly of eugenol and chavicol (55-65%), major portion of balance consists of terpenes (alpha-pinene, myrcene and dipentene) small quantities of citro, nerol, cineol and other terpenoids have also been found (FCTXAV 11,855,73). Yellow or brown liquid; aromatic odor, pungent, spicy taste. Sol in alc, glacial acetic acid.

SYNS: BAY LEAF OIL ◇ BOIS d'INDE ◇ LAUREL LEAF OIL ◇ MYRCIA OIL ◇ MYRICIA OIL ◇ OIL of BAY ◇ OIL of MYRCIA

USE IN FOOD:

Purpose: Flavoring agent.

Where Used: Meat, soups, stews.

Regulations: FDA - 21CFR 182.10, 182.20. GRAS when used at a level not in excess of the amount reasonably required to accomplish the intended effect.

SAFETY PROFILE: Moderately toxic by ingestion. When heated to decomposition it emits acrid smoke.

TOXICITY DATA and CODEN

orl-rat LD50: 1800 mg/kg FCTXAV 11,855,73

BAU000 CAS: 8012-89-3
BEESWAX

PROP: Yellow to brownish-yellow, soft to brittle wax. Mp: 62-65°, d: 0.95-0.96. Sol in chloroform, ether, fixed oils; sltly sol in alc.

SYNS: BEESWAX, WHITE ◇ BEESWAX, YELLOW

USE IN FOOD:

Purpose: Candy glaze, candy polish, flavoring agent, miscellaneous and general-purpose food chemical, surface-finishing agent.

Where Used: Candy (hard), candy (soft), chewing gum, confections, frostings.

Regulations: FDA - 21CFR 184.1973. GRAS with a limitation of 0.065 percent in chewing gum, 0.005 percent in confections and frostings, 0.04 percent in hard candy, 0.1 percent in soft candy, 0.002 percent in all other foods when used in accordance with good manufacturing practice.

SAFETY PROFILE: A mild allergen. Combustible when heated.

BAV750 CAS: 1302-78-9
BENTONITE

PROP: A clay containing appreciable amounts of the clay mineral montmorillonite; light yellow or green, cream, pink, gray to black. Insol in water and common organic solvents.

SYNS: ALBAGEL PREMIUM USP 4444 ◇ BENTONITE 2073 ◇ BENTONITE MAGMA ◇ HI-JEL ◇ IMVITE I.G.B.A. ◇ MAGBOND ◇ MONTMORILLONITE ◇ PANTHER CREEK BENTONITE ◇ SOUTHERN BENTONITE ◇ TIXOTON ◇ VOLCLAY ◇ VOLCLAY BENTONITE BC ◇ WILKINITE

USE IN FOOD:

Purpose: Colorant, pigment, stabilizer.

Where Used: Wine.

Regulations: GRAS when used in accordance with good manufacturing practice.

SAFETY PROFILE: Poison by intravenous route causing blood clotting. An experimental tumorigen.

TOXICITY DATA and CODEN

orl-mus TDLo:12000 g/kg/28W-C:ETA ANYAA9 57,678,54
ivn-rat LD50:35 mg/kg BSIBAC 44,1685,68

BAY500 CAS: 100-52-7
BENZALDEHYDE
DOT: 1989
mf: C_7H_6O mw: 106.13

PROP: Colorless liquid; burning taste with bitter almond odor. Mp: −26°, bp: 179°, flash p: 148°F, d: 1.041, autoign temp: 377°F, vap press: 1 mm @ 26.2°, vap d: 3.65, refr index: 1.544. Sltly sol in water; misc in alc, ether, oils.

SYNS: ALMOND ARTIFICIAL ESSENTIAL OIL ◇ ARTIFICIAL ALMOND OIL ◇ BENZENECARBALDEHYDE ◇ BENZENECARBONAL ◇ BENZOIC ALDEHYDE ◇ FEMA No. 2127 ◇ NCI-C56133

USE IN FOOD:

Purpose: Flavoring agent.

Where Used: Various.

Regulations: FDA - 21CFR 182.60. GRAS when used at a level not in excess of the amount reasonably required to accomplish the intended effect.

EPA Genetic Toxicology Program.

DOT Classification: Combustible Liquid; Label: None

SAFETY PROFILE: Poison by ingestion and intraperitoneal routes. Moderately toxic by subcutaneous route. An allergen. Acts as a feeble local anesthetic. Local contact may cause contact dermatitis. Causes central nervous system depression in small doses and convulsions in larger doses. A skin irritant. Mutagenic data. Combustible liquid. To fight fire, use water (may be used as a blanket), alcohol, foam, dry chemical. A strong reducing agent. Reacts violently with peroxyformic acid and other oxidizers.

TOXICITY DATA and CODEN

skn-rbt 500 mg/24H MOD FCTXAV 14,659,76
orl-rat LD50:1300 mg/kg FCTXAV 2,327,64
scu-rat LDLo:5000 mg/kg AIPTAK 27,163,22
scu-rbt LD50:5000 mg/kg FCTXAV 14,693,76

BBA000 CAS: 1708-39-0
BENZAL GLYCERYL ACETAL
mf: $C_{10}H_{12}O_3$ mw: 180.22

PROP: Colorless to pale yellow liquid; mild almond odor. D: 1.183-1.193, refr index: 1.535-1.541, flash p: 74°.

SYNS: BENZALDEHYDE GLYCERYL ACETAL (FCC) ◇ BENZYLIDENE GLYCEROL ◇ BUTYL PHENYL ACETATE ◇ FEMA No. 2209 ◇ 2-PHENYL-m-DIOXAN-5-OL

USE IN FOOD:

Purpose: Flavoring agent.

Where Used: Various.

Regulations: FDA - 21CFR 172.515. Use at a level not in excess of the amount reasonably required to accomplish the intended effect.

SAFETY PROFILE: Moderately toxic by ingestion and intraperitoneal routes. Mildly toxic by skin contact. Combustible liquid. When heated to decomposition it emits acrid smoke and irritating fumes.

TOXICITY DATA and CODEN

orl-rat LD50:3150 mg/kg FCTXAV 14,699,76
ipr-mus LD50:1296 mg/kg AIPTAK 85,474,51
skn-rbt LD50:5000 mg/kg FCTXAV 14,699,76

BBL500 CAS: 122-78-1
BENZENEACETALDEHYDE
mf: C_8H_8O mw: 120.16

PROP: Oily, colorless liquid which polymerizes and grows more viscous on standing; odor similar to lilac and hyacinth. Has been crystallized, mp: 33-34°, d:(25/25) 1.023-1.030, refr index: 1.525-1.545, bp: (10) 78°, n (20/D) 1.524-1.528, flash p: 68°. Sltly sol in water; sol in alc, ether, propylene glycol. One part is sol in two parts of 80% alc forming a clear solution.

SYNS: FEMA No. 2874 ◇ HYACINTHIN ◇ PAA ◇ PHENYLACETALDEHYDE (FCC) ◇ PHENYLACETIC ALDEHYDE ◇ PHENYLETHANAL ◇ α-TOLUALDEHYDE ◇ α-TOLUIC ALDEHYDE

USE IN FOOD:

Purpose: Flavoring agent.

Where Used: Bakery products, beverages (nonalcoholic), chewing gum, confections, gelatin desserts, ice cream, maraschino cherries, puddings.

Regulations: FDA - 21CFR 172.515. Use at a level not in excess of the amount reasonably required to accomplish the intended effect.

SAFETY PROFILE: Moderately toxic by ingestion. Human skin irritant. Combustible liquid. When heated to decomposition it emits acrid smoke and irritating fumes.

TOXICITY DATA and CODEN

skn-hmn 2%/48H FCTXAV 17,377,79
orl-rat LD50:1550 mg/kg FCTXAV 17,377,79

BBP750 CAS: 608-73-1
BENZENE HEXACHLORIDE
mf: $C_6H_6Cl_6$ mw: 290.82

PROP: Technical grade contains 68.7% α-BHC, 6.5% β-BHC and 13.5% γ-BHC (JPFCD2 14,305,79). White, crystalline powder. Mp: 113°, vap press: 0.0317 mm @ 20°.

SYNS: BHC (USDA) ◇ COMPOUND-666 ◇ DBH ◇ ENT 8,601 ◇ GAMMEXANE ◇ HCCH ◇ HEXA ◇ HEXACHLOR ◇ HEXACHLORAN ◇ HEXACHLOROCYCLOHEXANE ◇ 1,2,3,4,5,6-HEXACHLOROCYCLOHEXANE ◇ HEXYLAN

USE IN FOOD:

Purpose: Pesticide.

Where Used: Various.

Regulations: USDA CES Ranking: B-2 (1987).

IARC Cancer Review: Animal Sufficient Evidence IMEMDT 5,47,74

SAFETY PROFILE: Poison by ingestion and subcutaneous routes. Moderately toxic by skin contact. An experimental carcinogen, neoplastigen, and tumorigen by ingestion and skin contact. Human systemic effects by inhalation: headache, nausea or vomiting, and fever. Implicated in aplastic anemia. Experimental reproductive effects. Mutagenic data. Lindane is more toxic than DDT or dieldrin. When heated to decomposition it emits highly toxic fumes of phosgene, HCl and Cl^-. Potentially violent reaction with dimethylformamide + iron. When heated to decomposition it emits highly toxic fumes of phosgene, HCl and Cl^-.

A toxic organochlorine pesticide which is persistent in the environment and accumulates in mammalian tissue. For cattle, the oral LD50 <= 100 mg/kg. The various isomers have different actions; the γ (lindane) and α isomers are central nervous system stimulants, the principal symptom being convulsions. The β and Δ isomers are central nervous system depressants. The use of thermal vaporizers with lindane has caused acute poisoning by inhalation.

Dermatitis and perhaps other manifestations based on sensitivity represent a sort of chronic, though probably not systemic intoxication, which has been observed in humans.

The signs and symptoms of confirmed acute poisoning in humans have paralleled those of experimental animals. These signs and symptoms are: excitation, hyperirritability, loss of equilibrium, clonic-tonic convulsions, and later depression.

There is some evidence that the pulmonary edema and vascular collapse may be of neurogenic origin also. The symptoms in animals systemically poisoned by the γ isomer alone are essentially similar to those caused by mixtures, although the onset may be earlier. Workers acutely exposed to high air concentrations of lindane and its decomposition products show headache, nausea, and irritation of eyes, nose and throat.

In rare instances, urticaria has followed exposure to lindane vapor. Unlike the signs and symptoms already mentioned, this allergic manifestation occurs only in susceptible individuals, and usually only after a period of sensitization.

TOXICITY DATA and CODEN

mmo-omi 100 mg/L MILEDM 5,103,77
otr-rat-orl 875 mg/kg/7W-I CRNGDP 5,479,84
orl-rat TDLo:8100 mg/kg (90D male):REP APTOA6 52,12,83
orl-mus TDLo:9120 mg/kg (22W male):REP BECTA6 26,508,81
orl-mus TDLo:6720 mg/kg/80W-C:CAR JPFCD2 14(3),305,79
skn-mus TDLo:1600 mg/kg/80W-I:ETA JPFCD2 14(3),305,79
orl-mus TD :12600 mg/kg/30W-C:CAR JCROD7 99,143,81
ihl-man TCLo:400 μg/kg/3D:CNS,GIT,MET GISAAA 49(10),26,84
orl-rat LD50:100 mg/kg ATXKA8 22,115,66
skn-rat LD50:900 mg/kg 85DPAN -,-,71/76

BBQ500 CAS: 58-89-9
BENZENE HEXACHLORIDE-γ isomer

DOT: 2761
mf: $C_6H_6Cl_6$ mw: 290.82

SYNS: AALINDAN ◇ AFICIDE ◇ AGRISOL G-20 ◇ AGROCIDE ◇ AGRONEXIT ◇ AMEISENATOD ◇ AMEISENMITTEL MERCK ◇ APARSIN ◇ APHTIRIA ◇ APLIDAL ◇ ARBITEX ◇ BBH ◇ BEN-HEX ◇ BENTOX 10 ◇ γ-BENZENE HEXACHLORIDE ◇ BEXOL ◇ BHC ◇ γ-BHC ◇ CELANEX ◇ CHLORESENE ◇ CODECHINE ◇ DBH ◇ DETMOL-EXTRAKT ◇ DETOX 25 ◇ DEVORAN ◇ DOL GRANULE ◇ DRILL TOX-SPEZIAL AGLUKON ◇ ENT 7,796 ◇ ENTOMOXAN ◇ EXAGAMA ◇ FORLIN ◇ GALLOGAMA ◇ GAMACID ◇ GAMAPHEX ◇ GAMENE ◇ GAMISO ◇ GAMMA-COL ◇ GAMMAHEXA ◇ GAMMAHEXANE ◇ GAMMALIN ◇ GAMMOPAZ ◇ HCCH ◇ HCH ◇ γ-HCH ◇ HECLOTOX ◇ HEXACHLORAN ◇ γ-HEXACHLORAN ◇ γ-HEXACHLORANE ◇ γ-HEXACHLOROBENZENE ◇ 1-α,2-α,3-β,4-α,5-α,6-β-HEXACHLOROCYCLOHEXANE ◇ γ-HEXACHLOROCYCLOHEXANE ◇ 1,2,3,4,5,6-HEXACHLOROCYCLOHEXANE, γ-ISOMER ◇ HEXATOX ◇ HEXICIDE ◇ HGI ◇ INEXIT ◇ ISOTOX ◇ JACUTIN ◇ KOKOTINE ◇ KWELL ◇ LENDINE ◇ LENTOX ◇ LIDENAL ◇ LINDAGRAIN ◇ LINDANE (ACGIH, DOT, USDA) ◇ LINTOX ◇ MILBOL 49 ◇ MSZYCOL ◇ NCI-C00204 ◇ NEO-SCABICIDOL ◇ NEXIT ◇ NOVIGAM ◇ OVADZIAK ◇ PEDRACZAK ◇ QUELLADA ◇ RCRA WASTE NUMBER U129 ◇ SANG gamma ◇ STREUNEX ◇ TAP 85 ◇ VITON

USE IN FOOD:

Purpose: Pesticide.

Where Used: Various.

Regulations: USDA CES Ranking: A-2 (1986).

EPA Extremely Hazardous Substances List. EPA Genetic Toxicology Program. Community Right-To-Know List. NCI Carcinogenesis Bioassay (feed); No Evidence: mouse, rat NCITR* NCI-CG-TR-14,77; IARC Cancer Review: Animal Sufficient Evidence IMEMDT 5,47,74; IMEMDT 20,195,79

OSHA PEL: TWA 0.5 mg/m^3 (skin) ACGIH TLV: TWA 0.5 mg/m^3 (skin) DOT Classification: ORM-A; Label: None

SAFETY PROFILE: A human systemic poison by ingestion. Also a poison by ingestion, skin contact, intraperitoneal, intravenous, and intramuscular routes. An experimental neoplastigen and teratogen. Human systemic effects by ingestion: convulsions, dyspnea, and cyanosis. Other experimental animal reproductive effects. Mutagenic data. When heated to decomposition it emits toxic fumes of Cl^-, HCl, and phosgene.

TOXICITY DATA and CODEN

dns-ofs:lvr 45 μmol/L HKXUDL 4,268,84
msc-ham:lng 200 mg/L GISAAA 49(5),82,84
orl-rat TDLo:200 mg/kg (6-15D preg):TER TXCYAC 9,239,78
orl-mus TDLo:2730 mg/kg (13W male):REP FCTXAV 19,131,81
orl-rbt TDLo:260 mg/kg (6-18D preg):TER TXCYAC 9,239,78
orl-mus TDLo:25 g/kg/73W-C:NEO FCTXAV 11,433,73
orl-chd LDLo:180 mg/kg:CNS,PUL CMEP** -,1,56
orl-chd TDLo:111 mg/kg:CNS AEHLAU 25,374,72
orl-rat LD50:76 mg/kg SPEADM 74-1,-,74
skn-rat LD50:500 mg/kg WRPCA2 9,119,70

BCD500 CAS: 8030-30-6
BENZIN

DOT: 1255/1256/1271/2553

PROP: Dark straw-colored to colorless liquid. Bp: 149-216°, flash p: 107°F (CC), d: 0.862-0.892, autoign temp: 531°F. Sol in benzene, toluene, xylene, etc. Made from American coal oil and consists chiefly of pentane, hexane, and heptane (XPHPAW 255,43,40).

SYNS: AROMATIC SOLVENT ◇ COAL TAR NAPHTHA ◇ HI-FLASH NAPHTHAETHYLEN ◇ NAPHTA (DOT) ◇ NAPHTHA ◇ NAPHTHA DISTILLATE (DOT) ◇ NAPHTHA PETROLEUM (DOT) ◇ NAPHTHA, SOLVENT (DOT)

◇ PETROLEUM BENZIN ◇ PETROLEUM DISTILLATES (NAPHTHA) ◇ PETROLEUM ETHER (DOT) ◇ PETROLEUM NAPHTHA (DOT) ◇ PETROLEUM SPIRIT (DOT) ◇ SKELLY-SOLVE-F ◇ VM & P NAPHTHA

USE IN FOOD:

Purpose: Color diluent, protective coating, solvent.

Where Used: Eggs (shell), fruits (fresh), vegetables (fresh).

Regulations: FDA - 21CFR 73.1, 172.250. Use at a level not in excess of the amount reasonably required to accomplish the intended effect.

OSHA PEL: TWA 100 ppm ACGIH TLV: TWA 300 ppm; STEL 400 ppm DOT Classification: Flammable Liquid; Label: Flammable Liquid NIOSH REL: TWA 350 mg/m^3; CL 1800 mg/m^3/15M

SAFETY PROFILE: A human poison via intravenous route. Human systemic effects by intravenous route: dyspnea, respiratory stimulation, and other unspecified respiratory effects. Mildly toxic by inhalation. Can cause unconsciousness which may go into coma, stentorious breathing, and bluish tint to the skin. Recovery follows removal from exposure. In mild form, intoxication resembles drunkenness. On a chronic basis, no true poisoning; sometimes headache, lack of appetite, dizziness, sleeplessness, indigestion, and nausea. A common air contaminant. Flammable when exposed to heat or flame; can react with oxidizing materials. Keep containers tightly closed. Slight explosion hazard. To fight fire, use foam, CO_2, dry chemical.

TOXICITY DATA and CODEN

ihl-hmn LCLo:3 pph/5M TABIA2 3,231,33
ivn-man LDLo:27 mg/kg:PUL CTOXAO 16,335,80
ihl-rat LCLo:1600 ppm/6H CHINAG (17),1078,39

BCE500 CAS: 81-07-2
1,2-BENZISOTHIAZOL-3(2H)-ONE-1,1-DIOXIDE

mf: $C_7H_5NO_3S$ mw: 183.19

PROP: White crystals or powder; odorless with sweet taste. Mp: 228° (decomp), bp: sublimes. Sol in water, alc, chloroform, ether.

SYNS: ANHYDRO-o-SULFAMINE BENZOIC ACID ◇ 3-BENZISOTHIAZOLINONE-1,1-DIOXIDE ◇ o-BENZOIC SULPHIMIDE ◇ o-BENZOSULFIMIDE ◇ BENZOSULPHIMIDE ◇ BENZO-2-SULPHIMIDE ◇ o-BENZOYL SULFIMIDE ◇ o-BENZOYL SULPHIMIDE ◇ 1,2-DIHYDRO-2-KETOBENZISOSULFONAZOLE ◇ 1,2-DIHYDRO-2-KETOBENZISOSULPHONAZOLE ◇ 2,3-DIHYDRO-3-OXOBENZISOSULFONAZOLE ◇ 2,3-DIHYDRO-3-OXOBENZISOSULPHONAZOLE ◇ GARANTOSE ◇ GLUCID ◇ GLUSIDE ◇ HERMESETAS ◇ 3-HYDROXYBENZISOTHIAZOL-S,S-DIOXIDE ◇ INSOLUBLE SACCHARINE ◇ KANDISET ◇ NATREEN ◇ RCRA WASTE NUMBER U202 ◇ SACARINA ◇ SACCAHARIMIDE ◇ SACCHARINA ◇ SACCHARIN ACID ◇ SACCHARINE ◇ SACCHARINOL ◇ SACCHARINOSE ◇ SACCHAROL ◇ SAXIN ◇ SUCRE EDULCOR ◇ SUCRETTE ◇ o-SULFOBENZIMIDE ◇ o-SULFOBENZOIC ACID IMIDE ◇ 2-SULPHOBENZOIC IMIDE ◇ SYKOSE ◇ SYNCAL ◇ ZAHARINA

USE IN FOOD:

Purpose: Masticatory substance in chewing gum base, nonnutritive sweetener.

Where Used: Artificial sweetener, bacon, beverage mixes, beverages, chewing gum, desserts, fruit juice drinks, jam.

Regulations: FDA - 21CFR 180.37. Limitation of 12 mg per fluid ounce in beverages, fruit juice drinks, beverage mixes. Limitation of 20 mg per teaspoon of sugar sweetening equivalent and 230 mg per designated size in processed foods. USDA - 9CFR 318.7. Limitation of 0.01 percent in bacon.

IARC Cancer Review: Human Inadequate Evidence IMEMDT 22,111,80; Animal Sufficient Evidence IMEMDT 22,111,80. EPA Genetic Toxicology Program. Community Right-To-Know List.

SAFETY PROFILE: An experimental carcinogen, neoplastigen, tumorigen, and teratogen. A possible human carcinogen. Mild acute toxicity by ingestion. Experimental reproductive effects. Mutagenic data. When heated to decomposition it emits toxic NO_x and SO_x.

TOXICITY DATA and CODEN

cyt-smc 200 mg/L NATUAS 294,263,81
sce-ham:lng 100 mg/L BJCAAI 45,769,82
orl-mus TDLo:16800 mg/kg (MGN):REP TXCYAC 8,285,77
orl-mus TDLo:252 g/kg (MGN):REP DBTEAD 17,103,69
orl-mus TDLo:155 mg/kg (7D preg):TER IIZAAX 16,330,64
orl-rat TDLo:2008 g/kg/2Y-C:ETA JAPMA8 40,583,51
skn-mus TDLo:9600 mg/kg/10W-I:ETA BJCAAI 10,363,56

imp-mus TDLo:80 mg/kg:NEO BJCAAI 11,212,57
orl-mus LD50:17 g/kg EXPEAM 35,1364,79

BCJ250 CAS: 115-29-7
BENZOEPIN
DOT: 2761
mf: $C_9H_6Cl_6O_3S$ mw: 406.91

PROP: A mixture of 2 isomers. Brown crystals. Mp (α): 106°, mp (β): 212°, d: 1.745 @ 20°/20°. Nearly insol in water; sol in most organic solvents.

SYNS: BEOSIT ◇ BIO 5,462 ◇ CHLORTHIEPIN ◇ CRISULFAN ◇ CYCLODAN ◇ DEVISULPHAN ◇ ENDOCEL ◇ ENDOSOL ◇ ENDOSULFAN (ACGIH, DOT) ◇ ENDOSULPHAN ◇ ENSURE ◇ ENT 23,979 ◇ FMC 5462 ◇ 1,2,3,4,7,7-HEXACHLOROBICYCLO(2.2.1)HEPTEN-5,6-BIOXYMETHYLENESULFITE ◇ α,β-1,2,3,4,7,7-HEXACHLOROBICYCLO (2.2.1)-2-HEPTENE-5,6-BISOXYMETHYLENE SULFITE ◇ HEXACHLOROHEXAHYDROMETHANO 2,4,3-BENZODIOXATHIEPIN-3-OXIDE ◇ 6,7,8,9,10,10-HEXACHLORO-1,5,5a,6,9,9a-HEXAHYDRO-6,9-METHANO-2,4,3-BENZODIOXATHIEPIN-3-OXIDE ◇ 1,4,5,6,7,7-HEXACHLORO-5-NORBORNENE-2,3-DIMETHANOL CYCLIC SULFITE ◇ HILDAN ◇ HOE 2,671 ◇ INSECTOPHENE ◇ KOP-THIODAN ◇ MALIX ◇ NCI-C00566 ◇ NIA 5462 ◇ NIAGARA 5,462 ◇ OMS 570 ◇ RCRA WASTE NUMBER P050 ◇ SULFUROUS ACID, cyclic ester with 1,4,5,6,7,7-HEXACHLORO-5-NORBORNENE-2,3-DIMETHANOL ◇ THIFOR ◇ THIMUL ◇ THIODAN ◇ THIOFOR ◇ THIOMUL ◇ THIONEX ◇ THIOSULFAN ◇ THIOSULFAN TIONEL ◇ TIOVEL

USE IN FOOD:

Purpose: Insecticide.

Where Used: Tea (dried).

Regulations: FDA - 21CFR 193.170. Insecticide residue tolerance of 25 ppm in dried tea.

EPA Extremely Hazardous Substances List. NCI Carcinogenesis Bioassay (feed); No Evidence: mouse, rat NCITR* NCI-CG-TR-62,77.

ACGIH TLV: TWA 0.1 mg/m^3 (skin) DOT Classification: Poison B; Label: Poison

SAFETY PROFILE: Poison by ingestion, inhalation, skin contact, intraperitoneal, subcutaneous, and possibly other routes. An experimental tumorigen, neoplastigen, and teratogen. Other experimental reproductive effects. Human mutagenic data. A central nervous system stimulant producing convulsions. A highly toxic organochlorine pesticide which does not accumulate significantly in human tissue. Absorption is normally slow, but is increased by alcohols, oil, emulsifiers. When heated to decomposition it emits toxic fumes of Cl^- and SO_x.

TOXICITY DATA and CODEN

sln-dmg-orl 200 ppm/48H MUREAV 136,115,84
mmo-smc 100 mg/L TGANAK 18,455,84
sce-hmn:lym 1 μmol/L ARTODN 52,221,83
cyt-mus-unr 1 mg/kg TGANAK 16(1),45,82
cyt-ham-ipr 8 mg/kg ARTODN 58,152,85
orl-rat TDLo:45 mg/kg (6-14D preg):TER APTOA6 42,150,78
orl-rat TDLo:600 mg/kg (60D male):REP TOLED5 18(Suppl 1),139,83
orl-mus TDLo:330 mg/kg/78W-I:NEO NTIS** PB223-159
scu-mus TDLo:2 mg/kg:ETA NTIS** PB223-159
orl-rat LD50:18 mg/kg ARSIM* 20,9,66
ihl-rat LC50:80 mg/m^3/4H JOCMA7 9,35,67
skn-rat LD50:74 mg/kg WRPCA2 9,119,70
ipr-rat LD50:8 mg/kg RREVAH 22,1,68

BCL750 CAS: 65-85-0
BENZOIC ACID
mf: $C_7H_6O_2$ mw: 122.13

PROP: White crystalline powder. Mp: 121.7°, bp: 249°, flash p: 250°F (CC), d: 1.316, autoign temp: 1060°F, vap press: 1 mm @ 96.0° (sublimes), vap d: 4.21. Mod sol in water; sol in alc, ether, chloroform, fixed oils.

SYNS: ACIDE BENZOIQUE (FRENCH) ◇ BENZENECARBOXYLIC ACID ◇ BENZENEFORMIC ACID ◇ BENZENEMETHANOIC ACID ◇ BENZOATE ◇ BENZOESAEURE (GERMAN) ◇ BENZOIC ACID (DOT) ◇ CARBOXYBENZENE ◇ DRACYLIC ACID ◇ KYSELINA BENZOOVA (CZECH) ◇ PHENYL CARBOXYLIC ACID ◇ PHENYLFORMIC ACID ◇ RETARDER BA ◇ RETARDEX ◇ SALVO LIQUID ◇ SALVO POWDER ◇ TENN-PLAS

USE IN FOOD:

Purpose: Antimicrobial agent, flavoring agent, preservative.

Where Used: Cinnamon (naturally occurring), cloves (ripe, naturally occurring), cranberries (naturally occurring), plums (naturally occurring), prunes (naturally occurring).

Regulations: FDA - 21CFR 184.1021. GRAS with a limitation of 0.1 percent in foods when used in accordance with good manufacturing practice.

EPA Genetic Toxicology Program.

DOT Classification: Orm-E; Label: None

SAFETY PROFILE: Poison by subcutaneous route. Moderately toxic by ingestion and intraperitoneal routes. Human systemic effects by inhalation: dyspnea and allergic dermatitis. Severe eye irritant. A human skin irritant. Combustible when exposed to heat or flame; can react with oxidizing materials. The powder burns rapidly in oxygen. To fight fire, use water, CO_2, water spray or mist, dry chemical. When heated to decomposition it emits acrid smoke and irritating fumes.

TOXICITY DATA and CODEN

skn-hmn 22 mg/3D-I MOD 85DKA8 -,127,77
skn-rbt 500 mg/24H MLD BIOFX* 28-4/73
eye-rbt 100 mg SEV BIOFX* 28-4/73
mmo-esc 10 mmol/L ZBPIA9 112,226,59
dni-hmn:lym 5 mmol/L PNASA6 79,1171,82
orl-man LDLo:500 mg/kg FCTXAV 17,715,79
skn-hmn TDLo:6 mg/kg:PUL,SKN JOALAS 16,195,45
orl-rat LD50:2530 mg/kg MarJV# 29MAR77

BCM000 CAS: 120-51-4
BENZOIC ACID, BENZYL ESTER

mf: $C_{14}H_{12}O_2$ mw: 212.26

PROP: Found in Peru and Tolu Balsams, in Ylang-Ylang and in about 20 other essential oils (FCTXAV 11,1011,73). Colorless oily liquid; slt aromatic odor. Mp. 21°, bp: 324°, flash p: 298°F (CC), d: 1.116, refr index: 1.568, vap d: 7.3, autoign temp: 898°F. Misc with alc, chloroform, ether; ins in glycerin, water.

SYNS: ASCABIN ◇ ASCABIOL ◇ BENYLATE ◇ BENZOIC ACID, PHENYLMETHYL ESTER ◇ BENZYL ALCOHOL BENZOIC ESTER ◇ BENZYL BENZENECARBOXYLATE ◇ BENZYL BENZOATE (FCC) ◇ BENZYLETS ◇ BENZYL PHENYLFORMATE ◇ COLEBENZ ◇ FEMA No. 2138 ◇ NOVOSCABIN ◇ PERUSCABIN ◇ SCABANCA ◇ VANZOATE ◇ VENZONATE

USE IN FOOD:

Purpose: Flavoring agent.

Where Used: Various.

Regulations: FDA - 21CFR 172.515. Use at a level not in excess of the amount reasonably required to accomplish the intended effect.

SAFETY PROFILE: Moderately toxic by ingestion and skin contact. No data on chronic effects. Combustible liquid. Can react with oxidizing materials. To fight fire, use CO_2, water spray or mist, dry chemical. When heated to decomposition it emits acrid and irritating fumes and smoke.

TOXICITY DATA and CODEN

orl-rat LD50:500 mg/kg FMCHA2 -,C30,83
skn-rat LD50:4000 mg/kg JPETAB 93,26,48

BCP250 CAS: 119-53-9
BENZOIN

mf: $C_{14}H_{12}O_2$ mw: 212.26

SYNS: BENZOYLPHENYLCARBINOL ◇ BITTER ALMOND OIL CAMPHOR ◇ α-HYDROXYBENZYL PHENYL KETONE ◇ α-HYDROXY-α-PHENYLACETOPHENONE ◇ 2-HYDROXY-2-PHENYLACETOPHENONE ◇ NCI-C50011

USE IN FOOD:

Purpose: Color diluent.

Where Used: Fruit, vegetables.

Regulations: FDA - 21CFR 73.1. No residue.

NCI Carcinogenesis Bioassay (feed); No Evidence: mouse, rat NCITR* NCI-CG-TR-204,80.

SAFETY PROFILE: Mutagenic data. When heated to decomposition it emits acrid smoke and irritating fumes.

TOXICITY DATA and CODEN

mmo-sat 750 μg/plate PMRSDJ 5,187,85
dnd-rat:lvr 63700 μg/L PMRSDJ 5,353,85

BCS250 CAS: 119-61-9
BENZOPHENONE

mf: $C_{13}H_{10}O$ mw: 182.23

PROP: Rhombic, white crystals; persistent rose-like odor. mp (α): 49°, mp (β): 26°, mp (γ): 47°, bp: 305.4°, d (α): 1.0976 @ 50°/50°, d (β): 1.108 @ 23°/40°, vap press: 1 mm @ 108.2. Sol in fixed oils; sltly sol in propylene glycol; ins in glycerol.

SYNS: BENZOYLBENZENE ◇ DIPHENYL KETONE ◇ DIPHENYLMETHANONE ◇ FEMA No. 2134 ◇ α-OXODIPHENYLMETHANE ◇ PHENYL KETONE

USE IN FOOD:

Purpose: Flavoring agent.

Where Used: Various.

Regulations: FDA - 21CFR 172.515. Use at a level not in excess of the amount reasonably required to accomplish the intended effect.

SAFETY PROFILE: An experimental tumorigen. Moderately toxic by ingestion and intraperitoneal routes. Combustible when heated. Incompatible with oxidizers. When heated to decomposition it emits acrid and irritating fumes.

TOXICITY DATA and CODEN

skn-mus TDLo:8 g/kg/2Y-I:ETA TXAPA9 30,7,74
orl-mus LD50:2895 mg/kg JETOAS 9,99,76
ipr-mus LD50:727 mg/kg JETOAS 9,99,76

BDJ250 CAS: 2310-17-0
S-((3-BENZOXAZOLINYL-6-CHLORO-2-OXO)METHYL) O,O-DIETHYLPHOSPHORODITHIOATE

mf: $C_{12}H_{15}ClNO_4PS_2$ mw: 367.82

SYNS: AZOFENE ◇ BENZOPHOSPHATE ◇ BENZPHOS ◇ CHIPMAN 11974 ◇ S-(6-CHLORO-3-(MERCAPTO-METHYL)-2-BENZOXAZOLINONE)-O,O-DIETHYL PHOSPHORODITHIOATE ◇ 3-(6-CHLORO-2-OXOBENZOXAZOLIN-3-YL)METHYL-O,O-DIETHYL PHOSPHOROTHIOLOTHIONATE ◇ O,O-DIAETHYL-S-(6-CHLOR-2-OXO-BEN(b)-1,3-OXALIN-3-YL)-METHYL-DIT HIOPHOSPHAT (GERMAN) ◇ O,O-DIETHYL-S-((6-CHLOOR-2-OXO-BENZOXAZOLIN-3-YL)-METHYL)-DITHIO FOSFAAT (DUTCH) ◇ O,O-DIETHYL-S-(6-CHLOROBENZOXAZOLINYL-3-METHYL)DITHIOPHOSPHATE ◇ O,O-DIETHYL-S-((6-CHLORO-2-OXOBENZOXAZOLIN-3-YL)METHYL) PHOSPHORODITHIOATE ◇ O,O-DIETHYL-S-(6-CHLORO-2-OXO-BENZOXAZOLIN-3-YL)METHYL-PHOSPHORO THIOLOTHIONATE ◇ 3-DIETHYLDITHIOPHOSPHORYLMETHYL-6-CHLOROBENZOXAZOLONE-2 ◇ O,O-DIETIL-S-((6-CLORO-2-OXO-BENZOSSAZOLIN-3-IL)-METIL)-DITIOFOSFATO (ITALIAN) ◇ ENT 27,163 ◇ FOZALON ◇ NIA-9241 ◇ NIAGARA 9241 ◇ NPH-1091 ◇ PHASOLON ◇ PHOSALON ◇ PHOSALONE ◇ PHOZALON ◇ RHODIA RP 11974 ◇ RUBITOX ◇ ZOLON ◇ ZOLONE ◇ ZOLONE PM ◇ ZOOLON

USE IN FOOD:

Purpose: Insecticide.

Where Used: Animal feed, apple pomace (dried), citrus pulp (dried), grape pomace (dried), prunes (dried), raisins, tea (dried).

Regulations: FDA - 21CFR 193.340. Insecticide residue tolerance of 40 ppm in dried prunes, 20 ppm in raisins, 8 ppm in dried tea. 21CFR 561.300. Limitation of 85 ppm in dried apple pomace, 45 ppm in dried grape pomace, 12 ppm in dried citrus pulp when used for animal feed.

EPA: Farm Worker Field Reentry FEREAC 39,16888,74.

SAFETY PROFILE: Poison by ingestion, skin contact, and possibly other routes. A cholinesterase inhibitor. When heated to decomposition it emits very toxic fumes of Cl^-, NO_x, PO_x, and SO_x.

TOXICITY DATA and CODEN

orl-rat LD50:85 mg/kg KSKZAN 16(2),59,78
skn-rat LD50:390 mg/kg WRPCA2 9,119,70

BDS000 CAS: 94-36-0
BENZOYL PEROXIDE

DOT: 2085

mf: $C_{14}H_{10}O_4$ mw: 242.24

PROP: White, granular, tasteless, odorless powder. Mp: 103-106° (decomp), bp: decomposes explosively, autoign temp: 176°F. Sol in benzene, acetone, chloroform; sltly sol in alc; insol in water.

SYNS: ACETOXYL ◇ ACNEGEL ◇ AZTEC BPO ◇ BENOXYL ◇ BENZAC ◇ BENZAKNEW ◇ BENZOIC ACID, PEROXIDE ◇ BENZOPEROXIDE ◇ BENZOYL ◇ BENZOYLPEROXID (GERMAN) ◇ BENZOYLPEROXYDE (DUTCH) ◇ BENZOYL SUPEROXIDE ◇ BZF-60 ◇ CADET ◇ CADOX ◇ CLEARASIL BENZOYL PEROXIDE LOTION ◇ CLEARASIL BP ACNE TREATMENT ◇ CUTICURA ACNE CREAM ◇ DEBROXIDE ◇ DIBENZOYLPEROXID (GERMAN) ◇ DIBENZOYL PEROXIDE ◇ DIBENZOYLPEROXYDE (DUTCH) ◇ DIPHENYLGLYOXAL PEROXIDE ◇ DRY AND CLEAR ◇ EPI-CLEAR ◇ FOSTEX ◇ GAROX ◇ INCIDOL ◇ LOROXIDE ◇ LUCIDOL ◇ LUPERCO ◇ LUPEROX FL ◇ NAYPER B and BO ◇ NOROX BZP-250 ◇ NOVADELOX ◇ OXY-5 ◇ OXY-10 ◇ OXYLITE ◇ OXY WASH ◇ PANOXYL ◇ PEROSSIDO di BENZOILE (ITALIAN) ◇ PEROXYDE de BENZOYLE (FRENCH) ◇ PERSADOX ◇ QUINOLOR COMPOUND ◇ SULFOXYL ◇ SUPEROX ◇ THERADERM ◇ TOPEX ◇ VANOXIDE ◇ XERAC

USE IN FOOD:

Purpose: Bleaching agent.

Where Used: Cheese (asiago fresh), cheese (asiago medium), cheese (asiago old), cheese (asiago soft), cheese (blue), cheese (caciocavallo siciliano), cheese (emmentaler), cheese (gorgonzola), cheese (parmesan), cheese (provolone), cheese (reggiano), cheese (romano), cheese (Swiss), margarine, sausage casings, shortening, whey (annatto-colored).

Regulations: FDA - 21CFR 184.1157. GRAS when used in accordance with good manufacturing practice.

IARC Cancer Review: Animal Inadequate Evidence IMEMDT 36,267,85; Human Inadequate Evidence IMEMDT 36,267,85. EPA Genetic Toxicology Program. Community Right-To-Know List.

OSHA PEL: TWA 5 mg/m^3 ACGIH TLV: TWA 5 mg/m^3 DFG MAK: 5 mg/m^3 DOT Classification: Organic Peroxide; Label: Organic Peroxide NIOSH REL: TWA 5 mg/m^3

SAFETY PROFILE: Poison by ingestion and intraperitoneal routes. An experimental tumorigen. Can cause dermatitis, asthmatic effects, testicular atrophy, and vasodilation. An allergen and eye irritant. Human mutagenic data. Moderate fire hazard by spontaneous chemical reaction in contact with reducing agents. It ignites readily and burns rapidly. A powerful oxidizer. Dangerous explosion hazard; may explode spontaneously, when heated to above melting point, or when overheated under confinement. It is moderately sensitive to heat, shock, friction or contact with combustible materials. Explosive decomposition above the mp (103°) forms flammable products.

Explosive or violent reaction on contact with N,N-dimethylaniline; aniline; dimethyl sulfide; lithium tetrahydroaluminate; and N-bromosuccinimide + 4-toluic acid. Mixture with carbon tetrachloride + ethylene explodes at elevated temperatures and pressures. Reacts violently in contact with various organic or inorganic acids; alcohols; amines; metallic naphthenates; as well as with polymerization accelerators; i.e., dimethylaniline; and (CCl_4 + C_2H_4). Violent reaction with charcoal when heated above 50°. Decomposition produces dense white smoke of benzoic acid; phenyl benzoate; terphenyls; biphenyls; benzene and carbon dioxide. Vigorous reaction leading to ignition with methylmethacrylate; and vinyl acetate + ethyl acetate. To fight fire, use water spray, foam. All precautions must be taken to guard against fire and explosion hazards. Keep in a cool place; out of the direct rays of the sun; away from sparks, open flames and other sources of heat; avoid shock, rough handling, friction from grinding, etc. Isolated storage is required; keep away from possible contact with acids; alcohols; ethers; or other reducing agents or polymerization catalysts such as dimethylaniline. Complete instructions on storage and handling available from manufacturer.

TOXICITY DATA and CODEN

eye-rbt 500 mg/24H MLD 28ZPAK -,52,72
dnd-hmn:oth 100 μmol/L CNREA8 45,2522,85
oms-hmn:oth 56 μmol/L CNREA8 45,2522,85
skn-mus TDLo:24 g/kg/30W-I:ETA SCIEAS 213,1023,81
orl-rat LD50:7710 mg/kg 28ZPAK -,52,72
ipr-mus LDLo:250 mg/kg YKYUA6 31,855,80

BDX000 CAS: 140-11-4
BENZYL ACETATE

mf: $C_9H_{10}O_2$ mw: 150.19

PROP: Colorless liquid; sweet, floral fruity odor. Mp: −51.5°, bp: 213.5°, flash p: 216°F (CC), d: 1.06, autoign temp: 862°F, vap press: 1 mm @ 45°, vap d: 5.1, refr index: 1.501. Sol in alc, most fixed oils, propylene glycol; ins in glycerin, water @ 214°.

SYNS: ACETIC ACID BENZYL ESTER ◇ ACETIC ACID PHENYLMETHYL ESTER ◇ α-ACETOXYTOLUENE ◇ BENZYL ETHANOATE ◇ FEMA No. 2135 ◇ NCI-C06508

USE IN FOOD:

Purpose: Flavoring agent.

Where Used: Various.

Regulations: FDA - 21CFR 172.515. Use at a level not in excess of the amount reasonably required to accomplish the intended effect.

NTP Carcinogenesis Studies (gavage); Some Evidence: mouse, rat NTPTR* NTP-TR-250,86; IARC Cancer Review: Animal Limited Evidence IMEMDT 40,109,86.

SAFETY PROFILE: A poison by inhalation. Moderately toxic by ingestion and subcutaneous routes. Human systemic effects by inhalation: an antipsychotic, unspecified respiratory and urinary system effects. Combustible liquid. To fight fire, use alcohol foam, CO_2. When heated to decomposition it emits irritating fumes.

TOXICITY DATA and CODEN

ihl-hmn TCLo:50 ppm:CNS,PUL,KID TGNCDL 2,31,61
orl-rat LD50:2490 mg/kg FCTXAV 2,327,64
ihl-mus LCLo:1300 mg/m^3/22H AGGHAR 5,1,33
skn-cat LDLo:10 g/kg JPETAB 84,358,45

BDX500 CAS: 100-51-6
BENZYL ALCOHOL
mf: C_7H_8O mw: 108.15

PROP: Found in jasmine, hyacinth, ylang-ylang oils and at least two dozen other essential oils (FCTXAV 11,1011,73). Water-white liquid; faint, aromatic odor, sharp burning taste. Mp: −15.3°, bp: 205.7°, flash p: 213°F (CC), d: 1.042, autoign temp: 817°F, vap press: 1 mm @ 58.0°, vap d: 3.72, refr index: 1.540. Misc with alc, chloroform, ether, water @ 206°(decomp).

SYNS: BENZAL ALCOHOL ◇ BENZENECARBINOL ◇ BENZENEMETHANOL ◇ BENZOYL ALCOHOL ◇ FEMA No. 2137 ◇ HYDROXYTOLUENE ◇ α-HYDROXYTOLUENE ◇ NCI-C06111 ◇ PHENOLCARBINOL ◇ PHENYLCARBINOL ◇ PHENYLMETHANOL ◇ PHENYLMETHYL ALCOHOL ◇ α-TOLUENOL

USE IN FOOD:

Purpose: Flavoring agent.

Where Used: Various.

Regulations: FDA - 21CFR 172.515. Use at a level not in excess of the amount reasonably required to accomplish the intended effect.

EPA Genetic Toxicology Program.

SAFETY PROFILE: Poison by ingestion, intraperitoneal, intravenous, parenteral routes. Moderately toxic by inhalation, skin contact and subcutaneous routes. A moderate skin and severe eye irritant. Combustible liquid. Mixtures with sulfuric acid decompose explosively at 180°. Exothermic polymerization is catalyzed by HBr + iron when heated above 100°. To fight fire, use alcohol foam, CO_2, dry chemical. When heated to decomposition it emits acrid smoke and fumes.

TOXICITY DATA and CODEN

skn-rbt 10 mg/24H open MLD AMIHBC 4,119,51
eye-rbt 750 μg open SEV AMIHBC 4,119,51
orl-rat LD50:1230 mg/kg FCTXAV 2,327,64
ihl-rat LCLo:1000 ppm/8H AMIHBC 4,119,51
ipr-rat LD50:400 mg/kg NPIRI* 1,6,74
scu-rat LDLo:1700 mg/kg RMSRA6 15,561,1895

BED000 CAS: 103-37-7
BENZYL n-BUTYRATE
mf: $C_{11}H_{14}O_2$ mw: 178.25

PROP: Colorless liquid; floral plum-like odor. D: 1.006, refr index: 1.492, flash p: +100°. Sol in fixed oils; ins in glycerin, propylene glycol, water @ 239°.

SYNS: BENZYL n-BUTANOATE ◇ FEMA No. 2140

USE IN FOOD:

Purpose: Flavoring agent.

Where Used: Various.

Regulations: FDA - 21CFR 172.515. Use at a level not in excess of the amount reasonably required to accomplish the intended effect.

SAFETY PROFILE: Moderately toxic by ingestion. Combustible liquid. When heated to decomposition it emits acrid smoke and irritating fumes.

TOXICITY DATA and CODEN

orl-rat LD50:2330 mg/kg FCTXAV 2,327,64

BEG750 CAS: 103-41-3
BENZYL CINNAMATE
mf: $C_{16}H_{14}O_2$ mw: 238.30

PROP: Found in balsams of Peru, Tolu, Styrax, Copaiba and others (FCTXAV 11,1011,73). White crystals; aromatic odor. Mp: 39°, bp: 350.0°, vap press: 1 mm @ 173.8°, flash p: +100°. Sol in fixed oils; ins in glycerin, propylene glycol.

SYNS: BENZYL ALCOHOL CINNAMIC ESTER ◇ BENZYL γ-PHENYLACRYLATE ◇ CINNAMEIN ◇ trans-CINNAMIC ACID BENZYL ESTER ◇ FEMA No. 2142 ◇ 3-PHENYL-2-PROPENOIC ACID PHENYLMETHYL ESTER (9CI)

USE IN FOOD:

Purpose: Flavoring agent.

Where Used: Various.

Regulations: FDA - 21CFR 172.515. Use at a level not in excess of the amount reasonably required to accomplish the intended effect.

SAFETY PROFILE: Moderately toxic by ingestion. A mild allergen and skin irritant. Combustible liquid. When heated to decomposition it emits acrid smoke and irritating fumes.

TOXICITY DATA and CODEN

skn-rbt 500 mg MLD FCTXAV 11,1011,73
orl-rat LDLo:5530 mg/kg FCTXAV 2,327,64

BEM000 CAS: 139-07-1
BENZYLDIMETHYLDODECYL-AMMONIUM CHLORIDE
mf: $C_{21}H_{38}N{\cdot}Cl$ mw: 340.05

SYN: DODECYL DIMETHYL BENZYLAMMONIUM CHLORIDE

USE IN FOOD:

Purpose: Antimicrobial agent.

Where Used: Beets, sugarcane, sugarcane juice (raw).

Regulations: FDA - 21CFR 18172.165. Limitation of 0.25-1.0 ppm. 21CFR 173.320, 173.320. Limitation of 0.05 ppm based on weight of raw sugarcane or raw beets.

SAFETY PROFILE: A skin and eye irritant. When heated to decomposition it emits very toxic fumes of NO_x, NH_3 and Cl^-.

TOXICITY DATA and CODEN

skn-rbt 1 mg/24H OYYAA2 6,329,72
eye-rbt 1 mg OYYAA2 6,329,72

BEO250 CAS: 103-50-4
BENZYL ETHER

mf: $C_{14}H_{14}O$ mw: 198.28

PROP: Colorless to pale yellow liquid. Mp: 5°, bp: 298°, flash p: 275°F (CC), d: 1.039, vap d: 6.84, refr index: 1.557.

SYNS: BENZYL OXIDE (CZECH) ◇ DIBENZYLETHER (CZECH) ◇ FEMA No. 2371

USE IN FOOD:

Purpose: Flavoring agent.

Where Used: Various.

Regulations: FDA - 21CFR 172.515. Use at a level not in excess of the amount reasonably required to accomplish the intended effect.

SAFETY PROFILE: Moderately toxic by ingestion. Vapors are probably narcotic in high concentration. A skin and eye irritant. Combustible when exposed to heat or flame; can react with oxidizing materials. Moderate explosion hazard by spontaneous chemical reaction. To fight fire, use CO_2, dry chemical.

TOXICITY DATA and CODEN

skn-rbt 500 mg/24H MLD 28ZPAK -,38,72
eye-rbt 500 mg/24H MLD 28ZPAK -,38,72
orl-rat LD50:2500 mg/kg FCTXAV 16,637,78

BEP500 CAS: 10453-86-8
5-BENZYL-3-FURYL METHYL(±)-cis,trans-CHRYSANTHEMATE

mf: $C_{22}H_{26}O_3$ mw: 338.48

SYNS: BENZOFUROLINE ◇ BENZYFUROLINE ◇ (5-BENZYL-3-FURYL) METHYL-2,2-DIMETHYL-3-(2-METHYLPROPENYL)-CYCLOPROPANECARBOXYLATE ◇ CHRYSON ◇ CHRYSRON ◇ DIMETHYL-3-(2-METHYL-1-PROPENYL)CYCLOPROPANECARBOXYLATE ◇ ENT 27,474 ◇ FMC 17370 ◇ FOR-SYN ◇ NIA 17170 ◇ NRDC 104 ◇ NSC 195022 ◇ OMS-1206 ◇ PREMGARD ◇ PYNOSECT ◇ PYRETHERM ◇ RESMETHRIN ◇ RESMETRINA (PORTUGUESE) ◇ SBP-1382 ◇ S.B. PENICK 1382 ◇ SYNTHRIN

USE IN FOOD:

Purpose: Pesticide.

Where Used: Various.

Regulations: FDA - 21CFR 193.464. Insecticide residue tolerance of 3.0 ppm.

EPA Genetic Toxicology Program.

SAFETY PROFILE: Poison by inhalation, ingestion and intravenous routes. Moderately toxic by skin contact. When heated to decomposition it emits acrid and irritating fumes.

TOXICITY DATA and CODEN

orl-rat LD50:1347 mg/kg PCBPBS 2,308,72
ivn-rat LDLo:160 mg/kg BIOGAL 41(10),283,75
ihl-mus LD50:99 mg/kg BECTA6 19,113,78
skn-rbt LD50:2500 mg/kg SPEADM 78-1,9,78

BFD400
BENZYL PHENYLACETATE

mf: $C_{15}H_{14}O_2$ mw: 226.27

PROP: Colorless liquid; sweet, floral odor with honey undertone. D: 1.095-1.099, refr index: 1.553-1.558, flash p: +100°. Sol in alc, chloroform, ether.

SYN: FEMA No. 2149

USE IN FOOD:

Purpose: Flavoring agent.

Where Used: Various.

Regulations: FDA - 21CFR 172.515. Use at a level not in excess of the amount reasonably required to accomplish the intended effect.

SAFETY PROFILE: Combustible liquid. When heated to decomposition it emits acrid smoke and irritating fumes.

BFD800
BENZYL PROPIONATE

mf: $C_{10}H_{12}O_2$ mw: 164.20

PROP: Colorless liquid; sweet, floral fruity odor. D: 1.028-1.032, refr index: 1.496-1.500. Sol alc, most oils; sltly sol in propylene glycol; ins in glycerin, water.

SYN: FEMA No. 2150

USE IN FOOD:

Purpose: Flavoring agent.

Where Used: Various.

Regulations: FDA - 21CFR 172.515. Use at a level not in excess of the amount reasonably required to accomplish the intended effect.

SAFETY PROFILE: When heated to decomposition it emits acrid smoke and irritating fumes.

BFJ750 CAS: 118-58-1

BENZYL SALICYLATE

mf: $C_{14}H_{12}O_3$ mw: 228.26

PROP: Thick colorless liquid; pleasant odor. Bp: 208° @ 26 mm, d: 1.175 @ 20°, refr index: 1.579. Sol in fixed oils; ins in glycerin, propylene glycol.

SYNS: BENZYL-o-HYDROXYBENZOATE ◇ FEMA No. 2151

USE IN FOOD:

Purpose: Flavoring agent.

Where Used: Various.

Regulations: FDA - 21CFR 172.515. Use at a level not in excess of the amount reasonably required to accomplish the intended effect.

SAFETY PROFILE: Moderately toxic by ingestion. Combustible when exposed to heat or flame. When heated to decomposition it emits acrid smoke and irritating fumes. Incompatible with oxidizing materials.

TOXICITY DATA and CODEN

orl-rat LD50:2227 mg/kg FCTXAV 11,1029,73

BFN250 CAS: 1694-09-3

BENZYL VIOLET 3B

mf: $C_{39}H_{41}N_3O_6S_2 \cdot Na$ mw: 734.94

SYNS: ACID VIOLET ◇ A.F. VIOLET No 1 ◇ AIZEN FOOD VIOLET No 1 ◇ BENZYL VIOLET ◇ CALCOCID VIOLET 4BNS ◇ C.I. 42640 ◇ C.I. FOOD VIOLET 2 ◇ COOMASSIE VIOLET ◇ DISPERSED VIOLET 12197 ◇ FD & C VIOLET NO. 1 ◇ FORMYL VIOLET S4BN ◇ PERGACID VIOLET 2B ◇ SOLAR VIOLET 5BN ◇ WOOL VIOLET

USE IN FOOD:

Purpose: Color additive.

Where Used: Prohibited from foods.

Regulations: FDA - 21CFR 81.10. Provisional listing terminated.

IARC Cancer Review: Animal Sufficient Evidence IMEMDT 16,153,78. EPA Genetic Toxicology Program.

SAFETY PROFILE: An experimental carcinogen and tumorigen. Mutagenic data. When heated to decomposition it emits very toxic fumes of NO_x, NH_3, Na_2O and SO_x.

TOXICITY DATA and CODEN

mma-sat 320 μg/plate MUREAV 89,21,81
orl-rat TDLo:498 g/kg/28W-C:CAR JNCIAM 51,1337,73
scu-rat TDLo:9360 mg/kg/2Y-I:ETA FEPRA7 16,367,57

BFO000 CAS: 8007-75-8

BERGAMOT OIL RECTIFIED

PROP: Yellow-green liquid; agreeable odor. *Composition:* 1-linalyl acetate, 1-linalool, d-limonene, dipentene, bergaptene. By rectification of bergamot oil expressed, under vacuum, to remove completely the furocoumarins and other related nonvolatile residues; found in the fruit of citrus *Bergamia risso et poiteau (Fam. rutaceae)* (FCTXAV 11,1011,73). D: 0.875-0.880 @ 25°/25°. Misc with alc, glacial acetic acid; sol in fixed oils; ins in glycerin, propylene glycol.

SYNS: BERGAMOTTE OEL (GERMAN) ◇ OIL of BERGAMOT, COLDPRESSED ◇ OIL of BERGAMOT, RECTIFIED

USE IN FOOD:

Purpose: Flavoring agent.

Where Used: Bakery products, beverages (alcoholic), chewing gum, confections, gelatin desserts, ice cream, puddings.

Regulations: FDA - 21CFR 182.20. GRAS when used at a level not in excess of the amount reasonably required to accomplish the intended effect.

SAFETY PROFILE: Mildly toxic by ingestion. A mild skin irritant and allergen. Combustible. When heated to decomposition it emits acrid smoke and irritating fumes.

TOXICITY DATA and CODEN

skn-rbt 500 mg/24H MLD FCTXAV 11,1035,73
orl-rat LD50:11520 mg/kg PHARAT 14,435,59

BFW750 CAS: 128-37-0
BHT (FOOD GRADE)
mf: $C_{15}H_{24}O$ mw: 220.39

PROP: White, crystalline solid; faint characteristic odor. Bp: 265°, fp: 68°, flash p: 260°F (TOC), d: 1.048 @ 20°/4°, vap d: 7.6. Sol in alc; ins in water, propylene glycol.

SYNS: ADVASTAB 401 ◇ AGIDOL ◇ ANTIOXIDANT DBPC ◇ ANTIOXIDANT 29 ◇ AO 29 ◇ AO 4K ◇ 2,6-BIS(1,1-DIMETHYLETHYL)-4-METHYLPHENOL ◇ BUKS ◇ BUTYLATED HYDROXYTOLUENE ◇ BUTYLHYDROXYTOLUENE ◇ CAO 1 ◇ CAO 3 ◇ CATALIN CAO-3 ◇ CHEMANOX 11 ◇ DBMP ◇ DBPC (technical grade) ◇ DIBUTYLATED HYDROXYTOLUENE ◇ 2,6-DI-tert-BUTYL-p-CRESOL (ACGIH) ◇ 2,6-DI-tert-BUTYL-1-HYDROXY-4-METHYLBENZENE ◇ 3,5-DI-tert-BUTYL-4-HYDROXYTOLUENE ◇ 2,6-DI-terc. BUTYL-p-KRESOL (CZECH) ◇ 2,6-DI-tert-BUTYL-p-METHYLPHENOL ◇ 2,6-DI-tert-BUTYL-4-METHYLPHENOL ◇ FEMA No. 2184 ◇ 4-HYDROXY-3,5-DI-tert-BUTYLTOLUENE ◇ IMPRUVOL ◇ IONOL ◇ IONOL (antioxidant) ◇ METHYL DI-tert-BUTYLPHENOL ◇ 4-METHYL-2,6-DI-terc. BUTYLFENOL (CZECH) ◇ 4-METHYL-2,6-DI-tert-BUTYLPHENOL ◇ NCI-C03598 ◇ NONOX TBC ◇ PARABAR 441 ◇ SUSTANE ◇ TENOX BHT ◇ TOPANOL ◇ VANLUBE PCX

USE IN FOOD:

Purpose: Antioxidant.

Where Used: Beef patties (fresh), beef patties (pregrilled), beet sugar, cereals (dry breakfast), chewing gum, emulsion stabilizers for shortening, fats (rendered animal), margarine, meat (dried), meatballs (cooked or raw), oleomargarine, pizza toppings (cooked or raw), pork, potato flakes, potato granules, potato shreds (dehydrated), poultry, sausage (brown and serve), sausage (dry), sausage (fresh Italian), sweetpotato flakes, yeast.

Regulations: FDA - 21CFR 172.115. Limitation of total BHA and BHT 50 ppm in dehydrated potato shreds; 50 ppm in dry breakfast cereals; 200 ppm emulsion stabilizers for shortenings; 50 ppm potato flakes; 10 ppm in potato granules; 50 ppm in sweetpotato flakes. 21CFR 172.615. Limitation of 0.1 percent. 21CFR 173.340. Limitation of 0.1 percent of defoamer. 21CFR 181.24. Limitation of 0.005 percent migrating from food packages. 21CFR 182.3173. GRAS with a limitation of 5 percent of the fat and oil content including volatile oil content of the food when used in accordance with good manufacturing practice. USDA - 9CFR 318.7. Limitation of 0.003 percent in dry sausage. Limitation of 0.01 percent in rendered animal fat. Limitation of 0.02 percent individually or in combination with other antioxidants approved for use in margarine. 9CFR 381.147. Limitation of 0.01 percent in poultry based on fat content.

IARC Cancer Review: Animal Limited Evidence IMEDT 40,161,86. NCI Carcinogenesis Bioassay Completed; Results Negative (NCITR* NCI-CG-TR-150,79). EPA Genetic Toxicology Program.

ACGIH TLV: TLV 10 mg/m^3

SAFETY PROFILE: Poison by intraperitoneal and intravenous routes. Moderately toxic by ingestion. An experimental carcinogen and neoplastigen. Experimental reproductive effects. A human skin irritant. A skin and eye irritant in experimental animals. Combustible when exposed to heat or flame. It can react with oxidizing materials. To fight fire, use CO_2, dry chemical. When heated to decomposition it emits acrid smoke and fumes.

TOXICITY DATA and CODEN

skn-hmn 500 mg/48H MLD AMIHBC 5,311,52
skn-rbt 500 mg/48H MOD AMIHBC 5,311,52
eye-rbt 100 mg/24H MOD 28ZPAK -,57,72
dni-hmn:lym 20 μmol/L BBRCA9 80,963,78
dns-rat:lvr 100 pmol/L CRNGDP 5,1547,84
orl-rat TDLo:6 g/kg (13W male/13W pre-3W post):REP FCTOD7 24,1,86
orl-mus TDLo:43800 mg/kg (52D pre/1-21D preg):REP FCTXAV 3,371,65
orl-wmn TDLo:80 mg/kg NEJMAG 314,648,86
orl-rat LD50:890 mg/kg NEOLA4 24,253,77
ipr-mus LD50:138 mg/kg TXAPA9 61,475,81

BGD100 CAS: 58-85-5
BIOTIN
mf: $C_{10}H_{16}N_2O_3S$ mw: 244.31

PROP: White crystalline powder. Sltly sol in water, alc; ins in common organic solvents.

SYNS: d-BIOTIN ◇ cis-HEXAHYDRO-2-OXO-1H-YHIENO[3,4]IMIDAZOLE-4-VALERIC ACID

USE IN FOOD:

Purpose: Dietary supplement, nutrient.

Where Used: Various.

Regulations: FDA - 21CFR 182.5159, 182.8159. GRAS when used in accordance with good manufacturing practice.

SAFETY PROFILE: When heated to decomposition emits toxic fumes of NO_x, SO_x.

BGJ750 CAS: 132-27-4
2-BIPHENYLOL, SODIUM SALT
mf: $C_{12}H_9O \cdot Na$ mw: 192.20

SYNS: BACTROL ◇ D.C.S. ◇ DORVICIDE A ◇ DOWICIDE ◇ 2-HYDROXYDIPHENYL SODIUM ◇ MIL-DU-RID ◇ MYSTOX WFA ◇ NATRIPHENE ◇ OPP-Na ◇ ORPHENOL ◇ o-PHENYLPHENOL SODIUM SALT ◇ 2-PHENYLPHENOL SODIUM SALT ◇ PREVENTOL-ON ◇ SODIUM-2-HYDROXYDIPHENYL ◇ SODIUM-o-PHENYLPHENATE ◇ SODIUM-2-PHENYLPHENATE ◇ SODIUM-o-PHENYLPHENOLATE ◇ SODIUM-o-PHENYLPHENOXIDE ◇ SOPP ◇ STOMOLD B ◇ TOPANE

USE IN FOOD:

Purpose: Animal glue adjuvant.

Where Used: Packaging materials.

Regulations: FDA - 21CFR 178.3120. Use as an animal glue preservative only.

IARC Cancer Review: Animal Limited Evidence IMEMDT 30,329,83.

SAFETY PROFILE: Moderately toxic by ingestion. An experimental carcinogen, tumorigen, and teratogen. Other experimental reproductive effects. A human skin irritant. A severe skin irritant to experimental animals. When heated to decomposition it emits toxic fumes of Na_2O.

TOXICITY DATA and CODEN

skn-hmn 1 mg MccSB# 15JUN84
skn-rbt 50 mg/24H SEV MccSB# 15JUN84
mmo-asn 16 µmol/L PHYTAJ 66,217,76
sln-asn 52 µmol/L EVHPAZ 31,81,79
orl-mus TDLo:900 mg/kg (7-15D preg):REP TRENAF 29,89,78
orl-mus TDLo:72 g/kg (60D male):REP TRENAF 29,99,78
orl-rat LD50:656 mg/kg TRENAF 30(2),57,79

BGO750 CAS: 8001-88-5
BIRCH TAR OIL

PROP: Brown liquid; leather-like odor. D: 0.886-0.950. Sol in fixed oils; ins in glycerin, mineral oil, and propylene glycol. Found in the tar of the bark and wood of *Betula pendula* Roth (Fam. *Betulaceae*) and prepared by steam distillation of the tar obtained by dry distillation of the bark and wood (FCTXAV 11,1011,73).

SYN: BIRCH TAR OIL, RECTIFIED (FCC)

USE IN FOOD:

Purpose: Flavoring agent.

Where Used: Beer (birch).

Regulations: GRAS when used at a level not in excess of the amount reasonably required to accomplish the intended effect.

SAFETY PROFILE: A skin irritant. Moderately irritating to eyes and mucous membranes. A mild allergen. Combustible when exposed to heat or flame; can react with oxidizing materials.

TOXICITY DATA and CODEN

skn-rbt 500 mg/24H FCTXAV 11,1037,73

BHA750 CAS: 155-04-4
BIS(2-BENZOTHIAZOLYLTHIO)ZINC
mf: $C_{14}H_8N_2S_4 \cdot Zn$ mw: 397.85

SYNS: 2-BENZOTHIAZOLETHIOL, ZINC SALT (2:1) ◇ BIS(MERCAPTOBENZOTHIAZOLATO)ZINC ◇ HERMAT Zn-MBT ◇ 2-MERCAPTOBENZOTHIAZOLE ZINC SALT ◇ OXAF ◇ PENNAC ZT ◇ TISPERSE MB-58 ◇ USAF GY-7 ◇ VULKACIT ZM ◇ ZENITE ◇ ZENITE SPECIAL ◇ ZETAX ◇ ZINC-2-BENZOTHIAZOLETHIOLATE ◇ ZINC BENZOTHIAZOLYL MERCAPTIDE ◇ ZINC BENZOTHIAZOL-2-YLTHIOLATE ◇ ZINC BENZOTHIAZYL-2-MERCAPTIDE ◇ ZINC MERCAPTOBENZOTHIAZOLATE ◇ ZINC-2-MERCAPTOBENZOTHIAZOLE ◇ ZINC MERCAPTOBENZOTHIAZOLE SALT ◇ ZMBT ◇ ZnMB

USE IN FOOD:

Purpose: Animal glue adjuvant.

Where Used: Packaging materials.

Regulations: FDA - 21CFR 178.3120. Use as an animal glue preservative only.

Zinc compounds are on the Community Right-To-Know List.

SAFETY PROFILE: Poison by intraperitoneal route. Moderately toxic by ingestion and subcutaneous routes. An experimental carcinogen. When heated to decomposition it emits very toxic fumes of SO_x, NO_x, and ZnO.

TOXICITY DATA and CODEN

scu-mus TDLo:1000 mg/kg:CAR NTIS** PB223-159
orl-rat LD50:540 mg/kg VCTDC* 12/9/76
ipr-mus LD50:200 mg/kg NTIS** AD277-689

BHM000 CAS: 111-17-1
BIS(2-CARBOXYETHYL) SULFIDE
mf: $C_6H_{10}O_4S$ mw: 178.22

PROP: Mp: 134°. Very sol in alc, hot water, acetate; sltly sol in water.

SYNS: DIETHYL SULFIDE-2,2′-DICARBOXYLIC ACID ◇ KYSELINA-β,β′-THIODIPROPIONOVA (CZECH) ◇ 2-(2,3,5,6-TETRAMETHYLPHENOXY)PROPIONIC AICD ◇ TDPA ◇ 4-THIAHEPTANEDIOIC ACID ◇ THIODIPROPIONIC ACID ◇ β,β′-THIODIPROPIONIC ACID ◇ 3,3′-THIODIPROPIONIC ACID ◇ TYOX A

USE IN FOOD:

Purpose: Antioxidant.

Where Used: Packaging materials.

Regulations: FDA - 21CFR 181.24. Limitation of 0.005 percent migrating from food packages.

SAFETY PROFILE: A poison by intraperitoneal and intravenous routes. Moderately toxic by ingestion. A skin and eye irritant. When heated to decomposition it emits toxic fumes of SO_x.

TOXICITY DATA and CODEN

skn-rbt 500 mg/24H MLD 28ZPAK -,171,72
eye-rbt 20 mg/24H MOD 28ZPAK -,171,72
orl-rat LD50:3980 mg/kg 28ZPAK -,171,72

BIM250 CAS: 55-56-1
1,6-BIS(5-(p-CHLOROPHENYL) BIGUANIDINO)HEXANE

mf: $C_{22}H_{30}Cl_2N_{10}$ mw: 505.52

SYNS: 1,6-BIS(p-CHLOROPHENYLDIGUANIDO)HEXANE ◇ CHLORHEXIDIN (CZECH) ◇ CHLORHEXIDINE ◇ 1,6-DI(4′-CHLOROPHENYLDIGUANIDO)HEXANE ◇ 1,1′-HEXAMETHYLENEBIS(5-(p-CHLOROPHENYL)BIGUANIDE ◇ HIBITANE ◇ NOLVASAN ◇ ROTERSEPT ◇ STERIDO

USE IN FOOD:

Purpose: Animal drug.

Where Used: Beef.

Regulations: FDA - 21CFR 556.420. Tolerance of zero in uncooked tissues of calves.

SAFETY PROFILE: Mildly toxic by ingestion. Experimental reproductive effects. A human skin irritant. Mutagenic data. When heated to decomposition it emits very toxic fumes of Cl^- and NO_x.

TOXICITY DATA and CODEN

skn-hmn 1500 μg/3D-I MLD 85DKA8 -,127,77
mma-sat 400 nmol/L CBINA8 28,249,79
dnr-esc 7 μmol/disc CBINA8 28,249,79
orl-rat TDLo:1500 mg/kg (30D male):REP YACHDS 6,2599,78
orl-mus TDLo:1680 mg/kg (7D pre):REP MEXPAG 10,361,64
orl-rat LD50:9200 mg/kg YACHDS 6,2599,78

BIO750 CAS: 115-32-2
1,1-BIS(p-CHLOROPHENYL)-2,2,2-TRICHLOROETHANOL

DOT: 2761
mf: $C_{14}H_9Cl_5O$ mw: 370.48

PROP: Material used in cancer bioassay was 40-60% pure (NCITR* NCI-CG-TR-90,78).

SYNS: ACARIN ◇ 1,1-BIS(CHLOROPHENYL)-2,2,2-TRICHLOROETHANOL ◇ 1,1-BIS(4-CHLOROPHENYL)-2,2,2-TRICHLOROETHANOL ◇ CARBAX ◇ CEKUDIFOL ◇ 4-CHLORO-α-(4-CHLOROPHENYL)-α-(TRICHLOROMETHYL)BENZENEMETHANOL ◇ CPCA ◇ DICHLOROKELTHANE ◇ DECOFOL ◇ DI-(p-CHLOROPHENYL)TRICHLOROMETHYLCARBINOL ◇ 4,4′-DICHLORO-α-(TRICHLOROMETHYL)BENZHYDROL ◇ DICOFOL ◇ DTMC ◇ ENT 23,648 ◇ FW 293 ◇ HIFOL ◇ KELTANE ◇ p,p′-KELTHANE ◇ KELTHANE (DOT) ◇ KELTHANE DUST BASE ◇ KELTHANETHANOL ◇ MILBOL ◇ MITIGAN ◇ NCI-C00486 ◇ 2,2,2-TRICHLOOR-1,1-BIS(4-CHLOOR FENYL)-ETHANOL (DUTCH) ◇ 1,1,1-TRICHLOR-2,2-BIS(4-CHLORPHENYL)-AETHANOL (GERMAN) ◇ 2,2,2-TRICHLOR-1,1-BIS(4-CHLOR-PHENYL)-AETHANOL (GERMAN) ◇ 2,2,2-TRICHLORO-1,1-BIS (4-CHLOROPHENYL)-ETHANOL (FRENCH) ◇ 2,2,2-TRICHLORO-1,1-BIS(4-CLORO-FENIL)-ETANOLO (ITALIAN) ◇ 2,2,2-TRICHLORO-1,1-DI-(4-CHLOROPHENYL) ETHANOL

USE IN FOOD:

Purpose: Insecticide.

Where Used: Tea (dried).

Regulations: FDA - 21CFR 193.80. Insecticide residue tolerance of 45 ppm on dried tea.

IARC Cancer Review: Animal Limited Evidence IMEMDT 30,87,83. NCI Carcinogenesis Bioassay (feed); Clear Evidence: mouse NCITR* NCI-CG-TR-90,78; No Evidence: rat NCITR* NCI-CG-TR-90,78. Community Right-To-Know List.

DOT Classification: ORM-E; Label: None

SAFETY PROFILE: Poison by ingestion and skin contact. Moderately toxic by intraperitoneal route. An experimental carcinogen. Human mutagenic data. When heated to decomposition it emits toxic fumes of Cl^-.

TOXICITY DATA and CODEN

sce-hmn:lym 1 μmol/L ARTODN 52,221,83
orl-mus TDLo:17 g/kg/78W-C:CAR NCITR* NCI-CG-TR-90,78

orl-mus TD :35 g/kg/78W-C:CAR NCITR* NCI-CG-TR-90,78
orl-rat LD50:575 mg/kg WRPCA2 9,119,70
skn-rat LD50:100 mg/kg WRPCA2 9,119,70

BJK500 CAS: 137-30-4
BIS(DIMETHYLDITHIOCARBAMATO) ZINC

mf: $C_6H_{12}N_2S_4 \cdot Zn$ mw: 305.81

PROP: White powder. Mp: 248-250°; d: 1.65 @ 20°/20°.

SYNS: AAPROTECT ◇ AAVOLEX ◇ AAZIRA ◇ ACCELERATOR L ◇ ACETO ZDED ◇ ACETO ZDMD ◇ ALCOBAM ZM ◇ AMYL ZIMATE ◇ ANTENE ◇ BIS(DIMETHYLCARBAMODITHIOATO-S,S')ZINC ◇ BIS(DIMETHYLDITHIOCARBAMATE de ZINC) (FRENCH) ◇ BIS(N,N-DIMETIL-DITIOCARBAMMATO) DI ZINCO (ITALIAN) ◇ CARBAMIC ACID, DIMETHYLDITHIO-, ZINC SALT (2:) ◇ CARBAZINC ◇ CIRAM ◇ CORONA COROZATE ◇ COROZATE ◇ CUMAN ◇ CUMAN L ◇ CYMATE ◇ DIMETHYLCARBAMODITHIOIC ACID, ZINC COMPLEX ◇ DIMETHYLCARBAMODITHIOIC ACID, ZINC SALT ◇ DIMETHYLDITHIOCARBAMATE ZINC SALT ◇ DIMETHYLDITHIOCARBAMIC ACID, ZINC SALT ◇ DRUPINA 90 ◇ EPTAC 1 ◇ ENT 988 ◇ FUCLASIN ◇ FUCLASIN ULTRA ◇ FUKLASIN ◇ FUNGOSTOP ◇ HERMAT ZDM ◇ HEXAZIR ◇ KARBAM WHITE ◇ METHASAN ◇ METHAZATE ◇ METHYL ZIMATE ◇ METHYL ZINEB ◇ METHYL ZIRAM ◇ MEXENE ◇ MEZENE ◇ MILBAM ◇ MILBAN ◇ MOLURAME ◇ MYCRONIL ◇ NCI-C50442 ◇ ORCHARD BRAND ZIRAM ◇ POMARSOL Z FORTE ◇ PRODARAM ◇ RHODIACID ◇ SOXINAL PZ ◇ SOXINOL PZ ◇ TRICARBAMIX Z ◇ TSIMAT ◇ TSIRAM (RUSSIAN) ◇ USAF P-2 ◇ VANCIDE MZ-96 ◇ VULCACURE ◇ VULKACITE L ◇ Z 75 ◇ ZARLATE ◇ Z-C SPRAY ◇ ZERLATE ◇ ZIMATE ◇ ZIMATE METHYL ◇ ZINC BIS(DIMETHYLDITHIOCARBAMATE) ◇ ZINC BIS(DIMETHYLDITHIOCARBAMOYL)DISULPHIDE ◇ ZINC BIS(DIMETHYLTHIOCARBAMOYL)DISULFIDE ◇ ZINC DIMETHYLDITHIOCARBAMATE ◇ ZINC N,N-DIMETHYLDITHIOCARBAMATE ◇ ZINCMATE ◇ ZINK-BIS (N,N-DIMETHYL-DITHIOCARBAMAAT) (DUTCH) ◇ ZINK-BIS(N,N-DIMETHYL-DITHIOCARBAMAT) (GERMAN) ◇ ZINKCARBAMATE ◇ ZINK-(N,N-DIMETHYL-DITHIOCARBAMAT) (GERMAN) ◇ ZIRAM ◇ ZIRAM TECHNICAL ◇ ZIRAMVIS ◇ ZIRASAN ◇ ZIRBERK ◇ ZIREX 90 ◇ ZIRIDE ◇ ZIRTHANE ◇ ZITOX

USE IN FOOD:

Purpose: Animal glue adjuvant.

Where Used: Packaging materials.

Regulations: FDA - 21CFR 178.3120. Use as an animal glue preservative only.

IARC Cancer Review: Animal Inadequate Evidence IMEMDT 12,259,76; NTP Carcinogenesis Bioassay (feed); Clear Evidence: mouse, rat NTPTR* NTP-TR-238,83. EPA Genetic Toxicology Program. Zinc and its compounds are on the Community Right-To-Know List.

SAFETY PROFILE: Poison by ingestion, intraperitoneal, and intravenous routes. Moderately toxic by inhalation. Mutagenic data. An experimental carcinogen and tumorigen. Severe irritant to eyes, nose, and throat. When heated to decomposition it emits very toxic fumes of NO_x and SO_x.

TOXICITY DATA and CODEN

mmo-sat 5 μg/plate MUREAV 68,313,79
dnd-esc 1 μmol/L ARTODN 46,277,80
orl-rat TDLo:250 mg/kg (6-15D preg):REP EESADV 7,531,83
orl-rat TDLo:500 mg/kg (6-15D preg):REP EESADV 7,531,83
orl-mus TDLo:840 mg/kg/13W-I:ETA GISAAA 37(9),25,72
orl-rat LD50:1400 mg/kg FMCHA2 -,C259,80
ipr-rat LD50:23 mg/kg JAPMA8 41,662,52
ihl-rat LD50:1230 mg/kg EQSFAP 3,618,75

BJP000 CAS: 122-34-9
2,4-BIS(ETHYLAMINO)-6-CHLORO-s-TRIAZINE

mf: $C_7H_{12}ClN_5$ mw: 201.69

SYNS: A 2079 ◇ AKTINIT S ◇ AQUAZINE ◇ BATAZINA ◇ 2,4-BIS(AETHYLAMINO)-6-CHLOR-1,3,5-TRIAZIN (GERMAN) ◇ BITEMOL ◇ BITEMOL S 50 ◇ CAT (HERBICIDE) ◇ CDT ◇ CEKUSAN ◇ CEKUZINA-S ◇ CET ◇ 1-CHLORO-3,5-BISETHYLAMINO-2,4,6-TRIAZINE ◇ 2-CHLORO-4,6-BIS(ETHYLAMINO)-s-TRIAZINE ◇ 2-CHLORO-4,6-BIS (ETHYLAMINO)-1,3,5-TRIAZINE ◇ FRAMED ◇ G 27692 ◇ GEIGY 27,692 ◇ GESARAN ◇ GESATOP ◇ GESATOP 50 ◇ H 1803 ◇ HERBAZIN ◇ HERBAZIN 50 ◇ HERBEX ◇ HERBOXY ◇ HUNGAZIN DT ◇ PREMAZINE ◇ PRIMATOL S ◇ PRINCEP ◇ PRINTOP ◇ RADOCON ◇ RADOKOR ◇ SIMADEX ◇ SIMANEX ◇ SIMAZIN ◇ SIMAZINE (USDA) ◇ SIMAZINE 80W ◇ SYMAZINE ◇ TAFAZINE ◇ TAFAZINE 50-W ◇ TAPHAZINE ◇ TRIAZINE A 384 ◇ W 6658 ◇ ZEAPUR

USE IN FOOD:

Purpose: Herbicide.

Where Used: Animal feed, molasses, potable water, sugarcane byproducts, sugarcane syrups.

Regulations: FDA - 21CFR 193.400. Herbicide residue tolerance of 1 ppm sugarcane byproducts, molasses, and syrup, 0.01 ppm in potable water. 21CFR 561.350. Limitation of 1 ppm sugarcane byproduct molasses when used for animal feed. USDA CES Ranking: C-3 (1988).

EPA Genetic Toxicology Program.

SAFETY PROFILE: Poison by intravenous route. Mutagenic data. An experimental tumorigen. A skin and eye irritant. When heated to decomposition it emits very toxic fumes of Cl^- and NO_x.

TOXICITY DATA and CODEN

skn-rbt 500 mg open MLD CIGET* -,-,77
eye-rbt 80 mg MOD CIGET* -,-,77
sln-dmg-par 396 μmol/L JTEHD6 3,691,77
ihl-rat TCLo: 17 mg/m^3/2H (7-14D preg): REP NTIS** PB277,077
scu-rat TDLo: 16 g/kg/61W-I: ETA VOONAW 16(1),82,70
orl-rat LD50: 5000 mg/kg RREVAH 10,97,65

BJS000 CAS: 117-81-7
BIS(2-ETHYLHEXYL)PHTHALATE

mf: $C_{24}H_{38}O_4$ mw: 390.62

SYNS: BEHP ◇ BIS(2-ETHYLHEXYL)-1,2-BENZENEDICARBOXYLATE ◇ BISOFLEX 81 ◇ BISOFLEX DOP ◇ COMPOUND 889 ◇ DAF 68 ◇ DEHP ◇ DI(2-ETHYLHEXYL)ORTHOPHTHALATE ◇ DI(2-ETHYLHEXYL)PHTHALATE ◇ DIOCTYL PHTHALATE ◇ DI-sec-OCTYL PHTHALATE (ACGIH) ◇ DOP ◇ EHTYLHEXYL PHTHALATE ◇ ERGOPLAST FDO ◇ 2-ETHYLHEXYL PHTHALATE ◇ EVIPLAST 80 ◇ EVIPLAST 81 ◇ FLEXIMEL ◇ FLEXOL DOP ◇ FLEXOL PLASTICIZER DOP ◇ GOOD-RITE GP 264 ◇ HATCOL DOP ◇ HERCOFLEX 260 ◇ KODAFLEX DOP ◇ MOLLAN O ◇ NCI-C52733 ◇ NUOPLAZ DOP ◇ OCTOIL ◇ OCTYL PHTHALATE ◇ PALATINOL AH ◇ PHTHALIC ACID DIOCTYL ESTER ◇ PITTSBURGH PX-138 ◇ PLATINOL AH ◇ PLATINOL DOP ◇ RC PLASTICIZER DOP ◇ RCRA WASTE NUMBER U028 ◇ REOMOL DOP ◇ REOMOL D 79P ◇ SICOL 150 ◇ STAFLEX DOP ◇ TRUFLEX DOP ◇ VESTINOL AH ◇ VINICIZER 80 ◇ WITCIZER 312

USE IN FOOD:

Purpose: Plasticizer.

Where Used: Packaging materials.

Regulations: FDA - 21CFR 181.27. Limited to foods of high water content.

IARC Cancer Review: Human Inadequate Evidence IMEMDT 29,269,82; Animal Sufficient Evidence IMEMDT 29,269,82. NTP Carcinogenesis Bioassay (feed); Clear Evidence: mouse, rat NTPTR* NTP-TR-217,82. EPA Genetic Toxicology Program. Community Right-To-Know List.

OSHA PEL: TWA 5 mg/m^3; STEL 10 mg/m^3 ACGIH TLV: TWA 5 mg/m^3; STEL 10 mg/m^3 NIOSH REL: (DEHP) Reduce to lowest feasible level

SAFETY PROFILE: Poison by intravenous route. Suspected human carcinogen and an experimental teratogen. Affects the human gastrointestinal tract. A mild skin and eye irritant. When heated to decomposition it emits acrid smoke.

TOXICITY DATA and CODEN

skn-rbt 500 mg/24H MLD 28ZPAK -,48,72
eye-rbt 500 mg/24H MLD 28ZPAK -,48,72
dns-rat: lvr 500 μmol/L PMRSDJ 5,371,85
sln-ham: lvr 50 mg/L PMRSDJ 5,397,85
orl-rat TDLo: 35 mg/kg (14D pre): REP FCTXAV 15,389,77
orl-mus TDLo: 50 mg/kg (7D preg): REP EVHPAZ 45,71,82
orl-rat TDLo: 216 g/kg/2Y-C: CAR NTPTR* NTP-TR-217,82
orl-mus TDLo: 260 g/kg/2Y-C: CAR NTPTR* NTP-TR-217,82
orl-man TDLo: 143 mg/kg: GIT JIHTAB 27,130,45
orl-rat LD50: 30600 mg/kg EVHPAZ 3,131,73
ipr-rat LD50: 30700 mg/kg JIHTAB 27,130,45

BKE500 CAS: 120-40-1
N,N-BIS(2-HYDROXYETHYL)DODECANAMIDE

mf: $C_{16}H_{33}NO_3$ mw: 287.50

PROP: Solid. Mp: 36°

SYNS: BIS(2-HYDROXYETHYL)LAURAMIDE ◇ N,N-BIS(HYDROXYETHYL)LAURAMIDE ◇ N,N-BIS(β-HYDROXYETHYL)LAURAMIDE ◇ N,N-BIS(2-HYDROXYETHYL)LAURAMIDE ◇ CLINDROL 101CG ◇ CLINDROL SUPERAMIDE 100L ◇ COCO DIETHANOLAMIDE ◇ COCONUT OIL AMIDE OF DIETHANOLAMINE ◇ COMPERLAN LD ◇ CONDENSATE PL ◇ CRILLON L.D.E. ◇ DIETHANOLLAURAMIDE ◇ N,N-DIETHANOLLAURAMIDE ◇ N,N-DIETHANOLLAURIC ACID AMIDE ◇ EMID 6511 ◇ EMID 6541 ◇ ETHYLAN MLD ◇ HETAMIDE ML ◇ LAURAMIDE DEA ◇ LAURIC ACID DIETHANOLAMIDE ◇ LAURIC DIETHANOLAMIDE ◇ LAUROYL DIETHANOLAMIDE ◇ LAURYL DIETHANOLAMIDE ◇ LDA ◇ LDE ◇ MONAMID 150-LW ◇ NCI-C55323 ◇ NINOL AA-62 EXTRA ◇ NINOL 4821 ◇ NINOL AA62

◇ ONYXOL 345 ◇ REWOMID DLMS ◇ RICHAMIDE 6310 ◇ ROLAMID CD ◇ STANDAMIDD LD ◇ STEINAMID DL 203 S ◇ SUPER AMIDE L-9A ◇ SYNOTOL L-60 ◇ UNAMIDE J-56 ◇ VARAMID ML 1

USE IN FOOD:

Purpose: Antistatic agent.

Where Used: Packaging materials.

Regulations: FDA - 21CFR 178.3130. Limitation of 0.5 percent in polyethylene containers.

SAFETY PROFILE: Moderately toxic by ingestion. When heated to decomposition it emits toxic fumes of NO_x.

TOXICITY DATA and CODEN

orl-rat LD50:2700 mg/kg JSCCA5 13,469,62

BKL250 CAS: 7287-19-6
2,4-BIS(ISOPROPYLAMINO)-6-METHYLMERCAPTO-s-TRIAZINE

mf: $C_{10}H_{19}N_5S$ mw: 241.40

SYNS: 4,6-BIS(ISOPROPYLAMINO)-2-METHYLMERCAPTO-s-TRIAZINE ◇ 2,4-BIS(ISOPROPYLAMINO)-6-METHYLTHIO-s-TRIAZINE ◇ 2,4-BIS(ISOPROPYLAMINO)-6-METHYLTHIO-1,3,5-TRIAZINE ◇ N,N′-BIS(1-METHYLETHYL)-6-METHYL-THIO-1,3,5-TRIAZINE-2,4-DIAMINE ◇ CAPAROL ◇ G 34161 ◇ GESAGARD ◇ MERKAZIN ◇ 2-METHYLMERCAPTO-4,6-BIS(ISOPROPYLAMINO)-s-TRIAZINE ◇ 2-METHYLTHIO-4,6-BIS(ISOPROPYLAMINO)-s-TRIAZINE ◇ POLISIN ◇ PRIMATOL Q ◇ PROMETREX ◇ PROMETRIN ◇ PROMETRYN ◇ PROMETRYNE (USDA) ◇ SELEKTIN ◇ SESAGARD

USE IN FOOD:

Purpose: Herbicide.

Where Used: Celery, corn (fodder), corn (fresh), pigeon peas.

Regulations: USDA CES Ranking: C-3 (1985).

SAFETY PROFILE: Moderately toxic by ingestion. An eye irritant. When heated to decomposition it emits very toxic fumes of NO_x and SO_x.

TOXICITY DATA and CODEN

eye-rbt 80 mg MLD CIGET* -,-,77
orl-rat LD50:2100 mg/kg HYSAAV 34(1-3),433,69

BLC000 CAS: 868-18-8
BISODIUM TARTRATE

mf: $C_4H_4O_6 \cdot 2Na$ mw: 194.06

PROP: Transparent crystals; colorless and odorless. Sol in water.

SYNS: 2,3-DIHYDROXY-(R-(R*,R*))-BUTANEDIOIC ACID DISODIUM SALT (9CI) ◇ DISODIUM TARTRATE ◇ DISODIUM l-(+)-TARTRATE ◇ SODIUM TARTRATE (FCC) ◇ SODIUM l-(+)-TARTRATE

USE IN FOOD:

Purpose: Emulsifier, pH control agent, sequestrant.

Where Used: Fats, jams, jellies, margarine, meat products, oils, oleomargarine, sausage casings.

Regulations: FDA - 21CFR 184.1801. GRAS when used in accordance with good manufacturing practice. USDA - 9CFR 318.7. Sufficient for purpose.

SAFETY PROFILE: Moderately toxic by ingestion. When heated to decomposition it emits acrid smoke and irritating fumes.

TOXICITY DATA and CODEN

orl-mus LDLo:3686 mg/kg JAPMA8 31,12,42

BLC250 CAS: 10380-28-6
BIS(8-OXYQUINOLINE)COPPER

mf: $C_{18}H_{12}CuN_2O_2$ mw: 351.86

PROP: Yellow-green powder.

SYNS: BIOQUIN ◇ BIOQUIN 1 ◇ BIS(8-QUINOLINATO)COPPER ◇ BIS(8-QUINOLINOLATO)COPPER ◇ BIS(8-QUINOLINOLATO-N(1),O(8))-COPPER ◇ CELLU-QUIN ◇ COPPER-8 ◇ COPPER HYDROXYQUINOLATE ◇ COPPER-8-HYDROXYQUINOLATE ◇ COPPER-8-HYDROXYQUINOLINATE ◇ COPPER-8-HYDROXYQUINOLINE ◇ COPPER OXINATE ◇ COPPER (2+) OXINATE ◇ COPPER OXINE ◇ COPPER OXYQUINOLATE ◇ COPPER OXYQUINOLINE ◇ COPPER QUINOLATE ◇ COPPER-8-QUINOLATE ◇ COPPER QUINOLINOLATE ◇ COPPER-8-QUINOLINOLATE ◇ COPPER-8-QUINOLINOL ◇ CUNILATE ◇ CUNILATE 2472 ◇ CUPRIC-8-HYDROXYQUINOLATE ◇ CUPRIC-8-QUINOLINOLATE ◇ DOKIRIN ◇ FRUITDO ◇ 8-HYDROXYQUINOLINE COPPER COMPLEX ◇ MILMER ◇ OXIME COPPER ◇ OXINE COPPER ◇ OXINE CUIVRE ◇ OXYQUINOLINOLEATE de CUIVRE (FRENCH) ◇ QUINONDO

USE IN FOOD:

Purpose: Preservative for wood.

Where Used: Packaging materials.

Regulations: FDA - 21CFR 178.3800.

IARC Cancer Review: Animal Inadequate Evidence IMEMDT 15,103,77. Copper and its compounds are on the Community Right-To-Know List.

SAFETY PROFILE: Poison by intraperitoneal route. An experimental tumorigen. Mutagenic data. When heated to decomposition it emits toxic fumes of NO_x.

TOXICITY DATA and CODEN

mma-sat 5 μg/plate MUREAV 116,185,83
scu-mus TDLo: 156 mg/kg/39W-I: ETA JNCIAM 24,109,60
ipr-mus LD50: 67 mg/kg TXAPA9 5,599,63

BLU000 CAS: 13356-08-6
BIS(TRIS(β,β-DIMETHYLPHENETHYL) TIN)OXIDE

mf: $C_{60}H_{78}OSn_2$ mw: 1052.76

SYNS: BENDEX ◇ BIS(TRIS(2-METHYL-2-PHENYLPROPYL)TIN)OXIDE ◇ DI(TRI-(2,2-DIMETHYL-2-PHENYLETHYL)TIN)OXIDE ◇ ENT 27,738 ◇ FENBUTATIN OXIDE ◇ HEXAKIS(β,β-DIMETHYLPHENETHYL)DISTANNOXANE ◇ HEXAKIS(2-METHYL-2-PHENYLPROPYL)DISTANNOXANE ◇ SD 14114 ◇ SHELL SD-14114 ◇ TORQUE ◇ VENDEX

USE IN FOOD:

Purpose: Insecticide.

Where Used: Animal feed, apple pomace (dried), citrus pulp (dried), grape pomace (dried), prunes (dried), raisin waste, raisins.

Regulations: FDA - 21CFR 193.236. Insecticide residue tolerance of 20.0 ppm in prunes, dried, 20 ppm in raisins. 21CFR 561.255. Limitation of 75.0 ppm in dried apple pomace, 35.0 ppm in dried citrus pulp, 100 ppm in dried grape pomace, 20 ppm in raisin waste when used for animal feed.

OSHA PEL: TWA 0.1 mg/m^3 ACGIH TLV: TWA 0.1(Sn)/m^3 (skin) NIOSH REL: TWA 0.1 mg(Sn)/m^3

SAFETY PROFILE: Moderately toxic by ingestion and skin contact. When heated to decomposition it emits acrid smoke and irritating fumes.

TOXICITY DATA and CODEN

orl-rat LD50: 2630 mg/kg 85ARAE 1,17,77
skn-rat LD50: 1000 mg/kg TIUSAD 110,6,76

BLV500 CAS: 8013-76-1
BITTER ALMOND OIL

PROP: Volatile oil from dried ripe kernels of bitter almonds or from other kernels containing amygdalin, such as apricots, cherries, plums, and especially peaches. Colorless liquid; strong almond odor. Bp: 179°, d: 1.045-1.070 @ 15°. Sltly sol in water; sol in fixed oils and propylene glycol; insol in glycerin.

SYNS: ALMOND OIL BITTER, FFPA (FCC) ◇ OIL, BITTER ALMOND

USE IN FOOD:

Purpose: Flavoring agent.

Where Used: Baked goods, beverages (alcoholic), beverages (nonalcoholic), cakes, chewing gum, confections, gelatin desserts, ice cream, maraschino cherries, pastries, puddings.

Regulations: FDA - 21CFR 182.20. GRAS when used at a level not in excess of the amount reasonably required to accomplish the intended effect. Must be treated and redistilled to remove hydrocyanic acid.

SAFETY PROFILE: A human poison by ingestion. Moderately toxic by skin contact. A skin irritant. When heated to decomposition it emits toxic fumes of CN^-.

TOXICITY DATA and CODEN

skn-rbt 500 mg/24H MOD FCTXAV 17,705,79
orl-hmn LDLo: 107 mg/kg FCTXAV 17,705,79
orl-rat LD50: 960 mg/kg FCTXAV 17,705,79
skn-rbt LD50: 1220 mg/kg FCTXAV 17,705,79

BLW000
BIXA ORELLANA

PROP: From solvent extraction of *Bixa orellana* L. seeds (JAPMA8 49,218,60). Yellow red solutions or powder.

SYNS: ACHIOTE ◇ ANNATTO EXTRACT (FCC)

USE IN FOOD:

Purpose: Color additive.

Where Used: Ink (food marking), oleomargarine, poultry, sausage casings, shortening.

Regulations: FDA - 21CFR 73.30. Use in accordance with good manufacturing practice. USDA - 9CFR 318.7, 9CFR 381.147. Sufficient for purpose.

SAFETY PROFILE: Moderately toxic by intraperitoneal route. Human systemic effects by skin contact. When heated to decomposition it emits acrid smoke and irritating fumes.

TOXICITY DATA and CODEN

orl-hmn TDLo: 357 mg/kg: SKN ARTODN (Suppl. 1),141,78
ipr-mus LD50: 700 mg/kg JAPMA8 49,218,60

BLW250 CAS: 8006-82-4
BLACK PEPPER OIL

PROP: From steam distillation of dried fruit of *Piper nigrum L.* (Fam. *Piperaceae*). Main constituents include α- and β-pinene, β-caryophyllene, l-limonene, d-hydrocarveol, piperidine and piperrine (FCTXAV 16,637,78). A colorless to greenish liquid; odor and taste of pepper. Sol in fixed oils, mineral oil, propylene glycol; sltly sol in glycerin.

USE IN FOOD:

Purpose: Flavoring agent.

Where Used: Meat, salads, soups, vegetables.

Regulations: FDA - 21CFR 182.20. GRAS when used at a level not in excess of the amount reasonably required to accomplish the intended effect.

SAFETY PROFILE: A moderate skin irritant. Mutagenic data. When heated to decomposition it emits acrid smoke and irritating fumes.

TOXICITY DATA and CODEN

skn-rbt 500 mg/24H MOD FCTXAV 16,637,78
dnr-bcs 20 mg/disc TOFOD5 8,91,85

BMA550
BOIS de ROSE OIL

PROP: From steam distillation of chipped wood of *Aniba rosaeodora* var. *amazonica* Ducke, (Fam. *Lauraceae*). Colorless to pale yellow liquid; slt pleasant floral odor. Sol in fixed oils, propylene glycol, mineral oil; sltly sol in glycerin.

SYN: LIGNALOE OIL

USE IN FOOD:

Purpose: Flavoring agent.

Where Used: Bakery products, beverages (nonalcoholic), chewing gum, confections, gelatin desserts, ice cream products, puddings.

Regulations: FDA - 21CFR 182.20. GRAS when used at a level not in excess of the amount reasonably required to accomplish the intended effect.

SAFETY PROFILE: When heated to decomposition it emits acrid smoke and irritating fumes.

BMD100 CAS: 76-49-3
BORNYL ACETATE
mf: $C_{12}H_{20}O_2$ mw: 196.29

PROP: Colorless liquid or white crystalline solid; sweet, piney odor. D: 0.981-0.985, refr index: 1.462. flash p: 89°. Sol in alc, fixed oils; sltly sol in water; insol in glycerin, propylene glycol @ 226°.

SYNS: l-BORNYL ACETATE ◇ FEMA No. 2159

USE IN FOOD:

Purpose: Flavoring agent.

Where Used: Various.

Regulations: FDA - 21CFR 172.515. Use at a level not in excess of the amount reasonably required to accomplish the intended effect.

SAFETY PROFILE: Combustible liquid. When heated to decomposition it emits acrid smoke and irritating fumes.

BMN775
BROMELIN

PROP: From pineapples *Ananas comosus* and *Ananas bracteatus* L. White to tan amorphous powder. Sol in water; insol in alc, chloroform, ether.

SYNS: ANANASE ◇ BROMELAIN ◇ EXTRANASE ◇ INFLAMEN ◇ TRAUMANASE

USE IN FOOD:

Purpose: Chillproofing of beer, enzyme, meat tenderizing, preparation of precooked cereals, processing aid, tissue softening agent.

Where Used: Beer, bread, cereals (precooked), meat (raw cuts), poultry, wine.

Regulations: USDA - 9CFR 318.7, 381.147. Solutions consisting of water and approved proteolytic enzymes applied or injected into raw meat cuts shall not result in a gain of more than 3 percent above the weight of the untreated product. BATF - 27CFR 240.1051. GRAS when used in accordance with good manufacturing practice.

SAFETY PROFILE: When heated to decomposition it emits acrid smoke and irritating fumes.

BMO500 CAS: 7758-01-2
BROMIC ACID, POTASSIUM SALT
DOT: 1484
mf: $BrO_3 \cdot K$ mw: 167.01

PROP: White crystals or crystalline powder. Mp: 350° (approx), decomp @ 370°, d: 3.27 @ 17.5°. Sol in water; sltly sol in alc.

SYN: POTASSIUM BROMATE (DOT, FCC)

USE IN FOOD:

Purpose: Dough conditioner, maturing agent.

Where Used: Baked goods, beverages (fermented malt), confectionery products.

Regulations: FDA - 21CFR 172.730. Limitation of 75 ppm of bromate in treated malt, 25 ppm in fermented malt beverage.

IARC Cancer Review: Animal Sufficient Evidence IMEMDT 40,207,86.

DOT Classification: Oxidizer; Label: Oxidizer

SAFETY PROFILE: A poison by ingestion. An experimental carcinogen. A powerful oxidizer. An irritant to skin, eyes, and mucous membranes. Mutagenic data. Mixtures with sulfur may ignite. Violent reaction with Al; Al + dinitrotoluene @ 290°; As; C; Cu; $Pb(C_2H_3O_2)_2$; metal sulfides; organic matter; P; S. Aqueous solutions react violently with selenium. When heated to decomposition it emits very toxic fumes of Br^- and K_2O.

TOXICITY DATA and CODEN

mma-sat 1 mg/plate AMONDS 3,253,80
cyt-rat-ipr 500 μmol/kg MUREAV 147,274,85
cyt-rat-orl 3 mmol/kg MUREAV 147,274,85
orl-rat TDLo:9625 mg/kg/2Y-C:CAR ESKHA5 100,93,82
orl-rat TDLo:38500 mg/kg/2Y-C:CAR GANNA2 73,335,82
orl-rat LD50:321 mg/kg ESKHA5 100,93,82

BMO825
BROMINATED VEGETABLE (SOYBEAN) OIL

PROP: Pale yellow to dark brown viscous, oily liquid; bland or fruity odor and bland taste. Sol in alc, chloroform, ether, hexane, fixed oils; insol in water.

SYN: VEGETABLE (SOYBEAN) OIL, brominated

USE IN FOOD:

Purpose: Beverage stabilizer, flavoring agent.

Where Used: Beverages (fruit flavored).

Regulations: FDA - 21CFR 180.30. Limitation of 15 ppm in fruit flavored beverages.

SAFETY PROFILE: Experimental reproductive effects. When heated to decomposition it emits toxic fumes of Br^-.

TOXICITY DATA and CODEN

orl-rat TDLo:9 g/kg (2W male/2W pre-14D post):REP TJADAB 28,309,83
orl-rat TDLo:27 g/kg (2W male/2W pre-1D post):REP TJADAB 28,309,83

BNA750 CAS: 41198-08-7
O-(4-BROMO-2-CHLOROPHENYL)-O-ETHYL-S-PROPYL PHOSPHOROTHIOATE

mf: $C_{11}H_{15}BrClO_3PS$ mw: 373.65

SYNS: CGA 15324 ◇ CURACRON ◇ POLYCRON ◇ PROFENOFOS ◇ SELECRON

USE IN FOOD:

Purpose: Insecticide.

Where Used: Animal feed, cottonseed hulls, soapstock.

Regulations: FDA - 21CFR 561.53. Limitation of 6.0 ppm in cottonseed hulls, 15.0 ppm in soapstock when used for animal feed.

SAFETY PROFILE: Poison by ingestion and skin contact. An insecticide. When heated to decomposition it emits very toxic SO_x, PO_x, Br^-, and Cl^-.

TOXICITY DATA and CODEN

orl-rat LD50:400 mg/kg SPEADM 78-1,35,78
skn-rat LD50:300 mg/kg CIGET* -,-,77

BNM500 CAS: 74-83-9
BROMO METHANE

DOT: 1062
mf: CH_3Br mw: 94.95

PROP: Colorless, transparent, volatile liquid or gas; burning taste, chloroform-like odor. Bp: 3.56°, lel: 13.5%, uel: 14.5%, fp: −93°, flash p: none, d: 1.732 @ 0°/0°, autoign temp: 998°F, vap d: 3.27, vap press: 1824 mm @ 25°.

SYNS: BROM-METHAN (GERMAN) ◇ BROMO-O-GAS ◇ BROMOMETANO (ITALIAN) ◇ BROMURE de METHYLE (FRENCH) ◇ BROMURO di METILE (ITALIAN) ◇ BROOMMETHAAN (DUTCH) ◇ DAWSON 100 ◇ DOWFUME ◇ DOWFUME MC-2 SOIL FUMIGANT ◇ EDCO ◇ EMBAFUME ◇ FUMIGANT-1 (OBS.) ◇ HALON 1001 ◇ ISCOBROME ◇ KAYAFUME ◇ MB ◇ MBX ◇ MEBR ◇ METAFUME ◇ METHOGAS ◇ METHYLBROMID (GERMAN) ◇ METHYL BROMIDE (ACGIH, USDA) ◇ METYLU BROMEK (POLISH) ◇ MONOBROMOMETHANE ◇ PESTMASTER (OBS.) ◇ PROFUME (OBS.) ◇ R 40B1 ◇ RCRA WASTE NUMBER U029 ◇ ROTOX ◇ TERABOL ◇ TERR-O-GAS 100 ◇ ZYTOX

USE IN FOOD:

Purpose: Fumigant.

Where Used: Animal feed, apples, barley (milling fractions), cereal grains, corn (milling fractions), corn grits, cracked rice, fava beans, fermented malt beverages, flour, grain sorghum (milo, milling fractions), kiwi fruit, lentils, macadamia nuts, oats (milling fractions), pistachio nuts, rice (milling fractions), rye (milling fractions), sweet potatoes, wheat (milling fractions).

Regulations: FDA - 21CFR 193.225. Fumigant residue tolerance of 125 ppm as Br in cereal grain. 21CFR 193.230. Pesticide residue tolerance of 25 ppm as Br in fermented malt beverages. 21CFR 561.260. Limitation of 400 ppm as Br in dog food. Limitation of 125 ppm as Br in barley, corn, grain sorghum, oats, rice, rye, and wheat milling fractions when used for animal feed. USDA CES Ranking: B-4 (1986).

IARC Cancer Review: Human Inadequate Evidence IMEMDT 41,187,86; Animal Limited Evidence IMEMDT 41,187,86. Community Right-To-Know List. EPA Extremely Hazardous Substances List.

OSHA PEL: 5 ppm (skin) ACGIH TLV: TWA 5 ppm (skin) DOT Classification: Poison A; Label: Poison Gas NIOSH REL: Reduce to lowest level

SAFETY PROFILE: A human poison by inhalation. An experimental carcinogen by ingestion. Human systemic effects by inhalation: anorexia, nausea or vomiting. Corrosive to skin; can produce severe burns. Human mutagenic data. A powerful fumigant gas which is one of the most toxic of the common organic halides. It is hemotoxic and narcotic with delayed action. The effects are cumulative and damaging to nervous system, kidneys, and lung. Central nervous system effects include blurred vision, mental confusion, numbness, tremors, and speech defects.

Methyl bromide is reported to be eight times more toxic on inhalation than ethyl bromide. Moreover, because of its greater volatility, it is a much more frequent cause of poisoning. Death following acute poisoning is usually caused by its irritant effect on the lungs. In chronic poisoning, death is due to injury to the central nervous system. Fatal poisoning has always resulted from exposure to relatively high concentrations of methyl bromide vapors (from 8,600 to 60,000 ppm). Nonfatal poisoning has resulted from exposure to concentrations as low as 100-500 ppm. In addition to injury to the lung and central nervous system, the kidneys may be damaged with development of albuminuria and, in fatal cases, cloudy swelling and/or tubular degeneration. The liver may be enlarged. There are no characteristic blood changes.

Mixtures of 10-15 percent with air may be ignited with difficulty. Moderately explosive when exposed to sparks or flame. Forms explosive mixtures with air within narrow limits at atmospheric pressure, with wider limits at higher pressure. The explosive sensitivity of mixtures with air may be increased by the presence of aluminum; magnesium; zinc; or their alloys. Incompatible with metals; dimethyl sulfoxide; ethylene oxide. To fight fire, use foam, water, CO_2, dry chemical. When heated to decomposition it emits toxic fumes of Br^-.

TOXICITY DATA and CODEN

mmo-sat 400 ppm DHEFDK FDA-78-1046,78
sln-dmg-ihl 150 mg/m^3/6H MUREAV 113,272,83
orl-rat TDLo:3250 mg/kg/13W-I:CAR TXAPA9 72,262,84
ihl-man LCLo:60000 ppm/2H BJIMAG 2,24,45
ihl-chd LCLo:1 mg/m^3/2H NHOZAX 23,241,69
ihl-hmn TCLo:35 ppm:GIT INMEAF 11,575,42
orl-rat LD50:214 mg/kg TXAPA9 72,262,84
ihl-rat LC50:302 ppm/8H TXAPA9 81,183,85

BOO632 CAS: 5486-03-3
BUQUINOLATE

mf: $C_{20}H_{27}NO_5$ mw: 361.42

PROP: Crystals. Mp: 288-291°.

SYNS: BONAID ◇ ETHYL-6,7-DIISOBUTOXY-4-HYDROXYQUINOLINE-3-CARBOXYLATE ◇ 4-HYDROXY-6,7-BIS(2-METHYLPROPOXY)-3-QUINOLINECARBOXYLIC ACID ETHER ESTER ◇ 4-HYDROXY-6,7-DIISOBUTOXY-3-QUINOLINECARBOXYLIC ACID ETHYL ESTER

USE IN FOOD:

Purpose: Animal drug.

Where Used: Chicken, eggs.

Regulations: FDA - 21CFR 556.90. Limitation of 0.4 ppm in uncooked liver, kidney, and skin of chicken and 0.1 ppm in uncooked chicken muscle. Limitation of 0.5 ppm in egg yoke, 0.2 ppm in whole eggs. 21CFR 558.105.

SAFETY PROFILE: When heated to decomposition it emits acrid smoke and irritating fumes.

BOR500 CAS: 106-97-8
n-BUTANE
DOT: 1011
mf: C_4H_{10} mw: 58.14

PROP: Colorless gas; faint disagreeable odor. Bp: −0.5°, fp: −138°, lel: 1.9%, uel: 8.5%, flash p: −76°F (CC), d: 0.599, autoign temp: 761°F, vap press: 2 atm @ 18.8°, vap d: 2.046.

SYNS: BUTANE (ACGIH, DOT) ◇ BUTANEN (DUTCH) ◇ BUTANI (ITALIAN) ◇ DIETHYL ◇ METHYLETHYLMETHANE

USE IN FOOD:

Purpose: Aerating agent, gas, propellant.

Where Used: Various.

Regulations: FDA - 21CFR 184.1165. GRAS when used in accordance with good manufacturing practice.

ACGIH TLV: TWA 800 ppm DFG MAK: 1000 ppm (2350 mg/m^3) DOT Classification: Flammable Gas; Label: Flammable Gas

SAFETY PROFILE: Mildly toxic via inhalation. Causes drowsiness. An asphyxiant. Very dangerous fire hazard when exposed to heat, flame, or oxidizers. Highly explosive when exposed to flame, or when mixed with [$Ni(CO)_4$ + O_2]. To fight fire, stop flow of gas. When heated to decomposition it emits acrid smoke and fumes.

TOXICITY DATA and CODEN

ihl-rat LC50:658 g/m^3/4H FATOAO 30,102,67
ihl-mus LC50:680 g/m^3/2H FATOAO 30,102,67

BOS500 CAS: 107-88-0
1,3-BUTANEDIOL
mf: $C_4H_{10}O_2$ mw: 90.14

PROP: Viscous liquid. Bp: 207.5°, fp: <−50°, flash p: 250°F, d: 1.006 @ 20°/20°, autoign temp: 741°F, vap press: 0.06 mm @ 20°, vap d: 3.2.

SYNS: 1,3-BUTANDIOL (GERMAN) ◇ BUTANE-1,3-DIOL ◇ β-BUTYLENE GLYCOL ◇ 1,3-BUTYLENE GLYCOL (FCC) ◇ 1,3-DIHYDROXYBUTANE ◇ METHYLTRIMETHYLENE GLYCOL

USE IN FOOD:

Purpose: Flavoring agent, solvent for flavoring agents.

Where Used: Various.

Regulations: FDA - 21CFR 173.220. FDA, 184.1278 GRAS when used in the minimum amount required to accomplish the intended effect.

SAFETY PROFILE: Mildly toxic by ingestion and subcutaneous routes. An eye irritant. Combustible when exposed to heat or flame. Incompatible with oxidizing materials. To fight fire, use foam, alcohol foam, CO_2, dry chemical. When heated to decomposition it emits acrid smoke and irritating fumes.

TOXICITY DATA and CODEN

eye-rbt 505 mg AJOPAA 29,1363,46
orl-rat LD50:23 g/kg AMIHBC 4,119,51
scu-rat LD50:20 g/kg NPIRI* 1,14,74

BOT500 CAS: 431-03-8
2,3-BUTANEDIONE
DOT: 2346
mf: $C_4H_6O_2$ mw: 86.10

PROP: Greenish-yellow liquid; strong odor. Bp: 88°, flash p: 80°F, d: 0.9904 @ 15°/15°, refr index: 1.393-1.397, vap d: 3.00.Misc in alc, fixed oils, propylene glycol; sol in glycerin, water.

SYNS: BIACETYL ◇ DIACETYL (FCC) ◇ 2,3-DIKETOBUTANE ◇ DIMETHYL DIKETONE ◇ DIMETHYLGLYOXAL ◇ FEMA No. 2370

USE IN FOOD:

Purpose: Flavoring agent.

Where Used: Margarine, oleomargarine.

Regulations: FDA - 21CFR 184.1278. GRAS when used in accordance with good manufacturing practice. USDA - 9CFR 318.7. Sufficient for purpose. DOT Classification: Flammable Liquid; Label: Flammable Liquid

SAFETY PROFILE: A poison by intraperitoneal route. Moderately toxic by ingestion. A skin irritant. Human mutagenic data. Flammable liquid. Dangerous fire hazard when exposed to heat or flame. To fight fire, use alcohol foam, CO_2, dry chemical. When heated to decomposition it emits acrid smoke and fumes.

TOXICITY DATA and CODEN

skn-rbt 500 mg/24H MOD FCTXAV 17(Suppl.),695,79
mmo-sat 1 mg/plate MUREAV 67,367,79
oms-hmn:emb 20 mg/L BEXBAN 74,828,72

ipr-rat LD50: 400 mg/kg FCTXAV 7,571,69
orl-gpg LD50: 990 mg/kg FCTXAV 2,327,64

BOV000 CAS: 96-48-0
4-BUTANOLIDE
mf: $C_4H_6O_2$ mw: 86.10

PROP: Colorless liquid; mild caramel odor. Mp: −44°, bp: 206°, flash p: 209°F (OC), d: 1.124 @ 25°/4°, refr index: 1.434-1.454 @ 25°, vap d: 3.0.

SYNS: γ-6480 ◇ γ-BL ◇ BLO ◇ BLON ◇ BUTYRIC ACID LACTONE ◇ γ-BUTYROLACTONE (FCC) ◇ BUTYRYL LACTONE ◇ α-BUTYROLACTONE ◇ 4-DEOXYTETRONIC ACID ◇ DIHYDRO-2(3H)-FURANONE ◇ FEMA No. 3291 ◇ γ-HYDROXYBUTYROLACTONE ◇ 4-HYDROXYBUTANOIC ACID LACTONE ◇ γ-HYDROXYBUTYRIC ACID CYCLIC ESTER ◇ 4-HYDROXYBUTYRIC ACID γ-LACTONE ◇ NCI-C55878 ◇ TETRAHYDRO-2-FURANONE

USE IN FOOD:
Purpose: Flavoring agent.
Where Used: Candy, milk (soy).
Regulations: GRAS when used at a level not in excess of the amount reasonably required to accomplish the intended effect.

IARC Cancer Review: Animal No Evidence IMEMDT 11,231,76. EPA Genetic Toxicology Program.

SAFETY PROFILE: Moderately toxic by ingestion, intravenous and intraperitoneal routes. An experimental tumorigen by skin contact. Experimental reproductive effects. Mutagenic data. Less acutely toxic than β-propiolactone.
Combustible when exposed to heat or flame; can react with oxidizing materials. To fight fire, use foam, alcohol foam, CO_2, dry chemical. Potentially explosive reaction with butanol + 2,4-dichlorophenol + sodium hydroxide. When heated to decomposition it emits acrid and irritating fumes.

TOXICITY DATA and CODEN

dnd-bcs 20 μL/disc PMRSDJ 1,175,81
otr-ham:kdy 25 mg/L PMRSDJ 1,638,81
orl-rat TDLo: 25 g/kg (20D male): REP ARANDR 10,239,83
skn-mus TDLo: 50 g/kg/42W-I: ETA JNCIAM 31,41,63
orl-rat LD50: 1800 mg/kg 85GMAT -,31,82
ipr-rat LD50: 1000 mg/kg AITEAT 13,70,65

BOV250 CAS: 78-93-3
2-BUTANONE
DOT: 1193/1232
mf: C_4H_8O mw: 72.12

PROP: Colorless liquid; acetone-like odor. Bp: 79.57°, fp: −85.9°, lel: 1.8%, uel: 11.5%, flash p: 22°F (TOC), d: 0.80615 @ 20°/20°, vap press: 71.2 mm @ 20°, autoign temp: 960°F, vap d: 2.42, ULC: 85-90. Misc with alc, ether, fixed oils; water.

SYNS: AETHYLMETHYLKETON (GERMAN) ◇ BUTANONE 2 (FRENCH) ◇ ETHYL METHYL CETONE (FRENCH) ◇ ETHYLMETHYLKETON (DUTCH) ◇ ETHYL METHYL KETONE (DOT) ◇ FEMA No. 2170 ◇ MEK ◇ METHYL ACETONE (DOT) ◇ METHYL ETHYL KETONE (ACGIH) ◇ METILETILCHETONE (ITALIAN) ◇ METYLOETYLOKETON (POLISH) ◇ RCRA WASTE NUMBER U159

USE IN FOOD:
Purpose: Flavoring agent.
Where Used: Various.
Regulations: FDA - 21CFR 172.515. Use at a level not in excess of the amount reasonably required to accomplish the intended effect.

Community Right-To-Know List. EPA Genetic Toxicology Program.

OSHA PEL: TWA 200 ppm; STEL 300 ppm ACGIH TLV: TWA 200 ppm; STEL 300 ppm DFG MAK: 200 ppm (590 mg/m^3) DOT Classification: Flammable Liquid; Label: Flammable Liquid NIOSH REL: (Ketones) TWA 590 mg/m^3

SAFETY PROFILE: Moderately toxic by ingestion, skin contact and intraperitoneal routes. Human systemic effects by inhalation: conjunctiva irritation and unspecified effects on the nose and respiratory system. An experimental teratogen. Experimental reproductive effects. A strong irritant. Human eye irritation @ 350 ppm. Affects peripheral nervous system and central nervous system.
Highly flammable liquid. Reaction with hydrogen peroxide + nitric acid forms a heat- and shock- sensitive explosive product. Ignition on contact with potassium tert-butoxide. Mixture with 2-propanol will produce explosive peroxides during storage. Vigorous reaction with chloroform + alkali. Incompatible with chlorosulfonic acid; oleum. To fight fire, use alcohol foam, CO_2, dry chemical. When heated to decomposition it emits acrid smoke and fumes.

TOXICITY DATA and CODEN

eye-hmn 350 ppm JIHTAB 25,282,43
skn-rbt 500 mg/24H MOD JIHTAB 25,282,43
sln-smc 33800 ppm MUREAV 149,339,85
ihl-rat TCLo:3000 ppm/7H (6-15D preg):REP TXAPA9 28,452,74
ihl-rat TCLo:1000 ppm/(6-15D preg):TER TXAPA9 28,452,74
ihl-hmn TCLo:100 ppm/5M:IRR JIHTAB 25,282,43
orl-rat LD50:2737 mg/kg TXAPA9 19,699,71
ihl-rat LCLo:2000 ppm/4H JIHTAB 31,343,49

BOV700
BUTAN-3-ONE-2-YL BUTYRATE

mf: $C_8H_{14}O_3$ mw: 158.19

PROP: White to slightly yellow liquid; red berry odor. D: 0.972-0.992, refr index: 1.408-1.429. Sol in alc, propylene glycol, most oils; insol in water.

SYN: FEMA No. 3332

USE IN FOOD:

Purpose: Flavoring agent.

Where Used: Various.

Regulations: GRAS when used at a level not in excess of the amount reasonably required to accomplish the intended effect.

SAFETY PROFILE: When heated to decomposition it emits acrid smoke and irritating fumes.

BPU750 CAS: 123-86-4
n-BUTYL ACETATE

DOT: 1123

mf: $C_6H_{12}O_2$ mw: 116.18

PROP: Colorless liquid; strong fruity odor. Bp: 126°, fp: −73.5°, ULC: 50-60, lel: 1.4%, uel: 7.5°, flash p: 72°F, d: 0.88 @ 20°/20°, refr index: 1.393-1.396, autoign temp: 797°F, vap press: 15 mm @ 25°. Misc with alc, ether, propylene glycol; sltly sol in water.

SYNS: ACETATE de BUTYLE (FRENCH) ◇ ACETIC ACID n-BUTYL ESTER ◇ BUTILE (ACETATI di) (ITALIAN) ◇ BUTYLACETAT (GERMAN) ◇ BUTYL ACETATE ◇ 1-BUTYL ACETATE ◇ BUTYLACETATEN (DUTCH) ◇ BUTYLE (ACETATE de) (FRENCH) ◇ BUTYL ETHANOATE ◇ FEMA No. 2174 ◇ OCTAN n-BUTYLU (POLISH)

USE IN FOOD:

Purpose: Flavoring agent.

Where Used: Various.

Regulations: FDA - 21CFR 172.515. Use at a level not in excess of the amount reasonably required to accomplish the intended effect.

OSHA PEL: TWA 150 ppm; STEL 200 ppm ACGIH TLV: TWA 150 ppm; STEL 200 ppm DFG MAK: 200 ppm (950 mg/m^3) DOT Classification: Flammable Liquid; Label: Flammable Liquid

SAFETY PROFILE: Moderately toxic by intraperitoneal route. Mildly toxic by inhalation and ingestion. An experimental teratogen. A skin and severe eye irritant. Human systemic effects by inhalation: conjunctiva irritation, unspecified nasal and respiratory system effects. A mild allergen. High concentrations are irritating to eyes and respiratory tract and cause narcosis. Evidence of chronic systemic toxicity is inconclusive.

Flammable liquid. Moderately explosive when exposed to flame. Ignites on contact with potassium-tert-butoxide. To fight fire, use alcohol foam, CO_2, dry chemical. When heated to decomposition it emits acrid and irritating fumes.

TOXICITY DATA and CODEN

eye-hmn 300 ppm JIHTAB 25,282,43
skn-rbt 500 mg/24H MOD FCTXAV 17,509,79
eye-rbt 20 mg SEV AMIHBC 10,61,54
ihl-hmn TCLo:200 ppm:NOSE,EYE,PUL JIHTAB 25,282,43
ihl-rat TCLo:1500 ppm/7H (7-16D preg):TER NTIS** PB83-258038
orl-rat LD50:14 g/kg AMIHBC 10,61,54
ihl-rat LC50:2000 ppm/4H NPIRI* 1,7,74

BPW500 CAS: 71-36-3
n-BUTYL ALCOHOL

DOT: 1120

mf: $C_4H_{10}O$ mw: 74.14

PROP: Colorless liquid; vinous odor. Bp: 117.5°, ULC: 40, lel: 1.4%, uel: 11.2%, fp: −88.9°, flash p: 95-100°F, d: 0.80978 @ 20°/4°, autoign temp: 689°F, vap press: 5.5 mm @ 20°, vap d: 2.55. Misc in alc, ether, organic solvents; sltly sol in water

SYNS: ALCOOL BUTYLIQUE (FRENCH) ◇ BUTANOL (FRENCH) ◇ 1-BUTANOL ◇ n-BUTANOL ◇ BUTAN-1-OL ◇ BUTANOL (DOT) ◇ BUTANOLEN (DUTCH) ◇ BUTANOLO (ITALIAN) ◇ BUTYL ALCOHOL (DOT) ◇ BUTYL HYDROXIDE ◇ BUTYLOWY ALKOHOL (POLISH) ◇ BUTYRIC or NOR-

MAL PRIMARY BUTYL ALCOHOL ◇ CCS 203 ◇ FEMA No. 2178 ◇ 1-HYDROXYBUTANE ◇ METHYLOLPROPANE ◇ PROPYLCARBINOL ◇ PROPYLMETHANOL ◇ RCRA WASTE NUMBER U031

USE IN FOOD:

Purpose: Color diluent, flavoring agent.

Where Used: Confectionery, food supplements in tablet form, gum.

Regulations: FDA - 21CFR 73.1. Limitation of no residue. 21CFR 172.515. Use at a level not in excess of the amount reasonably required to accomplish the intended effect.

Community Right-To-Know List. EPA Genetic Toxicology Program.

OSHA PEL: CL 50 ppm (skin) ACGIH TLV: CL 50 ppm (skin) DFG MAK: 100 ppm (300 mg/m^3) DOT Classification: Flammable or Combustible Liquid; Label: Flammable Liquid

SAFETY PROFILE: A poison by intravenous route. Moderately toxic by skin contact, ingestion, subcutaneous, intraperitoneal, and possibly other routes. Human systemic effects by inhalation: conjunctiva irritation, unspecified respiratory system and nasal effects. A skin and severe eye irritant. Though animal experiments have shown the butyl alcohols to possess toxic properties, they have produced few cases of poisoning in industry probably because of their low volatility. The use of normal butyl alcohol is reported to have resulted in irritation of the eyes, with corneal inflammation, slight headache and dizziness, slight irritation of the nose and throat, and dermatitis about the fingernails and along the side of the fingers. Keratitis has also been reported. Mutagenic data.

Flammable liquid. Moderately explosive when exposed to flame. Incompatible with Al; chromium trioxide; oxidizing materials. To fight fire, use water spray, alcohol foam, CO_2, dry chemical. When heated to decomposition it emits acrid smoke and fumes.

TOXICITY DATA and CODEN

eye-hmn 50 ppm JIHTAB 25,282,43
skn-rbt 405 mg/24H MOD BIOFX* 2-5/69
eye-rbt 750 µg/24H SEV 28ZPAK -,35,72
cyt-smc 10 mmol/tube HEREAY 33,457,47
ihl-hmn TCLo:25 ppm:IRR JIHTAB 25,282,43
orl-rat LD50:790 mg/kg SAMJAF 43,795,69
ihl-rat LC50:8000 ppm/4H NPIRI* 1,10,74

BPY000 CAS: 13952-84-6
sec-BUTYLAMINE
DOT: 1125
mf: $C_4H_{11}N$ mw: 73.16

PROP: Liquid. Mp: −104°, bp: 63°, flash p: 15°F, d: 0.724 @ 20°.

SYNS: 2-AB ◇ 2-AMINOBUTANE ◇ BUTAFUME ◇ 2-BUTANAMINE ◇ DECCOTANE ◇ FRUCOTE ◇ 1-METHYLPROPYLAMINE ◇ TUTANE

USE IN FOOD:

Purpose: Fungicide.

Where Used: Animal feed, citrus molasses, dried citrus pulp.

Regulations: FDA - 21CFR 561.60. Limitation of 90 ppm in citrus molasses and dried citrus pulp when used for cattle feed

DFG MAK: 5 ppm (15 mg/m^3) DOT Classification: Flammable Liquid

SAFETY PROFILE: A poison by ingestion. A powerful irritant. Moderately toxic by skin contact. Dangerous fire hazard when exposed to heat or flame. To fight fire, use alcohol foam, water spray or mist, dry chemical. Incompatible with oxidizing materials. When heated to decomposition it emits toxic fumes of NO_x.

TOXICITY DATA and CODEN

orl-rat LD50:152 mg/kg TXAPA9 63,150,82
skn-rbt LD50:2500 mg/kg 28ZEAL 5,33,76

BQI000 CAS: 25013-16-5
BUTYLATED HYDROXYANISOLE
mf: $C_{11}H_{16}O_2$ mw: 180.27

PROP: White waxy solid; faint characteristic odor. Mp: 48-63°. Sol in alc, propylene glycol; insol in water.

SYNS: ANTRANCINE 12 ◇ BHA (FCC) ◇ BUTYLHYDROXYANISOLE ◇ tert-BUTYLHYDROXYANISOLE ◇ tert-BUTYL-4-HYDROXYANISOLE ◇ 2(3)-tert-BUTYL-4-HYDROXYANISOLE ◇ BUTYLOHYDROKSYANIZOL (POLISH) ◇ EMBANOX ◇ FEMA No. 2183 ◇ NIPANTIOX 1-F ◇ PREMERGE PLUS ◇ SUSTANE ◇ SUSTANE 1-F ◇ TENOX BHA ◇ VERTAC

USE IN FOOD:

Purpose: Antioxidant, preservative.

Where Used: Beef patties (fresh), beef patties (pregrilled), beet sugar, beverages (dry mixes), beverages prepared from dry mixes, cereals (dry breakfast), chewing gum, desserts (dry mixes), desserts prepared from dry mixes, dry yeast (active), emulsion stabilizers for shortening, fats (rendered animal), fruit (dry diced glazed), margarine, meat (dried), meatballs (cooked or raw), oleomargarine, pizza toppings (cooked or raw), pork, potato flakes, potato granules, potato shreds (dehydrated), poultry, rice, sausage (brown and serve), sausage (dry), sausage (fresh Italian), sweetpotato flakes.

Regulations: FDA - 21CFR 172.110, 172.515. Limitation of total BHA and BHT of 50 ppm in dehydrated potato shreds; 1,000 ppm (BHA only) active dry yeast; 1 ppm (BHA only) in beverages and desserts prepared from dry mixes; 50 ppm in dry breakfast cereals; 32 ppm (BHA only) in dry diced glazed fruit; 90 ppm (BHA only) in dry mixes for beverages and desserts; 200 ppm in emulsion stabilizers for shortenings; 50 ppm in potato flakes; 10 ppm in potato granules; 50 ppm in sweetpotato flakes. 21CFR 172.515. Limitation of 0.5 percent of the volatile oil content of the flavoring substance. 21CFR 172.615. Limitation of 0.1 percent. 21CFR 173.340. Limitation of 0.1 percent of defoamer. 21CFR 181.24. Limitation of 0.005 percent migrating from food packages. 21CFR 182.3169. GRAS with a limitation of 5 percent of the fat and oil content including volatile oil content of the food when used in accordance with good manufacturing practice. USDA - 9CFR 318.7. Limitation of 0.003 percent in dry sausage, 0.006 in combination in dry sausage. Limitation of 0.01 percent in rendered animal fat, 0.02 percent in combination in rendered animal fat. Limitation of 0.02 percent individually or in combination with other antioxidants approved for use in margarine. 9CFR 381.147. Limitation of 0.01 percent in poultry based on fat content.

IARC Cancer Review: Animal Sufficient Evidence IMEMDT 40,123,86. EPA Genetic Toxicology Program.

SAFETY PROFILE: An experimental carcinogen, neoplastigen, and tumorigen. Moderately toxic by ingestion and intraperitoneal routes. Experimental reproductive effects. Mutagenic data. When heated to decomposition it emits acrid and irritating fumes.

TOXICITY DATA and CODEN

mmo-omi 12500 μg/L FMLED7 14,183,82
sce-ham:fbr 100 μmol/L JNCIAM 58,1635,77
orl-rat TDLo:30 g/kg (2W male/2W pre-2W post):REP NTOTDY 3,321,81
orl-mus TDLo:12600 mg/kg (1-21D preg):REP FEPRA7 31,596,72
orl-rat TDLo:728 g/kg/2Y-C:CAR GANNA2 73,332,82
orl-ham TDLo:202 g/kg/24W-C:NEO GANNA2 74,459,83
orl-rat TD :874 g/kg/2Y-C:CAR GANNA2 73,332,82
orl-rat TD :182 g/kg/2Y-C:ETA GANNA2 73,332,82
ipr-rat LD50:2200 mg/kg TOLED5 27,15,85
orl-rbt LD50:2100 mg/kg JAOCA7 54,239,77

BQI050
BUTYLATED HYDROXYMETHYLPHENOL

mf: $C_{15}H_{24}O_2$ mw: 236.35

PROP: White crystalline powder. Mp: 140-141°. Sol in alc; insol in water, propylene glycol.

SYN: 4-HYDROXYMETHYL-2,6-DI-tert-BUTYLPHENOL

USE IN FOOD:

Purpose: Antioxidant.

Where Used: Packaging materials.

Regulations: FDA - 21CFR 172.150. Limitation of 0.02 percent of the oil or fat content of the food, including the volatile oil content. 21CFR 178.2550. Limitation of 0.5 mg/sq in. of food-contact surface.

SAFETY PROFILE: When heated to decomposition it emits acrid smoke and irritating fumes.

BQM500 CAS: 109-21-7
n-BUTYL n-BUTANOATE

mf: $C_8H_{16}O_2$ mw: 144.24

PROP: Colorless liquid; pineapple odor. Bp: 166°, flash p: 128°F (OC), d: 0.67-0.871, refr index: 1.405, vap d: 5.0. Misc with alc, ether, vegetable oils; sltly sol in propylene glycol, water.

SYNS: BUTYL BUTYRATE (FCC) ◇ n-BUTYL BUTYRATE ◇ n-BUTYL n-BUTYRATE ◇ FEMA No. 2186

USE IN FOOD:

Purpose: Flavoring agent.

Where Used: Various.

Regulations: FDA - 21CFR 172.515. Use at a level not in excess of the amount reasonably required to accomplish the intended effect.

SAFETY PROFILE: Moderately toxic via intraperitoneal route. Mildly toxic by ingestion. Moderately irritating to eyes, skin, and mucous membranes by inhalation. Narcotic in high concentrations. Combustible liquid. To fight fire, use alcohol foam, foam, CO_2, dry chemical. Incompatible with oxidizing materials. When heated to decomposition it emits acrid and irritating fumes.

TOXICITY DATA and CODEN

skn-rbt 500 mg/24H MOD FCTXAV 17,521,79
ipr-rat LD50:2300 mg/kg FCTXAV 17,521,79
orl-rbt LD50:9520 mg/kg IMSUAI 41,31,72

BQP000 CAS: 7492-70-8
BUTYL BUTYROLLACTATE

mf: $C_{11}H_{20}O_4$ mw: 216.28

PROP: Colorless liquid; butter, creamlike odor. D: 0.970, refr index: 1.420, flash p: +100°. Misc with alc, fixed oils; sol in propylene glycol; insol in water.

SYNS: BUTANOIC ACID-2-BUTOXY-1-METHYL-2-OXOETHYL ESTER (9CI) ◇ BUTYL BUTYRYL LACTATE ◇ BUTYRIC ACID ESTER WITH BUTYL LACTATE ◇ FEMA No. 2190 ◇ LACTIC ACID, BUTYL ESTER, BUTYRATE

USE IN FOOD:

Purpose: Flavoring agent.

Where Used: Baked goods, candy.

Regulations: FDA - 21CFR 172.515. Use at a level not in excess of the amount reasonably required to accomplish the intended effect.

SAFETY PROFILE: A skin irritant. Combustible liquid. When heated to decomposition it emits acrid smoke and irritating fumes.

TOXICITY DATA and CODEN

skn-rbt 500 mg/24H FCTXAV 17,241,79

BQP750 CAS: 85-70-1
BUTYL CARBOBUTOXYMETHYL PHTHALATE

mf: $C_{18}H_{24}O_6$ mw: 336.42

SYNS: BUTYL PHTHALATE BUTYL GLYCOLATE ◇ BUTYL PHTHALYL BUTYL GLYCOLATE ◇ DIBUTYL-o-(o-CARBOXYBENZOYL) GLYCOLATE ◇ DIBUTYL-o-CARBOXYBENZOYLOXYACETATE ◇ SANTICIZIER B-16

USE IN FOOD:

Purpose: Plasticizer.

Where Used: Packaging materials.

Regulations: FDA - 21CFR 181.27.

SAFETY PROFILE: Mildly toxic via intraperitoneal route. An experimental teratogen. Other experimental reproductive effects. Mutagenic data. An eye irritant. When heated to decomposition it emits acrid and irritating fumes.

TOXICITY DATA and CODEN

eye-rbt 500 mg AJOPAA 29,1363,46
cyt-ham:fbr 125 mg/L/24H MUREAV 48,337,77
ipr-rat TDLo:2296 mg/kg (5-15D preg):REP JPMSAE 61,51,72
ipr-rat TDLo:689 mg/kg (5-15D preg):TER JPMSAE 61,51,72
orl-rat LD50:7 g/kg EVHPAZ 3,131,73
ipr-rat LD50:6889 mg/kg JPMSAE 61,51,72

BQQ500 CAS: 111-76-2
BUTYL CELLOSOLVE

DOT: 2369

mf: $C_6H_{14}O_2$ mw: 118.20

PROP: Clear, mobile liquid; pleasant odor. Bp: 168.4-170.2°, fp: −74.8°, flash p: 160°F (COC), d: 0.9012 @ 20°/20°, vap press: 300 mm @ 140°.

SYNS: BUCS ◇ BUTOKSYETYLOWY ALKOHOL (POLISH) ◇ 2-BUTOSSI-ETANOLO (ITALIAN) ◇ 2-BUTOXY-AETHANOL (GERMAN) ◇ BUTOXYETHANOL ◇ 2-BUTOXY-1-ETHANOL ◇ n-BUTOXYETHANOL ◇ 2-BUTOXYETHANOL (ACGIH) ◇ o-BUTYL ETHYLENE GLYCOL ◇ BUTYL GLYCOL ◇ BUTYLGLYCOL (FRENCH, GERMAN) ◇ BUTYL OXITOL ◇ DOWANOL EB ◇ EKTASOLVE EB ◇ ETHYLENE GLYCOL-n-BUTYL ETHER ◇ ETHYLENE GLYCOL MONOBUTYL ETHER (DOT) ◇ GAFCOL EB ◇ GLYCOL BUTYL ETHER ◇ GLYCOL ETHER EB ◇ GLYCOL ETHER EB ACETATE ◇ GLYCOL MONOBUTYL ETHER ◇ JEFFERSOL EB ◇ MONOBUTYL GLYCOL ETHER ◇ 3-OXA-1-HEPTANOL ◇ POLY-SOLV EB

USE IN FOOD:

Purpose: Flume wash water additive.

Where Used: Sugar beets.

Regulations: FDA - 21CFR 173.315. Limitation of 1 ppm in wash water.

Glycol ethers are on the Community Right-To-Know List.

OSHA PEL: TWA 25 ppm (skin) ACGIH TLV: TWA 25 ppm (skin) DOT Classification: Poison B; Label: St. Andrews Cross Flammable Liquid

SAFETY PROFILE: Poison via ingestion, skin contact, intraperitoneal and intravenous routes. Moderately toxic via inhalation and subcutaneous routes. Human systemic effects by inhalation: nausea or vomiting, headache, nose tumors, unspecified eye effects. An experimental teratogen. Other experimental reproductive effects. A skin and eye irritant. Flammable liquid when exposed to heat or flame. To fight fire, use foam, CO_2, dry chemical. Incompatible with oxidizing materials; heat; flame. When heated to decomposition it emits acrid smoke and irritating fumes.

TOXICITY DATA and CODEN

skn-rbt 500 mg open MLD UCDS**
eye-rbt 18 mg AJOPAA 29,1363,46
ihl-rat TCLo:200 ppm/6H (6-15D preg):REP EVHPAZ 57,47,84
ihl-rat TCLo:25 ppm/6H (6-15D preg):TER EVHPAZ 57,47,84
ihl-rbt TCLo:100 ppm/6H (6-18D preg):TER EVHPAZ 57,47,84
ihl-hmn TCLo:195 ppm/8H:GIT AMIHAB 14,114,56
ihl-hmn TCLo:100 ppm:NOSE,EYE,CNS NPIRI* 1,50,74
orl-rat LD50:1480 mg/kg JIHTAB 23,259,41
ihl-rat LC50:450 ppm/4H TXAPA9 68,405,83
ipr-rat LD50:220 mg/kg 85GMAT -,67,82

BRM500 CAS: 1948-33-0
tert-BUTYLHYDROQUINONE

mf: $C_{10}H_{14}O_2$ mw: 166.24

PROP: White crystalline solid; characteristic odor. Mp: 126.5-128.5°. Sol in alc, ether; insol in water.

SYNS: MONO-tert-BUTYLHYDROQUINONE ◇ MTBHQ ◇ SUSTANE ◇ TBHQ (FCC) ◇ TENOX TBHQ

USE IN FOOD:

Purpose: Antioxidant.

Where Used: Beef patties (fresh), beef patties (pregrilled), dry cereals, edible fats, fats (rendered animal), margarine, meat (dried), meatballs (cooked or raw), oleomargarine, pizza toppings (cooked or raw), pork, potato chips, poultry, sausage (brown and serve), sausage (dry), sausage (fresh Italian), vegetable oils.

Regulations: FDA - 21CFR 172.185. Limitation of 0.02 percent of the oil or fat content of the food, including the volatile oil content. USDA - 9CFR 318.7. Limitation of 0.003 percent in dry sausage, 0.006 percent in combination only with BHA and BHT. Limitation of 0.01 percent in rendered animal fat, 0.02 percent in combination only with BHA and/or BHT. Limitation of 0.02 percent individually or in combination with other antioxidants approved for use in margarine. 9CFR 381.147. Limitation of 0.01 percent in poultry based on fat content.

SAFETY PROFILE: Poison by intraperitoneal route. Moderately toxic by ingestion. Mutagenic data. When heated to decomposition it emits acrid smoke and irritating fumes.

TOXICITY DATA and CODEN

cyt-mus-ipr 200 mg/kg FCTOD7 22,459,84
orl-rat LD50:700 mg/kg JAOCA7 52,53,75
ipr-rat LD50:300 mg/kg JAOCA7 52,53,75

BRQ350
BUTYL ISOBUTYRATE

mf: $C_8H_{16}O_2$ mw: 44.44

PROP: Colorless liquid; apple-pineapple odor. D: 0.859-0.864, refr index: 1.401, flash p: 45°. Misc with alc, ether, fixed oils; insol in glycerin, propylene glycol, water @ 166°.

SYN: FEMA No. 2188

USE IN FOOD:

Purpose: Flavoring agent.

Where Used: Various.

Regulations: FDA - 21CFR 172.515. Use at a level not in excess of the amount reasonably required to accomplish the intended effect.

SAFETY PROFILE: Combustible liquid. When heated to decomposition it emits acrid smoke and irritating fumes.

BSH100 CAS: 87-18-3
p-tert-BUTYLPHENYL SALICYLATE

mf: $C_{17}H_{18}O_3$ mw: 270.13

USE IN FOOD:

Purpose: Plasticizer.

Where Used: Packaging materials.

Regulations: FDA - 21CFR 181.27. Use in accordance with good manufacturing practice.

SAFETY PROFILE: When heated to decomposition it emits acrid smoke and irritating fumes.

BSH250 CAS: 78-48-8
BUTYL PHOSPHOROTRITHIOATE
mf: $C_{12}H_{27}OPS_3$ mw: 314.54

PROP: Liquid. Bp: 150° @ 0.3 mm. Insol in water; sol in aliphatic, aromatic and chlorinated hydrocarbons.

SYNS: B-1,776 ◇ BUTIFOS ◇ BUTIPHOS ◇ CHEMAGRO 1,776 ◇ CHEMAGRO B-1776 ◇ DEF ◇ DEF DEFOLIANT ◇ DE-GREEN ◇ E-Z-OFF D ◇ FOS-FALL "A" ◇ ORTHO PHOSPHATE DEFOLIANT ◇ S,S,S-TRIBUTYL PHOSPHOROTRITHIOATE ◇ S,S,S-TRIBUTYL TRITHIOPHOSPHATE

USE IN FOOD:

Purpose: Defoliant.

Where Used: Animal feed, cottonseed hulls.

Regulations: FDA - 21CFR 561.390. Limitation of 6 ppm in cottonseed hulls when used for animal feed.

SAFETY PROFILE: A poison by ingestion, skin contact, intraperitoneal and possibly other routes. Experimental reproductive effects. Animal experiments show an anti-cholinesterase effect.

TOXICITY DATA and CODEN

orl-rat TDLo: 216 mg/kg (8W pre): REP MZUZA8 (2),48,80
orl-rat LD50: 150 mg/kg TXAPA9 14,515,69
skn-rat LD50: 168 mg/kg WRPCA2 9,119,70
ipr-rat LD50: 210 mg/kg 34ZIAG -,199,69
ihl-mus LCLo: 3804 mg/m^3/1H 34ZIAG -,199,69

BSL600 CAS: 123-95-5
BUTYL STEARATE
mf: $C_{22}H_{44}O_2$ mw: 340.57

PROP: Crystals for alcohol, propanol, or ether.

SYN: OCTADECANOIC ACID, BUTYL ESTER

USE IN FOOD:

Purpose: Plasticizer.

Where Used: Packaging materials.

Regulations: FDA - 21CFR 181.27. Use in accordance with good manufacturing practice.

SAFETY PROFILE: When heated to decomposition it emits acrid smoke and irritating fumes.

BSU250 CAS: 123-72-8
n-BUTYRALDEHYDE
DOT: 1129
mf: C_4H_8O mw: 72.12

PROP: Colorless, mobile liquid; pungent, nutty odor. Mp: −100°, bp: 74.7°, flash p: 20°F (CC), (−6°), d: 0.902 @ 20°/4°, autoign temp: 446°F, lel: 2.5%, uel: 12.5%, vap d: 2.5, D: 0.797-0.802. Sol in water; misc with ether @ 74.8°.

SYNS: ALDEHYDE BUTYRIQUE (FRENCH) ◇ ALDEIDE BUTIRRICA (ITALIAN) ◇ BUTAL ◇ BUTALDEHYDE ◇ BUTALYDE ◇ BUTANAL ◇ n-BUTANAL (CZECH) ◇ BUTYRALDEHYD (GERMAN) ◇ BUTYRALDEHYDE (CZECH) ◇ n-BUTYL ALDEHYDE ◇ BUTYRAL ◇ BUTYRIC ALDEHYDE ◇ FEMA No. 2219 ◇ NCI-C56291

USE IN FOOD:

Purpose: Flavoring agent

Where Used: Various.

Regulations: FDA - 21CFR 172.515. Use at a level not in excess of the amount reasonably required to accomplish the intended effect.

Community Right-To-Know List.

DOT Classification: Flammable Liquid; Label: Flammable Liquid

SAFETY PROFILE: Moderately toxic by ingestion, inhalation, skin contact, intraperitoneal and subcutaneous routes. Severe skin and eye irritant. Human immunological effects by inhalation: delayed hypersensitivity. Highly flammable liquid. To fight fire, use foam, CO_2, dry chemical. Incompatible with oxidizing materials. Reacts vigorously with chlorosulfonic acid; HNO_3; oleum; H_2SO_4. When heated to decomposition it emits acrid smoke and fumes.

TOXICITY DATA and CODEN

skn-rbt 500 mg/24H SEV 28ZPAK -,40,72
eye-rbt 75 µg open SEV AMIHBC 4,119,51
spm-mus-ipr 30 mg/kg MUREAV 39,317,77
ihl-hmn TCLo: 580 mg/m^3: IMM BMJOAE 2,913,56
orl-rat LD50: 2490 mg/kg 28ZPAK -,40,72
ihl-rat LCLo: 8000 ppm/4H AMIHBC 4,119,51
ipr-rat LD50: 800 mg/kg FCTXAV 17,731,79

BSW000 CAS: 107-92-6
n-BUTYRIC ACID
DOT: 2820
mf: $C_4H_8O_2$ mw: 88.12

PROP: Colorless liquid; strong, rancid butter odor. Mp: −7.9°, bp: 163.5°, flash p: 161°F, fp: −5.5°, d: 0.9590 @ 20°/20°, refr index: 1.397, autoign temp: 846°F, vap press: 0.43 mm @ 20°, vap d: 3.04, lel: 2.0%, uel: 10.0%.

SYNS: BUTANOIC ACID ◇ BUTTERSAEURE (GERMAN) ◇ ETHYLACETIC ACID ◇ FEMA No. 2221 ◇ 1-PROPANECARBOXYLIC ACID ◇ PROPYLFORMIC ACID

USE IN FOOD:

Purpose: Flavoring agent

Where Used: Various.

Regulations: FDA - 21CFR 182.60. GRAS when used at a level not in excess of the amount reasonably required to accomplish the intended effect.

DOT Classification: Corrosive Material; Label: Corrosive

SAFETY PROFILE: Moderately toxic by ingestion, skin contact, subcutaneous, intraperitoneal and intravenous routes. Human mutagenic data. Severe skin and eye irritant. A corrosive material. Combustible liquid. Could react with oxidizing materials. Incandescent reaction with chromium trioxide above 100°. To fight fire, use alcohol foam, CO_2, dry chemical. When heated to decomposition it emits acrid smoke and irritating fumes.

TOXICITY DATA and CODEN

skn-rbt 10 mg/24H open SEV AMIHBC 10,61,54
eye-rbt 250 μg open SEV AMIHBC 10,61,54
dnd-hmn:hla 3 mmol/L CELLB5 12,855,77
dni-hmn:lym 4 mmol/L HAONDL 2,381,84
orl-rat LD50:2940 mg/kg AMIHBC 4,119,51
ipr-mus LD50:3180 mg/kg JPPMAB 21,85,69
skn-rbt LD50:530 mg/kg UCDS** 4/10/68

BSW500 CAS: 539-90-2
BUTYRIC ACID ISOBUTYL ESTER

mf: $C_8H_{16}O_2$ mw: 144.24

PROP: Colorless liquid; apple-pineapple odor. D: 0.858-0863, refr index: 1.402. Sol in alc, fixed oils; sltly sol in water; insol in glycerin.

SYNS: FEMA No. 2187 ◇ ISOBUTYL BUTANOATE ◇ ISOBUTYL BUTYRATE (FCC) ◇ 2-METHYLPROPYL BUTYRATE

USE IN FOOD:

Purpose: Flavoring agent.

Where Used: Various.

Regulations: FDA - 21CFR 172.515. Use at a level not in excess of the amount reasonably required to accomplish the intended effect.

SAFETY PROFILE: Mildly toxic by ingestion and intraduodenal routes. A skin irritant. When heated to decomposition it emits acrid smoke and irritating fumes.

TOXICITY DATA and CODEN

skn-rbt 500 mg/24H MLD FCTXAV 17,833,79
idu-rbt LD50:9500 mg/kg FCTXAV 17(Suppl),695,79
orl-rbt LD50:9520 mg/kg IMSUAI 41,31,72

C

CAD000 CAS: 7440-43-9
CADMIUM

mf: Cd mw: 112.40

PROP: Hexagonal crystals, silver-white, malleable metal. Mp: 320.9°, bp: 767 ± 2°, d: 8.642, vap press: 1 mm @ 394°.

SYNS: C.I. 77180 ◇ COLLOIDAL CADMIUM ◇ KADMIUM (GERMAN)

USE IN FOOD:

Purpose: Contaminant.

Where Used: Animal feed.

Regulations: USDA CES Ranking: B-4 (1985).

IARC Cancer Review: Animal Sufficient Evidence IMEMDT 11,39,76; IMEMDT 2,74,73. Cadmium and its compounds are on the Community Right-To-Know List. EPA Genetic Toxicology Program.

OSHA PEL: TWA 0.1 mg(Cd)/m^3; CL 0.6 mg(Cd)/m^3 (fume) ACGIH TLV: TWA 0.01 mg(Cd)/m^3 (dust), Human carcinogen DFG BAT: Blood 1.5 μg/dL; Urine 15 μg/dL NIOSH REL: (Cadmium) Reduce to lowest feasible level

SAFETY PROFILE: A human poison by inhalation and possibly other routes. Poison experimentally by ingestion, inhalation, intraperitoneal, subcutaneous, intramuscular, and intravenous routes. In humans inhalation causes an excess of protein in the urine. An experimental carcinogen, tumorigen, neoplastigen, and teratogen. Experimental reproductive effects. Mutagenic data. The dust ignites spontaneously in air and is flammable and explosive when exposed to heat, flame, or by chemical reaction with oxidizing agents; metals; HN_3; Zn; Se; and Te. Explodes on contact with hydrazoic acid. Violent or explosive reaction when heated with ammonium nitrate. Vigorous reaction when heated with nitryl fluoride. When heated temperature it emits toxic fumes of Cd.

TOXICITY DATA and CODEN

cyt-ham:ovr 1 μmol/L CGCGBR 26,251,80
orl-rat TDLo:220 mg/kg (1-22D preg): REP,TER TOLED5 11,233,82
ivn-rat TDLo:1250 μg/kg (14D preg): REP,TER JJATDK 1,264,81
ims-rat TDLo:40 mg/kg/4W-I:CAR JEPTDQ 1(1),51,77
ims-rat TD :70 mg/kg:ETA BJCAAI 18,124,64
ihl-man TCLo:88 μg/m^3/8.6Y:KID AEHLAU 28,147,74
ihl-hmn LCLo:39 mg/m^3/20M AIHAAP 31,180,70
unk-man LDLo:15 mg/kg 85DCAI 2,73,70
orl-rat LD50:225 mg/kg TXAPA9 41,667,77

CAK500 CAS: 58-08-2
CAFFEINE

mf: $C_8H_{10}N_4O_2$ mw: 194.22

PROP: White, fleecy masses; odorless with bitter taste. Mp: 236.8°. Sol in water, alc, chloroform, ether.

SYNS: CAFFEIN ◇ COFFEIN (GERMAN) ◇ COFFEINE ◇ 3,7-DIHYDRO-1,3,7-TRIMETHYL-1H-PURINE-2,6-DIONE ◇ ELDIATRIC C ◇ FEMA No. 2224 ◇ GUARANINE ◇ KOFFEIN (GERMAN) ◇ METHYLTHEOBROMIDE ◇ 1-METHYLTHEOBROMINE ◇ 7-METHYLTHEOPHYLLINE ◇ NCI-C02733 ◇ NO-DOZ ◇ ORGANEX ◇ THEIN ◇ THEINE ◇ 1,3,7-TRIMETHYL-2,6-DIOXOPURINE ◇ 1,3,7-TRIMETHYLXANTHINE

USE IN FOOD:

Purpose: Flavoring agent, stimulant.

Where Used: Beverages (cola), beverages (orange).

Regulations: FDA - 21CFR 182.1180. Limitation of 0.02 percent. GRAS when used at a level not in excess of the amount reasonably required to accomplish the intended effect.

EPA Genetic Toxicology Program.

SAFETY PROFILE: A human poison by ingestion. An experimental poison by ingestion, subcutaneous, intraperitoneal, intramuscular, rectal, and intravenous routes. Human systemic effects by ingestion, intravenous and intramuscular routes include: convulsions, muscle spasms, tremors, poor muscle coordination, somnolence, nausea or vomiting, and blood pressure increase. A human teratogen causing developmental abnormalities of the craniofacial and musculoskeletal systems, pregnancy termination (abortion) and stillbirth. Human maternal effects include an unspecified effect on labor or childbirth. Human mutagenic data. An experimental teratogen. Large doses (above 1.0 gram) cause palpitation, excitement, insomnia, dizziness, headache, and vomiting. Continued excessive use of caffeine in tea or coffee may lead to digestive disturbances, constipation, palpitations, shortness of breath, and depressed mental states. It is also implicated in cardiac disorders under those conditions. When heated to decomposition it emits toxic fumes of NO_x.

TOXICITY DATA and CODEN

dns-hmn:oth 1 mmol/L BIOJAU 35,665,81
dni-hmn:oth 4 mmol/L BIOJAU 35,665,81
orl-wmn TDLo:6750 mg/kg (1-39W preg): REP,TER LANCAO 1,1415,81
orl-wmn TDLo:3276 mg/kg (1-39W preg):REP POMDAS 62(3),64,77
orl-wmn TDLo:1092 mg/kg (1-91D preg):REP POMDAS 62,(3),64,77
orl-rat TDLo: 627 mg/kg (1-22D preg):REP EXPEAM 36,1105,80
ivn-hmn TDLo:7 mg/kg:CNS APTOA6 15,331,59
orl-inf TDLo:14700 μg/kg:CNS CLBIAS 10,148,77
ivn-inf TDLo:68 mg/kg:CNS AJDCAI 134,495,80
ims-inf TDLo:36 mg/kg:CNS AJDCAI 134,495,80
ivn-wmn LDLo:57 mg/kg:CNS,CVS APTOA6 15,331,59
orl-hmn LDLo:192 mg/kg JNDRAK 5,252,65
orl-chd LDLo:320 mg/kg FNSCA6 3,275,74
orl-wmn LDLo:1 g/kg:GIT BIATDR -,6,73
orl-rat LD50:192 mg/kg JNDRAK 5,252,65

CAL000 CAS: 470-82-6
CAJEPUTOL

mf: $C_{10}H_{18}O$ mw: 154.28

PROP: Colorless liquid characteristic odor with pungent, cooling taste. D: 0.921-0.924, refr index: 1.455-1.460, flash p: 122°F. Sol in alc, fixed oils glycerin, propylene glycol.

SYNS: 1,8-CINEOL ◇ CINEOLE ◇ 1,8-CINEOLE ◇ 1,8-EPOXY-p-MENTHANE ◇ EUCALYPTOL (FCC) ◇ EUCALYPTOLE ◇ FEMA No. 2465 ◇ LIMONENE OXIDE ◇ NCI-C56575 ◇ 1,8-OXIDO-p-MENTHANE ◇ 1,3,3-TRIMETHYL-2-OXABICYCLO(2.2.2)OCTANE

USE IN FOOD:

Purpose: Flavoring agent.

Where Used: Bakery products, beverages (nonalcoholic), chewing gum, confections, ice cream products.

Regulations: FDA - 21CFR 172.515. Use at a level not in excess of the amount reasonably required to accomplish the intended effect.

SAFETY PROFILE: Poison by subcutaneous and intramuscular routes. Moderately toxic by ingestion. Experimental reproductive effects. Combustible liquid. When heated to decomposition it emits acrid smoke and fumes.

TOXICITY DATA and CODEN

scu-rat TDLo:2 g/kg (19-22D preg):REP BCPCA6 22,543,73
orl-rat LD50:2480 mg/kg FCTXAV 2,327,64
scu-mus LDLo:50 mg/kg TFAKA4 1,134,55
ims-mus LD50:100 mg/kg JSICAZ 21,342,62

CAL750 CAS: 62-54-4
CALCIUM ACETATE

mf: $C_4H_6O_4{\cdot}Ca$ mw: 158.18

PROP: Fine white, bulky powder. Sol in water; sltly sol in alc.

SYNS: ACETATE OF LIME ◇ BROWN ACETATE ◇ CALCIUM DIACETATE ◇ GRAY ACETATE ◇ LIME ACETATE ◇ LIME PYROLIGNITE ◇ SORBO-CALCIAN ◇ SORBO-CALCION ◇ TELTOZAN ◇ VINEGAR SALTS

USE IN FOOD:

Purpose: Firming agent, pH control agent, processing aid, sequestrant, stabilizer, texturizer, thickener.

Where Used: Baked goods, cake mixes, fillings, gelatins, packaging materials, puddings, sweet sauces, syrups, toppings.

Regulations: FDA - 21CFR 181.29, 182.6190, 184.1185. GRAS with a limitation of 0.2 percent in baked goods, 0.2 percent in gelatins, puddings, and fillings, 0.15 percent in sweet sauces, toppings, and syrups, 0.0001 percent in all other foods when used in accordance with good manufacturing practice.

SAFETY PROFILE: Poison by intravenous route. Mutagenic data. When heated to decomposition it emits acrid smoke and fumes.

TOXICITY DATA and CODEN

dns-rat-rat 1290 μmol/kg/5D-I CRNGDP 6,1819,85
ivn-rat LDLo:147 mg/kg JPETAB 71,1,41
ivn-mus LD50:52 mg/kg JLCMAK 29,809,44

CAM200 CAS: 9005-35-0
CALCIUM ALGINATE

mf: $[(C_6H_7O_6)_2Ca]_n$ mw: 195.16

PROP: White to yellow, granular powder. Insol in water, organic solvents.

SYN: ALGIN

USE IN FOOD:

Purpose: Emulsifier, stabilizer, thickener.

Where Used: Alcoholic beverages, baked goods, confections, egg products, fats, frostings, fruits (fabricated), gelatins, gravies, jams, jellies, oils, puddings, sauces, sweet sauces.

Regulations: FDA - 21CFR 184.1187. GRAS with a limitation of 0.002 percent in baked goods, 0.4 percent in alcoholic beverages, 0.4 percent in confections and frostings, 0.6 percent in egg products, 0.5 percent in fats and oils, 0.25 percent in gelatins and puddings, 0.4 percent in gravies and sauces, 0.5 percent in jams and jellies, 0.5 percent in sweet sauces, 0.3 percent in all other foods when used in accordance with good manufacturing practice.

SAFETY PROFILE: When heated to decomposition it emits acrid smoke and irritating fumes.

CAM600 CAS: 5743-27-1
CALCIUM ASCORBATE
mf: $C_{12}H_{14}CaO_{12}\cdot 2H_2O$ mw: 426.35

PROP: White crystalline powder; odorless. Sol in water; sltly sol in alc; insol in ether.

USE IN FOOD:

Purpose: Antioxidant, preservative.

Where Used: Various.

Regulations: FDA - 21CFR 182.3189. GRAS when used in accordance with good manufacturing practice.

SAFETY PROFILE: When heated to decomposition it emits acrid smoke and irritating fumes.

CAM675
CALCIUM BENZOATE
mf: $C_{14}H_{10}O_4\cdot 3H_2O$ mw: 374.26

PROP: Orthorhombic crystals or powder. D: 1.44. Sol in water.

USE IN FOOD:

Purpose: Preservative.

Where Used: Margarine, oleomargarine.

Regulations: USDA - 9CFR 318.7. Limitation of 0.1 percent, or if used in combination with sorbic acid and its salts, 0.2 percent (expressed as the acids in the weight of the finished food.)

SAFETY PROFILE: Combustible when exposed to heat or flame. When heated to decomposition it emits acrid smoke and irritating fumes.

CAN400
CALCIUM BROMATE
mf: $Ca(BrO_3)_2\cdot H_2O$ mw: 313.90

PROP: White crystalline powder. Very sol in water.

USE IN FOOD:

Purpose: Dough conditioner, maturing agent.

Where Used: Baked goods.

Regulations: GRAS when used in accordance with good manufacturing practice.

SAFETY PROFILE: A nuisance dust.

CAO000 CAS: 1317-65-3
CALCIUM CARBONATE
mf: $CO_3\cdot Ca$ mw: 100.09

PROP: White microcrystalline powder. Mp: 825° (α), 1339° (β) @ 102.5 atm; d: 2.7-2.95. Found in nature as the minerals limestone, marble, aragonite, calcite, and vaterite. Odorless, tasteless powder or crystals. Two crystalline forms are of commercial importance: Aragonite, orthorhombic, mp: 825° (decomp), d: 2.83, formed at temperatures above 30°; calcite, hexagonal-rhombohedral, mp: 1339° (102.5 atm), d: 2.711, formed at temperatures below 30°. At about 825° is decomp into CaO and CO_2. Practically insol in water, alc; sol in dil acids.

SYNS: AGRICULTURAL LIMESTONE ◇ AGSTONE ◇ ARAGONITE ◇ ATOMIT ◇ BELL MINE PULVERIZED LIMESTONE ◇ CALCITE ◇ CALCIUM CARBONATE (ACGIH) ◇ CARBONIC ACID, CALCIUM SALT (1:1) ◇ CHALK ◇ DOLOMITE ◇ FRANKLIN ◇ LIMESTONE (FCC) ◇ LITHOGRAPHIC STONE ◇ MARBLE ◇ NATURAL CALCIUM CARBONATE ◇ PORTLAND STONE ◇ SOHNHOFEN STONE ◇ VATERITE

USE IN FOOD:

Purpose: Alkali, dietary supplement, dough conditioner, firming agent, modifier for chewing gum, nutrient, pH control agent, release agent for chewing gum, stabilizer, texturizer for chewing gum, yeast food.

Where Used: Baking powder, chewing gum, desserts (dry mixes), dough, packaging materials, wine.

Regulations: FDA - 21CFR 181.29, 182.5191, 184.1191, 184.1409. GRAS when used in accordance with good manufacturing practice. BATF - 27CFR 240.1051. Limitation of 30 pounds/1000 gallons of wine. The natural or fixed acids shall not be reduced below 5 g/L.

ACGIH TLV: TWA 10 mg/m^3 (total dust)

SAFETY PROFILE: A severe eye and moderate skin irritant. Ignites on contact with F_2. Incompatible with acids; alum; ammonium salts; (Mg + H_2). Calcium carbonate is a common air contaminant.

TOXICITY DATA and CODEN

skn-rbt 500 mg/24H MOD 28ZPAK -,267,72
eye-rbt 750 μg/24H SEV 28ZPAK -,267,72
orl-rat LD50:6450 mg/kg 28ZPAK -,267,72

CAO750 CAS: 10043-52-4
CALCIUM CHLORIDE
mf: $CaCl_2$ mw: 110.98

PROP: Cubic, colorless, deliquescent crystals. Mp: 772°, bp: >1600°, d: 2.512 @ 25°. Sol in water, alc.

SYNS: CALCIUM CHLORIDE, ANHYDROUS ◇ CALPLUS ◇ CALTAC ◇ DOWFLAKE ◇ LIQUIDOW ◇ PELADOW ◇ SNOMELT ◇ SUPERFLAKE ANHYDROUS

USE IN FOOD:

Purpose: Anticaking agent, antimicrobial agent, curing agent, firming agent, flavor enhancer, humectant, nutrient supplement, pH control agent, pickling agent, processing aid, sequestrant, stabilizer, surface-active agent, synergist, texturizer, thickener.

Where Used: Apple slices, baked goods, beverage bases (nonalcoholic), beverages (nonalcoholic), cheese, coffee, condiments, dairy product analogs, fruits (processed), fruit juices, gravies, jams (commercial), jellies (commercial), meat (raw cuts), meat products, milk (evaporated), pickles, plant protein products, potatoes (canned), poultry (raw cuts), relishes, sauces, tea, tomatoes (canned), vegetable juices (processed).

Regulations: FDA - 21CFR 184.1193. GRAS with a limitation of 0.3 percent in baked goods and dairy product analogs, 0.22 percent in nonalcoholic beverages and beverage bases, 0.2 percent in cheese and processed fruit and fruit juices, 0.32 percent in coffee and tea, 0.4 percent in condiments and relishes, 0.2 percent in gravies and sauces, 0.1 percent in commercial jams and jellies, 0.25 percent in meat products, 0.2 percent in plant protein products, 0.4 percent in processed vegetable juices, 0.05 percent in all other foods when used in accordance with good manufacturing practice. USDA - 9CFR 318.7, 381.147. Limitation of not more than 3 percent of a 0.8 molar solution. Solutions consisting of water and approved proteolytic enzymes applied or injected into raw meat cuts shall not result in a gain of more than 3 percent above the weight of the untreated product.

EPA Genetic Toxicology Program.

SAFETY PROFILE: Poison by intravenous, intramuscular, intraperitoneal, and subcutaneous routes. Moderately toxic by ingestion. An experimental tumorigen. Mutagenic data. Reacts violently with (B_2O_3 + CaO); BrF_3. Reaction with zinc releases explosive hydrogen gas. Catalyzes exothermic polymerization of methyl vinyl ether. Exothermic reaction with water. When heated to decomposition it emits toxic fumes of Cl^-.

TOXICITY DATA and CODEN

sln-smc 200 mmol/L MUTAEX 1,21,86
dns-rat-ipr 2500 μmol/kg JOENAK 65,45,75
orl-rat TDLo:112 g/kg/20W-C:ETA AJCAA7 23,550,35
orl-rat LD50:1000 mg/kg CNJMAQ 12,216,48
ipr-rat LD50:264 mg/kg OYYAA2 14,963,77
scu-rat LD50:2630 mg/kg OYYAA2 14,963,77

CAP850 CAS: 813-94-5
CALCIUM CITRATE
mf: $Ca_3(C_6H_5O_7)_2{\cdot}4H_2O$ mw: 570.50

PROP: Fine white powder. Sltly sol in water; insol in alc.

USE IN FOOD:

Purpose: Buffer, dietary supplement, firming agent, nutrient, sequestrant.

Where Used: Beans (lima), flour, peppers.

Regulations: FDA - 21CFR 182.1195, 182.5195, 182.6195, 182.8195. GRAS when used in accordance with good manufacturing practice.

SAFETY PROFILE: When heated to decomposition it emits acrid smoke and irritating fumes.

CAR000 CAS: 139-06-0
CALCIUM CYCLOHEXYLSULPHAMATE
mf: $C_{12}H_{24}N_2O_6S_2{\cdot}Ca$ mw: 396.58

PROP: White, crystalline powder; almost odorless; freely sol in water; practically insol in alc, benzene, chloroform, and ether.

SYNS: CALCIUM CYCLAMATE ◇ CALCIUM CYCLOHEXANESULFAMATE ◇ CALCIUM CYCLOHEXANE SULPHAMATE ◇ CALCIUM CYCLOHEXYLSULFAMATE ◇ CYCLAMATE CALCIUM ◇ CYCLAMATE, CALCIUM SALT ◇ CYCLAN ◇ CYCLOHEXANESULFAMIC ACID, CALCIUM SALT ◇ CYCLOHEXYLSULPHAMIC ACID, CALCIUM SALT ◇ CYLAN ◇ DIETIL ◇ KALZIUMZYKLAMATE (GERMAN) ◇ SUCARYL CALCIUM

USE IN FOOD:

Purpose: Nonnutritive sweetener.

Where Used: Prohibited from foods.

Regulations: FDA - 21CFR 189.135. Prohibited from direct addition or use in human food.

IARC Cancer Review: Animal Limited Evidence IMEMDT 22,55,80. EPA Genetic Toxicology Program.

SAFETY PROFILE: Poison by ingestion and intravenous routes. An experimental tumorigen and neoplastigen. Experimental reproductive effects. Human mutagenic data. When heated to decomposition it emits very toxic fumes of SO_x and NO_x.

TOXICITY DATA and CODEN

sln-dmg-orl 5 mmol/L DRISAA 46,114,71
dni-hmn:lng 100 mg/L JCLBA3 47,30a,70
orl-rat TDLo:55 mg/kg (1-22D preg):REP AJCNAC 23,782,70
orl-rat TDLo:3465 g/kg/88W-C:NEO JNCIAM 49,751,72
scu-rat TDLo:45 g/kg/66W-I:ETA FCTXAV 9,463,71
orl-rat LDLo:10 mg/kg CLDND*
ivn-mus LDLo:3500 μg/kg CLDND*

CAR775
CALCIUM DISODIUM EDTA

mf: $C_{10}H_{12}CaN_2Na_2O_8 \cdot 2H_2O$ mw: 410.30

PROP: White crystalline powder; hygroscopic with a faint salt taste. Sol in water.

SYNS: CALCIUM DISODIUM EDETATE ◇ CALCIUM DISODIUM ETHYLENEDIAMINETETRAACETATE ◇ CALCIUM DISODIUM (ETHYLENEDINITRILO)TETRAACETATE

USE IN FOOD:

Purpose: Hog scald agent, preservative, sequestrant.

Where Used: Beverages (distilled alcoholic), beverages (fermented malt), cabbage (pickled), clams (cooked canned), corn (canned), crabmeat (cooked canned), cucumbers (pickled), dressings (nonstandardized), egg product (that is hard-cooked and consists, in a cylindrical shape, of egg white with an inner core of egg yoke), French dressing, hog carcasses, lima beans (dried, cooked canned), margarine, mayonnaise, mushrooms, oleomargarine, pecan pie filling, pinto beans (processed dry), potato salad, potatoes (canned white), salad dressings, sauces, shrimp (cooked canned), soft drinks (canned carbonated), spice extractives in soluble carriers, spread (sandwich), spreads (artificially colored and lemon-flavored or orange-flavored).

Regulations: FDA - 21CFR 172.120. Limitation of 220 ppm in cabbage, pickled; 33 ppm in carbonated soft drinks, canned; 110 ppm in white potatoes, canned; 340 ppm in clams, cooked canned; 275 ppm in crabmeat, cooked canned; 220 ppm in cucumbers, pickled; 25 ppm in distilled alcoholic beverages; 75 ppm in dressings, nonstandardized; 310 ppm in lima beans, dried, cooked canned; 200 ppm (by weight of egg yolk) in egg product that is hard-cooked and consists, in a cylindrical shape, of egg white with an inner core of egg yoke; 25 ppm in fermented malt beverages; 75 ppm in French dressing; 75 ppm in mayonnaise; 200 ppm in mushrooms, cooked canned; 75 ppm in oleomargarine; 100 ppm in pecan pie filling; 100 ppm in potato salad; 800 ppm in pinto beans, processed dry; 75 ppm in salad dressing; 100 ppm in sandwich spread; 75 ppm in sauces; 250 ppm in shrimp, cooked canned; 60 ppm in spice extractives in soluble carriers; 100 ppm in spreads, artificially colored and lemon-flavored or orange-flavored. Limitation in combination with disodium EDTA of 75 ppm in dressings, nonstandardized; 75 ppm in French dressing; 75 ppm in mayonnaise; salad dressings; 100 ppm in sandwich spread; 75 ppm in sauces. USDA - 9CFR 318.7. Sufficient for purpose. Limitation of 75 ppm in finished oleomargarine or margarine.

SAFETY PROFILE: When heated to decomposition it emits toxic fumes of NO_x.

CAS750 CAS: 299-28-5
CALCIUM GLUCONATE

mf: $C_{12}H_{22}O_{14} \cdot Ca$ mw: 430.42

PROP: White, fluffy powder or granules; odorless and tasteless. Sol in hot water; less sol in cold water; insol in alc, acetic acid, and other organic solvents. Mp: loses H_2O @ 120°.

SYN: GLUCONATE de CALCIUM (FRENCH)

USE IN FOOD:

Purpose: Firming agent, formulation aid, sequestrant, stabilizer, texturizer, thickener.

Where Used: Baked goods, dairy product analogs, gelatins, gels, puddings, sugar substitutes.

Regulations: FDA - 21CFR 184.1199. GRAS with a limitation of 1.75 percent in baked goods, 0.4 percent in dairy product analogs, 4.5 percent in gelatins and puddings, 0.01 percent in sugar substitutes when used in accordance with good manufacturing practice.

SAFETY PROFILE: Moderately toxic by subcutaneous, intraperitoneal, and intravenous routes. Human systemic effects in infants by intramuscular route: dermatitis and fever. When heated to decomposition it emits acrid smoke and fumes.

TOXICITY DATA and CODEN

ims-inf TDLo: 143 mg/kg: SKN,MET JAMAAP 129,347,45
ims-inf LDLo: 10 g/kg JAMAAP 129,347,45
ivn-rat LD50: 950 mg/kg NIIRDN 6,226,82
ipr-mus LD50: 2200 mg/kg JDGRAX 15(1-2),121,84

CAS800 CAS: 27214-00-2
CALCIUM GLYCEROPHOSPHATE

mf: $C_3H_7CaO_6P$ mw: 210.14

PROP: Fine white hygroscopic powder. Sol in water; insol in alc.

USE IN FOOD:

Purpose: Dietary supplement, nutrient, stabilizer.

Where Used: Baking powder.

Regulations: FDA - 21CFR 181.29, 182.5201, 182.8201. GRAS when used in accordance with good manufacturing practice.

SAFETY PROFILE: When heated to decomposition emits toxic fumes of PO_x.

CAS825
CALCIUM HEXAMETAPHOSPHATE

USE IN FOOD:

Purpose: Sequestrant

Where Used: Various.

Regulations: FDA - 21CFR 182.6203. GRAS when used in accordance with good manufacturing practice.

SAFETY PROFILE: A nuisance dust.

CAT210 CAS: 7789-77-7
CALCIUM HYDROGEN PHOSPHATE

mf: $CaHPO_42H_2O$ mw: 136.06

USE IN FOOD:

Purpose: Stabilizer.

Where Used: Packaging materials.

Regulations: FDA - 21CFR 181.29. Use in accordance with good manufacturing practice.

SAFETY PROFILE: When heated to decomposition it emits acrid smoke and irritating fumes.

CAT250 CAS: 1305-62-0
CALCIUM HYDROXIDE

mf: CaH_2O_2 mw: 74.10

PROP: Rhombic, trigonal, colorless crystals or white power; sltly bitter taste. Mp: loses H_2O @ 580°, bp: decomp, d: 2.343. Sol in water, glycerin; insol in alc.

SYNS: BELL MINE ◇ CALCIUM HYDRATE ◇ HYDRATED LIME ◇ KEMIKAL ◇ LIME WATER ◇ SLAKED LIME

USE IN FOOD:

Purpose: Buffer, firming agent, hog scald agent, miscellaneous and general-purpose food chemical, neutralizing agent.

Where Used: Hog carcasses, tripe.

Regulations: FDA - 21CFR 184.1205. GRAS when used in accordance with good manufacturing practice. USDA - 9CFR 318.7. Sufficient for purpose.

ACGIH TLV: TWA 5 mg/m^3

SAFETY PROFILE: Mildly toxic by ingestion. A severe eye irritant. A skin, mucous membrane and respiratory system irritant. Mutagenic data. Causes dermatitis. Dust is considered to be a significant industrial hazard. A common air contaminant. Violent reaction with maleic anhydride; nitroethane; nitromethane; nitroparaffins; nitropropane; phosphorus. Reaction with polychlorinated phenols + potassium nitrate forms extremely toxic products.

TOXICITY DATA and CODEN

eye-rbt 10 mg SEV TXAPA9 55,501,80
cyt-rat/ast 1200 mg/kg GANNA2 54,155,62
orl-mus LD50:7300 mg/kg YKYUA6 32,1477,81

CAT500 CAS: 7789-80-2
CALCIUM IODATE
mf: $Ca(IO_3)_2 \cdot H_2O$ mw: 407.90

PROP: White powder. Sltly sol in water; insol in alc.

USE IN FOOD:

Purpose: Dough conditioner, maturing agent.

Where Used: Bread.

Regulations: FDA - 21CFR 184.1206. GRAS with a limitation of 0.0075 percent in flour when used in accordance with good manufacturing practice.

SAFETY PROFILE: A nuisance dust.

CAT600 CAS: 814-80-2
CALCIUM LACTATE
mf: $C_6H_{10}CaO_6 \cdot xH_2O$ mw: 218.22

PROP: White crystalline powder with up to 5 H_2O. Sol in water; insol in alc.

USE IN FOOD:

Purpose: Buffer, dough conditioner, firming agent, flavor enhancer, flavoring agent, leavening agent, nutrient supplement, stabilizer, thickener, yeast food.

Where Used: Bread, cake (angel food), fruits (canned), meat food sticks, meringues, milk (dry powder), sausage, sausage (imitation), vegetables (canned), whipped toppings.

Regulations: FDA - 21CFR 184.1207. GRAS except for infant foods and infant formulas when used in accordance with good manufacturing practice. USDA - 9CFR 318.7. Limitation of 0.6 percent in product formulation.

SAFETY PROFILE: When heated to decomposition it emits acrid smoke and irritating fumes.

CAT650 CAS: 5001-51-4
CALCIUM LACTOBIONATE
mf: $C_{24}H_{42}CaO_{24}$ mw: 754.66

PROP: White powder. Mp: 120° (decomp). Sol in water; insol in alc, ether.

SYN: CALCIUM 4-(β-d-GALACTOSIDO)-d-GLUCONATE

USE IN FOOD:

Purpose: Firming agent.

Where Used: Pudding mixes (dry).

Regulations: FDA - 21CFR 172.720. Use at a level not in excess of the amount reasonably required to accomplish the intended effect.

SAFETY PROFILE: When heated to decomposition it emits acrid smoke and irritating fumes.

CAT675
CALCIUM LIGNOSULFONATE

USE IN FOOD:

Purpose: Dispersing agent for pesticides.

Where Used: Bananas.

Regulations: FDA - 21CFR 172.715. Use at a level not in excess of the amount reasonably required to accomplish the intended effect.

SAFETY PROFILE: When heated to decomposition it emits acrid smoke and irritating fumes.

CAU300 CAS: 142-17-6
CALCIUM OLEATE
mf: $C_{36}H_{66}CaO_4$ mw: 602.97

PROP: Pale yellow transparent solid.

SYNS: 9-OCTADECENOIC ACID CALCIUM SALT ◇ OLEIC ACID CALCIUM SALT

USE IN FOOD:

Purpose: Stabilizer.

Where Used: Packaging materials.

Regulations: FDA - 21CFR 181.29. Use in accordance with good manufacturing practice.

SAFETY PROFILE: When heated to decomposition it emits acrid smoke and irritating fumes.

CAU500 CAS: 1305-78-8
CALCIUM OXIDE
DOT: 1910
mf: CaO mw: 56.08

PROP: Cubic, white crystals. Mp: 2580°, d: 3.37, bp: 2850°. Sol in water, glycerin; insol in alc.

SYNS: BURNT LIME ◇ CALCIA ◇ CALX ◇ LIME ◇ LIME, BURNED ◇ LIME, UNSLAKED (DOT) ◇ OXYDE de CALCIUM (FRENCH) ◇ QUICKLIME (DOT) ◇ WAPNIOWY TLENEK (POLISH)

USE IN FOOD:

Purpose: Alkali, dietary supplement, dough conditioner, hog scald agent, nutrient, poultry scald agent, yeast food.

Where Used: Hog carcasses, poultry, tripe.

Regulations: FDA - 21CFR 182.5210, 184.1210. GRAS when used in accordance with good manufacturing practice. USDA - 9CFR 318.7, 381.147. Sufficient for purpose.

OSHA PEL: TWA 5 mg/m^3 ACGIH TLV: TWA 2 mg/m^3 DFG MAK: 5 mg/m^3 DOT Classification: ORM-B; Label: None

SAFETY PROFILE: A caustic and irritating material. A common air contaminant. A powerful caustic to living tissue. The powdered oxide may react explosively with water. Mixtures with ethanol may ignite if heated and thus can cause an air-vapor explosion. Violent reaction with (B_2O_3 + $CaCl_2$); interhalogens (e.g., BF_3; ClF_3); F_2; HF; P_2O_5 + heat; water. Incandescent reaction with liquid HF. Incompatible with phosphorus(V) oxide.

CAU750 CAS: 137-08-6
CALCIUM-d-PANTOTHENATE

mf: $C_{19}H_{34}N_2O_{10}$•Ca mw: 490.63

PROP: White, sltly hygroscopic powder; odorless; bitter taste. Mp: 170-172°, decomp @ 195-196°. Sol in water and glycerin; insol in alc, chloroform, and ether.

SYNS: CALCIUM d(+)-N-(α,γ-DIHYDROXY-β,β-DIMETHYLBUTYRYL)-β-ALANINATE ◇ CALCIUM PANTOTHENATE (FCC) ◇ CALCIUM PANTOTHENATE ◇ d-CALCIUM PANTOTHENATE ◇ CALPANATE ◇ DEXTRO CALCIUM PANTOTHENATE ◇ N-(2,4-DIHYDROXY-3,3-DIMETHYLBUTYRYL)-β-ALANINE CALCIUM ◇ PANCAL ◇ PANTHOJECT ◇ PANTHOLIN ◇ PANTOTHENATE CALCIUM ◇ PANTOTHENIC ACID, CALCIUM SALT ◇ (+)-PANTOTHENIC ACID, CALCIUM SALT ◇ VITAMIN B-5

USE IN FOOD:

Purpose: Dietary supplement, nutrient.

Where Used: Various.

Regulations: FDA - 21CFR 182.5212, 184.1212. GRAS when used in accordance with good manufacturing practice.

SAFETY PROFILE: Moderately toxic by intraperitoneal, subcutaneous, and intravenous routes. Mildly toxic by ingestion. A vitamin. When heated to decomposition it emits toxic fumes of NO_x.

TOXICITY DATA and CODEN

ipr-rat LD50:820 mg/kg PSEBAA 45,311,40
scu-rat LD50:3400 mg/kg PSEBAA 45,311,40
ivn-rat LD50:830 mg/kg NIIRDN 6,599,82
orl-mus LD50:10 g/kg NIIRDN 6,599,82

CAU780
CALCIUM PANTOTHENATE, CALCIUM CHLORIDE DOUBLE SALT

mf: $C_{19}H_{34}N_2O_{10}$•Ca_2Cl_2 mw: 601.61

PROP: White, sltly hygroscopic powder; odorless with bitter taste. Sol in water and glycerin; insol in alc, chloroform, and ether.

USE IN FOOD:

Purpose: Dietary supplement, nutrient.

Where Used: Various.

Regulations: FDA - 21CFR 172.330. d- or dl-form. 21CFR 182.5212. GRAS when used in accordance with good manufacturing practice.

SAFETY PROFILE: Moderately toxic by intraperitoneal, subcutaneous, and intravenous routes. Mildly toxic by ingestion. A vitamin. When heated to decomposition it emits toxic fumes of NO_x.

CAV500 CAS: 1305-79-9
CALCIUM PEROXIDE

DOT: 1457

mf: CaO_2 mw: 72.08

PROP: Yellow crystals or powder or white crystals, decomposes in air. Mp: decomp @ 275°. insol in water; sol in acids, forming hydrogen peroxide.

SYNS: CALCIUM DIOXIDE ◇ CALCIUM SUPEROXIDE

USE IN FOOD:

Purpose: Dough conditioner, oxidizing agent.

Where Used: Bakery products.

Regulations: GRAS when used in accordance with good manufacturing practice.

DOT Classification: Oxidizer; Label: Oxidizer

SAFETY PROFILE: Irritating in concentrated form. Will react with moisture to form slaked lime. Flammable if hot and mixed with finely divided combustible material. Mixtures with oxidizable materials can also be ignited by grinding and are explosion hazards. A strong alkali. An oxidizer. Mixtures with polysulfide polymers may ignite.

CAW100 CAS: 7757-93-9
CALCIUM PHOSPHATE, DIBASIC

mf: $CaHPO_4$•$2H_2O$ mw: 172.09

PROP: White powder. Sol in dilute acid; insol in water, alc.

SYN: DICALCIUM PHOSPHATE

USE IN FOOD:

Purpose: Dietary supplement, dough conditioner, nutrient, stabilizer, yeast food.

Where Used: Baked goods, cereal products, dessert gels, packaging materials.

Regulations: FDA - 21CFR 181.29, 182.1217, 182.5212, 182.8217. GRAS when used in accordance with good manufacturing practice.

SAFETY PROFILE: Skin and eye irritant. A nuisance dust.

CAW110 CAS: 7758-23-8
CALCIUM PHOSPHATE, MONOBASIC
mf: $Ca(H_2PO_4)_2$ mw: 234.05

PROP: White crystals or granular powder. Sltly sol in water; insol in alc.

SYNS: ACID CALCIUM PHOSPHATE ◇ CALCIUM BIPHOSPHATE ◇ MONOCALCIUM PHOSPHATE

USE IN FOOD:

Purpose: Buffer, dietary supplement, dough conditioner, firming agent, leavening agent, nutrient, sequestrant, stabilizer, yeast food.

Where Used: Cereals, dough, fruit jellies, packaging materials, preserves.

Regulations: FDA - 21CFR 181.29, 182.1217, 182.5212, 182.6215, 182.8217. GRAS when used in accordance with good manufacturing practice.

SAFETY PROFILE: A nuisance dust.

CAW120 CAS: 12167-74-7
CALCIUM PHOSPHATE, TRIBASIC
mf: $10CaO \cdot 3P_2O_5 \cdot H_2O$ mw: 1004.64

PROP: White powder. Sol in dilute HCl; insol in water, alc.

SYNS: PERCIPITATED CALCIUM PHOSPHATE ◇ TRICALCIUM PHOSPHATE

USE IN FOOD:

Purpose: Anticaking agent, buffer, dietary supplement, fat rendering aid, nutrient, stabilizer.

Where Used: Cereals, desserts, fats (rendered animal), flour, lard, packaging materials, table salt, vinegar (dry).

Regulations: FDA - 21CFR 181.29, 182.1217, 182.5212, 182.8217. GRAS when used in accordance with good manufacturing practice. USDA - 9CFR 318.7. Sufficient for purpose.

SAFETY PROFILE: Skin and eye irritant. A nuisance dust.

CAW400 CAS: 4075-81-4
CALCIUM PROPIONATE
mf: $C_6H_{10}CaO_4$ mw: 186.22

PROP: White crystals; faint odor of propionic acid. Sol in water.

USE IN FOOD:

Purpose: Antimicrobial agent, mold and rope inhibitor, preservative.

Where Used: Cheese, confections, dough (fresh pie), fillings, frostings, gelatins, jams, jellies, pizza crust, puddings.

Regulations: FDA - 21CFR 181.23, 184.1221. GRAS when used in accordance with good manufacturing practice. USDA - 9CFR 318.7. Limitation of 0.32 percent alone or in combination with sodium propionate based on the weight of the flour brace used. 9CFR 381.147. Limitation of 0.3 percent alone or in combination with sodium propionate based on the weight of the flour used in fresh pie dough.

SAFETY PROFILE: When heated to decomposition it emits acrid smoke and irritating fumes.

CAW450 CAS: 7790-76-3
CALCIUM PYROPHOSPHATE
mf: $Ca_2P_2O_7$ mw: 254.10

PROP: Fine white powder. Sol in dilute HCl; insol in water.

USE IN FOOD:

Purpose: Buffer, dietary supplement, neutralizing agent, nutrient.

Where Used: Various.

Regulations: FDA - 21CFR 182.5223, 182.8223. GRAS when used in accordance with good manufacturing practice.

SAFETY PROFILE: A nuisance dust.

CAW500 CAS: 9007-13-0
CALCIUM RESINATE
DOT: 1313/1314
mf: $Ca(C_{44}H_{62}O_4)_2$ mw: 1349.50

PROP: Yellowish-white, amorphous powder or lumps.

SYNS: CALCIUM RESINATE, FUSED (DOT) ◇ CALCIUM RESINATE, technically pure (DOT) ◇ LIMED ROSIN

USE IN FOOD:

Purpose: Color diluent.

Where Used: Eggs (shell).

Regulations: FDA - 21CFR 73.1. No residue.

DOT Classification: Flammable Solid; Label: Flammable Solid; Flammable Solid; Label: Flammable Solid, fused

SAFETY PROFILE: Flammable solid when heated; can react with oxidizing materials. When heated to decomposition it emits acrid smoke and fumes.

CAW525
CALCIUM RICINOLEATE

USE IN FOOD:

Purpose: Stabilizer.

Where Used: Packaging materials.

Regulations: FDA - 21CFR 181.29. Use in accordance with good manufacturing practice.

SAFETY PROFILE: When heated to decomposition it emits acrid smoke and irritating fumes.

CAW600
CALCIUM SACCHARIN

mf: $C_{14}H_8CaN_2O_6S_2 \cdot 3.5H_2O$ mw: 467.48

PROP: White crystalline powder; faint aromatic odor. Sol in water.

SYN: 1,2-BENZISOTHIAZOLIN-3-ONE 1,1-DIOXIDE CALCIUM SALT

USE IN FOOD:

Purpose: Nonnutritive sweetener.

Where Used: Bacon, bakery products (nonstandardized), beverage mixes, beverages, chewing gum, fruit juice drinks, jam desserts, vitamin tablets (chewable).

Regulations: FDA - 21CFR 180.37. Limitation of 12 mg per fluid ounce of beverages; 12 mg per fluid ounce of fruit juice drinks; 12 mg per fluid ounce of beverage mixes, 30 mg per serving in processed foods. USDA - 9CFR 318.7. Limitation of 0.01 percent in bacon.

SAFETY PROFILE: When heated to decomposition emits toxic fumes of NO_x.

CAW850 CAS: 1344-95-2
CALCIUM SILICATE

PROP: Varing proportions of CaO and SiO_2. White powder. insol in water.

USE IN FOOD:

Purpose: Anticaking agent, filter aid.

Where Used: Baking powder, chips (fabricated), table salt.

Regulations: FDA - 21CFR 172.410, 182.2227. GRAS with a limitation to 2 percent by weight in table salt, 5 percent by weight in baking powder.

SAFETY PROFILE: A nuisance dust.

CAX275
CALCIUM SORBATE

PROP: Solid. Sltly sol in water.

USE IN FOOD:

Purpose: Mold retardant, preservative.

Where Used: Cheese, margarine, oleomargarine, packaging materials.

Regulations: FDA - 21CFR 182.3189. GRAS when used in accordance with good manufacturing practice. USDA - 9CFR 318.7. Limitation of 0.1 percent individually, of if used in combination with its salts or benzoic acid or its salts, 0.2 percent (expressed as the acids in the weight of the finished product. Not allowed in cooked sausage.

SAFETY PROFILE: When heated to decomposition it emits acrid smoke and irritating fumes.

CAX350 AS: 1592-23-0
CALCIUM STEARATE

PROP: Variable proportions of calcium stearate and calcium palmitate. Fine white powder; slt characteristic odor. insol in water, alc, ether.

USE IN FOOD:

Purpose: Anticaking agent, binder, emulsifier, flavoring agent, lubricant, release agent, stabilizer, thickener.

Where Used: Beet sugar, candy (pressed), garlic salt, meat tenderizer, molasses (dry), salad dressing mix, vanilla, yeast.

Regulations: FDA - 21CFR 172.863, 173.340. Must conform to FDA specifications for fats or fatty acids derived from edible oils. 21CFR 181.29. Use at a level not in excess of the amount reasonably required to accomplish the intended effect. 21CFR 184.1229. GRAS when used in accordance with good manufacturing practice.

SAFETY PROFILE: When heated to decomposition it emits acrid smoke and irritating fumes.

CAX375
CALCIUM STEAROYL LACTATE

PROP: Cream-colored powder; caramel odor. Sltly sol in hot water.

SYN: CALCIUM STEAROYL-2-LACTATE

USE IN FOOD:

Purpose: Dough conditioner, stabilizer, whipping agent.

Where Used: Bakery products (yeast-leavened), coffee whiteners, egg white (dried), egg white (liquid and frozen), margarine (low fat), potatoes (dehydrated), puddings, vegetable toppings (whipped).

Regulations: FDA - 21CFR 172.844. Limitation of 0.5 parts for each 100 parts of flour in yeast-leavened bakery products; 0.05 percent in egg white, liquid and frozen, and egg white, dried; 0.3 percent in whipped vegetable topping; 0.5 percent in dehydrated potatoes. Must conform to FDA specifications for fats or fatty acids derived from edible oils.

SAFETY PROFILE: When heated to decomposition it emits acrid smoke and irritating fumes.

CAX500 CAS: 7778-18-9
CALCIUM SULFATE

mf: $CaSO_4$ mw: 136.14

PROP: Pure anhydrous, white powder or odorless crystals. D: 2.964; mp: 1450°.

USE IN FOOD:

Purpose: Anticaking agent, color, coloring agent, dietary supplement, dough conditioner, dough strengthener, drying agent, firming agent, flour treating agent, formulation aid, leavening agent, nutrient supplement, pH control agent, processing aid, sequestrant, stabilizer, synergist, texturizer, thickener, yeast food.

Where Used: Baked goods, canned potatoes, canned tomatoes, carrots (canned), confections, frostings, frozen dairy dessert mixes, frozen dairy desserts, gelatins, grain products, ice cream (soft serve), lima beans (canned), pasta, peppers (canned), puddings, wine (sherry).

Regulations: FDA - 21CFR 184.1230. GRAS with a limitation of 1.3 percent in baked goods, 3.0 percent in confections and frostings, 0.5 percent in frozen dairy desserts and mixes, 0.4 percent in gelatins and puddings, 0.5 percent in grain products and pastas, 0.35 percent in processed vegetables, 0.07 percent in all other foods when used in accordance with good manufacturing practice. BATF - 27CFR 240.1051. Limitation of 16.69 pounds/1000 gallons.

SAFETY PROFILE: Reacts violently with aluminum when heated. Mixtures with diazomethane react exothermically and eventually explode. Mixtures with phosphorus ignite at high temperatures. When heated to decomposition it emits toxic fumes of SO_x.

CBA500 CAS: 79-92-5
CAMPHENE

DOT: 9011

mf: $C_{10}H_{16}$ mw: 136.26

PROP: Colorless cubic crystals; oily odor. Mp: 50-51°, bp: 159°, d: 0.842 @ 54°/4°, refr index: 1.452 @ 55°. Sol in alc; misc in fixed oils; insol in water.

SYN: FEMA No. 2229

USE IN FOOD:

Purpose: Flavoring agent.

Where Used: Various.

Regulations: FDA - 21CFR 172.515. Use at a level not in excess of the amount reasonably required to accomplish the intended effect.

DOT Classification: ORM-A; Label: None

SAFETY PROFILE: Mutagenic data. Combustible; yields flammable vapors when heated and can react with oxidizing materials. To fight fire, use water spray, foam, fog, CO_2. When heated to decomposition it emits acrid smoke and irritating fumes.

TOXICITY DATA and CODEN

bfa-rat/sat 2500 mg/kg NUCADQ 1,10,79

CBC100
CANANGA OIL

PROP: From flowers of the tree *Cananga odorata* f. et Thoms., (Fam. *Anonaceae*). Yellow liquid; harsh floral odor. Sol in fixed oils, mineral oil; insol in glycerin, propylene glycol.

USE IN FOOD:

Purpose: Flavoring agent.

Where Used: Bakery products, beverages (non-alcoholic), confections, ice cream products.

Regulations: FDA - 21CFR 182.20. GRAS when used at a level not in excess of the amount reasonably required to accomplish the intended effect.

SAFETY PROFILE: When heated to decomposition it emits acrid smoke and irritating fumes.

CBC175 CAS: 8006-44-8
CANDELILLA WAX

PROP: From the leaves of *Euphorbia antisyphilitica*. A hard, brown wax. D: 0.983. Sol in chloroform, toluene; insol in water.

USE IN FOOD:

Purpose: Lubricant, masticatory substance in chewing gum base, surface-finishing agent.

Where Used: Candy (hard), chewing gum.

Regulations: FDA - 21CFR 184.1976. GRAS when used in accordance with good manufacturing practice.

SAFETY PROFILE: When heated to decomposition it emits acrid smoke and irritating fumes.

CBC400
CANDIDIA GUILLIERMONDII

PROP: Derived from *Candidia guilliermondii* fam. *Cryptococcaceae*.

SYN: ATCC No. 20474

USE IN FOOD:

Purpose: Production aid.

Where Used: Citric acid.

Regulations: FDA - 21CFR 173.160. Use at a level not in excess of the amount reasonably required to accomplish the intended effect.

SAFETY PROFILE: When heated to decomposition it emits acrid smoke and irritating fumes.

CBC425
CANDIDA LIPOLYTICA

PROP: Derived from *Candida lipolytica* fam. *Cryptococcaceae*.

USE IN FOOD:

Purpose: Production aid.

Where Used: Citric acid.

Regulations: FDA - 21CFR 173.165. Use at a level not in excess of the amount reasonably required to accomplish the intended effect.

SAFETY PROFILE: When heated to decomposition it emits acrid smoke and irritating fumes.

CBE800 CAS: 514-78-3
CANTHAXANTHIN
mf: $C_{40}H_{52}O_2$ mw: 564.80

PROP: Dark crystalline powder. Sol in chloroform; very sltly sol in acetone; insol in water.

SYNS: CANTHA ◇ β-CAROTENE-4,4′-DIONE ◇ 4,4′-DIKETO-β-CAROTENE

USE IN FOOD:

Purpose: Color additive.

Where Used: Carbonated soda, feed (broiler chicken), salad dressings, spaghetti sauce.

Regulations: FDA - 21CFR 73.75. Limitation of 30 mg per pound of solid or pint of liquid food, 4.41 mg/kg (4 g/ton) of complete feed.

SAFETY PROFILE: When heated to decomposition it emits acrid smoke and irritating fumes.

CBG000 CAS: 133-06-2
CAPTAN
DOT: 9099
mf: $C_9H_8Cl_3NO_2S$ mw: 300.59

PROP: Odorless crystals. Insol in water; sol in benzene and chloroform.

SYNS: AACAPTAN ◇ AGROSOL S ◇ AGROX 2-WAY and 3-WAY ◇ AMERCIDE ◇ BANGTON ◇ BEAN SEED PROTECTANT ◇ CAPTAF ◇ CAPTANCAPTENEET 26,538 ◇ CAPTANE ◇ CAPTAN-STREPTOMYCIN 7.5-0.1 POTATO SEED PIECE PROTECTANT ◇ CAPTEX ◇ ENT 26,538 ◇ ESSO FUNGICIDE 406 ◇ FLIT 406 ◇ FUNGUS BAN TYPE II ◇ GLYODEX 3722 ◇ GRANOX PPM ◇ GUSTAFSON CAPTAN 30-DD ◇ HEXACAP ◇ KAPTAN ◇ LE CAPTANE (FRENCH) ◇ MALIPUR ◇ MERPAN ◇ MICRO-CHECK 12 ◇ NCI-CO0077 ◇ NERACID ◇ ORTHOCIDE ◇ OSOCIDE ◇ SR406 ◇ STAUFFER CAPTAN ◇ 3a,4,7,7a-TETRAHYDRO-N-(TRICHLOROMETHANESULPHENYL)PHTHALIMIDE ◇ 3a,4,7,7a-TETRAHYDRO-2-((TRICHLOROMETHYL)THIO)-1H-ISOINDOLE-1,3(2H)-DIONE ◇ 1,2,3,6-TETRAHYDRO-N-(TRICHLOROMETHYLTHIO)PHTHALIMIDE ◇ N-(TRICHLOR-METHYLTHIO)-PHTHALIMID (GERMAN) ◇ N-TRICHLOROMETHYLMERCAPTO-4-CYCLOHEXENE-1,2-DICARBOXIMIDE ◇ N-(TRICHLOROMETHYLMERCAPTO)-Δ^4-TETRAHYDROPHTHALIMIDE ◇ N-TRICHLOROMETHYL-

THIOCYCLOHEX-4-ENE-1,2-DICARBOXIMIDE ◇ N-TRICHLOROMETHYLTHIO-cis-Δ^4-CYCLOHEXENE-1,2-DICARBOXIMIDE ◇ N-((TRICHLOROMETHYL)THIO)-4-CYCLOHEXENE-1,2-DICARBOXIMIDE ◇ TRICHLOROMETHYLTHIO-1,2,5,6-TETRAHYDROPHTHALAMIDE ◇ N-((TRICHLOROMETHYL)THIO)TETRAHYDROPHTHALIMIDE ◇ N-TRICHLOROMETHYLTHIO-3A,4,7,7A-TETRAHYDROPHTHALIMIDE ◇ VANCIDE 89 ◇ VANGARD K ◇ VANICIDE ◇ VONDCAPTAN

USE IN FOOD:

Purpose: Fungicide.

Where Used: Almonds, animal feed, apples, beans, beef, beets, broccoli, cabbage, carrots, corn, garlic, kale, lettuce, peaches, peas, pork, potatoes, raisins, spinach, strawberries.

Regulations: FDA - 21CFR 193.40. Fungicide residue tolerance of 50 ppm in raisins. 21CFR 561.65. Limitation of 100 ppm on detreated corn seed for cattle and hog feed. USDA CES Ranking: B-3 (1987).

NCI Carcinogenesis Bioassay (feed); Clear Evidence: mouse NCITR* NCI-CG-TR-15,77; No Evidence: rat NCITR* NCI-CG-TR-15,77; IARC Cancer Review: Animal Limited Evidence IMEMDT 30,295,83. EPA Genetic Toxicology Program. Community Right-To-Know List.

ACGIH TLV: TWA 5 mg/m^3 DOT Classification: ORM-E; Label; None

SAFETY PROFILE: Moderately toxic to humans by ingestion. Moderately toxic experimentally by ingestion, inhalation, and intraperitoneal routes. An experimental tumorigen, neoplastigen, and teratogen. A suspected carcinogen. Experimental reproductive effects. Human mutagenic data. A fungicide. When heated to decomposition it emits toxic fumes of Cl^-, SO_x and NO_x.

TOXICITY DATA and CODEN

mmo-sat 310 ng/plate MUREAV 130,79,84
dns-hmn:fbr 1 μmol/L MUREAV 42,161,77
cyt-hmn:lng 10 mg/L ANYAA9 160,344,69
orl-rat TDLo:500 mg/kg (5D male):REP,TER FCTXAV 10,353,72
orl-mus TDLo:250 mg/kg (5D male):REP TXAPA9 23,277,72
orl-mus TDLo:1075 g/kg/80W-C:NEO NCITR* NCI-CG-TR-15,77
orl-mus TD :540 g/kg/80W-C:ETA NCITR* NCI-CG-TR-15,77
orl-hmn LDLo:1071 mg/kg 34ZIAG -,151,69
ihl-mus LC50:5000 mg/m^3/2H TXAPA9 45,320,78

CBG125 CAS: 8028-89-5
CARAMEL

PROP: Dark brown to black liquid or solid; burnt sugar odor, pleasant bitter taste. Sol in water (colloidal).

SYN: CARAMEL COLOR

USE IN FOOD:

Purpose: Color additive.

Where Used: Baked goods, colas, root beer.

Regulations: FDA - 21CFR 73.85, 182.1235. GRAS when used in accordance with good manufacturing practice.

SAFETY PROFILE: Mutagenic data. When heated to decomposition it emits acrid smoke and irritating fumes.

TOXICITY DATA and CODEN

mma-sat 50 mg/plate FCTOD7 22,623,84
cyt-ham:fbr 8 g/L FCTOD7 22,623,84

CBG500 CAS: 8000-42-8
CARAWAY OIL

PROP: The main constituent of caraway oil is 1-carvone; found in the fruits of *Carum carvi* L. (Fam. *Umbelliferae*). (FCTXAV 11, 1011,73). Colorless liquid; odor and taste of caraway.

SYNS: KUEMMEL OIL (GERMAN) ◇ OIL of CARAWAY

USE IN FOOD:

Purpose: Flavoring agent.

Where Used: Bakery products, beverages (nonalcoholic), condiments, ice cream products.

Regulations: FDA - 21CFR 182.10, 182.20. GRAS when used at a level not in excess of the amount reasonably required to accomplish the intended effect.

SAFETY PROFILE: Moderately toxic by ingestion and skin contact. A skin irritant. Mutagenic data. When heated to decomposition it emits acrid smoke and irritating fumes.

TOXICITY DATA and CODEN

skn-rbt 500 mg/24H FCTXAV 11,1051,73
mmo-sat 5 μg/plate KEKHB8 (9),11,79
orl-rat LD50:3500 mg/kg FCTXAV 11,1051,73
skn-rbt LD50:1780 mg/kg FCTXAV 11,1051,73

CBJ000 CAS: 121-59-5
N-CARBAMOYLARSANILIC ACID
mf: $C_7H_9AsN_2O_4$ mw: 260.10

PROP: White, nearly odorless powder; slt acid taste; sol in alc and water. Mp: 174°.

SYNS: AMABEVAN ◇ AMEBAN ◇ AMEBARSONE ◇ AMIBIARSON ◇ AMINARSON ◇ AMINARSONE ◇ AMINOARSON ◇ (4-((AMINOCARBONYL)AMINO) PHENYL)ARSONIC ACID ◇ ARSAMBIDE ◇ p-ARSONO-PHENYLUREA ◇ p-CARBAMIDOBENZENEARSONIC ACID ◇ p-CARBAMINO PHENYL ARSONIC ACID ◇ CARBAMINO-PHENYL-p-ARSONIC ACID ◇ 4-CARBAMYLAMINOPHE-NYLARSONIC ACID ◇ N-CARBAMYL ARSANILIC ACID ◇ CARBARSONE (USDA) ◇ CARBASONE ◇ FENARSONE ◇ HISTOCARB ◇ LEUCARSONE ◇ p-UREIDOBENZENEAR-SONIC ACID ◇ 4-UREIDO-1-PHENYLARSONIC ACID

USE IN FOOD:

Purpose: Animal feed drug.

Where Used: Animal feed.

Regulations: FDA - 21CFR 558.120. USDA CES Ranking: C-2 (1987).

Arsenic and its compounds are on the Community Right-To-Know List.

OSHA PEL: TWA 10 μmg/m^3 ACGIH TLV: TWA 0.2 mg(As)/m^3

SAFETY PROFILE: Poison by ingestion. Moderately toxic by intraperitoneal route. An experimental tumorigen. When heated to decomposition it emits very toxic fumes of As and NO_x.

TOXICITY DATA and CODEN

orl-rat TDLo:5000 mg/kg:ETA CNREA8 26,619,66
orl-rat LD50:510 mg/kg MEIEDD 10,246,83
ipr-rat LDLo:1000 mg/kg JPETAB 80,393,44

CBM500 CAS: 116-06-3
CARBANOLATE
mf: $C_7H_{14}N_2O_2S$ mw: 190.29

PROP: A solid material.

SYNS: ALDECARB ◇ ALDICARB (USDA) ◇ ALDICARBE (FRENCH) ◇ AMBUSH ◇ ENT 27,093 ◇ 2-METHYL-2-(METHYLTHIO)PROPANAL-O-((METHYLAMINO)CARBO-NYL)OXIME ◇ 2-METHYL-2-(METHYLTHIO)PROPIONAL-DEHYDE-O-(METHYLCARBAMOYL)OXIME ◇ 2-METHYL-2-(METHYLTHIO)PROPIONALDEHYDE OXIME ◇ 2-METHYL-2-METHYLTHIO-PROPIONAL-DEHYD-O-(N-METHYL-CARBAMOYL)-OXIM (GERMAN) ◇ 2-METIL-2-TIOMETIL-PROPIONALDEID-O-(N-METIL-CARBAMOIL)-OSSIMA (ITALIAN) ◇ NCI-C08640 ◇ OMS-771 ◇ RCRA WASTE NUMBER P070 ◇ TEMIC ◇ TEMIK ◇ TEMIK G10 ◇ UC-21149

USE IN FOOD:

Purpose: Insecticide, nematocide.

Where Used: Animal feed, bananas, beans, citrus fruit, coffee, hops (dried), peanuts, pecans, potatoes, sorghum, soybeans, sugar beets, sugarcane, sweet potatoes.

Regulations: FDA - 21CFR 193.15. Insecticide residue tolerance of 50 ppm on dried hops. 21CFR 561.30. Limitation of 0.6 ppm in citrus pulp, dried. Limitation of 0.3 ppm in cottonseed hulls, 0.5 ppm in sorghum bran when used for animal feed. USDA CES Ranking: A-4 (1986).

NCI Carcinogenesis Bioassay (feed); No Evidence: mouse, rat NCITR* NCI-CG-TR-136,79. EPA Extremely Hazardous Substances List.

SAFETY PROFILE: Deadly poison by ingestion, skin contact, subcutaneous and possibly other routes. Human mutagenic data. A powerful systemic poison, pesticide, nematocide, acaricide. In 1985 over 150 people in California exhibited toxic effects from eating watermelons contaminated with aldicarb. When heated to decomposition it emits very toxic fumes of NO_x and SO_x.

TOXICITY DATA and CODEN

sce-hmn:lym 10 mg/L MUREAV 138,175,84
otr-rat:emb 117 μg/plate JJATDK 1,190,81
orl-rat LD50:650 μg/kg TXAPA9 14,515,69
skn-rat LD50:2500 μg/kg TXAPA9 14,515,69

CBM750 CAS: 63-25-2
CARBARYL
mf: $C_{12}H_{11}NO_2$ mw: 201.24

PROP: White crystals. Mp: 142°, d: 1.232 @ 20°/20°.

SYNS: CARBATOX-60 ◇ CRAG SEVIN ◇ ENT 23,969 ◇ EXPERIMENTAL INSECTICIDE 7744 ◇ KARBARYL (POLISH) ◇ N-METHYLCARBAMATE de 1-NAPHTYLE (FRENCH) ◇ METHYLCARBAMATE-1-NAPHTHALENOL ◇ METHYL-CARBAMATE-1-NAPHTHOL ◇ METHYLCARBAMIC ACID-1-NAPHTHYL ESTER ◇ N-METHYL-1-NAFTYL-CARBA-MAAT (DUTCH) ◇ N-METHYL-1-NAPHTHYL-CARBAMAT

(GERMAN) ◇ N-METHYL-α-NAPHTHYLCARBAMATE ◇ N-METHYL-1-NAPHTHYL CARBAMATE ◇ N-METHYL-α-NAPHTHYLURETHAN ◇ N-METIL-1-NAFTIL-CARBAMMATO (ITALIAN) ◇ α-NAFTYL-N-METHYLKARBAMAT (CZECH) ◇ 1-NAPHTHOL-N-METHYLCARBAMATE ◇ 1-NAPHTHYL METHYLCARBAMATE ◇ α-NAPHTHYL N-METHYLCARBAMATE ◇ 1-NAPHTHYL-N-METHYL-CARBAMATE ◇ SEVIN

USE IN FOOD:

Purpose: Insecticide.

Where Used: Animal feed, pineapples.

Regulations: FDA - 21CFR 561.66. Limitation of 20 ppm in pineapple bran when used for animal feed. USDA CES Ranking: B-2 (1988).

IARC Cancer Review: Animal Inadequate Evidence IMEMDT 12,37,76. Community Right-To-Know List.

OSHA PEL: TWA 5 mg/m^3 ACGIH TLV: TWA 5 mg/m^3 DOT Classification: ORM-A; Label: None NIOSH REL: TWA 5 mg/m^3

SAFETY PROFILE: Poison by ingestion, intravenous, intraperitoneal, and possibly other routes. Human mutagenic data. An experimental carcinogen, teratogen, and tumorigen. Experimental reproductive effects. An eye and severe skin irritant. Absorbed by all routes, although skin absorption is slow. No accumulation in tissue. Symptoms include blurred vision, headache, stomach ache, vomiting. Symptoms similar to but less severe than those due to parathion. A reversible cholinesterase inhibitor. When heated to decomposition it emits toxic fumes of NO_x.

TOXICITY DATA and CODEN

skn-rbt 12 mg/24H SEV JAFCAU 9,30,61
eye-rbt 500 mg/24H MOD 28ZPAK -,164,72
mmo-sat 250 μg/plate RPZHAW 30,81,79
cyt-hmn:emb 40 μg/kg ZDVKAP 20(4),14,77
orl-rat TDLo:50 mg/kg (9 or 10D preg): REP,TER SAKNAH 39,471,65
orl-dog TDLo:388 mg/kg (preg):REP,TER TXAPA9 13,392,68
orl-rat TDLo:5700 mg/kg/95W-I:CAR VPITAR 29,71,70
imp-rat TDLo:80 mg/kg:ETA VPITAR 29,71,70
orl-rat LD50:250 mg/kg BWHOA6 44(1-3),241,71
ihl-rat LD50:721 mg/kg EQSFAP 3,618,75
skn-rat LD50:4000 mg/kg 85DPAN -,-,71/76
ipr-rat LD50:48 mg/kg PSEBAA 114,509,63

CBP000 CAS: 121-75-5
CARBETHOXY MALATHION

DOT: 2783
mf: $C_{10}H_{19}O_6PS_2$ mw: 330.38

PROP: Brown to yellow liquid; characteristic odor. D: 1.23 @ 25°/4°, mp: 2.9°, bp: 156° @ 0.7 mm. Miscible in organic solvents, sltly water-sol.

SYNS: S-(1,2-BIS(AETHOXY-CARBONYL)-AETHYL)-O,O-DIMETHYL-DITHIOPHASPHAT (GERMAN) ◇ S-(1,2-BIS (ETHOXY-CARBONYL)-ETHYL)-O,O-DIMETHYL-DITHIOFOSFAAT (DUTCH) ◇ S-1,2-BIS(ETHOXYCARBONYL) ETHYL-O,O-DIMETHYL THIOPHOSPHATE ◇ S-(1,2-BIS (ETOSSI-CARBONIL)-ETIL)-O,O-DIMETIL-DITIOFOSFATO (ITALIAN) ◇ CALMATHION ◇ CARBETOVUR ◇ CARBETOX ◇ CARBOFOS ◇ CARBOPHOS ◇ CELTHIGN ◇ CHEMATHION ◇ CIMEXAN ◇ COMPOUND 4049 ◇ CYTHION ◇ DETMOL MA ◇ DETMOL MA 96% ◇ S-(1,2-DI(ETHOXYCARBONYL)ETHYL DIMETHYL PHOSPHOROTHIOLOTHIONATE ◇ DIETHYL MERCAPTOSUCCINATE-S-ESTER with O,O-DIMETHYLPHOSPHORODITHIOATE ◇ O,O-DIMETHYL-S-1,2-DIKARBETOXYLETHYLDITIOFOSFAT (CZECH) ◇ O,O-DIMETHYL-DITHIOPHOSPHATE DIETHYLMERCAPTOSUCCINATE ◇ DITHIOPHOSPHATE de O,O-DIMETHYLE et de S-(1,2-DICARBOETHOXYETHYLE) (FRENCH) ◇ EMMATOS EXTRA ◇ ETHIOLACAR ◇ ETIOL ◇ EXTERMATHION ◇ FORMAL ◇ FORTHION ◇ FOSFOTHION ◇ FOSFOTION ◇ FYFANON ◇ HILTHION ◇ KARBOFOS ◇ KOP-THION ◇ KYPFOS ◇ MALACIDE ◇ MALAFOR ◇ MALAGRAN ◇ MALAKILL ◇ MALAMAR ◇ MALAPHELE ◇ MALAPHOS ◇ MALASOL ◇ MALASPRAY ◇ MALATHION ◇ MALATHION (ACGIH,DOT) ◇ MALATHIOZOO ◇ MALATHON ◇ MALATION (POLISH) ◇ MALATOL ◇ MALATOX ◇ MALDISON ◇ MALMED ◇ MALPHOS ◇ MALTOX ◇ MALTOX MLT ◇ MERCAPTOSUCCINIC ACID DIETHYL ESTER ◇ MERCAPTOTHION ◇ MERCAPTOTION (SPANISH) ◇ OLEOPHOSPHOTHION ◇ ORTHO MALATHION ◇ PHOSPHOTHION ◇ PRIODERM ◇ SADOFOS ◇ SADOPHOS ◇ SIPTOX I ◇ SUMITOX ◇ TAK ◇ VEGFRU MALATOX ◇ VETIOL ◇ ZITHIOL

USE IN FOOD:

Purpose: Insecticide.

Where Used: Animal feed, citrus pulp (dehydrated), grapes, nonmedicated cattle feed concentrate blocks, packaging materials, safflower oil.

Regulations: FDA - 21CFR 193.260. Insecticide residue tolerance of 12 ppm in grapes, 0.6 ppm in safflower oil. 21CFR 561.270. Limitation of 50 ppm in dehydrated citrus pulp, 10 ppm in nonmedicated cattle feed concentrate blocks when used for animal feed.

IARC Cancer Review: Animal No Evidence IMEMDT 30,103,83; NCI Carcinogenesis Bioassay (feed); No Evidence: mouse, rat NCITR* NCI-CG-TR-24,78; No Evidence: rat NCITR* NCI-CG-TR-192,79. EPA Genetic Toxicology Program.

OSHA PEL: TWA 10 mg/m^3 Total Dust; 5 mg/m^3 Respirable Dust (skin) ACGIH TLV: TWA 10 mg/m^3 (skin) DOT Classification: ORM-A; Label: None NIOSH REL: (Malathion) TWA 15 mg/m^3

SAFETY PROFILE: A human poison by ingestion. An experimental poison by ingestion, inhalation, intraperitoneal, intravenous, intraarterial, subcutaneous, and possibly other routes. Human systemic effects by ingestion: coma, blood pressure depression, and difficulty in breathing. Human mutagenic data. Has caused allergic sensitization of the skin. An organic phosphate cholinesterase inhibitor-type insecticide. When heated to decomposition it emits toxic fumes of PO_x and SO_x.

TOXICITY DATA and CODEN

mmo-sat 10 mg/L TGANAK 15(3),68,81
mmo-bcs 1 nmol/plate MSERDS 5,93,81
orl-rat TDLo:5550 mg/kg (91D pre/1-20D preg):TER JTEHD6 14,267,84
orl-rat TDLo:191 mg/kg (9D preg):REP JFMAAQ 54,452,67
orl-rat TDLo:283 mg/kg (9D preg):TER JFMAAQ 54,452,67
orl-man LDLo:471 mg/kg:CNS,CVS,PUL ATXKA8 23,11,67
orl-wmn LDLo:246 mg/kg AEHLAU 33,240,78
orl-rat LD50:370 mg/kg JFMAAQ 54,452,67
ihl-rat LCLo:1200 $\mu g/m^3$/4H 85GMAT -,56,82
skn-rat LD50:4444 mg/kg CMEP** -,1,56

CBS400
CARBOHYDRASE, ASPERGILLUS

PROP: From fermentation of *Aspergillus oryzae* var. Tan amorphous powder or liquid. Sol in water.

USE IN FOOD:

Purpose: Production aid, tenderizing agent.

Where Used: Alcoholic beverages, ale, bakery products, beer, dairy products, meat (raw cuts), poultry, starch syrups.

Regulations: USDA - 9CFR 318.7. Solutions consisting of water and approved proteolytic enzymes applied or injected into raw meat cuts shall not result in a gain of more than 3 percent above the weight of the untreated product.

SAFETY PROFILE: When heated to decomposition it emits acrid smoke and irritating fumes.

CBS405
CARBOHYDRASE and CELLILASE

PROP: Derived from *Aspergillus niger*.

SYN: CELLILASE and CARBOHYDRASE

USE IN FOOD:

Purpose: Tissue release agent.

Where Used: Clams, shrimp.

Regulations: FDA - 21CFR 173.120. Use at a level not in excess of the amount reasonably required to accomplish the intended effect.

SAFETY PROFILE: When heated to decomposition it emits acrid smoke and irritating fumes.

CBS410
CARBOHYDRASE and PROTEASE, MIXED

PROP: From controlled fermentation of *Bacillus licheniformus* var. Brown amorphous powders or liquid. Sol in water; insol in alc, chloroform, ether.

USE IN FOOD:

Purpose: Enzyme.

Where Used: Beer, beverages (alcoholic), candy, dextrose, fish meal, nutritive sweeteners, protein hydrolyzates, starch syrups.

Regulations: FDA - 21CFR 184.1027. GRAS when used in accordance with good manufacturing practice.

SAFETY PROFILE: When heated to decomposition it emits acrid smoke and irritating fumes.

CBS415
CARBOHYDRASE, RHIZOPUS

PROP: Derived from *Rhizopus oryzae*.

USE IN FOOD:

Purpose: Production aid.

Where Used: Dextrose.

Regulations: FDA - 21CFR 173.130. Must be refrigerated from production to use.

SAFETY PROFILE: When heated to decomposition it emits acrid smoke and irritating fumes.

CBT250 CAS: 4564-87-8
CARBOMYCIN
mf: $C_{42}H_{67}NO_{16}$ mw: 842.10

SYNS: CARBOMYCIN A ◇ DELTAMYCIN A ◇ 9-DEOXY-12,13-EPOXY-9-OXOLEUCOMYCIN V 3-ACETATE 4^B-(3-METHYLBUTANOATE) ◇ M-4209 ◇ MAGNAMYCIN ◇ MAGNAMYCIN A

USE IN FOOD:

Purpose: Animal drug.

Where Used: Chicken.

Regulations: FDA - 21CFR 556.110. Tolerance of zero in cooked edible tissues of chicken.

SAFETY PROFILE: Poison by subcutaneous route. Moderately toxic by intravenous and intramuscular routes. When heated to decomposition it emits toxic fumes of NO_x.

TOXICITY DATA and CODEN

scu-mus LD50:295 mg/kg ANTCAO 3,55,53
ivn-mus LD50:550 mg/kg MEIEDD 10,250,83
ims-mus LD50:1000 mg/kg ANTCAO 3,55,53

CBT750 CAS: 1333-86-4
CARBON BLACK

PROP: A generic term applied to a family of high-purity colloidal carbons commercially produced by carefully controlled pyrolysis of gaseous or liquid hydrocarbons. Carbon blacks, including commercial colloidal carbons such as furnace blacks, lamp blacks and acetylene blacks, usually contain less than several tenths percent of extractable organic matter and less than one percent ash.

SYNS: ACETYLENE BLACK ◇ CHANNEL BLACK ◇ FURNACE BLACK ◇ LAMP BLACK

USE IN FOOD:

Purpose: Color additive.

Where Used: Various.

Regulations: FDA - 21CFR 81.10. Provisional listing terminated.

IARC Cancer Review: Human Inadequate Evidence IMEMDT 33,35,84; Animal Inadequate Evidence IMEMDT 33,35,84

OSHA PEL: TWA 3.5 mg/m^3 ACGIH TLV: TWA 3.5 mg/m^3 NIOSH REL: TWA 3.5 mg/m^3

SAFETY PROFILE: Mildly toxic by ingestion, inhalation, and skin contact. A nuisance dust in high concentrations. While it is true that the tiny particulates of carbon black contain some molecules of carcinogenic materials, the carcinogens are apparently held tightly and are not eluted by hot or cold water, gastric juices, or blood plasma.

CBU250 CAS: 124-38-9
CARBON DIOXIDE
DOT: 1013/1845/2187
mf: CO_2 mw: 44.01

PROP: Colorless, odorless gas. Mp: subl @ −78.5° (−56.6° @ 5.2 atm), vap d: 1.53.

SYNS: ANHYDRIDE CARBONIQUE (FRENCH) ◇ CARBONIC ACID GAS ◇ CARBONIC ANHYDRIDE ◇ KOHLENDIOXYD (GERMAN) ◇ KOHLENSAURE (GERMAN)

USE IN FOOD:

Purpose: Aerating agent, carbonation, cooling agent, gas, leavening agent, modified atmospheres for pest control, pH control agent, processing aid, propellant.

Where Used: Beverages (carbonated), fruit, meat, poultry, wine.

Regulations: FDA - 21CFR 184.1240. GRAS when used in accordance with good manufacturing practice. 21CFR 193.45. Approved for modified atmospheres for pest control. USDA - 9CFR 318.7, 381.147. Sufficient for purpose. BATF - 27CFR 240.1051. The carbon dioxide content of the finished wine shall not be increased during the transfer operation.

OSHA PEL: TWA 5000 ppm ACGIH TLV: TWA 5000 ppm; STEL 30000 ppm DFG MAK: 5000 ppm (9000 mg/m^3) DOT Classification: Nonflammable gas; Label: Nonflammable Gas; ORM-A; Label: None NIOSH REL: TWA 10000 ppm; CL 30000 ppm/10M

SAFETY PROFILE: An asphyxiant. An experimental teratogen. Other experimental reproductive effects. Contact of carbon dioxide snow with the skin can cause burns. Dusts of magnesium, zirconium, titanium, and some magnesium-aluminum alloys ignite and then explode in CO_2 atmospheres. Dusts of aluminum, chromium, and manganese ignite and then explode when heated in CO_2. Several bulk metals will burn in CO_2. Reacts vigorously with (Al + Na_2O_2); Cs_2O; $Mg(C_2H_5)_2$; Li; (Mg + Na_2O_2);

K; KHC; Na; Na_2C_2; NaK; Ti. CO_2 fire extinguishers can produce highly incendiary sparks of 5-15 mJ at 10-20 KV by electrostatic discharge. Incompatible with acrylaldehyde, aziridine, metal acetylides, sodium peroxide.

TOXICITY DATA and CODEN

ihl-rat TCLo:6 pph/24H (10D preg):TER CIRUAL 8,1218,60
ihl-rbt TCLo:13 pph/4H (9-12D preg):TER ZMOAAN 56,165,65
ihl-mus TCLo:55 pph/2H (3D male):REP JRPFA4 13,165,67
ihl-hmn LCLo:10 pph/1M AOHYA3 17,159,74
ihl-hmn LCLo:9 pph/5M TABIA2 3,231,33

CBV500 CAS: 75-15-0
CARBON DISULFIDE

DOT: 1131
mf: CS_2 mw: 76.13

PROP: Clear, colorless liquid; nearly odorless when pure. Mp: −110.8°, bp: 46.5°, lel: 1.3%, uel: 50%, flash p: −22°F (CC), d: 1.261 @ 20°/20°, autoign temp: 257°F, vap press: 400 mm @ 28°, vap d: 2.64.

SYNS: CARBON BISULFIDE (DOT) ◇ CARBON BISULPHIDE ◇ CARBON DISULPHIDE ◇ CARBONE (SUFURE de) (FRENCH) ◇ CARBONIO (SOLFURO di) (ITALIAN) ◇ CARBON SULFIDE ◇ CARBON SULPHIDE (DOT) ◇ DITHIOCARBONIC ANHYDRIDE ◇ KOHLENDISULFID (SCHWEFELKOHLENSTOFF) (GERMAN) ◇ KOOLSTOFDISULFIDE (ZWAVELKOOLSTOF) (DUTCH) ◇ NCI-C04591 ◇ RCRA WASTE NUMBER P022 ◇ SCHWEFELKOHLENSTOFF (GERMAN) ◇ SOLFURO di CARBONIO (ITALIAN) ◇ SULPHOCARBONIC ANHYDRIDE ◇ WEEVILTOX ◇ WEGLA DWUSIARCZEK (POLISH)

USE IN FOOD:

Purpose: Fumigant.

Where Used: Cereal grains.

Regulations: FDA - 21CFR 193.225, 193.230. Fumigant.

EPA Genetic Toxicology Program. Community Right-To-Know List. EPA Extremely Hazardous Substances List.

OSHA PEL: TWA 4 ppm; STEL 12 ACGIH TLV: TWA 10 ppm (skin) DFG MAK: 10 ppm (30 mg/m^3) DOT Classification: Flammable Liquid; Label: Flammable Liquid; IMO: Flammable Liquid; Label: Flammable Liquid, Poison NIOSH REL: TWA 1 ppm; CL 10 ppm/15M

SAFETY PROFILE: A human poison by ingestion and possibly other routes. Mildly toxic to humans by inhalation. An experimental poison by intraperitoneal route. Human reproductive effects on spermatogenesis by inhalation. An experimental teratogen. Other experimental reproductive effects. Human mutagenic data. The main toxic effect is on the central nervous system, acting as a narcotic and anesthetic in acute poisoning with death following from respiratory failure. In chronic poisoning, the effect on the nervous system is one of central and peripheral damage which may be permanent if the damage has been severe.

Flammable liquid. A dangerous fire hazard when exposed to heat, flame, sparks, friction, or oxidizing materials. Severe explosion hazard when exposed to heat or flame. Ignition and potentially explosive reaction when heated in contact with rust or iron. Mixtures with sodium or potassium-sodium alloys are powerful, shock-sensitive explosives. Explodes on contact with permanganic acid. Potentially explosive reaction with nitrogen oxide; chlorine (catalyzed by iron). Mixtures with dinitrogen tetraoxide are heat-, spark- and shock-sensitive explosives. Reacts with metal azides to produce shock- and heat-sensitive, explosive metal azidodithioformates. Aluminum powder ignites in CS_2 vapor. The vapor ignites on contact with fluorine. Reacts violently with azides; CsN_3; ClO; ethylamine diamine; ethylene imine; $Pb(N_3)_2$; LiN_3; (H_2SO_4 + permanganates); KN_3; RbN_3; NaN_3; phenylcopper-triphenylphosphine complexes. Incompatible with air, metals, oxidants. To fight fire, use water, CO_2, dry chemical, fog, mist. When heated to decomposition it emits highly toxic fumes of SO_x.

TOXICITY DATA and CODEN

mmo-sat 100 μL/plate NIOSH* 5AUG77
sce-hmn:lym 10200 μg/L BCTKAG 14,115,81
ihl-man TCLo:40 mg/m^3 (91W male):REP MELAAD 60,566,69
orl-rat TDLo:2 g/kg (6-15D preg):TER TOXID9 4,86,84
ihl-rat TCLo:10 mg/m^3/8H (1-22D preg):TER ATSUDG 4,252,80
orl-hmn LDLo:14 mg/kg 32ZWAA 8,225,74
ihl-hmn LCLo:4000 ppm/30M 29ZWAE -,118,68
ihl-hmn LCLo:2000 ppm/5M TABIA2 3,231,33
orl-rat LD50:3188 mg/kg GISAAA 31(1),13,66

CBY000 CAS: 56-23-5
CARBON TETRACHLORIDE
DOT: 1846
mf: CCl_4 mw: 153.81

PROP: Colorless liquid; heavy, ethereal odor. Mp: −22.6°, bp: 76.8°, fp: −22.9°, flash p: none, d: 1.597 @ 20°, vap press: 100 mm @ 23.0°.

SYNS: BENZINOFORM ◇ CARBONA ◇ CARBON CHLORIDE ◇ CARBON TET ◇ CZTEROCHLOREK WEGLA (POLISH) ◇ ENT 4,705 ◇ FASCIOLIN ◇ FLUKOIDS ◇ METHANE TETRACHLORIDE ◇ NECATORINA ◇ NECATORINE ◇ PERCHLOROMETHANE ◇ R 10 ◇ RCRA WASTE NUMBER U211 ◇ TETRACHLOORKOOLSTOF (DUTCH) ◇ TETRACHLOORMETAAN ◇ TETRACHLORKOHLENSTOFF, TETRA (GERMAN) ◇ TETRACHLORMETHAN (GERMAN) ◇ TETRACHLOROCARBON ◇ TETRACHLOROMETHANE ◇ TETRACHLORURE de CARBONE (FRENCH) ◇ TETRACLOROMETANO (ITALIAN) ◇ TETRACLORURO di CARBONIO (ITALIAN) ◇ TETRAFINOL ◇ TETRAFORM ◇ TETRASOL ◇ UNIVERM ◇ VERMOESTRICID

USE IN FOOD:

Purpose: Fumigant.

Where Used: Cereal grains.

Regulations: FDA - 21CFR 193.225, 193.230. Fumigant.

IARC Cancer Review: Animal Sufficient Evidence IMEMDT 20,371,79; IMEMDT 1,53,72; Human Inadequate Evidence IMEMDT 1, 53,72; Human Limited Evidence IMEMDT 20,371,79. Community Right-To-Know List. EPA Genetic Toxicology Program.

OSHA PEL: TWA 2 ppm ACGIH TLV: TWA 5 ppm; STEL 30 (skin); Suspected Carcinogen DFG MAK: 10 ppm (65 mg/m^3) DOT Classification: ORM-A; Label: None; Poison B; Label: Poison NIOSH REL: CL 2 ppm/ 60M

SAFETY PROFILE: A human poison by ingestion and possibly other routes. Poison by subcutaneous and intravenous routes. Mildly toxic by inhalation. An experimental carcinogen, neoplastigen, tumorigen, teratogen, and suspected human carcinogen. Human systemic effects by inhalation and ingestion: nausea or vomiting, pupillary constriction, coma, anti-psychotic effects, tremors, somnolence, anorexia, unspecified respiratory system and gastrointestinal system effects. An eye and skin irritant. Damages liver, kidneys, and lungs. Mutagenic data. A narcotic. Individual susceptibility varies widely. Contact dermatitis can result from skin contact.

Carbon tetrachloride has a narcotic action resembling that of chloroform, though not as strong. Following exposures to high concentrations, the victim may become unconscious, and if exposure is not terminated, death can follow from respiratory failure. The after-effects following recovery from narcosis are more serious than those of delayed chloroform poisoning, usually taking the form of damage to the kidneys, liver, and lungs. Exposure to lower concentrations, insufficient to produce unconsciousness, usually results in severe gastrointestinal upset and may progress to serious kidney and hepatic damage. The kidney lesion is an acute nephrosis; the liver involvement consists of an acute degeneration of the central portions of the lobules. When recovery takes place, there may be no permanent disability. Marked variation in individual susceptibility to carbon tetrachloride exists; some persons appear to be unaffected by exposures which seriously poison their fellow workers. Alcoholism and previous liver and kidney damage seem to render the individual more susceptible. Concentrations on the order of 1000 to 1500 ppm are sufficient to cause symptoms if exposure continues for several hours. Repeated daily exposure to such concentration may result in poisoning.

Though the common form of poisoning following industrial exposure is usually one of gastrointestinal upset, which may be followed by renal damage, other cases have been reported in which the central nervous system has been affected with production of polyneuritis, narrowing of the visual fields, and other neurological changes. Prolonged exposure to small amounts of carbon tetrachloride has also been reported as causing cirrhosis of the liver.

Locally, a dermatitis may be produced following long or repeated contact with the liquid. The skin oils are removed and the skin becomes red, cracked and dry. The effect of carbon tetrachloride on the eyes either as a vapor or as a liquid, is one of irritation with lacrimation and burning.

Industrial poisoning is usually acute with malaise, headache, nausea, dizziness, and confusion which may be followed by stupor and sometimes loss of consciousness. Symptoms of liver and kidney damage may follow later with development of dark urine, sometimes jaundice and liver enlargement, followed by scanty urine, albumenuria and renal casts; uremia may develop and cause death. Where exposure has been less acute, the symptoms are usually headache, dizziness, nausea, vomiting, epigastric distress, loss of appetite, and fatigue. Visual disturbances (blind spots, spots before the eyes, a visual "haze" and restriction of the visual fields), secondary anemia, and occasionally a slight jaundice may occur. Dermatitis may be noticed on the exposed parts.

Forms impact-sensitive explosive mixtures with particulates of many metals, e.g., aluminum (when ball milled or heated to 152° in a closed container); barium (bulk metal also reacts violently); beryllium; potassium (200 times more shock-sensitive than mercury fulminate); potassium-sodium alloy (more sensitive than potassium); lithium; sodium; zinc (burns readily). Also forms explosive mixtures with chlorine trifluoride; calcium hypochlorite (heat sensitive); calcium disilicide (friction and pressure sensitive); triethyldialuminum trichloride (heat sensitive); decaborane(14) (impact sensitive); dinitrogen tetraoxide. Violent or explosive reaction on contact with fluorine. Forms explosive mixtures with ethylene between 25-105° and 30-80 bar. Potentially explosive reaction on contact with boranes. 9:1 mixtures of methanol and CCl_4 react exothermically with aluminum, magnesium, or zinc. Potentially dangerous reaction with dimethyl formamide; 1,2,3,4,5,6-hexachlorocyclohexane; or dimethylacetamide when iron is present as a catalyst. CCl_4 has caused explosions when used as a fire extinguisher on wax fires and uranium fires. Incompatible with aluminum trichloride, dibenzoyl peroxide, potassium-tert-butoxide. Vigorous exothermic reaction with allyl alcohol; $Al(C_2H_5)_3$; (benzoyl peroxide + C_2H_4); BrF_3; diborane; disilane; liquid O_2; Pu; ($AgClO_4$ + HCl); potassium-tert-butoxide; tetraethylenepentamine; tetrasilane; trisilane; Zr. When heated to decomposition it emits toxic fumes of Cl^- and phosgene. It has been banned from household use by FDA.

TOXICITY DATA and CODEN

skn-rbt 4 mg MLD XEURAQ MDDC-1715
eye-rbt 2200 μg/30S MLD XEURAQ MDDC-1715
mmo-asn 5000 ppm MUREAV 147,288,85
sln-asn 5000 ppm MUREAV 147,288,85
orl-rat TDLo:2 g/kg (7-8D preg):REP 85DJA5 -,95,71
orl-rat TDLo:3 g/kg (14D preg):TER BEXBAN 82,1262,76
par-rat TDLo:2384 mg/kg (18D preg):TER BEXBAN 76,1467,73
orl-mus TDLo:4400 mg/kg/19W-I:NEO JNCIAM 20,431,58
par-mus TDLo:305 g/kg/30W-I:ETA BEXBAN 89,845,80
scu-rat TD :182 g/kg/70W-I:CAR JNCIAM 44,419,70
orl-hmn LDLo:43 mg/kg 32ZWAA 8,128,74
ihl-hmn TCLo:20 ppm:GIT 85CYAB 2,136,59
orl-wmn TDLo:1800 mg/kg:EYE,CNS TXMDAX 69,86,73
orl-man TDLo:1700 mg/kg:CNS,PUL,GIT SAMJAF 49,635,75
ihl-hmn TCLo:45 ppm/3D:CNS,GIT LANCAO 1,360,60
orl-rat LD50:2800 mg/kg TXAPA9 17,498,70
ihl-rat LC50:8000 ppm/4H NPIRI* 1,16,74
skn-rat LD50:5070 mg/kg SPEADM 74-1,-,74

CCC500 CAS: 5234-68-4
CARBOXINE

mf: $C_{12}H_{13}NO_2S$ mw: 235.32

SYNS: 5-CARBOXANILIDO-2,3-DIHYDRO-6-METHYL-1,4-OXATHIIN ◇ CARBOXIN (USDA) ◇ D 735 ◇ DCMO ◇ 2,3-DIHYDRO-5-CARBOXANILIDO-6-METHYL-1,4-OXATHIIN ◇ 5,6-DIHYDRO-2-METHYL-3-CARBOXANILIDO-1,4-OXATHIIN (GERMAN) ◇ 2,3-DIHYDRO-6-METHYL-1,4-OXATHIIN-5-CARBOXANILIDE ◇ 5,6-DIHYDRO-2-METHYL-1,4-OXATHIIN-3-CARBOXANILIDE ◇ 5,6-DIHYDRO-2-METHYL-N-PHENYL-1,4-OXATHIIN-3-CARBOXAMIDE ◇ F 735 ◇ FLO PRO V SEED PROTECTANT ◇ VITAVAX

USE IN FOOD:

Purpose: Fungicide.

Where Used: Barley (forage), beans (forage), corn (fodder), oats (forage), peanut (hulls), rice (straw), sorghum (fodder), wheat (forage).

Regulations: USDA CES Ranking: C-4 (1987).

SAFETY PROFILE: Poison by ingestion. Moderately toxic by skin contact and possibly other routes. When heated to decomposition it emits very toxic fumes of NO_x and SO_x.

TOXICITY DATA and CODEN

orl-rat LD50:430 mg/kg GTPZAB 23(2),55,79
skn-rat LD50:1050 mg/kg GTPZAB 23(2),55,79

CCJ625 CAS: 8000-66-6
CARDAMON OIL

PROP: From the seed of *Elettaria acrdamomun* (L.) Maton (Fam. *Zingiberazeae*). Colorless liquid; aromatic penetrating odor of cardamom, pungent taste. Misc with alc.

SYNS: CARDAMON ◇ OIL of CARDAMON

USE IN FOOD:

Purpose: Flavoring agent.

Where Used: Bakery products, beverages (alcoholic), beverages (nonalcoholic), bread, chewing gum, condiments, confections, cookies, desserts, ice cream products, marinades, meat, pickles, punches (hot fruit).

Regulations: FDA - 21CFR 182.10. GRAS when used at a level not in excess of the amount reasonably required to accomplish the intended effect.

SAFETY PROFILE: Mutagenic data. When heated to decomposition it emits acrid smoke and fumes.

TOXICITY DATA and CODEN

mmo-sat 2500 ng/plate KEKHB8 (9),11,79
dnr-bcs 19 mg/disc SKEZAP 25,378,84

CCK590 CAS: 1390-65-4
CARMINE

mf: $C_{22}H_{20}O_{13}$ mw: 492.39

PROP: An aqueous extract of cochineal obtained from the dried female insects *Dactylopius coccus costa (Cossus cacti* L.). Bright red crystals from water. Decomp @ 250°. Sol in water, alc, ether; insol in benzene, chloroform. Carmine is the aluminum or calcium-aluminum lake on aluminum hydroxide substrate of carminic acid.

SYNS: B ROSE LIQUID ◇ CARMINIC ACID

USE IN FOOD:

Purpose: Color additive.

Where Used: Various.

Regulations: FDA - 21CFR 73.100. Must be pasteurized to destroy *Salmonella*. Use in accordance with good manufacturing practice.

SAFETY PROFILE: When heated to decomposition it emits acrid smoke and irritating fumes.

CCK640 CAS: 8015-86-9
CARNAUBA WAX

PROP: From leaf leaves of Brazilian wax palm *Copernicia careferia* (Arruda) Mart. Hard, brittle light yellow to brown solid. D: 0.997; mp: 82-85°. Sol in chloroform; sltly sol in boiling alc; insol in water.

SYN: BRAZIL WAX

USE IN FOOD:

Purpose: Anticaking agent, candy glaze, candy polish, formulation aid, lubricant, release agent, surface-finishing agent.

Where Used: Baked goods, baking mixes, candy (soft), chewing gum, confections, frostings, fruit juices, fruit juices (processed), fruits (fresh), fruits (processed), gravies, sauces.

Regulations: FDA - 21CFR 184.1978. GRAS when used in accordance with good manufacturing practice.

SAFETY PROFILE: When heated to decomposition it emits acrid smoke and irritating fumes.

CCK685 CAS: 7235-40-7
β-CAROTENE

mf: $C_{40}H_{56}$ mw: 536.88

PROP: Red crystals or crystalline powder. Sol in carbon disulfide, benzene, chloroform; sltly sol in ether, hexane, veg. oil; insol in water.

SYN: CAROTENE

USE IN FOOD:

Purpose: Color additive, dietary supplement, nutrient.

Where Used: Beverages (orange), cheese, dairy products, ice cream, oleomargarine, puddings.

Regulations: FDA - 21CFR 73.95, 182.5245. GRAS when used in accordance with good manufacturing practice.

SAFETY PROFILE: When heated to decomposition it emits acrid smoke and irritating fumes.

CCK691
CAROTENE COCHINEAL

USE IN FOOD:

Purpose: Coloring agent.

Where Used: Casings, fats (rendered).

Regulations: USDA - 9CFR 318.7. Sufficient for purpose.

SAFETY PROFILE: When heated to decomposition it emits acrid smoke and irritating fumes.

CCL250 CAS: 9000-07-1
CARRAGEEN

PROP: A sulfated polysaccharide. Dried plant of seaweed *Chondrus crispus, Chondrus ocella-*

tus, Eucheuma cottonil, Eucheuma spinosum, Gigartina acicularis, Gigartina pistillata, Gigartina radula, Gigartina stellata. Yellow-white when powdered. Sol in water @ 80°; insol in organic solvents. Dried, bleached *Chondrus crispus* containing salts of sulfated polygalactose esters.

SYNS: 3,6-ANHYDRO-d-GALACTAN ◇ AUBYGEL GS ◇ AUBYGUM DM ◇ BURTONITE-V-40-E ◇ CARASTAY ◇ CARASTAY G ◇ CARRAGEENAN (FCC) ◇ CARRAGEENAN GUM ◇ CARRAGHEANIN ◇ CARRAGHEEN ◇ CARRAGHEENAN ◇ CHONDRUS ◇ CHONDRUS EXTRACT ◇ COLLOID 775 ◇ COREINE ◇ EUCHEUMA SPINOSUM GUM ◇ FLANOGEN ELA ◇ GALOZONE ◇ GELCARIN ◇ GELCARIN HMR ◇ GELOZONE ◇ GENU ◇ GENUGEL ◇ GENUGEL CJ ◇ GENUGOL RLV ◇ GENUVISCO J ◇ GUM CARRAGEENAN ◇ GUM CHON 2 ◇ GUM CHROND ◇ IRISH GUM ◇ IRISH MOSS EXTRACT ◇ IRISH MOSS GELOSE ◇ KILLEEN ◇ LYGOMME CDS ◇ PEARLPUSS ◇ PELLUGEL ◇ PENCOGEL ◇ PIG-WRACK ◇ SATIAGEL GS350 ◇ SATIAGUM 3 ◇ SATIAGUM STANDARD ◇ SEAKEM CARRAGEENIN ◇ SEATREM ◇ SELF ROCK MOSS ◇ VISCARIN

USE IN FOOD:

Purpose: Binder, emulsifier, extender, gelling agent, stabilizer, thickener.

Where Used: Dairy products, dessert gels (water), jelly (low calorie), meat (restructured), poultry.

Regulations: FDA - 21CFR 172.620, 172.623, 172.626. Limitation of 5 percent polysorbate 80 in carrageenan and 500 ppm in final product. 21CFR 182.7255. GRAS when used as a stabilizer in accordance with good manufacturing practice. USDA - 9CFR 318.7. Limitation of 1.5 percent in restructured meat food products. 9CFR 381.147. Sufficient for purpose.

IARC Cancer Review: Animal Limited Evidence IMEMDT 10,181,76.

SAFETY PROFILE: Poison by intravenous route. An experimental neoplastigen and tumorigen. A suspected carcinogen. When heated to decomposition it emits acrid smoke and fumes.

TOXICITY DATA and CODEN

orl-rat TDLo:2100 g/kg/40W-C:ETA CNREA8 38,4427,78
scu-rat TDLo:525 mg/kg/21W-I:NEO 13BYAH -,83,62
ivn-rbt LDLo:5 mg/kg JPPMAB 17,647,65

CCL750 CAS: 8015-88-1
CARROT SEED OIL

PROP: Distilled from the seeds of *Daucus carota* L. (Fam. *Umbelliferae*). (FCTXAV 14, 659,76). Light yellow to amber liquid; aromatic odor. Sol in fixed oils, mineral oil; insol in glycerin, propylene glycol.

USE IN FOOD:

Purpose: Flavoring agent.

Where Used: Various.

Regulations: FDA - 21CFR 182.10, 182.20. GRAS when used at a level not in excess of the amount reasonably required to accomplish the intended effect.

SAFETY PROFILE: A skin irritant. When heated to decomposition it emits acrid smoke and irritating fumes.

TOXICITY DATA and CODEN

skn-rbt 500 mg/24H MLD FCTXAV 14,659,76
skn-gpg 500 mg/24H MLD FCTXAV 14,659,76

CCM000 CAS: 499-75-2
CARVACROL

mf: $C_{10}H_{14}O$ mw: 150.24

PROP: Colorless to pale yellow liquid; spicy thymol odor. D: 0.974-0.980, refr index: 1.521-1.526, flash p: 212.°F. Sol in alc, ether; insol in water.

SYNS: 2-p-CYMENOL ◇ FEMA No. 2245 ◇ 2-HYDROXY-p-CYMENE ◇ ISOPROPYL-o-CRESOL ◇ 5-ISOPROPYL-2-METHYLPHENOL ◇ ISOTHYMOL ◇ 2-METHYL-5-ISOPROPYLPHENOL ◇ o-THYMOL

USE IN FOOD:

Purpose: Flavoring agent.

Where Used: Various.

Regulations: FDA - 21CFR 172.515. Use at a level not in excess of the amount reasonably required to accomplish the intended effect.

SAFETY PROFILE: Poison by ingestion and subcutaneous route. Moderately toxic by skin contact. A severe skin irritant. Combustible liquid. When heated to decomposition it emits acrid smoke and irritating fumes.

TOXICITY DATA and CODEN

skn-rbt 500 mg/24H SEV FCTXAV 17(suppl)695,79
orl-rat LD50:810 mg/kg FCTXAV 2,327,64

skn-rbt LDLo: 2700 mg/kg JAPMA8 38,366,49
scu-rbt LDLo: 1000 mg/kg HBTXAC 5,46,59

CCN000 CAS: 87-44-5
CARYOPHYLLENE

mf: $C_{15}H_{26}$ mw: 206.41

PROP: Found in oil of clove, cinnamon leaves and copaiba balsam and in minor quantities in various other essential oils, especially lavender; prepared by isolation from clove leaf oil, clove stem oil, cinnamon leaf oil or pine oil fractions (FCTXAV 11,1011,73). Colorless to sltly yellow oily liquid; clove odor. D: 0.897-0.910, refr index: 1.498-1.504, flash p: 206°F. Sol in alc, ether; insol in water.

SYNS: β-CARYOPHYLLENE (FCC) ◇ FEMA No. 2252 ◇ 8-METHYLENE-4,11,11-(TRIMETHYL)BICYCLO(7.2.0) UNDEC-4-ENE

USE IN FOOD:

Purpose: Flavoring agent.

Where Used: Various.

Regulations: FDA - 21CFR 172.515. Use at a level not in excess of the amount reasonably required to accomplish the intended effect.

SAFETY PROFILE: A skin irritant. Combustible liquid. When heated to decomposition it emits acrid smoke and irritating fumes.

TOXICITY DATA and CODEN

skn-rbt 500 mg/24H FCTXAV 11,1059,73

CCO500
CASCARILLA OIL

PROP: From steam distillation of bark of Croton *cascarilla* Benn. or *Croton eluteria Benn.* (Fam. *Euphorbiaceae*). Light yellow to brown liquid; spicy odor. Sol in fixed oils; insol in glycerin, propylene glycol.

SYN: SWEETWOOD BARK OIL

USE IN FOOD:

Purpose: Flavoring agent.

Where Used: Various.

Regulations: FDA - 21CFR 182.20. GRAS when used at a level not in excess of the amount reasonably required to accomplish the intended effect.

SAFETY PROFILE: When heated to decomposition it emits acrid smoke and irritating fumes.

CCO750 CAS: 8007-80-5
CASSIA OIL

PROP: Chief constituent is cinnamic aldehyde, found in the leaves and twigs of *Cinnamomum cassia blume* (FCTXAV 13,91,75). Yellow liquid; cinnamon odor, spicy burning taste. Sol in fixed oils, propylene glycol; insol in glycerin, mineral oil.

SYNS: ARTIFICIAL CINNAMON OIL ◇ CINNAMON BARK OIL ◇ CINNAMON BARK OIL, CEYLON TYPE (FCC) ◇ CINNAMON OIL ◇ KASSIA OEL (GERMAN) ◇ OIL of CASSIA ◇ OIL of CHINESE CINNAMON ◇ OIL of CINNAMON ◇ OIL of CINNAMON, CEYLON ◇ OILS, CINNAMON

USE IN FOOD:

Purpose: Flavoring agent.

Where Used: Various.

Regulations: FDA - 21CFR 182.10, 182.20. GRAS when used at a level not in excess of the amount reasonably required to accomplish the intended effect.

SAFETY PROFILE: Poison by skin contact. Moderately toxic by ingestion and intraperitoneal routes. A human skin irritant. Mutagenic data. When heated to decomposition it emits acrid smoke and irritating fumes.

TOXICITY DATA and CODEN

skn-hmn 100% FCTXAV 13,109,75
skn-rbt 500 mg/24H SEV FCTXAV 13,91,75
dnr-bcs 600 μg/disc TOFOD5 8,91,85
orl-rat LD50: 2800 mg/kg FCTXAV 13,91,75

CCP250 CAS: 8001-79-4
CASTOR OIL

PROP: From seeds of *Ricinus communis* L. (Fam. *Euphorbiaceae*). A colorless to pale yellow, viscous liquid; bland taste, characteristic odor. Mp: −12°, bp: 313°, flash p: 445°F (CC), d: 0.96, autoign temp: 840°F. Sol in alc; misc abs alc, glacial acetic acid, chloroform, ether.

SYNS: AROMATIC CASTOR OIL ◇ CASTOR OIL AROMATIC ◇ COSMETOL ◇ CRYSTAL O ◇ GOLD BOND ◇ NCI-C55163 ◇ NEOLOID ◇ OIL of PALMA CHRISTI ◇ PHORBYOL ◇ RICINUS OIL ◇ RICIRUS OIL ◇ TANGANTANGAN OIL

USE IN FOOD:

Purpose: Antisticking agent, component of protective coatings, drying oil, release agent.

Where Used: Candy (hard), vitamin and mineral tablets.

Regulations: FDA - 21CFR 73.1. Limitation of 500 ppm in finished food. 21CFR 172.876. Limitation of 500 ppm in hard candy. 21CFR 181.26. Use at a level not in excess of the amount reasonably required to accomplish the intended effect.

SAFETY PROFILE: Moderately toxic by ingestion. An allergen. An additive permitted in food for human consumption. An eye irritant. A purgative. Combustible when exposed to heat. Spontaneous heating may occur. To fight fire, use CO_2, dry chemical, fog, mist.

TOXICITY DATA and CODEN

eye-rbt 500 mg AJOPAA 29,1363,46

CCP525 CAS: 9001-05-2
CATALASE from MICROCOCCUS LYSODEIKTICUS

PROP: Derived from *Micrococcus lysodeikticus*.

SYNS: CAPERASE ◇ EQUILASE ◇ OPTIDASE

USE IN FOOD:

Purpose: Production aid.

Where Used: Cheese.

Regulations: FDA - 21CFR 173.135. Use at a level not in excess of the amount reasonably required to accomplish the intended effect.

SAFETY PROFILE: When heated to decomposition it emits acrid smoke and irritating fumes.

CCQ500 CAS: 8007-20-3
CEDAR LEAF OIL

PROP: Constituent is d-α-thujone, found in leaves of *Thuja occidentalis* L. (Fam. *Cupressaaceae*) (FCTXAV 12,807,74). Yellowish, volatile oil; strong sage odor. D: 0.910-0.920. Sol in fixed oils, mineral oil, propylene glycol; insol in glycerin.

SYNS: OIL of ARBOR VITAE ◇ OIL of CEDAR LEAF ◇ OIL of THUJA ◇ OIL of WHITE CEDAR ◇ OILS, CEDAR LEAF ◇ OIL THUJA ◇ THUJA OIL ◇ WHITE CEDAR OIL

USE IN FOOD:

Purpose: Flavoring agent.

Where Used: Various.

Regulations: FDA - 21CFR 172.310. Finished food must be thujone free. Use at a level not in excess of the amount reasonably required to accomplish the intended effect.

SAFETY PROFILE: Moderately toxic by ingestion. A skin irritant. Ingestion of large quantities causes hypertension, bradycardia, tachypnea, convulsions, death. When heated to decomposition it emits acrid smoke and fumes.

TOXICITY DATA and CODEN

skn-rbt 500 mg/24H MOD FCTXAV 12,807,74
orl-rat LD50:830 mg/kg FCTXAV 12,807,74

CCS575
CEFTIOFUR

USE IN FOOD:

Purpose: Animal drug.

Where Used: Beef.

Regulations: FDA - 21CFR 556.113. Limitation of 3 ppm in muscle, 9 ppm in kidney, 6 ppm in liver, and 12 ppm in fat of cattle.

SAFETY PROFILE: When heated to decomposition it emits acrid smoke and irritating fumes.

CCS660
CELERY SEED OIL

PROP: From steam distillation of fruit and seed of *Apium graveolens* L. Yellow to green-brown liquid; aromatic odor. D: 0.870-0.910. Sol in fixed oils, mineral oil; sltly sol in propylene glycol; insol in glycerin.

USE IN FOOD:

Purpose: Flavoring agent.

Where Used: Meats, salads, sauces, soups.

Regulations: FDA - 21CFR 182.10, 182.20. GRAS when used at a level not in excess of the amount reasonably required to accomplish the intended effect.

SAFETY PROFILE: When heated to decomposition it emits acrid smoke and irritating fumes.

CCU100
CELLULOSE, MICROCRYSTALLINE

PROP: Fine white crystalline powder from treatment of α-cellulose with mineral acids. Insol in water, most organic solvents.

SYN: CELLULOSE GEL

USE IN FOOD:

Purpose: Anticaking agent, binding agent, disintegrating agent, dispersing agent, tableting agent.

Where Used: Frozen desserts, shredded cheese, whipped toppings.

Regulations: GRAS when used in accordance with good manufacturing practice.

SAFETY PROFILE: A nuisance dust. When heated to decomposition it emits acrid smoke and irritating fumes.

CCU150 CAS: 9004-34-6
CELLULOSE, POWDERED

PROP: Fine white fibrous particles from treatment of bleached cellulose from wood or cotton. Insol in water, most organic solvents.

USE IN FOOD:

Purpose: Anticaking agent, binding agent, bulking agent, disintegrating agent, dispersing agent, filter aid, texturizing agent, thickening agent.

Where Used: Various.

Regulations: GRAS when used in accordance with good manufacturing practice.

SAFETY PROFILE: A nuisance dust. When heated to decomposition it emits acrid smoke and irritating fumes.

CCU250 CAS: 9004-70-0
CELLULOSE TETRANITRATE

mf: $C_{12}H_{16}(ONO_2)_4O_6$ mw: 504.3

PROP: White, amorphous solid. D: 1.66, flash p: 55°F.

SYNS: CELLOIDIN ◇ CELLULOSE NITRATE ◇ COLLODION COTTON ◇ COLLOXYLIN ◇ GUNCOTTON ◇ NITROCELLULOSE ◇ NITROCOTTON ◇ PYRALIN ◇ PYROXYLIN ◇ PYROXYLIN PLASTICS (DOT) ◇ PYROXYLIN PLASTIC SCRAP (DOT) ◇ SOLUBLE GUN COTTON ◇ XYLOIDIN

USE IN FOOD:

Purpose: Manufacture of paper and paperboard.

Where Used: Packaging materials.

Regulations: FDA - 21CFR 181.30. Use at a level not in excess of the amount reasonably required to accomplish the intended effect.

DOT Classification: Flammable Solid; Label: Flammable Solid

SAFETY PROFILE: Flammable solid. Highly dangerous fire hazard in the dry state when exposed to heat, flame, or powerful oxidizers. When wet with 35% of denatured ethanol it is about as hazardous as ethanol alone or gasoline. Dry cellulose tetranitrate burns rapidly with intense heat and ignites easily. Moderately dangerous explosion hazard. To fight fire, use copious volumes of water; alcohol foam. CO_2 is effective in extinguishing fires of nitrocellulose solvents.

CCX500 CAS: 21593-23-7
CEPHAPIRIN

mf: $C_{17}H_{17}N_3O_6S_2$ mw: 423.49

SYNS: CEFAPIRIN (GERMAN) ◇ 3-(HYDROXYMETHYL)-8-OXO-7-(2-(4-PYRIDYLTHIO)ACETAMIDO)-5-THIA-1-AZABICYCLO(4.2.0)OCT-2-ENE-2-CARBOXYLIC ACID, ACETATE (ESTER)

USE IN FOOD:

Purpose: Animal drug.

Where Used: Beef, milk.

Regulations: FDA - 21CFR 556.115. Limitation of 0.02 ppm in milk and 0.1 ppm in uncooked edible tissues of dairy cattle.

SAFETY PROFILE: Moderately toxic by intravenous route. When heated to decomposition it emits very toxic fumes of NO_x and SO_x.

TOXICITY DATA and CODEN

ivn-rat LD50:4580 mg/kg ARZNAD 29,424,79

CDH500 CAS: 8002-66-2
CHAMOMILE OIL

PROP: By steam distillation of the flowers and stalks of *Matrilaria chamomilla* L. (FCTXAV 12,807,74). Blue-yellowish-brown liquid; strong odor and bitter aromatic taste. Composed of amyl and butyl esters of angelic, tiglic acids, and butyric acid. D: 0.905-0.915 @ 15°/15°. Sol in fixed oils, propylene glycol; insol in mineral oil, glycerin.

SYNS: CAMOMILE OIL GERMAN ◇ CHAMOMILE-GERMAN OIL ◇ GERMAN CHAMOMILE OIL ◇ HUNGARIAN CHAMOMILE OIL

USE IN FOOD:

Purpose: Flavoring agent.

Where Used: Various.

Regulations: FDA - 21CFR 182.10. GRAS when used at a level not in excess of the amount reasonably required to accomplish the intended effect.

SAFETY PROFILE: A mild allergen. A skin irritant. When heated to decomposition it emits acrid and irritating fumes.

TOXICITY DATA and CODEN

skn-rbt 500 mg/24H MOD FCTXAV 12,851,74

CDH750 CAS: 8015-92-7
CHAMOMILE OIL (ROMAN)

PROP: Obtained by the steam distillation of the dried flowers of *Anthemis nobilis* L. (FCTXAV 12,807,74). Blue liquid, turning brownish-yellow; strong aromatic odor. Composition: Amyl and butyl esters of angelic and tiglic acids, butyric acid, etc. D: 0.905-0.915 @ 15°/15°. Sol in fixed oils, mineral oil, propylene glycol; insol in glycerin.

SYN: CAMOMILE OIL, ENGLISH TYPE (FCC)

USE IN FOOD:

Purpose: Flavoring agent.

Where Used: Various.

Regulations: FDA - 21CFR 182.10, 182.20. GRAS when used at a level not in excess of the amount reasonably required to accomplish the intended effect.

SAFETY PROFILE: A mild allergen. A skin irritant. Combustible when heated. When heated to decomposition it emits acrid smoke and irritating fumes.

TOXICITY DATA and CODEN

skn-rbt 500 mg/24H MOD FCTXAV 12,853,74

CDP250 CAS: 56-75-7
CHLORAMPHENICOL

mf: $C_{11}H_{12}Cl_2N_2O_5$ mw: 323.15

PROP: Crystalline. Mp: 151°. Sltly sol in water.

SYNS: ALFICETYN ◇ AMBOFEN ◇ AMPHENICOL ◇ AMPHICOL ◇ AMSECLOR ◇ ANACETIN ◇ AQUAMYCETIN ◇ AUSTRACIL ◇ AUSTRACOL ◇ BIOCETIN ◇ BIOPHENICOL ◇ CATILAN ◇ CHEMICETIN ◇ CHEMICETINA ◇ CHLOMIN ◇ CHLOMYCOL ◇ CHLORAMEX ◇ CHLORAMFICIN ◇ CHLORAMFILIN ◇ d-CHLORAMPHENICOL ◇ d-threo-CHLORAMPHENICOL ◇ CHLORAMSAAR ◇ CHLORASOL ◇ CHLORA-TABS ◇ CHLORICOL ◇ CHLORNITROMYCIN ◇ CHLOROCAPS ◇ CHLOROCID ◇ CHLOROCOL ◇ CHLOROJECT L ◇ CHLOROMAX ◇ CHLOROMYCETIN ◇ CHLORONITRIN ◇ CHLOROPTIC ◇ CHLOROVULES ◇ CIDOCETINE ◇ CIPLAMYCETIN ◇ CLORAMIDINA ◇ CLOROAMFENICOLO (ITALIAN) ◇ CLOROMISAN ◇ CLOROSINTEX ◇ COMYCETIN ◇ CYLPHENICOL ◇ DESPHEN ◇ DETREOMYCINE ◇ DEXTROMYCETIN ◇ d-(-)-threo-2-DICHLOROACETAMIDO-1-p-NITROPHENYL-1,3-PROPANEDIOL ◇ d-(-)-threo-2,2-DICHLORO-N-(β-HYDROXY-α-(HYDROXYMETHYL))-p-NITROPHENETHYLACETAMIDE ◇ DOCTAMICINA ◇ ECONOCHLOR ◇ EMBACETIN ◇ EMETREN ◇ ENICOL ◇ ENTEROMYCETIN ◇ ERBAPLAST ◇ ERTILEN ◇ FARMICETINA ◇ FENICOL ◇ GLOBENICOL ◇ HALOMYCETIN ◇ HORTFENICOL ◇ INTRAMYCETIN ◇ ISMICETINA ◇ ISOPHENICOL ◇ KAMAVER ◇ KEMICETINE ◇ LEUKOMYAN ◇ LEVOMYCETIN ◇ LOROMISIN ◇ MASTIPHEN ◇ MEDIAMYCETINE ◇ MICOCHLORINE ◇ MICROCETINA ◇ MYCHEL ◇ MYCINOL ◇ d-threo-1-(p-NITROPHENYL)-2-(DICHLOROACETYLAMINO)-1,3-PROPANEDIOL ◇ NORIMYCIN V ◇ NOVOCHLOROCAP ◇ NOVOMYCETIN ◇ NOVOPHENICOL ◇ OFTALENT ◇ OLEOMYCETIN ◇ OPTHOCHLOR ◇ OTOPHEN ◇ PANTOVERNIL ◇ PARAXIN ◇ PETNAMYCETIN ◇ QUEMICETINA ◇ RIVOMYCIN ◇ ROMPHENIL ◇ SEPTICOL ◇ SINTOMICETINA ◇ STANOMYCETIN ◇ SYNTHOMYCINE ◇ TEVCOCIN ◇ TIFOMYCINE ◇ TREOMICETINA ◇ UNIMYCETIN ◇ VETICOL

USE IN FOOD:

Purpose: Animal drug.

Where Used: Beef, pork, lamb.

Regulations: USDA CES Ranking: A-2 (1985).

IARC Cancer Review: Human Limited Evidence IMEMDT 10,85,76. EPA Genetic Toxicology Program.

SAFETY PROFILE: Poison by intraperitoneal, intravenous, and subcutaneous routes. Moderately toxic by ingestion. A human carcinogen which causes leukemia, aplastic anemia, and other bone marrow changes by ingestion. Human systemic effects by an unknown route: changes in plasma or blood volume, unspecified liver effects, and hemorrhaging. An experimental tumorigen and teratogen. Other experimental reproductive effects. Human mutagenic data. When heated to decomposition it emits very toxic fumes of NO_x and Cl^-.

TOXICITY DATA and CODEN

oms-bcs 10 mg/L MGGEAE 189,73,83
dni-hmn:bmr 1500 μmol/L 46GFA5 -,17,81
orl-rat TDLo:23 g/kg (1-21D preg):TER TJADAB 12,291,75
ipr-rat TDLo:250 mg/kg (3D preg):REP EXPEAM 35,1649,70
orl-mus TDLo:5500 mg/kg (5-15D preg):TER TXAPA9 19,667,71
orl-wmn TDLo:300 mg/kg/60W-I:CAR,BLD NEJMAG 277,1003,67
ipr-mus TDLo:2500 mg/kg/5W-I:ETA,BLD CNREA8 41,3478,81

orl-wmn TD :1680 mg/kg/6W-I:CAR,BLD NEJMAG 277,1003,67
orl-man TD :434 mg/kg/W-C:CAR,BLD ACHAAH 66,267,81
orl-inf TDLo:440 mg/kg:CNS,GIT,MET JAMAAP 234,149,75
orl-wmn LDLo:400 mg/kg JAMAAP 234,149,75
ims-inf TDLo:250 mg/kg/2D:CVS NEJMAG 262,787,60
orl-rat TDLo:2500 mg/kg FRPSAX 10,3,55

CDP700 CAS: 530-43-8
CHLORAMPHENICOL PALMITATE

mf: $C_{27}H_{42}Cl_2N_2O_6$ mw: 561.61

SYNS: CAP-P ◇ CAP-PALMITATE ◇ CHLORAMPHENICOL MONOPALMITATE ◇ DETREOPAL ◇ α-ESTER PALMITIC ACID with D-threo-(-)-2,2-DICHLORO-N-(β-HYDROXY-α-(HYDROXYMETHYL)-p-NITROPHENETHYL)ACETAMIDE

USE IN FOOD:

Purpose: Animal drug.

Where Used: Beef, pork, lamb.

Regulations: USDA CES Ranking: A-2 (1985).

SAFETY PROFILE: Moderately toxic by oral route. An experimental teratogen. An antibiotic. When heated to decomposition it emits very toxic fumes of NO_x and Cl^-.

TOXICITY DATA and CODEN

orl-rbt TDLo:1200 mg/kg (7-12D preg):TER VHAGAS 71,623,77
orl-mus LD50:2640 mg/kg NIIRDN 6,248,82

CDR750 CAS: 57-74-9
CHLORDANE

DOT: 2762

mf: $C_{10}H_6Cl_8$ mw: 409.76

PROP: Colorless to amber; odorless, viscous liquid. Bp: 175°, d: 1.57-1.63 @ 15.5°/15.5°.

SYNS: ASPON-CHLORDANE ◇ BELT ◇ CD 68 ◇ CHLOORDAAN (DUTCH) ◇ CHLORDAN ◇ γ-CHLORDAN ◇ CHLORDANE, LIQUID (DOT) ◇ CHLORINDAN ◇ CHLOR KIL ◇ CHLORODANE ◇ CHLORTOX ◇ CLORDAN (ITALIAN) ◇ CORODANE ◇ CORTILAN-NEU ◇ DICHLOROCHLORDENE ◇ DOWCHLOR ◇ ENT 9,932 ◇ ENT 25,552-X ◇ HCS 3260 ◇ KYPCHLOR ◇ M 140 ◇ M 410 ◇ NCI-C00099 ◇ NIRAN ◇ 1,2,4,5,6,7,8,8-OCTACHLOOR-3a,4,7,7a-TETRAHYDRO-4,7-endo-METHANO-INDAAN (DUTCH) ◇ OCTACHLOR ◇ OCTACHLORODIHYDRODICYCLOPENTADIENE ◇ 1,2,4,5,6,7,8,8-OCTACHLORO-2,3,3a,4,7,7a-HEXAHYDRO-4,7-METHANOINDENE ◇ 1,2,4,5,6,7,8,8-OCTACHLORO-2,3,3a,4,7,7a-HEXAHYDRO-4,7-METHANO-1H-INDENE ◇ 1,2,4,5,6,7,8,8-OCTACHLORO-3a,4,7,7a-HEXAHYDRO-4,7-METHYLENE INDANE ◇ OCTACHLORO-4,7-METHANOHYDROINDANE ◇ OCTACHLORO-4,7-METHANOTETRAHYDROINDANE ◇ 1,2,4,5,6,7,8,8-OCTACHLORO-4,7-METHANO-3a,4,7,7a-TETRAHYDROINDANE ◇ 1,2,4,5,6,7,8,8-OCTACHLORO-3a,4,7,7a-TETRAHYDRO-4,7-METHANOINDAN ◇ 1,2,4,5,6,7,8,8-OCTACHLORO-3a,4,7,7a-TETRAHYDRO-4,7-METHANOINDANE ◇ 1,2,4,5,6,7,10,10-OCTACHLORO-4,7,8,9-TETRAHYDRO-4,7-METHYLENEINDANE ◇ 1,2,4,5,6,7,8,8-OCTACHLOR-3a,4,7,7a-TETRAHYDRO-4,7-endo-METHANO-INDAN (GERMAN) ◇ OCTA-KLOR ◇ OKTATERR ◇ ORTHO-KLOR ◇ 1,2,4,5,6,7,8,8-OTTOCHLORO-3A,4,7,7A-TETRAIDRO-4,7-endo-METANO-INDANO (ITALIAN) ◇ RCRA WASTE NUMBER U036 ◇ SD 5532 ◇ SHELL SD-5532 ◇ SYNKLOR ◇ TAT CHLOR 4 ◇ TOPICHLOR 20 ◇ TOPICLOR ◇ TOPICLOR 20 ◇ TOXICHLOR ◇ VELSICOL 1068

USE IN FOOD:

Purpose: Pesticide.

Where Used: Various.

Regulations: USDA CES Ranking: A-2 (1987).

IARC Cancer Review: Human Inadequate Evidence IMEMDT 20,45,79; Animal Sufficient Evidence IMEMDT 20,45,79. NCI Carcinogenesis Bioassay (feed); Clear Evidence: mouse NCITR* NCI-CG-TR-8,77; No Evidence: rat NCITR* NCI-CG-TR-8,77. EPA Genetic Toxicology Program. Community Right-To-Know List. EPA Extremely Hazardous Substances List.

OSHA PEL: TWA 0.5 mg/m^3 (skin) ACGIH TLV: TWA 0.5 mg/m^3 (skin) DFG MAK: 0.5 mg/m^3 DOT Classification: Combustible Liquid; Label: None; Flammable Liquid; Label: Flammable Liquid

SAFETY PROFILE: Poison to humans by ingestion and possibly other routes. An experimental poison by ingestion, inhalation, intravenous, and intraperitoneal routes. Moderately toxic by skin contact. Human systemic effects by ingestion or skin contact: tremors, convulsions, excitement, ataxia (loss of muscle coordination), and gastritis. A suspected human carcinogen. An experimental carcinogen and teratogen. Other experimental reproductive effects. Human mutagenic data. Flammable or combustible liquid. It is no longer permitted for use as a termiticide in homes. When heated to decomposition it emits toxic fumes of Cl^-.

TOXICITY DATA and CODEN

sce-ofs-mul 54 pmol/L MUREAV 118,61,83
dns-hmn:fbr 1 μmol/L MUREAV 42,161,77
orl-mus TDLo:3360 μg/kg (1-21D preg):TER JEPTDQ 2(2),357,78
orl-mus TDLo:152 mg/kg (1-19D preg):TER TXAPA9 62,402,82
orl-mus TDLo:7 mg/kg (15-21D preg):REP BJPCBM 49,311,73
orl-mus TDLo:2020 mg/kg/80W-C:CAR NCITR* NCI-CG-TR-8,77
orl-mus TD :3780 mg/kg/80W-C:CAR NCITR* NIC-CG-TR-8,77
orl-hmn LDLo:29 mg/kg:LIV CMEP** -,1,56
orl-wmn LDLo:120 μg/kg:CNS,GIT CMEP** -,1,56
skn-hmn LDLo:428 mg/kg:CNS 34ZIAG -,648,69
orl-rat LD50:283 mg/kg DOEAAH 34,25,79

CDS125 CAS: 16672-87-0
CHLORETHEPHON

mf: $C_2H_6ClO_3P$ mw: 144.50

PROP: Very hygroscopic needles from benzene. Mp: 74-75°. Freely sol in water, methanol, acetone, ethylene glycol, propylene glycol; sltly sol in benzene, toluene; practically insol in petr ether.

SYNS: AMCHEM 68-250 ◇ BROMOFLOR ◇ CAMPOSAN ◇ CEP ◇ 2-CEPA ◇ CEPHA ◇ CEPHA 10LS ◇ 2-CHLORAETHYL-PHOSPHONSAEURE (GERMAN) ◇ 2-CHLORETHYL-PHOSPHONIC ACID ◇ 2-CHLOROETHANEPHOSPHONIC ACID ◇ ETHEFON ◇ ETHEL ◇ ETHEPHON ◇ ETHEVERSE ◇ ETHREL ◇ FLORDIMEX ◇ FLOREL ◇ G 996 ◇ KAMPOSAN ◇ ROLL-FRUCT ◇ TOMATHREL

USE IN FOOD:

Purpose: Plant growth regulator.

Where Used: Animal feed, barley (milling fractions, except flour), raisin waste, sugarcane molasses, wheat milling fractions (except flour).

Regulations: FDA - 21CFR 193.186. Plant growth regulator residue tolerance of 5.0 ppm in barley and wheat milling fractions (except flour), 1.5 ppm in sugarcane, molasses. 21CFR 561.225. Limitation of 65.0 ppm in raisin waste when used for animal feed.

EPA Genetic Toxicology Program.

SAFETY PROFILE: Moderately toxic by ingestion. Mildly toxic by skin contact. Caution: Spray formulations are quite acidic, about pH 1.0. May be irritating to exposed skin and eyes, or if inhaled. When heated to decomposition it emits toxic fumes of Cl^- and PO_x.

TOXICITY DATA and CODEN

orl-rat LD50:3400 mg/kg ZKMAAX 20,274,80
skn-rbt LD50:5730 mg/kg 85DPAN -,-,71/76

CDV750 CAS: 7782-50-5
CHLORINE

DOT: 1017

mf: Cl_2 mw: 70.90

PROP: Greenish-yellow gas, liquid, or rhombic crystals. Mp: −101°, bp: −34.5°, d: (liq) 1.47 @ 0° (3.65 atm), vap press: 4800 mm @ 20°, vap d: 2.49. Sol in water.

SYNS: BERTHOLITE ◇ CHLOOR (DUTCH) ◇ CHLOR (GERMAN) ◇ CHLORE (FRENCH) ◇ CHLORINE MOL. ◇ CLORO (ITALIAN) ◇ MOLECULAR CHLORINE

USE IN FOOD:

Purpose: Antimicrobial agent, bleaching agent, oxidizing agent.

Where Used: Various.

Regulations: GRAS when used in accordance with good manufacturing practice.

Community Right-To-Know List. EPA Extremely Hazardous Substances List.

OSHA PEL: TWA CL 1 ppm; STEL 1 ppm ACGIH TLV: TWA 0.5 ppm; STEL 1 ppm DFG MAK: 0.5 ppm (1.5 mg/m^3) DOT Classification: Nonflammable Gas; Label: Nonflammable Gas and Poison; Poison A; Label: Poison Gas NIOSH REL: (Chlorine) CL 0.5 ppm/15M

SAFETY PROFILE: Moderately toxic to humans by inhalation. Very irritating by inhalation. Human mutagenic data. Human respiratory system effects by inhalation: changes in the trachea or bronchi, emphysema, chronic pulmonary edema or congestion. A strong irritant to eyes and mucous membranes. Chlorine is extremely irritating to the mucous membranes of the eyes @ 3 ppm and the respiratory tract. Combines with moisture to liberate O_2 and forms HCl. Both these substances, if present in quantity, cause inflammation of the tissues with which they come in contact. A concentration of 3.5 ppm produces a detectable odor; 15 ppm causes immediate irritation of the throat. Concentrations of 50 ppm are dangerous for even short exposures; 1000 ppm may be fatal, even when exposure is brief. Because of its intensely irritating properties, severe industrial exposure seldom occurs, as the worker is forced to leave the exposure area before he can be seriously affected. In cases where this is impossible, the initial irritation of the eyes and mucous membranes of the nose and throat is followed by coughing, a feeling of suffocation, and later, pain and a feeling of constriction in the chest. If exposure has been severe, pulmonary edema may follow with rales being heard over the chest. It is a common air contaminant.

Explodes on contact with many chemicals under a variety of conditions.

TOXICITY DATA and CODEN

cyt-hmn:lym 20 ppm CBINA8 6,375,73
spm-mus-orl 20 mg/kg/5D-C ENMUDM 7,201,85
ihl-hmn LCLo:2530 mg/m^3/30M:PUL 28ZOAH -,150,37
ihl-rat LC50:293 ppm/1H NTIS** PB214-270

CEA000 CAS: 79-11-8
CHLOROACETIC ACID
mf: $C_2H_3ClO_2$ mw: 94.50
DOT: 1750/1751

PROP: Colorless crystals. Mp: (α) 63°, (β) 56°, (τ) 50°, bp: 189°, flash p: 259°F, d: 1.58 @ 20°/20°, vap d: 3.26.

SYNS: ACIDE CHLORACETIQUE (FRENCH) ◇ ACIDE MONOCHLORACETIQUE (FRENCH) ◇ ACIDOMONOCLOROACETICO (ITALIAN) ◇ CHLORACETIC ACID ◇ α-CHLOROACETIC ACID ◇ CHLOROACETIC ACID, LIQUID (DOT) ◇ CHLOROACETIC ACID, SOLID (DOT) ◇ CHLOROETHANOIC ACID ◇ MCA ◇ MONOCHLOORAZIJNZUUR (DUTCH) ◇ MONOCHLORACETIC ACID ◇ MONOCHLORESSIGSAEURE (GERMAN) ◇ MONOCHLOROACETIC ACID ◇ MONOCHLOROETHANOIC ACID ◇ NCI-C60231

USE IN FOOD:

Purpose: Packaging adhesives, preservative.

Where Used: Packaging materials, prohibited from alcoholic beverages, prohibited from foods.

Regulations: FDA - 21CFR 189.155. Prohibited from direct addition or use in human food. 21CFR 175.105. Limitation of 10 ppb migration level from package adhesive.

EPA Genetic Toxicology Program. EPA Extremely Hazardous Substances List. Community Right-To-Know List.

DOT Classification: Corrosive Material; Label: Corrosive, liquid, solution or solid

SAFETY PROFILE: Poison by ingestion, inhalation, subcutaneous, and intravenous route. An experimental tumorigen. A corrosive skin, eye, and mucous membrane irritant. Mutagenic data. Combustible liquid when exposed to heat or flame. To fight fire, use water spray, fog, mist, dry chemical, foam. When heated to decomposition it emits toxic fumes of Cl^-.

TOXICITY DATA and CODEN

mma-mus:lym 548 mg/L MUREAV 97,49,82
scu-mus TDLo:100 mg/kg:ETA NTIS** PB223-159
scu-mus TD :1300 mg/kg/65W-I:ETA JNCIAM 53,695,74
ihl-rat LC50:180 mg/m^3 GTPZAB 18(9),32,74
scu-rat LD50:5 mg/kg TXAPA9 22,303,72
orl-mus LD50:165 mg/kg JPETAB 86,336,46

CEU500 CAS: 928-51-8
4-CHLORO-1-BUTANOL
mf: C_4H_9ClO mw: 108.58

SYNS: 4-CHLORBUTAN-1-OL (GERMAN) ◇ 4-CHLORO-1-BUTANE-OL ◇ 4-CHLOROBUTANOL ◇ TETRAMETHYLENE CHLOROHYDRIN

USE IN FOOD:

Purpose: Animal drug.

Where Used: Milk.

Regulations: FDA - 21CFR 556.140. Limitation of zero in milk.

SAFETY PROFILE: An experimental neoplastigen. Moderately toxic by ingestion. Mutagenic data. When heated to decomposition it emits toxic fumes of Cl^-.

TOXICITY DATA and CODEN

mmo-sat 20 μmol/plate MUREAV 90,91,81
ipr-mus TDLo:3650 mg/kg/8W-I:NEO CNREA8 39,391,79
orl-mus LD50:990 mg/kg ZHYGAM 26,17,80

CFX000 CAS: 15972-60-8
2-CHLORO-2′,6′-DIETHYL-N-(METHOXYMETHYL)ACETANILIDE
mf: $C_{14}H_{20}ClNO_2$ mw: 269.80

SYNS: ALACHLOR (USDA) ◇ ALANEX ◇ ALOCHLOR ◇ CHLORESSIGSAEURE-N-(METHOXYMETHYL)-2,6-DIAETHYLANILID (GERMAN) ◇ 2-CHLORO-N-(2,6-DIETHYLPHENYL)-N-(METHOXYMETHYL)ACETAMIDE ◇ CP 50144

◇ LASSO ◇ LAZO ◇ METACHLOR ◇ METHACHLOR ◇ PILLARZO

USE IN FOOD:

Purpose: Herbicide.

Where Used: Various.

Regulations: USDA CES Ranking: A-2 (1985).

EPA Genetic Toxicology Program.

SAFETY PROFILE: Moderately toxic by ingestion, skin contact, and possibly other routes. Human mutagenic data. When heated to decomposition it emits very toxic fumes of Cl^- and NO_x.

TOXICITY DATA and CODEN

mmo-omi 90 mg/L JASIAB 104,571,85
mrc-smc 33 μg/plate MUREAV 136,233,84
orl-rat LD50:1200 mg/kg WRPCA2 9,119,70
skn-rbt LD50:3500 mg/kg GUCHAZ 6,3,73

CHX500 CAS: 74-87-3
CHLOROMETHANE

DOT: 1063
mf: CH_3Cl mw: 50.49

PROP: Colorless gas; ethereal odor and sweet taste. D: 0.918 @ 20°/4°, mp: −97°, bp: −23.7°, flash p: <32°F, lel: 8.1%, uel: 17%, autoign temp: 1170°F, vap d: 1.78. Sltly sol in water; miscible with chloroform, ether, glacial acetic acid, sol in alcohol.

SYNS: ARTIC ◇ CHLOOR-METHAAN (DUTCH) ◇ CHLOR-METHAN (GERMAN) ◇ CHLORURE de METHYLE (FRENCH) ◇ CLOROMETANO (ITALIAN) ◇ CLORURO di METILE (ITALIAN) ◇ METHYLCHLORID (GERMAN) ◇ METHYL CHLORIDE (ACGIH, DOT) ◇ METYLU CHLOREK (POLISH) ◇ MONOCHLOROMETHANE ◇ RCRA WASTE NUMBER U045

USE IN FOOD:

Purpose: Propellant.

Where Used: Various.

Regulations: FDA - 21CFR 193.300. Propellant for pesticides.

IARC Cancer Review: Human Inadequate Evidence IMEMDT 41,161,86; Animal Inadequate Evidence IMEMDT 41,161,86. EPA Genetic Toxicology Program.

OSHA PEL: TWA 50 ppm; STEL 100 ppm (skin) ACGIH TLV: TWA 50 ppm; STEL 100 ppm (skin) DFG MAK: 50 ppm (105 mg/m^3) DOT Classification: Flammable Gas; Label: Flammable Gas; IMO: Poison A; Label: Poison Gas and Flammable Gas

SAFETY PROFILE: A poison. Very mildly toxic by inhalation. An experimental teratogen. Other experimental reproductive effects. Human mutagenic data. Human systemic effects by inhalation: convulsions, nausea or vomiting, and unspecified effects on the eye.

Chloromethane has slight irritant properties and may be inhaled without noticeable discomfort. It has some narcotic action, but this effect is weaker than that of chloroform. Acute poisoning, characterized by the narcotic effect, is rare in industry. In exposures to high concentrations, dizziness, drowsiness, incoordination, confusion, nausea and vomiting, abdominal pains, hiccoughs, diplopia, and dimness of vision are followed by delirium, convulsions, and coma. Death may be immediate; however, if the exposure is not fatal, recovery is usually slow. Degenerative changes in the central nervous system are not uncommon. The liver, kidneys, and bone marrow may be affected, with resulting acute nephritis and anemia. Death may occur several days after exposure resulting from degenerative changes in the heart, liver, and especially the kidneys. Repeated exposure to low concentrations causes damage to the central nervous system and, less frequently, to the liver, kidneys, bone marrow and cardiovascular system. Hemorrhages into the lungs, intestinal tract, and dura have been reported. Sprayed on the skin, chloromethane produces anesthesia through freezing of the tissues as it evaporates.

Flammable gas. Very dangerous fire hazard when exposed to heat, flame or powerful oxidizers. Moderate explosion hazard when exposed to flame and sparks. Explodes on contact with interhalogens (e.g., bromine trifluoride; bromine pentafluoride); magnesium and alloys; potassium and alloys; sodium and alloys; zinc. Potentially explosive reaction with aluminum when heated to 152° in a sealed container. Mixtures with aluminum chloride + ethylene react exothermically and then explode when pressurized to above 30 bar. May ignite on contact with aluminum chloride or powdered aluminum. To fight fire, stop flow of gas and use CO_2, dry chemical or water spray. When heated to decomposition it emits highly toxic fumes of Cl^-.

TOXICITY DATA and CODEN

oms-hmn:lym 3 pph MUREAV 155,75,85
sce-hmn:lym 3 pph MUREAV 155,75,85
ihl-rat TCLo:2000 ppm/6H (5D male):REP JACTDZ 4(1),224,85
ihl-rat TCLo:1500 ppm/6H (7-19D preg):TER TJADAB 27,181,83
ihl-mus TCLo:500 ppm/6H (6-17D preg):TER TJADAB 27,197,83
ihl-hmn LCLo:20000 ppm/2H:EYE,CNS,GIT 34ZIAG -,386,69
ihl-rat LC50:152000 mg/m^3/30M FAVUAI 7,35,75

CIF775 CAS: 60177-39-1

CHLOROMETHYLATED AMINATED STRYENE-DIVINYLBENZENE RESIN

USE IN FOOD:

Purpose: Clarifying agent, decolorizing agent.

Where Used: Sugar liquor and juices.

Regulations: FDA - 21CFR 173.70. Limitation of 500 ppm by weight of sugar solids.

SAFETY PROFILE: When heated to decomposition it emits toxic fumes of Cl^- and NO_2.

CIK750 CAS: 321-54-0

3-CHLORO-4-METHYL-7-COUMARINYL DIETHYLPHOSPHATE

mf: $C_{14}H_{16}ClO_6P$ mw: 346.72

SYNS: COROXON ◇ COUMAPHOS-O-ANALOG ◇ COUMAPHOS OXYGEN ANALOG (USDA) ◇ O,O-DI(2-CHLOROETHYL)-7-(3-CHLORO-4-METHYLCOUMARINYL) PHOSPHATE ◇ O,O-DIETHYL-O-(3-CHLORO-4-METHYL-COUMARIN-7-YL)PHOSPHATE ◇ DIETHYL-3-CHLORO-4-METHYL-7-COUMARINYL PHOSPHATE ◇ PHOSPHORIC ACID, DIETHYL ESTER, with 3-CHLORO-7-HYDROXY-4-METHYLCOUMARIN

USE IN FOOD:

Purpose: Insecticide.

Where Used: Beef, eggs, goat, horse, lamb, pork, poultry.

Regulations: USDA CES Ranking: B-2 (1988).

SAFETY PROFILE: Deadly poison by ingestion. When heated to decomposition it emits very toxic fumes of PO_x and Cl^-.

TOXICITY DATA and CODEN

orl-ckn LD50:2200 μg/kg BCPCA6 16,1183,67

CJI500 CAS: 76-15-3

CHLOROPENTAFLUOROETHANE

DOT: 1020

mf: C_2ClF_5 mw: 154.47

PROP: Colorless gas. Bp: −39.3°, mp: −38°, d: 1.5678 @ -42°. Insol in water; sol in alc and ether.

SYNS: CHLOROPENTAFLUORETHANE (ACGIH, DOT) ◇ F-115 ◇ FLUOROCARBON-115 ◇ FREON 115 ◇ GENETRON 115 ◇ HALOCARBON 115 ◇ MONOCHLOROPENTAFLUOROETHANE (DOT)

USE IN FOOD:

Purpose: Aerating agent.

Where Used: Foamed food products, sprayed food products.

Regulations: FDA - 21CFR 173.345. Use at a level not in excess of the amount reasonably required to accomplish the intended effect.

ACGIH TLV: TWA 1000 ppm DOT Classification: Nonflammable Gas; Label: Nonflammable Gas

SAFETY PROFILE: Mildly toxic by inhalation. A nonflammable gas. When heated to decomposition it emits toxic fumes of F^- and Cl^-.

CJJ250 CAS: 6164-98-3

CHLOROPHENAMIDINE

mf: $C_{10}H_{13}ClN_2$ mw: 196.70

SYNS: ACARON ◇ BERMAT ◇ C 8514 ◇ CARZOL ◇ CDM ◇ CHLORDIMEFORM ◇ CHLORFENAMIDINE ◇ N′-(4-CHLORO-2-METHYLPHENYL)-N,N-DIMETHYLMETHANIMIDAMIDE ◇ CHLOROPHENAMADIN ◇ N′-(4-CHLORO-o-TOLYL)-N,N-DIMETHYLFORMAMIDINE ◇ CHLORPHENAMIDINE ◇ N′-(4-CHLOR-o-TOLYL)-N,N-DIMETHYLFORMAMIDIN (GERMAN) ◇ CIBA 8514 ◇ N,N-DIMETHYL-N′-(2-METHYL-4-CHLOROPHENYL)-FORMAMIDINE ◇ N,N-DIMETHYL-N′-(2-METHYL-4-CHLORPHENYL)-FORMADIN (GERMAN) ◇ ENT 27,335 ◇ ENT 27,567 ◇ EP-333 ◇ FUNDAL ◇ FUNDAL 500 ◇ FUNDEX ◇ GALECRON ◇ N′-(2-METHYL-4-CHLOROPHENYL)-N,N-DIMETHYLFORMAMIDINE ◇ N′-(2-METHYL-4-CHLORPHENYL)-FORMAMIDIN-HYDROCHLORID (GERMAN) ◇ NSC 190935 ◇ RS 141 ◇ SCHERING 36268 ◇ SN 36268 ◇ SPANON ◇ SPANONE

USE IN FOOD:

Purpose: Pesticide.

Where Used: Animal feed, apple pomace (dried), cottonseed hulls, prunes (dried).

Regulations: FDA - 21CFR 193.15. Insecticide residue tolerance of 15 ppm on dried prunes. 21CFR 561.80. Limitation of 25 ppm in dried apple pomace, 10 ppm in cottonseed hulls when used for animal feed.

EPA Genetic Toxicology Program.

SAFETY PROFILE: Poison by ingestion, skin contact, and intraperitoneal routes. An experimental carcinogen. Experimental reproductive effects. Human mutagenic data. An eye and skin irritant. When heated to decomposition it emits very toxic fumes of NO_x and Cl^-.

TOXICITY DATA and CODEN

skn-rbt 500 mg open MLD CIGET* 6/2/75
eye-rbt 100 mg MLD CIGET* 6/2/75
mmo-smc 5 ppm RSTUDV 6,161,76
dni-hmn:hla 1 mmol/L BECTA6 11,184,74
orl-rat TDLo:1800 μg/kg (5-22D preg):REP BECTA6 20,760,78
orl-mus TDLo:6552 mg/kg/78W-C:CAR CHYCDW 19,154,85
orl-rat LD50:160 mg/kg KSKZAN 16(2),59,78
skn-rat LD50:640 mg/kg FMCHA2 -,C51,83

CJO250 CAS: 43121-43-3
1-(4-CHLOROPHENOXY)-3,3-DIMETHYL-1-(1,2,4-TRIAZOL-1-YL)-2-BUTAN-2-ONE

mf: $C_{14}H_{16}ClN_3O_2$ mw: 293.78

SYNS: AMIRAL ◇ BAY 6681 F ◇ BAYLETON ◇ BAY-MEB-6447 ◇ 1-((tert-BUTYLCARBONYL-4-CHLOROPHENOXY)METHYL)-1H-1,2,4-TRIAZOLE ◇ 1-(4-CHLOROPHENOXY)-3,3-DIMETHYL-1-(1H-1,2,4-TRIAZOL-1-YL)-2-BUTANONE ◇ MEB 6447 ◇ TRIADIMEFON

USE IN FOOD:

Purpose: Fungicide.

Where Used: Barley (milled fractions, except flour), wheat (milled fractions, except flour).

Regulations: FDA - 21CFR 193.83. Fungicide residue tolerance of 4.0 ppm in barley and wheat, milled fractions (except flour).

SAFETY PROFILE: Poison by ingestion. When heated to decomposition it emits very toxic fumes of Cl^- and NO_x.

TOXICITY DATA and CODEN

orl-rat LD50:400 mg/kg FMCHA2 -,C27,83
orl-mus LD50:363 mg/kg MEIEDD 10,1372,83
orl-rbt LD50:500 mg/kg 85DPAN -,-,71/76

CJV250 CAS: 35367-38-5
1-(4-CHLOROPHENYL)-3-(2,6-DIFLUOROBENZOYL)UREA

mf: $C_{14}H_9ClF_2N_2O_2$ mw: 310.70

SYNS: N-(((4-CHLOROPHENYL)AMINO)CARBONYL)-2,6-DIFLUOROBENZAMIDE ◇ DIFLUBENZURON ◇ DIFLURON ◇ DIMILIN ◇ DU 112307 ◇ ENT 29,054 ◇ OMS 1804 ◇ PDD 6040I ◇ PH 60-40 ◇ PHILIPS-DUPHAR PH 60-40 ◇ TH 6040 ◇ THOMPSON-HAYWARD TH6040

USE IN FOOD:

Purpose: Insecticide.

Where Used: Animal feed, soybean hulls, soybean soapstock.

Regulations: FDA - 21CFR 561.420. Limitation of 0.5 ppm in soybean hulls, 0.1 ppm in soybean soapstock when used for animal feed.

EPA Genetic Toxicology Program.

SAFETY PROFILE: Moderately toxic by skin contact. Mildly toxic by ingestion and possibly other routes. Mutagenic data. When heated to decomposition it emits very toxic fumes of Cl^-, F^- and NO_x.

TOXICITY DATA and CODEN

mmo-sat 1 mg/plate PCBPBS 10,174,79
cyt-mus-unr 500 mg/kg TGANAK 16(1),45,82
orl-mus LD50:4640 mg/kg SPEADM 78-1,-,78
skn-rbt LD50:2000 mg/kg SPEADM 74-1,-,74

CKM000 CAS: 116-29-0
p-CHLOROPHENYL-2,4,5-TRICHLOROPHENYL SULFONE

mf: $C_{12}H_6Cl_4O_2S$ mw: 356.04

PROP: Crystals. Mp: 147°. Nearly water-insol.

SYNS: AKARITOX ◇ AREDION ◇ 4-CHLOROPHENYL-2,4,5-TRICHLOROPHENYL SULFONE ◇ p-CHLOROPHENYL-2,4,5-TRICHLOROPHENYL SULPHONE ◇ DUPHAR ◇ ENT 23,737 ◇ FMC 5488 ◇ MITION ◇ NIA 5488 ◇ POLACARITOX ◇ ROZTOZOL ◇ SULFONE-2,4,4′,5-TETRACHLORODIPHENYL ◇ TEDION ◇ TEDION V-18 ◇ 2,4,4′,5-TETRACHLOOR-DIFENYL-SULFON (DUTCH) ◇ 2,4,4′,5-TETRACHLOR-DIPHENYL-SULFON (GERMAN) ◇ 2,4,4′,5-TETRACHLORODIPHENYL SULFONE ◇ 2,4,5,4′-TETRACHLORODIPHENYLSULPHONE ◇ 2,4,4′,5-TETRACLORO-DIFENIL-SOLFONE (ITALIAN) ◇ TETRADICHLONE ◇ TETRADIFON ◇ TETRADIPHON ◇ TETRAFIDON ◇ 1,2,4-TRICHLORO-5-((4-CHLOROPHENYL)SULFONYL)-BENZENE ◇ V-18

USE IN FOOD:

Purpose: Pesticide.

Where Used: Figs (dried), hops (dried), tea (dried).

Regulations: FDA - 21CFR 193.420. Pesticide residue tolerance of 120 ppm in dried hops, 10 ppm in dried figs, 8 ppm dried tea.

EPA Genetic Toxicology Program.

SAFETY PROFILE: Moderately toxic by ingestion. Mildly toxic by skin contact. An experimental teratogen. When heated to decomposition it emits highly toxic fumes of Cl^- and SO_x.

TOXICITY DATA and CODEN

scu-mus TDLo: 1953 mg/kg (6-14D preg): TER NTIS** PB223,-160
orl-rat LD50: 566 mg/kg WRPCA2 9,119,70

CKN000 CAS: 1406-65-1
CHLOROPHYLL

PROP: Dark green solution.

SYNS: BIOPHYLL ◇ CHLOROPHYL, GREEN ◇ C.I. 1956 ◇ DAROTOL ◇ DEODOPHYLL ◇ E 140 ◇ GREEN CHLOROPHYL ◇ L-GRUEN No. 1 (GERMAN) ◇ No. 1249 ◇ No. 1403 ◇ No. 75810

USE IN FOOD:

Purpose: Coloring agent.

Where Used: Casings, fats (rendered), oleomargarine, shortening.

Regulations: USDA - 9CFR 318.7. Sufficient for purpose.

SAFETY PROFILE: Poison by intravenous and intraperitoneal routes. When heated to decomposition it emits toxic fumes of NO_x.

TOXICITY DATA and CODEN

ipr-mus LD50: 400 mg/kg ARZNAD 4,19,54
ivn-mus LD50: 285 mg/kg ARZNAD 4,19,54

CMA750 CAS: 57-62-5
CHLORTETRACYCLINE

mf: $C_{22}H_{23}ClN_2O_8$ mw: 478.92

PROP: Golden yellow crystals. Mp: 168-169°. Sltly sol in water; very sol in aq soln pH 7.65; freely sol in the "cellosolves," dioxane, "Carbitol;" sol in methanol, ethanol, butanol, acetone, ethyl acetate, and benzene; insol in ether and petroleum ether.

SYNS: ACRONIZE ◇ AUREOCINA ◇ AUREOMYCIN ◇ AUREOMYCIN A-377 ◇ AUREOMYKOIN ◇ BIOMITSIN ◇ BIOMYCIN ◇ 7-CHLORO-4-(DIMETHYLAMINO)-1,4,11,12a-OCTAHYDRO-2-NAPHTHACENECARBOXAMIDE ◇ 7-CHLOROTETRACYCLINE ◇ CHRYSOMYKINE ◇ CTC ◇ DUOMYCIN ◇ FLAMYCIN

USE IN FOOD:

Purpose: Animal drug, animal feed drug.

Where Used: Animal feed, beef, chicken, duck, eggs, lamb, milk, pork, turkey.

Regulations: FDA - 21CFR 556.150. Limitation of 4 ppm in uncooked kidney, 1 ppm in uncooked muscle, liver, fat, and skin of chickens, turkeys, and ducks. Limitation of zero in eggs. Limitation of 4 ppm in uncooked kidney, 2 ppm in uncooked liver, 1 ppm in uncooked muscle, and 0.2 ppm in uncooked fat of swine. Limitation of 4 ppm in uncooked liver and kidney, 1 ppm in uncooked muscle and fat of calves. Limitation of 0.1 ppm in uncooked kidney, liver, and muscle of beef cattle and nonlactating dairy cows. Limitation of zero in uncooked fat and milk of beef cattle and nonlactating dairy cows. Limitation of 1 ppm in uncooked kidney, 0.5 ppm in uncooked liver, 0.1 ppm in uncooked muscle of sheep. 21CFR 558.128.

SAFETY PROFILE: Poison by intravenous, subcutaneous, intracerebral, and intraperitoneal routes. Moderately toxic by ingestion. Experimental reproductive effects. When heated to decomposition it emits toxic fumes of Cl^- and NO_x.

TOXICITY DATA and CODEN

scu-mus TDLo: 372 mg/kg (1-6D preg): REP ASPHAK 23,481,69
ipr-rat LDLo: 335 mg/kg CLDND*
ivn-rat LD50: 118 mg/kg ANYAA9 51,182,48
orl-mus LD50: 2500 mg/kg ARZNAD 5,1,55

CMC750 CAS: 67-97-0
CHOLECALCIFEROL

mf: $C_{27}H_{44}O$ mw: 384.71

PROP: White crystals; odorless. insol in water; sol in alc, chloroform, fatty oils.

SYNS: COLECALCIFEROL ◇ 7-DEHYDROCHOLESTROL, ACTIVATED ◇ DELSTEROL ◇ DEPARAL ◇ D3-VIGANTOL ◇ OLEOVITAMIN D3 ◇ RICKETON ◇ 9,10-SECOCHOLESTA-5,7,10(19)-TRIEN-3-β-OL ◇ TRIVITAN ◇ VIGORSAN ◇ VITAMIN D3 ◇ VITINC DAN-DEE-3

USE IN FOOD:

Purpose: Dietary supplement, nutrient.

Where Used: Cereals (breakfast), grain products, margarine, milk, milk products, pasta.

Regulations: FDA - 21CFR 166.110, 182.5953, 184.1950. GRAS with a limitation of 350 IU/100 grams in breakfast cereals, 90 IU/100 grams in grain products and pastas, 42 IU/100 grams in milk, 89 IU/100 grams in milk products when used in accordance with good manufacturing practice.

SAFETY PROFILE: Poison by ingestion. An experimental teratogen. When heated to decomposition it emits acrid smoke and irritating fumes.

TOXICITY DATA and CODEN

scu-rat TDLo:90 mg/kg (12-20D preg):TER FOMOAJ 29,333,70
orl-rat LD50:42 mg/kg TXAPA9 43,125,78

CMF300
CHOLINE BITARTRATE
mf: $C_9H_{19}NO_7$ mw: 253.25

PROP: White crystalline powder; acetic taste. Sol in water; sltly sol in alc; insol in ether, chloroform, benzene.

SYN: (2-HYDROXYETHYL)TRIMETHYLAMMONIUM BITARTRATE

USE IN FOOD:

Purpose: Dietary supplement, nutrient.

Where Used: Various.

Regulations: FDA - 21CFR 182.5250, 182.8250. GRAS when used in accordance with good manufacturing practice.

SAFETY PROFILE: When heated to decomposition emits toxic fumes of NO_x.

CMF750 CAS: 67-48-1
CHOLINE HYDROCHLORIDE
mf: $C_5H_{14}NO{\bullet}Cl$ mw: 139.65

PROP: Colorless to white hygroscopic crystals; slt odor of trimethylamine. Sol in water and alc.

SYNS: BIOCOLINA ◇ CHLORIDE de CHOLINE (FRENCH) ◇ CHOLINE CHLORHYDRATE ◇ CHOLINE CHLORIDE (FCC) ◇ CHOLINIUM CHLORIDE ◇ HEPACHOLINE ◇ (2-HYDROXYETHYL)TRIMETHYLAMMONIUM CHLORIDE ◇ LIPOTRIL

USE IN FOOD:

Purpose: Fungicide.

Where Used: Various.

Regulations: USDA CES Ranking: B-3 (1987).

NCI Carcinogenesis Bioassay (feed); Clear Evidence: mouse NCITR* NCI-CG-TR-15,77; No Evidence: rat NCITR* NCI-CG-TR-15,77; IARC Cancer Review: Animal Limited Evidence IMEMDT 30,295,83. EPA Genetic Toxicology Program. Community Right-To-Know List.

ACGIH TLV: TWA 5 mg/m^3 DOT Classification: ORM-E; Label; None

SAFETY PROFILE: Moderately toxic to humans by ingestion. Moderately toxic experimentally by ingestion, inhalation, and intraperitoneal routes. An experimental tumorigen, neoplastigen, and teratogen. A suspected carcinogen. Experimental reproductive effects. Human mutagenic data. When heated to decomposition it emits toxic fumes of Cl^-, SO_x and NO_x.

TOXICITY DATA and CODEN

cyt-ham:ovr 500 μg/L ENMUDM 7,1,85
sce-ham:ovr 500 μg/L ENMUDM 7,1,85
orl-mus LD50:3900 :1075 g/kg/80W-C:NEO NCITR* NCI-CG-TR-15,77
orl-mus TD :540 g/kg/80W-C:ETA NCITR* NCI-CG-TR-15,77
orl-hmn LDLo:1071 mg/kg 34ZIAG -,151,69
orl-rat LD50:9 g/kg ARSIM* 20,6,66
ihl-mus LC50:5000 mg/m^3/2H TXAPA9 45,320,78

CMP969 CAS: 104-55-2
CINNAMALDEHYDE
mf: C_9H_8O mw: 132.17

PROP: Found in Ceylon and Chinese cinnamon oils. Yellowish, oily liquid; strong odor of cinnamon. D: 1.048-1.052, mp: −7.5°, bp: 246.0° (some decomp), d: 1.048-1.052 @ 25°/25°, refr index: 1.619-1.623, flash p: 248°F. Very sltly sol in water; misc with alc, ether, chloroform, fixed oils.

SYNS: BENZYLIDENEACETALDEHYDE ◇ CASSIA ALDEHYDE ◇ CINNAMAL ◇ CINNAMYL ALDEHYDE ◇ CINNIMIC ALDEHYDE ◇ FEMA No. 2286 ◇ NCI-C56111 ◇ PHENYLACROLEIN ◇ 3-PHENYLACROLEIN ◇ 3-PHENYLPROPENAL ◇ 3-PHENYL-2-PROPENAL ◇ ZIMTALDEHYDE

USE IN FOOD:

Purpose: Flavoring agent.

Where Used: Bakery products, beverages (nonalcoholic), chewing gum, condiments, confections, ice cream products, meat.

Regulations: FDA - 21CFR 182.60. GRAS when used at a level not in excess of the amount reasonably required to accomplish the intended effect.

SAFETY PROFILE: Poison by intravenous and parenteral routes. Moderately toxic by ingestion and intraperitoneal routes. A severe human skin irritant. Mutagenic data. Combustible liquid. May ignite after a delay period in contact with NaOH. When heated to decomposition it emits acrid smoke and fumes.

TOXICITY DATA and CODEN

skn-hmn 40 mg/48H SEV FCTXAV 17,253,79
mma-sat 500 μg/plate FCTOD7 22,623,84
sln-dmg-par 2 pph ENMUDM 7,677,85
orl-rat LD50:2220 mg/kg FCTXAV 2,327,64
ipr-mus LD50:610 mg/kg YKKZAJ 92,135,72

CMP975 CAS: 621-82-9
CINNAMIC ACID
mf: $C_9H_8O_2$ mw: 148.17

PROP: Occurs free and partly esterified in storax, balsam Peru or Tolu, oil of cinnamon, coca leaves. White monoclinic crystals; honey floral odor. D: (4/4) 1.2475, mp: 133°, bp: 300°, flash p: +212°F. One gram dissolves in about 2000 mL water at 25° (more sol in hot water), in 6 mL alc, 5 mL methanol, 12 mL chloroform. Freely sol in benzene, ether, acetone, glacial acetic acid, carbon disulfide, fixed oils.

SYNS: FEMA No. 2288 ◇ PHENYLACRYLIC ACID ◇ tert-β-PHENYLACRYLIC ACID ◇ 3-PHENYLACRYLIC ACID ◇ 3-PHENYLPROPENOIC ACID ◇ 3-PHENYL-2-PROPENOIC ACID ◇ ZIMTSAEURE (GERMAN)

USE IN FOOD:

Purpose: Flavoring agent.

Where Used: Various.

Regulations: FDA - 21CFR 172.515. Use at a level not in excess of the amount reasonably required to accomplish the intended effect.

SAFETY PROFILE: Poison by intravenous and intraperitoneal routes. Moderately toxic by ingestion. A skin irritant. Combustible liquid. When heated to decomposition it emits acrid smoke and fumes.

TOXICITY DATA and CODEN

skn-rbt 500 mg/24H MLD FCTXAV 16,687,78
orl-rat LD50:2500 mg/kg FCTXAV 16,687,78
ipr-rat LD50:1600 mg/kg BCFAAI 112,53,73

CMQ500
CINNAMON LEAF OIL

PROP: Extracted by steam distillation of leaves from *Cinnamomum zeylanicum* Nees. Light to dark brown liquid; spicy cinnamon, clove odor and taste. Sol in fixed oils, propylene glycol, mineral oil; insol in glycerin.

SYNS: CINNAMON LEAF OIL, Ceylon ◇ CINNAMON LEAF OIL, Seychelles

USE IN FOOD:

Purpose: Flavoring agent.

Where Used: Bakery products, beverages (nonalcoholic), chewing gum, condiments, ice cream, meat, pickles.

Regulations: FDA - 21CFR 182.10, 182.20. GRAS when used at a level not in excess of the amount reasonably required to accomplish the intended effect.

SAFETY PROFILE: When heated to decomposition it emits acrid smoke and irritating fumes.

CMQ730 CAS: 103-54-8
CINNAMYL ACETATE
mf: $C_{11}H_{12}O_2$ mw: 176.23

PROP: Colorless liquid; sweet floral odor. D: 1.047-1.051, refr index: 1.539-1.543, flash p: 244°F. Misc with chloroform, ether, fixed oils; insol in glycerin, water @ 264°.

SYNS: ACETIC ACID, CINNAMYL ESTER ◇ FEMA No. 2293 ◇ γ-PHENYLALLYL ACETATE ◇ 3-PHENYL-2-PROPEN-1-YL ACETATE

USE IN FOOD:

Purpose: Flavoring agent.

Where Used: Various.

Regulations: FDA - 21CFR 172.515. Use at a level not in excess of the amount reasonably required to accomplish the intended effect.

SAFETY PROFILE: Moderately toxic by ingestion and intraperitoneal routes. Combustible liquid. When heated to decomposition it emits acrid smoke and fumes.

TOXICITY DATA and CODEN

orl-rat LD50:3300 mg/kg FCTXAV 11,1065,73
ipr-mus LD50:1200 mg/kg PHMCAA 3,62,61

CMQ740 CAS: 104-54-1
CINNAMYL ALCOHOL
mf: $C_9H_{10}O$ mw: 134.19

PROP: Occurs (in the esterified form) in storax and in balsam Peru, cinnamon leaves, hyacinth oil. Needles or crystalline mass; odor of hyacinth. Mp: 33°, d: 1.0397, bp: 250.0°, n (20/D) 1.58190. Sol in water, glycerol, propylene glycol. Freely sol in alc, ether, other common organic solvents.

SYNS: CINNAMIC ALCOHOL CINNAMYL ALCOHOL, SYNTHETIC ◇ FEMA No. 2294 ◇ γ-PHENYLALLYL ALCOHOL ◇ 3-PHENYLALLYL ALCOHOL ◇ 3-PHENYL-2-PROPEN-1-OL ◇ STYRONE ◇ STYRYL CARBINOL

USE IN FOOD:

Purpose: Flavoring agent.

Where Used: Various.

Regulations: FDA - 21CFR 172.515. Use at a level not in excess of the amount reasonably required to accomplish the intended effect.

SAFETY PROFILE: Moderately toxic by ingestion. A skin irritant. When heated to decomposition it emits acrid smoke and fumes.

TOXICITY DATA and CODEN

skn-rbt 500 mg/24H MOD FCTXAV 12,855,74
orl-rat LD50:2000 mg/kg FCTXAV 12,855,74

CMR500 CAS: 104-65-4
CINNAMYL FORMATE

mf: $C_{10}H_{10}O_2$ mw: 162.20

PROP: Colorless liquid; balsamic odor. D: 1.077-1.082, refr index: 1.550-1.556, flash p: 212°F. Misc with alc, chloroform, ether, fixed oils; insol in water @ 250°.

SYNS: CINNAMYL ALCOHOL, FORMATE ◇ CINNAMYL METHANOATE ◇ FEMA No. 2299 ◇ FORMIC ACID, CINNAMYL ESTER ◇ 3-PHENYL-2-PROPEN-1-YL FORMATE

USE IN FOOD:

Purpose: Flavoring agent.

Where Used: Various.

Regulations: FDA - 21CFR 172.515. Use at a level not in excess of the amount reasonably required to accomplish the intended effect.

SAFETY PROFILE: Moderately toxic by ingestion. Combustible liquid. When heated to decomposition it emits acrid smoke and irritating fumes.

TOXICITY DATA and CODEN

orl-rat LD50:2900 mg/kg FCTXAV 14,659,76

CMR800
CINNAMYL ISOVALERATE

mf: $C_{14}H_{18}O_2$ mw: 218.30

PROP: Colorless to sltly yellow liquid; spicy, floral, fruity odor. D: 0.991-0.996, refr index: 1.518-1.524, flash p: +212°F. Misc in alc, chloroform, ether, most oils; insol in glycerin, propylene glycol, water @ 313°.

SYN: FEMA No. 2302

USE IN FOOD:

Purpose: Flavoring agent.

Where Used: Various.

Regulations: FDA - 21CFR 172.515. Use at a level not in excess of the amount reasonably required to accomplish the intended effect.

SAFETY PROFILE: Combustible liquid. When heated to decomposition it emits acrid smoke and irritating fumes.

CMR850
CINNAMYL PROPIONATE

mf: $C_{12}H_{14}O_2$ mw: 190.24

PROP: Colorless to pale yellow liquid; spicy, fruity, balsamic odor. D: 1.029-1.033, refr index: 1.523-1.537, flash p: +212°F. Misc in alc, chloroform, ether, most oils; insol in glycerin, propylene glycol, water @ 289°.

SYN: FEMA No. 2301

USE IN FOOD:

Purpose: Flavoring agent.

Where Used: Various.

Regulations: FDA - 21CFR 172.515. Use at a level not in excess of the amount reasonably required to accomplish the intended effect.

SAFETY PROFILE: Combustible liquid. When heated to decomposition it emits acrid smoke and irritating fumes.

CMS750 CAS: 77-92-9
CITRIC ACID

mf: $C_6H_8O_7$ mw: 192.14

PROP: Colorless, odorless crystals (crystals are monoclinic holohedra and crystallize from hot concd aq soln); acid taste. Mp: 153° (anhydrous form), bp: decomp; d: 1.665, flash p: +212°F. Sol in water, alc, ether.

SYNS: ACILETTEN ◇ ANHYDROUS CITRIC ACID ◇ CITRETTEN ◇ CITRIC ACID, ANHYDROUS ◇ CITRO ◇ FEMA No. 2306 ◇ 2-HYDROXY-1,2,3-PROPANETRICAR-

BOXYLIC ACID ◇ β-HYDROXYTRICARBALLYLIC ACID ◇ KYSELINA CITRONOVA (CZECH)

USE IN FOOD:

Purpose: Acidifier, curing accelerator, dispersing agent, flavoring agent, sequestrant, synergist for antioxidants.

Where Used: Beef (cured), chili con carne, cured comminuted meat food product, fats (poultry), fruits (frozen), lard, meat (dried), pork (cured), pork (fresh), potato sticks, potatoes (instant), poultry, sausage (dry), sausage (fresh pork), shortening, wheat chips, wine.

Regulations: FDA - 21CFR 182.1033, 182.6033. GRAS when used in accordance with good manufacturing practice. USDA - 9CFR 318.7. Sufficient for purpose. Limitation of 10 percent solution to spray surfaces of cured cuts of meat. Not to exceed 500 ppm or 1.8 mg/sq in. of surface ascorbic acid, erythorbic acid, or sodium ascorbate, singly, or in combination; and/or not to exceed either 250 ppm or 0.9 mg/sq in. of surface of citric acid or sodium citrate, singly, or in combination on fresh pork cuts. Limitation of 0.01 percent alone or in combination with antioxidants in lard, shortening, fresh pork sausage, and dried meats. Limitation of 0.003 percent in dry sausage in combination with antioxidants. USDA - 9CFR 381.147. May be used to replace up to 50 percent of the ascorbic acid or sodium ascorbate in poultry. Limitation of 0.01 percent alone or in combination with antioxidants in poultry fats. BATF - 27CFR 240.1051. Limitation of 5.8 pounds/1000 gallons.

SAFETY PROFILE: Poison by intravenous route. Moderately toxic by subcutaneous and intraperitoneal routes. Mildly toxic by ingestion. A severe eye and moderate skin irritant. An irritating organic acid, some allergenic properties. Combustible liquid. Potentially explosive reaction with metal nitrates. When heated to decomposition it emits acrid smoke and fumes.

TOXICITY DATA and CODEN

skn-rbt 500 mg/24H MOD 28ZPAK -,105,72
eye-rbt 750 μg/24H SEV 28ZPAK -,105,72
orl-rat LD50:6730 mg/kg 28ZPAK -,105,72
ipr-rat LD50:883 mg/kg JPETAB 94,65,48

CMS845 CAS: 106-23-0
CITRONELLAL

mf: $C_{10}H_{18}O$ mw: 154.25

PROP: Colorless to sltly yellow liquid; intense lemon-citronnella-rose odor. D: 0.850-0.860, refr index: 1.446-1.456, flash p: 170°F. Sol in alc, most oils; sltly sol in propylene glycol; insol glycerin, water.

SYNS: 3,7-DIMETHYL-6-OCTENAL ◇ FEMA No. 2307

USE IN FOOD:

Purpose: Flavoring agent.

Where Used: Bakery products, beverages (nonalcoholic), chewing gum, confections, gelatin desserts, ice cream, meat, puddings.

Regulations: FDA - 21CFR 172.515. Use at a level not in excess of the amount reasonably required to accomplish the intended effect.

SAFETY PROFILE: Combustible liquid. When heated to decomposition it emits acrid smoke and irritating fumes.

CMS850 CAS: 107-75-5
CITRONELLAL HYDRATE

mf: $C_{10}H_{20}O_2$ mw: 172.30

PROP: Colorless liquid; sweet, floral, lily odor. D: 0.918-0.923, refr index: 1.447-1.450, flash p: +212°F. Sol in fixed oils, propylene glycol; insol in glycerin.

SYNS: CYCLALIA ◇ CYCLOSIA ◇ 3,7-DIMETHYL-7-HYDROXYOCTANAL ◇ FEMA No. 2583 ◇ FIXOL ◇ HYDROXYCITRONELLAL (FCC) ◇ 7-HYDROXYCITRONELLAL ◇ 7-HYDROXY-3,7-DIMETHYLOCTAN-1-AL ◇ 7-HYDROXY-3,7-DIMETHYL OCTANAL ◇ LAURINE ◇ LILYL ALDEHYDE ◇ MUSUET SYNTHETIC ◇ MUSUETTINE PRINCIPLE ◇ PHIXIA

USE IN FOOD:

Purpose: Flavoring agent.

Where Used: Various.

Regulations: FDA - 21CFR 172.515. Use at a level not in excess of the amount reasonably required to accomplish the intended effect.

SAFETY PROFILE: A skin irritant. Combustible liquid. When heated to decomposition it emits acrid smoke and irritating fumes.

TOXICITY DATA and CODEN

skn-rbt 500 mg/24H FCTXAV 12,921,74

CMT250 CAS: 106-22-9
CITRONELLOL

mf: $C_{10}H_{20}O$ mw: 156.30

PROP: Colorless oily liquid; rose odor. D: 0.850-0.860, refr index: 1.454-1.462, flash p: 215°F. Sol in fixed oils, propylene glycol; sltly sol in water; insol in glycerin @ 225°.

SYNS: CEPHROL ◇ 2,6-DIMETHYL-2-OCTEN-8-OL ◇ 3,7-DIMETHYL-6-OCTEN-1-OL ◇ FEMA No. 2309 ◇ FEMA No. 2980 ◇ RHODINOL ◇ RODINOL

USE IN FOOD:

Purpose: Flavoring agent.

Where Used: Bakery products, beverages (nonalcoholic), chewing gum, confections, gelatin desserts, ice cream products, puddings.

Regulations: FDA - 21CFR 172.515. Use at a level not in excess of the amount reasonably required to accomplish the intended effect.

SAFETY PROFILE: Poison by intravenous route. Moderately toxic by ingestion, skin contact, and intramuscular routes. Combustible liquid. When heated to decomposition it emits acrid smoke and irritating fumes.

TOXICITY DATA and CODEN

orl-rat LD50:3450 mg/kg FCTXAV 13,757,75
ivn-mus LDLo:100 mg/kg CBCCT* 5,139,53
ims-mus LD50:4000 mg/kg JSICAZ 21,342,62
skn-rbt LD50:2650 mg/kg FCTXAV 13,757,75

CMT600
CITRONELLYL BUTYRATE

mf: $C_{14}H_{26}O_2$ mw: 226.36

PROP: Colorless liquid; strong, fruity-rosy odor. D: 0.873-0.883; refr index: 1.444-1.448, flash p: +212°F. Misc in alc, ether, chloroform, most oils; insol water @ 245°.

SYNS: 3,7-DIMETHYL-6-OCTEN-1-YL BUTYRATE ◇ FEMA No. 2312

USE IN FOOD:

Purpose: Flavoring agent.

Where Used: Various.

Regulations: FDA - 21CFR 172.515. Use at a level not in excess of the amount reasonably required to accomplish the intended effect.

SAFETY PROFILE: Combustible liquid. When heated to decomposition it emits acrid smoke and irritating fumes.

CMT750 CAS: 105-85-1
CITRONELLYL FORMATE

mf: $C_{11}H_{20}O_2$ mw: 184.31

PROP: Colorless liquid; strong, fruity odor. D: 0.890-0.93, refr index: 1.443-1.452, flash p: 197°F. Sol in alc, fixed oils; sltly sol in propylene glycol; insol in glycerin, water @ 235°.

SYNS: 3,7-DIMETHYL-6-OCTEN-1-OL FORMATE ◇ 2,6-DIMETHYL-2-OCTEN-8-YL FORMATE ◇ 3,7-DIMETHYL-6-OCTEN-1-YL FORMATE ◇ FEMA No. 2314 ◇ FORMIC ACID, CITRONELLYL ESTER ◇ FORMIC ACID-3,7-DIMETHYL-6-OCTEN-1-YL ESTER

USE IN FOOD:

Purpose: Flavoring agent.

Where Used: Various.

Regulations: FDA - 21CFR 172.515. Use at a level not in excess of the amount reasonably required to accomplish the intended effect.

SAFETY PROFILE: Mildly toxic by ingestion. A human skin irritant. Combustible liquid. When heated to decomposition it emits acrid smoke and irritating fumes.

TOXICITY DATA and CODEN

skn-hmn 20 mg/48H MLD FCTXAV 11,1073,73
skn-rbt 500 mg/24H FCTXAV 11,1073,73
orl-rat LD50:8400 mg/kg FCTXAV 11,1011,73

CMT900
CITRONELLYL ISOBUTYRATE

mf: $C_{14}H_{26}O_2$ mw: 226.36

PROP: Colorless liquid; rosy-fruity odor. D: 0.870-0.880, refr index: 1.440-1.448, flash p: +212°F. Misc in alc, chloroform, ether, most oils; insol in water @ 249°.

SYNS: 3,7-DIMETHYL-6-OCTEN-1-YL ISOBUTYRATE ◇ FEMA No. 2313

USE IN FOOD:

Purpose: Flavoring agent.

Where Used: Various.

Regulations: FDA - 21CFR 172.515. Use at a level not in excess of the amount reasonably required to accomplish the intended effect.

SAFETY PROFILE: Combustible liquid. When heated to decomposition it emits acrid smoke and irritating fumes.

CMU100
CITRONELLYL PROPIONATE

mf: $C_{13}H_{24}O_2$ mw: 212.33

PROP: Colorless liquid; fruity-rosy odor. D: 0.877-0.886, refr index: 1.443-1.449, flash p:

+212°F. Misc in alc, most oils; insol in water @ 242°.

SYN: FEMA No. 2316

USE IN FOOD:

Purpose: Flavoring agent.

Where Used: Baked goods, beverages, candy, ice cream.

Regulations: FDA - 21CFR 172.515. Use in accordance with good manufacturing practice.

SAFETY PROFILE: Combustible liquid. When heated to decomposition it emits acrid smoke and irritating fumes.

CMU900
CLARY OIL

PROP: From steam distillation of flowering tops and leaves of *Salvia sclarea* L. (Fam. *Labiatae*). Pale yellow liquid; herbaceous odor. Sol in fixed oils, mineral oil; insol in glycerin, propylene glycol.

SYNS: CLARY SAGE OIL ◇ OIL of MUSCATEL

USE IN FOOD:

Purpose: Flavoring agent.

Where Used: Various.

Regulations: FDA - 21CFR 182.10, 182.20. GRAS when used at a level not in excess of the amount reasonably required to accomplish the intended effect.

SAFETY PROFILE: When heated to decomposition it emits acrid smoke and irritating fumes.

CMX895 CAS: 40665-92-7
CLOPROSTENOL
mf: $C_{22}H_{29}ClO_6$ mw: 424.96

SYNS: ESTRUMATE ◇ ICI 80996 ◇ racemic-ICI 80,996 ◇ I.C.I. LTD. COMPOUND NUMBER 80996

USE IN FOOD:

Purpose: Animal drug.

Where Used: Beef.

Regulations: USDA CES Ranking: B-4 (1988).

SAFETY PROFILE: Experimental reproductive effects. When heated to decomposition it emits toxic fumes of Cl^-.

TOXICITY DATA and CODEN

ipr-rat TDLo:900 ng/kg (20D preg):REP NATUAS 250,330,74

orl-ham TDLo:60 μg/kg (4-6D preg):REP PRGLBA 10,5,75

CMX896
CLOPYRALID

SYN: 3,6-DICHLORO-2-PYRIDINECARBOXYLIC ACID

USE IN FOOD:

Purpose: Herbicide.

Where Used: Animal feed, barley (milled fractions, except flour), oats (milled fractions, except flour), wheat (milled fractions, except flour).

Regulations: FDA - 21CFR 193.472. Limitation of 12.0 ppm in barley, milled fractions (except flour); in oats, milled fractions (except flour); in wheat, milled fractions (except flour). 21CFR 561.439. Limitation as above when used for animal feed.

SAFETY PROFILE: When heated to decomposition it emits acrid smoke and irritating fumes.

CMX920
CLORSULON

USE IN FOOD:

Purpose: Animal drug.

Where Used: Beef.

Regulations: FDA - 21CFR 556.163. Limitation of 1 ppm in kidney of cattle.

SAFETY PROFILE: When heated to decomposition emits toxic fumes of Cl^-.

CMY100 CAS: 8015-97-2
CLOVE LEAF OIL MADAGASCAR

PROP: From steam distillation of leaves of *Eugenis caryophyllata* Thunberg (*Eugenia aromatica* L. Baill.) (Fam. *Myrtaceae*). Pale yellow liquid. Ref. index: 1.527-1.538 ZBJ000 20°. Sol in propylene glycol, fixed oils; insol in glycerin, mineral oil.

SYNS: CLOVE LEAF OIL ◇ OILS, CLOVE LEAF

USE IN FOOD:

Purpose: Flavoring agent.

Where Used: Bakery products, beverages (alcoholic), beverages (nonalcoholic), cakes, chewing gum, condiments, confections, cookies, fruit punches, fruits (spiced), gelatin desserts, ice cream, marinades, meat, meat sauces, pickles, puddings, relishes, sauces.

Regulations: FDA - 21CFR 184.1257. GRAS when used at a level not in excess of the amount reasonably required to accomplish the intended effect.

SAFETY PROFILE: Moderately toxic by ingestion and skin contact. A severe skin irritant. When heated to decomposition it emits acrid smoke and fumes.

TOXICITY DATA and CODEN

skn-mus 100 % FCTXAV 16,695,78
skn-rbt 500 mg/24H SEV FCTXAV 16,695,78
orl-rat LD50: 1370 mg/kg FCTXAV 16,695,78
skn-rbt LD50: 1200 mg/kg FCTXAV 16,695,78

CNA500 CAS: 6147-53-1
COBALT ACETATE TETRAHYDRATE

mf: $C_4H_6O_4 \cdot Co \cdot 4H_2O$ mw: 249.11

SYNS: ACETIC ACID, COBALT(2+) SALT, TETRAHYDRATE ◇ COBALT DIACETATE TETRAHYDRATE ◇ COBALTOUS ACETATE TETRAHYDRATE ◇ OCTAN KOBALTNATY (CZECH)

USE IN FOOD:

Purpose: Foam stabilizer.

Where Used: Prohibited from foods and malt beverages (fermented).

Regulations: FDA - 21CFR 189.120. Prohibited from direct addition or use in human food.

Cobalt and its compounds are on the Community Right-To-Know List.

OSHA PEL: TWA 0.1 mg(Co)/m^3 (fume and dust) ACGIH TLV: TWA 0.05 mg(Co)/m^3 NIOSH REL: TWA (To Cobalt) TWA 0.1 mg/m^3

SAFETY PROFILE: Moderately toxic by ingestion. A skin and eye irritant. Human mutagenic data. When heated to decomposition it emits acrid smoke and irritating fumes.

TOXICITY DATA and CODEN

skn-rbt 500 mg/24H MOD 28ZPAK -,21,72
eye-rbt 500 mg/24H MLD 28ZPAK -,21,72
cyt-hmn: lym 600 μg/L CYGEDX 12(3),46,78
orl-rat LD50: 708 mg/kg FCTOD7 20,311,82

CNB450
COBALT CAPRYLATE

USE IN FOOD:

Purpose: Drying agent.

Where Used: Packaging materials.

Regulations: FDA - 21CFR 181.25. Use in accordance with good manufacturing practice.

SAFETY PROFILE: When heated to decomposition it emits acrid smoke and irritating fumes.

CNB599 CAS: 7646-79-9
COBALT(II) CHLORIDE

mf: Cl_2Co mw: 129.83

PROP: Blue powder. Mp: 724°, bp: 1049°, d: 3.348.

SYNS: COBALT DICHLORIDE ◇ COBALT MURIATE ◇ COBALTOUS CHLORIDE ◇ COBALTOUS DICHLORIDE ◇ KOBALT CHLORID (GERMAN)

USE IN FOOD:

Purpose: Foam stabilizer.

Where Used: Prohibited from malt beverages (fermented).

Regulations: FDA - 21CFR 189.120. Prohibited from direct addition or use in human food.

EPA Genetic Toxicology Program. Cobalt and its compounds are on the Community Right-To-Know List.

OSHA PEL: TWA 0.1 mg(Co)/m^3 (fume and dust) ACGIH TLV: TWA 0.05 mg(Co)/m^3 NIOSH REL: (To Cobalt) TWA 0.1 mg/m^3

SAFETY PROFILE: Poison experimentally by ingestion, skin contact, intraperitoneal, intravenous, and subcutaneous routes. Moderately toxic to humans by ingestion. An experimental carcinogen and teratogen. Human systemic effects by ingestion: anorexia, goiter (increased thyroid size), and weight loss. Experimental reproductive effects. Human mutagenic data. Incompatible with metals (e.g., sodium and potassium). When heated to decomposition it emits toxic fumes of Cl^-.

TOXICITY DATA and CODEN

mrc-bcs 50 mmol/L MUREAV 77,109,80
mmo-smc 100 mmol/L CPBTAL 33,1571,85
orl-rat TDLo: 11 mg/kg (1-22D preg): REP GISAAA 45(2)6,80
ipr-rat TDLo: 30 g/kg (15-16D preg): TER TJADAB 29(3),23A,84
ipr-mus TDLo: 25 mg/kg (10D preg): TER JPMSAE 58,766,69
scu-rat TDLo: 400 mg/kg: CAR LBANAX 11,43,77

orl-chd TDLo:48 mg/kg:CNS,END,MET JAMAAP 157,117,55
orl-chd LDLo:1500 mg/kg 34ZIAG -,182,69
orl-rat LD50:80 mg/kg HYSAAV 36,277,71

CNC235
COBALT LINOLEATE

USE IN FOOD:

Purpose: Drying agent.

Where Used: Packaging materials.

Regulations: FDA - 21CFR 181.25. Use in accordance with good manufacturing practice.

SAFETY PROFILE: When heated to decomposition it emits acrid smoke and irritating fumes.

CNE125 CAS: 10124-43-3
COBALT(II) SULFATE (1:1)

mf: $O_4S \cdot Co$ mw: 154.99

PROP: Red to lavender dimorphic, orthorhombic crystals. D: 3.71. Stable to 708°. Dissolves slowly in boiling water.

SYNS: COBALTOUS SULFATE ◇ COBALT SULFATE ◇ COBALT SULFATE (1:1) ◇ COBALT (2+) SULFATE ◇ COBALT(II) SULPHATE ◇ SULFURIC ACID, COBALT(2+) SALT (1:1)

USE IN FOOD:

Purpose: Boiler water additive, foam stabilizer.

Where Used: Prohibited from malt beverages (fermented).

Regulations: FDA - 21CFR 189.120. Prohibited from direct addition or use in human food. 21CFR 173.310. Catalyst in boiler water.

Cobalt and its compounds are on the Community Right-To-Know List. EPA Genetic Toxicology Program.

OSHA PEL: TWA 0.1 mg(Co)/m^3 (fume and dust) ACGIH TLV: TWA 0.05 mg(Co)/m^3

NIOSH REL: TWA 0.1 mg/m^3

SAFETY PROFILE: Poison by intravenous and intraperitoneal routes. Moderately toxic by ingestion. When heated to decomposition it emits toxic fumes of SO_x.

TOXICITY DATA and CODEN

orl-rat LD50:424 mg/kg FCTOD7 20,311,82
ipr-mus LD50:143 mg/kg COREAF 256,1043,63
ivn-dog LDLo:20 mg/kg HBAMAK 4,1289,35

CNE240
COBALT TALLATE

USE IN FOOD:

Purpose: Drying agent.

Where Used: Packaging materials.

Regulations: FDA - 21CFR 181.25. Use in accordance with good manufacturing practice.

SAFETY PROFILE: When heated to decomposition it emits acrid smoke and irritating fumes.

CNF250 CAS: 104-61-0
COCONUT ALDEHYDE

mf: $C_9H_{16}O_2$ mw: 156.25

PROP: Colorless to sltly yellow liquid; coconut odor. D: 0.958-0.966, refr index: 1.446-1.450, flash p: +212°F. Sol in alc, fixed oils, propylene glycol; insol in water.

SYNS: ALDEHYDE C-18 ◇ γ-N-AMYLBUTYROLACTONE ◇ FEMA No. 2781 ◇ 4-HYDROXYNONANOIC ACID, γ-LACTONE ◇ γ-NONALACTONE (FCC) ◇ 1,4-NONALOLIDE ◇ PRUNOLIDE

USE IN FOOD:

Purpose: Flavoring agent.

Where Used: Baked goods, candy, gelatins, ice cream, puddings.

Regulations: FDA - 21CFR 172.515. Use at a level not in excess of the amount reasonably required to accomplish the intended effect.

SAFETY PROFILE: Moderately toxic by ingestion. A skin irritant. Combustible liquid. When heated to decomposition it emits acrid smoke and irritating fumes.

TOXICITY DATA and CODEN

skn-rbt 500 mg/24H MLD FCTXAV 13,681,75
orl-rat LD50:6600 mg/kg FCTXAV 13,681,75

CNG825
COGNAC OIL

PROP: From steam distillation of wine lees. Green to blue-green liquid; cognal odor. Sol in fixed oils, mineral oil; very sltly sol in propylene glycol; insol in glycerin.

SYNS: COGNAC OIL, WHITE ◇ COGNAC OIL, GREEN ◇ ETHYL OENANTHATE ◇ WINE YEAST OIL

USE IN FOOD:

Purpose: Flavoring agent.

Where Used: Various.

Regulations: FDA - 21CFR 182.50. GRAS when used at a level not in excess of the amount reasonably required to accomplish the intended effect.

SAFETY PROFILE: When heated to decomposition it emits acrid smoke and irritating fumes.

CNH792 CAS: 8001-61-4
COPAIBA OIL

PROP: From steam distillation of South American *Copaifera* L. (Fam. *Leguminosae*) balsam. Colorless to yellow liquid; characteristic odor, aromatic, slightly bitter taste. D: 0.880-0.907; ref. index: 1.493-1.500 @ 20°. Sol in alc, fixed oils, mineral oil.

USE IN FOOD:

Purpose: Flavoring agent.

Where Used: Various.

Regulations: FDA - 21CFR 172.510. Use at a level not in excess of the amount reasonably required to accomplish the intended effect.

SAFETY PROFILE: When heated to decomposition it emits acrid smoke and irritating fumes.

CNI000 CAS: 7440-50-8
COPPER
af: Cu aw: 63.54

PROP: A metal with a distinct reddish color. Mp: 1083°, bp: 2324°, d: 8.92, vap press: 1 mm @ 1628°.

SYNS: ALLBRI NATURAL COPPER ◇ ANAC 110 ◇ ARWOOD COPPER ◇ BRONZE POWDER ◇ CDA 101 ◇ CDA 102 ◇ CDA 110 ◇ CDA 122 ◇ C.I. 77400 ◇ C.I. PIGMENT METAL 2 ◇ COPPER-AIRBORNE ◇ COPPER BRONZE ◇ COPPER-MILLED ◇ COPPER SLAG-AIRBORNE ◇ COPPER SLAG-MILLED ◇ 1721 GOLD ◇ GOLD BRONZE ◇ KAFAR COPPER ◇ M1 (COPPER) ◇ M2 (COPPER) ◇ OFHC Cu ◇ RANEY COPPER

USE IN FOOD:

Purpose: Herbicide.

Where Used: Potable water.

Regulations: FDA - 21CFR 193.90. Herbicides residue tolerance of 1 ppm in potable water.

Copper and its compounds are on the Community Right-To-Know List.

ACGIH TLV: TWA (dust, mist) 1 mg(Cu)/m^3; (fume) 0.2 mg/m^3 DFG MAK: (dust) 1 mg/m^3; (fume) 0.1 mg/m^3

SAFETY PROFILE: An experimental tumorigen and teratogen. Other experimental reproductive effects. Human systemic effects by ingestion: nausea and vomiting. Liquid copper explodes on contact with water. Potentially explosive reaction with actylenic compounds; 3-bromopropyne; ethylene oxide; lead azide; and ammonium nitrate. Ignites on contact with chlorine; chlorine trifluoride; fluorine (above 121°); and hydrazinium nitrate (above 70°). Reacts violently with C_2H_2; bromates; chlorates; iodates; (Cl_2 + OF_2); dimethyl sulfoxide + trichloroacetic acid; ethylene oxide; H_2O_2; hydrazine mononitrate; hydrazoic acid; H_2S + air, $Pb(N_3)_2$; K_2O_2; NaN_3; Na_2O_2; sulfuric acid. Incandescent reaction with potassium dioxide. Incompatible with 1-bromo-2-propyne.

TOXICITY DATA and CODEN

orl-rat TDLo: 152 mg/kg (22W pre): TER GISAAA 45(3),8,80
iut-rat TDLo: 250 μg/kg (1D pre): REP IJEBA6 19,1124,81
ipl-rat TDLo: 100 mg/kg: ETA AIHAAP 41,836,80
orl-hmn TDLo: 120 μg/kg: GIT PHRPA6 73,910,58

CNM100 CAS: 527-09-3
COPPER GLUCONATE
mf: $C_{12}H_{22}CuO_{14}$ mw: 453.84

PROP: Fine light blue powder. Sol in water; sltly sol in alc.

USE IN FOOD:

Purpose: Dietary supplement, nutrient, synergist.

Where Used: Various.

Regulations: FDA - 21CFR 182.5260, 184.1260. GRAS when used in accordance with current good manufacturing practice.

SAFETY PROFILE: When heated to decomposition it emits acrid smoke and irritating fumes.

CNP250 CAS: 7758-98-7
COPPER(II) SULFATE (1:1)
DOT: 9109
mf: $O_4S \cdot Cu$ mw: 159.60

PROP: Blue crystals or blue, crystalline granules or powder. D: 2.284.

SYNS: BCS COPPER FUNGICIDE ◇ BLUE COPPER ◇ BLUE STONE ◇ BLUE VITRIOL ◇ COPPER MONOSULFATE ◇ COPPER SULFATE ◇ CP BASIC SULFATE ◇ CUPRIC SULFATE ◇ KUPPERSULFAT (GERMAN)

◇ ROMAN VITRIOL ◇ SULFATE de CUIVRE (FRENCH) ◇ SULFURIC ACID, COPPER(2+) SALT (1:1) ◇ TNCS 53 ◇ TRINAGLE

USE IN FOOD:

Purpose: Nutrient supplement, processing aid.

Where Used: Brandy, distilling spirits, infant formula, wine.

Regulations: FDA - 21CFR 184.1261. GRAS when used in accordance with good manufacturing practice. BATF - 27CFR 240.1051. Limitation of 0.2 ppm of copper in wine. GRAS when used in accordance with good manufacturing practice. 27CFR 240.1051a. Limitation of 0.2 ppm of copper in brandy and distilling spirits.

Copper and its compounds are on the Community Right-To-Know List. EPA Genetic Toxicology Program.

OSHA PEL: TWA 1 mg(Cu)/m^3 ACGIH TLV: TWA 1 mg(Cu)/m^3 DOT Classification: ORM-E; label: None

SAFETY PROFILE: A human poison by ingestion. An experimental poison by ingestion, subcutaneous, parenteral, intravenous, and intraperitoneal routes. An experimental tumorigen. Human systemic effects by ingestion: gastritis, diarrhea, nausea or vomiting, damage to kidney tubules, and hemolysis. Mutagenic data. Reacts violently with hydroxylamine; magnesium. When heated to decomposition it emits toxic fumes of SO_x.

TOXICITY DATA and CODEN

otr-ham:emb 80 μmol/L CNREA8 39,193,79
mmo-bcs 400 μmol/L AMAHA5 21,297,74
ipr-rat TDLo:7500 μg/kg (3D preg):REP BECTA6 25,702,80
scu-mus TDLo:12768 μg/kg (30D male):REP JRPFA4 7,21,64
par-ckn TDLo:10 mg/kg:ETA BEXBAN 9,519,40
orl-man LDLo:857 mg/kg:GIT ATXKA8 17,20,58
orl-chd TDLo:200 mg/kg:KID,BLD AJDCAI 131,149,77
orl-hmn LDLo:50 mg/kg JAMAAP 235,801,76
orl-hmn TDLo:11 mg/kg:GIT LANCAO 2,700,60
orl-rat LD50:300 mg/kg 36SBA8 1,507,77

CNR000 CAS: 8001-31-8
COPRA (OIL)

DOT: 1363

PROP: From the kernel of the fruit of the coconut palm *Cocos nucifera*. Fatty solid or liquid; sweet, nutty taste. Mp: 21-27°.

SYNS: COCONUT BUTTER ◇ COCONUT MEAL PELLETS, containing 6-13% moisture and no more than 10% residual fat (DOT) ◇ COCONUT OIL (FCC) ◇ COCONUT PALM OIL ◇ COPRA (DOT) ◇ COPRA PELLETS (DOT) ◇ FREE COCONUT OIL

USE IN FOOD:

Purpose: Coating agent, emulsifying agent, formulation aid, texturizer.

Where Used: Baked goods, candy, desserts, margarine.

Regulations: GRAS when used in accordance with good manufacturing practice.

DOT Classification: ORM-C; Label: None; Flammable Solid; Label: None

SAFETY PROFILE: Flammable solid when exposed to heat or flame. May spontaneously heat and ignite if stored wet and hot.

CNR735 CAS: 8008-52-4
CORIANDER OIL

PROP: From steam distillation of ripe fruit of *Coriandrum sativum* L. (Fam. *Umbelliferae*). Colorless liquid; characteristic odor and taste. D: 0.863-0.875, refr index: 1.462 @ 20°.

SYNS: OIL of CORIANDER ◇ OILS, CORIANDER

USE IN FOOD:

Purpose: Flavoring agent.

Where Used: Curry powder, meat, sausage.

Regulations: FDA - 21CFR 182.10, 182.20. GRAS when used at a level not in excess of the amount reasonably required to accomplish the intended effect.

SAFETY PROFILE: Moderately toxic by ingestion. Mutagenic data. A skin irritant. When heated to decomposition it emits acrid smoke and fumes.

TOXICITY DATA and CODEN

skn-rbt 500 mg/24H FCTXAV 11,1077,73
dnr-bcs 10 mg/disc TOFOD5 8,91,85
orl-rat LD50:4130 mg/kg FCTXAV 11,1077,73

CNR850
CORN ENDOSPERM OIL

PROP: Reddish-brown liquid.

USE IN FOOD:

Purpose: Color additive.

Where Used: Chicken feed.

Regulations: FDA - 21CFR 73.315. Feed must be supplemented sufficiently with xanthophyll and associated carotenoids to accomplish the intended effect.

SAFETY PROFILE: When heated to decomposition it emits acrid smoke and irritating fumes.

CNR980 CAS: 66071-96-3
CORN GLUTEN

SYN: CORN GLUTEN MEAL

USE IN FOOD:

Purpose: Nutrient supplement.

Where Used: Various.

Regulations: FDA - 21CFR 184.1321. GRAS when used in accordance with good manufacturing practice.

SAFETY PROFILE: When heated to decomposition it emits acrid smoke and irritating fumes.

CNS000 CAS: 8001-30-7
CORN OIL

PROP: Light yellow, clear, oily liquid; faint characteristic odor. Mp: −10°, flash p: 490°F (CC), d: 0.92, autoign temp: 740°F. From wet milling of *Zea mays* (85DIA2 2,70,77).

USE IN FOOD:

Purpose: Coating agent, emulsifying agent, formulation aid, texturizer.

Where Used: Bakery products, margarine, mayonnaise, salad oil.

Regulations: GRAS when used in accordance with good manufacturing practice.

SAFETY PROFILE: Human skin irritant. An experimental teratogen. May be an allergen. Combustible liquid when exposed to heat or flame. Dangerous spontaneous heating may occur during storage if leaks impregnate rags, waste, etc. To fight fire, use CO_2, dry chemical.

TOXICITY DATA and CODEN

skn-hmn 300 mg/3D-I MLD 85DKA8 -,127,77
orl-rat TDLo: 12500 mg/kg (15-19D preg): TER TJADAB 27,75A,83
orl-mus TDLo: 30 g/kg (15-17D preg): TER TJADAB 29(2),57A,84

CNS100
CORN SILK and CORN SILK EXTRACT

PROP: Extracted from the fresh styles and stigmas of *Zea mays L.*

USE IN FOOD:

Purpose: Flavoring agent.

Where Used: Baked goods, baking mixes, beverages (nonalcoholic), candy (soft), frozen dairy desserts.

Regulations: FDA - 21CFR 184.1262. GRAS with a limitation of 30 ppm in baked goods and baking mixes. Limitation of 20 ppm in nonalcoholic beverages, 10 ppm in frozen dairy desserts, 20 ppm in soft candy, 4 ppm in all other foods.

SAFETY PROFILE: When heated to decomposition it emits acrid smoke and irritating fumes.

CNT400
COSTUS ROOT OIL

PROP: From the dried roots of *Saussurea lappa* Clarke (Fam. *Compositae*). Light yellow to brown liquid; persistent odor. D: 0.995-1.039, refr index: 1.515 @ 20°. Sol in fixed oils, mineral oil; insol in glycerin, propylene glycol.

USE IN FOOD:

Purpose: Flavoring agent.

Where Used: Various.

Regulations: FDA - 21CFR 172.510. Use at a level not in excess of the amount reasonably required to accomplish the intended effect.

SAFETY PROFILE: When heated to decomposition it emits acrid smoke and irritating fumes.

CNT950
COTTONSEED, MODIFIED PRODUCTS

PROP: Extracted from decorticated, partially defatted, cooked, ground cottonseed kernels.

USE IN FOOD:

Purpose: Processing aid.

Where Used: Candy (hard).

Regulations: FDA - 21CFR 172.894. Use at a level not in excess of the amount reasonably required to accomplish the intended effect.

SAFETY PROFILE: When heated to decomposition it emits acrid smoke and irritating fumes.

CNU000 CAS: 8001-29-4
COTTONSEED OIL (UNHYDROGENATED)

PROP: Oily, pale yellow, nearly odorless liquid from seeds of species of *Gossypium hirsutum*.

Flash p: 486°F (CC), fp: 0° to 5°, d: 0.915-0.921 @ 25°/25°, autoign temp: 650°F.

SYNS: DEODORIZED WINTERIZED COTTONSEED OIL ◇ NCI-C50168

USE IN FOOD:

Purpose: Coating agent, emulsifying agent, formulation aid, texturizer.

Where Used: Cooking oil, margarine, salad oil, shortening.

Regulations: GRAS when used in accordance with good manufacturing practice.

SAFETY PROFILE: An experimental tumorigen and teratogen. An allergen. Combustible liquid when exposed to heat or flame. However, if allowed to impregnate rags or oily waste, it can become a dangerous hazard due to spontaneous heating. To fight fire, use CO_2, dry chemical.

TOXICITY DATA and CODEN

ipr-rat TDLo:2256 mg/kg (5-15D preg):TER JDREAF 51,1632,72

ipr-rat TDLo:10 g/kg (5-15D preg):REP JPMSAE 61,51,72

orl-mus TDLo:2940 g/kg/35W-C:ETA LPDSAP 17,115,82

CNU750 CAS: 56-72-4

COUMAPHOS

DOT: 2783

mf: $C_{14}H_{16}ClO_5PS$ mw: 362.78

SYNS: AGRIDIP ◇ ASUNTHOL ◇ BAYER 21/199 ◇ BAYMIX 50 ◇ 3-CHLORO-7-HYDROXY-4-METHYL-COUMARIN-O,O-DIETHYL PHOSPHOROTHIOATE ◇ 3-CHLORO-7-HYDROXY-4-METHYL-COUMARIN-O-ESTER WITH-O,O-DIETHYL PHOSPHOROTHIOATE ◇ O-3-CHLORO-4-METHYL-7-COUMARINYL-O,O-DIETHYL PHOSPHOROTHIOATE ◇ 3-CHLORO-4-METHYL-7-COUMARINYL DIETHYL PHOSPHOROTHIOATE ◇ 3-CHLORO-4-METHYL-7-HYDROXYCOUMARIN DIETHYL THIOPHOSPHORIC ACID ESTER ◇ 3-CHLORO-4-METHYLUMBELLIFERONE-O-ESTER WITH-O,O-DIETHYL PHOSPHOROTHIOATE ◇ CUMAFOS (DUTCH) ◇ O,O-DIAETHYL-O-(3-CHLOR-4-METHYL-CUMARIN-7-YL)-MONOTHIOPHOSPHAT (GERMAN) ◇ O,O-DIETHYL-O-(3-CHLOOR-4-METHYL-CUMARIN-7-YL)MONOTHIOFOSFAAT (DUTCH) ◇ O,O-DIETHYL-O-(3-CHLORO-4-METHYL-7-COUMARINYL)PHOSPHOROTHIOATE ◇ O,O-DIETHYL-O-(3-CHLORO-4-METHYLCOUMARINYL-7) THIOPHOSPHATE ◇ O,O-DIETHYL-O-(3-CHLORO-4-METHYL-2-OXO-2H-BENZOPYRAN-7-YL)PHOSPHOROTHIOATE ◇ O,O-DIETHYL-3-CHLORO-4-METHYL-7-UMBELLIFERONE THIOPHOSPHATE ◇ O,O-DIETHYL-O-(3-CHLORO-4-METHYLUMBELLIFERYL)PHOSPHOROTHIOATE ◇ DIETHYL-3-CHLORO-4-METHYLUMBELLIFERYL THIONOPHOSPHATE ◇ DIETHYL THIOPHOSPHORIC ACIDESTER OF 3-CHLORO-4-METHYL-7-HYDROXYCOUMARIN ◇ O,O-DIETIL-O-(3-CLORO-4-METIL-CUMARIN-7-IL-MONOTIOFOSFATO) (ITALIAN) ◇ DIOLICE ◇ ENT 17,956 ◇ MELDONE ◇ NCI-C08662 ◇ THIOPHOSPHATE de O,O-DIETHYLE et de O-(3-CHLORO-4-METHYL-7-COUMARINYLE) (FRENCH)

USE IN FOOD:

Purpose: Animal feed drug.

Where Used: Animal feed.

Regulations: FDA - 21CFR 558.185. USDA CES Ranking: B-2 (1988).

NCI Carcinogenesis Bioassay (feed); No Evidence: mouse, rat NCITR* NCI-CG-TR-96,79. EPA Extremely Hazardous Substances List.

DOT Classification: Poison B; Label: Poison; Poison B; Label: Poison, liquid

SAFETY PROFILE: Poison by ingestion, skin contact, inhalation, intraperitoneal and possibly other routes. Mutagenic data. When heated to decomposition, it emits very toxic fumes of SO_x, PO_x, and Cl^-.

TOXICITY DATA and CODEN

otr-rat:emb 1400 ng/plate JJATDK 1,190,81

orl-rat LD50:13 mg/kg DOEAAH 35,25,79

ihl-rat LC50:303 mg/m3 VHTODE 24,87,82

skn-rat LD50:860 mg/kg WRPCA2 9,119,70

CNV000 CAS: 91-64-5

COUMARIN

mf: $C_9H_6O_2$ mw: 146.15

PROP: Crystals; fragrant, pleasant odor; burning taste. Mp: 70°, bp: 291.0°, vap press: 1 mm @ 106.0°.

SYNS: 2H-1-BENZOPYRAN-2-ONE ◇ 1,2-BENZOPYRONE ◇ cis-o-COUMARINIC ACID LACTONE ◇ COUMARINIC ANHYDRIDE ◇ o-HYDROXYCINNAMIC ACID LACTONE ◇ o-HYDROXYZIMTSAURE-LACTON (GERMAN) ◇ NCI-C60297 ◇ 2-OXO-1,2-BENZOPYRAN ◇ RATTEX ◇ TONKA BEAN CAMPHOR

USE IN FOOD:

Purpose: Flavoring agent.

Where Used: Prohibited from foods.

Regulations: FDA - 21CFR 189.130. Prohibited from direct addition or use in human food.

IARC Cancer Review: Animal Limited Evidence IMEMDT 10,113,76. EPA Genetic Toxicology Program.

SAFETY PROFILE: Poison by ingestion, intraperitoneal, and subcutaneous routes. An experimental carcinogen, tumorigen, and teratogen. Mutagenic data. Combustible when exposed to heat or flame. When heated to decomposition it emits acrid smoke and fumes.

TOXICITY DATA and CODEN

mma-sat 1 mg/plate ENMUDM 5(Suppl 1),3,83
dnd-mam:Lym 20 mmol/L PNASA6 48,686,62
orl-mus TDLo:3600 mg/kg (6-17D preg):TER ARZNAD 17,97,67
orl-rat TD :200 g/kg/2Y-C:ETA TXCYAC 1,93,73
orl-rat LD50:293 mg/kg FCTXAV 12,385,74
scu-mus LD50:242 mg/kg YKKZAJ 83,1124,63

CNV100
COUMARONE-INDENE RESIN

USE IN FOOD:

Purpose: Protective coating.

Where Used: Grapefruit, lemons, limes, oranges, tangelos, tangerines.

Regulations: FDA - 21CFR 172.215. Limitation of 200 ppm on a fresh-weight basis.

SAFETY PROFILE: When heated to decomposition it emits acrid smoke and irritating fumes.

COE175
CUBEB OIL

PROP: From steam distillation of mature, unripe fruit of *piper cubeba* L. (Fam. *Piperaceae*). Colorless to light green liquid; spicy odor, slt acrid taste. D: 0.898-0.928, refr index: 1.492-1.502 @ 20°. Sol in fixed oils, mineral oil; insol in glycerin, propylene glycol.

USE IN FOOD:

Purpose: Flavoring agent.

Where Used: Various.

Regulations: FDA - 21CFR 172.510. Use at a level not in excess of the amount reasonably required to accomplish the intended effect.

SAFETY PROFILE: When heated to decomposition it emits acrid smoke and irritating fumes.

COE500 CAS: 122-03-2
CUMALDEHYDE

mf: $C_{10}H_{12}O$ mw: 148.22

PROP: Found in at least 50 essential oils such as cumin, eucalyptus species, cinnamon, boldo and rue, and as main constituent of oil of *Pectis papposa harn* and *gray* (FCTXAV 12,385,74). Colorless to pale yellow liquid; pungent odor of cumin. D: 0.976-0.980, refr index: 1.529-1.534, flash p: 199°F. Sol in alc, ether; insol in water.

SYNS: p-CUMIC ALDEHYDE ◇ CUMINALDEHYDE ◇ CUMINIC ALDEHYDE (FCC) ◇ CUMINYL ALDEHYDE ◇ FEMA No. 2341 ◇ p-ISOPROPYLBENZALDEHYDE ◇ 4-ISOPROPYLBENZALDEHYDE ◇ p-ISOPROPYLBENZENECARBOXALDEHYDE ◇ 4-(1-METHYLETHYL)-BENZALDEHYDE (9CI)

USE IN FOOD:

Purpose: Flavoring agent.

Where Used: Various.

Regulations: FDA - 21CFR 172.515. Use at a level not in excess of the amount reasonably required to accomplish the intended effect.

SAFETY PROFILE: Moderately toxic by ingestion and skin contact. A skin irritant. Combustible liquid. When heated to decomposition it emits acrid smoke and irritating fumes.

TOXICITY DATA and CODEN

skn-rbt 500 mg/24H FCTXAV 12,395,74
orl-rat LD50:1390 mg/kg FCTXAV 2,327,64
skn-rbt LD50:2800 mg/kg FCTXAV 12,395,74

COF000 CAS: 93-53-8
CUMENE ALDEHYDE

mf: $C_9H_{10}O$ mw: 134.19

PROP: Colorless liquid; floral odor. D: 0.998-1.006, refr index: 1.515-1.520, flash p: 156°F. Sol in fixed oils; sltly sol in propylene glycol; insol in glycerin.

SYNS: FEMA No. 2886 ◇ α-FORMYLETHYLBENZENE ◇ HYACINTHAL ◇ HYDRATROPA ALDEHYDE ◇ HYDRATROPIC ALDEHYDE ◇ α-METHYL PHENYLACETALDEHYDE ◇ α-METHYL-α-TOLUIC ALDEHYDE ◇ 2-PHENYLPROPANAL ◇ α-PHENYLPROPIONALDEHYDE ◇ 2-PHENYLPROPIONALDEHYDE (FCC)

USE IN FOOD:

Purpose: Flavoring agent.

Where Used: Various.

Regulations: FDA - 21CFR 172.515. Use at a level not in excess of the amount reasonably required to accomplish the intended effect.

SAFETY PROFILE: Moderately toxic by ingestion. Combustible liquid. When heated to decomposition it emits acrid smoke and irritating fumes.

TOXICITY DATA and CODEN

orl-rat LD50:2800 mg/kg FCTXAV 2,327,64

COF325 CAS: 8014-13-9
CUMIN OIL

PROP: From steam distillation of *Cuminum cyminum* L. Light yellow to brown liquid; strong odor. D: 0.905-0.925, refr index: 1.501 @ 20°. Sol in fixed oils, mineral oil; very sol in glycerin, propylene glycol.

SYNS: CUMMIN ◇ OILS, CUMIN

USE IN FOOD:

Purpose: Flavoring agent.

Where Used: Cheese, meat, relishes, soups.

Regulations: FDA - 21CFR 182.10, 182.20. GRAS when used at a level not in excess of the amount reasonably required to accomplish the intended effect.

SAFETY PROFILE: Moderately toxic by ingestion and skin contact. A skin irritant. Mutagenic data. When heated to decomposition it emits acrid smoke and irritating fumes.

TOXICITY DATA and CODEN

skn-rbt 500 mg/24H MOD FCTXAV 12,869,74
mmo-sat 100 μL/plate NUCADQ 1,10,79
bfa-rat/sat 2500 mg/kg NUCADQ 1,10,79
orl-rat LD50:2500 mg/kg FCTXAV 12,869,74
skn-rbt LD50:3560 mg/kg FCTXAV 12,869,74

COF680 CAS: 7681-65-4
CUPROUS IODIDE

mf: CuI mw: 190.46

PROP: White crystalline powder.

SYNS: COPPER IODIDE ◇ COPPER(I) IODIDE

USE IN FOOD:

Purpose: Dietary supplement.

Where Used: Table salt.

Regulations: FDA - 21CFR 184.1265. GRAS with a limitation of 0.01 percent in table salt.

SAFETY PROFILE: When heated to decomposition it emits toxic fumes of I^-.

COG000 CAS: 8024-37-1
CURCUMIN

SYNS: C.I. 75300 ◇ CURCUMA OIL ◇ CURCUMINE ◇ NCI-C60015 ◇ TURMERIC OIL ◇ TURMERIC OLEORESIN

USE IN FOOD:

Purpose: Color additive.

Where Used: Various.

Regulations: FDA - 21CFR 73.615. Use in accordance with good manufacturing practice.

SAFETY PROFILE: Moderately toxic by intraperitoneal route. A skin irritant. Mutagenic data. When heated to decomposition it emits acrid smoke and irritating fumes.

TOXICITY DATA and CODEN

skn-rbt 500 mg/24H MLD FCTOD7 21,839,83
cyt-ham:lng 20 mg/L GMCRDC 27,95,81
ipr-mus LD50:1500 mg/kg IJMRAQ 64,601,76

COH500 CAS: 420-04-2
CYANAMIDE

mf: CH_2N_2 mw: 42.05

PROP: Deliquescent crystals. Mp: 45°, bp: 260°, flash p: 285°F, d: 1.282, vap d: 1.45.

SYNS: AMIDOCYANOGEN ◇ CARBAMONITRILE ◇ CARBIMIDE ◇ CYANOAMINE ◇ CYANOGENAMIDE ◇ CYANOGEN NITRIDE ◇ HYDROGEN CYANAMIDE ◇ USAF EK-1995

USE IN FOOD:

Purpose: Fumigant.

Where Used: Bacon (uncooked), cereal flours, cereals that are cooked before being eaten, cocoa, ham (uncooked), sausage (uncooked).

Regulations: FDA - 21CFR 193.240. Fumigant residue tolerance of 125 ppm in cereal flours, 90 ppm in cereals that are cooked before being eaten, 50 ppm in uncooked bacon, ham, and sausage, 200 ppm in cocoa.

Cyanide and its compounds are on the Community Right-To-Know List.

OSHA PEL: TWA 2 mg/m^3 ACGIH TLV: TWA 2 mg/m^3

SAFETY PROFILE: Poison by ingestion, inhalation, and intraperitoneal route. Moderately toxic by skin contact. A severe eye irritant. Combustible when exposed to heat or flame. To fight fire, use CO_2, dry chemical. Thermally unstable. Contact with moisture (water), acids, or alkalies may cause a violent reaction above 40°. Concentrated aqueous solutions may undergo explosive polymerization. Mixture with 1,2-phenylenediamine salts may cause explosive polymerization. When heated to decomposition or on contact with acid or acid fumes, it emits toxic fumes of CN^- and NO_x.

TOXICITY DATA and CODEN

eye-rbt 100 mg SEV 34ZIAG -,190,69
orl-rat LD50:125 mg/kg MEIEDD 10,383,83
ipr-rat LDLo:200 mg/kg PSEBAA 54,254,43
ipr-mus LD50:200 mg/kg NTIS** AD277-689
skn-rbt LD50:590 mg/kg 34ZIAG -,190,69

CON750
CYANO(4-FLUORO-3-PHENOXYPHENYL)METHYL-3-(2,2-DICHLOROETHENYL)-2,2-DIMETHYL-CYCLOPROPANECARBOXYLATE

USE IN FOOD:

Purpose: Insecticide.

Where Used: Animal feed, cottonseed hulls, cottonseed oil, tomato pomace (dried), tomato pomace (wet), tomato products (concentrated).

Regulations: FDA - 21CFR 193.98. Limitation of 2.0 ppm in cottonseed oil. Temporary tolerance of 0.5 ppm in concentrated tomato products. 21CFR 561.96. Limitation of 2.0 ppm in cotton seed hulls, tomato pomace (wet), 5.0 ppm in tomato pomace (dry) when used for animal feed.

SAFETY PROFILE: When heated to decomposition emits toxic fumes of CN^-, F^-, and Cl^-.

COQ380
(+)CYANO(3-PHENOXYPHENYL)METHYL(±)-1-(DIFLUOROMETHOXY)-α-(1-METHYLETHYL)BENZENEACETATE

SYN: FLUCYTHRINATE

USE IN FOOD:

Purpose: Insecticide.

Where Used: Animal feed, apple pomace (dried), apples, cottonseed oil.

Regulations: FDA - 21CFR 193.99. Limitation of 0.2 ppm in cottonseed oil. 21CFR 561.435. Limitation of 10.0 ppm in dried apple pomace when used for animal feed.

SAFETY PROFILE: When heated to decomposition emits toxic fumes of CN^-, F^-.

COU500 CAS: 103-95-7
CYCLAMEN ALDEHYDE

mf: $C_{13}H_{18}O$ mw: 190.31

PROP: Colorless liquid; strong, floral odor. D: 0.946-0.952, refr index: 1.503-1.508. Sol in fixed oils; insol in propylene glycol, glycerin.

SYNS: ALDEHYDE B ◇ CYCLAMAL ◇ FEMA No. 2743 ◇ p-ISOPROPYL-α-METHYLHYDROCINNAMIC ALDEHYDE ◇ p-ISOPROPYL-α-METHYLPHENYLPROPYL ALDEHYDE ◇ α-METHYL-p-ISOPROPYLHYDROCINNAMALDEHYDE ◇ 2-METHYL-3-(p-ISOPROPYLPHENYL)PROPIONALDEHYDE

USE IN FOOD:

Purpose: Flavoring agent.

Where Used: Various.

Regulations: GRAS when used at a level not in excess of the amount reasonably required to accomplish the intended effect.

SAFETY PROFILE: Moderately toxic by ingestion. A human skin irritant. When heated to decomposition it emits acrid smoke and irritating fumes.

TOXICITY DATA and CODEN

skn-hmn 15 mg/48H MLD FCTXAV 12,385,74
orl-rat LD50:3810 mg/kg FCTXAV 2,327,64

CPB000 CAS: 110-82-7
CYCLOHEXANE

DOT: 1145

mf: C_6H_{12} mw: 84.18

PROP: Colorless, mobile liquid; pungent odor. Mp: 6.5°, bp: 80.7°, fp: 4.6°, flash: p: 1.4°F, ULC: 90-95, lel: 1.3%, uel: 8.4%, d: 0.7791 @ 20°/4°, autoign temp: 473°F, vap press: 100 mm @ 60.8°, vap d: 2.90.

SYNS: CICLOESANO (ITALIAN) ◇ CYCLOHEXAAN (DUTCH) ◇ CYCLOHEXAN (GERMAN) ◇ CYKLOHEKSAN (POLISH) ◇ HEXAHYDROBENZENE ◇ HEXAMETHYLENE ◇ HEXANAPHTHENE ◇ RCRA WASTE NUMBER U056

USE IN FOOD:

Purpose: Color diluent.

Where Used: Various.

Regulations: FDA - 21CFR 73.

Community Right-To-Know List.

OSHA PEL: TWA 300 ppm ACGIH TLV: TWA 300 ppm DOT Classification: Flammable Liquid; Label: Flammable Liquid

SAFETY PROFILE: Poison by intravenous route. Moderately toxic by ingestion. A systemic irritant by inhalation and ingestion. A skin irritant. Mutagenic data. Flammable liquid. Dangerous fire hazard when exposed to heat or flame; can react with oxidizing materials. Moderate explosion hazard in the form of vapor when exposed to flame. When mixed hot with liquid dinitrogen tetraoxide an explosion resulted. To fight fire, use foam, CO_2, dry chemical, spray, fog. When heated to decomposition it emits acrid smoke and fumes.

TOXICITY DATA and CODEN

skn-rbt 1548 mg/2D-I JIHTAB 25,199,43
dnd-esc 10 μmol/L MUREAV 89,95,81
orl-rat LD50:29820 mg/kg JIHTAB 25,415,43
ivn-rbt LDLo:77 mg/kg JPMRAB 3,1,28

CPF000 CAS: 622-45-7
CYCLOHEXYL ACETATE

DOT: 2243
mf: $C_8H_{14}O_2$ mw: 142.22

PROP: Pale yellow liquid; fruity odor. Bp: 177°, d: 0.996, vap d: 4.9, flash p: 136°F, autoign temp: 633°F.

SYNS: CYCLOHEXANOL ACETATE ◇ CYCLOHEXANOLAZETAT (GERMAN) ◇ CYCLOHEXANYL ACETATE

USE IN FOOD:

Purpose: Flavoring agent.

Where Used: Baked goods, beverages, candy, ice cream.

Regulations: FDA - 21CFR 172.515. Use at a level not in excess of the amount reasonably required to accomplish the intended effect.

DOT Classification: Flammable or Combustible Liquid; Label: Flammable Liquid

SAFETY PROFILE: Moderately toxic by subcutaneous route. Mildly toxic by ingestion and skin contact. Human systemic effects by inhalation: conjunctiva irritation and unspecified respiratory system changes. A systemic irritant to humans. Flammable when exposed to heat or flame. When heated to decomposition it emits acrid smoke and irritating fumes.

TOXICITY DATA and CODEN

skn-rbt 500 mg/24H MOD FCTXAV 17(suppl),695,79
ihl-hmn TCLo:3000 mg/m^3/45M:IRR AHYGAJ 78,260,13
orl-rat LD50:6730 mg/kg TXAPA9 28,313,74
skn-rbt LD50:10 g/kg TXAPA9 28,313,74

CPF500 CAS: 108-91-8
CYCLOHEXYLAMINE

DOT: 2357
mf: $C_6H_{13}N$ mw: 99.20

PROP: Liquid; strong, fishy odor. Mp: −17.7°, bp: 134.5°, flash p: 69.8°F, d: 0.865 @ 25°/25°, autoign temp: 560°F, vap d. 3.42.

SYNS: AMINOCYCLOHEXANE ◇ AMINOHEXAHYDROBENZENE ◇ CHA ◇ CYCLOHEXANAMINE ◇ HEXAHYDROANILINE ◇ HEXAHYDROBENZENAMINE

USE IN FOOD:

Purpose: Boiler water additive.

Where Used: Various.

Regulations: FDA - 21CFR 173.310. Limitation of 10 ppm in steam and excluding use of such steam in contact with milk and milk products.

IARC Cancer Review: Animal No Evidence IMEMDT 22,55,80. EPA Extremely Hazardous Substances List. EPA Genetic Toxicology Program.

ACGIH TLV: TWA 10 ppm (skin) DFG MAK: 10 ppm (40 mg/m^3) DOT Classification: Flammable Liquid; Label: Flammable Liquid, Corrosive; Flammable or Combustible Liquid; Label: Flammable, Corrosive

SAFETY PROFILE: A poison by ingestion, skin contact, and intraperitoneal routes. Moderately toxic by subcutaneous and parenteral routes. An experimental teratogen. Other experimental reproductive effects. Severe human skin irritant. Can cause dermatitis; convulsions. Human mutagenic data. Flammable or combustible liquid. Dangerous fire hazard when exposed to heat, flame, or oxidizers. To fight fire, use alcohol foam, CO_2, dry chemical. When heated to decomposition it emits toxic fumes of NO_x.

TOXICITY DATA and CODEN

skn-hmn 125 mg/48H SEV AMIHBC 5,311,52
cyt-hmn:leu 10 μmol/L/5H MUREAV 39,1,76
hma-mus/leu 450 mg/kg/3D MUREAV 31,5,75
orl-rat TDLo:5600 mg/kg (4W male):REP FCTXAV 19,291,81
orl-mus TDLo:600 mg/kg (6-11D preg):TER SEIJBO 11,51,71

orl-mus TDLo: 120 mg/kg (6-11D preg): TER SEIJBO 11,51,71
orl-rat LD50: 156 mg/kg SKEZAP 14,542,73
skn-rbt LD50: 277 mg/kg AIHAAP 30,470,69

CPQ625 CAS: 100-88-9
N-CYCLOHEXYLSULPHAMIC ACID
mf: $C_6H_{13}NO_3S$ mw: 179.26

PROP: Crystals; sweet-sour taste. Mp: 169-170°. Fairly strong acid. Very sparingly soluble in water. Slowly hydrolyzed by hot water.

SYNS: CYCLAMATE ◇ CYCLAMIC ACID ◇ CYCLOHEXANESULPHAMIC ACID ◇ CYCLOHEXYLAMIDOSULPHURIC ACID ◇ CYCLOHEXYLAMINESULPHONIC ACID ◇ CYCLOHEXYLSULFAMIC ACID (9CI) ◇ CYCLOHEXYLSULPHAMIC ACID ◇ HEXAMIC ACID ◇ SUCARYL ◇ SUCARYL ACID

USE IN FOOD:

Purpose: Nonnutritive sweetener.

Where Used: Prohibited from foods.

Regulations: FDA - 21CFR 189.135. Prohibited from direct addition or use in human food.

SAFETY PROFILE: Poison by intravenous route. Mildly toxic by ingestion. A human carcinogen by ingestion (bladder tumors and hematuria). When heated to decomposition it emits toxic fumes of SO_x and NO_x.

TOXICITY DATA and CODEN

orl-man TDLo: 22 g/kg/77W-C: CAR,KID JOURAA 118,258,77
orl-man TD : 131 g/kg/5Y-C: CAR,KID JOURAA 118,258,77
orl-man TD : 164 g/kg/6Y-C: CAR,KID JOURAA 118,258,77
orl-rat LD50: 12 g/kg AJMSA9 225,551,53

CPS000 CAS: 115-25-3
CYCLOOCTAFLUOROBUTANE
DOT: 1976
mf: C_4F_8 mw: 200.03

PROP: Colorless, odorless gas. Bp: −6.04°, mp: −41.4°, d (liquid): 1.513 @ −70°F.

SYNS: FC-C 318 ◇ FREON C-318 ◇ HALOCARBON C-138 ◇ OCTAFLUOROCYCLOBUTANE (DOT) ◇ PERFLUOROCYCLOBUTANE ◇ PROPELLANT C318 ◇ R-C 318

USE IN FOOD:

Purpose: Aerating agent, propellant.

Where Used: Foamed food products, sprayed food products.

Regulations: FDA - 21CFR 173.360.

EPA Genetic Toxicology Program.

DOT Classification: Nonflammable Gas; Label: Nonflammable Gas

SAFETY PROFILE: Mildly toxic by ingestion and inhalation. Can cause slight transient effects at high concentrations. No anesthesia or central nervous system effects. Nonflammable Gas. Mutagenic data. When heated to decomposition it emits highly toxic fumes of F^-.

TOXICITY DATA and CODEN

sln-dmg-ihl 99 pph/10M ENVRAL 7,275,74

CQI000 CAS: 99-87-6
p-CYMENE
DOT: 2046
mf: $C_{10}H_{14}$ mw: 134.24

PROP: Colorless to pale yellow liquid; odorless. Mp: −68.2°, bp: 176°, lel: 0.7%, @ 100°, ULC: 30-35, flash p: 117°F (CC), d: 0.853, refr index: 1.489, autoign temp: 817°F, vap d: 4.62, vap press: 1 mm @ 17.3°, flash p: (technical) 127°F, uel (technical): 5.6%. Found in nearly 100 volatile oils including lemongrass, sage, thyme, coriander, star anise, and cinnamon (FCTXAV 12,385,74). Sol in alc, ether, acetone, benzene.

SYNS: CAMPHOGEN ◇ CYMENE ◇ CYMOL ◇ DOLCYMENE ◇ FEMA No. 2356 ◇ 4-ISOPROPYL-1-METHYLBENZENE ◇ p-ISOPROPYLTOLUENE ◇ p-METHYL-CUMENE ◇ p-METHYLISOPROPYL BENZENE ◇ 1-METHYL-4-ISOPROPYLBENZENE ◇ PARACYMENE ◇ PARACYMOL

USE IN FOOD:

Purpose: Flavoring agent.

Where Used: Various.

Regulations: FDA - 21CFR 172.515. Use at a level not in excess of the amount reasonably required to accomplish the intended effect.

DOT Classification: Flammable or Combustible Liquid; Label: Flammable Liquid

SAFETY PROFILE: Mildly toxic by ingestion. Humans sustain central nervous system effects at low doses. Mutagenic data. A skin irritant. Flammable or combustible liquid. Explosion Hazard: Slight in the form of vapor. To fight fire, use foam, CO_2, dry chemical. When heated to decomposition it emits acrid smoke and fumes.

TOXICITY DATA and CODEN

skn-rbt 500 mg/24H MOD FCTXAV 12,401,74
cyt-smc 200 μmol/tube HEREAY 33,457,47
orl-rat LD50:4750 mg/kg FCTXAV 2,327,64

CQK250 CAS: 52-89-1
l-CYSTEIN HYDROCHLORIDE
mf: $C_3H_7NO_2S \cdot ClH$ mw: 157.63

PROP: White crystalline powder; characteristic acetic taste. Mp: 175° (decomp). Sol in water, alc.

SYNS: CYSTEINE CHLORHYDRATE ◇ CYSTEINE HYDROCHLORIDE ◇ l-CYSTEINE HYDROCHLORIDE ◇ l-CYSTEINE MONOHYDROCHLORIDE (FCC)

USE IN FOOD:

Purpose: Dietary supplement, dough conditioner, nutrient.

Where Used: Baked goods (yeast leavened), baking mixes.

Regulations: FDA - 21CFR 172.320. Limitation 2.3 percent by weight. 21CFR 184.1271. GRAS with a limitation of 0.009 parts per 100 parts of flour

SAFETY PROFILE: Moderately toxic by intraperitoneal, intravenous, and possibly other routes. Mutagenic data. Used as a nutrient and dough conditioner. When heated to decomposition it emits very toxic fumes of NO_x, SO_x and Cl^-.

TOXICITY DATA and CODEN

mma-sat 20 mg/plate FCTOD7 22,623,84
cyt-ham:fbr 2 g/L FCTOD7 22,623,84
ipr-mus LD50:1250 mg/kg NTIS** AD691-490

CQK325 CAS: 56-89-3
l-CYSTINE
mf: $C_6H_{12}N_2O_4S_2$ mw: 240.30

PROP: Naturally occurring levorotatory form. Colorless to white hexagonal tablets from water. Decomp 260-261°. Sltly sol in water, alc. d-Cystine: Crystals. Sltly sol in water. dl-Cystine, the synthetic racemic form: Crystals. Sltly sol in water. meso-Cystine, the internally compensated form: Crystals. Sltly sol in water.

SYNS: CYSTEINE DISULFIDE ◇ CYSTIN ◇ (-)-CYSTINE ◇ CYSTINE ACID ◇ DICYSTEINE ◇ β,β′-DITHIODIALANINE ◇ GELUCYSTINE ◇ OXIDIZED l-CYSTEINE

USE IN FOOD:

Purpose: Dietary supplement, dough strengthener, nutrient.

Where Used: Baked goods (yeast leavened), baking mixes.

Regulations: FDA - 21CFR 172.320. Limitation of 2.3 percent by weight. 21CFR 184.1271. GRAS with a limitation of 0.009 parts per 100 parts of flour when used in accordance with good manufacturing practice.

SAFETY PROFILE: Experimental reproductive effects. When heated to decomposition it emits toxic fumes of PO_x and SO_x.

TOXICITY DATA and CODEN

orl-rat TDLo:9300 mg/kg (93D male):REP OYYAA2 15,199,78
orl-rat LDLo:25 g/kg OYYAA2 15,199,78

D

DAE450 CAS: 25152-84-5

trans,trans-2,4-DECADIENAL

mf: $C_{10}H_{16}O$ mw: 152.23

PROP: Yellow liquid; chicken fat odor. D: 0.806-0.876, refr index: 1.514-1.516, flash p: +212°F. Sol in alc, fixed oils; insol in water @ 104°.

SYNS: FEMA No. 3135 ◇ HEPTENYL ACROLEIN

USE IN FOOD:

Purpose: Flavoring agent.

Where Used: Various.

Regulations: FDA - 21CFR 172.515. Use at a level not in excess of the amount reasonably required to accomplish the intended effect.

SAFETY PROFILE: Combustible liquid. When heated to decomposition it emits acrid smoke and irritating fumes.

DAF200 CAS: 705-86-2

Δ-DECALACTONE

mf: $C_{10}H_{18}O_2$ mw: 170.28

PROP: Colorless liquid; coconut, fruity odor, butterlike on dilution. Refr index: 1.456-1.459. Very sol in alc, propylene glycol; insol in water @ 281°.

SYNS: AMYL-Δ-VALEROLACTONE ◇ DECANOLIDE-1,5 ◇ FEMA No. 2361

USE IN FOOD:

Purpose: Flavoring agent.

Where Used: Oleomargarine.

Regulations: FDA - 21CFR 172.515. Limitation of 10 ppm in oleomargarine.

SAFETY PROFILE: A skin and eye irritant. When heated to decomposition it emits acrid smoke and irritating fumes.

TOXICITY DATA and CODEN

skn-rbt 500 mg/24H MLD FCTXAV 14,659,76
eye-rbt 100 mg MLD NTIS** AD-A053-896

DAG000 CAS: 112-31-2

1-DECANAL

mf: $C_{10}H_{20}O$ mw: 156.30

PROP: Found in over 50 sources including citrus oils, citronella and lemongrass (FCTXAV 11,477,73). Colorless to light yellow liquid; floral, fatty odor. D: 0.830 @ 15°/4°, bp: 208°, flash p: 185°F. Sol in 80% alcohol, fixed oils, volatile oils and mineral oils; insol in water and glycerol.

SYNS: ALDEHYDE C10 ◇ C-10 ALDEHYDE ◇ CAPRALDEHYDE ◇ 1-DECYL ALDEHYDE ◇ FEMA No. 2362

USE IN FOOD:

Purpose: Flavoring agent.

Where Used: Various.

Regulations: FDA - 21CFR 182.60. Use at a level not in excess of the amount reasonably required to accomplish the intended effect.

SAFETY PROFILE: Moderately toxic by ingestion. Mildly toxic by skin contact. A severe skin irritant. Combustible liquid. When heated to decomposition it emits acrid smoke and irritating fumes.

TOXICITY DATA and CODEN

skn-rbt 14372 μg/24H open SEV AIHAAP 23,95,62
orl-rat LD50:3730 mg/kg FCTXAV 11,477,73
skn-rbt LD50:5040 mg/kg FCTXAV 11,477,73

DAG200 CAS: 112-31-2

1-DECANAL (MIXED ISOMERS)

SYN: FEMA No. 2362

USE IN FOOD:

Purpose: Flavoring agent.

Where Used: Various.

Regulations: FDA - 21CFR 182.60. Use at a level not in excess of the amount reasonably required to accomplish the intended effect.

SAFETY PROFILE: Moderately toxic by ingestion. A severe skin irritant. When heated to decomposition it emits acrid smoke and fumes.

TOXICITY DATA and CODEN

skn-rbt 14372 μg/24H open SEV AIHAAP 23,95,62
orl-rat LD50:3730 mg/kg AIHAAP 23,95,62

DAH400 CAS: 334-48-5
DECANOIC ACID
mf: $C_{10}H_{20}O_2$ mw: 172.30

PROP: White crystals; unpleasant odor. D: 0.8782 @ 50°/4°, bp: 270°, mp: 31.4°. Sol in most organic solvents and in dilute nitric acid; insol in water.

SYNS: CAPRIC ACID ◇ n-CAPRIC ACID ◇ CAPRINIC ACID ◇ CAPRYNIC ACID ◇ n-DECANOIC ACID ◇ n-DECOIC ACID ◇ DECYLIC ACID ◇ n-DECYLIC ACID ◇ HEXACID 1095 ◇ NEO-FAT 10 ◇ 1-NONANECARBOXYLIC ACID

USE IN FOOD:

Purpose: Component in the manufacture of other food-grade additives, defoaming agent, lubricant.

Where Used: Various.

Regulations: FDA - 21CFR 172.860. Use in accordance with good manufacturing practice.

SAFETY PROFILE: Poison by intravenous route. Mutagenic data. A skin irritant. When heated to decomposition it emits acrid smoke and irritating fumes.

TOXICITY DATA and CODEN

skn-rbt 500 mg/24H MOD FCTXAV 17,735,79
sln-smc 14500 ppb ANYAA9 407,186,83
ivn-mus LD50:129 mg/kg APTOA6 18,141,61

DAI350 CAS: 3913-71-1
2-DECENAL
mf: $C_{10}H_{18}O$ mw: 154.28

PROP: Sltly yellow liquid; orange odor. D: 0.836-0.846, refr index: 1.452-1.457. Sol in alc, fixed oils; insol in water.

SYNS: trans-2-DECEN-1-AL ◇ DECENALDEHYDE ◇ FEMA No. 2366

USE IN FOOD:

Purpose: Flavoring agent.

Where Used: Various.

Regulations: FDA - 21CFR 172.515. Use at a level not in excess of the amount reasonably required to accomplish the intended effect.

SAFETY PROFILE: Moderately toxic by skin contact. Mildly toxic by ingestion. A severe skin irritant. When heated to decomposition it emits acrid smoke and fumes.

TOXICITY DATA and CODEN

skn-rbt 500 mg/24H SEV FCTXAV 17,761,79
orl-rat LD50:5000 mg/kg FCTXAV 17,761,79
skn-rbt LD50:3400 mg/kg FCTXAV 17,761,79

DAI360
cis-4-DECENAL
mf: $C_{10}H_{18}O$ mw: 154.28

PROP: Colorless to slightly yellow liquid; fatty, orangelike odor. D: 0.847, refr index: 1.442-1.444, Sol in alc, fixed oils; insol in water.

SYNS:

cis-4-DECEN-1-AL (FCC) ◇ FEMA No. 3264

USE IN FOOD:

Purpose: Flavoring agent.

Where Used: Various.

Regulations: GRAS when used at a level not in excess of the amount reasonably required to accomplish the intended effect.

SAFETY PROFILE: When heated to decomposition it emits acrid smoke and irritating fumes.

DAI495 CAS: 18507-89-6
DECOQUINATE
mf: $C_{24}H_{35}NO_5$ mw: 417.53

SYNS: DECCOX ◇ 6-DECYLOXY-7-ETHOXY-4-HYDROXY-3-QUINOLINECARBOXYLID ACID ETHYL ESTER ◇ ETHYL 6-(N-DECYLOXY)-7-ETHOXY-4-HYDROXYQUINOLINE-3-CARBOXYLATE

USE IN FOOD:

Purpose: Animal drug.

Where Used: Beef, chicken, goat.

Regulations: FDA - 21CFR 556.170. Limitation of 2 ppm in uncooked edible tissues other than muscle and 1 ppm in skeletal muscle of chickens, cattle, and goats. 21CFR 558.195. USDA CES Ranking: Z-1 (1986).

SAFETY PROFILE: When heated to decomposition it emits acrid smoke and irritating fumes.

DAI600 CAS: 112-30-1
DECYL ALCOHOL
mf: $C_{10}H_{22}O$ mw: 158.32

PROP: Found in sweet orange and a few other essential oils (FCTXAV 11,95,73). Colorless, viscous, refractive liquid; floral fruity odor. Mp: 7°, bp: 232.9°, flash p: 180°F (OC), d: 0.8297 @ 20°/4°, refr index: 1.435-1.439, vap press: 1 mm @ 69.5°, vap d: 5.3. Sol in alc, ether, mineral oil, propylene glycol, fixed oils; insol glycerin water @ 233°.

SYNS: AGENT 504 ◇ ALCOHOL C-10 ◇ ANTAK ◇ C 10 ALCOHOL ◇ CAPRIC ALCOHOL ◇ CAPRINIC ALCOHOL ◇ DECANAL DIMETHYL ACETAL ◇ DECANOL ◇ n-DECANOL ◇ 1-DECANOL (FCC) ◇ n-DECATYL ALCOHOL ◇ n-DECYL ALCOHOL ◇ DECYLIC ALCOHOL ◇ DYTOL S-91 ◇ EPAL 10 ◇ FEMA No. 2365 ◇ LOROL 22 ◇ NONYLCARBINOL ◇ PRIMARY DECYL ALCOHOL ◇ ROYALTAC ◇ SIPOL L10

USE IN FOOD:

Purpose: Flavoring agent, intermediate.

Where Used: Baked goods, beverages, candy, ice cream.

Regulations: FDA - 21CFR 172.515, 172.864. Use at a level not in excess of the amount reasonably required to accomplish the intended effect.

SAFETY PROFILE: Moderately toxic by skin contact. Mildly toxic by ingestion, inhalation and possibly other routes. An experimental tumorigen. A severe human skin and eye irritant. Combustible when exposed to heat or flame; can react with oxidizing materials. To fight fire, use foam, CO_2, dry chemical. When heated to decomposition it emits acrid smoke and irritating fumes.

TOXICITY DATA and CODEN

skn-hmn 75 mg/3D-I SEV 85DKA8 -,127,77
skn-rbt 2600 mg/kg/24H MOD AIHAAP 34,493,73
eye-rbt 83 mg SEV AIHAAP 34,493,73
skn-mus TDLo: 12 g/kg/25W-I: ETA TXAPA9 9,70,66
orl-rat LD50: 4720 mg/kg AIHAAP 34,493,73
ihl-mus LC50: 4 g/m^3/2H 85GMAT -,42,82
skn-rbt LD50: 3560 mg/kg FCTXAV 11,95,73

DAJ000 CAS: 1322-98-1
DECYL BENZENE SODIUM SULFONATE

mf: $C_{16}H_{25}O_3S \cdot Na$ mw: 320.46

SYNS: SODIUM DECYLBENZENESULFONAMIDE ◇ SODIUM DECYLBENZENESULFONATE

USE IN FOOD:

Purpose: Defoaming agent, dispersing adjuvant.

Where Used: Citrus fruit (fresh).

Regulations: FDA - 21CFR 172.210.

SAFETY PROFILE: Poison by intravenous route. Moderately toxic by ingestion. A severe eye irritant. When heated to decomposition it emits toxic fumes of SO_x.

TOXICITY DATA and CODEN

eye-rbt 450 mg SEV AROPAW 40,668,48
orl-mus LD50: 2000 mg/kg PSTGAW 3,1,45
ivn-mus LD50: 115 mg/kg JAPMA8 38,428,49

DAP000 CAS: 301-12-2
DEMETON-O-METHYL SULFOXIDE

mf: $C_6H_{15}O_3PS_2$ mw: 230.30

SYNS: BAY 21097 ◇ DEMETON-S-METHYL-SULFOXID (GERMAN) ◇ DEMETON-S-METHYL SULFOXIDE ◇ DEMETON-METHYL SULPHOXIDE ◇ O,O-DIMETHYL-S-(2-AETHYLSULFINYL-AETHYL)-THIOLPHOSPHAT (GERMAN) ◇ O,O-DIMETHYL-S-(2-ETHTHIONYLETHYL) PHOSPHOROTHIOATE ◇ DIMETHYL-S-(2-ETHTHIONYLETHYL) THIOPHOSPHATE ◇ O,O-DIMETHYL-S-(2-ETHYLSULFINYL-ETHYL)-MONOTHIOFOSFAAT (DUTCH) ◇ O,O-DIMETHYL-S-(2-(ETHYLSULFINYL)ETHYL) PHOSPHOROTHIOATE ◇ O,O-DIMETHYL-S-(2-ETHYLSULFINYL)ETHYL THIOPHOSPHATE ◇ O,O-DIMETHYL-S-ETHYLSULPHINYLETHYL PHOSPHOROTHIOLATE ◇ O,O-DIMETHYL-S-(3-OXO-3-THIA-PENTYL)-MONOTHIOPHOSPHAT (GERMAN) ◇ O,O-DIMETIL-S-(2-ETIL-SOLFINIL-ETIL)-MONOTIOFOSFATO (ITALIAN) ◇ ENT 24,964 ◇ S-(2-(ETHYLSULFINYL) ETHYL)-O,O-DIMETHYL PHOSPHOROTHIOATE ◇ ISOMETHYLSYSTOX SULFOXIDE ◇ METAISOSYSTOXSULFOXIDE ◇ METASYSTEMOX ◇ METASYSTOX-R ◇ METHYL DEMETON-O-SULFOXIDE ◇ METILMERCAPTOFOSOKSID ◇ OXYDEMETONMETHYL ◇ OXYDEMETON-METILE (ITALIAN) ◇ R 2170 ◇ THIOPHOSPHATE de O,O-DIMETHYLE et de S-2-ETHYLSULFINYLETHYLE (FRENCH)

USE IN FOOD:

Purpose: Insecticide.

Where Used: Animal feed, sorghum (milled fractions, except flour).

Regulations: FDA - 21CFR 561.234. Limitation of 2.0 ppm in when used for animal feed.

EPA Genetic Toxicology Program.

SAFETY PROFILE: Poison by ingestion, skin contact, intravenous, intraperitoneal, and possibly other routes. Human mutagenic data. When heated to decomposition it emits very toxic fumes of PO_x and SO_x.

TOXICITY DATA and CODEN

mmo-sat 50 μg/plate MUREAV 124,97,83
sce-hmn: lym 20 mg/L MUREAV 124,97,83
orl-rat LD50: 30 mg/kg AEPPAE 234,352,58
skn-rat LD50: 100 mg/kg WRPCA2 9,119,70

ipr-rat LD50:20 mg/kg GUCHAZ 6,385,73
ivn-rat LD50:47 mg/kg BIJOAK 67,187,57

DAP200 CAS: 126-75-0
DEMETON-S
mf: $C_8H_{19}O_3PS_2$ mw: 258.36

SYNS: O,O-DIAETHYL-S-(2-AETHYLTHIO-AETHYL)-MONOTHIOPHOSPHAT (GERMAN) ◇ DIAETHYLTHIO-PHOSPHORSAEUREESTER des AETHYLTHIOGLY-KOL (GERMAN) ◇ DIETHYL-S-(2-ETHIOETHYL)THIO-PHOSPHATE ◇ O,O-DIETHYL-S-(2-ETHTHIOETHYL) PHOSPHOROTHIOATE ◇ O,O-DIETHYL-S-ETHYL-2-ETHYLMERCAPTOPHOSPHOROTHIOLATE ◇ O,O-DIETHYL-S-(2-ETHYLTHIO-ETHYL)-MONOTHIOFOSFAAT (DUTCH) ◇ O,O-DIETHYL-S-2-(ETHYLTHIO)ETHYL PHOS-PHOROTHIOATE ◇ O,O-DIETHYL-S-(2-(ETHYLTHIO) ETHYL) PHOSPHOROTHIOLATE (USDA) ◇ O,O-DIETIL-S-(2-ETILTIO-ETIL)-MONOTIOFOSFATO (ITALIAN) ◇ O,O-DIETYL-S-2-ETYLMERKAPTOETYLTIOFOSFAT (CZECH) ◇ 2-(ETHYLTHIO)-ETHANETHIOL S-ESTER with O,O-DIETHYL PHOSPHOROTHIOATE ◇ ISODEMETON ◇ IZOSYSTOX (CZECH) ◇ PO-SYSTOX ◇ THIOLDEMETON ◇ THIOL SYSTOX ◇ THIOPHOSPHATE de O,O-DIETHYLE et de S-(2-ETHYLTHIO-ETHYLE) (FRENCH)

USE IN FOOD:

Purpose: Pesticide.

Where Used: Animal feed, sugar beet pulp (dehydrated).

Regulations: FDA - 21CFR 561.130. Limitation of 5 ppm in dehydrated sugar beet pulp when used for animal feed. USDA CES Ranking: A-2 (1988).

SAFETY PROFILE: Poison by ingestion, intraperitoneal and subcutaneous routes. When heated to decomposition it emits very toxic fumes of PO_x and SO_x.

TOXICITY DATA and CODEN

orl-rat LD50:1500 µg/kg AEPPAE 217,144,53
ipr-rat LD50:1500 µg/kg AMIHAB 13,606,56

DAQ400 CAS: 83-44-3
DEOXYCHOLATIC ACID
mf: $C_{24}H_{40}O_4$ mw: 392.64

PROP: A white crystalline powder. Mp: 178°. Sol in alc, acetone; sltly sol in ether, chloroform; insol in water.

SYNS: CHOLEIC ACID ◇ CHOLEREBIC ◇ CHOLOREBIC ◇ DEGALOL ◇ DEOXYCHOLIC ACID (FCC) ◇ 7-α-DEOXY-CHOLIC ACID ◇ DESOXYCHOLIC ACID ◇ DESOXYCHOL-SAEURE (GERMAN) ◇ 3,12-DIHYDROXYCHOLANIC ACID ◇ 3-α,12-α-DIHYDROXYCHOLANIC ACID ◇ 3-α,12-α-DIHY-DROXY-5-β-CHOLAN-24-OIC ACID ◇ 3-α,12-α-DIHY-DROXY-5-β-CHOLANOIC ACID ◇ 3-α,12-α-DIHYDROXY-CHOLANSAEURE (GERMAN) ◇ DROXOLAN ◇ 17-β-(1-METHYL-3-CARBOXYPROPYL)-ETIOCHOLANE-3-α, 12-α-DIOL ◇ PYROCHOL ◇ SEPTOCHOL

USE IN FOOD:

Purpose: Emulsifier.

Where Used: Egg white (dried).

Regulations: GRAS when used at a level not in excess of the amount reasonably required to accomplish the intended effect.

SAFETY PROFILE: Poison by intraperitoneal route. Moderately toxic by ingestion and intravenous routes. An experimental tumorigen. Mutagenic data. When heated to decomposition it emits acrid smoke and irritating fumes.

TOXICITY DATA and CODEN

mmo-sat 20 mg/L MUREAV 158,45,85
sln-smc 100 mg/L CRNGDP 5,447,84
skn-mus TDLo:2700 mg/kg/10W-I:ETA BJCAAI 10,363,56
scu-mus TDLo:1120 mg/kg/22W-I:ETA NATUAS 145,627,40
orl-rat LD50:1 g/kg NAIZAM 33,71,82

DBD800 CAS: 9004-53-9
DEXTRINS
mf: $(C_6H_{10}O_5)_n \cdot xH_2O$

PROP: An intermediate product formed by the hydrolysis of starches. It describes a class of substances. Yellow or white powder or granules. Sol in water; insol in alc and ether, colloidal in properties.

SYNS: ARTIFICIAL GUM ◇ DEXTRANS ◇ STARCH GUM ◇ TAPIOCA ◇ VEGETABLE GUM

USE IN FOOD:

Purpose: Binder, colloidal stabilizer, extender, formulation aid, processing aid, surface-finishing agent, thickener.

Where Used: Baked goods, beverages (dry mix), confectionery products, egg roll, food-contact surfaces, gravies, pie fillings, poultry, puddings, soups.

Regulations: FDA - 21CFR 184.1277, 186.1275. GRAS when used in accordance with good manufacturing practice. USDA - 9CFR 318.7, 9CFR 381.147. Sufficient for purpose.

SAFETY PROFILE: Mildly toxic by intravenous route. When heated to decomposition it emits acrid smoke and irritating fumes.

DBH700
DIACETYL TARTARIC ACID ESTERS of MONO- and DIGLYCERIDES

PROP: Vary from sticky, viscous liquid to waxy solid; faint acid odor. Sol in oil, methanol, acetone, acetic acid, water.

USE IN FOOD:

Purpose: Emulsifier.

Purpose: Emulsifier, emulsifier salt, flavoring agent.

Where Used: Baked goods, baking mixes, beverages (nonalcoholic), chocolate couverture, coffee whiteners, confections, dairy product analogs, fats, frostings, margarine, oils, oleomargarine.

Regulations: FDA - 21CFR 184.1101. GRAS when used in accordance with good manufacturing practice. USDA - 9CFR 318.7. Sufficient for purpose, 0.5 percent in oleomargarine or margarine. 9CFR 381.147. Sufficient for purpose in poultry.

SAFETY PROFILE: When heated to decomposition it emits acrid smoke and irritating fumes.

DBI099 CAS: 10311-84-9
DIALIFOR

mf: $C_{14}H_{17}NO_4PS_2$ mw: 393.86

SYNS: S-(2-CHLORO-1-(1,3-DIHYDRO-1,3-DIOXO-2H-ISOINDOL-2-YL)ETHYL)-O,O-DIETHYL PHOSPHORODITHIOATE ◇ S-(2-CHLORO-1-PHTHALIMIDOETHYL)-O,O-DIETHYL PHOSPHORODITHIOATE ◇ O,O-DIETHYL-S-(2-CHLORO-1-PHTHALIMIDOETHYL)PHOSPHORODITHIOATE ◇ ENT 27,320 ◇ HERCULES 14503 ◇ PHOSPHORODITHIOIC ACID-S-(2-CHLORO-1-(1,3-DIHYDRO-1,3-DIOXO-2H-ISOINDOL-2-YL)ETHYL-O,O-DIETHYL ESTER ◇ PHOSPHORODITHIOIC ACID-S-(2-CHLORO-1-PHTHALIMIDOETHYL)-O,)-DIETHYL ESTER ◇ TORAK

USE IN FOOD:

Purpose: Insecticide.

Where Used: Animal feed, apple pomace (dried), citrus pulp (dried), grape pomace (dried), raisin waste, raisins.

Regulations: FDA - 21CFR 193.130. Insecticide residue tolerance of 2 ppm in raisins. 21CFR 561.140. Limitation of 40 ppm in dried apple pomace, 20 ppm in dried grape pomace, 15 ppm in dried citrus pulp, 10 ppm in raisin waste when used for animal feed.

SAFETY PROFILE: Poison by ingestion and skin contact. Experimental reproductive effects. When heated to decomposition it emits toxic fumes of SO_x, PO_x, and NO_x.

TOXICITY DATA and CODEN

orl-ham TDLO:100 mg/kg (8D preg):REP TXAPA9 16,24,70
orl-ham TDLO:200 mg/kg (7D preg):REP TXAPA9 16,24,70
orl-rat LD50:5 mg/kg BESAAT 15,122,69

DCJ800 CAS: 68855-54-9
DIATOMACEOUS EARTH

PROP: Composed of skeletons of small aquatic plants related to algae and contains as much as 88% amorphous silica (DTLVS* 4,120,80). White to buff colored solid. Insol in water; sol in hydrofluoric acid.

SYNS: D.E. ◇ DIATOMACEOUS SILICA ◇ DIATOMITE ◇ INFUSORIAL EARTH ◇ KIESELGUHR

USE IN FOOD:

Purpose: Fat refining, filter aid, insecticide.

Where Used: Fats (rendered).

Regulations: FDA - 21CFR 193.135, 561.145. Insecticide. USDA - 9CFR 318.7. Sufficient for purpose.

OSHA PEL: 80 $mg/m^3/\%SiO_2$ ACGIH TLV: TWA 10 mg/m^3 (dust)

SAFETY PROFILE: The dust may cause fibrosis of the lungs. Roasting or calcining at high temperatures produces cristobalite, and tridymite, thus increasing the fibrogenicity of the material.

DCL000 CAS: 333-41-5
DIAZIDE

DOT: 2783

mf: $C_{12}H_{21}N_2O_3PS$ mw: 304.38

PROP: Liquid with faint ester-like odor. Bp: 84° @ 0.002 mm, d: 1.116 @ 20°/4°. Miscible in organic solvents.

SYNS: ALFA-TOX ◇ BASUDIN ◇ BASUDIN 10 G ◇ BAZUDEN ◇ DAZZEL ◇ O,O-DIAETHYL-O-(2-ISOPROPYL-4-METHYL-PYRIMIDIN-6-YL)-MONOTHIOPHOSPHAT (GERMAN) ◇ O,O-DIAETHYL-O-(2-ISOPROPYL-4-METHYL)-

6-PYRIMIDYL-THIONOPHOSPHAT (GERMAN) ◇ DIANON ◇ DIATERR-FOS ◇ DIAZAJET ◇ DIAZATOL ◇ DIAZINON (ACGIH, DOT) ◇ DIAZINONE ◇ DIAZITOL ◇ DIAZOL ◇ O,O-DIETHYL-O-(2-ISOPROPYL-4-METHYL-PYRIMIDIN-6-YL)MONOTHIOFOSFAAT (DUTCH) ◇ O,O-DIETHYL-O-(2-ISOPROPYL-4-METHYL-6-PYRIMIDINYL)PHOSPHORO-THIOATE ◇ O,O-DIETHYL-O-(2-ISOPROPYL-6-METHYL-4-PYRIMIDINYL) PHOSPHOROTHIOATE ◇ DIETHYL 4-(2-ISO-PROPYL-6-METHYLPYRIMIDINYL)PHOSPHOROTHIONATE ◇ O,O-DIETHYL-O-(2-ISOPROPYL-4-METHYL-6-PYRIMI-DYL)PHOSPHOROTHIOATE ◇ O,O-DIETHYL-O-(2-ISO-PROPYL-4-METHYL-6-PYRIMIDYL) THIONOPHOSPHATE ◇ O,O-DIETHYL-2-ISOPROPYL-4-METHYLPYRIMIDYL-6-THIOPHOSPHATE ◇ O,O-DIETHYL-O-6-METHYL-2-ISOPRO-PYL-4-PYRIMIDINYL PHOSPHOROTHIOATE ◇ O,O-DIETIL-O-(2-ISOPROPIL-4-METIL-PIRIMIDIN-6-IL)-MONOTIOFOS-FATO (ITALIAN) ◇ DIMPYLATE ◇ DIPOFENE ◇ DIZINON ◇ DYZOL ◇ ENT 19,507 ◇ G 301 ◇ G-24480 ◇ GARDENTOX ◇ GEIGY 24480 ◇ O-2-ISOPROPYL-4-METHYLPYRIMIDYL-O,O-DIETHYL PHOSPHOROTHIOATE ◇ ISOPROPYL-METHYLPYRIMIDYL DIETHYL THIOPHOSPHATE ◇ KAYAZINON ◇ KAYAZOL ◇ NCI-C08673 ◇ NEDCIDOL ◇ NEOCIDOL ◇ NIPSAN ◇ NUCIDOL ◇ SAROLEX ◇ SPECTRACIDE ◇ THIOPHOSPHATE de O,O-DIETHYLE et de o-2-ISOPROPYL-4-METHYL-6-PYRIMIDYLE (FRENCH)

USE IN FOOD:

Purpose: Insecticide.

Where Used: Animal feed.

Regulations: FDA - 21CFR 193.142. Insecticide to be used in 1 percent spray and 2 percent dust formulations. 21CFR 561.415. Animal feed.

NCI Carcinogenesis Bioassay (feed); No Evidence: mouse, rat NCITR* NCI-CG-TR-137,79. EPA Genetic Toxicology Program.

ACGIH TLV: TWA 0.1 mg/m^3 DOT Classification: ORM-A; Label: None

SAFETY PROFILE: Poison by ingestion, skin contact, subcutaneous, intravenous, intraperitoneal, and possibly other routes. Mildly toxic by inhalation. Human systemic effects by ingestion: changes in motor activity, muscle weakness, and sweating. An experimental teratogen. Experimental reproductive effects. A skin and severe eye irritant. Human mutagenic data. When heated to decomposition it emits very toxic fumes of NO_x, PO_x and SO_x.

TOXICITY DATA and CODEN

skn-rbt 500 mg open MOD CIGET* -,-,77
eye-rbt 100 mg SEV CIGET* -,-,77
cyt-hmn:lym 500 μg/L TSITAQ 18,1490,76
cyt-ham:lng 100 mg/L/27H MUREAV 66,277,79
orl-rat TDLo:63500 μg/kg (10D preg):REP JFMAAQ 54,452,67
orl-mus TDLo:3960 μg/kg (1-22D preg):TER JTEHD6 3,989,77
orl-hmn TDLo:214 mg/kg:CNS,SKN CTOXAO 12,435,78
orl-rat LD50:66 mg/kg DOEAAH 35,25,79
ihl-rat LC50:3500 mg/m^3/4H FMCHA2 -,C75,83
skn-rat LD50:180 mg/kg PMJMAQ -,156,57

DDM000 CAS: 10222-01-2
α,α-DIBROMO-α-CYANOACETAMIDE
mf: $C_3H_2Br_2N_2O$ mw: 241.89

SYNS: DBNPA ◇ DIBROMOCYANOACETAMIDE ◇ 2,2-DIBROMO-3-NITRILOPROPIONAMIDE

USE IN FOOD:

Purpose: Antimicrobial agent.

Where Used: Beets, sugarcane.

Regulations: FDA - 21CFR 173.320. Limitation of not more than 10 ppm and not less than 2 ppm based on weight of raw sugarcane or raw beets.

Cyanide and its compounds are on the Community Right-To-Know List.

SAFETY PROFILE: Poison by ingestion and intravenous routes. A severe skin and eye irritant. An antimicrobial agent. When heated to decomposition it emits very toxic fumes of Br^- and NO_x.

TOXICITY DATA and CODEN

skn-rbt 500 mg SEV PHMCAA 15,226,73
eye-rbt 100 mg SEV PHMCAA 15,226,73
ivn-mus LD50:10 mg/kg CSLNX* NX#07898
orl-mam LD50:118 mg/kg PHMCAA 15,226,73

DDV600 CAS: 77-58-7
DIBUTYLBIS(LAUROYLOXY)-STANNANE
mf: $C_{32}H_{64}O_4Sn$ mw: 631.65

PROP: Pale yellow liquid to colorless solid (when pure). Mp: 23°, bp: non-distillable @ 10 mm, flash p: 455°F (OC), d: 1.066 @ 20°/20°, vap d: 21.8.

SYNS: BIS(DODECANOYLOXY)DI-n-BUTYLSTANNANE ◇ BIS(LAUROYLOXY)DIBUTYLSTANNANE ◇ BIS(LAUROYLOXY)DI(n-BUTYL)STANNANE ◇ BUTYNORATE ◇ DBTL ◇ DIBUTYLBIS(LAUROYLOXY)TIN ◇ DI-n-BUTYL-

TIN DI(DODECANOATE) ◇ DIBUTYLTIN DILAURATE (USDA) ◇ DIBUTYLTIN LAURATE ◇ DIBUTYL-ZINN-DILAURAT (GERMAN) ◇ FOMREZ SUL-4 ◇ LAUDRAN DI-n-BUTYLCINICITY (CZECH) ◇ LAURIC ACID, DIBUTYLSTANNYLENE deriv. ◇ LAURIC ACID, DIBUTYLSTANNYLENE SALT ◇ STABILIZER D-22 ◇ THERM CHEK 820 ◇ TIN DIBUTYL DILAURATE ◇ TINOSTAT

USE IN FOOD:

Purpose: Pesticide.

Where Used: Various

Regulations: USDA CES Ranking: A-1 (1988).

OSHA PEL: TWA 0.1 mg(Sn)/m^3 ACGIH TLV: TWA 0.1 mg(Sn)m^3 (skin) NIOSH REL: (Organotin Compounds) TWA 0.1 mg(Sn)/m^3

SAFETY PROFILE: Poison by ingestion and intraperitoneal routes. A skin and eye irritant. Avoid the vapor produced by heating. Combustible when exposed to heat or flame; reacts with oxidizers. When heated to decomposition it emits acrid smoke and fumes.

TOXICITY DATA and CODEN

skn-rbt 500 mg/24H MLD 28ZPAK -,230,72
eye-rbt 100 mg/24H MOD 28ZPAK -,230,72
orl-rat LD50:175 mg/kg ARZNAD 10,44,60
ipr-rat LDLo:85 mg/kg BJPCAL 10,16,55

DEH600 CAS: 109-43-3
DIBUTYL SEBACATE

mf: $C_{18}H_{34}O_4$ mw: 314.52

PROP: Clear liquid. Bp: 180° @ 3 mm, fp: −11°, flash p: 353°F (COC), d: 0.936 @ 20°/20°, vap d: 10.8.

SYNS: BIS(n-BUTYL)SEBACATE ◇ DECANEDIOIC ACID, DIBUTYL ESTER ◇ DI-n-BUTYL SEBACATE ◇ KODAFLEX DBS ◇ MONOPLEX DBS ◇ POLYCIZER DBS ◇ PX 404 ◇ SEBACIC ACID, DIBUTYL ESTER ◇ STAFLEX DBS

USE IN FOOD:

Purpose: Plasticizer.

Where Used: Packaging materials.

Regulations: FDA - 21CFR 181.27.

SAFETY PROFILE: Mildly toxic by ingestion. Experimental reproductive effects. Combustible liquid when exposed to heat or flame; can react with oxidizing materials. To fight fire, use CO_2, dry chemical. When heated to decomposition it emits acrid smoke and fumes.

TOXICITY DATA and CODEN

orl-rat TDLo:418 g/kg (10W male/10D pre): REP AMIHBC 7,310,53
orl-rat LD50:16 g/kg NPIRI* 2,22,75

DFA600 CAS: 75-71-8
DICHLORODIFLUOROMETHANE

DOT: 1028
mf: CCl_2F_2 mw: 120.91

PROP: Colorless, almost odorless gas. Mp: −158°, bp: −29°, vap press: 5 atm @ 16.1°.

SYNS: ALGOFRENE TYPE 2 ◇ ARCTON 6 ◇ DIFLUORODICHLOROMETHANE ◇ DWUCHLORODWUFLUOROMETAN (POLISH) ◇ ELECTRO-CF 12 ◇ ESKIMON 12 ◇ F 12 ◇ FC 12 ◇ FLUOROCARBON-12 ◇ FREON F-12 ◇ FRIGEN 12 ◇ GENETRON 12 ◇ HALON ◇ ISCEON 122 ◇ ISOTRON 12 ◇ KAISER CHEMICALS 12 ◇ LEDON 12 ◇ PROPELLANT 12 ◇ RCRA WASTE NUMBER U075 ◇ R 12 (DOT) ◇ REFREGERANT 12 ◇ UCON 12 ◇ UCON 12/HALOCARBON 12

USE IN FOOD:

Purpose: Direct-contact freezing agent.

Where Used: Various.

Regulations: FDA - 21CFR 173.355. Use in accordance with good manufacturing practice.

EPA Genetic Toxicology Program.

OSHA PEL: TWA 1000 ppm ACGIH TLV: TWA 1000 ppm DFG MAK: 1000 ppm (5000 mg/m^3) DOT Classification: Nonflammable Gas; Label: Nonflammable gas

SAFETY PROFILE: Human systemic effects by inhalation: conjunctiva irritation, fibrosing alveolitis, and liver changes. Narcotic in high concentrations. Nonflammable Gas. Can react violently with Al. When heated to decomposition it emits highly toxic fumes of phosgene, Cl^-, and F^-.

TOXICITY DATA and CODEN

ihl-hmn TCLo:200000 ppm/30M: EYE,PUL,LIV EJTXAZ 9,385,76
ihl-rat LC50:80 pph/30M EJTXAZ 9,385,76

DFF900 CAS: 107-06-2
1,2-DICHLOROETHANE

DOT: 1184
mf: $C_2H_4Cl_2$ mw: 98.96

PROP: Colorless, clear liquid; pleasant odor, sweet taste. Bp: 83.5°, ULC: 60-70, lel: 6.2%, uel: 15.9%, fp: −35.7°, flash p: 56°F, d: 1.257 @ 20°/4°, autoign temp: 775°F, vap press: 100 mm @ 29.4°, vap d: 3.35, refr index: 1.445 @ 20°. Sol in alc, ether, acetone, carbon tetrachloride; sltly sol in water.

SYNS: AETHYLENCHLORID (GERMAN) ◇ BICHLORURE d'ETHYLENE (FRENCH) ◇ BORER SOL ◇ BROCIDE ◇ CHLORURE d'ETHYLENE (FRENCH) ◇ CLORURO di ETHENE (ITALIAN) ◇ 1,2-DCE ◇ DESTRUXOL BORER-SOL ◇ 1,2-DICHLOORETHAAN (DUTCH) ◇ 1,2-DICHLORAETHAN (GERMAN) ◇ DICHLOREMULSION ◇ DI-CHLOR-MULSION ◇ DICHLORO-1,2-ETHANE (FRENCH) ◇ α,β-DICHLOROETHANE ◇ sym-DICHLOROETHANE ◇ DICHLOROETHYLENE ◇ 1,2-DICLOROETANO (ITALIAN) ◇ DUTCH LIQUID ◇ DUTCH OIL ◇ EDC ◇ ENT 1,656 ◇ ETHANE DICHLORIDE ◇ ETHYLEENDICHLORIDE (DUTCH) ◇ ETHYLENE CHLORIDE ◇ ETHYLENE DICHLORIDE (ACGIH, DOT, FCC) ◇ 1,2-ETHYLENE DICHLORIDE ◇ GLYCOL DICHLORIDE ◇ NCI-C00511 ◇ RCRA WASTE NUMBER U077

USE IN FOOD:

Purpose: Extraction solvent, flume wash water additive, pesticide.

Where Used: Spice oleoresins, sugar beets.

Regulations: FDA - 21CFR 173.230. Limitation of 30 ppm in spice oleoresins, limit of 30 ppm for all chlorinated solvent residues. 21CFR 173.315. Limitation of 0.1 ppm in wash water.

IARC Cancer Review: Human Limited Evidence IMEMDT 20,429,79; Animal Sufficient Evidence IMEMDT 20,429,79. NCI Carcinogenesis Bioassay (gavage); Clear Evidence: mouse-rat NCITR* NCI-CG-TR-55,78. EPA Genetic Toxicology Program.

OSHA PEL: TWA 1 ppm; STEL 2 ppm ACGIH TLV: TWA 10 ppm DOT Classification: Flammable Liquid; Label: Flammable Liquid; IMO: Flammable Liquid; Label: Flammable Liquid, Poison NIOSH REL: TWA 1 ppm; CL 2 ppm/15M

SAFETY PROFILE: A human poison by ingestion. Poison experimentally by intravenous and subcutaneous routes. Moderately toxic by inhalation, skin contact, and intraperitoneal routes. An experimental carcinogen, neoplastigen, tumorigen and teratogen. Human systemic effects by ingestion and inhalation: flaccid paralysis without anesthesia (usually neuromuscular blockade), somnolence, cough, jaundice, nausea or vomiting, hypermotility, diarrhea, ulceration or bleeding from the stomach, fatty liver degeneration, change in cardiac rate, cyanosis and coma. An experimental transplacentral carcinogen. It may also cause dermatitis, edema of the lungs, toxic effects on the kidneys, and severe corneal effects. A strong narcotic. Experimental reproductive effects. A skin and severe eye irritant, and strong local irritant. Its smell and irritant effects warn of its presence at relatively safe concentrations. Human mutagenic data.

Flammable liquid. A dangerous fire hazard if exposed to heat, flame or oxidizers. Moderately explosive in the form of vapor when exposed to flame. Violent reaction with Al; N_2O_4; NH_3; dimethylaminopropylamine. Can react vigorously with oxidizing materials and emit vinyl chloride and HCl. To fight fire, use water, foam, CO_2, dry chemicals. When heated to decomposition it emits highly toxic fumes of Cl^- and phosgene.

TOXICITY DATA and CODEN

skn-rbt 600 mg open MLD UCDS** 3/23/70
eye-rbt 63 mg SEV UCDS** 3/23/70
mmo-sat 40 μmol/plate CBINA8 20,1,78
msc-hmn:lym 100 mg/L MUREAV 142,133,85
ihl-rat TCLo:300 ppm/7H (6-15D preg):REP BANRDU 5,149,80
orl-rat TDLo:5286 mg/kg/69W-I:CAR BANRDU 5,35,80
ihl-rat TCLo:5 ppm/7H/78W-I:ETA BANRDU 5,3,80
orl-mus TDLo:3536 mg/kg/78W-I:CAR BANRDU 5,35,80
ihl-hmn TCLo:4000 ppm/H:CNS,PNS,GIT PCOC** -,500,66
orl-hmn TDLo:428 mg/kg:GIT,CNS,PUL SOMEAU 22,132,58
orl-man TDLo:892 mg/kg:GIT,LIV WILEAR 28,983,75
orl-hmn LDLo:286 mg/kg:GIT,LIV CLCEAL 86,203,47
orl-rat LD50:670 mg/kg FMCHA2 -,C99,83
ihl-rat LC50:1000 ppm/7H AMIHBC 4,482,51

DFH600 CAS: 321-55-1
O,O-DI(2-CHLOROETHYL)-O-(3-CHLORO-4-METHYLCOUMARIN-7-YL) PHOSPHATE

mf: $C_{14}H_{14}Cl_3O_6P$ mw: 415.60

SYNS: O,O-BIS(2-CHLOROETHYL)-O-(3-CHLORO-4-METHYL-7-COUMARINYL) PHOSPHATE ◇ 2-CHLOROE-

THANOL HYDROGEN PHOSPHATE ESTER with 3-CHLORO-7-HYDROXY-4-METHYLCOUMARIN ◇ 2-CHLOROETHANOL PHOSPHATE DIESTER ESTER with 3-CHLORO-7-HYDROXY-4-METHYLCOUMARIN ◇ 3-CHLORO-7-HYDROXY-4-METHYLCOUMARIN BIS(2-CHLOROETHYL)PHOSPHATE ◇ 3-CHLORO-4-METHYL-UMBELLIFERONE BIS(2-CHLOROETHYL)PHOSPHATE ◇ DI-(2-CHLOROETHYL)-3-CHLORO-4-METHYL-7-COUMARINYL PHOSPHATE ◇ DI-(2-CHLOROETHYL)-3-CHLORO-4-METHYLCOUMARIN-7-YL PHOSPHATE ◇ EUSTIDIL ◇ GALLOXON ◇ GALOXANE ◇ 96H60 ◇ HALOXON ◇ HELMIRANE ◇ HELMIRON ◇ HELMIRONE ◇ LOXON ◇ LUXON ◇ LXON

USE IN FOOD:

Purpose: Animal drug.

Where Used: Beef.

Regulations: FDA - 21CFR 556.310. Tolerance of 0.1 ppm in edible tissues of cattle.

SAFETY PROFILE: Moderately toxic by ingestion and intraperitoneal routes. Human mutagenic data. When heated to decomposition it emits very toxic fumes of PO_x and Cl^-.

TOXICITY DATA and CODEN

dni-hmn:oth 10 mg/L JTEHD6 10,143,82
orl-rat LD50:900 mg/kg FAZMAE 17,108,73

DFV400 CAS: 50-65-7
2′,5-DICHLORO-4′-NITROSALICYLANILIDE

mf: $C_{13}H_8Cl_2N_2O_4$ mw: 327.13

SYNS: BAY 2353 ◇ BAYER 73 ◇ BAYER 2353 ◇ BAYLUSCID ◇ CHEMAGRO 2353 ◇ 5-CHLORO-N-(2-CHLORO-4-NITROPHENYL)-2-HYDROXYBENZAMIDE ◇ 5-CHLORO-2′-CHLORO-4′-NITROSALICYLANILIDE ◇ 2-CHLORO-4-NITROPHENYLAMIDE-6-CHLOROSALICYLIC ACID ◇ N-(2-CHLORO-4-NITROPHENYL)-5-CHLOROSALICYLAMIDE ◇ CLONITRALID ◇ 2′,5-DICHLOR-4′-NITRO-SALIZYLSAEUREANILID (GERMAN) ◇ DICHLOSALE ◇ ENT 25,823 ◇ FENASAL ◇ HL 2447 ◇ 2-HYDROXY-5-CHLORO-N-(2-CHLORO-4-NITROPHENYL)BENZAMIDE ◇ IOMESAN ◇ IOMEZAN ◇ NICLOSAMIDE ◇ PHENASAL ◇ VERMITIN ◇ YOMESAN

USE IN FOOD:

Purpose: Animal feed drug, pesticide.

Where Used: Animal feed.

Regulations: FDA - 21CFR 558.367.

SAFETY PROFILE: Poison by intraperitoneal and intravenous routes. Moderately toxic by ingestion. Human mutagenic data. When heated to decomposition it emits very toxic fumes of Cl^- and NO_x.

TOXICITY DATA and CODEN

mma-sat 5 μg/plate MUREAV 117,79,83
cyt-hmn:lym 6 mL/L MUREAV 173,81,86
orl-rat LDLo:10 g/kg ZTMPA5 13,1,62
ipr-rat LD50:250 mg/kg ZTMPA5 13,1,62

DFY600 CAS: 94-75-7
DICHLOROPHENOXYACETIC ACID

DOT: 2765

mf: $C_8H_6Cl_2O_3$ mw: 221.04

PROP: White powder. Mp: 141°; bp: 160° @ 0.4 mm; vap d: 7.63.

SYNS: ACIDE-2,4-DICHLORO PHENOXYACETIQUE (FRENCH) ◇ ACIDO (2,4-DICLORO-FENOSSI)-ACETICO (ITALIAN) ◇ AGROTECT ◇ AMIDOX ◇ AMOXONE ◇ AQUA-KLEEN ◇ BH 2,4-D ◇ CHIPCO TURF HERBICIDE 'D' ◇ CHLOROXONE ◇ CROP RIDER ◇ CROTILIN ◇ 2,4-D (ACGIH, DOT, USDA) ◇ D 50 ◇ DACAMINE ◇ 2,4-D ACID ◇ DEBROUSSAILLANT 600 ◇ DECAMINE ◇ DED-WEED ◇ DED-WEED LV-69 ◇ DESORMONE ◇ (2,4-DICHLOOR-FENOXY)-AZIJNZUUR (DUTCH) ◇ 2,4-DICHLOROPHENOXYACETIC ACID (DOT) ◇ 2,4-DICHLORPHENOXYACETIC ACID ◇ (2,4-DICHLOR-PHENOXY)-ESSIGSAEURE (GERMAN) ◇ DICOPUR ◇ DICOTOX ◇ DINOXOL ◇ DMA-4 ◇ DORMONE ◇ 2,4-DWUCHLOROFENOKSYOCTOSY KWAS (POLISH) ◇ EMULSAMINE BK ◇ EMULSAMINE E-3 ◇ ENT 8,538 ◇ ENVERT 171 ◇ ENVERT DT ◇ ESTERON ◇ ESTERON 99 ◇ ESTERON 76 BE ◇ ESTERON BRUSH KILLER ◇ ESTERON 99 CONCENTRATE ◇ ESTERONE FOUR ◇ ESTERON 44 WEED KILLER ◇ FARMCO ◇ FERNESTA ◇ FERNIMINE ◇ FERNOXONE ◇ FOREDEX 75 ◇ FORMOLA 40 ◇ HEDONAL (The herbicide) ◇ HERBIDAL ◇ IPANER ◇ KROTILINE ◇ LAWN-KEEP ◇ MACRONDRAY ◇ MIRACLE ◇ MONOSAN ◇ MOXONE ◇ NETAGRONE 600 ◇ NSC 423 ◇ PENNAMINE ◇ PHENOX ◇ PIELIK ◇ PLANOTOX ◇ PLANTGARD ◇ RCRA WASTE NUMBER U240 ◇ RHODIA ◇ SALVO ◇ SPRITZ-HORMIN/2,4-D ◇ SUPER D WEEDONE ◇ SUPERORMONE CONCENTRE ◇ TRANSAMINE ◇ TRIBUTON ◇ TRINOXOL ◇ U 46 ◇ U-5043 ◇ U 46DP ◇ VERGEMASTER ◇ VERTON D ◇ VIDON 638 ◇ VISKO-RHAP DRIFT HERBICIDES ◇ WEED-AG-BAR ◇ WEEDAR-64 ◇ WEED-B-GON ◇ WEEDEZ WONDER BAR ◇ WEEDONE LV4 ◇ WEED TOX ◇ WEEDTROL

USE IN FOOD:

Purpose: Herbicide.

Where Used: Barley (milled fractions, except flour), oats (milled fractions, except flour), potable water, rye (milled fractions), sugarcane bagasse, sugarcane molasses, wheat (milled fractions, except flour).

Regulations: FDA - 21CFR 193.100. Herbicide residue tolerance of 5 ppm in sugarcane molasses, 2 ppm in milled fractions (except flour) of barley, oats, wheat, 0.1 ppm in potable water. 21CFR 561.100. Limitation of 5 ppm in sugarcane bagasse and sugarcane molasses. Limitation of 2 ppm in milled fractions from barley, oats, rye, and wheat used as or converted into animal feed. CES Ranking: B-2 (1987).

IARC Cancer Review: Human Limited Evidence IMEMDT 41,357,86; Animal Inadequate Evidence IMEMDT 15,111,77; Human Inadequate Evidence IMEMDT 15,111,77. EPA Genetic Toxicology Program. Community Right-To-Know List.

OSHA PEL: TWA 10 mg/m^3 ACGIH TLV: TWA 10 mg/m^3 DFG MAK: 10 mg/m^3 DOT Classification: ORM-A; Label: None

SAFETY PROFILE: Poison by ingestion, intravenous, and intraperitoneal routes. Moderately toxic by skin contact. An experimental carcinogen and teratogen. A suspected human carcinogen. Human systemic effects by ingestion: somnolence, convulsions, coma, and nausea or vomiting. Can cause liver and kidney injury. A skin and severe eye irritant. Human mutagenic data. Experimental reproductive effects. When heated to decomposition it emits toxic fumes of Cl^-.

TOXICITY DATA and CODEN

skn-rbt 500 mg/24H MLD 28ZPAK -,279,72
eye-rbt 750 μg/24H SEV 28ZPAK -,279,72
sce-hmn:lym 10 mg/L JOHEA8 73,224,82
cyt-rat-ipr 2500 μg/kg RABIDH 32,265,83
orl-rat TDLo:220 μg/kg (1-22D preg):TER GISAAA 50(10),76,85
orl-mus TDLo:900 mg/kg (6-14D preg):TER NTIS** PB223-160
orl-ham TDLo:200 mg/kg (7-11D preg):REP BECTA6 6,599,67
orl-hmn LDLo:80 mg/kg:GIT,CNS ARPAAQ 94,270,72
orl-man LDLo:93 mg/kg:CNS PAREAQ 14,225,52
orl-rat LD50:370 mg/kg FMCHA2 -,C68,83
skn-rat LD50:1500 mg/kg WRPCA2 9,119,70
ipr-mus LDLo:125 mg/kg TXAPA9 23,288,72

DGC400 CAS: 330-54-1
3-(3,4-DICHLOROPHENYL)-1,1-DIMETHYLUREA

DOT: 2767
mf: $C_9H_{10}Cl_2N_2O$ mw: 233.11

PROP: Crystals. Mp: 159°. Sltly sol in water, hydrocarbon solvents.

SYNS: AF 101 ◇ CEKIURON ◇ CRISURON ◇ DAILON ◇ DCMU ◇ DIATER ◇ 3-(3,4-DICHLOOR-FENYL)-1,1-DIMETHYLUREUM (DUTCH) ◇ DICHLORFENIDIM ◇ 3-(3,4-DICHLOROPHENOL)-1,1-DIMETHYLUREA ◇ N'-(3,4-DICHLOROPHENYL)-N,N-DIMETHYLUREA ◇ 1-(3,4-DICHLOROPHENYL)-3,3-DIMETHYLUREE (FRENCH) ◇ 3-(3,4-DICHLOR-PHENYL)-1,1-DIMETHYL-HARNSTOFF (GERMAN) ◇ 3-(3,4-DICLORO-FENYL)-1,1-DIMETIL-UREA (ITALIAN) ◇ 1,1-DIMETHYL-3-(3,4-DICHLOROPHENYL) UREA ◇ DI-ON ◇ DIREX 4L ◇ DIUREX ◇ DIUROL ◇ DIURON (ACGIH, DOT) ◇ DIURON 4L ◇ DMU ◇ DREXEL ◇ DREXEL DIURON 4L ◇ DURAN ◇ DYNEX ◇ FARMCO DIURON ◇ HERBATOX ◇ HW 920 ◇ KARMEX ◇ KARMEX DIURON HERBICIDE ◇ KARMEX DW ◇ MARMER ◇ SUP'R FLO ◇ TELVAR ◇ TELVAR DIURON WEED KILLER ◇ UNIDRON ◇ USAF P-7 ◇ USAF XR-42 ◇ VONDURON

USE IN FOOD:

Purpose: Herbicide.

Where Used: Animal feed.

Regulations: FDA - 21CFR 561.220. Limitation of 4 ppm in dried citrus pulp when used for livestock feed.

EPA Genetic Toxicology Program. Chlorophenol compounds are on The Community Right-To-Know List.

ACGIH TLV: TWA 10 mg/m^3 DOT Classification: ORM-E; Label: None

SAFETY PROFILE: An experimental tumorigen and teratogen. Moderately toxic by ingestion, intraperitoneal and possibly other routes. An experimental teratogen. Mutagenic data. When heated to decomposition emits highly toxic fumes of Cl^- and NO_x.

TOXICITY DATA and CODEN

mma-sat 3 μg/plate MUREAV 58,353,78
dni-mus-orl 1 g/kg MUREAV 58,353,78
orl-rat TDLo:360 mg/kg (6-15D preg):TER GISAAA 49(8),83,84
scu-mus TDLo:1935 mg/kg (6-14D preg):TER NTIS** PB223-160
orl-mus TDLo:153 g/kg/78W-I:ETA NTIS** PB223-159

orl-rat LD50: 1017 mg/kg JAFCAU 18,1104,70
ipr-mus LDLo: 500 mg/kg NTIS** AD277-689

DGI000 CAS: 709-98-8
DICHLOROPROPIONANILIDE

mf: $C_9H_9Cl_2NO$ mw: 218.09

PROP: Light brown solid (pure); liquid (technical grade). Mp (pure): 85-89°, bp (technical grade): 91-95°.

SYNS: BAY 30130 ◇ CHEM RICE ◇ CRYSTAL PROPANIL-4 ◇ DCPA ◇ N-(3,4-DICHLOROPHENYL)PROPANAMIDE ◇ N-(3,4-DICHLOROPHENYL)PROPIONAMIDE ◇ 3,4-DICHLOROPROPIONANILIDE ◇ 3′,4′-DICHLOROPROPIONANILIDE ◇ DIPRAM ◇ DPA ◇ FARMCO PROPANIL ◇ FW 734 ◇ GRASCIDE ◇ HERBAX TECHNICAL ◇ MONTROSE PROPANIL ◇ PROPANEX ◇ PROPANID ◇ PROPANIDE ◇ PROPANIL ◇ PROPIONIC ACID-3,4-DICHLOROANILIDE ◇ PROP-JOB ◇ RISELECT ◇ ROGUE ◇ ROSANIL ◇ S 10165 ◇ STAM ◇ STAM F 34 ◇ STAM LV 10 ◇ STAM M-4 ◇ STAMPEDE ◇ STAMPEDE 3E ◇ STAM SUPERNOX ◇ STREL ◇ SUPERNOX ◇ SURCOPUR ◇ SURPUR ◇ VERTAC

USE IN FOOD:

Purpose: Herbicide.

Where Used: Animal feed, rice, rice bran, rice hulls, rice polishings.

Regulations: FDA - 21CFR 561.150. Limitation of 10 ppm in rice bran, rice hulls, and rice polishings when used for animal feed.

EPA Genetic Toxicology Program.

SAFETY PROFILE: Poison by ingestion. Moderately toxic by an unspecified route. Mildly toxic by skin contact. Mutagenic data. When heated to decomposition it emits very toxic fumes of Cl^- and NO_x.

TOXICITY DATA and CODEN

cyt-mus-unr 100 mg/kg TGANAK 14(6),41,80
cyt-mus-orl 100 mg/kg CYGEDX 14(6),38,80
orl-rat LD50: 560 mg/kg WRPCA2 9,119,70
skn-rbt LD50: 4830 mg/kg FMCHA2 -,C197,83

DGI400 CAS: 75-99-0
2,2-DICHLOROPROPIONIC ACID

DOT: 1760
mf: $C_3H_4Cl_2O_2$ mw: 142.97

PROP: White to tan powder.

SYNS: BASFAPON ◇ BASFAPON B ◇ BASFAPON/BASFAPON N ◇ BASINEX ◇ BH DALAPON ◇ CRISAPON ◇ DALAPON (USDA) ◇ DALAPON 85 ◇ DED-WEED ◇ DEVIPON ◇ α-DICHLOROPROPIONIC ACID ◇ α,α-DICHLOROPROPIONIC ACID ◇ DOWPON ◇ DOWPON M ◇ GRAMEVIN ◇ KENAPON ◇ LIROPON ◇ PROPROP ◇ RADAPON ◇ REVENGE ◇ UNIPON

USE IN FOOD:

Purpose: Herbicide.

Where Used: Animal feed, citrus pulp (dehydrated), potable water.

Regulations: FDA - 21CFR 193.105. Herbicide residue tolerance of 0.2 ppm in potable water. 21CFR 561.110. Limitation of 20 ppm in dehydrated citrus pulp when used for animal feed. USDA CES Ranking: A-3 (1985).

EPA Genetic Toxicology Program.

ACGIH TLV: TWA 1 ppm DOT Classification: Corrosive Material; Label: Corrosive

SAFETY PROFILE: Moderately toxic by ingestion. Corrosive. A skin irritant. Mutagenic data. When heated to decomposition it emits toxic fumes of Cl^-.

TOXICITY DATA and CODEN

skn-rbt 100 μg/24H open AIHAAP 23,95,62
mmo-omi 500 ppm IJEBA6 11,114,73
orl-rat LD50: 970 mg/kg FMCHA2 -,C69,83

DGI600 CAS: 127-20-8
α,α-DICHLOROPROPIONIC ACID SODIUM SALT

mf: $C_3H_3Cl_2O_2 \cdot Na$ mw: 164.95

SYNS: BASFAPON B ◇ DALAPON ◇ DALAPON SODIUM ◇ DALAPON SODIUM SALT ◇ 2,2-DICHLOROPROPIONIC ACID, SODIUM SALT ◇ DOWPON ◇ 2,2-DPA ◇ GRAMEVIN ◇ NATRIUMSALZ DER 2,2-DICHLORPROPIONSAURE ◇ RADAPON ◇ SODIUM DALAPON ◇ SODIUM-α,α-DICHLOROPROPIONATE ◇ SODIUM-2,2-DICHLOROPROPIONATE ◇ UNIPON

USE IN FOOD:

Purpose: Herbicide.

Where Used: Potable water.

Regulations: FDA - 21CFR 193.105. Herbicide residue tolerance of 0.2 ppm in potable water.

EPA Genetic Toxicology Program.

SAFETY PROFILE: Moderately toxic by ingestion and possibly other routes. Mutagenic data. When heated to decomposition it emits toxic fumes of Na_2O and Cl^-.

TOXICITY DATA and CODEN

cyt-mus-unr 200 mg/kg TGANAK 16(1),45,82
orl-rat LD50:3860 mg/kg WRPCA2 9,119,70

DHI000 CAS: 97-96-1
DIETHYL ACETALDEHYDE
DOT: 1178
mf: $C_6H_{12}O$ mw: 100.18

PROP: Colorless liquid; ungent odor. Bp: 116.8°, flash p: 70°F (OC), fp: −89°, d: 0.808-0.814, vap press: 13.7 mm @ 20°, vap d: 3.45, lel: 1.2%, uel: 7.7%. Misc in alc, ether; sltly sol in water.

SYNS: ALDEHYDE-2-ETHYLBUTYRIQUE (FRENCH) ◇ 2-ETHYLBUTANAL ◇ 2-ETHYLBUTRIC ALDEHYDE ◇ ETHYL BUTYRALDEHYDE ◇ α-ETHYLBUTYRALDEHYDE ◇ ETHYL BUTYRALDEHYDE (DOT) ◇ 2-ETHYLBUTYRALDEHYDE (DOT,FCC) ◇ FEMA No. 2426

USE IN FOOD:

Purpose: Flavoring agent.

Where Used: Various.

Regulations: FDA - 21CFR 172.515. Use at a level not in excess of the amount reasonably required to accomplish the intended effect.

DOT Classification: Flammable Liquid; Label: Flammable Liquid

SAFETY PROFILE: Moderately toxic by ingestion. Mildly toxic by inhalation. A skin irritant. Flammable liquid. Can react vigorously with oxidizing materials. To fight fire, use alcohol foam, CO_2, dry chemical. When heated to decomposition it emits acrid smoke and fumes.

TOXICITY DATA and CODEN

skn-rbt 500 mg open MLD UCDS** 12/14/71
orl-rat LD50:3980 mg/kg AMIHBC 4,119,51
ihl-rat LCLo:8000 ppm/4H AMIHBC 4,119,51

DHI400 CAS: 88-09-5
DIETHYLACETIC ACID
mf: $C_6H_{12}O_2$ mw: 116.18

PROP: Colorless, volatile liquid; rancid odor. Mp: −93°, bp: 121.0°, flash p: 78°F (CC), d: 0.917, vap press: 10 mm @ 15.3°, vap d: 4.0, autoign temp: 865°F. Misc in alc, ether, water.

SYNS: 2-ETHYL BUTANOIC ACID ◇ α-ETHYLBUTYRIC ACID ◇ 2-ETHYLBUTYRIC ACID (FCC) ◇ FEMA No. 2429 ◇ 3-PENTANECARBOXYLIC ACID

USE IN FOOD:

Purpose: Flavoring agent.

Where Used: Various.

Regulations: FDA - 21CFR 172.515. Use at a level not in excess of the amount reasonably required to accomplish the intended effect.

SAFETY PROFILE: Moderately toxic by ingestion and skin contact. An irritant to skin and mucous membranes. A severe eye irritant. Narcotic in high concentrations. Flammable liquid. To fight fire, use CO_2, dry chemical, alcohol foam. When heated to decomposition it emits acrid smoke and fumes.

TOXICITY DATA and CODEN

skn-rbt 10 mg/24H open MLD AMIHBC 10,61,54
eye-rbt 250 μg open SEV AMIHBC 10,61,54
orl-rat LD50:2200 mg/kg AMIHBC 10,61,54
skn-rbt LD50:520 mg/kg AMIHBC 10,61,54

DIN800 CAS: 29232-93-7
2-DIETHYLAMINO-6-METHYLPYRIMIDIN-4-YL DIMETHYL PHOSPHOROTHIONATE
mf: $C_{11}H_{20}N_3O_3PS$ mw: 305.37

SYNS: ACTELIC ◇ ACTELLIC ◇ ACTELLIFOG ◇ BLEX ◇ O-(2-(DIETHYLAMINO)-6-METHYL-4-PYRIMIDINYL)-O,O-DIMETHYL PHOSPHOROTHIOATE ◇ O-(2-DIETHYLAMINO-6-METHYLPYRIMIDIN-4-YL)-O,O-DIMETHYL PHOSPHOROTHIOATE ◇ ENT 27,699GC ◇ METHYL PIRIMIPHOS ◇ PIRIMIFOS-METHYL ◇ PLANT PROTECTION PP511 ◇ PP511 ◇ PYRIMIDINE PHOSPHATE ◇ PYRIMIPHOS METHYL ◇ SILOSAN

USE IN FOOD:

Purpose: Insecticide.

Where Used: Animal feed, corn milling fractions, corn oil, sorghum milling fractions.

Regulations: FDA - 21CFR 193.468. Insecticide residue tolerance of 40 ppm in corn milling fractions (except flour), 88 ppm in corn oil, 40 ppm in sorghum milling fractions (except flour). 21CFR 561.432. Animal feeds.

SAFETY PROFILE: Moderately toxic by ingestion and possibly other routes. Mutagenic data. When heated to decomposition it emits very toxic fumes of NO_x, PO_x and SO_x.

TOXICITY DATA and CODEN

mmo-sat 5 μL/plate MUREAV 28,405,75
cyt-mus-unr 500 mg/kg TGANAK 16(1),45,82
orl-rat LD50:1250 mg/kg SSCMBX 20,33,83

DIZ100 CAS: 1609-47-8
DIETHYL DICARBONATE
mf: $C_6H_{10}O_5$ mw: 162.16

PROP: Viscous liquid; fruity odor. D: 1.12, visc (20°): 1.97 cp. Soluble in alcohols, esters, ketones, hydrocarbons.

SYNS: BAYCOVIN ◇ DEPC ◇ DICARBONIC ACID DIETHYL ESTER ◇ DIETHYL ESTER of PYROCARBONIC ACID ◇ DIETHYL OXYDIFORMATE ◇ DIETHYL PYROCARBONATE ◇ DIETHYL PYROCARBONIC ACID ◇ DKD ◇ ETHYL PYROCARBONATE ◇ OXYDIFORMIC ACID DIETHYL ESTER ◇ PIREF ◇ PYROCARBONATE d'ETHYLE (FRENCH) ◇ PYROCARBONIC ACID, DIETHYL ESTER ◇ PYROKOHLENSAEURE DIAETHYL ESTER (GERMAN)

USE IN FOOD:

Purpose: Ferment inhibitor, fungicide.

Where Used: Prohibited from alcoholic beverages, prohibited from foods.

Regulations: FDA - 21CFR 189.140. Prohibited from direct addition or use in human food. Legal for use in wine in other countries.

SAFETY PROFILE: Poison by ingestion and intraperitoneal routes. Concentrated DEPC is irritating to eyes, mucous membranes and skin. When heated to decomposition it emits acrid smoke and fumes.

TOXICITY DATA and CODEN

orl-rat LD50:850 mg/kg FAONAU 51A,69,72
ipr-rat LD50:100 mg/kg ZLUFAR 114,292,61
orl-mus LD50:2027 mg/kg ZLUFAR 139,287,69

DJH600 CAS: 100-37-8
N,N-DIETHYLETHANOLAMINE
DOT: 2686
mf: $C_6H_{15}NO$ mw: 117.22

PROP: Colorless, hygroscopic liquid. Bp: 162°, flash p: 140°F (OC), d: 0.8851 @ 20°/20°, vap press: 1.4 mm @ 20°, vap d: 4.03.

SYNS: DEAE ◇ DIAETHYLAMINOAETHANOL (GERMAN) ◇ DIETHYLAMINOETHANOL ◇ β-DIETHYLAMINOETHANOL ◇ N-DIETHYLAMINOETHANOL ◇ 2-(DIETHYLAMINO)ETHANOL ◇ 2-N-DIETHYLAMINOETHANOL ◇ 2-DIETHYLAMINOETHANOL (ACGIH) ◇ DIETHYLAMINOETHANOL (DOT) ◇ β-DIETHYLAMINOETHYL ALCOHOL ◇ DIETHYLETHANOLAMINE ◇ N,N-DIETHYL-N-(β-HYDROXYETHYL)AMINE ◇ 2-HYDROXYTRIETHYLAMINE

USE IN FOOD:

Purpose: Boiler water additive.

Where Used: Various.

Regulations: FDA - 21CFR 173.310. Limitation of 15 ppm in steam and excluding use of such steam in contact with milk and milk products.

OSHA PEL: TWA 10 ppm (skin) ACGIH TLV: TWA 10 ppm (skin) DOT Classification: Flammable or Combustible Liquid; Label: Flammable Liquid

SAFETY PROFILE: Poison by intraperitoneal and intravenous routes. Moderately toxic by ingestion, skin contact, subcutaneous, intramuscular, and possibly other routes. Human systemic effects by inhalation: nausea or vomiting. A skin and severe eye skin irritant. Combustible liquid. Flammable when exposed to heat or flame; can react with oxidizing materials. To fight fire, use alcohol foam, CO_2, dry chemical. When heated to decomposition it emits toxic fumes of NO_x.

TOXICITY DATA and CODEN

skn-rbt 500 mg open MLD UCDS** 6/11/63
eye-rbt 5 mg SEV UCDS** 6/11/63
ihl-hmn TCLo:200 ppm:GIT 34ZIAG -,216,69
orl-rat LD50:1300 mg/kg JIHTAB 26,269,44
ihl-rat LCLo:4500 mg/m³/4H GTPZAB 14(11),52,70
ipr-rat LD50:1220 mg/kg TXAPA9 12,486,68

DJL000 CAS: 577-11-7
DI-(2-ETHYLHEXYL) SODIUM SULFOSUCCINATE
mf: $C_{20}H_{38}O_7S \cdot Na$ mw: 445.63

PROP: White, waxlike, plastic solid; octyl alcohol odor. Sol in hexane, glycerin, alc; sltly sol in water.

SYNS: AEROSOL GPG ◇ ALCOPOL O ◇ ALPHASOL OT ◇ BEROL 478 ◇ BIS(ETHYLHEXYL) ESTER of SODIUM SULFOSUCCINIC ACID ◇ BIS(2-ETHYLHEXYL)SODIUM SULFOSUCCINATE ◇ BIS(2-ETHYLHEXYL)-S-SODIUM SULFOSUCCINATE ◇ 1,4-BIS(2-ETHYLHEXYL) SODIUM SULFOSUCCINATE ◇ 1,4-BIS(2-ETHYLHEXYL)SULFOBUTANEDIOIC ACID ESTER, SODIUM SALT ◇ CELANOL DOS 75 ◇ CLESTOL ◇ COLACE ◇ COMPLEMIX ◇ CONSTONATE ◇ COPROL ◇ DEFILIN ◇ DIOCTLYN ◇ DIOCTYLAL ◇ DIOCTYL ESTER of SODIUM SULFOSUCCINATE ◇ DIOCTYL ESTER of SODIUM SULFOSUCCINIC ACID ◇ DIOCTYL-MEDO FORTE ◇ DIOCTYL SODIUM SULFOSUCCINATE (FCC) ◇ DIOCTYL SULFOSUCCINATE SODIUM SALT ◇ DIOMEDICONE ◇ DIOSUCCIN ◇ DIOTILAN ◇ DIOVAC ◇ DOCUSATE SODIUM ◇ DOXINATE ◇ DOXOL ◇ DSS ◇ DULSIVAC ◇ DUOSOL ◇ 2-ETHYL-

HEXYL SULFOSUCCINATE SODIUM ◇ HUMIFEN WT 27G ◇ KONLAX ◇ KOSATE ◇ LAXINATE ◇ MANOXAL OT ◇ MERVAMINE ◇ MODANE SOFT ◇ MOLATOC ◇ MOLCER ◇ MOLOFAC ◇ MONAWET MD 70E ◇ NEKAL WT-27 ◇ NEVAX ◇ NIKKOL OTP 70 ◇ NORVAL ◇ OBSTON ◇ RAPISOL ◇ REGUTOL ◇ REQUTOL ◇ REVAC ◇ SANMORIN OT 70 ◇ SBO ◇ SOBITAL ◇ SODIUM BIS(2-ETHYLHEXYL) SULFOSUCCINATE ◇ SODIUM DI-(2-ETHYLHEXYL) SULFOSUCCINATE ◇ SODIUM DIOCTYL SULFOSUCCINATE ◇ SODIUM DIOCTYL SULPHOSUCCINATE ◇ SODIUM-2-ETHYLHEXYLSULFOSUCCINATE ◇ SODIUM SULFODI-(2-ETHYLHEXYL)SULFOSUCCINATE ◇ SOFTIL ◇ SOLIWAX ◇ SOLUSOL-75% ◇ SOLUSOL-100% ◇ SULFIMEL DOS ◇ TEX WET 1001 ◇ TRITON GR-5 ◇ VATSOL OT ◇ VELMOL ◇ WAXSOL ◇ WETAID SR

USE IN FOOD:

Purpose: Emulsifier, hog scald agent, poultry scald agent, processing aid, wetting agent.

Where Used: Beverage mixes (dry), cocoa, eggs (shell), fruit juice drinks, gelatin dessert (dry), gum, hog carcasses, milk, molasses, poultry.

Regulations: FDA - 21CFR 73.1. Limitation of 9 ppm in finished food. With cocoa (21CFR 163.117, 172.520, and 172.810): limitation of 75 ppm in finished beverage. 21CFR 172.810. Limitation of 15 ppm in gelatin dessert, 10 ppm in finished beverage or fruit drinks, 25 ppm in molasses; 0.5 percent in gums or hydrophilic colloids; 10 percent in fumaric acid-acidulated fruit juice drinks. USDA - 9CFR 318.7, 381.147. Sufficient for purpose.

SAFETY PROFILE: Poison by intravenous route. Moderately toxic by ingestion and intraperitoneal routes. A skin and severe eye irritant. When heated to decomposition it emits toxic fumes of SO_x and Na_2O.

TOXICITY DATA and CODEN

eye-rbt 250 μg MLD AROPAW 34,99,45
eye-rbt 1% SEV JAPMA8 38,428,49
orl-rat LD50:1900 mg/kg JSCCA5 13,469,62
ipr-rat LD50:590 mg/kg BCTKAG 7,161,74

DJX000 CAS: 84-66-2
DIETHYL-o-PHTHALATE

mf: $C_{12}H_{14}O_4$ mw: 222.26

PROP: Clear, colorless liquid. Mp: −40.5°, bp: 302°, flash p: 325°F (OC), d: 1.110, vap d: 7.66.

SYNS: ANOZOL ◇ 1,2-BENZENEDICARBOXYLIC ACID, DIETHYL ESTER ◇ DIETHYL PHTHALATE (ACGIH) ◇ ESTOL 1550 ◇ ETHYL PHTHALATE ◇ NCI-C60048 ◇ NEANTINE ◇ PALATINOL A ◇ PHTHALIC ACID, DIETHYL ESTER ◇ PHTHALOL ◇ PHTHALSAEUREDIAETHYLESTER (GERMAN) ◇ PLACIDOL E ◇ RCRA WASTE NUMBER U088 ◇ SOLVANOL

USE IN FOOD:

Purpose: Plasticizer.

Where Used: Packaging materials.

Regulations: FDA - 21CFR 181.27.

ACGIH TLV: TWA 5 mg/m^3

SAFETY PROFILE: Poison by intravenous route. Moderately toxic by ingestion, subcutaneous, and intraperitoneal routes. Human systemic effects by inhalation: lachrimation, respiratory obstruction, and other unspecified respiratory system effects. An eye irritant and systemic irritant by inhalation. An experimental teratogen. Other experimental reproductive effects. Narcotic in high concentrations. Combustible when exposed to heat or flame. To fight fire, use water spray, mist, foam. When heated to decomposition it emits acrid smoke and irritating fumes.

TOXICITY DATA and CODEN

eye-rbt 112 mg JPETAB 82,377,44
mmo-sat 200 μg/plate JTEHD6 16,61,85
orl-rat TDLo:53480 mg/kg (14D male):REP FCTXAV 16,415,78
ipr-rat TDLo:506 mg/kg (5-15D preg):TER JPMSAE 61,51,72
ihl-hmn TCLo:1000 mg/m^3:EYE,PUL AGGHAR 5,1,33
orl-rat LD50:8600 mg/kg GTPZAB 24(3),25,80
ipr-rat LD50:5058 mg/kg JPMSAE 61,51,72

DJY600 CAS: 110-40-7
DIETHYL SEBACATE

mf: $C_{14}H_{26}O_4$ mw: 258.40

PROP: Colorless to sltly yellow liquid; faint fruity odor. D: 0.960-0.965, refr index: 1.435. Misc with alc, ether, other organic solvents, fixed oils; insol in water @ 302°.

SYNS: DIETHYL DECANEDIOATE ◇ DIETHYL-1,10-DECANEDIOATE ◇ ETHYL SEBACATE ◇ FEMA No. 2376 ◇ SEBACIC ACID, DIETHYL ESTER

USE IN FOOD:

Purpose: Flavoring agent.

Where Used: Various.

Regulations: FDA - 21CFR 172.515. Use at a level not in excess of the amount reasonably required to accomplish the intended effect.

SAFETY PROFILE: Mildly toxic by ingestion. A skin irritant. When heated to decomposition it emits acrid smoke and irritating fumes.

TOXICITY DATA and CODEN

skn-rbt 500 mg/24H MLD FCTXAV 16,637,78
orl-rat LD50: 14470 mg/kg FCTXAV 2,327,64

DKV150 CAS: 619-01-2
DIHYDROCARVEOL
mf: $C_{10}H_{18}O$ mw: 154.28

PROP: Colorless, oily liquid; spearmint odor. D: 0.921-0.926, refr index: 1.477-1.481, flash p: +153°F. Sol in alc, fixed oils; insol in water.

SYNS: 1,6-DIHYDROCARVEOL ◇ FEMA No. 2379 ◇ 8-p-MENTHEN-2-OL ◇ 6-METHYL-3-ISOPROPYLCYCLOHEXANOL

USE IN FOOD:

Purpose: Flavoring agent.

Where Used: Various.

Regulations: FDA - 21CFR 172.515. Use at a level not in excess of the amount reasonably required to accomplish the intended effect.

SAFETY PROFILE: A moderate skin and eye irritant. A combustible liquid. When heated to decomposition it emits acrid smoke and irritating fumes.

TOXICITY DATA and CODEN

skn-rbt 500 mg/24H MOD FCTXAV 17,771,79

DKV175 CAS: 7764-50-3
d-DIHYDROCARVONE
mf: $C_{10}H_{16}O$ mw: 152.26

PROP: Colorless liquid; spearmint-like odor. D: 0.923-0.928, refr index: 1.470-1.474. Sol in alc, fixed oils; insol in water.

SYNS: FEMA No. 3565 ◇ 8-p-MENTHEN-2-ONE ◇ p-MENTH-8-EN-2-ONE ◇ d-2-METHYL-5-(1-METHYLENENYL)-CYCLOHEXANONE

USE IN FOOD:

Purpose: Flavoring agent.

Where Used: Various.

Regulations: FDA - 21CFR 172.515. Use at a level not in excess of the amount reasonably required to accomplish the intended effect.

SAFETY PROFILE: Moderately toxic by subcutaneous route. When heated to decomposition it emits acrid smoke and irritating fumes.

TOXICITY DATA and CODEN

scu-mus LD50:2900 mg/kg FCTXAV 18,665,80

DMC600 CAS: 123-33-1
1,2-DIHYDROPYRIDAZINE-3,6-DIONE
mf: $C_4H_4N_2O_2$ mw: 112.10

PROP: Crystals. Mp: > 300°. Sol in water and alc.

SYNS: 1,2-DIHYDRO-3,6-PYRIDAZINEDIONE ◇ ENT 18,870 ◇ 6-HYDROXY-3(2H)-PYRIDAZINONE ◇ MALEIC ACID HYDRAZIDE ◇ MALEIC HYDRAZIDE ◇ N,N-MALEOYLHYDRAZINE ◇ 1,2,3,6-TETRAHYDRO-3,6-DIOXOPYRIDAZINE

USE IN FOOD:

Purpose: Pesticide.

Where Used: Potato chips.

Regulations: FDA - 21CFR 193.270. Pesticide residue tolerance of 160 ppm in potato chips.

IARC Cancer Review: Animal Inadequate Evidence IMEMDT 4,173,74.

SAFETY PROFILE: Moderately toxic by ingestion. An experimental tumorigen. Mutagenic data. Can cause chronic liver damage and acute central nervous system effects. When heated to decomposition emits highly toxic fumes of NO_x.

TOXICITY DATA and CODEN

cyt-grh-orl 5 mg CYTOAN 37,345,72
mma-sat 50 µL/plate MUREAV 66,247,79
scu-rat TDLo:2600 mg/kg/65W-I:ETA BJCAAI 19,392,65
orl-rat LD50:3800 mg/kg WRPCA2 9,119,70

DME000 CAS: 128-46-1
DIHYDROSTREPTOMYCIN
mf: $C_{21}H_{41}N_7O_{12}$ mw: 583.69

SYNS: DHMS ◇ DST

USE IN FOOD:

Purpose: Animal drug.

Where Used: Beef, milk.

Regulations: FDA - 21CFR 556.200. Tolerance of zero in uncooked edible tissues of calves, milk.

EPA Genetic Toxicology Program.

SAFETY PROFILE: Poison by intravenous and intramuscular routes. Moderately toxic by subcutaneous and intraperitoneal routes. Human teratogenic effects by unspecified route: developmental abnormalities of the eye and ear. An experimental teratogen. Other experimental reproductive effects. Mutagenic data. When heated to decomposition it emits toxic fumes of NO_x.

TOXICITY DATA and CODEN

cyt-mus-par 100 mg/kg NULSAK 2,161,71
unr-wmn TDLo:260 mg/kg (19-22W preg): TER SJRDAH 50,61,69
scu-rat TDLo:600 mg/kg (6-10D preg):TER OSDIAF 14,107,65
scu-rat LD50:1100 mg/kg ARZNAD 12,597,62
ivn-mus LD50:200 mg/kg 85GDA2 1,96,80

DMW000 CAS: 7361-61-7
5,6-DIHYDRO-2-(2,6-XYLIDINO)-4H-1,3-THIAZINE

mf: $C_{12}H_{16}N_2S$ mw: 220.36

SYNS: BAY 1470 ◇ BAY VA 1470 ◇ N-(5,6-DIHYDRO-4H-1,3-THIAZINYL)-2,6-XYLIDINE ◇ 2-(2,6-DIMETHYLANILINO)-5,6-DIHYDRO-4H-1,3-THIAZINE ◇ 2-(2,6-DIMETHYLPHENYLAMINO)-4H-5,6-DIHYDRO-1,3-THIAZINE ◇ N-(2,6-DIMETHYLPHENYL)-5,6-DIHYDRO-4H-1,3-THIAZIN-2-AMINE ◇ N-(2,6-DIMETHYLPHENYL)-5,6-DIHYDRO-4H-1,3-THIAZINE-2-AMINE (9CI) ◇ ROMPUN ◇ WH 7286 ◇ XYLZIN ◇ XYLAZINE (USDA)

USE IN FOOD:

Purpose: Animal drug.

Where Used: Meat.

Regulations: USDA CES Ranking: Z-4 (1986).

SAFETY PROFILE: Poison by ingestion, subcutaneous and intravenous routes. A veterinary sedative, analgesic, muscle relaxant. When heated to decomposition it emits very toxic fumes of NO_x and SO_x.

TOXICITY DATA and CODEN

orl-rat LD50:130 mg/kg DTTIAF 75,565,68
scu-mus LD50:121 mg/kg DTTIAF 75,565,68

DNH125 CAS: 141-04-8
DIISOBUTYL ADIPATE

mf: $C_{14}H_{26}O_4$ mw: 258.40

SYNS: DIBA ◇ FTAFLEX DIBA ◇ ISOBUTYL ADIPATE

USE IN FOOD:

Purpose: Plasticizer.

Where Used: Packaging materials.

Regulations: FDA - 21CFR 181.27.

SAFETY PROFILE: Moderately toxic by intraperitoneal route. Mildly toxic by ingestion. An experimental teratogen. Experimental reproductive effects. When heated to decomposition it emits acrid smoke and fumes.

TOXICITY DATA and CODEN

ipr-rat TDLo:1190 mg/kg (5-15D preg):REP JPMSAE 62,1596,73
ipr-rat TDLo:595 mg/kg (5-15D preg):TER JPMSAE 62,1596,73
ipr-rat LD50:5950 mg/kg JPMSAE 62,1596,73
orl-gpg LD50:12300 mg/kg GWXXBX #2703360

DNU390
DILL SEED OIL, AMERICAN TYPE

PROP: From steam distillation of the salks, leaves, and seeds of *Anethum graveolens* L. Yellowish liquid. D: 0.884-0.900, refr index: 1.480 @ 20°. Sol in propylene glycol; insol in glycerin.

SYNS: DILL OIL ◇ DILL HERB OIL, AMERICAN TYPE

USE IN FOOD:

Purpose: Flavoring agent.

Where Used: Dips, meat, sauces, spreads.

Regulations: FDA - 21CFR 184.1282. GRAS when used at a level not in excess of the amount reasonably required to accomplish the intended effect.

SAFETY PROFILE: When heated to decomposition it emits acrid smoke and irritating fumes.

DNU392
DILL SEED OIL, INDIAN TYPE

PROP: From steam distillation of the dried ripe fruit of *Anethum sowa* D.C. (Fam. *Umbelliferae* (FCC III). Yellowish liquid; harsh caraway odor and taste. D: 0.925-0.980, refr index: 1.486 @ 20°. Sol in fixed oils, mineral oil; sltly sol in propylene glycol; insol in glycerin.

SYNS: DILL OIL, INDIAN TYPE ◇ DILL SEED OIL, INDIAN

USE IN FOOD:

Purpose: Flavoring agent.

Where Used: Dips, meat, sauces, spreads.

Regulations: FDA - 21CFR 184.1282. GRAS when used at a level not in excess of the amount reasonably required to accomplish the intended effect.

SAFETY PROFILE: When heated to decomposition it emits acrid smoke and irritating fumes.

DNU400 CAS: 8006-75-5
DILL WEED OIL

PROP: From steam distillation of the dried ripe fruit of *Anethum graveolens* L. (Fam. *Umbelliferae*. Yellowish liquid; caraway odor and taste. D: 0.890-0.915, refr index: 1.4836 @ 20°. Sol in fixed oils, mineral oil; propylene glycol; insol in glycerin.

SYNS: DILL FRUIT OIL ◇ DILL HERB OIL ◇ DILL OIL ◇ DILL SEED OIL ◇ DILL SEED OIL, EUROPEAN TYPE (FCC)

USE IN FOOD:

Purpose: Flavoring agent.

Where Used: Various.

Regulations: FDA - 21CFR 172.510. Use at a level not in excess of the amount reasonably required to accomplish the intended effect.

SAFETY PROFILE: Mildly toxic by ingestion. A skin irritant. Mutagenic data. When heated to decomposition it emits acrid smoke and fumes.

TOXICITY DATA and CODEN

skn-rbt 500 mg/24H MOD FCTOD7 20 (Suppl),673,82
mma-sat 1 mg/plate JOPHDQ 3,236,80
orl-rat LD50:4040 mg/kg FCTXAV 14,659,76

DOB400 CAS: 72-43-5
DIMETHOXY-DDT
DOT: 2761
mf: $C_{16}H_{15}Cl_3O_2$ mw: 345.66

PROP: Crystals. Mp: 78°, vap d: 12.

SYNS: 2,2-BIS(p-ANISYL)-1,1,1-TRICHLOROETHANE ◇ 1,1-BIS(p-METHOXYPHENYL)-2,2,2-TRICHLOROETHANE ◇ 2,2-BIS(p-METHOXYPHENYL)-1,1,1-TRICHLOROETHANE ◇ CHEMFORM ◇ DIANISYLTRICHLORETHANE ◇ 2,2-DI-p-ANISYL-1,1,1-TRICHLOROETHANE ◇ p,p'-DIMETHOXYDIPHENYLTRICHLOROETHANE ◇ DIMETHOXY-DT ◇ 2,2-DI-(p-METHOXYPHENYL)-1,1,1-TRICHLOROETHANE ◇ DI(p-METHOXYPHENYL)-TRICHLOROMETHYL METHANE ◇ DMDT ◇ p,p'-DMDT ◇ ENT 1,716 ◇ MARALATE ◇ MARLATE ◇ METHOXCIDE ◇ METHOXO ◇ METHOXYCHLOR (ACGIH, DOT, USDA) ◇ p,p'-METHOXYCHLOR ◇ METHOXY-DDT ◇ METOKSYCHLOR (POLISH) ◇ METOX ◇ MOXIE ◇ NCI-C00497 ◇ RCRA WASTE NUMBER U247 ◇ 1,1,1-TRICHLOR-2,2-BIS(4-METHOXY-PHENYL)-AETHAN (GERMAN) ◇ 1,1,1-TRICHLORO-2,2-BIS(p-ANISYL)ETHANE ◇ 1,1,1-TRICHLORO-2,2-BIS (p-METHOXYPHENOL)ETHANOL ◇ 1,1,1-TRICHLORO-2,2-BIS(p-METHOXYPHENYL)ETHANE ◇ 1,1,1-TRICHLORO-2,2-DI(4-METHOXYPHENYL)ETHANE ◇ 1,1'-(2,2,2-TRICHLOROETHYLIDENE)BIS(4-METHOXYBENZENE)

USE IN FOOD:

Purpose: Insecticide.

Where Used: Various.

Regulations: USDA CES Ranking: D-4 (1987).

IARC Cancer Review: Animal No Evidence IMEMDT 20,259,79; Animal Inadequate Evidence IMEMDT 5,193,74. NCI Carcinogenesis Bioassay (feed); No Evidence: mouse, rat NCITR* NCI-CG-TR-35,78. EPA Genetic Toxicology Program. Community Right-To-Know List.

OSHA PEL: TWA 10 mg/m^3 Total Dust; 5 mg/m^3 Respirable Dust ACGIH TLV: TWA 10 mg/m^3 DOT Classification: ORM-E; Label: None

SAFETY PROFILE: Moderately toxic by ingestion, intraperitoneal and skin contact. An experimental carcinogen, tumorigen, and teratogen. Human systemic effects by skin contact: somnolence. Experimental reproductive effects. Mutagenic data. When heated to decomposition emits highly toxic fumes of Cl^-.

TOXICITY DATA and CODEN

spm-rat-orl 28 g/kg/10W-C PSEBAA 176,187,84
otr-mus:fbr 2 mg/L JJIND8 67,1303,81
orl-rat TDLo:66 g/kg (33D male):REP JPETAB 138,126,62
orl-rat TDLo:2 g/kg (6-15D preg):TER TXAPA9 45,435,78
orl-rat TDLo:18200 mg/kg/2Y-C:CAR,TER EVHPAZ 36,205,80
orl-mus TDLo:56700 mg/kg/90W-C: CAR,TER JCROD7 93,173,79
orl-dog TDLo:383 g/kg/3Y-C:ETA EVHPAZ 36,205,80
orl-hmn LDLo:6430 mg/kg PCOC** -,705,66
skn-hmn TDLo:2414 mg/kg:CNS PCOC** -,705,66
orl-rat LD50:5000 mg/kg JPETAB 99,140,50

DOK200 CAS: 6358-53-8

1-((2,5-DIMETHOXYPHENYL)AZO)-2-NAPHTHOL

mf: $C_{18}H_{16}N_2O_3$ mw: 308.36

PROP: Mp: 156°. Sltly water-sol; mod sol in alc.

SYNS: C.I. 12156 ◇ C.I. SOLVENT RED 80 ◇ CITRUS RED NO. 2 ◇ 2,5-DIMETHOXYBENZENEAZO-β-NAPHTHOL ◇ 1-((2,5-DIMETHOXYPHENYL)AZO)-2-NAPHTHALENOL ◇ 2,5-DIMETHOXY-1-(PHENYLAZO)-2-NAPHTHOL ◇ 1-(1-(2,5-DIMETHOXYPHENYL)AZO)-2-NAPHTHOL ◇ 1-(2,5-DIMETHYLOXYPHENYLAZO)-2-NAPHTHOL

USE IN FOOD:

Purpose: Color.

Where Used: Oranges.

Regulations: FDA - 21CFR 74.302. Limitation of 2.0 ppm by weight calculated on the basis of the whole fruit.

IARC Cancer Review: Animal Sufficient Evidence IMEMDT 8,101,75. EPA Genetic Toxicology Program.

SAFETY PROFILE: An experimental carcinogen. Mutagenic data. When heated to decomposition it emits toxic fumes of NO_x.

TOXICITY DATA and CODEN

mmo-sat 500 μg/plate MUREAV 56,249,78
scu-mus TDLo:20 g/kg/80W-C:CAR FCTXAV 4,493,66
imp-mus TDLo:80 mg/kg:CAR BJCAAI 22,825,68

DOL600 CAS: 6923-22-4

3-(DIMETHOXYPHOSPHINYLOXY)N-METHYL-cis-CROTONAMIDE

mf: $C_7H_{14}NO_5P$ mw: 223.19

PROP: A reddish-brown solid; mild ester odor. Bp: 125°.

SYNS: APADRIN ◇ BILOBRAN ◇ CRISODRIN ◇ CRISODIN ◇ O,O-DIMETHYL-O-(2-N-METHYLCARBAMOYL-1-METHYL-VINYL)-FOSFAAT (DUTCH) ◇ O,O-DIMETHYL-O-(2-N-METHYLCARBAMOYL-1-METHYL)-VINYL-PHOSPHAT (GERMAN) ◇ O,O-DIMETHYL-O-(2-N-METHYL-CARBAMOYL-1-METHYL-VINYL) PHOSPHATE ◇ DIMETHYL-1-METHYL-2-(METHYLCARBAMOYL)VINYL-PHOSPHATE, cis ◇ (E)-DIMETHYL 1-METHYL-3-(METHYLAMINO)-3-OXO-1-PROPENYL PHOSPHATE ◇ DIMETHYL PHOSPHATE ESTER OF 3-HYDROXY-ETHYL-cis-CROTON-AMIDE ◇ DIMETHYL PHOSPHATE OF 3-HYDROXY-N-METHYL-cis-CROTONAMINE ◇ O,O-DIMETIL-O-(2-N-METILCARBAMOIL-1-METIL-VINIL)-FOSFATO (ITALIAN) ◇ ENT 27,129 ◇ HAZODRIN ◇ 3-HYDROXY-N-METHYL-cis-CROTONAMIDE DIMETHYL PHOSPHATE ◇ cis-1-METHYL-2-METHYL CARBAMOYL VINYL PHOSPHATE ◇ MONOCIL 40 ◇ MONOCRON ◇ MONOCROTOPHOS (ACGIH) ◇ NUVACRON ◇ PHOSPHATE de DIMETHYLE et de 2-METHYLCARBAMOYL1-METHYL VINYLE (FRENCH) ◇ PHOSPHORIC ACID, DIMETHYL ESTER, ESTER WITH cis-3-HYDROXY-N-METHYLCROTONAMIDE ◇ PLANTDRIN ◇ PILLARDRIN ◇ SHELL SD 9129 ◇ SUSVIN ◇ ULVAIR

USE IN FOOD:

Purpose: Insecticide.

Where Used: Tomato products (concentrated).

Regulations: FDA - 21CFR 193.151. Insecticide residue tolerance of 2 ppm in concentrated tomato products.

EPA Genetic Toxicology Program. EPA Extremely Hazardous Substances List.

ACGIH TLV: TWA 0.25 mg/m^3 (skin)

SAFETY PROFILE: Poison by ingestion, inhalation, skin contact, intraperitoneal, subcutaneous, and intravenous routes. Mutagenic data. Use may be restricted. When heated to decomposition it emits very toxic NO_x and PO_x.

TOXICITY DATA and CODEN

mmo-sat 500 μg/plate NTIS** PB84-138973
mmo-esc 5 μL/plate MUREAV 28,405,75
skn-rat LD50:112 mg/kg WRPCA2 9,119,70
orl-rat LD50:8 mg/kg FMCHA2 -,C161,83
ihl-rat LC50:63 mg/m^3/4H EGESAQ 24,173,80
ipr-rat LDLo:20 mg/kg BECTA6 19,47,78
scu-rat LD50:6964 μg/kg BJPCBM 40,124,70
ivn-rat LD50:9200 μg/kg NTIS** PB277-077

DOP600 CAS: 30560-19-1

O,S-DIMETHYLACETYLPHOSPHORO-AMIDOTHIOATE

mf: $C_4H_{10}NO_3PS$ mw: 183.18

SYNS: ACEPHAT (GERMAN) ◇ ACEPHATE ◇ ACETYL-PHOSPHORAMIDOTHIOIC ACID-O,S-DIMETHYL ESTER ◇ CHEVRON RE 12,420 ◇ ENT 27,822 ◇ ORTHENE ◇ ORTHENE-755 ◇ ORTHO 12420 ◇ ORTRAN ◇ ORTRIL ◇ RE 12420 ◇ 75 SP

USE IN FOOD:

Purpose: Insecticide.

Where Used: Animal feed.

Regulations: FDA - 21CFR 193.10. Pesticide residue tolerance of 0.02 ppm in food. 21CFR 561.20. Limitation of 8 ppm in cottonseed meal, 4 ppm in cottonseed hulls and soybean meal when used for animal feed.

EPA Genetic Toxicology Program.

SAFETY PROFILE: Poison by ingestion. Moderately toxic by skin contact and inhalation. Human mutagenic data. When heated to decomposition it emits very toxic fumes of NO_x, PO_x and SO_x.

TOXICITY DATA and CODEN

mmo-sat 3 mg/plate NTIS** PB80-133226
mrc-smc 50000 ppm NTIS** PB80-133226
orl-rat LD50:700 mg/kg MEIEDD 10,5,83
ihl-mus LCLo:2200 mg/m^3/5H TXAPA9 45,232,78

DOR500 CAS: 25988-97-0
DIMETHYLAMINE-EPICHLOROHYDRIN COPOLYMER

SYN: EPICHLOROHYDRIN-DIMETHYLAMINE COPOLYMER

USE IN FOOD:

Purpose: Decolorizing agent, flocculent.

Where Used: Sugar liquor and juices.

Regulations: FDA - 21CFR 173.60. Limitation of 150 ppm by weight of sugar solids.

SAFETY PROFILE: When heated to decomposition it emits acrid smoke and irritating fumes.

DQD400 CAS: 1596-84-5
DIMETHYLAMINOSUCCINAMIC ACID

mf: $C_6H_{12}N_2O_3$ mw: 160.20

SYNS: ALAR ◇ ALAR-85 ◇ AMINOZIDE ◇ B 995 ◇ BERNSTEINSAEURE-2,2-DIMETHYLHYDRAZID (GERMAN) ◇ B-NINE ◇ BUTANEDIOIC ACID MONO(2,2-DIMETHYLHYDRAZIDE) ◇ DAMINOZIDE (USDA) ◇ DIMAS ◇ N-DIMETHYL AMINO-β-CARBAMYL PROPIONIC ACID ◇ N-(DIMETHYLAMINO)SUCCINAMIC ACID ◇ N-DIMETHYLAMINO-SUCCINAMIDSAEURE (GERMAN) ◇ DMASA ◇ DMSA ◇ KYLAR ◇ NCI-C03827 ◇ SADH ◇ SUCCINIC ACID-2,2-DIMETHYLHYDRAZIDE ◇ SUCCINIC-1,1-DIMETHYL HYDRAZIDE

USE IN FOOD:

Purpose: Plant growth regulator.

Where Used: Animal feed, apples, cherries, grapes, nectarines, peaches, peanut meal, peanuts, pears, tomato pomace (dried), tomato products (concentrated).

Regulations: FDA - 21CFR 193.410. Plant growth regulator residue tolerance of 3 ppm in concentrated tomato products. 21CFR 561.360. Limitation of 90 ppm in peanut meal, 10 ppm in dried tomato pomace when used as animal feed. USDA CES Ranking: B-3 (1985).

EPA Genetic Toxicology Program. NCI Carcinogenesis Bioassay (feed); Clear Evidence: mouse, rat NCITR* NCI-CG-TR-83,78.

SAFETY PROFILE: Poison by intraperitoneal route. Moderately toxic by ingestion and possibly other routes. An experimental carcinogen, tumorigen, and teratogen. When heated to decomposition it emits toxic fumes of NO_x.

TOXICITY DATA and CODEN

orl-rat TDLo:182 g/kg/2Y-C:CAR,TER NCITR* NCI-CG-TR-83,78
orl-mus TDLo:2600 g/kg/62W-C:CAR CNREA8 37,3497,77
orl-mus TD :873 g/kg/2Y-C:ETA NCITR* NCI-CG-TR-83,78
orl-rat LD50:8400 mg/kg FMCHA2 -,D10,80

DQM600 CAS: 22781-23-3
2,2-DIMETHYL-1,3-BENZODIOXOL-4-OL METHYLCARBAMATE

mf: $C_{11}H_{13}NO_4$ mw: 223.25

SYNS: BENCARBATE ◇ BENDIOCARB ◇ BICAM ULV ◇ 2,2-DIMETHYL-1,3-BENZDIOXOL-4-YL-N-METHYLCARBAMATE ◇ 2,2-DIMETHYLBENZO-1,3-DIOXOL-4-YL METHYLCARBAMATE ◇ 2,2-DIMETHYL-4-(N-METHYLAMINOCARBOXYLATO)-1,3-BENZODIOXOLE ◇ 2,2-DIMETHYL-4-(N-METHYLCARBAMATO)-1,3-BENZODIOXOLE ◇ DYCARB ◇ FICAM ◇ GARVOX ◇ 2,3-ISOPROPYLIDENEDIOXYPHENYL METHYLCARBAMATE ◇ MC6897 ◇ METHYLCARBAMIC ACID-2,3-(ISOPROPYLIDENEDIOXY)PHENYL ESTER ◇ MULTAMAT ◇ NIOMIL ◇ ROTATE ◇ TATTOO ◇ TURCAM

USE IN FOOD:

Purpose: Insecticide.

Where Used: Animal feed, corn oil.

Regulations: FDA - 21CFR 193.152. Insecticide residue tolerance of 0.1 ppm in corn oil. (Expired 1/27/1983) 21CFR 561.191. Limited to spot and crack treatment.

EPA Genetic Toxicology Program.

SAFETY PROFILE: Poison by ingestion. Moderately toxic by skin contact. When heated to decomposition it emits toxic fumes of NO_x.

TOXICITY DATA and CODEN

orl-rat LD50:40 mg/kg FMCHA2 -,C29,83
skn-rat LD50:1000 mg/kg GUCHAZ 6,31,73

DQQ200 CAS: 100-86-7
DIMETHYL BENZYL CARBINOL
mf: $C_{10}H_{14}O$ mw: 150.24

PROP: White crystalline solid; floral odor. D: 0.972-0.977, flash p: 198°F. Sol in fixed oils mineral oil, propylene glycol; insol in glycerin.

SYNS: BENZYL DIMETHYL CARBINOL ◇ α,α-DIMETHYLPHENETHYL ALCOHOL ◇ 1,1-DIMETHYL-2-PHENYLETHANOL ◇ DMBC ◇ FEMA No. 2393

USE IN FOOD:

Purpose: Flavoring agent.

Where Used: Various.

Regulations: FDA - 21CFR 172.515. Use at a level not in excess of the amount reasonably required to accomplish the intended effect.

SAFETY PROFILE: Moderately toxic by ingestion. Combustible liquid. When heated to decomposition it emits acrid smoke and irritating fumes.

TOXICITY DATA and CODEN

orl-rat LD50:1280 mg/kg FCTXAV 2,327,64

DQQ375
DIMETHYL BENZYL CARBINYL ACETATE
mf: $C_{12}H_{16}O_2$ mw: 192.26

PROP: Colorless liquid to solid at room temp; floral, fruity odor. D: 0.995-1.002, refr index: 1.490-1.495, flash p: +212°F. Sol in fixed oils; sltly sol in propylene glycol; insol in water.

SYNS: α,α-DIMETHYLPHENETHYL ACETATE ◇ DIMETHYL BENZYL CARBINYL ACETATE ◇ FEMA No. 2392

USE IN FOOD:

Purpose: Flavoring agent.

Where Used: Various.

Regulations: FDA - 21CFR 172.515. Use at a level not in excess of the amount reasonably required to accomplish the intended effect.

SAFETY PROFILE: Combustible liquid. When heated to decomposition it emits acrid smoke and irritating fumes.

DQQ380
DIMETHYL BENZYL CARBINYL BUTYRATE
mf: $C_{14}H_{20}O_2$ mw: 220.31

PROP: Colorless liquid; prunelike odor. D: 0.960-0.981, refr index: 1.473-1.493 @ 25°, flash p: +151°F. Sol in alc, fixed oils; insol in water, propylene glycol.

SYNS: α,α-DIMETHYLPHENRTHYL BUTYRATE ◇ FEMA No. 2394

USE IN FOOD:

Purpose: Flavoring agent.

Where Used: Various.

Regulations: FDA - 21CFR 172.515. Use at a level not in excess of the amount reasonably required to accomplish the intended effect.

SAFETY PROFILE: Combustible liquid. When heated to decomposition it emits acrid smoke and irritating fumes.

DRI500
DIMETHYL DIALKYL AMMONIUM CHLORIDE

USE IN FOOD:

Purpose: Colorizing agent.

Where Used: Sugar liquors.

Regulations: FDA - 21CFR 172.712. Limitation of 700 ppm by weight of the sugar solids.

SAFETY PROFILE: When heated to decomposition it emits acrid smoke and irritating fumes.

DRJ600 CAS: 300-76-5
O,O-DIMETHYL-O-(1,2-DIBROMO-2,2-DICHLOROETHYL)PHOSPHATE
DOT: 2783
mf: $C_4H_7Br_2Cl_2O_4P$ mw: 380.80

PROP: Slightly sol in aliphatic hydrocarbons, very sol in aromatic hydrocarbons. Mp: 27.0°.

SYNS: ARTHODIBROM ◇ BROMCHLOPHOS ◇ BROMEX ◇ DIBROM ◇ O-(1,2-DIBROM-2,2-DICHLORAETHYL)-O,O-DIMETHYL-PHOSPHAT (GERMAN) ◇ 1,2-DIBROMO-2,2-DICHLOROETHYL DIMETHYL PHOSPHATE ◇ O-(1,2-DIBROMO-2,2-DICLORO-ETIL)-O,O-DIMETIL-FOSTATO (ITALIAN) ◇ O-(1,2-DIBROOM-2,2-DICHLOOR-ETHYL)-O,O-DIMETHYL-FOSFAAT (DUTCH) ◇ DIMETHYL 1,2-DIBROMO-2,2-DICHLORETHYL PHOSPHATE ◇ O,O-DIMETHYL-O-2,2-DICHLORO-1,2-DIBROMOETHYL PHOSPHATE ◇ ENT 24,988 ◇ HIBROM ◇ NALED (ACGIH, USDA)

◇ ORTHO 4355 ◇ ORTHODIBROM ◇ ORTHODIBROMO ◇ PHOSPHATE de O,O-DIMETHLE et de O-(1,2-DIBROMO-2,2-DICHLORETHYLE) (FRENCH) ◇ RE-4355

USE IN FOOD:

Purpose: Insecticide.

Where Used: Various.

Regulations: USDA CES Ranking: B-4 (1987).

OSHA PEL: TWA 3 mg/m^3 ACGIH TLV: TWA 3 mg/m^3 (skin)

SAFETY PROFILE: Poison by ingestion and inhalation. Moderately toxic by skin contact. A human skin irritant and a severe experimental skin irritant. Mutagenic data. An insecticide of the cholinesterase inhibitor type, and a nonsystemic acaricide. When heated to decomposition it emits very toxic fumes of Br^-, Cl^- and PO_x.

TOXICITY DATA and CODEN

skn-man 42 mg/21D-I open TXAPA9 21,369,72
skn-rbt 500 mg/24H SEV TXAPA9 21,369,72
mma-sat 50 μg/plate JAFCAU 29,268,81
mmo-bcs 50 μg/plate JAFCAU 29,268,81
orl-rat LD50:250 mg/kg TXAPA9 14,515,69
ihl-rat LD50:7700 μg/kg BECTA6 19,113,78
skn-rat LD50:800 mg/kg WRPCA2 9,119,70

DRJ850 CAS: 4525-33-1
DIMETHYL DICARBONATE

mf: $(CH_3OCO)_2O$ mw: 134.09

USE IN FOOD:

Purpose: Fungicide, yeast inhibitor.

Where Used: Wine.

Regulations: FDA - 21CFR 172.133. Limitation of 200 ppm in wine.

SAFETY PROFILE: When heated to decomposition it emits acrid smoke and irritating fumes.

DRK200 CAS: 62-73-7
DIMETHYL DICHLOROVINYL PHOSPHATE

DOT: 2783

mf: $C_4H_7Cl_2O_4P$ mw: 220.98

PROP: Liquid. Bp: 120° @ 14 mm, bp: 77° @ 1 mm. Sltly sol in water and glycerin; miscible with aromatic and chlorinated hydrocarbon solvents and alcohol.

SYNS: APAVAP ◇ ASTROBOT ◇ ATGARD ◇ BAY 19149 ◇ BENFOS ◇ BIBESOL ◇ BREVINYL ◇ CANOGARD ◇ CEKUSAN ◇ CHLORVINPHOS ◇ CYANOPHOS ◇ CYPONA ◇ DDVF ◇ DEDEVAP ◇ DERIBAN ◇ DERRIBANTE ◇ DEVIKOL ◇ (2,2-DICHLOOR-VINYL)-DIMETHYL-FOSFAAT (DUTCH) ◇ DICHLOORVO (DUTCH) ◇ DICHLORFOS (POLISH) ◇ 2,2-DICHLOROETHENOL DIMETHYL PHOSPHATE ◇ 2,2-DICHLOROETHENYL DIMETHYL PHOSPHATE ◇ 2,2-DICHLOROETHENYL PHOSPHORIC ACID DIMETHYL ESTER ◇ DICHLOROPHOS (ACGIH, USDA) ◇ DICHLOROVAS ◇ (2,2-DICHLORO-VINIL)DIMETILFOSFATO (ITALIAN) ◇ 2,2-DICHLOROVINYL ALCOHOL, DIMETHYL PHOSPHATE ◇ 2,2-DICHLOROVINYL DIMETHYL PHOSPHATE ◇ 2,2-DICHLOROVINYL DIMETHYL PHOSPHORIC ACID ESTER ◇ DICHLOROVOS ◇ DICHLORPHOS ◇ (2,2-DICHLOR-VINYL)-DIMETHYL-PHOSPHAT (GERMAN) ◇ O-(2,2-DICHLORVINYL)-O,O-DIMETHYLPHOSPHAT (GERMAN) ◇ DICHLORVOS (ACGIH, DOT) ◇ DIMETHYL-2,2-DICHLOROETHENYL PHOSPHATE ◇ DIMETHYL-2,2-DICHLOROVINYL PHOSPHATE ◇ O,O-DIMETHYL DICHLOROVINYL PHOSPHATE ◇ O,O-DIMETHYL-O-2,2-DICHLOROVINYL PHOSPHATE ◇ O,O-DIMETHYL-O-(2,2-DICHLOR-VINYL)-PHOSPHAT (GERMAN) ◇ DIVIPAN ◇ DQUIGARD ◇ DUO-KILL ◇ DURAVOS ◇ ENT 20,738 ◇ EQUIGEL ◇ ESTROSEL ◇ ESTROSOL ◇ FECAMA ◇ FLY-DIE ◇ FLY FIGHTER ◇ HERKAL ◇ KRECALVIN ◇ LINDAN ◇ MAFU ◇ MARVEX ◇ MOPARI ◇ NCI-C00113 ◇ NERKOL ◇ NOGOS ◇ NO-PEST ◇ NO-PEST STRIP ◇ NSC-6738 ◇ NUVA ◇ OKO ◇ OMS 14 ◇ PHOSPHATE de DIMETHYLE et de 2,2-DICHLOROVINYLE (FRENCH) ◇ PHOSPHORIC ACID-2,2-DICHLOROETHENYL DIMETHYL ESTER ◇ PHOSVIT ◇ SD-1750 ◇ SZKLARNIAK ◇ TAP 9VP ◇ TASK ◇ TASK TABS ◇ TENAC ◇ TETRAVOS ◇ VAPONA ◇ VAPONITE ◇ VERDICAN ◇ VERDIPOR ◇ VINYLOFOS ◇ VINYLOPHOS

USE IN FOOD:

Purpose: Animal drug, animal feed drug, insecticide.

Where Used: Animal feed, cereals, cookies (packaged), crackers (packaged), figs (dried), flour, pork, sugar.

Regulations: FDA - 21CFR 193.140. Insecticide residue tolerance of 0.5 ppm on packaged or bagged nonperishable processed food, 0.5 ppm in dried figs. 21CFR 556.180. Tolerance of 0.1 ppm in edible tissue of swine. 21CFR 558.205. USDA CES Ranking: B-4 (1987).

IARC Cancer Review: Animal Inadequate Evidence IMEMDT 20,97,79. NCI Carcinogenesis Bioassay (feed); No Evidence: mouse, rat NCITR* NCI-CG-TR-10,77. EPA Genetic Toxicology Program. Community Right-To-Know List. EPA Extremely Hazardous Substances List.

OSHA PEL: TWA 1 mg/m^3 (skin) ACGIH TLV: TWA 0.1 ppm DOT Classification: Poison B; Label: Poison; Poison B; Label: Poison

SAFETY PROFILE: Poison by ingestion, inhalation, skin contact, subcutaneous, intravenous, intraperitoneal, and possibly other routes. An experimental teratogen. Experimental reproductive effects. Human mutagenic data. A cholinesterase inhibitor. No neurotoxicity has been observed. It is very rapidly metabolized and excreted. When heated to decomposition it emits very toxic fumes of Cl^- and PO_x.

TOXICITY DATA and CODEN

dnd-hmn:lym 62 mg/L TUMOAB 66,425,80
dni-hmn:lym 62 mg/L TUMOAB 66,425,80
orl-rat TDLo:39200 μg/kg (14-21D preg):REP NUPOBT 15,255,77
ipr-rat TDLo:15 mg/kg (11D preg):TER AEHLAU 16,805,68
ihl-rbt TCLo:4 mg/m^3/23H (1-28D preg):TER ATXKA8 30,29,72
orl-rat LD50:25 mg/kg DOEAAH 35,25,79
ihl-rat LC50:15 mg/m^3/4H GISAAA 33(12),35,68
skn-rat LD50:70400 μg/kg APYPAY 32,507,81
ipr-rat LDLo:40 mg/kg BECTA6 19,47,78
scu-rat LD50:10800 μg/kg APYPAY 32,507,81

DSD775 CAS: 106-72-9
2,6-DIMETHYL-5-HEPTENAL
mf: $C_9H_{16}O$ mw: 140.23

PROP: Pale yellow liquid; melon odor. D: 0.852-0.858, refr index: 1.443-1.448

SYN: FEMA No. 2497

USE IN FOOD:

Purpose: Flavoring agent.

Where Used: Various.

Regulations: FDA - 21CFR 172.515. Use at a level not in excess of the amount reasonably required to accomplish the intended effect.

SAFETY PROFILE: Skin and eye irritant. When heated to decomposition it emits acrid smoke and irritating fumes.

DSO200 CAS: 23422-53-9
N,N-DIMETHYL-N′-(((METHYLAMINO)CARBONYL)OXY)-PHENYLMETHANIMIDAMIDE MONOHYDROCHLORIDE
mf: $C_{11}H_{15}N_3O_2 \cdot ClH$ mw: 257.75

SYNS: CARZOL SP ◇ DICARZOL ◇ m-(((DIMETHYLAMINO)METHYLENE)AMINO)PHENYLMETHYL CARBAMATE,HYDROCHLORIDE ◇ 3-DIMETHYLAMINOMETHYLENEIMINOPHENYL-N-METHYLCARBAMATE, HYDROCHLORIDE ◇ ENT 27,566 ◇ EP-332 ◇ FORMETANATE HYDROCHLORIDE ◇ MORTON EP332 ◇ NOR-AM EP 332 ◇ SCHERING 36056 ◇ SN 36056

USE IN FOOD:

Purpose: Insecticide.

Where Used: Animal feed, citrus molasses, grapefruit, lemons, limes, oranges, tangerines.

Regulations: FDA - 21CFR 561.250. Limitation of 10 ppm in citrus molasses when used for animal feed.

EPA Extremely Hazardous Substances List.

SAFETY PROFILE: Poison by ingestion. Mildly toxic by skin contact. When heated to decomposition it emits very toxic fumes of NO_x and HCl.

TOXICITY DATA and CODEN

orl-mus LD50:18 mg/kg 28ZEAL 5,118,76
skn-rbt LD50:10200 mg/kg 28ZEAL 5,118,76

DSP400 CAS: 60-51-5
O,O-DIMETHYL METHYLCARBAMOYLMETHYL PHOSPHORODITHIOATE
mf: $C_5H_{12}NO_3PS_2$ mw: 229.27

SYNS: AC-12682 ◇ AMERICAN CYANAMID 12880 ◇ BI-58 ◇ CEKUTHOATE ◇ CL 12880 ◇ CYGON ◇ CYGON INSECTICIDE ◇ DAPHENE ◇ DE-FEND ◇ DEMOS-L40 ◇ DEVIGON ◇ DIMATE 267 ◇ DIMETATE ◇ DIMETHOAAT (DUTCH) ◇ DIMETHOATE (USDA) ◇ DIMETHOAT (GERMAN) ◇ DIMETHOAT TECHNISCH 95% ◇ DIMETHOGEN ◇ O,O-DIMETHYLDITHIOPHOSPHORYLACETIC ACID-N-MONOMETHYLAMIDE SALT ◇ O,O-DIMETHYL-DITHIOPHOSPHORYLESSIGSAEURE MONOMETHYLAMID (GERMAN) ◇ O,O-DIMETHYL-S-(2-(METHYLAMINO)-2-OXOETHYL) PHOSPHORODITHIOATE ◇ O,O-DIMETHYL-S-(N-METHYL-CARBAMOYL)-METHYL-DITHIOFOSFAAT (DUTCH) ◇ (O,O-DIMETHYL-S-(N-METHYL-CARBAMOYL-METHYL)-DITHIOPHOSPHAT) (GERMAN) ◇ O,O-DIMETHYL-S-(N-METHYLCARBAMOYLMETHYL) DITHIOPHOSPHATE ◇ O,O-DIMETHYL-S-(N-METHYLCARBAMOYLMETHYL) PHOSPHORODITHIOATE ◇ O,O-DIMETHYL-S-(N-METHYLCARBAMYLMETHYL) THIOTHIONOPHOSPHATE ◇ O,O-DIMETHYL-S-(N-MONOMETHYL)-CARBAMYL METHYLDITHIOPHOSPHATE ◇ O,O-DIMETHYL-S-(2-OXO-3-AZA-BUTYL)-DITHIOPHOSPHAT (GERMAN) ◇ O,O-DIMETIL-S-(N-METIL-

CARBAMOIL-METIL)-DITIOFOSFATO (ITALIAN) ◇ DIMETON ◇ DIMEVUR ◇ DITHIOPHOSPHATE de O,O-DIMETHYLE et de S(-N-METHYLCARBAMOYL-METHYLE) (FRENCH) ◇ EI-12880 ◇ ENT 24,650 ◇ EXPERIMENTAL INSECTICIDE 12,880 ◇ FERKETHION ◇ FORTION NM ◇ FOSFAMID ◇ FOSFOTOX ◇ FOSTION MM ◇ L-395 ◇ LURGO ◇ S-METHYLCARBAMOYLMETHYL-O,O-DIMETHYL PHOSPHORODITHIOATE ◇ N-MONOMETHYLAMIDE OF O,O-DIMETHYLDITHIOPHOSPHORYLACETIC ACID ◇ NC-262 ◇ NCI-C00135 ◇ PERFECTHION ◇ PHOSPHAMID ◇ PHOSPHORODITHIOIC ACID-O,O-DIMETHYL-S-(2-(METHYLAMINO)-2-OXOETHYL) ESTER ◇ RACUSAN ◇ RCRA WASTE NUMBER P044 ◇ REBELATE ◇ ROGODIAL ◇ ROGOR ◇ ROXION U.A. ◇ SINORATOX ◇ TRIMETION

USE IN FOOD:

Purpose: Insecticide.

Where Used: Animal feed, citrus pulp (dried).

Regulations: FDA - 21CFR 561.170. Limitation of 5 ppm dried citrus pulp when used for animal feed. USDA CES Ranking: B-3 (1986).

NCI Carcinogenesis Bioassay (feed); No Evidence: mouse, rat NCITR* NCI-CG-TR-4,77. EPA Genetic Toxicology Program. EPA Extremely Hazardous Substances List.

SAFETY PROFILE: Poison by ingestion, skin contact, intraperitoneal, subcutaneous, and possibly other routes. Moderately toxic by intravenous route. An experimental carcinogen and teratogen. Experimental reproductive effects. Human mutagenic data. When heated to decomposition it emits very toxic fumes of NO_x, PO_x, and SO_x.

TOXICITY DATA and CODEN

mma-sat 500 μg/plate JTEHD6 16,403,85
mmo-omi 100 mg/L TGANAK 15(3),68,81
orl-rat TDLo:120 mg/kg (6-15D preg):TER BECTA6 22,522,79
orl-mus TDLo:1050 mg/kg (MGN):REP TXAPA9 26,29,73
orl-rat TDLo:256 mg/kg/4W-I:CAR ARGEAR 41,311,73
ims-rat TDLo:176 mg/kg/6W-I:CAR ARGEAR 41,311,73
orl-hmn LD50:30 mg/kg GUCHAZ 6,209,73
skn-rat LD50:353 mg/kg BJIMAG 26,59,69
ipr-rat LD50:100 mg/kg BJIMAG 21,52,64
ipr-mus LD50:45 mg/kg BJIMAG 21,52,64
scu-mus LD50:60 mg/kg BJIMAG 21,52,64

DSP600 CAS: 23135-22-0
N′,N′-DIMETHYL-N-((METHYLCARBAMOYL)OXY)-1-METHYLTHIOOXAMIMIDIC ACID

mf: $C_7H_{13}N_3O_3S$ mw: 219.29

SYNS: D-1410 ◇ 2-(DIMETHYLAMINO)-N-(((METHYLAMINO)CARBONYL)OXY)-2-OXOETHANIMIDOTHIOIC ACID METHYL ESTER ◇ 2-DIMETHYLAMINO-1-(METHYLTHIO)GLYOXAL-o-METHYLCARBAMOYLMONOXIME ◇ N,N-DIMETHYL-α-METHYLCARBAMOYLOXYIMINO-α (METHYLTHIO)ACETAMIDE ◇ N′,N′-DIMETHYL-N-((METHYLCARBAMOYL)OXY)-1-THIOOXAMIMIDIC ACID METHYL ESTER ◇ DPX 1410 ◇ INSECTICIDE-NEMATICIDE 1410 ◇ METHYL-2-(DIMETHYLAMINO)-N-(((METHYLAMINO)CARBONYL)OXY)-2-OXOETHANIMIDOTHIOATE ◇ METHYL-1-(DIMETHYLCARBAMOYL)-N-(METHYLCARBAMOYLOXY)THIOFORMIMIDATE ◇ S-METHYL-1-(DIMETHYLCARBAMOYL)-N-((METHYLCARBAMOYL)OXY)THIOFORMIMIDATE ◇ METHYL-N′,N′-DIMETHYL-N-((METHYLCARBAMOYL)OXY)-1-THIOOXAMIMIDATE ◇ OXAMYL ◇ THIOXAMYL ◇ VYDATE ◇ VYDATE L INSECTICIDE/NEMATICIDE ◇ VYDATE L OXAMYL INSECTICIDE/NEMATOCIDE

USE IN FOOD:

Purpose: Insecticide.

Where Used: Animal feed, pineapple bran, pineapples.

Regulations: FDA - 21CFR 561.285. Limitation of 6 ppm in pineapple bran when used for animal feed.

EPA Extremely Hazardous Substances List.

SAFETY PROFILE: Poison by ingestion and inhalation. Moderately toxic by skin contact. When heated to decomposition it emits very toxic fumes of NO_x and SO_x.

TOXICITY DATA and CODEN

orl-rat LD50:5 mg/kg JAFCAU 26,550,78
ihl-rat LC50:170 mg/m^3/1H 85DPAN -,-,71/76
skn-rbt LD50:740 mg/kg SPEADM 78-1,61,78

DSQ000 CAS: 122-14-5
DIMETHYL-3-METHYL-4-NITROPHENYLPHOSPHOROTHIONATE

mf: $C_9H_{12}NO_5PS$ mw: 277.25

SYNS: ACCOTHION ◇ ACEOTHION ◇ AGRIA 1050 ◇ AGRIYA 1050 ◇ AGROTHION ◇ AMERICAN CYANAMID CL-47,300 ◇ ARBOGAL ◇ BAY 41831 ◇ BAYER 41831 ◇ BAYER S 5660 ◇ CEKUTROTHION ◇ CL 47300

◇ CP 47114 ◇ CYFEN ◇ CYTEL ◇ CYTEN ◇ O,O-DIMETHYL-O-(3-METHYL-4-NITROFENYL)-MONOTHIOFOSFAAT (DUTCH) ◇ O,O-DIMETHYL-O-(3-METHYL-4-NITRO-PHENYL)-MONOTHIOPHOSPHAT (GERMAN) ◇ O,O-DIMETHYL-O-(3-METHYL-4-NITROPHENYL) PHOSPHOROTHIOATE ◇ O,O-DIMETHYL-O-(3-METHYL-4-NITROPHENYL) THIOPHOSPHATE ◇ O,O-DIMETHYL-O-(3-METHYL) PHOSPHOROTHIOATE ◇ O,O-DIMETHYL-O-(4-NITRO-3-METHYLPHENYL)THIOPHOSPHATE ◇ O,O-DIMETHYL-O-4-NITRO-m-TOLYL PHOSPHOROTHIOATE ◇ O,O-DIMETIL-O-(3-METIL-4-NITRO-FENIL)-MONOTIOFOSFATO (ITALIAN) ◇ EI 47300 ◇ ENT 25,715 ◇ FALITHION ◇ FENITOX ◇ FENITROTHION ◇ FENITROTION (HUNGARIAN) ◇ FOLETHION ◇ H-35-F 87 (BVM) ◇ 8057HC ◇ KOTION ◇ MEP (Pesticide) ◇ METATHIONE ◇ METATION ◇ METHYLNITROPHOS ◇ MONSANTO CP 47114 ◇ NITROPHOS ◇ NOVATHION ◇ NUVANOL ◇ OLEOSUMIFENE ◇ OMS 43 ◇ OVADOFOS ◇ PENNWALT C-4852 ◇ PHENITROTHION ◇ S 112A ◇ S 5660 ◇ SUMITHIAN ◇ THIOPHOSPHATE de O,O-DIMETHYLE et de O-(3-METHYL-4-NITROPHENYLE) (FRENCH) ◇ VERTHION

USE IN FOOD:

Purpose: Insecticide.

Where Used: Wheat gluten.

Regulations: FDA - 21CFR 193.156. Insecticide residue tolerance of 30 ppm in wheat gluten. Experimental use tolerance of 0.5 ppm in foods. (Expired 6/17/1983)

EPA Genetic Toxicology Program. EPA Extremely Hazardous Substances List.

SAFETY PROFILE: Poison by ingestion, inhalation, intravenous, intraperitoneal, and possibly other routes. Moderately toxic by skin contact, intratracheal, and subcutaneous routes. Human systemic effects by ingestion: hypermotility, diarrhea, nausea or vomiting, and dyspnea. Mutagenic data. When heated to decomposition it emits very toxic fumes of NO_x, PO_x and SO_x.

TOXICITY DATA and CODEN

mmo-sat 500 μg/plate MUREAV 116,185,83
cyt-ham:lng 200 mg/L GMCRDC 27,95,81
orl-wmn TDLo:800 mg/kg:GIT,PUL ARTODN 56,136,84
orl-rat LD50:250 mg/kg TXAPA9 21,315,72
ihl-rat LC50:378 mg/m^3/4H EGESAQ 24,173,80
skn-rat LD50:750 mg/kg SRTCDF -,101,77
ipr-rat LD50:300 mg/kg TXAPA9 63,91,82
ivn-rat LD50:33 mg/kg ABCHA6 27,669,63

DTC600 CAS: 122-19-0
DIMETHYLOCTADECYLBENZYLAMMONIUM CHLORIDE

mf: $C_{27}H_{50}N{\bullet}Cl$ mw: 424.23

SYNS: AMMONYX 4 ◇ AMMONYX CA SPECIAL ◇ ARQUAD DM18B-90 ◇ BARQUAT SB-25 ◇ BENZYLDIMETHYLSTEARYLAMMONIUM CHLORIDE ◇ BENZYLSTEARYLDIMETHYLAMMONIUM CHLORIDE ◇ CARSOQUAT SDQ-25 ◇ DEHYQUART STC-25 ◇ DIMETHYLBENZYLOCTADECYLAMMONIUM CHLORIDE ◇ INTEXAN SB-85 ◇ J SOFT C 4 ◇ KATAMINE AB ◇ NISSAN CATION S2-100 ◇ N-OCTADECYL-N-BENZYL-N,N-DIMETHYLAMMONIUMCHLORIDE ◇ OCTADECYLDIMETHYLBENZYLAMMONIUM CHLORIDE ◇ ORTHOSAN MB ◇ QUATERNOL 1 ◇ STEARALKONIUM CHLORIDE ◇ STEARYLDIMETHYLBENZYLAMMONIUM CHLORIDE ◇ STEBAC ◇ TALLOW BENZYL DIMETHYLAMMONIUM CHLORIDE ◇ TRITON X-40 ◇ VARISOFT SDC

USE IN FOOD:

Purpose: Antimicrobial agent.

Where Used: Beets, sugarcane, sugarcane juice (raw).

Regulations: FDA - 21CFR 172.165. Limitation of 1.5-6.0 ppm. 21CFR 173.320. Limitation of 0.05 ppm based on weight of raw sugarcane or raw beets.

SAFETY PROFILE: Poison by intraperitoneal route. Moderately toxic by ingestion. A human skin irritant and severe experimental eye irritant. When heated to decomposition it emits very toxic fumes of NO_x, NH_3 and Cl^-.

TOXICITY DATA and CODEN

skn-hmn 3 mg/3D-I MLD 85DKA8 -,127,77
skn-man 125 mg/2D MLD PSTGAW 20,16,53
skn-rbt 1 mg/24H OYYAA2 6,329,72
eye-rbt 200 μg SEV PSTGAW 20,16,53
orl-rat LD50:1250 mg/kg JACTDZ 1(2),57,82

DTC800 CAS: 5392-40-5
3,7-DIMETHYL-2,6-OCTADIENAL

mf: $C_{10}H_{16}O$ mw: 152.26

PROP: Mobile, pale yellow liquid; strong lemon odor. D: 0.891-0.897 @ 15°, refr index: 1.486-1.490, flash p: 198°F. Sol in 5 volumes of 60% alcohol; sol in all proportions of benzyl benzoate, diethyl phthalate, glycerin, propylene glycol, mineral oil, fixed oils and 95% alc; insol in water.

SYNS: BUTOBEN ◇ BUTYL p-HYDROXYBENZOATE ◇ CITRAL (FCC) ◇ FEMA No. 2203 ◇ NCI-C56348 ◇ NERAL

USE IN FOOD:

Purpose: Flavoring agent.

Where Used: Baked goods, candy, ice cream.

Regulations: FDA - 21CFR 172.515, 182.60. Use at a level not in excess of the amount reasonably required to accomplish the intended effect.

SAFETY PROFILE: Moderately toxic by intraperitoneal route. Mildly toxic by ingestion. Experimental reproductive effects. A human and experimental skin irritant. Combustible liquid. When heated to decomposition it emits acrid smoke and irritating fumes.

TOXICITY DATA and CODEN

skn-hmn 40 mg/24H MLD FCTXAV 17,259,79
skn-rbt 500 mg/24H MOD FCTXAV 17,259,79
skn-rat TDLo: 28 g/kg (60D pre): REP FCTXAV 18,547,80
ipr-rat TDLo: 1800 mg/kg (22W pre): REP FCTXAV 18,547,80
orl-rat LD50: 4960 mg/kg FCTXAV 2,327,64
ipr-rat LD50: 460 mg/kg JRPFA4 55,347,79

DTD000 CAS: 106-24-1

3,7-DIMETHYL-(E)-2,6-OCTADIEN-1-OL

mf: $C_{10}H_{18}O$ mw: 154.28

PROP: Colorless to pale yellow, oily liquid; pleasant geranium odor. D: 0.870-0.890 @ 15°, refr index: 1.469-1.478, mp: 15°, bp: 230°, flash p: 214°F. Sol in fixed oils, propylene glycol; sltly sol in water; insol in glycerin @ 230°.

SYNS: 2,6-DIMETHYL-trans-2,6-OCTADIEN-8-OL ◇ 3,7-DIMETHYL-trans-2,6-OCTADIEN-1-OL ◇ FEMA No. 2507 ◇ GERANIOL (FCC) ◇ GERANIOL ALCOHOL ◇ GERANIOL EXTRA ◇ GERANYL ALCOHOL ◇ GUANIOL ◇ LEMONOL

USE IN FOOD:

Purpose: Flavoring agent.

Where Used: Various.

Regulations: FDA - 21CFR 182.60. Use at a level not in excess of the amount reasonably required to accomplish the intended effect.

SAFETY PROFILE: Poison by intravenous route. Moderately toxic by ingestion and intramuscular routes. Combustible liquid. When heated to decomposition it emits acrid smoke and irritating fumes.

TOXICITY DATA and CODEN

orl-rat LD50: 3600 mg/kg FCTXAV 2,327,64
ivn-rbt LDLo: 50 mg/kg NYKZAU 58,394,62

DTD200 CAS: 106-25-2

2-cis-3,7-DIMETHYL-2,6-OCTADIEN-1-OL

mf: $C_{10}H_{18}O$ mw: 154.28

PROP: Colorless liquid; sweet, rose odor. D: 0.875-0.880, refr index: 1.467-1.478. Sol in alc, chloroform, ether, water @ 227°.

SYNS: 3,7-DIMETHYL-(Z)-2,6-OCTADIEN-1-OL ◇ FEMA No. 2770 ◇ NEROL (FCC)

USE IN FOOD:

Purpose: Flavoring agent.

Where Used: Various.

Regulations: FDA - 21CFR 172.515. Use at a level not in excess of the amount reasonably required to accomplish the intended effect.

SAFETY PROFILE: Moderately toxic by intramuscular route. Mildly toxic by ingestion. A skin irritant. When heated to decomposition it emits acrid smoke and irritating fumes.

TOXICITY DATA and CODEN

skn-rbt 500 mg/24H MOD FCTXAV 14,623,76
orl-rat LD50: 4500 mg/kg FCTXAV 14,623,76

DTD600 CAS: 115-95-7

3,7-DIMETHYL-1,6-OCTADIEN-3-OL ACETATE

mf: $C_{12}H_{20}O_2$ mw: 196.32

PROP: Clear, colorless, oily liquid; odor of bergamot. Bp: 108-110°, d: 0.898-0.914, flash p: 185°F. Sol in alc, ether, diethyl phthalate, benzyl benzoate, mineral oil, fixed oils; sltly sol in propylene glycol; insol in water, glycerin.

SYNS: ACETIC ACID LINALOOL ESTER ◇ BERGAMIOL ◇ 3,7-DIMETHYL-1,6-OCTADIEN-3-YL ACETATE ◇ FEMA No. 2636 ◇ LICAREOL ACETATE ◇ LINALOL ACETATE ◇ LINALOOL ACETATE ◇ LINALYL ACETATE (FCC)

USE IN FOOD:

Purpose: Flavoring agent.

Where Used: Various.

Regulations: FDA - 21CFR 182.60. Use at a level not in excess of the amount reasonably required to accomplish the intended effect.

SAFETY PROFILE: Mildly toxic by ingestion. Combustible liquid. When heated to decomposition it emits acrid smoke and irritating fumes.

TOXICITY DATA and CODEN

orl-rat LD50:14550 mg/kg FCTXAV 2,327,64

DTD800 CAS: 105-87-3

trans-3,7-DIMETHYL-2,6-OCTADIEN-1-OL ACETATE

mf: $C_{12}H_{20}O_2$ mw: 196.32

PROP: Colorless, sweet, clear liquid; odor of lavender. D: 0.907-0.918 @ 15°, refr index: 1.458-1.464, bp: 128-129° @ 16 mm, flash p: 219°F. Sol in alc, fixed oils, ether; sltly sol in propylene glycol; insol in water and glycerol.

SYNS: ACETIC ACID GERANIOL ESTER ◇ 3,7-DIMETHYL-2-trans-6-OCTADIENYL ACETATE ◇ trans-3,7-DIMETHYL-2,6-OCTADIEN-1-YL ACETATE ◇ trans-2,6-DIMETHYL-2,6-OCTADIEN-8-YL ETHANOATE ◇ FEMA No. 2509 ◇ GERANIOL ACETATE ◇ GERANYL ACETATE (FCC) ◇ NCI-C54728

USE IN FOOD:

Purpose: Flavoring agent.

Where Used: Various.

Regulations: FDA - 21CFR 182.60. Use at a level not in excess of the amount reasonably required to accomplish the intended effect.

SAFETY PROFILE: Mildly toxic by ingestion. Combustible liquid. When heated to decomposition it emits acrid smoke and irritating fumes.

TOXICITY DATA and CODEN

orl-rat LD50:6330 mg/kg FCTXAV 2,327,64

DTE600 CAS: 106-21-8

DIMETHYLOCTANOL

mf: $C_{10}H_{22}O$ mw: 158.32

PROP: Colorless liquid; sweet, rose odor. D: 0.26-0.842, refr index: 1.435. Sol in fixed oils, propylene glycol; insol in glycerin.

SYNS: DIHYDROCITRONELLOL ◇ 2,6-DIMETHYL-8-OCTANOL ◇ 3,7-DIMETHYL-1-OCTANOL (FCC) ◇ FEMA No. 2391 ◇ GERANIOL TETRAHYDRIDE ◇ PELARGOL ◇ PERHYDROGERANIOL ◇ TETRAHYDROGERANIOL

USE IN FOOD:

Purpose: Flavoring agent.

Where Used: Bakery products, beverages (nonalcoholic), chewing gum, confections, ice cream products, pickles.

Regulations: FDA - 21CFR 172.515. Use at a level not in excess of the amount reasonably required to accomplish the intended effect.

SAFETY PROFILE: Moderately toxic by skin contact. A skin irritant. When heated to decomposition it emits acrid smoke and irritating fumes.

TOXICITY DATA and CODEN

skn-rbt 500 mg/24H FCTXAV 13,517,75
skn-rbt LD50:2400 mg/kg FCTXAV 12,535,74

DTF400 CAS: 141-25-3

2,6-DIMETHYL-1-OCTEN-8-OL

mf: $C_{10}H_{20}O$ mw: 156.30

PROP: Flash p: +212°F.

SYNS: α-CITRONELLOL ◇ 3,7-DIMETHYL-7-OCTEN-1-OL ◇ FEMA No. 2981 ◇ RHODINOL (FCC)

USE IN FOOD:

Purpose: Flavoring agent.

Where Used: Various.

Regulations: FDA - 21CFR 172.515. Use at a level not in excess of the amount reasonably required to accomplish the intended effect.

SAFETY PROFILE: Moderately toxic by intramuscular route. When heated to decomposition it emits acrid smoke and irritating fumes.

TOXICITY DATA and CODEN

ims-mus LD50:4000 mg/kg JSICAZ 21,342,62

DTR850

DIMETHYLPOLYSILOXANE

mf: $[(CH_3)_2SiO—]$

PROP: Clear, colorless viscous liquid. D: 0.96, refr index: 1.400. Sol in hydrocarbon solvents; insol in water.

SYNS: DIMETHYL SILICONE ◇ POLYDIMETHYLSILOXANE

USE IN FOOD:

Purpose: Defoaming agent, hog scald agent, poultry scald agent, release agent.

Where Used: Chewing gum, gelatins, gelatin (dry), hog carcasses, poultry, salt, sugar, wine.

Regulations: FDA - 21CFR 173.340. Limitation of 10 ppm in food, except zero tolerance in milk, 110 ppm in dry gelatin, 250 ppm in salt. 21CFR 181.28. Use at a level not in excess of the amount reasonably required to accomplish the intended effect. USDA - 9CFR 318.7, 381.147. Sufficient for purpose.

SAFETY PROFILE: Combustible liquid. When heated to decomposition it emits acrid smoke and irritating fumes.

DTU400 CAS: 5910-89-4
2,3-DIMETHYLPYRAZINE
mf: $C_6H_8N_2$ mw: 108.16

PROP: Colorless liquid; nutty cocoa odor. D: 1.000-1.022 @ 20°, refr index: 1.506-1.509, flash p: 147°F (OC), d: 0.99, vap d: 3.72, bp: 182.2°. Misc with water, organic solvents.

SYNS: 2,3-DIMETHYL-1,4-DIAZINE ◇ FEMA No. 3271

USE IN FOOD:

Purpose: Flavoring agent.

Where Used: Various.

Regulations: GRAS when used at a level not in excess of the amount reasonably required to accomplish the intended effect.

SAFETY PROFILE: Moderately toxic by ingestion and intraperitoneal routes. Combustible liquid. When heated to decomposition it emits toxic fumes of NO_x.

TOXICITY DATA and CODEN

orl-rat LD50:613 mg/kg DCTODJ 3,249,80
ipr-mus LD50:1390 mg/kg TXAPA9 17,244,70

DTU600 CAS: 123-32-0
2,5-DIMETHYLPYRAZINE
mf: $C_6H_8N_2$ mw: 108.16

PROP: Colorless liquid; potato taste. D: 0.980-1.000, refr index: 1.497-1.501, flash p: 147°F (OC), d: 0.99, vap d: 3.72, bp: 182.2°. Misc with water, organic solvents.

SYNS: 2,5-DIMETHYL-1,4-DIAZINE ◇ FEMA No. 3272

USE IN FOOD:

Purpose: Flavoring agent.

Where Used: Various.

Regulations: GRAS when used at a level not in excess of the amount reasonably required to accomplish the intended effect.

SAFETY PROFILE: Moderately toxic by ingestion and intraperitoneal routes. Mutagenic data. Combustible liquid when exposed to heat, open flame, spark, oxidizers. To fight fire use water spray, mist, dry chemical, CO_2, foam. When heated to decomposition it emits toxic fumes of NO_x.

TOXICITY DATA and CODEN

mmo-smc 3300 μg/L FCTXAV 18,581,80
cyt-ham:ovr 2500 μg/L FCTXAV 18,581,80
orl-rat LD50:1020 mg/kg DCTODJ 3,249,80
ipr-mus LD50:1350 mg/kg TXAPA9 17,244,70

DTU800 CAS: 108-50-9
2,6-DIMETHYLPYRAZINE
mf: $C_6H_8N_2$ mw: 108.16

PROP: White to yellow crystals; nutty, coffee odor. Mp: 48°, d: .965 @ 50°. Sol in water, organic solvents @ 155°.

SYN: FEMA No. 3273

USE IN FOOD:

Purpose: Flavoring agent.

Where Used: Various.

Regulations: GRAS when used at a level not in excess of the amount reasonably required to accomplish the intended effect.

SAFETY PROFILE: Moderately toxic by intraperitoneal route. When heated to decomposition it emits toxic fumes of NO_x.

TOXICITY DATA and CODEN

ipr-mus LD50:1080 mg/kg TXAPA9 17,244,70

DTV300 CAS: 625-84-3
2,5-DIMETHYLPYRROLE
mf: C_6H_9N mw: 95.15

PROP: Colorless to yellow oily liquid. D: 0.935-0.945 @ 20°, refr index: 1.503- 1.506. Very sol in alc, ether; very sltly sol in water.

SYN: FEMA No. 7071

USE IN FOOD:

Purpose: Flavoring agent.

Where Used: Various.

Regulations: GRAS when used at a level not in excess of the amount reasonably required to accomplish the intended effect.

SAFETY PROFILE: When heated to decomposition emits toxic fumes of NO_x.

DUJ200 CAS: 5598-13-0
O,O-DIMETHYL-O-(3,5,6-TRICHLORO-2-PYRIDYL)PHOSPHOROTHIOATE
mf: $C_7H_7Cl_3NO_3PS$ mw: 322.53

SYNS: CHLORPYRIFOS-METHYL ◇ DOWCO 217 ◇ DURSBAN METHYL ◇ ENT 27,520 ◇ METHYL CHLOR-

PYRIFOS ◇ METHYL DURSBAN ◇ NOLTRAN ◇ NSC 60380 ◇ OMS-1155 ◇ RELDAN ◇ ZERTELL

USE IN FOOD:

Purpose: Insecticide.

Where Used: Animal feed, barley (milling fractions, except flour), oats (milling fractions, except flour), rice (milling fractions, except flour), sorghum (milling fractions, except flour), wheat (milling fractions, except flour).

Regulations: FDA - 21CFR 193.471. Insecticide residue tolerance of 90 ppm in barley milling fractions (except flour), 130 ppm in oats milling fractions (except flour), 90 ppm in sorghum milling fractions (except flour), 30 ppm in rice milling fractions (except flour), 30 ppm in wheat milling fractions (except flour). 21CFR 561.437. Limitation of as above when used for animal feed.

SAFETY PROFILE: Poison by ingestion. Moderately toxic by intraperitoneal route. A skin irritant. When heated to decomposition it emits very toxic fumes of Cl^-, NO_x, PO_x, and SO_x.

TOXICITY DATA and CODEN

skn-rbt 500 mg/24H MLD TXAPA9 21,369,72
orl-rat LD50:941 mg/kg BESAAT 15,123,69
scu-rat LD50:6900 mg/kg YKYUA6 30,409,79

DUK800 CAS: 2164-17-2
1,1-DIMETHYL-3-(α,α,α-TRIFLUORO-m-TOLYL) UREA

mf: $C_{10}H_{11}F_3N_2O$ mw: 232.23

SYNS: C 2059 ◇ CIBA 2059 ◇ COTORAN ◇ COTORAN MULTI 50WP ◇ COTTONEX ◇ 1,1-DIMETHYL-3-(3-TRIFLUOROMETHYLPHENYL)UREA ◇ N,N-DIMETHYL-N′-(3-TRIFLUOROMETHYLPHENYL)UREA ◇ FLUOMETURON ◇ HERBICIDE C-2059 ◇ LANEX ◇ NCI-C08695 ◇ PAKHTARAN ◇ 3-(5-TRIFLUORMETHYLPHENYL)-,1-DIMETHYLHARNSTOFF (GERMAN) ◇ N-(m-TRIFLUOROMETHYLPHENYL)-N′,N′-DIMETHYLUREA ◇ N-(3-TRIFLUOROMETHYLPHENYL)-N′-N′-DIMETHYLUREA ◇ 3-(m-TRIFLUOROMETHYLPHENYL)-1,1-DIMETHYLUREA

USE IN FOOD:

Purpose: Herbicide.

Where Used: Animal feed.

Regulations: FDA - 21CFR 561.240. Limitation of 0.2 ppm in sugarcane bagasse when used for animal feed.

EPA Genetic Toxicology Program. IARC Cancer Review: Animal Inadequate Evidence IMEMDT 30,245,83. NCI Carcinogenesis Bioassay (feed); No Evidence: rat NCITR* NCI-CG-TR-195,80; Equivocal Evidence: mouse NCITR* NCI-CG-TR-195,80.

SAFETY PROFILE: Moderately toxic by ingestion, intraperitoneal, and possibly other routes. An experimental carcinogen. Mutagenic data. When heated to decomposition it emits very toxic fumes of F^- and NO_x.

TOXICITY DATA and CODEN

mma-sat 1 μg/plate MUREAV 58,353,78
dni-mus-orl 1 g/kg MUREAV 58,353,78
orl-mus TDLo:87 g/kg/2Y-C:CAR NCITR* NCI-CG-TR-195,80
orl-rat LD50:6416 mg/kg PESTD5 17,351,76
ipr-rat LD50:685 mg/kg PESTD5 17,351,76

DUQ150
3,5-DINITROBENZAMIDE

USE IN FOOD:

Purpose: Animal drug.

Where Used: Chicken.

Regulations: FDA - 21CFR 556.220. Limitation of zero in uncooked edible tissues of chickens.

SAFETY PROFILE: When heated to decomposition emits toxic fumes of NO_x.

DUV600 CAS: 1582-09-8
2,6-DINITRO-N,N-DIPROPYL-4-(TRIFLUOROMETHYL)BENZENAMINE

mf: $C_{13}H_{16}F_3N_3O_4$ mw: 335.32

PROP: Technical product contains 84-88 ppm diproplynitrosoamine (NCITR* NCI-CG-TR-34,78).

SYNS: AGREFLAN ◇ AGRIFLAN 24 ◇ CRISALIN ◇ DIGERMIN ◇ 2,6-DINITRO-N,N-DI-N-PROPYL-α,α,α-TRIFLURO-p-TOLUIDINE ◇ 2,6-DINITRO-4-TRIFLUORMETHYL-N,N-DIPROPYLANILIN (GERMAN) ◇ 4-(DI-N-PROPYLAMINO)-3,5-DINITRO-1-TRIFLUOROMETHYLBENZENE ◇ N,N-DI-N-PROPYL-2,6-DINITRO-4-TRIFLUOROMETHYLANILINE ◇ N,N-DIPROPYL-4-TRIFLUOROMETHYL-2,6-DINITROANILINE ◇ ELANCOLAN ◇ L-36352 ◇ LILLY 36,352 ◇ NCI-C00442 ◇ NITRAN ◇ OLITREF ◇ SU SEGURO CARPIDOR ◇ TREFANOCIDE ◇ TREFICON ◇ TREFLAM ◇ TREFLAN ◇ TREFLANOCIDE ELANCOLAN ◇ TRIFLUORALIN (USDA) ◇ TRIFLURALIN ◇ TRIFLURALINE ◇ α,α,α-TRIFLUORO-2,6-DINITRO-N,N-DIPROPYL-p-TOLUIDINE ◇ TRIFUREX ◇ TRIKEPIN ◇ TRIM

USE IN FOOD:

Purpose: Herbicide.

Where Used: Barley, carrots, peppermint oil, soybeans, spearmint oil, wheat.

Regulations: FDA - 21CFR 193.440. Herbicide residue tolerance of 2 ppm in peppermint oil and spearmint oil. USDA CES Ranking: C-4 (1986).

NCI Carcinogenesis Bioassay (feed); Clear Evidence: mouse NCITR* NCI-CG-TR-34,78; No Evidence: rat NCITR* NCI-CG-TR-34,78. EPA Genetic Toxicology Program. Community Right-To-Know List.

SAFETY PROFILE: Moderately toxic by ingestion and intraperitoneal routes. An experimental carcinogen, tumorigen, and teratogen. Experimental reproductive effects. Human mutagenic data. When heated to decomposition it emits very toxic fumes of F^- and NO_x.

TOXICITY DATA and CODEN

mma-sat 1 mg/plate ENMUDM 8(Suppl 7),1,86
sce-hmn:lym 1 mg/L BSIBAC 60,2149,84
orl-mus TDLo:10 mg/kg (6-15D preg):TER TJADAB 15,15A,77
orl-mus TDLo:10 g/kg (6-15D preg):REP TJADAB 23,33,81
ipr-mus TDLo:200 mg/kg (1D male):TER EESADV 4,263,80
orl-mus TDLo:180 g/kg/78W-C:CAR NCITR* NCI-CG-TR-34,78
ipr-mus TDLo:2600 μg/kg/39D-I:ETA PATHAB 73,707,81
orl-mus TD :340 g/kg/78W-C:CAR NCITR* NCI-CG-TR-34,78
orl-mus LD50:5000 mg/kg GUCHAZ 6,524,73
ipr-mus LDLo:1500 mg/kg BECTA6 20,554,78

DVG400 CAS: 148-01-6
3,5-DINITRO-o-TOLUAMIDE

mf: $C_8H_7N_3O_5$ mw: 225.18

PROP: Yellowish solid. Mp: 177°. Very sltly sol in water; sol in acetone, acetonitrile, and dimethyl formamide.

SYNS: COCCIDINE A ◇ COCCIDOT ◇ DINITOLMID ◇ DINITOLMIDE ◇ DINITOLMIDE (ACGIH) ◇ D.O.T. ◇ 2-METHYL-3,5-DINITROBENZAMIDE ◇ ZOALENE ◇ ZOAMIX

USE IN FOOD:

Purpose: Animal drug.

Where Used: Chicken, turkey.

Regulations: FDA - 21CFR 556.770. Limitation of 6 ppm in liver and kidney, 3 ppm in muscle of chickens; 3 ppm muscle and liver of turkeys. 21CFR 558.680.

ACGIH TLV: TWA 5 mg/m^3; STEL 10 mg/m^3

SAFETY PROFILE: Poison by intravenous route. Moderately toxic by ingestion. Mutagenic data. A strong exothermic reaction above 248° has caused industrial explosions. When heated to decomposition it emits toxic fumes of NO_x.

TOXICITY DATA and CODEN

mmo-sat 500 μg/plate MUREAV 77,21,80
mmo-esc 500 μg/plate MUREAV 77,21,80
orl-rat LD50:600 mg/kg 29ZVAB -,537,69

DVQ709 CAS: 78-34-2
DIOXATHION

mf: $C_{12}H_{26}O_6P_2S_4$ mw: 456.56

PROP: Nonvolatile, stable solid. Nonflammable. Insol in water.

SYNS: BIS(DITHIOPHOSPHATE de O,O-DIETHYLE) de S,S'-(1,4-DIOXANNE-2,3-DIYLE) (FRENCH) ◇ 1,4-DIOSSAN-2,3-DIYL-BIS(O,O-DIETIL-DITIOFOSFATO) (ITALIAN) ◇ 1,4-DIOXAAN-2,3-DIYL-BIS(O,O-DIETHYL-DITHIOFOSFAAT) (DUTCH) ◇ 2,3-p-DIOXANDITHIOL S,S-BIS(O,O-DIETHYL PHOSPHORODITHIOATE) ◇ 1,4-DIOXAN-2,3-DIYL-BIS(O,O-DIAETHYL-DITHIOPHOSPHAT) (GERMAN) ◇ 1,4-DIOXAN-2,3-DIYL-BIS(O,O-DIETHYLPHOSPHOROTHIOLOTHIONATE) ◇ 1,4-DIOXAN-2,3-DIYL-O,O,O',O'-TETRAETHYL DI(PHOSPHOROMITHIOATE) ◇ 2,3-p-DIOXANE-S,S-BIS(O,O-DIETHYLPHOSPHOROITHIOATE) ◇ p-DIOXANE-2,3-DITHIOL-S,S-DIESTER WITH O,O-DIETHYL PHOSPHORODITHIOATE ◇ p-DIOXANE-2,3-DIYL ETHYL PHOSPHORODITHIOATE ◇ ENT 22,897 ◇ NCI-C00395 ◇ PHOSPHORODITHIOIC ACID-S,S'-1,4-DIOXANE-2,3-DIYL O,O,O',O'-TETRAETHYL ESTER

USE IN FOOD:

Purpose: Pesticide.

Where Used: Animal feed, citrus pulp (dehydrated).

Regulations: FDA - 21CFR 561.210. Limitation of 18 ppm in dehydrated citrus pulp when used for cattle feed.

NCI Carcinogenesis Bioassay (feed); No Evidence: mouse, rat NCITR* NCI-CG-TR-125,78.

ACGIH TLV: TWA 0.2 mg/m^3

SAFETY PROFILE: Poison by ingestion, inhalation, skin contact, and intraperitoneal routes.

A cholinesterase inhibitor. When heated to decomposition it emits very toxic fumes of PO_x and SO_x.

TOXICITY DATA and CODEN

skn-rat LD50:63 mg/kg TXAPA9 5,605,63
orl-rat LD50:20 mg/kg WRPCA2 9,119,70
ihl-rat LC50:1398 mg/m^3/1H TXAPA9 5,605,63

DVX800 CAS: 122-39-4
DIPHENYLAMINE
mf: $C_{12}H_{11}N$ mw: 169.24

PROP: Crystals; floral odor. Mp: 52.9°, bp: 302.0°, flash p: 307°F (CC), d: 1.16, autoign temp: 1173°F, vap press: 1 mm @ 108.3°, vap d: 5.82. Sol in benzene, ether, and carbon disulfide.

SYNS: ANILINOBENZENE ◇ BIG DIPPER ◇ C.I. 10355 ◇ DFA ◇ N,N-DIPHENYLAMINE ◇ DPA ◇ NO SCALD ◇ N-PHENYLANILINE ◇ N-PHENYLBENEZENAMINE ◇ SCALDIP

USE IN FOOD:

Purpose: Insecticide.

Where Used: Various.

Regulations: USDA CES Ranking: B-4 (1985).

EPA Genetic Toxicology Program.

ACGIH TLV: TWA 10 mg/m^3

SAFETY PROFILE: Poison by ingestion. An experimental teratogen. Experimental reproductive effects. Action similar to aniline but less severe. Combustible when exposed to heat or flame. Can react violently with hexachloromelamine; trichloromelamine. Can react with oxidizing materials. To fight fire, use CO_2, dry chemical. When heated to decomposition it emits highly toxic fumes of NO_x.

TOXICITY DATA and CODEN

orl-rat TDLo:7500 mg/kg (17-22D preg):TER PEREBL 4,448,70
orl-rat LDLo:3000 mg/kg CNREA8 26,619,66

DWB800 CAS: 1241-94-7
DIPHENYL-2-ETHYLHEXYL PHOSPHATE
mf: $C_{20}H_{27}O_4P$ mw: 362.44

SYNS: 2-ETHYL-1-HEXANOL ESTER with DIPHENYL PHOSPHATE ◇ 2-ETHYLHEXYL DIPHENYL ESTER PHOSPHORIC ACID ◇ 2-ETHYLHEXYL DIPHENYLPHOSPHATE ◇ SANTICIZER 141 (MONSANTO)

USE IN FOOD:

Purpose: Plasticizer.

Where Used: Packaging materials.

Regulations: FDA - 21CFR 181.27.

SAFETY PROFILE: Poison by intravenous route. When heated to decomposition it emits toxic fumes of PO_x.

TOXICITY DATA and CODEN

ivn-rbt LDLo:272 mg/kg AMIHBC 8,170,53

DWQ000 CAS: 7727-21-1
DIPOTASSIUM PERSULFATE
DOT: 1492
mf: $H_2O_8S_2 \cdot 2K$ mw: 272.34

PROP: White, odorless crystals. Mp: decomp @ 100°, d: 2.477.

SYNS: ANTHION ◇ PEROXYDISULFURIC ACID DIPOTASSIUM SALT ◇ POTASSIUM PEROXYDISULFATE ◇ POTASSIUM PEROXYDISULPHATE ◇ POTASSIUM PERSULFATE (ACGIH, DOT)

USE IN FOOD:

Purpose: Defoaming agent, dispersing adjuvant, poultry scald agent.

Where Used: Citrus fruit (fresh), poultry.

Regulations: FDA - 21CFR 172.210. Use in accordance with good manufacturing practice. USDA - 9CFR 381.147. Sufficient for purpose.

ACGIH TLV: TWA 5 mg(S_2O_8)/m^3 DOT Classification: Oxidizer; Label: Oxidizer

SAFETY PROFILE: Moderately toxic. An irritant and allergen. A powerful oxidizer. Flammable when exposed to heat or by chemical reaction. Can react with reducing materials. It liberates oxygen above 100° when dry or @ about 50° when in solution. When heated to decomposition it emits highly toxic fumes of SO_x and K_2O.

DWX800 CAS: 85-00-7
DIQUAT DIBROMIDE
DOT: 2781
mf: $C_{12}H_{12}N_2 \cdot 2Br$ mw: 344.08

PROP: Yellow crystals. Mp: 355°. Sol in water.

SYNS: AQUACIDE ◇ DEIQUAT ◇ DEXTRONE ◇ 9,10-DIHYDRO-8a,10,-DIAZONIAPHENANTHRENE DIBROMIDE ◇ 9,10-DIHYDRO-8a,10a-DIAZONIAPHENANTHRENE(1,1′-ETHYLENE-2,2′-BIPYRI-

DYLIUM)DIBROMIDE ◇ 5,6-DIHYDRO-DIPYRIDO (1,2a;2,1c)PYRAZINIUM DIBROMIDE ◇ 6,7-DIHYDRO-PYRIDO(1,2a;2',1'-C)PYRAZINEDIUM DIBROMIDE ◇ DIQUAT (ACGIH,DOT) ◇ 1,1'-ETHYLENE-2,2'-BIPYRIDYLIUM DIBROMIDE ◇ ETHYLENE DIPYRIDYLIUM DIBROMIDE ◇ 1,1-ETHYLENE 2,2-DIPYRIDYLIUM DIBROMIDE ◇ 1,1'-ETHYLENE-2,2'-DIPYRIDYLIUM DIBROMIDE ◇ FB/2 ◇ FEGLOX ◇ PREEGLONE ◇ REGLON ◇ REGLONE ◇ WEEDTRINE-D

USE IN FOOD:

Purpose: Herbicide.

Where Used: Animal feed, potable water, potato chips, potato wastes (dried), potatoes (processed).

Regulations: FDA - 21CFR 193.160. Herbicide residue tolerance of 0.01 ppm in potable water, 0.5 ppm in processed potatoes including potato chips. 21CFR 561.215. Limitation of 1.0 ppm in dried potato wastes when used for animal feed.

EPA Genetic Toxicology Program.

ACGIH TLV: TWA 0.5 mg/m^3 DOT Classification: ORM-E; Label: None

SAFETY PROFILE: Poison by ingestion, subcutaneous, intravenous, intraperitoneal, and possibly other routes. An experimental teratogen. Experimental reproductive effects. Human mutagenic data. A skin and eye irritant. When heated to decomposition it emits very toxic fumes of NO_x and Br^-.

TOXICITY DATA and CODEN

skn-rbt 400 mg/kg/20D MLD BJIMAG 27,51,70
eye-rbt 10 mg MLD BJIMAG 27,51,70
mmo-sat 100 nmol/plate TOLED5 3,169,79
dnr-sat 10 μg/plate MUREAV 68,183,79
ipr-rat TDLo: 7 mg/kg (7D preg): TER 26UZAB 6,257,68
ipr-mus TDLo: 10800 μg/kg (9-12D preg): TER BECTA6 25,513,80
orl-rat LD50: 120 mg/kg PRKHDK 1,31,75

DXC400 CAS: 144-33-2
DISODIUM CITRATE

mf: $C_6H_6O_7 \cdot 2Na$ mw: 236.10

PROP: White crystals or granular powder; odorless. Mp: loses water @ 150°, bp: decomp @ red heat. Sol in water; insol in alc.

SYNS: DISODIUM HYDROGEN CITRATE ◇ NATRIUM CITRICUM (GERMAN) ◇ SODIUM CITRATE (FCC)

USE IN FOOD:

Purpose: Buffer, nutrient for cultured buttermilk, sequestrant.

Where Used: Beef (cured), beverages (carbonated), cream (nondairy coffee whiteners), cured comminuted meat food product, margarine, milk (evaporated), oleomargarine, pork (cured), pork (fresh).

Regulations: FDA - 21CFR 182.1751, 182.6751. GRAS when used in accordance with good manufacturing practice. USDA - 9CFR 318.7. Sufficient for purpose. Limitation of 10 percent solution to spray surfaces of cured cuts of meat. Not to exceed 500 ppm or 1.8 mg/sq in. of surface ascorbic acid, erythorbic acid, or sodium ascorbate, singly, or in combination; and/or not to exceed either 250 ppm or 0.9 mg/sq in. of surface of citric acid or sodium citrate, singly, or in combination on fresh pork cuts. Sufficient for purpose in oleomargarine or margarine.

SAFETY PROFILE: Poison by intravenous route. Moderately toxic by intraperitoneal and subcutaneous routes. When heated to decomposition it emits toxic fumes of Na_2O.

TOXICITY DATA and CODEN

ipr-rat LD50: 1724 mg/kg JPETAB 94,65,48
scu-mus LD50: 2580 mg/kg ARZNAD 15,852,65

DXD200 CAS: 142-59-6
DISODIUM ETHYLENE-1,2-BISDITHIOCARBAMATE

mf: $C_4H_6N_2S_4 \cdot 2Na$ mw: 256.34

PROP: Crystals. Sol in water.

SYNS: CARBON D ◇ CHEM BAM ◇ DINATRIUM-AETHYLENBISDITHIOCARBAMAT (GERMAN) ◇ DINATRIUM-(N,N'-AETHYLEN-BIS(DITHIOCARBAMAT)) (GERMAN) ◇ DINATRIUM-(N,N'-ETHYLEEN-BIS(DITHIOCARBAMAAT)) (DUTCH) ◇ DISODIUM ETHYLENEBIS(DITHIOCARBAMATE) ◇ DITHANE A-40 ◇ DITHANE D-14 ◇ DSE ◇ 1,2-ETHANEDIYLBISCARBAMODITHIOIC ACID DISODIUM SALT ◇ N,N'-ETHYLENE BIS(DITHIOCARBAMATE de SODIUM) (FRENCH) ◇ ETHYLENEBIS(DITHIOCARBAMATE) DISODIUM SALT ◇ ETHYLENEBIS(DITHIOCARBAMIC ACID) DISODIUM SALT ◇ N,N'-ETILEN-BIS(DITIOCARBAMMATO) DI SODIO (ITALIAN) ◇ NABAM ◇ NABAME (FRENCH) ◇ PARZATE ◇ SPRING-BAK

USE IN FOOD:

Purpose: Antimicrobial agent.

Where Used: Beets, sugarcane.

Regulations: FDA - 21CFR 173.320. Limitation of 3.0 ppm based on weight of raw sugarcane or raw beets.

EPA Genetic Toxicology Program.

SAFETY PROFILE: Poison by ingestion. Moderately toxic by intraperitoneal route. An experimental teratogen. Experimental reproductive effects. Mutagenic data. When heated to decomposition it emits very toxic fumes of NO_x, Na_2O and SO_x.

TOXICITY DATA and CODEN

mmo-omi 1000 ppm MMAPAP 50,233,73
scu-mus TDLo: 194 mg/kg (6-14D preg): TER NTIS** Pb 223,-,160
scu-mus TDLo: 418 mg/kg (6-14D preg): REP NTIS** PB223-160
orl-rat LD50: 395 mg/kg FEPRA7 11,391,52
ipr-rat LD50: 500 mg/kg 85DPAN -,-,71/76

DXE400 CAS: 860-22-0
DISODIUM INDIGO-5,5-DISULFONATE

mf: $C_{16}H_{10}N_2O_8S_2 \cdot 2Na$ mw: 468.38

PROP: Blue-brown to red-brown powder. Sol in water, conc sulfuric acid; sltly sol in alc.

SYNS: ACID BLUE W ◇ ACID LEATHER BLUE IC ◇ A.F. BLUE No. 2 ◇ AIRDALE BLUE IN ◇ AMACID BRILLIANT BLUE ◇ ANILINE CARMINE POWDER ◇ ATUL INDIGO CARMINE ◇ 1311 BLUE ◇ 12070 BLUE ◇ BUCACID INDIGOTINE B ◇ CANACERT INDIGO CARMINE ◇ CARMINE BLUE (BIOLOGICAL STAIN) ◇ C.I. 73015 ◇ C.I. 7581 ◇ C.I. ACID BLUE 74 ◇ C.I. FOOD BLUE 1 ◇ DISODIUM SALT of 1-INDIGOTIN-S,S′-DISULPHONIC ACID ◇ DOLKWAL INDIGO CARMINE ◇ E 132 ◇ F D & C BLUE No. 2 (FCC) ◇ GRAPE BLUE A GEIGY ◇ INDIGO CARMINE ◇ INDIGO CARMINE (BIOLOGICAL STAIN) ◇ INDIGO CARMINE DISODIUM SALT ◇ INDIGO EXTRACT ◇ INDIGO-KARMIN (GERMAN) ◇ 5,5′-INDIGOTIN DISULFONIC ACID ◇ INDIGOTINE ◇ INDIGOTINE DISODIUM SALT ◇ INTENSE BLUE ◇ L-BLAU 2 (GERMAN) ◇ MAPLE INDIGO CARMINE ◇ SACHSISCHBLAU ◇ SCHULTZ Nr. 1309 (GERMAN) ◇ SODIUM 5,5′-INDIGOTIDISULFONATE ◇ SOLUBLE INDIGO ◇ USACERT BLUE No.2

USE IN FOOD:

Purpose: Color additive.

Where Used: Various.

Regulations: FDA - 21CFR 74.102, 81.1. Use in accordance with good manufacturing practice.

EPA Genetic Toxicology Program.

SAFETY PROFILE: Poison by intravenous route. Moderately toxic by ingestion and subcutaneous routes. An experimental neoplastigen. Mutagenic data. When heated to decomposition it emits very toxic fumes of SO_x, NO_x, and Na_2O.

TOXICITY DATA and CODEN

cyt-ham: fbr 12 g/L FCTOD7 22,623,84
cyt-ham: ovr 20 μmol/L/5H-C ENMUDM 1,27,79
scu-rat TDLo: 9 g/kg/2Y-I: NEO TXAPA9 8,29,66
orl-rat LD50: 2 g/kg SCPHA4 47,39,79
ivn-rat LD50: 93 mg/kg NIIRDN 6,83,82

DXE500 CAS: 4691-65-0
DISODIUM INOSINATE

mf: $C_{10}H_{13}N_4O_8P \cdot 2Na$ mw: 394.22

PROP: Colorless to white crystals; characteristic taste. Sol in water; sltly sol in alc; insol in ether.

SYNS: DISODIUM IMP ◇ DISODIUM-5′-INOSINATE ◇ DISODIUM INOSINE-5′-MONOPHOSPHATE ◇ DISODIUM INOSINE-5′-PHOSPHATE ◇ IMP DISODIUM SALT ◇ 5′-IMP DISODIUM SALT ◇ IMP SODIUM SALT ◇ INOSINE-5′-MONOPHOSPHATE DISODIUM ◇ INOSIN-5′-MONOPHOSPHATE DISODIUM ◇ SODIUM INOSINATE ◇ SODIUM-5′-INOSINATE

USE IN FOOD:

Purpose: Flavor enhancer.

Where Used: Ham, meats (cured), poultry, sausage.

Regulations: FDA - 21CFR 172.535. Must contain no more than 150 ppm soluble barium. USDA - 9CFR 318.7, 381.147. Sufficient for purpose.

SAFETY PROFILE: Moderately toxic by several routes. An experimental teratogen. Mutagenic data. When heated to decomposition it emits toxic fumes of PO_x, NO_x and Na_2O.

TOXICITY DATA and CODEN

cyt-ham: fbr 1 g/L FCTOD7 22,623,84
ipr-mus TDLo: 250 mg/kg (10D preg): TER JJPAAZ 22,201,72
ipr-mus TDLo: 500 mg/kg (13D preg): TER JJPAAZ 22,201,72
orl-rat LD50: 15900 mg/kg AJINO* -,-,73
ipr-rat LD50: 4850 mg/kg AJINO* -,-,73
scu-rat LD50: 3900 mg/kg AJINO* -,-,73

DXF800 CAS: 7758-16-9
DISODIUM PYROPHOSPHATE

mf: $H_2O_7P_2 \cdot Na_2$ mw: 221.94

PROP: White, crystalline powder. D: 1.862, mp: 220° (decomp). Sol in water.

SYNS: DINATRIUMPYROPHOSPHAT (GERMAN) ◇ DIPHOSPHORIC ACID, DISODIUM SALT ◇ DISODIUM DIHYDROGEN PYROPHOSPHATE ◇ DISODIUM DIPHOSPHATE ◇ SODIUM ACID PYROPHOSPHATE (FCC) ◇ SODIUM PYROPHOSPHATE

USE IN FOOD:

Purpose: Buffer, cooked out juices retention agent, hog scald agent, leavening agent, poultry scald agent, sequestrant.

Where Used: Biscuits, bologna, bologna (garlic), doughnuts, fish products (canned), frankfurters, hog carcasses, knockwurst, meat products, potatoes (processed), poultry, poultry food products, vienna, wieners.

Regulations: FDA - 21CFR 182.1087, 182.6787. GRAS when used in accordance with good manufacturing practice. USDA - 9CFR 318.7. Limitation of 8 ounces in 100 pounds of meat or 0.5 percent of final product. In meat food products, where allowed, limitation of 5 percent of phosphate in pickle at 10 percent pump level, 0.5 percent of phosphate in product (only clear solution may be injected into product. 9CFR 381.147. Limitation of 0.5 percent in total poultry product.

SAFETY PROFILE: Poison by intravenous route. Moderately toxic by ingestion and subcutaneous routes. An irritant to skin, eyes, and mucous membranes. When heated to decomposition it emits toxic fumes of PO_x and Na_2O.

TOXICITY DATA and CODEN

orl-mus LD50:2650 mg/kg ARZNAD 7,445,57
scu-mus LD50:480 mg/kg ARZNAD 7,445,57
ivn-mus LD50:59 mg/kg ARZNAD 7,445,57

DXG650
DISTEARYL THIODIPROPIONATE

USE IN FOOD:

Purpose: Antioxidant.

Where Used: Packaging materials.

Regulations: FDA - 21CFR 181.24. Limitation of 0.005 percent migrating from food packages.

SAFETY PROFILE: When heated to decomposition emits toxic fumes of SO_x.

DXQ750
DIVINYLBENZENE COPOLYMER

USE IN FOOD:

Purpose: Removal of organic substances from aqueous foods.

Where Used: Various.

Regulations: FDA - 21CFR 173.65. The aqueous food stream contacting the polymer must be maintained at 175°F or less for foods of Types I, II, and VI-B (excluding carbonated beverages) described in 21CFR 176.170(c).

SAFETY PROFILE: When heated to decomposition it emits acrid smoke and irritating fumes.

DXS700
Δ-DODECALACTONE

mf: $C_{12}H_{22}O_2$ mw: 198.31

PROP: Colorless to yellow liquid; coconut-fruity odor. Refr index: 1.458-1.461, flash p: +151°F. Very sol in alc, propylene glycol, veg. oil; insol in water.

SYN: FEMA No. 2401

USE IN FOOD:

Purpose: Flavoring agent.

Where Used: Oleomargarine.

Regulations: FDA - 21CFR 172.515. Limitation of 10 ppm in oleomargarine.

SAFETY PROFILE: Combustible liquid. When heated to decomposition it emits acrid smoke and irritating fumes.

DXT000 CAS: 112-54-9
1-DODECANAL

mf: $C_{12}H_{24}O$ mw: 184.36

PROP: Reported in pine-needle, lime, sweet-orange, and a dozen other essential oils (FCTXAV 11,477,73). Colorless to light yellow liquid; fatty odor. D: 0.826-0.836, refr index: 1.433-1.439, flash p: 180°F. Sol in alc, fixed oils, propylene glycol; insol in glycerin, water.

SYNS: C-12 ALDEHYDE, LAURIC ◇ 1-DODECYL ALDEHYDE ◇ DUODECYLIC ALDEHYDE ◇ FEMA No. 2615 ◇ LAURYL ALDEHYDE (FCC)

USE IN FOOD:

Purpose: Flavoring agent.

Where Used: Various.

Regulations: FDA - 21CFR 172.515. Use at a level not in excess of the amount reasonably required to accomplish the intended effect.

SAFETY PROFILE: Mildly toxic by ingestion. A human and experimental skin irritant. Combustible liquid. When heated to decomposition it emits acrid smoke and irritating fumes.

TOXICITY DATA and CODEN

skn-hmn 5 mg/48H MLD FCTXAV 11,1079,73
skn-rbt 500 mg/24H MOD FCTXAV 11,1079,73
orl-rat LD50: 23 g/kg FCTXAV 11,483,73

DXU300
trans-2-DODECEN-1-AL
mf: $C_{12}H_{22}O$ mw: 182.31

PROP: Slightly yellow liquid; fatty, citruslike odor. D: 0.839-0.049, refr index: 1.462. Sol in alc, fixed oils; insol in water.

SYN: FEMA No. 2402

USE IN FOOD:

Purpose: Flavoring agent.

Where Used: Various.

Regulations: FDA - 21CFR 172.515. Use at a level not in excess of the amount reasonably required to accomplish the intended effect.

SAFETY PROFILE: When heated to decomposition it emits acrid smoke and irritating fumes.

DXV600 CAS: 112-53-8
DODECYL ALCOHOL
mf: $C_{12}H_{26}O$ mw: 186.38

PROP: Colorless solid, liquid above 21°; floral odor. D: 0.830-0.836, refr index: 1.440-1.444, mp: 24°, bp: 259°, flash p: 260°F, autoign temp: 527°F. Sol in 2 parts of 70% alc, fixed oils, propylene glycol; insol in water, glycerin.

SYNS: ALCOHOL C-12 ◇ ALFOL 12 ◇ CACHALOT L-50 ◇ CO 12 ◇ CO-1214 ◇ n-DODECANOL ◇ 1-DODECANOL ◇ n-DODECYL ALCOHOL ◇ DUODECYL ALCOHOL ◇ DYTOL J-68 ◇ EPAL 12 ◇ FEMA No. 2617 ◇ LAURIC ALCOHOL ◇ LAURINIC ALCOHOL ◇ LAURYL 24 ◇ LAURYL ALCOHOL (FCC) ◇ n-LAURYL ALCOHOL, PRIMARY ◇ LOROL ◇ MA-1214 ◇ SIPOL L12

USE IN FOOD:

Purpose: Flavoring agent, intermediate.

Where Used: Various.

Regulations: FDA - 21CFR 172.515, 172.864. Use at a level not in excess of the amount reasonably required to accomplish the intended effect.

SAFETY PROFILE: Moderately toxic by intraperitoneal route. Mildly toxic by ingestion. An experimental tumorigen. A severe human skin irritant. Combustible when exposed to heat or flame; can react with oxidizing materials. To fight fire, use dry chemical, CO_2. When heated to decomposition it emits acrid smoke and irritating fumes.

TOXICITY DATA and CODEN

skn-hmn 75 mg/3D-I SEV 85DKA8 -,127,77
skn-mus TDLo: 19 g/kg/39W-I: ETA TXAPA9 9,70,66
orl-rat LD50: 12800 mg/kg FCTXAV 11,95,73

DXW200 CAS: 25155-30-0
DODECYL BENZENE SODIUM SULFONATE
DOT: 9146
mf: $C_{18}H_{29}O_3S \cdot Na$ mw: 348.52

PROP: White to light yellow flakes, granules or powder.

SYNS: AA-9 ◇ ABESON NAM ◇ BIO-SOFT D-40 ◇ CALSOFT F-90 ◇ CONCO AAS-35 ◇ CONOCO C-50 ◇ DETERGENT HD-90 ◇ DODECYLBENZENESULFONIC ACID SODIUM SALT ◇ DODECYLBENZENESULPHONATE, SODIUM SALT ◇ DODECYLBENZENSULFONAN SODNY (CZECH) ◇ MERCOL 25 ◇ NACCANOL NR ◇ NECCANOL SW ◇ PILOT HD-90 ◇ PILOT SF-40 ◇ RICHONATE 1850 ◇ SANTOMERSE 3 ◇ SODIUM DODECYLBENZENESULFONATE (DOT) ◇ SODIUM DODECYLBENZENESULFONATE, DRY ◇ SODIUM LAURYLBENZENESULFONATE ◇ SOLAR 40 ◇ SOL SODOWA KWASU LAURYLOBENZENOSULFONOWEGO (POLISH) ◇ SULFAPOL ◇ SULFAPOLU (POLISH) ◇ SULFRAMIN 85 ◇ SULFRAMIN 40 FLAKES ◇ SULFRAMIN 40 GRANULAR ◇ SULFRAMIN 1238 SLURRY ◇ p-1',1',4',4'-TETRAMETHYLOKTYLBENZENSULFONAN SODNY (CZECH) ◇ ULTRAWET K

USE IN FOOD:

Purpose: Animal glue adjuvant, denuding agent, lye peeling agent, poultry scald agent, washing water agent.

Where Used: Fruits, hog carcasses, packaging materials, poultry, vegetables.

Regulations: FDA - 21CFR 173.315. Limitation of 0.2 percent in wash water. 21CFR 178.3120. Use at a level not in excess of the amount reasonably required to accomplish the intended effect. USDA - 9CFR 318.7, 381.147. Sufficient for purpose.

DOT Classification: ORM-E; Label: None

SAFETY PROFILE: Poison by intravenous route. Moderately toxic by ingestion. A skin and severe eye irritant. When heated to decomposition it emits toxic fumes of Na_2O.

TOXICITY DATA and CODEN

skn-rbt 500 mg/24H MOD 28ZPAK -,195,72
eye-rbt 250 μg/24H SEV 28ZPAK -,195,72
orl-rat LD50:1260 mg/kg FSASAX 63,938,61

DXX200 CAS: 1166-52-5
DODECYL GALLATE
mf: $C_{19}H_{30}O_5$ mw: 338.49

SYNS: LAURYL GALLATE ◇ NIPAGALLIN LA ◇ PROGALLIN LA

USE IN FOOD:

Purpose: Antioxidant.

Where Used: Cream cheese, fats, margarine, oils, oleomargarine, potatoes (instant mashed).

Regulations: USDA - 9CFR 318.7. Limitation of 0.02 percent individually or in combination with other antioxidants approved for use in margarine.

SAFETY PROFILE: Moderately toxic by ingestion. When heated to decomposition it emits acrid smoke and irritating fumes.

TOXICITY DATA and CODEN

orl-mus LD50:1600 mg/kg 14CYAT 2,1897,63

DXZ000 CAS: 7631-98-3
N-DODECYLSARCOSINE SODIUM SALT
mf: $C_{15}H_{30}NO_2 \cdot Na$ mw: 279.45

SYN: SODIUM-N-LAURYL SARCOSINE

USE IN FOOD:

Purpose: Antifogging agent, antistatic agent.

Where Used: Packaging materials.

Regulations: FDA - 21CFR 178.3130. Use at a level not in excess of the amount reasonably required to accomplish the intended effect.

SAFETY PROFILE: Poison by intravenous route. When heated to decomposition it emits toxic fumes of NO_x and Na_2O.

TOXICITY DATA and CODEN

ivn-mus LD50:180 mg/kg CSLNX* NX#00171

DYE000 CAS: 2921-88-2
DOWCO 179
DOT: 2783
mf: $C_9H_{11}Cl_3NO_3PS$ mw: 350.59

SYNS: BRODAN ◇ CHLORPYRIFOS (ACGIH, DOT, USDA) ◇ O,O-DIAETHYL-O-3,5,6-TRICHLOR-2-PYRIDYLMONOTHIOPHOSPHAT (GERMAN) ◇ O,O-DIETHYL-O-3,5,6-TRICHLORO-2-PYRIDYL PHOSPHOROTHIOATE ◇ DURSBAN ◇ DURSBAN F ◇ ENT 27,311 ◇ ERADEX ◇ LORSBAN ◇ OMS-0971 ◇ PYRINEX ◇ 3,5,6-TRICHLORO-2-PYRIDINOL-O-ESTER WITH O,O-DIETHYL PHOSPHOROTHIOATE

USE IN FOOD:

Purpose: Insecticide.

Where Used: Animal feed, apple pomace (dried), beet sugar molasses, beet sugar pulp (dried), citrus oil, citrus pulp (dried), corn oil, corn soapstock, grape pomace (dried), peanut oil, soybean milling fractions, sunflower seed hulls, wheat (milling fractions, except flour).

Regulations: FDA - 21CFR 193.85. Insecticide residue tolerance of 25.0 ppm in citrus oil, 3.0 ppm in corn oil, 10.0 in corn oil, 1.5 ppm peanut oil, 3.0 ppm wheat, milling fractions (except flour). 21CFR 561.98. Limitation of 12.0 ppm in dried apple pomace, 15.0 ppm in beet sugar molasses, 5.0 ppm in dried beet sugar pulp, 5.0 ppm in dried citrus pulp, 1.0 ppm in corn soapstock, 2.0 ppm in dried grape pomace, 1.5 ppm in soybean milling fractions, and 0.5 ppm in sunflower seed hulls when used for animal feed. USDA CES Ranking: B-4 (1986).

EPA Genetic Toxicology Program.

ACGIH TLV: TWA 0.2 mg/m^3 DOT Classification: ORM-A; Label: None

SAFETY PROFILE: Poison by ingestion, skin contact, inhalation, subcutaneous, and possibly other routes. An experimental teratogen. Mutagenic data. When heated to decomposition it emits very toxic fumes of Cl^-, NO_x, PO_x, and SO_x.

TOXICITY DATA and CODEN

cyt-dmg-orl 50 ppb/3S ENMUDM 5,835,83
orl-mus TDLo:10 mg/kg (6-15D preg):TER TXAPA9 54,31,80
orl-mus TDLo:250 mg/kg (6-15D preg):TER TXAPA9 54,31,80
orl-rat LD50:97 mg/kg FMCHA2 -,D183,80
ihl-rat LD50:78 mg/kg BECTA6 19,113,78
skn-rat LD50:202 mg/kg TXAPA9 14,515,69

E

EAR000 CAS: 145-73-3

ENDOTHAL

mf: $C_8H_{10}O_5$ mw: 186.18

PROP: Solid. Mp: 144°. Sol in water.

SYNS: AQUATHOL ◇ ENDOTHALL ◇ ENDOTHAL TECHNICAL ◇ 3,6-ENDOOXOHEXAHYDROPHTHALIC ACID ◇ 3,6-ENDOXOHEXAHYDROPHTHALIC ACID ◇ 3,6-endo-EPOXY-1,2-CYCLOHEXANEDICARBOXYLIC ACID ◇ HEXAHYDRO-3,6-endo-OXYPHTHALIC ACID ◇ HYDOUT ◇ HYDROTHAL-47 ◇ 7-OXABICYCLO(2.2.1)HEPTANE-2,3-DICARBOXYLIC ACID ◇ RCRA WASTE NUMBER P088 ◇ TRI-ENDOTHAL

USE IN FOOD:

Purpose: Herbicide.

Where Used: Potable water.

Regulations: FDA - 21CFR 193.180. Herbicide residue tolerance of 0.2 ppm potable water.

SAFETY PROFILE: Poison by ingestion. Very irritating to skin, eyes, and mucus membranes. Causes diarrhea. When heated to decomposition it emits acrid smoke and fumes.

TOXICITY DATA and CODEN

orl-rat LD50:38 mg/kg PCOC** -,471,66

EAT500 CAS: 72-20-8

ENDRIN

DOT: 2761

mf: $C_{12}H_8Cl_6O$ mw: 380.90

PROP: White crystals. Mp: decomp @ 200°.

SYNS: COMPOUND 269 ◇ ENDREX ◇ ENDRIN (ACGIH, DOT) ◇ ENDRINE (FRENCH) ◇ ENT 17,251 ◇ EXPERIMENTAL INSECTICIDE 269 ◇ HEXACHLOROEPOXYOCTAHYDRO-endo,endo-DIMETHANONAPHTHALENE ◇ 3,4,5,6,9,9-HEXACHLORO-1a,2,2a,3,6,6a,7,7a-OCTAHYDRO-2,7:3,6-DIMETHANONAPHTH(2,3-b)OXIRENE ◇ HEXADRIN ◇ MENDRIN ◇ NCI-C00157 ◇ NENDRIN ◇ RCRA WASTE NUMBER P051

USE IN FOOD:

Purpose: Insecticide.

Where Used: Various.

Regulations: USDA CES Ranking: A-3 (1986).

IARC Cancer Review: Animal Inadequate Evidence IMEMDT 5,157,74; Human Inadequate Evidence IMEMDT 5,157,74. NCI Carcinogenesis Bioassay (feed); No Evidence: mouse, rat NCITR* NCI-CG-TR-12,79. EPA Genetic Toxicology Program. EPA Extremely Hazardous Substances List.

OSHA PEL: TWA 0.1 mg/m^3 (skin) ACGIH TLV: TWA 0.1 mg/m^3 (skin) DFG MAK: 0.1 mg/m^3 DOT Classification: Poison B; Label: Poison; Poison B; Label: Poison, liquid

SAFETY PROFILE: Poison by ingestion, skin contact, intravenous and possibly other routes. A suspected human carcinogen. An experimental teratogen. Experimental reproductive effects. Mutagenic data. A central nervous system stimulant. Highly toxic to birds, fish and humans. Many cases of fatal poisoning have been attributed to it. Does not accumulate in human tissue. In humans, ingestion of 1 mg/kg has caused symptoms.

A dangerous fire hazard. Mixtures with parathion dissolve very exothermically in petroleum solvents and may cause an air-vapor explosion.

TOXICITY DATA and CODEN

sce-ofs-mul 54 pmol/L MUREAV 118,61,83
cyt-rat-par 1 mg/kg BECTA6 9,65,73
orl-rat TDLo:2320 μg/kg (4D pre):TER OYYAA2 6,673,72
orl-mus TDLo:10 mg/kg (8-12D preg):REP JTEHD6 10,541,82
orl-mus TDLo:2500 μg/kg (9D preg):TER TJADAB 9,11,74
orl-rat LD50:3 mg/kg WRPCA2 9,119,70
skn-rat LD50:12 mg/kg SPEADM 78-1,13,78

EBH525

EPOXIDIZED SOYBEAN OIL

PROP: Iodine number maximum of 6, oxirane oxygen minimum of 6.0 percent.

USE IN FOOD:

Purpose: Plasticizer.

Where Used: Packaging materials.

Regulations: FDA - 21CFR 181.27. Use in accordance with good manufacturing practice.

SAFETY PROFILE: When heated to decomposition it emits acrid smoke and irritating fumes.

EBW500 CAS: 1024-57-3
EPOXYHEPTACHLOR
mf: $C_{10}H_5Cl_7O$ mw: 389.30

SYNS: ENT 25,584 ◇ HCE ◇ HEPTACHLOR EPOXIDE (USDA) ◇ 1,4,5,6,7,8,8-HEPTACHLORO-2,3-EPOXY-2,3,-3a,4,7,7a-HEXAHYDRO-4,7-METHANOINDENE ◇ 1,4,5,-6,7,8,8-HEPTACHLORO-2,3-EPOXY-3a,4,7,7a-TETRAHY-DRO-4,7-METHANOINDAN ◇ 2,3,4,5,6,7,7-HEPTACHLORO-1a,1b,5,5a,6,6a-HEXAHYDRO-2,5-METHANO-2H-INDENO(1,2-b)OXIRENE ◇ VELSICOL 53-CS-17

USE IN FOOD:
Purpose: Pesticide.
Where Used: Various.
Regulations: USDA CES Ranking: A-1 (1987).

IARC Cancer Review: Human Inadequate Evidence IMEMDT 20,129,79; Animal Inadequate Evidence IMEMDT 5,173,74; Animal Limited Evidence IMEMDT 20,129,79. EPA Genetic Toxicology Program.

SAFETY PROFILE: Poison by ingestion and intravenous routes. An experimental carcinogen. A suspected human carcinogen. Human mutagenic data. When heated to decomposition it emits toxic fumes of Cl^-.

TOXICITY DATA and CODEN

mma-hmn:fbr 10 μmol/L MUREAV 42,161,77
orl-mus TDLo:580 mg/kg/69W-C:CAR ARTODN 38,163,77
orl-mus TD :876 mg/kg/2Y-C:CAR ECEBDI 45,147,77
orl-rat LD50:47 mg/kg ARSIM* 20,27,66

ECF500 CAS: 75-56-9
EPOXYPROPANE
mf: C_3H_6O mw: 58.09
DOT: 1280

PROP: Colorless liquid; ethereal odor. Bp: 33.9°, lel: 2.8%, uel: 37%, fp: −104.4°, flash p: −35°F (TOC), d: 0.8304 @ 20°/20°, vap press: 400 mm @ 17.8°, vap d: 2.0. Sol in water, alc, and ether.

SYNS: 1,2-EPOXYPROPANE ◇ 2,3-EPOXYPROPANE ◇ METHYL ETHYLENE OXIDE ◇ METHYL OXIRANE ◇ NCI-C50099 ◇ OXYDE de PROPYLENE (FRENCH) ◇ PROPENE OXIDE ◇ PROPYLENE EPOXIDE ◇ PROPYLENE OXIDE (ACGIH,DOT) ◇ 1,2-PROPYLENE OXIDE

USE IN FOOD:
Purpose: Fumigant.
Where Used: Cocoa, glace fruit, gum, nutmeats (processed, except peanuts), prunes (dried), spices (processed), starch.
Regulations: FDA - 21CFR 193.380. Fumigant residue tolerance of 300 ppm in cocoa, 700 ppm in glace fruit, 300 ppm in gums, 300 ppm in processed nutmeats (except peanuts), 700 ppm in dried prunes, 300 ppm in processed spices, 300 ppm in starch.

IARC Cancer Review: Human Inadequate Evidence IMEMDT 36,227,85; Animal Sufficient Evidence IMEMDT 36,227,85; Animal Limited Evidence IMEMDT 11,191,76. Carcinogenesis Studies (inhalation); Some Evidence: rat NTPTR* NTP-TR-267,85; Clear Evidence: mouse NTPTR* NTP-TR-267,85. EPA Genetic Toxicology Program. Community Right-To-Know List. EPA Extremely Hazardous Substances List.

OSHA PEL: TWA 20 ppm ACGIH TLV: TWA 20 ppm DOT Classification: Flammable Liquid; Label: Flammable Liquid

SAFETY PROFILE: Poison by intraperitoneal route. Moderately toxic by ingestion, inhalation, and skin contact. An experimental carcinogen, neoplastigen, and tumorigen by ingestion, inhalation and subcutaneous routes. A suspected human carcinogen. Experimental reproductive effects. Human mutagenic data. A severe skin and eye irritant.

Flammable liquid. A very dangerous fire and explosion hazard when exposed to heat or flame. Explosive reaction with epoxy resin and sodium hydroxide. Forms explosive mixtures with oxygen. Reacts with ethylene oxide + polyhydric alcohol to form the thermally unstable polyether alcohol. Incompatible with NH_4OH; chlorosulfonic acid; HCl; HF; HNO_3; oleum; H_2SO_4. Dangerous; can react vigorously with oxidizing materials. Keep away from heat and open flame. To fight fire, use alcohol foam, CO_2, dry chemical. When heated to decomposition it emits acrid smoke and fumes.

TOXICITY DATA and CODEN

skn-rbt 50 mg/6M SEV AMIHAB 13,228,56
eye-rbt 5 mg SEV AJOPAA 29,1363,46
mmo-sat 350 μg/plate ABCHA6 47,2461,83
mmo-omi 25 mmol/L MUREAV 73,1,80
ihl-rat TCLo:500 ppm/7H (7-16D preg):TER NTIS** PB83-258038

ihl-rat TCLo:500 ppm/7H (15D pre/1-16D preg):REP SWEHDO 9,94,83
orl-rat TDLo:10798 mg/kg/2Y-I:CAR BJCAAI 46,924,82
ihl-rat TCLo:100 ppm/7H/2Y-I:NEO TXAPA9 76,69,84
scu-rat TDLo:1500 mg/kg/46W-I:ETA ANYAA9 68,750,58
scu-mus TDLo:272 mg/kg/95W-I:CAR ZHPMAT 174,383,81
orl-rat LD50:520 mg/kg GISAAA 46(7),76,81
ihl-rat LCLo:4000 ppm/4H AIHAAP 30,470,69
ipr-rat LD50:364 mg/kg TXAPA9 52,422,80

EDA000 CAS: 50-14-6
ERGOCALCIFEROL
mf: $C_{28}H_{44}O$ mw: 396.72

PROP: White crystals; odorless. Mp: 115-118°. Insol in water; sol in alc, chloroform, ether, and fatty oils.

SYNS: d-ARTHIN ◇ CALCIFEROL ◇ CALCIFERON 2 ◇ CONDACAPS ◇ CONDOCAPS ◇ CONDOL ◇ CRTRON ◇ CRYSTALLINA ◇ DARAL ◇ DAVITAMON D ◇ DAVITIN ◇ DECAPS ◇ DEE-OSTEROL ◇ DEE-RON ◇ DEE-RONAL ◇ DEE-ROUAL ◇ DELTALIN ◇ DERATOL ◇ DETALUP ◇ DIACTOL ◇ DIVIT URTO ◇ DORAL ◇ DRISDOL ◇ ERGORONE ◇ ERGOSTEROL, ACTIVATED ◇ ERGOSTEROL, IRRADIATED ◇ ERTRON ◇ FORTODYL ◇ GELTABS ◇ HI-DERATOL ◇ INFRON ◇ IRRADIATED ERGOSTA-5,7,22-TRIEN-3-β-OL ◇ METADEE ◇ MULSIFEROL ◇ MYKOSTIN ◇ OLEOVITAMIN D ◇ OSTELIN ◇ RADIOSTOL ◇ RADSTERIN ◇ 9,10,SECOERGOSTA-5,-7,10(19),22-TETRAEN-3-β-OL ◇ SHOCK-FEROL ◇ STEROGYL ◇ VIGANTOL ◇ VIOSTEROL ◇ VITAMIN D2 (FCC) ◇ VITAVEL-D

USE IN FOOD:

Purpose: Dietary supplement, nutrient.

Where Used: Cereals (breakfast), grain products, margarine, milk, milk products, pasta.

Regulations: FDA - 21CFR 166.110, 182.5950, 182.5953, 184.1950. GRAS with a limitation of 350 IU/100 grams in breakfast cereals, 90 IU/100 grams in grain products and pastas, 42 IU/100 grams in milk, 89 IU/100 grams in milk products when used in accordance with good manufacturing practice.

EPA Extremely Hazardous Substances List.

SAFETY PROFILE: Poison by ingestion, intraperitoneal, intravenous and intramuscular routes. An experimental teratogen. Human systemic effects by ingestion: anorexia, nausea or vomiting, and weight loss. Experimental reproductive effects. When heated to decomposition it emits acrid smoke and irritating fumes.

TOXICITY DATA and CODEN

orl-rat TDLo:45 mg/kg (13-21D preg):TER ARPAAQ 73,371,62
orl-rat TDLo:33750 μg/kg (9D pre):REP IJMDAI 2,14,66
scu-rat TDLo:17500 μg/kg (1-7D preg):TER FESTAS 12,343,61
orl-wmn TDLo 12600 mg/kg/72W: CNS,GIT,MET LANCAO 1,1164,80
orl-rat LDLo:5 mg/kg NIIRDN 6,128,82

EDE600
ERYTHORBIC ACID

PROP: White or sltly yellow crystals or powder. Mp: 164-171° (decomp). Sol in water, alc; sltly sol in glycerin.

SYN: d-ARABOASCORBIC ACID

USE IN FOOD:

Purpose: Antioxidant, curing accelerator, preservative.

Where Used: Bananas (frozen), beef (cured), cured comminuted meat food product, pork (cured), pork (fresh), poultry.

Regulations: FDA - 21CFR 182.3041. GRAS when used in accordance with good manufacturing practice. USDA - 9CFR 318.7, 9CFR 381.147. Limitation of 75 ounces to 100 gallons of pickle at 10 percent pump level, 0.75 ounces to 100 pounds of meat. Not to exceed 500 ppm or 1.8 mg/sq in. of surface ascorbic acid, erythorbic acid or sodium ascorbate, singly, or in combination; and/or not to exceed either 250 ppm or 0.9 mg/sq in. of surface of citric acid or sodium citrate, singly, or in combination on fresh pork cuts.

SAFETY PROFILE: When heated to decomposition it emits acrid smoke and irritating fumes.

EDH500 CAS: 114-07-8
ERYTHROMYCIN
mf: $C_{37}H_{67}NO_{13}$ mw: 734.05

PROP: White or slightly yellow, crystalline powder; odorless; freely sol in alc, chloroform, and ether; very sltly sol in water. Mp: 133-138°.

SYNS: DOTYCIN ◇ EM ◇ E-MYCIN ◇ ERYCIN ◇ ERYTHROCIN ◇ ERYTHROGRAN ◇ ERYTHROGUENT

◇ ERYTHROMYCIN A ◇ ILOTYCIN ◇ PANTOMICINA ◇ PROPIOCINE ◇ ROBIMYCIN

USE IN FOOD:

Purpose: Animal drug.

Where Used: Beef, chicken, eggs, pork, turkey.

Regulations: FDA - 21CFR 556.230. Tolerance of 0.1 ppm in uncooked edible tissues of swine, zero in uncooked edible tissues of beef cattle and milk, 0.025 ppm in uncooked eggs, 0.125 ppm in uncooked edible residue of chickens and turkeys.

EPA Genetic Toxicology Program.

SAFETY PROFILE: Poison by intravenous, and intramuscular routes. Moderately toxic by ingestion, intraperitoneal, and subcutaneous routes. An experimental teratogen. Mutagenic data. When heated to decomposition it emits toxic fumes of NO_x.

TOXICITY DATA and CODEN

dnr-esc 20 μL/disc MUREAV 97,1,82
mrc-smc 1 g/L MUREAV 12,357,71
orl-rat TDLo:6 g/kg (10-15D preg):TER JJANAX 25,187,72
orl-mus TDLo:12 g/kg (8-13D preg):TER JJANAX 25,193,72
orl-rat LD50:9272 mg/kg AMPMAR 39,259,78
scu-rat LDLo:427 mg/kg CLDND*

EDK200
ERYTHROMYCIN THIOCYANATE

USE IN FOOD:

Purpose: Animal feed drug.

Where Used: Animal feed.

Regulations: FDA - 21CFR 558.248. Use at a level not in excess of the amount reasonably required to accomplish the intended effect.

SAFETY PROFILE: When heated to decomposition it emits acrid smoke and irritating fumes.

EDN100
ESTERASE-LIPASE

PROP: Derived from *Mucor miehei.*

USE IN FOOD:

Purpose: Flavor enhancer.

Where Used: Cheese, fats, milk products, oils.

Regulations: FDA - 21CFR 173.140. Use at a level not in excess of the amount reasonably required to accomplish the intended effect.

SAFETY PROFILE: When heated to decomposition it emits acrid smoke and irritating fumes.

EDP000 CAS: 50-50-0
ESTRADIOL-3-BENZOATE

mf: $C_{25}H_{28}O_3$ mw: 376.53

PROP: White or slightly yellow to brownish crystalline powder, odorless. Almost insol in water; sol in alcohol, acetone, and dioxane; sparingly sol in vegetable oils; sltly sol in ether. Mp: 191-196°

SYNS: BENOVOCYLIN ◇ BENZHORMOVARINE ◇ BENZOATE d'OESTRADIOL (FRENCH) ◇ BENZOESTROFOL ◇ BENZOFOLINE ◇ BENZO-GYNOESTRYL ◇ BENZOIC ACID ESTRADIOL ◇ DIFFOLLISTEROL ◇ DIFOLLICULINE ◇ DIHYDROESTRIN BENZOATE ◇ DIHYDROFOLLICULIN BENZOATE ◇ DIMENFORMON BENZOATE ◇ DIMENFORMONE ◇ DIOGYN B ◇ EBZ ◇ ESTON-B ◇ ESTRADIOL BENZOATE ◇ ESTRADIOL-17-β-BENZOATE ◇ ESTRADIOL-17-β-3-BENZOATE ◇ β-ESTRADIOL BENZOATE ◇ β-ESTRADIOL-3-BENZOATE ◇ 17-β-ESTRADIOL BENZOATE ◇ 17-β-ESTRADIOL-3-BENZOATE ◇ ESTRADIOL MONOBENZOATE ◇ 17-β-ESTRADIOL MONOBENZOATE ◇ ESTRA-1,3,5(10)-TRIENE-3,17-DIOL (17-β)-3-BENZOATE ◇ ESTRA-1,3,5(10)-TRIENE-3,17-β-DIOL, 3-BENZOATE ◇ 1,3,5(10)-ESTRATRIENE-3,17-β-DIOL 3-BENZOATE ◇ FEMESTRONE ◇ FOLLICORMON ◇ FOLLIDRIN ◇ GRAAFINA ◇ de GRAAFINA ◇ GYNECORMONE ◇ GYNFORMONE ◇ HIDROESTRON ◇ HORMOGYNON ◇ HYDROXYESTRIN BENZOATE ◇ MEE ◇ ODB ◇ OESTRADIOL BENZOATE ◇ OESTRADIOL-3-BENZOATE ◇ β-OESTRADIOL BENZOATE ◇ β-OESTRADIOL 3-BENZOATE ◇ 17-β-OESTRADIOL 3-BENZOATE ◇ OESTRADIOL MONOBENZOATE ◇ OESTRAFORM (BDH) ◇ 1,3,5(10)-OESTRATRIENE-3,17-β-DIOL 3-BENZOATE ◇ OVAHORMON BENZOATE ◇ OVASTEROL-B ◇ OVEX ◇ OVOCYCLIN BENZOATE ◇ OVOCYCLIN M ◇ OVOCYCLIN-MB ◇ PRIMOGYN B ◇ PRIMOGYN BOLEOSUM ◇ PRIMOGYN I ◇ PROGYNON B ◇ PROGYNON BENZOATE ◇ RECTHORMONE OESTRADIOL ◇ SOLESTRO ◇ UNISTRADIOL

USE IN FOOD:

Purpose: Animal drug.

Where Used: Beef, lamb.

Regulations: FDA - 21CFR 556.240. Tolerance of 120 parts per trillion (ppt) in muscle, 480 ppt in fat, 360 ppt in kidney, 240 ppt in liver of heifers, steers, and calves. Tolerance of 120 ppt in muscle, 600 ppt in fat, kidney, and liver of lambs.

IARC Cancer Review: Animal Sufficient Evidence IMEMDT 21,279,79.

SAFETY PROFILE: An experimental carcinogen, tumorigen, and teratogen. Human reproductive effects by intramuscular route: menstrual cycle changes and disorders. Experimental reproductive effects. Mutagenic data. A steroid. When heated to decomposition it emits acrid smoke and irritating fumes.

TOXICITY DATA and CODEN

dni-rat-scu 10 μg/kg JOENAK 65,45,75
ims-wmn TDLo: 1 mg/kg (5D pre): REP FESTAS 26,405,75
orl-rat TDLo: 1063 μg/kg (42D pre-21D post): REP JAFCAU 22,969,74
ims-rat TDLo: 2500 μg/kg (13D preg): TER ACENA7 33,520,60
scu-mus TDLo: 28 mg/kg/(12-16D preg): TER PSEBAA 97,809,58
imp-rat TDLo: 25 mg/kg: NEO EJCAAH 13,1437,77
scu-mus TDLo: 24 mg/kg/36W-I: CAR CNREA8 8,337,48
par-mus TDLo: 38 mg/kg/57W-I: ETA CNREA8 1,359,41
scu-mus TD : 38 mg/kg/39W-I: CAR CNREA8 8,337,48

EDR400
ESTRADIOL MONOPAMITATE

USE IN FOOD:

Purpose: Animal drug.

Where Used: Chicken.

Regulations: FDA - 21CFR 556.250. Limitation of zero in chickens.

SAFETY PROFILE: When heated to decomposition it emits acrid smoke and irritating fumes.

EDS500 CAS: 8016-88-4
ESTRAGON OIL

PROP: From steam distillation of leaves, stems, and flowers from *Artemesia dracunculus* L. Pale yellow to amber liquid; spicy licorice and sweet basil odor. Sol in fixed oils, mineral oil; insol in propylene glycol, glycerin.

SYN: TARRAGON OIL (FCC)

USE IN FOOD:

Purpose: Flavoring agent.

Where Used: Various.

Regulations: FDA - 21CFR 182.10, 182.20. GRAS when used at a level not in excess of the amount reasonably required to accomplish the intended effect.

SAFETY PROFILE: Moderately toxic by ingestion. A skin irritant. When heated to decomposition it emits acrid smoke and irritating fumes.

TOXICITY DATA and CODEN

skn-rbt 500 mg/24H FCTXAV 12,709,74
orl-rat LD50: 1900 mg/kg FCTXAV 12,709,74

EEA500 CAS: 107-15-3
1,2-ETHANEDIAMINE

DOT: 1604
mf: $C_2H_8N_2$ mw: 60.12

PROP: Volatile, colorless, hygroscopic liquid; ammonia-like odor. Mp: 8.5°, bp: 117.2°, flash p: 110°F (CC), d: 0.8994 @ 20°/4°, vap press: 10.7 mm @ 20°, vap d: 2.07, autoign temp: 725°F.

SYNS: AETHALDIAMIN (GERMAN) ◇ AETHYLENEDIAMIN (GERMAN) ◇ 1,2-DIAMINOAETHAN (GERMAN) ◇ 1,2-DIAMINO-ETHAAN (DUTCH) ◇ 1,2-DIAMINOETHANE ◇ 1,2-DIAMINO-ETHANO (ITALIAN) ◇ DIMETHYLENEDIAMINE ◇ ETHYLEENDIAMINE (DUTCH) ◇ ETHYLENEDIAMINE ◇ 1,2-ETHYLENEDIAMINE ◇ ETHYLENE-DIAMINE (FRENCH) ◇ NCI-C60402

USE IN FOOD:

Purpose: Animal drug, animal glue adjuvant, flume wash water additive.

Where Used: Milk, packaging materials, sugar beets.

Regulations: FDA - 21CFR 173.315. Limitation of 0.1 ppm in wash water. 21CFR 178.3120. Use at a level not in excess of the amount reasonably required to accomplish the intended effect. 21CFR 556.270. Tolerance of zero in milk.

EPA Extremely Hazardous Substances List.

OSHA PEL: TWA 10 ppm ACGIH TLV: TWA 10 ppm DOT Classification: Corrosive Material; Label: Corrosive; Corrosive Material; Label: Corrosive, Flammable Liquid

SAFETY PROFILE: An irritant poison in humans by inhalation. Experimental poison by inhalation, intraperitoneal, subcutaneous, and intravenous routes. Moderately toxic by ingestion and skin contact. Corrosive. A severe skin and eye irritant. An allergen and sensitizer. Mutagenic data. Flammable when exposed to heat, flame or oxidizers. Can react violently with acetic acid; acetic anhydride; acrolein; acrylic acid; acrylonitrile; allyl chloride; CS_2; chlorosulfonic acid; epichlorohydrin; ethylene chlorohydrin; HCl; mesityl oxide; HNO_3; oleum; $AgClO_4$; H_2SO_4; β-propiolactone; vinyl acetate. To fight fire, use CO_2, dry chemical, alcohol foam. When heated to decomposition it emits toxic fumes of NO_x and NH_3.

TOXICITY DATA and CODEN

skn-rbt 10 mg/24H open SEV AMIHBC 4,119,51
eye-rbt 675 μg SEV AJOPAA 29,1363,46
mmo-sat 33 μg/plate ENMUDM 5(Suppl 1),3,83
mma-sat 1 mg/plate ENMUDM 5(Suppl 1),3,83
ihl-hmn TCLo:200 ppm:PNS AMIHBC 9,223,54
orl-rat LD50:500 mg/kg 85GMAT -,66,82
ihl-rat LCLo:4000 ppm/8H AMIHBC 4,119,51

EEK100 CAS: 59-06-3
ETHOPABATE

mf: $C_{12}H_{15}NO_4$ mw: 237.25

PROP: Odorless white to pink crystals. Sol in methanol, ethanol, acetone, acetonitrile, isopropanol, p-dioxane, ethyl acetate, methylene chloride.

SYNS: 4-ACETAMIDO-2-ETHOLXBENZOIC ACID METHYL ESTER ◇ 2-ETHOXY-4-ACETAMIDOBENZOID ACID METHYL ESTER ◇ ETHYL PABATE ◇ METHYL 4-ACETAMIDO-2-ETHOXYBENZOATE

USE IN FOOD:

Purpose: Animal drug.

Where Used: Chickens.

Regulations: FDA - 21CFR 556.260. Limitation of 1.5 ppm in uncooked liver and kidney, 0.5 ppm in uncooked muscle of chickens.

SAFETY PROFILE: When heated to decomposition it emits acrid smoke and irritating fumes.

EET550
2-[1-(ETHOXYIMINO)BUTYL]-5-[2-(ETHYLTHIO)PROPYL]-3-HYDROXY-2-CYCLOHEXENE-1-ONE

USE IN FOOD:

Purpose: Herbicide.

Where Used: Animal feed, cottonseed soapstock, flaxseed meal, peanut soapstock, potato pomace (dry), sugarbeet molasses, sunflower meal, tomato products (concentrated).

Regulations: FDA - 21CFR 193.479. Limitation of 24.0 ppm in tomato products, concentrated. 21CFR 561.430. Limitation of 15 ppm in cottonseed soapstock, 7 ppm in flaxseed meal, 75.0 ppm in peanut soapstock, 0.5 ppm in sugarbeet molasses, 20.0 ppm in sunflower meal, 12.0 ppm in dry potato pomace when used for animal feed.

SAFETY PROFILE: When heated to decomposition it emits acrid smoke and irritating fumes.

EEU100
ETHOXYLATED MONO- and DIGLYCERIDES

PROP: Mix of stearate, palmitate, and lesser amounts of myristate partial esters of glycerin condensed with approx. 20 moles of ethylene oxide per mole of α-monoglyceride reaction mixtures. (FCC III) Pale, sltly yellow, oily liquid; mildly bitter taste. Sol in water, alc, xylene; sltly sol in mineral oil, vegetable oil.

SYNS: POLYGLYCERATE (60) ◇ POLYOXYETHYLENE (20) MONO- and DIGLYCERIDES of FATTY ACIDS

USE IN FOOD:

Purpose: Dough conditioner, emulsifier.

Where Used: Bakery products (yeast-leavened), cake mixes, cakes, desserts (frozen), icing mixes, icings, margarine, milk or cream substitutes for beverage coffee, peanut butter, puddings, whipped toppings (vegetable oil).

Regulations: FDA - 21CFR 172.834. Limitation of 0.5 percent in flour for yeast-leavened bakery products; 0.5 percent of dry ingredients for cakes and cake mixes; 0.45 percent in whipped vegetable oil toppings; 0.5 percent in icings and icing mixes; 0.2 percent in frozen desserts; 0.4 percent in milk or cream substitutes for beverage coffee. Must conform to FDA specifications for fats or fatty acids derived from edible oils.

SAFETY PROFILE: When heated to decomposition it emits acrid smoke and irritating fumes.

EFE000 CAS: 150-69-6
4-ETHOXYPHENYLUREA

mf: $C_9H_{12}N_2O_2$ mw: 180.23

PROP: Needle-like crystals. Mp: 174°.

SYNS: p-AETHOXYPHYLHARNSTOFF (GERMAN) ◇ DULCINE ◇ N-(4-ETHOXYPHENYL)UREA ◇ p-ETHOXYPHENYLUREA ◇ NCI-C02073 ◇ PHENETHYLCARBAMID (GERMAN) ◇ p-PHENETOLCARBAMID (GERMAN) ◇ p-PHENETOLCARBAMIDE ◇ p-PHENETOLECARBAMIDE ◇ p-PHENETYLUREA ◇ SUCROL ◇ SUESSTOFF ◇ VALZIN

USE IN FOOD:

Purpose: Nonnutritive sweetener.

Where Used: Prohibited from foods.

Regulations: FDA - 21CFR 189.145. Prohibited from direct addition or use in human food.

IARC Cancer Review: Animal Inadequate Evidence IMEMDT 12,97,76.

SAFETY PROFILE: Human poison by ingestion. Moderately toxic experimentally by ingestion. An experimental tumorigen. Human systemic effects by ingestion: somnolence, hallucinations, distorted perceptions and changes in motor activity. In adults 20 grams to 40 grams produces dizziness, nausea, methemoglobinemia, cyanosis, hypotension. When heated to decomposition it emits toxic fumes of NO_x.

TOXICITY DATA and CODEN

orl-rat TDLo:232 g/kg/59W-C:ETA JAPMA8 40,583,51
orl-wmn TDLo:600 mg/kg:CNS MEKLA7 43,105,48
orl-chd LDLo:400 mg/kg MEKLA7 43,105,48
orl-rat LD50:1900 mg/kg NCIMR* NIH-71-E-2144

EFR000 CAS: 141-78-6
ETHYL ACETATE
DOT: 1173
mf: $C_4H_8O_2$ mw: 88.12

PROP: Colorless liquid; fragrant odor. Mp: −83.6°, bp: 77.15°, ULC: 85-90, lel: 2.2%, uel: 11%, flash p: 24°F, d: 0.8946 @ 25°, autoign temp: 800°F, vap press: 100 mm @ 27.0°, vap d: 3.04. Misc with alc, ether, glycerin, volatile oils, water @ 54°.

SYNS: ACETIC ETHER ◇ ACETIDIN ◇ ACETOXYETHANE ◇ AETHYLACETAT (GERMAN) ◇ ESSIGESTER (GERMAN) ◇ ETHYLACETAAT (DUTCH) ◇ ETHYL ACETIC ESTER ◇ ETHYLE (ACETATE d') (FRENCH) ◇ ETHYL ETHANOATE ◇ ETILE (ACETATO di) (ITALIAN) ◇ FEMA No. 2414 ◇ OCTAN ETYLU (POLISH) ◇ RCRA WASTE NUMBER U112 ◇ VINEGAR NAPHTHA

USE IN FOOD:

Purpose: Color diluent, flavoring agent, solvent.

Where Used: Coffee (decaffeinated), fruits, tea (decaffeinated), vegetables.

Regulations: FDA - 21CFR 73.1, 173.228, 182.60. Use at a level not in excess of the amount reasonably required to accomplish the intended effect.

EPA Genetic Toxicology Program.

OSHA PEL: TWA 400 ppm ACGIH TLV: TWA 400 ppm DFG MAK: 400 ppm (1400 mg/m^3) DOT Classification: Flammable Liquid; Label: Flammable Liquid

SAFETY PROFILE: Poison by inhalation. Moderately toxic by intraperitoneal and subcutaneous routes. Mildly toxic by ingestion. Human systemic effects by inhalation: olfactory changes, conjunctiva irritation, and pulmonary changes. Human eye irritant. Mutagenic data. Irritating to mucous surfaces, particularly the eyes, gums and respiratory passages, and is also mildly narcotic. On repeated or prolonged exposures, it causes conjunctival irritation and corneal clouding. It can cause dermatitis. High concentrations have a narcotic effect and can cause congestion of the liver and kidneys. Chronic poisoning has been described as producing anemia, leucocytosis (transient increase in the white blood cell count), and cloudy swelling, and fatty degeneration of the viscera.

Highly flammable liquid. A very dangerous fire hazard when exposed to heat or flame; can react vigorously with oxidizing materials. Moderate explosion hazard when exposed to flame. Potentially explosive reaction with lithium tetrahydroaluminate. Ignites on contact with potassium tert-butoxide. Violent reaction with chlorosulfonic acid; ($LiAlH_2$ + 2-chloromethyl furan); oleum. To fight fire, use CO_2, dry chemical or alcohol foam. When heated to decomposition it emits acrid smoke and irritating fumes.

TOXICITY DATA and CODEN

eye-hmn 400 ppm JIHTAB 25,282,43
sln-smc 24400 ppm MUREAV 149,339,85
cyt-ham:fbr 9 g/L FCTOD7 22,623,84
ihl-hmn TCLo:400 ppm:NOSE,EYE,PUL JIHTAB 25,282,43
orl-rat LD50:5620 mg/kg YKYUA6 32,1241,81
ihl-rat LC50:1600 ppm/8H 14CYAT 2,1879,63

EFS000 CAS: 141-97-9
ETHYL ACETYL ACETATE
mf: $C_6H_{10}O_3$ mw: 130.16

PROP: Colorless liquid; fruity odor. Bp: 180.8°, fp: −45°, flash p: 185°F (COC), autoign temp: 563°F, d: 1.0261 @ 20°/20°, refr index: 1.418, vap press: 1 mm @ 28.5°, vap d: 4.48. Misc with alc, ether, ethyl acetate, water.

SYNS: ACETOACETIC ACID, ETHYL ESTER ◇ ACETOACETIC ESTER ◇ ACTIVE ACETYL ACETATE ◇ DIACETIC ETHER ◇ EAA ◇ ETHYL ACETOACETATE (FCC) ◇ ETHYL ACETYLACETONATE ◇ ETHYL BENZYL ACETOACETATE ◇ ETHYL-3-OXOBUTANOATE ◇ ETHYL-3-OXOBUTYRATE ◇ FEMA No. 2415 ◇ 3-OXOBUTANOIC ACID ETHYL ESTER

USE IN FOOD:

Purpose: Flavoring agent.

Where Used: Various.

Regulations: FDA - 21CFR 172.515. Use at a level not in excess of the amount reasonably required to accomplish the intended effect.

SAFETY PROFILE: Moderately toxic by ingestion. A skin and eye irritant. Combustible liquid when exposed to heat or flame; can react with oxidizing materials. Explosive reaction when heated with Zn + trimbromoneopentyl alcohol or 2,2,2-tris(bromomethyl)ethanol. To fight fire, use alcohol foam, CO_2, dry chemical. When heated to decomposition it emits acrid smoke and irritating fumes.

TOXICITY DATA and CODEN

skn-rbt 510 mg open MLD UCDS** 3/12/69
eye-rbt 23 mg AJOPAA 29,1363,46
orl-rat LD50:3980 mg/kg JIHTAB 31,60,49

EFT000 CAS: 140-88-5
ETHYL ACRYLATE

DOT: 1917

mf: $C_5H_8O_2$ mw: 100.13

PROP: Colorless liquid; acrid penetrating odor. Bp: 99.8°, fp: <−72°: lel: 1.8%, flash p: 60°F (OC), 48.2°F d: 0.916-0.919, vap press, 29.3 mm @ 20°, vap d: 3.45. Misc with alc, ether; sltly sol in water.

SYNS: ACRYLATE d'ETHYLE (FRENCH) ◇ ACRYLIC ACID ETHYL ESTER ◇ ACRYLSAEUREAETHYLESTER (GERMAN) ◇ AETHYLACRYLAT (GERMAN) ◇ ETHOXYCARBONYLETHYLENE ◇ ETHYLACRYLAAT (DUTCH) ◇ ETHYLAKRYLAT (CZECH) ◇ ETHYL PROPENOATE ◇ ETHYL-2-PROPENOATE ◇ ETIL ACRILATO (ITALIAN) ◇ ETILACRILATULUI (ROMANIAN) ◇ FEMA No. 2418 ◇ NCI-C50384 ◇ 2-PROPENOIC ACID, ETHYL ESTER ◇ RCRA WASTE NUMBER U113

USE IN FOOD:

Purpose: Flavoring agent.

Where Used: Various.

Regulations: FDA - 21CFR 172.515. Use at a level not in excess of the amount reasonably required to accomplish the intended effect.

IARC Cancer Review: Animal Sufficient Evidence IMEMDT 39,81,86; Animal Inadequate Evidence IMEMDT 19,47,79; Human Inadequate Evidence IMEMDT 19,47,79. NTP Carcinogenesis Studies (gavage); Clear Evidence: mouse, rat NTPTR* NTP-TR-259,86. Community Right-To-Know List.

OSHA PEL: TWA 5 ppm; STEL 25 ppm (skin) ACGIH TLV: TWA 5 ppm; STEL 25 ppm (skin) DFG MAK: 5 ppm (20 mg/m^3) DOT Classification: Flammable Liquid; Label: Flammable Liquid

SAFETY PROFILE: Poison by ingestion and inhalation. Moderately toxic by skin contact and intraperitoneal routes. A suspected human carcinogen. An experimental carcinogen. Human systemic effects by inhalation: eye, olfactory and pulmonary changes. A skin and eye irritant. Characterized in its terminal stages by dyspnea, cyanosis, and convulsive movements. It caused severe local irritation of the gastro-enteric tract; and toxic degenerative changes of cardiac, hepatic, renal, and splenic tissues were observed. It gave no evidence of cumulative effects. When applied to the intact skin of rabbits, the ethyl ester caused marked local irritation, erythema, edema, thickening, and vascular damage. Animals subjected to a fairly high concentrations of these esters suffered irritation of the mucous membranes of the eyes, nose, and mouth as well as lethargy, dyspnea, and convulsive movements.

Flammable liquid. A very dangerous fire hazard when exposed to heat or flame; can react vigorously with oxidizing materials. Violent reaction with chlorosulfonic acid. To fight fire, use CO_2, dry chemical or alcohol foam. When heated to decomposition it emits acrid smoke and irritating fumes.

TOXICITY DATA and CODEN

eye-mky 1204 ppm/15H-I JIHTAB 31,317,49
skn-rbt 500 mg open MLD UCDS** 12/14/71
eye-rbt 45 mg MLD UCDS** 12/14/71
mma-mus:lym 20 mg/L ENMUDM 8(Suppl 6),4,86
cyt-ham:lng 9800 µg/L GMCRDC 27,95,81
ihl-hmn TCLo:50 ppm:NOSE,EYE,PUL 34ZIAG -,75,69
orl-rat LD50:800 mg/kg BCTKAG 12,405,79

ihl-rat LC50:2180 ppm/4H:NOSE,EYE,PUL JTEHD6 16,811,85
skn-rat LDLo:1800 mg/kg PJPPAA 32,223,80

EFU000 CAS: 64-17-5
ETHYL ALCOHOL
DOT: 1170
mf: C_2H_6O mw: 46.08

PROP: Clear, colorless liquid; fragrant odor and burning taste. Bp: 78.32°, ULC: 70, lel: 3.3%, uel: 19% @ 60°, fp: $<-130°$, flash p: 55.6°F, d: 0.7893 @ 20°/4°, autoign temp: 793°F, vap press: 40 mm @ 19°, vap d: 1.59, refr index: 1.364. Misc in water, alc, chloroform, and ether.

SYNS: ABSOLUTE ETHANOL ◇ AETHANOL (GERMAN) ◇ AETHYLALKOHOL (GERMAN) ◇ ALCOHOL ◇ ALCOHOL, ANHYDROUS ◇ ALCOHOL, DEHYDRATED ◇ ALCOOL ETHYLIQUE (FRENCH) ◇ ALCOOL ETILICO (ITALIAN) ◇ ALGRAIN ◇ ALKOHOL (GERMAN) ◇ ALKOHOLU ETYLOWEGO (POLISH) ◇ ANHYDROL ◇ COLOGNE SPIRIT ◇ COLOGNE SPIRITS (ALCOHOL) (DOT) ◇ ETANOLO (ITALIAN) ◇ ETHANOL ◇ ETHANOL 200 PROOF ◇ ETHANOL SOLUTION (DOT) ◇ ETHYLALCOHOL (DUTCH) ◇ ETHYL ALCOHOL ANHYDROUS ◇ ETHYL HYDRATE ◇ ETHYL HYDROXIDE ◇ ETYLOWY ALKOHOL (POLISH) ◇ FERMENTATION ALCOHOL ◇ GRAIN ALCOHOL ◇ JAYSOL ◇ JAYSOL S ◇ METHYLCARBINOL ◇ MOLASSES ALCOHOL ◇ NCI-C03134 ◇ POTATO ALCOHOL ◇ SD ALCOHOL 23-HYDROGEN ◇ SPIRITS OF WINE ◇ SPIRT ◇ TECSOL

USE IN FOOD:

Purpose: Antimicrobial agent, extraction solvent, vehicle.

Where Used: Pizza crust.

Regulations: FDA - 21CFR 184.1293. GRAS with a limitation of 2.0 percent in pizza crusts when used in accordance with good manufacturing practice.

EPA Genetic Toxicology Program.

OSHA PEL: TWA 1000 ppm ACGIH TLV: TWA 1000 ppm DFG MAK: 1,000 ppm (1,900 mg/m^3) DOT Classification: Flammable Liquid; Label: Flammable Liquid; Flammable or Combustible Liquid; Label: Flammable Liquid

SAFETY PROFILE: Moderately toxic to humans by ingestion. Moderately toxic experimentally by intravenous and intraperitoneal routes. Mildly toxic by inhalation and skin contact. An experimental tumorigen and teratogen. Human systemic effects by ingestion and subcutaneous route: sleep disorders, hallucinations, distorted perceptions, convulsions, motor activity changes, ataxia, coma, antipsychotic, headache, pulmonary changes, alteration in gastric secretion, nausea or vomiting, other gastrointestinal changes, menstrual cycle changes and body temperature decrease. Human reproductive effects by ingestion, intravenous, and intrauterine routes: changes in female fertility index. Effects on newborn include: changes in apgar score, neonatal measures or effects and drug dependence. Experimental reproductive effects. Human mutagenic data. An eye and severe skin irritant.

The systemic effect of ethanol differs from that of methanol. Ethanol is rapidly oxidized in the body to carbon dioxide and water, and in contrast to methanol, no cumulative effect occurs. Though ethanol possesses narcotic properties, concentrations sufficient to produce this effect are not reached in industry. Concentrations below 1,000 ppm usually produce no signs of intoxication. Exposure to concentrations over 1,000 ppm may cause headache, irritation of the eyes, nose and throat, and, if continued for an hour, drowsiness and lassitude, loss of appetite and inability to concentrate. There is no concrete evidence that repeated exposure to ethanol vapor results in cirrhosis of the liver. Ingestion of large doses can cause alcohol poisoning. Repeated ingestions can lead to alcoholism. It is a central nervous system depressant.

Flammable liquid when exposed to heat or flame; can react vigorously with oxidizers. To fight fire, use alcohol foam, CO_2, dry chemical. Ignites and then explodes on contact with acetic anhydride + sodium hydrogen sulfate. Reacts violently with many chemicals.

TOXICITY DATA and CODEN

skn-rbt 500 mg/24H SEV 28ZPAK -,34,72
eye-rbt 100 mg/24H MOD 28ZPAK -,34,72
mmo-esc 140 g/L MUREAV 130,97,84
cyt-mus-orl 40 g/kg NATUAS 302,258,83
orl-wmn TDLo:41 g/kg (41W preg):REP AJDCAI 129,1075,75
ivn-wmn TDLo:8 g/kg (32W preg):REP AJDCAI 134,419,80
orl-rat TDLo:4 g/kg (13D preg):TER CYGEDX 15,23,81
orl-mus TDLo:5800 mg/kg (7D preg):TER TJADAB 27,231,83

orl-mus TDLo:320 mg/kg/50W-I:ETA CALEDQ 13,345,81
orl-man TDLo:50 mg/kg:GIT JPETAB 56,117,36
orl-man TDLo:1430 μg/kg:CNS JPETAB 197,488,76
orl-wmn TDLo:256 g/kg/12W:CNS,END JAMAAP 238,2143,77
orl-chd LDLo:2000 mg/kg ATXKA8 17,183,58
orl-hmn LDLo:1400 mg/kg:CNS,GIT NPIRI* 1,44,74
scu-inf LDLo:19440 mg/kg:CNS,MET AJCPAI 5,466,35
orl-rat LD50:7060 mg/kg TXAPA9 16,718,70
ihl-rat LC50:20000 ppm/10H NPIRI* 1,44,74

EGM000 CAS: 87-25-2
ETHYL ANTHRANILATE
mf: $C_9H_{11}NO_2$ mw: 165.21

PROP: Colorless to amber liquid; floral, orange blossom odor. D: 1.115-1.120, refr index: 1.5631.566, flash p: +151°F. Sol in alc, fixed oils, propylene glycol.

SYNS: o-AMINOBENZOIC ACID, ETHYL ESTER ◇ ETHYL-o-AMINOBENZOATE ◇ FEMA No. 2421

USE IN FOOD:
Purpose: Flavoring agent.
Where Used: Various.
Regulations: FDA - 21CFR 172.515. Use at a level not in excess of the amount reasonably required to accomplish the intended effect.

SAFETY PROFILE: Moderately toxic by ingestion. A skin irritant. Combustible liquid. When heated to decomposition it emits toxic fumes of NO_x.

TOXICITY DATA and CODEN
skn-rbt 500 mg/24H MOD FCTXAV 14,759,76
orl-rat LD50:3750 mg/kg FCTXAV 14,759,76

EGR000 CAS: 93-89-0
ETHYL BENZOATE
mf: $C_9H_{10}O_2$ mw: 150.19

PROP: Colorless liquid; heavy fruity odor. Mp: −34.6°, bp: 213.4°, flash p: >204°F, d: 1.048 @ 20°/20°, refr index: 1.502-1.506, vap press: 1 mm @ 44.0°, vap d: 5.17, autoign temp: 914°F. Sol alc, fixed oils, propylene glycol; insol in glycerin, water @ 212°; misc in petroleum, chloroform, ether.

SYNS: BENZOIC ETHER ◇ ESSENCE OF NIOBE ◇ FEMA No. 2422

USE IN FOOD:
Purpose: Flavoring agent.
Where Used: Various.
Regulations: FDA - 21CFR 172.515. Use at a level not in excess of the amount reasonably required to accomplish the intended effect.

SAFETY PROFILE: Moderately toxic by ingestion. Mildly toxic by skin contact. A skin and eye irritant. Combustible liquid when exposed to heat or flame; can react with oxidizing materials. To fight fire, use foam, CO_2, dry chemical. When heated to decomposition it emits acrid smoke and irritating fumes.

TOXICITY DATA and CODEN
skn-rbt 10 mg/24H open MLD AMIHBC 10,61,54
eye-rbt 500 mg open AMIHBC 10,61,54
orl-rat LD50:2100 mg/kg JPETAB 84,358,45
skn-cat LDLo:10 g/kg JPETAB 84,358,45

EHE000 CAS: 105-54-4
ETHYL n-BUTYRATE
DOT: 1180
mf: $C_6H_{12}O_2$ mw: 116.18

PROP: Colorless liquid; banana-pineapple odor. D: 0.874, refr index: 1.391, mp: −100.8°, bp: 121.6°, flash p: 79°F. Sol in water, fixed oils, propylene glycol; misc in alc and ether; insol in glycerin @ 121°.

SYNS: BUTANOIC ACID ETHYL ESTER ◇ BUTYRIC ETHER ◇ ETHYL BUTANOATE ◇ ETHYL BUTYRATE (DOT,FCC) ◇ FEMA No. 2427

USE IN FOOD:
Purpose: Flavoring agent.
Where Used: Various.
Regulations: FDA - 21CFR 182.60. Use at a level not in excess of the amount reasonably required to accomplish the intended effect.

DOT Classification: Flammable or Combustible Liquid; Label: Flammable Liquid

SAFETY PROFILE: Mildly toxic by ingestion. A skin irritant. Flammable liquid when exposed to heat or flame; can react vigorously with oxidizing materials. When heated to decomposition it emits acrid smoke and irritating fumes.

TOXICITY DATA and CODEN
skn-rbt 500 mg/24H MOD FCTXAV 12,703,74
orl-rat LD50:13 g/kg FCTXAV 2,327,64

EHE500 CAS: 110-38-3
ETHYL CAPRATE
mf: $C_{12}H_{24}O_2$ mw: 200.36

PROP: Colorless liquid; oily, brandy odor. Bp: 243°, d: 0.863, refr index: 1.424, vap d: 6.9, flash p: +212°F. Sol in fixed oils; insol in glycerin, propylene glycol @ 243°.

SYNS: CAPRIC ACID ETHYL ESTER ◇ DECANOIC ACID, ETHYL ESTER ◇ ETHYL CAPRINATE ◇ ETHYL DECANOATE (FCC) ◇ ETHYL DECYLATE ◇ FEMA No. 2432

USE IN FOOD:

Purpose: Flavoring agent.

Where Used: Various.

Regulations: FDA - 21CFR 172.515. Use at a level not in excess of the amount reasonably required to accomplish the intended effect.

SAFETY PROFILE: A skin irritant. Combustible liquid when exposed to heat or flame; can react with oxidizing materials. When heated to decomposition it emits acrid smoke and irritating fumes.

TOXICITY DATA and CODEN

skn-rbt 500 mg/24H MLD FCTXAV 16,733,78

EHF000 CAS: 123-66-0
ETHYL CAPROATE
DOT: 1177
mf: $C_8H_{16}O_2$ mw: 144.24

PROP: Colorless liquid; mild wine odor. Bp: 163°, flash p: 130°F (OC), d: 0.867-0.871, refr index: 1.406-1.409, vap d: 5.0. Sol in fixed oils; sltly sol in propylene glycol; insol in glycerin @ 166.

SYNS: ETHYL BUTYLACETATE (DOT) ◇ ETHYL HEXANOATE (FCC) ◇ FEMA No. 2439

USE IN FOOD:

Purpose: Flavoring agent.

Where Used: Various.

Regulations: FDA - 21CFR 172.515. Use at a level not in excess of the amount reasonably required to accomplish the intended effect.

DOT Classification: Flammable or Combustible Liquid; Label: Flammable Liquid

SAFETY PROFILE: A skin irritant. Flammable or combustible when exposed to heat or flame; can react with oxidizing materials. When heated to decomposition it emits acrid smoke and irritating fumes. To fight fire, use CO_2, foam, dry chemical.

TOXICITY DATA and CODEN

skn-rbt 500 mg/24H MOD FCTXAV 14,659,76

EHG100 CAS: 9004-57-3
ETHYL CELLULOSE

PROP: Ethyl ether of cellulose. White to light tan powder. Sol in some organic solvents; insol in water, glycerin, propylene glycol.

USE IN FOOD:

Purpose: Binder and filler in dry vitamin preparations, color diluent, fixative in flavoring compounds, protective coating component for vitamin and mineral tablets.

Where Used: Confectionery, eggs (shell), food supplements in tablet form, gum.

Regulations: FDA - 21CFR 73.1, 172.868. Use at a level not in excess of the amount reasonably required to accomplish the intended effect.

SAFETY PROFILE: When heated to decomposition it emits acrid smoke and irritating fumes.

EHN000 CAS: 103-36-6
ETHYL-trans-CINNAMATE
mf: $C_{11}H_{12}O_2$ mw: 176.23

PROP: Nearly colorless, oily liquid; faint cinnamon odor. D: 1.049 @ 20°/4°, refr index: 1.558-1.561, bp: 271°, mp: 9°, flash p: +212°F. Misc in alc, ether, fixed oils; insol in glycerin, water @ 272°.

SYNS: ETHYL CINNAMATE (FCC) ◇ ETHYL-β-PHENYLACRYLATE ◇ ETHYL-3-PHENYLPROPENOATE ◇ FEMA No. 2430

USE IN FOOD:

Purpose: Flavoring agent.

Where Used: Various.

Regulations: FDA - 21CFR 172.515. Use at a level not in excess of the amount reasonably required to accomplish the intended effect.

SAFETY PROFILE: Moderately toxic by ingestion. Combustible liquid. When heated to decomposition it emits acrid smoke and irritating fumes.

TOXICITY DATA and CODEN

orl-rat LD50:4000 mg/kg VPITAR 33(5),48,74

EIK000 CAS: 78-78-4
ETHYLDIMETHYLMETHANE
DOT: 1265
mf: C_5H_{12} mw: 72.17

PROP: Colorless liquid with pleasant odor. Bp: 27.8°, fp: −160.5°, flash p: < −60°F (CC), vap press: 595 mm @ 21.1°, vap d: 2.48, lel: 1.4%, uel: 7.6%.

SYNS: ISOAMYLHYDRIDE ◇ ISOPENTANE ◇ 2-METHYLBUTANE

USE IN FOOD:

Purpose: Adjuvant in foamed plastic, blowing agent.

Where Used: Packaging materials.

Regulations: FDA - 21CFR 178.3010.

DOT Classification: Flammable Liquid; Label: Flammable Liquid NIOSH REL: (Alkanes) TWA 350 mg/m^3

SAFETY PROFILE: Probably mildly toxic and narcotic by inhalation. Flammable Liquid. A very dangerous fire and explosion hazard when exposed to heat, flame or oxidizers. Keep away from sparks, heat or open flame; can react with oxidizing materials. To fight fire, use foam, CO_2, dry chemical. When heated to decomposition it emits acrid smoke and irritating fumes.

EIL000
2-ETHYL-3,5(6)-DIMETHYLPYRAZINE
mf: $C_8H_{12}N_2$ mw: 136.20

PROP: Colorless to sltly yellow liquid; roasted cocoa odor. D: 0.950-0.980, refr index: 1.500, flash p: 154°F. Sol in water, organic solvents.

SYN: FEMA No. 3149

USE IN FOOD:

Purpose: Flavoring agent.

Where Used: Various.

Regulations: GRAS when used at a level not in excess of the amount reasonably required to accomplish the intended effect.

SAFETY PROFILE: Combustible liquid. When heated to decomposition emits toxic fumes of NO_x.

EIQ500
ETHYLENEBIS(DITHIOCARBAMATO)-MANGANESE and ZINC ACETATE (50:1))

SYNS: MANEB PLUS ZINC ACETATE (50:1) ◇ ZINC ACETATE PLUS MANEB (1:50)

USE IN FOOD:

Purpose: Fungicide.

Where Used: Animal feed, barley bran, barley flour, oat bran, oat flour, raisins, rye bran, rye flour, wheat bran, wheat flour.

Regulations: FDA - 21CFR 193.460. Fungicide residue tolerance of 28 ppm in raisins; 20 ppm in barley bran, oat bran, rye bran, wheat bran; 1 ppm in barley flour, oat flour, rye flour, wheat flour. 21CFR 561.410. Limitation of 20 ppm in milled animal feed fractions of barley, oats, rye, wheat.

Manganese and its compounds, as well as Zinc and its compounds, are on the Community Right-To-Know List.

SAFETY PROFILE: Experimental reproductive effects. When heated to decomposition it emits very toxic fumes of NO_x, ZnO and SO_x.

TOXICITY DATA and CODEN

orl-rat TDLo:765 mg/kg (11D preg):REP TJADAB 14,171,76

EIV000 CAS: 64-02-8
N,N′-ETHYLENEDIAMINEDIACETIC ACID TETRASODIUM SALT
mf: $C_{10}H_{12}N_2O_8 \cdot 4Na$ mw: 380.20

SYNS: AQUAMOLLIN ◇ CALSOL ◇ CELON E ◇ CELON H ◇ CELON IS ◇ CHEELOX BF ◇ CHEELOX BR-33 ◇ CHELON 100 ◇ CHEMCOLOX 200 ◇ COMPLEXONE ◇ CONIGON BC ◇ DISTOL 8 ◇ EDATHANIL TETRASODIUM ◇ EDETATE SODIUM ◇ EDETIC ACID TETRASODIUM SALT ◇ EDTA, SODIUM SALT ◇ EDTA TETRASODIUM SALT ◇ ENDRATE TETRASODIUM ◇ N,N′-1,2-ETHANEDIYLBIS(N-(CARBOXYMETHYL)GLYCINE TETRASODIUM SALT ◇ ETHYLENEBIS(IMINODIACETIC ACID) TETRASODIUM SALT ◇ ETHYLENEDIAMINETETRAACETIC ACID, TETRASODIUM SALT ◇ HAMP-ENE 100 ◇ HAMP-ENE 215 ◇ HAMP-ENE 220 ◇ HAMP-ENE Na4 ◇ IRGALON ◇ KALEX ◇ KEPMPLEX 100 ◇ KOMPLXON ◇ METAQUEST C ◇ NERVANAID B LIQUID ◇ NERVANID B ◇ NULLAPON B ◇ NULLAPON BF-78 ◇ NULLAPON BFC CONC ◇ PERMA KLEER 50 CRYSTALS ◇ PERMA KLEER TETRA CP ◇ QUESTEX 4 ◇ SEQUESTRENE 30A ◇ SEQUESTRENE Na 4 ◇ SEQUESTRENE ST ◇ SODIUM EDETATE ◇ SODIUM

EDTA ◇ SODIUM ETHYLENEDIAMINETETRAACETATE ◇ SODIUM ETHYLENEDIAMINETETRAACETIC ACID ◇ SODIUM SALT of ETHYLENEDIAMINETETRAACETIC ACID ◇ SYNTES 12A ◇ SYNTRON B ◇ TETRACEMIN ◇ TETRASODIUM EDTA ◇ TETRASODIUM ETHYLENEDIAMINETETRAACETATE ◇ TETRASODIUM ETHYLENEDIAMINETETRACETATE ◇ TETRASODIUM (ETHYLENEDINITRILO)TETRAACETATE ◇ TETRASODIUM SALT of EDTA ◇ TETRASODIUM SALT of ETHYLENEDIAMINETETRACETICACID ◇ TETRINE ◇ TRILON B ◇ TST ◇ TYCLAROSOL ◇ VERSENE 100 ◇ VERSENE POWDER ◇ WARKEELATE PS-43

USE IN FOOD:

Purpose: Flume wash water additive, poultry scald agent.

Where Used: Beets (sugar), poultry.

Regulations: FDA - 21CFR 173.315. Limitation of 0.1 ppm in wash water. USDA - 9CFR 381.147. Sufficient for purpose.

SAFETY PROFILE: Poison by intraperitoneal route. A skin and eye irritant. When heated to decomposition it emits toxic fumes of NO_x and Na_2O.

TOXICITY DATA and CODEN

skn-rbt 500 mg/24H MOD 28ZPAK -,306,72
eye-rbt 100 mg/24H MOD 28ZPAK -,306,72
ipr-mus LD50:330 mg/kg REPMBN 10,391,62

EIX500 CAS: 139-33-3
ETHYLENEDIAMINETETRAACETIC ACID, DISODIUM SALT

mf: $C_{10}H_{14}N_2O_8 \cdot 2Na$ mw: 336.24

PROP: White crystalline powder. Sol in water.

SYNS: CHELADRATE ◇ CHELAPLEX III ◇ CHELATON III ◇ COMPLEXON III ◇ d'E.D.T.A. DISODIQUE (FRENCH) ◇ DISODIUM DIACID ETHYLENEDIAMINETETRAACETATE ◇ DISODIUM DIHYDROGEN ETHYLENEDIAMINETETRAACETATE ◇ DISODIUM DIHYDROGEN (ETHYLENEDINITRILO)TETRAACETATE ◇ DISODIUM EDATHAMIL ◇ DISODIUM EDETATE ◇ DISODIUM EDTA (FCC) ◇ DISODIUM ETHYLENEDIAMINETETRAACETATE ◇ DISODIUM ETHYLENEDIAMINETETRAACETIC ACID ◇ DISODIUM (ETHYLENEDINITRILO) TETRAACETATE ◇ DISODIUM (ETHYLENEDINITRILO) TETRAACETIC ACID ◇ DISODIUM SALT of EDTA ◇ DISODIUM SEQUESTRENE ◇ DISODIUM TETRACEMATE ◇ DISODIUM VERSENATE ◇ DISODIUM VERSENE ◇ EDATHAMIL DISODIUM ◇ EDETATE DISODIUM ◇ EDTA, DISODIUM SALT ◇ ENDRATE DISODIUM ◇ N,N'-1,2-ETHANEDIYLBIS(N-(CARBOXYMETHYL)GLYCINE) DISODIUM SALT ◇ ETHYLENEBIS(IMINODIACETIC ACID) DISODIUM SALT ◇ ETHYLENEDIAMINETETRAACETATE DISODIUM SALT ◇ (ETHYLENEDINITRILO)TETRAACETIC ACID DISODIUM SALT ◇ F 1 (complexon) ◇ KIRESUTO B ◇ METAQUEST B ◇ PERMA KLEER 50 CRYSTALS DISODIUM SALT ◇ SELEKTON B 2 ◇ SEQUESTRENE SODIUM 2 ◇ SODIUM VERSENATE ◇ TETRACEMATE DISODIUM ◇ TITRIPLEX III ◇ TRILON BD ◇ TRIPLEX III ◇ VERESENE DISODIUM SALT ◇ VERSENE SODIUM 2

USE IN FOOD:

Purpose: Hog scald agent, preservative, sequestrant, stabilizer.

Where Used: Black-eyed peas (canned), cereal products (ready-to-eat, containing dried bananas), chickpeas (canned cooked), dressings (nonstandardized), French dressing, gefilte fish balls or patties in packing medium, hog carcasses, kidney beans (canned), mayonnaise, nonnutritive sweeteners, potatoes (white, frozen including cut potatoes), salad dressings, sauces, sausage (cooked), spread (sandwich), strawberry pie filling (canned).

Regulations: FDA - 21CFR 172.135. Limitation of 150 ppm in aqueous multivitamin preparations; 145 ppm in black-eyed peas, canned; 165 ppm in chickpeas, canned cooked; 165 ppm in kidney beans, canned; 500 ppm in strawberry pie filling, canned; 36 ppm in sausage, cooked; 75 ppm in dressings, nonstandardized; 75 ppm in French dressing; 100 ppm in white potatoes, frozen including cut potatoes; 50 ppm in gefilte fish balls or patties in packing medium; 75 ppm in mayonnaise; 315 ppm in cereal products, read-to-eat, containing dried bananas; 75 ppm in salad dressings; 100 ppm in sandwich spread; 75 ppm in sauces; 0.1% of dry product in nonnutritive sweeteners (21CFR 180.37). Limitation in conjunction with calcium disodium EDTA of 75 ppm in dressings, nonstandardized; 75 ppm in French dressing; 75 ppm in mayonnaise; 75 ppm in salad dressings; 100 ppm in sandwich spread; 75 ppm in sauces. USDA - 9CFR 318.7. Sufficient for purpose.

EPA Genetic Toxicology Program.

SAFETY PROFILE: Poison by intravenous route. Moderately toxic by ingestion. An experimental teratogen. Experimental reproductive effects. Mutagenic data. The calcium disodium salt of EDTA is used as a chelating agent in treating lead poisoning. Also used as an anticoagulant. When heated to decomposition it emits toxic fumes of NO_x and Na_2O.

TOXICITY DATA and CODEN

cyt-grh-par 1 mmol/L CISCB7 16,18,74
orl-rat TDLo:31429 mg/kg (1-22D preg):REP SCIEAS 173,62,72
orl-rat TDLo:12857 mg/kg (7-15D preg):TER SCIEAS 173,62,72
orl-rat LD50:2000 mg/kg FEPRA7 27,465,68

EIY500 CAS: 106-93-4
1,2-ETHYLENE DIBROMIDE
DOT: 1605
mf: $C_2H_4Br_2$ mw: 187.88

PROP: Colorless, heavy liquid; sweet odor. Bp: 131.4°, fp: 9.3°, flash p: none, d: 2.172 @ 25°/25°, 2.1707 @ 25°/4°, vap press: 17.4 mm @ 30°, vap d: 6.48.

SYNS: AETHYLENBROMID (GERMAN) ◇ BROMOFUME ◇ BROMURO di ETILE (ITALIAN) ◇ CELMIDE ◇ DBE ◇ 1,2-DIBROMAETHAN (GERMAN) ◇ 1,2-DIBROMOETANO (ITALIAN) ◇ DIBROMOETHANE ◇ α,β-DIBROMOETHANE ◇ sym-DIBROMOETHANE ◇ DIBROMURE d'ETHYLENE (FRENCH) ◇ 1,2-DIBROOMETHAAN (DUTCH) ◇ DOWFUME 40 ◇ DOWFUME EDB ◇ DOWFUME W-8 ◇ DOWFUME W-85 ◇ DOWFUME W-90 ◇ DOWFUME W-100 ◇ DWUBROMOETAN (POLISH) ◇ EDB ◇ EDB-85 ◇ E-D-BEE ◇ ENT 15,349 ◇ ETHYLENE BROMIDE ◇ ETHYLENE DIBROMIDE (ACGIH, DOT, USDA) ◇ FUMO-GAS ◇ GLYCOL BROMIDE ◇ GLYCOL DIBROMIDE ◇ ISCOBROME D ◇ KOPFUME ◇ NCI-C00522 ◇ NEPHIS ◇ PESTMASTER ◇ PESTMASTER EDB-85 ◇ RCRA WASTE NUMBER U067 ◇ SOILBROM-40 ◇ SOILBROM-85 ◇ SOILBROM-90 ◇ SOILBROM-100 ◇ SOILBROME-85 ◇ SOILBROM-90EC ◇ SOILFUME ◇ UNIFUME

USE IN FOOD:

Purpose: Fumigant.

Where Used: Cereal grains, corn grits, cracked rice, fermented malt beverages.

Regulations: FDA - 21CFR 193.225. Fumigant. residue tolerance of 125 ppm as Br in cereal grain. 21CFR 193.230. Pesticide residue tolerance of 25 ppm as Br in fermented malt beverages. USDA CES Ranking: A-4 (1986).

Community Right-To-Know List. IARC Cancer Review: Animal Sufficient Evidence IMEMDT 15,195,77. NCI Carcinogenesis Bioassay (gavage); Clear Evidence: mouse, rat NCITR* NCI-CG-TR-86,78; NTP Carcinogenesis Bioassay (inhalation); Clear Evidence: mouse, rat NTPTR* NTP-TR-210,82. EPA Genetic Toxicology Program.

OSHA PEL: TWA 20 ppm; CL 30 ppm; Pk 50 ppm/5M/8H ACGIH TLV: Suspected carcinogen; No TWA DOT Classification: ORM-A; Label:None; Poison B; Label: Poison NIOSH REL: (EDB) 0.045 ppm; CL 1 mg/m^3/15M

SAFETY PROFILE: Human poison by ingestion. Experimental poison by ingestion, skin contact, intraperitoneal and possibly other routes. Moderately toxic by inhalation and rectal routes. An experimental carcinogen, neoplastigen, and teratogen. Experimental reproductive effects. Human mutagenic data. A severe human and experimental skin irritant. An experimental eye irritant. Implicated in worker sterility. When heated to decomposition it emits toxic fumes of Br^-.

TOXICITY DATA and CODEN

skn-hmn 1538 mg/24H SEV AGGHAR 8,591,38
skn-rbt 1%/14D SEV AMIHBC 6,158,52
eye-rbt 1% AMIHBC 6,158,52
mmo-sat 1 μg/plate ENMUDM 7(Suppl 5),1,85
mma-esc 333 μg/plate ENMUDM 7(Suppl 5),1,85
orl-rat TDLo:50 mg/kg (5D male):TER MUREAV 77,71,80
ihl-rat TCLo:66670 ppb/4H (3-20D preg):REP NTOTDY 5,579,83
ihl-mus TCLo:38 ppm/23H (6-15D preg):TER TXAPA9 45,347,78
orl-rat TDLo:540 mg/kg /78W-C:CAR BANRDU 5,279,80
ihl-rat TCLo:10 ppm/6H/2Y-I:CAR NCITR* NCI-CG-TR-201,82
orl-wmn LDLo:90 mg/kg 34ZIAG -,257,69
orl-rat LD50:108 mg/kg SPEADM 74-1,-,74
ihl-rat LCLo:400 ppm/2H AMIHBC 6,158,52
skn-rat LD50:300 mg/kg 85DPAN -,-,71/76

EJH500 CAS: 109-86-4
ETHYLENE GLYCOL METHYL ETHER
DOT: 1188
mf: $C_3H_8O_2$ mw: 76.11

PROP: Colorless liquid; mild, agreeable odor. Misc in water, alc, ether, benzene. Bp: 124.5°, fp: −86.5°, flash p: 115°F (OC), lel: 2.5%, uel: 14%, d: 0.9660 @ 20°/4°, autoign temp: 545°F, vap press: 6.2 mm @ 20°, vap d: 2.62.

SYNS: AETHYLENGLYKOL-MONOMETHYLAETHER (GERMAN) ◇ DOWANOL EM ◇ EGM ◇ EGME ◇ ETHER MONOMETHYLIQUE de l'ETHYLENE-GLYCOL (FRENCH)

◇ ETHYLENE GLYCOL MONOMETHYL ETHER (DOT) ◇ GLYCOL ETHER EM ◇ GLYCOLMETHYL ETHER ◇ GLYCOL MONOMETHYL ETHER ◇ JEFFERSOL EM ◇ MECS ◇ 2-METHOXY-AETHANOL (GERMAN) ◇ 2-METHOXYETHANOL (ACGIH) ◇ METHOXYHYDROXYETHANE ◇ METHYL CELLOSOLVE (DOT) ◇ METHYL ETHOXOL ◇ METHYL GLYCOL ◇ METHYLGLYKOL (GERMAN) ◇ METHYL OXITOL ◇ METIL CELLOSOLVE (ITALIAN) ◇ METOKSYETYLOWY ALKOHOL (POLISH) ◇ 2-METOSSIETANOLO (ITALIAN) ◇ MONOMETHYL ETHER of ETHYLENE GLYCOL ◇ POLY-SOLV EM ◇ PRIST

USE IN FOOD:

Purpose: Binder, color diluent, extender.

Where Used: Confectionery, food supplements in tablet form, gum, poultry.

Regulations: FDA - 21CFR 73.1. No residue. USDA - 9CFR 381.147. Limitation of 0.15 percent in poultry.

Community Right-To-Know List.

OSHA PEL: TWA 25 ppm (skin) ACGIH TLV: TWA 5 ppm (skin) DOT Classification: Combustible Liquid; Label: None; Flammable or Combustible Liquid; Label: Flammable Liquid NIOSH REL: TWA (Glycol Ethers): Reduce to lowest level

SAFETY PROFILE: Moderately toxic to humans by ingestion. Moderately toxic experimentally by ingestion, inhalation, skin contact, intraperitoneal and intravenous routes. Human systemic effects by inhalation: change in motor activity, tremors, and convulsions. An experimental teratogen. Experimental reproductive effects. Mutagenic data. A skin and eye irritant. When used under conditions which do not require the application of heat, this material probably presents little hazard to health. The blood picture may resemble that produced by exposure to benzene. The persons affected had been exposed to vapors of methyl "Cellosolve", ethanol and methanol, ethyl acetate and petroleum naphtha.

Flammable or combustible when exposed to heat or flame. A moderate explosion hazard. Can react with oxidizing materials to form explosive peroxides. To fight fire, use alcohol foam, CO_2, dry chemical. When heated to decomposition it emits acrid smoke and irritating fumes.

TOXICITY DATA and CODEN

skn-rbt 483 mg/24H MLD TXAPA9 19,276,71
eye-rbt 97 mg TXAPA9 19,276,71
spm-rat-orl 500 mg/kg ENMUDM 6,390,84
spm-mus-orl 500 mg/kg ENMUDM 6,390,84
orl-rat TDLo: 175 mg/kg (7-13D preg): TER TJADAB 32,33,85
orl-rat TDLo: 700 mg/kg (9-15D preg): REP TJADAB 31,58A,85
ipr-rat TDLo: 190 mg/kg (12D preg): TER JJATDK 4,35,84
ihl-hmn TCLo: 25 ppm: CNS JIHTAB 20,134,38
orl-hmn LDLo: 3380 mg/kg JIHTAB 28,267,46
orl-rat LD50: 2460 mg/kg JIHTAB 23,259,41
ihl-rat LC50: 1500 ppm/7H NPIRI* 1,57,74

EJN500 CAS: 75-21-8
ETHYLENE OXIDE

DOT: 1040
mf: C_2H_4O mw: 44.06

PROP: Colorless gas at room temperature. Mp: −111.3°, bp: 10.7°, ULC: 100, lel: 3.0%, uel: 100%, flash p: −4°F, d: 0.8711 @ 20°/20°, autoign temp: 804°F, vap press: 1095 mm @ 20°, vap d: 1.52. Misc in water and alc; very sol in ether.

SYNS: AETHYLENOXID (GERMAN) ◇ AMPROLENE ◇ ANPROLENE ◇ ANPROLINE ◇ DIHYDROOXIRENE ◇ DIMETHYLENE OXIDE ◇ ENT 26,263 ◇ E.O. ◇ 1,2-EPOXYAETHAN (GERMAN) ◇ EPOXYETHANE ◇ 1,2-EPOXYETHANE ◇ ETHENE OXIDE ◇ ETHYLEENOXIDE (DUTCH) ◇ ETHYLENE (OXYDE d') (FRENCH) ◇ ETILENE (OSSIDO di) (ITALIAN) ◇ ETO ◇ ETYLENU TLENEK (POLISH) ◇ FEMA No. 2433 ◇ MERPOL ◇ NCI-C50088 ◇ OXACYCLOPROPANE ◇ OXANE ◇ OXIDOETHANE ◇ α,β-OXIDOETHANE ◇ OXIRAAN (DUTCH) ◇ OXIRANE ◇ OXYFUME ◇ OXYFUME 12 ◇ RCRA WASTE NUMBER U115 ◇ STERILIZING GAS ETHYLENE OXIDE 100% ◇ T-GAS

USE IN FOOD:

Purpose: Fumigant.

Where Used: Spices (ground).

Regulations: FDA - 21CFR 193.200. Fumigant residue tolerance of 50 ppm in ground spices.

IARC Cancer Review: Animal Inadequate Evidence IMEMDT 11,157,76; Human Inadequate Evidence IMEMDT 36,189,85; Animal Sufficient Evidence IMEMDT 36,189,85. Community Right-To-Know List. EPA Extremely Hazardous Substances List. EPA Genetic Toxicology Program.

OSHA PEL: TWA 1 ppm ACGIH TLV: TWA 1 ppm; Suspected Carcinogen DOT Classification: Flammable Liquid; Label: Flammable Liquid; Flammable Gas; Label: Poison Gas and

Flammable Gas NIOSH REL: (Oxirane) TWA 0.1 ppm; CL 5 ppm/10M/D

SAFETY PROFILE: Poison by ingestion, intraperitoneal, subcutaneous, intravenous, and possibly other routes. Moderately toxic by inhalation. A suspected human carcinogen. An experimental carcinogen, tumorigen, neoplastigen, and teratogen. Human systemic effects by inhalation: convulsions, nausea, vomiting, olfactory and pulmonary changes. Experimental reproductive effects. Mutagenic data. A human skin irritant and experimental eye irritant. An irritant to mucous membranes of respiratory tract. High concentrations can cause pulmonary edema.

Highly flammable liquid or gas. Severe explosion hazard when exposed to flame. To fight fire, use alcohol foam, CO_2, dry chemical. Violent polymerization occurs on contact with many chemicals. Explosive reaction with glycerol at 200°. Rapid compression of the vapor with air causes explosions. When heated to decomposition it emits acrid smoke and irritating fumes.

TOXICITY DATA and CODEN

skn-hmn 1%/7S AMIHBC 2,549,50
eye-rbt 18 mg/6H MOD BUYRAI 31,25,77
dns-hmn:leu 4 mmol/L CBINA8 47,265,83
sce-hmn:lym 4 pph TCMUD8 6,15,86
ihl-rat TCLo:100 ppm/6H (6-15D preg):TER TXAPA9 48,A84,79
ihl-rat TCLo:150 ppm/7H (7-16D preg):TER NTIS** PB83-258038
ivn-rbt TDLo:324 mg/kg (6-14D preg):REP NTIS** PB83-242016
orl-rat TDLo:1186 mg/kg/2Y-I:CAR BJCAAI 46,924,82
ihl-rat TCLo:33 ppm/6H/2Y-I:CAR TXAPA9 75,105,84
ihl-hmn TCLo:12500 ppm/10S:NOSE JOHYAY 32,409,32
ihl-wmn TCLo:500 ppm/2M:CNS,GIT,PUL DICPBB 15,384,81
orl-rat LD50:72 mg/kg SPEADM 78-1,17,78
ihl-rat LC50:800 ppm/4H 34ZIAG -,258,69
scu-rat LD50:187 mg/kg GISAAA 48(1),23,83

EJO025
ETHYLENE OXIDE POLYMER

USE IN FOOD:

Purpose: Foam stabilizer.

Where Used: Beverages (fermented malt).

Regulations: FDA - 21CFR 172.770. Limitation of 300 ppm in fermented malt beverages.

SAFETY PROFILE: When heated to decomposition it emits acrid smoke and irritating fumes.

EKF000 CAS: 298-04-4
O,O-ETHYL-S-2(ETHYLTHIO)ETHYL PHOSPHORODITHIOATE

DOT: 2783

mf: $C_8H_{19}O_2PS_3$ mw: 274.42

SYNS: BAYER 19639 ◇ O,O-DIAETHYL-S-(2-AETHYLTHIO-AETHYL)-DITHIOPHOSPHAT (GERMAN) ◇ O,O-DIAETHYL-S-(3-THIA-PENTYL)-DITHIOPHOSPHAT (GERMAN) ◇ O,O-DIETHYL-S-(2-ETHTHIOETHYL) PHOSPHORODITHIOATE ◇ O,O-DIETHYL-S-(2-ETHTHIOETHYL) THIOTHIONOPHOSPHATE ◇ O,O-DIETHYL-S-(2-ETHYLMERCAPTOETHYL) DITHIOPHOSPHATE ◇ O,O-DIETHYL-S-(2-ETHYLTHIO-ETHYL)-DITHIOFOSFAAT (DUTCH) ◇ O,O-DIETHYL-2-ETHYLTHIOETHYL PHOSPHORODITHIOATE ◇ O,O-DIETHYL-S-2-(ETHYLTHIO)ETHYL PHOSPHORODITHIOATE ◇ O,O-DIETIL-S-(2-ETILTIO-ETIL)-DITIOFOSFATO (ITALIAN) ◇ DIMAZ ◇ DISULFATON ◇ DISULFOTON (ACGIH, DOT) ◇ DI-SYSTON ◇ DISYSTOX ◇ DITHIODEMETON ◇ DITHIOPHOSPHATE de O,O-DIETHYLE et de S-(2-ETHYLTHIO-ETHYLE) (FRENCH) ◇ DITHIOSYSTOX ◇ ENT 23,437 ◇ S-2-(ETHYLTHIO)ETHYL O,O-DIETHYL ESTER of PHOSPHORODITHIOIC ACID ◇ FRUMIN AL ◇ M-74 ◇ RCRA WASTE NUMBER P039 ◇ S 276 ◇ SOLVIREX ◇ THIODEMETON ◇ THIODEMETRON

USE IN FOOD:

Purpose: Insecticide.

Where Used: Animal feed, pineapple bran, pineapples, sugar beet pulp (dehydrated).

Regulations: FDA - 21CFR 561.160. Limitation of 5 ppm in dehydrated sugar beet pulp and pineapple bran when used for animal feed.

EPA Extremely Hazardous Substances List. EPA Genetic Toxicology Program.

ACGIH TLV: TWA 0.1 mg/m^3 DOT Classification: Poison B; Label: Poison; Poison B; Label: Poison, dry or liquid

SAFETY PROFILE: Poison by ingestion, inhalation, skin contact, intraperitoneal, intravenous, and possibly other routes. Human mutagenic data. When heated to decomposition it emits very toxic SO_x and PO_x.

TOXICITY DATA and CODEN

mmo-sat 5 μL/plate MUREAV 28,405,75
msc-mus:lym 40 mg/L NTIS** PB84-138973

orl-rat LD50:2 mg/kg FMCHA2 -,C85,83
ihl-rat LC50:200 mg/m^3 85GYAZ -,25,71
skn-rat LD50:6 mg/kg TXAPA9 14,515,69

EKF575
2-ETHYL FENCHOL
mf: $C_{12}H_{22}O$ mw: 182.30

PROP: Pale yellow liquid; camphor, earthy odor. D: 0.946-0.967, refr index: 1.470-1.491. Sol in alc, propylene glycol, fixed oils; insol in water.

SYN: FEMA No. 3491

USE IN FOOD:

Purpose: Flavoring agent.

Where Used: Various.

Regulations: GRAS when used at a level not in excess of the amount reasonably required to accomplish the intended effect.

SAFETY PROFILE: When heated to decomposition it emits acrid smoke and irritating fumes.

EKL000 CAS: 109-94-4
ETHYL FORMATE
DOT: 1190
mf: $C_3H_6O_2$ mw: 74.09

PROP: Colorless liquid; sharp, rum-like odor. Mp: −79°, bp: 54.3°, lel: 2.7%, uel: 13.5%, flash p: −4°F (CC), d: 0.9236 @ 20°/20°, refr index: 1.359, autoign temp: 851°F, vap press: 100 mm @ 5.4°, vap d: 2.55. Sol in fixed oils, propylene glycol, water (decomp); sltly sol in mineral oil; insol in glycerin @ 54°.

SYNS: AETHYLFORMIAT (GERMAN) ◇ AREGINAL ◇ ETHYLE (FORMIATE d') (FRENCH) ◇ ETHYLFORMIAAT (DUTCH) ◇ ETHYL FORMIC ESTER ◇ ETHYL METHANOATE ◇ ETILE (FORMIATO di) (ITALIAN) ◇ FEMA No. 2434 ◇ FORMIC ACID, ETHYL ESTER ◇ FORMIC ETHER ◇ MROWCZAN ETYLU (POLISH)

USE IN FOOD:

Purpose: Flavoring agent, insecticide.

Where Used: Baked goods, candy (hard), candy (soft), chewing gum, fillings, frozen dairy desserts, gelatins, puddings, raisins, Zante currents.

Regulations: FDA - 21CFR 172.515, 184.1295. GRAS with a limitation of 0.05 percent in baked goods, 0.04 percent in chewing gum, hard candy, soft candy, 0.02 percent in frozen dairy desserts; 0.03 percent in gelatins, puddings, and fillings; 0.1 percent in all other foods when used in accordance with good manufacturing practice. 21CFR 193.210. Insecticide residue tolerance of 250 ppm in raisins and Zante currents.

OSHA PEL: TWA 100 ppm ACGIH TLV: TWA 100 ppm DOT Classification: Flammable Liquid; Label: Flammable Liquid

SAFETY PROFILE: Moderately toxic by ingestion and subcutaneous routes. Mildly toxic by skin contact and inhalation. An experimental tumorigen. A powerful inhalation irritant in humans. A skin and eye irritant. Highly flammable liquid. A very dangerous fire and explosion hazard when exposed to heat, flame or oxidizers. To fight fire, use alcohol foam, spray, mist, dry chemical. When heated to decomposition it emits acrid smoke and irritating fumes.

TOXICITY DATA and CODEN

skn-rbt 460 mg open MLD UCDS** 4/10/68
eye-rbt 20 mg open AMIHBC 10,61,54
skn-mus TDLo:110 g/kg/9W-I:ETA BJCAAI 9,177,55
orl-rat LD50:1850 mg/kg FCTXAV 2,327,64
ihl-rat LCLo:8000 ppm/4H AMIHBC 10,61,54

EKN050 CAS: 106-30-9
ETHYL HEPTANOATE
mf: $C_9H_{18}O_2$ mw: 158.24

PROP: Colorless liquid; wine-brandy odor. D: 0.867-0.872, refr index: 1.411, flash p: 149°F. Misc in alc, chloroform, fixed oils; sltly sol in propylene glycol.

SYNS: ETHYL HEPTOATE ◇ FEMA No. 2437

USE IN FOOD:

Purpose: Flavoring agent.

Where Used: Various.

Regulations: FDA - 21CFR 172.515. Use at a level not in excess of the amount reasonably required to accomplish the intended effect.

SAFETY PROFILE: Combustible liquid. When heated to decomposition it emits acrid smoke and irritating fumes.

ELS000 CAS: 97-62-1
ETHYL ISOBUTYRATE
mf: $C_6H_{12}O_2$ mw: 116.18

PROP: Colorless, volatile liquid; fruity, aromatic odor. Mp: −88°, bp: 110-111°, d: 0.862, vap press: 40 mm @ 33.8°, vap d: 4.01, refr index: 1.385, flash p: <64.4°F.

SYNS: ETHYL ISOBUTANOATE ◇ ETHYLISOBUTYRATE (DOT) ◇ ETHYL-2-METHYLPROPANOATE ◇ ETHYL-2-METHYLPROPIONATE ◇ FEMA No. 2428 ◇ ISOBUTYRIC ACID, ETHYL ESTER ◇ 2-METHYLPROPIONIC ACID, ETHYL ESTER

USE IN FOOD:

Purpose: Flavoring agent.

Where Used: Baked goods, beverages, candy.

Regulations: FDA - 21CFR 172.515. Use at a level not in excess of the amount reasonably required to accomplish the intended effect.

DOT Classification: Flammable Liquid; Label: Flammable Liquid

SAFETY PROFILE: Moderately toxic by intraperitoneal route. A skin irritant. Flammable Liquid. A very dangerous fire hazard when exposed to heat or flame; can react vigorously with oxidizing materials. To fight fire, use foam, CO_2, dry chemical. When heated to decomposition it emits acrid smoke and irritating fumes.

TOXICITY DATA and CODEN

skn-rbt 500 mg/24H MOD FCTXAV 16,741,78
ipr-mus LD50:800 mg/kg FCTXAV 16,741,78

ELY700 CAS: 106-33-2
ETHYL LAURATE

mf: $C_{14}H_{28}O_2$ mw: 228.37

PROP: Colorless, oily liquid; fruity-floral odor. D: 0.858, refr index: 1.430, flash p: +212°F. Misc in alc, chloroform, ether; insol in water @ 269°.

SYNS: ETHYL DODECANOATE ◇ FEMA No. 2441

USE IN FOOD:

Purpose: Flavoring agent.

Where Used: Various.

Regulations: FDA - 21CFR 172.515. Use at a level not in excess of the amount reasonably required to accomplish the intended effect.

SAFETY PROFILE: Combustible liquid. When heated to decomposition it emits acrid smoke and irritating fumes.

EMA500 CAS: 105-53-3
ETHYL MALONATE

mf: $C_7H_{12}O_4$ mw: 160.19

PROP: Clear, colorless liquid; fruit-like odor. Bp: 198.9°, fp: −49.8°, flash p: 200°F (OC), d: 1.055 @ 25°/25°, refr index: 1.413-1.416, vap press: 1 mm @ 40.0°, vap d: 5.52. Sol in fixed oils, propylene glycol; sltly sol in alc, water; insol in glycerin, mineral oil @ 200°.

SYNS: CARBETHOXYACETIC ESTER ◇ DICARBETHOXYMETHANE ◇ DIETHYL MALONATE (FCC) ◇ DIETHYL PROPANEDIOATE ◇ FEMA No. 2375 ◇ MALONIC ACID, DIETHYL ESTER ◇ MALONIC ESTER ◇ METHANEDICARBOXYLIC ACID, DIETHYL ESTER ◇ PROPANEDIOIC ACID, DIETHYL ESTER

USE IN FOOD:

Purpose: Flavoring agent.

Where Used: Various.

Regulations: FDA - 21CFR 172.515. Use at a level not in excess of the amount reasonably required to accomplish the intended effect.

SAFETY PROFILE: Mildly toxic by ingestion. A skin irritant. Combustible liquid when exposed to heat or flame; can react with oxidizing materials. To fight fire use water to blanket fire, foam, CO_2, dry chemical. When heated to decomposition it emits acrid smoke and irritating fumes.

TOXICITY DATA and CODEN

skn-rbt 500 mg/24H MLD FCTXAV 14,745,76
orl-rat LD50:15 g/kg AIHAAP 30,470,69

EMA600 CAS: 4940-11-8
ETHYL MALTOL

mf: $C_7H_8O_3$ mw: 140.15

PROP: White crystalline powder; sweet fruity taste. Mp: 90°. Sol in water, alc, propylene glycol, chloroform.

SYNS: 2-ETHYL-3-HYDROXY-4H-PYRAN-4-ONE ◇ 2-ETHYL PYROMECONIC ACID ◇ 3-HYDROXY-2-ETHYL-4-PYRONE

USE IN FOOD:

Purpose: Flavoring agent, processing aid.

Where Used: Chocolate, desserts, wine.

Regulations: FDA - 21CFR 172.515. Use at a level not in excess of the amount reasonably required to accomplish the intended effect. BATF - 27CFR 240.1051. Limitation of 100 mg/L.

SAFETY PROFILE: Moderately toxic by ingestion and subcutaneous routes. Mutagenic data.

When heated to decomposition it emits acrid smoke and irritating fumes.

TOXICITY DATA and CODEN

mmo-sat 1 mg/plate MUREAV 67,367,79
orl-rat LD50:1150 mg/kg TXAPA9 15,604,69
scu-mus LD50:910 mg/kg CPBTAL 22,1008,74

EMP600 CAS: 7452-79-1
ETHYL 2-METHYLBUTYRATE
mf: $C_7H_{14}O_2$ mw: 130.19

PROP: Colorless liquid; strong, apple-like odor. D: 0.861-0.866, refr index: 1.396, flash p: +153°F. Sol in alc, propylene glycol; misc in fixed oils; very sltly sol in water.

SYN: FEMA No. 2443

USE IN FOOD:

Purpose: Flavoring agent.

Where Used: Various.

Regulations: FDA - 21CFR 172.515. Use at a level not in excess of the amount reasonably required to accomplish the intended effect.

SAFETY PROFILE: Combustible liquid. When heated to decomposition it emits acrid smoke and irritating fumes.

EMS500 CAS: 563-12-2
ETHYL METHYLENE PHOSPHORODITHIOATE
DOT: 2783
mf: $C_9H_{22}O_4P_2S_4$ mw: 384.49

PROP: Liquid. Mp: −13°, d: 1.220 @ 20°/4°. Sltly sol in water; sol in xylene, chloroform, acetone.

SYNS: AC 3422 ◇ BIS(S-(DIETHOXYPHOSPHINOTHIOYL)MERCAPTO) METHANE ◇ BLADAN ◇ DIETHION ◇ EMBATHION ◇ ENT 24,105 ◇ ETHANOX ◇ ETHIOL ◇ ETHION (ACGIH,DOT) ◇ ETHODAN ◇ FMC-1240 ◇ FOSFONO 50 ◇ HYLEMOX ◇ ITOPAZ ◇ KWIT ◇ METHANEDITHIOL-S,S-DIESTER with O,O-DIETHYL ESTER PHOSPHORODITHIOIC ACID ◇ METHYLEEN-S,S′-BIS(O,O-DIETHYL-DITHIOFOSFAAT) (DUTCH) ◇ S,S′-METHYLEN-BIS(O,O-DIAETHYL-DITHIOPHOSPHAT) (GERMAN) ◇ METHYLENE-S,S′-BIS(O,O-DIAETHYL-DITHIOPHOSPHAT) (GERMAN) ◇ S,S′-METHYLENE O,O,O′,O′-TETRAETHYL PHOSPHORODITHIOATE ◇ NIAGARA 1240 ◇ NIALATE ◇ PHOSPHOTOX E ◇ RHODIACIDE ◇ RHODOCIDE ◇ RODOCID ◇ RP 8167 ◇ SOPRATHION ◇ O,O,O′,O′-TETRAAETHYL-BIS(DITHIOPHOSPHAT) (GERMAN) ◇ O,O,O′,O′-TETRAETHYL S,S′-METHYLENEBISPHOSPHORDITHIOATE ◇ O,O,O′,O′-TETRAETHYL-S,S′-METHYLENEBISPHOSPHORODITHIOATE ◇ TETRAETHYL S,S′-METHYLENE BIS(PHOSPHOROTHIOLOTHIONATE) ◇ O,O,O′,O′-TETRAETHYL S,S′-METHYLENE DI(PHOSPHORODITHIOATE) ◇ VEGFRU FOSMITE

USE IN FOOD:

Purpose: Insecticide.

Where Used: Animal feed, citrus pulp (dried), raisins, tea (dried).

Regulations: FDA - 21CFR 193.190. Insecticide residue tolerance of 10 ppm in dried tea, 4 ppm in raisins. 21CFR 561.230. Limitation of 10 ppm in dried citrus pulp when used for animal feed.

EPA: Farm Worker Field Reentry FEREAC 39,16888,74. EPA Genetic Toxicology Program. EPA Extremely Hazardous Substances List.

ACGIH TLV: TWA 0.4 mg/m^3 (skin) DOT Classification: Poison B; Label: Poison

SAFETY PROFILE: Poison by ingestion, skin contact, intraperitoneal and possibly other routes. Human systemic effects by ingestion: flaccid paralysis without anesthesia, motor activity changes, fever and inhibition of cholinesterase. When heated to decomposition it emits highly toxic fumes of SO_x and PO_x.

TOXICITY DATA and CODEN

orl-inf TDLo:15700 μg/kg:CNS,PNS,MET TXMDAX 63,71,67
orl-hmn TDLo:100 μg/kg:BIO TXAPA9 22,286,72
orl-rat LD50:13 mg/kg PHJOAV 185,361,60
skn-rat LD50:62 mg/kg TXAPA9 14,515,69

ENC000 CAS: 77-83-8
ETHYL METHYLPHENYLGLYCIDATE
mf: $C_{12}H_{14}O_3$ mw: 206.26

PROP: Colorless to yellowish liquid; strawberry-like odor. D: 1.086-1.112, refr index: 1.504-1.513, flash p: 273°F. Sol in fixed oils, propylene glycol; insol in glycerin.

SYNS: C-16 ALDEHYDE ◇ EMPG ◇ α-β-EPOXY-β-METHYLHYDROCINNAMIC ACID, ETHYL ESTER ◇ ETHYL α,β-EPOXY-β-METHYLHYDROCINNAMATE ◇ ETHYL 2,3-EPOXY-3-METHYL-3-PHENYLPROPIONATE ◇ ETHYL ESTER of 2,3-EPOXY-3-PHENYLBUTANOIC ACID ◇ FEMA No. 2444 ◇ FRAESEOL ◇ 3-METHYL-3-

PHENYLGLYCIDIC ACID ETHYL ESTER ◇ STRAWBERRY ALDEHYDE

USE IN FOOD:

Purpose: Flavoring agent.

Where Used: Beverages, candy, ice cream.

Regulations: FDA - 21CFR 182.60. Use at a level not in excess of the amount reasonably required to accomplish the intended effect.

SAFETY PROFILE: Mildly toxic by ingestion. Combustible liquid. When heated to decomposition it emits acrid smoke and irritating fumes.

TOXICITY DATA and CODEN

orl-rat LD50:5470 mg/kg FCTXAV 2,327,64

ENF200 CAS: 15707-23-0
2-ETHYL-3-METHYLPYRAZINE
mf: $C_7H_{10}N_2$ mw: 122.17

PROP: Colorless to sltly yellow liquid; strong, raw potato odor. D: 0.980-0.999 @ 20°, refr index: 1.502. Sol in water, organic solvents.

SYN: FEMA No. 3155

USE IN FOOD:

Purpose: Flavoring agent.

Where Used: Various.

Regulations: GRAS when used at a level not in excess of the amount reasonably required to accomplish the intended effect.

SAFETY PROFILE: When heated to decomposition it emits acrid smoke and irritating fumes.

ENI000 CAS: 35400-43-2
O-ETHYL-O-(4-(METHYLTHIO)PHENYL) S-PROPYL PHOSPHORODITHIOATE
mf: $C_{12}H_{19}O_2PS_3$ mw: 322.46

SYNS: BAY-NTN-9306 ◇ BOLSTAR ◇ O-ETHYL-O-(4-METHYLMERCAPTO)PHENYL)-S-N-PROPYLPHOSPHOROTHIONOTHIOLATE ◇ O-ETHYL-O-(4-(METHYLTHIO)PHENYL)PHOSPHORODITHIOIC ACID-S-PROPYL ESTER ◇ HELOTHION ◇ SULPROFOS (ACGIH)

USE IN FOOD:

Purpose: Insecticide.

Where Used: Animal feed, cottonseed hulls, cottonseed oil, soybean hulls.

Regulations: FDA - 21CFR 193.212. Insecticide residue tolerance of 1 ppm in cottonseed oil. 21CFR 561.233. Limitation of 10.0 ppm in cottonseed hulls, 1.0 ppm in soybean hulls when used for animal feed.

ACGIH TLV: TWA 1 mg/m^3

SAFETY PROFILE: Poison by ingestion. Moderately toxic by skin contact. When heated to decomposition it emits very toxic fumes of PO_x and SO_x.

TOXICITY DATA and CODEN

orl-rat LD50:65 mg/kg 85ARAE 1,205,77
skn-rbt LD50:820 mg/kg FMCHA2 -,C34,83

ENL850 CAS: 124-06-1
ETHYL MYRISTATE
mf: $C_{16}H_{23}O_2$ mw: 256.42

PROP: Colorless to pale yellow liquid; waxy odor. D: 0.857, refr index: 1.434, flash p: +212°F.

SYN: FEMA No. 2445

USE IN FOOD:

Purpose: Flavoring agent.

Where Used: Various.

Regulations: FDA - 21CFR 172.515. Use at a level not in excess of the amount reasonably required to accomplish the intended effect.

SAFETY PROFILE: Combustible liquid. When heated to decomposition it emits acrid smoke and irritating fumes.

ENW000 CAS: 123-29-5
ETHYL NONANOATE
mf: $C_{11}H_{22}O_2$ mw: 186.33

PROP: Colorless liquid; fruity, cognac odor. D: 0.863-0867, refr index: 1.420, flash p: 185°F. Misc with alc, propylene glycol; insol in water.

SYNS: ETHYL NONYLATE ◇ ETHYL PELARGONATE ◇ FEMA No. 2447 ◇ NONANOIC ACID, ETHYL ESTER ◇ WINE ETHER

USE IN FOOD:

Purpose: Flavoring agent.

Where Used: Beverages, beverages (alcoholic), candy, ice cream.

Regulations: FDA - 21CFR 172.515. Use at a level not in excess of the amount reasonably required to accomplish the intended effect.

SAFETY PROFILE: Mildly toxic by ingestion. A skin irritant. Combustible liquid. When heated to decomposition it emits acrid smoke and irritating fumes.

TOXICITY DATA and CODEN

skn-rbt 500 mg/24H MOD FCTXAV 16,747,78
orl-gpg LD50:24 g/kg FCTXAV 2,327,64

ENY000 CAS: 106-32-1
ETHYL OCTANOATE
mf: $C_{10}H_{20}O_2$ mw: 172.30

PROP: Colorless liquid; wine-brandy fruit odor. D: 0.865-0.869, refr index: 1.417, flash p: 185°F. Sol in fixed oils; sltly sol in propylene glycol; insol in glycerin, water @ 209°.

SYNS: ETHYL CAPRYLATE ◇ ETHYL OCTYLATE ◇ FEMA No. 2449 ◇ OCTANOIC ACID, ETHYL ESTER

USE IN FOOD:

Purpose: Flavoring agent.

Where Used: Various.

Regulations: FDA - 21CFR 172.515. Use at a level not in excess of the amount reasonably required to accomplish the intended effect.

SAFETY PROFILE: Mildly toxic by ingestion. A skin irritant. Combustible liquid. When heated to decomposition it emits acrid smoke and irritating fumes.

TOXICITY DATA and CODEN

skn-rbt 500 mg/24H MOD FCTXAV 14,763,76
orl-rat LD50:25960 mg/kg FCTXAV 14,763,76

EOB050
ETHYL OXYHYDRATE

PROP: Colorless liquid; sharp rum-like odor. Misc in alc, glycerin, propylene glycol.

SYNS: FEMA No. 2996 ◇ RUM ETHER ◇ SALICYLALDEHYDE

USE IN FOOD:

Purpose: Flavoring agent.

Where Used: Beverages, candy, ice cream.

Regulations: FDA - 21CFR 172.515. Use at a level not in excess of the amount reasonably required to accomplish the intended effect.

SAFETY PROFILE: When heated to decomposition it emits acrid smoke and irritating fumes.

EOH000 CAS: 101-97-3
ETHYL PHENYLACETATE
mf: $C_{10}H_{12}O_2$ mw: 164.22

PROP: Colorless liquid; sweet, honey odor. Bp: 227°, d: 1.033 @ 20°, refr index: 1.496-1.500, vap d: 5.67, flash p: +100 C. Sol in fixed oils; insol in glycerin, propylene glycol, water.

SYNS: BENZENEACETIC ACID, ETHYL ESTER (9CI) ◇ ETHYL BENZENEACETATE ◇ ETHYL PHENACETATE ◇ ETHYL-2-PHENYLETHANOATE ◇ ETHYL-α-TOLUATE ◇ FEMA No. 2452 ◇ PHENYLACETIC ACID, ETHYL ESTER ◇ α-TOLUIC ACID, ETHYL ESTER

USE IN FOOD:

Purpose: Flavoring agent.

Where Used: Various.

Regulations: FDA - 21CFR 172.515. Use at a level not in excess of the amount reasonably required to accomplish the intended effect.

SAFETY PROFILE: Moderately toxic by ingestion. Combustible liquid. When heated to decomposition it emits acrid smoke and irritating fumes.

TOXICITY DATA and CODEN

orl-rat LD50:3300 mg/kg FCTXAV 13,99,75

EOK600 CAS: 121-39-1
ETHYL PHENYLGLYCIDATE
mf: $C_{11}H_{12}O_3$ mw: 192.23

PROP: Colorless liquid; strong strawberry odor. D: 1.120, refr index: 1.516-1.521. Sol in alc, chloroform, ether; insol in water.

SYNS: ETHYL-α,β-EPOXYHYDROCINNAMATE ◇ ETHYL-α,β-EPOXY-α-PHENYLPROPIONATE ◇ ETHYL-3-PHENYLGLYCIDATE ◇ FEMA No. 2454

USE IN FOOD:

Purpose: Flavoring agent.

Where Used: Various.

Regulations: FDA - 21CFR 172.515. Use at a level not in excess of the amount reasonably required to accomplish the intended effect.

SAFETY PROFILE: Moderately toxic by ingestion. Mutagenic data. When heated to decomposition it emits acrid smoke and irritating fumes.

TOXICITY DATA and CODEN

mmo-sat 1 g/L MUREAV 89,269,81
msc-ham:ovr 103 mg/L MUREAV 138,1,84
orl-rat LD50:2300 mg/kg FCTXAV 13,101,75

EOR525
ETHYLPHTHALYL ETHYL GLYCOLATE

USE IN FOOD:

Purpose: Plasticizer.

Where Used: Packaging materials.

Regulations: FDA - 21CFR 181.27. Use in accordance with good manufacturing practice.

SAFETY PROFILE: When heated to decomposition it emits acrid smoke and irritating fumes.

EPB500 CAS: 105-37-3
ETHYL PROPIONATE

DOT: 1195

mf: $C_5H_{10}O_2$ mw: 102.15

PROP: Colorless liquid; fruity, rum odor. Bp: 210°F, mp: −72.6°, flash p: 54°F (CC), d: 0.886, refr index: 1.383, autoign temp: 824°F, vap press: 40 mm @ 27.2°, vap d: 3.52, lel: 1.9%, uel: 11%. Misc with alc, ether, propylene glycol; sol in fixed oils; 1 mL in 42 mL water @ 99°.

SYNS: FEMA No. 2456 ◇ PROPIONATE d'ETHYLE (FRENCH) ◇ PROPIONIC ACID, ETHYL ESTER ◇ PROPIONIC ETHER

TOXICITY DATA and CODEN

skn-rbt 500 mg/24H MOD FCTXAV 16,749,78
ipr-rat LD50: 1200 mg/kg FCTXAV 16,749,78
ipr-mus LD50: 1300 mg/kg 14CYAT 2,1879,63
orl-rbt LD50: 3500 mg/kg 14CYAT 2,1879,63

SAFETY PROFILE: Moderately toxic by ingestion and intraperitoneal routes. A skin irritant. Flammable Liquid. A very dangerous fire and explosion hazard when exposed to heat or flame; can react vigorously with oxidizing materials. To fight fire, use foam, CO_2, dry chemical. When heated to decomposition it emits acrid smoke and irritating fumes.

EQF000 CAS: 121-32-4
ETHYL VANILLIN

mf: $C_9H_{10}O_3$ mw: 166.19

PROP: Fine, crystalline needles; vanilla odor. Mp: 76.5°, flash p: +212°F. Sol in alc, chloroform, ether, propylene glycol; sltly sol in water.

SYNS: BOURBONAL ◇ ETHAVAN ◇ ETHOVAN ◇ 3-ETHOXY-4-HYDROXYBENZALDEHYDE ◇ ETHYLPROTAL ◇ FEMA No. 2464 ◇ 4-HYDROXY-3-ETHOXYBENZALDEHYDE ◇ PROTOCATECHUIC ALDEHYDE ETHYL ETHER ◇ QUANTROVANIL ◇ VANILLAL ◇ VANIROM

USE IN FOOD:

Purpose: Flavoring agent.

Where Used: Baked goods, beverages, ice cream, sauces.

Regulations: FDA - 21CFR 182.60. Use at a level not in excess of the amount reasonably required to accomplish the intended effect.

DOT Classification: Flammable Liquid; Label: Flammable Liquid

SAFETY PROFILE: Moderately toxic by ingestion, intraperitoneal, subcutaneous, and intravenous routes. A human skin irritant. Mutagenic data. Combustible liquid. When heated to decomposition it emits acrid smoke and irritating fumes.

TOXICITY DATA and CODEN

skn-hmn 10 mg/48H MLD FCTXAV 13,103,75
cyt-ham: fbr 250 mg/L FCTOD7 22,623,84
scu-rat LDLo: 1800 mg/kg JAPMA8 29,425,40

EQQ000 CAS: 8000-48-4
EUCALYPTUS OIL

PROP: From steam distillation of leaves of *Eucalyptus globulus* Labillardiere. Chief constituent is eucalyptol (FCTXAV 13,19,75). Colorless to pale-yellow liquid; spicy odor and taste. Composition: eucalyptol, aldehydes, d-pinene. Mp: −15.4° (approx). D: 0.905-0.925 @ 25°/25°.

SYNS: DINKUM OIL ◇ EUKALYPTUS OEL (GERMAN) ◇ OIL of EUCALYPTUS

USE IN FOOD:

Purpose: Flavoring agent.

Where Used: Bakery products, beverages (alcoholic), beverages (nonalcoholic), confections, ice cream.

Regulations: FDA - 21CFR 172.510. Use at a level not in excess of the amount reasonably required to accomplish the intended effect.

SAFETY PROFILE: A human poison by ingestion. Human systemic effects by ingestion: ciliary eye spasms, somnolence and respiratory depression. A skin irritant. When heated to decomposition it emits acrid smoke and irritating fumes.

TOXICITY DATA and CODEN

skn-rbt 500 mg/24H MOD FCTXAV 13,91,75
orl-chd TDLo: 218 mg/kg: EYE,CNS,PUL ADCHAK 28,475,53

orl-man LDLo:375 mg/kg ADCHAK 28,475,53
orl-rat LD50:2480 mg/kg FCTXAV 13,91,75

EQR500 CAS: 97-53-0
EUGENOL
mf: $C_{10}H_{12}O_2$ mw: 164.22

PROP: Colorless or yellowish liquid; pungent, clove odor. D: 1.064-1.070, refr index: 1.540, bp: 253.5°, flash p: 219°F. Sol in alc, chloroform, ether, volatile oils; very sltly sol in water.

SYNS: 4-ALLYLGUAIACOL ◇ 4-ALLYL-1-HYDROXY-2-METHOXYBENZENE ◇ 4-ALLYL-2-METHOXYPHENOL ◇ CARYOPHYLLIC ACID ◇ EUGENIC ACID ◇ Fa 100 ◇ FEMA No. 2467 ◇ 1-HYDROXY-2-METHOXY-4-ALLYLBENZENE ◇ 4-HYDROXY-3-METHOXYALLYLBENZENE ◇ 1-HYDROXY-2-METHOXY-4-PROP-2-ENYLBENZENE ◇ 2-METHOXY-4-ALLYLPHENOL ◇ 2-METHOXY-4-PROP-2-ENYLPHENOL ◇ 2-METHOXY-4-(2-PROPENYL)PHENOL ◇ 2-METOKSY-4-ALLILOFENOL (POLISH) ◇ NCI-C50453 ◇ SYNTHETIC EUGENOL

USE IN FOOD:

Purpose: Flavoring agent.

Where Used: Baked goods, beverages (nonalcoholic), candy, chewing gum, condiments, confections, gelatin desserts, ice cream, meat, puddings, spice oils.

Regulations: FDA - 21CFR 184.1257. GRAS when used at a level not in excess of the amount reasonably required to accomplish the intended effect.

IARC Cancer Review: Animal Limited Evidence IMEMDT 36,75,85. NTP Carcinogenesis Studies (feed); Equivocal Evidence: mouse NTPTR* NTP-TR-223,83; No Evidence: rat NTPTR* NTP-TR-223,83. EPA Genetic Toxicology Program.

SAFETY PROFILE: Moderately toxic by ingestion, intraperitoneal, and subcutaneous routes. An experimental carcinogen and tumorigen. Human mutagenic data. A human skin irritant. Combustible liquid. When heated to decomposition it emits acrid smoke and irritating fumes.

TOXICITY DATA and CODEN

skn-hmn 40 mg/48H MLD FCTXAV 13,545,75
mma-sat 50 μg/plate NTIS** AD-A116-715
oms-ham:ovr 400 mg/L CALEDQ 14,251,81
orl-mus TDLo:37080 mg/kg/2Y-I:ETA NTPTR* NTP-TR-223,82
orl-rat LD50:1930 mg/kg PSEBAA 73,148,50
ipr-rat LDLo:800 mg/kg RMSRA6 16,449,1896

EQS000 CAS: 93-28-7
EUGENOL ACETATE
mf: $C_{12}H_{14}O_3$ mw: 206.26

PROP: Solid or pale yellow liquid; mild clove odor. D: 1.87, mp: 29-30°, bp: 281.2°, flash p: +151°F. Insol in water; sol in alc and ether.

SYNS: ACETEUGENOL ◇ 1-ACETOXY-2-METHOXY-4-ALLYLBENZENE ◇ ACETYLEUGENOL ◇ 4-ALLYL-2-METHOXYPHENOL ACETATE ◇ 4-ALLYL-2-METHOXYPHENYL ACETATE ◇ 1,3,4-EUGENOL ACETATE ◇ EUGENYL ACETATE ◇ FEMA No. 2469

USE IN FOOD:

Purpose: Flavoring agent.

Where Used: Various.

Regulations: FDA - 21CFR 172.515. Use at a level not in excess of the amount reasonably required to accomplish the intended effect.

SAFETY PROFILE: Moderately toxic by ingestion. A skin irritant. Combustible liquid. When heated to decomposition it emits acrid smoke and irritating fumes.

TOXICITY DATA and CODEN

skn-rbt 500 mg/24H MOD FCTXAV 12,807,74
orl-rat LD50:1670 mg/kg FCTXAV 2,327,64

F

FAE000 CAS: 38444-45-9
F D & C BLUE No. 1
mf: $C_{37}H_{36}N_2O_9S_3 \cdot 2Na$ mw: 794.91

PROP: Dark purple to bronze powder. Sol in water, ether, conc sulfuric acid.

SYNS: ACID SKY BLUE A ◇ AIZEN FOOD BLUE NO. 2 ◇ 1206 BLUE ◇ BRILLIANT BLUE FCD NO. 1 ◇ BRILLIANT BLUE FCF ◇ CANACERT BRILLIANT BLUE FCF ◇ C.I. 42090 ◇ C.I. ACID BLUE 9, DISODIUM SALT ◇ C.I. FOOD BLUE 2 ◇ COGILOR BLUE 512.12 ◇ COSMETIC BLUE LAKE ◇ D & C BLUE NO. 4 ◇ DISPERSED BLUE 12195 ◇ DOLKWAL BRILLIANT BLUE ◇ EDICOL BLUE CL 2 ◇ ERIOGLAUCINE G ◇ FDC BLUE NO. 1 ◇ FENAZO BLUE XI ◇ FOOD BLUE 2 ◇ FOOD BLUE DYE NO. 1 ◇ HEXACOL BRILLIANT BLUE A ◇ INTRACID PURE BLUE L ◇ MERANTINE BLUE EG ◇ USACERT BLUE NO. 1

USE IN FOOD:

Purpose: Color additive.

Where Used: Baked goods, candy, confections.

Regulations: FDA - 21CFR 74.101. Use in accordance with good manufacturing practice.

IARC Cancer Review: Animal Sufficient Evidence IMEMDT 16,171,78.

SAFETY PROFILE: An experimental neoplastigen. Mutagenic data. When heated to decomposition it emits very toxic fumes of NO_x, Na_2O and SO_x.

TOXICITY DATA and CODEN

cyt-ham:lng 4400 mg/L GMCRDC 27,95,81
scu-rat TDLo:5.5 g/kg/97W-I:NEO ZEKBAI 64,287,61

FAG000 CAS: 2353-45-9
F D & C GREEN NO. 3
mf: $C_{37}H_{36}N_2O_{10}S_3 \cdot 2Na$ mw: 810.91

PROP: Red to brown-violet powder. Sol in water, conc sulfuric acid.

SYNS: AIZEN FOOD GREEN NO. 3 ◇ C.I. 42053 ◇ C.I. FOOD GREEN 3 ◇ FAST GREEN FCF ◇ 1724 GREEN ◇ SOLID GREEN FCF

USE IN FOOD:

Purpose: Color additive.

Where Used: Beverages, cereals, desserts, soft drinks.

Regulations: FDA - 21CFR 74.203. Use in accordance with good manufacturing practice.

IARC Cancer Review: Animal Sufficient Evidence IMEMDT 16,187,78. EPA Genetic Toxicology Program.

SAFETY PROFILE: An experimental neoplastigen. Mutagenic data. When heated to decomposition it emits very toxic fumes of NO_x and SO_x.

TOXICITY DATA and CODEN

mma-sat 10 mg/plate FCTOD7 22,623,84
cyt-ham:fbr 4 g/L FCTOD7 22,623,84
scu-rat TDLo:5925 mg/kg/48W-I:NEO JNCIAM 24,769,60

FAG050 CAS: 25956-17-6
F D & C RED No. 40
mf: $C_{18}H_{14}N_2O_8S_2Na_2$ mw: 496.42

PROP: Red powder. Sol in water; sltly sol in abs alc.

SYNS: ALLURA RED AC ◇ C.I. 16035

USE IN FOOD:

Purpose: Color additive.

Where Used: Beverages, candy, cereals, confections, desserts, ice cream.

Regulations: FDA - 21CFR 74.340. Use in accordance with good manufacturing practice.

SAFETY PROFILE: Experimental reproductive effects. When heated to decomposition it emits very toxic fumes of NO_x and SO_x.

TOXICITY DATA and CODEN

orl-rat TDLo:38500 mg/kg(14D male/14D pre):REP TXCYAC 28,207,83
orl-rat TDLo:83500 mg/kg(2W male/2W pre-2W post):REP TXCYAC 28,207,83

FAG150 CAS: 2783-94-0
F D & C YELLOW No. 6
mf: $C_{16}H_{10}N_2O_7S_2Na_2$ mw: 452.36

PROP: Orange powder. Sol in water, conc sulfuric acid; sltly sol in abs alc.

SYNS: ACID YELLOW TRA ◇ AIZEN FOOD YELLOW No. 5 ◇ CANACERT SUNSET YELLOW FCF ◇ C.I. 15985 ◇ GELBORANGE-S (GERMAN) ◇ SUNSET YELLOW FCF

USE IN FOOD:

Purpose: Color additive.

Where Used: Bakery goods, beverages, butter, candy, cereals, cheese, confections, desserts, ice cream.

Regulations: FDA - 21CFR 74.706, 81.1. Use in accordance with good manufacturing practice.

IARC Cancer Review: Animal Inadequate Evidence IMEMDT 8,257,75.

SAFETY PROFILE: Moderately toxic by intraperitoneal route. When heated to decomposition it emits very toxic fumes of NO_x and SO_x.

TOXICITY DATA and CODEN

cyt-ham:lng 2 g/l GMCRDC 27,95,81
ipr-rat LD50:4600 mg/kg:CNS,GIT FCTXAV 5,747,67

FAG759 CAS: 52-85-7
FAMPHUR
mf: $C_{10}H_{16}NO_5PS_2$ mw: 325.36

PROP: Crystalline powder. Mp: 55°. Very sol in chloroform and carbon tetrachloride; sltly sol in water.

SYNS: AMERICAN CYANAMID-38023 ◇ AMERICAN CYANAMID CL-38,023 ◇ BO-ANA ◇ CL-38023 ◇ CYFLEE ◇ O-(4-((DIMETHYLAMINO)SULFONYL)PHENYL) O,O-DIMETHYL PHOSPHOROTHIOATE ◇ O,O-DIMETHYL-O-(p-(N,N-DIMETHYLSULFAMOYL)PHENYL)PHOSPHOROTHIOATE ◇ DOVIP ◇ ENT 25,644 ◇ FAMFOS ◇ FAMOPHOS ◇ FAMOPHOS WARBEX ◇ FAMPHOS ◇ FANFOS ◇ RCRA WASTE NUMBER P097 ◇ WARBEX

USE IN FOOD:

Purpose: Animal feed drug.

Where Used: Animal feed.

Regulations: FDA - 21CFR 558.254.

SAFETY PROFILE: Poison by ingestion and intramuscular routes. Moderately toxic by skin contact. A cholinesterase inhibitor. When heated to decomposition it emits toxic fumes of NO_x and SO_x.

TOXICITY DATA and CODEN

orl-rat LD50:35 mg/kg TXAPA9 21,315,72
skn-rbt LD50:1460 mg/kg WRPCA2 9,119,70

FAG875 CAS: 4602-84-0
FARNESOL
mf: $C_{15}H_{26}O$ mw: 222.41

PROP: trans-Farnesol: Light yellow liquid; mild oily odor. Bp: 111°, n (25/D) 1.4872. Commercial farnesol: Bp: 110-113°, d: (20/4) 0.8871, n (20/D) 1.4870., refr index: 1.487-1.492. Insol in water @ 263°.

SYNS: FARNESYL ALCOHOL ◇ FEMA No. 2478 ◇ 3,7,11-TRIMETHYL-2,6,10-DODECATRIEN-1-OL

USE IN FOOD:

Purpose: Flavoring agent.

Where Used: Various.

Regulations: FDA - 21CFR 172.515. Use at a level not in excess of the amount reasonably required to accomplish the intended effect.

SAFETY PROFILE: Moderately toxic by intraperitoneal route. Mildly toxic by ingestion. Mutagenic data. When heated to decomposition it emits acrid smoke and irritating fumes.

TOXICITY DATA and CODEN

dni-oin:ovr 100 μmol/L ABCHA6 43,1285,79
orl-rat LD50:6000 mg/kg THERAP 27,893,72
ipr-mus LD50:443 mg/kg THERAP 27,893,72

FAG900
FATTY ACIDS

PROP: Consists of capric, caprylic, lauric, myristic, oleic, palmitic, and stearic acids manufactured from fats and oils derived from edible sources.

USE IN FOOD:

Purpose: Component in the manufacture of other food-grade additives, defoaming agent, lubricant.

Where Used: Various.

Regulations: FDA - 21CFR 172.860. Use in accordance with good manufacturing practice.

SAFETY PROFILE: When heated to decomposition it emits acrid smoke and irritating fumes.

FAK000 CAS: 22224-92-6
FENAMIPHOS
mf: $C_{13}H_{22}NO_3PS$ mw: 303.39

SYNS: O-AETHYL-O-(3-METHYL-4-METHYLTHIOPHENYL)-ISOPROPYLAMIDO-PHOSPHORSAEURE ESTER (GERMAN) ◇ BAY 68138 ◇ ENT 27,572 ◇ ETHYL-3-METHYL-4-(METHYLTHIO)PHENYL(1-METHYLETHYL)PHOSPHORAMIDATE ◇ ETHYL-4-(METHYLTHIO)-m-TOLYL ISOPROPYL PHOSPHOR AMIDATE ◇ ISOPROPYLAMINO-O-ETHYL-(4-METHYLMERCAPTO-3-METHYLPHENYL)PHOSPHATE

◇ 1-(METHYLETHYL)-ETHYL 3-METHYL-4-(METHYLTHIO) PHENYL PHOSPHORAMIDATE ◇ NEMACUR ◇ NSC 195106 ◇ PHANAMIPHOS

USE IN FOOD:

Purpose: Nematocide.

Where Used: Animal feed, apple pomace (dried), apples, citrus molasses, citrus oil, citrus pulp (dried), grape pomace, grapes, pineapple bran, pineapples, raisin waste, raisins.

Regulations: FDA - 21CFR 193.463. Nematocide residue tolerence of 25.0 ppm in citrus oil, 0.3 ppm in raisins. 21CFR 561.232. Limitation of 5.0 ppm in dried apple pomace, 2.5 ppm in citrus molasses, 2.5 ppm in dried citrus pulp, 1.0 ppm in grape pomace, 10.0 ppm in pineapple bran, and 3.0 ppm in raisin waste when used for animal feed.

EPA Extremely Hazardous Substances List.

ACGIH TLV: TWA 0.1 mg/m^3 (skin)

SAFETY PROFILE: Poison by ingestion, inhalation, and skin contact. When heated to decomposition it emits very toxic fumes of NO_x, PO_x and SO_x.

TOXICITY DATA and CODEN

orl-rat LD50: 8 mg/kg BESAAT 15,116,69
ihl-rat LC50: 91 mg/m^3/4H 85DPAN -,-,71/76
skn-rat LD50: 500 mg/kg 85DPAN -,0,71/76

FAK100 CAS: 60168-88-9
FENARIMOL

mf: $C_{17}H_{12}Cl_2N_2O$ mw: 331.21

PROP: White, odorless crystals. Mp: 117-119°. Practically insol in water; sol in most organic solvents.

SYNS: BLOC ◇ (2-CHLOROPHENYL)-α-(4-CHLOROPHENYL)-5-PYRIMIDINEMETHANOL ◇ α-(2-CHLOROPHENYL)-α-(4-CHLOROPHENYL)-5-PYRIMIDINEMETHANOL ◇ EL 222 ◇ RIMIDIN ◇ RUBIGAN

USE IN FOOD:

Purpose: Fungicide.

Where Used: Animal feed, apple pomace (dried), apple pomace (wet), apples.

Regulations: FDA - 21CFR 561.438. Limitation of 0.2 ppm in wet and dry apple pomace when used for animal feed.

SAFETY PROFILE: Moderately toxic by ingestion. Mutagenic data. When heated to decomposition it emits toxic fumes of Cl^- and NO_x.

TOXICITY DATA and CODEN

sln-asn 6 mg/L MUREAV 79,169,80
cyt-mus-orl 450 mg/kg DBANAD 36,1351,83
orl-rat LD50: 2500 mg/kg FMCHA2 -,C209,83

FAL100 CAS: 43210-67-9
FENBENDAZOLE

mf: $C_{15}H_{13}N_3O_2S$ mw: 299.37

SYNS: FENBENDAZOL ◇ HOE 881 ◇ PANACUR ◇ (5-(PHENYLTHIO)-2-BENZIMIDAZOLECARBAMIC ACID, METHYL ESTER

USE IN FOOD:

Purpose: Animal drug, animal feed drug.

Where Used: Animal feed, beef, pork.

Regulations: FDA - 21CFR 556.275. Tolerance of 0.8 ppm (marker residue) in liver of cattle. Tolerance of 5 ppm in swine muscle, 15 ppm in swine liver, 20 ppm in swine kidney, and 20 ppm in swine skin and fat. 21CFR 558.258. USDA CES Ranking: B-3 (1987).

SAFETY PROFILE: Human mutagenic data. When heated to decomposition it emits toxic fumes of SO_x and NO_x.

TOXICITY DATA and CODEN

oth:hmn:lym 100 mg/L ENMUDM 2,67,80

FAP000 CAS: 8006-84-6
FENNEL OIL

PROP: From steam distillation of *Foeniculum vulgare* Miller (Fam. *Umbelliferae*) (FCTXAV 12,807,74). Colorless to pale yellow liquid; odor and taste of fennel.

SYNS: BITTER FENNEL OIL ◇ FENCHEL OEL (GERMAN) ◇ OIL of FENNEL

USE IN FOOD:

Purpose: Flavoring agent.

Where Used: Various.

Regulations: FDA - 21CFR 182.10, 182.20. GRAS when used at a level not in excess of the amount reasonably required to accomplish the intended effect.

SAFETY PROFILE: Moderately toxic by ingestion. Mutagenic data. A severe skin irritant. When heated to decomposition it emits acrid smoke and irritating fumes.

TOXICITY DATA and CODEN

skn-mus 100 % SEV FCTXAV 12,879,74
skn-rbt 500 mg/24H FCTXAV 14,309,76

mma-sat 2500 μg/plate JAFCAU 30,563,82
orl-rat LD50:3120 mg/kg PHARAT 14,435,59

FAQ500 CAS: 69381-94-8
FENPROSTALENE

mf: $C_{23}H_{30}O_6$ mw: 402.49

SYNS: BOVILENE ◇ 7-[3,5-DIHYDROXY-2-(3-HYDROXY-4-PHENOXY-1-BUTENYL)CYCLOPENTYL]-4,5-HEPTADIENOIC ACID METHYL ESTER ◇ SYNCHROCEPT B

USE IN FOOD:

Purpose: Animal drug.

Where Used: Beef.

Regulations: FDA - 21CFR 556.277. Limitation of 10 ppb in muscle, 20 ppb in liver, 30 ppb in kidney, 40 ppb in fat, 100 ppb in the injection site in cattle.

SAFETY PROFILE: When heated to decomposition it emits acrid smoke and irritating fumes.

FAQ999 CAS: 55-38-9
FENTHION

mf: $C_{10}H_{15}O_3PS_2$ mw: 278.34

SYNS: BAY 29493 ◇ BAYCID ◇ BAYER 9007 ◇ BAYTEX ◇ O,O-DIMETHYL-O-4-(METHYLMERCAPTO)-3-METHYLPHENYL PHOSPHOROTHIOATE ◇ O,O-DIMETHYL-p-4-(METHYLMERCAPTO)-3-METHYLPHENYL THIOPHOSPHATE ◇ O,O-DIMETHYL-O-(3-METHYL-4-METHYLMERCAPTOPHENYL)PHOSPHOROTHIOATE ◇ O,O-DIMETHYL-O-(3-METHYL-4-METHYLTHIO-FENYL)-MONOTHIOFOSFAAT (DUTCH) ◇ O,O-DIMETHYL-O-(3-METHYL-4-METHYLTHIOPHENYL)-MONOTHIOPHOSPHAT (GERMAN) ◇ O,O-DIMETHYL-O-3-METHYL-4-METHYLTHIOPHENYL PHOSPHOROTHIOATE ◇ O,O-DIMETHYL-O-(3-METHYL-4-METHYLTHIO-PHENYL)-THIONOPHOSPHAT (GERMAN) ◇ O,O-DIMETHYL-O-(4-METHYLTHIO-3-METHYLPHENYL) PHOSPHOROTHIOATE ◇ O,O-DIMETHYL-O-(4-(METHYLTHIO)-m-TOLYL) PHOSPHOROTHIOATE ◇ O,O-DIMETIL-O-(3-METIL-4-METILTIO-FENIL)-MONOTIOFOSFATO (ITALIAN) ◇ DMTP ◇ ENT 25,540 ◇ ENTEX ◇ LEBAYCID ◇ MERCAPTOPHOS ◇ 4-METHYLMERCAPTO-3-METHYLPHENYL DIMETHYL THIOPHOSPHATE ◇ MPP ◇ NCI-C08651 ◇ OMS 2 ◇ PHOSPHOROTHIOIC ACID-O,O-DIMETHYL-O-(3-METHYL-4-METHYLTHIOPHENYLE) (FRENCH) ◇ QUELETOX ◇ S 1752 ◇ SPOTTON ◇ TALODEX ◇ THIOPHOSPHATE de O,O-DIMETHYLE et de O-(3-METHYL-4-METHYLTHIOPHENYLE) (FRENCH) ◇ TIGUVON

USE IN FOOD:

Purpose: Insecticide.

Where Used: Fish, meat, sauces.

Regulations: USDA CES Ranking: C-3 (1985).

NCI Carcinogenesis Bioassay Completed; Results Negative: Rat (NCITR* NCI-CG-TR-103,79);Indefinite: Mouse (NCITR* NCI-CG-TR-103,79). EPA Genetic Toxicology Program.

ACGIH TLV: TWA 0.2 mg/m^3 (skin)

SAFETY PROFILE: A human poison by an unspecified route. Poison experimentally by ingestion, skin contact, intraperitoneal, intravenous and intramuscular routes. Moderately toxic by inhalation. An experimental tumorigen and teratogen. Experimental reproductive effects. Mutagenic data. When heated to decomposition it emits very toxic fumes of PO_x and SO_x.

TOXICITY DATA and CODEN

mma-sat 333 μg/plate ENMUDM 8(Suppl 7),1,86
sce-ham:lng 40 mg/L ENMUDM 4,621,82
orl-mus TDLo:1050 mg/kg (MGN):REP TXAPA9 26,29,73
ipr-mus TDLo:40 mg/kg (11D preg):TER TXAPA9 24,324,73
orl-mus TDLo:1730 mg/kg/103W-C:ETA NCITR* NCI-CG-TR-103,79
unr-hmn LD50:50 mg/kg DTLVS* 4,191,80
orl-rat LD50:180 mg/kg KSKZAN 16(2),59,78
ihl-rat LCLo:1 g/m^3/2H 85GYAZ -,27,71
skn-rat LD50:330 mg/kg TXAPA9 2,88,60

FAR100 CAS: 51630-58-1
FENVALERATE

mf: $C_{25}H_{22}ClNO_3$ mw: 419.93

PROP: Clear, yellow, viscous liquid at 23°. D: 1.17, n (20/D) 1.5533. Solubility at 20° (g/L): acetone, >450; chloroform, >450; methanol, >450; hexane, 77. Insol in water. Decomp gradually between 150-300°.

SYNS: BELMARK ◇ α-CYANO-3-PHENOXYBENZYL-2-(4-CHLOROPHENYL)ISOVALERATE PYDRIN ◇ α-CYANO-3-PHENOXYBENZYL-2-(4-CHLOROPHENYL)-3-METHYLBUTYRATE ◇ CYANO(3-PHENOXYPHENYL)METHYL 4-CHLORO-α-(1-METHYLETHYL)BENZENEACETATE ◇ ECTRIN ◇ PHENVALERATE ◇ PYDRIN ◇ S 5602 ◇ SANMARTON ◇ SD 43775 ◇ SUMICIDIN ◇ SUMIFLY ◇ SUMIPOWER ◇ WL 43775

USE IN FOOD:

Purpose: Insecticide.

Where Used: Animal feed, apple pomace (dried), apples, soybean hulls, soybeans, sugarcane bagasse, sunflower hulls, sunflower seeds, tomato pomace (dried), tomatoes.

Regulations: FDA - 21CFR 193.97. Insecticide residue tolerence of 0.05 ppm on all food items. 21CFR 561.97. Limitation of 20 ppm in dried apple pomace, 10 ppm in dried tomato pomace, 1.0 ppm in soybean hulls, 20.0 ppm in sugarcane bagasse, 2.0 ppm in sunflower hulls when used for animal feed.

Cyanide and its compounds are on the Community Right-To-Know List.

SAFETY PROFILE: Poison by ingestion, intravenous and intracerebral routes. Moderately toxic by skin contact. Experimental reproductive effects. Highly toxic to fish and bees. Corrosive, causes eye damage. A skin irritant. When heated to decomposition it emits toxic fumes of Cl^-, NO_x and CN^-.

TOXICITY DATA and CODEN

orl-dog TDLo:1138 mg/kg (26W pre):REP FAATDF 4,577,84
orl-rat LD50:451 mg/kg FMCHA2 -,C104,83
ivn-rat LDLo:50 mg/kg ARTODN 45,325,80
skn-rbt LD50:2500 mg/kg FMCHA2 -,C104,83

FAS700 CAS: 1333-00-2
FERRIC AMMONIUM CITRATE

PROP: A complex salt of undermined structure. Transparent green scales, granules, powder, or crystals; ammoniacal odor, mild iron-metallic taste. Sol in water; insol in alc. Deliquescent.

SYNS: FERRIC AMMONIUM CITRATE, GREEN ◇ IRON(III) AMMONIUM CITRATE

USE IN FOOD:

Purpose: Anticaking agent, dietary supplement, nutrient.

Where Used: Various.

Regulations: FDA - 21CFR 172.430. Limitation of 25 ppm in finished salt. 21CFR 184.1296. GRAS when used in accordance with good manufacturing practice.

SAFETY PROFILE: When heated to decomposition it emits acrid smoke and irritating fumes.

FAU000 CAS: 7705-08-0
FERRIC CHLORIDE
DOT: 1773/2582
mf: Cl_3Fe mw: 162.20

PROP: Black-brown solid. Mp: 292°, bp: 319.0°, d: 2.90 @ 25°, vap press: 1 mm @ 194.0°.

SYNS: CHLORURE PERRIQUE (FRENCH) ◇ FERRIC CHLORIDE, SOLID, ANHYDROUS (DOT) ◇ FERRIC CHLORIDE, SOLID (DOT) ◇ FLORES MARTIS ◇ IRON CHLORIDE ◇ IRON(III) CHLORIDE ◇ IRON CHLORIDE, SOLID (DOT) ◇ IRON SESQUICHLORIDE, SOLID (DOT) ◇ IRON TRICHLORIDE ◇ PERCHLORURE de FER (FRENCH)

USE IN FOOD:

Purpose: Flavoring agent.

Where Used: Various.

Regulations: FDA - 21CFR 184.1298 GRAS when used in accordance with good manufacturing practice.

EPA Genetic Toxicology Program.

ACGIH TLV: TWA 1 mg(Fe)/m^3 DOT Classification: ORM-B, Label: None, anhydrous; Corrosive Material; Label: Corrosive

SAFETY PROFILE: Poison by intravenous route. Moderately toxic by ingestion. Experimental reproductive effects. Corrosive. Probably an eye, skin, and mucous membrane irritant. Reacts with water to produce toxic and corrosive fumes. Catalyzes potentially explosive polymerization of ethylene oxide; chlorine + monomers (e.g., styrene). Forms shock-sensitive explosive mixtures with some metals (e.g., potassium; sodium). Violent reaction with allyl chloride. When heated to decomposition it emits highly toxic fumes of HCl.

TOXICITY DATA and CODEN

itt-rat TDLo:12976 μg/kg (1D male):REP JRPFA4 7,21,64
ivg-rat TDLo:29 mg/kg (1D pre):REP CCPTAY 4,91,71
orl-rat LD50:1872 mg/kg GISAAA 48(9),71,83

FAW100 CAS: 2338-05-8
FERRIC CITRATE
mf: $C_6H_5FeO_7$ mw: 244.95

PROP: White or red crystals; odorless with slt metallic taste. Sol in water.

SYN: IRON(III) CITRATE

USE IN FOOD:

Purpose: Nutrient supplement.

Where Used: Various.

Regulations: FDA - 21CFR 184.1298 GRAS when used in accordance with good manufacturing practice.

SAFETY PROFILE: When heated to decomposition it emits acrid smoke and irritating fumes.

FAZ500 CAS: 10045-86-0
FERRIC PHOSPHATE
mf: $FePO_4 \cdot xH_2O$ mw: 150.82

PROP: Yellowish to buff powder; odorless. Sol in mineral acids; insol in water, acetic acid.

SYNS: FERRIC ORTHOPHOSPHATE ◇ IRON PHOSPHATE

USE IN FOOD:

Purpose: Dietary supplement, nutrient supplement.

Where Used: Egg substitutes (frozen), pasta products, rice products.

Regulations: FDA - 21CFR 182.5301, 182.8301, 184.1301. GRAS when used in accordance with good manufacturing practice.

SAFETY PROFILE: When heated to decomposition it emits very toxic fumes of PO_x.

FAZ525 CAS: 10058-44-3
FERRIC PYROPHOSPHATE
mf: $Fe_4(P_2O_7)_3 \cdot xH_2O$ mw: 745.22

PROP: Tan to yellow powder. Sol in mineral acids; insol in water.

SYN: IRON PYROPHOSPHATE

USE IN FOOD:

Purpose: Dietary supplement, nutrient supplement.

Where Used: Various.

Regulations: FDA - 21CFR 182.5304, 182.8304, 184.1304. GRAS when used in accordance with good manufacturing practice.

SAFETY PROFILE: When heated to decomposition it emits very toxic fumes of PO_x.

FBA000 CAS: 10028-22-5
FERRIC SULFATE
DOT: 9121
mf: $Fe_2O_{12}S_3$ mw: 399.88
PROP: Yellow solid.

SYNS: DIIRON TRISULFATE ◇ IRON PERSULFATE ◇ IRON SESQUISULFATE ◇ IRON SULFATE (2:3) ◇ IRON(III) SULFATE ◇ IRON TERSULFATE ◇ SULFURIC ACID, IRON (3^+) SALT (3:2)

USE IN FOOD:

Purpose: Flavoring agent.

Where Used: Various.

Regulations: FDA - 21CFR 184.1307. GRAS when used in accordance with good manufacturing practice.

ACGIH TLV: TWA 1 mg/m^3; STEL 2 mg/m^3 DOT Classification: ORM-E

SAFETY PROFILE: When heated to decomposition it emits toxic fumes of SO_x and Fe^-.

FBC100 CAS: 1336-80-7
FERROCHOLINATE
mf: $C_6H_{10}FeO_{10} \cdot C_5H_{14}NO$ mw: 402.21

PROP: Greenish-brown, reddish-brown or brown amorphous solid with glistening surface upon fracture. Sol in water, acids, and alkalies.

SYNS: CHELAFER ◇ CHEL-IRON ◇ FERRIC CHOLINE CITRATE ◇ FERROLIP ◇ IRON CHOLINE CITRATE COMPLEX

USE IN FOOD:

Purpose: Nutrient supplement.

Where Used: Various.

Regulations: FDA - 21CFR 172.370. Use at a level not in excess of the amount reasonably required to accomplish the intended effect.

SAFETY PROFILE: Poison by intravenous and intraperitoneal routes. Mildly toxic by ingestion. When heated to decomposition it emits toxic fumes of NO_x.

TOXICITY DATA and CODEN

orl-mus LD50:5500 mg/kg AJMSA9 241,296,61
ipr-mus LD50:151 mg/kg AJMSA9 241,296,61
ivn-mus LD50:210 mg/kg AJMSA9 241,296,61

FBH050 CAS: 14536-17-5
FERROUS ASCORBATE

PROP: Blue-violet solid.

SYN: IRON(II) ASCORBATE

USE IN FOOD:

Purpose: Nutrient supplement.

Where Used: Various.

Regulations: FDA - 21CFR 184.1307a GRAS when used in accordance with good manufacturing practice.

SAFETY PROFILE: A nuisance dust.

FBH100 CAS: 563-71-3
FERROUS CARBONATE
mf: $CFeO_3$ mw: 115.86

PROP: White solid; odorless.

SYN: IRON(II) CARBONATE

USE IN FOOD:

Purpose: Nutrient supplement.

Where Used: Various.

Regulations: FDA - 21CFR 184.1307b GRAS when used in accordance with good manufacturing practice.

SAFETY PROFILE: A nuisance dust.

FBJ075 CAS: 23383-11-1
FERROUS CITRATE
mf: $C_6H_6FeO_7$ mw: 245.96

PROP: White crystals or sltly colored powder.

SYN: IRON(II) CITRATE

USE IN FOOD:

Purpose: Nutrient supplement.

Where Used: Various.

Regulations: FDA - 21CFR 184.1307c GRAS when used in accordance with good manufacturing practice.

SAFETY PROFILE: When heated to decomposition it emits acrid smoke and irritating fumes.

FBJ100 CAS: 141-01-5
FERROUS FUMARATE
mf: $C_4H_2O_4$•Fe mw: 169.91

PROP: Reddish-orange to reddish-brown granular powder; odorless, almost tasteless. D: 2.435. Solubility at 25° in water: 0.14 g/100 mL; in alcohol <0.01 g/100 mL. Solubility in acid is limited by liberation of fumaric acid.

SYNS: CPIRON ◇ ERCO-FER ◇ ERCOFERRO ◇ FEOSTAT ◇ FEROTON ◇ FERROFUME ◇ FERRONAT ◇ FERRONE ◇ FERROTEMP ◇ FERRUM ◇ FERSAMAL ◇ FIRON ◇ FUMAFER ◇ FUMAR-F ◇ FUMIRON ◇ GALFER ◇ HEMOTON ◇ IRCON ◇ IRON FUMARATE ◇ METERFER ◇ METERFOLIC ◇ ONE-IRON ◇ PALAFER ◇ TOLERON ◇ TOLFERAIN ◇ TOLIFER

USE IN FOOD:

Purpose: Dietary supplement, nutrient supplement.

Where Used: Cereals, waffles (frozen).

Regulations: FDA - 21CFR 172.350. Limited as a source of iron in foods for special dietary purposes. 21CFR 184.1307d GRAS when used in accordance with good manufacturing practice.

ACGIH TLV: TWA 1 mg/(Fe)/m^3

SAFETY PROFILE: Poison by intraperitoneal route. Moderately toxic by ingestion and subcutaneous routes. When heated to decomposition it emits acrid smoke and irritating fumes.

TOXICITY DATA and CODEN

orl-rat LD50:3850 mg/kg NIIRDN 6,683,82
ipr-rat LD50:185 mg/kg NIIRDN 6,683,82
scu-rat LD50:500 mg/kg NIIRDN 6,683,82

FBK000 CAS: 6047-12-7
FERROUS GLUCONATE
mf: $C_{12}H_{22}O_{14}$•Fe•H_2O mw: 482.17

PROP: Yellowish-gray or pale greenish-yellow, fine powder or granules with slt odor of burned sugar. Sol in water and glycerin; insol in alc.

SYNS: FERGON ◇ FERGON PREPARATIONS ◇ FERLUCON ◇ FERRONICUM ◇ GLUCO-FERRUM ◇ IROMIN ◇ IRON GLUCONATE ◇ IROX (GADOR) ◇ NIONATE ◇ RAY-GLUCIRON

USE IN FOOD:

Purpose: Color additive, dietary supplement, nutrient supplement.

Where Used: Olives (ripe), vitamin pills.

Regulations: FDA - 21CFR 73.160, 182.5308, 182.8308, 184.1308. GRAS when used in accordance with good manufacturing practice.

ACGIH TLV: TWA 1 mg(Fe)/m^3

SAFETY PROFILE: Poison by intraperitoneal and intravenous routes. Moderately toxic by ingestion. An experimental tumorigen and teratogen. Human systemic effects by ingestion: hypermotility, diarrhea, nausea, and vomiting. When heated to decomposition it emits acrid smoke and irritating fumes.

TOXICITY DATA and CODEN

scu-mus TDLo:2600 mg/kg/13W-I:ETA,TER JNCIAM 24,109,60
orl-chd TDLo:162 mg/kg:GIT JAMAAP 218,1179,71
orl-rat LD50:2237 mg/kg NTIS** UR-3490-168

FBO000 CAS: 7782-63-0
FERROUS SULFATE HEPTAHYDRATE
mf: $O_4S \cdot Fe \cdot 7H_2O$ mw: 278.05

PROP: Pale blue green monoclinic crystals or granules; odorless with a salt taste. D: 2.99-3.08. Sol in water; insol in alc.

SYNS: COPPERAS ◇ FEOSOL ◇ FER-IN-SOL ◇ FERO-GRADUMET ◇ FERROUS SULFATE (FCC) ◇ FESOFOR ◇ FESOTYME ◇ GREEN VITROL ◇ HAEMOFORT ◇ IRONATE ◇ IRON(II) SULFATE (1:1), HEPTAHYDRATE ◇ IRON VITROL ◇ IROSUL ◇ MOL-IRON ◇ PRESFERSUL ◇ SULFERROUS

USE IN FOOD:

Purpose: Clarifying agent, dietary supplement, nutrient supplement, processing aid, stabilizer.

Where Used: Baking mixes, cereals, infant foods, pasta products, wine.

Regulations: FDA - 21CFR 182.5315, 182, 8315, 184.1315. GRAS when used in accordance with good manufacturing practice. BATF - 27CFR 240.1051. Limitation of 3 ounces/1000 gallons of wine.

ACGIH TLV: TWA 1 mg(Fe)/m^3

SAFETY PROFILE: Poison by intravenous, intraperitoneal, and subcutaneous routes. Moderately toxic by ingestion. Mutagenic data. When heated to decomposition it emits toxic fumes of SO_x.

TOXICITY DATA and CODEN

mmo-esc 30 μmol/L CIWYAO 49,144,50
orl-rat LDLo: 1389 mg/kg EQSSDX 1,1,75

FBS000 CAS: 9001-33-6
FICIN

PROP: A proteolytic enzyme in the crude latex of the fig tree *Ficus* (JPETAB 71,20,41). White powder. Very sol in water.

SYNS: DEBRICIN ◇ FICUS PROTEASE ◇ FICUS PROTEINASE ◇ HIGUEROXYL DELABARRE ◇ TL 367

USE IN FOOD:

Purpose: Chillproofing of beer, enzyme, meat tenderizing, preparation of precooked cereals, processing aid, tenderizing agent, tissue softening agent.

Where Used: Beer, cereals (precooked), meat (raw cuts), poultry, wine.

Regulations: USDA - 9CFR 318.7, 381.147. Solutions consisting of water and approved proteolytic enzymes applied or injected into raw meat cuts shall not result in a gain of more than 3 percent above the weight of the untreated product. BATF - 27CFR 240.1051. GRAS when used in accordance with good manufacturing practice.

SAFETY PROFILE: Poison by inhalation and intravenous routes. Mildly toxic by ingestion. When heated to decomposition it emits toxic fumes.

TOXICITY DATA and CODEN

orl-rat LD50: 10 g/kg MEIEDD 10,585,83
ihl-mus LCLo: 290 mg/m^3/10M NDRC** NDCrc-132,Sept,42
ihl-rbt LCLo: 290 mg/m^3/10M NDRC** NDCrc-132,Sept,42

FBU850
FIR NEEDLE OIL, CANADIAN TYPE

PROP: Found in the needles and twigs of *Abies balsamea* L. Mill (Fam. *Pinaceae*) (FCTXAV 13,449,75). Colorless to faintly yellow liquid; pleasant odor. Sol in fixed oils, mineral oil; sltly sol in propylene glycol; insol in glycerin.

SYN: BALSAM FIR OIL

USE IN FOOD:

Purpose: Flavoring agent.

Where Used: Various.

Regulations: FDA - 21CFR 172.210. Use at a level not in excess of the amount reasonably required to accomplish the intended effect.

SAFETY PROFILE: When heated to decomposition it emits acrid smoke and irritating fumes.

FBV000 CAS: 8021-29-2
FIR NEEDLE OIL, SIBERIAN

PROP: Found in the needles and twigs of *Abies sibirica* Ledeb. (Fam. *Pinaceae*) (FCTXAV 13,449,75). Colorless to faintly yellow liquid. Sol in fixed oils, mineral oil; insol in glycerin, propylene glycol.

SYN: PINE NEEDLE OIL

USE IN FOOD:

Purpose: Flavoring agent.

Where Used: Various.

Regulations: FDA - 21CFR 172.210. Use at a level not in excess of the amount reasonably required to accomplish the intended effect.

SAFETY PROFILE: Mildly toxic by ingestion. A human and experimental skin irritant. When heated to decomposition it emits acrid smoke and irritating fumes.

TOXICITY DATA and CODEN

skn-hmn 12500 μg/48H FCTXAV 13,450,75
skn-rbt 500 mg/24H MOD FCTXAV 13,450,75
orl-rat LD50: 10200 mg/kg FCTXAV 13,450,75

FDA885 CAS: 69806-50-4
FLUAZIFOP-BUTYL

mf: $C_{19}H_{20}F_3NO_4$ mw: 383.38

SYN: (±)-2-[4-[[5-(TRIFLUOROMETHYL)-2-PYRIDINYL] OXY]PHENOXY]PROPANOIC ACID BUTYL ESTER

USE IN FOOD:

Purpose: Herbicide.

Where Used: Cottonseed oil, soybean oil.

Regulations: FDA - 21CFR 193.466. Limitation of 0.2 ppm in cottonseed oil, 2.0 ppm in soybean oil.

SAFETY PROFILE: When heated to decomposition emits toxic fumes of F^-, NO_x.

FLU000 CAS: 1649-18-9
4′-FLUORO-4-(4-(2-PYRIDYL)-1-PIPERAZINYL)BUTYROPHENONE

mf: $C_{19}H_{22}FN_3O$ mw: 327.44

SYNS: AZAPERONE (USDA) ◇ AZEPERONE ◇ EUCALMYL ◇ FLUOPERIDOL ◇ 1-(3-(4-FLUOROBENZOYL)PROPYL)-4-(2-PYRIDYL)PIPERAZINE ◇ 1-(4-FLUOROPHENYL)-4-(4-(2-PYRIDINYL)-1-PIPERAZINYL)-1-BUTANONE ◇ R 1929 ◇ STRESNIL ◇ SUICALM

USE IN FOOD:

Purpose: Animal drug.

Where Used: Meat.

Regulations: USDA CES Ranking: B-4 (1986).

SAFETY PROFILE: Poison by ingestion, intravenous, intraperitoneal and subcutaneous routes. When heated to decomposition it emits very toxic fumes of F^- and NO_x.

TOXICITY DATA and CODEN

orl-rat LD50: 245 mg/kg ARZNAD 24,1798,74
scu-rat LD50: 450 mg/kg ARZNAD 24,1798,74

FMT000 CAS: 59-30-3
FOLIC ACID

mf: $C_{19}H_{19}N_7O_6$ mw: 441.45

PROP: A member of the vitamin B complex. Orange-yellow needles or platelets; odorless. Sol in dilute alkali hydroxide and carbonate solns; sltly sol in water; insol in lipid solvents, acetone, alc, chloroform, ether.

SYNS: l-N-(p-(((-2-AMINO-4-HYDROXY-6-PTERIDINYL)METHYL)AMINO)BENZOYL)GLUTAMIC ACID ◇ FOLACIN ◇ FOLATE ◇ FOLCYSTEINE ◇ NSC 3073 ◇ PTEGLU ◇ PTEROYLGLUTAMIC ACID ◇ PTEROYL-l-GLUTAMIC ACID ◇ PTEROYLMONOGLUTAMIC ACID ◇ PTEROYL-l-MONOGLUTAMIC ACID ◇ USAF CB-13 ◇ VITAMIN Bc ◇ VITAMIN M

USE IN FOOD:

Purpose: Dietary supplement, nutrient.

Where Used: Green vegetables, liver, nuts.

Regulations: FDA - 21CFR 172.345. Limitation of daily intake 0.4 mg for foods unlabeled with reference to age, 0.1 mg for infants, 0.3 mg for children under 4 years old, 0.4 mg for adults or children over 4 years old, and 0.8 mg for pregnant or lactating women.

SAFETY PROFILE: Poison by intraperitoneal and intravenous routes. An experimental teratogen. Experimental reproductive effects. Mutagenic data. When heated to decomposition it emits toxic fumes of NO_x.

TOXICITY DATA and CODEN

dns-rat-ipr 150 mg/kg CNREA8 29,136,69
dns-mus-ipr 250 mg/kg CNREA8 29,136,69
par-rat TDLo: 150 mg/kg (10D preg): TER FEPRA7 23,292,64
ivn-rat LD50: 500 mg/kg NIIRDN 6,869,82
ipr-mus LD50: 85 mg/kg EXPEAM 41,72,85

FMU100
FOOD STARCH, MODIFIED

PROP: White powders; tasteless and odorless. Insol in water, alc, ether, chloroform.

USE IN FOOD:

Purpose: Binder, colloidal stabilizer, thickener.

Where Used: Desserts, gravies, pie fillings, sauces.

Regulations: FDA - 21CFR 172.892, 178.3520. Use at a level not in excess of the amount reasonably required to accomplish the intended effect.

SAFETY PROFILE: When heated to decomposition it emits acrid smoke and irritating fumes.

FMV000 CAS: 50-00-0
FORMALDEHYDE
DOT: 1198/2209
mf: CH_2O mw: 30.03

PROP: Clear, water-white, very sltly acid gas or liquid; pungent odor. Pure formaldehyde is not available commercially because of its tendency to polymerize. It is sold as aqueous solutions containing from 37 percent to 50% formaldehyde by weight and varying amounts of methanol. Some alcoholic solns are used industrially and the physical properties and hazards may be greatly influenced by the solvent. Lel: 7.0%, uel: 73.0%, autoign temp: 806°F, d: 1.0, bp: −3°F, flash p: (37% methanol-free): 185°F, flash p: (15% methanol-free): 122°F.

SYNS: ALDEHYDE FORMIQUE (FRENCH) ◇ ALDEIDE FORMICA (ITALIAN) ◇ BFV ◇ FA ◇ FANNOFORM ◇ FORMALDEHYD (CZECH, POLISH) ◇ FORMALDEHYDE, solution (DOT) ◇ FORMALIN ◇ FORMALIN 40 ◇ FORMALIN (DOT) ◇ FORMALINA (ITALIAN) ◇ FORMALINE (GERMAN) ◇ FORMALIN-LOESUNGEN (GERMAN) ◇ FORMALITH ◇ FORMIC ALDEHYDE ◇ FORMOL ◇ FYDE ◇ HOCH ◇ IVALON ◇ KARSAN ◇ LYSOFORM ◇ METHANAL ◇ METHYL ALDEHYDE ◇ METHYLENE GLYCOL ◇ METHYLENE OXIDE ◇ MORBOCID ◇ NCI-C02799 ◇ OPLOSSINGEN (DUTCH) ◇ OXOMETHANE ◇ OXYMETHYLENE ◇ PARAFORM ◇ POLYOXYMETHYLENE GLYCOLS ◇ RCRA WASTE NUMBER U122 ◇ SUPERLYSOFORM

USE IN FOOD:

Purpose: Animal glue adjuvant, preservative.

Where Used: Cottonseed, packaging materials.

Regulations: FDA - 21CFR 173.340. Limitation of 1.0 percent of the dimethylpolysiloxane content. 21CFR 178.3120. Use as a preservative only at a level not in excess of the amount reasonably required to accomplish the intended effect.

IARC Cancer Review: Human Inadequate Evidence IMEMDT 29,345,82; Animal Sufficient Evidence IMEMDT 29,345,82. EPA Genetic Toxicology Program.

OSHA PEL: TWA 1 ppm; STEL 2 ppm ACGIH TLV: TWA 1 ppm; Suspected Carcinogen DOT Classification: Combustible Liquid; Label: None; ORM-A; Label: None; Flammable or Combustible Liquid; Label: Flammable Liquid NIOSH REL: (Formaldehyde) Limit to lowest feasible level.

SAFETY PROFILE: Human poison by ingestion. Experimental poison by ingestion, skin contact, inhalation, intravenous, intraperitoneal, and subcutaneous routes. Moderately toxic by an unspecified route. A suspected human carcinogen. An experimental carcinogen, tumorigen, and teratogen. Human systemic effects by inhalation: lacrimation, olfactory changes, aggression and pulmonary changes. Experimental reproductive effects. Human mutagenic data. A human skin and eye irritant. A severe experimental eye and skin irritant. If swallowed it causes violent vomiting and diarrhea which can lead to collapse. Frequent or prolonged exposure can cause hypersensitivity leading to contact dermatitis, possibly of an eczematoid nature. An air concentration of 20 ppm is quickly irritating to eyes.

Combustible liquid when exposed to heat or flame; can react vigorously with oxidizers. A moderate explosion hazard when exposed to heat or flame. The gas is a more dangerous fire hazard than the vapor. Should formaldehyde be involved in a fire, irritating gaseous formaldehyde may be evolved. When aqueous formaldehyde solutions are heated above their flash points, a potential for an explosion hazard exists. High formaldehyde concentration or methanol content lowers the flash point. Reacts with NO_x at about 180°; the reaction becomes explosive. Also reacts violently with perchloric acid + aniline; performic acid; nitromethane; magnesium carbonate; H_2O_2. Moderately dangerous; because of irritating vapor which may exist in toxic concentrations locally if storage tank is ruptured. To fight fire stop flow of gas (for pure form); alcohol foam for 37% methanol-free form. When heated to decomposition it emits acrid smoke and fumes.

TOXICITY DATA and CODEN

skn-hmn 150 μg/3D-I MLD 85DKA8 -,127,77
eye-hmn 4 ppm/5M IAPWAR 4,79,61
eye-hmn 1 ppm/6M nse MLD AIHAAP 44,463,83
skn-rbt 500 mg/24H SEV 28ZPAK -,40,72
eye-rbt 10 mg SEV TXAPA9 55,501,80
mma-sat 5 μL/plate BIMADU 6,129,85
dnd-hmn:fbr 100 μmol/L ENMUDM 7,267,85
orl-rat TDLo:200 mg/kg (1D male):REP TJADAB 26(3),14A,82
ihl-rat TCLo:1 mg/m^3/24H (1-22D preg):TER HYSAAV 34(5),266,69
ipr-mus TDLo:240 mg/kg (7-14D preg):TER TJADAB 30(1),34A,84

ihl-rat TCLo: 14300 ppb/6H/2Y-I: CAR CNREA8 43,4382,83
scu-rat TDLo: 1170 mg/kg/65W-I: ETA GANNA2 45,451,54
ihl-rat TC: 15 ppm/6H/78W-I: CAR CNREA8 49,3398,80
ihl-hmn TCLo: 17 mg/m^3/30M: EYE,PUL JAMAAP 165,1908,57
ihl-man TCLo: 300 $\mu g/m^3$: NOSE,CNS GTPZAB 12(7),20,68
orl-wmn LDLo: 108 mg/kg 29ZWAE -,328,68
orl-rat LD50: 800 mg/kg JIHTAB 23,259,41
ihl-rat LC50: 590 mg/m^3 GISAAA 41(6),103,76

FNA000 CAS: 64-18-6
FORMIC ACID
DOT: 1779
mf: CH_2O_2 mw: 46.03

PROP: Colorless, fuming liquid; pungent, penetrating odor. Bp: 100.8°, fp: 8.2°, flash p: 156°F (OC), d: 1.2267 @ 15°/4°, 1.220 @ 20°/4°, autoign temp: 1114°F, vap press: 40 mm @ 24.0°, vap d: 1.59, flash p:(90% soln): 122°F, autoign temp (90% soln): 813°F, lel (90% soln) = 18%, uel (90% soln) = 57%. Misc in water, alc, glycerin, ether.

SYNS: ACIDE FORMIQUE (FRENCH) ◇ ACIDO FORMICO (ITALIAN) ◇ AMEISENSAEURE (GERMAN) ◇ AMINIC ACID ◇ FORMYLIC ACID ◇ HYDROGEN CARBOXYLIC ACID ◇ KWAS METANIOWY (POLISH) ◇ METHANOIC ACID ◇ MIERENZUUR (DUTCH) ◇ RCRA WASTE NUMBER U123

USE IN FOOD:

Purpose: Flavoring adjuvant, paper manufacturing aid.

Where Used: Food packaging.

Regulations: FDA - 21CFR 186.1316. GRAS as an indirect additive when used in accordance with good manufacturing practice.

OSHA PEL: TWA 5 ppm ACGIH TLV: TWA 5 ppm DOT Classification: Corrosive Material; Label: Corrosive; Corrosive Material; Label: Corrosive, solution

SAFETY PROFILE: Poison by intravenous route. Moderately toxic by ingestion and intraperitoneal routes. Mildly toxic by inhalation. Mutagenic data. Corrosive. A skin and severe eye irritant. Flammable liquid when exposed to heat or flame; can react vigorously with oxidizing materials. Explosive reaction with furfuryl alcohol; H_2O_2; $Tl(NO_3)_3 \cdot 3H_2O$; nitromethane; P_2O_5. To fight fire, use CO_2, dry chemical, alcohol foam. When heated to decomposition it emits acrid smoke and irritating fumes.

TOXICITY DATA and CODEN

skn-rbt 610 mg open MLD UCDS** 5/8/68
eye-rbt 122 mg SEV UCDS** 5/8/68
mmo-esc 70 ppm/3H AMNTA4 85,119,51
cyt-nml: oth 100 mmol/L CAANAT 56,712,72
orl-rat LD50: 1100 mg/kg GTPZAB 23(12),49,79
ihl-rat LC50: 15 g/m^3/15M GTPZAB 23(12),49,79

FOI000 CAS: 6804-07-5
2-FORMYLQUINOXALINE-1,4-DIOXIDE CARBOMETHOXYHYDRAZONE
mf: $C_{11}H_{10}N_4O_4$ mw: 262.25

SYNS: CARBADOX (USDA) ◇ FORTIGRO ◇ GS 6244 ◇ MECADOX ◇ (2-QUINOXALINYLMETHYLENE)-HYDRAZINECARBOXYLIC ACID METHYL ESTER-N,N'-DIOXIDE

USE IN FOOD:

Purpose: Animal drug, animal feed drug.

Where Used: Animal feed, pork.

Regulations: FDA - 21CFR 556.100. Tolerance of zero in swine. 21CFR 558.115. USDA CES Ranking: A-3 (1987).

SAFETY PROFILE: Human mutagenic data. When heated to decomposition it emits toxic fumes of NO_x.

TOXICITY DATA and CODEN

dns-hmn: oth 50 mg/L JTEHD6 10,143,82
sce-ham: lng 50 mg/L MUREAV 139,199,84

FOU000 CAS: 110-17-8
FUMARIC ACID
DOT: 9126
mf: $C_4H_4O_4$ mw: 116.08

PROP: White crystals; odorless. Mp: 287°, d: 1.635 @ 20°/4°. Bp: 290°. Sol in water, ether; very sltly sol in chloroform.

SYNS: ALLOMALEIC ACID ◇ BOLETIC ACID ◇ trans-BUTENEDIOIC ACID ◇ (E)-BUTENEDIOIC ACID ◇ trans-1,2-ETHYLENEDICARBOXYLIC ACID ◇ (E)1,2-ETHYLENEDICARBOXYLIC ACID ◇ KYSELINA FUMAROVA (CZECH) ◇ LICHENIC ACID ◇ NSC-2752 ◇ U-1149 ◇ USAF EK-P-583

USE IN FOOD:

Purpose: Acidifier, curing accelerator, flavoring agent.

Where Used: Beverage mixes (dry), candy, desserts, pie fillings, poultry, wine.

Regulations: FDA - 21CFR 172.350. Use at a level not in excess of the amount reasonably required to accomplish the intended effect. USDA - 9CFR 318.7, 9CFR 381.147. Limitation of 0.065 percent of the weight of the meat. BATF - 27CFR 240.1051. Limitation of 25 pounds/1000 gallons of wine. The fumaric acid content of the finished wine shall not exceed 3.0 g/L.

DOT Classification: ORM-E; Label: None

SAFETY PROFILE: Poison by intraperitoneal route. Mildly toxic by ingestion and skin contact. A skin and eye irritant. Combustible when exposed to heat or flame; can react vigorously with oxidizing materials. When heated to decomposition it emits acrid smoke and irritating fumes.

TOXICITY DATA and CODEN

skn-rbt 500 mg/24H MLD 28ZPAK -,51,72
eye-rbt 100 mg/24H MOD 28ZPAK -,51,72
orl-rat LD50:10700 mg/kg TXAPA9 42,417,77

FPB875 CAS: 35554-44-0
FUNGAFLOR

mf: $C_{14}H_{14}Cl_2N_2O$ mw: 297.20

PROP: Solidified oil. Sltly sol in organic solvents; poorly sol in water.

SYNS: (±)-1-(β-(ALLYLOXY)-2,4-DICHLOROPHENETHYL)IMIDAZOLE ◇ ENILOCONAZOL (SP) ◇ 1-(2-(2,4-DICHLORPHENYL)-2-PROPENYLOXY)AETHYL)-1H-IMIDAZOLE ◇ 1-(2-(2,4-DICHLOROPHENYL)-2-(2-PROPENYLOXY) ETHYL)-1H-IMIDAZOLE ◇ IMAVEROL ◇ IMAZALIL ◇ R 23979

USE IN FOOD:

Purpose: Fungicide.

Where Used: Animal feed, citrus oil, citrus pulp (dried).

Regulations: FDA - 21CFR 193.467. Fungicide residue tolerence of 25.0 ppm in citrus oil. 21CFR 561.429. Limitation of 25.0 in dried citrus pulp when used for animal feed.

SAFETY PROFILE: Poison by ingestion and intraperitoneal routes. Experimental reproductive effects. A skin and eye irritant. When heated to decomposition it emits toxic fumes of Cl^- and NO_x.

TOXICITY DATA and CODEN

skn-rbt 640 mg/kg/24H MLD ARZNAD 31,309,81
eye-rbt 49 mg MOD ARZNAD 31,309,81
orl-rat TDLo:2240 mg/kg (16-22D preg/21D post):REP ARZNAD 31,309,81
orl-rat TDLo:560 mg/kg (16-22D preg/21D post):REP ARZNAD 31,309,81
orl-rat LD50:227 mg/kg ARZNAD 31,309,81
ipr-rat LD50:155 mg/kg ARZNAD 31,309,81

FPE000 CAS: 1563-66-2
FURADAN

DOT: 2757

mf: $C_{12}H_{15}NO_3$ mw: 221.28

PROP: White, crystalline solid; odorless. Mp: 105-152°, d: 1.180 @ 20°/20°, vap press: 2 × 10^{-5} mm @ 33°. Sltly water-sol.

SYNS: BAY 70143 ◇ CARBOFURAN (ACGIH, DOT, USDA) ◇ CURATERR ◇ D 1221 ◇ 2,3-DIHYDRO-2,2-DIMETHYLBENZOFURANYL-7-N-METHYLCARBAMATE ◇ 2,3-DIHYDRO-2,2-DIMETHYL-7-BENZOFURANYL METHYLCARBAMATE ◇ 2,2-DIMETHYL-7-COUMARANYL N-METHYLCARBAMATE ◇ 2,2-DIMETHYL-2,3-DIHYDROBENZOFURAN-7-YL ESTER, METHYLCARBAMIC ACID ◇ 2,2-DIMETHYL-2,3-DIHYDRO-7-BENZOFURANYL-N-METHYLCARBAMATE ◇ ENT 27,164 ◇ FMC 10242 ◇ FURODAN ◇ METHYL CARBAMIC ACID 2,3-DIHYDRO-2,2-DIMETHYL-7-BENZOFURANYL ESTER ◇ NIA 10242 ◇ NIAGRA 10242 ◇ YALTOX

USE IN FOOD:

Purpose: Insecticide.

Where Used: Animal feed, grape pomace, peanut soapstock fatty acids, raisins, raisin waste, soybean soapstock fatty acids, sunflower seed hulls, sunflower seed meal, sunflower seed soapstock.

Regulations: FDA - 21CFR 561.67. Limitation of 2.0 ppm in grape pomace, 24.0 ppm in peanut soapstock fatty acids, 6.0 ppm in rasin waste and soybean soapstock fatty acids, 1.4 ppm in sunflower seed hulls and meal, 3.0 ppm in sunflower seed soapstock when used for animal feed. 21CFR 193.43. Insecticide residue tolerence of 2 ppm in raisins. USDA CES Ranking: C-3 (1986).

EPA Extremely Hazardous Substances List. EPA Genetic Toxicology Program.

ACGIH TLV: TWA 0.1 mg/m^3 DOT Classification: Poison B; Label: Poison; Poison B; Label: Poison, liquid

SAFETY PROFILE: Poison by inhalation, ingestion, skin contact, intravenous, and possibly other routes. An experimental teratogen. Experimental reproductive effects. Human mutagenic data. When heated to decomposition it emits toxic fumes of NO_x.

TOXICITY DATA and CODEN

mmo-sat 10 mg/plate MUREAV 116,185,83
msc-ham: lng 1 mmol/L ENVRAL 29,48,82
orl-mus TDLo: 210 μg/kg (1-12D preg): REP JEPTDQ 2(2),357,78
orl-mus TDLo: 10500 μg/kg (1-21D preg): TER JEPTDQ 4(5-6),53,80
orl-rat LD50: 5300 μg/kg ARSIM* 20,16,66
ihl-rat LC50: 85 mg/m^3 JOCMA7 12,16,70
skn-rat LD50: 120 mg/kg WRPCA2 9,119,70

FPG000 CAS: 98-01-1
2-FURALDEHYDE

DOT: 1199
mf: $C_5H_4O_2$ mw: 96.09

PROP: Colorless-yellowish liquid; almond-like odor. Bp: 161.7° @ 764 mm, lel: 2.1%, uel: 19.3%, flash p: 140°F (CC), d: 1.154-1.158, refr index: 1.522-1.528, autoign temp: 600°F, vap d: 3.31. Sol in water; misc with alc.

SYNS: ARTIFICIAL ANT OIL ◇ FEMA No. 2489 ◇ FURAL ◇ FURALE ◇ 2-FURANALDEHYDE ◇ 2-FURANCARBONAL ◇ 2-FURANCARBOXALDEHYDE ◇ FURFURAL (ACGIH,DOT,FCC) ◇ 2-FURFURAL ◇ FURFURALDEHYDE ◇ FURFURALE (ITALIAN) ◇ FURFUROL ◇ FURFUROLE ◇ 2-FURIL-METANALE (ITALIAN) ◇ FUROLE ◇ α-FUROLE ◇ 2-FURYL-METHANAL ◇ NCI-C56177 ◇ PYROMUCIC ALDEHYDE ◇ RCRA WASTE NUMBER U125

USE IN FOOD:

Purpose: Flavoring agent.

Where Used: Various.

Regulations: GRAS when used at a level not in excess of the amount reasonably required to accomplish the intended effect.

EPA Genetic Toxicology Program.

OSHA PEL: TWA 2 ppm (skin) ACGIH TLV: TWA 2 ppm (skin) DOT Classification: Combustible Liquid; Label: None; Flammable or Combustible Liquid; Label: Flammable Liquid

SAFETY PROFILE: Poison by ingestion, intraperitoneal, subcutaneous, intravenous and intramuscular routes. Moderately toxic by inhalation and skin contact. Human mutagenic data. A skin and eye irritant. Mutagenic data. The liquid is dangerous to the eyes. The vapor is irritating to mucous membranes and is a central nervous system poison. However, its low volatility reduces its toxicity effect. Ingestion of furfural has produced cirrhosis of the liver in rats. In industry there is a tendency to minimize the danger of acute effects resulting from exposure to it. This is particularly true because of its low volatility.

Combustible liquid when exposed to heat or flame; can react with oxidizing materials. Moderate explosion hazard when exposed to heat or flame or by chemical reaction. An exothermic polymerization of almost explosive violence can occur upon contact with strong mineral acids or alkalies. Keep away from heat and open flames. Mixture with sodium hydrogen carbonate ignites spontaneously. To fight fire, use alcohol foam, CO_2, dry chemical. When heated to decomposition it emits acrid smoke and irritating fumes.

TOXICITY DATA and CODEN

skn-rbt 500 mg/24H MOD 28ZPAK -,139,72
eye-rbt 20 mg/24H MOD 28ZPAK -,139,72
mma-sat 7 uL/plate MUREAV 58,205,78
cyt-ham: ovr 2500 μmol/L CALEDQ 13,89,81
ihl-hmn TCLo: 310 μg/m^3 GISAAA 26(6),3,61
orl-rat LD50: 65 mg/kg BCTKAG 13,371,80
ihl-rat LCLo: 153 ppm/4H 28ZPAK -,139,72

FPQ000 CAS: 9000-21-9
FURCELLERAN GUM

PROP: Vegetable gum from *Furcellaria fastigiata* (Fam. *Rodophyceae*) available as an odorless white powder. Sol in warm water.

SYN: BURTONITE 44

USE IN FOOD:

Purpose: Emulsifier, stabilizer, thickener.

Where Used: Flans, gelled meat products, jams, jellies, milk puddings.

Regulations: FDA - 21CFR 172.655. Use at a level not in excess of the amount reasonably required to accomplish the intended effect.

SAFETY PROFILE: Moderately toxic by ingestion. When heated to decomposition it emits acrid smoke and fumes.

TOXICITY DATA and CODEN

orl-rat LD50: 5000 mg/kg FDRLI* 124,-,76

FQT000 CAS: 8013-75-0
FUSEL OIL
DOT: 1201

PROP: Colorless to pale yellow liquid; odorless. D: 0.807-0.813, refr index: 1.405-1.410. Composition of grain fusel oil is methanol, ethanol, acetaldehyde, and other alcohols (ARGEAR 33,49,69).

SYNS: FEMA No. 2497 ◇ FUSELOEL (GERMAN) ◇ FUSEL OIL, REFINED (FCC) ◇ HUILE de FUSEL (FRENCH)

USE IN FOOD:

Purpose: Flavoring agent.

Where Used: Various.

Regulations: FDA - 21CFR 172.515. Use at a level not in excess of the amount reasonably required to accomplish the intended effect.

DOT Classification: Combustible Liquid; Label: None; Flammable Liquid; Label: Flammable Liquid

SAFETY PROFILE: Mutagenic data. Suspected of containing carcinogens. Flammable liquid when exposed to heat or flame; can react vigorously with oxidizing materials. When heated to decomposition it emits acrid smoke and fumes.

TOXICITY DATA and CODEN

mmo-esc 7000 ppm ARGEAR 33,49,69
dlt-mus-scu 12500 mg/kg/5D-C ARGEAR 33,49,69

G

GAV050
α-GALACTOSIDASE

PROP: Derived from *Mortierella vinaceae* var. *raffinoseutilizer*.

SYN: ATCC No. 20034

USE IN FOOD:

Purpose: Production aid.

Where Used: Sugar (beet).

Regulations: FDA - 21CFR 173.145. Use at a level not in excess of the amount reasonably required to accomplish the intended effect.

SAFETY PROFILE: When heated to decomposition it emits acrid smoke and irritating fumes.

GBU800
GARLIC OIL

PROP: From steam distillation of *Allium sativum* L. (Fam. *Liliaceae*). Clear to yellow liquid; strong odor and taste of garlic. Sol in fixed oils, mineral oil; insol in glycerin, alc, propylene glycol.

USE IN FOOD:

Purpose: Flavoring agent.

Where Used: Meat, sauces, vegetables.

Regulations: FDA - 21CFR 184.1317. GRAS when used at a level not in excess of the amount reasonably required to accomplish the intended effect.

SAFETY PROFILE: An eye irritant. When heated to decomposition it emits acrid smoke and irritating fumes.

GCS000 CAS: 1405-41-0
GENTAMYCIN SULFATE

SYNS: GARAMYCIN ◇ GENOPTIC ◇ GENOPTIC S.O.P. ◇ GM SULFATE ◇ NSC-82261 ◇ SCH 9724

USE IN FOOD:

Purpose: Animal drug.

Where Used: Pork, turkey.

Regulations: FDA - 21CFR 556.300. Limitation of 0.1 ppm in turkey, 0.1 ppm in swine muscle. USDA CES Ranking: B-2 (1986).

SAFETY PROFILE: Poison by intravenous, intraperitoneal, and intramuscular routes. Moderately toxic by subcutaneous route. Mutagenic data. When heated to decomposition it emits very toxic fumes of SO_x.

TOXICITY DATA and CODEN

dnd-esc 5 mg/L MUREAV 89,95,81
spm-rat-unr 9600 μg/kg/8D JOURAA 112,348,74
ipr-rat LD50:630 mg/kg JJANAX 26,221,73
ivn-rat LD50:96 mg/kg JZKEDZ 8,219,82

GCY000 CAS: 105-86-2
GERANIOL FORMATE

mf: $C_{11}H_{18}O_2$ mw: 182.29

PROP: Colorless to pale yellow liquid; rose odor. D: 0.906-0.920, refr index: 1.457-1.466, flash p: 205°F. Sol in alc, fixed oils; insol in glycerin, propylene glycol, water @ 216°.

SYNS: trans-3,7-DIMETHYL-2,6-OCTADIEN-1-OL FORMATE ◇ trans-3,7-DIMETHYL-2,6-OCTADIEN-1-YL FORMATE ◇ 3,7-DIMETHYL-2,6-OCTADIENYL ESTER FORMIC ACID (E) ◇ FEMA No. 2514 ◇ FORMIC ACID, GERANIOL ESTER ◇ GERANYL FORMATE (FCC)

USE IN FOOD:

Purpose: Flavoring agent.

Where Used: Various.

Regulations: FDA - 21CFR 172.515. Use at a level not in excess of the amount reasonably required to accomplish the intended effect.

SAFETY PROFILE: A human skin irritant and an experimental eye irritant. Combustible liquid. When heated to decomposition it emits acrid smoke and irritating fumes.

TOXICITY DATA and CODEN

skn-hmn 10 mg/48H MLD FCTXAV 12,893,74
eye-rbt 100 mg/24H MLD FCTXAV 12,893,74

GDA000 CAS: 8000-46-2
GERANIUM OIL ALGERIAN TYPE

PROP: From steam distillation of leaves from *Pelargonium graveolens* l'Her (Fam. *Geraniaceae*). Containsol geraniol and geranyl tiglate (FCTXAV 14,659,76). Yellow liquid; odor of rose and geraniol. D: 0.886-0.898, refr index: 1.454-1.472 @ 20°. Sol in fixed oils, mineral oil; insol in glycerin.

SYNS: GERANIUM OIL ◇ OIL of GERANIUM ◇ OIL of PELARGONIUM ◇ OIL of ROSE GERANIUM ◇ OIL ROSE GERANIUM ALGERIAN ◇ PELARGONIUM OIL ◇ ROSE GERANIUM OIL ALGERIAN

USE IN FOOD:

Purpose: Flavoring agent.

Where Used: Bakery products, beverages (nonalcoholic), chewing gum, confections, gelatin desserts, ice cream, puddings.

Regulations: FDA - 21CFR 182.10, 182.20. GRAS when used at a level not in excess of the amount reasonably required to accomplish the intended effect.

SAFETY PROFILE: A skin irritant. When heated to decomposition it emits acrid smoke and irritating fumes.

TOXICITY DATA and CODEN

skn-rbt 500 mg/24H MLD FCTXAV 14,781,76

GDE800
GERANYL BENZOATE

mf: $C_{17}H_{22}O_2$ mw: 258.36

PROP: Sltly yellow liquid; floral odor resembling ylang ylang oil. D: 0.978-0.984, refr index: 1.513-1.518, flash p: +212°F. Misc in alc, chloroform; insolin water @ 305°.

SYNS: 3,7-DIMETHYL-2,6-OCTADIEN-1-YL BENZOATE ◇ FEMA No. 2511

USE IN FOOD:

Purpose: Flavoring agent.

Where Used: Various.

Regulations: FDA - 21CFR 172.515. Use at a level not in excess of the amount reasonably required to accomplish the intended effect.

SAFETY PROFILE: Combustible liquid. When heated to decomposition it emits acrid smoke and irritating fumes.

GDE825
GERANYL BUTYRATE

mf: $C_{14}H_{24}O_2$ mw: 224.34

PROP: Colorless to pale yellow liquid; fruity, roselike odor. D: 0.889-0.904, refr index: 1.455-1.462, flash p: +199°F. Sol in alc, fixed oils; insol in glycerin, propylene glycol, water @ 253°.

SYNS: 3,7-DIMETHYL-2,6-OCTADIENE-1-YL BUTYRATE ◇ FEMA No. 2512

USE IN FOOD:

Purpose: Flavoring agent.

Where Used: Various.

Regulations: FDA - 21CFR 172.515. Use at a level not in excess of the amount reasonably required to accomplish the intended effect.

SAFETY PROFILE: Combustible liquid. When heated to decomposition it emits acrid smoke and irritating fumes.

GDK000 CAS: 109-20-6
GERANYL ISOVALERATE

mf: $C_{15}H_{26}O_2$ mw: 238.41

SYNS: trans-3,7-DIMETHYL-2,6-OCTADIENYL ISOPENTANOATE ◇ (E)-ISOVALERIC ACID-3,7-DIMETHYL-2,6-OCTADIENYL ESTER ◇ (E)-3-METHYLBUTYRIC ACID-3,7-DIMETHYL-2,6-OCTADIENYL ESTER

USE IN FOOD:

Purpose: Flavoring agent.

Where Used: Baked goods, beverages, candy, ice cream.

Regulations: FDA - 21CFR 172.515. Use at a level not in excess of the amount reasonably required to accomplish the intended effect.

SAFETY PROFILE: A skin irritant. When heated to decomposition it emits acrid smoke and irritating fumes.

TOXICITY DATA and CODEN

skn-rbt 500 mg/24H MLD FCTXAV 14,785,76

GDM400
GERANYL PHENYLACETATE

mf: $C_{18}H_{24}O_2$ mw: 272.39

PROP: Yellow liquid; honey-rose odor. D: 0.971-0.978, refr index: 1.507-1.511, flash p: +212°F. Misc in alc, chloroform, ether; insol in water.

SYNS: 3,7-DIMETHYL-2,6-OCTADIEN-1-YL PHENYLACETATE ◇ FEMA No. 2516

USE IN FOOD:

Purpose: Flavoring agent.

Where Used: Various.

Regulations: FDA - 21CFR 172.515. Use at a level not in excess of the amount reasonably required to accomplish the intended effect.

SAFETY PROFILE: Combustible liquid. When heated to decomposition it emits acrid smoke and irritating fumes.

GDM450
GERANYL PROPIONATE

mf: $C_{13}H_{22}O_2$ mw: 210.32

PROP: Colorless liquid; rosy, fruity odor. D: 0.896-0.913, refr index: 1.456-1.464, flash p: +212°F. Sol in alc, fixed oils; insol in glycerin, propylene glycol, water @ 253°.

SYNS: 3,7-DIMETHYL-2,6-OCTADADIEN-1-YL PROPIONATE ◇ FEMA No. 2517

USE IN FOOD:

Purpose: Flavoring agent.

Where Used: Various.

Regulations: FDA - 21CFR 172.515. Use at a level not in excess of the amount reasonably required to accomplish the intended effect.

SAFETY PROFILE: Combustible liquid. When heated to decomposition it emits acrid smoke and irritating fumes.

GEM000 CAS: 77-06-5
GIBBERELLIC ACID

mf: $C_{19}H_{22}O_6$ mw: 346.41

PROP: A plant growth-promoting hormone. White crystals or crystalline powder. Mp: 233-235°. Sltly sol in water, ether; sol in methanol, ethanol, acetone, aqueous solns of sodium bicarbonate and sodium acetate; moderately sol in ethyl acetate.

SYNS: BERELEX ◇ BRELLIN ◇ CEKUGIB ◇ FLORALTONE ◇ GA ◇ GIBBERELLIN ◇ GIBBREL ◇ GIB-SOL ◇ GIB-TABS ◇ GROCEL ◇ NCI-C55823 ◇ PRO-GIBB ◇ 2,4a,7-TRIHYDROXY-1-METHYL-8-METHYLENEGIBB-3-ENE-1,10-CARBOXYLIC ACID 1-4-LACTONE

USE IN FOOD:

Purpose: Enzyme activator.

Where Used: Fermented malt beverages.

Regulations: FDA - 21CFR 172.725. Limitation of 2 ppm in treated barley malt, 0.5 ppm in finished beverage.

EPA Genetic Toxicology Program.

SAFETY PROFILE: Mildly toxic by ingestion. An experimental tumorigen. Mutagenic data. When heated to decomposition it emits acrid smoke and irritating fumes.

TOXICITY DATA and CODEN

dnd-sal:spr 1 mmol/L PYTCAS 11,3135,72
dnd-mam:lym 1 mmol/L PYTCAS 11,3135,72
orl-mus TDLo:142 g/kg/78W-I:ETA NTIS** PB223-159
orl-rat LD50:6300 mg/kg 85ARAE 3,43,76/77

GEQ000 CAS: 8007-08-7
GINGER OIL

PROP: From steam distillation of ground rhizomes of *Zingiber officinale* Roscoe (Fam. *Zingiberaceae*) (FCTXAV 12,807,74). Yellow liquid; odor of ginger. D: 0.870-0.882, refr index: 1.488 @ 20°. Sol in fixed oils, mineral oil, alc; insol in glycerin, propylene glycol.

USE IN FOOD:

Purpose: Flavoring agent.

Where Used: Baked goods, beverages, desserts, meat, relishes, sauces.

Regulations: FDA - 21CFR 182.10, 182.20. GRAS when used at a level not in excess of the amount reasonably required to accomplish the intended effect.

SAFETY PROFILE: A skin irritant. Mutagenic data. When heated to decomposition it emits acrid smoke and irritating fumes.

TOXICITY DATA and CODEN

skn-rbt 500 mg/24H MOD FCTXAV 12,901,74
dnr-bcs 5 μL/disc TOFOD5 8,91,85

GEY000 CAS: 50-70-4
GLUCITOL

mf: $C_6H_{14}O_6$ mw: 182.20

PROP: White crystalline powder; odorless with sweet taste. D: 1.47 @ −5°, mp: 93° (metastable form); 97.5°, (stable form), bp: 105°. Sol in water; sltly sol in methanol, ethanol, acetic acid, phenol, and acetamide; almost insol in other organic solvents.

SYNS: CHOLAXINE ◇ DIAKARMON ◇ d-GLUCITOL ◇ GULITOL ◇ l-GULITOL ◇ KARION ◇ NIVITIN ◇ SIONIT ◇ SIONON ◇ SORBICOLAN ◇ SORBITE ◇ SORBITOL (FCC) ◇ d-SORBITOL ◇ SORBO ◇ SORBOL ◇ SORBOSTYL ◇ SORVILANDE

USE IN FOOD:

Purpose: Anticaking agent, curing agent, drying agent, emulsifier, firming agent, flavoring agent, formulation aid, free-flow agent, humectant, lubricant, nutritive sweetener, pickling agent, release agent, sequestrant, stabilizer, surface-finishing agent, texturizing agent, thickener.

Where Used: Baked goods, baking mixes, beverages (low calorie), candy (hard), candy (soft), chewing gum, chocolate, cough drops, frank, frankfurter (labeled), frozen dairy desserts, furter, jams (commercial nonstandardized), jellies (commercial nonstandardized), knockwurst, sausage (cooked), shredded coconut, wieners.

Regulations: FDA - 21CFR 184.1835. GRAS with a limitation of 99 percent in hard candy and cough drops, 75 percent in chewing gum, 98 percent in soft candy, 30 percent in nonstandardized jams and jellies, commercial, 30 percent in baked goods and baking mixes, 17 percent in frozen dairy desserts, 12 percent in all other foods when used in accordance with good manufacturing practice. USDA - 9CFR 318.7. Limitation of not more than 2 percent of the weight of the formula, excluding the formula weight of water or ice; not permitted in combination with corn syrup, and/or corn syrup solids.

EPA Genetic Toxicology Program.

SAFETY PROFILE: Mildly toxic by ingestion. Human systemic effects by ingestion: hypermotility and diarrhea. When heated to decomposition it emits acrid smoke and irritating fumes.

TOXICITY DATA and CODEN

orl-wmn TDLo:1700 mg/kg/D:GIT AJDDAL 23,568,78
orl-rat LD50:15900 mg/kg FAONAU 53A,500,74
ivn-rat LD50:7100 mg/kg FAONAU 53A,500,74

GFA200 CAS: 90-80-2
GLUCONO Δ-LACTONE

mf: $C_6H_{10}O_6$ mw: 178.14

PROP: White crystalline powder. MP: decomp @ 153°. Sol in water, sltly sol in alc.

USE IN FOOD:

Purpose: Acidifier, binder, curing agent, leavening agent, pH control agent, pickling agent, sequestrant.

Where Used: Cured comminuted meat food product, dessert mixes, frankfurters, genoa salami, sausages.

Regulations: FDA - 21CFR 184.1318. GRAS when used in accordance with good manufacturing practice. USDA - 9CFR 318.7. Limitation of 8 ounces in 100 pounds of meat. Limitation of 6 ounces to 100 pounds of genoa salami.

SAFETY PROFILE: When heated to decomposition it emits acrid smoke and irritating fumes.

GFG000 CAS: 50-99-7
d-GLUCOSE

mf: $C_6H_{12}O_6$ mw: 180.18

PROP: Colorless crystals or white crystalline or granular powder; odorless with sweet taste. D: 1.544, mp: 146°. Sol in water; sltly sol in alc. α Form monohydrate, crystals from water. Mp: 83°. α Form: anhydrous crystals from hot ethanol or water: mp: 146°. Very sparingly sol in abs alc, ether, acetone; sol in hot glacial acetic acid, pyridine, aniline. β Form: crystals from hot H_2O + ethanol, from dil acetic acid or from pyridine; mp: 148-155°.

SYNS: ANHYDROUS DEXTROSE ◇ CARTOSE ◇ CERELOSE ◇ CORN SUGAR ◇ DEXTROPUR ◇ DEXTROSE (FCC) ◇ DEXTROSE, ANHYDROUS ◇ DEXTROSOL ◇ GLUCOLIN ◇ GLUCOSE ◇ d-GLUCOSE, ANHYDROUS ◇ GLUCOSE LIQUID ◇ GRAPE SUGAR ◇ SIRUP

USE IN FOOD:

Purpose: Formulation aid, humectant, nutritive sweetener, processing aid, texturizing agent.

Where Used: Bakery products, confections, ham (chopped or processed), hamburger, ice cream, luncheon meat, meat loaf, poultry, sausage.

Regulations: USDA - 9CFR 318.7, 381.147. Sufficient for purpose.

EPA Genetic Toxicology Program.

SAFETY PROFILE: Mildly toxic by ingestion. Experimental reproductive effects. Mutagenic data. Potentially explosive reaction with potassium nitrate + sodium peroxide when heated in a sealed container. Mixtures with alkali release carbon monoxide when heated. When heated to decomposition it emits acrid smoke and irritating fumes.

TOXICITY DATA and CODEN

mmo-sat 25 mg/plate NARHAD 12,2127,84
oms-omi 1 mol/L ARMKA7 91,305,73
ipr-rat TDLo:300 g/kg (30D pre):REP OYYAA2 6,251,72
orl-rat LD50:25800 mg/kg 85AIAL -,39,73

GFG050
GLUCOSE ISOMERASE ENZYME PREPARATIONS, INSOLUBLE

SYN: INSOLUBLE GLUCOSE ISOMERASE ENZYME PREPARATIONS

USE IN FOOD:

Purpose: Enzyme.

Where Used: Corn syrup (high fructose).

Regulations: FDA - 21CFR 184.1372. GRAS when used in accordance with good manufacturing practice.

SAFETY PROFILE: When heated to decomposition it emits acrid smoke and irritating fumes.

GFO000 CAS: 56-86-0
l-GLUTAMIC ACID

mf: $C_5H_9NO_4$ mw: 147.15

PROP: A nonessential amino acid present in all complete proteins. White crystals or crystalline powder. Mp (dl form): 194°, d (dl form): 1.4601 @ 20°/4°, mp (l form): 224-225°, d (l form): 1.538 @ 20°/4°. Sltly sol in water.

SYNS: α-AMINOGLUTARIC ACID ◇ l-2-AMINOGLUTARIC ACID ◇ 2-AMINOPENTANEDIOIC ACID ◇ 1-AMINOPROPANE-1,3-DICARBOXYLIC ACID ◇ GLUSATE ◇ GLUTACID ◇ GLUTAMIC ACID ◇ α-GLUTAMIC ACID ◇ d-GLUTAMIENSUUR ◇ GLUTAMINIC ACID ◇ l-GLUTAMINIC ACID ◇ GLUTAMINOL ◇ GLUTATON

USE IN FOOD:

Purpose: Dietary supplement, nutrient, salt substitute.

Where Used: Various.

Regulations: FDA - 21CFR 182.1045. GRAS when used in accordance with good manufacturing practice.

SAFETY PROFILE: Human systemic effects by ingestion and intravenous route: headache and nausea or vomiting. When heated to decomposition it emits toxic fumes of NO_x.

TOXICITY DATA and CODEN

orl-hmn TDLo:71 mg/kg:CNS SCIEAS 163,826,69

ivn-hmn TDLo:117 mg/kg:GIT AJMSA9 214,281,47

GFO025 CAS: 138-15-8
l-GLUTAMIC ACID HYDROCHLORIDE

mf: $C_5H_9NO_4HCL$ mw: 183.59

PROP: White crystals or crystalline powder. Sol in water; insol in alc, ether.

SYNS: α-AMINOGLUTARIC ACID HYDROCHLORIDE ◇ l-2-AMINOGLUTARIC ACID HYDROCHLORIDE ◇ 2-AMINOPENTANEDIOIC ACID HYDROCHLORIDE ◇ 1-AMINOPROPANE-1,3-DICARBOXYLIC ACID HYDROCHLORIDE ◇ GLUTAMIC ACID HYDROCHLORIDE ◇ α-GLUTAMIC ACID HYDROCHLORIDE ◇ GLUTAMINIC ACID HYDROCHLORIDE ◇ l-GLUTAMINIC ACID HYDROCHLORIDE

USE IN FOOD:

Purpose: Dietary supplement, flavoring agent, nutrient, salt substitute.

Where Used: Various.

Regulations: FDA - 21CFR 172.320. Limitation 12.2 percent by weight. 21CFR 182.1047. GRAS when used as a salt substitute in accordance with good manufacturing practice.

SAFETY PROFILE: A general-purpose food additive. When heated to decomposition it emits toxic fumes of NO_x and Cl^-.

GFO050 CAS: 56-85-9
GLUTAMINE

mf: $C_5H_{10}N_2O_3$ mw: 146.17

PROP: l-Form (natural): Fine opaque needles from water or dil ethanol. Decomp 185-186°. Sol in water; practically insol in methanol, ethanol, ether, benzene, acetone, ethyl acetate, chloroform. dl-Form: prisms from dil acetone. Mp: 185-186°.

SYNS: 2-AMINOGLUTARAMIC ACID ◇ l-2-AMINOGLUTARAMIDIC ACID ◇ CEBROGEN ◇ GLUMIN ◇ GLUTAMIC ACID AMIDE ◇ GLUTAMIC ACID-5-AMIDE ◇ γ-GLUTAMINE ◇ l-GLUTAMINE (9CI, FCC) ◇ LEVOGLUTAMID ◇ LEVOGLUTAMIDE ◇ STIMULINA

USE IN FOOD:

Purpose: Dietary supplement, nutrient.

Where Used: Various.

Regulations: FDA - 21CFR 172.320. Limitation 12.4 percent by weight.

SAFETY PROFILE: Mildly toxic by ingestion. Experimental reproductive effects. When heated to decomposition it emits toxic fumes of NO_x.

TOXICITY DATA and CODEN

orl-rat TDLo:300 g/kg (30D male):REP KSRNAM 8,902,74

orl-rat LD50:7500 mg/kg NIIRDN 6,228,82

GFQ000 CAS: 111-30-8
GLUTARALDEHYDE
mf: $C_5H_8O_2$ mw: 100.13

SYNS: CIDEX ◇ GLUTARAL ◇ GLUTARALDEHYD (CZECH) ◇ GLUTARDIALDEHYDE ◇ GLUTARIC DIALDEHYDE ◇ NCI-C55425 ◇ 1,5-PENTANEDIAL ◇ 1,5-PENTANEDIONE ◇ POTENTIATED ACID GLUTARALDEHYDE ◇ SONACIDE

USE IN FOOD:
Purpose: Fixing agent.
Where Used: Enzymes.
Regulations: FDA - 21CFR 173.357. Wash fixed enzyme preparation to remove fixing materials.

ACGIH TLV: CL 0.2 ppm

SAFETY PROFILE: Poison by ingestion, intravenous and intraperitoneal routes. Moderately toxic by inhalation, skin contact, and subcutaneous routes. An experimental teratogen. Experimental reproductive effects. Mutagenic data. A severe human skin irritant. A severe eye and human skin irritant. When heated to decomposition it emits acrid smoke and irritating fumes.

TOXICITY DATA and CODEN

skn-hmn 6 mg/3D-I SEV 85DKA8 -,127,77
skn-rbt 500 mg/24H SEV 28ZPAK -,42,72
eye-rbt 250 μg/24H SEV 28ZPAK -,42,72
oms-nml:oth 50 mmol/L MUREAV 148,25,85
cyt-ham:ovr 160 μg/L ENMUDM 7,1,85
orl-rat TDLo:875 mg/kg (35D male):REP OYYAA2 12,11,76
orl-mus TDLo:8 g/kg (6-15D preg):TER TJADAB 22,51,80
orl-rat LD50:134 mg/kg OYYAA2 19,503,80
ihl-rat LC50:5000 ppm/4H 28ZPAK -,42,72

GGA000 CAS: 56-81-5
GLYCERIN
mf: $C_3H_8O_3$ mw: 92.11

PROP: Colorless or pale yellow liquid; odorless, syrupy, sweet and warm taste. Mp: 17.9 (solidifies at a much lower temp), bp: 290°, ULC: 10-20, flash p: 320°F, d: 1.260 @ 20°/4°, autoign temp: 698°F, vap press: 0.0025 mm @ 50°, vap d: 3.17. Misc with water, alc; insol in chloroform, ether, oils.

SYNS: GLYCERIN, ANHYDROUS ◇ GLYCERINE ◇ GLYCERIN, SYNTHETIC ◇ GLYCERITOL ◇ GLYCEROL ◇ GLYCYL ALCOHOL ◇ GROCOLENE ◇ MOON ◇ 1,2,3-PROPANETRIOL ◇ STAR ◇ SUPEROL ◇ SYNTHETIC GLYCERIN ◇ 90 TECHNICAL GLYCERINE ◇ TRIHYDROXYPROPANE ◇ 1,2,3-TRIHYDROXYPROPANE

USE IN FOOD:
Purpose: Bodying agent, humectant, plasticizer, solvent.
Where Used: Baked goods, candy, marshmallows, packaging materials.
Regulations: FDA - 21CFR 172.866, 178.3500. Use at a level not in excess of the amount reasonably required to accomplish the intended effect. 21CFR 182.1320. GRAS when used in accordance with good manufacturing practice.

ACGIH TLV: TWA 10 mg/m^3 (mist)

SAFETY PROFILE: Poison by subcutaneous route. Mildly toxic by ingestion. Human systemic effects by ingestion: headache and nausea or vomiting. Experimental reproductive effects. Human mutagenic data. A skin and eye irritant. In the form of mist it is a nuisance particulate and inhalation irritant.

Combustible liquid when exposed to heat, flame, or powerful oxidizers. Mixtures with hydrogen peroxide are highly explosive. Ignites on contact with potassium permanganate, calcium hypochlorite. Mixture with nitric acid + sulfuric acid forms the explosive glyceryl nitrate. Mixture with perchloric acid + lead oxide forms explosive perchlorate esters. To fight fire, use alcohol foam, CO_2, dry chemical. When heated to decomposition it emits acrid smoke and fumes.

TOXICITY DATA and CODEN

skn-rbt 500 mg/24H MLD 28ZPAK -,37,72
eye-rbt 500 mg/24H MLD 28ZPAK -,37,72
dni-hmn:lym 200 mmol/L PNASA6 79,1171,82
itt-rat TDLo:280 mg/kg (2D male):REP CCPTAY 29,291,84
itt-rat TDLo:1600 mg/kg (1D male):REP CCPTAY 29,291,84
orl-hmn TDLo:1428 mg/kg:CNS,GIT 34ZIAG -,288,69
orl-rat LD50:12600 mg/kg FEPRA7 4,142,45

GGA850
GLYCEROL ESTER of PARTIALLY DIMERIZED ROSIN

PROP: Hard, pale amber-colored resin. Sol in acetone, benzene; insol in water.

USE IN FOOD:

Purpose: Color diluent, masticatory substance in chewing gum base.

Where Used: Chewing gum, confectionery, eggs (shell), food supplements in tablet form, fruits, gum, vegetables.

Regulations: FDA - 21CFR 73.1, 172.615. Use in amounts not to exceed those required to produce the intended physical or other technical effect.

SAFETY PROFILE: When heated to decomposition it emits acrid smoke and irritating fumes.

GGA860
GLYCEROL ESTER of PARTIALLY HYDROGENATED WOOD ROSIN

PROP: Medium hard, pale amber resin. Sol in acetone, benzene; insol in water, alc.

USE IN FOOD:

Purpose: Masticatory substance in chewing gum base.

Where Used: Chewing gum.

Regulations: FDA - 21CFR 172.615. Use in amounts not to exceed those required to produce the intended physical or other technical effect.

SAFETY PROFILE: When heated to decomposition it emits acrid smoke and irritating fumes.

GGA865
GLYCEROL ESTER of POLYMERIZED ROSIN

PROP: Hard, pale amber resin. Sol in acetone, benzene; insol in water, alc.

USE IN FOOD:

Purpose: Masticatory substance in chewing gum base.

Where Used: Chewing gum.

Regulations: FDA - 21CFR 172.615. Use in amounts not to exceed those required to produce the intended physical or other technical effect.

SAFETY PROFILE: When heated to decomposition it emits acrid smoke and irritating fumes.

GGA870
GLYCEROL ESTER of TALL OIL ROSIN

PROP: Pale amber resin. Sol in acetone, benzene; insol in water.

USE IN FOOD:

Purpose: Masticatory substance in chewing gum base.

Where Used: Chewing gum.

Regulations: FDA - 21CFR 172.615. Use in amounts not to exceed those required to produce the intended physical or other technical effect.

SAFETY PROFILE: When heated to decomposition it emits acrid smoke and irritating fumes.

GGA875
GLYCEROL ESTER of WOOD ROSIN

PROP: Hard, pale amber resin. Sol in acetone, benzene; insol in water.

USE IN FOOD:

Purpose: Beverage stabilizer, masticatory substance in chewing gum base.

Where Used: Chewing gum, citrus beverages.

Regulations: FDA - 21CFR 172.615. Use in amounts not to exceed those required to produce the intended physical or other technical effect. 21CFR 172.735. Limitation of 100 ppm in the finished beverage.

SAFETY PROFILE: When heated to decomposition it emits acrid smoke and irritating fumes.

GGA885
GLYCEROL-LACTO OLEATE

USE IN FOOD:

Purpose: Emulsifier.

Where Used: Fats (rendered animal).

Regulations: USDA - 9CFR 318.7, 9CFR 381.147. Sufficient for purpose.

SAFETY PROFILE: When heated to decomposition it emits acrid smoke and irritating fumes.

GGA900
GLYCEROL-LACTO STEARATE

USE IN FOOD:

Purpose: Emulsifier.

Where Used: Cake mixes, chocolate coatings, fats (rendered animal), shortening, whipped vegetable toppings.

Regulations: USDA - 9CFR 318.7, 9CFR 381.147. Sufficient for purpose.

SAFETY PROFILE: When heated to decomposition it emits acrid smoke and irritating fumes.

GGA910
GLYCEROL-LACTO PALMITATE

USE IN FOOD:

Purpose: Emulsifier.

Where Used: Cakes, fats (rendered animal), whipped topping.

Regulations: FDA - 21CFR 172.852. Use in accordance with good manufacturing practice. USDA - 9CFR 318.7, 9CFR 381.147. Sufficient for purpose.

SAFETY PROFILE: When heated to decomposition it emits acrid smoke and irritating fumes.

GGA925
GLYCEROL MONOOLEATE

USE IN FOOD:

Purpose: Defoamer, dispersing agent, emulsifier, plasticizer.

Where Used: Coffee whiteners, packaging materials, vegetable oil.

Regulations: FDA - 21CFR 181.27, 182.4505. Use in accordance with good manufacturing practice.

SAFETY PROFILE: When heated to decomposition it emits acrid smoke and irritating fumes.

GGQ100
GLYCERYL BEHENATE

USE IN FOOD:

Purpose: Formulation aid.

Where Used: Excipient formulations in tablets.

Regulations: FDA - 21CFR 184.1328. Use in accordance with good manufacturing practice.

SAFETY PROFILE: When heated to decomposition it emits acrid smoke and irritating fumes.

GGR200 CAS: 25496-72-4
GLYCERYL MONOOLEATE

USE IN FOOD:

Purpose: Flavor adjuvant, flavoring agent, solvent, vehicle.

Where Used: Baking mixes, beverages (nonalcoholic), beverages bases, chewing gum, meat products.

Regulations: FDA - 21CFR 184.1323. GRAS when used in accordance with good manufacturing practice.

SAFETY PROFILE: When heated to decomposition it emits acrid smoke and irritating fumes.

GGS600 CAS: 139-43-5
GLYCERYL TRI(12-ACETOXYSTEARATE)

USE IN FOOD:

Purpose: Polymer adjuvant.

Where Used: Packaging materials.

Regulations: FDA - 21CFR 178.3505. Limitation, in combination with calcium carbonate, of 1 weight-percent of the total mixture.

SAFETY PROFILE: When heated to decomposition it emits acrid smoke and irritating fumes.

GGU000 CAS: 555-43-1
GLYCERYL TRISTEARATE

USE IN FOOD:

Purpose: Crystallization accelerator, fermentation aid, formulation aid, fractionation aid, lubricant, release agent, surface-finishing agent, winterization agent.

Where Used: Chocolate (imitation), cocoa, confections, fats, oils.

Regulations: FDA - 21CFR 172.811. Limitation of 1 percent in cocoa and imitation chocolate. Limitation of 0.5 percent as a formulation aid, lubricant, release agent, surface finishing agent in food. Limitation of 3.0 percent in confections. Limitation of 1.0 percent as a formulation aid in fats and oils. Limitation of 0.5 percent as a winterization agent and fractionation aid in fats and oils.

SAFETY PROFILE: When heated to decomposition it emits acrid smoke and irritating fumes.

GHA000 CAS: 56-40-6
GLYCINE

mf: $C_2H_5NO_2$ mw: 75.08

PROP: The simplest amino acid and the principal amino acid in sugar cane. White crystals; odorless, sweet taste. Mp: 232-236° (decomp), d: 1.1607. Sol in water; insol in alcohol and ether.

SYNS: AMINOACETIC ACID ◇ GLYCOLIXIR ◇ HAMPSHIRE GLYCINE

USE IN FOOD:

Purpose: Dietary supplement, nutrient.

Where Used: Beverage bases, beverages, fats (rendered animal).

Regulations: FDA - 21CFR 170.50. GRAS for animal feed only (21CFR 582.5049). 21CFR 172.320. Limitation 3.5 percent by weight. 21CFR 172.812. Limitation of 0.2 percent in finished beverage. USDA - 9CFR 318.7. Limitation of 0.01 percent in rendered animal fat.

SAFETY PROFILE: Moderately toxic by intravenous route. Mildly toxic by ingestion. When heated to decomposition it emits toxic fumes of NO_x.

TOXICITY DATA and CODEN

orl-rat LD50:7930 mg/kg YACHDS 5,1502,77
scu-rat LD50:5200 mg/kg YACHDS 5,1502,77

GIA000 CAS: 36734-19-7
GLYCOPHEN
mf: $C_{13}H_{13}Cl_2N_3O_3$ mw: 330.19

SYNS: CHIPCO 26019 ◇ 3-(3,5-DICHLOROPHENYL)-N-(1-METHYLETHYL)-2,4-DIOXO-1-IMIDAZOLIDINECARBOXAMIDE ◇ GLYCOPHENE ◇ IPRODIONE ◇ 1-ISOPROPYL CARBAMOYL-3-(3,5-DICHLOROPHENYL)-HYDANTOIN ◇ LFA 2043 ◇ MRC 910 ◇ PROMIDIONE ◇ ROP 500 F ◇ ROVRAL ◇ RP 26019

USE IN FOOD:

Purpose: Fungicide.

Where Used: Animal feed, ginseng (dried), grape pomace (dried), raisin waste, raisins, soapstock.

Regulations: FDA - 21CFR 193.253. Fungicide residue tolerance of 300 ppm in raisins, 4.0 ppm in dried ginseng. 21CFR 561.263. Limitation of 225.0 ppm in dried grape pomace, 300.0 ppm in raisin waste, 10.0 ppm in soapstock when used for animal feed.

SAFETY PROFILE: Moderately toxic by ingestion. When heated to decomposition it emits very toxic fumes of NO_x and Cl^-.

TOXICITY DATA and CODEN

orl-rat LD50:4400 mg/kg FMCHA2 -,C132,83

GJS300
GRAPE COLOR EXTRACT

USE IN FOOD:

Purpose: Color additive.

Where Used: Nonbeverage food.

Regulations: FDA - 21CFR 73.169. Use in accordance with good manufacturing practice.

SAFETY PROFILE: When heated to decomposition it emits acrid smoke and irritating fumes.

GJU000 CAS: 8016-20-4
GRAPEFRUIT OIL

PROP: From the fresh peel of *Citrus paradisi* Macfayden (*Citrus decumana* L.). Yellow liquid. Sol in fixed oils, mineral oil; sltly sol in propylene glycol; insol in glycerin.

SYN: GRAPEFRUIT OIL, COLDPRESSED ◇ GRAPEFRUIT OIL, EXPRESSED ◇ OIL of GRAPEFRUIT ◇ OIL of SHADDOCK

USE IN FOOD:

Purpose: Flavoring agent.

Where Used: Bakery products, beverages (nonalcoholic), chewing gum, confections, gelatin desserts, ice cream products, puddings, syrups.

Regulations: FDA - 21CFR 182.20. GRAS when used at a level not in excess of the amount reasonably required to accomplish the intended effect.

SAFETY PROFILE: An experimental tumorigen. Mutagenic data. A skin irritant. When heated to decomposition it emits acrid smoke and irritating fumes.

TOXICITY DATA and CODEN

skn-rbt 500 mg/24H MLD FCTXAV 12,743,74
dnr-bcs 20 mg/disc TOFOD5 8,91,85
skn-mus TDLo:280 g/kg/33W-I:ETA JNCIAM 24,1389,60

GJU100
GRAPE SKIN EXTRACT

PROP: Red to purple powder or liquid concentrate.

SYN: ENOCIANINA

USE IN FOOD:

Purpose: Color.

Where Used: Beverages (alcoholic), beverages (carbonated), beverages (still).

Regulations: FDA - 21CFR 73.170. Use in accordance with good manufacturing practice.

SAFETY PROFILE: When heated to decomposition it emits acrid smoke and irritating fumes.

GLS800 CAS: 5550-12-9
GUANYLIC ACID SODIUM SALT
mf: $C_{10}H_{14}N_5O_8P\cdot 2Na$ mw: 409.24

PROP: Colorless to white crystals; characteristic taste. Sol in water; sltly sol in alc; insol in ether.

SYNS: DISODIUM GMP ◇ DISODIUM GUANYLATE (FCC) ◇ DISODIUM-5'-GMP ◇ DISODIUM-5'-GUANYLATE ◇ GMP DISODIUM SALT ◇ 5'-GMP DISODIUM SALT ◇ GMP SODIUM SALT ◇ SODIUM GMP ◇ SODIUM GUANOSINE-5'-MONOPHOSPHATE ◇ SODIUM GUANYLATE ◇ SODIUM-5'-GUANYLATE

USE IN FOOD:

Purpose: Flavor enhancer.

Where Used: Canned foods, poultry, sauces, snack items, soups.

Regulations: FDA - 21CFR 172.530. Use at a level not in excess of the amount reasonably required to accomplish the intended effect. USDA - 9CFR 318.7, 381.147. Sufficient for purpose.

SAFETY PROFILE: Moderately toxic by intraperitoneal, subcutaneous and intravenous routes. Mildly toxic by ingestion. Mutagenic data. When heated to decomposition it emits toxic fumes of PO_x, NO_x and Na_2O.

TOXICITY DATA and CODEN

cyt-ham:fbr 1 g/L FCTOD7 22,623,84
orl-mus LD50:15 g/kg AJINO* -,-,73
ipr-mus LD50:5010 mg/kg AJINO* -,-,73
scu-mus LD50:5050 mg/kg AJINO* -,-,73
ivn-mus LD50:3580 mg/kg AJINO* -,-,73

GLU000 CAS: 9000-30-0
GUAR GUM

PROP: Yellowish-white powder, dispersible in hot or cold water, obtained from the ground endosperms of *Cyanopsis tetragonoloan* L. Taub (Fam. *Leguminosae*). White powder; odorless. Sol in water; insol in oils, grease, hydrocarbons, ketones, esters.

SYNS: A-20D ◇ BURTONITE V-7-E ◇ CYAMOPSIS GUM ◇ DEALCA TP1 ◇ DECORPA ◇ GALACTASOL ◇ GENDRIV 162 ◇ GUAR ◇ GUAR FLOUR ◇ GUM CYAMOPSIS ◇ GUM GUAR ◇ INDALCA AG ◇ JAGUAR NO. 124 ◇ JAGUAR GUM A-20-D ◇ JAGUAR PLUS ◇ LYCOID DR ◇ NCI-C50395 ◇ REGONOL ◇ REIN GUARIN ◇ SUPERCOL U POWDER ◇ SYNGUM D 46D ◇ UNI-GUAR

USE IN FOOD:

Purpose: Emulsifier, firming agent, formulation aid, stabilizer, thickener.

Where Used: Baked goods, baking mixes, beverages, cereals (breakfast), cheese, dairy product analogs, fats, gravies, ice cream, jams, jellies, milk products, oils, sauces, soup mixes, soups, sweet sauces, syrups, toppings, vegetable juices, vegetables (processed).

Regulations: FDA - 21CFR 184.1339. GRAS with a limitation of 0.35 percent in baked goods and baking mixes, 1.2 percent in breakfast cereals, 0.8 percent in cheese, 1.0 percent in dairy product analogs, 2.0 percent in fats and oils; 1.2 percent in gravies and sauces, 1.0 percent in jams and jellies, 0.6 percent in milk products, 2.0 percent in processed vegetables and vegetable juices, 0.8 percent in soups and soup mixes, 1.0 percent in sweet sauces, toppings, and syrups, 0.5 percent in all other foods when used in accordance with good manufacturing practice.

NTP Carcinogenesis Bioassay (feed); No Evidence: mouse, rat NTPTR* NTP-TR-229,82. EPA Genetic Toxicology Program.

SAFETY PROFILE: Mildly toxic by ingestion. When heated to decomposition it emits acrid smoke and irritating fumes.

TOXICITY DATA and CODEN

orl-rat LD50:7060 mg/kg FCTXAV 19,287,81

GLW100 CAS: 9000-29-7
GUM GUAIAC

PROP: From wood of *guajacum officinale* L. or *Guajacum sanctum* L. (Fam. *Zygophyllaceae*). Brown solid; balsamic odor, sltly acrid taste. Sol in alc, ether, chloroform, solns of alkalies; sltly sol in carbon disulfide, benzene.

SYN: GUAIAC GUM

USE IN FOOD:

Purpose: Antioxidant, preservative.

Where Used: Fats (rendered animal).

Regulations: FDA - 21CFR 181.24. Limitation of 0.005 percent migrating from food packages. USDA - 9CFR 318.7. Limitation of 0.01 percent in rendered animal fat.

SAFETY PROFILE: Moderately toxic by ingestion. When heated to decomposition it emits acrid smoke and irritating fumes.

TOXICITY DATA and CODEN

orl-gpg LD50:1120 mg/kg AFREAW 3,197,51

GLY000 CAS: 9000-28-6
GUM GHATTI

PROP: The gummy exudation from the stem of *Anogeissus latifolia*. Colorless to pale yellow tears; almost odorless. Sltly sol in water.

SYN: INDIAN GUM

USE IN FOOD:

Purpose: Emulsifier.

Where Used: Beverage mixes, beverages, buttered syrup, oils.

Regulations: FDA - 21CFR 184.1333. GRAS with a limitation of 0.2 percent in beverages and beverage mixes, 0.1 percent in all other foods when used in accordance with good manufacturing practice.

SAFETY PROFILE: Mildly toxic by ingestion. When heated to decomposition it emits acrid smoke and irritating fumes.

TOXICITY DATA and CODEN

orl-rat LD50:17 g/kg FDRLI* 124,-,76

H

HAF600
HALOFUGINONE HYDROBROMIDE
mf: $C_{16}H_{18}Br_2ClN_3O_3$ mw: 495.612

PROP: Crystals.

SYNS: 7-BROMO-6-CHLOROFEBRIFUGINE HYDROBROMIDE ◇ 7-BROMO-6-CHLORO-3-[3-(3-HYDROXY-2-PIPERDINYL)-2-OXOPROPYL]-4(3H)-QUINAZOLINONE HYDROBROMIDE

USE IN FOOD:

Purpose: Animal drug.

Where Used: Chicken.

Regulations: FDA - 21CFR 556.308. Limitation of 0.1 ppm in liver of chickens. 21CFR 558.265. USDA CES Ranking: B-1 (1988).

SAFETY PROFILE: When heated to decomposition it emits acrid smoke and irritating fumes.

HAM500 CAS: 7440-59-7
HELIUM
DOT: 1046/1963
mf: He mw: 4.00

PROP: Colorless, odorless, tasteless, inert gas. Mp: −272.2° @ 26 atm, bp: −268.9°, d: (gas): 0.1785 g/L @ 0°, d: (liquid): 0.147 @ −270.8°.

SYNS: HELIUM, COMPRESSED (DOT) ◇ HELIUM, REFRIGERATED LIQUID (DOT)

USE IN FOOD:

Purpose: Processing aid.

Where Used: Various.

Regulations: FDA - 21CFR 184.1355. GRAS when used in accordance with good manufacturing practice.

DOT Classification: Nonflammable Gas; Label: Nonflammable Gas

SAFETY PROFILE: A simple asphyxiant. Nonflammable Gas.

HAR000 CAS: 76-44-8
HEPTACHLOR
DOT: 2761
mf: $C_{10}H_5Cl_7$ mw: 373.30

PROP: Crystals. Mp: 96°. Nearly insol in water; sol in organic solvents.

SYNS: AGROCERES ◇ 3-CHLOROCHLORDENE ◇ DRINOX ◇ E 3314 ◇ ENT 15,152 ◇ EPTACLORO (ITALIAN) ◇ 1,4,5,6,7,8,8-EPTACLORO-3a,4,7,7a-TETRAIDRO-4,7-endo-METANO-INDENE (ITALIAN) ◇ GPKh ◇ H-34 ◇ HEPTACHLOOR (DUTCH) ◇ 1,4,5,6,7,8,8-HEPTACHLOOR-3a,4,7,7a-TETRAHYDRO-4,7-endo-METHANO-INDEEN (DUTCH) ◇ HEPTACHLOR (ACGIH, DOT) ◇ HEPTACHLORE (FRENCH) ◇ 3,4,5,6,7,8,8-HEPTACHLORODICYCLOPENTADIENE ◇ 3,4,5,6,7,8,8a-HEPTACHLORODICYCLOPENTADIENE ◇ 1,4,5,6,7,10,10-HEPTACHLORO-4,7,8,9,-TETRAHYDRO-4,7-ENDOMETHYLENEINDENE ◇ 1,4,5,6,7,8,8-HEPTACHLORO-3a,4,7,7a-TETRAHYDRO-4,7-ENDOMETHANOINDENE ◇ 1,4,5,6,7,8,8a-HEPTACHLORO-3a,4,7,7a-TETRAHYDRO-4,7-METHANOINDANE ◇ 1,4,5,6,7,8,8-HEPTACHLORO-3a,4,7,7a-TETRAHYDRO-4,7-METHANOINDENE ◇ 1(3a),4,5,6,7,8,8-HEPTACHLORO-3a(1),4,7,7a-TETRAHYDRO-4,7-METHANOINDENE ◇ 1,4,5,6,7,8,8-HEPTACHLORO-3a,4,7,7a-TETRAHYDRO-4,7-METHANOL-1H-INDENE ◇ 1,4,5,6,7,8,8-HEPTACHLORO-3a,4,7,7,7a-TETRAHYDRO-4,7-METHYLENE INDENE ◇ 1,4,5,6,7,10,10-HEPTACHLORO-4,7,8,9-TETRAHYDRO-4,7-METHYLENEINDENE ◇ 1,4,5,6,7,8,8-HEPTACHLOR-3a,4,7,7,7a-TETRAHYDRO-4,7-endo-METHANO-INDEN (GERMAN) ◇ HEPTAGRAN ◇ HEPTAMUL ◇ NCI-C00180 ◇ RCRA WASTE NUMBER P059 ◇ RHODIACHLOR ◇ VELSICOL 104

USE IN FOOD:

Purpose: Pesticide.

Where Used: Various.

Regulations: USDA CES Ranking: A-1 (1987).

IARC Cancer Review: Human Inadequate Evidence IMEMDT 20,129,79; Animal Inadequate Evidence IMEMDT 5,173,74; Animal Sufficient Evidence IMEMDT 20,129,79. NCI Carcinogenesis Bioassay (feed) Clear Evidence: Mouse (NCITR* NCI-CG-TR-9,77); Results negative: Rat (NCITR* NCI-CG-TR-9,77). EPA Genetic Toxicology Program. Community Right-To-Know List.

OSHA PEL: TWA 500 $\mu g/m^3$ (skin) ACGIH TLV: TWA 0.5 mg/m^3 (skin) DOT Classification: ORM-E; Label; None

SAFETY PROFILE: A poison by ingestion, skin contact, intraperitoneal, intravenous, and possibly other routes. An experimental carcinogen.

Human mutagenic data. Acute exposure and chronic doses have caused liver damage. In man, a dose of 1-3 grams can cause serious symptoms, especially where liver impairment is the case. Acute symptoms include tremors, convulsions, kidney damage, respiratory collapse, and death. Dangerous; when heated to decomposition it emits toxic fumes of Cl^-.

NOTE: The EPA has canceled registration of pesticides containing heptachlor with the exception of its use by subsurface ground insertion external to the dwelling for termite control.

TOXICITY DATA and CODEN

mma-hmn:fbr 100 μmol/L MUREAV 42,161,77
cyt-rat-orl 60 μg/kg 34LXAP -,555,76
dlt-rat-orl 60 μg/kg 34LXAP -,555,76
cyt-mus-ipr 5200 μg/kg SOGEBZ 2,80,66
orl-mus TDLo:403 mg/kg/80W-C:CAR NCITR* NCI-CG-TR-9,77
orl-mus TD :930 mg/kg/80W-C:CAR NCITR* NCI-CG-TR-9,77
orl-rat LD50:40 mg/kg PHJOAV 185,361,60
skn-rat LD50:119 mg/kg SPEADM 78-1,12,78
ipr-rat LD50:27 mg/kg FCTXAV 11,63,73
orl-mus LD50:68 mg/kg SPEADM 78-1,12,78
ipr-mus LD50:130 mg/kg SOGEBZ 2,80,66
ivn-mus LDLo:20 mg/kg JPETAB 107,266,53
skn-rbt LD50:2000 mg/kg AFDOAQ 16,3,52
orl-gpg LD50:116 mg/kg PCOC** -,576,66
orl-ham LD50:100 mg/kg EJTXAZ 7,159,74
unr-mam LD50:60 mg/kg 30ZDA9 -,59,71

HAV450 CAS: 5910-85-0
2,4-HEPTADIENAL

mf: $C_7H_{10}O$ mw: 110.17

PROP: Slightly yellow liquid; green odor. Refr index: 1.478-1.480, flash p: 140°F. Sol in alc, fixed oils, water.

SYNS: FEMA No. 3164 ◇ HEPTADIENAL-2,4 ◇ trans,trans-2,4-HEPTADIENAL ◇ 2,4-HEPTDIENAL

USE IN FOOD:

Purpose: Flavoring agent.

Where Used: Various.

Regulations: FDA - 21CFR 172.515. Use at a level not in excess of the amount reasonably required to accomplish the intended effect.

SAFETY PROFILE: Poison by skin contact. Moderately toxic by ingestion. A severe skin irritant. Combustible liquid. When heated to decomposition it emits acrid smoke and fumes.

TOXICITY DATA and CODEN

skn-rbt 500 mg SEV FCTOD7 21,855,83
orl-rat LD50:1150 mg/kg FCTOD7 21,855,83
skn-rbt LD50:313 mg/kg FCTOD7 21,855,83

HBA550
γ-HEPTALACTONE

mf: $C_7H_{12}O_2$ mw: 128.17

PROP: Colorless, sltly oily liquid; coconut, sweet, malty, caramel odor. D: 0.997-1.004 @ 20°, refr index: 1.439-1.445. Misc in alc, fixed oils; very sltly sol in water.

SYNS: FEMA No. 2539 ◇ HEPTANOLIDE-1,4

USE IN FOOD:

Purpose: Flavoring agent.

Where Used: Various.

Regulations: FDA - 21CFR 172.515. Use at a level not in excess of the amount reasonably required to accomplish the intended effect.

SAFETY PROFILE: When heated to decomposition it emits acrid smoke and irritating fumes.

HBB500 CAS: 111-71-7
HEPTANAL

mf: $C_7H_{14}O$ mw: 114.18

PROP: Colorless liquid; penetrating, fruity odor. D: 0.814-0.819, refr index: 1.412-1.420, mp: −43.3°, bp: 152.8°, flash p: 93°F. Sol in alc, ether, fixed oils; sltly sol in water @ 153°; misc in alc, ether.

SYNS: ENANTHAL ◇ ENANTHALDEHYDE ◇ ENANTHOLE ◇ FEMA No. 2540 ◇ HEPTALDEHYDE ◇ OENANTHALDEHYDE ◇ OENANTHOL

USE IN FOOD:

Purpose: Flavoring agent.

Where Used: Various.

Regulations: FDA - 21CFR 172.515. Use at a level not in excess of the amount reasonably required to accomplish the intended effect.

SAFETY PROFILE: Mildly toxic by ingestion. Flammable liquid. When heated to decomposition it emits acrid smoke.

TOXICITY DATA and CODEN

orl-rat LD50:14 g/kg FDRLI* 123,-,76

HBG000 CAS: 110-43-0

2-HEPTANONE

DOT: 1110

mf: $C_7H_{14}O$ mw: 114.21

PROP: Colorless, mobile liquid; penetrating, fruity odor. Bp: 151.5°, flash p: 120°F (OC), autoign temp: 991°F, vap d: 3.94, d: 0.8197 @ 15°/4°. Very sltly sol in water; sol in alc and ether.

SYNS: AMYL-METHYL-CETONE (FRENCH) ◇ n-AMYL METHYL KETONE ◇ AMYL METHYL KETONE (DOT) ◇ FEMA No. 2544 ◇ METHYL-AMYL-CETONE (FRENCH) ◇ METHYL n-AMYL KETONE (ACGIH) ◇ METHYL AMYL KETONE (DOT) ◇ METHYL PENTYL KETONE

USE IN FOOD:

Purpose: Flavoring agent.

Where Used: Various.

Regulations: FDA - 21CFR 172.515. Use at a level not in excess of the amount reasonably required to accomplish the intended effect.

OSHA PEL: TWA 100 ppm ACGIH TLV: TWA 50 ppm DOT Classification: Combustible Liquid; Label: None; IMO: Flammable or Combustible Liquid; Label: Flammable Liquid

NIOSH REL: (Ketones) TWA 465 mg/m^3

SAFETY PROFILE: Moderately toxic by ingestion. Mildly toxic by inhalation and skin contact. A skin irritant. Combustible liquid when exposed to heat or flame; can react with oxidizing materials. To fight fire, use foam, CO_2, dry chemical. When heated to decomposition it emits acrid smoke and fumes.

TOXICITY DATA and CODEN

skn-rbt 14 mg/24H open MLD AIHAAP 23,95,62
orl-rat LD50:1670 mg/kg UCDS** 8/11/58
ihl-rat LCLo:4000 ppm/4H AIHAAP 23,95,62

HBG500 CAS: 106-35-4

3-HEPTANONE

mf: $C_7H_{14}O$ mw: 114.21

PROP: Clear mobile liquid; fatty odor. Mp: −36.7°, bp: 148°, flash p: 115°F (OC), d: 0.8198 @ 20°/20°, vap d: 3.93. Misc with alc, ether, water @ 149°.

SYNS: AETHYLBUTYLKETON (GERMAN) ◇ n-BUTYL ETHYL KETONE ◇ EPTAN-3-ONE (ITALIAN) ◇ ETHYLBUTYLCETONE (FRENCH) ◇ ETHYLBUTYLKETON (DUTCH) ◇ ETHYL BUTYL KETONE (ACGIH) ◇ ETILBUTILCHETONE (ITALIAN) ◇ FEMA No. 2545 ◇ HEPTAN-3-ON (DUTCH, GERMAN) ◇ HEPTAN-3-ONE

USE IN FOOD:

Purpose: Flavoring agent.

Where Used: Various.

Regulations: FDA - 21CFR 172.515. Use at a level not in excess of the amount reasonably required to accomplish the intended effect.

OSHA PEL: TWA 50 ppm ACGIH TLV: TWA 50 ppm

SAFETY PROFILE: Moderately toxic by ingestion and inhalation. A skin and eye irritant. Combustible liquid. Can react with oxidizing materials. To fight fire, use foam, CO_2, dry chemical.

TOXICITY DATA and CODEN

skn-rbt 500 mg/24H MOD FCTXAV 16,731,78
eye-rbt 100 mg MLD FCTXAV 16,731,78
orl-rat LD50:2760 mg/kg JIHTAB 31,60,49
ihl-rat LCLo:2000 ppm/4H JIHTAB 31,343,49

HBI800

cis-4-HEPTEN-1-AL

mf: $C_7H_{12}O$ mw: 112.17

PROP: Sltly yellow liquid; fatty odor. Refr index: 1.432-1.436, flash p: 68°F. Sol in alc, fixed oils; insol in water.

SYNS: FEMA No. 3289 ◇ 4-HEPTENAL ◇ n-PROPYLIDENE BUTYRALDEHYDE

USE IN FOOD:

Purpose: Flavoring agent.

Where Used: Various.

Regulations: FDA - 21CFR 172.515. Use at a level not in excess of the amount reasonably required to accomplish the intended effect.

SAFETY PROFILE: Flammable liquid. When heated to decomposition it emits acrid smoke and irritating fumes.

HBL500 CAS: 111-70-6

HEPTYL ALCOHOL

mf: $C_7H_{16}O$ mw: 116.23

PROP: Colorless liquid; citrus odor. Mp: −34.6°, bp: 175.8°, d: 0.824 @ 20°/4°, refr index: 1.423-1.427, flash p: 160°F. Misc in alc, fixed oils, ether; sltly sol in water @ 175°.

SYNS: l'ALCOOL n-HEPTYLIQUE PRIMAIRE (FRENCH) ◇ ENANTHIC ALCOHOL ◇ FEMA No. 2548 ◇ n-HEPTANOL

◇ 1-HEPTANOL ◇ n-HEPTANOL-1 (FRENCH) ◇ 1-HYDROXY-HEPTANE

USE IN FOOD:

Purpose: Flavoring agent.

Where Used: Various.

Regulations: FDA - 21CFR 172.515. Use at a level not in excess of the amount reasonably required to accomplish the intended effect.

SAFETY PROFILE: Moderately toxic by ingestion and skin contact. Mildly toxic by inhalation. Combustible liquid. Can react with oxidizing materials. When heated to decomposition it emits acrid smoke and fumes.

TOXICITY DATA and CODEN

orl-rat LD50:500 mg/kg AMPMAR 35,501,74
orl-mus LD50:1500 mg/kg GISAAA 31,16,66
ihl-mus LC50:6600 mg/kg 85GMAT -,72,82

HBO500 CAS: 112-23-2
HEPTYL FORMATE
mf: $C_8H_{16}O_2$ mw: 144.24

SYNS: FORMIC ACID, HEPTYL ESTER ◇ HEPTANOL, FORMATE ◇ n-HEPTYL METHANOATE

USE IN FOOD:

Purpose: Flavoring agent.

Where Used: Baked goods, beverages, candy, ice cream.

Regulations: FDA - 21CFR 172.515. Use at a level not in excess of the amount reasonably required to accomplish the intended effect.

SAFETY PROFILE: A skin irritant. When heated to decomposition it emits acrid smoke and fumes.

TOXICITY DATA and CODEN

skn-rbt 500 mg/24H MOD FCTXAV 16,771,78

HBP300
HEPTYLPARABEN
mf: $C_{14}H_{20}O_3$ mw: 236.31

PROP: Small colorless crystals or white crystalline powder; odorless, burning taste. Mp: 48-51°. Sol in alc, ether; very sltly sol in water.

SYN: n-HEPTYL p-HYDROXYBENZOATE

USE IN FOOD:

Purpose: Antioxidant, preservative.

Where Used: Beer, beverages (fermented malt), fruit drinks (noncarbonated), soft drinks (noncarbonated).

Regulations: FDA - 21CFR 172.145. Limitation 12 ppm in fermented malt beverages, 20 ppm in noncarbonated soft drinks and fruit drinks. BATF - 27CFR 240.1051. Limitation of 12 ppm in wine.

SAFETY PROFILE: When heated to decomposition it emits acrid smoke and irritating fumes.

HCP000 CAS: 36653-82-4
1-HEXADECANOL
mf: $C_{16}H_{34}O$ mw: 242.50

PROP: Solid or leaf-like crystals. Mp: 49.3°, bp: 190° @ 15 mm, d: 0.8176 @ 50°/4°. Insol in H_2O; sol in alc, chloroform, ether.

SYNS: ADOL ◇ ALCOHOL C-16 ◇ ATALCO C ◇ CACHALOT C-50 ◇ CETAFFINE ◇ CETAL ◇ CETALOL CA ◇ CETYL ALCOHOL ◇ CETYLIC ALCOHOL ◇ CETYLOL ◇ CO-1670 ◇ CRODACOL-CAS ◇ CYCLAL CETYL ALCOHOL ◇ DYTOL F-11 ◇ EPAL 16NF ◇ ETHAL ◇ ETHOL ◇ HEXADECANOL ◇ n-HEXADECANOL ◇ HEXADECAN-1-OL ◇ HEXADECYL ALCOHOL ◇ n-HEXADECYL ALCOHOL ◇ LOROL 24 ◇ LOXANOL K ◇ PALMITYL ALCOHOL ◇ PRODUCT 308

USE IN FOOD:

Purpose: Color diluent, intermediate.

Where Used: Confectionery, food supplements in tablet form, gum.

Regulations: FDA - 21CFR 73.1. No residue. 21CFR 172.864. Use at a level not in excess of the amount reasonably required to accomplish the intended effect.

SAFETY PROFILE: Moderately toxic by ingestion and intraperitoneal routes. An eye and human skin irritant. Flammable when exposed to heat or flame; can react with oxidizing materials. To fight fire, use foam, CO_2, dry chemical. When heated to decomposition it emits acrid smoke and fumes.

TOXICITY DATA and CODEN

skn-hmn 75 mg/3D-I MLD 85DKA8 -,127,77
skn-rbt 2600 mg/kg/24H MLD AIHAAP 34,493,73
eye-rbt 82 mg MLD AIHAAP 34,493,73
orl-rat LD50:6400 mg/kg FCTXAV 16,683,78

HEM000 CAS: 66-25-1
1-HEXANAL
DOT: 1207
mf: $C_6H_{12}O$ mw: 100.18

PROP: Colorless liquid; powerful fatty-green odor. Reported in about a dozen essential oils (FCTXAV 11,95,73). Mp: −56.3°, bp: 128.7°, flash p: 90°F (OC), d: 0.808-0.812, refr index: 1.402-1.407, vap press: 8.6 mm @ 20°, vap d: 3.45. Sol in alc, fixed oils, propylene glycol; very sltly sol in water.

SYNS: ALDEHYDE C-6 ◇ CAPROALDEHYDE ◇ CAPROIC ALDEHYDE ◇ CAPRONALDEHYDE ◇ n-CAPROYLALDEHYDE ◇ FEMA No. 2557 ◇ HEXALDEHYDE (DOT) ◇ HEXANAL

USE IN FOOD:

Purpose: Flavoring agent.

Where Used: Various.

Regulations: FDA - 21CFR 172.515. Use at a level not in excess of the amount reasonably required to accomplish the intended effect.

DOT Classification: Flammable or Combustible Liquid; Label: Flammable Liquid

SAFETY PROFILE: Mildly toxic by ingestion and inhalation. An irritant to skin and eyes. Flammable liquid. A dangerous fire hazard when exposed to heat or flame; can react vigorously with oxidizing materials. When heated to decomposition it emits acrid smoke and fumes.

TOXICITY DATA and CODEN

skn-rbt 10 mg/24H open MLD AMIHBC 10,61,54
eye-rbt 100 mg/24H MLD FCTXAV 11,95,73
orl-rat LD50:4890 mg/kg AMIHBC 10,61,54
ihl-rat LCLo:2000 ppm/4H AMIHBC 10,61,54

HEN000 CAS: 110-54-3
n-HEXANE
DOT: 1208
mf: C_6H_{14} mw: 86.20

PROP: Colorless clear liquid; faint odor. Bp: 69°, ULC: 90-95, lel: 1.2%, uel: 7.5%, fp: −95.6°, flash p: −9.4°F, d: 0.6603 @ 20°/4°, autoign temp: 437°F, vap press: 100 mm @ 15.8°, vap d: 2.97. Insol in water; misc in chloroform, ether, alc. Very volatile liquid.

SYNS: ESANI (ITALIAN) ◇ GETTYSOLVE-B ◇ HEKSAN (POLISH) ◇ HEXANE (DOT) ◇ HEXANEN (DUTCH) ◇ HEXANES (FCC) ◇ NCI-C60571

USE IN FOOD:

Purpose: Extraction solvent.

Where Used: Hops extract, spice oleoresins.

Regulations: FDA - 21CFR 173.270. 25 ppm in spice oleoresins, 2.2 percent in hops extracts.

OSHA PEL: TWA 50 ppm ACGIH TLV: TWA 50 ppm DOT Classification: Flammable Liquid; Label: Flammable Liquid NIOSH REL: TWA (Alkanes) 350 mg/m^3

SAFETY PROFILE: Slightly toxic by ingestion and inhalation. Human systemic effects by inhalation: hallucinations. Experimental teratogenic and reproductive effects. Mutagenic data. An eye irritant. Can cause a motor neuropathy in exposed workers. May be irritating to respiratory tract and narcotic in high concentrations. Inhalation of 5000 ppm for 1/6-hour produces marked vertigo; 2500-1000 ppm for 12 hours produces drowsiness, fatigue, loss of appetite, paresthesia in distal extremities; 2500-500 ppm produces muscle weakness, cold pulsation in extremities, blurred vision, headache, anorexia, and onset of polyneuropathy. 2000 ppm for 1/6-hour produces no symptoms. 1000-500 ppm for 3-6 months produces fatigue, loss of appetite, distal paresthesia. Dangerous if abused.

Flammable liquid. A very dangerous fire and explosion hazard when exposed to heat or flame; can react vigorously with oxidizing materials. Mixtures with dinitrogen tetraoxide may explode at 28°. To fight fire, use CO_2, dry chemical. When heated to decomposition it emits acrid smoke and fumes.

TOXICITY DATA and CODEN

eye-rbt 10 mg MLD TXAPA9 55,501,80
cyt-ham:fbr 500 mg/L FCTOD7 22,623,84
ihl-rat TCLo:1 pph/6H (65D male):REP FAATDF 4,191,84
orl-mus TDLo:238 g/kg (6-15D preg):TER DCTODJ 3,393,80
ihl-hmn TCLo:5000 ppm/10M:CNS BMRII* 2979,-,29
orl-rat LD50:28710 mg/kg TXAPA9 19,699,71

HEU000 CAS: 142-62-1
HEXANOIC ACID
DOT: 1706
mf: $C_6H_{12}O_2$ mw: 116.18

PROP: Oily, colorless liquid; odor of Limburger cheese. Bp: 205.0°, fp: −3.4°, flash p: 215°F (COC), d: 0.9295 @ 20°/20°, refr index: 1.415-1.418, vap press: 0.18 mm @ 20°, vap d: 4.0, autoign temp: 716°F. Very sol in ether, fixed oils; sltly sol in water.

SYNS: BUTYLACETIC ACID ◇ CAPROIC ACID ◇ n-CAPROIC ACID ◇ CAPRONIC ACID ◇ FEMA No. 2559 ◇ HEXACID 698 ◇ n-HEXANOIC ACID ◇ n-HEXOIC ACID ◇ PENTIFORMIC ACID ◇ PENTYLFORMIC ACID

USE IN FOOD:

Purpose: Flavoring agent.

Where Used: Various.

Regulations: FDA - 21CFR 172.515. Use at a level not in excess of the amount reasonably required to accomplish the intended effect.

DOT Classification: Corrosive Material; Label: Corrosive

SAFETY PROFILE: Moderately toxic by ingestion, skin contact, intraperitoneal, and subcutaneous routes. Mutagenic data. Corrosive. A skin and severe eye irritant. Combustible when exposed to heat or flame; can react with oxidizing materials. To fight fire, use CO_2, dry chemical, fog, mist. When heated to decomposition it emits acrid smoke and fumes.

TOXICITY DATA and CODEN

skn-rbt 10 mg/24H open MLD AMIHBC 10,61,54
eye-rbt 695 μg SEV AJOPAA 29,1363,46
oms-nml:oth 10 mmol/L CHROAU 40,1,73
cyt-nml:oth 10 mmol/L CHROAU 40,1,73
orl-rat LD50:3000 mg/kg JIHTAB 26,269,44
ihl-mus LC50:4100 mg/m^3/2H 85GMAT -,32,82

HFA300 CAS: 51235-04-2
HEXAZINONE
mf: $C_{12}H_{20}N_4O_2$ mw: 252.36

SYNS: 3-CYCLOHEXYL-6-(DIMETHYLAMINO)-1-METHYL-s-TRIAZINE-2,4(1H,3H)-DIONE ◇ 3-CYCLOHEXYL-6-(DIMETHYLAMINO)-1-METHYL-1,3,5-TRIAZINE-2,4(1H,3H)-DIONE ◇ DPX 3674 ◇ VELPAR ◇ VELPAR WEED KILLER

USE IN FOOD:

Purpose: Herbicide.

Where Used: Various.

Regulations: USDA CES Ranking: D-4 (1985).

SAFETY PROFILE: Moderately toxic by ingestion and intraperitoneal routes. Mildly toxic by skin contact. Experimental reproductive effects. An eye irritant. When heated to decomposition it emits toxic fumes of NO_x.

TOXICITY DATA and CODEN

eye-rbt 48 mg MOD FAATDF 4,603,84
orl-rat TDLo:51700 mg/kg (94D male/94D pre):REP FAATDF 4,960,84
orl-rat LD50:1690 mg/kg 85ARAE 2,135,77
skn-rat LD50:5278 mg/kg FMCHA2 -,C126,83

HFA525
trans-2-HEXEN-1-AL
mf: $C_6H_{10}O$ mw: 98.15

PROP: Pale yellow liquid; fruity, vegetable odor. D: 0.841-0.848, refr index: 1.445-1.449, flash p: 100°F. Sol in alc, propylene glycol, fixed oils; very sltly sol in water.

SYN: FEMA No. 2560

USE IN FOOD:

Purpose: Flavoring agent.

Where Used: Various.

Regulations: FDA - 21CFR 172.515. Use at a level not in excess of the amount reasonably required to accomplish the intended effect.

SAFETY PROFILE: Flammable liquid. When heated to decomposition it emits acrid smoke and irritating fumes.

HFD500 CAS: 928-95-0
2-HEXEN-1-OL, (E)-
mf: $C_6H_{12}O$ mw: 100.18

PROP: Colorless liquid; fruity-green odor. D: 0.836-0.841, refr index: 0.437-1.442, flash p: 129°F. Sol in alc, propylene glycol, fixed oils; very sltly sol in water.

SYNS: FEMA No. 2562 ◇ trans-2-HEXENOL ◇ 2-HEXENOL ◇ trans-2-HEXEN-1-OL (FCC)

USE IN FOOD:

Purpose: Flavoring agent.

Where Used: Various.

Regulations: FDA - 21CFR 172.515. Use at a level not in excess of the amount reasonably required to accomplish the intended effect.

SAFETY PROFILE: Moderately toxic by ingestion. Mildly toxic by skin contact. A skin irritant. Combustible liquid. When heated to decomposition it emits acrid smoke and fumes.

TOXICITY DATA and CODEN

skn-rbt 500 mg/24H FCTXAV 12,911,74
orl-rat LD50:3500 mg/kg FCTXAV 12,911,74
skn-rbt LD50:4500 mg/kg FCTXAV 12,911,74

HFE000 CAS: 928-96-1
cis-3-HEXENOL
mf: $C_6H_{12}O$ mw: 100.18

PROP: Colorless liquid; powerful grassy-green odor. D: 0.846-0.850, refr index: 1.43-1.441, bp: 137°, flash p: 111°F. Sol in alc, propylene glycol, fixed oils; very sltly sol in water.

SYNS: BLATTERALKOHOL ◇ FEMA No. 2563 ◇ β-γ-HEXENOL ◇ cis-3-HEXEN-1-OL (FCC) ◇ LEAF ALCOHOL

USE IN FOOD:

Purpose: Flavoring agent.

Where Used: Various.

Regulations: FDA - 21CFR 172.515. Use at a level not in excess of the amount reasonably required to accomplish the intended effect.

SAFETY PROFILE: A poison by intraperitoneal route. Mildly toxic by ingestion. Combustible liquid. When heated to decomposition it emits acrid smoke and fumes.

TOXICITY DATA and CODEN

orl-rat LD50:4700 mg/kg FCTXAV 12,909,74
ipr-rat LD50:600 mg/kg FCTXAV 7,451,69

HFE550
cis-3-HEXENYL 2-METHYLBUTYRATE

mf: $C_{11}H_{20}O_2$ mw: 184.28

PROP: Colorless liquid; powerful, fruity odor like unripe apples. D: 0.876-0.880, refr index: 1.430, flash p: 153°F. Sol in alc, fixed oils; insol in water.

SYN: FEMA No. 3497

USE IN FOOD:

Purpose: Flavoring agent.

Where Used: Various.

Regulations: FDA - 21CFR 172.515. Use at a level not in excess of the amount reasonably required to accomplish the intended effect.

SAFETY PROFILE: Combustible liquid. When heated to decomposition it emits acrid smoke and irritating fumes.

HFG500 CAS: 108-10-1
HEXONE

DOT: 1245
mf: $C_6H_{12}O$ mw: 100.18

PROP: Colorless mobile liquid; fruity, ethereal odor. Bp: 118°, lel: 1.4%, uel: 7.5%, flash p: 62.6°F, d: 0.796-0.799, fp: −80.2°, autoign temp: 858°F, vap press: 16 mm @ 20°. Misc with alc, ether; sol in alc.

SYNS: FEMA No. 2731 ◇ HEXON (CZECH) ◇ ISOBUTYLMETHYLKETON (CZECH) ◇ ISOBUTYL METHYL KETONE ◇ ISOPROPYLACETONE ◇ METHYL-ISOBUTYL-CETONE (FRENCH) ◇ METHYLISOBUTYLKETON (DUTCH, GERMAN) ◇ METHYL ISOBUTYL KETONE (ACGIH,DOT) ◇ METYLOIZOBUTYLOKETON (POLISH) ◇ 4-METHYL-PENTAN-2-ON (DUTCH, GERMAN) ◇ 4-METHYL-2-PENTANON (CZECH) ◇ 2-METHYL-4-PENTANONE ◇ 4-METHYL-2-PENTANONE (FCC) ◇ METILISOBUTILCHETONE (ITALIAN) ◇ 4-METILPENTAN-2-ONE (ITALIAN) ◇ MIBK ◇ MIK ◇ RCRA WASTE NUMBER U161 ◇ SHELL MIBK

USE IN FOOD:

Purpose: Flavoring agent.

Where Used: Various.

Regulations: FDA - 21CFR 172.515. Use at a level not in excess of the amount reasonably required to accomplish the intended effect.

Community Right-To-Know List.

OSHA PEL: TWA 50 ppm; STEL 75 ppm ACGIH TLV: TWA 50 ppm; STEL 75 ppm DOT Classification: Flammable Liquid; Label: Flammable Liquid NIOSH REL: (Ketones) TWA 200 mg/m^3

SAFETY PROFILE: A poison by intraperitoneal route. Moderately toxic by ingestion. Mildly toxic by inhalation. Very irritating to the skin, eyes and mucous membranes. A human systemic irritant by inhalation. Narcotic in high concentration. Flammable liquid when exposed to heat, flame or oxidizers. Ignites on contact with potassium-tert-butoxide. Moderately explosive in the form of vapor when exposed to heat or flame. May form explosive peroxides upon exposure to air. Can react vigorously with reducing materials. To fight fire, use alcohol foam, CO_2, dry chemical.

TOXICITY DATA and CODEN

eye-hmn 200 ppm/15M JIHTAB 28,262,46
skn-rbt 500 mg/24H MLD 28ZPAK -,42,72
eye-rbt 40 mg SEV UCDS** 4/25/58
orl-rat LD50:2080 mg/kg UCDS** 4/25/58
ihl-rat LC50:8000 ppm/4H 28ZPAK -,42,72

HFI500 CAS: 142-92-7
HEXYL ACETATE

mf: $C_8H_{16}O_2$ mw: 144.24

PROP: Colorless liquid; fruity odor. D: 0.878, mp: −60.9°, bp: 171.5°, refr index: 1.407, flash p: 109°F. Insol in water; very sol in alc and ether.

SYNS: ACETIC ACID HEXYL ESTER ◇ FEMA No. 2565 ◇ n-HEXYL ACETATE (FCC) ◇ 1-HEXYL ACETATE ◇ HEXYL ALCOHOL, ACETATE ◇ HEXYL ETHANOATE

USE IN FOOD:

Purpose: Flavoring agent.

Where Used: Various.

Regulations: FDA - 21CFR 172.515. Use at a level not in excess of the amount reasonably required to accomplish the intended effect.

SAFETY PROFILE: Mildly toxic by ingestion. Combustible liquid. When heated to decomposition it emits acrid smoke and fumes.

TOXICITY DATA and CODEN

orl-rat LD50:42 g/kg TXAPA9 28,313,74

HFJ500 CAS: 111-27-3
n-HEXYL ALCOHOL

mf: $C_6H_{14}O$ mw: 102.20

PROP: Colorless liquid. Bp: 157.2°, fp: −44.6°, flash p: 145°F, d: 0.816-0.821, vap press: 1 mm @ 24.4°, vap d: 3.52. Misc in alc, ether; sltly sol in water.

SYNS: AMYLCARBINOL ◇ CAPROYL ALCOHOL ◇ EPAL 6 ◇ FEMA No. 2567 ◇ HEXANOL ◇ n-HEXANOL (DOT) ◇ 1-HEXANOL ◇ HEXYL ALCOHOL ◇ 1-HYDROXYHEXANE ◇ PENTYLCARBINOL

USE IN FOOD:

Purpose: Flavoring agent, intermediate.

Where Used: Various.

Regulations: FDA - 21CFR 172.864, 172.515. Use at a level not in excess of the amount reasonably required to accomplish the intended effect.

DOT Classification: Flammable or Combustible Liquid; Label: Flammable Liquid

SAFETY PROFILE: Poison by intravenous route. Moderately toxic by ingestion and skin contact. A skin and severe eye irritant. Combustible liquid. Can react with oxidizing materials. To fight fire, use alcohol foam, CO_2, dry chemical.

TOXICITY DATA and CODEN

skn-rbt 410 mg open MLD UCDS** 4/21/67
eye-rbt 250 μg open SEV AMIHBC 4,119,51
orl-rat LD50:720 mg/kg SAMJAF 43,795,69

HFM600
HEXYL-2-BUTENOATE

mf: $C_{10}H_{18}O_2$ mw: 170.24

PROP: Colorless liquid; fruity odor. D: 0.880, refr index: 1.428-1.449. Sol in alc, fixed oils; insol in water, propylene glycol.

SYNS: FEMA No. 3354 ◇ HEXYL CROTONATE

USE IN FOOD:

Purpose: Flavoring agent.

Where Used: Various.

Regulations: GRAS when used at a level not in excess of the amount reasonably required to accomplish the intended effect.

SAFETY PROFILE: When heated to decomposition it emits acrid smoke and irritating fumes.

HFO500 CAS: 101-86-0
HEXYL CINNAMALDEHYDE

mf: $C_{15}H_{20}O$ mw: 216.35

PROP: Pale yellow liquid; jasmine odor. D 0.953-0.959, refr index: 1.548-1.552. Sol in fixed oils; insol in propylene glycol, glycerin.

SYNS: FEMA No. 2569 ◇ α-HEXYLCINNAMALDEHYDE (FCC) ◇ HEXYL CINNAMIC ALDEHYDE ◇ α-HEXYLCINNAMIC ALDEHYDE ◇ α-n-HEXYL-β-PHENYLACROLEIN ◇ 2-(PHENYLMETHYLENE)OCTANOL

USE IN FOOD:

Purpose: Flavoring agent.

Where Used: Various.

Regulations: FDA - 21CFR 172.515. Use at a level not in excess of the amount reasonably required to accomplish the intended effect.

SAFETY PROFILE: Moderately toxic by ingestion. A skin irritant. When heated to decomposition it emits acrid smoke and fumes.

TOXICITY DATA and CODEN

skn-rbt 500 mg/24H MOD FCTXAV 12,915,74
orl-rat LD50:3100 mg/kg FCTXAV 12,915,74

HFQ600
HEXYL ISOVALERATE

mf: $C_{11}H_{22}O_2$ mw: 186.30

PROP: Colorless liquid; pungent, fruity odor. D: 0.853, refr index: 1.417. Sol in alc, fixed oils; insol in water.

SYN: FEMA No. 3500

USE IN FOOD:

Purpose: Flavoring agent.

Where Used: Various.

Regulations: FDA - 21CFR 172.515. Use at a level not in excess of the amount reasonably required to accomplish the intended effect.

SAFETY PROFILE: When heated to decomposition it emits acrid smoke and irritating fumes.

HFR200
HEXYL 2-METHYLBUTYRATE

mf: $C_{11}H_{22}O_2$ mw: 186.30

PROP: Colorless liquid; strong, fresh-green, fruity odor. D: 0.854, refr index: 1.416-1.421, flash p: 122°F. Sol in alc, fixed oils; insol in water.

SYN: FEMA No. 3499

USE IN FOOD:

Purpose: Flavoring agent.

Where Used: Various.

Regulations: FDA - 21CFR 172.515. Use at a level not in excess of the amount reasonably required to accomplish the intended effect.

SAFETY PROFILE: Combustible liquid. When heated to decomposition it emits acrid smoke and irritating fumes.

HGB100
HIGH-FRUCTOSE CORN SYRUP

PROP: Water-white to light yellow viscous liquid; sweet taste. Misc with water.

SYN: CORN SYRUP, HIGH-FRUCTOSE

USE IN FOOD:

Purpose: Flavoring agent, nutritive sweetener.

Where Used: Beverages (carbonated), candy, confections, desserts (frozen), drinks (dairy), fruits (canned), ham (chopped or processed), hamburger, ice cream, luncheon meat, meat loaf, poultry, sausage.

Regulations: FDA - 21CFR 182.1866. GRAS when used in accordance with good manufacturing practice. Insoluble glucose isomerase used in preparation must comply with FDA regulations in 21CFR 184.1372. USDA - 9CFR 318.7. Limitation of 2.0 percent. 9CFR 381.147. Sufficient for purpose.

SAFETY PROFILE: When heated to decomposition it emits acrid smoke and irritating fumes.

HGE700 CAS: 71-00-1
HISTIDINE

mf: $C_6H_9N_3O_2$ mw: 155.18

PROP: l-Histidine, the natural form. White needles, plates, or crystalline powder; sltly bitter taste. Decomp 287° (softens at 277°). Solubility in water at 25°: 41.9 g/L. Sol in water; very sltly sol in alc; insol in ether.

SYNS: l-α-AMINO-4(OR 5)-IMIDAZOLEPROPIONIC ACID ◇ GLYOXALINE-5-ALANINE ◇ l-HISTIDINE (FCC)

USE IN FOOD:

Purpose: Dietary supplement, nutrient.

Where Used: Various.

Regulations: FDA - 21CFR 172.310. Limitation 2.4 percent by weight.

SAFETY PROFILE: Experimental reproductive effects. Human mutagenic data. When heated to decomposition it emits toxic fumes of NO_x.

TOXICITY DATA and CODEN

dni-hmn:oth 1 mmol/L JIDEAE 65,400,75
cyt-ham:lng 2500 ppm TOLED5 28,117,85
ipr-rat TDLo:7 g/kg (35D pre):REP OYYAA2 17,807,79

HGE800
HISTIDINE MONOHYDROCHLORIDE

mf: $C_6H_9N_3O_2 \cdot HCl \cdot H_2O$ mw: 209.63

PROP: White needles, plates, or crystalline powder; sltly bitter taste. Decomp 250°. Sol in water; insol in alc, ether.

SYNS: l-α-AMINO-4(OR 5)-IMIDAZOLEPROPIONIC ACID MONOHYDROCHLORIDE ◇ GLYOXALINE-5-ALANINE MONOHYDROCHLORIDE

USE IN FOOD:

Purpose: Dietary supplement, nutrient.

Where Used: Various.

Regulations: FDA - 21CFR 172.310. Limitation 2.4 percent by weight.

SAFETY PROFILE: When heated to decomposition it emits toxic fumes of NO_x.

HGK750
HOP EXTRACT, MODIFIED

PROP: An extract of hops by a variety of organic solvent extractions.

SYN: MODIFIED HOP EXTRACT.

USE IN FOOD:

Purpose: Flavoring agent.

Where Used: Beer.

Regulations: FDA - 21CFR 172.560. Use in accordance with good manufacturing practice.

SAFETY PROFILE: When heated to decomposition it emits acrid smoke and irritating fumes.

HGK800
HOPS OIL

PROP: From steam distillation of cones from female *Humulus lupulus* L. or *Humulus americanus* Nutt. (Fam. *Moraceae*). Yellow liquid; aromatic odor. D: 0.825-0.926, refr index: 1.470-1.494 @ 20°. Sol in fixed oils, mineral oil; insol in glycerin, propylene glycol.

USE IN FOOD:

Purpose: Flavoring agent.

Where Used: Various.

Regulations: FDA - 21CFR 182.20. GRAS when used at a level not in excess of the amount reasonably required to accomplish the intended effect.

SAFETY PROFILE: When heated to decomposition it emits acrid smoke and irritating fumes.

HGS000 CAS: 302-01-2
HYDRAZINE

DOT: 2029/2030
mf: H_4N_2 mw: 32.06

PROP: Colorless, oily, fuming liquid or white crystals. Mp: 1.4°, bp: 113.5°, flash p: 100°F (OC), d: 1.1011 @ 15° (liquid), autoign temp: can vary from 74°F in contact with iron rust, 270°F in contact with black iron, 313°F in contact with stainless steel, 518°F in contact with glass. Vap d: 1.1; lel: 4.7%, uel: 100%.

SYNS: ANHYDROUS HYDRAZINE (DOT) ◇ DIAMIDE ◇ DIAMINE ◇ HYDRAZINE, ANHYDROUS (DOT) ◇ HYDRAZINE, AQUEOUS SOLUTION (DOT) ◇ HYDRAZINE BASE ◇ HYDRAZYNA (POLISH) ◇ RCRA WASTE NUMBER U133

USE IN FOOD:

Purpose: Boiler water additive.

Where Used: Various.

Regulations: FDA - 21CFR 173.310d. Zero tolerance in steam.

IARC Cancer Review: Animal Sufficient Evidence IMEMDT 4,127,74. EPA Extremely Hazardous Substances List. Community Right-To-Know List. Genetic Toxicology Program.

OSHA PEL: TWA 0.1 ppm (skin) ACGIH TLV: TWA 0.1 ppm (skin); Suspected Carcinogen DOT Classification: Flammable Liquid; Label: Flammable Liquid and Poison; Corrosive Material; Label: Corrosive, aqueous solution NIOSH REL: (Hydrazines) CL 0.04 mg/m^3/2H

SAFETY PROFILE: A poison by ingestion, skin contact, intraperitoneal, intravenous, and possibly other routes. Moderately toxic by inhalation. An experimental carcinogen, neoplastigen, and tumorigen of lungs, nervous system, liver, kidney, hematopoietic organs, breast, and subcutaneous tissue. An experimental teratogen. Other experimental reproductive effects. Human mutagenic data. A powerful reducing agent which is corrosive to the eyes, skin, and mucous membranes. May cause skin sensitization as well as systemic poisoning. Hydrazine and some of its derivatives may cause damage to the liver and destruction of red blood cells.

Flammable liquid. A very dangerous fire hazard when exposed to heat, flame or oxidizing agents. Severe explosion hazard when exposed to heat or flame or by chemical reaction. Potentially explosive reactions with many chemicals. The vapor will burn without air. It is a powerful explosive. It is very sensitive and must not be used without full and complete instructions from the manufacturer for handling, storage and disposal. Dangerous; when heated to decomposition it emits highly toxic fumes of NO_x and NH_3.

TOXICITY DATA and CODEN

mmo-omi 70 μg/L MUREAV 173,233,86
sce-ham:lng 1 mmol/L HUGEDQ 54,155,80
ihl-rat TCLo:4 mg/m^3/2H (7-20D preg):REP GISAAA 39(10),23,74
ipr-rat TDLo:50 mg/kg (6-15D preg):TER APTOD9 19,A21,80
ipr-mus TDLo:80 mg/kg (6-9D preg):TER NTIS** AD-A084-023
ihl-rat TCLo:1 ppm/6H/1Y-I:ETA PAACA3 21,74,80
orl-mus TDLo:1951 mg/kg/2Y-C:NEO IJCNAW 9,109,72
ipr-mus TDLo:400 mg/kg/5W-I:CAR UICMAI 7,180,67
orl-rat LD50:60 mg/kg MEPAAX 24,71,73
ihl-rat LC50:570 ppm/4H AMIHAB 12,609,55

HHL000 CAS: 7647-01-0
HYDROCHLORIC ACID

DOT: 1050/1789/2186
mf: ClH mw: 36.46

PROP: Colorless, fuming gas or colorless, fuming liquid; strongly corrosive with pungent odor. Mp: −114.3°, bp: −84.8°, d: 1.639 g/L (gas) @ 0°, 1.194 @ −26° (liquid), vap press: 4.0 atm @ 17.8°. Misc with water, alc.

SYNS: ACIDE CHLORHYDRIQUE (FRENCH) ◇ ACIDO CLORIDRICO (ITALIAN) ◇ CHLOORWATERSTOF (DUTCH) ◇ CHLOROHYDRIC ACID ◇ CHLOROWODOR (POLISH) ◇ CHLORWASSERSTOFF (GERMAN) ◇ HYDROCHLORIC ACID, ANHYDROUS (DOT) ◇ HYDROCHLORIC ACID, SOLUTION, INHIBITED (DOT) ◇ HYDROCHLORIDE ◇ HYDROGEN CHLORIDE (ACGIH, DOT) ◇ HYDROGEN CHLORIDE, ANHYDROUS (DOT) ◇ HYDROGEN CHLORIDE, REFRIGERATED LIQUID (DOT) ◇ MURIATIC ACID (DOT) ◇ SPIRITS OF SALT (DOT)

USE IN FOOD:

Purpose: Acid, buffer, neutralizing agent.

Where Used: Various.

Regulations: FDA - 21CFR 182.1057. GRAS when used as a buffer and neutralizing agent in accordance with good manufacturing practice. Must be free of contamination introduced from by-product manufacturing process.

EPA Extremely Hazardous Substances List. Community Right-To-Know List. EPA Genetic Toxicology Program.

OSHA PEL: CL 5 ppm ACGIH TLV: CL 5 ppm DOT Classification: Corrosive Material; Label: Corrosive, solution; Nonflammable Gas; Label: Nonflammable Gas; IMO: Nonflammable Gas; Label: Nonflammable Gas, Corrosive

SAFETY PROFILE: A human poison by an unspecified route. Mildly toxic to humans by inhalation. Moderately toxic experimentally by ingestion. A corrosive irritant to the skin, eyes and mucous membranes. Mutagenic data. An experimental teratogen. A concentration of 35 ppm causes irritation of the throat after short exposure. In general, hydrochloric acid causes little trouble in industry other than from accidental splashes and burns. It is a common air contaminant and is heavily used in industry and homes.

Nonflammable Gas. Explosive reaction with many chemicals. Potentially dangerous reaction with sulfuric acid releases HCl gas. When heated to decomposition it emits toxic fumes of Cl^-.

TOXICITY DATA and CODEN

eye-rbt 100 mg rns MLD TXCYAC 23,281,82
dnr-esc 25 µg/well ENMUDM 3,429,81
cyt-grh-par 20 mg NULSAK 9,119,66
ihl-rat TCLo:450 mg/m³/1H (1D pre):TER AKGIAO 53(6),69,77
ihl-hmn LCLo:1300 ppm/30M 29ZWAE -,207,68
ihl-rat LC50:3124 ppm/1H AMRL** TR-74-78,74

HHP000 CAS: 104-53-0
HYDROCINNAMALDEHYDE

mf: $C_9H_{10}O$ mw: 134.19

PROP: Colorless to sltly yellow liquid; strong floral, hyacinth odor. Bp: 221-224°, d: 1.010-1.020, refr index: 1.520-1.532, flash p: 203°F. Misc with alc, ether; insol in water.

SYNS: BENZENEPROPANAL ◇ BENZYLACETALDEHYDE ◇ DIHYDROCINNAMALDEHYDE ◇ FEMA No. 2887 ◇ HYDROCINNAMIC ALDEHYDE ◇ 3-PHENYLPROPANAL ◇ 3-PHENYL-1-PROPANAL ◇ 3-PHENYLPROPIONALDEHYDE (FCC) ◇ β-PHENYLPROPIONALDEHYDE ◇ 3-PHENYLPROPYL ALDEHYDE

USE IN FOOD:

Purpose: Flavoring agent

Where Used: Various.

Regulations: FDA - 21CFR 172.515. Use at a level not in excess of the amount reasonably required to accomplish the intended effect.

SAFETY PROFILE: A poison by intravenous route. A human skin irritant. Combustible liquid. When heated to decomposition it emits acrid smoke and fumes.

TOXICITY DATA and CODEN

skn-hmn 100 % FCTXAV 12,967,74
ivn-mus LD50:56 mg/kg CSLNX* NX#05219

HHP050 CAS: 122-97-4
HYDROCINNAMIC ALCOHOL

mf: $C_9H_{12}O$ mw: 136.21

PROP: Colorless sltly viscous liquid; sweet, hyacinth-mignonette odor. D: .998-1.002, refr index: 1.524-1.528, flash p: 228°F. Sol in fixed oils, propylene glycol; insol in glycerin.

SYNS: 3-BENZENEPROPANOL ◇ FEMA No. 2885 ◇ HYDROCINNAMYL ALCOHOL ◇ (3-HYDROXYPROPYL) BENZENE ◇ γ-PHENYLPROPANOL ◇ 3-PHENYLPROPANOL ◇ 3-PHENYL-1-PROPANOL (FCC) ◇ PHENYLPROPYL ALCOHOL ◇ γ-PHENYLPROPYL ALCOHOL ◇ 3-PHENYLPROPYL ALCOHOL

USE IN FOOD:

Purpose: Flavoring agent

Where Used: Various.

Regulations: FDA - 21CFR 172.515. Use at a level not in excess of the amount reasonably required to accomplish the intended effect.

SAFETY PROFILE: Moderately toxic by ingestion. Mildly toxic by skin contact. A skin irritant. Combustible liquid. When heated to decomposition it emits toxic fumes.

TOXICITY DATA and CODEN

skn-rbt 500 mg/24H MOD FCTXAV 17,893,79
orl-rat LD50:2300 mg/kg FCTXAV 17,893,79
skn-rbt LD50:5000 mg/kg FCTXAV 17,893,79

HHP500 CAS: 122-72-5
HYDROCINNAMYL ACETATE

mf: $C_{11}H_{14}O_2$ mw: 178.25

PROP: Colorless liquid; spicy, floral odor. D: 1.012, refr index: 1.494, flash p: +212°F. Sol in alc; insol in water.

SYNS: FEMA No. 2890 ◇ 3-PHENYL-1-PROPANOL ACETATE ◇ PHENYLPROPYL ACETATE ◇ 3-PHENYLPROPYL ACETATE (FCC) ◇ 3-PHENYL-1-PROPYL ACETATE

USE IN FOOD:

Purpose: Flavoring agent

Where Used: Various.

Regulations: FDA - 21CFR 172.515. Use at a level not in excess of the amount reasonably required to accomplish the intended effect.

SAFETY PROFILE: Mildly toxic by ingestion. Combustible liquid. When heated to decomposition it emits acrid smoke and fumes.

TOXICITY DATA and CODEN

orl-rat LD50:4700 mg/kg FCTXAV 12,965,74

HHR000 CAS: 125-04-2
HYDROCORTISONE SODIUM SUCCINATE

mf: $C_{25}H_{35}O_9 \cdot Na$ mw: 502.59

PROP: White, odorless, hygroscopic, amorphous solid. Mp: 169-171°. Very sol in water and alc; insol in chloroform, very sltly sol in acetone.

SYNS: A-HYDROCORT ◇ BUCCALSONE ◇ CORLAN ◇ CORTISOL HEMISUCCINATE SODIUM SALT ◇ CORTISOL SODIUM HEMISUCCINATE ◇ CORTISOL SODIUM SUCCINATE ◇ CORTISOL-21-SODIUM SUCCINATE ◇ CORTISOL SUCCINATE, SODIUM SALT ◇ EL-CORTELAN SOLUBLE ◇ EMI-CORLIN ◇ FLEBOCORTID ◇ HYCORACE ◇ HYDROCORTISONE-21-SODIUM SUCCINATE ◇ 21-(HYDROGEN SUCCINATE)CORTISOL, MONOSODIUM SALT ◇ INTRACORT ◇ NORDICORT ◇ ORALSONE ◇ SODIUM HYDROCORTISONE SUCCINATE ◇ SODIUM HYDROCORTISONE-21-SUCCINATE ◇ SOLU-CORTEF ◇ SOLU-GLYC ◇ U 4905

USE IN FOOD:

Purpose: Animal drug.

Where Used: Milk.

Regulations: FDA - 21CFR 556.320. Limitation of 10 ppb in milk.

SAFETY PROFILE: Moderately toxic by intraperitoneal route. Experimental teratogenic and reproductive effects. When heated to decomposition it emits toxic fumes of Na_2O.

TOXICITY DATA and CODEN

ipr-rat TDLo:1800 mg/kg (9-14D preg):TER KSRNAM 4,2969,70
ipr-rat TDLo:1500 mg/kg (9-14D preg):REP KSRNAM 4,2969,70
scu-rbt TDLo:30 mg/kg (28-30D preg):TER KRMJAC 27,51,80
ipr-rat LD50:1320 mg/kg NIIRDN 6,625,82

HHR500 CAS: 119-84-6
HYDROCOUMARIN

mf: $C_9H_8O_2$ mw: 148.17

PROP: Colorless to pale yellow liquid; coconut odor. D: 1.186, refr index: 1.555, flash p: 266°F.

SYNS: 1,2-BENZODIHYDROPYRONE (FCC) ◇ 2-CHROMANONE ◇ DIHYDROCOUMARIN ◇ 3,4-DIHYDROCOUMARIN ◇ o-HYDROXY-HYDROCINNAMIC ACID-Δ-LACTONE ◇ FEMA No. 2381 ◇ MELILOTIN ◇ MELILOTOL ◇ NCI-C55890 ◇ 2-OXOCHROMAN ◇ USAF DO-12

USE IN FOOD:

Purpose: Flavoring agent.

Where Used: Various.

Regulations: GRAS when used at a level not in excess of the amount reasonably required to accomplish the intended effect.

SAFETY PROFILE: A poison by intraperitoneal route. Moderately toxic by ingestion. A skin irritant. Combustible liquid. When heated to decomposition it emits acrid smoke and fumes.

TOXICITY DATA and CODEN

skn-rbt 500 mg/24H MOD FCTXAV 12,521,74
orl-rat LD50: 1460 mg/kg FCTXAV 2,327,64

HHW560 CAS: 8016-14-6
HYDROGENATED FISH OIL

PROP: Oil. Mp: >32°.

USE IN FOOD:

Purpose: Constituent of cotton and cotton fabrics.

Where Used: Packaging materials.

Regulations: FDA - 21CFR 186.1551. GRAS when used in accordance with good manufacturing practice.

SAFETY PROFILE: When heated to decomposition it emits acrid smoke and irritating fumes.

HHW575
HYDROGENATED SPERM OIL

USE IN FOOD:

Purpose: Release agent in bakery pans.

Where Used: Baked goods.

Regulations: FDA - 21CFR 173.275. Use at a level not in excess of the amount reasonably required to accomplish the intended effect.

SAFETY PROFILE: When heated to decomposition it emits acrid smoke and irritating fumes.

I

IAQ000 CAS: 96-45-7
2-IMIDAZOLIDINETHIONE
mf: $C_3H_6N_2S$ mw: 102.17

PROP: White crystals. Water solubility: 9 g/ 100 mL @ 30°. Often occurs as a main degradation product of the metal salts of ethylene bisdithiocarbamic acid.

SYNS: 4,5-DIHYDROIMIDAZOLE-2(3H)-THIONE ◇ ETHYLENE THIOUREA ◇ N,N'-ETHYLENETHIOUREA ◇ 1,3-ETHYLENE-2-THIOUREA ◇ l'ETHYLENE THIOUREE (FRENCH) ◇ ETU ◇ 2-MERCAPTOIMIDAZOLINE ◇ 2-MERKAPTOIMIDAZOLIN (CZECH) ◇ NA-22 ◇ NCI-C03372 ◇ PENNAC CRA ◇ RCRA WASTE NUMBER U116 ◇ RODANIN S-62 (CZECH) ◇ SODIUM-22 NEOPRENE ACCELERATOR ◇ 2-THIOL-DIHYDROGLYOXALINE ◇ USAF EL-62 ◇ VULKACIT NPV/C2 ◇ WARECURE C

USE IN FOOD:

Purpose: Used in rubber articles.

Where Used: Prohibited from foods.

Regulations: FDA - 21CFR 189.250. Prohibited from indirect addition or use in human food.

IARC Cancer Review: Animal Sufficient Evidence IMEMDT 7,45,74. Community Right-To-Know List. EPA Genetic Toxicology Program.

NIOSH REL: (ETU) Use encapsulated form; minimize exposure.

SAFETY PROFILE: Poison by ingestion and intraperitoneal routes. An experimental carcinogen. Experimental teratogenic and reproductive effects. Mutagenic data. An eye irritant. When heated to decomposition it emits very toxic fumes of NO_x and SO_x.

TOXICITY DATA and CODEN

eye-rbt 500 mg/24H MLD 28ZPAK -,167,72
mma-sat 3333 μg/plate ENMUDM 8(Suppl 7),1,86
otr-ham:kdy 80 μg/L BJCAAI 37,873,78
orl-rat TDLo:60 mg/kg (13D preg):TER TJADAB 24,131,81
ihl-rat TCLo:120 mg/m^3/3H (7-14D preg):REP NTIS** PB277-077
skn-rat TDLo:100 mg/kg (12-13D preg):TER TXAPA9 41,35,77
orl-rat TDLo:5306 mg/kg/77W-C:CAR JNCIAM 49,583,72
orl-rat TD :11466 mg/kg/78W-C:CAR JJIND8 67,75,81
orl-rat LD50:265 mg/kg 28ZPAK -,167,72
ipr-mus LD50:200 mg/kg NTIS** AD277-689

ICM000 CAS: 120-72-9
INDOLE
mf: C_8H_7N mw: 117.16

PROP: Colorless to yellowish scales; intense fecal odor. Mp: 52°, bp: 253°; volatile with steam. Sol in hot water, alc, ether, petroleum ether; insol in mineral oil, glycerin.

SYNS: 1-AZAINDENE ◇ 1-BENZAZOLE ◇ BENZOPYRROLE ◇ 2,3-BENZOPYRROLE ◇ FEMA No. 2593 ◇ INDOL (GERMAN) ◇ KETOLE

USE IN FOOD:

Purpose: Flavoring agent.

Where Used: Various.

Regulations: FDA - 21CFR 172.515. Use at a level not in excess of the amount reasonably required to accomplish the intended effect.

SAFETY PROFILE: A poison by intraperitoneal and subcutaneous routes. Moderately toxic by ingestion and skin contact. An experimental carcinogen and tumorigen. When heated to decomposition it emits toxic fumes of NO_x.

TOXICITY DATA and CODEN

scu-mus TDLo:1000 mg/kg/25W-I:CAR KLWOAZ 35,504,57
scu-mus TD :2000 mg/kg/20W-I:ETA AICCA6 19,660,63
orl-rat LD50:1000 mg/kg AIHAAP 23,95,62

IDE300 CAS: 87-89-8
INOSITOL
mf: $C_6H_{12}O_6$ mw: 180.16

PROP: White crystals or crystalline powder; odorless with a sweet taste. Sol in water; insol in ether, chloroform.

SYNS: cis-1,2,3,5-trans-4,6-CYCLOHEXANEHEXOL ◇ i-INOSITOL ◇ meso-INOSITOL

USE IN FOOD:

Purpose: Dietary supplement, nutrient.

Where Used: Various.

Regulations: FDA - 21CFR 182.5370, 184.1370. GRAS when used in accordance with good manufacturing practice.

SAFETY PROFILE: When heated to decomposition it emits acrid smoke and irritating fumes.

IDH200 CAS: 8013-17-0

INVERT SUGAR

PROP: Hygroscopic liquid; sweet taste. Very sol in water, glycerin, glycols; sltly sol in acetone, alc.

SYN: INVERT SUGAR SYRUP

USE IN FOOD:

Purpose: Nutritive sweetener.

Where Used: Candy, icings, soft drinks.

Regulations: GRAS when used in accordance with good manufacturing practice.

SAFETY PROFILE: When heated to decomposition it emits acrid smoke and irritating fumes.

IDL100

IODINATED CASEIN

USE IN FOOD:

Purpose: Animal feed drug.

Where Used: Animal feed.

Regulations: FDA - 21CFR 558.295.

SAFETY PROFILE: When heated to decomposition it emits acrid smoke and irritating fumes.

IFW000 CAS: 127-41-3

α-IONONE

mf: $C_{13}H_{20}O$ mw: 192.33

PROP: Colorless oil; woody, violet odor. D: 0.930, refr index: 1.497-1.502, bp: 136.1. Sol in alc, fixed oils propylene glycol; sltly sol in water; misc in ether; insol in glycerin.

SYNS: α-CYCLOCITRYLIDENEACETONE ◇ FEMA No. 2594 ◇ 4-(2,6,6-TRIMETHYL-2-CYCLOHEXEN-1-YL)-3-BUTEN-2-ONE

USE IN FOOD:

Purpose: Flavoring agent.

Where Used: Various.

Regulations: FDA - 21CFR 172.515. Use at a level not in excess of the amount reasonably required to accomplish the intended effect.

SAFETY PROFILE: Mildly toxic by ingestion. When heated to decomposition it emits acrid smoke and fumes.

TOXICITY DATA and CODEN

orl-rat LD50:4590 mg/kg FCTXAV 2,327,64

IFX000 CAS: 14901-07-6

β-IONONE

mf: $C_{13}H_{20}O$ mw: 192.33

PROP: Colorless oil; woody odor. D: 0.944, refr index: 1.517-1.522, bp: 140°, flash p: +234°F. Sol in alc, fixed oils, propylene glycol; sltly sol in water; misc in ether; insol in glycerin.

SYNS: β-CYCLOCITRYLIDENEACETONE ◇ FEMA No. 2595 ◇ 4-(2,6,6-TRIMETHYL-1-CYCLOHEXEN-1-YL)-3-BUTEN-2-ONE

USE IN FOOD:

Purpose: Flavoring agent.

Where Used: Various.

Regulations: FDA - 21CFR 172.515. Use at a level not in excess of the amount reasonably required to accomplish the intended effect.

SAFETY PROFILE: Mildly toxic by ingestion. Combustible liquid. When heated to decomposition it emits acrid smoke and fumes.

TOXICITY DATA and CODEN

orl-rat LD50:4590 mg/kg FCTXAV 2,327,64

IGH000 CAS: 14885-29-1

IPROPRAN

mf: $C_7H_{11}N_3O_2$ mw: 169.21

SYNS: IPRONIDAZOLE (USDA) ◇ 2-ISOPROPYL-1-METHYL-5-NITROIMIDAZOLE ◇ 1-METHYL-2-(1-METHYLETHYL)-5-NITRO-1H-IMIDAZOLE ◇ RO 7-1554

USE IN FOOD:

Purpose: Animal drug, animal feed drug.

Where Used: Animal feed, turkey.

Regulations: FDA - 21CFR 556.340. Limitation of zero in turkey. 21CFR 558.305. USDA CES Ranking: Z-4 (1986).

SAFETY PROFILE: Moderately toxic by ingestion. Mutagenic data. Used as an antiprotozoal and antimicrobial agent. When heated to decomposition it emits toxic fumes of NO_x.

TOXICITY DATA and CODEN

mmo-sat 1 μmol/L TCMUD8 3,429,83
mmo-smc 5 ppm MUREAV 86,243,81
orl-trk LD50:640 mg/kg POSCAL 49,92,70

IGK800 CAS: 7439-89-6
IRON
mf: Fe mw: 55.85

PROP: From decomposition of iron pentacarbonyl: dark grey powder. From electrodeposition: lusterless, gray black powder. From chemical reduction: gray-black powder.

SYNS: ANCOR EN 80/150 ◇ ARMCO IRON ◇ CARBONYL IRON ◇ IRON, CARBONYL (FCC) ◇ IRON, ELECTROLYTIC ◇ IRON, ELEMENTAL ◇ IRON, REDUCED (FCC)

USE IN FOOD:

Purpose: Dietary supplement, nutrient supplement.

Where Used: Baked goods, cereal products, flour, pasta.

Regulations: FDA - 21CFR 182.5375, 182.8375, 184.1375. GRAS when used in accordance with good manufacturing practice.

SAFETY PROFILE: Poison by intraperitoneal route. Iron is potentially toxic in all forms and by all routes of exposure. The inhalation of large amounts of iron dust results in iron pneumoconiosis (arc welder's lung). Chronic exposure to excess levels of iron (> 50-100 mg Fe/day) can result in pathological deposition of iron in the body tissues, the symptoms of which are fibrosis of the pancreas, diabetes mellitus, and liver cirrhosis.

As with other metals, it becomes more reactive as it is more finely divided. Ultrafine iron powder is pyrophoric and potentially explosive.

TOXICITY DATA and CODEN

ipr-rbt LDLo:20 mg/kg NTIS** PB158-508

IGQ050
IRON CAPRYLATE

USE IN FOOD:

Purpose: Drying agent.

Where Used: Packaging materials.

Regulations: FDA - 21CFR 181.25. Use in accordance with good manufacturing practice.

SAFETY PROFILE: When heated to decomposition it emits acrid smoke and irritating fumes.

IHA050
IRON LINOLEATE

USE IN FOOD:

Purpose: Drying agent.

Where Used: Packaging materials.

Regulations: FDA - 21CFR 181.25. Use in accordance with good manufacturing practice.

SAFETY PROFILE: When heated to decomposition it emits acrid smoke and irritating fumes.

IHB700
IRON NAPHTHENATE

USE IN FOOD:

Purpose: Drying agent.

Where Used: Packaging materials.

Regulations: FDA - 21CFR 181.25. Use in accordance with good manufacturing practice.

SAFETY PROFILE: When heated to decomposition it emits acrid smoke and irritating fumes.

IHD000 CAS: 1309-37-1
IRON(III) OXIDE
mf: Fe_2O_3 mw: 159.70

SYNS: ANCHRED STANDARD ◇ ANHYDROUS IRON OXIDE ◇ ANHYDROUS OXIDE of IRON ◇ ARMENIAN BOLE ◇ BAUXITE RESIDUE ◇ BLACK OXIDE of IRON ◇ BLENDED RED OXIDES of IRON ◇ BURNTISLAND RED ◇ BURNT SIENNA ◇ BURNT UMBER ◇ CALCOTONE RED ◇ CAPUT MORTUUM ◇ C.I. 77491 ◇ C.I. PIGMENT RED 101 ◇ COLCOTHAR ◇ COLLOIDAL FERRIC OXIDE ◇ CROCUS MARTIS ADSTRINGENS ◇ DEANOX ◇ EISENOXYD ◇ ENGLISH RED ◇ FERRIC OXIDE ◇ FERRUGO ◇ INDIAN RED ◇ IRON OXIDE (ACGIH) ◇ IRON OXIDE RED ◇ IRON SESQUIOXIDE ◇ JEWELER'S ROUGE ◇ LEVANOX RED 130A ◇ LIGHT RED ◇ MANUFACTURED IRON OXIDES ◇ MARS BROWN ◇ MARS RED ◇ NATURAL IRON OXIDES ◇ NATURAL RED OXIDE ◇ OCHRE ◇ PRUSSIAN BROWN ◇ RADDLE ◇ 11554 RED ◇ RED IRON OXIDE ◇ RED OCHRE ◇ ROUGE ◇ RUBIGO ◇ SIENNA ◇ SPECULAR IRON ◇ STONE RED ◇ SUPRA ◇ SYNTHETIC IRON OXIDE ◇ VENETIAN RED ◇ VITRIOL RED ◇ VOGEL'S IRON RED ◇ YELLOW FERRIC OXIDE ◇ YELLOW OXIDE OF IRON

USE IN FOOD:

Purpose: Color additive, constituent of paperboard.

Where Used: Cat food, dog food, packaging materials.

Regulations: FDA - 21CFR 73.200. Limitation of 0.25 percent. 21CFR 186.1300, 186.1374.

GRAS when used in accordance with good manufacturing practice.

IARC Cancer Review: Human Limited Evidence IMEMDT 1,29,72; Animal No Evidence IMEMDT 1,29,72.

ACGIH TLV: TWA 5 mg(Fe)/m^3 (vapor, dust)

SAFETY PROFILE: A poison by subcutaneous route. A suspected human carcinogen. An experimental tumorigen. Catalyzes the potentially explosive polymerization of ethylene oxide.

TOXICITY DATA and CODEN

scu-rat TDLo: 135 mg/kg: ETA PBPHAW 14,47,78
ipr-rat LD50: 5500 mg/kg GTPZAB 26(4),23,82

IHM000 CAS: 7720-78-7
IRON(II) SULFATE (1:1)

DOT: 9125
mf: $O_4S \cdot Fe$ mw: 151.91

PROP: Grayish white to buff powder. Slowly sol in water; insol in alc.

SYNS: COPPERAS ◇ DURETTER ◇ DUROFERON ◇ EXSICCATED FERROUS SULFATE ◇ EXSICCATED FERROUS SULPHATE ◇ FEOSOL ◇ FEOSPAN ◇ FER-IN-SOL ◇ FERO-GRADUMET ◇ FERRALYN ◇ FERRO-GRADUMET ◇ FERROSULFAT (GERMAN) ◇ FERROSULFATE ◇ FERRO-THERON ◇ FERROUS SULFATE (DOT, FCC) ◇ FERSOLATE ◇ GREEN VITRIOL ◇ IRON MONOSULFATE ◇ IRON PROTOSULFATE ◇ IRON VITRIOL ◇ IROSPAN ◇ IROSUL ◇ SLOW-FE ◇ SULFERROUS ◇ SULFURIC ACID, IRON(2^+) SALT (1:1)

USE IN FOOD:

Purpose: Dietary supplement, nutrient.

Where Used: Various.

Regulations: FDA - 21CFR 182.5315, 182,8315. GRAS when used in accordance with good manufacturing practice.

EPA Genetic Toxicology Program.

ACGIH TLV: TWA 1 mg/(Fe)/m^3 DOT Classification: ORM-E; Label: None

SAFETY PROFILE: A human poison by ingestion. Moderately toxic to humans by an unspecified route. An experimental poison by ingestion, intraduodenal, intraperitoneal, intravenous, and subcutaneous routes. An experimental tumorigen. Human systemic effects by ingestion: aggression, somnolence, brain recording changes, diarrhea, nausea or vomiting, bleeding from the stomach, coma. A systemic toxin. Experimental reproductive effects. Mutagenic data. When heated to decomposition it emits toxic fumes of SO_x.

TOXICITY DATA and CODEN

dnd-omi 2 µmol/L BBRCA9 77,1150,77
mmo-smc 100 mmol/L MUREAV 117,149,83
orl-rat TDLo: 7200 mg/kg: TER OYYAA2 17,483,79
itt-rat TDLo: 12153 mg/kg (1D male): REP JRPFA4 7,21,64
scu-mus TDLo: 1600 mg/kg/16W-I: ETA JNCIAM 24,109,60
orl-wmn TDLo: 10560 µg/kg: GIT JAMAAP 236,2320,76
orl-chd TDLo: 20 mg/kg: BRN,CNS JOPDAB 94,147,79
orl-chd TDLo: 150 mg/kg: CNS,GIT NEJMAG 273,1124,65
orl-wmn TDLo: 60 mg/kg: CNS,GIT JAMAAP 229,1333,74
orl-rat LD50: 319 mg/kg JOPDAB 69,663,66

IHN075
IRON TALLATE

USE IN FOOD:

Purpose: Drying agent.

Where Used: Packaging materials.

Regulations: FDA - 21CFR 181.25. Use in accordance with good manufacturing practice.

SAFETY PROFILE: When heated to decomposition it emits acrid smoke and irritating fumes.

IHP400 CAS: 106-27-4
ISOAMYL BUTYRATE

mf: $C_9H_{18}O_2$ mw: 158.24

PROP: Colorless liquid; fruity odor. D: 0.860, refr index: 1.409-1.414, flash p: 149°F. Sol in alc, fixed oils; insol in glycerin, propylene glycol, water @ 179°.

SYNS: AMYL BUTYRATE ◇ FEMA No. 2060

USE IN FOOD:

Purpose: Flavoring agent.

Where Used: Baked goods, dessert gels, puddings.

Regulations: FDA - 21CFR 172.515. Use at a level not in excess of the amount reasonably required to accomplish the intended effect.

SAFETY PROFILE: Combustible liquid. When heated to decomposition it emits acrid smoke and irritating fumes.

IHS000 CAS: 110-45-2
ISOAMYL FORMATE
DOT: 1109
mf: $C_6H_{12}O_2$ mw: 116.18

PROP: Clear liquid; fruity odor. Bp: 123.3°, d: 0.877 @ 20°, refr index: 1.396, vap press: 10 mm @ 17.1°, flash p: 127°F. Misc with alc, ether, propylene glycol; very sltly sol in water; insol in glycerin.

SYNS: FEMA No. 2069 ◇ FORMIC ACID, ISOPENTYL ESTER ◇ ISOAMYL METHANOATE ◇ ISOPENTYL ALCOHOL, FORMATE ◇ ISOPENTYL FORMATE ◇ 3-METHYLBUTYL FORMATE

USE IN FOOD:

Purpose: Flavoring agent.

Where Used: Candy, dessert gels, ice cream, puddings.

Regulations: FDA - 21CFR 172.515. Use at a level not in excess of the amount reasonably required to accomplish the intended effect.

DOT Classification: Flammable or Combustible Liquid; Label: Flammable Liquid.

SAFETY PROFILE: Moderately toxic by ingestion. A skin irritant. This material is very irritating and can cause narcosis. Combustible liquid. Can react with oxidizing materials. When heated to decomposition it emits acrid smoke and fumes.

TOXICITY DATA and CODEN

skn-rbt 500 mg/24H MOD FCTXAV 17,829,79
orl-rat LD50:9840 mg/kg FCTXAV 2,327,64

IHU100
ISOAMYL HEXANOATE
mf: $C_{11}H_{22}O_2$ mw: 186.29

PROP: Colorless liquid; fruity odor. D: 0.858-0.863, refr index: 1.418-1.422, flash p: 190°F. Sol in alc, fixed oils; insol in glycerin, propylene glycol, water @ 222°.

SYNS: AMYL HEXANOATE ◇ FEMA No. 2075 ◇ ISOAMYL CAOPROATE ◇ PENTYL HEXANOATE

USE IN FOOD:

Purpose: Flavoring agent.

Where Used: Candy, desserts, ice cream.

Regulations: FDA - 21CFR 172.515. Use at a level not in excess of the amount reasonably required to accomplish the intended effect.

SAFETY PROFILE: Combustible liquid. When heated to decomposition it emits acrid smoke and irritating fumes.

IHX600
ISOBORNYL ACETATE
mf: $C_{12}H_{20}O_2$ mw: 196.29

PROP: Colorless liquid; camphoraceous, piney, balsamic odor. D: 0.980, refr index: 1.462, flash p: +212°F. Sol in alc, fixed oils; sltly sol in propylene glycol; insol in water @ 227°.

SYN: FEMA No. 2160

USE IN FOOD:

Purpose: Flavoring agent.

Where Used: Various.

Regulations: FDA - 21CFR 172.515. Use at a level not in excess of the amount reasonably required to accomplish the intended effect.

SAFETY PROFILE: Combustible liquid. When heated to decomposition it emits acrid smoke and irritating fumes.

IIJ000 CAS: 110-19-0
ISOBUTYL ACETATE
DOT: 1213/1123
mf: $C_6H_{12}O_2$ mw: 116.18

PROP: Colorless, neutral liquid; fruit-like odor. Mp: −98.9°, bp: 118°, flash p: 64°F (CC) (18°), d: 0.8685 @ 15°, refr index: 1.389, vap press: 10 mm @ 12.8°, autoign temp: 793°F, vap d: 4.0, lel: 2.4%, uel: 10.5%. Very sol in alc, fixed oils, propylene glycol; sltly sol in water.

SYNS: ACETATE d'ISOBUTYLE (FRENCH) ◇ ACETIC ACID, ISOBUTYL ESTER ◇ ACETIC ACID-2-METHYLPROPYL ESTER ◇ FEMA No. 2175 ◇ 2-METHYLPROPYL ACETATE ◇ 2-METHYL-1-PROPYL ACETATE ◇ β-METHYLPROPYL ETHANOATE

USE IN FOOD:

Purpose: Flavoring agent.

Where Used: Various.

Regulations: FDA - 21CFR 172.515. Use at a level not in excess of the amount reasonably required to accomplish the intended effect.

OSHA PEL: TWA 150 ppm ACGIH TLV: TWA 150 ppm; STEL 187 ppm DOT Classification: Flammable Liquid; Label: Flammable Liquid

SAFETY PROFILE: Mildly toxic by ingestion and inhalation. A skin and eye irritant. Upon absorption by the body it can hydrolyze to acetic acid and isobutanol. Highly flammable liquid. A very dangerous fire and moderate explosion hazard when exposed to heat, flame or oxidizers. To fight fire, use alcohol foam, CO_2, dry chemical. When heated to decomposition it emits acrid smoke and fumes.

TOXICITY DATA and CODEN

skn-rbt 500 mg/24H MOD FCTXAV 16,637,78
eye-rbt 500 mg/24H MOD FCTXAV 16,637,78
orl-rat LD50:13400 mg/kg NPIRI* 1,8,74
ihl-rat LCLo:8000 ppm/4H AIHAAP 23,95,62

IIL000 CAS: 78-83-1
ISOBUTYL ALCOHOL
DOT: 1212
mf: $C_4H_{10}O$ mw: 74.14

PROP: Clear mobile liquid; sweet odor. Bp: 107.90°, flash p: 82°F, ULC: 40-45, lel: 1.2%, uel: 10.9% @ 212°F, fp: −108°, d: 0.800, autoign temp: 800°F, vap press: 10 mm @ 21.7°, vap d: 2.55. Sltly sol in water; misc with alc, ether.

SYNS: ALCOOL ISOBUTYLIQUE (FRENCH) ◇ FEMA No. 2179 ◇ FERMENTATION BUTYL ALCOHOL ◇ 1-HYDROXYMETHYLPROPANE ◇ ISOBUTANOL (DOT) ◇ ISOBUTYLALKOHOL (CZECH) ◇ ISOPROPYLCARBINOL ◇ 2-METHYL PROPANOL ◇ 2-METHYL-1-PROPANOL ◇ 2-METHYLPROPAN-1-OL ◇ 2-METHYLPROPYL ALCOHOL ◇ RCRA WASTE NUMBER U140

USE IN FOOD:

Purpose: Color diluent, flavoring agent.

Where Used: Confectionery, food supplements in tablet form, gum.

Regulations: FDA - 21CFR 73.1, 172.515. Use in accordance with good manufacturing practice. 21CFR 73.1. Limitation of no residue.

OSHA PEL: TWA 50 ppm ACGIH TLV: TWA 50 ppm DOT Classification: Flammable or Combustible Liquid; Label: Flammable Liquid

SAFETY PROFILE: Poison by intravenous and intraperitoneal routes. Moderately toxic by ingestion and skin contact. Mildly toxic by inhalation. An experimental carcinogen and tumorigen. A severe skin and eye irritant. Mutagenic data. Flammable liquid. Dangerous fire hazard when exposed to heat or flame. Moderately explosive in the form of vapor when exposed to heat, flame or oxidizers. Keep away from heat and open flame. To fight fire, use alcohol foam, CO_2, dry chemical. When heated to decomposition it emits acrid smoke and fumes.

TOXICITY DATA and CODEN

skn-rbt 500 mg/24H SEV 28ZPAK -,35,72
eye-rbt 2 mg open SEV AMIHBC 10,61,54
mmo-esc 25000 ppm ABMGAJ 23,843,69
cyt-smc 20 mmol/tube HEREAY 33,457,47
orl-rat TDLo:29 g/kg/I:ETA ARGEAR 45,19,75
scu-rat TDLo:9 g/kg/I:CAR ARGEAR 45,19,75
orl-rat LD50:2460 mg/kg AMIHBC 10,61,54
ihl-rat LCLo:8000 ppm/4H AMIHBC 10,61,54

IIN300
ISOBUTYL-2-BUTENOATE
mf: $C_8H_{14}O_2$ mw: 142.19

PROP: Colorless liquid; powerful fruity odor. D: 0.880, refr index: 1.426-1.430. Sol in alc, propylene glycol, fixed oils; sltly sol in water.

SYN: FEMA No. 3432

USE IN FOOD:

Purpose: Flavoring agent.

Where Used: Various.

Regulations: GRAS when used at a level not in excess of the amount reasonably required to accomplish the intended effect.

SAFETY PROFILE: When heated to decomposition it emits acrid smoke and irritating fumes.

IIQ000 CAS: 122-67-8
ISOBUTYL CINNAMATE
mf: $C_{13}H_{16}O_2$ mw: 204.29

PROP: Colorless liquid; sweet, fruity odor. D: 1.001, refr index: 1.539-1.541, flash p: +212°F. Misc with alc, chloroform, ether, fixed oils; insol in water.

SYNS: CINNAMIC ACID, ISOBUTYL ESTER ◇ FEMA No. 2193 ◇ LABDANOL ◇ 3-PHENYL-2-PROPENOIC ACID, 2-METHYLPROPYL ESTER

USE IN FOOD:

Purpose: Flavoring agent.

Where Used: Baked goods, beverages, candy, ice cream.

Regulations: FDA - 21CFR 172.515. Use at a level not in excess of the amount reasonably required to accomplish the intended effect.

SAFETY PROFILE: A skin irritant. Combustible liquid. When heated to decomposition it emits acrid smoke and fumes.

TOXICITY DATA and CODEN

skn-rbt 500 mg/24H MLD FCTXAV 14,799,76

IIQ500
ISOBUTYLENE-ISOPRENE COPOLYMER

SYN: BUTYL RUBBER

USE IN FOOD:

Purpose: Masticatory substance in chewing gum base.

Where Used: Chewing gum.

Regulations: FDA - 21CFR 172.615. Use in amounts not to exceed those required to produce the intended physical or other technical effect.

SAFETY PROFILE: When heated to decomposition it emits acrid smoke and irritating fumes.

IIR000 CAS: 542-55-2
ISOBUTYL FORMATE

DOT: 2393
mf: $C_5H_{10}O_2$ mw: 102.15

PROP: Liquid. D: 0.885 @ 20°/4°, mp: −95.3°, bp: 98.2°, flash p: <70°F, autoign temp: 608°F, lel: 2.0%, uel: 8%. Sol in water @ 22°; misc in alc and ether.

SYNS: FORMIC ACID, ISOBUTYL ESTER ◇ TETRYL FORMATE

USE IN FOOD:

Purpose: Flavoring agent.

Where Used: Baked goods, beverages, candy, ice cream.

Regulations: FDA - 21CFR 172.515. Use at a level not in excess of the amount reasonably required to accomplish the intended effect.

DOT Classification: Flammable Liquid; Label: Flammable Liquid

SAFETY PROFILE: Moderately toxic by ingestion. A very dangerous fire hazard when exposed to heat, open flame, or oxidizers. A moderate explosion hazard when exposed to heat or flame. To fight fire, use water spray, foam, CO_2, dry chemical. When heated to decomposition it emits acrid smoke and fumes.

TOXICITY DATA and CODEN

orl-rbt LD50:3064 mg/kg IMSUAI 41,31,72

IJF400
ISOBUTYL PHENYLACETATE

mf: $C_{12}H_{16}O_2$ mw: 192.23

PROP: Colorless liquid; rose, honey-like odor. D: 0.984-0.988, refr index: 1.486, flash p: 241°F. Sol in alc, fixed oils; insol in glycerin, propylene glycol, water.

SYN: FEMA No. 2210

USE IN FOOD:

Purpose: Flavoring agent.

Where Used: Various.

Regulations: FDA - 21CFR 172.515. Use at a level not in excess of the amount reasonably required to accomplish the intended effect.

SAFETY PROFILE: Combustible liquid. When heated to decomposition it emits acrid smoke and irritating fumes.

IJN000 CAS: 87-19-4
ISOBUTYL SALICYLATE

mf: $C_{11}H_{14}O_3$ mw: 194.25

PROP: Colorless liquid; orchid odor. D: 1.062-1.066, refr index: 1.507, flash p: 250°F. Sol in fixed oils; insol in glycerin, propylene glycol.

SYNS: FEMA No. 2213 ◇ ISOBUTYL-o-HYDROXYBENZOATE ◇ SALICYLIC ACID, ISOBUTYL ESTER

USE IN FOOD:

Purpose: Flavoring agent.

Where Used: Various.

Regulations: FDA - 21CFR 172.515. Use at a level not in excess of the amount reasonably required to accomplish the intended effect.

SAFETY PROFILE: Moderately toxic by ingestion. Combustible liquid. When heated to decomposition it emits acrid smoke and fumes.

TOXICITY DATA and CODEN

orl-rat LD50:1560 mg/kg FCTXAV 13,681,75

IJS000 CAS: 78-84-2
ISOBUTYRALDEHYDE

DOT: 2045
mf: C_4H_8O mw: 72.12

PROP: Transparent, colorless, highly refractive liquid; pungent odor. Mp: −65°, bp: 64°, flash p: −40°F (CC), d: 0.783-0.788, autoign temp: 434°F, lel: 1.6%, uel: 10.6%, vap d: 2.5. Sol in water; misc in alc, ether, benzene, carbon disulfide, acetone, toluene, chloroform.

SYNS: FEMA No. 2220 ◇ ISOBUTANAL ◇ ISOBUTYRALDEHYD (CZECH) ◇ ISOBUTYLALDEHYDE ◇ ISOBUTYL ALDEHYDE (DOT) ◇ ISOBUTYRIC ALDEHYDE ◇ 2-METHYLPROPANAL ◇ 2-METHYL-1-PROPANAL ◇ 2-METHYLPROPIONALDEHYDE ◇ NCI-C60968 ◇ VALINE ALDEHDYE

USE IN FOOD:

Purpose: Flavoring agent.

Where Used: Various.

Regulations: FDA - 21CFR 172.515. Use at a level not in excess of the amount reasonably required to accomplish the intended effect.

Community Right-To-Know List.

DOT Classification: Flammable Liquid; Label: Flammable Liquid

SAFETY PROFILE: Moderately toxic by ingestion. Mildly toxic by skin contact and inhalation. A severe skin and eye irritant. A highly flammable liquid. A very dangerous fire hazard when exposed to heat, flame or oxidizers. Moderately explosive in the form of vapor when exposed to heat or flame. Can react vigorously with reducing materials. When heated to decomposition it emits acrid smoke and fumes. To fight fire, use dry chemical, CO_2, mist, foam.

TOXICITY DATA and CODEN

skn-rbt 500 mg/24H SEV 28ZPAK -,41,72
eye-rbt 20 mg open SEV AMIHBC 10,61,54
eye-rbt 100 mg/24H MOD 28ZPAK -,41,72
orl-rat LD50:2810 mg/kg 28ZPAK -,41,72
ihl-rat LCLo:8000 ppm/4H AMIHBC 10,61,54

IJU000 CAS: 79-31-2

ISOBUTYRIC ACID

DOT: 2529

mf: $C_4H_8O_2$ mw: 88.12

PROP: Colorless liquid; pungent odor of rancid butter. Mp: −47°, bp: 154.5°, flash p: 132°F (TOC), d: 0.949 @ 20°/4°, refr index: 1.392, vap press: 1 mm @ 14.7°, vap d: 3.04, autoign temp: 935°F. Misc with alc, chloroform and ether. Misc with alc, fixed oils, glycerin, propylene glycol; insol in water.

SYNS: DIMETHYLACETIC ACID ◇ FEMA No. 2222 ◇ ISOPROPYLFORMIC ACID ◇ α-METHYLPROPIONIC ACID ◇ 2-METHYLPROPANOIC ACID ◇ 2-METHYLPROPIONIC ACID

USE IN FOOD:

Purpose: Flavoring agent.

Where Used: Various.

Regulations: FDA - 21CFR 172.515. Use at a level not in excess of the amount reasonably required to accomplish the intended effect.

DOT Classification: Corrosive Material; Label: Corrosive; Flammable or Combustible Liquid; Label: Flammable Liquid

SAFETY PROFILE: A poison by ingestion. Moderately toxic by skin contact. A corrosive irritant to the eyes, skin and mucous membranes. Flammable liquid when exposed to heat or flame; can react with oxidizing materials. To fight fire, use alcohol foam, CO_2, dry chemical. When heated to decomposition it emits acrid smoke and fumes.

TOXICITY DATA and CODEN

skn-rbt 139 μg/24H open AIHAAP 23,95,62
orl-rat LD50:280 mg/kg AIHAAP 23,95,62
skn-rbt LD50:500 mg/kg AIHAAP 23,95,62

IJV000 CAS: 103-28-6

ISOBUTYRIC ACID, BENZYL ESTER

mf: $C_{11}H_{14}O_2$ mw: 178.25

PROP: Colorless liquid; floral, jasmine odor. D: 1.001-1.005, refr index: 1.489, flash p: 212°F. Sol in alc, fixed oils; sltly sol in propylene glycol; insol in glycerin @ 229°.

SYNS: BENZYL ISOBUTYRATE (FCC) ◇ BENZYL-2-METHYL PROPIONATE ◇ FEMA No. 2141

USE IN FOOD:

Purpose: Flavoring agent.

Where Used: Various.

Regulations: FDA - 21CFR 172.515. Use at a level not in excess of the amount reasonably required to accomplish the intended effect.

SAFETY PROFILE: Moderately toxic by ingestion. Combustible liquid. When heated to decomposition it emits acrid smoke and fumes.

TOXICITY DATA and CODEN

orl-rat LD50:2850 mg/kg FCTXAV 11,1023,73

IKQ000 CAS: 97-54-1

ISOEUGENOL

mf: $C_{10}H_{12}O_2$ mw: 164.22

PROP: Pale yellow oil; carnation odor. D: 1.079-1.085, refr index: 1.572-1.577, mp: −10°, bp: 266°. cis Form: liquid, bp: 133° @ 11 mm, d: 1.088 @ 20°/4°. trans Form: crystals, mp: 33°; bp: 140° @ 12 mm; d: 1.087 @ 20°/4°, flash p: +212°F. Sol in fixed oils, propylene glycol; very sltly sol in water; misc in alc and ether; insol in glycerin.

SYNS: FEMA No. 2468 ◇ 1-HYDROXY-2-METHOXY-4-PROPENYLBENZENE ◇ 4-HYDROXY-3-METHOXY-1-PROPENYLBENZENE ◇ 2-METHOXY-4-PROPENYLPHENOL ◇ NCI-C60979 ◇ 4-PROPENYLGUAIACOL

USE IN FOOD:

Purpose: Flavoring agent.

Where Used: Various.

Regulations: FDA - 21CFR 172.515. Use at a level not in excess of the amount reasonably required to accomplish the intended effect.

SAFETY PROFILE: Moderately toxic by ingestion. Human mutagenic data. Combustible liquid. When heated to decomposition it emits acrid smoke and fumes.

TOXICITY DATA and CODEN

sce-hmn:lym 250 μmol/L MUREAV 169,129,86
orl-rat LD50:1560 mg/kg TXAPA9 6,378,64

IKR000 CAS: 93-16-3
1,3,4-ISOEUGENOL METHYL ETHER

mf: $C_{11}H_{14}O_2$ mw: 178.25

PROP: Colorless to pale yellow liquid; clove-carnation odor. D: 1.047, refr index: 1.566, flash p: +212°F. Sol in fixed oils; insol in glycerin, propylene glycol.

SYNS: 1,2-DIMETHOXY-4-PROPENYLBENZENE ◇ FEMA No. 2476 ◇ ISOEUGENYL METHYL ETHER ◇ ISOHOMOGENOL ◇ METHYL ISOEUGENOL (FCC) ◇ 4-PROPENYL VERATROLE

USE IN FOOD:

Purpose: Flavoring agent.

Where Used: Various.

Regulations: FDA - 21CFR 172.515. Use at a level not in excess of the amount reasonably required to accomplish the intended effect.

SAFETY PROFILE: Poison by intravenous route. Moderately toxic by intraperitoneal route. A skin irritant. Combustible liquid. When heated to decomposition it emits acrid smoke and fumes.

TOXICITY DATA and CODEN

skn-rbt 500 mg/24H FCTXAV 13,865,75
ipr-mus LD50:570 mg/kg AIPTAK 199,226,72

IKX000 CAS: 73-32-5
ISOLEUCINE

mf: $C_6H_{13}NO_2$ mw: 131.17

PROP: An essential amino acid; many isomeric forms. White crystalline powder; bitter taste. Mp: (dl): 292° (decomp), (l): 283-284° (decomp). Sltly sol in water; nearly insol in alc; insol in ether.

SYNS: 2-AMINO-3-METHYLPENTANOIC ACID ◇ α-AMINO-β-METHYLVALERIC ACID ◇ l-ISOLEUCINE (FCC)

USE IN FOOD:

Purpose: Dietary supplement, nutrient.

Where Used: Various.

Regulations: FDA - 21CFR 172.310. Limitation 6.6 percent by weight.

SAFETY PROFILE: Mildly toxic by intraperitoneal route. A nutrient and/or dietary supplement food additive. When heated to decomposition it emits toxic fumes of NO_x.

TOXICITY DATA and CODEN

ipr-rat LD50:6822 mg/kg ABBIA4 58,253,55

IKX010 CAS: 443-79-8
dl-ISOLEUCINE

mf: $C_6H_{13}NO_2$ mw: 131.17

PROP: White crystalline powder; sltly bitter taste. Mp: 292° (decomp). Sol in water; insol in alc, ether.

SYN: dl-2-AMINO-3-METHYLVALERIC ACID

USE IN FOOD:

Purpose: Dietary supplement, nutrient.

Where Used: Various.

Regulations: FDA - 21CFR 172.310. Limitation 6.6 percent by weight.

SAFETY PROFILE: When heated to decomposition emits toxic fumes of NO_x.

ILR100 CAS: 27554-26-3
ISOOCTYL PHTHALATE

mf: $C_{24}H_{38}O_4$ mw: 390.62

SYNS: 1,2-BENZENEDICARBOXYLIC ACID, DIISOOCTYL ESTER ◇ BIS(6-METHYLHEPTYL)ESTER OF PHTHALIC ACID ◇ CORFLEX 880 ◇ DIISOOCTYL PHTHALATE ◇ FLEXOL PLASTICIZER DIP ◇ HEXAPLAS M/O

USE IN FOOD:

Purpose: Plasticizer.

Where Used: Packaging materials.

Regulations: FDA - 21CFR 181.27. Limited to foods of high water content.

SAFETY PROFILE: Moderately toxic by ingestion. Mildly toxic by skin contact. A skin irritant. When heated to decomposition it emits acrid smoke and irritating fumes.

TOXICITY DATA and CODEN

skn-rbt 500 mg open MLD UCDS** 6/11/65
orl-rat LD50:22 g/kg EVHPAZ 3,61,73

ILR150
ISOPARAFFINIC PETROLEUM HYDROCARBONS, SYNTHETIC

SYN: SYNTHETIC ISOPARAFFINIC PETROLEUM HYDROCARBONS

USE IN FOOD:

Purpose: Coating agent, float, froth-flotation cleaning, insecticide formulations component.

Where Used: Eggs, fruits, pickles, vegetables, vinegar, wine.

Regulations: FDA - 21CFR 172.882, 561.365. Use at a level not in excess of the amount reasonably required to accomplish the intended effect.

SAFETY PROFILE: When heated to decomposition it emits acrid smoke and irritating fumes.

ILV000 CAS: 123-92-2
ISOPENTYL ALCOHOL ACETATE

mf: $C_7H_{14}O_2$ mw: 130.21

PROP: Colorless liquid; banana-like odor. Bp: 142.0°, ULC: 55-60, lel: 1% @ 212°F, uel: 7.5%, flash p: 77°F, d: 0.876, refr index: 1.400, autoign temp: 680°F, vap d: 4.49. Misc in alc, ether, ethyl acetate, fixed oils; sltly sol in water; insol in glycerin, propylene glycol.

SYNS: ACETIC ACID, ISOPENTYL ESTER ◇ BANANA OIL ◇ FEMA No. 2055 ◇ ISOAMYL ACETATE (ACGIH, FCC) ◇ ISOAMYL ETHANOATE ◇ ISOPENTYL ACETATE ◇ 3-METHYLBUTYL ACETATE ◇ 3-METHYL-1-BUTYL ACETATE ◇ 3-METHYLBUTYL ETHANOATE ◇ PEAR OIL

USE IN FOOD:

Purpose: Flavoring agent.

Where Used: Beverages, candy, ice cream.

Regulations: FDA - 21CFR 172.515. Use at a level not in excess of the amount reasonably required to accomplish the intended effect.

OSHA PEL: TWA 100 ppm ACGIH TLV: TWA 100 ppm; STEL 125 ppm

SAFETY PROFILE: Mildly toxic by ingestion, inhalation and subcutaneous routes. Exposure to concentrations of about 1,000 ppm for 1 hour can cause headache, fatigue, pulmonary irritation and serious toxicity effects. Highly flammable liquid. When exposed to heat or flame; can react vigorously with reducing materials. Moderately explosive in the form of vapor when exposed to heat or flame. To fight fire, use alcohol foam, CO_2, dry chemical. When heated to decomposition it emits acrid smoke and fumes.

TOXICITY DATA and CODEN

orl-rat LD50:16600 mg/kg YKYUA6 32,1241,81
ihl-cat LCLo:35000 mg/m^3 AGGHAR 5,1,33

IME000 CAS: 87-20-7
ISOPENTYL SALICYLATE

mf: $C_{12}H_{16}O_3$ mw: 208.28

PROP: Coorless liquid; pleasant odor. D: 1.047, refr index: 1.503-1.509, flash p: 271°F. Misc with alc, chloroform, ether, fixed oils; insol in glycerin, propylene glycol, water.

SYNS: FEMA No. 2084 ◇ ISOAMYL o-HYDROXYBENZOATE ◇ ISOAMYL SALICYLATE (FCC) ◇ ISOPENTYL-2-HYDROXYPHENYL METHANOATE ◇ 3-METHYLBUTYL 2-HYDROXYBENZOATE ◇ SALICYLIC ACID, ISOPENTYL ESTER

USE IN FOOD:

Purpose: Flavoring agent.

Where Used: Various.

Regulations: FDA - 21CFR 172.515. Use at a level not in excess of the amount reasonably required to accomplish the intended effect.

SAFETY PROFILE: Moderately toxic by intravenous route. Experimental reproductive effects. Combustible liquid. When heated to decomposition it emits acrid smoke and fumes.

TOXICITY DATA and CODEN

orl-rat TDLo:12600 mg/kg (42D male):REP FCTXAV 13,185,75
ivn-dog LD50:500 mg/kg 14CYAT 2,1847,63

INJ000 CAS: 67-63-0
ISOPROPYL ALCOHOL

DOT: 1219

mf: C_3H_8O mw: 60.11

PROP: Clear, colorless liquid; slt odor, sltly bitter taste. Mp: −88.5 to −89.5°, bp: 82.5°, lel: 2.5%, uel: 12%, flash p: 53°F (CC), d: 0.7854 @ 20°/4°, refr index: 1.377 @ 20°, vap d: 2.07, ULC: 70. fp: −89.5°; autoign temp: 852°F. Misc with water, alc, ether, chloroform; insol in salt solns.

SYNS: ALCOOL ISOPROPILICO (ITALIAN) ◇ ALCOOL ISOPROPYLIQUE (FRENCH) ◇ DIMETHYLCARBINOL ◇ ISOHOL ◇ ISOPROPANOL (DOT) ◇ ISO-PROPYLALKOHOL (GERMAN) ◇ LUTOSOL ◇ PETROHOL ◇ PROPAN-2-OL ◇ 2-PROPANOL ◇ i-PROPANOL (GERMAN) ◇ sec-PROPYL ALCOHOL (DOT) ◇ i-PROPYLALKOHOL (GERMAN) ◇ SPECTRAR

USE IN FOOD:

Purpose: Color diluent, extraction agent.

Where Used: Beet sugar, confectionery, food supplements in tablet form, gum, hops extract, lemon oil, spice oleoresins, yeast.

Regulations: FDA - 21CFR 73.1. No residue. 21CFR 173.240. Limitation of 50 ppm in spice oleoresins, 6 ppm in lemon oil, 2 percent in hops extract. 21CFR 173.340.

The isopropyl alcohol strong acid manufacturing process is on the Community Right-To-Know List. EPA Genetic Toxicology Program.

OSHA PEL: TWA 400 ppm; STEL 500 ppm ACGIH TLV: TWA 400 ppm; STEL 500 ppm DOT Classification: Flammable Liquid; Label: Flammable Liquid NIOSH REL: (Isopropyl Alcohol) TWA 400 ppm; CL 800 ppm/15M

SAFETY PROFILE: Poison by ingestion and subcutaneous routes. Moderately toxic to humans by an unspecified route. Moderately toxic experimentally by intravenous and intraperitoneal routes. Mildly toxic by skin contact. Human systemic effects by ingestion or inhalation: flushing, pulse rate decrease, blood pressure lowering, anesthesia, narcosis, headache, dizziness, mental depression, hallucinations, distorted perceptions, dyspnea, respiratory depression, nausea or vomiting, coma. Experimental teratogenic and reproductive effects. Mutagenic data. An eye and skin irritant. An FDA over-the-counter drug.

There is some evidence that humans can acquire a slight tolerance to this material. It acts very much like ethanol in regard to absorption, metabolism and elimination but with a stronger narcotic action.

Flammable liquid. A very dangerous fire hazard when exposed to heat, flame or oxidizers. Moderately explosive when exposed to heat or flame. Reacts with air to form dangerous peroxides. Can react vigorously with oxidizing materials. To fight fire, use CO_2, dry chemical, alcohol foam. When heated to decomposition it emits acrid smoke and fumes.

TOXICITY DATA and CODEN

skn-rbt 500 mg MLD NTIS** AD-A106-944
eye-rbt 16 mg AJOPAA 29,1363,46
cyt-smc 200 mmol/tube HEREAY 33,457,47
orl-rat TDLo: 11340 mg/kg (45D pre): REP GISAAA 43(1),8,78
orl-rat TDLo: 32400 μg/kg (26W pre): TER GISAAA 43(1),8,78
orl-rat TDLo: 6480 mg/kg (26W male/26W pre): REP GISAAA 43(1),8,78
orl-man TDLo: 14432 mg/kg: CNS,CVS,PUL NEJMAG 277,699,67
orl-hmn TDLo: 223 mg/kg: CNS,CVS JLCMAK 12,326,27
orl-hmn LDLo: 3570 mg/kg: CNS,PUL,GIT 34ZIAG -,339,69
orl-rat LD50: 5045 mg/kg GISAAA 43(1),8,78
ihl-rat LCLo: 12000 ppm/8H IAEC** 17JUN74

IOO222
ISOPROPYL CITRATE

USE IN FOOD:

Purpose: Preservative, sequestrant.

Where Used: Oleomargarine, vegetable oils.

Regulations: FDA - 21CFR 182.6386. GRAS with a limitation of 0.02 percent when used in accordance with good manufacturing practice. USDA - 9CFR 318.7. Limitation of 0.02 percent in oleomargarine.

SAFETY PROFILE: When heated to decomposition it emits acrid smoke and irritating fumes.

IRY000 CAS: 94-86-0
ISOSAFROEUGENOL

mf: $C_{11}H_{14}O_2$ mw: 178.25

PROP: White crystalline powder; vanilla odor. Flash p: +212°F. Sol fixed oils; insol in water.

SYNS: 6-ETHOXY-m-ANOL ◇ 1-ETHOXY-2-HYDROXY-4-PROPENYLBENZENE ◇ FEMA No. 2922 ◇ HYDROXY METHYL ANETHOL ◇ PROPENYLGUAETHOL (FCC)

USE IN FOOD:

Purpose: Flavoring agent.

Where Used: Various.

Regulations: FDA - 21CFR 172.515. Use at a level not in excess of the amount reasonably required to accomplish the intended effect.

SAFETY PROFILE: Moderately toxic by ingestion. Combustible liquid. When heated to decomposition it emits acrid smoke and fumes.

TOXICITY DATA and CODEN

orl-rat LD50:2400 mg/kg AFDOAQ 15,82,51

ISR000 CAS: 62-56-6
ISOTHIOUREA
DOT: 2877
mf: CH_4N_2S mw: 76.13

PROP: White powder or crystals. Mp: 177°, bp: decomp, d: 1.405.

SYNS: PSEUDOTHIOUREA ◇ RCRA WASTE NUMBER U219 ◇ SULOUREA ◇ THIOCARBAMIDE ◇ β-THIOPSEUDOUREA ◇ THIOUREA (DOT) ◇ 2-THIOUREA ◇ THU ◇ TSIZP 34 ◇ USAF EK-497

USE IN FOOD:

Purpose: Antimycotic agent.

Where Used: Citrus fruit.

Regulations: FDA - 21CFR 189.190. Prohibited from direct addition or use in human food.

IARC Cancer Review: Animal Sufficient Evidence IMEMDT 7,95,74. EPA Genetic Toxicology Program.

DOT Classification: Poison B; Label: St. Andrews Cross

SAFETY PROFILE: A human poison by an unspecified route. An experimental poison by ingestion and intraperitoneal routes. An experimental carcinogen, neoplastigen, and tumorigen. Human mutagenic data. Human systemic effects by ingestion: hemorrhage, granulocytopenia (reduction in number of granulocytes), and changes in cell count (unspecified). May cause depression of bone marrow with anemia, leukopenia and thrombocytopenia. May also cause allergic skin eruptions. Causes hepatic tumors upon chronic administration. Experimental teratogenic and reproductive effects. May react violently with acrolein. Incompatible with acrylaldehyde; H_2O_2; HNO_3. When heated to decomposition it emits very toxic fumes of NO_x and SO_x.

TOXICITY DATA and CODEN

mmo-sat 150 μg/plate ABCHA6 44,3017,80
dnd-esc 20 μmol/L MUREAV 89,95,81
orl-rat TDLo:1 g/kg (12D preg):REP TJADAB 23,335,81
orl-rat TDLo:1400 mg/kg (16-22D preg):TER AMASA4 32,271,75
orl-rat TDLo:4800 mg/kg (17-22D preg/10D post):TER TJADAB 31,57A,85
orl-rat TDLo:78 g/kg/56W-C:CAR CNREA8 17,302,57
mul-rat TDLo:151 g/kg/52W-I:CAR CNREA8 17,302,57
orl-wmn TDLo:1660 mg/kg/5W:BLD LANCAO 246,179,44
orl-rat LD50:125 mg/kg HBTXAC 5,177,59
ipr-rat LD50:436 mg/kg FCTXAV 3,597,65

ISU000 CAS: 503-74-2
ISOVALERIC ACID
DOT: 1760
mf: $C_5H_{10}O_2$ mw: 102.15

PROP: Colorless liquid; acid taste, disagreeable rancid-cheese odor. Solidifies @ −37°, d: 0.931 @ 20°/4°, refr index: 1.403, mp: −34.5° (−50°), bp: 175-177°. Sol in water @ 16°; misc in alc, chloroform, ether.

SYNS: DELPHINIC ACID ◇ FEMA No. 3102 ◇ ISOPENTANOIC ACID (DOT) ◇ ISOPROPYLACETIC ACID ◇ ISOVALERIANIC AICD ◇ 3-METHYLBUTANOIC ACID ◇ β-METHYLBUTYRIC ACID ◇ 3-METHYLBUTYRIC ACID

USE IN FOOD:

Purpose: Flavoring agent.

Where Used: Various.

Regulations: FDA - 21CFR 172.515. Use at a level not in excess of the amount reasonably required to accomplish the intended effect.

DOT Classification: Corrosive Material; Label: Corrosive

SAFETY PROFILE: A poison by skin contact. Moderately toxic by ingestion and intravenous routes. A corrosive skin and eye irritant. When heated to decomposition it emits acrid smoke and fumes.

TOXICITY DATA and CODEN

skn-rbt 470 mg open MOD UCDS** 1/31/72
eye-rbt 940 μg MLD UCDS** 1/31/72
orl-rat LD50:2000 mg/kg UCDS** 1/31/72

ISV000 CAS: 2835-39-4
ISOVALERIC ACID, ALLYL ESTER
mf: $C_8H_{14}O_2$ mw: 142.22

SYNS: ALLYL ISOVALERATE ◇ ALLYL ISOVALERIANATE ◇ ALLYL 3-METHYLBUTYRATE ◇ FEMA No. 2045 ◇ 3-METHYLBUTANOIC ACID, 2-PROPENYL ESTER ◇ 3-METHYLBUTYRIC ACID, ALLYL ESTER ◇ NCI-C54717 ◇ 2-PROPENYL ISOVALERATE ◇ 2-PROPENYL 3-METHYLBUTANOATE

USE IN FOOD:

Purpose: Flavoring agent.

Where Used: Various.

Regulations: FDA - 21CFR 172.515. Use at a level not in excess of the amount reasonably required to accomplish the intended effect.

IARC Cancer Review: Animal Limited Evidence IMEMDT 36,69,85. NTP Carcinogenesis Studies (gavage); Clear Evidence: mouse, rat NTPTR* NTP-TR-253,83.

SAFETY PROFILE: A poison by ingestion. Moderately toxic by skin contact. An experimental carcinogen and tumorigen. A skin irritant. When heated to decomposition it emits acrid smoke and fumes.

TOXICITY DATA and CODEN

skn-rbt 500 mg/24H MOD FCTXAV 17,703,79
orl-rat TDLo:31930 mg/kg/2Y-I:CAR NTPTR* NTP-TR-253,83
orl-mus TDLo:31930 mg/kg/2Y-I:CAR NTPTR* NTP-TR-253,83
orl-rat TD :15065 mg/kg/2Y-I:ETA NTPTR* NTP-TR-253,83
orl-rat LD50:230 mg/kg FCTXAV 17,703,79
skn-rbt LD50:560 mg/kg FCTXAV 17,703,79

ISW000 CAS: 103-38-8
ISOVALERIC ACID, BENZYL ESTER
mf: $C_{12}H_{16}O_2$ mw: 192.28

PROP: Colorless liquid; fruity apple odor. D: 0.985-0.9911, refr index: 1.486, flash p: +212°F. Sol in alc, fixed oils; sltly sol in propylene glycol; insol in glycerin, water @ 246°.

SYNS: BENZYL ISOVALERATE (FCC) ◇ BENZYL-3-METHYLBUTANOATE ◇ BENZYL-3-METHYL BUTYRATE ◇ FEMA No. 2152 ◇ ISOPENTANOIC ACID, PHENYLMETHYL ESTER ◇ ISOPROPYL ACETIC ACID, BENZYL ESTER ◇ 3-METHYLBUTANOIC ACID, PHENYLETHYL ESTER

USE IN FOOD:

Purpose: Flavoring agent.

Where Used: Various.

Regulations: FDA - 21CFR 172.515. Use at a level not in excess of the amount reasonably required to accomplish the intended effect.

SAFETY PROFILE: A skin irritant. Combustible liquid. When heated to decomposition it emits acrid smoke and fumes.

TOXICITY DATA and CODEN

skn-rbt 500 mg/24H MLD FCTXAV 12,829,74

ISX000 CAS: 109-19-3
ISOVALERIC ACID, BUTYL ESTER
mf: $C_9H_{18}O_2$ mw: 158.27

PROP: Colorless to pale yellow liquid; fruity odor. Vap d: 5.45, bp: 150°, d: 0.851-0.857, refr index: 1.407. Misc with alc, fixed oils; sltly sol in propylene glycol; insol in water.

SYNS: n-BUTYL ISOPENTANOATE ◇ n-BUTYL ISOVALERATE ◇ 1-BUTYL ISOVALERATE ◇ BUTYL ISOVALERIANATE ◇ BUTYL 3-METHYLBUTYRATE ◇ FEMA No. 2218 ◇ 3-METHYLBUTANOIC ACID, BUTYL ESTER

USE IN FOOD:

Purpose: Flavoring agent.

Where Used: Various.

Regulations: FDA - 21CFR 172.515. Use at a level not in excess of the amount reasonably required to accomplish the intended effect.

EPA Extremely Hazardous Substances List.

SAFETY PROFILE: Mildly toxic by ingestion. A skin irritant. Flammable when exposed to heat, flame, sparks, and oxidizers. To fight fire, use alcohol foam, dry chemical, spray, mist, fog. When heated to decomposition it emits acrid smoke and fumes.

TOXICITY DATA and CODEN

skn-rbt 500 mg/24H MLD FCTXAV 18,659,80
orl-rbt LD50:8200 mg/kg FCTXAV 18,659,80

ISY000 CAS: 108-64-5
ISOVALERIC ACID, ETHYL ESTER
mf: $C_7H_{14}O_2$ mw: 130.21

PROP: Colorless, oily liquid; apple odor. Flash p: 77°F, d: 0.868 @ 20°/20°, refr index: 1.395-1.399, bp: 135°, mp: −99°. Misc with alc, fixed oils, benzene, ether; sol in propylene glycol; sltly sol in water @ 135°.

SYNS: ETHYL ISOVALERATE (FCC) ◇ FEMA No. 2463 ◇ 3-METHYLBUTANOIC ACID, ETHYL ESTER ◇ 3-METHYLBUTYRIC ACID, ETHYL ESTER

USE IN FOOD:

Purpose: Flavoring agent.

Where Used: Various.

Regulations: FDA - 21CFR 172.515. Use at a level not in excess of the amount reasonably required to accomplish the intended effect.

SAFETY PROFILE: Moderately toxic by intraperitoneal route. Mildly toxic by ingestion. A skin irritant. Flammable liquid when exposed to heat, flame or sparks. When heated to decomposition it emits acrid smoke and fumes.

TOXICITY DATA and CODEN

skn-rbt 500 mg/24H MLD FCTXAV 16,743,78
ipr-rat LD50:1200 mg/kg FCTXAV 16,743,78
orl-rbt LD50:7031 mg/kg IMSUAI 41,31,72

ISZ000 CAS: 35154-45-1
(Z)-ISOVALERIC ACID-3-HEXENYL
mf: $C_{11}H_{20}O_2$ mw: 184.31

PROP: Colorless liquid; sweet, apple odor. D: 0.869-0.874, refr index: 1439-1.435. Sol in alc, propylene glycol, fixed oils; insol in water.

SYNS: AI3-35966 ◇ FEMA No. 3498 ◇ cis-3-HEXENYL ISOVALERATE (FCC)

USE IN FOOD:

Purpose: Flavoring agent.

Where Used: Various.

Regulations: FDA - 21CFR 172.515. Use at a level not in excess of the amount reasonably required to accomplish the intended effect.

SAFETY PROFILE: A skin irritant. When heated to decomposition it emits acrid smoke and fumes.

TOXICITY DATA and CODEN

skn-rbt 500 mg/24H MLD NTIS** AD-A053-884

ITB000 CAS: 659-70-1
ISOVALERIC ACID, ISOPENTYL ESTER
mf: $C_{10}H_{20}O_2$ mw: 172.30

PROP: Colorless liquid; apple odor. D: 0.851-0.857, refr index: 1.411, flash p: 162°F. Misc in alc, fixed oils; sltly sol in propylene glycol; insol in water.

SYNS: FEMA No. 2085 ◇ ISOAMYL ISOVALERATE (FCC) ◇ ISOPENTYL ISOVALERATE

USE IN FOOD:

Purpose: Flavoring agent.

Where Used: Various.

Regulations: FDA - 21CFR 172.515. Use at a level not in excess of the amount reasonably required to accomplish the intended effect.

SAFETY PROFILE: Mildly toxic by ingestion. A skin irritant. Combustible liquid. When heated to decomposition it emits acrid smoke and fumes.

TOXICITY DATA and CODEN

skn-rbt 500 mg/24H MOD FCTXAV 16,789,78
orl-rbt LD50:13956 mg/kg IMSUAI 41,31,72

ITD875 CAS: 70288-86-7
IVERMECTIN

SYNS: 22,23-DIHYDROAVERMECTIN B1 ◇ HYVERMECTIN ◇ MK 933

USE IN FOOD:

Purpose: Animal drug.

Where Used: Beef, pork, reindeer.

Regulations: FDA - 21CFR 556.334. Limitation of 15 ppb in cattle and reindeer liver, 20 ppb in swine liver. USDA CES Ranking: B-1 (1986).

SAFETY PROFILE: Poison by subcutaneous route. When heated to decomposition emits toxic fumes of NO_x.

TOXICITY DATA and CODEN

scu-ctl LDLo:8 mg/kg SCIEAS 221,823,83

J

JEA000 CAS: 8012-91-7
JUNIPER BERRY OIL

PROP: A volatile oil. Principal constituents include d-pinene, camphene, 1-terpineol-4 and other oxygenated constituents. From steam distillation of the fruit of *Juniperus communis* L. (Fam. *Cupressaceae*). (FCTXAV 14,307,76). Colorless to faint green-yellow liquid; aromatic bitter taste. Sol in fixed oils, mineral oil; insol in glycerin, propylene glycol.

SYNS: OIL of JUNIPER BERRY ◇ WACHOLDERBEER OEL (GERMAN)

USE IN FOOD:

Purpose: Flavoring agent.

Where Used: Various.

Regulations: FDA - 21CFR 182.20. GRAS when used at a level not in excess of the amount reasonably required to accomplish the intended effect.

SAFETY PROFILE: Mildly toxic by ingestion. A human skin irritant. An allergen. A systemic irritant. If taken internally, a severe kidney irritation similar to that caused by turpentine may result. When heated to decomposition it emits acrid smoke and fumes.

TOXICITY DATA and CODEN

skn-hmn 100% FCTXAV 14,333,76
skn-rbt 500 mg/24H MOD FCTXAV 14,333,76
orl-rat LD50:6280 mg/kg PHARAT 14,435,59

K

KAJ000 CAS: 40596-69-8

KABAT

mf: $C_{19}H_{34}O_3$ mw: 310.53

PROP: Amber liquid. Bp: 100°. Solubility in water: 1.39 ppm. Sol in most organic solvents.

SYNS: ALTOSID ◇ ALTOSID IGR ◇ ALTOSID SR 10 ◇ ENT 70,460 ◇ ISOPROPYL(2E,4E)-11-METHOXY-3,7,11-TRIMETHYL-2,4-DODECADIENOATE ◇ MANTA ◇ METHOPRENE ◇ (E,E)-11-METHOXY-3,7,11-TRIMETHYL-2,4-DODECANDIENOATE ◇ ZR 515

USE IN FOOD:

Purpose: Insect growth regulator.

Where Used: Animal feed, apples (dried), apricots (dried), barley cereal, beef, corn cereal, corn meal, grits, hominy, macaroni, oat cereal, peaches (dried), potable water, prunes (dried), raisins, rice cereal, rye cereal, spices, wheat cereal, wheat flour.

Regulations: FDA - 21CFR 193.285. Insect growth regulator residue tolerance of 10 ppm in raisins, wheat flour, macaroni, rice cereal, rye cereal, barley cereal, wheat cereal, corn cereal, corn meal, grits, hominy, oat cereal, spices, dried apples, dried apricots, dried peaches, dried prunes. Exempt from tolerance in potable water. 21CFR 561.282. Limitation of 22.7 to 45.4 mg/100 pounds of cattle body weight per month in the form of mineral and/or protein blocks when used for animal feed.

SAFETY PROFILE: Moderately toxic by skin contact. Mildly toxic by ingestion. Mutagenic data. When heated to decomposition it emits acrid smoke and fumes.

TOXICITY DATA and CODEN

dni-oin:ovr 100 μmol/L ABCHA6 43,1285,79
oms-oin:ovr 100 μmol/L ABCHA6 43,1285,79
orl-dog LD50:5000 mg/kg EVHPAZ 14,119,76
skn-rbt LD50:3000 mg/kg EVHPAZ 14,119,76

KBB600 CAS: 1332-58-7

KAOLIN

PROP: Fine white to light yellow powder; earth taste. Insol in ether, alc, dil acids, and alkali solutions.

SYN: CHINA CLAY

USE IN FOOD:

Purpose: Anticaking agent, clarifying agent, paper manufacturing aid.

Where Used: Food packaging, wine.

Regulations: FDA - 21CFR 182.2727, 182.2729, 184.1155, 186.1256. GRAS as an indirect additive when used in accordance with good manufacturing practice. BATF - 27CFR 240.1051. The sodium content of the wine shall not be increased.

SAFETY PROFILE: A nuisance dust.

KBK000 CAS: 9000-36-6

KARAYA GUM

PROP: Dried exudate of the tree, *Sterculia ureus* Roxburgh (Fam. *Sterculiaceae*). Fine, white powder; slt odor of acetic acid. insol in alc; swells in water to a gel.

USE IN FOOD:

Purpose: Emulsifier, formulation aid, stabilizer, thickener.

Where Used: Baked goods, candy (soft), frozen dairy desserts, milk products, toppings.

Regulations: FDA - 21CFR 184.1349. GRAS with a limitation of 0.3 percent in frozen dairy desserts, 0.02 percent in milk products, 0.9 percent in soft candy, 0.002 percent in all other foods when used in accordance with good manufacturing practice.

SAFETY PROFILE: Very mildly toxic by ingestion. A mild allergen.

TOXICITY DATA and CODEN

orl-rat LDLo:30 g/kg FOREAE 13,29,48

KDK700

KELP

PROP: Dehydrated seaweed, dark green to brown; salty, characteristic taste. From *Macrocystis pyrifera, Laminaria digitata, Laminaria saccharina, and Laminaria cloustoni.*

USE IN FOOD:

Purpose: Dietary supplement (iodine).

Where Used: Various.

Regulations: FDA - 21CFR 172.365. Limitation of total iodine to 225 μg per day without reference to age, 45 μg for infants, 105 μg for children under 4 years old, 225 μg for adults and children 4 or more years old, and 300 μg for pregnant or lactating women.

SAFETY PROFILE: When heated to decomposition it emits acrid smoke and irritating fumes.

L

LAC000 CAS: 8016-26-0

LABDANUM OIL

PROP: Main constituents are acetophenone, 1,5,5-trimethyl-6-cyclohexanone and ladaniol found in the gum of the shrub *Cistus ladaniferus* L. (Fam. *Cistaceae*). Prepared by steam distillation of the crude gum. Yellow, viscous liquid; powerful balsamic odor. D: 0.905-0.993, refr index: 1.492-1.507 @ 20°, flash p: 187°F. Sol in fixed oils, mineral oil; insol in glycerin, propylene glycol.

SYN: OIL of LABDANUM

USE IN FOOD:

Purpose: Flavoring agent.

Where Used: Various.

Regulations: FDA - 21CFR 172.515. Use at a level not in excess of the amount reasonably required to accomplish the intended effect.

SAFETY PROFILE: Mildly toxic by ingestion. A skin irritant. Combustible liquid. When heated to decomposition it emits acrid smoke and fumes.

TOXICITY DATA and CODEN

skn-rbt 500 mg/24H MOD FCTXAV 14,307,76
orl-rat LD50:8980 mg/kg FCTXAV 14,307,76

LAE350

LACTASE ENZYME PREPARATIONS FROM KLUYVEROMYCES LACTIS

PROP: Derived from *Kluyveromyces lactis*.

USE IN FOOD:

Purpose: Enzyme.

Where Used: Milk.

Regulations: FDA - 21CFR 184.1388. GRAS when used in accordance with good manufacturing practice.

SAFETY PROFILE: When heated to decomposition it emits acrid smoke and irritating fumes.

LAE400

LACTATED MONO-DIGLYCERIDES

PROP: Soft to hard waxy solid. Dispersible in hot water; moderately sol in hot isopropanol, xylene, cottonseed oil.

USE IN FOOD:

Purpose: Emulsifier, stabilizer.

Where Used: Margarine, oleomargarine.

Regulations: USDA - 9CFR 318.7. Sufficient for purpose, 0.5 percent in oleomargarine or margarine.

SAFETY PROFILE: When heated to decomposition it emits acrid smoke and irritating fumes.

LAG000 CAS: 50-21-5

LACTIC ACID

mf: $C_3H_6O_3$ mw: 90.09

PROP: Yellow to colorless crystals or syrupy 50% liquid. Mp: 16.8°, bp: 122° @ 15 mm, d: 1.249 @ 15°. Volatile with superheated steam; sol in H_2O, alc, furfurol; sltly sol in ether; insol in chloroform, petr ether, carbon disulfide. Misc in water, (alc + ether).

SYNS: ACETONIC ACID ◇ ETHYLIDENELACTIC ACID ◇ 1-HYDROXYETHANECARBOXYLIC ACID ◇ 2-HYDROXYPROPANOIC ACID ◇ 2-HYDROXYPROPIONIC ACID ◇ α-HYDROXYPROPIONIC ACID ◇ KYSELINA MLECNA (CZECH) ◇ dl-LACTIC ACID ◇ MILCHSAURE (GERMAN) ◇ MILK ACID ◇ ORDINARY LACTIC ACID ◇ racemic LACTIC ACID

USE IN FOOD:

Purpose: Acid, antimicrobial agent, curing agent, flavor enhancer, flavoring agent, pH control agent, pickling agent, solvent, vehicle.

Where Used: Cheese spreads, egg (dry powder), olives, poultry, salad dressing mix, wine.

Regulations: FDA - 21CFR 184.1061. GRAS when used at a level not in excess of the amount reasonably required to accomplish the intended effect. USDA - 9CFR 318.7, 9CFR 381.147. Sufficient for purpose. BATF - 27CFR 240.1051. GRAS when used in accordance with good manufacturing practice.

SAFETY PROFILE: Moderately toxic by ingestion and rectal routes. Mutagenic data. A severe skin and eye irritant. An FDA over the counter drug. Mixtures with nitric acid + hydrofluoric acid may react vigorously and are storage hazards. When heated to decomposition it emits acrid smoke and irritating fumes.

TOXICITY DATA and CODEN

skn-rbt 500 mg/24H SEV 28ZPAK -,105,72
eye-rbt 750 μg/24H SEV 28ZPAK -,105,72
mmo-esc 210 ppm/3H AMNTA4 85,119,51
orl-rat LD50:3730 mg/kg JIHTAB 23,259,41

LAJ000 CAS: 97-64-3
LACTIC ACID, ETHYL ESTER
DOT: 1192
mf: $C_5H_{10}O_3$ mw: 118.15

PROP: Colorless liquid; mild odor. Bp: 154°, ULC: 30-35%, lel: 1.55% @ 212°F, flash p: 115°F (CC), flash p (technical): 131°F, d: 1.029-1.032, refr index: 1.410-1.420, autoign temp: 752°F, vap d: 4.07. Very sol in alc, ether, chloroform, water.

SYNS: ACTYLOL ◇ ACYTOL ◇ ETHYL-α-HYDROXYPROPIONATE ◇ ETHYL 2-HYDROXYPROPIONATE ◇ ETHYL LACTATE (DOT,FCC) ◇ FEMA No. 2440 ◇ LACTATE d'ETHYLE (FRENCH) ◇ SOLACTOL

USE IN FOOD:

Purpose: Flavoring agent.

Where Used: Various.

Regulations: FDA - 21CFR 172.515. Use at a level not in excess of the amount reasonably required to accomplish the intended effect.

DOT Classification: Flammable or Combustible Liquid; Label: Flammable Liquid; Combustible Liquid; Label: None

SAFETY PROFILE: Moderately toxic by ingestion, intraperitoneal, subcutaneous and intravenous routes. Flammable or combustible liquid when exposed to heat or flame; can react with oxidizing materials. Slight explosion hazard in the form of vapor when exposed to flame. To fight fire, use foam, CO_2, dry chemical. When heated to decomposition it emits acrid smoke and irritating fumes.

TOXICITY DATA and CODEN

orl-mus LD50:2500 mg/kg JPETAB 65,89,39
scu-mus LD50:2500 mg/kg JPETAB 65,89,39

LAL000 CAS: 5905-52-2
LACTIC ACID, IRON(2+) SALT (2:1)
mf: $C_6H_{10}O_6$•Fe mw: 234.01

PROP: Greenish-white crystals; slt peculiar odor. Moderately sol in water; sltly sol in alc.

SYNS: FERROUS LACTATE ◇ IRON(2+) LACTATE

USE IN FOOD:

Purpose: Dietary supplement, nutrient supplement.

Where Used: Various.

Regulations: FDA - 21CFR 182.5311, 182.8311, 184.1311. GRAS when used in accordance with good manufacturing practice.

ACGIH TLV: TWA 1 mg(Fe)/m^3

SAFETY PROFILE: Poison by ingestion. An experimental tumorigen. When heated to decomposition it emits acrid smoke and irritating fumes.

TOXICITY DATA and CODEN

scu-mus TDLo:4200 mg/kg/21W-I:ETA JNCIAM 24,109,60
orl-mus LD50:147 mg/kg JPMSAE 54-1211,65

LAM000 CAS: 72-17-3
LACTIC ACID, MONOSODIUM SALT
mf: $C_3H_5O_3$•Na mw: 112.07

PROP: Hygroscopic solid; slt salt taste.

SYNS: 2-HYDROXYPROPANOIC ACID MONOSODIUM SALT ◇ LACOLIN ◇ LACTIC ACID SODIUM SALT ◇ PER-GLYCERIN ◇ SODIUM LACTATE

USE IN FOOD:

Purpose: Cooked out juices retention agent, corrosion preventative, denuding agent, emulsifier, flavor enhancer, flavoring agent, hog scald agent, humectant, lye peeling agent, pH control agent, washing agent.

Where Used: Biscuits, fruits, hog carcasses, meat products, nuts, sponge cake, Swiss roll, tripe, vegetables, water (bottled), water (canned).

Regulations: FDA - 21CFR 184.1768. GRAS when used in accordance with good manufacturing practice. Not authorized for infant foods and infant formulas. USDA - 9CFR 318.7. In meat food products, where allowed, limitation of 5 percent of phosphate in pickle at 10 percent pump level, 0.5 percent of phosphate in product (only clear solution may be injected into product.

SAFETY PROFILE: Moderately toxic by intraperitoneal route. An eye irritant. When heated to decomposition it emits toxic fumes of Na_2O.

TOXICITY DATA and CODEN

eye-rbt 100 mg MLD FCTOD7 20,573,82
ipr-rat LD50:2000 mg/kg FAONAU 40,146,67

LAR400
LACTYLATED FATTY ACID ESTERS of GLYCEROL and PROPYLENE GLYCOL

PROP: Soft to hard waxy solid. Dispersible in hot water; moderately sol in hot isopropanol, benzene, chloroform, soybean oil.

SYN: PROPYLENE GLYCOL LACTOSTEARATE

USE IN FOOD:

Purpose: Emulsifier, plasticizer, stabilizer, surface-active agent, whipping agent.

Where Used: Cake mixes, coffee whiteners, icings, toppings.

Regulations: FDA - 21CFR 172.850. When heated to decomposition it emits acrid smoke and irritating fumes. Must conform to FDA specifications for fats or fatty acids derived from edible oils.

SAFETY PROFILE: When heated to decomposition it emits acrid smoke and irritating fumes.

LAR800
LACTYLIC ESTERS of FATTY ACIDS

PROP: Hard, waxy solid to liquid. Dispersible in hot water; sol in organic solvents, vegetable oil.

USE IN FOOD:

Purpose: Emulsifier, plasticizer, surface-active agent.

Where Used: Baked goods, bakery mixes, frozen desserts, fruit juices (dehydrated), fruits (dehydrated), milk or cream substitutes for beverage coffee, pancake mixes, pudding mixes, rice (precooked instant), shortening (liquid), vegetable juices (dehydrated), vegetables (dehydrated).

Regulations: FDA - 21CFR 172.848. Use at a level not in excess of the amount reasonably required to accomplish the intended effect. Must conform to 21CFR 172.860(b), 172.862 regulations for fats or fatty acids from edible sources.

SAFETY PROFILE: When heated to decomposition it emits acrid smoke and irritating fumes.

LAU550
LANOLIN, ANHYDROUS

PROP: Yellow-white semisolid. Insol in water; sol in chloroform, ether.

SYN: WOOL FAT

USE IN FOOD:

Purpose: Masticatory substance in chewing gum base.

Where Used: Chewing gum.

Regulations: FDA - 21CFR 172.615. Use in amounts not to exceed those required to produce the intended physical or other technical effect.

SAFETY PROFILE: When heated to decomposition it emits acrid smoke and irritating fumes.

LBE300
LARD (UNHYDROGENATED)

PROP: Whitish fat rendered from pork fat. Mp: 42°

SYNS: BLEACHED LARD ◇ BLEACHED-DEODORIZED LARD

USE IN FOOD:

Purpose: Coating agent, emulsifying agent, formulation aid, texturizer.

Where Used: Cake mixes.

Regulations: GRAS when used in accordance with good manufacturing practice.

SAFETY PROFILE: When heated to decomposition it emits acrid smoke and irritating fumes.

LBF500 CAS: 11054-70-9
LASALOCID
mf: $C_{35}H_{54}O_8$ mw: 602.89

SYN: ANTIBIOTIC X 537

USE IN FOOD:

Purpose: Animal drug.

Where Used: Beef, chicken, lamb.

Regulations: FDA - 21CFR 556.347. Tolerance of 0.3 ppm in chicken skin with adhering fat (marker residue), 0.7 ppm in cattle liver (marker residue), and 1.2 ppm in sheep muscle (safe concentration). 21CFR 558.311.

SAFETY PROFILE: Poison by ingestion and intraperitoneal routes. An eye and skin irritant. When heated to decomposition it emits acrid smoke and irritating fumes.

TOXICITY DATA and CODEN

skn-rbt 100 mg/24H MLD DCTODJ 8,451,85
eye-rbt 50 mg MOD DCTODJ 8,451,85

LBK000 CAS: 8006-78-8
LAUREL LEAF OIL

PROP: Main constituent is cineole. From steam distillation of the leaves of *Laurus nobilis* L. (Fam. *Lauraceae)*. Yellow liquid; aromatic and spicy odor. D: 0.905-0.929, refr index: 1.465 at 20°. Sol in fixed oils, mineral oil, propylene glycol; insolin glycerin.

SYNS: BAY LEAF OIL ◇ OIL of LAUREL LEAF

USE IN FOOD:

Purpose: Flavoring agent.

Where Used: Various.

Regulations: FDA - 21CFR 182.20. GRAS when used at a level not in excess of the amount reasonably required to accomplish the intended effect.

SAFETY PROFILE: Moderately toxic by ingestion. A skin irritant. When heated to decomposition it emits acrid smoke and irritating fumes.

TOXICITY DATA and CODEN

skn-rbt 500 mg/24H MOD FCTXAV 14,337,76
orl-rat LD50:3950 mg/kg FCTXAV 14,337,76

LBL000 CAS: 143-07-7
LAURIC ACID
mf: $C_{12}H_{24}O_2$ mw: 200.36

PROP: Colorless, needle-like crystals; slt odor of bay oil. Mp: 48°, bp: 225° @ 100 mm, d: 0.883, vap press: 1 mm @ 121.0°. Insol in water; sol in chloroform, benzene, alc, ether, and petroleum ether.

SYNS: DODECANOIC ACID ◇ DODECOIC ACID ◇ DUODECYLIC ACID ◇ HYDROFOL ACID 1255 ◇ HYSTRENE 9512 ◇ LAUROSTEARIC ACID ◇ NEO-FAT 12 ◇ NINOL AA62 EXTRA ◇ 1-UNDECANECARBOXYLIC ACID ◇ WECOLINE 1295

USE IN FOOD:

Purpose: Component in the manufacture of other food-grade additives, defoaming agent, lubricant.

Where Used: Coconut oil, vegetable fats.

Regulations: FDA - 21CFR 172.860. Use in accordance with good manufacturing practice.

SAFETY PROFILE: Poison by intravenous route. Mildly toxic by ingestion. An experimental neoplastigen. Mutagenic data. Combustible when exposed to heat or flame; can react with oxidizing materials. When heated to decomposition it emits acrid smoke and irritating fumes.

TOXICITY DATA and CODEN

cyt-smc 10 mg/L NATUAS 294,263,81
skn-mus TDLo:108 g/kg/15W-I:NEO APMIAL 46,51,59
orl-rat LD50:12 g/kg FDRLI* 123,-,76

LCA000 CAS: 8022-15-9
LAVANDIN OIL

PROP: Main constituent is Linalool. Prepared by steam distillation of the flowering stalks of the plants *Lavanoula hybrida reverchon*, Lavandula abrialis (Fam. *Labiatae*), *Lavandula officinalis*, or *Lavandula latifolia*. Yellow liquid; camphoraceous odor of lavender. D: 0.885, refr index: 1.460 @ 20°. Sol in fixed oils, propylene glycol, mineral oil; insol in glycerin.

SYNS: OIL of LAVANDIN, ABRIAL TYPE

USE IN FOOD:

Purpose: Flavoring agent.

Where Used: Various.

Regulations: FDA - 21CFR 182.20. GRAS when used at a level not in excess of the amount reasonably required to accomplish the intended effect.

SAFETY PROFILE: A skin irritant. When heated to decomposition it emits acrid smoke and irritating fumes.

TOXICITY DATA and CODEN

skn-rbt 500 mg/24H MLD FCTXAV 14,443,76

LCD000 CAS: 8000-28-0
LAVENDER OIL

PROP: Found in the flowers of *Lavandula officinalis* Chaix et Villars (*Lavabdula vera* De Candolle (Fam. *Labiatae*). The main constituent is linalyl acetate. A colorless to yellow liquid; characteristic odor and taste of lavender flowers. D: 0.875, refr index: 1.459-1.470 @ 20°.

SYNS: LAVENDEL OEL (GERMAN) ◇ OIL of LAVENDER

USE IN FOOD:

Purpose: Flavoring agent.

Where Used: Bakery products, beverages (nonalcoholic), chewing gum, confections, ice cream products.

Regulations: FDA - 21CFR 182.10, 182.20. GRAS when used at a level not in excess of the amount reasonably required to accomplish the intended effect.

SAFETY PROFILE: Mildly toxic by ingestion. A skin irritant. When heated to decomposition it emits acrid smoke and irritating fumes.

TOXICITY DATA and CODEN

skn-rbt 500 mg/24H MLD FCTXAV 14,451,76
orl-rat LD50:9040 mg/kg PHARAT 14,435,59

LCF000 CAS: 7439-92-1
LEAD

mf: Pb mw: 207.19

PROP: Bluish-gray, soft metal. Mp: 327.43°, bp: 1740°, d: 11.34 @ 20°/4°. vap press: 1 mm @ 973°.

SYNS: C.I. 77575 ◇ C.I. PIGMENT METAL 4 ◇ GLOVER ◇ LEAD FLAKE ◇ LEAD S2 ◇ OLOW (POLISH) ◇ OMAHA ◇ OMAHA & GRANT ◇ SI ◇ SO

USE IN FOOD:

Purpose: Contaminant.

Where Used: Animal feed

Regulations: USDA CES Ranking: B-4 (1985).

IARC Cancer Review: Animal Inadequate Evidence IMEMDT 23,325,80. Lead and its compounds are on the Community Right-To-Know List. EPA Genetic Toxicology Program.

OSHA PEL: TWA 0.05 mg(Pb)/m^3 ACGIH TLV: TWA 0.15 mg(Pb)/m^3 NIOSH REL: TWA (Inorganic Lead) 0.10 mg(Pb)/m^3

SAFETY PROFILE: Poison by ingestion. Moderately toxic by intraperitoneal route. It is a suspected carcinogen of the lungs and kidneys. Human systemic effects by ingestion and inhalation: loss of appetite, anemia, malaise, insomnia, headache, irritability, muscle and joint pains, tremors, flaccid paralysis without anesthesia, hallucinations and distorted perceptions, muscle weakness, gastritis and liver changes. The major organ systems affected are the nervous system, blood system, and kidneys. Lead encephalopathy is accompanied by severe cerebral edema, increase in cerebral spinal fluid pressure, proliferation and swelling of endothelial cells in capillaries and arterioles, proliferation of glial cells, neuronal degeneration and areas of focal cortical necrosis in fatal cases.

Experimental evidence now suggests that blood levels of lead below 10 μg/dl can have the effect of diminishing the IQ scores of children. Low levels of lead impair neurotransmission and immune system function and may increase systolic blood pressure. Reversible kidney damage can occur from acute exposure. Chronic exposure can lead to irreversible vascular schlerosis, tubular cell atrophy, interstitial fibrosis, and glomerular sclerosis. Severe toxicity can cause sterility, abortion and neonatal mortality and morbidity. An experimental teratogen. Experimental reproductive effects. Human mutagenic data. Very heavy intoxication can sometimes be detected by formation of a dark line on the gum margins, the so-called "lead line."

TOXICITY DATA and CODEN

cyt-hmn-unr 50 μg/m^3 MUREAV 147,301,85
cyt-rat-ihl 23 μg/m^3/16W GTPZAB 26(10),38,82
orl-rat TDLo: 1140 mg/kg (14D pre-21D post): REP PHMCAA 20,201,78
orl-rat TDLo: 1100 mg/kg (1-22D preg):TER FEPRA7 37,895,78
ihl-rat TCLo: 10 mg/m^3/24H (1-21D preg):TER ZHPMAT 165,294,77
orl-wmn TDLo:450 mg/kg/6Y:PNS:CNS JAMAAP 237,2627,77
ihl-hmn TCLo: 10 μg/m^3:GIT:LIV VRDEA5 (5),107,81
ipr-rat LDLo: 1000 mg/kg EQSSDX 1,1,75

LEF180
LECITHIN

PROP: A complex mixture from soybeans and other plants. Light yellow to brown semisolid; slt nutlike odor, bland taste.

USE IN FOOD:

Purpose: Antioxidant, emulsifier.

Where Used: Baked goods, beverage powders, cocoa powder, fat (griddling), fillings, meat products, oleomargarine, poultry, shortening.

Regulations: FDA - 21CFR 184.1400. GRAS when used in accordance with good manufacturing practice. USDA - 9CFR 318.7. Limitation of 0.5 percent in oleomargarine, sufficient for purpose in other foods. 9CFR 381.147. Sufficient for purpose

SAFETY PROFILE: When heated to decomposition it emits acrid smoke and irritating fumes.

LEG000
LEMONGRASS OIL EAST INDIAN

PROP: From steam distillation of the freshly cut and partially dried grasses of *Cymbopogon flexuosus* and *Andropogon nardus var. flexuosus*. The main constituent is citral. Dark yellow to brown-red liquid; heavy lemon odor. D: 0.894-0.902, refr index: 1.483. Sol in mineral oil, propylene glycol, alc; insol in water, glycerin.

SYNS: BRITISH EAST INDIAN LEMONGRASS OIL ◇ COCHIN ◇ EAST INDIAN LEMONGRASS OIL ◇ LEMONGRAS OEL (GERMAN) ◇ OIL of LEMONGRASS, EAST INDIAN

USE IN FOOD:

Purpose: Flavoring agent.

Where Used: Bakery products, beverages (nonalcoholic), chewing gum, confections, gelatin desserts, ice cream, puddings.

Regulations: FDA - 21CFR 182.20. GRAS when used at a level not in excess of the amount reasonably required to accomplish the intended effect.

SAFETY PROFILE: Mildly toxic by ingestion. A skin irritant. When heated to decomposition it emits acrid smoke and irritating fumes.

TOXICITY DATA and CODEN

skn-mus 100% MLD FCTXAV 14,455,76
orl-rat LD50:5600 mg/kg FCTXAV 14(5),443,76

LEH000 CAS: 8007-02-1
LEMONGRASS OIL WEST INDIAN

PROP: Main constituent is citral. From steam distillation of freshly cut and partially dried grasses of *Cymbopogon citratus* (STAPF) and *Andropogon nardus var. ceriferus* (Hack). Light yellow to brown liquid; light lemon odor. D: 0.869-0.894, refr index: 1.483. Sol in mineral oil, propylene glycol; insol in water,

SYNS: GUATEMALA LEMONGRASS OIL ◇ MADAGASCAR LEMONGRASS OIL ◇ OIL of LEMONGRASS, WEST INDIAN ◇ WEST INDIAN LEMONGRASS OIL

USE IN FOOD:

Purpose: Flavoring agent.

Where Used: Various.

Regulations: FDA - 21CFR 182.20. GRAS when used at a level not in excess of the amount reasonably required to accomplish the intended effect.

SAFETY PROFILE: A skin irritant. When heated to decomposition it emits acrid smoke and irritating fumes.

TOXICITY DATA and CODEN

skn-mus 100% MLD FCTXAV 14,443,76

LEI000 CAS: 8008-56-8
LEMON OIL

PROP: Expressed from the peel of the fruit of *Citrus limon* L. Burmann filius (Fam. *Rutaceae*). Pale yellow liquid; taste and odor of lemon peel. D: 0.849, refr index: 1.473 @ 20°. Misc with dehydrated alc, glacial acetic acid.

SYNS: CEDRO OIL ◇ LEMON OIL, COLDPRESSED (FCC) ◇ LEMON OIL, EXPRESSED ◇ OIL of LEMON ◇ ZITRONEN OEL (GERMAN)

USE IN FOOD:

Purpose: Flavoring agent.

Where Used: Bakery products, beverages (nonalcoholic), cereals, chewing gum, condiments, confections, gelatin desserts, ice cream, meat, puddings, syrups.

Regulations: FDA - 21CFR 182.20. GRAS when used at a level not in excess of the amount reasonably required to accomplish the intended effect. Do not use if terebinthine odor can be detected.

SAFETY PROFILE: Moderately toxic by ingestion. An experimental tumorigen. A skin irritant. When heated to decomposition it emits acrid smoke and irritating fumes.

TOXICITY DATA and CODEN

skn-mus 100% MLD FCTXAV 12,703,74
skn-mus TDLo:280 g/kg/33W-I:ETA JNCIAM 24,1389,60
orl-rat LD50:2840 mg/kg PHARAT 14,435,59

LEI025
LEMON OIL, DESERT TYPE, COLDPRESSED

PROP: Expressed without heat from the peel of the fruit of *Citrus limon* L. Burmann filius (Fam. *Rutaceae*). Pale yellow liquid; taste and odor of lemon peel. D: 0.846, refr index: 1.473 @ 20°. Misc with dehydrated alc, glacial acetic acid.

SYN: OIL of LEMON, DESERT TYPE, COLDPRESSED

USE IN FOOD:

Purpose: Flavoring agent.

Where Used: Various.

Regulations: FDA - 21CFR 182.20. GRAS Do not use if terebinthine odor can be detected. Use at a level not in excess of the amount reasonably required to accomplish the intended effect.

SAFETY PROFILE: A skin irritant. When heated to decomposition it emits acrid smoke and irritating fumes.

LEI030
LEMON OIL, DISTILLED

PROP: From distillation of fresh peel from *Citrus limon* L. Burmann filius (Fam. *Rutaceae*). Pale yellow liquid; taste and odor of fresh lemon peel. D: 0.842, refr index: 1.470 @ 20°. Misc with dehydrated alc, glacial acetic acid.

SYN: OIL of LEMON, DISTILLED

USE IN FOOD:

Purpose: Flavoring agent.

Where Used: Various.

Regulations: FDA - 21CFR 182.20. GRAS when used at a level not in excess of the amount reasonably required to accomplish the intended effect.

SAFETY PROFILE: A skin irritant. When heated to decomposition it emits acrid smoke and irritating fumes.

LER000 CAS: 328-38-1
dl-LEUCINE

mf: $C_6H_{13}NO_2$ mw: 131.17

PROP: dl Form (synthetic form): Leaflets from water; odorless with sweet taste. Mp: 290 (decomp). Sol in water, sltly sol in alc; insol in ether.

SYN: dl-2-AMINO-4-METHYLVALERIC ACID

USE IN FOOD:

Purpose: Dietary supplement, nutrient.

Where Used: Various.

Regulations: FDA - 21CFR 172.320. Limitation of 8.8 percent.

SAFETY PROFILE: Mildly toxic by intraperitoneal route. When heated to decomposition it emits toxic fumes of NO_x.

TOXICITY DATA and CODEN

ipr-rat LD50:6429 mg/kg ABBIA4 64,319,56

LES000 CAS: 61-90-5
l-LEUCINE

mf: $C_6H_{13}NO_2$ mw: 131.20

PROP: An essential amino acid; occurs in isomeric forms. White crystals. Mp (dl): 332° with decomp, mp (l): 295°, d: 1.239 @ 18°/4°. l Form (natural): glistening, hexagonal plates from aq alc. D: 1.291 @ 18°, subl @ 145-148°, decomp @ 293-295°. Sol in water, sltly sol in alc; insol in ether.

SYNS: α-AMINOISOCAPROIC ACID ◇ 2-AMINO-4-METHYLPENTANOIC ACID ◇ 2-AMINO-4-METHYLVALERIC ACID ◇ l-2-AMINO-4-METHYLVALERIC ACID ◇ α-AMINO-γ-METHYLVALERIC ACID ◇ LEUCIN (GERMAN) ◇ LEUCINE ◇ 4-METHYLNORVALINE

USE IN FOOD:

Purpose: Dietary supplement, nutrient.

Where Used: Various.

Regulations: FDA - 21CFR 172.320. Limitation of 8.8 percent.

SAFETY PROFILE: Moderately toxic by subcutaneous route. An experimental teratogen. Experimental reproductive effects. When heated to decomposition it emits toxic fumes of NO_x.

TOXICITY DATA and CODEN

orl-rat TDLo:138 g/kg (5-15D preg):TER JONUAI 74,93,61

ipr-rat TDLo:60 mg/kg/(6-9D preg):TER NATWAY 56,37,69

ipr-rat LD50:5379 mg/kg ABBIA4 58,253,55

LFA000 CAS: 6649-23-6
LEVAMISOLE

mf: $C_{11}H_{12}N_2S$ mw: 204.31

SYNS: 6-PHENYL-2,3,5,6-TETRAHYDROIMIDAZO(2,1-b)THIAZOLE ◇ 2,3,5,6-TETRAHYDRO-6-PHENYLIMIDAZO(2,1-b)THIAZOLE

USE IN FOOD:

Purpose: Animal drug, animal feed drug.

Where Used: Animal feed, beef, lamb, pork.

Regulations: FDA - 21CFR 556.350. Limitation of 0.1 ppm in cattle, sheep, and swine. 21CFR 558.315. (hydrochloride equivalent). USDA CES Ranking: C-2 (1985).

SAFETY PROFILE: Poison by ingestion, intravenous, intraperitoneal and subcutaneous routes. Human systemic effects by ingestion: coma, skin dermatitis and irritation, and fever. When heated to decomposition it emits very toxic fumes of NO_x and SO_x.

TOXICITY DATA and CODEN

orl-chd TDLo:40 mg/kg/8D:CNS,SKN,MET JOPDAB 93,304,78
orl-rat LD50:345 mg/kg DRUGAY 20,89,80

LFI000 CAS: 7660-25-5
LEVULOSE

mf: $C_6H_{12}O_6$ mw: 180.18

PROP: White, hygroscopic crystals or crystalline powder; odorless with sweet taste. D: 1.6. Sol in methanol, ethanol, water.

SYNS: FRUCTOSE (FCC) ◇ FRUIT SUGAR ◇ FRUTABS ◇ LAEVORAL ◇ LAEVOSAN ◇ LEVUGEN

USE IN FOOD:

Purpose: Formulation aid, nutritive sweetener, processing oil.

Where Used: Baked goods, beverages (low calorie).

Regulations: GRAS when used in accordance with good manufacturing practice.

SAFETY PROFILE: An experimental tumorigen. When heated to decomposition it emits acrid smoke and fumes.

TOXICITY DATA and CODEN

scu-mus TDLo:5000 mg/kg:ETA GANNA2 46,371,55

LFN300 CAS: 8008-94-4
LICORICE ROOT EXTRACT

SYNS: GLYCYRRHIZA ◇ GLYCYRRHIZAE (LATIN) ◇ GLYCYRRHIZA EXTRACT ◇ GLYCYRRHIZINA ◇ KANZO (JAPANESE) ◇ LICORICE ◇ LICORICE EXTRACT ◇ LICORICE ROOT

USE IN FOOD:

Purpose: Flavor enhancer, flavoring agent, surface-active agent.

Where Used: Bacon, baked goods, beverages (alcoholic), beverages (nonalcoholic), candy (hard), candy (soft), chewing gum, cocktail mixes, herbs, ice cream, imitation whipped products, plant protein products, seasonings, soft drinks, syrups, vitamin or mineral dietary supplements.

Regulations: FDA - 21CFR 184.1408. GRAS with a limitation of (as glycyrrhizn) 0.05 percent in baked goods, 0.1 percent in alcoholic beverages, 0.15 percent in nonalcoholic beverages, 1.1 percent in chewing gum, 16.0 percent in hard candy, 0.15 percent in herbs and seasonings, 0.15 percent in plant protein products, 3.1 percent in soft candy, 0.5 percent in vitamin or mineral dietary supplements, 0.1 percent in all other foods except sugar substitutes when used in accordance with good manufacturing practice. Not permitted to be used as a nonnutritive sweetener in sugar substitutes.

SAFETY PROFILE: Moderately toxic by intraperitoneal and subcutaneous routes. Mildly toxic by ingestion. Mutagenic data. When heated to decomposition it emits acrid smoke and irritating fumes.

TOXICITY DATA and CODEN

dnr-bcs 100 g/L MUREAV 97,81,82
orl-rat LD50:14200 mg/kg OYYAA2 14,535,77

LFU000 CAS: 5989-27-5
d-LIMONENE

mf: $C_{10}H_{16}$ mw: 136.26

PROP: Colorless liquid; citrus odor. Bp: 175.5-176°, d: 0.8402 @ 25°/4°, refr index: 1.471. Misc with alc, fixed oils; sltly sol in glycerin; insol in propylene glycol, water.

SYNS: FEMA No. 2633 ◇ (+)-4-ISOPROPENYL-1-METHYLCYCLOHEXENE ◇ d-(+)-LIMONENE ◇ (+)-R-LIMONENE ◇ d-p-MENTHA-1,8-DIENE ◇ p-MENTHA-1,8-DIENE ◇ (R)-1-METHYL-4-(1-METHYLETHENYL)-CYCLOHEXENE ◇ NCI-C55572

USE IN FOOD:

Purpose: Flavoring agent.

Where Used: Various.

Regulations: FDA - 21CFR 182.60. Use at a level not in excess of the amount reasonably required to accomplish the intended effect.

SAFETY PROFILE: Poison by intravenous route. Moderately toxic by intraperitoneal and intraduodenal routes. Mildly toxic by ingestion. An experimental tumorigen and teratogen. Experimental reproductive effects. When heated to decomposition it emits acrid smoke and irritating fumes.

TOXICITY DATA and CODEN

orl-rat TDLo:20083 mg/kg (9-15D preg):TER OYYAA2 10,179,75
orl-rat TDLo:252 g/kg (26W male):REP OYYAA2 9,403,75
orl-mus TDLo:3546 mg/kg (7-12D preg):TER OYYAA2 13,863,77
orl-mus TDLo:67 g/kg/39W-I:ETA JNCIAM 35,771,65
orl-rat LD50:4400 mg/kg NIIRDN 6,887,82

LFX000 CAS: 78-70-6
LINALOOL
mf: $C_{10}H_{18}O$ mw: 154.28

PROP: Colorless liquid; odor similar to that of bergamot oil and French lavender. D: 0.858-0.868 @ 25°, refr index: 1.461, bp: 195-199°, flash p: 172°F. Sol in alc, ether, fixed oils, propylene glycol; insol in glycerin.

SYNS: ALLO-OCIMENOL ◇ 2,6-DIMETHYL-2,7-OCTADIENE-6-OL ◇ 2,6-DIMETHYLOCTA-2,7-DIEN-6-OL ◇ 3,7-DIMETHYLOCTA-1,6-DIEN-3-OL ◇ 3,7-DIMETHYL-1,6-OCTADIEN-3-OL ◇ FEMA No. 2635 ◇ LINALOL ◇ LINALYL ALCOHOL

USE IN FOOD:

Purpose: Flavoring agent.

Where Used: Various.

Regulations: FDA - 21CFR 182.60. Use at a level not in excess of the amount reasonably required to accomplish the intended effect.

SAFETY PROFILE: Moderately toxic by ingestion. Mildly toxic by skin contact. A skin irritant. When heated to decomposition it emits acrid smoke and irritating fumes.

TOXICITY DATA and CODEN

skn-rbt 500 mg/24H MLD FCTXAV 14,673,76
orl-rat LD50:2790 mg/kg FCTXAV 2,327,64

LFZ000 CAS: 126-64-7
LINALYL BENZOATE
mf: $C_{17}H_{22}O_2$ mw: 258.39

PROP: Found in the essential oils of Ylang-Ylang and Tuberose (FCTXAV 14,443,76). Yellow to brown-yellow liquid; tuberose odor. D: 0.980-0.999, refr index: 1.505-1.520, flash p: 208°F. Sol in chloroform, alc, ether; insol in water.

SYNS: 3,7-DIMETHYL-1,6-OCTADIEN-3-OL BENZOATE ◇ 3,7-DIMETHYL-1,6-OCTADIEN-3-YL BENZOATE ◇ 1,5-DIMETHYL-1-VINYL-4-HEXEN-1-OL BENZOATE ◇ 1,5-DIMETHYL-1-VINYL-4-HEXEN-1-YL BENZOATE ◇ FEMA No. 2638

USE IN FOOD:

Purpose: Flavoring agent.

Where Used: Various.

Regulations: FDA - 21CFR 172.515. Use at a level not in excess of the amount reasonably required to accomplish the intended effect.

SAFETY PROFILE: A skin irritant. Combustible liquid. When heated to decomposition it emits acrid smoke and irritating fumes.

TOXICITY DATA and CODEN

skn-rbt 500 mg/24H MLD FCTXAV 14,461,76

LGA050
LINALYL FORMATE
mf: $C_{11}H_{18}O_2$ mw: 182.26

PROP: Colorless liquid; citrus, herbaceous odor. D: 0.910-0.918, refr index: 1.453-1.458, flash p: 189°F. Sol in alc, fixed oils; sltly sol in propylene glycol, water; insol in glycerin @ 202°.

SYNS: 3,7-DIMETHYL-1,6-OCTADIEN-3-YL FORMATE ◇ FEMA No. 2642

USE IN FOOD:

Purpose: Flavoring agent.

Where Used: Various.

Regulations: FDA - 21CFR 172.515. Use at a level not in excess of the amount reasonably required to accomplish the intended effect.

SAFETY PROFILE: Combustible liquid. When heated to decomposition it emits acrid smoke and irritating fumes.

LGB000 CAS: 78-35-3
LINALYL ISOBUTYRATE
mf: $C_{14}H_{24}O_2$ mw: 224.38

PROP: Colorless liquid; fresh, rosy odor. D: 0.882-0.888, refr index: 1.446-1.451, flash p: +212°F. Misc with alc, chloroform, ether; insol in water @ 20°.

SYNS: 3,7-DIMETHYL-1,6-OCTADIEN-3-OL ISOBUTYRATE ◇ 3,7-DIMETHYL-1,6-OCTADIEN-3-YL ISOBUTYRATE ◇ 1,5-DIMETHYL-1-VINYL-4-HEXENYL ESTER, ISOBUTYRIC ACID ◇ FEMA No. 2640 ◇ LINALOOL ISOBUTYRATE

USE IN FOOD:

Purpose: Flavoring agent.

Where Used: Various.

Regulations: FDA - 21CFR 172.515. Use at a level not in excess of the amount reasonably required to accomplish the intended effect.

SAFETY PROFILE: Combustible liquid. Mildly toxic by ingestion. When heated to decomposition it emits acrid smoke and irritating fumes.

TOXICITY DATA and CODEN

orl-mus LD50:15100 mg/kg FCTXAV 2,327,64

LGC100
LINALYL PROPIONATE
mf: $C_{13}H_{22}O_2$ mw: 210.32

PROP: Colorless liquid; fresh, pear odor. D: 0.893-0.902, refr index: 1.449-1.454, flash p: 189°F. Sol in alc, fixed oils; insol in glycerin @ 226°.

SYN: FEMA No. 2645

USE IN FOOD:

Purpose: Flavoring agent.

Where Used: Various.

Regulations: FDA - 21CFR 172.515. Use at a level not in excess of the amount reasonably required to accomplish the intended effect.

SAFETY PROFILE: Combustible liquid. When heated to decomposition it emits acrid smoke and irritating fumes.

LGD000 CAS: 154-21-2
LINCOMYCIN
mf: $C_{18}H_{34}N_2O_6S$ mw: 406.60

PROP: Sol in methanol, ethanol, butanol, isopropanol, ethyl acetate, n-butyl acetate, amylacetate, etc. Moderately sol in water.

SYNS: ALBIOTIC ◇ LINCOCIN ◇ LINCOLCINA ◇ LINCOLNENSIN ◇ LINCOMYCINE (FRENCH) ◇ NSC-70731 ◇ U-10149

USE IN FOOD:

Purpose: Animal drug, animal feed drug.

Where Used: Animal feed, chicken, pork.

Regulations: FDA - 21CFR 556.360. Tolerance of 0.15 ppm in milk, 0.1 ppm in edible tissues of chicken and swine. 21CFR 558.325.

EPA Genetic Toxicology Program.

SAFETY PROFILE: Poison by intramuscular route. Moderately toxic by ingestion and intraperitoneal routes. When heated to decomposition it emits very toxic fumes of SO_x and NO_x.

TOXICITY DATA and CODEN

orl-rat LD50:1000 mg/kg 85ERAY 1,186,78
scu-rat LD50:9780 mg/kg TXAPA9 18,185,71

LGF900
LINOLEAMIDE

SYN: LINOLEIC ACID AMIDE

USE IN FOOD:

Purpose: Release agent.

Where Used: Packaging materials.

Regulations: FDA - 21CFR 181.28. Use in accordance with good manufacturing practice.

SAFETY PROFILE: When heated to decomposition it emits acrid smoke and irritating fumes.

LGG000 CAS: 60-33-3
LINOLEIC ACID
mf: $C_{18}H_{32}O_2$ mw: 280.50

PROP: Colorless oil, easily oxidized by air. D: 0.9038 @ 18°/4°, mp: −12°, bp: 230° @ 16 mm. Sol in ether and ethanol; misc with dimethyl formamide, fat solvents, oils.

SYNS: LEINOLEIC ACID ◇ 9,12-LINOLEIC ACID ◇ cis,cis-9,12-OCTADECADIENOIC ACID ◇ cis-9,cis-12-OCTADECADIENOIC ACID ◇ 9,12-OCTADECADIENOIC ACID

USE IN FOOD:

Purpose: Dietary supplement, flavoring agent.

Where Used: Infant formula.

Regulations: FDA - 21CFR 182.5065. GRAS when used in accordance with good manufacturing practice and free from chickedema factor. 21CFR 184.1065. GRAS when used in accordance with good manufacturing practice, but 21CFR 412(g) governs use in infant formula.

SAFETY PROFILE: A human skin irritant. Ingestion can cause nausea and vomiting. When heated to decomposition it emits acrid smoke and irritating fumes.

TOXICITY DATA and CODEN

skn-hmn 75 mg/3D-I MOD 85DKA8 -,127,77

LGK000 CAS: 8001-26-1
LINSEED OIL

PROP: Yellowish liquid, peculiar odor, bland taste. Sltly sol in alc; misc with chloroform, ether, petr ether, carbon disulfide, oil, turpentine. Bp: 343°, mp: −19°, d: 0.93, flash p: (raw oil) 432°F (CC), flash p: (boiled) 403°F (CC), autoign temp: 650°F. From seed of *Linum usitatissimum*.

SYNS: GROCO ◇ L-310

USE IN FOOD:

Purpose: Drying oil.

Where Used: Packaging materials.

Regulations: FDA - 21CFR 181.26.

SAFETY PROFILE: An allergen and skin irritant to humans. Combustible liquid when exposed to heat or flame; can react with oxidizing materials. Subject to spontaneous heating. Violent reaction with Cl_2. To fight fire, use CO_2, dry chemical.

TOXICITY DATA and CODEN

skn-hmn 300 mg/3D-I MOD 85DKA8 -,127,77

LIA000 CAS: 9000-40-2
LOCUST BEAN GUM

PROP: From the ground endosperms of *Ceratonia ailiqua* (L.) Taub. (Fam. *Leguminosae*). White powder; odorless and tasteless but acquires a leguminous taste when boiled in water. A galactomannan polysaccharide. Mw: 310,000 (approx). Insol in most organic solvents.

SYNS: ALGAROBA ◇ CAROB BEAN GUM ◇ CAROB FLOUR ◇ NCI-C50419 ◇ ST. JOHN'S BREAD ◇ SUPERCOL

USE IN FOOD:

Purpose: Emulsifier, stabilizer, thickener.

Where Used: Baked goods, beverage bases (nonalcoholic), beverages, candy, cheese, fillings, gelatins, ice cream, jams, jellies, pies, puddings, soups.

Regulations: FDA - 21CFR 184.1343. GRAS with a limitation of 0.15 percent in baked goods, 0.25 percent in beverages and beverage bases, nonalcoholic, 0.8 percent in cheese, 0.75 percent in gelatins, puddings, and fillings, 0.75 percent in jams and jellies, 0.5 percent in all other foods when used in accordance with good manufacturing practice.

NTP Carcinogenesis Bioassay (feed); No Evidence: mouse, rat NTPTR* NTP-TR-221,82.

SAFETY PROFILE: Mildly toxic by ingestion. When heated to decomposition it emits acrid smoke and irritating fumes.

TOXICITY DATA and CODEN

orl-rat LD50:13 g/kg FDRLI* 124,-,76

LII000 CAS: 8016-31-7
LOVAGE OIL

PROP: The constituents include d-α-terpineol, butyl dihydrophthalides, butyl tetrahydrophthalides, coumarin, aldehydes and acetic and isovaleric acid. From steam distillation of fresh root of *Levisticum officinale* L. Koch syn. *Angelica levisticum*, Baillon (Fam. *Umbelliferae*). Yellow to green to brown liquid; strong odor and taste. D: 1.034-1.057, refr index: 1.536-1.554 @ 20°. Sol in fixed oils; sltly sol in mineral oil; insol in glycerin, propylene glycol.

USE IN FOOD:

Purpose: Flavoring agent.

Where Used: Various.

Regulations: FDA - 21CFR 182.510. Use at a level not in excess of the amount reasonably required to accomplish the intended effect.

SAFETY PROFILE: Moderately toxic by ingestion. A skin irritant. When heated to decomposition it emits acrid smoke and irritating fumes.

TOXICITY DATA and CODEN

skn-rbt 500 mg/24H MOD FCTXAV 16,813,78
orl-mus LD50:3400 mg/kg FCTXAV 16,637,78

LJO000 CAS: 657-27-2
l-LYSINE MONOHYDROCHLORIDE
mf: $C_6H_{14}N_2O_2 \cdot ClH$ mw: 182.68

PROP: White powder. Mp: 235-236°. Sol in water; insol in alc and ether. Crystals from dil ethanol. Mp: 263-264° (decomp) when anhydrous.

SYNS: 2,6-DIAMINOHEXANOIC ACID HYDROCHLORIDE ◇ l-LYSINE HYDROCHLORIDE ◇ LYSINE MONOHYDROCHLORIDE

USE IN FOOD:

Purpose: Dietary supplement, nutrient.

Where Used: Various.

Regulations: FDA - 21CFR 172.320. Limitation of 6.4 percent expressed as the free amino acid.

SAFETY PROFILE: Mildly toxic by ingestion. When heated to decomposition it emits very toxic fumes of HCl and NO_x.

TOXICITY DATA and CODEN

orl-rat LD50:10 g/kg JPMSAE 62,49,73
ipr-rat LD50:4019 mg/kg ABBIA4 58,253,55

M

MAD500 CAS: 546-93-0
MAGNESIUM(II) CARBONATE (1:1)
mf: $CO_3 \cdot Mg$ mw: 84.32

PROP: Very light, white powder; odorless. D: 3.04; decomp @ 350°. Sol in acids; insol in water and alc.

SYNS: CARBONATE MAGNESIUM ◇ CARBONIC ACID, MAGNESIUM SALT ◇ C.I. 77713 ◇ DCI LIGHT MAGNESIUM CARBONATE ◇ HYDROMAGNESITE ◇ MAGMASTER ◇ MAGNESIA ALBA ◇ MAGNESITE ◇ MAGNESIUM CARBONATE ◇ MAGNESIUM CARBONATE, PRECIPITATED ◇ STAN-MAG MAGNESIUM CARBONATE

USE IN FOOD:

Purpose: Alkali, anticaking agent, carrier, color-retention agent, drying agent.

Where Used: Dry mixes, table salt.

Regulations: GRAS when used in accordance with good manufacturing practice.

ACGIH TLV: TWA 10 mg/m^3

SAFETY PROFILE: Incompatible with formaldehyde. When heated to decomposition it emits acrid smoke and irritating fumes.

MAE250 CAS: 7786-30-3
MAGNESIUM CHLORIDE
mf: Cl_2Mg mw: 95.21

PROP: Thin, white to opaque, gray granules and/or flakes, deliquescent. Mp: 708° (712° with rapid heating), bp: 1412°, d: 2.325. Sol in H_2O evolving much heat, alc.

SYN: DUS-TOP

USE IN FOOD:

Purpose: Color-retention agent, firming agent, flavoring agent, tissue softening agent.

Where Used: Meat (raw cuts), poultry (raw cuts).

Regulations: FDA - 21CFR 182.5446, 184.1426. GRAS when used in accordance with good manufacturing practice. USDA - 9CFR 318.7, 381.147. Limitation of not more than 3 percent of a 0.8 molar solution. Solutions consisting of water and approved proteolytic enzymes applied or injected into raw meat cuts shall not result in a gain of more than 3 percent above the weight of the untreated product.

EPA Genetic Toxicology Program.

SAFETY PROFILE: Poison by intraperitoneal and intravenous routes. Moderately toxic by ingestion and subcutaneous routes. Human mutagenic data. In humid environments it causes steel to rust very rapidly. When heated to decomposition it emits toxic fumes of Cl^-.

TOXICITY DATA and CODEN

mmo-omi 8000 ppm APMBAY 6,45,58
cyt-hmn:hla 2 mmol/L JCLLAX 78,217,71
orl-rat LD50:2800 mg/kg JPETAB 35,1,29

MAG100
MAGNESIUM GLYCEROPHOSPHATE

USE IN FOOD:

Purpose: Plasticizer.

Where Used: Packaging materials.

Regulations: FDA - 21CFR 181.27. Use in accordance with good manufacturing practice.

SAFETY PROFILE: When heated to decomposition it emits acrid smoke and irritating fumes.

MAG550
MAGNESIUM HYDROGEN PHOSPHATE

USE IN FOOD:

Purpose: Plasticizer.

Where Used: Packaging materials.

Regulations: FDA - 21CFR 181.27. Use in accordance with good manufacturing practice.

SAFETY PROFILE: When heated to decomposition it emits acrid smoke and irritating fumes.

MAG750 CAS: 1309-42-8
MAGNESIUM HYDROXIDE
mf: H_2MgO_2 mw: 58.33

PROP: White powder, odorless. D: 2.36, mp: decomp @ 350°. Sol in solns of ammonium salts and dilute acids; almost insol in water and alc.

SYNS: MAGNESIA MAGMA ◇ MAGNESIUM HYDRATE ◇ MILK OF MAGNESIA

USE IN FOOD:

Purpose: Alkali, color-retention agent, drying agent, pH control agent, processing aid.

Where Used: Various.

Regulations: FDA - 21CFR 184.1428. GRAS when used in accordance with good manufacturing practice.

SAFETY PROFILE: Incompatible with maleic anhydride; P.

MAH500 CAS: 1309-48-4
MAGNESIUM OXIDE
mf: MgO mw: 40.31

PROP: White, bulky very fine powder; odorless. Mp: 2500-2800°, d: 3.65-3.75. Very sltly sol in water; sol in dil acids; insol in alc.

SYNS: CALCINED BRUCITE ◇ CALCINED MAGNESIA ◇ CALCINED MAGNESITE ◇ MAGNESIA ◇ MAGNESIA USTA ◇ MAGNEZU TLENEK (POLISH) ◇ SEAWATER MAGNESIA

USE IN FOOD:

Purpose: Alkali, anticaking agent, firming agent, free-flow agent, lubricant, neutralizing agent, nutrient agent, pH control agent, release agent.

Where Used: Various.

Regulations: FDA - 21CFR 182.5431, 184.1431. GRAS when used in accordance with good manufacturing practice.

OSHA PEL: TWA 10 mg/m^3 Total Dust; 5 mg/m^3 Respirable Dust ACGIH TLV: TWA 10 mg/m^3 (fume) DFG MAK: 6 mg/m^3 (fume)

SAFETY PROFILE: An experimental tumorigen. Inhalation of the fumes can produce a febrile reaction and leukocytosis in humans. Violent reaction or ignition with interhalogens (e.g., bromine pentafluoride; chlorine trifluoride). Incandescent reaction with phosphorus pentachloride.

TOXICITY DATA and CODEN

itr-ham TDLo:480 mg/kg/30W-I:ETA CNREA8 33,2209,73

ihl-hmn TCLo:400 mg/m^3 DTLVS* 3,147,71

MAH775 CAS: 7782-75-4
MAGNESIUM PHOSPHATE, DIBASIC
mf: $MgPHO_4 \cdot 3H_2O$ mw: 174.33

PROP: White crystalline powder. Sltly sol in water; insol in alc; sol in dil acid.

SYN: DIMAGNESIUM PHOSPHATE

USE IN FOOD:

Purpose: Dietary supplement, nutrient, pH control agent, stabilizer.

Where Used: Various.

Regulations: 21CFR 181.29, 182.5434, 184.1434. GRAS when used in accordance with good manufacturing practice.

SAFETY PROFILE: A nuisance dust.

MAH780
MAGNESIUM PHOSPHATE, TRIBASIC
mf: $Mg_3(PO_4)_2 \cdot xH_2O$ mw: 262.86

PROP: White crystalline powder; odorless. Sol in dil mineral acids; insol in water.

SYN: TRIMAGNESIUM PHOSPHATE

USE IN FOOD:

Purpose: Dietary supplement, nutrient, pH control agent, stabilizer.

Where Used: Various.

Regulations: 21CFR 181.29, 182.5434, 184.1434. GRAS when used in accordance with good manufacturing practice.

SAFETY PROFILE: A nuisance dust.

MAI000 CAS: 12057-74-8
MAGNESIUM PHOSPHIDE
DOT: 2011
mf: Mg_3P_2 mw: 134.87

SYNS: FOSFURI di MAGNESIO (ITALIAN) ◇ MAGNESIUM-FOSFIDE (DUTCH) ◇ PHOSPHURE de MAGNESIUM (FRENCH)

USE IN FOOD:

Purpose: Fumigant.

Where Used: Animal feed, processed foods.

Regulations: FDA - 21CFR 193.255. Fumigant residue tolerance of 0.01 phosphine in processed foods. 21CFR 561.268. Limitation of 0.1 ppm in animal feeds.

DOT Classification: Flammable Solid; Label: Dangerous When Wet, Poison

SAFETY PROFILE: A poison. Moderately toxic by inhalation. Flammable when exposed to heat, flame or oxidizing materials. Ignites when heated in chlorine, bromine or iodine vapors. Incandescent reaction with nitric acid. Reacts with water to evolve flammable phosphine gas. When heated to decomposition it emits toxic fumes of PO_x and phosphine.

TOXICITY DATA and CODEN

ihl-rat LCLo: 580 ppm/1H ZGSHAM 25,279,33

MAJ000 CAS: 1343-90-4
MAGNESIUM SILICATE HYDRATE
mf: $Mg_2O_8Si_3 \cdot H_2O$ mw: 278.91

PROP: Fine white powder; odorless and tasteless. Insol in water, alc.

USE IN FOOD:

Purpose: Anticaking agent, filter aid.

Where Used: Table salt.

Regulations: FDA - 21CFR 182.2437. GRAS with limitation of 2 percent in table salt.

SAFETY PROFILE: A human skin irritant.

TOXICITY DATA and CODEN

skn-hmn 300 μg/3D-I MLD 85DKA8 -,127,77

MAJ030 CAS: 557-04-0
MAGNESIUM STEARATE

PROP: Fine white bulky powder; faint, characteristic odor. Insol in water, alc, ether.

USE IN FOOD:

Purpose: Anticaking agent, binder, emulsifier, lubricant, nutrient agent, processing aid, release agent, stabilizer.

Where Used: Candy, gum, mint, sugarless gum.

Regulations: FDA - 21CFR 172.863, 173.340. Must conform to FDA specifications for salts of fats or fatty acids derived from edible oils. 21CFR 181.29. Use at a level not in excess of the amount reasonably required to accomplish the intended effect. 21CFR 184.1440. GRAS when used in accordance with good manufacturing practice.

SAFETY PROFILE: When heated to decomposition it emits acrid smoke and toxic fumes.

MAJ250 CAS: 7487-88-9
MAGNESIUM SULFATE (1:1)
mf: $O_4S \cdot Mg$ mw: 120.37

PROP: Opaque needles or granular crystalline powder; odorless with cooling, bitter, salt taste. Sol in water; slowly sol in glycerin; sltly sol in alc.

SYNS: EPSOM SALTS ◇ MAGNESIUM SULPHATE

USE IN FOOD:

Purpose: Dietary supplement, flavor enhancer, nutrient, processing aid.

Where Used: Various.

Regulations: FDA - 21CFR 182.5443, 184.1443. GRAS when used at a level not in excess of the amount reasonably required to accomplish the intended effect.

SAFETY PROFILE: Moderately toxic by ingestion, intraperitoneal, and subcutaneous routes. An experimental teratogen. Potentially explosive reaction when heated with ethoxyethynyl alcohols (e.g., 1-ethoxy-3-methyl-1-butyn-3-ol). When heated to decomposition it emits toxic fumes of SO_x.

TOXICITY DATA and CODEN

ipr-rat TDLo: 750 mg/kg (17-21D preg): TER GEPHDP 12,25,81
orl-mus LDLo: 5000 mg/kg HBAMAK 4,1364,35

MAN000 CAS: 6915-15-7
MALIC ACID
mf: $C_4H_6O_5$ mw: 134.10

PROP: White or colorless crystals; acid taste. Exhibits isomeric forms (dl, l and d). D (dl): 1.601, d (d or l): 1.595 @ 20°/40; mp (dl): 128°, mp (d or l): 100°; bp (dl): 150°, bp (d or l): 140° (decomp). Very sol in water and alc; sltly sol in ether.

SYNS: HYDROXYSUCCINIC ACID ◇ KYSELINA JABLECNA

USE IN FOOD:

Purpose: Acidifier, flavor enhancer, flavoring agent, pH control agent, synergist for antioxidants.

Where Used: Beverages (dry mix), candy (hard), candy (soft), chewing gum, fats (chicken), fillings, fruit juices, fruits (processed), gelatins, jams, jellies, lard, nonalcoholic beverages, puddings, shortening, soft drinks, wine.

Regulations: FDA - 21CFR 184.1069. GRAS with a limitation of 3.4 percent in non-alcoholic beverages, 3.0 percent in chewing gum, 0.8 percent in gelatins, puddings, and fillings, 6.9 percent in hard candy, 2.6 percent in jams and jellies, 3.5 percent in processed fruits and fruit juices, 3.0 percent in soft candy, 0.7 percent in all other foods when used in accordance with good manufacturing practice. USDA - 9CFR 318.7. Limitation of 0.01 percent on basis of total, weight in combination with antioxidants in lard and shortening. BATF - 27CFR 240.1051. GRAS when used in accordance with good manufacturing practice.

SAFETY PROFILE: Moderately toxic by ingestion. A skin and severe eye irritant. When heated to decomposition it emits acrid smoke and irritating fumes.

TOXICITY DATA and CODEN

skn-rbt 500 mg/24H MOD 28ZPAK -,105,72
eye-rbt 750 μg/24H SEV 28ZPAK -,105,72
orl-rat LDLo: 1600 mg/kg 14CYAT 2,1813,63

MAO300 CAS: 9050-36-6
MALTODEXTRIN

mf: $(C_6H_{10}O_5)_n$

PROP: White powder or solution from partial hydrolysis of corn starch.

USE IN FOOD:

Purpose: Bodying agent, bulking agent, carrier, crystallization inhibitor, texturizer.

Where Used: Candy, crackers, puddings.

Regulations: FDA - 21CFR 184.1444. GRAS when used in accordance with good manufacturing practice.

SAFETY PROFILE: When heated to decomposition it emits acrid smoke and irritating fumes.

MAO525
MALT SYRUP

PROP: Derived from barley (*Hordeum vulgare* L.). Brown liquid; sweet taste. Sol in water.

SYN: MALT EXTRACT

USE IN FOOD:

Purpose: Adjuvant, flavoring agent.

Where Used: Cured meat, poultry.

Regulations: FDA - 21CFR 184.1445. GRAS when used in accordance with good manufacturing practice. USDA - 9CFR 318.7. Limitation of 2.5 percent in cured meats. 9CFR 381.147. Sufficient for purpose.

SAFETY PROFILE: When heated to decomposition it emits acrid smoke and irritating fumes.

MAO900
MANDARIN OIL, COLDPRESSED

PROP: From expression of peel of *Citrus reticulata* Blanco var. *Mandarin*. Clear orange to brown-orange liquid; orange odor. D: 0.846. Sol in fixed oils, mineral oil; slt sol in propylene glycol; insol in glycerin.

USE IN FOOD:

Purpose: Flavoring agent.

Where Used: Various.

Regulations: FDA - 21CFR 182.20. GRAS when used at a level not in excess of the amount reasonably required to accomplish the intended effect.

SAFETY PROFILE: When heated to decomposition it emits acrid smoke and irritating fumes.

MAQ790
MANGANESE CAPRYLATE

USE IN FOOD:

Purpose: Drying agent.

Where Used: Packaging materials.

Regulations: FDA - 21CFR 181.25. Use in accordance with good manufacturing practice.

SAFETY PROFILE: When heated to decomposition it emits acrid smoke and irritating fumes.

MAR000 CAS: 7773-01-5
MANGANESE(II) CHLORIDE (1:2)

mf: Cl_2Mn mw: 125.84

PROP: Cubic, deliquescent, pink crystals. Mp: 650°, bp: 1190°, d: 2.977 @ 25°. Sol in water.

SYNS: MANGANESE DICHLORIDE ◇ MANGANOUS CHLORIDE

USE IN FOOD:

Purpose: Dietary supplement, nutrient.

Where Used: Various.

Regulations: GRAS when used in accordance with good manufacturing practice.

Manganese and its compounds are on the Community Right-To-Know List. EPA Genetic Toxicology Program.

ACGIH TLV: 5 mg(Mn)/m^3

SAFETY PROFILE: Poison by intraperitoneal, subcutaneous, intramuscular, intravenous, parenteral, and possibly other routes. Moderately toxic by ingestion. Mutagenic data. An experimental carcinogen and teratogen. Experimental reproductive effects. Reacts violently with potassium or sodium. When heated to decomposition it emits toxic fumes of Cl^-.

TOXICITY DATA and CODEN

mmo-esc 5 µmol/L MUREAV 126,9,84
msc-mus:lym 40 mg/L JTEHD6 9,367,82
orl-rat TDLo:106 mg/kg (30W pre):REP GISAAA 49(11),80,84
ipr-pig TDLo:4581 mg/kg (12-16W preg):TER DABBBA 33,2872,72
ivn-ham TDLo:30 mg/kg (8D preg):REP ADTEAS 5,51,72
ipr-mus TDLo:2080 mg/kg/26W-I:CAR FEPRA7 23,393,64
scu-mus TDLo:2080 mg/kg/26W-I:CAR FEPRA7 23,393,64
orl-mus LD50:1715 mg/kg TOLED5 7,221,81

MAR260 CAS: 10024-66-5
MANGANESE CITRATE
mf: $Mn_3(C_6H_5O_7)_2$ mw: 543.02

PROP: Pale orange or pinkish-white powder.

USE IN FOOD:

Purpose: Nutrient supplement, sequestrant.

Where Used: Baked goods, beverages (nonalcoholic), dairy product analogs, fish products, infant formula, meat products, milk products, poultry products.

Regulations: FDA - 21CFR 184.1449. GRAS when used in accordance with good manufacturing practice.

SAFETY PROFILE: When heated to decomposition it emits acrid smoke and irritating fumes.

MAS800 CAS: 6485-39-8
MANGANESE GLUCONATE
mf: $C_{12}H_{22}MnO_{14}•2H_2O$ mw: 481.27

PROP: Slightly pink powder. Sol in hot water; very sltly sol in alc.

USE IN FOOD:

Purpose: Dietary supplement, nutrient.

Where Used: Baked goods, beverages (nonalcoholic), dairy product analogs, fish products, meat products, milk products, poultry products.

Regulations: FDA - 21CFR 182.5452, 184.1452. GRAS when used in accordance with good manufacturing practice.

SAFETY PROFILE: When heated to decomposition emits toxic fumes of manganese.

MAS810
MANGANESE GLYCEROPHOSPHATE
mf: $C_3H_7MnO_6P•xH_2O$ mw: 225.00

PROP: White or pink powder; odorless and tasteless. Sol in citric acid solution; sltly sol in water; insol in alc.

USE IN FOOD:

Purpose: Dietary supplement, nutrient.

Where Used: Various.

Regulations: FDA - 21CFR 182.5455, 182.8455. GRAS when used in accordance with good manufacturing practice.

SAFETY PROFILE: When heated to decomposition emits toxic fumes of manganese.

MAS815 CAS: 10043-84-2
MANGANESE HYPOPHOSPHITE
mf: $Mn(PH_2O_2)_2•xH_2O$ mw: 184.91

PROP: Pink granular or crystalline powder; odorless and tasteless. Sol in water, alc.

USE IN FOOD:

Purpose: Dietary supplement, nutrient.

Where Used: Various.

Regulations: FDA - 21CFR 182.8458. GRAS when used in accordance with good manufacturing practice.

SAFETY PROFILE: When heated to decomposition emits toxic fumes of manganese.

MAS818
MANGANESE LINOLEATE

USE IN FOOD:

Purpose: Drying agent.

Where Used: Packaging materials.

Regulations: FDA - 21CFR 181.25. Use in accordance with good manufacturing practice.

SAFETY PROFILE: When heated to decomposition it emits acrid smoke and irritating fumes.

MAS820
MANGANESE NAPHTHENATE

USE IN FOOD:

Purpose: Drying agent.

Where Used: Packaging materials.

Regulations: FDA - 21CFR 181.25. Use in accordance with good manufacturing practice.

SAFETY PROFILE: When heated to decomposition it emits acrid smoke and irritating fumes.

MAU250 CAS: 7785-87-7
MANGANESE(II) SULFATE (1:1)

mf: $O_4S \cdot Mn$ mw: 151.00

PROP: Pink granular powder; odorless. Mp: 700°, bp: decomp @ 850°. d: 3.25. Very sol in water; more so in boiling H_2O; insol in alc.

SYNS: MANGANOUS SULFATE ◇ MAN-GRO ◇ NCI-C61143 ◇ SORBA-SPRAY Mn ◇ SULFURIC ACID, MANGANESE(2+) SALT

USE IN FOOD:

Purpose: Dietary supplement, nutrient.

Where Used: Beverages (nonalcoholic), dairy product analogs, fish products, meat products, milk products, poultry products.

Regulations: FDA - 21CFR 182.5461, 184.1461. GRAS when used in accordance with good manufacturing practice.

Manganese and its compounds are on the Community Right-To-Know List. EPA Genetic Toxicology Program.

ACGIH TLV: 5 mg(Mn)/m^3

SAFETY PROFILE: Poison by intraperitoneal route. An experimental neoplastigen. Mutagenic data. When heated to decomposition it emits toxic fumes of SO_x and manganese.

TOXICITY DATA and CODEN

dnr-bcs 50 mmol/L MUREAV 31,185,75
dni-smc 10 mmol/L MGGEAE 151,69,77
ipr-mus TDLo:660 mg/kg/8W-I:NEO CNREA8 36,1744,76
ipr-mus LD50:332 mg/kg COREAF 256,1043,63

MAV100
MANGANESE TALLATE

USE IN FOOD:

Purpose: Drying agent.

Where Used: Packaging materials.

Regulations: FDA - 21CFR 181.25. Use in accordance with good manufacturing practice.

SAFETY PROFILE: When heated to decomposition it emits acrid smoke and irritating fumes.

MAX875
MARJORAM OIL

PROP: From steam distillation of the herb *Marjoram hortensis* L. (Fam. *Labiatae*). Yellow to green-yellow liquid; spicy odor. Sol in fixed oils, mineral oil, partly sol in propylene glycol; insol in glycerin.

USE IN FOOD:

Purpose: Flavoring agent.

Where Used: Fish, meat, sauces, soups.

Regulations: FDA - 21CFR 182.10. GRAS when used at a level not in excess of the amount reasonably required to accomplish the intended effect.

SAFETY PROFILE: When heated to decomposition it emits acrid smoke and irritating fumes.

MBU500 CAS: 8015-01-8
MARJORAM OIL, SPANISH

PROP: Main constituent is cineole. From steam distillation of the flowering plant material from the shrub *Thymus mastichina* L. (Fam. *Labiatae)* (FCTXAV 14,443,76). Faintly yellow liquid. D: 0.904-0.920, refr index: 1.463 @ 20°. Sol in fixed oils; insol in glycerin, propylene glycol, mineral oil.

SYNS: OIL of MARJORAM, SPANISH ◇ SPANISH MARJORAM OIL

USE IN FOOD:

Purpose: Flavoring agent.

Where Used: Various.

Regulations: FDA - 21CFR 182.10. GRAS when used at a level not in excess of the amount reasonably required to accomplish the intended effect.

SAFETY PROFILE: A skin irritant. When heated to decomposition it emits acrid smoke and irritating fumes.

TOXICITY DATA and CODEN

skn-rbt 500 mg/24H MLD FCTXAV 14,467,76

MCB000 CAS: 108-78-1
MELAMINE

mf: $C_3H_6N_6$ mw: 126.15

PROP: Monoclinic, colorless prisms. Mp: <250°, bp: sublimes, d: 1.573 @ 250°, vap press: 50 mm @ 315°, vap d: 4.34. Sltly sol in water, very sltly sol in hot alc; insol in ether.

SYNS: AERO ◇ CYANURAMIDE ◇ CYANUROTRIAMIDE ◇ CYANUROTRIAMINE ◇ CYMEL ◇ NCI-C50715 ◇ 2,4,6-TRIAMINO-s-TRIAZINE

USE IN FOOD:

Purpose: Manufacture of paper and paperboard.

Where Used: Packaging materials.

Regulations: FDA - 21CFR 181.30.

IARC Cancer Review: Animal Inadequate Evidence IMEMDT 39,333,86. NTP Carcinogenesis Bioassay (feed); No Evidence: mouse NTPTR* NTP-TR-245,83; (feed); Clear Evidence: rat NTPTR* NTP-TR-245,83. Community Right-To-Know List.

SAFETY PROFILE: An experimental carcinogen and tumorigen. Moderately toxic by ingestion and intraperitoneal routes. An eye, skin, and mucous membrane irritant. Causes dermatitis in humans. Mutagenic data. When heated to decomposition it emits toxic fumes of NO_x and CN^-.

TOXICITY DATA and CODEN

eye-rbt 500 mg/24H MLD 28ZPAK -,153,72
mnt-mus-orl 1 g/kg ENMUDM 4,342,82
orl-rat TDLo: 195 g/kg/2Y-C:CAR TXAPA9 72,292,84
orl-rat TD : 197 g/kg/2Y-C:CAR NTPTR* NTP-TR-245,83
orl-rat LD50: 3161 mg/kg TXAPA9 72,292,84

MCB380 CAS: 2919-66-6
MELENGESTROL ACETATE

mf: $C_{25}H_{32}O_4$ mw: 396.57

SYNS: 17-(ACETYLOXY)-6-METHYL-16-METHYLENE-PREGNA-4,6-DIENE-3,20-DIONE (9CI) ◇ 6-DEHYDRO-16-METHYLENE-6-METHYL-17-ACETOXYPROGESTERONE ◇ 17-HYDROXY-6-METHYL-16-METHYLENEPREGNA-4,6-DIENE-3,20-DIONE, ACETATE ◇ MGA ◇ MGA 100 (STEROID)

USE IN FOOD:

Purpose: Animal drug, animal feed drug.

Where Used: Animal feed, beef.

Regulations: FDA - 21CFR 556.380. Limitation of zero in cattle. 21CFR 558.342.

SAFETY PROFILE: Experimental reproductive effects. When heated to decomposition it yields acrid smoke and irritating fumes.

TOXICITY DATA and CODEN

orl-rat TDLo: 40 mg/kg (10D male): REP FESTAS 15,419,64
scu-rbt TDLo: 120 μg/kg (12D pre): REP PSEBAA 143,681,73

MCB625
MENTHA ARVENSIS, OIL

PROP: From *Mentha arvensis var. piperascens* Holmes (forma piperascens Malinvaud) (Fam. *Cabiatae*) (CCPTAY 24,559,81). Colorless to yellow liquid, minty odor. D: 0.888-0.908, refr index: 1.458 @ 20°.Sol in fixed oils, mineral oil, propylene glycol; insol in glycerin.

SYNS: CORNMINT OIL, PARTIALLY DEMENTHOLIZED ◇ MENTHA ARVENSIS OIL, PARTIALLY DEMENTHOLIZED (FCC)

USE IN FOOD:

Purpose: Flavoring agent.

Where Used: Various.

Regulations: GRAS when used at a level not in excess of the amount reasonably required to accomplish the intended effect.

SAFETY PROFILE: Experimental reproductive effects. When heated to decomposition it emits acrid smoke and irritating fumes.

TOXICITY DATA and CODEN

scu-rat TDLo: 50 mg/kg (5D male): REP CCPTAY 24,559,81
scu-rat TDLo: 20 mg/kg (9-10D preg): REP CCPTAY 24,559,81

MCB750 CAS: 99-85-4
p-MENTHA-1,4-DIENE

mf: $C_{10}H_{16}$ mw: 136.26

PROP: Colorless liquid; citrus odor. D: 0.841, refr index: 1.4731.477. Sol in alc, fixed oils; insol in water.

SYNS: FEMA No. 3559 ◇ 1-METHYL-4-ISOPROPYLCYCLOHEXADIENE-1,4 ◇ γ-TERPINENE (FCC)

USE IN FOOD:

Purpose: Flavoring agent.

Where Used: Various.

Regulations: FDA - 21CFR 172.515. Use at a level not in excess of the amount reasonably required to accomplish the intended effect.

SAFETY PROFILE: Moderately toxic by ingestion. A skin irritant. When heated to decomposition it emits acrid smoke and irritating fumes.

TOXICITY DATA and CODEN

skn-rbt 500 mg MOD FCTXAV 14,659,76
orl-rat LD50:3650 mg/kg FCTXAV 14,875,76

MCC000 CAS: 99-83-2
p-MENTHA-1,5-DIENE
mf: $C_{10}H_{16}$ mw: 136.26

PROP: Colorless to sltly yellow liquid; mint odor. D: 0.835-0.865, refr index: 1.471-1.477, flash p: 120°F. Sol in alc; insol in water.

SYNS: α-FELLANDRENE ◇ FEMA No. 2856 ◇ 4-ISOPROPYL-1-METHYL-1,5-CYCLOHEXADIENE ◇ 5-ISOPROPYL-2-METHYL-1,3-CYCLOHEXADIENE ◇ 2-METHYL-5-ISOPROPYL-1,3-CYCLOHEXADIENE ◇ α-PHELLANDRENE (FCC)

USE IN FOOD:

Purpose: Flavoring agent.

Where Used: Various.

Regulations: FDA - 21CFR 172.515. Use at a level not in excess of the amount reasonably required to accomplish the intended effect.

SAFETY PROFILE: Mildly toxic by ingestion. A severe human skin irritant. Incompatible with air. Combustible liquid. When heated to decomposition it emits acrid smoke and irritating fumes.

TOXICITY DATA and CODEN

skn-man 100% SEV FCTXAV 16,843,78
orl-rat LD50:5700 mg/kg FCTXAV 16,843,78

MCC500 CAS: 5989-54-8
(S)-(-)-p-MENTHA-1,8-DIENE
mf: $C_{10}H_{16}$ mw: 136.26

PROP: Colorless liquid; light odor. D: 0.837-0.841, refr index: .469-1.473. Misc in alc, fixed oils; insol in water.

SYNS: 1-LIMONENE ◇ (-)-LIMONENE (FCC) ◇ 1-METHYL-4-(1-METHYLETHENYL)-(S)-CYCLOHEXENE

USE IN FOOD:

Purpose: Flavoring agent.

Where Used: Various.

Regulations: GRAS when used at a level not in excess of the amount reasonably required to accomplish the intended effect.

SAFETY PROFILE: A skin irritant. When heated to decomposition it emits acrid smoke and irritating fumes.

TOXICITY DATA and CODEN

skn-rbt 500 mg/24H MOD FCTXAV 16,809,78

MCD379 CAS: 2244-16-8
d-p-MENTHA-6,8,(9)-DIEN-2-ONE
mf: $C_{10}H_{14}O$ mw: 150.24

PROP: Colorless liquid; caraway odor. D: 0.956-0.960, refr index: 1.96-1.499. Sol in propylene glycol, fixed oils; misc in alc; insol in glycerin.

SYNS: (+)-CARVONE ◇ d-CARVVONE (FCC) ◇ d(+)-CARVONE ◇ (S)-CARVONE ◇ (S)-(+)-CARVONE ◇ FEMA No. 2249 ◇ d-1-METHYL-4-ISOPROPENYL-6-CYCLOHEXEN-2-ONE ◇ (S)-2-METHYL-5-(1-METHYLETHENYL)-2-CYCLOHEXEN-1-ONE

USE IN FOOD:

Purpose: Flavoring agent.

Where Used: Bakery products, beverages (alcoholic), beverages (nonalcoholic), chewing gum, condiments, confections, ice cream.

Regulations: FDA - 21CFR 182.60. Use at a level not in excess of the amount reasonably required to accomplish the intended effect.

SAFETY PROFILE: Poison by ingestion and skin contact. A skin irritant. When heated to decomposition it emits acrid smoke and irritating fumes.

TOXICITY DATA and CODEN

skn-rbt 500 mg/24H MLD FCTXAV 16,673,78
orl-rat LD50:3710 μg/kg FCTXAV 16,673,78
skn-rbt LD50:4 mg/kg FCTXAV 16,673,78

MCD500 CAS: 6485-40-1
1-6,8(9)-p-MENTHADIEN-2-ONE
mf: $C_{10}H_{14}O$ mw: 150.22

PROP: Colorless liquid; spearmint odor. D: 0.956-0.960, refr index: 1.495-1.499. Sol in propylene glycol, fixed oils; misc in alc; insol in glycerin.

SYNS: (−)-CARVONE ◇ 1-CARVONE ◇ l(−)-CARVONE (FCC) ◇ (R)-CARVONE ◇ FEMA No. 2249 ◇ (R)-(−)-p-MENTHA-6,8-DIEN-2-ONE ◇ 1-1-METHYL-4-ISOPROPENYL-6-CYCLOHEXEN-2-ONE ◇ (R)-2-METHYL-5-(1-METHYLETHENYL)-2-CYCLOHEXEN-1-ONE (9CI)

USE IN FOOD:

Purpose: Flavoring agent.

Where Used: Various.

Regulations: FDA - 21CFR 172.515. Use at a level not in excess of the amount reasonably required to accomplish the intended effect.

SAFETY PROFILE: Moderately toxic by ingestion. When heated to decomposition it emits acrid smoke and irritating fumes.

TOXICITY DATA and CODEN

orl-rat LD50: 1640 mg/kg FCTXAV 11,1057,73

MCE250 CAS: 1074-95-9
p-MENTHAN-3-ONE racemic

mf: $C_{10}H_{18}O$ mw: 154.28

PROP: Several stereoisomers found in nature; 1-menthone found in essential oils of Russian and American peppermint, Geranium, *Andropogon fragrans*, *Mentha timija*, *Mentha arvensis* and others; d-menthone found in essential oils of *Barosma pulchellum*, *Nepeta japonica maxim* and others; d-isomenthone isolated from *Micromeriabissinica benth.*, *Pelargonium tometosum jacquin,* and others; 1-isomenthone identified in *Reunion geranium*, *Pelargonium capitatum* and others (FCTXAV 14,443,76). Flash p: 156°F.

SYNS: FEMA No. 2667 ◇ 2-ISOPROPYL-5-METHYL-CYCLOHEXAN-1-ONE, racemic ◇ MENTHONE, racemic

USE IN FOOD:

Purpose: Flavoring agent.

Where Used: Various.

Regulations: FDA - 21CFR 172.515. Use at a level not in excess of the amount reasonably required to accomplish the intended effect.

SAFETY PROFILE: Moderately toxic by ingestion. A skin irritant. Combustible liquid. When heated to decomposition it emits acrid smoke and irritating fumes.

TOXICITY DATA and CODEN

skn-rbt 500 mg/24H MLD FCTXAV 14,443,76
orl-rat LD50: 2180 mg/kg FCTXAV 14(5),443,76

MCE750 CAS: 7786-67-6
p-MENTH-8-EN-3-OL

mf: $C_{10}H_{18}O$ mw: 154.28

PROP: Colorless liquid; mint odor. D: 0.904-0.913, refr index: 1.470-1.475. Misc in alc, ether, fixed oils; sltly sol in water.

SYNS: FEMA No. 2962 ◇ ISOPULEGOL (FCC) ◇ 8(9)-p-MENTHEN-3-OL ◇ 1-METHYL-4-ISOPROPENYLCYCLOHEXAN-3-OL

USE IN FOOD:

Purpose: Flavoring agent.

Where Used: Bakery products, beverages (nonalcoholic), chewing gum, confections, gelatin desserts, ice cream, puddings, syrups.

Regulations: FDA - 21CFR 172.515. Use at a level not in excess of the amount reasonably required to accomplish the intended effect.

SAFETY PROFILE: Moderately toxic by ingestion and skin contact. When heated to decomposition it emits acrid smoke and irritating fumes.

TOXICITY DATA and CODEN

orl-rat LD50: 1030 mg/kg FCTXAV 13,681,75
skn-rbt LD50: 3000 mg/kg FCTXAV 13,681,75

MCF750 CAS: 89-78-1
MENTHOL

mf: $C_{10}H_{20}O$ mw: 156.26

PROP: Hexagonal crystals or granules; peppermint taste and odor. D: 0.890 @ 15°/15°, vap press: 1 mm @ 56.0°, vap d: 5.38, mp: 41-43°, bp: 212°, flash p: +199°F. Very sol in alc, chloroform, ether, petr ether, glacial acetic acid, liquid petrolatum; sltly sol in water.

SYNS: FEMA No. 2665 ◇ HEXAHYDROTHYMOL ◇ 2-ISOPROPYL-5-METHYL-CYCLOHEXANOL ◇ p-MENTHAN-3-OL ◇ 1-MENTHOL ◇ 5-METHYL-2-(1-METHYLETHYL)CYCLOHEXANOL ◇ PEPPERMINT CAMPHOR

USE IN FOOD:

Purpose: Flavoring agent.

Where Used: Various.

Regulations: FDA - 21CFR 172.515. Use at a level not in excess of the amount reasonably required to accomplish the intended effect.

SAFETY PROFILE: Poison by intravenous route. Moderately toxic by ingestion and intraperitoneal routes. A severe eye irritant. Incompatible with phenol; β-naphthol; resorcinol or thymol in trituration; potassium permanganate; chromium trioxide; pyrogallol. Combustible liquid. When heated to decomposition it emits acrid smoke and irritating fumes.

TOXICITY DATA and CODEN

eye-rbt 750 μg SEV AJOPAA 29,1363,46
orl-rat LD50: 3180 mg/kg FCTXAV 2,327,64

MCG000 CAS: 15356-70-4
dl-MENTHOL

mf: $C_{10}H_{20}O$ mw: 156.30

SYNS: FEMA No. 2665 ◇ 4-ISOPROPYL-1-METHYLCYCLOHEXAN-3-OL ◇ dl-3-p-MENTHANOL ◇ 3-p-MENTHOL

◇ MENTHOL racemic ◇ MENTHOL racemique (FRENCH) ◇ 5-METHYL-2-(1-METHYLETHYL)-CYCLOHEXANOL (1-α,2-β,5-α)

USE IN FOOD:

Purpose: Flavoring agent.

Where Used: Various.

Regulations: FDA - 21CFR 172.515. Use at a level not in excess of the amount reasonably required to accomplish the intended effect.

NCI Carcinogenesis Bioassay (feed); No Evidence: mouse, rat NCITR* NCI-GC-TR-98,79.

SAFETY PROFILE: Moderately toxic by ingestion, intraperitoneal, and subcutaneous routes. A skin irritant. When heated to decomposition it emits acrid smoke and irritating fumes.

TOXICITY DATA and CODEN

skn-rbt 500 mg/24H MLD FCTXAV 14,443,76
orl-rat LD50:2900 mg/kg FAONAU 44A,59,67

MCG250 CAS: 2216-51-5
l-MENTHOL
mf: $C_{10}H_{20}O$ mw: 156.30

PROP: Found in high concentrations in oils of Peppermint (*Mentha Piperita)* and Japanese Mint Oil (*Mentha Arvensis)*, and in lower concentrations in Reunion Geranium Oil and in a large number of essential oils; prepared by isolation from *Mentha arvensis* Oils (FCTXAV 14,443,76).

SYNS: FEMA No. 2665 ◇ (-)-MENTHYL ALCOHOL ◇ (1R-(1-α,2-β,5-α))-5-METHYL-2-(1-METHYLETHYL)CYCLOHEXANOL ◇ U.S.P. MENTHOL

USE IN FOOD:

Purpose: Flavoring agent.

Where Used: Various.

Regulations: FDA - 21CFR 172.515. Use at a level not in excess of the amount reasonably required to accomplish the intended effect.

SAFETY PROFILE: Poison by intravenous route. Moderately toxic by ingestion, intraperitoneal, and subcutaneous routes. When heated to decomposition it emits acrid smoke and irritating fumes.

TOXICITY DATA and CODEN

orl-rat LD50:3300 mg/kg FAONAU 44A,59,67
ipr-rat LD50:700 mg/kg JPPMAB 35,110,83

MCG275 CAS: 89-80-5
MENTHONE
mf: $C_{10}H_{18}O$ mw: 154.28

PROP: Colorless liquid; mint odor. D: 0.888-0.895, refr index: 1.448-1.453. Sol in alc, fixed oils; very sltly sol in water.

SYNS: FEMA No. 2667 ◇ l-p-MENTHAN-3-ONE ◇ l-MENTHONE (FCC) ◇ p-MENTHONE ◇ trans-MENTHONE ◇ trans-5-METHYL-2-(1-METHYLETHYL)-CYCLOHEXANONE

USE IN FOOD:

Purpose: Flavoring agent.

Where Used: Various.

Regulations: FDA - 21CFR 172.515. Use at a level not in excess of the amount reasonably required to accomplish the intended effect.

SAFETY PROFILE: Moderately toxic by ingestion, intravenous and subcutaneous routes. Mutagenic data. When heated to decomposition it emits acrid smoke and irritating fumes.

TOXICITY DATA and CODEN

mmo-sat 6400 ng/plate MUREAV 138,17,84
mma-sat 32 μg/plate MUREAV 138,17,84
orl-rat LD50:500 mg/kg FRXXBL #2448856

MCG500 CAS: 16409-45-3
dl-MENTHYL ACETATE
mf: $C_{12}H_{22}O_2$ mw: 198.34

PROP: Colorless liquid; characteristic minty odor. D: 0.919 @ 20°/4°, refr index: 0.443-1.450, bp: 227°, flash p: 197°F. Sltly sol in water, glycerin; misc with alc, ether, propylene glycol, fixed oils.

SYNS: FEMA No. 2668 ◇ MENTHOL, ACETATE (8CI) ◇ MENTHYL ACETATE ◇ MENTHYL ACETATE racemic ◇ p-MENTH-3-YL ESTER-dl-ACETIC ACID

USE IN FOOD:

Purpose: Flavoring agent.

Where Used: Various.

Regulations: FDA - 21CFR 172.515. Use at a level not in excess of the amount reasonably required to accomplish the intended effect.

SAFETY PROFILE: Mildly toxic by ingestion. A skin irritant. Combustible liquid. When heated to decomposition it emits acrid smoke and irritating fumes.

TOXICITY DATA and CODEN

skn-rbt 500 mg/24H MLD FCTXAV 14,479,76
orl-rat LD50:7620 mg/kg FCTXAV 14,477,76

MCG750 CAS: 2623-23-6
1-p-MENTH-3-YL ACETATE
mf: $C_{12}H_{22}O$ mw: 182.34

PROP: Colorless liquid; minty odor. D: 0.919-0.924, refr index: 1.443-1.447. Sol in alc, propylene glycol, fixed oils; sltly sol in water, glycerin.

SYNS: FEMA No. 2668 ◇ l-2-ISOPROPYL-5-METHYL-CYCLOHEXAN-1-OL ACETATE ◇ (−)-MENTHYL ACETATE ◇ l-MENTHYL ACETATE (FCC) ◇ l-p-MENTH-3-YL ACETATE ◇ (R-(1α,2β,5α))-5-METHYL-2-(1-METHYLETHYL)-CYCLOHEXANOL ACETATE (9CI)

USE IN FOOD:

Purpose: Flavoring agent.

Where Used: Various.

Regulations: FDA - 21CFR 172.515. Use at a level not in excess of the amount reasonably required to accomplish the intended effect.

SAFETY PROFILE: A skin irritant. When heated to decomposition it emits acrid smoke and irritating fumes.

TOXICITY DATA and CODEN

skn-rbt 500 mg/24H MLD FCTXAV 14,477,76

MDM100 CAS: 57837-19-1
METALAXYL
mf: $C_{15}H_{21}NO_4$ mw: 279.35

PROP: White crystals.

SYN: N-(2,6-DIMETHYLPHENYL)-N-(METHOXYACETYL) ALANINE METHYL ESTER

USE IN FOOD:

Purpose: Fungicide.

Where Used: Animal feed, apple pomace (dried), apple pomace (wet), citrus molasses, citrus oil, citrus pulp, hops (dried), hops (spent), legume vegetable cannery wastes, peanut meal, peanut soapstock, potato chips, potato wastes (dried processed), potatoes (processed), soybean hulls, soybean meal, soybean soapstock, sugar beet molasses, tomato pomace (dried), tomato pomace (wet), tomatoes (processed), wheat (milling fractions).

Regulations: FDA - 21CFR 193.277. Limitation of 7.0 ppm in citrus oil, 4.0 ppm in processed potatoes including potato chips, 3.0 ppm in processed tomatoes, and 1.0 ppm in milling fractions of wheat. Limitation of 50 ppm in dried hops. (Expired 10/28/1988) 21CFR 561.273. Limitation of 2.0 ppm in dry apple pomace, 0.4 ppm in wet apple pomace, 7.0 ppm in citrus molasses, 7.0 ppm in citrus pulp, 2.0 ppm in dry hops, 5.0 ppm in legume vegetable cannery waste, 1.0 ppm in peanut meal, 2.0 in peanut soapstock, 4.0 ppm in dried processed potato waste, 2.0 ppm in soybean hulls or meal or soapstock, 1.0 ppm in sugar beet molasses, 20.0 ppm in tomato pomace (dry and wet) when used for animal feed. Limitation of 50 ppm in hops, spent. (Expired 10/28/1988)

SAFETY PROFILE: When heated to decomposition emits toxic fumes of NO_x.

MDN525
METHACRYLIC ACID-DIVINYLBENZENE COPOLYMER

USE IN FOOD:

Purpose: Carrier of vitamin B_{12}.

Where Used: Special dietary foods.

Regulations: FDA - 21CFR 172.775. Use in accordance with good manufacturing practice.

SAFETY PROFILE: When heated to decomposition it emits acrid smoke and irritating fumes.

MDR000 CAS: 75-09-2
METHANE DICHLORIDE
DOT: 1593
mf: CH_2Cl_2 mw: 84.93

PROP: Colorless, volatile liquid; odor of chloroform. Bp: 39.8°, lel: 15.5% in O_2, uel: 66.4% in O_2, fp: −96.7°, d: 1.326 @ 20°/4°, autoign temp: 1139°F, vap press: 380 mm @ 22°, vap d: 2.93, refr index: 1.424 @ 20L. Sol in water; misc with alcohol, acetone, chloroform, ether, carbon tetrachloride.

SYNS: AEROTHENE MM ◇ CHLORURE de METHYLENE (FRENCH) ◇ DCM ◇ DICHLOROMETHANE (DOT) ◇ FREON 30 ◇ METHYLENE BICHLORIDE ◇ METHYLENE CHLORIDE (ACGIH, DOT, FCC, USDA) ◇ METHYLENE DICHLORIDE ◇ METYLENU CHLOREK (POLISH) ◇ NCI-C50102 ◇ RCRA WASTE NUMBER U080 ◇ SOLMETHINE

USE IN FOOD:

Purpose: Color diluent, extraction solvent.

Where Used: Coffee (decaffeinated), fruits, hops extract, spice oleoresins, vegetables.

Regulations: FDA - 21CFR 73.1. No residue. 21CFR 173.255. Limitation of 30 ppm in spice oleoresins, 2.2 percent in hops extract, 10 ppm in decaffeinated coffee. USDA CES Ranking: A-2 (1986).

IARC Cancer Review: Human Inadequate Evidence IMEMDT 41,43,86; Animal Inadequate Evidence IMEMDT 20,449,79; Animal Sufficient Evidence IMEMDT 41,43,86. NTP Carcinogenesis Studies (inhalation); Clear Evidence: mouse, rat NTPTR* NTP-TR-306,86. EPA Genetic Toxicology Program. Community Right-To-Know List.

OSHA PEL: TWA 500 ppm; CL 1000 ppm; Pk 2000/5M/2H ACGIH TLV: TWA 50 ppm, Suspected Carcinogen DOT Classification: Poison B; Label: St. Andrews Cross NIOSH REL: (To Methylene Chloride) TWA 75 ppm; Pk 500 ppm/15M

SAFETY PROFILE: Poison by intravenous route. Moderately toxic by ingestion, subcutaneous and intraperitoneal routes. Mildly toxic by inhalation. An experimental carcinogen and tumorigen. Human systemic effects by ingestion and inhalation: paresthesia, somnolence, altered sleep time, convulsions, euphoria, and change in cardiac rate. An experimental teratogen. Experimental reproductive effects. An eye and severe skin irritant. Human mutagenic data.

It is flammable in the range of 12-19 percent in air but ignition is difficult. It will not form explosive mixtures with air at ordinary temperatures. Mixtures in air with methanol vapor are flammable. It will form explosive mixtures with an atmosphere having a high oxygen content; in liquid O_2; N_2O_4; K; Na; NaK. Explosive in the form of vapor when exposed to heat or flame. It can be decomposed by contact with hot surfaces and open flame, and then yield toxic fumes which are irritating and give warning of their presence. When heated to decomposition it emits highly toxic fumes of phosgene and Cl^-.

TOXICITY DATA and CODEN

skn-rbt 810 mg/24H SEV JETOAS 9,171,76
eye-rbt 162 mg MOD JETOAS 9,171,76
dni-hmn:fbr 5000 ppm/1H-C MUREAV 81,203,81
cyt-ham:ovr 5 g/L MUREAV 116,361,83
ihl-rat TCLo:4500 ppm/24H (1-17D preg):REP TXAPA9 52,29,80
ihl-mus TCLo:1250 ppm/7H (6-15D preg): REP TXAPA9 32,84,75
ihl-rat TCLo:3500 ppm/6H/2Y-I:CAR FAATDF 4,30,84
ihl-mus TCLo:2000 ppm/5H/2Y-C:CAR NTPTR* NTP-TR-306,86
orl-hmn LDLo:357 mg/kg:PNS,CNS 34ZIAG -,390,69
ihl-hmn TCLo:500 ppm/1Y-I:CNS,CVS ABHYAE 43,1123,68
ihl-hmn TCLo:500 ppm/8H:CNS SCIEAS 176,295,72
orl-rat LD50:2136 mg/kg PPGDS* JAN81
ihl-rat LC50:88000 mg/m^3/30M FAVUAI 7,35,75

MDS250 CAS: 67-56-1
METHANOL

DOT: 1230
mf: CH_4O mw: 32.05

PROP: Clear, colorless, very mobile liquid; slt alcoholic odor when pure; crude material may have a repulsive pungent odor. Mp: 64.8°, lel: 6.0%, uel: 36.5%, ULC: 70, fp: −97.8°, d: 0.7915 @ 20°/4°, autoign temp: 878°F, vap press: 100 mm @ 21.2°, vap d: 1.11. Misc in water, ethanol, ether, benzene; ketones and most other organic solvents.

SYNS: ALCOOL METHYLIQUE (FRENCH) ◇ ALCOOL METILICO (ITALIAN) ◇ CARBINOL ◇ COLONIAL SPIRIT ◇ COLUMBIAN SPIRITS (DOT) ◇ METANOLO (ITALIAN) ◇ METHANOL (DOT) ◇ METHYL ALCOHOL (ACGIH, DOT, FCC) ◇ METHYLALKOHOL (GERMAN) ◇ METHYL HYDROXIDE ◇ METHYLOL ◇ METYLOWY ALKOHOL (POLISH) ◇ MONOHYDROXYMETHANE ◇ PYROXYLIC SPIRIT ◇ RCRA WASTE NUMBER U154 ◇ WOOD ALCOHOL (DOT) ◇ WOOD NAPHTHA ◇ WOOD SPIRIT

USE IN FOOD:

Purpose: Extraction solvent.

Where Used: Hops extract, spice oleoresins.

Regulations: FDA - 21CFR 173.250. Limitation of 50 ppm in spice oleoresins, 2.2 percent in hops extracts.

Community Right-To-Know List. EPA Genetic Toxicology Program.

OSHA PEL: TWA 200 ppm; STEL 250 ppm (skin) ACGIH TLV: TWA 200 ppm; STEL 250 ppm (skin) DFG MAK: 200 ppm (260 mg/m^3) DOT Classification: Flammable Liquid; Label: Flammable Liquid, Poison NIOSH REL: TWA 200 ppm; CL 800 ppm/15M

SAFETY PROFILE: A human poison by ingestion. Poison experimentally by skin contact. Moderately toxic experimentally by intravenous and intraperitoneal routes. Mildly toxic by inhalation. Human systemic effects by ingestion and inhalation: optic nerve neuropathy, visual field changes, lacrimation, headache, cough, dyspnea, other respiratory effects, nausea or vomiting. An experimental teratogen. Experimental reproductive effects. An eye and skin irritant. Human mutagenic data. A narcotic.

Flammable liquid. Dangerous fire hazard when exposed to heat, flame, or oxidizers. Explosive in the form of vapor when exposed to heat or flame. Explosive reaction with chloroform + heat; diethyl zinc. Dangerous; can react vigorously with oxidizing materials. To fight fire, use alcohol foam. When heated to decomposition it emits acrid smoke and irritating fumes.

TOXICITY DATA and CODEN

skn-rbt 500 mg/24H MOD 28ZPAK -,33,72
eye-rbt 40 mg MOD UCDS** 3/24/70
mmo-smc 12 pph GENRA8 37,173,81
dnd-rat-orl 10 µmol/kg ENMUDM 4,317,82
orl-rat TDLo:7500 mg/kg (17-19D preg):REP TOXID9 1,32,81
ihl-rat TCLo:20000 ppm/7H (7-15D preg):TER FAATDF 5,727,85
ihl-rat TCLo:10000 ppm/7H (7-15D preg):TER FAATDF 5,727,85
orl-hmn LDLo:428 mg/kg:CNS,PUL NPIRI* 1,74,74
orl-hmn LDLo:143 mg/kg:EYE,PUL,GIT 34ZIAG -,382,69
ihl-hmn TCLo:86000 mg/m^3:EYE,PUL AGGHAR 5,1,33
ihl-hmn TCLo:300 ppm:EYE,CNS,PUL NPIRI* 1,74,74
orl-rat LD50:5628 mg/kg GTPZAB 19(11),27,75
ihl-rat LC50:64000 ppm/4H NPIRI* 1,74,74

MDT750 CAS: 63-68-3
l-METHIONINE
mf: $C_5H_{11}NO_2S$ mw: 149.23

PROP: White, crystalline powder or platelets; faint odor. Mp: 281° (decomp), d: 1.340. Sol in water, dilute acids, and alkalies; insol in abs alc, alc, benzene, acetone, ether.

SYNS: l-α-AMINO-γ-METHYLMERCAPTOBUTYRIC ACID ◇ 2-AMINO-4-(METHYLTHIO)BUTYRIC ACID ◇ l(−)-AMINO-γ-METHYLTHIOBUTYRIC ACID ◇ CYMETHION ◇ LIQUIMETH ◇ METHIONINE ◇ l-(-)-METHIONINE ◇ l-γ-METHYLTHIO-α-AMINOBUTYRIC ACID

USE IN FOOD:

Purpose: Dietary supplement, nutrient.

Where Used: Various.

Regulations: FDA - 21CFR 172.310. Limitation 3.1 percent by weight.

EPA Genetic Toxicology Program.

SAFETY PROFILE: Mildly toxic by ingestion and intraperitoneal routes. Human mutagenic data. An experimental teratogen. Experimental reproductive effects. When heated to decomposition it emits very toxic fumes of NO_x and SO_x.

TOXICITY DATA and CODEN

mmo-esc 100 mg/L PMRSDJ 1,376,81
dnr-smc 500 mg/L PMRSDJ 1,502,81
dnd-hmn:oth 2 µmol/L BBACAQ 696,15,82
orl-rat TDLo:26100 mg/kg (10-20D preg):TER DTTIAF 82,457,75
orl-rat TDLo:14720 mg/kg (10-20D preg):REP DTTIAF 82,457,75
orl-rat LD50:36 g/kg GISAAA 48(6),20,83

MDU600 CAS: 16752-77-5
METHOMYL
mf: $C_5H_{10}N_2O_2S$ mw: 162.23

PROP: White, crystalline solid; slt sulfurous odor. Moderately water-sol, mp: 79°.

SYNS: DU PONT INSECTICIDE 1179 ◇ ENT 27,341 ◇ INSECTICIDE 1,179 ◇ LANNATE L ◇ MESOMILE ◇ METHYL N-((METHYLAMINO)CARBONYL)OXY)ETHANIMIDO)THIOATE ◇ METHYL-N-((METHYLCARBAMOYL)OXY)THIOACETIMIDATE ◇ S-METHYL N-[(METHYLCARBAMOYLOOXY]THIOACETIMIDATE ◇ 2-METHYLTHIO-ACETALDEHYD-O-(METHYLCARBAMOYL)-OXIM (GERMAN) ◇ 2-METHYLTHIO-PROPIONALDEHYD-O-(METHYLCARBAMOYL)-OXIM (GERMAN) ◇ METOMIL (ITALIAN) ◇ NU-BAIT II ◇ NUDRIN ◇ RCRA WASTE NUMBER P066 ◇ 3-THIABUTAN-2-ONE, O-(METHYLCARBAMOYL)OXIME ◇ WL 18236

USE IN FOOD:

Purpose: Insecticide.

Where Used: Hops (dried).

Regulations: FDA - 21CFR 193.475. Limitation of 7 ppm in dried hops. (Expires 1/12/1990)

EPA Genetic Toxicology Program. EPA Extremely Hazardous Substances List.

ACGIH TLV: TWA 2.5 mg/m^3

SAFETY PROFILE: Poison by ingestion, inhalation and subcutaneous routes. Mildly toxic by skin contact. When heated to decomposition it emits very toxic fumes of NO_x and SO_x.

TOXICITY DATA and CODEN

orl-rat LD50: 17 mg/kg GUCHAZ 6,336,73
ihl-rat LC50: 77 ppm TXAPA9 40,1,77

MDW750 CAS: 100-06-1
4′-METHOXYACETOPHENONE
mf: $C_9H_{10}O_2$ mw: 150.19

PROP: Colorless to pale yellow fused solid; hawthorn odor. Flash p: +212°F. Sol in fixed oils, propylene glycol; misc in glycerin.

SYNS: ACETANISOLE (FCC) ◇ p-ACETYLANISOLE ◇ 4-ACETYLANISOLE ◇ BANANOTE ◇ FEMA No. 2005 ◇ LINARODIN ◇ p-METHOXYACETOPHENONE ◇ p-METHOXYPHENYL METHYL KETONE ◇ 4-METHOXYPHENYL METHYL KETONE ◇ NOVATONE

USE IN FOOD:

Purpose: Flavoring agent.

Where Used: Various.

Regulations: FDA - 21CFR 172.515. Use at a level not in excess of the amount reasonably required to accomplish the intended effect.

SAFETY PROFILE: Moderately toxic by ingestion. Human systemic effects by inhalation: pulse rate increased without fall in blood pressure and blood pressure elevation. A skin irritant. Combustible liquid. When heated to decomposition it emits acrid smoke and irritating fumes.

TOXICITY DATA and CODEN

skn-rbt 500 mg/24H MOD FCTXAV 12,807,74
ihl-hmn TCLo: 1700 μg/m³/39W-I:CVS GISAAA 50(4),86,85
orl-rat LD50: 1720 mg/kg FCTXAV 12,807,74

MED500 CAS: 105-13-5
p-METHOXYBENZYL ALCOHOL
mf: $C_8H_{10}O_2$ mw: 138.18

PROP: Needles or colorless liquid; floral odor. D: 1.113 @ 15°/15°, refr index: 1.543, mp: 25°, bp: 258.8°, flash p: +210°F. Insol in water; sol in alc and ether fixed oils; sltly sol glycerin.

SYNS: ANISE ALCOHOL ◇ ANISIC ALCOHOL ◇ p-ANISOL ALCOHOL ◇ ANISYL ALCOHOL (FCC) ◇ FEMA No. 2099 ◇ 4-METHOXYBENZENEMETHANOL ◇ 4-METHOXYBENZYL ALCOHOL

USE IN FOOD:

Purpose: Flavoring agent.

Where Used: Various.

Regulations: FDA - 21CFR 172.515. Use at a level not in excess of the amount reasonably required to accomplish the intended effect.

SAFETY PROFILE: Moderately toxic by ingestion. A skin irritant. Combustible liquid. When heated to decomposition it emits acrid smoke and irritating fumes.

TOXICITY DATA and CODEN

skn-rbt 500 mg/24H MOD FCTXAV 12,825,74
orl-rat LD50: 1200 mg/kg JPETAB 93,26,48

MEL500 CAS: 1918-00-9
2-METHOXY-3,6-DICHLOROBENZOIC ACID
DOT: 2769
mf: $C_8H_6Cl_2O_3$ mw: 221.04

SYNS: ACIDO (3,6-DICLORO-2-METOSSI)-BENZOICO (ITALIAN) ◇ BANEX ◇ BANLEN ◇ BANVEL ◇ BANVEL HERBICIDE ◇ BRUSH BUSTER ◇ COMPOUND B DICAMBA ◇ DIANAT (RUSSIAN) ◇ DIANATE ◇ DICAMBA (DOT) ◇ 3,6-DICHLOOR-2-METHOXY-BENZOEIZUUR (DUTCH) ◇ 3,6-DICHLOR-3-METHOXY-BENZOESAEURE (GERMAN) ◇ 3,6-DICHLORO-o-ANISIC ACID ◇ 2,5-DICHLORO-6-METHOXYBENZOIC ACID ◇ 3,6-DICHLORO-2-METHOXYBENZOIC ACID ◇ MDBA ◇ MEDIBEN ◇ VELSICOL COMPOUND "R" ◇ VELSICOL 58-CS-11

USE IN FOOD:

Purpose: Herbicide.

Where Used: Animal feed, sugarcane molasses.

Regulations: FDA - 21CFR 193.465. Herbicide residue tolerance of 2.0 ppm sugarcane molasses. 21CFR 561.427. Limitation of 2.0 ppm in sugarcane molasses when used for animal feed.

EPA Genetic Toxicology Program.

DOT Classification: ORM-E; Label: None

SAFETY PROFILE: Moderately toxic by ingestion and possibly other routes. Mutagenic data. When heated to decomposition it emits toxic fumes of Cl^-.

TOXICITY DATA and CODEN

mma-sat 53500 nmol/L MUREAV 136,233,84
cyt-mus-unr 500 mg/kg TGANAK 16(1),45,82
orl-rat LD50: 1040 mg/kg RREVAH 10,97,65

MEX350
2-METHOXY-3(5)-METHYLPYRAZINE
mf: $C_6H_8N_2O$ mw: 124.14

PROP: Colorless liquid; roasted, hazelnut odor. D: 1.000-1.090 @ 20°, refr index: 1.506, flash p: 131°F. Sol in water, organic solvents.

SYN: FEMA No. 3183

USE IN FOOD:

Purpose: Flavoring agent.

Where Used: Various.

Regulations: GRAS when used at a level not in excess of the amount reasonably required to accomplish the intended effect.

SAFETY PROFILE: Combustible liquid. When heated to decomposition emits toxic fumes of NO_x.

MFF580
4-p-METHOXYPHENYL-2-BUTANONE

PROP: Colorless to pale yellow liquid; sweet, floral odor. D: 1.042-1.048, refr index: 1.517-1.521, flash p: +212°F.

SYNS: ANISYLACETONE ◇ FEMA No. 2672

USE IN FOOD:

Purpose: Flavoring agent.

Where Used: Various.

Regulations: FDA - 21CFR 172.515. Use at a level not in excess of the amount reasonably required to accomplish the intended effect.

SAFETY PROFILE: Combustible liquid. When heated to decomposition it emits acrid smoke and irritating fumes.

MFN285 CAS: 3149-28-8
2-METHOXYPYRAZINE
mf: $C_5H_6N_2O$ mw: 110.12

PROP: Colorless to yellow liquid; nutty, cocoa-like odor. D: 1.110-1.140 @ 20°, refr index: 1.508. Sol in alc; insol in water @ 61°.

SYN: FEMA No. 3302

USE IN FOOD:

Purpose: Flavoring agent.

Where Used: Various.

Regulations: GRAS when used at a level not in excess of the amount reasonably required to accomplish the intended effect.

SAFETY PROFILE: Skin and eye irritant. When heated to decomposition emits toxic fumes of NO_x.

MFT500 CAS: 127-25-3
METHYL ABIETATE
mf: $C_{21}H_{32}O_2$ mw: 316.47

PROP: Colorless to thick yellow liquid; almost odorless. Flash p: 356°F (OC), vap d: 10.9. D: 1.040 @ 20°/20°, bp: 360-365° with decomp. Insol in water, misc in alc and ether, the usual organic solvents, and with aliphatic hydrocarbons. From the esterification of the resinous residue of turpentine (FCTXAV 12,807,74).

SYNS: ABIETIC ACID, METHYL ESTER ◇ METHYL ESTER OF WOOD ROSIN ◇ METHYL ESTER OF WOOD ROSIN, PARTIALLY HYDROGENATED (FCC)

USE IN FOOD:

Purpose: Masticatory substance in chewing gum base.

Where Used: Chewing gum.

Regulations: FDA - 21CFR 172.615. Use in amounts not to exceed those required to produce the intended physical or other technical effect.

SAFETY PROFILE: A skin irritant. Probably slightly toxic. Combustible liquid when exposed to heat or flame; can react with oxidizing materials. To fight fire, use CO_2, dry chemical. When heated to decomposition it emits acrid smoke and irritating fumes.

TOXICITY DATA and CODEN

skn-rbt 500 mg/24H MOD FCTXAV 12,807,74

MFW250 CAS: 122-00-9
4′-METHYL ACETOPHENONE
mf: $C_9H_{10}O$ mw: 134.19

PROP: Colorless liquid; fruity, actophenone odor. D: 0.996-1.004, refr index: 1.530-1.535, flash p: 198°F. Sol in fixed oils, propylene glycol insol in glycerin.

SYNS: p-ACETYLTOLUENE ◇ FEMA No. 2677 ◇ MELILOTAL ◇ p-METHYL ACETOPHENONE ◇ 1-METHYL-4-ACETYLBENZENE ◇ METHYL-p-TOLYL KETONE

USE IN FOOD:

Purpose: Flavoring agent.

Where Used: Various.

Regulations: FDA - 21CFR 172.515. Use at a level not in excess of the amount reasonably required to accomplish the intended effect.

SAFETY PROFILE: Moderately toxic by ingestion. A human skin irritant. Combustible liquid. When heated to decomposition it emits acrid smoke and irritating fumes.

TOXICITY DATA and CODEN

skn-hmn 100% FCTXAV 12,933,74
skn-rbt 500 mg/24H MLD FCTXAV 12,807,74
orl-rat LD50:1400 mg/kg FCTXAV 12,807,74

MFW500 CAS: 520-45-6
METHYLACETOPYRONONE
mf: $C_8H_8O_4$ mw: 168.16

PROP: White crystals or crystalline powder. Mp: 109°, bp: 269.0°, vap press: 1 mm @ 91.7°, vap d: 5.8. Moderately sol in water and organic solvents.

SYNS: 2-ACETYL-5-HYDROXY-3-OXO-4-HEXENOIC ACID Δ-LACTONE ◇ 3-ACETYL-6-METHYL-2,4-PYRANDIONE ◇ 3-ACETYL-6-METHYLPYRANDIONE-2,4 ◇ 3-ACETYL-6-METHYL-2H-PYRAN-2,4(3H)-DIONE ◇ DEHYDRACETIC ACID ◇ DEHYDROACETIC ACID (FCC) ◇ DHA ◇ DHS

USE IN FOOD:

Purpose: Preservative.

Where Used: Various.

Regulations: GRAS when used in accordance with good manufacturing practice.

SAFETY PROFILE: Poison by ingestion and intravenous routes. Moderately toxic by intraperitoneal route. An experimental tumorigen. Combustible when exposed to heat or flame. When heated to decomposition it emits acrid smoke and irritating fumes.

TOXICITY DATA and CODEN

scu-rat TDLo:592 mg/kg/37W-I:ETA BJCAAI 20,134,66
orl-rat LD50:500 mg/kg WRPCA2 9,119,70

MGP000 CAS: 104-93-8
p-METHYL ANISOLE
mf: $C_8H_{10}O$ mw: 122.18

PROP: Found in oil of Ylang-Ylang, Cananga, and others (FCTXAV 12,385,74). Colorless liquid; ylang-ylang odor. D: 0.996-0.970, refr index: 1.510-1.513, flash p: 144°F. Sol in fixed oils; insol in glycerin, propylene glycol.

SYNS: p-CRESOL METHYL ETHER ◇ p-CRESYL METHYL ETHER ◇ FEMA No. 2681 ◇ p-METHOXYTOLUENE ◇ 4-METHOXYTOLUENE ◇ 4-METHYL-1-METHOXYBENZENE ◇ 4-METHYLPHENOL METHYL ETHER ◇ METHYL-p-TOLYL ETHER ◇ p-TOLYL METHYL ETHER

USE IN FOOD:

Purpose: Flavoring agent.

Where Used: Various.

Regulations: FDA - 21CFR 172.515. Use at a level not in excess of the amount reasonably required to accomplish the intended effect.

SAFETY PROFILE: Moderately toxic by ingestion. A skin irritant. Combustible liquid. When heated to decomposition it emits acrid smoke and irritating fumes.

TOXICITY DATA and CODEN

skn-rbt 500 mg/24H closed MOD FCTXAV 12,385,74
orl-rat LD50:1920 mg/kg FCTXAV 12,385,74

MGQ250 CAS: 85-91-6
N-METHYLANTHRANILIC ACID, METHYL ESTER
mf: $C_9H_{11}NO_2$ mw: 165.21

PROP: Pale yellow liquid; grape-like odor. D: 1.126-1.132, refr index: 1.578-1.581, flash p: 196°F. Sol in fixed oils; sltly sol in propylene glycol; insol in water, glycerin.

SYNS: DIMETHYL ANTHRANILATE (FCC) ◇ FEMA No. 2718 ◇ 2-METHYLAMINO METHYL BENZOATE ◇ METHYL METHYLAMINOBENZOATE ◇ METHYL-N-METHYL ANTHRANILATE ◇ MMA

USE IN FOOD:

Purpose: Flavoring agent.

Where Used: Various.

Regulations: FDA - 21CFR 172.515. Use at a level not in excess of the amount reasonably required to accomplish the intended effect.

SAFETY PROFILE: Poison by intravenous route. Moderately toxic by ingestion. Combustible liquid. When heated to decomposition it emits toxic fumes of NO_x.

TOXICITY DATA and CODEN

orl-rat LDLo:3380 mg/kg FCTXAV 8,359,70

MHA500 CAS: 101-41-7
METHYL BENZENEACETATE
mf: $C_9H_{10}O_2$ mw: 150.19

PROP: Colorless liquid; honey, jasmine odor. D: 1.062, refr index: 1.503-1.509, vap d: 5.18, flash p: 192°F. Sol in alc, fixed oils; insol in glycerin, propylene glycol, water @ 215°.

SYNS: BENZENEACETIC ACID, METHYL ESTER ◇ FEMA No. 2733 ◇ METHYL PHENYLACETATE (FCC) ◇ METHYL-α-TOLUATE ◇ PHENYLACETIC ACID, METHYL ESTER

USE IN FOOD:

Purpose: Flavoring agent.

Where Used: Various.

Regulations: FDA - 21CFR 172.515. Use at a level not in excess of the amount reasonably required to accomplish the intended effect.

SAFETY PROFILE: Moderately toxic by ingestion and skin contact. A skin irritant. Combustible liquid. When heated to decomposition it emits acrid smoke and irritating fumes.

TOXICITY DATA and CODEN

skn-rbt 500 mg/24H FCTXAV 12,807,74
orl-rat LD50:2550 mg/kg FCTXAV 12,807,74
skn-rbt LD50:2400 mg/kg FCTXAV 12,807,74

MHA750 CAS: 93-58-3
METHYL BENZENECARBOXYLATE

mf: $C_8H_8O_2$ mw: 136.16

PROP: Colorless liquid; fragrant odor. Mp: −12.5°, bp: 199.6°, flash p: 181°F, d: 1.082-1.088, refr index: 1.515, vap press: 1 mm @ 39.0°, vap d: 4.69. Sol in alc, fixed oils, propylene glycol, water @ 30°; misc in alc, ether; insol in glycerin.

SYNS: FEMA No. 2683 ◇ METHYL BENZOATE (FCC) ◇ NIOBE OIL ◇ OIL of NIOBE

USE IN FOOD:

Purpose: Flavoring agent.

Where Used: Various.

Regulations: FDA - 21CFR 172.515. Use at a level not in excess of the amount reasonably required to accomplish the intended effect.

SAFETY PROFILE: Moderately toxic by ingestion. Mildly toxic by skin contact. A skin and eye irritant. Combustible liquid when exposed to heat or flame; can react with oxidizing materials. To fight fire, use foam, CO_2, dry chemical, water to blanket fire. When heated to decomposition it emits acrid smoke and irritating fumes.

TOXICITY DATA and CODEN

skn-rbt 10 mg/24H MLD AMIHBC 10,61,54
eye-rbt 500 mg AMIHBC 10,61,54
orl-rat LD50:1350 mg/kg FCTXAV 2,327,64

MHL000 CAS: 31431-39-7
METHYL-5-BENZOYL BENZIMIDAZOLE-2-CARBAMATE

mf: $C_{16}H_{13}N_3O_3$ mw: 295.32

SYNS: N-2 (5-BENZOYL-BENZIMIDAZOLE) CARBAMATE de METHYLE (FRENCH) ◇ 5-BENZOYL-2-BENZIMIDAZOLE-CARBAMIC ACID METHYL ESTER ◇ N-(BENZOYL-5-BENZIMIDAZOLYL)-2, CARBAMATE de METHYLE (FRENCH) ◇ MBDZ ◇ MEBENDAZOLE (USDA) ◇ OVITELMIN ◇ PANTELMIN ◇ R 17635 ◇ TELMIN ◇ VERMIRAX ◇ VERMOX

USE IN FOOD:

Purpose: Animal drug.

Where Used: Various.

Regulations: USDA CES Ranking: B-4 (1986).

EPA Genetic Toxicology Program.

SAFETY PROFILE: Poison by ingestion. Moderately toxic by intraperitoneal route. Human mutagenic data. Experimental reproductive effects. When heated to decomposition it emits toxic fumes of NO_x.

TOXICITY DATA and CODEN

mma-sat 600 nmol/plate CNREA8 38,4478,78
bfa-mus/sat 500 mg/kg CNREA8 38,4478,78
orl-rat TDLo:78400 μg/kg (8-15D preg):REP THERAP 31,505,76
orl-rat LDLo:320 mg/kg TXAPA9 24,371,73

MHM100
METHYLBENZYL ACETATE

mf: $C_{10}H_{12}O_2$ mw: 164.20

PROP: Colorless liquid; sweet, nutty odor. D: 1.030-1.035, refr index: 1.501. Sol in fixed oils; sltly sol in propylene glycol; insol in glycerin.

SYN: TOLYL ACETATE

USE IN FOOD:

Purpose: Flavoring agent.

Where Used: Various.

Regulations: FDA - 21CFR 172.515. Use at a level not in excess of the amount reasonably required to accomplish the intended effect.

SAFETY PROFILE: When heated to decomposition it emits acrid smoke and irritating fumes.

MHV250 CAS: 94-46-2
1-(3-METHYL)BUTYL BENZOATE
mf: $C_{12}H_{16}O$ mw: 176.28

PROP: Colorless to pale yellow liquid; pungent, fruity odor. D: 0.986-0.992, refr index: 1.492, flash p: +212°F.

SYNS: AMYL BENZOATE ◇ BENZOIC ACID, 1-(3-METHYL)BUTYL ESTER ◇ FEMA No. 2058 ◇ ISOAMYL BENZOATE (FCC) ◇ ISOPENTYL BENZOATE

USE IN FOOD:

Purpose: Flavoring agent.

Where Used: Various.

Regulations: FDA - 21CFR 172.515. Use at a level not in excess of the amount reasonably required to accomplish the intended effect.

SAFETY PROFILE: Mildly toxic by ingestion. A skin irritant. Combustible liquid. When heated to decomposition it emits acrid smoke and irritating fumes.

TOXICITY DATA and CODEN

skn-rbt 500 mg/24H MLD FCTXAV 11,1079,73
orl-rat LD50:6330 mg/kg FCTXAV 11,477,73

MHV500 CAS: 17804-35-2
METHYL-1-(BUTYLCARBAMOYL)-2-BENZIMIDAZOLYLCARBAMATE
mf: $C_{14}H_{18}N_4O_3$ mw: 290.36

SYNS: ARILATE ◇ BBC ◇ BENLATE 50 ◇ BENOMYL (ACGIH, USDA) ◇ BENOMYL 50W ◇ BNM ◇ 1-(BUTYLCARBAMOYL)-2-BENZIMIDAZOLECARBAMIC ACID, METHYL ESTER ◇ 1-(BUTYLCARBAMOYL)-2-BENZIMIDAZOL-METHYLCARBAMAT (GERMAN) ◇ 1-(N-BUTYLCARBAMOYL)-2-(METHOXY-CARBOXAMIDO)-BENZIMIDAZOL (GERMAN) ◇ DU PONT 1991 ◇ FUNDASOL ◇ FUNGICIDE 1991 ◇ MBC ◇ TERSAN 1991

USE IN FOOD:

Purpose: Fungicide.

Where Used: Apples, apricots, bananas, cherries, mangoes, nectarines, peaches, pears, pineapples, plums, raisins, tomato products (concentrated).

Regulations: FDA - 21CFR 193.30. Fungicide residue tolerance of 50 ppm in raisins and concentrated tomato products. 21CFR 561.50. Limitation of 125 ppm in dried grape pomace and raisin waste, 70 ppm in dried apple pomace, 50 ppm in dried citrus pulp, 20 ppm in rice hulls. USDA CES Ranking: B-3 (1986).

EPA Genetic Toxicology Program.

ACGIH TLV: TWA 10 mg/m^3

SAFETY PROFILE: Poison by ingestion. Mildly toxic by inhalation. An experimental teratogen. Experimental reproductive effects. Human mutagenic data. A human skin irritant. When heated to decomposition it emits toxic fumes of NO_x.

TOXICITY DATA and CODEN

skn-man 0.1% MLD LANCAO 2,1252,80
sln-smc 123 ppm ANYAA9 407,186,83
sln-hmn:lym 10 mg/L MUREAV 121,139,83
orl-rat TDLo:250 mg/kg (12D preg):REP BEXBAN 83,247,77
orl-rat TDLo:936 mg/kg (7-22D preg/15D post):TER TXAPA9 62,44,82
orl-mus TDLo:250 mg/kg (11D preg):TER JHEMA2 24,295,80
orl-rat LD50:10 g/kg JHEMA2 24,295,80
ihl-rat LD50:9920 mg/kg EQSFAP 3,618,75

MHW260
2-METHYLBUTYL ISOVALERATE
mf: $C_{10}H_{20}O_2$ mw: 172.27

PROP: Colorless liquid; herbaceous, fruity odor. D: 0.852, refr index: 1.413. Sol in alc, fixed oils; insol in water.

SYNS: FEMA No. 2753 ◇ 2-METHYLBUTYL-3-METHYLBUTANOATE

USE IN FOOD:

Purpose: Flavoring agent.

Where Used: Various.

Regulations: FDA - 21CFR 172.515. Use at a level not in excess of the amount reasonably required to accomplish the intended effect.

SAFETY PROFILE: When heated to decomposition it emits acrid smoke and irritating fumes.

MIF760 CAS: 9004-67-5
METHYL CELLULOSE

PROP: White, fibrous powders. Sol in water, some organic solvents.

USE IN FOOD:

Purpose: Binder, bodying agent, bulking agent, emulsifier, film former, stabilizer, thickener.

Where Used: Baked goods, fruit pie fillings, meat patties, vegetable patties.

Regulations: FDA - 21CFR 182.1480. GRAS when used in accordance with good manufacturing practice. USDA - 9CFR 318.7. Limitation of 0.15 percent in meat and vegetable products.

SAFETY PROFILE: When heated to decomposition it emits acrid smoke and irritating fumes.

MII500 CAS: 2971-90-6
METHYLCHLOROPINDOL
mf: $C_7H_7Cl_2NO$ mw: 192.05

SYNS: COCCIDIOSTAT C ◇ CLOPIDOL (ACGIH) ◇ COYDEN ◇ 3,5-DICHLORO-2,6-DIMETHYL-4-PYRIDINOL ◇ METHYLCHLORPINDOL ◇ METILCLORPINDOL

TOXICITY DATA and CODEN

USE IN FOOD:

Purpose: Animal drug, animal feed drug.

Where Used: Beef, cereal grains, chicken, fruits, goat, lamb, milk, pork, turkey, vegetables.

Regulations: FDA - 21CFR 556.160. Limitation of 0.2 ppm in cereal grains, vegetables, and fruits. Limitation of 15 ppm in uncooked liver and kidney, 5 ppm in uncooked muscle of chickens and turkeys. Limitation of 3 ppm in uncooked kidney, 1.5 ppm in uncooked liver, 0.2 ppm in uncooked muscle of cattle, sheep, and goats. Limitation of 0.2 ppm in swine. Limitation of 0.02 ppm in milk. 21CFR 558.175.

ACGIH TLV: TWA 10 mg/m^3

SAFETY PROFILE: A nuisance dust. When heated to decomposition it emits very toxic fumes of Cl^- and NO_x.

MIO000 CAS: 101-39-3
α-METHYLCINNAMALDEHYDE
mf: $C_{10}H_{10}O$ mw: 146.20

PROP: Yellow liquid; cinnamon odor. D: 1.035-1.039, refr index: 1.602-1.607, flash p: 174°F. Sol in fixed oils, propylene glycol; insol in glycerin.

SYNS: FEMA No. 2697 ◇ METHYL CINNAMIC ALDEHYDE ◇ α-METHYLCINNAMIC ALDEHYDE ◇ α-METHYLCINNIMAL ◇ 2-METHYL-3-PHENYL-2-PROPENAL

USE IN FOOD:

Purpose: Flavoring agent.

Where Used: Various.

Regulations: FDA - 21CFR 172.515. Use at a level not in excess of the amount reasonably required to accomplish the intended effect.

SAFETY PROFILE: Moderately toxic by ingestion. A skin irritant. Combustible liquid. When heated to decomposition it emits acrid smoke and irritating fumes.

TOXICITY DATA and CODEN

skn-gpg 5%/2W MLD ADVEA4 58,121,78
orl-rat LD50:2050 mg/kg FCTXAV 13,681,75

MIO500 CAS: 103-26-4
METHYL CINNAMATE
mf: $C_{10}H_{10}O_2$ mw: 162.20

PROP: White to sltly yellow crystals; fruity odor. D: 1.042 @ 36/0°, mp: 33.4°, bp: 263°, flash p: +212°F. Very sol in alc, ether; sol in fixed oils, glycerin, propylene glycol; insol in water.

SYNS: FEMA No. 2698 ◇ METHYL CINNAMYLATE ◇ METHYL-3-PHENYLPROPENOATE ◇ 3-PHENYL-2-PROPENOIC ACID METHYL ESTER (9CI)

USE IN FOOD:

Purpose: Flavoring agent.

Where Used: Various.

Regulations: FDA - 21CFR 172.515. Use at a level not in excess of the amount reasonably required to accomplish the intended effect.

SAFETY PROFILE: Moderately toxic by ingestion. Combustible liquid. When heated to decomposition it emits acrid smoke and irritating fumes.

TOXICITY DATA and CODEN

orl-rat LD50:2610 mg/kg FCTXAV 13,681,75

MIP750 CAS: 92-48-8
6-METHYLCOUMARIN
mf: $C_{10}H_8O_2$ mw: 160.18

PROP: White needles from benzene; coconut odor. Mp: 73-76, flash p: +153°F. Sol in alc and benzene.

SYNS: FEMA No. 2690 ◇ 6-MC ◇ 6-METHYL-2H-1-BENZOPYRAN-2-ONE ◇ 6-METHYLBENZOPYRONE ◇ 6-METHYL-1,2-BENZOPYRONE ◇ 6-METHYLCOUMARINIC ANHYDRIDE ◇ NCI-C55812 ◇ TONCARINE

USE IN FOOD:

Purpose: Flavoring agent.

Where Used: Various.

Regulations: FDA - 21CFR 172.515. Use at a level not in excess of the amount reasonably required to accomplish the intended effect.

EPA Genetic Toxicology Program.

SAFETY PROFILE: Poison by subcutaneous route. Moderately toxic by ingestion. A skin irritant. Mutagenic data. Combustible liquid. When heated to decomposition it emits acrid smoke and irritating fumes.

TOXICITY DATA and CODEN

skn-rbt 500 mg/24H MLD FCTXAV 14,605,76
mma-sat 3 μmol/plate FCTOD7 21,707,83
orl-rat LD50:1680 mg/kg FCTXAV 14,605,76

MJY500 CAS: 25057-89-0
3-(1-METHYLETHYL)-1H-2,1,3-BENZOTHIAZAIN-4(3H)-ONE-2,2-DIOXIDE
mf: $C_{10}H_{12}N_2O_3S$ mw: 240.30

SYNS: BAS 351-H ◇ BASAGRAN ◇ BENDIOXIDE ◇ BENTAZON ◇ 3-ISOPROPYL-2,1,3-BENZOTHIADIAZINON-(4)-2,2-DIOXID (GERMAN) ◇ 3-ISOPROPYL-1H-2,1,3-BENZOTHIADIAZIN-4(3H)-ONE-2,2-DIOXIDE

USE IN FOOD:

Purpose: Herbicide.

Where Used: Animal feed, spent mint hay.

Regulations: FDA - 21CFR 561.51. Limitation of 4 ppm on spent mint hay when used for animal feed.

EPA Genetic Toxicology Program.

SAFETY PROFILE: Moderately toxic by ingestion and skin contact. When heated to decomposition it emits very toxic fumes of SO_x and NO_x.

TOXICITY DATA and CODEN

orl-rat LD50:1100 mg/kg GUCHAZ 6,36,73
skn-rat LD50:2500 mg/kg 85DPAN -,-,71/76

MJY550
METHYL ETHYL CELLULOSE

PROP: White fibrous solid or powder. Disperses in water.

USE IN FOOD:

Purpose: Emulsifier, foaming agent, stabilizer.

Where Used: Meringues, whipped toppings.

Regulations: FDA - 21CFR 172.872. Use at a level not in excess of the amount reasonably required to accomplish the intended effect.

SAFETY PROFILE: When heated to decomposition it emits acrid smoke and irritating fumes.

MKA000 CAS: 31218-83-4
(E)-1-METHYLETHYL-3-(((ETHYLAMINO)METHOXYPHOSPHINOTHIOYL)OXY-2-BUTENOATE
mf: $C_{10}H_{20}NO_4PS$ mw: 281.34

SYNS: BLOTIC ◇ ENT 27,989 ◇ (3)-O-2-ISOPROPOXYCARBONYL-1-METHYLVINYL-O-METHYL ETHYLPHOSPHORAMIDOTHIOATE ◇ PROPETAMPHOS ◇ SAFROTIN ◇ SAN 52 139 I ◇ SANDOZ 52139 ◇ VEL 4283

USE IN FOOD:

Purpose: Insecticide.

Where Used: Animal feed.

Regulations: FDA - 21CFR 193.375. Insecticide residue tolerance of 0.1 ppm in food. 21CFR 561.434. Limitation of 0.1 ppm in animal feeds.

SAFETY PROFILE: Poison by ingestion. Moderately toxic by skin contact. When heated to decomposition it emits very toxic fumes of PO_x, SO_x and NO_x.

TOXICITY DATA and CODEN

orl-rat LD50:75 mg/kg 85ARAE 1,118,77
skn-rat LD50:2300 mg/kg FMCHA2 -,D256,80

MKG750 CAS: 107-31-3
METHYL FORMATE
DOT: 1243
mf: $C_2H_4O_2$ mw: 60.06

PROP: Colorless liquid; agreeable odor. Mp: −99.8°, bp: 31.5°, lel: 5.9%, uel: 20%, flash p: −2.2°F, d: 0.98149 @ 15°/4°, 0.975 @ 20°/4°, autoign temp: 869°F, vap press: 400 mm @ 16°/0°, vap d: 2.07. Solidifies at about 100°. Moderately sol in water, methyl alcohol; misc in alc.

SYNS: FORMIATE de METHYLE (FRENCH) ◇ METHYLE (FORMIATE de) (FRENCH) ◇ METHYL FORMATE (DOT) ◇ METHYLFORMIAAT (DUTCH) ◇ METHYLFORMIAT (GERMAN) ◇ METHYL METHANOATE ◇ METIL (FORMIATO di) (ITALIAN)

USE IN FOOD:

Purpose: Fumigant.

Where Used: Raisins, Zante currants (dried).

Regulations: FDA - 21CFR 193.310. Fumigant residue tolerance of 250 ppm as formic acid in raisins and dried Zante currants.

OSHA PEL: TWA 100 ppm; STEL 150 ppm ACGIH TLV: TWA 100 ppm; STEL 150 ppm DOT Classification: Flammable Liquid; Label: Flammable Liquid

SAFETY PROFILE: Moderately toxic by ingestion. Inhalation of vapor can cause irritation to nasal passages and conjunctiva, optic neuritis, narcosis, retching, and death from pulmonary irritation. Flammable liquid. Very dangerous fire hazard when exposed to heat or flame; can react vigorously with oxidizing materials. Explosive in the form of vapor when exposed to heat or flame. To fight fire, use alcohol foam, CO_2, dry chemical. When heated to decomposition it emits acrid smoke and irritating fumes.

TOXICITY DATA and CODEN

orl-rbt LD50:1622 mg/kg IMSUAI 41,31,72
ihl-gpg LCLo:10000 ppm 14CYAT -,1855,63

MKK000 CAS: 409-02-9
6-METHYL-5-HEPTEN-2-ONE
mf: $C_8H_{14}O$ mw: 126.22

PROP: Sltly yellow liquid; citrus-lemongrass odor. D: 0.846-0.851, refr index: .438-1.442, mp: −67.1, bp: 173-174°, flash p: 122°F. Insol in water; misc in alc, ether, chloroform.

SYN: FEMA No. 2707 ◇ METHYL HEPTENONE

USE IN FOOD:

Purpose: Flavoring agent.

Where Used: Various.

Regulations: FDA - 21CFR 172.515. Use at a level not in excess of the amount reasonably required to accomplish the intended effect.

SAFETY PROFILE: Moderately toxic by ingestion. A skin irritant. Combustible liquid. When heated to decomposition it emits acrid smoke and irritating fumes.

TOXICITY DATA and CODEN

skn-rbt 500 mg/24H FCTXAV 13,681,75
orl-rat LD50:3500 mg/kg FCTXAV 13,859,75

MLA250 CAS: 99-86-5
1-METHYL-4-ISOPROPYLCYCLOHEXADIENE-1,3
mf: $C_{10}H_{16}$ mw: 136.26

PROP: Colorless liquid; lemon odor. D: 0.834 @ 20°/4°, refr index: 1.475-1.480, bp: 181.5°. Insol in water; misc in alc, ether, fixed oils.

SYNS: FEMA No. 3558 ◇ p-MENTHA-1,3-DIENE ◇ 1-METHYL-4-ISOPROPYL-1,3-CYCLOHEXADIENE ◇ α-TERPINENE (FCC)

USE IN FOOD:

Purpose: Flavoring agent.

Where Used: Various.

Regulations: FDA - 21CFR 172.515. Use at a level not in excess of the amount reasonably required to accomplish the intended effect.

SAFETY PROFILE: Moderately toxic by ingestion. When heated to decomposition it emits acrid smoke and irritating fumes.

TOXICITY DATA and CODEN

orl-rat LD50:1680 mg/kg FCTXAV 14,873,76

MLA300
5-METHYL-2-ISOPROPYL-2-HEXENAL
mf: $C_{10}H_{18}O$ mw: 154.25

PROP: Sltly yellow liquid; herbaxeous, woody, fruity, chocolate odor. D: 0.845-0.860, refr index: 1.448. Sol in alc, fixed oils; insol in water, propylene glycol.

SYN: FEMA No. 3406

USE IN FOOD:

Purpose: Flavoring agent.

Where Used: Various.

Regulations: GRAS when used at a level not in excess of the amount reasonably required to accomplish the intended effect.

SAFETY PROFILE: When heated to decomposition it emits acrid smoke and irritating fumes.

MLL600 CAS: 53955-81-0
METHYL 2-METHYLBUTYRATE
mf: $C_6H_{12}O_2$ mw: 116.16

PROP: Colorless liquid; sweet, fruity, apple-like odor. D: 0.879, refr index: 1.393-1.397, flash p: 91°F. Sol in alc, fixed oils; insol in water.

SYNS: FEMA No. 2719 ◇ METHYL 2-METHYLBUTANOATE

USE IN FOOD:

Purpose: Flavoring agent.

Where Used: Various.

Regulations: FDA - 21CFR 172.515. Use at a level not in excess of the amount reasonably required to accomplish the intended effect.

SAFETY PROFILE: Flammable liquid. When heated to decomposition it emits acrid smoke and irritating fumes.

MND275 CAS: 111-12-6
METHYL 2-OCTYNOATE
mf: $C_9H_{14}O_2$ mw: 154.23

PROP: Colorless to sltly yellow liquid; powerful, unpleasant odor; violet odor when diluted. D: 0.919, refr index: 1.446, flash p: +212°F. Sol in fixed oils; sltly sol in propylene glycol; insol in glycerin

SYNS: FEMA No. 2729 ◇ FOLIONE ◇ METHYL HEPTINE CARBONATE ◇ METHYL 2-OCTINATE

USE IN FOOD:

Purpose: Flavoring agent.

Where Used: Various.

Regulations: FDA - 21CFR 172.515. Use at a level not in excess of the amount reasonably required to accomplish the intended effect.

SAFETY PROFILE: Moderately toxic by ingestion and skin contact. A moderate skin and eye irritant. A combustible liquid. When heated to decomposition it emits acrid smoke and irritating fumes.

TOXICITY DATA and CODEN

skn-rbt 500 mg/24H MOD FCTXAV 17,375,79
orl-rat LD50:1530 mg/kg FCTXAV 17,375,79
skn-rbt LD50:3300 mg/kg FCTXAV 17,375,79

MNG750 CAS: 118-71-8
2-METHYL-3-OXY-γ-PYRONE
mf: $C_6H_6O_3$ mw: 126.12

PROP: White crystalline powder; caramel-butterscotch odor. Sol in water, alc, glycerin, propylene glycol.

SYNS: CORPS PRALINE ◇ 3-HYDROXY-2-METHYL-4H-PYRAN-4-ONE ◇ 3-HYDROXY-2-METHYL-γ-PYRONE ◇ 3-HYDROXY-2-METHYL-4-PYRONE ◇ LARIXIC ACID ◇ LARIXINIC ACID ◇ MALTOL ◇ 2-METHYL-3-HYDROXY-4-PYRONE ◇ 2-METHYL PYROMECONIC ACID ◇ PALATONE ◇ TALMON ◇ VETOL

USE IN FOOD:

Purpose: Flavoring agent, processing aid.

Where Used: Baked goods, flour, wine.

Regulations: FDA - 21CFR 172.515. Use at a level not in excess of the amount reasonably required to accomplish the intended effect. BATF - 27CFR 240.1051. Limitation of 250 mg/L in wine.

SAFETY PROFILE: Moderately toxic by ingestion, intraperitoneal and subcutaneous routes. A skin irritant. Human mutagenic data. When heated to decomposition it emits acrid smoke and irritating fumes.

TOXICITY DATA and CODEN

skn-rbt 500 mg/24H MOD FCTXAV 13,841,75
mmo-sat 1 mg/plate MUREAV 67,367,79
mma-sat 3333 μg/plate ENMUDM 8(Suppl 7),1,86
orl-rat LD50:2330 mg/kg FCTXAV 13,681,75

MNR250 CAS: 140-39-6
4-METHYLPHENYL ACETATE
mf: $C_9H_{10}O_2$ mw: 150.19

PROP: Colorless liquid; strong floral odor. D: 1.044 @ 16°, refr index: 1.499-1.502, bp: decomp @ 360°, mp: 220°, vap d: 5.18, flash p: 203°F. Sol in fixed oils, propylene glycol, misc in alc and ether; insol in water, glycerin.

SYNS: ACETIC ACID-4-METHYLPHENYL ESTER ◇ p-ACETOXYTOLUENE ◇ 4-ACETOXYTOLUENE ◇ p-CRESOL ACETATE ◇ p-CRESYL ACETATE (FCC) ◇ FEMA No. 3073 ◇ 4-METHYLBENZOIC ACID METHYL ESTER ◇ p-METHYLPHENYL ACETATE ◇ NARCEOL ◇ PARACRESYL ACETATE ◇ p-TOLYL ACETATE ◇ p-TOLYL ETHANOATE

USE IN FOOD:

Purpose: Flavoring agent.

Where Used: Various.

Regulations: FDA - 21CFR 172.515. Use at a level not in excess of the amount reasonably required to accomplish the intended effect.

SAFETY PROFILE: Moderately toxic by ingestion and skin contact. Combustible liquid. When heated to decomposition it emits toxic smoke and irritating fumes.

TOXICITY DATA and CODEN

orl-rat LD50:1900 mg/kg FCTXAV 12,391,74
skn-rbt LD50:2100 mg/kg FCTXAV 12,391,74

MNT075
METHYL PHENYLCARBINYL ACETATE
mf: $C_{10}H_{12}O_2$ mw: 164.20

PROP: Colorless liquid; gardenia odor. D: 1.023, refr index: 1.493-1.497, flash p: 176°F. Sol in fixed oils, glycerin; insol in water.

SYNS: FEMA No. 2684 ◇ α-PHENYL ETHYL ACETATE

USE IN FOOD:

Purpose: Flavoring agent.

Where Used: Various.

Regulations: FDA - 21CFR 172.515. Use at a level not in excess of the amount reasonably required to accomplish the intended effect.

SAFETY PROFILE: Combustible liquid. When heated to decomposition it emits acrid smoke and irritating fumes.

MOR500 CAS: 83-43-2
METHYLPREDNISOLONE

mf: $C_{22}H_{30}O_5$ mw: 374.52

PROP: Crystals. Mp: 228-237°.

SYNS: MEDROL ◇ MEDROL DOSEPAK ◇ MEDRONE ◇ Δ^1-6-α-METHYLHYDROCORTISONE ◇ 6-α-METHYLPREDNISOLONE ◇ METRISONE ◇ NSC-19987 ◇ 11-β,17,21-TRIHYDROXY-6-α-METHYLPREGNA-1,4-DIENE-3,20-DIONE ◇ 11-β,17-α,21-TRIHYDROXY-6-α-METHYL-1,4-PREGNADIENE-3,20-DIONE ◇ URBASON ◇ URBASONE ◇ WYACORT

USE IN FOOD:

Purpose: Animal drug.

Where Used: Milk.

Regulations: FDA - 21CFR 556.400. Limitation of 10 ppb in milk.

SAFETY PROFILE: Moderately toxic by intraperitoneal route. When heated to decomposition it emits acrid smoke and irritating fumes.

TOXICITY DATA and CODEN

ipr-mus LD50:2292 mg/kg NIIRDN 6,832,82

MOW750 CAS: 109-08-0
2-METHYLPYRAZINE

mf: $C_5H_6N_2$ mw: 94.13

PROP: Liquid; nutty, cocoa odor. Mp: −29°, bp: 133° @ 737 mm, flash p: 122°F (COC), d: 1.0224 @ 25°/25°, refr index: 1.504, vap d: 3.2. Misc with water, alc, acetone, fixed oils.

SYN: FEMA No. 3309

USE IN FOOD:

Purpose: Flavoring agent.

Where Used: Various.

Regulations: GRAS when used at a level not in excess of the amount reasonably required to accomplish the intended effect.

SAFETY PROFILE: Moderately toxic by ingestion and intraperitoneal routes. Mutagenic data. Combustible liquid when exposed to heat, flame or oxidizers. Can react with oxidizing materials. To fight fire, use water spray, foam, dry chemical, CO_2. When heated to decomposition it emits highly toxic fumes of NO_x.

TOXICITY DATA and CODEN

mmo-smc 8500 μg/L FCTXAV 18,581,80
cyt-ham:ovr 2500 μg/L FCTXAV 18,581,80
orl-rat LD50:1800 mg/kg DCTODJ 3,249,80

MPI000 CAS: 119-36-8
METHYL SALICYLATE

mf: $C_8H_8O_3$ mw: 152.16

PROP: From steam distillation of leaves from *Gaultheria procumbens* L. (Fam. *Ericacaae*) or from the bark of *Betula lenta* L. (Fam. *Betulaceae*). Colorless, yellowish or reddish oily liquid; odor and taste of wintergreen. Mp: −8.6°, bp: 223.3°, ULC: 20-25, flash p: 214°F (CC), fp: −1.2°, d: 1.1840 @ 25°/25°, refr index: 1.535, autoign temp: 850°F, vap press: 1 mm @ 54.0°, vap d: 5.24. Sltly sol in water @ 222° (decomp); sol in chloroform, ether, alc, glacial acetic acid.

SYNS: ACIDE ANISIQUE (FRENCH) ◇ ACIDE METHYL-o-BENZOIQUE (FRENCH) ◇ o-ANISIC ACID ◇ BETULA OIL ◇ FEMA No. 2745 ◇ GAULTHERIA OIL, ARTIFICIAL ◇ o-HYDROXYBENZOIC ACID, METHYL ESTER ◇ 2-HYDROXYBENZOIC ACID METHYL ESTER ◇ o-METHOXYBENZOIC ACID ◇ 2-METHOXYBENZOIC ACID ◇ METHYL-o-HYDROXYBENZOATE ◇ METYLESTER KYSELINY SALICYLOVE (CZECH) ◇ NATURAL WINTERGREEN OIL ◇ OIL of WINTERGREEN ◇ SALICYLIC ACID, METHYL ESTER ◇ SWEET BIRCH OIL ◇ SYNTHETIC WINTERGREEN OIL ◇ TEABERRY OIL ◇ WINTERGREEN OIL (FCC) ◇ WINTERGREEN OIL, SYNTHETIC

USE IN FOOD:

Purpose: Flavoring agent.

Where Used: Baked goods, beverages, candy, chewing gum.

Regulations: GRAS when used at a level not in excess of the amount reasonably required to accomplish the intended effect.

SAFETY PROFILE: Human poison by ingestion. Moderately toxic experimentally by ingestion, intraperitoneal, intravenous, and subcutaneous routes. An experimental teratogen. Human systemic effects by ingestion: flaccid paralysis without anesthesia, general anesthesia, dyspnea, and nausea or vomiting. Experimental reproductive effects. A severe skin and eye irritant. Ingestion of relatively small amounts has caused severe poisoning and death. Combustible liquid when exposed to heat or flame; can react with oxidizing materials. To fight fire, use CO_2, dry chemical. When heated to decomposition it emits acrid smoke and irritating fumes.

TOXICITY DATA and CODEN

skn-rbt 500 mg/24H MOD FCTXAV 16,637,78
eye-rbt 500 mg/24H SEV 28ZPAK -,106,72
orl-rat TDLo:36450 mg/kg (MGN):REP TXAPA9 18,755,71
scu-rat TDLo:500 mg/kg (10D preg):TER AJPAA4 35,315,59
skn-ham TDLo:5250 mg/kg (7D preg):TER TJADAB 28,421,83
orl-chd LDLo:228 mg/kg:PUL,GIT AJDCAI 69,37,45
orl-chd LDLo:700 mg/kg:PNS,CNS,PUL ADCHAK 28,475,53
orl-hmn LDLo:506 mg/kg MEIEDD 10,876,83
orl-rat LD50:887 mg/kg FCTXAV 2,327,64

MQI550 CAS: 110-41-8
2-METHYLUNDECANAL
mf: $C_{12}H_{24}O$ mw: 184.32

PROP: Colorless to sltly yellow liquid; fatty odor. D: 0.822-0.830, refr index: 1.431. Sol in alc, fixed oils, propylene glycol; insol in glycerin.

SYNS: ALDEHYDE C-12 MNA ◇ FEMA No. 2749 ◇ METHYL n-NONYL ACETALDEHYDE

USE IN FOOD:

Purpose: Flavoring agent.

Where Used: Various.

Regulations: FDA - 21CFR 172.515. Use at a level not in excess of the amount reasonably required to accomplish the intended effect.

SAFETY PROFILE: When heated to decomposition it emits acrid smoke and irritating fumes.

MQR200 CAS: 1178-29-6
METOSERPATE HYDROCHLORIDE
mf: $C_{24}H_{32}N_2O_5 \cdot ClH$ mw: 465.04

SYNS: METHYL-18-EPIRESERPATE METHYL ETHER HYDROCHLORIDE ◇ PACITRAN ◇ SU-9064 ◇ SU 8842 HYDROCHLORIDE

USE IN FOOD:

Purpose: Animal drug.

Where Used: Chicken.

Regulations: FDA - 21CFR 556.410. Limitation of 0.02 ppm in chicken.

SAFETY PROFILE: Poison by ingestion and intravenous routes. When heated to decomposition it emits toxic fumes of NO_x and HCl.

TOXICITY DATA and CODEN

orl-rat LD50:182 mg/kg JPETAB 138,78,62
ivn-rat LD50:21 mg/kg 27ZQAG -,112,72

MQS225 CAS: 3704-09-4
MIBOLERONE
mf: $C_{20}H_{30}O_2$ mw: 302.50

PROP: Crystalline solid. Solubility in deionized water: 0.0454 mg/mL @ 37°.

SYNS: CHEQUE ◇ (7-α,17-β)-17-HYDROXY-7,17-DIMETHYL-ESTR-4-EN-3-ONE (9CI) ◇ 17-β-HYDROXY-7-α,17-DIMETHYLESTR-4-EN-3-ONE ◇ MATENON ◇ MIBOLERON ◇ U 10997

USE IN FOOD:

Purpose: Animal feed drug.

Where Used: Dog food.

Regulations: FDA - 21CFR 558.348.

SAFETY PROFILE: An experimental neoplastigen and teratogen. Experimental reproductive effects. When heated to decomposition it emits acrid smoke and irritating fumes.

TOXICITY DATA and CODEN

scu-mus TDLo:800 μg/kg (10D pre):REP STEDAM 9,235,67
orl-dog TDLo:196 μg/kg (6-61D preg/42D post):TER AJVRAH 39,837,78
orl-dog TDLo:8985 μg/kg/9.6Y-I:NEO TOPADD 13,177,85

MQU075
MILK-CLOTTING ENZYME from BACILLUS CEREUS

PROP: Derived from *Bacillus cereus*, fam. *Bacillaceae*.

USE IN FOOD:

Purpose: Milk clotting.

Where Used: Cheese.

Regulations: FDA - 21CFR 173.150. Use at a level not in excess of the amount reasonably required to accomplish the intended effect.

SAFETY PROFILE: When heated to decomposition it emits acrid smoke and irritating fumes.

MQU100
MILK-CLOTTING ENZYME from ENDOTHIA PARASITICA

PROP: Derived from *Edothia parasitica*, fam. *Diaporthacessae*.

USE IN FOOD:

Purpose: Milk clotting.

Where Used: Cheese.

Regulations: FDA - 21CFR 173.150. Use at a level not in excess of the amount reasonably required to accomplish the intended effect.

SAFETY PROFILE: When heated to decomposition it emits acrid smoke and irritating fumes.

MQU120
MILK-CLOTTING ENZYME from MUCOR MIEHEI

PROP: Derived from *Mucor miehei* Cooney et Emerson, fam. *Mucoraceae*.

USE IN FOOD:

Purpose: Milk clotting.

Where Used: Cheese.

Regulations: FDA - 21CFR 173.150. Use at a level not in excess of the amount reasonably required to accomplish the intended effect.

SAFETY PROFILE: When heated to decomposition it emits acrid smoke and irritating fumes.

MQU125
MILK-CLOTTING ENZYME from MUCOR PUSILLUS

PROP: Derived from *Mucor pusillus* Lindt, fam. *Mucoraceae*.

USE IN FOOD:

Purpose: Milk clotting.

Where Used: Cheese.

Regulations: FDA - 21CFR 173.150. Use at a level not in excess of the amount reasonably required to accomplish the intended effect.

SAFETY PROFILE: When heated to decomposition it emits acrid smoke and irritating fumes.

MQV750 CAS: 8012-95-1
MINERAL OIL

PROP: Colorless, oily liquid; practically tasteless and odorless. D: 0.83-0.86 (light), 0.875-0.905 (heavy), flash p: 444°F (OC), ULC: 10-20. Insol in water and alc; sol in benzene, chloroform, and ether. A mixture of liquid hydrocarbons from petroleum.

SYNS: ADEPSINE OIL ◇ ALBOLINE ◇ BAYOL F ◇ BLANDLUBE ◇ CRYSTOSOL ◇ DRAKEOL ◇ FONOLINE ◇ GLYMOL ◇ KAYDOL ◇ KONDREMUL ◇ MINERAL OIL, WHITE (FCC) ◇ MOLOL ◇ NEO-CULTOL ◇ NUJOL ◇ PAROL ◇ PAROLEINE ◇ PARRAFIN OIL ◇ PENETECK ◇ PENRECO ◇ PERFECTA ◇ PETROGALAR ◇ PETROLATUM, LIQUID ◇ PRIMOL 335 ◇ PROTOPET ◇ SAXOL ◇ TECH PET F ◇ WHITE MINERAL OIL

USE IN FOOD:

Purpose: Binder, defoaming agent, fermentation aid, lubricant, protective coating, release agent.

Where Used: Bakery products, beet sugar, confectionery, egg white solids, fruit (raw), fruits (dehydrated), grain, meat (frozen), pickles, potatoes (sliced), sorbic acid, starch (molding), vegetables (dehydrated), vegetables (raw), vinegar, wine, yeast.

Regulations: FDA - 21CFR 172.878. Limitation of 0.6 percent in capsules or tablets containing concentrations of flavoring, spices, condiments, nutrients, or food for dietary use. Limitation of 0.15 percent in bakery products, 0.02 percent in dehydrated fruits and vegetables, 0.1 percent in egg white solids, 0.095 percent in frozen meat, 0.3 percent in molding starch, 0.15 percent in yeast, 0.25 percent in sorbic acid, 0.2 percent in confectionery, 0.02 percent in grain. 21CFR 173.340. Limitation of 0.008 percent in wash water for sliced potatoes, 150 ppm in yeast. 21CFR 175.105, 176.200, 176.210, 177.2260, 177.2600, 177.2800, 178.3570, 178.3620, and 178.3910. May contain an antioxidant permitted by FDA in an amount not greater than required to produce it intended effect.

SAFETY PROFILE: A human carcinogen by inhalation which produces gastrointestinal tumors. A human teratogen by inhalation which causes testicular tumors in the fetus. Inhalation of vapor or particulates can cause aspiration pneumonia. An eye irritant. Combustible liquid when exposed to heat or flame. To fight fire, use dry chemical, CO_2, foam. When heated to decomposition it emits acrid smoke and fumes.

TOXICITY DATA and CODEN

eye-rbt 250 mg/5D MLD AMIHAB 14,265,56

ihl-man TCLo:5 mg/m^3/5Y-I:CAR,GIT,TER JOCMA7 23,333,81

skn-mus TDLo: 332 g/mg/20W-I: ETA ANYAA9 132,439,65
ipr-mus TD : 50 g/kg/9W-I: ETA IJCNAW 6,422,70
ipr-mus TD : 72 g/kg/26W-I: ETA JOIMA3 92,747,62

MQY750 CAS: 101-14-4
MOCA

mf: $C_{13}H_{12}Cl_2N_2$ mw: 267.17

SYNS: BIS AMINE ◇ CURALIN M ◇ CURENE 442 ◇ CYANASET ◇ DI(-4-AMINO-3-CHLOROPHENYL)METHANE ◇ DI-(4-AMINO-3-CLOROFENIL)METANO (ITALIAN) ◇ 4,4′-DIAMINO-3,3′-DICHLORODIPHENYLMETHANE ◇ 3,3′-DICHLOR-4,4′-DIAMINODIPHENYLMETHAN (GERMAN) ◇ 3,3′-DICHLORO-4,4′-DIAMINODIPHENYLMETHANE ◇ 3,3′-DICLORO-4,4′-DIAMINODIFENILMETANO (ITALIAN) ◇ MBOCA ◇ 4,4′-METHYLENE(BIS)-CHLOROANILINE ◇ METHYLENE-4,4′-BIS(o-CHLOROANILINE) ◇ p,p′-METHYLENEBIS(α-CHLOROANILINE) ◇ 4,4′-METHYLENEBIS(o-CHLOROANILINE) ◇ p,p′-METHYLENEBIS(o-CHLOROANILINE) ◇ 4,4′-METHYLENEBIS(2-CHLOROANILINE) ◇ 4,4′-METHYLENEBIS-2-CHLOROBENZENAMINE ◇ METHYLENE-BIS-ORTHOCHLOROANILINE ◇ 4,4-METILENE-BIS-o-CLOROANILINA (ITALIAN) ◇ RCRA WASTE NUMBER U158

USE IN FOOD:

Purpose: Packaging adhesives.

Where Used: Packaging materials.

Regulations: FDA - 21CFR 189.280. Prohibited from indirect addition to human food from food-contact surfaces.

IARC Cancer Review: Animal Sufficient Evidence IMEMDT 4,65,74. EPA Genetic Toxicology Program. Community Right-To-Know List.

ACGIH TLV: TWA 0.02 ppm (skin); Suspected Carcinogen NIOSH REL: (MOCA) TWA 0.003 mg/m^3; avoid skin contact

SAFETY PROFILE: Poison by intraperitoneal route. Moderately toxic by ingestion. An experimental carcinogen and tumorigen. Mutagenic data. When heated to decomposition it emits very toxic fumes of Cl^- and NO_x.

TOXICITY DATA and CODEN

otr-mus: fbr 10 μg/L JJIND8 67,1303,81
dns-ham: lvr 10 μmol/L TXAPA9 58,231,81
orl-rat TDLo: 4050 mg/kg/77W-C: CAR JEPTDQ 2(1),149,78
scu-rat TDLo: 25 g/kg/89W-C: CAR NATWAY 58,578,71
orl-rat TD : 27 g/kg/79W-C: ETA NATWAY 58,578,71
orl-rat LD50: 2100 mg/kg KCRZAE 26(9),28,67

MRA075
MODIFIED POLYACRYLAMIDE RESINS

PROP: Produced by copolymerization of acrylamide with not more than 5-mole percent of β-methacrylyloxyethyl trimethylammonium methyl sulfate.

SYN: POLYACRYLAMIDE RESINS, MODIFIED

USE IN FOOD:

Purpose: Flocculent.

Where Used: Sugar liquor or juice.

Regulations: FDA - 21CFR 173.10. Limitation of 5 ppm by weight of the juice.

SAFETY PROFILE: When heated to decomposition it emits acrid smoke and irritating fumes.

MRA250 CAS: 11015-37-5
MOENOMYCIN

PROP: Produced by *Streptomyces roseoflavus* (85ERAY 1,740,78).

SYNS: BAMBERMYCIN ◇ FLAVOMYCIN ◇ FLAVOPHOSPHOLIPOL ◇ MENOMYCIN ◇ MOENOMYCIN A

USE IN FOOD:

Purpose: Animal feed drug.

Where Used: Animal feed.

Regulations: FDA - 21CFR 558.95.

SAFETY PROFILE: Poison by intravenous route. Moderately toxic by subcutaneous route. When heated to decomposition it emits acrid smoke and irritating fumes.

TOXICITY DATA and CODEN

scu-mus LD50: 500 mg/kg 85ERAY 1,740,78
ivn-mus LD50: 200 mg/kg

MRE225 CAS: 17090-79-8
MONENSIC ACID

mf: $C_{36}H_{62}O_{11}$ mw: 670.98

PROP: Crystals. Mp: 103-105° (monohydrate). Very stable under alkaline conditions. Sltly sol in water; more sol in hydrocarbons; very sol in other organic solvents.

SYNS: A 3823A ◇ ELANCOBAN ◇ MONELAN ◇ MONENSIN (USDA) ◇ MONENSIN A

USE IN FOOD:

Purpose: Animal drug, animal feed drug.

Where Used: Animal feed, beef, chicken, turkey.

Regulations: FDA - 21CFR 556.420. Limitation of 0.05 ppm in cattle, 1.5 ppm in muscle of chicken or turkey. 21CFR 558.355. USDA CES Ranking: B-3 (1985).

SAFETY PROFILE: Poison by ingestion and intraperitoneal routes. An eye and skin irritant. When heated to decomposition it emits acrid smoke and irritating fumes.

TOXICITY DATA and CODEN

skn-rbt 100 mg/24H MLD DCTODJ 8,451,85
eye-rbt 50 mg MOD DCTODJ 8,451,85
orl-rat LD50:100 mg/kg DCTODJ 8,451,85

MRF000 CAS: 7558-63-6
MONOAMMONIUM GLUTAMATE

mf: $C_5H_9NO_4 \cdot H_3N$ mw: 164.19

PROP: White crystalline powder; odorless. Sol in water; insol in common organic solvents.

SYNS: AMMONIUMGLUTAMINAT (GERMAN) ◇ MAG ◇ MONOAMMONIUM l-GLUTAMATE

USE IN FOOD:

Purpose: Flavor enhancer, salt substitute.

Where Used: Meat, poultry.

Regulations: FDA - 21CFR 182.1500. GRAS when used in accordance with good manufacturing practice. USDA - 9CFR 318.7, 381.147. Sufficient for purpose.

SAFETY PROFILE: Moderately toxic by intraperitoneal route. When heated to decomposition it emits toxic fumes including NO_x and NH_3.

TOXICITY DATA and CODEN

ipr-rat LD50:1000 mg/kg HSZPAZ 300,97,55

MRH215
MONO- and DIGLYCERIDES

PROP: Yellow liquids to ivory-colored plastics to hard solids; bland odor and taste. Sol in alc, ethyl acetate, chloroform, other chlorinated hydrocarbons; insol in water.

USE IN FOOD:

Purpose: Dough strengthener, emulsifier, emulsifier salt, flavoring agent, formulation aid, lubricant, release agent, softener, solvent, stabilizer, surface-active agent, surface-finishing agent, texturizer, thickener, vehicle.

Where Used: Baked goods (yeast raised), cakes, caramel, chewing gum, coffee whiteners, fats (rendered animal), frozen desserts, fudge, fudge sauces, lard, oleomargarine, peanut butter, poultry, shortening, whipped topping.

Regulations: FDA - 21CFR 172.863, 184.1505. GRAS when used in accordance with good manufacturing practice. USDA - 9CFR 318.7. Sufficient for purpose in lard and shortening, 0.5 percent in oleomargarine. 9CFR 381.147. Sufficient for purpose.

SAFETY PROFILE: When heated to decomposition it emits acrid smoke and irritating fumes.

MRH218
MONO- and DIGLYCERIDES, MONOSODIUM PHOSPHATE DERIVATIVES

SYN: MONOSODIUM PHOSPHATE DERIVATIVES of MONO- and DIGLYCERIDES

USE IN FOOD:

Purpose: Emulsifier, emulsifier salt, lubricant, release agent, surface-active agent.

Where Used: Candy (soft), dairy product analogs.

Regulations: FDA - 184.1521. GRAS when used in accordance with good manufacturing practice.

SAFETY PROFILE: When heated to decomposition it emits acrid smoke and irritating fumes.

MRH225
MONO-, DI-, and TRIPOTASSIUM CITRATE

USE IN FOOD:

Purpose: Stabilizer.

Where Used: Packaging materials.

Regulations: FDA - 21CFR 181.29. Use in accordance with good manufacturing practice.

SAFETY PROFILE: When heated to decomposition it emits acrid smoke and irritating fumes.

MRH230
MONO-, DI-, and TRISODIUM CITRATE

USE IN FOOD:

Purpose: Curing accelerator, stabilizer.

Where Used: Packaging materials, poultry.

Regulations: FDA - 21CFR 181.29. Use in accordance with good manufacturing practice. 9CFR 381.147. May be used to replace up to 50 percent of the ascorbic acid or sodium ascorbate in poultry.a

SAFETY PROFILE: When heated to decomposition it emits acrid smoke and irritating fumes.

MRH235
MONO-, DI-, and TRISTEARYL CITRATE

USE IN FOOD:

Purpose: Flavor preservative, plasticizer, sequestrant.

Where Used: Oleomargarine, packaging materials.

Regulations: FDA - 21CFR 181.27. Use in accordance with good manufacturing practice. 21CFR 182.6851. Limitation of 0.15 percent as a sequestrant.

SAFETY PROFILE: When heated to decomposition it emits acrid smoke and irritating fumes.

MRH500 CAS: 141-43-5
MONOETHANOLAMINE

DOT: 2491
mf: C_2H_7NO mw: 61.10

PROP: Colorless liquid; ammoniacal odor. Hygroscopic, bp: 170.5°, fp: 10.5°, flash p: 200°F (OC), d: 1.0180 @ 20°/4°, vap press: 6 mm @ 60°, vap d: 2.11. Misc in water and alc; sltly sol in benzene; sol in chloroform.

SYNS: AETHANOLAMIN (GERMAN) ◇ 2-AMINOAETHANOL (GERMAN) ◇ 2-AMINOETANOLO (ITALIAN) ◇ 2-AMINOETHANOL ◇ β-AMINOETHYL ALCOHOL ◇ COLAMINE ◇ ETANOLAMINA (ITALIAN) ◇ ETHANOLAMINE (ACGIH, DOT) ◇ β-ETHANOLAMINE ◇ ETHANOLAMINE SOLUTION (DOT) ◇ ETHYLOLAMINE ◇ GLYCINOL ◇ 2-HYDROXYETHYLAMINE ◇ β-HYDROXYETHYLAMINE ◇ MEA ◇ MONOAETHANOLAMIN (GERMAN) ◇ OLAMINE ◇ THIOFACO M-50 ◇ USAF EK-1597

USE IN FOOD:

Purpose: Animal glue adjuvant, flume wash water additive.

Where Used: Packaging materials, sugar beets.

Regulations: FDA - 21CFR 173.315. Limitation of 0.3 ppm in wash water. 21CFR 178.3120. Use at a level not in excess of the amount reasonably required to accomplish the intended effect.

OSHA PEL: TWA 3 ppm ACGIH TLV: TWA 3 ppm; STEL 6 ppm DOT Classification: Corrosive Material; Label: Corrosive

SAFETY PROFILE: Poison by intraperitoneal route. Moderately toxic by ingestion, skin contact, subcutaneous, intravenous, and intramuscular routes. A corrosive irritant to skin, eyes, and mucous membranes. Flammable when exposed to heat or flame. A powerful, reactive base. To fight fire, use foam, alcohol foam, dry chemical. When heated to decomposition it emits toxic fumes of NO_x.

TOXICITY DATA and CODEN

skn-rbt 505 mg open MOD UCDS** 1/13/73
eye-rbt 763 μg SEV AJOPAA 29,1363,46
orl-rat LD50:2140 mg/kg 34ZIAG -,706,69
skn-rat LD50:1500 mg/kg GTPZAB 23(9),55,79

MRI300
MONOGLYCERIDE CITRATE

PROP: Soft, white-colored, lard-like, waxy solid; bland odor and taste. Sol in fat solvents, alc; insol in water.

USE IN FOOD:

Purpose: Antioxidant, synergist for antioxidants.

Where Used: Fats, fats (poultry), lard, margarine, oils, oleomargarine, sausage (fresh pork), meat (dried), shortening.

Regulations: FDA - 21CFR 172.832. Limitation of 200 ppm of the combined weight of the oil or fat and the additive. USDA - 9CFR 318.7, 381.147. Sufficient for purpose. Limitation of 0.5 percent in oleomargarine or margarine. Limitation of 0.02 percent in lard, shortening, poultry fats, fresh pork sausage, dried meats.

SAFETY PROFILE: When heated to decomposition it emits acrid smoke and irritating fumes.

MRI785
MONOISOPROPYL CITRATE

USE IN FOOD:

Purpose: Plasticizer, sequestrant, synergist for antioxidants.

Where Used: Fats (poultry), lard, meat (dried), oleomargarine, packaging materials, sausage (fresh pork), shortening.

Regulations: FDA - 21CFR 181.27, 182.6511. GRAS when used in accordance with good manufacturing practice. USDA - 9CFR 318.7. Limitation of 0.02 percent. 9CFR 381.147. Limitation of 0.01 percent in poultry fats.

SAFETY PROFILE: When heated to decomposition it emits acrid smoke and irritating fumes.

MRK500 CAS: 19473-49-5
MONOPOTASSIUM GLUTAMATE
mf: $C_5H_8NO_4 \cdot K$ mw: 185.24

PROP: White, free-flowing, hygroscopic crystalline powder; practically odorless. Freely sol in water, sltly sol in alc.

SYNS: l-GLUTAMIC ACID, MONOPOTASSIUM SALT ◇ MONOPOTASSIUM l-GLUTAMATE (FCC) ◇ MPG ◇ POTASSIUM GLUTAMATE ◇ POTASSIUM GLUTAMINATE

USE IN FOOD:

Purpose: Flavor enhancer, salt substitute.

Where Used: Meat.

Regulations: FDA - 21CFR 182.1516. GRAS when used at a level not in excess of the amount reasonably required to accomplish the intended effect.

EPA Genetic Toxicology Program.

SAFETY PROFILE: Mildly toxic by ingestion and possibly other routes. Human systemic effects by ingestion: headache. When heated to decomposition it emits toxic fumes of K_2O and NO_x.

TOXICITY DATA and CODEN

orl-hmn TDLo:57 mg/kg:CNS SCIEAS 163,826,69
orl-mus LD50:4500 mg/kg FATOAO 42,274,79

MRL500 CAS: 142-47-2
MONOSODIUM GLUTAMATE
mf: $C_5H_9NO_4 \cdot Na$ mw: 170.14

PROP: White or almost white crystals or powder; slt peptone-like odor, meal-like taste. Very sol in water; sltly sol in alc.

SYNS: ACCENT ◇ AJINOMOTO ◇ CHINESE SEASONING ◇ GLUTACYL ◇ GLUTAMIC ACID, SODIUM SALT ◇ GLUTAMMATO MONOSODICO (ITALIAN) ◇ GLUTAVENE ◇ MONOSODIOGLUTAMMATO (ITALIAN) ◇ MONOSODIUM-l-GLUTAMATE (FCC) ◇ α-MONOSODIUM GLUTAMATE ◇ MSG ◇ NATRIUMGLUTAMINAT (GERMAN) ◇ RL-50 ◇ SODIUM GLUTAMATE ◇ SODIUM l-GLUTAMATE ◇ l(+) SODIUM GLUTAMATE ◇ VETSIN ◇ ZEST

USE IN FOOD:

Purpose: Flavor enhancer.

Where Used: Meat, poultry, sauces, soups.

Regulations: FDA - 21CFR 182. GRAS when used at a level not in excess of the amount reasonably required to accomplish the intended effect. USDA - 9CFR 318.7, 381.147. Sufficient for purpose.

EPA Genetic Toxicology Program.

SAFETY PROFILE: Moderately toxic by intravenous route. Mildly toxic by ingestion and other routes. An experimental teratogen. Human systemic effects by ingestion and intravenous routes: somnolence, hallucinations and distorted perceptions, headache, dyspnea, nausea or vomiting. Experimental reproductive effects. When heated to decomposition it emits toxic fumes of NO_x and Na_2O.

TOXICITY DATA and CODEN

orl-rat TDLo:1315 g/kg (MGN):TER EXPEAM 28,260,72
orl-rat TDLo:48 g/kg (14D pre-21D post):REP TXAPA9 50,267,79
scu-mus TDLo:5104 mg/kg (17D preg):TER SEIJBO 14,77,74
orl-wmn TDLo:50 mg/kg:PUL NEJMAG 305,1154,81
orl-hmn TDLo:43 mg/kg:CNS,GIT HYSAAV 36(9),364,71
orl-rat LD50:17 g/kg FRPPAO 27,19,72

MRL750 CAS: 2163-80-6
MONOSODIUM METHYLARSONATE
mf: $CH_4AsO_3 \cdot Na$ mw: 161.96

SYNS: ANSAR 170 ◇ ARSONATE LIQUID ◇ ASAZOL ◇ BUENO ◇ DACONATE 6 ◇ DAL-E-RAD ◇ HERB-ALL ◇ HERBAN M ◇ MERGE ◇ MESAMATE ◇ MESAMATE CONCENTRATE ◇ METHYLARSENIC ACID, SODIUM SALT ◇ MONATE ◇ MONOSODIUM ACID METHANEARSONATE ◇ MONOSODIUM ACID METHARSONATE ◇ MONOSODIUM METHANEARSONATE ◇ MONOSODIUM METHANEARSONIC ACID ◇ MSMA ◇ NCI-C60071 ◇ PHYBAN ◇ SILVISAR 550 ◇ SODIUM ACID METHANEARSONATE ◇ SODIUM METHANEARSONATE ◇ TARGET MSMA ◇ TRANS-VERT ◇ WEED 108 ◇ WEED-E-RAD ◇ WEED-HOE

USE IN FOOD:

Purpose: Herbicide.

Where Used: Animal feed, cottonseed hulls.

Regulations: FDA - 21CFR 561.280. Limitation of 0.9 ppm as As_2O_3 in cottonseed hulls when used for animal feed.

Arsenic and its compounds are on the Community Right-To-Know List. EPA Genetic Toxicology Program.

OSHA PEL: TWA 0.5 mg(As)/m^3 ACGIH TLV: TWA 0.2 mg(As)/m^3

SAFETY PROFILE: Poison by unspecified route. Moderately toxic by ingestion. A skin and eye irritant. When heated to decomposition it emits toxic fumes of As and Na_2O.

TOXICITY DATA and CODEN

skn-rbt 54 mg open MLD CIGET* -,-,77
eye-rbt 34 mg MLD CIGET* -,-,77
orl-rat LD50:700 mg/kg FMCHA2 -,C163,83

MRN260 CAS: 26155-31-7
MORANTEL TARTRATE
mf: $C_{12}H_{16}N_2S{\bullet}C_4H_6O_6$ mw: 370.46

SYNS: BANMINTH II ◇ MORANTREL TARTRATE

USE IN FOOD:

Purpose: Animal drug.

Where Used: Beef, milk.

Regulations: FDA - 21CFR 556.425. Limitation of 0.70 ppm of N-methyl-1,3-propanediamine in liver of cattle. Limitation of 0.4 ppm in milk. 21CFR 558.360.

SAFETY PROFILE: When heated to decomposition it emits acrid smoke and irritating fumes.

TOXICITY DATA and CODEN

orl-rat LD50:926 mg/kg AUVJA2 46,297,70

MRP750 CAS: 110-91-8
MORPHOLINE
DOT: 2054/1760
mf: C_4H_9NO mw: 87.14

PROP: Colorless, hygroscopic oil; amine odor. Bp: 128.9°, fp: −7.5°, flash p: 100°F (OC), autoign temp: 590°F, vap press: 10 mm @ 23°, vap d: 3.00, mp: −4.9°, d: 1.007 @ 20°/4°. Volatile with steam; misc with H_2O evolving some heat; misc with acetone, benzene, ether, castor oil, methanol, ethanol, ethylene, glycol, linseed oil, turpentine, pine oil. Immiscible with concentrated NaOH solns.

SYNS: DIETHYLENEIMIDE OXIDE ◇ DIETHYLENE IMIDOXIDE ◇ DIETHYLENE OXIMIDE ◇ DIETHYLENIMIDE OXIDE ◇ MORPHOLINE, AQUEOUS MIXTURE (DOT) ◇ 1-OXA-4-AZACYCLOHEXANE ◇ TETRAHYDRO-p-ISOXAZINE ◇ TETRAHYDRO-1,4-ISOXAZINE ◇ TETRAHYDRO-1,4-OXAZINE ◇ TETRAHYDRO-2H-1,4-OXAZINE

USE IN FOOD:

Purpose: Boiler water additive, protective coating.

Where Used: Fruits (fresh), vegetables (fresh).

Regulations: FDA - 21CFR 172.235. Limited use as one of the salts of fatty acids meeting the requirements of 21CFR 172.860. FDA - 21CFR 173.310. Limitation of 10 ppm in steam and excluding use of such steam in contact with milk and milk products.

EPA Genetic Toxicology Program.

OSHA PEL: TWA 20 ppm (skin); STEL 30 ppm (skin) ACGIH TLV: TWA 20 ppm; STEL 30 ppm (skin) DFG MAK: 20 ppm (70 mg/m^3) DOT Classification: Flammable Liquid; Label: Flammable Liquid; Corrosive Material; Label: Corrosive, aqueous solution.

SAFETY PROFILE: Moderately toxic by ingestion, inhalation, skin contact, intraperitoneal and possibly other routes. An experimental neoplastigen. Mutagenic data. A corrosive irritant to skin, eyes, and mucous membranes. Can cause kidney damage. Flammable liquid. A very dangerous fire hazard when exposed to flame, heat, or oxidizers; can react with oxidizing materials. To fight fire, use alcohol foam, CO_2, dry chemical. Mixtures with nitromethane are explosive. May ignite spontaneously in contact with cellulose nitrate of high surface area. When heated to decomposition it emits highly toxic fumes of NO_x.

TOXICITY DATA and CODEN

skn-rbt 995 mg/24H SEV BIOFX* 10-4/70
eye-rbt 2 mg SEV AJOPAA 29,1363,46
otr-mus:lym 1 μL/L ENMUDM 4,390,82
orl-mus TDLo:2560 mg/kg/Y-C:NEO GISAAA 44(8),15,79
orl-rat LD50:1050 mg/kg UCDS** 4/21/67
ihl-rat LC50:8000 ppm/8H NPIRI* 1,85,74

MRW775
MYCLOBUTANIL

SYN: α-BUTYL-α(4-CHLOROPHENYL)-1H-1,2,4-THIAZOLE-1-PROPANENITRILE

USE IN FOOD:

Purpose: Fungicide.

Where Used: Raisins.

Regulations: FDA - 21CFR 193.477. Limitation of 5 ppm in raisins. (Expires 2/28/1988)

SAFETY PROFILE: When heated to decomposition emits toxic fumes of NO_x, SO_x, Cl^-.

MRZ150 CAS: 123-35-3
MYRCENE
mf: $C_{10}H_{16}$ mw: 136.26

PROP: Colorless to pale yellow liquid; sweet, balsamic odor. D: 0.789, refr index: 1.466-1.471, flash p: 99°F. Sol in alc, fixed oils; insol in water.

SYNS: FEMA No. 2762 ◇ 3-METHYLENE-7-METHYL-1,6-OCTADIENE ◇ 7-METHYL-3-METHYLENE-1,6-OCTADIENE

USE IN FOOD:

Purpose: Flavoring agent.

Where Used: Various.

Regulations: FDA - 21CFR 172.515. Use at a level not in excess of the amount reasonably required to accomplish the intended effect.

SAFETY PROFILE: A moderate skin and eye irritant. Flammable liquid. When heated to decomposition it emits acrid smoke and irritating fumes.

TOXICITY DATA and CODEN

skn-rbt 500 mg/24H MOD FCTXAV 14,615,76

MSA250 CAS: 544-63-8
MYRISTIC ACID
mf: $C_{14}H_{28}O_2$ mw: 228.36

PROP: White or faintly yellow crystals from methanol. Mp: 58.5°, bp: 250.5° @ 100 mm, d: 0.8622 @ 54°/4°. Sol in abs alc, methanol, ether, petroleum ether, benzene, chloroform; insol in water.

SYNS: CRODACID ◇ EMERY 655 ◇ HYDROFOL ACID 1495 ◇ HYSTRENE 9014 ◇ TETRADECANOIC ACID ◇ n-TETRADECOIC ACID ◇ 1-TRIDECANECARBOXYLIC ACID ◇ UNIVOL U 316S

USE IN FOOD:

Purpose: Component in the manufacture of other food-grade additives, defoaming agent, lubricant.

Where Used: Various.

Regulations: FDA - 21CFR 172.860. Use in accordance with good manufacturing practice.

SAFETY PROFILE: Poison by intravenous route. Mutagenic data. A human skin irritant. When heated to decomposition it emits acrid smoke and irritating fumes.

TOXICITY DATA and CODEN

skn-hmn 75 mg/3D-I MOD 85DKA8 -,127,77
sln-smc 2500 ppb ANYAA9 407,186,83
ivn-mus LD50:43 mg/kg APTOA6 18,141,61

MSB775 CAS: 9000-45-7
MYRRH OIL

PROP: From steam distillation of myrrh gum from *Commiphora* (Fam. *Burseraceae*). Light brown to green liquid; characteristic odor. Sol in fixed oils, sltly sol in mineral oil; insol in glycerin, propylene glycol.

USE IN FOOD:

Purpose: Flavoring agent.

Where Used: Various.

Regulations: FDA - 21CFR 172.510. Use at a level not in excess of the amount reasonably required to accomplish the intended effect.

SAFETY PROFILE: When heated to decomposition it emits acrid smoke and irritating fumes.

N

NAR500 CAS: 61789-51-3
NAPHTHENIC ACID, COBALT SALT
DOT: 2001

PROP: Brown, amorphous powder or bluish-red solid. Flash p: 120°F, d: 0.9, autoign temp: 529°F. Water-insol; sol in oil, alc, ether. Contains 6% cobalt (AMIHAB 12,477,55).

SYNS: COBALT NAPHTHENATE, POWDER (DOT) ◇ NAPHTHENATE de COBALT (FRENCH)

USE IN FOOD:

Purpose: Drying agent.

Where Used: Packaging materials.

Regulations: FDA - 21CFR 181.25.

Cobalt and its compounds are on the Community Right-To-Know List.

OSHA PEL: TWA 0.1 mg(Co)/m^3 (fume and dust) DOT Classification: Flammable Solid; Label: Flammable Solid NIOSH REL: (Cobalt) TWA 0.1 mg/m^3

SAFETY PROFILE: Moderately toxic by ingestion. Flammable when exposed to heat or flame. When heated to decomposition it emits acrid smoke and irritating fumes.

TOXICITY DATA and CODEN

orl-rat LD50:3900 mg/kg AMIHAB 12,477,55

NBO600 CAS: 55134-13-9
NARASIN
mf: $C_{43}H_{72}O_{11}$ mw: 765.05

PROP: Crystals. Mp: 98-100°. Sol in alcs, acetone, chloroform, ethyl acetate. Insol in water.

SYNS: 4-METHYLSALINOMYCIN ◇ MONTEBAN

USE IN FOOD:

Purpose: Animal drug.

Where Used: Chicken.

Regulations: FDA - 21CFR 556.428. Limitation of 0.6 ppm in muscle, 1.8 ppm in liver, 1.3 ppm in skin and fat of chickens. 21CFR 558.363.

SAFETY PROFILE: When heated to decomposition it emits acrid smoke and irritating fumes.

NBR000 CAS: 500-38-9
NDGA
mf: $C_{18}H_{22}O_4$ mw: 302.40

PROP: Crystals from acetic acid. Mp: 184-185°. Sol in methanol, ethanol, and ether; sltly sol in hot water and chloroform; nearly insol in benzene and petroleum ether.

SYNS: 1,4-BIS(3,4-DIHYDROXYPHENYL)-2,3-DIMETHYLBUTANE ◇ DIHYDRONORGUAIARETIC ACID ◇ β,γ-DIMETHYL-α,Δ-BIS(3,4-DIHYDROXYPHENYL)BUTANE ◇ 4,4′-(2,3-DIMETHYLTETRAMETHYLENE)DIPYROCATECHOL ◇ NORDIHYDROGUAIARETIC ACID ◇ NORDIHYDROGUAIRARETIC ACID

USE IN FOOD:

Purpose: Antioxidant.

Where Used: Prohibited from foods.

Regulations: 21CFR 181.24. Limitation of 0.005 percent migrating from food packages. 21CFR 189.165. Prohibited from direct addition or use in human food.

SAFETY PROFILE: Moderately toxic by intraperitoneal route. When heated to decomposition it emits acrid smoke and irritating fumes.

TOXICITY DATA and CODEN

orl-rat LD50:2 g/kg JAOCA7 54,239,77

NBU500 CAS: 16595-80-5
NEMICIDE
mf: $C_{11}H_{12}N_2S \cdot ClH$ mw: 240.77

SYNS: CITARIN L ◇ DECARIS ◇ IMIDAZO(2,1-β)THIAZOLE MONOHYDROCHLORIDE ◇ KW-2-LE-T ◇ LEVAMISOLE ◇ LEVAMISOLE HYDROCHLORIDE (USDA) ◇ LEV HYDROCHLORIDE ◇ LEVOMYSOL HYDROCHLORIDE ◇ NIRATIC HYDROCHLORIDE ◇ NIRATIC-PURON HYDROCHLORIDE ◇ NSC-177023 ◇ R-12,564 ◇ RIPERCOL-L ◇ SOLASKIL ◇ STIMAMIZOL HYDROCHLORIDE ◇ (−)-2,3,5,6-TETRAHYDRO-6-PHENYLIMIDAZO(2,1-b)THIAZOLE HYDROCHLORIDE ◇ l-(−)-2,3,5,6-TETRAHYDRO-6-PHENYL-IMIDAZO(2,1-B)THIAZOLE HYDROCHLORIDE ◇ 1-TETRAMISOLE HYDROCHLORIDE ◇ TRAMISOL ◇ TRAMISOLE ◇ WORM-CHEK

USE IN FOOD:

Purpose: Animal drug, animal feed drug.

Where Used: Animal feed, beef, lamb, pork.

Regulations: FDA - 21CFR 556.350. Limitation of 0.1 ppm in cattle, sheep, and swine. 21CFR 558.315. (hydrochloride equivalent). USDA CES Ranking: C-2 (1985).

SAFETY PROFILE: Poison by ingestion, intraperitoneal, subcutaneous, intravenous, and intramuscular routes. An experimental teratogen. Human systemic effects by ingestion: thrombocytopenia. Experimental reproductive effects. Mutagenic data. When heated to decomposition it emits very toxic fumes of NO_x, SO_x and HCl.

TOXICITY DATA and CODEN

dns-mus-unr 10 mg/kg CCROBU 59,531,75
orl-rat TDLo:330 mg/kg (7-17D preg):TER YACHDS 10,3155,82
orl-rat TDLo:480 mg/kg (19-22D preg/4D post):REP YACHDS 10,3155,82
orl-rat TDLo:2125 mg/kg (19-22D preg/21D post):REP YACHDS 10,3155,82
orl-wmn TDLo:180 mg/kg/36D:BLD BMJOAE 2(6086),555,77
orl-rat LD50:1170 mg/kg CHTHBK 14(4),244,69

NCG000 CAS: 1405-10-3
NEOMYCIN SULFATE

SYNS: BIOSOL VETERINARY ◇ FRADIOMYCIN SULFATE ◇ LIDAMYCIN CREME ◇ MYCAIFRADIN SULFATE ◇ MYCIFRADIN-N ◇ MYCIGIENT ◇ NEOBIOTIC ◇ NEO-MANTLE CREME ◇ NEOMIX ◇ NEOMYCINE SULFATE ◇ NEOMYCIN SULPHATE ◇ OTOBIOTIC ◇ QUINTESS-N ◇ USAF CB-19

USE IN FOOD:

Purpose: Animal drug.

Where Used: Beef, milk.

Regulations: FDA - 21CFR 556.430. Limitation of 0.25 ppm in calves, 0.15 ppm in milk. USDA CES Ranking: B-3 (1986).

EPA Genetic Toxicology Program.

SAFETY PROFILE: Poison by intraperitoneal, intramuscular, intravenous, and subcutaneous routes. Human systemic effects by ingestion: somnolence, hallucinations and distorted perceptions and anorexia. A human skin irritant. When heated to decomposition it emits very toxic fumes of SO_x.

TOXICITY DATA and CODEN

skn-hmn 6 mg/3D-I MLD 85DKA8 -,127,77
orl-wmn TDLo:12600 mg/kg/7D:CNS ARSUAX 111,822,76
scu-mus LD50:190 mg/kg ANTBAL 18,444,73

NCN600 CAS: 13997-19-8
NEQUINATE

mf: $C_{22}H_{23}NO_4$ mw: 365.43

PROP: Crystals. Mp: 287-288°.

SYNS: 7-(BENZYLOXY)-6-N-BUTYL-1,4-DIHYDRO-4-OXO-3-QUINOLINECARBOXYLIC ACID METHYL ESTER ◇ 7-(BENZYLOXY)-6-N-BUTYL-4-HYDROXY-3-QUINOLINECARBOXYLIC ACID METHYL ◇ 3-METHOXYCARBONYL-6-N-BUTYL-7-BENZYLOXY-4-OXOQUINOLINE ◇ STATYL

USE IN FOOD:

Purpose: Animal drug.

Where Used: Chicken.

Regulations: FDA - 21CFR 556.440. Limitation of 0.1 ppm in chickens. 21CFR 558.365.

SAFETY PROFILE: When heated to decomposition it emits acrid smoke and irritating fumes.

NCN700 CAS: 7212-44-4
NEROLIDOL

mf: $C_{15}H_{26}O$ mw: 222.37

PROP: Colorless to straw colored liquid; faint floral, rose-like odor. D: 0.870-0.880, refr index: 1.478-1.483. Sol in fixed oils, propylene glycol; insol in glycerin.

SYNS: FEMA No. 2772 ◇ 3,7,11-TRIMETNYL-1,6,10-DODECATRIEN-3-OL

USE IN FOOD:

Purpose: Flavoring agent.

Where Used: Various.

Regulations: FDA - 21CFR 172.515. Use at a level not in excess of the amount reasonably required to accomplish the intended effect.

SAFETY PROFILE: When heated to decomposition it emits acrid smoke and irritating fumes.

NCR000 CAS: 98-92-0
NIACINAMIDE

mf: $C_6H_6N_2O$ mw: 122.14

PROP: Colorless needles or white crystalline powder; odorless with a bitter taste. Mp: 129°, d: 1.40. Very sol in water, ether, glycerin.

SYNS: ACID AMIDE ◇ AMIDE PP ◇ AMINICOTIN ◇ AMIXICOTYN ◇ AMNICOTIN ◇ AUSTROVIT PP

◇ BENICOT ◇ DELONIN AMIDE ◇ DIPEGYL ◇ DIPIGYL ◇ ENDOBION ◇ FACTOR PP ◇ HANSAMID ◇ INOVITAN PP ◇ NAM ◇ NANDERVIT-N ◇ NIACEVIT ◇ NIAMIDE ◇ NICAMIDE ◇ NICAMINA ◇ NICAMINDON ◇ NICASIR ◇ NICOBION ◇ NICOFORT ◇ NICOGEN ◇ NICOMIDOL ◇ NICOSAN 2 ◇ NICOTA ◇ NICOTAMIDE ◇ NICOTILAMIDE ◇ NICOTILILAMIDO ◇ NICOTINE ACID AMIDE ◇ NICOTINIC ACID AMIDE ◇ NICOTINIC AMIDE ◇ NICOTINSAUREAMID (GERMAN) ◇ NICOTOL ◇ NICOTYLAMIDE ◇ NICOVEL ◇ NICOVIT ◇ NICOVITOL ◇ NICOZYMIN ◇ NIKO-TAMIN ◇ NIKOTINSAEUREAMID (GERMAN) ◇ NIOCINAMIDE ◇ NIOZYMIN ◇ PELMIN ◇ PELMINE ◇ PELONIN AMIDE ◇ PP-FACTOR ◇ PYRIDINE-3-CARBOXYLIC ACID AMIDE ◇ 3-PYRIDINECARBOXYLIC ACID AMIDE ◇ VI-NICOTYL ◇ VI-NICTYL ◇ VITAMIN B3 ◇ VITAMIN PP ◇ WITAMINA PP

USE IN FOOD:

Purpose: Dietary supplement, nutrient.

Where Used: Various.

Regulations: FDA - 21CFR 182.5535, 184.1535. GRAS when used in accordance with good manufacturing practice.

SAFETY PROFILE: Moderately toxic by ingestion, intravenous, intraperitoneal, and subcutaneous routes. Mutagenic data. When heated to decomposition it emits toxic fumes of NO_x.

TOXICITY DATA and CODEN

dni-rat:lvr 20 mmol/L JJIND8 69,1353,82
scu-rat LD50:1680 mg/kg PSEBAA 62,19,46
orl-mus LD50:2500 mg/kg NIIRDN 6,545,82

NCR025
NIACINAMIDE ASCORBATE

PROP: Lemon yellow-colored powder; very slt odor. Mp: 141-145°. Sol in water, alc; sltly sol in chloroform, ether, glycerin; insol in benzene.

SYN: NIACINAMIDE ASCORBIC ACID COMPLEX

USE IN FOOD:

Purpose: Dietary supplement, nutrient.

Where Used: Multivitamin preparations.

Regulations: FDA - 21CFR 172.315.

SAFETY PROFILE: When heated to decomposition it emits acrid smoke and irritating fumes.

NCW100 CAS: 330-95-0
NICARBAZIN

mf: $C_{19}H_{18}N_6O_6$ mw: 426.38

PROP: Crystals. Decomp 265-275°. Insol in water.

SYNS: N,N-BIS(4-NITRPHENYL)UREA, compd with 4,6-DIMETHYL-2(1H)-PYRIMIDINONE (1:1) ◇ 4,4′-DINITROCARBANILIDE compd with 4,6-DIMETHYL-2-PYRIMIDINOL (1:1) ◇ NICARB ◇ NICOXIN ◇ NICRAZIN

USE IN FOOD:

Purpose: Animal drug.

Where Used: Chicken.

Regulations: FDA - 21CFR 556.445. Limitation of 4 ppm in uncooked muscle, liver, skin, and neck of chickens. 21CFR 558.366.

SAFETY PROFILE: When heated to decomposition it emits acrid smoke and irritating fumes.

NCW500 CAS: 7440-02-0
NICKEL

af: Ni aw: 58.71

PROP: A silvery-white, hard, malleable and ductile metal. D: 8.90 @ 25°, vap press: 1 mm @ 1810°. Crystallizes as metallic cubes. Mp: 1455°, bp: 2730°. Stable in air at room temp.

SYNS: C.I. 77775 ◇ Ni 270 ◇ NICKEL 270 ◇ NICKEL (DUST) ◇ NICKEL (ITALIAN) ◇ NICKEL PARTICLES ◇ NICKEL SPONGE ◇ Ni 0901-S ◇ Ni 4303T ◇ NP 2 ◇ RANEY ALLOY ◇ RANEY NICKEL

USE IN FOOD:

Purpose: Catalyst for the transesterification of fats.

Where Used: Fats (rendered animal), oils.

Regulations: FDA - 21CFR 184.1537 GRAS when used in accordance with good manufacturing practice. USDA - 9CFR 318. Must be eliminated during processing.

IARC Cancer Review: Animal Inadequate Evidence IMEMDT 2,126,73; Animal Sufficient Evidence IMEMDT 11,75,76. Community Right-To-Know List. EPA Extremely Hazardous Substances List.

OSHA PEL: TWA 1 mg(Ni)/m^3 ACGIH TLV: TWA 1 mg/m^3 NIOSH REL: (Inorganic Nickel) TWA 0.015 mg(Ni)/m^3

SAFETY PROFILE: Poison by ingestion, intratracheal, intraperitoneal, subcutaneous, and intravenous routes. An experimental carcinogen, neoplastigen, tumorigen, and teratogen. Experimental reproductive effects. Ingestion of soluble salts causes nausea, vomiting, diarrhea. Mutagenic data. Hypersensitivity to nickel is common and can cause allergic contact dermatitis, pulmonary asthma, conjunctivitis, and inflammatory reactions around nickel-containing medical implants and prostheses. Powders may ignite spontaneously in air. Incompatible with oxidants (e.g., bromine pentafluoride; peroxyformic acid; potassium perchlorate; chlorine; nitryl fluoride; ammonium nitrate).

TOXICITY DATA and CODEN

otr-ham:kdy 400 mg/L IAPUDO 53,193,84
orl-rat TDLo:158 mg/kg (MGN):TER AEHLAU 23,102,71
ims-rat TDLo:56 mg/kg:CAR IAPUDO 53,127,84
ipl-rat TDLo:100 mg/kg/21W-I:ETA PWPSA8 16,150,73
imp-rat TDLo:250 mg/kg:CAR JNCIAM 16,55,55
ims-rat TD :200 mg/kg/21W-I:NEO PWPSA8 14,68,71
ims-rat TD :1 g/kg/17W-I:CAR PAACA3 9,28,68
ivn-dog LDLo:10 mg/kg 14CYAT 2,1120,63
orl-gpg LDLo:5 mg/kg AMPMAR 25,247,64

NDT000 CAS: 59-67-6
NICOTINIC ACID

mf: $C_6H_5NO_2$ mw: 123.12

PROP: The anti-pellagra vitamin. Colorless needles or white crystalline powder; slt odor. Mp: 236°, subl above mp, d: 1.473. Sol in water and boiling alc; insol in most lipid solvents. Nonhygroscopic and stable in air.

SYNS: ACIDE NICOTINIQUE (FRENCH) ◇ ACIDUM NICOTINICUM ◇ AKOTIN ◇ ANTI-PELLAGRA VITAMIN ◇ APELAGRIN ◇ BIONIC ◇ 3-CARBOXYPYRIDINE ◇ DASKIL ◇ DAVITAMON PP ◇ DIREKTAN ◇ EFACIN ◇ NAH ◇ NAOTIN ◇ NIACIN (FCC) ◇ NICACID ◇ NICAMIN ◇ NICANGIN ◇ NICO ◇ NICO-400 ◇ NICOBID ◇ NICOCAP ◇ NICOCIDIN ◇ NICOCRISINA ◇ NICODAN ◇ NICODELMINE ◇ NICOLAR ◇ NICONACID ◇ NICONAT ◇ NICONAZID ◇ NICOROL ◇ NICOSIDE ◇ NICO-SPAN ◇ NICOSYL ◇ NICOTAMIN ◇ NICOTENE ◇ NICOTIL ◇ NICOTINE ACID ◇ NICOTINIPCA ◇ NICOTINOYL HYDRAZINE ◇ NICOTINSAURE (GERMAN) ◇ NICOVASAN ◇ NICOVASEN ◇ NICOVEL ◇ NICYL ◇ NIPELLEN ◇ PELLAGRAMIN ◇ PELLAGRA PREVENTIVE FACTOR ◇ PELLAGRIN ◇ PELONIN ◇ PEVITON ◇ PP FACTOR ◇ P.P. FACTOR-PELLAGRA PREVENTIVE FACTOR ◇ PYRIDINE-3-CARBONIC ACID ◇ PYRIDINE-β-CARBOXYLIC ACID ◇ PYRIDINE-3-CARBOXYLIC ACID ◇ 3-PYRIDINECARBOXYLIC ACID ◇ PYRIDINE-CARBOXYLIQUE-3 (FRENCH) ◇ S115 ◇ SK-NIACIN ◇ TINIC ◇ VITAPLEX N ◇ WAMPOCAP

USE IN FOOD:

Purpose: Dietary supplement, nutrient.

Where Used: Various.

Regulations: FDA - 21CFR 182.5530, 184.1530. GRAS when used in accordance with good manufacturing practice. May not be used on fresh meat.

SAFETY PROFILE: Poison by intraperitoneal route. Moderately toxic by ingestion, intravenous and subcutaneous routes. An experimental carcinogen. When heated to decomposition it emits toxic fumes of NO_x.

TOXICITY DATA and CODEN

orl-mus TDLo:174 g/kg/94W-C:CAR ONCOBS 38,106,81
orl-rat LD50:7000 mg/kg NIIRDN 6,544,82

NEB050 CAS: 1414-45-5
NISIN PREPARATION

mf: $C_{143}H_{230}N_{42}O_{37}S_7$ mw: 3354.25

PROP: Crystals from ethanol. Derived from *Streptoccus lactus* Lancefield Group N.

USE IN FOOD:

Purpose: Preservative, antimicrobial agent.

Where Used: Cheese spreads (pasteurized).

Regulations: FDA - 21CFR 184.1538(b)(d). Limitation of 250 ppm Nisin in the finished product.

SAFETY PROFILE: When heated to decomposition it emits acrid smoke and irritating fumes.

NEJ000
NITRITES

PROP: Salts of nitrous acid.

USE IN FOOD:

Purpose: Curing agent.

Where Used: Bacon, meat (cured), meat products, poultry products.

Regulations: FDA - 21CFR 170.60, 172.170, 172.175. USDA - 9CFR 318.7. The use of nitrites, nitrates, or combination shall not result in more than 200 ppm of nitrite, calculated as sodium nitrate in finished product.

SAFETY PROFILE: Large amounts taken by mouth may produce nausea, vomiting, cyanosis (due to methemoglobin formation), collapse, and coma. Repeated small doses cause a fall in blood pressure, rapid pulse, headache, and visual disturbances. They have been implicated in an increased incidence of cancer. They may react with organic amines in the body to form carcinogenic nitrosamines. Organic nitrites are used to treat angina pectoris. Fire hazards are variable. They are generally powerful oxidizers. On contact with readily oxidized materials, a violent reaction such as a fire or explosion may ensue. Explosion hazards are also variable. Organic nitrites may decompose violently in contact with NH_4; salts; cyanide; KCN. Dangerous; shock may explode them; can react vigorously with reducing materials. When heated to decomposition they emit highly toxic fumes of NO_x.

NGE500 CAS: 59-87-0
NITROFURAZONE
mf: $C_6H_6N_4O_4$ mw: 198.16

PROP: Odorless, lemon-yellow crystals; bitter aftertaste. Darkens upon prolonged exposure to light. Decomp @ 236-240°. Very sltly sol in water; sltly sol in alc, propylene glycol; sol in alkaline solns; insol in ether.

SYNS: ALDOMYCIN ◇ ALFUCIN ◇ AMIFUR ◇ BABROCID ◇ BIOFUREA ◇ CHEMOFURAN ◇ COCAFURIN ◇ COXISTAT ◇ DERMOFURAL ◇ DYNAZONE ◇ ELDEZOL ◇ FEDACIN ◇ FLAVAZONE ◇ FRACINE ◇ FURACILLIN ◇ FURACINETTEN ◇ FURACOCCID ◇ FURACORT ◇ FURACYCLINE ◇ FURALDON ◇ FURAN-OFTENO ◇ FURAPLAST ◇ FURASEPTYL ◇ FURAZONE ◇ FURESOL ◇ FURFURIN ◇ FUVACILLIN ◇ HEMOFURAN ◇ IBIOFURAL ◇ MAMMEX ◇ MONOFURACIN ◇ NCI-C56064 ◇ NEFCO ◇ NF ◇ NIFUZON ◇ 5-NITROFURALDEHYDE SEMICARBAZIDE ◇ 6-NITROFURALDEHYDE SEMICARBAZIDE ◇ 5-NITRO-2-FURALDEHYDE SEMICARBAZONE ◇ 5-NITROFURAN-2-ALDEHYDE SEMICARBAZONE ◇ 5-NITRO-2-FURANCARBOXALDEHYDE SEMICARBAZONE ◇ 2((5-NITRO-2-FURANYL)METHYLENE)HYDRAZINECARBOXAMIDE ◇ 5-NITROFURFURAL SEMICARBAZONE ◇ (5-NITRO-2-FURFURYLIDENEAMINO)UREA ◇ NITROZONE ◇ NSC-2100 ◇ OTOFURAN ◇ SANFURAN ◇ SPRAY-DERMIS ◇ SPRAY-FORAL ◇ U-6421 ◇ USAF EA-4 ◇ VABROCID ◇ VADROCID ◇ VETERINARY NITROFURAZONE ◇ YATROCIN

USE IN FOOD:

Purpose: Animal feed drug.

Where Used: Animal feed, pork, poultry.

Regulations: FDA - 21CFR 558.370. Use at a level not in excess of the amount reasonably required to accomplish the intended effect.

IARC Cancer Review: Animal Inadequate Evidence IMEMDT 7,171,74. EPA Genetic Toxicology Program.

SAFETY PROFILE: Poison by ingestion and intraperitoneal routes. Moderately toxic by subcutaneous route. An experimental neoplastigen, tumorigen, teratogen. Experimental reproductive effects. A human sensitizer. Human mutagenic data. When heated to decomposition it emits toxic fumes of NO_x.

TOXICITY DATA and CODEN

dnd-esc 1 mg/L MUREAV 107,1,83
dnr-bcs 10 μg/disc KSRNAM 17,70,83
orl-rat TDLo:700 mg/kg (7D male):REP APJAAG 14,261,64
scu-mus TDLo:300 mg/kg (10D preg):TER TJADAB 12,206,75
scu-mus TDLo:300 mg/kg (10D preg):TER SEIJBO 15,234,75
orl-rat TDLo:31 g/kg/45W-C:NEO FEPRA7 25,419,66
orl-rat TD :4500 mg/kg/27D-I:ETA CNREA8 28,924,68
orl-rat LD50:590 mg/kg JAMAAP 133,299,47

NGG500 CAS: 67-45-8
3-((5-NITROFURFURYLIDENE)AMINO)-2-OXAZOLIDONE
mf: $C_8H_7N_3O_5$ mw: 225.18

SYNS: BIFURON ◇ CORIZIUM ◇ DIAFURON ◇ ENTEROTOXON ◇ FURAXONE ◇ FURAZOL ◇ FURAZOLIDON ◇ FURAZOLIDONE (USDA) ◇ FURAZON ◇ FURIDON ◇ FUROVAG ◇ FUROX ◇ FUROXAL ◇ FUROXANE ◇ FUROXONE SWINE MIX ◇ FUROZOLIDINE ◇ GIARDIL ◇ GIARLAM ◇ MEDARON ◇ NEFTIN ◇ NG-180 ◇ NICOLEN ◇ NIFULIDONE ◇ NIFURAN ◇ 3-(((5-NITRO-2-FURANYL)METHYLENE)AMINO)-2-OXAZOLIDINONE ◇ NITROFURAZOLIDONE ◇ NITROFURAZOLIDONUM ◇ 3-(5′-NITROFURFURALAMINO)-2-OXAZOLIDONE ◇ NITROFUROXON ◇ N-(5-NITRO-2-FURFURYLIDENE)-3-AMINOOXAZOLIDINE-2-ONE ◇ N-(5-NITRO-2-FURFURYLIDENE)-3-AMINO-2-OXAZOLIDONE ◇ 3-((5-NITROFURYLIDENE)AMINO)-2-OXAZOLIDONE ◇ 5-NITRO-N-(2-OXO-3-OXAZOLIDINYL)-2-FURANMETHANIMINE ◇ PURADIN ◇ ROPTAZOL ◇ SCLAVENTEROL ◇ TIKOFURAN ◇ TOPAZONE ◇ TRICHOFURON ◇ TRICOFURON ◇ USAF EA-1 ◇ VIOFURAGYN

USE IN FOOD:

Purpose: Animal drug, animal feed drug.

Where Used: Animal feed, pork.

Regulations: FDA - 21CFR 556.290. Tolerance of zero in cooked edible tissues of swine. 21CFR 558.262. USDA CES Ranking: A-1 (1987).

IARC Cancer Review: Animal Inadequate Evidence IMEMDT 31,141,83. EPA Genetic Toxicology Program.

SAFETY PROFILE: Poison by ingestion and intraperitoneal routes. Human systemic effects by ingestion: dyspnea, respiratory depression and rosinophillis. Experimental reproductive effects. Human mutagenic data. When heated to decomposition it emits toxic fumes of NO_x.

TOXICITY DATA and CODEN

pic-esc 2 mg/L MUREAV 156,69,85
mmo-omi 7 mg/L CBINA8 45,315,83
orl-rat TDLo:700 mg/kg (7D male):REP APJAAG 14,261,64
orl-mus TDLo:1 g/kg (1D preg):REP JOENAK 15,355,57
orl-man TDLo:11 mg/kg:PUL,BLD ARDSBL 105,823,72
orl-rat LD50:2336 mg/kg TXAPA9 18,185,71

NGP500 CAS: 7727-37-9
NITROGEN

DOT: 1066/1977
af: N_2 aw: 28.02

PROP: Colorless gas, colorless liquid or cubic crystals at low temp. Mp: −210.0°, d: 1.2506 g/L @ 0°, d (liquid): 0.808 @ −195.8°. Condenses to a liquid; sltly sol in water; sol in liquid ammonia, alc.

SYNS: NITROGEN, COMPRESSED (DOT) ◇ NITROGEN GAS ◇ NITROGEN, REFRIGERATED LIQUID (DOT)

USE IN FOOD:

Purpose: Aerating agent, gas, modified atmospheres for insect control, oxygen exclusion, propellant.

Where Used: Fruit, poultry, various food in sealed containers, wine.

Regulations: FDA - 21CFR 184.1540. 21CFR 193.323. Modified atmospheres for insect control. USDA - 9CFR 318.7, 381.147. Sufficient for purpose.

DOT Classification: Nonflammable Gas; Label: Nonflammable Gas

SAFETY PROFILE: Low toxicity. In high concentrations it is a simple asphyxiant. The release of nitrogen from solution in the blood, with formation of small bubbles, is the cause of most of the symptoms and changes found in compressed air illness (caisson disease). It is a narcotic at high concentration and high pressure. Both the narcotic effects and the bends are hazards of compressed air atmospheres such as found in underwater diving. Nonflammable Gas. Can react violently with lithium; neodymium; titanium under the proper conditions.

NGU000 CAS: 10024-97-2
NITROGEN OXIDE

DOT: 1070/2201
mf: N_2O mw: 44.02

PROP: Colorless gas, liquid or cubic crystals; slt sweet odor. Mp: −90.8°, bp: −88.49°, d: 1.977 g/L (liquid 1.226 @ −89°).

SYNS: DINITROGEN MONOXIDE ◇ FACTITIOUS AIR ◇ HYPONITROUS ACID ANHYDRIDE ◇ LAUGHING GAS ◇ NITROUS OXIDE (DOT) ◇ NITROUS OXIDE, COMPRESSED (DOT) ◇ NITROUS OXIDE, REFRIGERATED LIQUID (DOT)

USE IN FOOD:

Purpose: Aerating agent, gas, propellant.

Where Used: Dairy product analogs, wine.

Regulations: FDA - 21CFR 184.1545 GRAS when used in accordance with good manufacturing practice. BATF - 27CFR 240.1051. GRAS when used in accordance with good manufacturing practice.

EPA Genetic Toxicology Program.

DOT Classification: Nonflammable Gas; Label: Nonflammable Gas, Oxidizer, compressed
NIOSH REL: (Waste Anesthetic Gases and Vapors) TWA 25 ppm

SAFETY PROFILE: Moderately toxic by inhalation. Human systemic effects by inhalation: general anesthetic, decreased pulse rate without blood pressure fall and body temperature decrease. An experimental teratogen. Experimental reproductive effects. Mutagenic data. An asphyxiant. Does not burn but is flammable by chemical reaction and supports combustion. Moderate explosion hazard; it can form an explosive mixture with air. Also self-explodes at high temperatures.

TOXICITY DATA and CODEN

sln-dmg-ihl 99 pph/6M-C ENVRAL 7,286,74
dni-rat-ihl 75000 ppm/24H AACRAT 62,738,83
ihl-rat TCLo:5000 ppm/6H (1-19D preg):TER BJANAD 55,67,83
ihl-rat TCLo:20 pph/8H (28D male):REP ANESAV 44,104,76

ihl-ham TCLo:95 pph/24H (7D preg):TER CJPPA3 57,1229,79
ihl-hmn TDLo:24 mg/kg/2H:CNS,CVS,MET BJANAD 35,631,63
ihl-rat LC50:1068 mg/m^3/4H TPKVAL 15,53,79

NHP100
NITROMIDE and SULFANITRAN

USE IN FOOD:
Purpose: Animal feed drug.
Where Used: Animal feed.
Regulations: FDA - 21CFR 558.376.

SAFETY PROFILE: When heated to decomposition it emits acrid smoke and irritating fumes.

NIJ500 CAS: 98-72-6
4-NITROPHENYLARSONIC ACID
mf: $C_6H_6AsNO_5$ mw: 247.05

SYNS: NITARSONE ◇ 4-NITROBENZENEARSONIC ACID ◇ p-NITROPHENYLARSONIC ACID ◇ RAS-26

USE IN FOOD:
Purpose: Animal feed drug.
Where Used: Animal feed.
Regulations: FDA - 21CFR 558.369. Use at a level not in excess of the amount reasonably required to accomplish the intended effect.

Arsenic and its compounds are on the Community Right-To-Know List.

OSHA PEL: TWA 0.5 mg(As)/m^3 ACGIH TLV: TWA 0.2 mg(As)m^3

SAFETY PROFILE: Poison by ingestion and intravenous routes. When heated to decomposition it emits very toxic fumes of NO_x and As.

TOXICITY DATA and CODEN
orl-rat LDLo:100 mg/kg NCNSA6 5,13,53

NIY525 CAS: 553-79-7
5-NITRO-2-n-PROPOXYANILINE
mf: $C_9H_{12}N_2O_3$ mw: 196.20

SYN: P-4000

USE IN FOOD:
Purpose: Artificial sweetener.
Where Used: Prohibited from foods.
Regulations: FDA - 21CFR 189.175. Prohibited from foods.

SAFETY PROFILE: When heated to decomposition emits toxic fumes of NO_x.

NMV760 CAS: 557-48-2
trans,cis-2,6-NONADIENAL
mf: $C_9H_{14}O$ mw: 138.23

PROP: Slightly yellow liquid; powerful, violet, cucumber odor. D: 0.850-0.870, refr index: 1.470. Sol in alc, fixed oils; insol in water.

SYNS: CUCUMBER ALDEHYDE ◇ FEMA No. 3317 ◇ trans-2,cis-6-NONADIENAL ◇ 2,6-NONADIENAL ◇ VIOLET LEAF ALDEHYDE

USE IN FOOD:
Purpose: Flavoring agent.
Where Used: Various.
Regulations: FDA - 21CFR 172.515. Use at a level not in excess of the amount reasonably required to accomplish the intended effect.

SAFETY PROFILE: A moderate skin irritant. When heated to decomposition it emits acrid smoke and irritating fumes.

TOXICITY DATA and CODEN
skn-rbt 500 mg/24H MOD FCTOD7 20,769,82

NMV775 CAS: 5910-87-2
trans,trans-2,4-NONADIENAL
mf: $C_9H_{14}O$ mw: 138.21

PROP: Slightly yellow liquid; strong, fatty, floral odor. D: 0.850-0.870, refr index: 1.522. Sol in alc, fixed oils; insol in water.

SYN: FEMA No. 3212

USE IN FOOD:
Purpose: Flavoring agent.
Where Used: Various.
Regulations: FDA - 21CFR 172.515. Use at a level not in excess of the amount reasonably required to accomplish the intended effect.

SAFETY PROFILE: When heated to decomposition it emits acrid smoke and irritating fumes.

NMV780
trans,cis-2,6-NONADIENOL
mf: $C_9H_{16}O$ mw: 140.22

PROP: White to yellow liquid; powerful, vegetable odor. D: 0.860-0.880, refr index: 1.464. Insol in water.

SYNS: CUCUMBER ALCOHOL ◇ FEMA No. 2780

USE IN FOOD:
Purpose: Flavoring agent.

Where Used: Various.

Regulations: FDA - 21CFR 172.515. Use at a level not in excess of the amount reasonably required to accomplish the intended effect.

SAFETY PROFILE: When heated to decomposition it emits acrid smoke and irritating fumes.

NMV790
Δ-NONALACTONE
mf: $C_9H_{16}O_2$ mw: 156.25

SYNS: ALDEHYDE C-18 ◇ Δ-N-AMYLBUTYROLACTONE ◇ FEMA No. 3356 ◇ 5-HYDROXYNONANOIC ACID, LACTONE ◇ 1,5-NONALOLIDE

PROP: Colorless to pale yellow liquid; coconut odor. D: 0.980, refr index: 1.452.

USE IN FOOD:

Purpose: Flavoring agent.

Where Used: Various.

Regulations: FDA - 21CFR 172.515. Use at a level not in excess of the amount reasonably required to accomplish the intended effect.

SAFETY PROFILE: When heated to decomposition it emits acrid smoke and irritating fumes.

NMW500 CAS: 124-19-6
1-NONANAL
mf: $C_9H_{18}O$ mw: 142.27

PROP: Found in at least 20 essential oils, including rose and citrus oils and several species of pine oil (FCTXAV 11, 95,73). Colorless to light yellow liquid; citrus-rose odor. D: 0.820-0.830, refr index: 1.422-1.429, flash p: 162°F. Sol in alc; fixed oils, propylene glycol; insol in glycerin.

SYNS: ALDEHYDE C-9 ◇ C-9 ALDEHYDE ◇ FEMA No. 2782 ◇ NCI-C61018 ◇ 1-NONALDEHYDE ◇ 1-NONYL ALDEHYDE ◇ PELARGONIC ALDEHYDE

USE IN FOOD:

Purpose: Flavoring agent.

Where Used: Various.

Regulations: FDA - 21CFR 172.515. Use at a level not in excess of the amount reasonably required to accomplish the intended effect.

SAFETY PROFILE: A severe skin irritant. Combustible liquid. When heated to decomposition it emits acrid smoke and irritating fumes.

TOXICITY DATA and CODEN

skn-rbt 500 mg/24H SEV FCTXAV 11,1079,73

NNA300 CAS: 2463-53-8
2-NONENAL
mf: $C_9H_{16}O$ mw: 140.25

PROP: White to sltly yellow liquid; fatty, violet odor. D: 0.850-0.870, refr index: 1.457. Sol in alc, fixed oils; insol in water.

SYNS: FEMA No. 3213 ◇ HEPTYLIDENE ALDEHYDE ◇ β-HEXYLACROLEIN ◇ 2-NONEN-1-AL ◇ α-NONENYL ALDEHYDE ◇ trans-2-NONENAL (FCC)

USE IN FOOD:

Purpose: Flavoring agent.

Where Used: Various.

Regulations: GRAS when used at a level not in excess of the amount reasonably required to accomplish the intended effect.

SAFETY PROFILE: Moderately toxic by skin contact. Mildly toxic by ingestion. A severe skin irritant. When heated to decomposition it emits acrid smoke and irritating fumes.

TOXICITY DATA and CODEN

skn-rbt 500 mg/24H SEV FCTOD7 20(Suppl),775,82
orl-rat LD50:5 g/kg FCTOD7 20(Suppl),775,82
skn-rbt LD50:3700 mg/kg FCTOD7 20(Suppl),775,82

NNA530
cis-6-NONEN-1-OL
mf: $C_9H_{18}O$ mw: 142.23

PROP: White to slightly yellow liquid; powerful, melonlike odor. D: 0.850-0.870, refr index: 1.448. Insol in water.

SYN: FEMA No. 3465

USE IN FOOD:

Purpose: Flavoring agent.

Where Used: Various.

Regulations: GRAS when used at a level not in excess of the amount reasonably required to accomplish the intended effect.

SAFETY PROFILE: When heated to decomposition it emits acrid smoke and irritating fumes.

NNA532
trans-2-NONEN-1-OL
mf: $C_9H_{18}O$ mw: 142.23

PROP: White liquid; fatty, violet odor. D: 0.830-0.850, refr index: 1.444-1.448. Insol in water.

SYN: FEMA No. 3379

USE IN FOOD:

Purpose: Flavoring agent.

Where Used: Various.

Regulations: GRAS when used at a level not in excess of the amount reasonably required to accomplish the intended effect.

SAFETY PROFILE: When heated to decomposition it emits acrid smoke and irritating fumes.

NNB400 CAS: 143-13-5
NONYL ACETATE
mf: $C_{11}H_{22}O_2$ mw: 186.29

PROP: Colorless liquid; fruity odor. D: 0.864, refr index: 1.422, flash p: +153°F. Sol in alc, ether; insol in water.

SYN: FEMA No. 2788

USE IN FOOD:

Purpose: Flavoring agent.

Where Used: Various.

Regulations: FDA - 21CFR 172.515. Use at a level not in excess of the amount reasonably required to accomplish the intended effect.

SAFETY PROFILE: Combustible liquid. When heated to decomposition it emits acrid smoke and irritating fumes.

NNB500 CAS: 143-08-8
n-NONYL ALCOHOL
mf: $C_9H_{20}O$ mw: 144.29

PROP: Colorless liquid; rose-citrus odor. D: 0.827 @ 20°/4°, refr index: 1.43-1.435, mp: −5°, bp: 213.5°, flash p: 169°F. Insol in water; misc in alc, ether, chloroform.

SYNS: ALCOHOL C-9 ◇ FEMA No. 2789 ◇ NONALOL ◇ 1-NONANOL ◇ NONAN-1-OL ◇ NONYL ALCOHOL ◇ OCTYL CARBINOL ◇ PELARGONIC ALCOHOL

USE IN FOOD:

Purpose: Flavoring agent.

Where Used: Various.

Regulations: FDA - 21CFR 172.515. Use at a level not in excess of the amount reasonably required to accomplish the intended effect.

SAFETY PROFILE: Moderately toxic by ingestion. Mildly toxic by skin contact and inhalation. Combustible liquid. When heated to decomposition it emits acrid smoke and irritating fumes.

TOXICITY DATA and CODEN

orl-rat LDLo:1400 mg/kg 14CYAT 2,1466,63
ihl-mus LC50:5500 mg/m^3/2H 85GMAT -,94,82
skn-rbt LD50:5660 mg/kg FCTXAV 11,95,73

NNQ100
NORFLURAZON

SYN: 4-CHLORO-5-(METHYLAMINO)-2-(α,α,α-TRI-FLUORO-m-TOLYL)-3(2H)-PYRIDAZINONE

USE IN FOOD:

Purpose: Herbicide.

Where Used: Animal feed, citrus pulp (dried), hops (dried).

Regulations: FDA - 21CFR 193.324. Limitation of 3.0 ppm in dried hops. 21CFR 561.283. Limitation of 0.4 ppm in dried citrus pulp when used for animal feed.

SAFETY PROFILE: When heated to decomposition emits toxic fumes of NO_x, F^-, and Cl^-.

NOB000 CAS: 1476-53-5
NOVOBIOCIN, MONOSODIUM SALT
mf: $C_{31}H_{35}N_2O_{11}$•Na mw: 634.67

SYNS: ALBAMYCIN ◇ ALBAMYCIN SODIUM ◇ CATHOMYCIN SODIUM ◇ CATHOMYCIN SODIUM LYOVAC ◇ INAMYCIN ◇ MONOSODIUM NOVOBIOCIN ◇ NOVOBIOCIN MONOSODIUM ◇ NOVOBIOCIN, SODIUM derivative ◇ SODIUM ALBAMYCIN ◇ SODIUM NOVOBIOCIN ◇ U-6591

USE IN FOOD:

Purpose: Animal drug, animal feed drug.

Where Used: Animal feed, beef, chicken, duck, milk, turkey.

Regulations: FDA - 21CFR 556.460. Limitation of 0.1 ppm in milk, 1 ppm in cattle, chickens, turkeys, and ducks. 21CFR 558.415.

SAFETY PROFILE: Poison by intraperitoneal and subcutaneous routes. Moderately toxic by ingestion and subcutaneous routes. When heated to decomposition it emits toxic fumes of NO_x and Na_2O.

TOXICITY DATA and CODEN

ipr-rat LD50:400 mg/kg TXAPA9 24,37,73
orl-mus LD50:962 mg/kg ANTCAO 6,226,56

NOG500 CAS: 8008-45-5
NUTMEG OIL, EAST INDIAN

PROP: Major components are α- and β-pinene, camphene, myristicin, dipentene and sabanene.

Found in fruit of *Myristica fragrans Houttuyn* (Fam. *Myristicaceae)*. Prepared by steam distillation of dried nutmeg (FCTXAV 14,601,76). Colorless to pale yellow liquid; odor and taste of nutmeg. East Indian: d: 0.880-0.910, refr index: 1.474-1.488; West Indian: d: 0.854-0.880, refr index: 1.469-1.476 @20°. Sol in fixed oils, mineral oil; sltly sol in cold alc; very sol in hot alc, chloroform, ether; insol in glycerin, propylene glycol.

SYNS: MYRISTICA OIL ◇ NUTMEG OIL ◇ OIL of MYRISTICA ◇ OIL of NUTMEG

USE IN FOOD:

Purpose: Flavoring agent.

Where Used: Cakes, eggnog, fruit, puddings.

Regulations: FDA - 21CFR 182.10, 182.20. GRAS when used at a level not in excess of the amount reasonably required to accomplish the intended effect.

SAFETY PROFILE: Moderately toxic by ingestion. Experimental reproductive effects. Mutagenic data. A skin irritant. When heated to decomposition it emits acrid smoke and irritating fumes.

TOXICITY DATA and CODEN

skn-rbt 500 mg/24H MOD FCTXAV 14,631,76
dnr-bcs 20 mg/disc TOFOD5 8,91,85
orl-mus TDLo:2400 mg/kg (40D male):REP TOLED5 7,239,81
orl-mus TDLo:4 g/kg (40D male):REP TOLED5 7,239,81
orl-rat LD50:2620 mg/kg FCTXAV 14,631,76

NOH500 CAS: 1400-61-9
NYSTATIN
mf: $C_{46}H_{83}NO_{18}$ mw: 938.30

PROP: Yellow to light-tan powder; odor suggestive of cereals. Mp: decomp > 160°. Sparingly sol in methanol and ethanol; very sltly sol in water; insol in chloroform, ether and benzene.

SYNS: BIOFANAL ◇ CANDEX ◇ CANDIO-HERMAL ◇ DIASTATIN ◇ MORONAL ◇ MYCOSTATIN ◇ MYCOSTATIN 20 ◇ NILSTAT ◇ NYSTAN ◇ NYSTATINE ◇ NYSTAVESCENT ◇ O-V STATIN

USE IN FOOD:

Purpose: Animal drug, animal feed drug.

Where Used: Animal feed, eggs, pork, poultry.

Regulations: FDA - 21CFR 556.470. Limitation of zero in eggs, swine, poultry. 21CFR 558.430.

EPA Genetic Toxicology Program.

SAFETY PROFILE: Poison by intraperitoneal and intravenous routes. Moderately toxic by subcutaneous route. Mildly toxic by ingestion. An experimental teratogen. Experimental reproductive effects. Mutagenic data. When heated to decomposition it emits toxic fumes of NO_x.

TOXICITY DATA and CODEN

cyt-mus-par 50 mg/kg EXPEAM 33,306,77
orl-rat TDLo:100 mg/kg (9D preg):TER ANTBAL 20,45,75
orl-mus LD50:8000 mg/kg PMDCAY 14,105,77

O

OAV000 CAS: 31566-31-1

OCTADECANOIC ACID, MONOESTER with 1,2,3-PROPANETRIOL

mf: $C_{21}H_{42}O_4$ mw: 358.63

PROP: Pure white- or cream-colored, wax-like solid; faint odor. Mp: 58-59°, d: 0.97. Sol in (hot) alc, oils, and hydrocarbons.

SYNS: ABRACOL S.L.G ◇ ADMUL ◇ ADVAWAX 140 ◇ ALDO HMS ◇ ALDO-28 ◇ ARLACEL 161 ◇ ARMOSTAT 801 ◇ ATMOS 150 ◇ ATMUL 67 ◇ CEFATIN ◇ CELINHOL -A ◇ CERASYNT 1000-D ◇ CERASYNT S ◇ CITOMULGAN M ◇ CYCLOCHEM GMS ◇ DERMAGINE ◇ DISTEARIN ◇ DREWMULSE TP ◇ DRUMULSE AA ◇ EMCOL CA ◇ EMEREST 2400 ◇ EMCOL MSK ◇ EMUL P.7 ◇ ESTOL 603 ◇ GLYCERIN MONOSTEARATE ◇ GLYCEROL MONOSTEARATE ◇ GLYCERYL MONOSTEARATE ◇ GROCOR 5500 ◇ HODAG GMS ◇ IMWITOR 191 ◇ KESSCO 40 ◇ LIPO GMS 410 ◇ MONELGIN ◇ MONOSTEARIN ◇ OGEEN 515 ◇ ORBON ◇ PROTACHEM GMS ◇ SEDETINE ◇ STARFOL GMS 450 ◇ STEARIC ACID, MONOESTER WITH GLYCEROL ◇ STEARIC MONOGLYCERIDE ◇ TEGIN ◇ UNIMATE GMS ◇ USAF KE-7 ◇ WITCONOL MS

USE IN FOOD:

Purpose: Coating agent, emulsifier, lubricant, solvent, texture modifying agent.

Where Used: Baked goods, cake shortening, desserts, fruits, ice cream, nuts, peanut butter, puddings, shortening, whipped toppings.

Regulations: FDA - 21CFR 184.1324. GRAS when used in accordance with good manufacturing practice.

SAFETY PROFILE: Poison by intraperitoneal route. When heated to decomposition it emits acrid smoke and irritating fumes.

TOXICITY DATA and CODEN

ipr-mus LD50:200 mg/kg NTIS** AD277-689

OAX000 CAS: 112-92-5

1-OCTADECANOL

mf: $C_{18}H_{38}O$ mw: 270.56

PROP: Colorless solid or flakes. Mp: 58°, bp: 202° @ 10 mm, d: 0.8124 @ 59°/4°.

SYNS: ADOL ◇ ADOL 68 ◇ ATALCO S ◇ CO-1895 ◇ CO-1897 ◇ CRODACOL-S ◇ DECYL OCTYL ALCOHOL ◇ DYTOL E-46 ◇ LOROL 28 ◇ OCTADECANOL ◇ n-OCTADECANOL ◇ OCTA DECYL ALCOHOL ◇ n-OCTADECYL ALCOHOL ◇ POLAAX ◇ SIPOL S ◇ SIPONOL S ◇ STEAROL ◇ STEARYL ALCOHOL ◇ STERAFFINE ◇ USP XIII STEARYL ALCOHOL

USE IN FOOD:

Purpose: Intermediate.

Where Used: Various.

Regulations: FDA - 21CFR 172.864. Use at a level not in excess of the amount reasonably required to accomplish the intended effect.

SAFETY PROFILE: Mildly toxic by ingestion. An experimental neoplastigen. Flammable when exposed to heat or flame; can react with oxidizing materials. To fight fire, use foam, CO_2, dry chemical. When heated to decomposition it emits acrid smoke and irritating fumes.

TOXICITY DATA and CODEN

imp-mus TDLo:1000 mg/kg:NEO CNREA8 26,105,66

orl-rat LD50:20 g/kg 37ASAA 1,722,28

OBC000 CAS: 124-30-1

OCTADECYLAMINE

mf: $C_{18}H_{39}N$ mw: 269.58

SYNS: N-OCTADECYLAMINE ◇ OKTADECYLAMIN (CZECH) ◇ STEARYLAMINE

USE IN FOOD:

Purpose: Boiler water additive.

Where Used: Various.

Regulations: FDA - 21CFR 173.310. Limitation of 3 ppm in steam and excluding use of such steam in contact with milk and milk products.

SAFETY PROFILE: Poison by intraperitoneal route. A skin irritant. When heated to decomposition it emits toxic fumes of NO_x.

TOXICITY DATA and CODEN

skn-rbt 500 mg/24H MOD 28ZPAK -,63,72

ipr-mus LD50:250 mg/kg NTIS** AD691-490

OCE000 CAS: 104-50-7

γ-OCTALACTONE

mf: $C_8H_{14}O_2$ mw: 142.22

PROP: Colorless to pale yellow liquid; coconut odor. D: 0.970-0.980, refr index: 1.443-1.447. Sol in alc; sltly sol in water.

SYNS: γ-n-BUTYL-γ-BUTYROLACTONE ◇ FEMA No. 2798 ◇ 5-HYDROXYOCTANOIC ACID LACTONE ◇ OCTANOLIDE-1,4 ◇ TETRAHYDRO-6-PROPYL-2H-PYRAN-2-ONE

USE IN FOOD:

Purpose: Flavoring agent.

Where Used: Baked goods, candy, ice cream.

Regulations: FDA - 21CFR 172.515. Use at a level not in excess of the amount reasonably required to accomplish the intended effect.

SAFETY PROFILE: Mildly toxic by ingestion. A skin irritant. When heated to decomposition it emits acrid smoke and irritating fumes.

TOXICITY DATA and CODEN

skn-rbt 500 mg/24H MOD FCTXAV 14,821,76
orl-rat LD50:4400 mg/kg FCTXAV 14,821,76

OCO000 CAS: 124-13-0
1-OCTANAL
mf: $C_8H_{16}O$ mw: 128.24

PROP: Found in about 20 essential oils, including a number of citrus oils (FCTXAV 11,95,73). Colorless to light yellow liquid; fatty-orange odor. Bp: 163.4, flash p: 125°F (CC), d: 0.821 @ 20°/4°, refr index: 1.417-1.425, vap d: 4.41. Sol in alc, fixed oils, propylene glycol; insol in glycerin.

SYNS: ALDEHYDE C-8 ◇ C-8 ALDEHYDE ◇ FEMA No. 2797 ◇ OCTANALDEHYDE ◇ n-OCTYL ALDEHYDE

USE IN FOOD:

Purpose: Flavoring agent.

Where Used: Various.

Regulations: FDA - 21CFR 172.515. Use at a level not in excess of the amount reasonably required to accomplish the intended effect.

SAFETY PROFILE: Mildly toxic by ingestion and skin contact. A skin and eye irritant. Combustible when exposed to heat or flame; can react with oxidizing materials. To fight fire, use foam, CO_2, dry chemical.

TOXICITY DATA and CODEN

skn-rbt 500 mg/24H MLD FCTXAV 11,1079,73
eye-rbt 100 mg MLD FCTXAV 11,1079,73
orl-rat LD50:5630 mg/kg FCTXAV 11,95,73
skn-rbt LD50:6350 mg/kg FCTXAV 11,95,73

OCY000 CAS: 124-07-2
OCTANOIC ACID
mf: $C_8H_{16}O_2$ mw: 144.24

PROP: Colorless, oily liquid; unpleasant odor, burning rancid taste. D: 0.91 @ 20°, bp: 240°, mp: 17°. Slt sol in water; sol in most organic solvents.

SYNS: C-8 ACID ◇ CAPRYLIC ACID ◇ n-CAPRYLIC ACID ◇ 1-HEPTANECARBOXYLIC ACID ◇ HEXACID 898 ◇ NEO-FAT 8 ◇ OCTIC ACID ◇ n-OCTOIC ACID ◇ n-OCTYLIC ACID

USE IN FOOD:

Purpose: Adjuvant, antimicrobial agent, component in the manufacture of other food-grade additives, defoaming agent, flavoring agent, lubricant.

Where Used: Baked goods, candy (soft), cheese, fats, frozen dairy desserts, gelatins, meat products, oils, packaging materials, puddings, snack foods.

Regulations: FDA - 21CFR 172.860, 184.1025. GRAS with a limitation of 0.013 percent in baked goods, 0.04 percent in cheese, 0.005 percent in fats and oils, 0.005 percent in frozen dairy desserts, 0.005 percent in gelatin and puddings, 0.005 percent in meat products, 0.005 percent in soft candies, 0.016 percent in snack foods, 0.001 percent in other food categories when used in accordance with good manufacturing practice. 21CFR 186.1085. GRAS as an indirect additive.

SAFETY PROFILE: Moderately toxic by intravenous route. Mildly toxic by ingestion. Mutagenic data. A skin irritant. Yields irritating vapors which can cause coughing. When heated to decomposition it emits acrid smoke and irritating fumes.

TOXICITY DATA and CODEN

skn-rbt 500 mg/24H MOD FCTXAV 19,237,81
sln-smc 5 ppm ANYAA9 407,186,83
oms-nml:oth 10 mmol/L CHROAU 40,1,73
orl-rat LD50:10080 mg/kg FCTXAV 2,327,64

OCY100 CAS: 20296-29-1
3-OCTANOL
mf: $C_8H_{18}O$ mw: 130.28

PROP: Colorless liquid; strong, nutty odor. D: 0.816-0.821, refr index: 1.425. Sol in alc, fixed oils; insol in water.

SYNS: AMYLETHYLCARBINOL ◇ ETHYLAMYLCARBINOL ◇ ETHYL-n-AMYLCARBINOL ◇ FEMA No. 3581 ◇ OCTANOL-3

USE IN FOOD:

Purpose: Flavoring agent.

Where Used: Various.

Regulations: FDA - 21CFR 172.515. Use at a level not in excess of the amount reasonably required to accomplish the intended effect.

SAFETY PROFILE: A moderate skin and eye irritant. When heated to decomposition it emits acrid smoke and irritating fumes.

TOXICITY DATA and CODEN

skn-rbt 500 mg/24H MOD FCTXAV 17,881,79

ODG000 CAS: 111-13-7
2-OCTANONE

mf: $C_8H_{16}O$ mw: 128.24

PROP: Colorless liquid; pleasant apple odor. D: 0.813-0.818, refr index: 1.414-1.418, mp: −20.9°, bp: 173.5°, vap d: 4.4, flash p: 160°F. Sltly sol in water; sol in alc, hydrocarbons, ether, esters.

SYNS: FEMA No. 2802 ◇ METHYL HEXYL KETONE (FCC)

USE IN FOOD:

Purpose: Flavoring agent.

Where Used: Various.

Regulations: FDA - 21CFR 172.515. Use at a level not in excess of the amount reasonably required to accomplish the intended effect.

SAFETY PROFILE: Moderately toxic by an unspecified route. A skin irritant. Combustible liquid when exposed to heat, flame or oxidizers. To fight fire, use foam, alcohol foam. When heated to decomposition it emits acrid smoke and irritating fumes.

TOXICITY DATA and CODEN

skn-rbt 500 mg/24H MLD FCTXAV 13,681,75
unr-mus LD50:1600 mg/kg JMCMAR 19,1257,76

ODQ800
trans-2-OCTEN-1-AL

mf: $C_8H_{14}O$ mw: 126.20

PROP: Slightly yellow liquid; green odor. D: 0.830-0.850, refr index: 1.421-1.424. Sol in alc, fixed oils; sltly sol in water.

SYN: FEMA No. 3215

USE IN FOOD:

Purpose: Flavoring agent.

Where Used: Various.

Regulations: GRAS when used at a level not in excess of the amount reasonably required to accomplish the intended effect.

SAFETY PROFILE: When heated to decomposition it emits acrid smoke and irritating fumes.

ODW025
cis-3-OCTEN-1-OL

mf: $C_8H_{16}O$ mw: 128.22

PROP: White to yellowish liquid; musty, mushroom odor. D: 0.830-0.850, refr index: 1.440. Insol in water.

SYN: FEMA No. 3467

USE IN FOOD:

Purpose: Flavoring agent.

Where Used: Various.

Regulations: FDA - 21CFR 172.515. Use at a level not in excess of the amount reasonably required to accomplish the intended effect.

SAFETY PROFILE: When heated to decomposition it emits acrid smoke and irritating fumes.

ODW030
1-OCTEN-3-YL ACETATE

mf: $C_{10}H_{18}O_2$ mw: 170.24

PROP: Colorless liquid; metallic, mushroom odor. D: 0.865-0.886, refr index: 1.414-1.434 @ 25°. Sol in fixed oils; insol in water, propylene glycol.

SYNS: FEMA No. 3587 ◇ PINOCARVEOL

USE IN FOOD:

Purpose: Flavoring agent.

Where Used: Various.

Regulations: FDA - 21CFR 172.515. Use at a level not in excess of the amount reasonably required to accomplish the intended effect.

SAFETY PROFILE: When heated to decomposition it emits acrid smoke and irritating fumes.

ODW040
1-OCTEN-3-YL BUTYRATE

mf: $C_{12}H_{22}O_2$ mw: 198.31

PROP: Colorless liquid; metallic, mushroom odor. D: 0.859-0.880, refr index: 1.416-1.437

@ 25°. Sol in alc, fixed oils; sltly sol in propylene glycol; insol in water.

SYN: FEMA No. 3612

USE IN FOOD:

Purpose: Flavoring agent.

Where Used: Various.

Regulations: FDA - 21CFR 172.515. Use at a level not in excess of the amount reasonably required to accomplish the intended effect.

SAFETY PROFILE: When heated to decomposition it emits acrid smoke and irritating fumes.

OEG000 CAS: 112-14-1
1-OCTYL ACETATE
mf: $C_{10}H_{20}O_2$ mw: 172.30

PROP: Colorless liquid; orange-jasmine odor. D: 0.865, refr index: 1.418-1.421, mp: −38.5°, bp: 210°, flash p: 190°F. Insol in water; misc with alc, ether, fixed oils.

SYNS: ACETATE C-8 ◇ ACETIC ACID, OCTYL ESTER ◇ CAPRYLYL ACETATE ◇ FEMA No. 2806 ◇ 1-OCTANOL ACETATE ◇ n-OCTANYL ACETATE ◇ OCTYL ACETATE ◇ n-OCTYL ACETATE ◇ OCTYL ALCOHOL ACETATE

USE IN FOOD:

Purpose: Flavoring agent.

Where Used: Various.

Regulations: FDA - 21CFR 172.515. Use at a level not in excess of the amount reasonably required to accomplish the intended effect.

SAFETY PROFILE: Moderately toxic by ingestion. A skin irritant. Combustible liquid. When heated to decomposition it emits acrid smoke and irritating fumes.

TOXICITY DATA and CODEN

skn-rbt 500 mg/24H MLD FCTXAV 12,807,74
orl-rat LD50:3000 mg/kg AMIHBC 10,61,54

OEG100
3-OCTYL ACETATE
mf: $C_{10}H_{20}O_2$ mw: 172.27

PROP: Colorless liquid; rosy, minty odor. D: 0.856-0.860, refr index: 1.414, fp: 190°. Sol in alc, propylene glycol, fixed oils; sltly sol in water.

SYN: FEMA No. 3583

USE IN FOOD:

Purpose: Flavoring agent.

Where Used: Various.

Regulations: FDA - 21CFR 172.515. Use at a level not in excess of the amount reasonably required to accomplish the intended effect.

SAFETY PROFILE: Combustible liquid. When heated to decomposition it emits acrid smoke and irritating fumes.

OEI000 CAS: 111-87-5
OCTYL ALCOHOL
mf: $C_8H_{18}O$ mw: 130.26

PROP: Colorless liquid. D: 0.827 @ 20° 16/4°, mp: −16.7°, bp: 194.5°, flash p: 178°F. Sol in water; misc in alc, ether, and chloroform. Found in several citrus oils and at least 10 other natural sources. (FCTXAV 11,95,73).

SYNS: ALCOHOL C-8 ◇ ALFOL 8 ◇ CAPRYL ALCOHOL ◇ CAPRYLIC ALCOHOL ◇ DYTOL M-83 ◇ EPAL 8 ◇ FEMA No. 2800 ◇ HEPTYL CARBINOL ◇ 1-HYDROXYOCTANE ◇ LOROL 20 ◇ OCTANOL ◇ n-OCTANOL ◇ 1-OCTANOL (FCC) ◇ OCTILIN ◇ OCTYL ALCOHOL, NORMAL-PRIMARY ◇ PRIMARY OCTYL ALCOHOL ◇ SIPOL L8

USE IN FOOD:

Purpose: Flavoring agent, intermediate, solvent.

Where Used: Beverages, candy, gelatin desserts, ice cream, pudding mixes.

Regulations: FDA - 21CFR 172.230, 172.515. Only for encapsulating lemon oil, distilled lime oil, orange oil, peppermint oil, and spearmint oil. 21CFR 172.864, 173.280. Use at a level not in excess of the amount reasonably required to accomplish the intended effect.

SAFETY PROFILE: Poison by intravenous route. Moderately toxic by ingestion. Mutagenic data. A skin irritant. Combustible liquid when exposed to heat or flame; can react with oxidizing materials. To fight fire, use water foam, fog, alcohol foam, dry chemical, CO_2.

TOXICITY DATA and CODEN

skn-rbt 500 mg/24H MLD FCTXAV 11,1079,73
cyt-smc 2 mmol/tube HEREAY 33,457,47
orl-mus LD50:1790 mg/kg HYSAAV 31,310,66

OES000 CAS: 113-48-4
N-OCTYL BICYCLOHEPTENE DICARBOXIMIDE
mf: $C_{17}H_{25}NO_2$ mw: 275.43

SYNS: BICYCLO(2.2.1)HEPTENE-2-DICARBOXYLIC ACID, 2-ETHYLHEXYLIMIDE ◇ ENDOMETHYLENETETRA

HYDROPHTHALIC ACID, N-2-ETHYLHEXYL IMIDE ◇ ENT 8,184 ◇ N-(2-ETHYLHEXYL)BICYCLO-(2,2,1)-HEPT-5-ENE-2,3-DICARBOXIMIDE ◇ N-2-ETHYLHEXYLIMIDE ENDOMETHYLENETETRAHYDROPHTHALIC ACID ◇ N-(2-ETHYLHEXYL)-5-NORBORNENE-2,3-DICARBOXIMIDE ◇ 2-(2-ETHYLHEXYL)-3a,4,7,7a-TETRAHYDRO-4,7-METHANO-1H-ISOINDOLE-1,3(2H)-DIONE ◇ MGK-264 ◇ OCTACIDE 264 ◇ N-OCTYLBICYCLO-(2.2.1)-5-HEPTENE-2,3-DICARBOXIMIDE ◇ PYRODONE ◇ SYNERGIST 264 ◇ VAN DYK 264

USE IN FOOD:

Purpose: Pesticide.

Where Used: Various.

Regulations: FDA - 21CFR 193.320. Pesticide residue tolerance of 10 ppm.

SAFETY PROFILE: Moderately toxic by ingestion, skin contact and intraperitoneal routes. Experimental reproductive effects. Large doses can cause central nervous system stimulation followed by depression. When heated to decomposition it emits toxic fumes of NO_x.

TOXICITY DATA and CODEN

orl-rat TDLo:346 mg/kg (MGN):REP TXAPA9 9,555,66
orl-rat LD50:2800 mg/kg FMCHA2 -,C157,83
skn-rat LD50:470 mg/kg 30ZDA9 -,139,71

OEY100
OCTYL FORMATE

mf: $C_9H_{18}O_2$ mw: 158.24

PROP: Colorless liquid; fruity odor. D: 0.869, refr index: 1.418. Sol in fixed oils, propylene glycol; insol in glycerin.

SYN: FEMA No. 2809

USE IN FOOD:

Purpose: Flavoring agent.

Where Used: Various.

Regulations: FDA - 21CFR 172.515. Use at a level not in excess of the amount reasonably required to accomplish the intended effect.

SAFETY PROFILE: When heated to decomposition it emits acrid smoke and irritating fumes.

OFA000 CAS: 1034-01-1
OCTYL GALLATE

mf: $C_{15}H_{22}O_5$ mw: 282.37

USE IN FOOD:

Purpose: Antioxidant.

Where Used: Margarine, oleomargarine.

Regulations: USDA - 9CFR 318.7. Limitation of 0.02 percent individually or in combination with other antioxidants approved for use in margarine.

SAFETY PROFILE: Mildly toxic by ingestion. When heated to decomposition it emits acrid smoke and irritating fumes.

TOXICITY DATA and CODEN

orl-rat LD50:4700 mg/kg FOMAAB 26,99,51

OGI200
ODORLESS LIGHT PETROLEUM HYDROCARBONS

PROP: Liquid; faint odor. Bp: 300-650°.

USE IN FOOD:

Purpose: Coating agent, defoamer, float, froth-flotation cleaning, insecticide formulations component.

Where Used: Beet sugar, eggs, fruits, pickles, vegetables, vinegar, wine.

Regulations: FDA - 21CFR 172.884. Use at a level not in excess of the amount reasonably required to accomplish the intended effect.

SAFETY PROFILE: When heated to decomposition it emits acrid smoke and irritating fumes.

OGK000 CAS: 8015-79-0
OIL of CALAMUS

PROP: Extract of *Acorus calamus L., araceae.* Containing: asarone, eugenol; esters of acetic and heptylic acids. Volatile oil. Yellow to yellowish-brown liquid (viscid); aromatic odor, bitter taste. D: 0.960-0.9707 @ 20°/20°. Very sltly sol in water, misc with alc. Keep well closed, cool, and protected from light.

SYNS: CALAMUS OIL ◇ KALMUS OEL (GERMAN) ◇ OIL of SWEET FLAG

USE IN FOOD:

Purpose: Flavoring agent.

Where Used: Prohibited from foods.

Regulations: FDA - 21CFR 189.110. Prohibited from direct addition or use in human food.

SAFETY PROFILE: Poison by intraperitoneal route. Moderately toxic by ingestion. An experimental tumorigen. When heated to decomposition it emits acrid smoke and irritating fumes.

TOXICITY DATA and CODEN

orl-rat TDLo: 10 g/kg/59W-C: ETA PAACA3 8,24,67
orl-rat LD50: 777 mg/kg FCTXAV 2,327,64

OGM800
OIL of LIME OIL, COLDPRESSED

PROP: Expressed from the peel of *Citrus aurantofolia* Swingle (Mexican type) or *Citrus latifolia* (Tahitian type). Yellow to brown-green liquid. Sol in fixed oils, mineral oil; insol glycerin, propylene glycol.

USE IN FOOD:

Purpose: Flavoring agent.

Where Used: Various.

Regulations: FDA - 21CFR 182.20. GRAS when used at a level not in excess of the amount reasonably required to accomplish the intended effect.

SAFETY PROFILE: When heated to decomposition it emits acrid smoke and irritating fumes.

OGO000 CAS: 8008-26-2
OIL of LIME, DISTILLED

PROP: From distillation of juice or crushed fruit of *Citrus aurantofolia* Swingle. Colorless to green-yellow liquid. Sol in fixed oils, mineral oil; insol glycerin, propylene glycol.

SYNS: DISTILLED LIME OIL ◇ LIME OIL ◇ LIME OIL, DISTILLED (FCC) ◇ OILS, LIME

USE IN FOOD:

Purpose: Flavoring agent.

Where Used: Bakery products, beverages (nonalcoholic), chewing gum, condiments, confections, gelatin desserts, ice cream products, puddings.

Regulations: FDA - 21CFR 182.20. GRAS when used at a level not in excess of the amount reasonably required to accomplish the intended effect.

SAFETY PROFILE: An experimental tumorigen. Mutagenic data. A skin irritant. When heated to decomposition it emits acrid smoke and irritating fumes.

TOXICITY DATA and CODEN

skn-rbt 500 mg/24H MLD FCTXAV 12,729,74
dnr-bcs 20 mg/disc TOFOD5 8,91,85
orl-mus TDLo: 67 g/kg/39W-I: ETA JNCIAM 35,771,65

OGW000 CAS: 8007-12-3
OIL of NUTMEG, EXPRESSED

PROP: From steam distillation of dried arillode of the ripe seed of *Myristica fragrans* Houtt. (Fam. *Myristicaceae*). Colorless to pale yellow liquid; odor and taste of nutmeg. East Indian: d: 0.880-0.930, refr index: 1.474-1.488; West Indian: d: 0.854-0.880, refr index: 1.469-1.480 @20°. Sol in fixed oils, mineral oil; sltly sol in cold alc; very sol in hot alc, chloroform, ether; insol in glycerin, propylene glycol.

SYNS: MACE OIL (FCC) ◇ NCI-C56484 ◇ OIL of MACE ◇ OILS, MACE

USE IN FOOD:

Purpose: Flavoring agent.

Where Used: Bread, cakes, chocolate pudding, fruit salad.

Regulations: FDA - 21CFR 182.10, 182.20. GRAS when used at a level not in excess of the amount reasonably required to accomplish the intended effect.

EPA Genetic Toxicology Program.

SAFETY PROFILE: Moderately toxic by ingestion. A skin irritant. Human ingestion causes symptoms similar to volatile oil of nutmeg. When heated to decomposition it emits acrid smoke and irritating fumes.

TOXICITY DATA and CODEN

skn-rbt 500 mg/24H MOD FCTXAV 17,851,79
orl-rat LD50: 3640 mg/kg FCTXAV 2,327,64

OGY000 CAS: 8008-57-9
OIL of ORANGE

PROP: Yellow to deep-orange liquid; characteristic orange taste and odor. D: 0.842-0.846 @ 25°/25°, refr index: 1.472 @ 20°. Sol in 2 vols 90% alc, in 1 vol glacial acetic acid; sltly sol in water; misc with abs alc, carbon disulfide. Keep well closed, cool and protected from light. Oil expressed from the peel of *Citrus sinensis* L. Osbeck (Fam. *Rutaceae*) (BJCAAI 13, 92,59).

SYNS: NEAT OIL of SWEET ORANGE ◇ OIL of SWEET ORANGE ◇ ORANGE OIL ◇ ORANGE OIL, COLDPRESSED (FCC) ◇ SWEET ORANGE OIL

USE IN FOOD:

Purpose: Flavoring agent.

Where Used: Bakery products, beverages (nonalcoholic), chewing gum, condiments, ice cream products.

Regulations: FDA - 21CFR 182.20. GRAS when used at a level not in excess of the amount reasonably required to accomplish the intended effect.Do not use if terebinthine odor can be detected.

SAFETY PROFILE: An experimental neoplastigen. A skin irritant. When heated to decomposition it emits acrid smoke and irritating fumes.

TOXICITY DATA and CODEN

skn-rbt 500 mg/24H MOD FCTXAV 12,733,74
orl-mus TDLo:67 g/kg/40W-I:NEO JNCIAM 35,771,65

OGY010
ORANGE OIL, BITTER, COLDPRESSED

PROP: Oil expressed from the peel of *Citrus aurantium* L. Osbeck (Fam. *Rutaceae*). Pale yellow to yellow-brown liquid; characteristic orange odor and bitter taste. D: 0.845-0.851, refr index: 1.472 @ 20°. Misc in abs alc, in 1 vol glacial acetic acid; sol in fixed oils, mineral oil; insol in glycerin.

SYNS: OIL of BITTER ORANGE ◇ ORANGE OIL ◇ BITTER ORANGE OIL

USE IN FOOD:

Purpose: Flavoring agent.

Where Used: Various.

Regulations: FDA - 21CFR 182.20. GRAS when used at a level not in excess of the amount reasonably required to accomplish the intended effect.

SAFETY PROFILE: When heated to decomposition it emits acrid smoke and irritating fumes.

OGY020
ORANGE OIL, DISTILLED

PROP: From steam distillation of fresh peel of *Citrus sinensis* L. Osbeck (Fam. *Rutaceae*). Colorless to pale yellow liquid; odor of fresh orange peel. Sol in fixed oils, mineral oil, alc; insol in glycerin, propylene glycol.

USE IN FOOD:

Purpose: Flavoring agent.

Where Used: Various.

Regulations: FDA - 21CFR 182.20. GRAS when used at a level not in excess of the amount reasonably required to accomplish the intended effect.

SAFETY PROFILE: When heated to decomposition it emits acrid smoke and irritating fumes.

OHG000 CAS: 115-71-9
OIL of SANDALWOOD, EAST INDIAN

mf: $C_{15}H_{24}O$ mw: 220.39

PROP: From steam distillation of the ground dried wood of *Santalus album L.* (FCTXAV 12,807,74). Colorless to sltly yellow viscous liquid; sandalwood odor. D: 0.965-0.973, refr index: 1.505. Very sol in alc, fixed oils, propylene glycol; in in water, glycerin.

SYNS: 5-(2,3-DIMETHYLTRICYCLO(2.2.1.$0^{2,6}$)HEPT-3-YL)-2-METHYL-2-PENTEN-1-OL ◇ FEMA No. 3006 ◇ SANDALWOOD OIL, EAST INDIAN ◇ α-SANTALOL (FCC)

USE IN FOOD:

Purpose: Flavoring agent.

Where Used: Various.

Regulations: FDA - 21CFR 172.510. Use at a level not in excess of the amount reasonably required to accomplish the intended effect.

SAFETY PROFILE: Moderately toxic by ingestion. A skin irritant. When heated to decomposition it emits acrid smoke and irritating fumes.

TOXICITY DATA and CODEN

skn-mus 500 mg MLD FCTXAV 12,807,74
orl-rat LD50:3800 mg/kg FCTXAV 12,807,74

OHM600
OLEAMIDE

SYN: OLEIC ACID AMIDE

USE IN FOOD:

Purpose: Release agent.

Where Used: Packaging materials.

Regulations: FDA - 21CFR 181.28. Use in accordance with good manufacturing practice.

SAFETY PROFILE: When heated to decomposition it emits acrid smoke and irritating fumes.

OHO000 CAS: 6696-47-5
OLEANDOMYCIN HYDROCHLORIDE

mf: $C_{35}H_{61}NO_{12}\cdot ClH$ mw: 724.43

PROP: Long needles from ethyl acetate. Mp: 134-135°. Very sol in water.

SYN: OLEANDOMYCIN MONOHYDROCHLORIDE

USE IN FOOD:

Purpose: Animal drug, animal feed drug.

Where Used: Animal feed, chicken, pork, turkey.

Regulations: FDA - 21CFR 556.480. Limitation of zero in chickens, turkeys, and swine. 21CFR 558.435.

SAFETY PROFILE: Poison by intravenous route. Moderately toxic by ingestion and subcutaneous routes. When heated to decomposition it emits very toxic fumes of NO_x and HCl.

TOXICITY DATA and CODEN

orl-mus LD50:4000 mg/kg ANTCAO 7,419,57

OHO200 CAS: 7060-74-4
OLEANDOMYCIN PHOSPHATE

mf: $C_{35}H_{61}NO_{12}\cdot H_3O_4P$ mw: 785.97

SYN: MATROMYCIN

USE IN FOOD:

Purpose: Animal drug, animal feed drug.

Where Used: Animal feed, chicken, pork, turkey.

Regulations: FDA - 21CFR 556.480. Limitation of zero in chickens, turkeys, and swine. 21CFR 558.435.

SAFETY PROFILE: Poison by intravenous route. Moderately toxic by ingestion and subcutaneous routes. When heated to decomposition it emits toxic fumes of PO_x and NO_x.

TOXICITY DATA and CODEN

orl-mus LD50:4 g/kg NIIRDN 6,164,82

OHU000 CAS: 112-80-1
OLEIC ACID

mf: $C_{18}H_{34}O_2$ mw: 282.52

PROP: Colorless liquid; odorless when pure. Mp: 6°, bp: 360.0°, flash p: 372°F (CC), d: 0.895 @ 25°/25°, autoign temp: 685°F, vap press: 1 mm @ 176.5°, bp: 286° @ 100 mm. Insol in water; misc in alc and ether.

SYNS: CENTURY CD FATTY ACID ◇ EMERSOL 210 ◇ EMERSOL 220 WHITE OLEIC ACID ◇ GLYCON RO ◇ GROCO 2 ◇ HY-PHI 1055 ◇ INDUSTRENE 105 ◇ K 52 ◇ l'ACIDE OLEIQUE (FRENCH) ◇ METAUPON ◇ NEO-FAT 90-04 ◇ NEO-FAT 92-04 ◇ cis-Δ^9-OCTADECENOIC ACID ◇ cis-OCTADEC-9-ENOIC ACID ◇ cis-9-OCTADECENOIC ACID ◇ 9,10-OCTADECENOIC ACID ◇ PAMOLYN ◇ RED OIL ◇ TEGO-OLEIC 130 ◇ VOPCOLENE 27 ◇ WECOLINE OO ◇ WOCHEM NO. 320

USE IN FOOD:

Purpose: Binder, coatings, component in the manufacture of other food-grade additives, defoaming agent, flume wash water additive, lubricant.

Where Used: Beet sugar, citrus fruit (fresh), sugar beets, yeast.

Regulations: FDA - 21CFR 172.860, 172.862. From tall oil fatty acids: Must conform to 21CFR 172.862 specifications for fats or fatty acids derived from edible oils. Use at a level not in excess of the amount reasonably required to accomplish the intended effect. 21CFR 173.315. Limitation of 0.1 ppm in wash water. 21CFR 173.340, 172.862.

SAFETY PROFILE: Poison by intravenous route. Mildly toxic by ingestion. An experimental tumorigen. Mutagenic data. A human and experimental skin irritant. Combustible when exposed to heat or flame. To fight fire, use CO_2, dry chemical. Potentially dangerous reaction with perchloric acid + heat. When heated to decomposition it emits acrid smoke and irritating fumes.

TOXICITY DATA and CODEN

skn-hmn 15 mg/3D-I MOD 85DKA8 -,127,77
skn-rbt 500 mg open MLD UCDS** 11/29/63
cyt-smc 100 mg/L NATUAS 294,263,81
dns-mus-rec 35 mg/kg CALEDQ 23,253,84
scu-rbt TDLo:390 mg/kg/17W-I:ETA CRSBAW 137,760,43
orl-rat LD50:74 g/kg UCDS** 11/29/63

OHY000 CAS: 143-18-0
OLEIC ACID, POTASSIUM SALT

mf: $C_{18}H_{34}O_2\cdot K$ mw: 321.62

SYNS: POTASSIUM cis-9-OCTADECENOIC ACID ◇ POTASSIUM OLEATE

USE IN FOOD:

Purpose: Anticaking agent, binder, emulsifier, stabilizer.

Where Used: Packaging materials, various foods.

Regulations: FDA - 21CFR 172.863, 181.29. Use in accordance with good manufacturing practice.

SAFETY PROFILE: An eye irritant. When heated to decomposition it emits toxic fumes of K_2O.

TOXICITY DATA and CODEN

eye-rbt 12 mg/48H JANCA2 56,905,73

OIA000 CAS: 143-19-1
OLEIC ACID, SODIUM SALT
mf: $C_{18}H_{33}O_2 \cdot Na$ mw: 304.50

PROP: White powder; slt tallow odor. Mp: 232-235°.

SYNS: EUNATROL ◇ SODIUM OLEATE

USE IN FOOD:

Purpose: Stabilizer, anticaking agent, binder, emulsifier, paper manufacturing aid.

Where Used: Food packaging, various foods.

Regulations: FDA - 21CFR 172.863, 181.29. Use in accordance with good manufacturing practice. 21CFR 186.1770. GRAS as an indirect additive when used in accordance with good manufacturing practice.

SAFETY PROFILE: Poison by intravenous route. Combustible when exposed to heat or flame. When heated to decomposition it emits toxic fumes of Na_2O.

TOXICITY DATA and CODEN

ivn-mus LD50:152 mg/kg RPOBAR 2,327,70

OIM000 CAS: 8050-07-5
OLIBANUM GUM

PROP: Contains 3-8% volatile oil (pinene, dipentene, etc.), 60% resins, 20% gum (polysaccharide fraction) and 6-8% bassorin (FCTXAV 16,637,78). A gum from the trees *Boswellia carterii* Birdw. and other *Boswellia* species (Fam. *Burseraceae*).

SYN: FRANKINCENSE GUM

USE IN FOOD:

Purpose: Flavoring agent.

Where Used: Various.

Regulations: FDA - 21CFR 172.510. Use at a level not in excess of the amount reasonably required to accomplish the intended effect.

SAFETY PROFILE: A skin irritant. When heated to decomposition it emits acrid smoke and irritating fumes.

TOXICITY DATA and CODEN

skn-rbt 500 mg/24H MOD FCTXAV 16,837,78

OIM025
OLIBANUM OIL

PROP: Distilled from a gum from the trees *Boswellia carterii* Birdw. and other *Boswellia* species (Fam. *Burseraceae*). Pale liquid; pleasant balsamic odor. D: 0.862-0.889, refr index: 1.465-1.482 @ 20°. Sol in fixed oils, mineral oil; insol in glycerin, propylene glycol.

SYN: FRANKINCENSE OIL

USE IN FOOD:

Purpose: Flavoring agent.

Where Used: Various.

Regulations: FDA - 21CFR 172.510. Use at a level not in excess of the amount reasonably required to accomplish the intended effect.

SAFETY PROFILE: When heated to decomposition it emits acrid smoke and irritating fumes.

OJD200
ONION OIL

PROP: From steam distillation of bulbs of *Allium ceoa* L. (Fam. *Lillaceae*). Clear amber liquid; strong pungent odor and taste of onion. Sol in fixed oils, mineral oil, alc; insol in glycerin, propylene glycol.

SYN: OIL of ONION

USE IN FOOD:

Purpose: Flavoring agent.

Where Used: Various.

Regulations: FDA - 21CFR 182.20. GRAS when used at a level not in excess of the amount reasonably required to accomplish the intended effect.

SAFETY PROFILE: Skin irritant. When heated to decomposition it emits acrid smoke and irritating fumes.

OJK325 CAS: 15139-76-1
ORANGE B
mf: $C_{22}H_{16}N_4Na_2O_9S_2$ mw: 590.50

PROP: Dull orange crystals.

SYN: 1-(4-SULFOPHENYL)-3-ETHYLCARBOXY-4-(4-SULFONAPHTHYLAZO)-5-HYDROXYPYRAZOLE

USE IN FOOD:

Purpose: Color additive.

Where Used: Frankfurters, sausages.

Regulations: FDA - 21CFR 74.250. Limitation of 150 ppm in frankfurters and sausages.

SAFETY PROFILE: When heated to decomposition emits toxic fumes of SO_x.

OJO000 CAS: 8007-11-2
ORIGANUM OIL

PROP: Main constituent is carvacrol. From steam distillation of the herb *Thymus capitatus* Hoffm. et Link (FCTXAV 12,807,74). Yellow to dark red brown liquid; pungent spicy odor of thyme oil. D: 0.935-0.960, refr index: 1.502 @ 20°. Sol in fixed oil, propylene glycol, mineral oil; insol in glycerin.

SYN: OIL of ORIGANUM

USE IN FOOD:

Purpose: Flavoring agent.

Where Used: Various.

Regulations: FDA - 21CFR 182.20. GRAS when used at a level not in excess of the amount reasonably required to accomplish the intended effect.

SAFETY PROFILE: Poison by skin contact. Moderately toxic by ingestion. A severe skin irritant. When heated to decomposition it emits acrid smoke and irritating fumes.

TOXICITY DATA and CODEN

skn-mus 100%:SEV FCTXAV 12,945,74
orl-rat LD50:1850 mg/kg FCTXAV 12,945,74
skn-rbt LD50:320 mg/kg FCTXAV 12,945,74

OJO100
ORMETOPRIM

USE IN FOOD:

Purpose: Animal drug.

Where Used: Catfish, chicken, duck, salmonids, turkey.

Regulations: FDA - 21CFR 556.490. Limitation of 0.1 ppm in chickens, turkeys, ducks, salmonids, and catfish.

SAFETY PROFILE: When heated to decomposition it emits acrid smoke and irritating fumes.

OJW100
ORRIS ROOT OIL

PROP: From steam distillation of peeled, dried, aged rhizomes of *Iris pallida* L. (Fam. *Iridaceae*). Light yellow to brown solid at room temp. Mp: 38-50°. Sol in fixed oils, mineral oil, propylene glycol; insol in glycerin.

USE IN FOOD:

Purpose: Flavoring agent.

Where Used: Various.

Regulations: FDA - 21CFR 172.510. Use at a level not in excess of the amount reasonably required to accomplish the intended effect.

SAFETY PROFILE: When heated to decomposition it emits acrid smoke and irritating fumes.

OJY100 CAS: 19044-88-3
ORYZALIN

mf: $C_{12}H_{18}N_4O_6S$ mw: 346.36

SYNS: 3,5-DINITRO-N^4,N^4-DIPROPYLSULFANILAMIDE ◇ DIRIMAL ◇ RYZELAN ◇ SURFLAN

USE IN FOOD:

Purpose: Herbicide.

Where Used: Peppermint, spearmint.

Regulations: FDA - 21CFR 193.462. Limitation of 0.1 ppm in peppermint and spearmint oil.

SAFETY PROFILE: When heated to decomposition emits toxic fumes of NO_x, SO_x.

OQU100 CAS: 42874-03-3
OXYFLUORFEN

mf: $C_{15}H_{11}ClF_3NO_4$ mw: 361.72

PROP: Orange crystal solid. Sol in water and most solids.

SYN: 2-CHLORO-1-(3-ETHOXY-4-NITROPHENOXY)-4-(TRIFLUOROMETHYL)BENZENE

USE IN FOOD:

Purpose: Herbicide.

Where Used: Cottonseed oil, mint oil, soybean oil.

Regulations: FDA - 21CFR 193.325. Limitation of 0.25 ppm in cottonseed oil, mint oil, soybean oil.

SAFETY PROFILE: When heated to decomposition emits toxic fumes of Cl^-, F^-, and NO_2.

ORS100
OXYSTEARIN

PROP: Mixture of the glycerides of partially oxidized stearic and other fatty acids. Tan to light brown waxy solid; bland taste. refr index: 1.465. Sol in ether, solvent hexane, chloroform.

USE IN FOOD:

Purpose: Crystallization inhibitor in cooking oils, crystallization inhibitor in salad oil, defoaming agent, sequestrant.

Where Used: Beet sugar, cooking oil, salad oil, vegetable oils, yeast.

Regulations: FDA - 21CFR 172.818. Limitation of 0.125 percent of the combined weight of the oils or shortening. Must conform to FDA specifications for fats or fatty acids derived from edible oils. 21CFR 173.340.

SAFETY PROFILE: When heated to decomposition it emits acrid smoke and irritating fumes.

ORW000 CAS: 10028-15-6
OZONE

af: O_3 aw: 48.00

PROP: Unstable colorless gas or dark blue liquid; characteristic odor. Mp: −193°, bp: −111.9°, d (gas): 2.144 g/L, 1.71 @ −183°. D: (liquid) 1.614 g/mL @ −195.4°.

SYNS: OZON (POLISH) ◇ TRIATOMIC OXYGEN

USE IN FOOD:

Purpose: Antimicrobial agent.

Where Used: Bottled water.

Regulations: FDA - 21CFR 184.1563 Limitation of 0.4 mg/L of bottled water.

EPA Genetic Toxicology Program.

OSHA PEL: TWA 0.1 ppm; STEL 0.3 ppm ACGIH TLV: TWA CL 0.1 ppm DFG MAK: 0.1 ppm (0.2 mg/m^3)

SAFETY PROFILE: A human poison by inhalation. An experimental neoplastigen, tumorigen, and teratogen. Human systemic effects by inhalation: visual field changes, eye lacrimation, headache, decreased pulse rate with fall in blood pressure, blood pressure decrease, skin dermatitis, cough, dyspnea, respiratory stimulation and other pulmonary changes. Experimental reproductive effects. Human mutagenic data. A skin, eye, upper respiratory system and mucous membrane irritant.

Concentration of 0.015 ppm of ozone in air produces a barely detectable odor. Concentrations of 1 ppm produce a disagreeable sulfurlike odor and may cause headache and irritation of eyes and the upper respiratory tract; symptoms disappear after leaving the exposure.

A powerful and highly reactive oxidizing agent. A severe explosion hazard in liquid form when shocked, exposed to heat or flame, or in concentrated form by chemical reaction with powerful reducing agents. Incompatible with rubber; dinitrogen tetraoxide.

TOXICITY DATA and CODEN

eye-rbt 2 ppm/4H JPCAAC 10,17,60
mmo-esc 100 ppb/20M MEHYDY 4,165,78
dnr-esc 50 ppm/30M BBRCA9 77,220,77
ihl-rat TCLo:1040 ppt/24H (6-9D preg):TER TXAPA9 48,19,79
ihl-rat TCLo:1500 ppb/24H (17-20D preg): REP TOLED5 5,3,80
ihl-mus TCLo:5 ppm/2H/75D-I:NEO AEHLAU 20,16,70
ihl-mus TC:608 μg/m^3/24W-I:ETA JJIND8 75,771,85
ihl-hmn TCLo:100 ppm/1M:SKN,PUL NEACA9 19,686,41
ihl-man TCLo:1860 ppb/75M:EYE,CVS,PUL AEHLAU 10,517,65
ihl-hmn TCLo:1 ppm:PUL AEHLAU 10,295,65
ihl-hmn TCLo:200 ppb/3H:EYE,PUL 32ZWAA 8,182,74
ihl-rat LC50:4800 ppm/4H AMIHAB 15,181,57

P

PAE000 CAS: 8014-19-5
PALMAROSA OIL

PROP: From steam distillation of the grass *Cymbopogon Martini* Stapf. Var. Motia, mainly *Geraniol* (FCTXAV 12,807,74). Yellow oily liquid. D: 0.879-0.892, refr index: 1.473 @ 20°. Sol in fixed oils, propylene glycol, mineral oil; insol in glycerin.

SYN: GERANIUM OIL, EAST INDIAN TYPE ◇ GERANIUM OIL, TURKISH TYPE ◇ OIL of PALMAROSA

USE IN FOOD:

Purpose: Flavoring agent.

Where Used: Various.

Regulations: FDA - 21CFR 182.20. GRAS when used at a level not in excess of the amount reasonably required to accomplish the intended effect.

SAFETY PROFILE: A skin irritant. When heated to decomposition it emits acrid smoke and irritating fumes.

TOXICITY DATA and CODEN

skn-rbt 500 mg/24H MOD FCTXAV 12,947,74

PAE240
PALMITAMIDE

SYN: PALMITIC ACID AMIDE

USE IN FOOD:

Purpose: Release agent.

Where Used: Packaging materials.

Regulations: FDA - 21CFR 181.28. Use in accordance with good manufacturing practice.

SAFETY PROFILE: When heated to decomposition it emits acrid smoke and irritating fumes.

PAE250 CAS: 57-10-3
PALMITIC ACID

mf: $C_{17}H_{32}O_2$ mw: 256.48

PROP: Colorless plates or white crystalline powder; slt characteristic odor and taste. D: 0.849 @ 70°/4°, mp: 63-64°, bp: 271.5° @ 100 mm. Insol in water; very sltly sol in petr ether; sol in absolute ether, chloroform.

SYNS: CETYLIC ACID ◇ EMERSOL 140 ◇ EMERSOL 143 ◇ HEXADECANOIC ACID ◇ n-HEXADECOIC ACID ◇ HEXADECYLIC ACID ◇ HYDROFOL ◇ HYSTRENE 8016 ◇ INDUSTRENE 4516 ◇ 1-PENTADECANECARBOXYLIC ACID

USE IN FOOD:

Purpose: Component in the manufacture of other food-grade additives, defoaming agent, lubricant.

Where Used: Various.

Regulations: FDA - 21CFR 172.860. Use in accordance with good manufacturing practice. Must conform to FDA specifications for fats or fatty acids derived from edible oils.

SAFETY PROFILE: A poison by intravenous route. An experimental neoplastigen. A human skin irritant. When heated to decomposition it emits acrid smoke and irritating fumes.

TOXICITY DATA and CODEN

skn-hmn 75 mg/3D-I MLD 85DKA8 -,127,77
imp-mus TDLo:1000 mg/kg:NEO CNREA8 26,105,66
ivn-mus LD50:57 mg/kg APTOA6 18,141,61

PAE275
PALM KERNEL OIL (UNHYDROGENATED)

PROP: From the kernal of the fruit of the oil palm *Elaeis guineensis*. A fatty solid; characteristic sweet nutty flavor.

USE IN FOOD:

Purpose: Coating agent, emulsifying agent, formulation aid, texturizer.

Where Used: Confectionery products, margarine.

Regulations: GRAS when used in accordance with good manufacturing practice.

SAFETY PROFILE: When heated to decomposition it emits acrid smoke and irritating fumes.

PAE300
PALM OIL (UNHYDROGENATED)

PROP: From the pulp of the fruit of the oil palm *Elaeis guineensis*. A deep orange-red fatty semisolid @ 21-27°; characteristic sweet nutty flavor.

USE IN FOOD:

Purpose: Coating agent, emulsifying agent, formulation aid, texturizer.

Where Used: Margarine, shortening.

Regulations: GRAS when used in accordance with good manufacturing practice.

SAFETY PROFILE: When heated to decomposition it emits acrid smoke and irritating fumes.

PAG200 CAS: 81-13-0
d-PANTHENOL
mf: $C_9H_{19}NO_4$ mw: 205.29

PROP: Viscous, somewhat hygroscopic liquid; sltly bitter taste. D: (20/20) 1.2, bp: 118-120°, easily decomp on distillation. Freely sol in water, alc, methanol, ether; sltly sol in glycerin. Natural pH about 9.5.

SYNS: ALCOPAN-250 ◇ BEPANTHEN ◇ BEPANTHENE ◇ BEPANTOL ◇ COZYME ◇ DEXPANTHENOL (FCC) ◇ d-(+)-2,4-DIHYDROXY-N-(3-HYDROXYPROPYL)-3,3-DIMETHYLBUTYRAMIDE ◇ D-P-A INJECTION ◇ ILOPAN ◇ MOTILYN ◇ PANADON ◇ PANTHENOL ◇ d(+)-PANTHENOL (FCC) ◇ PANTHODERM ◇ PANTOL ◇ PANTOTHENOL ◇ d-PANTOTHENOL ◇ PANTOTHENYL ALCOHOL ◇ d-PANTOTHENYL ALCOHOL ◇ d(+)-PANTOTHENYL ALCOHOL ◇ THENALTON ◇ ZENTINIC

USE IN FOOD:

Purpose: Dietary supplement, nutrient.

Where Used: Various.

Regulations: FDA - 21CFR 182.5580. GRAS when used in accordance with good manufacturing practice.

SAFETY PROFILE: Moderately toxic by intravenous route. When heated to decomposition it emits toxic fumes of NO_x.

TOXICITY DATA and CODEN

ipr-mus LD50:9 g/kg FRPSAX 14,43,59

PAG500 CAS: 9001-73-4
PAPAIN

PROP: White to gray, sltly hygroscopic powder. Sol in water and glycerin; insol in other common organic solvents. The most thermostatic enzyme known, digests protein. Isolated from the latex of the green fruit and leaves of *Carcia papaya L.* (IJMRAQ 67,499,78).

SYNS: ARBUZ ◇ CAROID ◇ NEMATOLYT ◇ PAPAYOTIN ◇ SUMMETRIN ◇ TROMASIN ◇ VEGETABLE PEPSIN ◇ VELARDON ◇ VERMIZYM

USE IN FOOD:

Purpose: Chillproofing of beer, enzyme, meat tenderizing, preparation of precooked cereals, processing aid, tissue softening agent.

Where Used: Meat (raw cuts), poultry, wine.

Regulations: FDA - 21CFR 184.1585. GRAS when used in accordance with good manufacturing practice. USDA - 9CFR 318.7, 381.147. Solutions consisting of water and approved proteolytic enzymes applied or injected into raw meat cuts shall not result in a gain of more than 3 percent above the weight of the untreated product. BATF - 27CFR 240.1051. GRAS when used in accordance with good manufacturing practice.

SAFETY PROFILE: Experimental teratogenic and reproductive effects. An allergen. When heated to decomposition it emits toxic fumes of NO_x.

TOXICITY DATA and CODEN

orl-rat TDLo:750 mg/kg (8-17D preg):TER IJEBA6 18,953,80

ipr-rat TDLo:375 mg/kg (8D preg):TER IJMRAQ 72,300,80

ipr-rbt TDLo:500 mg/kg (17D preg):REP JAINAA 28,6,79

PAH275
PAPRIKA

PROP: Ground dried pod of mild capsicum *Capsicum annuum* L.

USE IN FOOD:

Purpose: Color additive.

Where Used: Chorizo sausage.

Regulations: FDA - 21CFR 73.340. Use in accordance with good manufacturing practice. USDA - 9CFR 318.7. Not allowed in meat except for chorizo sausage and meat products allowed by 9 CFR 319.

SAFETY PROFILE: When heated to decomposition it emits acrid smoke and irritating fumes.

PAH280
PAPRIKA OLEORESIN

PROP: Derived from organic solvent extraction of ground dried pod of mild capsicum *Capsicum annuum* L.

USE IN FOOD:

Purpose: Color additive.

Where Used: Chorizo sausage, condiment mixtures, salad dressings.

Regulations: FDA - 21CFR 73.345. Use in accordance with good manufacturing practice. USDA - 9CFR 318.7. Not allowed in meat except for chorizo sausage and meat products allowed by 9 CFR 319.

SAFETY PROFILE: When heated to decomposition it emits acrid smoke and irritating fumes.

PAH750 CAS: 8002-74-2
PARAFFIN

PROP: Colorless or white, translucent wax; odorless. D: approx 0.90, mp: 50-57°. Insol in water, alc; sol in benzene, chloroform, ether, carbon disulfide, oils; misc with fats.

SYNS: PARAFFIN WAX ◇ PARAFFIN WAX FUME (ACGIH)

USE IN FOOD:

Purpose: Adhesive component, coatings, masticatory substance in chewing gum base.

Where Used: Chewing gum.

Regulations: FDA - 21CFR 172.615, 175.105, 175.250, 178.3800. Use in amounts not to exceed those required to produce the intended physical or other technical effect.

ACGIH TLV: TWA 2 mg/m^3 (fume)

SAFETY PROFILE: The semi-refined, fully refined, and the crude paraffins are experimental tumorigens by the implant route. Many paraffin waxes contain carcinogens.

TOXICITY DATA and CODEN

imp-rat TDLo: 120 mg/kg: ETA CNREA8 33,1225,73
imp-mus TD : 660 mg/kg: ETA CALEDQ 6,21,79

PAI000 CAS: 30525-89-4
PARAFORMALDEHYDE

mf: $(CH_2O)_n$

PROP: White crystals; odor of formaldehyde. Flash p: 158°F, autoign temp: 572°F. Sltly sol in cold water; moderately sol in hot water yielding formaldehyde.

SYNS: FLO-MOR ◇ FORMAGENE ◇ PARAFORSN ◇ TRIFORMOL ◇ TRIOXYMETHYLENE

USE IN FOOD:

Purpose: Insecticide.

Where Used: Maple syrup.

Regulations: FDA - 21CFR 193.330. Pesticide residue tolerance of 2 ppm of formaldehyde in maple syrup. DOT Classification: ORM-A; Label: None; DOT-IMO: Flammable Solid; Label: None

SAFETY PROFILE: Moderately toxic by ingestion. A severe eye and skin irritant. Mutagenic data. Flammable when exposed to heat or flame; can react with oxidizing materials. To fight fire, use alcohol foam, CO_2, dry chemical. Incompatible with liquid oxygen. Dangerous; when heated to decomposition it emits toxic formaldehyde gas.

TOXICITY DATA and CODEN

skn-rbt 500 mg/2H SEV BIOFX* 28-4/73
eye-rbt 100 mg SEV BIOFX* 28-4/73
otr-rat: emb 2200 ng/plate JJATDK 1,190,81
orl-rat LD50: 800 mg/kg 28ZEAL 4,308,69
skn-rbt LDLo: 10000 mg/kg BIOFX* 28-4/73

PAI990 CAS: 4685-14-7
PARAQUAT

mf: $C_{12}H_{14}N_2$ mw: 186.28

SYNS: DIMETHYL VIOLOGEN ◇ GRAMOXONE S ◇ METHYL VIOLOGEN (2+) ◇ PARAQUAT DICATION

USE IN FOOD:

Purpose: Defoliant, desiccant, herbicide.

Where Used: Animal feed, beef, goat, hops (dried), lamb, mint hay (spent), peanuts, pork, sunflower seed hulls.

Regulations: FDA - 21CFR 193.331. Pesticide residue tolerance of 0.2 ppm in dried hops. 21CFR 561.289. Limitation of 3.0 ppm in spent mint hay, 6.0 ppm in sunflower seed hulls when used for animal feed. USDA CES Ranking: A-4 (1986).

EPA Genetic Toxicology Program.

OSHA PEL: TWA 0.1 mg/m^3 Respirable Dust (skin) ACGIH TLV: TWA 0.1 mg/m^3

SAFETY PROFILE: Poison by ingestion and intraperitoneal routes. Mutagenic data. Causes ulceration of digestive tract, diarrhea, vomiting, renal damage, jaundice, edema, hemorrhage, fibrosis of lung, and death from anoxia may result. When heated to decomposition it emits toxic fumes of NO_x.

TOXICITY DATA and CODEN

mmo-omi 20 ppm MUREAV 138,39,84
orl-rat LD50: 150 mg/kg FMCHA2 -,C118,83
orl-mus LD50: 120 mg/kg GEPHDP 14,541,83

PAL750 CAS: 8000-68-8
PARSLEY OIL

PROP: From steam distillation of above ground parts (herb oil) or ripe seed (seed oil) of *Petroselinium sativum* Hoffm. (Fam. *Umbelligerae*). Yellow to light brown liquid; odor of parsley. D (herb oil): 0.908-0.940, (seed oil): 1.040; refr index (herb oil): 1.503-1.530 @ 20°, (seed oil): 1.513-1.522 @ 20°. Sol in fixed oils, mineral oil; sltly sol in propylene glycol; insol in glycerin.

SYNS: OIL of PARSLEY ◇ PARSLEY HERB OIL (FCC) ◇ PARSLEY SEED OIL (FCC) ◇ PETERSILIENSAMEN OEL (GERMAN)

USE IN FOOD:

Purpose: Flavoring agent.

Where Used: Various.

Regulations: FDA - 21CFR 182.10, 182.20. GRAS when used at a level not in excess of the amount reasonably required to accomplish the intended effect.

SAFETY PROFILE: Moderately toxic by ingestion. A human skin irritant. When heated to decomposition it emits acrid smoke and irritating fumes.

TOXICITY DATA and CODEN

skn-hmn 10 mg/48H MLD FCTXAV 13,681,75
skn-rbt 500 mg/24H MLD FCTXAV 13,681,75
orl-rat LD50:3300 mg/kg FCTOD7 21,871,83

PAO000 CAS: 8002-03-7
PEANUT OIL

PROP: Straw-yellow to greenish-yellow or nearly colorless oil; nutty odor and bland taste. Mp: 2.7°, flash p: 540°F, d: 0.92, autoign temp: 833°F. Misc with ether, petr ether, chloroform, carbon disulfide; sol in benzene, carbon tetrachloride, oils; very sltly sol in alc. From seed of *Arachis hypogaea* (85DIA2 2,201,77).

SYNS: ARACHIS OIL ◇ EARTHNUT OIL ◇ GROUNDNUT OIL ◇ INDIGENOUS PEANUT OIL ◇ KATCHUNG OIL ◇ PECAN SHELL POWDER

USE IN FOOD:

Purpose: Coating agent, emulsifying agent, formulation aid, texturizer.

Where Used: Salad oil.

Regulations: GRAS when used in accordance with good manufacturing practice.

SAFETY PROFILE: An experimental tumorigen. A human skin irritant and mild allergen. Mutagenic data. Combustible when exposed to heat or flame; can react with oxidizing materials. To fight fire, use CO_2, dry chemical. When heated to decomposition it emits acrid smoke and irritating fumes.

TOXICITY DATA and CODEN

skn-hmn 300 mg/3D-I MLD 85DKA8 -,127,77
mma-sat 10 μL/plate FCTXAV 18,467,80
orl-mus TDLo:952 g/kg/1Y-I:ETA IJMRAQ 61,422,73
skn-mus TD :2276 g/kg/81W-C:ETA IJCNAW 10,652,72

PAO150 CAS: 9000-69-5
PECTIN

PROP: From citrus peel, apple pomace, or beet pulp. Yellow-white powder. Sol in water; insol in alc.

USE IN FOOD:

Purpose: Emulsifier, gelling agent, stabilizer, thickener.

Where Used: Beverages, jams, jellies.

Regulations: FDA - 21CFR 184.184.1588. GRAS when used in accordance with good manufacturing practice.

SAFETY PROFILE: When heated to decomposition it emits acrid smoke and irritating fumes.

PAQ000 CAS: 1406-05-9
PENICILLIN

mf: $(CH_3)_2C_5H_3NSO(COOH)NHCOOR$ (bicyclic)

PROP: A group of isomeric and closely related antibiotic compounds with outstanding bacterial activity. An extract from *Penicillium notatum* (JPETAB 77,40,43). Different varieties of penicillin are produced by adding the proper precursors to the nutrient solution.

SYN: PENIZILLIN (GERMAN)

USE IN FOOD:

Purpose: Animal drug, animal feed drug.

Where Used: Animal feed, beef, chicken, eggs, lamb, milk, pheasants, pork, quail, turkey.

Regulations: FDA - 21CFR 556.510. Limitation of 0.05 ppm in cattle. Limitation of zero in chickens, pheasants, quail, swine, sheep, eggs, milk, foods in which such milk has been used. Limitation of 0.01 ppm in turkeys. 21CFR 558.460.

EPA Genetic Toxicology Program.

SAFETY PROFILE: Human reproductive effects by ingestion: abortion. Human systemic effects by intramuscular route: dermatitis. Experimental reproductive effects. Has been implicated in aplastic anemia. When heated to decomposition it emits very toxic fumes of NO_x and SO_x.

TOXICITY DATA and CODEN

orl-wmn TDLo: 72 mg/kg (13-15W preg): REP BSFDA3 57,534,50
ims-man TDLo: 12385 μg/kg/16D-I: SKN MJAUAJ 1,305,47
scu-mus LDLo: 3200 mg/kg JPETAB 77,70,43

PAR500 PENNYROYAL OIL

PROP: Chief constituent is d-pulegone. From steam distillation of *Mentha pulegium L.* (Fam. *Labiatae*) (FCTXAV 12,807,74). Light yellow liquid; mint odor. Sol in fixed oils, propylene glycol, mineral oil; insol in glycerin.

SYN: AMERICAN PENNYROYAL OIL

USE IN FOOD:

Purpose: Flavoring agent.

Where Used: Various.

Regulations: FDA - 21CFR 172.10. Use at a level not in excess of the amount reasonably required to accomplish the intended effect.

SAFETY PROFILE: Experimental poison by ingestion. A skin irritant. When heated to decomposition it emits acrid smoke and irritating fumes.

TOXICITY DATA and CODEN

skn-mus 100% MOD FCTXAV 12,949,74
orl-rat LD50: 400 mg/kg FCTXAV 12,949,74

PAX250 PENTACHLOROPHENOL CAS: 87-86-5

DOT: 2020
mf: C_6HCl_5O mw: 266.32

PROP: Dark-colored flakes and sublimed needle crystals; characteristic odor. Mp: 191°, bp: 310° (decomp), d: 1.978, vap press: 40 mm @ 211.2°. Sol in ether, benzene; very sol in alc; insol in water; sltly sol in cold petr ether.

SYNS: CHEM-TOL ◇ CHLOROPHEN ◇ CRYPTOGIL OL ◇ DOWCIDE 7 ◇ DUROTOX ◇ FUNGIFEN ◇ GLAZD PENTA ◇ GRUNDIER ARBEZOL ◇ LAUXTOL ◇ LIROPREM ◇ NCI-C54933 ◇ PCP ◇ PENCHLOROL ◇ PENTA ◇ PENTACHLOORFENOL (DUTCH) ◇ PENTACHLOROFENOL ◇ PENTACHLOROPHENATE ◇ 2,3,4,5,6-PENTACHLOROPHENOL ◇ PENTACHLOROPHENOL (GERMAN) ◇ PENTACLOROFENOLO (ITALIAN) ◇ PENTACON ◇ PENTA-KIL ◇ PENTASOL ◇ PENWAR ◇ PERATOX ◇ PERMACIDE ◇ PERMAGARD ◇ PERMASAN ◇ PERMATOX DP-2 ◇ PERMITE ◇ PRILTOX ◇ RCRA WASTE NUMBER U242 ◇ SANTOBRITE ◇ SANTOPHEN ◇ SINITUHO ◇ TERM-I-TROL ◇ THOMPSON'S WOOD FIX ◇ WEEDONE

USE IN FOOD:

Purpose: Preservative.

Where Used: Packaging materials.

Regulations: FDA - 21CFR 178.3800. Limitation of 50 ppm in treated wood. USDA CES Ranking: B-1 (1985).

IARC Cancer Review: Human Limited Evidence IMEMDT 41,319,86; Animal Inadequate Evidence IMEMDT 20,303,79. EPA Extremely Hazardous Substances List. Chlorophenol compounds are on the Community Right-To-Know List. EPA Genetic Toxicology Program.

OSHA PEL: TWA 500 $\mu g/m^3$ (skin) ACGIH TLV: TWA 0.5 mg/m^3 (skin) DFG MAK: 0.05 ppm (0.5 mg/m^3) DOT Classification: ORM-E; Label: None

SAFETY PROFILE: Human poison by ingestion. Poison experimentally by ingestion, skin contact, intraperitoneal, and subcutaneous routes. A suspected human carcinogen. An experimental teratogen. A skin irritant. Mutagenic data. Acute poisoning is marked by weakness with changes in respiration, blood pressure, and urinary output. Also causes dermatitis, convulsions, and collapse. Chronic exposure can cause liver and kidney injury. Dangerous; when heated to decomposition it emits highly toxic fumes of Cl^-.

TOXICITY DATA and CODEN

skn-rbt 10 mg/24H open MLD AIHAAP 23,95,62
mma-sat 40 nmol/plate AIDZAC 10,305,82
orl-rat TDLo: 60 mg/kg/(9D preg): TER DABBBA 37,1184,76

scu-mus TDLo: 46 mg/kg: ETA NTIS** PB223-159
orl-hmn LDLo: 29 mg/kg 27ZXA3 -,256,63
orl-mus LD50: 117 mg/kg TOLED5 29,39,85

PBB800
PENTAERYTHRITOL ESTER of PARTIALLY HYDROGENATED WOOD ROSIN

PROP: Hard, amber-colored solid. Sol in acetone, benzene; insol in water.

USE IN FOOD:

Purpose: Coatings, masticatory substance in chewing gum base.

Where Used: Citrus fruit (fresh).

Regulations: FDA - 21CFR 172.210.

SAFETY PROFILE: When heated to decomposition it emits acrid smoke and irritating fumes.

PBB810
PENTAERYTHRITOL ESTER of WOOD ROSIN

PROP: Hard, amber-colored solid. Sol in acetone, benzene; insol in water.

USE IN FOOD:

Purpose: Masticatory substance in chewing gum base.

Where Used: Chewing gum.

Regulations: FDA - 21CFR 172.615. Use in amounts not to exceed those required to produce the intended physical or other technical effect.

SAFETY PROFILE: When heated to decomposition it emits acrid smoke and irritating fumes.

PBK250 CAS: 109-66-0
PENTANE

DOT: 1265
mf: C_5H_{12} mw: 72.17

PROP: Colorless liquid. Bp: 36.1°, flash p: <−40°F, fp: −129.8°, d: 0.626 @ 20°/4°, autoign temp: 588°F, vap press: 400 mm @ 18.5°, vap d: 2.48, lel: 1.5%, uel: 7.8%. Sol in water; misc in alc, ether, organic solvents.

SYNS: AMYL HYDRIDE (DOT) ◇ PENTAN (POLISH) ◇ PENTANEN (DUTCH) ◇ PENTANI (ITALIAN)

USE IN FOOD:

Purpose: Adjuvant in foamed plastic, blowing agent.

Where Used: Packaging materials.

Regulations: FDA - 21CFR 178.3010.

OSHA PEL: TWA 1000 ppm; STEL 750 ppm ACGIH TLV: TWA 600 ppm; STEL 750 ppm DFG MAK: 1000 ppm (2950 mg/m^3) DOT Classification: Flammable Liquid; Label: Flammable Liquid NIOSH REL: TWA 350 mg/m^3

SAFETY PROFILE: Moderately toxic by intravenous route. Narcotic in high concentration. The liquid can cause blisters on contact. Flammable liquid. Highly dangerous fire hazard when exposed to heat, flame, or oxidizers. Severe explosion hazard when exposed to heat or flame. Shock can shatter metal containers and release contents. To fight fire, use foam, CO_2, dry chemical. When heated to decomposition it emits acrid smoke and irritating fumes.

TOXICITY DATA and CODEN

ivn-mus LD50: 446 mg/kg JPMSAE 67,566,78

PBN250 CAS: 107-87-9
2-PENTANONE

DOT: 1249
mf: $C_5H_{10}O$ mw: 86.15

PROP: Water-white liquid; fruity, ethereal odor. D: 0.801-.0.806, vap d: 3.0, bp: 216°F, flash p: 45°F, autoign temp: 941°F, lel: 1.5%, uel: 8.2%. Sltly sol in water; misc with alc, ether.

SYNS: ETHYL ACETONE ◇ FEMA No. 2842 ◇ METHYLPROPYL-CETONE (FRENCH) ◇ METHYL-n-PROPYL KETONE ◇ METHYL PROPYL KETONE (ACGIH, DOT) ◇ METYLOPROPYLOKETON (POLISH) ◇ MPK

USE IN FOOD:

Purpose: Flavoring agent.

Where Used: Various.

Regulations: FDA - 21CFR 172.515. Use at a level not in excess of the amount reasonably required to accomplish the intended effect.

OSHA PEL: TWA 200 ppm; STEL 250 ppm ACGIH TLV: TWA 200 ppm; STEL 250 ppm DFG MAK: 200 ppm (700 mg/m^3) DOT Classification: Label: Flammable Liquid NIOSH REL: TWA 530 mg/m^3

SAFETY PROFILE: Moderately toxic by ingestion and intraperitoneal routes. Mildly toxic by skin contact and inhalation. Human systemic effects by inhalation: headache, nausea, irritation of the respiratory passages, eyes, and skin. A skin irritant. Mutagenic data. Highly flammable liquid. A very dangerous fire hazard when exposed to heat or flame; can react vigorously with oxidizing materials. An explosion hazard in the form of vapor when exposed to heat or flame. To fight fire, use alcohol foam. When heated to decomposition it emits acrid smoke and irritating fumes.

TOXICITY DATA and CODEN

skn-rbt 405 mg open MLD UCDS** 7/20/67
sln-smc 13600 ppm MUREAV 149,339,85
ihl-hmn TCLo:1500 ppm:EYE,CNS,GIT NPIRI* 1,83,74
orl-rat LD50:3730 mg/kg AIHAAP 23,95,62
ihl-rat LCLo:2000 ppm/4H AIHAAP 23,95,62

PCB250 CAS: 8006-90-4
PEPPERMINT OIL

PROP: From steam distillation of *Mentha piperita* L. (Fam. *Labiatae*). Colorless to pale yellow liquid; strong odor and taste of peppermint. D: 0.896-0.908 @ 25°/25°, refr index: 1.459 @ 20°.

SYN: PFEFFERMINZ OEL (GERMAN)

USE IN FOOD:

Purpose: Flavoring agent.

Where Used: Bakery products, beverages (alcoholic), beverages (nonalcoholic), chewing gum, confections, fruit cocktails, gelatin desserts, ice cream, meat, puddings, sauces, toppings.

Regulations: FDA - 21CFR 182.10, 182.20. GRAS when used at a level not in excess of the amount reasonably required to accomplish the intended effect.

SAFETY PROFILE: Moderately toxic by ingestion and intraperitoneal routes. An allergen. Mutagenic data. When heated to decomposition it emits acrid smoke and irritating fumes.

TOXICITY DATA and CODEN

dnr-bcs 5 μL/disc TOFOD5 8,91,85
orl-rat LD50:2426 mg/kg JPMSAE 54,1071,65
ipr-rat LD50:819 mg/kg JPMSAE 54,1071,65

PCJ400
PERLITE

PROP: Average density of 0.13. Expands when finely ground and heated. Natural glass, amorphous mineral consisting of fused sodium potassium aluminum silicate, containing <1% quartz.

USE IN FOOD:

Purpose: Filter aid.

Where Used: Various.

Regulations: GRAS when used in accordance with good manufacturing practice.

ACGIH TLV: TWA 10 mg/m^3 of total dust.

SAFETY PROFILE: A nuisance dust.

PCR100
PETITGRAIN OIL, PARAGUAY TYPE

PROP: From steam distillation of the leaves of *Citrus aurantium* L. subspecies *amara*. Yellow to brown liquid; harsh bitter odor. D: 0.878-0.889, refr index: 1.455 @ 20°. Sol in fixed oils, mineral oil, propylene glycol; insol in glycerin.

USE IN FOOD:

Purpose: Flavoring agent.

Where Used: Bakery products, beverages (nonalcoholic), chewing gum, condiments, gelatin desserts, ice cream, puddings.

Regulations: FDA - 21CFR 182.20. GRAS when used at a level not in excess of the amount reasonably required to accomplish the intended effect.

SAFETY PROFILE: When heated to decomposition it emits acrid smoke and irritating fumes.

PCR200
PETROLATUM

PROP: White to amber solid. D: 0.815-0.880, mp: 38-60°. Insol in water, alc; sol ether, solvent hexane, oil, benzene, carbon disulfide, chloroform, turpentine.

SYNS: WHITE PETROLATUM ◇ YELLOW PETROLATUM

USE IN FOOD:

Purpose: Defoaming agent, lubricant, polishing agent, protective coating, release agent, sealing agent.

Where Used: Bakery products, beet sugar, confectionery, egg white solids, fruits (dehydrated), vegetables (dehydrated), yeast.

Regulations: FDA - 21CFR 172.880. Limitation of 0.15 percent in bakery products, 0.2 percent in confectionery, 0.02 percent in dehydrated fruits and vegetables, 0.1 percent in egg white solids. 21CFR 172.884, 173.340. 21CFR 175.105, 175.176, 177.2600, 177.2800. May contain FDA approved antioxidants (21CFR 409) in amounts no greater than required to produce the intended effect.

SAFETY PROFILE: When heated to decomposition it emits acrid smoke and irritating fumes.

PCT600
PETROLEUM WAX

PROP: Translucent; tasteless and odorless wax. Mp: 48-93°. Insol in water; very sltly sol in organic solvents.

SYNS: MICROCRYSTALLINE WAX ◇ PETROLEUM WAX, SYNTHETIC (FCC) ◇ REFINED PETROLEUM WAX

USE IN FOOD:

Purpose: Defoaming agent, masticatory substance in chewing gum base, protective coating.

Where Used: Beet sugar, cheese, chewing gum, fruit (raw), fruit defoamer, pizza (frozen), vegetables (raw), yeast.

Regulations: FDA - 21CFR 172.230, 172.615. Must conform to 21CFR 172.886. Limitation of 1,050 ppm of poly(alkylacrylate) as an antioxidant. 21CFR 172.888. Use at a level not in excess of the amount reasonably required to accomplish the intended effect. 21CFR 173.340, 178.3710, 178.3720. May contain FDA approved antioxidants (21CFR 409) in amounts no greater than required to produce the intended effect.

SAFETY PROFILE: When heated to decomposition it emits acrid smoke and irritating fumes.

PDD750 CAS: 60-12-8
PHENETHYL ALCOHOL

mf: $C_8H_{10}O$ mw: 122.18

PROP: Colorless liquid; floral odor of roses. Mp: −27°, bp: 220°, flash p: 216°F, d: 1.0245 @ 15°, vap d: 4.21. Misc with alc, ether; sol in fixed oils, glycerin, propylene glycol.

SYNS: BENZYL CARBINOL ◇ FEMA No. 2858 ◇ PHENETHANOL ◇ β-PHENETHYL ALCOHOL ◇ 2-PHENETHYL ALCOHOL ◇ β-PHENYLETHANOL ◇ 2-PHENYLETHANOL ◇ β-PHENYLETHYL ALCOHOL ◇ 2-PHENYLETHYL ALCOHOL

USE IN FOOD:

Purpose: Flavoring agent.

Where Used: Various.

Regulations: FDA - 21CFR 172.515. Use at a level not in excess of the amount reasonably required to accomplish the intended effect.

SAFETY PROFILE: Poison by ingestion and intraperitoneal routes. Moderately toxic by skin contact. A skin and eye irritant. Experimental teratogenic effects. Causes severe central nervous system injury to experimental animals. Combustible when exposed to heat or flame; can react with oxidizing materials. To fight fire, use CO_2, dry chemical. When heated to decomposition it emits acrid smoke and irritating fumes.

TOXICITY DATA and CODEN

eye-rbt 12 g/10M MLD ARZNAD 9,349,59
skn-gpg 100 mg MLD FCTXAV 13,903,75
orl-rat TDLo:43 mg/kg (6-15D preg):TER JTEHD6 12,235,83
orl-rat TDLo:430 mg/kg (6-15D preg):TER JTEHD6 12,235,83
orl-rat LD50:1790 mg/kg FCTXAV 2,327,64

PDE000 CAS: 98-85-1
α-PHENETHYL ALCOHOL

DOT: 2937
mf: $C_8H_{10}O$ mw: 122.18

PROP: Colorless liquid; hyacinth odor. Bp: 204°, fp: 21.4°, d: 1.015 @ 20°/20°, refr index: 1.525, vap press: 0.1 mm @ 20°, vap d: 4.21, flash p: 205°F (OC). Sol in fixed oils, propylene glycol; very sol in glycerin.

SYNS: FEMA No. 2685 ◇ α-METHYLBENZYL ALCOHOL (FCC) ◇ METHYLPHENYLCARBINOL ◇ METHYPHENYLMETHANOL ◇ NCI-C55685 ◇ 1-PHENYLETHANOL ◇ PHENYLMETHYLCARBINOL ◇ STYRALLYL ALCOHOL ◇ STYRALYL ALCOHOL

USE IN FOOD:

Purpose: Flavoring agent.

Where Used: Various.

Regulations: FDA - 21CFR 172.515. Use at a level not in excess of the amount reasonably required to accomplish the intended effect.

EPA Genetic Toxicology Program.

DOT Classification: Poison B; Label: St. Andrews Cross

SAFETY PROFILE: Poison by ingestion and subcutaneous routes. Moderately toxic by skin contact. A skin and severe eye irritant. Combustible when exposed to heat or flame; can react with oxidizing materials. To fight fire, use alcohol foam, foam, CO_2, dry chemical.

TOXICITY DATA and CODEN

skn-rbt 500 mg/24H MOD FCTXAV 12,995,74
eye-rbt 2 mg SEV AJOPAA 29,1363,46
orl-rat LD50:400 mg/kg JIHTAB 26,269,44

PDF750 CAS: 103-48-0
PHENETHYL ISOBUTYRATE
mf: $C_{12}H_{16}O_2$ mw: 192.28

PROP: Colorless to light yellow liquid; fruity, rosy odor. D: 0.9871.486-1.490, flash p: +212°F. Sol in alc, fixed oils; insol in water @ 230°.

SYNS: BENZYLCARBINOL ISOBUTYRATE ◇ BENZYLCARBINYL ISOBUTYRATE ◇ FEMA No. 2862 ◇ PHENYLETHYL ISOBUTYRATE ◇ β-PHENYLETHYL ISOBUTYRATE ◇ 2-PHENYLETHYL ISOBUTYRATE ◇ 2-PHENYLETHYL-2-METHYLPROPIONATE

USE IN FOOD:

Purpose: Flavoring agent.

Where Used: Baked goods, beverages, candy, ice cream.

Regulations: FDA - 21CFR 172.515. Use at a level not in excess of the amount reasonably required to accomplish the intended effect.

SAFETY PROFILE: Mildly toxic by ingestion. Combustible liquid. When heated to decomposition it emits acrid smoke and irritating fumes.

TOXICITY DATA and CODEN

orl-rat LD50:5200 mg/kg FCTXAV 16,637,78

PDF775 CAS: 140-26-1
PHENETHYL ISOVALERATE
mf: $C_{13}H_{18}O_2$ mw: 206.31

PROP: Colorless to sltly yellow liquid; fruity, rosy odor. D: 0.973, refr index: 1.484, flash p: +212°F. Sol in alc, fixed oils; insol in water @ 263°.

SYNS: FEMA No. 2871 ◇ 3-METHYL-BUTANOIC ACID 2-PHENYLETHYL ESTER ◇ PHENETHYL ESTER ISOVALERIC ACID ◇ PHENYLETHYL ISOVALERATE ◇ β-PHENYLETHYL ISOVALERATE ◇ 2-PHENYLETHYL-3-METHYLBUTIRATE

USE IN FOOD:

Purpose: Flavoring agent.

Where Used: Various.

Regulations: FDA - 21CFR 172.515. Use at a level not in excess of the amount reasonably required to accomplish the intended effect.

SAFETY PROFILE: Mildly toxic by ingestion. Combustible liquid. When heated to decomposition it emits acrid smoke and irritating fumes.

TOXICITY DATA and CODEN

orl-rat LD50:6220 mg/kg VPITAR 33(5),48,74

PDF790
2-PHENETHYL 2-METHYLBUTYRATE
mf: $C_{13}H_{18}O_2$ mw: 206.28

PROP: Colorless liquid; floral, fruity odor. D: 0.973, refr index: 1.484, flash p: +212°F. Sol in alc, fixed oils; insol in water.

SYN: FEMA No. 3632

USE IN FOOD:

Purpose: Flavoring agent.

Where Used: Various.

Regulations: FDA - 21CFR 172.515. Use at a level not in excess of the amount reasonably required to accomplish the intended effect.

SAFETY PROFILE: Combustible liquid. When heated to decomposition it emits acrid smoke and irritating fumes.

PDI000 CAS: 102-20-5
PHENETHYL PHENYLACETATE
mf: $C_{16}H_{16}O_2$ mw: 240.32

PROP: Colorless to sltly yellow liquid above 26°; rosy, hyacinth odor. D: 1.079-1.082, flash p: +212°F. Sol in alc; insol in water.

SYNS: BENZENEACETIC ACID, 2-PHENYLETHYL ESTER ◇ BENZYLCARBINYL-α-TOLUATE ◇ FEMA No. 2866 ◇ PHENYLACETIC ACID, PHENETHYL ESTER ◇ β-PHENYLETHYL PHENYLACETATE ◇ 2-PHENYLETHYL PHENYLACETATE ◇ 2-PHENYLETHYL-α-TOLUATE

USE IN FOOD:

Purpose: Flavoring agent.

Where Used: Various.

Regulations: FDA - 21CFR 172.515. Use at a level not in excess of the amount reasonably required to accomplish the intended effect.

SAFETY PROFILE: Moderately toxic by ingestion. Combustible liquid. When heated to decomposition it emits acrid smoke and irritating fumes.

TOXICITY DATA and CODEN

orl-mus LD50:3190 mg/kg VPITAR 33(5),48,74

PDK200
PHENETHYL SALICYLATE

mf: $C_{15}H_{14}O_3$ mw: 242.27

PROP: White crystals; balsamic odor. Solidification point: 41°, flash p: +212°F. Sol in alc; insol in water.

SYN: FEMA No. 2868

USE IN FOOD:

Purpose: Flavoring agent.

Where Used: Various.

Regulations: FDA - 21CFR 172.515. Use at a level not in excess of the amount reasonably required to accomplish the intended effect.

SAFETY PROFILE: Combustible liquid. When heated to decomposition it emits acrid smoke and irritating fumes.

PDS900
PHENOXYETHYL ISOBUTYRATE

mf: $C_{12}H_{16}O_3$ mw: 208.26

PROP: Colorless liquid; honey, roselike odor. D: 1.044, refr index: 1.492, flash p: +212°F. Misc in alc, chloroform, ether; insol in water.

SYN: FEMA No. 2873

USE IN FOOD:

Purpose: Flavoring agent.

Where Used: Various.

Regulations: FDA - 21CFR 172.515. Use at a level not in excess of the amount reasonably required to accomplish the intended effect.

SAFETY PROFILE: Combustible liquid. When heated to decomposition it emits acrid smoke and irritating fumes.

PDX000 CAS: 101-48-4
PHENYLACETALDEHYDE DIMETHYL ACETAL

mf: $C_{10}H_{14}O_2$ mw: 166.24

PROP: Colorless liquid; strong odor. D: 1.000-1.006, refr index: 1.493, flash p: 194°F. Sol in fixed oils, propylene glycol; insol in glycerin.

SYNS: (2,2-DIMETHOXYETHYL)-BENZENE (9CI) ◇ 1,1-DIMETHOXY-2-PHENYLETHANE ◇ FEMA No. 2876 ◇ HYSCYLENE P ◇ PHENACETALDEHYDE DIMETHYL ACETAL ◇ α-TOLYL ALDEHYDE DIMETHYL ACETAL ◇ VIRIDINE

USE IN FOOD:

Purpose: Flavoring agent.

Where Used: Various.

Regulations: FDA - 21CFR 172.515. Use at a level not in excess of the amount reasonably required to accomplish the intended effect.

SAFETY PROFILE: Moderately toxic by ingestion. Combustible liquid. When heated to decomposition it emits acrid smoke and irritating fumes.

TOXICITY DATA and CODEN

orl-rat LD50:3500 mg/kg FCTXAV 13,681,75

PDY850 CAS: 103-82-2
PHENYLACETIC ACID

mf: $C_8H_8O_2$ mw: 136.16

PROP: Leaflets on distillation in vacuo; plates, tablets from petr ether; disagreeable odor of geranium. Mp: 76.5°, bp: 265.5°, d (77/4) 1.091, flash p: +212°F. Sltly sol in cold water; freely in hot water. Sol in alc, ether. Solubility @ 25° in chloroform (moles/L): 4.422; in carbon tetrachloride: 1.842; in acetylene tetrachloride: 4.513; in trichlorethylene: 3.299; in tetrachlorethylene: 1.558; in pentachloroethane: 3.252.

SYNS: BENZENACETIC ACID ◇ BENZENEACETIC ACID ◇ FEMA No. 2878 ◇ omega-PHENYLACETIC ACID ◇ α-TOLUIC ACID

USE IN FOOD:

Purpose: Flavoring agent.

Where Used: Various.

Regulations: FDA - 21CFR 172.515. Use at a level not in excess of the amount reasonably required to accomplish the intended effect.

SAFETY PROFILE: Moderately toxic by ingestion, subcutaneous and intraperitoneal routes. An experimental teratogen. Combustible liquid. When heated to decomposition it emits acrid smoke and irritating fumes.

TOXICITY DATA and CODEN

orl-rat TDLo:450 mg/kg (4D preg):TER VPITAR 32,50,73

orl-rat LD50:2250 mg/kg VPITAR 33(5),48,74

PEC500 CAS: 673-06-3
d-PHENYLALANINE
mf: $C_9H_{11}NO_2$ mw: 165.21

PROP: Needles from alc, white crystalline platlets. Mp: 104-105°; Sol in hot water; very sltly sol in alc; sltly sol petr ether.

SYNS: dl-α-AMINO-β-PHENYLPROPIONIC ACID ◇ NCI-C60195 ◇ d-β-PHENYLALANINE ◇ dl-PHENYLALANINE (FCC)

USE IN FOOD:

Purpose: Dietary supplement, nutrient.

Where Used: Various.

Regulations: FDA - 21CFR 172.310. Limitation 5.8 percent by weight.

SAFETY PROFILE: Mildly toxic by intraperitoneal route. Human systemic effects by ingestion: nausea, hypermotility, diarrhea. When heated to decomposition it emits toxic fumes of NO_x.

TOXICITY DATA and CODEN

orl-hmn TDLo: 500 mg/kg/5W-I:GIT JACTDZ 1(3),124,82
ipr-rat LD50: 5452 mg/kg ABBIA4 64,319,56

PEC750 CAS: 63-91-2
l-PHENYLALANINE
mf: $C_9H_{11}NO_2$ mw: 165.21

PROP: White crystals or crystalline powder; slt odor and bitter taste. Mp: decomp @ 275-283°. Sol in water; very sltly sol in alc, ether.

SYNS: (S)-α-AMINOBENZENEPROPANOIC ACID ◇ α-AMINOHYDROCINNAMIC ACID ◇ α-AMINO-β-PHENYLPROPIONIC ACID ◇ ANTIBIOTIC FN 1636 ◇ PAL ◇ PHENYLALANINE ◇ PHENYL-α-ALANINE ◇ (S)-PHENYLALANINE ◇ β-PHENYLALANINE ◇ l-β-PHENYLALANINE ◇ 3-PHENYLALANINE

USE IN FOOD:

Purpose: Dietary supplement, nutrient.

Where Used: Various.

Regulations: FDA - 21CFR 172.310. Limitation 5.8 percent by weight.

SAFETY PROFILE: Mildly toxic by intraperitoneal route. Experimental reproductive effects. When heated to decomposition it emits toxic fumes of NO_x.

TOXICITY DATA and CODEN

orl-rat TDLo: 220 g/kg (2W male/2W pre-3W post): REP NETOD7 1,79,79
orl-mky TDLo: 33600 mg/kg (1-24W preg): REP PEDIAU 42,27,68
ipr-rat LD50: 5287 mg/kg ABBIA4 58,253,55

PEJ750 CAS: 85-84-7
1-(PHENYLAZO)-2-NAPHTHYLAMINE
mf: $C_{16}H_{13}N_3$ mw: 247.32

SYNS: A.F YELLOW NO. 2 ◇ 1-BENZENE-AZO-β-NAPHTHYLAMINE ◇ 1-BENZENEAZO-2-NAPHTHYLAMINE ◇ CERISOL YELLOW AB ◇ C.I. 11380 ◇ C.I. FOOD YELLOW 10 ◇ C.I. SOLVENT YELLOW 5 ◇ DOLKWAL YELLOW AB ◇ EXT. D & C YELLOW NO. 9 ◇ FD & C YELLOW NO. 3 ◇ GRASAL YELLOW ◇ JAUNE AB ◇ OIL YELLOW A ◇ 1-(PHENYLAZO)-2-NAPHTHALENAMINE ◇ YELLOW AB ◇ YELLOW NO. 2

USE IN FOOD:

Purpose: Color additive.

Where Used: Various.

Regulations: FDA - 21CFR 81.10. Provisional listing terminated.

IARC Cancer Review: Animal No Evidence IMEMDT 8,279,75. EPA Genetic Toxicology Program.

SAFETY PROFILE: Moderately toxic by ingestion and subcutaneous routes. An experimental tumorigen. Mutagenic data. When heated to decomposition it emits toxic fumes of NO_x.

TOXICITY DATA and CODEN

dns-rat-orl 500 mg/kg ENMUDM 7,101,85
orl-rat TDLo: 8190 mg/kg/65W-C:ETA JPPMAB 7,591,55
orl-rbt LDLo: 1000 mg/kg JBCHA3 27,403,16
scu-rbt LDLo: 1000 mg/kg JBCHA3 27,403,16

PFB250 CAS: 103-45-7
2-PHENYLETHYL ACETATE
mf: $C_{10}H_{12}O_2$ mw: 164.22

PROP: Colorless liquid; sweet, rosy, honey odor. Mp: 164.2°, bp: 223.6°, fp: $<-20°$, flash p: 230°F, d: 1.032 @ 25°/25°, refr index: 1.497-1.501. Sol in alc, fixed oils, propylene glycol; insol in glycerin, water @ 232°.

SYNS: ACETIC ACID-2-PHENYLETHYL ESTER ◇ BENZYLCARBINYL ACETATE ◇ FEMA No. 2857 ◇ β-PHENETHYL ACETATE ◇ 2-PHENETHYL ACETATE ◇ β-PHENYLETHYL ACETATE

USE IN FOOD:

Purpose: Flavoring agent.

Where Used: Various.

Regulations: FDA - 21CFR 172.515. Use at a level not in excess of the amount reasonably required to accomplish the intended effect.

SAFETY PROFILE: Moderately toxic by ingestion. Mildly toxic by skin contact. Combustible when exposed to heat or flame; can react vigorously with oxidizing materials. To fight fire, use alcohol foam, CO_2 and dry chemical. When heated to decomposition it emits acrid smoke and irritating fumes.

TOXICITY DATA and CODEN

orl-rat LD50:3670 mg/kg VPITAR 33(5),48,74

PGA800
2-PHENYLPROPIONALDEHYDE DIMETHYL ACETAL

mf: $C_{11}H_{16}O_2$ mw: 180.25

PROP: Colorless to sltly yellow liquid; mushroom odor. D: 0.989-0.994, refr index: 1.492-1.497. Sol in alc, ether; insol in water.

SYNS: FEMA No. 2888 ◇ HYDRATROPIC ALDEHYDE DIMETHYL ACETAL

USE IN FOOD:

Purpose: Flavoring agent.

Where Used: Various.

Regulations: FDA - 21CFR 172.515. Use at a level not in excess of the amount reasonably required to accomplish the intended effect.

SAFETY PROFILE: When heated to decomposition it emits acrid smoke and irritating fumes.

PGS000 CAS: 298-02-2
PHORATE

mf: $C_7H_{17}O_2PS_3$ mw: 260.39

PROP: Liquid. Bp: 118-120° @ 0.8 mm, d: 1.156. @ 25°/4°. Insol in water; misc with carbon tetrachloride, dioxane, xylene.

SYNS: O,O-DIAETHYL-S-(AETHYLTHIO-METHYL)-DITHIOPHOSPHAT (GERMAN) ◇ O,O-DIETHYL-S-ETHYLMERCAPTOMETHYL DITHIOPHOSPHONATE ◇ O,O-DIETHYL-S-(ETHYLTHIO-METHYL)-DITHIOFOSFAAT (DUTCH) ◇ O,O-DIETHYL-S-ETHYLTHIOMETHYL DITHIOPHOSPHONATE ◇ O,O-DIETHYL-ETHYLTHIOMETHYL PHOSPHORODITHIOATE ◇ O,O-DIETHYL-S-(ETHYLTHIO)METHYL PHOSPHORODITHIOATE ◇ O,O-DIETHYL-S-ETHYLTHIOMETHYL THIOTHIONOPHOSPHATE ◇ O,O-DIETIL-S-(ETILTIO-METIL)-DITIOFOSFATO (ITALIAN) ◇ DITHIOPHOSPHATE de O,O-DIETHYLE et d'ETHYLTHIOMETHYLE (FRENCH) ◇ ENT 24,042 ◇ FORAAT (DUTCH) ◇ GRANUTOX ◇ L 11/6 ◇ PHORAT (GERMAN) ◇ PHORATE-10G ◇ RAMPART ◇ RCRA WASTE NUMBER P094 ◇ THIMET ◇ TIMET ◇ VEGFRU ◇ VERGFRU FORATOX

USE IN FOOD:

Purpose: Insecticide.

Where Used: Animal feed, sugar beet pulp (dried).

Regulations: FDA - 21CFR 561.290. Limitation of 1 ppm in dried sugar beet pulp when used for cattle feed.

EPA Extremely Hazardous Substances List. EPA Genetic Toxicology Program.

ACGIH TLV: TWA 0.05 mg/m^3 (skin)

SAFETY PROFILE: Poison by ingestion, skin contact, and intravenous routes. Experimental reproductive effects. Mutagenic data. A cholinesterase inhibitor. When heated to decomposition it emits toxic fumes of PO_x and SO_x.

TOXICITY DATA and CODEN

mnt-mus-ipr 750 μg/kg MUREAV 155,131,85
ipr-grb TDLo:7200 μg/kg (30D male):REP TXCYAC 18,133,80
orl-mus LD50:6590 μg/kg JPFCD2 B15,867,80

PHA500 CAS: 1071-83-6
N-(PHOSPHONOMETHYL)GLYCINE

mf: $C_3H_8NO_5P$ mw: 169.09

SYNS: GLYPHOSATE ◇ MON 0573

USE IN FOOD:

Purpose: Herbicide.

Where Used: Animal feed, citrus pulp (dried), molasses (sugarcane), olives (imported), palm oil, potable water, soybean oil, tea (dried), tea (instant).

Regulations: FDA - 21CFR 193.235. Herbicide residue tolerance of 30.0 ppm in molasses, sugarcane, 0.1 ppm in palm oil, 0.1 ppm in olives, imported, 1.0 ppm in tea, dried, 4.0 ppm in tea, instant. Herbicide residue tolerance of 0.1 ppm in potable water. (Expired 1/1/1983) 21CFR 561.253. Limitation of 0.4 ppm in dried citrus pulp, 20 ppm in soybean oil when used for animal feed.

SAFETY PROFILE: Poison by intraperitoneal route. Moderately toxic by ingestion. When heated to decomposition it emits very toxic fumes of NO_x and PO_x.

TOXICITY DATA and CODEN

orl-mus LD50:1568 mg/kg TXAPA9 45,319,78
ipr-rat LD50:238 mg/kg TXAPA9 45,319,78

PHB250 CAS: 7664-38-2
PHOSPHORIC ACID

DOT: 1805
mf: H_3O_4P mw: 98.00

PROP: Colorless liquid or rhombic crystals. Mp: 42.35°, loses $1/2H_2O$ @ 213°, fp: 42.4°, d: 1.864 @ 25°, vap press: 0.0285 mm @ 20°. Misc with water, alc.

SYNS: ACIDE PHOSPHORIQUE (FRENCH) ◇ ACIDO FOSFORICO (ITALIAN) ◇ FOSFORZUUROPLOSSINGEN (DUTCH) ◇ ORTHOPHOSPHORIC ACID ◇ PHOSPHORSAEURELOESUNGEN (GERMAN)

USE IN FOOD:

Purpose: Acid, sequestrant, synergist for antioxidants.

Where Used: Beverages, cheese, colas, fats (poultry), lard, margarine, oleomargarine, poultry, root beer, shortening.

Regulations: FDA - 21CFR 182.1073. GRAS when used in accordance with good manufacturing practice. USDA - 9CFR 318.7, 9CFR 381.147. Sufficient for purpose. Limitation of 0.01 percent in lard, shortening, and poultry fat.

Community Right-To-Know List. EPA Genetic Toxicology Program.

OSHA PEL: TWA 1 mg/m^3; STEL 3 mg/m^3
ACGIH TLV: TWA 1 mg/m^3; STEL 3 mg/m^3
DOT Classification: Corrosive Material; Label: Corrosive

SAFETY PROFILE: Human poison by an unspecified route. Moderately toxic by ingestion and skin contact. A corrosive irritant to eyes, skin, and mucous, membranes and a systemic irritant by inhalation. A strong acid. Mixtures with nitromethane are explosive. Reacts with chlorides + stainless steel to form explosive hydrogen gas. Potentially violent reaction with sodium tetrahydroborate. Dangerous; when heated to decomposition it emits toxic fumes of PO_x.

TOXICITY DATA and CODEN

skn-rbt 595 mg/24H SEV BIOFX* 17-4/70
eye-rbt 119 mg SEV BIOFX* 17-4/70
unr-man LDLo:220 mg/kg 85DCAI 2,73,70
orl-rat LD50:1530 mg/kg BIOFX* 17-4/70
skn-rbt LD50:2740 mg/kg BIOFX* 17-4/70

PHX250 CAS: 732-11-6
PHTHALIMIDOMETHYL-O,O-DIMETHYL PHOSPHORODITHIOATE

mf: $C_{11}H_{12}NO_4PS_2$ mw: 317.33

SYNS: APPA ◇ DECEMTHION P-6 ◇ (O,O-DIMETHYL-PHTHALIMIDIOMETHYL-DITHIOPHOSPHATE) ◇ O,O-DIMETHYL S-(N-PHTHALIMIDOMETHYL) DITHIOPHOSPHATE/S O,O-DIMETHYL S-PHTHALIMIDOMETHYL PHOSPHORODITHIOATE ◇ ENT 25,705 ◇ FTALOPHOS ◇ IMIDAN ◇ KEMOLATE ◇ N-(MERCAPTOMETHYL) PHTHALIMIDE S-(O,O-DIMETHYL PHOSPHORODITHIOATE) ◇ PERCOLATE ◇ PHOSMET ◇ PHOSPHORODITHIOIC ACID, S-((1,3-DIHYDRO-1,3-DIOXO-ISOINDOL-2-YL) METHYL) O,O-DIMETHYL ESTER ◇ PHTHALIMIDO-O,O-DIMETHYL PHOSPHORODITHIOATE ◇ PHTHALOPHOS ◇ PMP ◇ PROLATE ◇ R 1504 ◇ SMIDAN ◇ STAUFFER R 1504

USE IN FOOD:

Purpose: Insecticide.

Where Used: Cottonseed oil.

Regulations: FDA - 21CFR 193.275. Insecticide residue tolerance of 0.2 ppm in cottonseed oil.

EPA Extremely Hazardous Substances List.

SAFETY PROFILE: A human poison by ingestion. Poison experimentally by inhalation, ingestion, and possibly other routes. Moderately toxic by skin contact. Human systemic effects by inhalation: lacrimation, somnolence, and olfaction effects. Experimental teratogenic and reproductive effects. Mutagenic data. When heated to decomposition it emits very toxic fumes of NO_x, PO_x, and SO_x.

TOXICITY DATA and CODEN

cyt-mus-orl 20 mg/kg SOGEBZ 11,1534,75
mmo-sat 5 mg/plate MUREAV 116,185,83
orl-rat TDLo:30 mg/kg (13D preg):TER EVHPAZ 13,121,76
orl-rat TDLo:16500 μg/kg (1-22D preg):REP EVHPAZ 13,121,76
orl-mus TDLo:6200 μg/kg (12D preg):TER AXVMAW 34,791,80
orl-hmn LDLo:50 mg/kg SPEADM 74-1,-,74
ihl-hmn TCLo:2 mg/m^3/8H:NOSE,EYE,CNS HYSAAV 34,192,69
ihl-rat LC50:54 mg/m3/4H 85GMAT -,59,82
skn-rat LD50:1550 mg/kg WRPCA2 9,119,70

PIF750 CAS: 7681-93-8
PIMARICIN
mf: $C_{33}H_{47}NO_{13}$ mw: 665.81

PROP: An antibiotic produced by a strain of *Streptomyces chattanoogensis* (85ERAY 2,956,78).

SYNS: ANTIBIOTIC A-5283 ◇ CL 12,625 ◇ MYCOPHYT ◇ MYPROZINE ◇ NATACYN ◇ NATAMYCIN ◇ PIMAFUCIN ◇ TENNECETIN

USE IN FOOD:

Purpose: Fungicide, mold inhibitor.

Where Used: Cheese, wine.

Regulations: FDA - 21CFR 172.155. Limitation of 200-300 ppm solution applied to cuts in cheese.

SAFETY PROFILE: Poison by intravenous, intramuscular, subcutaneous, and intraperitoneal routes. Moderately toxic by ingestion. When heated to decomposition it emits toxic fumes of NO_x.

TOXICITY DATA and CODEN

orl-rat LD50:2730 mg/kg TXAPA9 8,97,66
ipr-rat LD50:85 mg/kg ANTCAO 9,406,59

PIG750 CAS: 8016-45-3
PIMENTA LEAF OIL

PROP: Main constituent is eugenol. From steam distillation of the shrub *Pimenta officinalis* Lindl. (Fam. *Myrtaceae*). (FCTXAV 12,807,74). Pale yellow to brown liquid; spicy odor. D: 1.037-1.050, refr index: 1.531 @ 20°. Sol in propylene glycol, fixed oils; insol in glycerin, mineral oil.

SYN: OIL of PIMENTA LEAF

USE IN FOOD:

Purpose: Flavoring agent.

Where Used: Various.

Regulations: FDA - 21CFR 182.20. GRAS when used at a level not in excess of the amount reasonably required to accomplish the intended effect.

SAFETY PROFILE: Moderately toxic by ingestion. A severe skin irritant. When heated to decomposition it emits acrid smoke and irritating fumes.

TOXICITY DATA and CODEN

skn-rbt 500 mg/24H SEV FCTXAV 12,807,74
orl-rat LD50:3600 mg/kg FCTXAV 12,807,74

PIH250 CAS: 80-56-8
2-PINENE
DOT: 2368
mf: $C_{10}H_{16}$ mw: 136.26

PROP: Liquid; odor of turpentine. Mp: −55°, bp: 155°, flash p: 91°F, d: 0.8592 @ 20°/4°, refr index: 1.464-1.468, vap press: 10 mm @ 37.3°, vap d: 4.7, autoign temp: 491°F. Insol in water; sol in alc, chloroform, ether, glacial acetic acid, fixed oils.

SYNS: ACINTENE A ◇ FEMA No. 2902 ◇ α-PINENE (FCC) ◇ 2,6,6-TRIMETHYLBICYCLO(3.1.1)-2-HEPT-2-ENE

USE IN FOOD:

Purpose: Flavoring agent.

Where Used: Various.

Regulations: FDA - 21CFR 172.515. Use at a level not in excess of the amount reasonably required to accomplish the intended effect.

DOT Classification: Flammable or Combustible Liquid; Label: Flammable Liquid

SAFETY PROFILE: A deadly poison by inhalation. Moderately toxic by ingestion. An eye, mucous membrane, and severe human skin irritant. Flammable liquid. A dangerous fire hazard when exposed to heat, flame, or oxidizing materials. To fight fire, use foam, CO_2, dry chemical. Explodes on contact with nitrosyl perchlorate.

TOXICITY DATA and CODEN

skn-man 100% SEV FCTXAV 16,637,78
orl-rat LD50:3700 mg/kg FCTXAV 16,637,78
ihl-rat LCLo:625 $\mu g/m^3$ FCTXAV 16,637,78

PIH500 CAS: 8000-26-8
PINE NEEDLE OIL, SCOTCH

PROP: Volatile oil from steam distillation of *Pinus sylvestris* L. (Fam. *Pinaceae*) constituted of dipentene, pinene, sylvestrene, cadinene and bornyl acetate. Yellow liquid; penetrating odor. Bp: 200-220°, flash p: 172°F (CC), d: 0.86, refr index: 1.473 @ 20°. Sol in fixed oils, mineral oil; sltly sol in propylene glycol; insol in glycerin.

SYNS: KIEFERNADEL OEL (GERMAN) ◇ SCOTCH PINE NEEDLE OIL

USE IN FOOD:

Purpose: Flavoring agent.

Where Used: Various.

Regulations: FDA - 21CFR 172.510. Use at a level not in excess of the amount reasonably required to accomplish the intended effect.

SAFETY PROFILE: Mildly toxic by ingestion. A weak allergen and a mild irritant. Flammable when exposed to heat or flame; can react vigorously with oxidizing materials. To fight fire, use foam, CO_2, dry chemical. When heated to decomposition it emits acrid smoke and irritating fumes.

TOXICITY DATA and CODEN

orl-rat LD50:6880 mg/kg PHARAT 14,435,59

PII000 CAS: 8000-26-8
PINUS PUMILIO OIL

PROP: From steam distillation of needles of *Pinus mugo* turra var. *pumilio* (Haenke) Zenari (Fam. *Pinaceae*) (FCTXAV 14,659,76). Colorless to yellow liquid; pleasant odor and a bitter, pungent taste. D: 0.853-0.871, refr index: 1.475 @ 20°.

SYNS: DWARF PINE NEEDLE OIL ◇ KNEE PINE OIL ◇ LATSCHENKIEFEROL ◇ OIL of MOUNTAIN PINE ◇ PINE NEEDLE OIL, DWARF (FCC) ◇ PINUS MONTANA OIL

USE IN FOOD:

Purpose: Flavoring agent.

Where Used: Various.

Regulations: FDA - 21CFR 172.510. Use at a level not in excess of the amount reasonably required to accomplish the intended effect.

SAFETY PROFILE: Mildly toxic by ingestion. A human skin irritant. When heated to decomposition it emits acrid smoke and irritating fumes.

TOXICITY DATA and CODEN

skn-hmn 12% FCTXAV 14,843,76
orl-rat LD50:6880 mg/kg PHARAT 14,435,59

PIW250 CAS: 120-57-0
PIPERONAL
mf: $C_8H_6O_3$ mw: 150.14

PROP: Colorless, lustrous crystals; floral odor. Mp: 37°, bp: 263°, vap press: 1 mm @ 87.0°. Very sol in alc, ether; sol in propylene glycol, fixed oils; insol water, glycerin.

SYNS: 3,4-BENZODIOXOLE-5-CARBOXALDEHYDE ◇ 3,4-DIHYDROXYBENZALDEHYDE METHYLENE KETAL ◇ DIOXYMETHYLENE-PROTOCATECHUIC ALDEHYDE ◇ FEMA No. 2911 ◇ HELIOTROPIN ◇ 3,4-METHYLENE-DIHYDROXYBENZALDEHYDE ◇ 3,4-METHYLENEDIOXYBENZALDEHYDE ◇ PIPERONALDEHYDE ◇ PIPERONYL ALDEHYDE ◇ PROTOCATECHUIC ALDEHYDE METHYLENE ETHER

USE IN FOOD:

Purpose: Flavoring agent.

Where Used: Various.

Regulations: FDA - 21CFR 182.60. Use at a level not in excess of the amount reasonably required to accomplish the intended effect.

SAFETY PROFILE: Moderately toxic by ingestion and intraperitoneal routes. Can cause central nervous system depression. A skin irritant. Combustible when exposed to heat or flame; can react with oxidizing materials.

TOXICITY DATA and CODEN

orl-rat LD50:2700 mg/kg TXAPA9 6,378,64
ipr-rat LDLo:1500 mg/kg RMSRA6 16,449,1896

PIX000 CAS: 326-61-4
PIPERONYL ACETATE
mf: $C_{10}H_{10}O_4$ mw: 194.20

PROP: Colorless to light yellow liquid; heliotrope odor.

SYNS: HELIOTROPYL ACETATE ◇ 3,4-METHYLENEDIOXYBENZYL ACETATE

USE IN FOOD:

Purpose: Flavoring agent.

Where Used: Baked goods, beverages, candy, ice cream.

Regulations: FDA - 21CFR 172.515. Use at a level not in excess of the amount reasonably required to accomplish the intended effect.

SAFETY PROFILE: Moderately toxic by ingestion. A skin irritant. When heated to decomposition it emits acrid smoke and irritating fumes.

TOXICITY DATA and CODEN

skn-rbt 500 mg/24H MLD FCTXAV 12,807,74
orl-rat LD50:2100 mg/kg FCTXAV 12,807,74

PIX250 CAS: 51-03-6
PIPERONYL BUTOXIDE
mf: $C_{19}H_{30}O_5$ mw: 338.49

PROP: Light brown liquid; mild odor. Bp: 180° @ 1 mm, flash p: 340°F, d: 1.04-1.07 @ 20°/20°. Misc with methanol, ethanol, benzene.

SYNS: BUTACIDE ◇ BUTOCIDE ◇ BUTOXIDE ◇ α-(2-(2-BUTOXYETHOXY)ETHOXY)-4,5-METHYLENEDIOXY-2-PROPYLTOLUENE ◇ α-(2-(2-n-BUTOXYETHOXY)-ETHOXY)-4,5-METHYLENEDIOXY-2-PROPYLTOLUENE ◇ 5-((2-(2-BUTOXYETHOXY)ETHOXY)METHYL)-6-PROPYL-1,3-BENZODIOXOLE ◇ BUTYL CARBITOL 6-PROPYLPIPERONYL ETHER ◇ BUTYL-CARBITYL (6-PROPYLPIPERONYL) ETHER ◇ ENT 14,250 ◇ FAC 5273 ◇ FMC 5273 ◇ 3,4-METHYLENDIOXY-6-PROPYLBENZYL-n-BUTYL-DIAETHYLENGLYKOLAETHER (GERMAN) ◇ (3,4-METHYLENEDIOXY-6-PROPYLBENZYL) (BUTYL) DIETHYLENE GLICOL ETHER ◇ 3,4-METHYLENEDIOXY-6-PROPYLBENZYL n-BUTYL DIETHYLENEGLYCOL ETHER ◇ NCI-C02813 ◇ NIA 5273 ◇ NUSYN-NOXFISH ◇ PB ◇ PRENTOX ◇ 6-(PROPYLPIPERONYL)-BUTYL CARBITYL ETHER ◇ 6-PROPYLPIPERONYL BUTYL DIETHYLENE GLYCOL ETHER ◇ 5-PROPYL-4-(2,5,8-TRIOXA-DODECYL)-1,3-BENZODIOXOL (GERMAN) ◇ PYBUTHRIN ◇ PYRENONE 606 ◇ SYNPREN-FISH

USE IN FOOD:

Purpose: Insecticide.

Where Used: Animal feed, dried foods, milled fractions derived from cereal grains, packaging materials.

Regulations: FDA - 21CFR 178.3730, 193.360. Insecticide residue tolerance of 10 ppm in milled fractions derived from cereal grains, dried foods. 21CFR 561.310. Limitation of 10 ppm when used for animal feed

IARC Cancer Review: Animal No Evidence IMEMDT 30,183,83. NCI Carcinogenesis Bioassay (feed); No Evidence: mouse, rat NCITR* NCI-CG-TR-120,79. Glycol ether compounds are on the Community Right-To-Know List.

SAFETY PROFILE: Poison by skin contact. Moderately toxic by ingestion and intraperitoneal routes. An experimental carcinogen and tumorigen. Experimental reproductive effects. Many glycol ether compounds have dangerous human reproductive effects. Mutagenic data. Combustible when exposed to heat or flame; can react with oxidizing materials. To fight fire, use foam, CO_2, dry chemical. When heated to decomposition it emits acrid smoke and irritating fumes.

TOXICITY DATA and CODEN

otr-ham:emb 500 μg/L CRNGDP 4,291,83
orl-rat TDLo: 3 g/kg (6-15D preg): REP FCTXAV 15,337,77
scu-mus TDLo: 9 g/kg (6-14D preg): REP NTIS** PB223-160
orl-mus TDLo: 33 g/kg/78W-I: CAR NTIS** PB223-159
scu-mus TDLo: 1000 mg/kg: ETA NTIS** PB223-159
orl-mus LD50: 3800 mg/kg BESAAT 15,85,69

PJH500 CAS: 9006-00-2

PLIOFILM

mf: $(C_3H_5Cl)_n$

SYNS: PERMASEAL ◇ RUBBER HYDROCHLORIDE ◇ RUBBER HYDROCHLORIDE POLYMER

USE IN FOOD:

Purpose: Manufacture of paper and paperboard.

Where Used: Packaging materials.

Regulations: FDA - 21CFR 181.30.

SAFETY PROFILE: An experimental tumorigen. When heated to decomposition it emits toxic fumes of Cl^-.

TOXICITY DATA and CODEN

imp-rat TDLo: 18 mg/kg: ETA CNREA8 15,333,55

PJK150 CAS: 9003-11-6

POLOXALENE

mf: $HO(CH_2CH_2O)_n[CH(CH_3)CH_2O]_n(CH_2CH_2O)_nH$

PROP: Liquid nonionic surfactant polymer.

SYNS: BIS[HYDROXYETHYLPOLY(ETHYLENEOXY)ETHYLPROPYLENEGLYCOL ◇ BLOAT GUARD ◇ DIPOLYOXYETHYLATEDPOLYPROPYLENEGLYCOL ETHER ◇ POLY(OXYETHYLENE)-POLY(OXYPROPYLENE)-POLY-(OXYETHYLENE) POLYMER ◇ THERABLOAT

USE IN FOOD:

Purpose: Animal feed drug.

Where Used: Animal feed.

Regulations: FDA - 21CFR 558.464.

SAFETY PROFILE: When heated to decomposition it emits acrid smoke and irritating fumes.

PJK151 CAS: 9003-11-6

POLOXALENE FREE-CHOICE LIQUID TYPE C FEED

mf: $HO(CH_2CH_2O)_n[CH(CH_3)CH_2O]_n(CH_2CH_2O)_nH$

PROP: Liquid nonionic surfactant polymer.

SYNS: BIS[HYDROXYETHYLPOLY(ETHYLENEOXY)ETHYLPROPYLENEGLYCOL ◇ BLOAT GUARD ◇ DIPOLYOXYETHYLATEDPOLYPROPYLENEGLYCOL ETHER ◇ POLY (OXYETHYLENE)-POLY(OXYPROPYLENE)-POLY(OXYETHYLENE) POLYMER ◇ THERABLOAT

USE IN FOOD:

Purpose: Animal feed drug.

Where Used: Animal feed.

Regulations: FDA - 21CFR 558.465.

SAFETY PROFILE: When heated to decomposition it emits acrid smoke and irritating fumes.

PJK200
POLOXAMER 331

PROP: Average molecular weight 3800. Colorless liquid. D: 1.02, refr index: 1.452. Very sltly sol in water; sol in alc; insol in propylene glycol, ethylene glycol.

SYNS: ETHYLENE OXIDE and PROPYLENE OXIDE BLOCK POLYMER ◇ PROPYLENE OXIDE and ETHYLENE OXIDE BLOCK POLYMER ◇ α-HYDRO-omega-HYDROXY-POLY-(OXYRTHYLENE)-POLY(OXYPROPYLENE)(51-57 MOLES)-POLY(OXYETHYLENE) BLOCK POLYMER

USE IN FOOD:

Purpose: Dough conditioner, foam control agent, solubilizing agent in flavor concentrates, stabilizing agent in flavor concentrates, surfactant, poultry scald agent.

Where Used: Dough, flavor concentrates, pork, poultry.

Regulations: FDA - 21CFR 172.808. Limitation of weight not to exceed weight of flavor concentrate, 0.05 percent in scald baths for poultry, 5 grams per hog in dehairing machines, 0.5 percent by weight of flour. 9CFR 381.147. Limitation of 0.05 percent by weight of poultry scald water.

SAFETY PROFILE: When heated to decomposition it emits acrid smoke and irritating fumes.

PJK350 CAS: 9003-05-8
POLYACRYLAMIDE

PROP: Contains not more than 0.2 percent of acrylamide monomer.

USE IN FOOD:

Purpose: Film former.

Where Used: Soft-shell gelatin capsules.

Regulations: FDA - 21CFR 172.255. Use at a level not in excess of the amount reasonably required to accomplish the intended effect.

SAFETY PROFILE: When heated to decomposition it emits acrid smoke and irritating fumes.

PJL750 CAS: 1336-36-3
POLYCHLORINATED BIPHENYLS

DOT: 2315

PROP: Bp: 340-375°, flash p: 383°F (COC), d: 1.44 @ 30°. A series of technical mixtures consisting of many isomers and compounds that vary from mobile oily liquids to white crystalline solids and hard noncrystalline resins. Technical products vary in composition, in the degree of chlorination and possibly according to batch (IARC** 7,262,74).

SYNS: AROCLOR ◇ CHLOPHEN ◇ CHLOREXTOL ◇ CHLORINATED BIPHENYL ◇ CHLORINATED DIPHENYL ◇ CHLORINATED DIPHENYLENE ◇ CHLORO BIPHENYL ◇ CHLORO-1,1-BIPHENYL ◇ CLOPHEN ◇ DYKANOL ◇ FENCLOR ◇ INERTEEN ◇ KANECHLOR ◇ MONTAR ◇ NOFLAMOL ◇ PCB (DOT, USDA) ◇ PHENOCHLOR ◇ POLYCHLORINATED BIPHENYL ◇ POLYCHLOROBIPHENYL ◇ PYRALENE ◇ PYRANOL ◇ SANTOTHERM ◇ SOVOL ◇ THERMINOL FR-1

USE IN FOOD:

Purpose: Insecticide.

Where Used: Various.

Regulations: USDA CES Ranking: A-4 (1985).

IARC Cancer Review: Human Limited Evidence IMEMDT 18,43,78. EPA Extremely Hazardous Substances List.

DOT Classification: ORM-E; Label: None

NIOSH REL: TWA (Polychlorinated Biphenyls) 0.001 mg/m^3

SAFETY PROFILE: Moderately toxic by ingestion. Some are poisons by other routes. Suspected human carcinogens. Experimental carcinogens and tumorigens. Experimental reproductive effects. Like the chlorinated naphthalenes, the chlorinated diphenyls have two distinct actions on the body, namely, a skin effect and a toxic action on the liver. The higher the chlorine content of the diphenyl compound, the more toxic is it liable to be. In persons who have suffered systemic intoxication, the usual signs and symptoms are nausea, vomiting, loss of weight, jaundice, edema and abdominal pain. Where the liver damage has been severe the patient may pass into coma and die. Combustible when exposed to heat or flame. When heated to decomposition they emit highly toxic fumes of Cl^-.

TOXICITY DATA and CODEN

orl-mam TDLo: 325 mg/kg (30D pre/1-36D preg): REP AMBOCX 6,239,77
orl-rat TDLo: 16800 mg/kg/2Y-C: ETA TOERD9 1,159,78
orl-mus TDLo: 1250 mg/kg/25W-I: CAR FCTOD7 21,688,83
orl-rat TD : 1250 mg/kg/25W-I: CAR FCTOD7 21,688,83
orl-mus LD50: 1900 mg/kg FKIZA4 60,544,69

PJQ425 CAS: 68424-04-4
POLYDEXTROSE

PROP: Off-white to light tan solid. Sol in water.

USE IN FOOD:

Purpose: Bulking agent, formulation aid, humectant, texturizer.

Where Used: Baked goods, baking mixes (fruit, custard, and pudding-filled pies), cakes, candy (hard), candy (soft), chewing gum, confections, cookies, fillings, frostings, frozen dairy desserts, frozen dairy dessert mixes, gelatins, puddings, salad dressings.

Regulations: FDA - 21CFR 172.841. Special labeling for single serving containing above 15 grams: "Sensitive individuals may experience a laxative effect from excessive consumption of this produce."

SAFETY PROFILE: When heated to decomposition it emits acrid smoke and irritating fumes.

PJQ430
POLYDEXTROSE SOLUTION

PROP: Clear, straw-colored liquid.

USE IN FOOD:

Purpose: Bulking agent, formulation aid, humectant, texturizer.

Where Used: Various.

Regulations: GRAS when used in accordance with good manufacturing practice.

SAFETY PROFILE: When heated to decomposition it emits acrid smoke and irritating fumes.

PJS750 CAS: 9002-88-4
POLYETHYLENE
mf: $(C_2H_4)_n$

PROP: Odorless. The high molecular weight compounds are tough, white leathery, resinous. D: 0.92 @ 20°/4°, mp: 85-110°. Sol in hot benzene; insol in water.

SYNS: AGILENE ◇ ALKATHENE ◇ BAKELITE DYNH ◇ DIOTHENE ◇ ETHENE POLYMER ◇ ETHYLENE HOMOPOLYMER ◇ ETHYLENE POLYMERS ◇ HOECHST PA 190 ◇ MICROTHENE ◇ POLYETHYLENE AS ◇ POLYWAX 1000 ◇ TENITE 800

USE IN FOOD:

Purpose: Masticatory substance in chewing gum base, protective coating.

Where Used: Avocados, bananas, beets, Brazil nuts, chestnuts, chewing gum, coconuts, eggplant, filberts, garlic, grapefruit, hazelnuts, lemons, limes, mangoes, muskmelons, onions, oranges, papaya, peas (in pods), pecans, pineapples, plantain, pumpkin, rutabaga, squash (acorn), sweetpotatoes, tangerines, turnips, walnuts, watermelon.

Regulations: FDA - 21CFR 172.260. Mild air oxidation resin. FDA - 21CFR 172.615.

IARC Cancer Review: Animal Sufficient Evidence IMEMDT 19,157,79; Human Inadequate Evidence IMEMDT 19,157,79.

SAFETY PROFILE: An experimental carcinogen and tumorigen by implant. Reacts violently with F_2. When heated to decomposition it emits acrid smoke and irritating fumes.

TOXICITY DATA and CODEN

imp-rat TDLo: 33 mg/kg: ETA CNREA8 15,333,55
imp-rat TD : 2120 mg/kg: ETA BJCAAI 23,401,69
imp-rat TD : 1476 mg/kg: ETA CORTBR 88,223,72

PJT000 CAS: 25322-68-3
POLYETHYLENE GLYCOL
mf: $H(OC_2H_4)_nOH$

PROP: Clear liquid or white solid. D: 1.110-1.140 @ 20°, mp: 4-10°, flash p: 471°F. Sol in organic solvents, aromatic hydrocarbons.

SYNS: ALKAPOL PEG-200 ◇ CARBOWAX ◇ α-HYDROXY-omega-HYDROXY-POLY(OXY-1,2-ETHANEDIYL) ◇ JEFFOX ◇ JORCHEM 400 ML ◇ LUTROL ◇ PEG ◇ PLURACOL P-410 ◇ POLY(ETHYLENE OXIDE) ◇ POLY-G SERIES ◇ POLYOX

USE IN FOOD:

Purpose: Binding agent, boiler water additive, coating agent, dispersing agent, flavoring adjuvant, lubricant, plasticizing agent.

Where Used: Beverages (carbonated), citrus fruit (fresh), nonnutritive sweeteners, sodium nitrite coating, tablets, vitamin or mineral preparations.

Regulations: FDA - 21CFR 172.820. Limitation of zero in milk. 21CFR 172.210, 173.340. Must comply with 21CFR 172.820. 21CFR 820. Limitation of zero in milk, otherwise use at a level not in excess of the amount reasonably required to accomplish the intended effect.

EPA Genetic Toxicology Program.

SAFETY PROFILE: Slightly toxic by ingestion. A skin and eye irritant. Combustible liquid when exposed to heat or flame. To fight fire, use water, foam, dry chemical. When heated to decomposition it emits acrid smoke and irritating fumes.

TOXICITY DATA and CODEN

skn-rbt 500 mg/24H MLD 28ZPAK -,255,72
eye-rbt 500 mg/24H MLD 28ZPAK -,255,72
orl-rat LD50:33750 mg/kg ARZNAD 26,1581,76

PJT200 CAS: 25322-68-3
POLYETHYLENE GLYCOL 200
mf: $H(OC_2H_4)_nOH$

PROP: Viscous, hydroscopic liquid with *n* about 4; slt characteristic odor. D (25°/25°) 1.127.

SYNS: CARBOWAX ◇ JEFFOX ◇ NYCOLINE ◇ PEG 200 ◇ PLURACOL E ◇ POLYAETHYLENGLYCOLE 200 (GERMAN) ◇ POLY-G ◇ POLYGLYCOL E ◇ SOLBASE

USE IN FOOD:

Purpose: Binding agent, coating agent, dispersing agent, flavoring adjuvant, lubricant, plasticizing agent.

Where Used: Various.

Regulations: FDA - 21CFR 172.820. Limitation of zero in milk. 21CFR 178.3750.

EPA Genetic Toxicology Program.

SAFETY PROFILE: Mildly toxic by ingestion. Caution: Solvent action on some plastics. When heated to decomposition it emits acrid smoke and irritating fumes.

TOXICITY DATA and CODEN

orl-rat LD50:28900 mg/kg ARZNAD 3,451,53
orl-mus LD50:38300 mg/kg ARZNAD 3,451,53
ipr-mus LD50:7500 mg/kg NTIS** AD628-313

PJT225 CAS: 25322-68-3
POLYETHYLENE GLYCOL 300
mf: $(C_6H_{11}NO)_n$

SYNS: POLYAETHYLENGLYKOLE 300 (GERMAN) ◇ PEG 300

USE IN FOOD:

Purpose: Binding agent, coating agent, dispersing agent, flavoring adjuvant, lubricant, plasticizing agent.

Where Used: Various.

Regulations: FDA - 21CFR 172.820. Limitation of zero in milk. 21CFR 178.3750.

EPA Genetic Toxicology Program.

SAFETY PROFILE: Mildly toxic by ingestion. When heated to decomposition it emits acrid smoke and irritating fumes.

TOXICITY DATA and CODEN

orl-rat LD50:27500 mg/kg ARZNAD 3,451,53
ipr-rat LD50:170000 mg/kg ARZNAD 3,451,53

PJT230 CAS: 25322-68-3
POLYETHYLENE GLYCOL 400
mf: $H(OC_2H_4)_nOH$

PROP: Liquid with *n* about 8.2 to 9.1. Mw: 380-420, d: 1.128, mp: 4-8°.

SYNS: PEG 400 ◇ POLYAETHYLENGLYKOLE 400 (GERMAN) ◇ POLY G 400

USE IN FOOD:

Purpose: Binding agent, coating agent, dispersing agent, flavoring adjuvant, lubricant, manufacture of paper and paperboard, plasticizing agent.

Where Used: Packaging materials.

Regulations: FDA - 21CFR 172.820. Limitation of zero in milk. 21CFR 178.3750, 181.30.

EPA Genetic Toxicology Program.

SAFETY PROFILE: Low toxicity by ingestion, intravenous, and intraperitoneal routes. When heated to decomposition it emits acrid smoke and irritating fumes.

TOXICITY DATA and CODEN

orl-mus LD50:28915 mg/kg PESTD5 17,351,76
ipr-mus LD50:9953 mg/kg PESTD5 17,351,76

PJT240 CAS: 25322-68-3
POLYETHYLENE GLYCOL 600
mf: $H(OC_2H_4)_nOH$

PROP: Liquid with *n* about 12.5 to 13.9.mw: 570-630, d: 1.128, mp: 20-25°.

SYNS: PEG 600 ◇ POLYAETHYLENGLYKOLE 600 (GERMAN)

USE IN FOOD:

Purpose: Binding agent, coating agent, dispersing agent, flavoring adjuvant, lubricant, plasticizing agent.

Where Used: Various.

Regulations: FDA - 21CFR 172.820. Limitation of zero in milk. 21CFR 178.3750.

EPA Genetic Toxicology Program.

SAFETY PROFILE: Low toxicity by ingestion. An eye irritant. When heated to decomposition it emits acrid smoke and irritating fumes.

TOXICITY DATA and CODEN

eye-rbt 100 mg MLD 34ZIAG -,747,69
orl-rat LD50:38100 mg/kg 34ZIAG -,747,69

PJT250 CAS: 25322-68-3
POLYETHYLENE GLYCOL 1000
mf: $H(OC_2H_4)_nOH$

SYNS: CARBOWAX 1000 ◇ MACROGOL 1000 ◇ PEG 1000 ◇ POLYAETHYLENGLYKOLE 1000 (GERMAN) ◇ POLYGLYCOL 1000 ◇ POLYGLYCOL E1000

USE IN FOOD:

Purpose: Binding agent, coating agent, dispersing agent, flavoring adjuvant, lubricant, plasticizing agent.

Where Used: Various.

Regulations: FDA - 21CFR 172.820. Limitation of zero in milk. 21CFR 178.3750.

EPA Genetic Toxicology Program.

SAFETY PROFILE: Moderately toxic by intraperitoneal and intravenous routes. Mildly toxic by ingestion. An experimental tumorigen. When heated to decomposition it emits acrid smoke and irritating fumes.

TOXICITY DATA and CODEN

ivg-mus TDLo:416 mg/kg/Y-I:ETA BJCAAI 15,252,61
orl-rat LD50:42 g/kg ARZNAD 3,451,53
ipr-rat LD50:15570 mg/kg ARZNAD 3,451,53

PJT500 CAS: 25322-68-3
POLYETHYLENE GLYCOL 1500
mf: $H(OC_2H_4)_nOH$

PROP: White, free-flowing powder. D: 1.15-1.21 @ 25°/25°, fp: 44-48°.

SYNS: CARBOWAX 1500 ◇ α-HYDRO-omega-HYDROXY-POLY(OXY-1,2-ETHANEDIYL) ◇ PEG 1500 ◇ POLYAETHYLENGLYKOLE 1500 (GERMAN) ◇ POLYOXYETHYLENE 1500

USE IN FOOD:

Purpose: Binding agent, coating agent, dispersing agent, flavoring adjuvant, lubricant, plasticizing agent.

Where Used: Various.

Regulations: FDA - 21CFR 172.820. Limitation of zero in milk. 21CFR 178.3750.

EPA Genetic Toxicology Program.

SAFETY PROFILE: Mildly toxic by ingestion. A human skin irritant. When heated to decomposition it emits acrid smoke and irritating fumes.

TOXICITY DATA and CODEN

skn-hmn 500 mg/48H JIDEAE 19,423,52
orl-rat LD50:44200 mg/kg ARZNAD 3,451,53
ipr-rat LD50:17700 mg/kg ARZNAD 3,451,53

PJT750 CAS: 25322-68-3
POLYETHYLENE GLYCOL 4000
mf: $H(OC_2H_4)_nOH$

PROP: White, free-flowing powder or white flakes. D: 1.20-1.21 @ 25°/25°Fp: 54-58°.

SYNS: CARBOWAX 4000 ◇ CARSONON PEG-4000 ◇ MACROGOL 4000 ◇ PEG 4000 ◇ POLYAETHYLENGLYKOLE 4000 (GERMAN) ◇ POLYGLYCOL 4000 ◇ POLYGLYCOL E-4000 ◇ POLYGLYCOL E-4000 USP ◇ POLYOXYETHYLENE (75)

USE IN FOOD:

Purpose: Binding agent, coating agent, dispersing agent, flavoring adjuvant, lubricant, plasticizing agent.

Where Used: Various.

Regulations: FDA - 21CFR 172.820. Limitation of zero in milk. 21CFR 178.3750.

EPA Genetic Toxicology Program.

SAFETY PROFILE: Mildly toxic by ingestion. A skin irritant. When heated to decomposition it emits acrid smoke and irritating fumes.

TOXICITY DATA and CODEN

skn-rbt 500 mg open MLD UCDS** 4/13/65
orl-rat LD50:50 g/kg 34ZIAG -,747,69

PJU000 CAS: 25322-68-3
POLYETHYLENE GLYCOL 6000
mf: $H(OC_2H_4)_nOH$

PROP: White, waxy solid. Mp: 58-62°, flash p: >887°F. Water-sol.

SYNS: CARBOWAX 6000 ◇ PEG 6000 ◇ POLYAETHYLENGLYKOLE 6000 (GERMAN)

USE IN FOOD:

Purpose: Binding agent, coating agent, color diluent, dispersing agent, flavoring adjuvant, lubricant, plasticizing agent.

Where Used: Eggs (shell).

Regulations: FDA - 21CFR 73.1, 172.820. Limitation of zero in milk. 21CFR 178.3750.

EPA Genetic Toxicology Program.

SAFETY PROFILE: Mildly toxic by ingestion. Mutagenic data. A skin irritant. Combustible when exposed to heat or flame. When heated to decomposition it emits acrid smoke and irritating fumes.

TOXICITY DATA and CODEN

skn-rbt 500 mg open MLD UCDS** 4/9/65
dnd-omi 100 g/L PNASA6 72,4288,75
cyt-ham:oth 50 pph DKBSAS 240,228,78
orl-rat LDLo:50 g/kg 34ZIAG -,747,69

PJX875
POLYGLYCEROL ESTERS of FATTY ACIDS

PROP: Yellow to amber oily viscous liquids; light tan to brown soft solids; tan to brown waxy solids. Dispersible in water; sol in organic solvents and oils.

USE IN FOOD:

Purpose: Cloud inhibitor, emulsifier.

Where Used: Cake mixes, confectionery products, fats (rendered animal), margarine, oleomargarine, salad oil, vegetable oils, whipped topping (dry mix).

Regulations: FDA - 21CFR 172.854. Use at a level not in excess of the amount reasonably required to accomplish the intended effect. Must conform to 21CFR 172.860(b), 172.862 specifications for fats or fatty acids derived from edible oils. USDA - 9CFR 318.7. Sufficient for purpose, 0.5 percent in oleomargarine or margarine.

SAFETY PROFILE: When heated to decomposition it emits acrid smoke and irritating fumes.

PJY800 CAS: 9003-27-4
POLYISOBUTYLENE

PROP: Soft to hard elastic light white solids; odorless and tasteless. Sol in benzene, diisobutylene.

USE IN FOOD:

Purpose: Masticatory substance in chewing gum base.

Where Used: Chewing gum.

Regulations: FDA - 21CFR 172.615. Use in amounts not to exceed those required to produce the intended physical or other technical effect.

SAFETY PROFILE: When heated to decomposition it emits acrid smoke and irritating fumes.

PJY850 CAS: 26099-09-2
POLYMALEIC ACID

USE IN FOOD:

Purpose: Production aid.

Where Used: Sugar liquor or juice.

Regulations: FDA - 21CFR 173.45. Limitation of 0.5 ppm by weight of the beet or cane sugar juice or liquor process stream.

SAFETY PROFILE: When heated to decomposition it emits acrid smoke and irritating fumes.

PJY855
POLYMALEIC ACID, SODIUM SALT

USE IN FOOD:

Purpose: Production aid.

Where Used: Sugar liquor or juice.

Regulations: FDA - 21CFR 173.45. Limitation of 0.5 ppm by weight of the beet or cane sugar juice or liquor process stream.

SAFETY PROFILE: When heated to decomposition it emits acrid smoke and irritating fumes.

PKG250 CAS: 9005-65-6
POLYOXYETHYLENE SORBITAN MONOOLEATE

PROP: Yellow to orange oily liquid; faint odor, bitter taste. Sol in water, alc, fixed oils, ethyl acetate, toluene; insol in mineral oil.

SYNS: ARMOTAN PMO-20 ◇ ATLOX 1087 ◇ CAPMUL POE-O ◇ CRILL 10 ◇ DREWMULSE POE-SMO ◇ DURFAX 80 ◇ EMSORB 6900 ◇ ETHOXYLATED SORBITAN MONOOLEATE ◇ GLYCOSPERSE O-20 ◇ HODAG SVO 9 ◇ LIPOSORB O-20 ◇ MONITAN ◇ MONTANOX 80 ◇ NCI-C60286 ◇ NIKKOL TO ◇ OLOTHORB ◇ POLYOXYETHYLENE SORBITAN OLEATE ◇ POLYSORBAN 80 ◇ POLYSORBATE 80 B.P.C. ◇ POLYSORBATE 80, U.S.P. ◇ PROTASORB O-20 ◇ ROMULGIN O ◇ SORBIMACROGOL OLEATE ◇ SORBITAL O 20 ◇ SORETHYTAN (20) MONOOLEATE ◇ SORLATE ◇ SVO 9 ◇ TWEEN 80

USE IN FOOD:

Purpose: Adjuvant, color diluent, dispersing agent, emulsifier, solubilizer, stabilizer.

Where Used: Baked goods, baking mixes, barbecue sauce, beet sugar, chewing gum, confectionery, cottage cheese, custard (frozen), dietary foods (special), dill oil in canned spiced green beans, edible oil (whipped topping), edible oils, fillings, food supplements in tablet form, frozen desserts (nonstandardized), fruit sherbet, gelatin dessert mix, ice cream, ice milk, icings, margarine, oleomargarine, pickle products, pickles, poultry, shortening, sodium chloride (coarse crystal), toppings, vitamin-mineral preparations with calcium caseinate and fat-soluble vitamins, vitamin-mineral preparations with calcium caseinate in the absence of fat-soluble vitamins, vitamins (fat-soluble with no calcium caseinate), yeast.

Regulations: FDA - 21CFR 73.1, 172.515, 172.840. Limitation of 0.1 percent in ice cream; 0.1 percent in frozen custard; 0.1 percent in ice milk; 0.1 percent in fruit sherbet; 0.1 percent in nonstandardized frozen desserts; 4 percent in yeast-defoamer formulations; 500 ppm in pickles and pickle products; 175 mg/day in vitamin-mineral preparations with calcium caseinate in the absence of fat-soluble vitamins; 475/day mg in vitamin-mineral preparations with calcium caseinate and fat-soluble vitamins; 300/day mg in fat-soluble vitamins with no calcium caseinate; 10 ppm in sodium chloride, coarse crystal; 360 mg/day in special dietary foods; 30 ppm in dill oil in canned spiced green beans; 1 percent in shortening and edible oils (Additional limitations when used in combination with polysorbate 60, polysorbate 65, and/or sorbitan monostearate.); 0.0175 percent in poultry scald water; 0.082 percent in gelatin dessert mix; 0.008 percent in cottage cheese (21CFR 133.128, 133.131); 0.005 percent in barbecue sauce. Must conform to FDA specifications for fats or fatty acids derived from edible oils. 21CFR 173.340. USDA - 9CFR 318.7, 381.147. Limitation of 1 percent alone. If used with polysorbate 60, the combined total shall not exceed 1 percent.

SAFETY PROFILE: Moderately toxic by intravenous route. Mildly toxic by ingestion. An experimental tumorigen. Experimental reproductive effects. Human mutagenic data. An eye irritant. When heated to decomposition it emits acrid smoke and irritating fumes.

TOXICITY DATA and CODEN

eye-rbt 150 mg MLD AROPAW 40,668,48
dni-hmn:lym 20 ppm BBRCA9 45,630,71
dni-mus:oth 20 ppm ENPBBC 5,84,75
orl-rat TDLo:635 g/kg (MGN):REP JONUAI 60,489,56
orl-rat TDLo:1270 g/kg (84D pre-21D post): REP JONUAI 60,489,56
scu-rat TDLo:10 g/kg/27W-I:ETA FCTXAV 9,463,71
orl-mus LD50:25 g/kg BCFAAI 101,173,82
ipr-mus LD50:8210 mg/kg ARZNAD 6,119,56

PKG750 CAS: 9005-67-8
POLYOXYETHYLENE SORBITAN MONOSTEARATE

mf: $C_{64}H_{126}O_{26}$ mw: 1311.90

PROP: Lemon to orange colored oily liquid; faint odor and bitter taste. Sol in water, aniline, ethyl acetate, toluene; insol in mineral oil, vegetable oil.

SYNS: CAPMUL ◇ LGYCOSPERSE S-20 ◇ LIPOSORB S-20 ◇ POLYOXYETHYLENE 20 SORBITAN MONOSTEARATE ◇ POLYSORBATE 60 (FCC) ◇ SORBITAN, MONOOCTADECANOATE, POLY(OXY-1,2-ETHANEDIYL) DERIVATIVES ◇ TWEEN 60

USE IN FOOD:

Purpose: Emulsifier, poultry scald agent, protective coating, stabilizer.

Where Used: Baked goods, bakery products (yeast-leavened), baking mixes, cacao products, cake mixes, cakes, chocolate coatings, confectionery coatings (nonstandardized), confectionery coatings (sugar-type), cream fillings, dressings, edible oil (whipped topping), fillings, fruit (raw), gelatin dessert mix (sugar-based), gelatin desserts (artificially sweetened), icings, milk or cream substitutes for beverage coffee, mixes (non-alcoholic) for alcoholic drinks, poultry, poultry fat (rendered), pudding mix (sugar-based), shortening (edible oil), soft drink mix (powdered), syrups (chocolate flavored), toppings, vegetables (raw).

Regulations: FDA - 21CFR 172.515, 172.836. Limitation of 0.4 percent of weight of whipped edible oil topping, 0.46 percent of cake or cake mix, 0.46 percent of icings, 0.5 percent of confectionery coatings, nonstandardized, 0.5 percent of cacao products, 0.2 percent of sugar-type confectionery coatings, 0.3 percent of dressing, 1 percent of shortening, 1 percent of edible oil, 0.4 percent in milk or cream substitutes in beverage coffee, 4.5 percent of nonalcoholic mixes for alcoholic drinks, 0.5 percent in yeast-leavened bakery products, 0.5 percent in gelatin desserts, artificially sweetened, 0.05 percent in chocolate flavored syrups, 4.5 percent in powdered soft drink mix, 0.5 percent in gelatin dessert mix, sugar-based, 0.5 percent in pudding mix, sugar-based. Additional limitations when used in combination with polysorbate 65 and/or sorbitan monostearate. 21CFR 172.878, 172.886. Use at a level not in excess of the amount reasonably required to accomplish the intended effect. Must conform to FDA specifications for fats or fatty acids derived from edible oils. 21CFR 173.340. USDA - 9CFR 318.7, 381.147. Limitation of 1 percent alone. If used with polysorbate 80, the combined total shall not exceed 1 percent. 9CFR 381.147. Limitation of 0.0175 percent in scald water.

SAFETY PROFILE: Moderately toxic by intravenous route. An experimental tumorigen. Experimental reproductive effects. When heated to decomposition it emits acrid smoke and irritating fumes.

TOXICITY DATA and CODEN

orl-rat TDLo:635 g/kg (MGN):REP JONUAI 60,489,56
orl-rat TDLo:1270 g/kg (84D pre-21D post): REP JONUAI 60,489,56
scu-rat TDLo:2100 mg/kg/7W-I:ETA 13BYAH -,83,62
skn-mus TDLo:168 g/kg/35W-I:ETA JNCIAM 25,607,60
ivn-rat LD50:1220 mg/kg FAONAU 53A,256,74

PKI500 CAS: 25322-69-4
POLYPROPYLENE GLYCOL

mf: $(C_3H_8O_2)_n$

PROP: Clear, colorless liquid. Mw: 400-2000, mp: does not crystallize, flash p: +390°F, d: 1.002-1.007. Sol in water, aliphatic ketones and alcs; insol in ether, aliphatic hydrocarbons.

SYNS: ALKAPOL PPG-1200 ◇ JEFFOX ◇ POLYPROPYLENGLYKOL (CZECH)

USE IN FOOD:

Purpose: Defoaming agent.

Where Used: Beet sugar, beverages, candy, coconut (shredded), icings, yeast.

Regulations: FDA - 21CFR 173.340. Use at a level not in excess of the amount reasonably required to accomplish the intended effect.

SAFETY PROFILE: Mildly toxic by ingestion. A skin and eye irritant. Combustible liquid when exposed to heat or flame; can react with oxidizing materials. To fight fire, use foam, CO_2, dry chemical. When heated to decomposition it emits acrid smoke and irritating fumes.

TOXICITY DATA and CODEN

skn-rbt 500 mg/24H MLD 28ZPAK -,255,72
eye-rbt 500 mg AJOPAA29,1363,46
orl-rat LD50:4190 mg/kg 28ZPAK -,255,72

PKL000 CAS: 9005-64-5
POLYSORBATE 20

PROP: Lemon to amber colored liquid; characteristic odor, bitter taste. Sol in water, alc, ethyl acetate, methanol, dioxane; insol in mineral oil, mineral spirits.

SYNS: GLYCOSPERSE L-20X ◇ POLYOXYETHYLENE (20) SORBITAN MONOLAURATE

USE IN FOOD:

Purpose: Emulsifier, flavoring agent, stabilizer.

Where Used: Various.

Regulations: FDA - 21CFR 172.515. Use at a level not in excess of the amount reasonably required to accomplish the intended effect.

SAFETY PROFILE: Moderately toxic by intravenous route. Mildly toxic by ingestion. A human skin irritant. When heated to decomposition it emits acrid smoke and irritating fumes.

TOXICITY DATA and CODEN

skn-hmn 15 mg/3D-I MLD 85DKA8 -,127,77
orl-rat LD50:37 g/kg FOREAE 21,348,56

PKL050 CAS:
POLYSORBATE 65

PROP: Tan, waxy solid; faint odor, bitter taste. Sol in mineral oil, vegetable oil, mineral spirits, acetone, ether, dioxane, alc, methanol; dispersible in water, carbon tetrachloride.

USE IN FOOD:

Purpose: Emulsifier, stabilizer.

Where Used: Cake fillings, cake mixes, cakes, frozen custard, frozen desserts, fruit sherbet, ice cream, ice milk, icings, milk or cream substitutes for beverage coffee, nonstandardized frozen desserts, whipped edible oil topping.

Regulations: FDA - 21CFR 172.838. Limitation of 0.1 percent in ice cream; 0.1 percent in frozen custard; 0.1 percent in ice milk; 0.1 percent in fruit sherbet; 0.1 percent in nonstandardized frozen desserts; 0.32 percent in cakes; 0.32 percent in cake mixes; 0.4 percent in whipped edible oil topping; 0.4 percent in milk or cream substitutes for beverage coffee; 0.32 percent in cake icings; 0.32 percent in cake fillings. Additional limitations when used in combination with polysorbate 60 and/or sorbitan monostearate. Must conform to FDA specifications for fats or fatty acids derived from edible oils. 21CFR 173.340.

SAFETY PROFILE: When heated to decomposition it emits acrid smoke and irritating fumes.

PKP750 CAS: 9002-89-5
POLYVINYL ALCOHOL

PROP: Colorless, amorphous powder. Mp: decomp over 200°, flash p: 175°F (OC), d: 1.329. Polymer of average molecular weight 120,000 (AMPLAO 67,589,59).

SYNS: ELVANOL ◇ ETHENOL HOMOPOLYMER (9CI) ◇ GELVATOLS ◇ GOHSENOLS ◇ POLY(VINYL ALCOHOL) ◇ VINYL ALCOHOL POLYMER

USE IN FOOD:

Purpose: Color diluent.

Where Used: Eggs (shell).

Regulations: FDA - 21CFR 73.1. No penetration allowed through shell.

IARC Cancer Review: Animal Limited Evidence IMEMDT 19,341,79; Human Inadequate Evidence IMEMDT 19,341,79.

SAFETY PROFILE: An experimental carcinogen and tumorigen. Flammable when exposed to heat or flame; can react with oxidizing materials. Slight explosion hazard in the form of dust when exposed to flame. To fight fire, use alcohol foam, CO_2, dry chemical. When heated to decomposition it emits acrid smoke and irritating fumes.

TOXICITY DATA and CODEN

scu-rat TDLo:2500 mg/kg:CAR AMPLAO 67,589,59
imp-rat TDLo:10 g/kg:ETA BJSUAM 52,49,65
imp-rat TD :3768 mg/kg:ETA EXPEAM 19,424,63

PKQ150
POLYVINYLPOLYPYRROLIDONE

PROP: White, hygroscopic powder; faint bland odor. Insol in water.

SYNS: CROSPOVIDONE ◇ PVPP ◇ 1-VINYL-2-PYRROLIDONE CROSSLINKED INSOLUBLE POLYMER

USE IN FOOD:

Purpose: Clarifying agent.

Where Used: Beverages, vinegar, wine.

Regulations: FDA - 21CFR 173.50. Must be removed by filtration. BATF - 27CFR 240.1051. Limitation of 6 pounds/1000 gallons in wine. Must be removed by filtration.

SAFETY PROFILE: When heated to decomposition it emits acrid smoke and irritating fumes.

PKQ250 CAS: 9003-39-8
POLY(1-VINYL-2-PYRROLIDINONE) HOMOPOLYMER

mf: $(C_6H_9ON)_n$

PROP: A free-flowing, white, amorphous powder. D: 1.23-1.29. Sol in water, chlorinated hydrocarbons, alc, amines, nitroparaffins, and lower molecular weight fatty acids.

SYNS: AGENT AT 717 ◇ ALBIGEN A ◇ ALDACOL Q ◇ AT 717 ◇ BOLINAN ◇ 1-ETHENYL-2-PYRROLIDINONE HOMOPOLYMER ◇ 1-ETHENYL-2-PYRROLIDINONE POLYMERS ◇ GANEX P 804 ◇ HEMODESIS ◇ HEMODEZ ◇ K25 (polymer) ◇ KOLLIDON ◇ LUVISKOL ◇ MPK 90

◇ NCI C60582 ◇ NEOCOMPENSAN ◇ PERAGAL ST ◇ PERISTON ◇ PLASDONE ◇ POLYCLAR L ◇ POLY(1-(2-OXO-1-PYRROLIDINYL)ETHYLENE) ◇ POLYVIDONE ◇ POLY(n-VINYLBUTYROLACTAM) ◇ POLYVINYLPYRROLIDONE ◇ POVIDONE (USP XIX) ◇ PROTAGENT ◇ PVP (FCC) ◇ SUBTOSAN ◇ VINISIL ◇ N-VINYLBUTYROLACTAM POLYMER ◇ N-VINYLPYRROLIDONE POLYMER

USE IN FOOD:

Purpose: Bodying agent, clarifying agent, color diluent, dispersing agent, stabilizer, tableting aid.

Where Used: Beer, citrus fruit (fresh), confectionery, flavor concentrates in tablet form, food supplements in tablet form, fruits, gum, nonnutritive sweeteners in concentrated liquid form, nonnutritive sweeteners in tablet form, vegetables, vinegar, vitamin and mineral concentrates in liquid form, vitamin and mineral concentrates in tablet form, wine.

Regulations: FDA - 21CFR 73.1, 172.210, 21CFR 173.55. Limitation of 10 ppm in beer, 40 ppm in vinegar, 60 ppm in wine. Use at a level not in excess of the amount reasonably required to accomplish the intended effect. BATF - 27CFR 240.1051. Limitation of 60 ppm in finished wine.

SAFETY PROFILE: Mildly toxic by intraperitoneal and intravenous routes. When heated to decomposition it emits toxic fumes of NO_x.

TOXICITY DATA and CODEN

ipr-mus LD50:12 g/kg FAONAU 53A,487,74
ivn-mky LDLo:5300 mg/kg NCIHL* NIH-69-2067,70

PKU600 CAS: 868-14-4
POTASSIUM ACID TARTRATE

mf: $C_4H_5KO_6$ mw: 188.18

PROP: Colorless crystals or white crystalline powder; acid taste. Sol in water, sltly sol in alc.

SYNS: CREAM of TARTER ◇ POTASSIUM BITARTRATE

USE IN FOOD:

Purpose: Acid, anticaking agent, antimicrobial agent, formulation aid, humectant, leavening agent, pH control agent, processing aid, stabilizer, surface-active agent, thickener.

Where Used: Baked goods, candy (hard), candy (soft), confections, crackers, frostings, gelatins, jams, jellies, margarine, oleomargarine, puddings, wine (grape).

Regulations: FDA - 21CFR 184.1077. GRAS when used in accordance with good manufacturing practice. USDA - 9CFR 318.7. Sufficient for purpose. BATF - 27CFR 240.1051. Limitation of 25 pounds/1000 gallons of grape wine.

SAFETY PROFILE: When heated to decomposition it emits acrid smoke and irritating fumes.

PKU700 CAS: 9005-36-1
POTASSIUM ALGINATE

mf: $(C_6H_7O_6K)_x$ mw: 214.22

PROP: White fibrous granular solid; odorless and tasteless. Sol in water; insol in alc, chloroform, ether.

SYN: ALGIN

USE IN FOOD:

Purpose: Emulsifier, stabilizer, thickener.

Where Used: Confections, frostings, fruit juices, fruits (processed), gelatins, puddings.

Regulations: FDA - 21CFR 184.1610. GRAS with a limitation of 0.1 percent in confections and frostings, 0.7 percent in gelatins and puddings, 0.25 percent in processed fruits and fruit juices, 0.01 percent in all other food when used in accordance with good manufacturing practice.

SAFETY PROFILE: When heated to decomposition it emits acrid smoke and irritating fumes.

PKW760 CAS: 582-25-2
POTASSIUM BENZOATE

mf: $C_7H_5O_2 \cdot K$ mw: 160.22

USE IN FOOD:

Purpose: Preservative.

Where Used: Margarine, oleomargarine, wine.

Regulations: USDA - 9CFR 318.7. Limitation of 0.1 percent, or if used in combination with sorbic acid and its salts, 0.2 percent (expressed as the acids in the weight of the finished food.) BATF - 27CFR 240.1051. Limitation of 0.1 percent in wine.

SAFETY PROFILE: Combustible when exposed to heat or flame. When heated to decomposition it emits acrid smoke and irritating fumes.

PKX100 CAS: 298-14-6
POTASSIUM BICARBONATE

mf: $KHCO_3$ mw: 100.12

PROP: Colorless, transparent, monoclinic prisms or white granular powder; odorless. Sol in water; insol in alc.

USE IN FOOD:

Purpose: Alkali, formulation aid, leavening agent, nutrient supplement, pH control agent, processing aid.

Where Used: Baked goods, margarine, oleomargarine.

Regulations: FDA - 21CFR 184.1613. GRAS when used in accordance with good manufacturing practice. USDA - 9CFR 318.7. Sufficient for purpose. BATF - 27CFR 240.1051. The natural or fixed acids shall not be reduced below 5 parts per thousand.

SAFETY PROFILE: A nuisance dust.

PKY500 CAS: 7758-02-3
POTASSIUM BROMIDE

mf: BrK mw: 119.01

PROP: Colorless, cubic, sltly hygroscopic crystals. Mp: 730°, bp: 1380°, d: 2.75 @ 25°, vap press: 1 mm @ 795°.

SYN: BROMIDE SALT OF POTASSIUM

USE IN FOOD:

Purpose: Lye peeling agent, washing water agent.

Where Used: Fruits, vegetables.

Regulations: FDA - 21CFR 173.315. Use at a level not in excess of the amount reasonably required to accomplish the intended effect.

SAFETY PROFILE: Large doses can cause central nervous system depression. Prolonged inhalation may cause skin eruptions. Mutagenic data. Violent reaction with BrF_3. When heated to decomposition it emits toxic fumes of K_2O and Br^-.

TOXICITY DATA and CODEN

hma-rat/ast 200 mg/kg GANNA2 54,155,63

PLA000 CAS: 584-08-7
POTASSIUM CARBONATE (2:1)

mf: CO_3•2K mw: 138.21

PROP: White, deliquescent, granular, translucent powder; odorless with alkaline taste. D: 2.428 @ 19°, mp: 891°, bp: decomposes. Sol in water; insol in alc.

SYNS: CARBONIC ACID, DIPOTASSIUM SALT ◇ KALIUMCARBONAT (GERMAN) ◇ K-GRAN ◇ PEARL ASH ◇ POTASH

USE IN FOOD:

Purpose: Alkali, boiler water additive, flavoring agent, nutrient supplement, pH control agent, processing aid.

Where Used: Margarine, oleomargarine, soups.

Regulations: FDA - 21CFR 173.310, 21CFR 184.1619. GRAS when used in accordance with good manufacturing practice. USDA - 9CFR 318.7. Sufficient for purpose. BATF - 27CFR 240.1051. The natural or fixed acids shall not be reduced below 5 parts per thousand.

SAFETY PROFILE: Poison by ingestion. A strong caustic. Incompatible with KCO; chlorine trifluoride; magnesium. When heated to decomposition it emits toxic fumes of K_2O.

TOXICITY DATA and CODEN

orl-rat LD50:1870 mg/kg AIHAAP 30,470,69
orl-bwd LD50:100 mg/kg AECTCV 12,355,83

PLA500 CAS: 7447-40-7
POTASSIUM CHLORIDE

mf: ClK mw: 74.55

PROP: Colorless or white crystals or powder; odorless with salty taste. D: 1.987, mp: 773° (sublimes @ 1500°). Sol in water; sltly sol in alc; insol in abs alc.

SYNS: CHLORID DRASELNY (CZECH) ◇ CHLOROPOTASSURIL ◇ DIPOTASSIUM DICHLORIDE ◇ EMPLETS POTASSIUM CHLORIDE ◇ ENSEAL ◇ KALITABS ◇ KAOCHLOR ◇ KAON-Cl ◇ KAY CIEL ◇ K-LOR ◇ KLOTRIX ◇ K-PRENDEDOME ◇ PFIKLOR ◇ POTASSIUM MONOCHLORIDE ◇ POTAVESCENT ◇ REKAWAN ◇ SLOW-K ◇ TRIPOTASSIUM TRICHLORIDE

USE IN FOOD:

Purpose: Dietary supplement, flavor enhancer, flavoring agent, gelling agent, nutrient, pH control agent, salt substitute, tissue softening agent, yeast food.

Where Used: Jelly (artificially sweetened), meat (raw cuts), poultry (raw cuts), preserves (artificially sweetened).

Regulations: FDA - 21CFR 182.5622, 184.1622. GRAS when used in accordance with good manufacturing practice. Preparations containing equal to or greater than 100 mg of potassium per tablet are drugs covered by 21CFR

201.306. USDA - 9CFR 318.7, 381.147. Limitation of not more than 3 percent of a 2.0 molar solution. A solution of the approved inorganic chlorides injected into or applied to raw meat cuts shall not result in a gain of more than 3 percent above the weight of the untreated product.

SAFETY PROFILE: A human poison by ingestion. Poison experimentally by ingestion, intravenous, and intraperitoneal routes. Moderately toxic by subcutaneous route. Human systemic effects by ingestion: nausea, blood clotting changes, cardiac arrythmias. An eye irritant. Mutagenic data. Explosive reaction with BrF_3; sulfuric acid + potassium permanganate. When heated to decomposition it emits toxic fumes of K_2O and Cl^-.

TOXICITY DATA and CODEN

eye-rbt 500 mg/24H MLD 28ZPAK -,8,72
mmo-sat 100 μg/plate NTPTB* APR 82
cyt-ham:lng 12 g/L FCTOD7 22,501,84
orl-wmn TDLo:60 mg/kg/D:GIT,BLD LANCAO 2,919,80
orl-inf LDLo:938 mg/kg/2D JAMAAP 240,1339,78
orl-man LDLo:20 mg/kg:CVS,GIT,BLD LANCAO 2,919,80
orl-rat LD50:2600 mg/kg 28ZPAK -,8,72

PLB500 CAS: 10141-00-1
POTASSIUM CHROMIC SULFATE

mf: $Cr \cdot 2H_2O_4S \cdot K$ mw: 287.26

SYNS: 0% BASICITY CHROME ALUM ◇ CHROME ALUM ◇ CHROME POTASH ALUM ◇ CHROMIC POTASSIUM SULFATE ◇ CHROMIC POTASSIUM SULPHATE ◇ CHROMIUM POTASSIUM SULFATE (1:1:2) ◇ CHROMIUM POTASSIUM SULPHATE ◇ CRYSTAL CHROME ALUM ◇ POTASSIUM CHROMIC SULPHATE ◇ POTASSIUM CHROMIUM ALUM ◇ POTASSIUM DISULPHATOCHROMATE(III) ◇ SULFURIC ACID, CHROMIUM (3+) POTASSIUM SALT (2:1:1)

Purpose: Animal glue adjuvant.

Where Used: Packaging materials.

Regulations: FDA - 21CFR 178.3120. Use only in glue as a colloidal flocculant added to the pulp suspension prior to the sheet-forming operation in the manufacture of paper and paper board.

Chromium and its compounds are on the Community Right-To-Know List. EPA Genetic Toxicology Program.

OSHA PEL: TWA 0.5 mg(Cr)/m^3 ACGIH TLV: TWA 0.5 mg(Cr)/m^3

SAFETY PROFILE: Chromate salts are carcinogens. Mutagenic data. When heated to decomposition it emits toxic fumes of K_2O.

TOXICITY DATA and CODEN

cyt-ham:ovr 1 mg/L CRNGDP 3,1331,82
sce-ham:ovr 1 mg/L CRNGDP 3,1331,82

PLB750 CAS: 866-84-2
POTASSIUM CITRATE

mf: $C_6H_5O_7 \cdot 3K$ mw: 306.41

PROP: Colorless transparent crystals or white powder; odorless with salty taste. D: 1.98, decomp when heated to 230°. Deliquescent, sol in water and glycerol; almost insol in alc.

SYNS: CITRIC ACID, TRIPOTASSIUM SALT ◇ TRIPOTASSIUM CITRATE MONOHYDRATE

USE IN FOOD:

Purpose: Miscellaneous and general-purpose buffer, pH control agent, sequestrant.

Where Used: Jelly (artificially sweetened), margarine, meat products, milk, oleomargarine, wine.

Regulations: FDA - 21CFR 182.1625, 182.6625. GRAS when used in accordance with good manufacturing practice. USDA - 9CFR 318.7. Sufficient for purpose. BATF - 27CFR 240.1051. Limitation of 25 pounds/1000 gallons of wine.

SAFETY PROFILE: Poison by intravenous route. When heated to decomposition it emits toxic fumes of K_2O.

TOXICITY DATA and CODEN

ivn-dog LD50:167 mg/kg AVERAG 44,555,37

PLG775
POTASSIUM GIBBERELLATE

mf: $C_{19}H_{21}KO_6$ mw: 384.47

PROP: White crystalline powder; odorless. Deliquescent, sol in water, alc, acetone.

USE IN FOOD:

Purpose: Enzyme activator.

Where Used: Fermented malt beverages.

Regulations: FDA - 21CFR 172.725. Limitation of 2 ppm in treated barley malt, 0.5 ppm in finished beverage.

SAFETY PROFILE: When heated to decomposition it emits acrid smoke and irritating fumes.

PLG800 CAS: 299-27-4
POTASSIUM GLUCONATE
mf: $C_6H_{12}O_7 \cdot K$ mw: 235.28

PROP: Yellowish-white crystals or powder; mild, sltly salty taste. Decomp at 180°. Freely sol in water, glycerin; practically insol in abs alc, ether, benzene, chloroform.

SYNS: d-GLUCONIC ACID, MONOPOTASSIUM SALT (9CI) ◇ GLUCONIC ACID POTASSIUM SALT ◇ GLUCONSAN K ◇ KALIUM-BETA ◇ KAON ◇ KAON ELIXIR ◇ KATORIN ◇ K-IAO ◇ POTALIUM ◇ POTASORAL ◇ POTASSIUM d-GLUCONATE ◇ POTASSURIL ◇ SIROKAL

USE IN FOOD:

Purpose: Denuding agent, dietary supplement, nutrient, sequestrant.

Where Used: Beverages (dry mix), cake mixes, desserts (dry mix), tripe.

Regulations: USDA - 9CFR 318.7. Sufficient for purpose.

SAFETY PROFILE: Moderately toxic by intraperitoneal route. Mildly toxic by ingestion. When heated to decomposition it emits toxic fumes of K_2O.

TOXICITY DATA and CODEN

orl-rat LD50:10380 mg/kg NIIRDN 6,226,82
ipr-rat LD50:2664 mg/kg KSRNAM 6,810,72

PLG810
POTASSIUM GLYCEROPHOSPHATE
mf: $C_3H_7K_2O_6P \cdot 3H_2O$ mw: 302.20

PROP: Pale yellow syrupy liquid. Sol in water.

USE IN FOOD:

Purpose: Dietary supplement, nutrient.

Where Used: Various.

Regulations: FDA - 21CFR 182.5628, 182.8628. GRAS when used in accordance with good manufacturing practice.

SAFETY PROFILE: When heated to decomposition it emits acrid smoke and irritating fumes.

PLJ500 CAS: 1310-58-3
POTASSIUM HYDROXIDE
DOT: 1813/1814
mf: HKO mw: 56.11

PROP: White, deliquescent pieces, lumps or sticks having crystalline fracture. Mp: 360°±7°, bp: 1320°, d: 2.044. Sol in water, alc.

SYNS: CAUSTIC POTASH ◇ CAUSTIC POTASH, dry, solid, flake, bead, or granular (DOT) ◇ CAUSTIC POTASH, liquid or solution (DOT) ◇ HYDROXYDE de POTASSIUM (FRENCH) ◇ KALIUMHYDROXID (GERMAN) ◇ KALIUMHYDROXYDE (DUTCH) ◇ LYE ◇ POTASSA ◇ POTASSE CAUSTIQUE (FRENCH) ◇ POTASSIO (IDROSSIDO di) (ITALIAN) ◇ POTASSIUM HYDRATE (DOT) ◇ POTASSIUM HYDROXIDE, dry, solid, flake, bead, or granular (DOT) ◇ POTASSIUM HYDROXIDE, liquid or solution (DOT) ◇ POTASSIUM (HYDROXYDE de) (FRENCH)

USE IN FOOD:

Purpose: Alkali, formulation aid, pH control agent, poultry scald agent, processing aid, stabilizer, thickener.

Where Used: Black olives, poultry.

Regulations: FDA - 21CFR 184.1631. GRAS when used in accordance with good manufacturing practice. USDA - 9CFR 381.147. Sufficient for purpose.

ACGIH TLV: CL 2 mg/m^3 DOT Classification: Corrosive Material; Label: Corrosive; Corrosive Material; Label: Corrosive, solution

SAFETY PROFILE: Poison by ingestion. An eye irritant and severe human skin irritant. Very corrosive to the eyes, skin, and mucous membranes. Mutagenic data. Ingestion may cause violent pain in throat and epigastrium, hematemesis, collapse. Stricture of esophagus may result if not immediately fatal. Above 84° it reacts with reducing sugars to form the poisonous carbon monoxide gas. Violent, exothermic reaction with water. When heated to decomposition it emits toxic fumes of K_2O.

TOXICITY DATA and CODEN

skn-hmn 50 mg/24H SEV TXAPA9 31,481,75
eye-rbt 1 mg/24H rns MOD TXAPA9 32,239,75
cyt-rat/ast 1800 mg/kg GANNA2 54,155,63
orl-rat LD50:365 mg/kg TXAPA9 32,239,75

PLJ750 CAS: 1310-58-3
POTASSIUM HYDROXIDE (solution)
DOT: 1813/1814
mf: HKO mw: 56.11

PROP: Clear liquid.

SYN: POTASSIUM HYDRATE (solution)

USE IN FOOD:

Purpose: Alkali, formulation aid, pH control agent, processing aid, stabilizer, thickener.

Where Used: Black olives.

Regulations: FDA - 21CFR 184.1631. GRAS when used in accordance with good manufacturing practice.

DOT Classification: Corrosive Material; Label: Corrosive

SAFETY PROFILE: Very corrosive to the eyes, skin, and mucous membranes. When heated to decomposition it emits toxic fumes of K_2O.

TOXICITY DATA and CODEN

skn-rbt 5 mg/24H MOD TXAPA9 32,239,75
eye-rbt 1 mg/24H rns MOD TXAPA9 32,239,75

PLK250 CAS: 7758-05-6
POTASSIUM IODATE
mf: $IO_3 \cdot K$ mw: 214.00

PROP: Colorless crystals or white crystalline powder. Mp: 560°, d: 3.89. Sol in water; insol in alc.

SYN: IODIC ACIODIC ACID, POTASSIUM SALT

USE IN FOOD:

Purpose: Dough conditioner, maturing agent.

Where Used: Baked goods, bread.

Regulations: FDA - 21CFR 184.1635. GRAS with a limitation of 0.0075 percent in flour when used in accordance with good manufacturing practice.

SAFETY PROFILE: Poison by ingestion and intraperitoneal routes. Violent reaction with organic matter. When heated to decomposition it emits very toxic fumes of I^- and K_2O.

TOXICITY DATA and CODEN

orl-mus LDLo: 531 mg/kg JPETAB 120,171,57
ipr-mus LD50: 136 mg/kg JPETAB 120,171,57

PLK500 CAS: 7681-11-0
POTASSIUM IODIDE
mf: IK mw: 166.00

PROP: Colorless or white granules. Mp: 723°, bp: 1420°, d: 3.13, vap press: 1 mm @ 745°. Sltly hygroscopic. Sol in water, glycerin, alc.

USE IN FOOD:

Purpose: Dietary supplement, nutrient supplement.

Where Used: Various.

Regulations: FDA - 21CFR 172.375. Limitation for daily intake of 225 μg for foods unlabeled with reference to age, 5 μg for infants, 105 μg for children under 4 years old, 225 μg for adults and children 4 years old or older, and 300 μg for pregnant or lactating women. 21CFR 184.1634. GRAS with a limitation of 0.01 percent in table salt when used in accordance with good manufacturing practice.

SAFETY PROFILE: Poison by intravenous route. Moderately toxic by ingestion and intraperitoneal routes. Human teratogenic effects by ingestion: developmental abnormalities of the endocrine system. Experimental teratogenic and reproductive effects. Mutagenic data. Explosive reaction with charcoal + ozone; trifluoroacetyl hypofluorite; fluorine perchlorate. Incompatible with oxidants; BrF_3; FClO; metallic salts; calomel. When heated to decomposition it emits very toxic fumes of K_2O and I^-.

TOXICITY DATA and CODEN

cyt-rat/ast 500 mg/kg GANNA2 54,155,63
orl-wmn TDLo: 3240 mg/kg (1-39W preg): TER ADCHAK 43,702,68
orl-rat TDLo: 10530 mg/kg (1-9D preg): TER JRPFA4 27,265,71
orl-rat TDLo: 822 mg/kg (2W male/2W pre-13D post): REP FCTOD7 22,963,84
orl-mus LDLo: 1862 mg/kg JPETAB 120,171,57

PLK650 CAS: 996-31-6
POTASSIUM LACTATE
mf: $C_3H_5O_3K$ mw: 128.17

USE IN FOOD:

Purpose: Flavoring adjuvant, flavor enhancer, flavoring agent, humectant, pH control agent.

Where Used: Various.

Regulations: FDA - 21CFR 184.1639. GRAS when used in accordance with good manufacturing practice. Not authorized for infant foods and infant formulas.

SAFETY PROFILE: When heated to decomposition it emits acrid smoke and irritating fumes.

PLL500 CAS: 7757-79-1
POTASSIUM NITRATE
DOT: 1486
mf: KNO_3 mw: 101.11

PROP: Transparent, colorless or white crystalline powder or crystals; odorless with a cooling, pungent, salty taste. Mp: 334°, bp: decomp @ 400°, d: 2.109 @ 16°. Sol in glycerol, water; mod sol in alc.

SYNS: KALIUMNITRAT (GERMAN) ◇ NITER ◇ NITRE ◇ NITRIC ACID, POTASSIUM SALT ◇ SALTPETER ◇ VICKNITE

USE IN FOOD:

Purpose: Antimicrobial agent, preservative.

Where Used: Cod roe, meat (cured), poultry.

Regulations: FDA - 21CFR 172.160. Limitation of 200 ppm in finished cod roe. 21CFR 181.33. USDA - 9CFR 318.7, 9CFR 381.147. Limitation of 7 pounds per 100 gallons of pickle; 3.5 ounces per 100 pound of meat, 2.75 ounces to 100 pounds chopped meat.

DOT Classification: Oxidizer; Label: Oxidizer

SAFETY PROFILE: Poison by intravenous route. Moderately toxic by ingestion. An experimental teratogen. Experimental reproductive effects. Mutagenic data. Ingestion of large quantities may cause gastroenteritis. Chronic exposure can cause anemia, nephritis, and methemoglobinemia. A powerful oxidizer. When heated to decomposition it emits very toxic fumes of NO_x and K_2O.

TOXICITY DATA and CODEN

mrc-esc 5 pph JGMIAN 8,45,53
orl-rat TDLo:22 g/kg (1-22D preg):TER JANSAG 15,1291,58
orl-rbt TDLo:6505 mg/kg (23-27D preg):REP SOVEA7 27,246,74
orl-gpg TDLo:1670 mg/kg:TER TXAPA9 12,179,68
orl-rat LD50:3750 mg/kg NYKZAU 81,469,83

PLM500 CAS: 7758-09-0
POTASSIUM NITRITE (1:1)

DOT: 1488
mf: $NO_2 \cdot K$ mw: 85.11

PROP: White or sltly yellowish, deliquescent prisms or sticks. Mp: 387°; bp: decomp, d: 1.915. Very sol in water; sltly sol in alc.

SYNS: NITROUS ACID, POTASSIUM SALT ◇ POTASSIUM NITRITE (DOT)

USE IN FOOD:

Purpose: Antimicrobial agent, color fixative in meat and meat products.

Where Used: Bacon, meat (cured), poultry.

Regulations: FDA - 21CFR 181.34. USDA - 9CFR 318.7. Sufficient for purpose. 9CFR 318.7, 9CFR 381.147. Limitation of 2 pounds per 100 gallons of pickle at 10 percent pump level; 1 ounce per 100 pounds of meat, 0.25 ounces to 100 pounds of chopped meat. The use of nitrites, nitrates, or combination shall not result in more than 200 ppm of nitrite, calculated as sodium nitrate in finished product.

DOT Classification: Oxidizer; Label: Oxidizer

SAFETY PROFILE: Poison by ingestion. Experimental teratogenic and reproductive effects. Nitrites have been implicated in an increased incidence of cancer. Mutagenic data. Flammable when exposed to heat or flame. A powerful oxidizing material. Slight explosion hazard when exposed to heat. Upon decomposition it emits toxic fumes of K_2O.

TOXICITY DATA and CODEN

mmo-omi 500 mmol/L JMOBAK 9,352,64
mmo-omi 500 mmol/L JOVIAM 7,673,71
orl-rat TDLo:30 g/kg (60D male/60D pre):REP DABBBA 28,3815,68
orl-gpg TDLo:137 g/kg (18W pre):REP TXAPA9 12,179,68
orl-gpg TDLo:201 g/kg (19W pre):TER TXAPA9 12,179,68
orl-rbt LD50:200 mg/kg SOVEA7 27,246,74

PLQ400
POTASSIUM PHOSPHATE, DIBASIC

mf: K_2HPO_4 mw: 174.18

PROP: Colorless or white granular solid. Deliquescent, sol in water; insol in alc.

SYNS: DIPOTASSIUM MONOPHOSPHATE ◇ DIPOTASSIUM PHOSPHATE

USE IN FOOD:

Purpose: Buffer, cooked out juices retention agent, sequestrant, yeast food.

Where Used: Cheese, coffee whiteners, meat products, poultry food products.

Regulations: FDA - 21CFR 182.6285. GRAS when used in accordance with good manufacturing practice. USDA - 9CFR 318.7. In meat food products, where allowed, limitation of 5 percent of phosphate in pickle at 10 percent pump level, 0.5 percent of phosphate in product (only clear solution may be injected into product). 9CFR 381.147. Limitation of 0.5 percent of total poultry product.

SAFETY PROFILE: A nuisance dust.

PLQ405
POTASSIUM PHOSPHATE, MONOBASIC
mf: KH_2PO_4 mw: 136.09

PROP: Colorless crystals or white crystalline powder; odorless. Sol in water; insol in alc.

SYNS: POTASSIUM BIPHOSPHATE ◇ POTASSIUM DIHYDROGEN PHOSPHATE ◇ MONOPOTASSIUM PHOSPHATE

USE IN FOOD:

Purpose: Buffer, cooked out juices retention agent, sequestrant, yeast food.

Where Used: Eggs (whole), meat products, milk, poultry food products.

Regulations: USDA - 9CFR 318.7. In meat food products, where allowed, limitation of 5 percent of phosphate in pickle at 10 percent pump level, 0.5 percent of phosphate in product (only clear solution may be injected into product. 9CFR 381.147. Limitation of 0.5 percent in total poultry product.

SAFETY PROFILE: A nuisance dust.

PLQ410
POTASSIUM PHOSPHATE, TRIBASIC
mf: K_3PO_4 mw: 212.27

PROP: White crystals. Hygroscopic, sol in water; insol in alc.

SYN: TRIPOTASSIUM PHOSPHATE

USE IN FOOD:

Purpose: Emulsifier.

Where Used: Various.

Regulations: GRAS when used in accordance with good manufacturing practice.

SAFETY PROFILE: A nuisance dust.

PLR125
POTASSIUM POLYMETAPHOSPHATE
mf: $(KPO_3)_x$

PROP: White powder; odorless. Insol in water; sol in dilute solutions of sodium salts.

SYNS: POTASSIUM KURROL'S SALT ◇ POTASSIUM METAPHOSPHATE

USE IN FOOD:

Purpose: Fat emulsifier, moisture-retaining agent.

Where Used: Various.

Regulations: GRAS when used in accordance with good manufacturing practice.

SAFETY PROFILE: A nuisance dust.

PLR200
POTASSIUM PYROPHOSPHATE
mf: $K_4P_2O_7$ mw: 330.34

PROP: Colorless crystals or white granular solid. Hygroscopic, sol in water; insol in alc.

SYN: TETRAPOTASSIUM PYROPHOSPHATE

USE IN FOOD:

Purpose: Cooked out juices retention agent, emulsifier, texturizer.

Where Used: Meat products, poultry food products.

Regulations: USDA - 9CFR 318.7. In meat food products, where allowed, limitation of 5 percent of phosphate in pickle at 10 percent pump level, 0.5 percent of phosphate in product (only clear solution may be injected into product. 9CFR 381.147. Limitation of 0.5 percent of total poultry product.

SAFETY PROFILE: A nuisance dust.

PLR250 CAS: 16731-55-8
POTASSIUM PYROSULFITE
DOT: 2693
mf: $O_5S_2{\cdot}K$ mw: 183.22

PROP: Monoclinic plates or white crystalline powder; sulfur dioxide odor. Mp: decomp; d: 2.3. Sol in water; insol in alc.

SYNS: POTASSIUM METABISULFITE (DOT, FCC) ◇ PYROSULFUROUS ACID, DIPOTASSIUM SALT

USE IN FOOD:

Purpose: Antioxidant, preservative, sterilizer.

Where Used: Fruits (fresh), meat, vegetables (fresh), wine.

Regulations: FDA - 21CFR 182.3637. GRAS when used in accordance with good manufacturing practice except that it is not used in meats or in foods recognized as sources of vitamin B_1 or on fruits and vegetables intended to be sold or served raw or to be presented to consumers as fresh. BATF - 27CFR 240.1051. As proscribed in 21CFR 4.22

EPA Genetic Toxicology Program.

DOT Classification: ORM-B; Label: None

SAFETY PROFILE: An experimental tumorigen. Experimental reproductive effects. A very irritating material. When heated to decomposition it emits toxic fumes of SO_x and K_2O.

TOXICITY DATA and CODEN

orl-rat TDLo:35 g/kg (49D pre/1-21D preg): REP CRSBAW 172,470,78
orl-mus TDLo:1440 g/kg/2Y-C:ETA EESADV 3,451,79
orl-mus TD :2880 g/kg/2Y-C:ETA EESADV 3,451,79

PLS750 CAS: 590-00-1
POTASSIUM SORBATE
mf: $C_6H_7O_2 \cdot K$ mw: 150.23

PROP: White crystals, crystalline powder, or pellets. Mp: 270° (decomp): d: 1.363 @ 25°/20°. Sol in alc, water.

SYNS: 2,4-HEXADIENOIC ACID POTASSIUM SALT ◇ SORBIC ACID, POTASSIUM SALT ◇ SORBISTAT-K ◇ SORBISTAT-POTASSIUM

USE IN FOOD:

Purpose: Mold retardant, preservative.

Where Used: Baked goods, beverages (carbonated), beverages (still), bread, cake batters, cake fillings, cake topping, cheese, cottage cheese (creamed), fish (smoked or salted), fruit juices (fresh), fruits (dried), margarine, oleomargarine, pickled goods, pie crusts, pie fillings, salad dressings, salads (fresh), sausage (dry), sea food cocktail, syrups (chocolate dairy), wine.

Regulations: FDA - 21CFR 182.3640. GRAS when used in accordance with good manufacturing practice. USDA - 9CFR 318.7. Limitation of 0.1 percent individually, or if used in combination with its salts or benzoic acid or its salts, 0.2 percent (expressed as the acids in the weight of the finished product. Limitation of 10 percent in water solution applied to casings after stuffing or casings may be dipped in a 10 percent solution prior to stuffing. Not allowed in cooked sausage. BATF - 27CFR 240.1051. Limitation of 300 mg/100 gallons of wine.

EPA Genetic Toxicology Program.

SAFETY PROFILE: Moderately toxic by intraperitoneal route. Mildly toxic by ingestion. Mutagenic data. When heated to decomposition it emits toxic fumes of K_2O.

TOXICITY DATA and CODEN

cyt-ham:lng 10 g/L ATSUDG (4),41,80
sce-ham:lng 10 g/L FCTOD7 22,501,84
orl-rat LD50:4920 mg/kg FAONAU 40,61,67
ipr-mus LD50:1300 mg/kg FAONAU 53A,121,74

PLS775 CAS: 593-29-3
POTASSIUM STEARATE
mf: $KC_{18}H_{35}O_2$ mw: 322.57

PROP: White powder usually has fatty odor.

SYN: STEARIC ACID POTASSIUM SALT.

USE IN FOOD:

Purpose: Anticaking agent, binder, emulsifier, stabilizer.

Where Used: Chewing gum, packaging materials.

Regulations: FDA - 21CFR 172.863. Must conform to FDA specifications for salts of fats or fatty acids derived from edible oils. 21CFR 181.29. Use in accordance with good manufacturing practice.

SAFETY PROFILE: When heated to decomposition it emits acrid smoke and irritating fumes.

PLT000 CAS: 7778-80-5
POTASSIUM SULFATE (2:1)
mf: $O_4S \cdot 2K$ mw: 174.26

PROP: Colorless to white, odorless crystals; bitter salty taste. D: 2.66, mp: 1067°. Sol in water; insol in alc.

SYN: SULFURIC ACID, DIPOTASSIUM SALT

USE IN FOOD:

Purpose: Miscellaneous and general-purpose food additive, water corrective.

Where Used: Beverages (nonalcoholic).

Regulations: FDA - 21CFR 184.1643. GRAS with a limitation of 0.015 percent in non-alcoholic beverages when used in accordance with good manufacturing practice.

SAFETY PROFILE: Moderately toxic to humans by ingestion. Moderately toxic experimentally by subcutaneous route. Swallowing large doses causes severe gastrointestinal tract effects. When heated to decomposition it emits toxic fumes of K_2O and SO_x.

TOXICITY DATA and CODEN

orl-wmn LDLo:800 mg/kg AEXPBL 21,169,1886
orl-rat LD50:6600 mg/kg GISAAA 50(7),24,85

PLT500 CAS: 10117-38-1
POTASSIUM SULFITE
mf: $O_3S \cdot 2K$ mw: 158.26

PROP: White crystals or granular powder; odorless. Sol in water; sltly sol in alc.

SYN: SULFUROUS ACID, DIPOTASSIUM SALT

USE IN FOOD:

Purpose: Antioxidant, preservative.

Where Used: Various.

Regulations: GRAS when used in accordance with good manufacturing practice.

SAFETY PROFILE: When heated to decomposition it emits toxic fumes of SO_x and K_2O.

PLW400
POTASSIUM TRIPOLYPHOSPHATE
mf: $K_5P_3O_{10}$ mw: 448.41

PROP: White granules or powder. Hygroscopic, sol in water.

SYNS: PENTAPOTASSIUM TRIPHOSPHATE ◇ POTASSIUM TRIPHOSPHATE

USE IN FOOD:

Purpose: Boiler water additive, cooked out juices retention agent, texturizer.

Where Used: Chewing gum, meat products, poultry food products.

Regulations: FDA - 21CFR 173.310, 182.1810. GRAS when used in accordance with good manufacturing practice. USDA - 9CFR 318.7. In meat food products, where allowed, limitation of 5 percent of phosphate in pickle at 10 percent pump level, 0.5 percent of phosphate in product (only clear solution may be injected into product). 9CFR 381.147. Limitation of 0.5 percent of total poultry product.

SAFETY PROFILE: A nuisance dust.

PLZ000 CAS: 53-03-2
PREDNISONE
mf: $C_{21}H_{26}O_5$ mw: 358.47

PROP: White, odorless, crystalline powder. Mp: 235° (with some decomp). Very sltly sol in water; sltly sol in alcohol, chloroform, methanol, and dioxane.

SYNS: ANCORTONE ◇ BICORTONE ◇ COLISONE ◇ CORTAN ◇ CORTANCYL ◇ Δ-CORTELAN ◇ CORTIDELT ◇ Δ-CORTISONE ◇ Δ^1-CORTISONE ◇ Δ-CORTONE ◇ COTONE ◇ DACORTIN ◇ DECORTANCYL ◇ DECORTIN ◇ DECORTISYL ◇ Δ-1-DEHYDROCORTISONE ◇ 1-DEHYDROCORTISONE ◇ DEKORTIN ◇ DELTACORTELAN ◇ DELTACORTISONE ◇ DELTACORTONE ◇ DELTA-DOME ◇ DELTISONE ◇ 17,21-DIHYDROXYPREGNA-1,4-DIENE-3,11,20-TRIONE ◇ ENCORTON ◇ HOSTACORTIN ◇ IN-SONE ◇ JUVASON ◇ LISACORT ◇ METACORTANDRACIN ◇ NCI-C04897 ◇ NSC 10023 ◇ ORASONE ◇ PARACORT ◇ PRECORT ◇ PREDNICEN-M ◇ PREDNILONGA ◇ PREDNISON ◇ PREDNIZON ◇ 1,4-PREGNADIENE-17-α,21-DIOL-3,11,20-TRIONE ◇ RECTODELT ◇ SERVISONE ◇ SK-PREDNISONE ◇ SUPERCORTIL ◇ U 6020 ◇ ULTRACORTEN ◇ WOJTAB ◇ ZENADRID (VETERINARY)

USE IN FOOD:

Purpose: Animal drug.

Where Used: Milk.

Regulations: FDA - 21CFR 556.530. Limitation of zero in milk.

IARC Cancer Review: Human Inadequate Evidence IMEMDT 26,293,81; Animal Inadequate Evidence IMEMDT 26,293,81. NCI Carcinogenesis Studies (ipr); No Evidence: mouse CANCAR 40,1935,77; (ipr); Equivocal Evidence: rat CANCAR 40,1935,77.

SAFETY PROFILE: Poison by intraperitoneal and subcutaneous routes. Moderately toxic by intramuscular route. An experimental tumorigen. Human systemic effects by ingestion and possibly other routes: sensory change involving peripheral nerves. Experimental reproductive effects. Mutagenic data. Has been implicated in aplastic anemia.

TOXICITY DATA and CODEN

mmo-sat 3333 μg/plate NTPTB*J JAN82
mma-sat 333 μg/plate ENMUDM 5(Suppl 1),3,83
scu-mus TDLo: 24 mg/kg (13-18D preg): REP PBBHAU 12,213,80
ipr-rat TDLo: 860 mg/kg/26W-I: ETA CANCAR 40S,1935,77
orl-man TDLo: 857 μg/kg: PNS NEURAI 36,729,86
ipr-mus LD50: 135 mg/kg NCISP* JAN86

PMA000 CAS: 50-24-8
PREDONIN
mf: $C_{21}H_{28}O_5$ mw: 360.49

SYNS: CODELCORTONE ◇ CO-HYDELTRA ◇ Δ^1-CORTISOL ◇ DECORTIN H ◇ Δ^1-DEHYDROCORTISOL ◇ Δ^1-DEHYDROHYDROCORTISONE ◇ 1-DEHYDROHYDROCORTISONE ◇ DELCORTOL ◇ DELTA-CORTEF ◇ DELTACORTENOL

◇ DELTACORTRIL ◇ DELTA F ◇ DELTA-STAB ◇ DEXA-CORTIDELT HOSTACORTIN H ◇ DI-ADRESON F ◇ DICORTOL ◇ DYDELTRONE ◇ FERNISOLONE ◇ HOSTACORTIN ◇ HYDELTRA ◇ HYDELTRONE ◇ Δ^1-HYDROCORTISONE ◇ HYDRODELTALONE ◇ HYDRODELTISONE ◇ HYDRORETROCORTIN ◇ METACORTANDRALONE ◇ METICORTELONE ◇ METI-DERM ◇ PARACORTOL ◇ PARACOTOL ◇ PRECORTANCYL ◇ PRECORTISYL ◇ PREDNE-DOME ◇ PREDNELAN ◇ PREDNIS ◇ PREDNISOLONE ◇ PREDONINE ◇ 1,4-PREGNADIENE-3,20-DIONE-11-β,17-α,21-TRIOL ◇ 1,4-PREGNADIENE-11-β,17-α,21-TRIOL-3,20-DIONE ◇ SCHERISOLON ◇ STERANE ◇ STEROLONE ◇ 11-β,17,21-TRIHYDROXY-PREGNA-1,4-DIENE-3,20-DIONE ◇ 11-β,17-α,21-TRIHYDROXYPREGNA-1,4-DIENE-3,20-DIONE ◇ 11-β,17-α,21-TRIHYDROXY-1,4-PREGNADIENE-3,20-DIONE ◇ ULACORT ◇ ULTRACORTENE-H

USE IN FOOD:

Purpose: Animal drug.

Where Used: Milk.

Regulations: FDA - 21CFR 556.520. Limitation of zero in milk.

EPA Genetic Toxicology Program.

SAFETY PROFILE: A poison by intravenous and subcutaneous routes. Moderately toxic by ingestion and intraperitoneal routes. Human teratogenic effects by an unspecified route: developmental abnormalities of the central nervous system; effects on embryo or fetus: fetal death, extra embryonic structures. Human reproductive effects by an unspecified route: stillbirth. An experimental teratogen. Experimental reproductive effects. Human mutagenic data. When heated to decomposition it emits acrid smoke and irritating fumes.

TOXICITY DATA and CODEN

dni-hmn-unr 6300 μg/kg/8W STBIBN 50,172,75
dni-hmn:lym 1 mg/L AJOGAH 127,151,77
unr-wmn TDLo:56 mg/kg (1-40W preg):TER LANCAO 1,117,68
orl-rat TDLo:250 mg/kg (5-15D preg):TER AJOGAH 92,234,65
scu-mus TDLo:8 mg/kg (12-13D preg):REP NTOTDY 4,289,82
ipr-rat LD50:2000 mg/kg ADTEAS 3,181,68
orl-mus LD50:1680 mg/kg ARZNAD 20,111,70

PMC325 CAS: 1912-24-9
PRIMATOL

mf: $C_8H_{14}ClN_5$ mw: 215.72

PROP: Crystals. Mp: 171-174°. Solubility at 25°: in water, 70 ppm; ether, 12,000 ppm; chloroform, 52,000 ppm; methanol, 18,000 ppm.

SYNS: A 361 ◇ AATREX ◇ AATREX 4L ◇ AATREX NINE-O ◇ AATREX 80W ◇ 2-AETHYLAMINO-4-CHLOR-6-ISOPROPYLAMINO-1,3,5-TRIAZIN (GERMAN) ◇ 2-AETHYLAMINO-4-ISOPROPYLAMINO-6-CHLOR-1,3,5-TRIAZIN (GERMAN) ◇ AKTIKON ◇ AKTIKON PK ◇ AKTINIT A ◇ AKTINIT PK ◇ ARGEZIN ◇ ATAZINAX ◇ ATRANEX ◇ ATRASINE ◇ ATRATOL A ◇ ATRAZIN ◇ ATRAZINE(ACGIH, USDA) ◇ ATRED ◇ ATREX ◇ CANDEX ◇ CEKUZINA-T ◇ 2-CHLORO-4-ETHYLAMINEISOPROPYLAMINE-s-TRIAZINE ◇ 1-CHLORO-3-ETHYLAMINO-5-ISOPROPYLAMINO-s-TRIAZINE ◇ 1-CHLORO-3-ETHYLAMINO-5-ISOPROPYLAMINO-2,4,6-TRIAZINE ◇ 2-CHLORO-4-ETHYLAMINO-6-ISOPROPYLAMINO-s-TRIAZINE ◇ 2-CHLORO-4-ETHYLAMINO-6-ISOPROPYLAMINO-1,3,5-TRIAZINE ◇ 6-CHLORO-N-ETHYL-N′-(1-METHYLETHYL)-1,3,5-TRIAZINE-2,4-DIAMINE (9CI) ◇ 2-CHLORO-4-(2-PROPYLAMINO)-6-ETHYLAMINO-s-TRIAZINE ◇ CRISATRINA ◇ CRISAZINE ◇ CYAZIN ◇ FARMCO ATRAZINE ◇ FENAMIN ◇ FENAMINE ◇ FENATROL ◇ G 30027 ◇ GEIGY 30,027 ◇ GESAPRIM ◇ GESOPRIM ◇ GRIFFEX ◇ HUNGAZIN ◇ HUNGAZIN PK ◇ INAKOR ◇ OLEOGESAPRIM ◇ PRIMATOL A ◇ PRIMAZE ◇ RADAZIN ◇ RADIZINE ◇ SHELL ATRAZINE HERBICIDE ◇ STRAZINE ◇ TRIAZINE A 1294 ◇ VECTAL ◇ VECTAL SC ◇ WEEDEX A ◇ WONUK ◇ ZEAZIN ◇ ZEAZINE

USE IN FOOD:

Purpose: Herbicide.

Where Used: Various.

Regulations: USDA CES Ranking: C-3 (1985).

EPA Genetic Toxicology Program.

ACGIH TLV: TWA 5 mg/m^3

SAFETY PROFILE: Poison by intraperitoneal route. Moderately toxic by ingestion. Mildly toxic by inhalation and skin contact. An experimental tumorigen. Human mutagenic data. Experimental reproductive effects. A skin and severe eye irritant. When heated to decomposition it emits toxic fumes of Cl^- and NO_x.

TOXICITY DATA and CODEN

skn-rbt 38 mg open MLD CIGET* -,-,77
eye-rbt 6320 μg SEV CIGET* -,-,77
sln-nsc 10 mg/L MUREAV 167,35,86
dns-hmn:fbr 3 mmol/L MUREAV 74,77,80
scu-rat TDLo:2400 mg/kg (3-9D preg):REP BECTA6 9,301,73

scu-mus TDLo:418 mg/kg (6-14D preg):REP NTIS** PB223-160
orl-mus TDLo:9000 mg/kg/78W-I:ETA NTIS** PB223-159
orl-rat LD50:1500 mg/kg NTOTDY 5,503,83
ihl-rat LC50:5200 mg/m^3/4H FMCHA2 -,C3,83

PMH500 CAS: 57-83-0
PROGESTERONE

mf: $C_{31}H_{30}O_2$ mw: 314.51

PROP: A female sex hormone. White, crystalline powder; odorless. D: 1.166 @ 23°, mp: 127-131°. Practically insol in water; sol in alc, acetone, and dioxane; sparingly sol in oils.

SYNS: CORLUTIN ◇ CORLUVITE ◇ CORPORIN ◇ CORPUS LUTEUM HORMONE ◇ CYCLOGEST ◇ Δ^4-PREGNENE-3,20-DIONE ◇ GLANDUCORPIN ◇ HORMOFLAVEINE ◇ HORMOLUTON ◇ LINGUSORBS ◇ LIPO-LUTIN ◇ LUCORTEUM SOL ◇ LUTEAL HORMONE ◇ LUTEOHORMONE ◇ LUTEOSAN ◇ LUTEX ◇ LUTOCYCLIN ◇ LUTROMONE ◇ NALUTRON ◇ NSC-9704 ◇ PERCUTACRINE ◇ PIAPONON ◇ 3,20-PREGNENE-4 ◇ PREGNENEDIONE ◇ PREGNENE-3,20-DIONE ◇ PREGN-4-ENE-3,20-DIONE ◇ 4-PREGNENE-3,20-DIONE ◇ PROGEKAN ◇ PROGESTEROL ◇ β-PROGESTERONE ◇ PROGESTERONUM ◇ PROGESTIN ◇ PROGESTONE ◇ PROLIDON ◇ SYNGESTERONE ◇ SYNOVEX S ◇ SYNTOLUTAN

USE IN FOOD:

Purpose: Animal drug.

Where Used: Beef, lamb.

Regulations: FDA - 21CFR 556.540. Limitation of no residue in excess of the following increments above the concentrations naturally present in untreated animals: 3 ppb in muscle, 12 ppb in fat, 9 ppb in kidney, 6 ppb in liver of steers and calves; 3 ppb in muscle, 15 ppb for fat, kidney, and liver of lambs.

IARC Cancer Review: Animal Limited Evidence IMEMDT 21,491,79; Animal Sufficient Evidence IMEMDT 6,135,74. EPA Genetic Toxicology Program.

SAFETY PROFILE: Poison by intravenous and intraperitoneal routes. An experimental carcinogen, neoplastigen, tumorigen, and teratogen. Human male reproductive effects by intramuscular route: changes in spermatogenesis, the prostate, seminal vesicle, Cowper's gland, and accessory glands; impotence, and breast development. Human female reproductive effects by ingestion, parenteral and intravaginal routes: fertility changes; menstrual cycle changes and disorders; uterus, cervix, and vagina changes. Human teratogenic effects by ingestion, parenteral and possibly other routes: developmental abnormalities of the urogenital system. Experimental reproductive effects. Human mutagenic data. When heated to decomposition it emits acrid smoke and irritating fumes.

TOXICITY DATA and CODEN

dni-hmn:lym 5 μmol/L PSEBAA 146,401,74
cyt-mus:emb 1 mg/L DANKAS 282,173,85
orl-wmn TDLo:200 mg/kg (20D pre):REP AJOGAH 85,427,63
orl-wmn TDLo:113 mg/kg (6-32W preg):TER JCEMAZ 19,1369,59
orl-wmn TDLo:100 mg/kg (20D pre):REP FESTAS 16,158,65
par-wmn TDLo:600 μg/kg (67-71D preg):TER MACPAJ 33,200,58
scu-mus TDLo:40 mg/kg:NEO BJCAAI 19,824,65
imp-dog TDLo:270 mg/kg/78W:ETA 36PYAS -,145,77
ivn-mus LDLo:100 mg/kg JMCMAR 11,117,68

PMH900
l-PROLINE

mf: $C_5H_9NO_2$ mw: 115.13

PROP: White crystals or crystalline powder; odorless with sweet taste. Very sol in water, alc; insol in ether.

SYN: l-2-PYRROLIDINECARBOXYLIC ACID

USE IN FOOD:

Purpose: Dietary supplement, nutrient.

Where Used: Various.

Regulations: FDA - 21CFR 172.310. Limitation 4.2 percent by weight.

SAFETY PROFILE: When heated to decomposition emits toxic fumes of NO_x.

PMJ750 CAS: 74-98-6
PROPANE

DOT: 1075/1978

mf: C_3H_8 mw: 44.11

PROP: Colorless gas. Bp: −42.1°, lel: 2.3%, uel: 9.5%, fp: −187.1°, flash p: −156°F, d: 0.5852 @ −44.5°/4°, autoign temp: 842°F, vap d: 1.56. Sol in water, alc, ether.

SYNS: DIMETHYLMETHANE ◇ PROPYL HYDRIDE

USE IN FOOD:

Purpose: Aerating agent, gas, propellant.

Where Used: Various.

Regulations: FDA - 21CFR 184.1655. GRAS when used in accordance with good manufacturing practice.

OSHA PEL: TWA 1000 ppm ACGIH TLV: Asphyxiant DOT Classification: Flammable Gas; Label: Flammable Gas

SAFETY PROFILE: Central nervous system effects at high concentrations. An asphyxiant. Flammable gas. Highly dangerous fire hazard when exposed to heat or flame; can react vigorously with oxidizers. Explosive in the form of vapor when exposed to heat or flame. To fight fire, stop flow of gas. When heated to decomposition it emits acrid smoke and irritating fumes.

PML000 CAS: 57-55-6
1,2-PROPANEDIOL
mf: $C_3H_8O_2$ mw: 76.11

PROP: Colorless viscous liquid; practically odorless. Bp: 188.2°, flash p: 210°F (OC), lel: 2.6%, uel: 12.6%, d: 1.0362 @ 25°/25°, autoign temp: 700°F, vap press: 0.08 mm @ 20°, vap d: 2.62, fp: −59°. Hygroscopic; misc with water, acetone, chloroform; sol in essential oils; immisc with fixed oils.

SYNS: 1,2-DIHYDROXYPROPANE ◇ DOWFROST ◇ METHYLETHYLENE GLYCOL ◇ METHYL GLYCOL ◇ MONOPROPYLENE GLYCOL ◇ PG 12 ◇ PROPANE-1,2-DIOL ◇ PROPYLENE GLYCOL (FCC) ◇ PROPYLENE GLYCOL USP ◇ α-PROPYLENEGLYCOL ◇ 1,2-PROPYLENE GLYCOL ◇ SIRLENE ◇ SOLAR WINTER BAN ◇ TRIMETHYL GLYCOL

USE IN FOOD:

Purpose: Anticaking agent, antioxidant, clarifying agent, dough strengthener, emulsifier, flavoring agent, formulation aid, hog scald agent, humectant, poultry scald agent, processing aid, solvent, stabilizer, surface-active agent, texturizer, thickener, vehicle, wetting agent.

Where Used: Beverages (alcoholic), confections, flavorings, frostings, frozen dairy products, hog carcasses, nut products, nuts, poultry, seasonings, wine.

Regulations: FDA - 21CFR 184.1666. GRAS with a limitation of 5 percent in alcoholic beverages, 24 percent in confections and frostings, 2.5 percent in frozen dairy products, 97 percent in seasonings and flavorings, 5 percent in nuts and nut products, 2.0 percent in all other foods when used in accordance with good manufacturing practice. USDA - 9CFR 318.7, 381.147. Sufficient for purpose. BATF - 27CFR 240.1051. Limitation of 40 ppm in wine.

EPA Genetic Toxicology Program.

SAFETY PROFILE: Experimental teratogenic and reproductive effects. An eye and human skin irritant. Human systemic effects by ingestion: general anesthesia, convulsions, changes in surface EEG. Mutagenic data. Combustible liquid when exposed to heat or flame; can react with oxidizing materials. To fight fire, use alcohol foam. When heated to decomposition it emits acrid smoke and irritating fumes.

TOXICITY DATA and CODEN

skn-hmn 104 mg/3D-I MOD 85DKA8 -,127,77
skn-man 10%/2D JIDEAE 19,423,52
eye-rbt 100 mg MLD FCTOD7 20,573,82
dni-mus-scu 8000 mg/kg APMUAN S274,304,81
cyt-mus-scu 8000 mg/kg APMUAN S274,304,81
ipr-mus TDLo:100 mg/kg (11D preg):REP KAIZAN 37,239,62
ipr-mus TDLo:100 mg/kg (15D preg):TER KAIZAN 37,239,62
orl-chd TDLo:79 g/kg/56W-I:CNS,BRN JOPDAB 93,515,78
orl-rat LD50:20 g/kg TXAPA9 45,362,78

PMN850 CAS: 139-40-2
PROPAZINE
mf: $C_9H_{16}ClN_7O_2$ mw: 229.75

SYNS: 2,4-BIS(ISOPROPYLAMINO)-6-CHLORO-s-TRIAZINE ◇ 2,4-BIS(PROPYLAMINO)-6-CHLOR-1,3,5-TRIAZIN (GERMAN) ◇ GESAMIL ◇ MILOGARD ◇ PLANTULIN ◇ PRIMATOL P ◇ PROPASIN ◇ PROZINEX

USE IN FOOD:

Purpose: Herbicide.

Where Used: Various.

Regulations: USDA CES Ranking: C-4 (1988).

SAFETY PROFILE: Moderately toxic by ingestion. An experimental tumorigen. Moderate eye irritation. When heated to decomposition it emits toxic fumes of NO_x and Cl^-.

TOXICITY DATA and CODEN

eye-rbt 400 mg open CIGET* -,-,77
orl-gpg TDLo:11 g/kg/78W-I:ETA NTIS** PB223-159
orl-mus LD50:3180 mg/kg 85GMAT -,35,82

PMQ750 CAS: 104-46-1
p-PROPENYLANISOLE
mf: $C_{10}H_{12}O$ mw: 148.22

PROP: Leaves from alc or light yellow liquid above 23°; sweet taste with anise odor. D: 0.991 @ 20°/20°, refr index: 1.557-1.561, mp: 22.5°, bp: 235.3°, flash p: 198°F. Very sltly sol in water; misc in abs alc, ether, chloroform.

SYNS: ACINTENE O ◇ ANETHOLE (FCC) ◇ ANISE CAMPHOR ◇ ARIZOLE ◇ FEMA No. 2086 ◇ ISOESTRAGOLE ◇ p-METHOXY-β-METHYLSTYRENE ◇ 1-(p-METHOXYPHENYL)PROPENE ◇ 1-METHOXY-4-PROPENYLBENZENE ◇ 4-METHOXYPROPENYLBENZENE ◇ MONASIRUP ◇ NAULI "GUM" ◇ OIL of ANISEED ◇ p-1-PROPENYLANISOLE ◇ 4-PROPENYLANISOLE ◇ p-PROPENYLPHENYL METHYL ETHER

USE IN FOOD:

Purpose: Flavoring agent.

Where Used: Various.

Regulations: FDA - 21CFR 182.60. Use at a level not in excess of the amount reasonably required to accomplish the intended effect.

SAFETY PROFILE: Poison by ingestion. An experimental tumorigen. Combustible liquid. When heated to decomposition it emits acrid smoke and irritating fumes.

TOXICITY DATA and CODEN

ipr-mus TDLo:2400 mg/kg/8W-I:ETA CNREA8 33,3069,73
orl-rat LD50:2090 mg/kg FCTXAV 2,327,64

PMS500 CAS: 1797-74-6
2-PROPENYL PHENYLACETATE
mf: $C_{11}H_{12}O_2$ mw: 176.23

PROP: Colorless to light yellow liquid; fruity odor of banana and honey.

SYNS: ALLYL PHENYLACETATE ◇ BENZENEACETIC ACID, 2-PROPENYL ESTER ◇ PHENYLACETIC ACID ALLYL ESTER

USE IN FOOD:

Purpose: Flavoring agent.

Where Used: Baked goods, candy.

Regulations: FDA - 21CFR 172.515. Use at a level not in excess of the amount reasonably required to accomplish the intended effect.

SAFETY PROFILE: Moderately toxic by ingestion. A human skin irritant. When heated to decomposition it emits acrid smoke and irritating fumes.

TOXICITY DATA and CODEN

skn-hmn 30 mg/48H FCTXAV 15,621,77
skn-rbt 310 mg/kg/24H MOD FCTXAV 15,621,77
orl-rat LD50:650 mg/kg FCTXAV 15,621,77

PMT750 CAS: 123-38-6
PROPIONALDEHYDE
DOT: 1275
mf: C_3H_6O mw: 58.09

PROP: Colorless, mobile liquid; suffocating odor. Mp: −81°, bp: 48°, flash p: 15-19°F (OC), d: 0.807 @ 20°/4°, lel: 2.9%, uel: 17%, vap d: 2.0, autoign temp: 405°F. Misc with alc, ether, water @ 49°.

SYNS: ALDEHYDE PROPIONIQUE (FRENCH) ◇ FEMA No. 2923 ◇ METHYLACETALDEHYDE ◇ NCI-C61029 ◇ PROPALDEHYDE ◇ PROPANAL ◇ PROPIONIC ALDEHYDE ◇ PROPYL ALDEHYDE ◇ PROPYLIC ALDEHYDE

USE IN FOOD:

Purpose: Flavoring agent.

Where Used: Various.

Regulations: FDA - 21CFR 172.515. Use at a level not in excess of the amount reasonably required to accomplish the intended effect.

Community Right-To-Know List.

DOT Classification: Flammable Liquid; Label: Flammable Liquid

SAFETY PROFILE: Moderately toxic by skin contact, ingestion and subcutaneous routes. Mildly toxic by inhalation. A skin and severe eye irritant. Flammable liquid. Dangerous fire hazard when exposed to heat or flame; reacts vigorously with oxidizers. To fight fire, use alcohol foam, CO_2, dry chemical. When heated to decomposition it emits acrid smoke and irritating fumes.

TOXICITY DATA and CODEN

skn-rbt 500 mg open MLD UCDS** 4/25/58
eye-rbt 41 mg SEV UCDS** 4/25/58
orl-rat LD50:1410 mg/kg AMIHBC 4,119,51
ihl-rat LCLo:8000 ppm/4H AMIHBC 4,119,51

PMU750 CAS: 79-09-4
PROPIONIC ACID
DOT: 1848
mf: $C_3H_6O_2$ mw: 74.09

PROP: Oily liquid; pungent, disagreeable, rancid odor. D: 0.998 @ 15°/4°, mp: −21.5°, bp: 141.1°, vap press: 10 mm @ 39.7°, vap d: 2.56, autoign temp: 955°F. Misc in water, alc, ether, chloroform.

SYNS: ACIDE PROPIONIQUE (FRENCH) ◇ CARBOXYETHANE ◇ ETHANECARBOXYLIC ACID ◇ ETHYLFORMIC ACID ◇ METACETONIC ACID ◇ METHYL ACETIC ACID ◇ PROPANOIC ACID ◇ PROPIONIC ACID, solution containing not less than 80% acid (DOT) ◇ PROPIONIC ACID GRAIN PRESERVER ◇ PROZOIN ◇ PSEUDOACETIC ACID ◇ SENTRY GRAIN PRESERVER ◇ TENOX P GRAIN PRESERVATIVE

USE IN FOOD:

Purpose: Antimicrobial agent, flavoring agent, mold and rope inhibitor, preservative.

Where Used: Various.

Regulations: FDA - 21CFR 172.515, 184.1091. GRAS when used at a level not in excess of the amount reasonably required to accomplish the intended effect.

ACGIH TLV: TWA 10 ppm; STEL 15 ppm DOT Classification: Corrosive Material; Label: Corrosive; Corrosive Material; Label: Corrosive, solution; Corrosive Material; Label: Corrosive, Flammable Liquid

SAFETY PROFILE: Poison by intraperitoneal route. Moderately toxic by ingestion, skin contact, and intravenous routes. A corrosive irritant to eyes, skin, and mucous membranes. Flammable liquid. Highly flammable when exposed to heat, flame, or oxidizers. To fight fire, use alcohol foam. When heated to decomposition it emits acrid smoke and irritating fumes.

TOXICITY DATA and CODEN

skn-rbt 495 mg open SEV UCDS** 3/24/70
eye-rbt 990 μg SEV UCDS** 3/24/70
orl-rat LD50:3500 mg/kg FMCHA2 -,C198,83

PNE250 CAS: 104-45-0
p-n-PROPYL ANISOLE
mf: $C_{10}H_{14}O$ mw: 150.24

PROP: Colorless to pale yellow liquid; anise odor. D: 0.940, refr index: 1.502-1.506, flash p: 185°F. Sol in fixed oils; insol in glycerin, propylene glycol.

SYNS: DIHYDROANETHOLE ◇ FEMA No. 2930 ◇ 1-METHOXY-4-PROPYLBENZENE ◇ 4-PROPYLANISOLE ◇ 4-n-PROPYLANISOLE

USE IN FOOD:

Purpose: Flavoring agent.

Where Used: Bakery products, beverages (nonalcoholic), confections, ice cream products.

Regulations: FDA - 21CFR 172.515. Use at a level not in excess of the amount reasonably required to accomplish the intended effect.

SAFETY PROFILE: Mildly toxic by ingestion. Mutagenic data. Combustible liquid. When heated to decomposition it emits acrid smoke and irritating fumes.

TOXICITY DATA and CODEN

sln-dmg-orl 5 mmol/L FCTOD7 21,707,83
orl-rat LD50:4400 mg/kg TXAPA9 6,378,64

PNJ750 CAS: 9005-37-2
PROPYLENE GLYCOL ALGINATE
mf: $(C_9H_{14}O_7)_8$ mw: 1873.6

PROP: White fibrous or granular powder; odorless and tasteless. Sol in water and dil organic acids.

SYNS: HYDROXY PROPYL ALGINATE ◇ KELCOLOID

USE IN FOOD:

Purpose: Emulsifier, stabilizer, thickener.

Where Used: Baked goods, beer, cheese, citrus fruit (raw), condiments, confections, dairy desserts (frozen), fats, flavorings, frostings, gelatins, gravies, ices (fruit and water), jams, jellies, oils, puddings, relishes, salad dressings, sauces (sweet), seasonings, syrups.

Regulations: FDA - 21CFR 172.210. Must comply with 21CFR 172.820. 21CFR 171.858. GRAS for foods in 21CFR 170.3(n). Limitation of 0.5 percent in frozen dairy desserts; 0.5 percent in fruit and water ices; 0.5 percent in frostings; 0.5 percent in confections; 0.5 percent in baked goods; 0.9 percent in cheeses; 1.1 percent in fats; 1.1 percent in oil; 0.6 percent in gelatin; 0.6 percent in puddings; 0.5 percent in gravies; 0.5 percent in sweet sauces; 0.4 percent in jams and jellies; 0.6 percent in condiments; 0.6 percent in relish; 1.7 percent in seasonings; 1.7 percent in flavors; 0.3 percent in surface agents in other foods. 21CFR 173.340.

SAFETY PROFILE: Mildly toxic by ingestion. When heated to decomposition it emits acrid smoke and irritating fumes.

TOXICITY DATA and CODEN

orl-rat LD50:7200 mg/kg FDRLI* 124,-,76

PNL225 PROPYLENE GLYCOL MONO- and DIESTERS

PROP: Clear liquid or white to yellow beads or flakes; bland odor and taste. Insol in water, sol in alc, ethyl acetate, chloroform.

SYNS: PROPYLENE GLYCOL MONO- and DIESTERS of FATTY ACIDS ◇ PROPYLENE GLYCOL MONOSTEARATE

USE IN FOOD:

Purpose: Emulsifier, stabilizer.

Where Used: Beet sugar, cake batters, cake icings, cake shortening, margarine, oils, oleomargarine, poultry fat (rendered), whipped toppings, yeast.

Regulations: FDA - 21CFR 172.856. Use at a level not in excess of the amount reasonably required to accomplish the intended effect. 21CFR 172.860, 172.862. Must conform to specifications for fats or fatty acids derived from edible oils. 21CFR 173.340. USDA - 9CFR 318.7. Limitation of 2.0 percent in oleomargarine or margarine. 9CFR 381.147. Sufficient for purpose.

SAFETY PROFILE: When heated to decomposition it emits acrid smoke and irritating fumes.

PNM750 n-PROPYL GALLATE

CAS: 121-79-9

mf: $C_{10}H_{12}O_5$ mw: 212.22

PROP: Odorless, fine, ivory powder or crystals; sltly bitter taste. Mp: 147-149°. Sltly sol in water, sol in alc, ether.

SYNS: GALLIC ACID, PROPYL ESTER ◇ NIPA 49 ◇ NIPAGALLIN P ◇ PROGALLIN P ◇ n-PROPYL ESTER of 3,4,5-TRIHYDROXYBENZOIC ACID ◇ PROPYL GALLATE ◇ n-PROPYL-3,4,5-TRIHYDROXYBENZOATE ◇ TENOX PG ◇ 3,4,5-TRIHYDROXYBENZENE-1-PROPYLCARBOXYLATE ◇ 3,4,5-TRIHYDROXYBENZOIC ACID, n-PROPYL ESTER

USE IN FOOD:

Purpose: Antioxidant.

Where Used: Beef patties (fresh), beef patties (pregrilled), chewing gum, fats (rendered animal), margarine, meat (dried), meat products, meatballs (cooked or raw), oils, oleomargarine, pizza toppings (cooked or raw), pork, potato sticks, poultry, sausage (brown and serve), sausage (dry), sausage (fresh Italian).

Regulations: FDA - 21CFR 172.615. Use in amounts not to exceed those required to produce the intended physical or other technical effect. 21CFR 181.24. Limitation of 0.005 percent migrating from food packages. 21CFR 184.1660. GRAS with a limitation of 0.02 percent in the fat and oil content, including the essential oil content when used in accordance with good manufacturing practice. USDA - 9CFR 318.7. Limitation of 0.003 percent in dry sausage. Limitation of 0.01 percent in rendered animal fat. Limitation of 0.02 percent individually or in combination with other antioxidants approved for use in margarine. 9CFR 381.147. Limitation of 0.01 percent in poultry based on fat content.

NTP Carcinogenesis Bioassay (feed); No Evidence: mouse, rat NTPTR* NTP-TR-240,82.

SAFETY PROFILE: Poison by ingestion and intraperitoneal routes. An experimental tumorigen. Experimental teratogenic and reproductive effects. Mutagenic data. Combustible when exposed to heat or flame; can react with oxidizing materials. When heated to decomposition it emits acrid smoke and irritating fumes.

TOXICITY DATA and CODEN

mmo-sat 200 μg/plate SYSWAE 12,41,79
cyt-ham:fbr 40 mg/L ESKHA5 96,55,78
orl-rat TDLo:45 g/kg (1-22D preg):TER SKEZAP 20,378,79
orl-rat TDLo:19 g/kg (1-22D preg):REP SKEZAP 20,378,79
orl-rat TDLo:2500 mg/kg (1-22D preg):REP AJANA2 110,29,62
orl-mus TDLo:168 g/kg/2Y-C:ETA NKEZA4 29,25,82
orl-rat LD50:2600 mg/kg VPITAR 18,24,59

POC750 PROSTAGLANDIN F2-α-THAM

CAS: 38562-01-5

mf: $C_{20}H_{34}NO_5 \cdot C_4H_{11}NO_3$ mw: 475.70

SYNS: 7-(3,5-DIHYDROXY-2-(3-HYDROXY-1-OCTENYL) CYCLOPENTYL)-5-HEPTENOIC ACID, THAM ◇ 7-(3,5-DI-HYDROXY-2-(3-HYDROXY-1-OCTENYL)CYCLOPENTYL)-5-HEPTENOIC ACID, TRIMETHAMINE SALT ◇ DINOPROST TROMETHAMINE (USDA) ◇ 583E ◇ LUTALYSE ◇ PGF2-α THAM ◇ PGF2-α TRIS SALT ◇ PGF2-α TROMETHAMINE ◇ PROSTAGLANDIN F2-α THAM SALT ◇ PROSTAGLANDIN F2a TROMETHAMINE ◇ THAM ◇ TROMETHAMINE PROSTAGLANDIN F2-α ◇ U-14

USE IN FOOD:

Purpose: Animal drug.

Where Used: Various.

Regulations: USDA CES Ranking: B-4 (1988).

SAFETY PROFILE: Poison by intraperitoneal, subcutaneous, intravenous, and intramuscular routes. Moderately toxic by ingestion. Human reproductive effects by intervaginal route: terminates pregnancy, effects on fertility. Experimental teratogenic and reproductive effects. When heated to decomposition it emits toxic fumes of NO_x.

TOXICITY DATA and CODEN

ivg-wmn TDLo:4 mg/kg (37D preg):REP PRGLBA 2,453,72
ivg-wmn TDLo:12 mg/kg (5W preg):REP AJOGAH 117,346,73
ipr-rat TDLo:300 μg/kg (9-14D preg):TER KSRNAM 7,652,73
orl-rat LD50:665 mg/kg YKYUA6 32,1129,81

POH750 CAS: 127-91-3
PSEUDOPINENE
mf: $C_{10}H_{16}$ mw: 136.26

PROP: Colorless liquid; pine odor. D: 0.864, refr index: 1.477, flash p: 88°F. Sol in fixed oils; insol in water, propylene glycol, glycerin

SYNS: 6,6-DIMETHYL-2-METHYLENEBICYCLO (3.1.1)HEPTANE ◇ FEMA No. 2903 ◇ NOPINEN ◇ NOPINENE ◇ β-PINENE (FCC) ◇ 2(10)-PINENE ◇ PSEUDOPINEN

USE IN FOOD:

Purpose: Flavoring agent.

Where Used: Various.

Regulations: FDA - 21CFR 172.515. Use at a level not in excess of the amount reasonably required to accomplish the intended effect.

SAFETY PROFILE: Mildly toxic by ingestion. A skin irritant. Flammable liquid. When heated to decomposition it emits acrid smoke and irritating fumes.

TOXICITY DATA and CODEN

skn-rbt 500 mg/24H MOD FCTXAV 16,859,78
orl-rat LD50:4700 mg/kg FCTXAV 16,859,78

POO000 CAS: 97-11-0
PYRETHRIN
mf: $C_{21}H_{28}O_3$ mw: 328.49

SYNS: 2-CYCLOPENTENYL-4-HYDROXY-3-METHYL-2-CYCLOPENTEN-1-ONE CHRYSANTHEMATE ◇ 3-(2-CYCLOPENTEN-1-YL)-2-METHYL-4-OXO-2-CYCLOPENTEN-1-YL CHRYSANTHEMUMATE ◇ 3-(2-CYCLOPENTENYL)-2-METHYL-4-OXO-2-CYCLOPENTENYL CHRYSANTHEMUM-MONOCARBOXYLATE ◇ CYCLOPENTENYLRETHONYL CHRYSANTHEMATE ◇ ENT 22,952

USE IN FOOD:

Purpose: Insecticide.

Where Used: Animal feed, dried foods, milled fractions derived from cereal grains, packaging materials.

Regulations: FDA - 21CFR 193.390. Insecticide residue tolerance of 1 ppm in milled fractions derived from cereal grains, dried foods. 21CFR 561.340. Limitation of 1 ppm when used for animal feed

SAFETY PROFILE: Moderately toxic by ingestion and possibly other routes. When heated to decomposition it emits acrid smoke and irritating fumes.

TOXICITY DATA and CODEN

orl-rat LD50:1410 mg/kg ARSIM* 20,7,66

PPK500 CAS: 58-56-0
PYRIDOXOL HYDROCHLORIDE
mf: $C_8H_{11}NO_3 \cdot ClH$ mw: 205.66

PROP: Commercial form of pyridoxine (Vitamin B_6). Colorless to white platelets or crystalline powder; odorless. Mp: 204-206° (decomp). Sol in water, alc, acetone; sltly sol in other organic solvents; insol in ether.

SYNS: ADERMINE HYDROCHLORIDE ◇ BECILAN ◇ BENADON ◇ CAMPOVITON 6 ◇ HEXABETALIN ◇ HEBABIONE HYDROCHLORIDE ◇ HEXAVIBEX ◇ HEXERMIN ◇ HEXOBION ◇ 3-HYDROXY-4,5-DIMETHYLOL-α-PICOLINE HYDROCHLORIDE ◇ 5-HYDROXY-6-METHYL-3,4-PYRIDINEDICARBINOL HYDROCHLORIDE ◇ 5-HYDROXY-6-METHYL-3,4-PYRIDINEDIMETHANOL HYDROCHLORIDE ◇ 2-METHYL-3-HYDROXY-4,5-BIS(HYDROXYMETHYL)PYRIDINE HYDROCHLORIDE ◇ PYRIDIPCA ◇ PYRIDOXINE HYDROCHLORIDE (FCC) ◇ PYRIDOXINIUM CHLORIDE ◇ PYRIDOXINUM HYDROCHLORICUM (HUNGARIAN) ◇ VITAMIN B6-HYDROCHLORIDE

USE IN FOOD:

Purpose: Dietary supplement, nutrient.

Where Used: Baked goods, beverage bases (nonalcoholic), beverages (nonalcoholic), cereal s (breakfast), dairy product analogs, meat products, milk products, plant protein products, snack foods.

Regulations: FDA - 21CFR 184.1676. GRAS when used in accordance with good manufacturing practice.

SAFETY PROFILE: Poison by intravenous route. Moderately toxic by ingestion, intramuscular, and subcutaneous routes. Human reproductive effects by ingestion and intramuscular routes: postpartum changes. Experimental teratogenic effects. Human mutagenic data. When heated to decomposition it emits very toxic fumes of NO_x and HCl.

TOXICITY DATA and CODEN

sce-hmn:lym 2 mg/L MUREAV 124,175,83
orl-wmn TDLo:2 mg/kg (38W preg):REP APSVAM 72,525,83
orl-rat TDLo:8040 μg/kg (14W pre-12D post): TER JONUAI 104,111,74
orl-rat LD50:4000 mg/kg ARZNAD 11,922,61

PPS250 CAS: 109-97-7
PYRROLE

mf: C_4H_5N mw: 67.10

PROP: Colorless liquid, darkens on standing; mild nutty odor. Fp: −24°, flash p: 102°F (TCC), d: 0.968 @ 20°/4°, refr index: 1.507, vap d: 2.31, bp: 130-131° @ 761 mm. Sltly sol in water; very sol in alc, fixed oils, benzene, ether; insol in alkali.

SYNS: 1-AZA-2,4-CYCLOPENTADIENE ◇ AZOLE ◇ DIVINYLENIMINE ◇ FEMA No. 3386 ◇ IMIDOLE ◇ MONOPYRROLE

USE IN FOOD:

Purpose: Flavoring agent.

Where Used: Various.

Regulations: GRAS when used at a level not in excess of the amount reasonably required to accomplish the intended effect.

SAFETY PROFILE: Poison by subcutaneous, intraperitoneal, and possibly other routes. Flammable when exposed to heat or flame; can react with oxidizing materials. To fight fire, use foam, CO_2, dry chemical. When heated to decomposition it emits highly toxic fumes of NO_x.

TOXICITY DATA and CODEN

scu-mus LD50:61 mg/kg 28ZEAL 4,335,69

Q

QIJ000 CAS: 60-93-5
QUININE DIHYDROCHLORIDE
mf: $C_{20}H_{24}N_2O_2 \cdot 2ClH$ mw: 397.38

PROP: White needles or crystalline powder; odorless with very bitter taste. Sol in water, alc, glycerin; sltly sol in chloroform; very sltly sol in ether.

SYNS: ACID QUININE HYDROCHLORIDE ◇ CHININDIHYDROCHLORID (GERMAN) ◇ 6′-METHOXYCINCHONAN-9-OL DIHYDROCHLORIDE ◇ QUININE BIMURIATE ◇ (−)-QUININE DIHYDROCHLORIDE

USE IN FOOD:

Purpose: Flavoring agent.

Where Used: Beverages (carbonated).

Regulations: FDA - 21CFR 172.575. Limitation of 83 ppm.

SAFETY PROFILE: Poison by intravenous and subcutaneous routes. Moderately toxic by ingestion. Mutagenic data. When heated to decomposition it emits very toxic fumes of NO_x and HCl.

TOXICITY DATA and CODEN

mma-sat 2800 nmol/plate MUREAV 66,33,79
orl-rat LD50:1392 mg/kg JPETAB 91,157,47
ivn-rat LD50:78 mg/kg JPETAB 91,157,47

QMA000 CAS: 804-63-7
QUININE SULFATE
mf: $C_{20}H_{24}N_2O_2 \cdot O_4S$ mw: 420.52

PROP: Fine white needlelike crystals; odorless with a very bitter taste. Sol in water, alc; sltly sol in chloroform.

SYNS: QUININE BISULFATE ◇ QUININE HYDROGEN SULFATE

USE IN FOOD:

Purpose: Flavoring agent.

Where Used: Beverages (bitter lemon), beverages (carbonated), quinine water, tonic water.

Regulations: FDA - 21CFR 172.575. Limitation of 83 ppm.

SAFETY PROFILE: Human poison by ingestion. Human systemic effects by ingestion: flaccid paralysis without anesthesia, visual field changes, tinnitus, motor activity changes and blood angranulocytosis. Experimental reproductive effects. Mutagenic data. When heated to decomposition it emits very toxic fumes of SO_x and NO_x.

TOXICITY DATA and CODEN

pic-esc 100 μg/plate CNREA8 43,2819,83
orl-rat TDLo:1425 mg/kg (14D pre-21D post): REP BNEOBV 36,273,79
orl-hmn TDLo:4300 μg/kg:BLD,PNS BMJOAE 1,605,77
orl-hmn TDLo:43 mg/kg:CNS,EYE,EAR JIMSAX 67,46,74
orl-wmn LDLo:220 mg/kg CTOXAO 7,129,74

R

RBF100 CAS: 26538-44-3
RALGRO
mf: $C_{18}H_{26}O_5$ mw: 322.44

SYNS: 6-(6,10-DIHYDROXYUNDECYL)-β-RESORCYLIC ACID-mu-LACTONE ◇ FRIDERON ◇ MK-188 ◇ P1496 ◇ RALABOL ◇ RALONE ◇ ZEARALANOL ◇ ZEARANOL ◇ ZERANOL (USDA)

USE IN FOOD:

Purpose: Animal drug.

Where Used: Beef, lamb.

Regulations: FDA - 21CFR 556.760. Limitation of zero in cattle and sheep. USDA CES Ranking: C-2 (1986).

SAFETY PROFILE: Experimental reproductive effects. When heated to decomposition it emits acrid smoke and irritating fumes.

TOXICITY DATA and CODEN

orl-rat TDLo:52 mg/kg (6-18D preg):REP TJADAB 25,37,82
scu-mus LD50:7 mg/kg (10-16D preg):REP TXAPA9 41,138,77

RBK200
RAPESEED OIL

PROP: Pale yellow liquid.

SYNS: FULLY HYDROGENATED RAPESEED OIL ◇ LOW ERUCIC ACID RAPESEED OIL ◇ RAPE SEED OIL ◇ SUPERGLYCERINATED FULLY HYDROGENATED RAPESEED OIL

USE IN FOOD:

Purpose: Emulsifier, stabilizer, thickener.

Where Used: Cake mixes, edible fats, peanut butter, shortening.

Regulations: FDA - 21CFR 184.1555. GRAS with a limitation of 2 percent in peanut butter of fully hydrogenated rapeseed oil. Limitation of 4 percent in shortening or 0.5 percent in total cake mix of superglycerinated fully hydrogenated rapeseed oil. Limitation of 2 percent of low erucic acid rapeseed in edible fats.

SAFETY PROFILE: When heated to decomposition it emits acrid smoke and irritating fumes.

RBU000 CAS: 5471-51-2
RASPBERRY KETONE
mf: $C_{10}H_{12}O_2$ mw: 164.22

PROP: White solid; raspberry odor. Mp: 81-86°, flash p: +212°F.

SYNS: FEMA No. 2588 ◇ FRAMBINONE ◇ 4-(4-HYDROXYPHENYL)-2-BUTANONE ◇ p-HYDROXYBENZYL ACETONE ◇ 1-(p-HYDROXYPHENYL)-3-BUTANONE ◇ 4-(p-HYDROXYPHENYL)-2-BUTANONE (FCC) ◇ OXYPHENALON ◇ RHEOSMIN

USE IN FOOD:

Purpose: Flavoring agent.

Where Used: Various.

Regulations: FDA - 21CFR 172.515. Use at a level not in excess of the amount reasonably required to accomplish the intended effect.

SAFETY PROFILE: Poison by intraperitoneal route. Moderately toxic by ingestion. Combustible liquid. When heated to decomposition it emits acrid smoke and irritating fumes.

TOXICITY DATA and CODEN

orl-rat LD50:1320 mg/kg FCTXAV 8,349,70
ipr-rat LD50:350 mg/kg FCTXAV 8,349,70

RCZ100 CAS: 9001-98-3
RENNET

SYN: BOVINE RENNET

USE IN FOOD:

Purpose: Binder, enzyme, extender, processing aid, stabilizer, thickener.

Where Used: Fillings, frozen dairy desserts, gelatins, loaves (nonspecific), milk products, poultry, puddings, sausage, sausage (imitation), soups, stews.

Regulations: FDA - 21CFR 184.1685. GRAS when used in accordance with good manufacturing practice. USDA - 9CFR 318.7. Rennet treated calcium reduced dried milk and calcium lactate, limitation of 3.5 percent in sausages (calcium lactate required at a rate of 10 percent of binder). Rennet treated calcium reduced dried milk and calcium lactate, sufficient for purpose in imitation sausages, nonspecific loaves, soups, and stews (calcium lactate required at a rate of 10 percent of binder). Rennet treated calcium reduced dried milk and calcium lactate, sufficient for purpose in imitation sausages, nonspecific loaves, soups, and stews (calcium lactate required at a rate of 25 percent of binder). Rennet treated sodium casenate and calcium lactate, sufficient for purpose in imitation sausages, nonspecific loaves, soups, and stews (calcium lactate required at a rate of 25 percent of binder). 9CFR 381.147. Rennet treated calcium reduced dried milk and calcium lactate, sufficient for purpose in poultry.

SAFETY PROFILE: When heated to decomposition it emits acrid smoke and irritating fumes.

REU000 CAS: 68-26-8
all-trans-RETINOL

mf: $C_{20}H_{30}O$ mw: 286.50

PROP: Light yellow to red oil; mild fishy odor. Very sol in chloroform, ether; sol in abs alcohol, vegetable oil; insol in glycerin, water.

SYNS: ACON ◇ AFAXIN ◇ AGIOLAN ◇ ALPHALIN ◇ ALPHASTEROL ◇ ANATOLA ◇ ANTI-INFECTIVE VITAMIN ◇ ANTIXEROPHTHALMIC VITAMIN ◇ AORAL ◇ APEXOL ◇ AQUASYNTH ◇ AVIBON ◇ AVITA ◇ AVITOL ◇ BIOSTEROL ◇ CHOCOLA A ◇ 3,7-DIMETHYL-9-(2,6,6-TRIMETHYL-1-CYCLOHEXEN-1-YL)-2,4,6,8-NONATETRAEN-1-OL ◇ DISATABS TABS ◇ DOFSOL ◇ EPITELIOL ◇ HI-A-VITA ◇ LARD FACTOR ◇ MYVPACK ◇ OLEOVITAMIN A ◇ OPHTHALAMIN ◇ PREPALIN ◇ RETINOL ◇ RETROVITAMIN A ◇ TESTAVOL ◇ VAFLOL ◇ VI-ALPHA ◇ VITAMIN A (FCC) ◇ VITAMIN A1 ◇ VITAMIN A1 ALCOHOL ◇ all-trans-VITAMIN A ALCOHOL ◇ VITAVEL-A ◇ VITPEX ◇ VOGAN ◇ VOGAN-NEU

USE IN FOOD:

Purpose: Dietary supplement, nutrient.

Where Used: Various.

Regulations: FDA - 21CFR 182.5930, 184.1930. GRAS when used in accordance with good manufacturing practice.

EPA Genetic Toxicology Program.

SAFETY PROFILE: Moderately toxic by ingestion. Human teratogenic effects by ingestion: developmental abnormalities of the craniofacial area and urogenital system. An experimental teratogen. Experimental reproductive effects. Human mutagenic data. When heated to decomposition it emits acrid smoke and irritating fumes.

TOXICITY DATA and CODEN

oms-hmn:lym 4 mg/L EJCODS 21,1089,85
sce-hmn:lym 4 mg/L EJCODS 21,1089,85
orl-wmn TDLo:68 mg/kg (1-39W preg):TER OBGNAS 43,750,74
orl-rat TDLo:76560 μg/kg (8-10D preg):TER NTOTDY 3,1,81
orl-rat TDLo:22 mg/kg (1-22D preg):REP AMPLAO 66,278,58
ipr-rat TDLo:60 mg/kg (10D preg):TER TJADAB 30(1),13A,84
orl-rat LD50:2000 mg/kg AVSUAR 74,29,75

REZ000 CAS: 127-47-9
RETINOL ACETATE

mf: $C_{22}H_{32}O_2$ mw: 328.54

SYNS: CRYSTALETS ◇ MYVAK ◇ MYVAX ◇ RETINYL ACETATE ◇ all-trans-RETINYL ACETATE ◇ VITAMIN A ACETATE ◇ trans-VITAMIN A ACETATE ◇ VITAMIN A ALCOHOL ACETATE

USE IN FOOD:

Purpose: Dietary supplement, nutrient.

Where Used: Various.

Regulations: FDA - 21CFR 182.5933, 184.1930. GRAS when used in accordance with good manufacturing practice.

SAFETY PROFILE: Moderately toxic by ingestion. An experimental neoplastigen and teratogen. Experimental reproductive effects. Mutagenic data. When heated to decomposition it emits acrid smoke and irritating fumes.

TOXICITY DATA and CODEN

dni-rat:mmr 3 μmol/L JJIND8 70,949,83
orl-rat TDLo:480 mg/kg (6-19D preg):REP TOXID9 4,84,84
orl-rat TDLo:310 mg/kg (10-12D preg):TER TJADAB 1,299,68
orl-rat TDLo:1377 mg/kg (15-19D preg):TER TJADAB 21,58A,80
orl-rat TDLo:51800 mg/kg/2Y-C:NEO JJIND8 74,715,85
orl-mus LDLo:1000 mg/kg APMIAL 70,398,67

RHA000 CAS: 141-11-7
RHODINYL ACETATE

mf: $C_{12}H_{22}O_2$ mw: 198.34

PROP: Mixture of acetates of geraniol and l-citronellol, found in geranium oil (FCTXAV 12,807,74). Colorless to sltly yellow liquid;

fresh rose odor. D: 0.895-0.908, refr index: 1.450-1.458. Sol alc, fixed oils; insol in glycerin, propylene glycol, water @ 237°.

SYNS: α-CITRONELLYL ACETATE ◇ 3,7-DIMETHYL-7-OCTEN-1-OL ACETATE ◇ FEMA No. 2981 ◇ RHODINOL ACETATE

USE IN FOOD:

Purpose: Flavoring agent.

Where Used: Various.

Regulations: FDA - 21CFR 172.515. Use at a level not in excess of the amount reasonably required to accomplish the intended effect.

SAFETY PROFILE: A skin irritant. When heated to decomposition it emits acrid smoke and irritating fumes.

TOXICITY DATA and CODEN

skn-rbt 500 mg/24H MLD FCTXAV 12,975,74

RHA500
RHODINYL FORMATE

mf: $C_{11}H_{20}O_2$ mw: 184.28

PROP: Colorless to slightly yellow liquid; leafy, rose-like odor. D: 0.901-0.908, refr index: 1.453-1.458. Sol in alc, fixed oils; insol in glycerin, propylene glycol, water @ 200°.

SYN: FEMA No. 2984

USE IN FOOD:

Purpose: Flavoring agent.

Where Used: Various.

Regulations: FDA - 21CFR 172.515. Use at a level not in excess of the amount reasonably required to accomplish the intended effect.

SAFETY PROFILE: When heated to decomposition it emits acrid smoke and irritating fumes.

RIF500 CAS: 130-40-5
RIBOFLAVIN 5′-PHOSPHATE SODIUM

mf: $C_{127}H_{20}N_4NaO_9P \cdot 2H_2O$ mw: 1817.56

PROP: Fine orange-yellow crystalline powder; slt odor. Hygroscopic; sol in water. Decomposed by light when in solution.

SYN: RIBOFLAVIN 5′-PHOSPHATE ESTER MONOSODIUM SALT

USE IN FOOD:

Purpose: Dietary supplement, nutrient supplement.

Where Used: Milk products.

Regulations: FDA - 21CFR 182.5697, 184.1697. GRAS when used in accordance with good manufacturing practice.

SAFETY PROFILE: When heated to decomposition emits toxic fumes of NO_x and NaO_2.

RIK000 CAS: 83-88-5
RIBOFLAVINE

mf: $C_{17}H_{20}N_4O_6$ mw: 376.37

PROP: Orange to yellow crystals; slt odor. Mp: 282° (decomp). Sltly sol in water, alc; insol in ether, chloroform.

SYNS: BEFLAVINE ◇ 6,7-DIMETHYL-9-d-RIBITYLISOALLOXAZINE ◇ 7,8-DIMETHYL-10-d-RIBITYLISOALLOXAZINE ◇ 7,8-DIMETHYL-10-(d-RIBO-2,3,4,5-TETRAHYDROXYPENTYL)ISOALLOXAZINE ◇ FLAVAXIN ◇ HYFLAVIN ◇ HYRE ◇ LACTOFLAVIN ◇ LACTOFLAVINE ◇ RIBIPCA ◇ RIBODERM ◇ RIBOFLAVIN ◇ RIBOFLAVINEQUINONE ◇ VITAMIN B2 ◇ VITAMIN G

USE IN FOOD:

Purpose: Color additive, dietary supplement, nutrient supplement.

Where Used: Various.

Regulations: FDA - 21CFR 73.450, 182.5695, 184.1695. GRAS when used in accordance with good manufacturing practice.

SAFETY PROFILE: Poison by intravenous route. Moderately toxic by intraperitoneal and subcutaneous routes. Mutagenic data. When heated to decomposition it emits toxic fumes of NO_x.

TOXICITY DATA and CODEN

cyt-ham:lng 300 mg/L GMCRDC 27,95,81
ipr-rat LD50:560 mg/kg JPETAB 76,75,42
scu-rat LD50:5000 mg/kg JPETAB 76,75,42

RJF800
RICE BRAN WAX

PROP: Tan to brown hard wax. Mp: 75°. Sol in chloroform, benzene; insol in water.

USE IN FOOD:

Purpose: Coating agent, masticatory substance in chewing gum base, release agent.

Where Used: Candy, chewing gum, fruits (fresh), vegetables (fresh).

Regulations: FDA - 21CFR 172.615. Use in amounts not to exceed those required to produce the intended physical or other technical effect. 21CFR 172.890. Limitation of 50 ppm in candy, 50 ppm in fresh fruits and vegetables, 2.5 percent in chewing gum. 21CFR 178.3860. Limitation of 1.0 percent in the polymer.

SAFETY PROFILE: When heated to decomposition it emits acrid smoke and irritating fumes.

RLK890 CAS: 25875-51-8
ROBENIDINE
mf: $C_{15}H_{13}Cl_2N_5$ mw: 334.23

PROP: Crystals from ethanol. Mp: 289-290°.

SYNS: 1,3-BIS((p-CHLOROBENZYLIDENE)AMINO) GUANIDINE ◇ CARBONIMIDIC DIHYDRAZIDE, BIS((4-CHLOROPHENYL)METHYLENE)- ◇ CHEMCOCCIDE ◇ CHEMOCCIDE ◇ CHIMCOCCIDE ◇ KHIMCOCCID ◇ KHIMCOECID ◇ KHIMKOKTSID ◇ KHIMKOKTSIDE

USE IN FOOD:

Purpose: Animal drug.

Where Used: Chicken.

Regulations: FDA - 21CFR 556.580. Limitation of 0.2 ppm in skin and fat of chickens and 0.1 ppm in other tissues of chickens. 21CFR 558.515. (hydrochloride).

SAFETY PROFILE: Moderately toxic by ingestion. When heated to decomposition it emits toxic fumes of Cl^- and NO_x.

TOXICITY DATA and CODEN

orl-rat LD50:1350 mg/kg VETNAL 57(9),53,81

RMA000 CAS: 50471-44-8
RONILAN
mf: $C_{12}H_9Cl_2NO_3$ mw: 286.12

SYNS: BAS 352 F ◇ 3-(3,5-DICHLOROPHENYL)-5-ETHENYL-5-METHYL-2,4-OXAZOLIDINEDIONE ◇ 3-(3,5-DICHLOROPHENYL)-5-METHYL-5-VINYL-2,4-OXAZOLIDINEDIONE ◇ VINCLOZOLIN (GERMAN)

USE IN FOOD:

Purpose: Insecticide.

Where Used: Animal feed, prunes, raisins.

Regulations: FDA - 21CFR 193.137. Insecticide residue tolerance of 75 ppm in prunes. Limitation of 30 ppm in raisins. (Expires 5/13/1988) 21CFR 561.440. Limitation of 42.0 ppm in dry grape pomace when used for animal feed. (Expires 5/13/1988)

SAFETY PROFILE: Mildly toxic by ingestion. Mutagenic data. When heated to decomposition it emits very toxic fumes of Cl^- and NO_x.

TOXICITY DATA and CODEN

mma-sat 100 mg/L ATSUDG (5),345,82
mma-ssp 100 mg/L ATSUDG (5),345,82
orl-gpg LD50:8000 mg/kg 85DPAN -,-,71/76

RMU000 CAS: 8000-25-7
ROSEMARY OIL

PROP: Constituents are α-pinene, camphene, and cineole. From steam distillation of flowering tops of *Rosmarinus officinalis* L. (Fam. *Labiatae*) (FCTXAV 12,807,74). Colorless to pale yellow liquid; odor of rosemary. D: 0.894-0.912, refr index: 1.464 @ 20°.

SYNS: ROSEMARIE OIL ◇ ROSMARIN OIL (GERMAN)

USE IN FOOD:

Purpose: Flavoring agent.

Where Used: Lamb, meat, poultry, soups.

Regulations: FDA - 21CFR 182.10, 182.20. GRAS when used at a level not in excess of the amount reasonably required to accomplish the intended effect.

SAFETY PROFILE: Mildly toxic by ingestion. A skin irritant. When heated to decomposition it emits acrid smoke and irritating fumes.

TOXICITY DATA and CODEN

skn-rbt 500 mg/24H MOD FCTXAV 12,977,74
orl-rat LD50:5000 mg/kg FCTXAV 12,977,74

RNA000 CAS: 8007-01-0
ROSE OIL

PROP: Volatile oil from steam distillation of fresh flowers of *Rosa gallica* L. and *Rosa Damascena* Mill. and varieties of these species (Fam. *Rosaceae*). Colorless to yellow liquid; odor and taste of rose. D: 0.848-0.863 @ 30°/15°, refr index: 1.457 @ 30°.

SYN: ROSEN OEL (GERMAN)

USE IN FOOD:

Purpose: Flavoring agent.

Where Used: Bakery products, beverages (nonalcoholic), condiments, confections, ice cream, meat.

Regulations: FDA - 21CFR 182.20. GRAS when used at a level not in excess of the amount reasonably required to accomplish the intended effect.

SAFETY PROFILE: Mildly toxic by ingestion. When heated to decomposition it emits acrid smoke and irritating fumes.

TOXICITY DATA and CODEN

orl-rat LD50:12560 mg/kg PHARAT 14,435,59

RQU750
RUE OIL and HERB

PROP: From steam distillation of fresh blossoming plants *Ruta graveolens* L., *Ruta montana* L., or *Ruta bracteosa* L. (Fam. *Rutaceae*). Yellow to amber liquid; fatty odor. Sol in fixed oils, mineral oil; insol in glycerin, propylene glycol.

USE IN FOOD:

Purpose: Flavoring agent.

Where Used: Baked goods, candy (soft), frozen dairy desserts and mixes.

Regulations: FDA - 21CFR 184.1698, 1699. GRAS with a limitation for rue herb: 2 ppm; for rue oil: 10 ppm in baked goods, 10 ppm in frozen dairy desserts and mixes, 10 ppm in soft candy, 4 ppm in all other food when used in accordance with good manufacturing practice.

SAFETY PROFILE: When heated to decomposition it emits acrid smoke and irritating fumes.

S

SAB500 CAS: 128-44-9

SACCHARIN

mf: $C_7H_4NO_3S \cdot Na$ mw: 205.17

PROP: White crystals or crystalline powder; odorless, very sweet taste. Sol in water, alc.

SYNS: ARTIFICIAL SWEETENING SUBSTANZ GENDORF 450 ◇ CRISTALLOSE ◇ CRYSTALLOSE ◇ DAGUTAN ◇ KRISTALLOSE ◇ MADHURIN ◇ ODA ◇ SACCHARINE SOLUBLE ◇ SACCHARINNATRIUM ◇ SACCHARIN, SODIUM ◇ SACCHARIN, SODIUM SALT ◇ SACCHARIN SOLUBLE ◇ SACCHAROIDUM NATRICUM ◇ SAXIN ◇ SODIUM 1,2 BENZISOTHIAZOLIN-3-ONE-1,1-DIOXIDE ◇ SODIUM o-BENZOSULFIMIDE ◇ SODIUM BENZOSULPHIMIDE ◇ SODIUM o-BENZOSULPHIMIDE ◇ SODIUM 2-BENZOSULPHIMIDE ◇ SODIUM SACCHARIDE ◇ SODIUM SACCHARIN (FCC) ◇ SODIUM SACCHARINATE ◇ SODIUM SACCHARINE ◇ SOLUBLE GLUSIDE ◇ SOLUBLE SACCHARIN ◇ SUCARIL ◇ SUCRA ◇ o-SULFONBENZOIC ACID IMIDE SODIUM SALT ◇ SULPHOBENZOIC IMIDE, SODIUM SALT ◇ SWEETA ◇ SYKOSE ◇ WILLOSETTEN

USE IN FOOD:

Purpose: Nonnutritive sweetener.

Where Used: Bacon, bakery products (nonstandardized), beverage mixes, beverages, chewing gum, desserts, fruit juice drinks, jam, relishes, vitamin tablets (chewable).

Regulations: FDA - 21CFR 180.37. Limitation of 12 mg per fluid ounce of beverages; 12 mg per fluid ounce of fruit juice drinks; 12 mg per fluid ounce of beverage mixes, 30 mg per serving in processed foods. USDA - 9CFR 318.7. Limitation of 0.01 percent in bacon.

IARC Cancer Review: Animal Sufficient Evidence IMEMDT 22,111,80. EPA Genetic Toxicology Program.

SAFETY PROFILE: Moderately toxic by ingestion and intraperitoneal routes. An experimental carcinogen, neoplastigen, tumorigen, and teratogen. A promoter. Experimental reproductive effects. Human mutagenic data. When heated to decomposition it emits very toxic fumes of SO_x, Na_2O and NO_x.

TOXICITY DATA and CODEN

cyt-hmn:leu 500 mg/L MUREAV 32,81,75
sce-hmn:leu 20 μmol/L ENMUDM 1,177,79
orl-mus TDLo:103 g/kg (30D male):TER IJMRAQ 60,599,72
orl-mus TDLo:2 g/kg (MGN):REP TOLED5 19,267,83
ipr-mus TDLo:2 g/kg (1D male):REP JEPTDQ 2(4),1047,79
orl-rat TDLo:1092 g/kg/1Y-C:CAR GANNA2 74,8,83
imp-mus TDLo:176 mg/kg:NEO SCIEAS 168,1238,70
orl-rat TDLo:112 g/kg/8W-C:NEO CNREA8 37,2943,77
orl-rat TD :224 g/kg/8W-C:NEO CNREA8 37,2943,77
orl-rat TD :1330 g/kg/95W-C:ETA CBINA8 11,225,75
orl-rat LD50:14200 mg/kg FCTXAV 6,313,68

SAC000 CAS: 8001-23-8

SAFFLOWER OIL

PROP: From *Carthanus tinctorius*, consists of triglycerides of linoleic acid (85DIA2 2, 287,77). Light yellow oil. D: 0.9211 @ 25°/25°. Sol in oil and fat solvents.

SYN: SAFFLOWER OIL (UNHYDROGENATED) (FCC)

USE IN FOOD:

Purpose: Coating agent, emulsifying agent, formulation aid, texturizer.

Where Used: Various.

Regulations: GRAS when used in accordance with good manufacturing practice.

SAFETY PROFILE: A human skin irritant. Ingestion of large doses can cause vomiting. When heated to decomposition it emits acrid smoke and irritating fumes.

TOXICITY DATA and CODEN

skn-hmn 300 mg/3D-I MLD 85DKA8 -,127,77

SAC100

SAFFRON

PROP: From the dried stigma of *Ceocus saffron* L.

USE IN FOOD:

Purpose: Coloring agent.

Where Used: Casings, fats (rendered).

Regulations: FDA - 21CFR 73.500. Sufficient for purpose. USDA - 9CFR 318.7. Sufficient for purpose.

SAFETY PROFILE: When heated to decomposition it emits acrid smoke and irritating fumes.

SAD000 CAS: 94-59-7
SAFROL

mf: $C_{10}H_{10}O_2$ mw: 162.20

PROP: Colorless liquid or crystals; sassafras odor. Mp: 11°, bp: 234.5°, d: 1.0960 @ 20°, vap press: 1 mm @ 63.8°. Insol in water; very sol in alc; misc with chloroform, ether.

SYNS: 5-ALLYL-1,3-BENZODIOXOLE ◇ ALLYLCATECHOL METHYLENE ETHER ◇ ALLYLDIOXYBENZENE METHYLENE ETHER ◇ 1-ALLYL-3,4-METHYLENEDIOXYBENZENE ◇ 4-ALLYL-1,2-METHYLENEDIOXYBENZENE ◇ m-ALLYLPYROCATECHIN METHYLENE ETHER ◇ 4-ALLYLPYROCATECHOL FORMALDEHYDE ACETAL ◇ ALLYLPYROCATECHOL METHYLENE ETHER ◇ 1,2-METHYLENEDIOXY-4-ALLYLBENZENE ◇ 3,4-METHYL ENEDIOXY-ALLYBENZENE ◇ 5-(2-PROPENYL)-1,3-BENZODIOXOLE ◇ RCA WASTE NUMBER U203 ◇ RHYUNO OIL ◇ SAFROLE ◇ SAFROLE MF ◇ SHIKIMOLE ◇ SHIKOMOL

USE IN FOOD:

Purpose: Flavoring agent.

Where Used: Prohibited from foods.

Regulations: FDA - 21CFR 189.180. Prohibited from direct addition or use in human food.

IARC Cancer Review: Animal Sufficient Evidence IMEMDT 10,231,76, IMEMDT 1, 169,72. Community Right-To-Know List. EPA Genetic Toxicology Program.

SAFETY PROFILE: Poison by intraperitoneal and intravenous routes. Moderately toxic by ingestion and subcutaneous routes. An experimental carcinogen and neoplastigen. Experimental reproductive effects. Human mutagenic data. A skin irritant. Combustible when exposed to heat or flame. When heated to decomposition it emits acrid smoke and irritating fumes.

TOXICITY DATA and CODEN

skn-rbt 500 mg/24H MOD FCTXAV 12,983,74
dns:hmn:hla 10 μL/L PMRSDJ 5,347,85
otr-mus:emb 100 mg/L PMRSDJ 5,659,85
ipr-mus TDLo:1 g/kg (5D male):REP PMRSDJ 1,712,81
orl-mus TDLo:480 mg/kg (12-18D preg):REP CNREA8 39,4378,79
orl-rat TDLo:200 g/kg/94W-C:CAR CNREA8 37,1883,77
orl-mus LDLo:22 g/kg/90W-I:CAR CNREA8 39,4378,79
orl-mus TD :210 g/kg/52W-C:NEO CNREA8 37,1883,77
orl-rat LD50:1950 mg/kg TXAPA9 7,18,65

SAE000 CAS: 8022-56-8
SAGE OIL

PROP: From steam distillation of plants from *Salvia lavandulaefolia* Vahl. or *Salvia hispanorium* Lag. (Fam. *Labiatae*) (FCTXAV 12,807,74). Colorless to yellow oil. D: 0.909-0.932, refr index: 1.468 @ 20°. Sol in fixed oils, glycerin, mineral oil, propylene glycol.

SYNS: SAGE OIL, SPANISH TYPE ◇ SALBEI OEL (GERMAN)

USE IN FOOD:

Purpose: Flavoring agent.

Where Used: Fish, pork, poultry, seasonings, soups.

Regulations: FDA - 21CFR 182.20. GRAS when used at a level not in excess of the amount reasonably required to accomplish the intended effect.

SAFETY PROFILE: Moderately toxic by ingestion. When heated to decomposition it emits acrid smoke and irritating fumes.

TOXICITY DATA and CODEN

orl-rat LD50:2600 mg/kg PHARAT 14,435,59

SAE500 CAS: 8022-56-8
SAGE OIL, DALMATIAN TYPE

PROP: Main constituent is thujone. From steam distillation of leaves from *Salvia officinalis* l. (FCTXAV 12,807,74). Yellow liquid; thujone odor and taste. D: 0.903-0.925, refr index: 1.457 @ 20°. Sol in fixed oils, mineral oil; sltly sol in propylene glycol; insol in glycerin.

SYNS: DALMATIAN SAGE OIL ◇ SAGE OIL ◇ SALBEI OEL (GERMAN)

USE IN FOOD:

Purpose: Flavoring agent.

Where Used: Fish, pork, poultry, seasonings, soups.

Regulations: FDA - 21CFR 182.10, 182.20. GRAS when used at a level not in excess of the amount reasonably required to accomplish the intended effect.

SAFETY PROFILE: Moderately toxic by ingestion. Mutagenic data. A human skin irritant. When heated to decomposition it emits acrid smoke and irritating fumes.

TOXICITY DATA and CODEN

skn-hmn 100% FCTXAV 12,987,74
skn-rbt 500 mg/24H MOD FCTXAV 12,987,74
dnr-bcs 10 mg/disc TOFOD5 8,91,85
orl-rat LD50:2600 mg/kg FCTXAV 12,987,74

SAI000 CAS: 69-72-7
SALICYLIC ACID
mf: $C_7H_6O_3$ mw: 138.13

PROP: D: 1.443 @ 20°/4°, mp: 158.3°, bp: 211° @ 20 mm ±. Sol in water, alc, ether.

SYNS: ACIDO SALICILICO (ITALIAN) ◇ o-HYDROXYBENZOIC ACID ◇ 2-HYDROXYBENZOIC ACID ◇ KERALYT ◇ ORTHOHYDROXYBENZOIC ACID ◇ RETARDER W ◇ SA ◇ SAX

USE IN FOOD:

Purpose: Animal drug, fungicide.

Where Used: Milk (prohibited), wine (prohibited).

Regulations: FDA - 21CFR 556.590. Limitation of zero in milk.

EPA Genetic Toxicology Program.

SAFETY PROFILE: Poison by ingestion, intravenous and intraperitoneal routes. Moderately toxic by subcutaneous route. An experimental teratogen. Human systemic effects by skin contact: ear tinnitus. Mutagenic data. A skin and severe eye irritant. Experimental reproductive effects. When heated to decomposition it emits acrid smoke and irritating fumes.

TOXICITY DATA and CODEN

skn-rbt 500 mg/24H MLD BIOFX* 21-3/71
eye-rbt 100 mg SEV BIOFX* 21-3/71
mmo-smc 1 mmol/L/3H MUREAV 60,291,79
dni-mus-orl 100 mg/kg MUREAV 46,305,77
orl-rat TDLo:40 mg/kg (20-21D preg):REP PRGLBA 4,93,73
orl-rat TDLo:700 mg/kg (8-14D preg):TER SKEZAP 14,549,73
orl-mus TDLo:1 g/kg (17D preg):TER APTOA6 35,107,74
skn-man TDLo:57 mg/kg:EAR JAMAAP 244,660,80
orl-rat LD50:891 mg/kg BIOFX* 21-3/71

SAL000 CAS: 118-61-6
SALICYLIC ETHYL ESTER
mf: $HO \cdot C_6H_4 \cdot CO_2 \cdot C_2H_5$ mw: 166.18

PROP: Colorless liquid; wintergreen odor. D: 1.127, refr index: 1.520, mp: 1.3°, bp: 233-234°. Sol in alc, ether, acetic acid, fixed oils; sltly sol in water, glycerin.

SYNS: ETHYL-o-HYDROXYBENZOATE ◇ ETHYL SALICYLATE (FCC) ◇ FEMA No. 2458 ◇ SALICYLIC ETHER

USE IN FOOD:

Purpose: Flavoring agent.

Where Used: Various.

Regulations: FDA - 21CFR 172.515. Use at a level not in excess of the amount reasonably required to accomplish the intended effect.

SAFETY PROFILE: Moderately toxic by ingestion and subcutaneous routes. A skin irritant. When heated to decomposition it emits acrid smoke and irritating fumes.

TOXICITY DATA and CODEN

skn-rbt 500 mg/24H MOD FCTXAV 16,637,78
orl-rat LD50:1320 mg/kg FCTXAV 16,637,78

SAN500 CAS: 53003-10-4
SALINOMYCIN
mf: $C_{42}H_{70}O_{11}$ mw: 751.02

PROP: Mp: 112.5-113.5°.

SYN: COXISTAC

USE IN FOOD:

Purpose: Animal feed drug.

Where Used: Animal feed.

Regulations: FDA - 21CFR 558.550.

SAFETY PROFILE: When heated to decomposition it emits acrid smoke and irritating fumes.

SAO550
SALTS of FATTY ACIDS

PROP: Consists of aluminum, calcium, magnesium, potassium, and sodium salts of capric, caprylic, lauric, myristic, oleic, palmitic, and stearic acids manufactured from fats and oils derived from edible sources.

USE IN FOOD:

Purpose: Anticaking agent, binder, emulsifier.

Where Used: Various foods.

Regulations: FDA - 21CFR 172.863. Must conform to FDA specifications for salts of fats or fatty acids derived from edible oils. Use in accordance with good manufacturing practice.

SAFETY PROFILE: When heated to decomposition it emits acrid smoke and irritating fumes.

SAU400
SANTALYL ACETATE

PROP: Mixture of α- and β-isomers from acetylation of santalol. Colorless to sltly yellow liquid; sandalwood odor. D: 0.980, refr index: 1.488-1.491, flash p: +212°F. Sol in alc; insol in water.

SYN: FEMA No. 3007

USE IN FOOD:

Purpose: Flavoring agent.

Where Used: Various.

Regulations: FDA - 21CFR 172.515. Use at a level not in excess of the amount reasonably required to accomplish the intended effect.

SAFETY PROFILE: Combustible liquid. When heated to decomposition it emits acrid smoke and irritating fumes.

SAV000 CAS: 91-53-2
SANTOQUINE

mf: $C_{14}H_{19}NO$ mw: 217.34

PROP: Clear, light yellow liquid. Mp: <0°, bp: 125° @ 2 mm, vap d: 7.48, d: 1.030 @ 25°, refr index: 1.57.

SYNS: 1,2-DIHYDRO-6-ETHOXY-2,2,4-TRIMETHYL!xo-QUINOLINE ◇ 1,2-DIHYDRO-2,2,4-TRIMETHYL-6-ETHOXY-QUINOLINE ◇ EMQ ◇ EQ ◇ 6-ETHOXY-1,2-DIHYDRO-2,2,4-TRIMETHYLQUINOLINE ◇ ETHOXYQUIN (FCC) ◇ ETHOXYQUINE ◇ 6-ETHOXY-2,2,4-TRIMETHYL-1,2-DI-HYDROQUINOLINE ◇ NIFLEX ◇ NIX-SCALD ◇ SANTO-FLEX A ◇ SANTOFLEX AW ◇ SANTOQUIN ◇ STOP-SCALD ◇ 2,2,4-TRIMETHYL-6-ETHOXY-1,2-DIHYDRO-QUINOLINE ◇ USAF B-24

USE IN FOOD:

Purpose: Antioxidant.

Where Used: Apples, chili powder, eggs, fats (uncooked, of meat from animals), milk, paprika, pears.

Regulations: FDA - 21CFR 172.140. Limitation of 100 ppm in chili powder; 100 ppm in paprika; 5 ppm in fat, uncooked, of meat from animals except poultry; 3 ppm in fat, uncooked, of meat from poultry; 0.5 ppm in eggs; zero tolerance in milk.

EPA Genetic Toxicology Program.

SAFETY PROFILE: Poison by intraperitoneal route. Moderately toxic by ingestion. Mutagenic data. Combustible when exposed to heat or flame; can react with oxidizing materials. When heated to decomposition it emits toxic fumes of NO_x.

TOXICITY DATA and CODEN

mma-sat 200 μg/plate PCBRD2 141,407,84
orl-rat LD50:800 mg/kg RCTEA4 45,627,72

SAY900
SASSAFRAS

PROP: A yellowish-reddish, volatile oil; pungent, aromatic odor and taste. D: 1.065-1.077 @ 25°/25°. Sol in alc, ether, chloroform, glacial acetic acid, CS_2. Safrole-free ethanol extract of *Sassafras albidum* root bark (JNCIAM 60,683,78).

SYN: SASSAFRAS ALBIDUM

USE IN FOOD:

Purpose: Flavoring agent.

Where Used: Bakery products, beverages (nonalcoholic), confections, gelatin desserts, puddings.

Regulations: FDA - 21CFR 172.580. Use at a level not in excess of the amount reasonably required to accomplish the intended effect.

SAFETY PROFILE: An experimental neoplastigen. A skin irritant. When heated to decomposition it emits acrid smoke and irritating fumes.

TOXICITY DATA and CODEN

skn-rbt 500 mg/24H MOD FCTOD7 20(Suppl),825,82
scu-rat TDLo:3540 mg/kg/59W-I:NEO JNCIAM 60,683,78

SBA000 CAS: 8016-68-0
SAVORY OIL (SUMMER VARIETY)

PROP: From steam distillation of *Saturiea hortensis* L. (Fam. *Labiatae*) (FCTXAV

14,659,76). Light yellow to dark brown liquid; spicy odor. D: 0.875-0.954, refr index: 1.486-1.505 @ 20°. Sol in fixed oils, mineral oil; insol in glycerin, propylene glycol.

USE IN FOOD:

Purpose: Flavoring agent.

Where Used: Salads, sauces, soups.

Regulations: FDA - 21CFR 182.10, 182.20. GRAS when used at a level not in excess of the amount reasonably required to accomplish the intended effect.

SAFETY PROFILE: Poison by skin contact. Moderately toxic by ingestion. A severe skin irritant. When heated to decomposition it emits acrid smoke and irritating fumes.

TOXICITY DATA and CODEN

skn-rbt 500 mg/24H SEV FCTXAV 14,659,76
orl-rat LD50:1370 mg/kg FCTXAV 14,659,76
skn-gpg LD50:340 mg/kg FCTXAV 14,659,76

SCA350 CAS: 302-84-1
dl-SERINE
mf: $C_3H_7NO_3$ mw: 105.09

PROP: White crystals or crystalline powder. Mp: 246° (decomp). Sol in water; insol in alc, ether.

USE IN FOOD:

Purpose: Dietary supplement, nutrient.

Where Used: Various.

Regulations: FDA - 21CFR 172.310. Limitation 8.4 percent by weight.

SAFETY PROFILE: When heated to decomposition emits toxic fumes of NO_x.

SCA355 CAS: 56-45-1
l-SERINE
mf: $C_3H_7NO_3$ mw: 105.10

PROP: White crystals or crystalline powder; odorless with sweet taste. Mp: 228° (decomp). Sol in water; insol in alc, ether.

SYN: l-2-AMINO-3-HYDROXYPROPANOIC ACID

USE IN FOOD:

Purpose: Dietary supplement, nutrient.

Where Used: Various.

Regulations: FDA - 21CFR 172.310. Limitation 8.4 percent by weight.

SAFETY PROFILE: When heated to decomposition emits toxic fumes of NO_x.

SCC700
SHELLAC, BLEACHED

PROP: From the resinous secretion, called lac, of the insect *Llaccifer (Tachardia) lacca* Kerr (Fam. *Coccidae*). Off white, amorphous, granular solid. Sol in alc; insol in water; sltly sol in acetone, ether.

SYNS: WHITE SHELLAC ◇ REGULAR BLEACHED SHELLAC

USE IN FOOD:

Purpose: Coating agent, color diluent, glaze, surface-finishing agent.

Where Used: Confectionery, food supplements in tablet form, gum.

Regulations: FDA - 21CFR 73.1.

SAFETY PROFILE: When heated to decomposition it emits acrid smoke and irritating fumes.

SCC705
SHELLAC, BLEACHED, WAX-FREE

PROP: From the resinous secretion, called lac, of the insect *Llaccifer (Tachardia) lacca* Kerr (Fam. *Coccidae*). light yellow, amorphous, granular solid. Sol in alc; insol in water; sltly sol in acetone, ether.

SYN: REFINED BLEACHED SHELLAC

USE IN FOOD:

Purpose: Coating agent, color diluent, glaze, surface-finishing agent.

Where Used: Confectionery, food supplements in tablet form, gum.

Regulations: FDA - 21CFR 73.1. No residue.

SAFETY PROFILE: When heated to decomposition it emits acrid smoke and irritating fumes.

SCH000 CAS: 7631-86-9
SILICA, AMORPHOUS FUMED
mf: O_2Si mw: 60.09

PROP: A finely powdered microcellular silica foam with minimum SiO_2 content of 89.5%. Insol in water; sol in hydrofluoric acid.

SYNS: ACTICEL ◇ AEROSIL ◇ AMORPHOUS SILICA DUST ◇ AQUAFIL ◇ CAB-O-GRIP II ◇ CAB-O-SIL ◇ CAB-O-SPERSE ◇ CATALOID ◇ COLLOIDAL SILICA

◇ COLLOIDAL SILICON DIOXIDE ◇ DAVISON SG-67 ◇ DICALITE ◇ DRI-DIE INSECTICIDE 67 ◇ ENT 25,550 ◇ FLO-GARD ◇ FOSSIL FLOUR ◇ FUMED SILICA ◇ FUMED SILICON DIOXIDE ◇ HI-SEL ◇ LO-VEL ◇ LUDOX ◇ NALCOAG ◇ NYACOL ◇ NYACOL 830 ◇ NYACOL 1430 ◇ SANTOCEL ◇ SG-67 ◇ SILICA AEROGEL ◇ SILICA, AMORPHOUS ◇ SILICIC ANHYDRIDE ◇ SILICON DIOXIDE (FCC) ◇ SILIKILL ◇ SYNTHETIC AMORPHOUS SILICA ◇ VULKASIL

USE IN FOOD:

Purpose: Anticaking agent, antifoaming agent, carrier, chillproofing agent in malt beverages, color diluent, conditioning agent, defoaming agent, ink (food marking), processing aid.

Where Used: Bacon (cured), baking powder, beer, coffee whiteners, egg yolk (dried), flour, fruits, gelatin desserts, pudding mixes, salt, soups (powdered), tortilla chips, vanilla powder, vegetables.

Regulations: FDA - 21CFR 73.1. Limitation of 2 percent of ink solids. 21CFR 172.230. Only for encapsulating lemon oil, distilled lime oil, orange oil, peppermint oil, and spearmint oil. 21CFR 172.480. Limitation of 2 percent by weight of the food, must be removed from beer by filtration. 21CFR 173.340. 21CFR 182.1711. GRAS when used as an anti-foaming agent in accordance with good manufacturing practice. USDA - 9CFR 318.7. Limitation of 4 percent in the dry tocophenol-containing bacon curing agent.

IARC Cancer Review: Animal Inadequate Evidence IMEMDT 42,209,88; Human Inadequate Evidence IMEMDT 42,209,88.

OSHA PEL: TWA 80 $mg/m^3/\%SiO_2$

SAFETY PROFILE: Poison by intraperitoneal, intravenous, and intratracheal routes. Moderately toxic by ingestion. Much less toxic than crystalline forms. Does not cause silicosis.

TOXICITY DATA and CODEN

orl-rat LD50:3160 mg/kg ARSIM* 20,9,66
ivn-rat LD50:15 mg/kg BSIBAC 44,1685,68

SCI000 CAS: 7631-86-9
SILICA, AMORPHOUS HYDRATED

mf: O_2Si mw: 60.09

SYNS: SILICA AEROGEL ◇ SILICA GEL ◇ SILICA XEROGEL ◇ SILICIC ACID

USE IN FOOD:

Purpose: Clarifying agent.

Where Used: Wine.

Regulations: BATF - 27CFR 240.1051. Limitation of 20 pounds collodial silicon dioxide at a 30 percent concentration in 1000 gallons of wine. Silicon dioxide shall be removed by filtration.

IARC Cancer Review: Animal Inadequate Evidence IMEMDT 42,209,88; Human Inadequate Evidence IMEMDT 42,209,88.

OSHA PEL: TWA 80 $mg/m^3/\%SiO_2$ ACGIH TLV: TWA 10 mg/m^3

SAFETY PROFILE: The pure unaltered form is considered nontoxic. Some deposits contain small amounts of crystalline quartz which is therefore fibrogenic. When diatomaceous earth is calcined (with or without fluxing agents) some silica is converted to cristobalite and is therefore fibrogenic. Tridymite has never been detected in calcined diatomaceous earth.

SEG500 CAS: 127-09-3
SODIUM ACETATE

mf: $C_2H_3O_2{\bullet}Na$ mw: 82.04

PROP: White granular powder. Autoign temp: 1125°F, d: 1.45, mp: 58°. Decomp @ higher temp. Sol in water, alc.

SYNS: ACETIC ACID, SODIUM SALT ◇ ANHYDROUS SODIUM ACETATE ◇ NATRIUMACETAT (GERMAN) ◇ SODIUM ACETATE, ANHYDROUS (FCC)

USE IN FOOD:

Purpose: Boiler water additive, flavoring agent, pH control agent.

Where Used: Candy (hard), candy (soft), cereals (breakfast), fats, grain products, jams, jellies, meat products, oils, pasta, snack foods, soup mixes, soups, sweet sauces.

Regulations: FDA - 21CFR 173.310, 184.1721. GRAS with a limitation of 0.007 percent in breakfast cereal, 0.5 percent in fats and oils, 0.6 percent in grain products pastas and snack foods, 0.15 percent in hard candy, 0.12 percent in jams and jellies, 0.12 percent in meat products, 0.2 percent in soft candy, 0.05 percent in soups and soup mixes, 0.05 percent in sweet sauces when used in accordance with good manufacturing practice.

EPA Genetic Toxicology Program.

SAFETY PROFILE: Poison by intravenous route. Moderately toxic by ingestion. A skin and eye irritant. When heated to decomposition it emits toxic fumes of Na_2O.

TOXICITY DATA and CODEN

skn-rbt 500 mg/24H MLD BIOFX* 19-3/71
eye-rbt 10 mg MLD BIOFX* 19-3/71
orl-rat LD50:3530 mg/kg FAONAU 40,127,67

SEG800 CAS: 7681-38-1
SODIUM ACID SULFATE (SOLID)

DOT: 1821/2837
mf: $HO_4S \cdot Na$ mw: 120.06

PROP: White crystals or granules. Mp: >315° (decomp), d: 2.435 @ 13°. Sol in water.

SYNS: GBS ◇ NITRE CAKE ◇ SODIUM ACID SULFATE ◇ SODIUM ACID SULFATE, SOLUTION (DOT) ◇ SODIUM BISULFATE, FUSED ◇ SODIUM BISULFATE, SOLID (DOT, FCC) ◇ SODIUM BISULFATE, SOLUTION (DOT) ◇ SODIUM HYDROGEN SULFATE, SOLID (DOT) ◇ SODIUM HYDROGEN SULFATE, SOLUTION (DOT) ◇ SODIUM PYROSULFATE ◇ SULFURIC ACID, MONOSODIUM SALT

USE IN FOOD:

Purpose: Acid, preservative.

Where Used: Fruit (dried), fruits (fresh), lemon juice, meat, vegetables (fresh).

Regulations: FDA - 21CFR 182.3739. GRAS when used in accordance with good manufacturing practice except that it is not used in meats or in foods recognized as sources of vitamin B_1 or on fruits and vegetables intended to be sold or served raw or to be presented to consumers as fresh.

DOT Classification: ORM-B; Label: None; Corrosive Material; Label: Corrosive, solid and solution

SAFETY PROFILE: A corrosive irritant to skin, eyes, and mucous membranes. Mutagenic data. Reacts with moisture to form sulfuric acid. Incompatible with calcium hypochlorite. When heated to decomposition it emits toxic fumes of SO_x and Na_2O.

TOXICITY DATA and CODEN

mmo-omi 1000 ppm POASAD 34,114,53

SEH000 CAS: 9005-38-3
SODIUM ALGINATE

mf: $(C_6H_7O_6Na)_n$ mw: 198.11

PROP: Colorless to slight yellow filamentous or granular solid or powder; odorless and tasteless. In water it forms a viscous colloidal soln; insol in ether, alc, chloroform.

SYNS: ALGIN ◇ ALGINATE KMF ◇ ALGIN (POLYSACCHARIDE) ◇ ALGIPON L-1168 ◇ AMNUCOL ◇ ANTIMIGRANT C 45 ◇ CECALGINE TBV ◇ COHASAL-1H ◇ DARID QH ◇ DARILOID QH ◇ DUCKALGIN ◇ HALLTEX ◇ K'-ALGILINE ◇ KELCO GEL LV ◇ KELCOSOL ◇ KELGIN ◇ KELGUM ◇ KELSET ◇ KELSIZE ◇ KELTEX ◇ KELTONE ◇ LAMITEX ◇ MANUCOL ◇ MANUCOL DM ◇ MANUTEX ◇ MEYPRALGIN R/LV ◇ MINUS ◇ MOSANON ◇ NOURALGINE ◇ OG 1 ◇ PECTALGINE ◇ PROCTIN ◇ PROTACELL 8 ◇ PROTANAL ◇ PROTATEK ◇ SNOW ALGIN H ◇ SODIUM POLYMANNURONATE ◇ STIPINE ◇ TAGAT ◇ TRAGAYA

USE IN FOOD:

Purpose: Boiler water additive, emulsifier, firming agent, flavor enhancer, formulation aid, processing aid, stabilizer, surface-active agent, texturizer, thickener.

Where Used: Candy (hard), condiments, confections, edible films, frostings, fruit juices, fruits (processed), gelatins, puddings, relishes, sauces, toppings.

Regulations: FDA - 21CFR 173.310, 21CFR 184.1724. GRAS with a limitation of 1.0 percent in condiments and relishes except pimento ribbon for stuffed olives, 6.0 percent in confections and frostings, 4.0 percent in gelatins, puddings, 10.0 percent in hard candy, 2.0 percent in processed fruits and fruit juices, 1.0 percent in all other foods when used in accordance with good manufacturing practice.

SAFETY PROFILE: Poison by intravenous and intraperitoneal routes. When heated to decomposition it emits toxic fumes of Na_2O.

TOXICITY DATA and CODEN

ivn-rat LD50:1000 mg/kg FAONAU 53A,382,74

SEH500
SODIUM n-ALKYLBENZENE SULFONATE

USE IN FOOD:

Purpose: Hog scald agent.

Where Used: Hog carcasses.

Regulations: USDA - 9CFR 318.7. Sufficient for purpose.

SAFETY PROFILE: When heated to decomposition it emits toxic fumes of SO_x and Na_2O.

SEM000 CAS: 1344-00-9
SODIUM ALUMINOSILICATE

PROP: Fine, white, amorphous powder or beads; odorless and tasteless. Insol in water, alc and other organic solvents.

SYNS: NCI-C55505 ◇ SODIUM SILICOALUMINATE

USE IN FOOD:

Purpose: Anticaking agent.

Where Used: Cake mixes, dry mixes, nondairy creamers, salt, sugar (powdered).

Regulations: FDA - 21CFR 182.2727. GRAS with limitation of 2 percent when used in accordance with good manufacturing practice.

SAFETY PROFILE: An irritant to skin, eyes, and mucous membranes. When heated to decomposition it emits toxic fumes of Na_2O.

SEM300
SODIUM ALUMINUM PHOSPHATE, ACIDIC
mf: $NaAl_3H_{14}(PO_4)_8 \cdot 4H_2O$ mw: 949.88

PROP: White powder; odorless. Insol in water; sol in hydrochloric acid.

SYN: SALP

USE IN FOOD:

Purpose: Leavening agent.

Where Used: Cakes, pancake mixes.

Regulations: FDA - 21CFR 182.1781. GRAS when used in accordance with good manufacturing practice.

SAFETY PROFILE: A nuisance dust.

SEM305
SODIUM ALUMINUM PHOSPHATE, BASIC

PROP: White powder; odorless. Insol in water; sol in hydrochloric acid.

SYN: KASAL

USE IN FOOD:

Purpose: Leavening agent.

Where Used: Cheese (processed).

Regulations: FDA - 21CFR 182.1781. GRAS when used in accordance with good manufacturing practice.

SAFETY PROFILE: A nuisance dust.

SEN000 CAS: 7782-92-5
SODIUM AMIDE
DOT: 1425
mf: H_2NNa mw: 39.02

PROP: White, crystalline powder. Mp: 210°, bp: 400°.

SYN: SODAMIDE

USE IN FOOD:

Purpose: Catalyst for the transesterification of fats.

Where Used: Fats (rendered animal).

Regulations: USDA - 9CFR 318.7. Must be eliminated during processing.

DOT Classification: Label: Flammable Solid and Dangerous When Wet

SAFETY PROFILE: An intense irritant to tissue, skin, and eyes. Flammable by chemical reaction. Ignites or explodes with heat or grinding. Explosive reaction with moisture; halocarbons oxidants; sodium nitrite; air. Can become explosive in storage. Will react with water or steam to produce heat and toxic and corrosive fumes of sodium hydroxide and ammonia. When heated to decomposition it emits highly toxic fumes of NH_3 and Na_2O.

SFB000 CAS: 532-32-1
SODIUM BENZOATE
mf: $C_7H_5O_2 \cdot Na$ mw: 144.11

PROP: White crystalline solid; odorless. Sol in water, alc.

SYNS: ANTIMOL ◇ BENZOATE OF SODA ◇ BENZOATE SODIUM ◇ BENZOESAEURE (NA-SALZ) (GERMAN) ◇ BENZOIC ACID, SODIUM SALT ◇ SOBENATE ◇ SODIUM BENZOIC ACID

USE IN FOOD:

Purpose: Antimicrobial agent, preservative.

Where Used: Distilling materials, margarine, oleomargarine.

Regulations: FDA - 21CFR 181.23, 184.1733. GRAS with a limitation of 0.1 percent when used in accordance with good manufacturing practice. USDA - 9CFR 318.7. Limitation of 0.1 percent, or if used in combination with sorbic acid and its salts, 0.2 percent (expressed as the acids in the weight of the finished food). BATF - 27CFR 240.1051. Limitation of 0.1 percent in distilling materials.

EPA Genetic Toxicology Program.

SAFETY PROFILE: Poison by subcutaneous and intravenous routes. Moderately toxic by ingestion, intramuscular and intraperitoneal routes. An experimental teratogen. Experimental reproductive effects. Mutagenic data. Larger doses of 8-10 grams by mouth may cause nausea and vomiting. Small doses have little or no effect. Combustible when exposed to heat or flame. When heated to decomposition it emits toxic fumes of Na_2O.

TOXICITY DATA and CODEN

cyt-ham:lng 1 g/L ATSUDG (4)41,80
cyt-ham:fbr 2 g/L/48H MUREAV 48,337,77
orl-rat TDLo:44 g/kg (1-22D preg):TER ESKHA5 96,47,78
orl-rat TDLo:44 g/kg (1-22D preg):REP ESKHA5 96,47,78
ipr-rat TDLo:3 g/kg (12-14D preg):TER TXAPA9 19,373,71
orl-rat LD50:4070 mg/kg JIHTAB 30,63,48

SFC500 CAS: 144-55-8
SODIUM BICARBONATE

mf: $NaHCO_3$ mw: 84.01

PROP: White crystalline powder. Sol in water; insol in alc.

SYN: BAKING SODA

USE IN FOOD:

Purpose: Alkali, cleaning agent, leavening agent, pH control agent, poultry scald agent.

Where Used: Baked goods, beverages (dry mix), fats (rendered), margarine, oleomargarine, pickles (cured), soups, poultry, vegetables.

Regulations: FDA - 21CFR 184.1736. GRAS when used in accordance with good manufacturing practice. USDA - 9CFR 318.7, 381.147. Sufficient for purpose.

SAFETY PROFILE: A nuisance dust.

SFE000 CAS: 7631-90-5
SODIUM BISULFITE (1:1)

DOT: 2693

mf: $HO_3S \cdot Na$ mw: 104.06

PROP: White, crystalline powder; odor of sulfur dioxide, disagreeable taste. D: 1.48, mp: decomp. Very sol in hot or cold water; sltly sol in alc.

SYNS: BISULFITE de SODIUM (FRENCH) ◇ HYDROGEN SULFITE SODIUM ◇ SODIUM ACID SULFITE ◇ SODIUM BISULFITE ◇ SODIUM BISULFITE (ACGIH) ◇ SODIUM BISULFITE, SOLID (DOT) ◇ SODIUM BISULFITE, SOLUTION (DOT) ◇ SODIUM HYDROGEN SULFITE ◇ SODIUM HYDROGEN SULFITE, SOLID (DOT) ◇ SODIUM HYDROGEN SULFITE, SOLUTION (DOT) ◇ SODIUM SULHYDRATE ◇ SULFUROUS ACID, MONOSODIUM SALT

USE IN FOOD:

Purpose: Preservative.

Where Used: Various.

Regulations: GRAS when used in accordance with good manufacturing practice.

EPA Genetic Toxicology Program.

ACGIH TLV: TWA 5 mg/m^3 DOT Classification: ORM-B; Label: None; Corrosive Material; Label: Corrosive

SAFETY PROFILE: Poison by intravenous and intraperitoneal routes. Moderately toxic by ingestion. A corrosive irritant to skin, eyes, and mucous membranes. Mutagenic data. An allergen. When heated to decomposition it emits toxic fumes of SO_x and Na_2O.

TOXICITY DATA and CODEN

mmo-sat 1 mmol/L TCMUD8 5,195,85
mmo-esc 1 mol/L BBRCA9 39,983,70
orl-rat LD50:2000 mg/kg DTLVS* 4,369,80

SFN700
SODIUM CALCIUM ALUMINOSILICATE, HYDRATED

USE IN FOOD:

Purpose: General purpose food additive.

Where Used: Various.

Regulations: FDA - 21CFR 182.2729. Limitation of 2 percent in accordance with good manufacturing practice.

SAFETY PROFILE: A nuisance dust.

SFO000 CAS: 497-19-8
SODIUM CARBONATE (2:1)

mf: $CO_3 \cdot 2Na$ mw: 105.99

PROP: White, odorless, small crystals or crystalline powder; alkali taste. Mp: 851°, bp: decomp, d: 2.509 @ 0°. Hygroscopic; sol in water.

SYNS: CARBONIC ACID, DISODIUM SALT ◇ CRYSTOL CARBONATE ◇ DISODIUM CARBONATE ◇ SODA ASH ◇ TRONA

USE IN FOOD:

Purpose: Alkali, antioxidant, boiler water additive, curing agent, fat rendering agent, flavoring agent, hog scald agent, pH control agent, pickling agent, poultry scald agent, processing aid.

Where Used: Baked goods, dessert gels (water), fats (rendered), hog carcasses, margarine, oleomargarine, poultry, puddings, sauces, soups (instant).

Regulations: FDA - 21CFR 173.310, 184.1742. GRAS when used in accordance with good manufacturing practice. USDA - 9CFR 318.7, 381.147. Sufficient for purpose.

EPA Genetic Toxicology Program.

SAFETY PROFILE: Poison by intraperitoneal route. Moderately toxic by inhalation and subcutaneous routes. Mildly toxic by ingestion. Experimental reproductive effects. A skin and eye irritant. When heated to decomposition it emits toxic fumes of Na_2O.

TOXICITY DATA and CODEN

skn-rbt 500 mg/24H MLD 28ZPAK -,7,72
eye-rbt 100 mg/24H MOD 28ZPAK -,8,72
iut-mus TDLo: 84800 ng/kg (4D preg): REP JRPFA4 63,365,81
orl-rat LD50: 4090 mg/kg 28ZPAK -,8,72
ihl-rat LC50: 2300 mg/m^3/2H ENVRAL 31,138,83

SFO500 CAS: 9004-32-4
SODIUM CARBOXYMETHYL CELLULOSE

PROP: A synthetic cellulose gum (the sodium salt of carboxy methyl cellulose not less than 99.5% on a dry weight basis, with maximum substitution of 0.95 carboxymethyl groups per anhydroglucose unit, and with a minimum viscosity of 25 centipoises for 2% weight aqueous solutions at 25°). Colorless, odorless, hygroscopic powder or granules. Insol in most organic solvents.

SYNS: AC-DI-SOL NF ◇ AQUAPLAST ◇ B10 ◇ BLANOSE BWM ◇ B 10 (POLYSACCHARIDE) ◇ CARBOXYMETHYL CELLULOSE ◇ CARBOXYMETHYL CELLULOSE, SODIUM ◇ CARBOXYMETHYL CELLULOSE, SODIUM SALT ◇ CARMETHOSE ◇ CELLOFAS ◇ CELLOGEL C ◇ CELLPRO ◇ CELLUFIX FF 100 ◇ CELLUGEL ◇ CELLULOSE GLYCOLIC ACID, SODIUM SALT ◇ CELLULOSE GUM ◇ CELLULOSE SODIUM GLYCOLATE ◇ CMC ◇ CM-CELLULOSE Na SALT ◇ CMC 7H ◇ CMC SODIUM SALT ◇ COLLOWELL ◇ COPAGEL PB 25 ◇ COURLOSE A 590 ◇ DAICEL 1150 ◇ FINE GUM HES ◇ GLIKOCEL TA ◇ KMTS 212 ◇ LOVOSA ◇ LUCEL (polysaccharide) ◇ MAJOL PLX ◇ MODOCOLL 1200 ◇ NACM-CELLULOSE SALT ◇ NYMCEL S ◇ POLYFIBRON 120 ◇ SANLOSE SN 20A ◇ SARCELL TEL ◇ S 75M ◇ SODIUM CELLULOSE GLYCOLATE ◇ SODIUM CMC ◇ SODIUM CM-CELLULOSE ◇ SODIUM SALT OF CARBOXYMETHYLCELLULOSE ◇ TYLOSE 666 ◇ UNISOL RH

USE IN FOOD:

Purpose: Binder, boiler water additive, extender, stabilizer, thickener.

Where Used: Pies (baked), poultry.

Regulations: FDA - 21CFR 173.310, 182.1745. GRAS when used in accordance with good manufacturing practice. USDA - 9CFR 318.7. Limitation of 1.5 percent. Must be added dry. 9CFR 381.147. Sufficient for purpose.

SAFETY PROFILE: Mildly toxic by ingestion. An experimental neoplastigen. Experimental reproductive effects. When heated to decomposition it emits toxic fumes of Na_2O.

TOXICITY DATA and CODEN

orl-rat TDLo: 140 mg/kg (14D male): REP OYYAA2 14,623,77
scu-rat TDLo: 1900 mg/kg/19W-I: NEO 13BYAH -,83,62
orl-rat LD50: 27000 mg/kg FOREAE 13,29,48

SFQ000 CAS: 9005-46-3
SODIUM CASEINATE

PROP: Coarse, white powder; odorless. Insol in water, alc.

SYNS: CASEIN AND CASEINATE SALTS (FCC) ◇ CASEIN-SODIUM ◇ CASEIN, SODIUM COMPLEX ◇ CASEINS, SODIUM COMPLEXES ◇ NUTROSE

USE IN FOOD:

Purpose: Binder, clarifying agent, emulsifier, extender, stabilizer.

Where Used: Bread, cereals, cheese (imitation), coffee whiteners, desserts, egg substitutes, loaves (nonspecific), meat (processed), poultry, sausage (imitation), soups, stews, whipped toppings, whipped toppings (vegetable oil), wine.

Regulations: FDA - 21CFR 182.1748. GRAS when used in accordance with good manufacturing practice. USDA - 9CFR 318.7, 9CFR 381.147. Sufficient for purpose in meat and poultry. BATF - 27CFR 240.1051. GRAS when used in accordance with good manufacturing practice.

SAFETY PROFILE: An experimental tumorigen. When heated to decomposition it emits toxic fumes of Na_2O.

TOXICITY DATA and CODEN

scu-mus TDLo:45 g/kg/15D-I:ETA JNCIAM 57,1367,76

SFS000 CAS: 7775-09-9

SODIUM CHLORATE

DOT: 2428/1495

mf: $ClO_3 \cdot Na$ mw: 106.44

PROP: Colorless, odorless crystals; cooling, saline taste. Mp: 248-261°, bp: decomp, d: 2.490 @ 15°.

SYNS: ASEX ◇ ATLACIDE ◇ ATRATOL ◇ B-HERBATOX ◇ CHLORATE OF SODA (DOT) ◇ CHLORATE SALT of SODIUM ◇ CHLORAX ◇ CHLORSAURE (GERMAN) ◇ DE-FOL-ATE ◇ DESOLET ◇ DREXEL DEFOL ◇ DROP LEAF ◇ EVAU-SUPER ◇ FALL ◇ GRAIN SORGHUM HARVEST-AID ◇ GRANEX O ◇ HARVEST-AID ◇ KLOREX ◇ KUSA-TOHRU ◇ KUSATOL ◇ NATRIUMCHLORAAT (DUTCH) ◇ NATRIUMCHLORAT (GERMAN) ◇ ORTHO C-1 DEFOLIANT & WEED KILLER ◇ OXYCIL ◇ RASIKAL ◇ SHED-A-LEAF ◇ SHED-A-LEAF 'L' ◇ SODA CHLORATE (DOT) ◇ SODIUM (CHLORATE de) (FRENCH) ◇ SODIO (CLORATO di) (ITALIAN) ◇ SODIUM CHLORATE, AQUEOUS SOLUTION (DOT) ◇ TRAVEX ◇ TUMBLEAF ◇ UNITED CHEMICAL DEFOLIANT NO. 1 ◇ VAL-DROP

USE IN FOOD:

Purpose: Animal glue adjuvant.

Where Used: Packaging materials.

Regulations: FDA - 21CFR 178.3120. Use at a level not in excess of the amount reasonably required to accomplish the intended effect.

DOT Classification: Oxidizer; Label: Oxidizer

SAFETY PROFILE: Human poison by unspecified routes. Moderately toxic experimentally by ingestion and intraperitoneal routes. Human systemic effects by ingestion: blood hemolysis with or without anemia, methemoglobinemia-carboxhemoglobinemia and pulmonary changes. Mutagenic data. A skin, mucous membrane and eye irritant. Damages the red blood cells of humans when ingested. Used as an herbicide.

A powerful oxidizer. It can explode on contact with flame or sparks (static discharge) and has caused many industrial explosions. May react explosively with agricultural materials (e.g., peat; powdered sulfur; sawdust). Can also react violently with nitrobenzene, paper, metal sulfides, dibasic organic acids, organic matter. When heated to decomposition it emits toxic fumes of Cl^- and Na_2O.

TOXICITY DATA and CODEN

skn-rbt 500 mg/24H MLD BIOFX* 24,3/71
eye-rbt 10 mg MLD BIOFX* 21-3/71
mma-sat 40 µmol/plate MUREAV 90,91,81
sln-dmg-orl 250 mmol/L MUREAV 90,91,81
orl-wmn TDLo:800 mg/kg:PUL,BLD 34ZIAG -,539,69
unr-hmn LDLo:214 mg/kg GUCHAZ 6,461,73
orl-rat LD50:1200 mg/kg PHJOAV 185,361,60

SFT000 CAS: 7647-14-5

SODIUM CHLORIDE

mf: ClNa mw: 58.44

PROP: Colorless, transparent crystals or white, crystalline powder. Mp: 801°, bp: 1413°, d: 2.165, vap press: 1 mm @ 865°. Sol in water, glycerin.

SYNS: COMMON SALT ◇ DENDRITIS ◇ EXTRA FINE 200 SALT ◇ EXTRA FINE 325 SALT ◇ HALITE ◇ H.G. BLENDING ◇ NATRIUMCHLORID (GERMAN) ◇ PUREX ◇ ROCK SALT ◇ SALINE ◇ SALT ◇ SEA SALT ◇ STERLING ◇ TABLE SALT ◇ TOP FLAKE ◇ USP SODIUM CHLORIDE ◇ WHITE CRYSTAL

USE IN FOOD:

Purpose: Chilling media, curing agent, dough conditioner, flavoring agent, intensifier, nutrient, preservative.

Where Used: Baked goods, butter, cheese, nuts (salted), poultry, sausage.

Regulations: USDA - GRAS when used in accordance with good manufacturing practice. USDA - 9CFR 381.147. Limitation of 700 pounds to 10,000 gallons of water for poultry chilling.

EPA Genetic Toxicology Program.

SAFETY PROFILE: Poison by intraperitoneal and intracervical routes. Moderately toxic by ingestion, intravenous and subcutaneous routes. An experimental teratogen. Human systemic effects by ingestion: blood pressure increase. Human reproductive effects by intraplacental route: terminates pregnancy. Experimental reproductive effects. Human mutagenic data. A skin and eye irritant. When bulk sodium chloride is heated to high temperature, a vapor is emitted which is irritating, particularly to the eyes. Ingestion of large amounts of sodium chloride can cause irritation of the stomach. Improper use of salt tablets may produce this effect. When heated to decomposition it emits toxic fumes of Cl^- and Na_2O.

TOXICITY DATA and CODEN

skn-rbt 50 mg/24H MLD BIOFX* 20-3/71
eye-rbt 100 mg/24H MOD 28ZPAK -,7,72
dni-hmn:fbr 125 mmol/L CNREA8 46,713,86
ipc-wmn TDLo:27 mg/kg (15W preg):REP AJOGAH 118,218,74
ipr-rat TDLo:1710 mg/kg (13D preg):TER SEIJBO 8,197,68
iut-rat TDLo:500 mg/kg (4D preg):REP BIREBV 21,47,79
orl-hmn TDLo:12357 mg/kg/23D-C:CVS AJDDAL 21,180,54
orl-rat LD50:3000 mg/kg TXAPA9 20,57,71

SFT500 CAS: 7758-19-2
SODIUM CHLORITE (DOT)
DOT: 1496
mf: $ClNaO_2$ mw: 90.44

PROP: White crystals or crystalline powder. Bp: decomp @ 180-200°.

SYN: TEXTILE

USE IN FOOD:

Purpose: Paper slimicide.

Where Used: Food packaging.

Regulations: FDA - 21CFR 186.1750. GRAS when used in accordance with good manufacturing practice.

DOT Classification: Oxidizer; Label: Oxidizer

SAFETY PROFILE: Poison by ingestion. An experimental teratogen. Experimental reproductive effects. Mutagenic data. May act as an irritant due to its oxidizing power. A powerful oxidizing agent; ignited by friction, heat or shock. An explosive sensitive to impact or heating to 200°. Potentially explosive reaction with acids, oils, organic matter, oxalic acid + water, zinc. Can react vigorously on contact with reducing materials. When heated to decomposition it emits highly toxic fumes of Cl^- and Na_2O.

TOXICITY DATA and CODEN

mma-sat 300 μg/plate FCTOD7 22,623,84
cyt-ham:fbr 20 mg/L FCTOD7 22,623,84
orl-rat TDLo:800 mg/kg (8-15D preg):TER EVHPAZ 46,25,82
orl-rat TDLo:16 g/kg (8-15D preg):REP EVHPAZ 46,25,82
ipr-rat TLDo:160 mg/kg (8-15D preg):TER EVHPAZ 46,25,82
orl-rat LD50:165 mg/kg YKYUA6 31,959,80

SFW000 CAS: 361-09-1
SODIUM CHOLATE
mf: $C_{24}H_{39}O_5{\bullet}Na$ mw: 430.62

SYNS: CHOLIC ACID, MONOSODIUM SALT ◇ CHOLIC ACID, SODIUM SALT ◇ DS-Na ◇ SODIUM CHOLIC ACID ◇ OX BILE EXTRACT ◇ PURIFIED OXGALL ◇ TRIHYDROXY-3,7,12-CHOLANATE de Na (FRENCH) ◇ (3-α,5-β,7-α,12-α)3,7,12-TRIHYDROXY-CHOLAN-24-OIC ACID, MONOSODIUM SALT

USE IN FOOD:

Purpose: Surfactant.

Where Used: Cheese.

Regulations: FDA - 21CFR 184.1560. GRAS with a limitation of 0.002 percent in cheese.

SAFETY PROFILE: Poison by intravenous route. When heated to decomposition it emits toxic fumes of Na_2O.

TOXICITY DATA and CODEN

ivn-mus LD50:200 mg/kg AIPTAK 90,18,52

SGC000 CAS: 139-05-9
SODIUM CYCLAMATE
mf: $C_6H_{12}NO_3S{\bullet}Na$ mw: 201.24

PROP: White, crystalline powder; practically odorless. Sol in water. Almost insol in alc, benzene, chloroform, ether.

SYNS: ASSUGRIN ◇ ASSUGRIN FEINUSS ◇ ASSUGRIN VOLLSUSS ◇ ASUGRYN ◇ CYCLAMATE ◇ CYCLAMATE SODIUM ◇ CYCLAMIC ACID SODIUM SALT ◇ CYCLOHEXANESULFAMIC ACID, MONOSODIUM SALT ◇ CYCLOHEXANESULPHAMIC ACID, MONOSODIUM SALT ◇ CYCLOHEXYL SULPHAMATE SODIUM ◇ DULZOR-ETAS ◇ HACHI-SUGAR ◇ IBIOSUC ◇ NATREEN ◇ NATRIUMZYKLAMATE (GERMAN) ◇ SODIUM CYCLOHEXANESULFAMATE ◇ SODIUM CYCLOHEXANESULPHAMATE ◇ SODIUM CYCLOHEXYL AMIDOSULPHATE ◇ SODIUM CYCLOHEXYL SULFAMATE ◇ SODIUM CYCLOHEXYL SULFAMIDATE ◇ SODIUM CYCLOHEXYL SULPHAMATE ◇ SODIUM SUCARYL ◇ SUCARYL SODIUM ◇ SUCCARIL ◇ SUCROSA ◇ SUESSETTE ◇ SUESTAMIN ◇ SUGARIN ◇ SUGARON

USE IN FOOD:

Purpose: Nonnutritive sweetener.

Where Used: Prohibited from foods.

Regulations: FDA - 21CFR 189.135. Prohibited from direct addition or use in human food.

IARC Cancer Review: Animal Limited Evidence IMEMDT 22,55,80. EPA Genetic Toxicology Program.

SAFETY PROFILE: Moderately toxic by intravenous and intraperitoneal routes. Mildly toxic by ingestion. An experimental neoplastigen, tumorigen, and teratogen. May be a carcinogen. Experimental reproductive effects. Human mutagenic data. When heated to decomposition it emits very toxic fumes of Na_2O, SO_x and NO_x.

TOXICITY DATA and CODEN

dni-hmn:hla 800 μg/L INHEAO 9,188,71
cyt-hmn:leu 10 μmol/L/5H MUREAV 39,1,76
orl-mus TDLo:180 mg/kg (7D preg):TER IIZAAX 16,330,64
orl-mus TDLo:420 g/kg (MGN):REP TXCYAC 8,285,77
orl-mus TDLo:5 g/kg (7D preg):TER ARZNAD 19,923,69
orl-rat TDLo:63 g/kg/9W-C:NEO CNREA8 37,2943,77
orl-rat TD :610 g/kg/87W-C:ETA CBINA8 11,255,75
orl-rat LD50:15250 mg/kg FCTXAV 6,313,68

SGD000 CAS: 4418-26-2
SODIUM DEHYDROACETIC ACID

mf: $C_8H_7O_4 \cdot Na$ mw: 190.14

PROP: White powder; odorless with slt characteristic taste. Mp: 109-111°. Sol in water, propylene glycol, glycerin.

SYNS: DEHYDROACETIC ACID, SODIUM SALT ◇ DHA-SODIUM ◇ HARVEN ◇ 4-HEXENOIC ACID, 2-ACETYL-5-HYDROXY-3-OXO, Δ-LACTONE, SODIUM derivative ◇ 3-(1-HYDROXYETHYLIDENE)-6-METHYL-2H-PYRAN-2,4(3H)-DIONE, SODIUM SALT ◇ SODIUM DEHYDROACETATE (FCC)

USE IN FOOD:

Purpose: Preservative.

Where Used: Squash (cut or peeled).

Regulations: FDA - 21CFR 172.130. Limitation of 65 ppm on prepared squash.

EPA Genetic Toxicology Program.

SAFETY PROFILE: Poison by intravenous route. Moderately toxic by ingestion. An experimental teratogen. Experimental reproductive effects. Mutagenic data. When heated to decomposition it emits toxic fumes of Na_2O.

TOXICITY DATA and CODEN

cyt-ham:fbr 30 g/L FCTOD7 22,623,84
cyt-ham:lng 2500 mg/L GMCRDC 27,95,81
orl-mus TDLo:500 mg/kg (6-15D preg):TER NKEZA4 27,91,80
orl-mus TDLo:2 g/kg (6-15D preg):TER NKEZA4 27,91,80
orl-mus LD50:1175 mg/kg LONZA# 02JUN80

SGE400 CAS: 126-96-5
SODIUM DIACETATE

mf: $C_4H_7NaO_4 \cdot xH_2O$

PROP: white crystalline powder; odor of acetic acid. Sol in water.

SYN: SODIUM HYDROGEN DIACETATE

USE IN FOOD:

Purpose: Antimicrobial agent, flavoring agent, mold and rope inhibitor, pH control agent, preservative, sequestrant.

Where Used: Baked goods, bread, candy (soft), fats, gravies, meat products, oils, sauces, snack foods, soups.

Regulations: FDA - 21CFR 184.1754. GRAS with a limitation of 0.4 percent in baked goods, 0.1 percent in fats and oils, 0.1 percent in meat products, 0.1 percent in soft candy, 0.25 percent in gravies and sauces, 0.05 percent in snack foods and soups and gravies when used in accordance with good manufacturing practice.

SAFETY PROFILE: When heated to decomposition it emits acrid smoke and irritating fumes.

SGL500 CAS: 7558-80-7
SODIUM DIHYDROGEN PHOSPHATE (1:2:1)

mf: $H_2O_4P \cdot Na$ mw: 119.98

PROP: White crystalline powder or granules; odorless. Hygroscopic; sol in water; insol in alc.

SYNS: MONOSODIUM DIHYDROGEN PHOSPHATE ◇ MONOSODIUM PHOSPHATE ◇ MONOSORB XP-4 ◇ PRIMARY SODIUM PHOSPHATE ◇ SODIUM ACID PHOSPHATE ◇ SODIUM BIPHOSPHATE ◇ SODIUM BIPHOSPHATE ANHYDROUS ◇ SODIUM PHOSPHATE, MONOBASIC (FCC)

USE IN FOOD:

Purpose: Buffer, dietary supplement, emulsifier, nutrient, poultry scald agent.

Where Used: Beverages (dry mix), cheese, meat products, poultry, poultry products, soft drinks.

Regulations: FDA - 21CFR 182.1751, 182.6085, 182.6778, 182.8890. GRAS when used in accordance with good manufacturing practice. USDA - 9CFR 318.7. In meat food products, where allowed, limitation of 5 percent of phosphate in pickle at 10 percent pump level, 0.5 percent of phosphate in product (only clear solution may be injected into product. 9CFR 381.147. Limitation of 0.5 percent of total poultry product. 9CFR 381.147. Sufficient for purpose.

SAFETY PROFILE: Poison by intramuscular route. Mildly toxic by ingestion. A human and experimental eye irritant. When heated to decomposition it emits toxic fumes of PO_x and Na_2O.

TOXICITY DATA and CODEN

eye-hmn 50 mg MLD ARZNAD 9,349,59
eye-rbt 150 mg MLD ARZNAD 9,349,59
orl-rat LD50: 8290 mg/kg 28ZPAK -,16,72

SGM500 CAS: 128-04-1
SODIUM N,N-DIMETHYLDITHIOCARBAMATE

mf: $C_3H_6NS_2 \cdot Na$ mw: 143.21

PROP: Crystals.

SYNS: ACETO SDD 40 ◇ ALCOBAM NM ◇ BROGDEX 555 ◇ CARBON S ◇ DIBAM ◇ DIMETHYLDITHIOCARBAMIC ACID, SODIUM SALT ◇ DMDK ◇ METHYL NAMATE ◇ SDDC ◇ SHARSTOP 204 ◇ STA-FRESH 615 ◇ STERISEAL LIQUID #40 ◇ THIOSTOP N ◇ VINSTOP ◇ VULNOPOL NM ◇ WING STOP B

USE IN FOOD:

Purpose: Antimicrobial agent.

Where Used: Beets, sugarcane.

Regulations: FDA - 21CFR 173.320(b)(2). Limitation of 3.0 ppm based on weight of raw sugarcane or raw beets.

SAFETY PROFILE: Moderately toxic by ingestion and intraperitoneal routes. Mutagenic data. When heated to decomposition it emits very toxic fumes of NO_x, SO_x and Na_2O.

TOXICITY DATA and CODEN

mmo-sat 50 μg/plate MUREAV 116,185,83
mma-sat 5 μg/plate PCBRD2 141,407,84
orl-rat LD50: 1000 mg/kg PCOC** -,1029,66

SGR700
SODIUM ERYTHORBATE

mf: $C_6H_7NaO_6 \cdot H_2O$ mw: 216.12

PROP: White crystalline powder; odorless. Sol in water.

USE IN FOOD:

Purpose: Antioxidant, curing accelerator, preservative.

Where Used: Baked goods, beef, beef (cured), beverages, bologna, cured comminuted meat food product, frankfurters, meat (cured), pork (cured), pork (fresh), potato salad, poultry.

Regulations: USDA - 9CFR 318.7, 9CFR 381.147. Limitation of 87.5 ounce per 100 gallons of pickle at 10 percent pump level; 7/8 ounce per 100 pounds meat, 10 percent to surfaces of cured cuts prior to packaging. Not to exceed 500 ppm or 1.8 mg/sq in. of surface ascorbic acid, erythorbic acid or sodium ascorbate, singly, or in combination; and/or not to exceed either 250 ppm or 0.9 mg/sq in. of surface of citric acid or sodium citrate, singly, or in combination on fresh pork cuts.

SAFETY PROFILE: When heated to decomposition it emits acrid smoke and irritating fumes.

SHE300
SODIUM FERRIC PYROPHOSPHATE

mf: $Na_8Fe_4(P_2O_7)_5 \cdot XH_2O$ mw: 1277.02

PROP: White to tan powder; odorless. Insol in water; sol in hydrochloric acid.

SYNS: FERRIC SODIUM PYROPHOSPHATE ◇ SODIUM IRON PYROPHOSPHATE

USE IN FOOD:

Purpose: Dietary supplement, nutrient.

Where Used: Various.

Regulations: FDA - 21CFR 182.5306, 182.8308. GRAS when used in accordance with good manufacturing practice.

SAFETY PROFILE: A nuisance dust.

SHE350 CAS: 13601-19-9
SODIUM FERROCYANIDE

mf: $Na_4Fe(CN)_6 \cdot 10H_2O$ mw: 484.06

PROP: Yellow crystals or crystalline powder. Sol in water; insol in most organic solvents.

SYN: YELLOW PRUSSIATE of SODA

USE IN FOOD:

Purpose: Anticaking agent for sodium chloride, processing aid.

Where Used: Sodium chloride, wine.

Regulations: FDA GRAS per FDA advisory of 6/22/82. Limitation of 13 ppm calculated as anhydrous sodium ferrocyanide. BATF - 27CFR 240.1051. Limitation of 1 ppm residue in finished wine.

SAFETY PROFILE: When heated to decomposition emits toxic fumes of CN^-.

SHJ000 CAS: 141-53-7

SODIUM FORMATE

mf: $CHO_2 \cdot Na$ mw: 68.01

PROP: White, deliquescent crystals. Mp: 253°, d: 1.92 @ 20°.

SYN: SALACHLOR

USE IN FOOD:

Purpose: Paper manufacturing aid.

Where Used: Food packaging.

Regulations: FDA - 21CFR 186.1756. GRAS when used in accordance with good manufacturing practice.

SAFETY PROFILE: Moderately toxic by ingestion, intravenous, subcutaneous and possibly other routes. Combustible when exposed to heat or flame. When heated to decomposition it emits toxic fumes of Na_2O.

TOXICITY DATA and CODEN

orl-mus LD50:11200 mg/kg ZERNAL 9,332,69

SHK800 CAS: 527-07-1

SODIUM GLUCONATE

mf: $C_6H_{12}O_7 \cdot Na$ mw: 219.17

PROP: White to tan granular or crystalline powder. Very sol in water, sltly sol in alc; insol in ether.

SYNS: GLONSEN ◇ GLUCONATO di SODIO (ITALIAN) ◇ GLUCONIC ACID SODIUM SALT ◇ MONOSODIUM GLUCONATE ◇ PASEXON 100T ◇ PMP SODIUM GLUCONATE ◇ SODIUM d-GLUCONATE

USE IN FOOD:

Purpose: Dietary supplement, nutrient, sequestrant.

Where Used: Various.

Regulations: FDA - 21CFR 182.6757. GRAS when used in accordance with good manufacturing practice.

SAFETY PROFILE: Low toxicity by intravenous route. When heated to decomposition it emits acrid smoke and irritating fumes.

TOXICITY DATA and CODEN

ivn-rbt LDLo:7630 mg/kg AFSPA2 68,1,39

SHM500 CAS: 10124-56-8

SODIUM HEXAMETAPHOSPHATE

mf: $O_{18}P_6 \cdot 6Na$ mw: 611.76

PROP: White powder or flakes. Sol in water.

SYNS: CALGON ◇ CHEMI-CHARL ◇ HEXAMETAPHOSPHATE, SODIUM SALT ◇ HMP ◇ MEDI-CALGON ◇ PHOSPHATE, SODIUM HEXAMETA ◇ POLYPHOS ◇ SHMP

USE IN FOOD:

Purpose: Hog scald agent, poultry scald agent.

Where Used: Hog carcasses, poultry.

Regulations: USDA - 9CFR 318.7, 381.147. Sufficient for purpose.

SAFETY PROFILE: Poison by intravenous route. Moderately toxic by intraperitoneal and subcutaneous routes. Mildly toxic by ingestion. When heated to decomposition it emits toxic fumes of PO_x and Na_2O.

TOXICITY DATA and CODEN

orl-mus LD50:7250 mg/kg ARZNAD 7,445,57

ipr-mus LD50:870 mg/kg REPMBN 10,391,62

SHS000 CAS: 1310-73-2

SODIUM HYDROXIDE

DOT: 1823/1824

mf: HNaO mw: 40.00

PROP: White, pieces, lumps or sticks. Mp: 318.4°, bp: 1390°, d: 2.120 @ 20°/4°, vap press: 1 mm @ 739°. Deliquescent, sol in water, alc.

SYNS: CAUSTIC SODA ◇ CAUSTIC SODA, BEAD (DOT) ◇ CAUSTIC SODA, DRY (DOT) ◇ CAUSTIC SODA, FLAKE (DOT) ◇ CAUSTIC SODA, GRANULAR (DOT) ◇ CAUSTIC SODA, LIQUID (DOT) ◇ CAUSTIC SODA, SOLID (DOT) ◇ CAUSTIC SODA, SOLUTION (DOT) ◇ HYDROXYDE de SODIUM (FRENCH) ◇ LEWIS-RED DEVIL LYE ◇ LYE (DOT) ◇ NATRIUMHYDROXID (GERMAN) ◇ NATRIUMHYDROXYDE (DUTCH) ◇ SODA LYE ◇ SODIO(IDROSSIDO di) (ITALIAN) ◇ SODIUM HYDRATE (DOT) ◇ SODIUM HYDROXIDE, BEAD (DOT) ◇ SODIUM HYDROXIDE, DRY (DOT) ◇ SODIUM HYDROXIDE, FLAKE (DOT) ◇ SODIUM HYDROXIDE, GRANULAR (DOT) ◇ SODIUM HYDROXIDE, SOLID (DOT) ◇ SODIUM(HYDROXYDE de) (FRENCH) ◇ WHITE CAUSTIC

USE IN FOOD:

Purpose: Alkali, boiler water additive, cooked out juices retention agent, hog scald agent, pH control agent, poultry scald agent, processing aid.

Where Used: Black olives, brandy, fats (rendered animal), hog carcasses, margarine, meat food products containing phosphates, oleomargarine, poultry food products containing phosphates, tripe, wine spirit.

Regulations: FDA - 21CFR 173.310, 184.1763. GRAS when used in accordance with good manufacturing practice. USDA - 9CFR 318.7. Sufficient for purpose in hog carcasses and tripe. May be used only in combination with phosphates in a ratio not to exceed one part sodium hydroxide to four parts phosphate. The combination shall not exceed 5 percent in pickle at 10 percent pump level, 0.5 percent in product. Sufficient for purpose in oleomargarine or margarine. 9CFR 381.147. May be used only in combination with phosphates in a ratio not to exceed one part sodium hydroxide to four parts phosphate. 9CFR 381.147. Sufficient for purpose. BATF - 27CFR 240.1051a. Finished brandy or wine spirit produced from the distilling material must be free of chemical residue resulting from treatment with sodium hydroxide.

EPA Genetic Toxicology Program.

OSHA PEL: CL 2 mg/m^3 ACGIH TLV: Cl 2 mg/m^3 DFG MAK: 2 mg/m^3 DOT Classification: Corrosive Material; Label: Corrosive NIOSH REL: (Sodium Hydroxide) CL 2 mg/m^3/15M

SAFETY PROFILE: Poison by intraperitoneal route. Moderately toxic by ingestion. Mutagenic data. A corrosive irritant to skin, eyes, and mucous membranes. This material, both solid and in solution, has a markedly corrosive action upon all body tissue causing burns and frequently deep ulceration, with ultimate scarring. Mists, vapors, and dusts of this compound cause small burns, and contact with the eyes rapidly causes severe damage to the delicate tissue. Ingestion causes very serious damage to the mucous membranes or other tissues with which contact is made. Inhalation of the dust or concentrated mist can cause damage to the upper respiratory tract and to lung tissue, depending upon the severity of the exposure.

A strong base. Caution: Under the proper conditions of temperature, pressure, and state of division, it can ignite or react violently with many organic compounds. Dangerous material to handle. When heated to decomposition it emits toxic fumes of Na_2O.

TOXICITY DATA and CODEN

eye-mky 1%/24H SEV TXAPA9 6,701,64
skn-rbt 500 mg/24H SEV 28ZPAK -,7,72
cyt-grh-par 20 mg NULSAK 9,119,66
ipr-mus LD50:40 mg/kg COREAF 257,791,63
orl-rbt LDLo:500 mg/kg AEPPAE 184,587,37

SHS500 CAS: 1310-73-2
SODIUM HYDROXIDE (liquid)

DOT: 1823/1824
mf: HNaO mw: 40.00

PROP: Clear to sltly turbid, colorless liquid.

SYNS: CAUSTIC SODA SOLUTION ◇ LYE SOLUTION ◇ SODA LYE ◇ SODIUM HYDRATE SOLUTION ◇ SODIUM HYDROXIDE SOLUTION (FCC) ◇ WHITE CAUSTIC SOLUTION

USE IN FOOD:

Purpose: Alkali, boiler water additive, hog scald agent, pH control agent, processing aid.

Where Used: Black olives, tripe.

Regulations: FDA - 21CFR 173.310, 184.1763. GRAS when used in accordance with good manufacturing practice. USDA - 9CFR 318.7. Sufficient for purpose.

Community Right-To-Know List.

DOT Classification: Corrosive Material; Label: Corrosive

SAFETY PROFILE: Poison by intraperitoneal route. Moderately toxic by ingestion. Mutagenic data. A corrosive irritant to skin, eyes, and mucous membranes. When heated to decomposition it emits toxic fumes of Na_2O.

TOXICITY DATA and CODEN

cyt-grh-par 20 μL NULSAK 9,119,66
orl-rbt LDLo:500 mg/kg AEPPAE 184,587,37

SHU500 CAS: 7681-52-9
SODIUM HYPOCHLORITE

DOT: 1791
mf: ClHO•Na mw: 75.45

PROP: Mp: decomp.

SYNS: ANTIFORMIN ◇ B-K LIQUID ◇ CARREL-DAKIN SOLUTION ◇ CHLOROS ◇ CHLOROX ◇ CLOROX ◇ DAKINS SOLUTION ◇ HYCLORITE ◇ MILTON ◇ SURCHLOR

USE IN FOOD:

Purpose: Lye peeling agent, washing water additive.

Where Used: Fruits, vegetables.

Regulations: FDA - 21CFR 173.315. Use at a level not in excess of the amount reasonably required to accomplish the intended effect.

EPA Genetic Toxicology Program.

DOT Classification: ORM-B; Label: None

SAFETY PROFILE: Human mutagenic data. An eye irritant. Corrosive and irritating by ingestion and inhalation. The anhydrous salt is highly explosive and sensitive to heat or friction. Reacts to form explosive products with amines. Solutions in water are storage hazards due to oxygen evolution. When heated to decomposition it emits toxic fumes of Na_2O and Cl^-.

TOXICITY DATA and CODEN

eye-rbt 10 mg MOD TXAPA9 55,501,80
mma-sat 1 mg/plate AMONDS 3,253,80
dnr-esc 20 nL/disc MUREAV 41,61,76

SHV000 CAS: 7681-53-0
SODIUM HYPOPHOSPHITE
mf: $H_2O_2P•Na$ mw: 87.98

PROP: Colorless, pearly, crystalline plates or white granular powder; bittersweet, saline taste. Deliquescent; sol in water; sltly sol in alc.

SYNS: NATRIUMHYPOPHOSPHIT (GERMAN) ◇ SODIUM PHOSPHINATE

USE IN FOOD:

Purpose: Antioxidant, preservative.

Where Used: Cod liver oil emulsions.

Regulations: FDA - 21CFR 184.1764. GRAS when used in accordance with good manufacturing practice.

SAFETY PROFILE: Poison by subcutaneous route. Moderately toxic by intraperitoneal route. Flammable when exposed to heat or flame. Aqueous solutions may explode on evaporation. Potentially explosive reaction with oxidants (e.g., chlorates; nitrates). Heat causes it to evolve phosphine. It can explode. When heated to decomposition it emits toxic fumes of PO_x and Na_2O.

TOXICITY DATA and CODEN

ipr-mus LD50:1584 mg/kg COREAF 257,791,63

SIG500 CAS: 2492-26-4
SODIUM 2-MERCAPTOBENZOTHIAZOLE
mf: $C_7H_4NS_2•Na$ mw: 189.23

SYNS: 2-MERCAPTOBENZOTHIAZOLE SODIUM DERIVATIVE ◇ 2-MERCAPTOBENZOTHIAZOLE SODIUM SALT

USE IN FOOD:

Purpose: Animal glue adjuvant.

Where Used: Packaging materials.

Regulations: FDA - 21CFR 178.3120. For use as a preservative only.

SAFETY PROFILE: Moderately toxic by ingestion. When heated to decomposition it emits very toxic fumes of NO_x, SO_x and Na_2O.

TOXICITY DATA and CODEN

orl-rat LD50:3120 mg/kg FMCHA2 -,D327,80

SII000 CAS: 7681-57-4
SODIUM METABISULFITE
DOT: 2693
mf: $O_5S_2•2Na$ mw: 190.10

PROP: Colorless crystals or white to yellowish powder; odor of sulfur dioxide. Sol in water; sltly sol in alc.

SYNS: DISODIUM PYROSULFITE ◇ SODIUM METABOSULPHITE ◇ SODIUM PYROSULFITE

USE IN FOOD:

Purpose: Antioxidant, boiler water additive, preservative.

Where Used: Fruit (dried), fruit (fresh), lemon drinks, maraschino cherries, meat, shrimp, vegetables (fresh).

Regulations: FDA - 21CFR 173.310, 182.3766. GRAS when used in accordance with good manufacturing practice except that it is not used in meats or in foods recognized as sources of vitamin B_1 or on fruits and vegetables intended to be sold or served raw or to be presented to consumers as fresh.

EPA Genetic Toxicology Program.

ACGIH TLV: TWA 5 mg/m^3 DOT Classification: ORM-B; Label: None

SAFETY PROFILE: Poison by intravenous route. Moderately toxic by parenteral route. Experimental reproductive effects. Mutagenic data. When heated to decomposition it emits toxic fumes of SO_x and Na_2O.

TOXICITY DATA and CODEN

cyt-ham:ovr 180 μg/L ENMUDM 7,1,85
sce-ham:ovr 200 μg/L ENMUDM 7,1,85

orl-rat TDLo: 20 g/kg (MGN): REP FCTXAV 10,291,72
orl-rat TDLo: 40 g/kg (MGN): REP FCTXAV 10,291,72
ivn-rat LD50: 115 mg/kg DTLVS* 4,371,80

SII500 CAS: 10361-03-2
SODIUM METAPHOSPHATE

mf: $O_3P \cdot Na$ mw: 101.96

PROP: Sodium metaphosphate exists as a number of different molecular species, some of which exhibit various crystalline forms. The vitreous sodium phosphates having a Na_2O/P_2O_3 mole ratio near unity are classified as sodium metaphosphates. The term also extends to short-chain vitreous compositions, the compounds of which exhibit the polyphosphate formula $Na_{n+2}P_nO_{3n+1}$ with *n* as low as 4-5. In such as $(NaPO_3)$, *n* may be a small integer <3 (cyclic molecules) or a large number (polymers). Amorphous white solids. Very sol in water.

SYNS: GRAHAM'S SALT ◇ METAFOS ◇ SODIUM HEXAMETAPHOSPHATE ◇ SODIUM POLYPHOSPHATES, GLASSY ◇ SODIUM TETRAPOLYPHOSPHATE

USE IN FOOD:

Purpose: Cooked out juices retention agent, emulsifier, sequestrant, texturizer.

Where Used: Cheese, dairy products, fish, lima beans, meat food products, milk, peanuts, peas (canned), poultry food products.

Regulations: FDA - 21CFR 182.6760. GRAS when used in accordance with good manufacturing practice. USDA - 9CFR 318.7. In meat food products, where allowed, limitation of 5 percent of phosphate in pickle at 10 percent pump level, 0.5 percent of phosphate in product (only clear solution may be injected into product). 9CFR 381.147. Limitation of 0.5 percent of total poultry product.

Reported in EPA TSCA Inventory.

SAFETY PROFILE: Moderately toxic by intraperitoneal route. When heated to decomposition it emits toxic fumes of Na_2O and PO_x.

TOXICITY DATA and CODEN

ipr-mus LD50: 830 mg/kg REPMBN 10,391,62

SIJ500 CAS: 124-41-4
SODIUM METHOXIDE

DOT: 1431
mf: $CH_3O \cdot Na$ mw: 54.03

PROP: White, amorphous, free-flowing powder. Decomp in air above 127°; decomp by water. Sol in methyl and ethyl alc, fats, esters.

SYNS: METHANOL, SODIUM SALT ◇ SODIUM METHYLATE (DOT, FCC) ◇ SODIUM METHYLATE, DRY (DOT)

USE IN FOOD:

Purpose: Catalyst for the transesterification of fats.

Where Used: Fats.

Regulations: USDA - 9CFR 318.7. Must be eliminated during processing.

DOT Classification: Flammable Solid; Label: Flammable Solid and Dangerous When Wet

SAFETY PROFILE: A corrosive and irritating material. It hydrolyzes into methanol and sodium hydroxide. May ignite spontaneously in moist air. Flammable when exposed to heat or flame. Ignites on contact with water. When heated to decomposition it emits toxic fumes of Na_2O.

SIM400
SODIUM MONO- and DIMETHYL NAPHTHALENE SULFONATE

USE IN FOOD:

Purpose: Anticaking agent, crystallization of sodium carbonate, hog scald agent, lye peeling agent.

Where Used: Fish (cured), meat (cured), pork, potable water.

Regulations: FDA - 21CFR. USDA - 9CFR 318.7. Sufficient for purpose. Limitation of 250 ppm in sodium carbonate; 0.1 percent in sodium nitrite.

SAFETY PROFILE: When heated to decomposition it emits toxic fumes of SO_x and Na_2O.

SIM500 CAS: 7558-79-4
SODIUM MONOHYDROGEN PHOSPHATE (2:1:1)

DOT: 9147
mf: $HO_4P \cdot 2Na$ mw: 141.96

PROP: Colorless, translucent crystals or white powder. Sol in water; very sltly sol in alc.

SYNS: DIBASIC SODIUM PHOSPHATE ◇ DISODIUM HYDROGEN PHOSPHATE ◇ DISODIUM MONOHYDROGEN PHOSPHATE ◇ DISODIUM ORTHOPHOSPHATE ◇ DISODIUM PHOSPHATE ◇ DISODIUM PHOSPHORIC ACID ◇ DSP ◇ EXSICCATED SODIUM PHOSPHATE ◇ NATRIUMPHOSPHAT (GERMAN) ◇ PHOSPHORIC ACID, DISODIUM

SALT ◇ SODA PHOSPHATE ◇ SODIUM HYDROGEN PHOSPHATE ◇ SODIUM PHOSPHATE, DIBASIC (DOT, FCC)

USE IN FOOD:

Purpose: Buffer, cooked out juices retention agent, dietary supplement, emulsifier, hog scald agent, nutrient, poultry scald agent, sequestrant, stabilizer, texturizer.

Where Used: Coffee whiteners, cream sauce, hog carcasses, meat food products, milk (evaporated), poultry, pudding (instant), whipped products.

Regulations: FDA - 21CFR 181.29, 21CFR 182.1778, 182.5778, 182.6290, 182.8890. GRAS when used in accordance with good manufacturing practice. USDA - 9CFR 318.7. Sufficient for purpose. In meat food products, where allowed, limitation of 5 percent of phosphate in pickle at 10 percent pump level, 0.5 percent of phosphate in product (only clear solution may be injected into product. 9CFR 381.147. Limitation of 0.5 percent of total poultry product.

DOT Classification: ORM-E; Label: None

SAFETY PROFILE: Poison by intravenous route. Moderately toxic by intraperitoneal, subcutaneous, and intramuscular routes. Mildly toxic by ingestion. A skin and eye irritant. When heated to decomposition it emits toxic fumes of PO_x and Na_2O.

TOXICITY DATA and CODEN

skn-rbt 500 mg/24H MLD 28ZPAK -,16,72
eye-rbt 500 mg/24H MLD 28ZPAK -,16,72
orl-rat LD50: 17 g/kg 28ZPAK -,16,72

SIO000 CAS: 7631-99-4

SODIUM(I) NITRATE (1:1)

DOT: 1498
mf: $NO_3 \cdot Na$ mw: 85.00

PROP: Colorless, transparent, odorless crystals; saline, sltly bitter taste. Mp: 306.8°, bp: decomp @ 380°, d: 2.261. Deliquescent in moist air; sol in water, sltly sol in alc.

SYNS: CHILE SALTPETER ◇ CUBIC NITER ◇ NITRATE de SODIUM (FRENCH) ◇ NITRATINE ◇ NITRIC ACID, SODIUM SALT ◇ SODA NITER ◇ SODIUM NITRATE (DOT)

USE IN FOOD:

Purpose: Antimicrobial agent, boiler water additive, preservative.

Where Used: Chub (smoked), meat, poultry, sablefish (smoked), salmon (smoked), shad (smoked).

Regulations: FDA - 21CFR 171.170. Limitation of [sodium nitrite] - 500 [200] ppm in sablefish, smoked cooked; 500 [200] ppm in salmon, smoked cooked; 500 [200] ppm in shad, smoked cooked; 500 [200] ppm in meat curing preparations for home use. 21CFR 172.177. Limitation of 200 ppm in chub, smoked. 21CFR 173.310, 181.33. USDA - 9CFR 318.7, 381.147. Limitation of 7 pounds per 100 gallons of pickle; 3.5 ounces per 100 pounds of meat, 2.75 ounces to 100 pounds of chopped meat.

EPA Genetic Toxicology Program.

DOT Classification: Oxidizer; Label: Oxidizer

SAFETY PROFILE: Poison by intravenous route. Moderately toxic by ingestion. An experimental tumorigen. Human mutagenic data. A powerful oxidizer. It will ignite with heat or friction. When heated to decomposition it emits toxic fumes of NO_x and Na_2O.

TOXICITY DATA and CODEN

mmo-omi 1000 ppm POASAD 34,114,53
dns-hmn:hla 6 mmol/L AEMBAP 177,269,84
orl-rat TDLo: 100 g/kg/2Y-C:ETA FCTOD7 22,715,84
orl-rbt LD50: 2680 mg/kg SOVEA7 27,246,74

SIP500 CAS: 5064-31-3

SODIUM NITRILOTRIACETATE

mf: $C_6H_6NO_6 \cdot 3Na$ mw: 257.10

SYNS: HAMPSHIRE NTA ◇ NITRILOTRIACETIC ACID, TRISODIUM SALT ◇ NTA ◇ TRISODIUM NITRILOTRIACETATE ◇ TRISODIUM NITRILOTRIACETIC ACID

USE IN FOOD:

Purpose: Boiler water additive.

Where Used: Various.

Regulations: FDA - 21CFR 173.310. Limitation of 5 ppm in steam and excluding use of such steam in contact with milk and milk products.

SAFETY PROFILE: Poison by intraperitoneal route. Moderately toxic by ingestion. An experimental neoplastigen. Experimental reproductive effects. Mutagenic data. When heated to decomposition it emits toxic fumes of NO_x and Na_2O.

TOXICITY DATA and CODEN

or-rat:emb 495 μg/plate JJATDK 1,190,81
orl-rat TDLo: 39 g/kg (8W male/8W pre-3W post):REP FCTXAV 9,509,71

orl-rat TDLo:70300 mg/kg/2Y-C:NEO JJIND8 66,869,81
orl-rat LD50:1100 mg/kg TXAPA9 18,398,71

SIQ500 CAS: 7632-00-0
SODIUM NITRITE
DOT: 1500
mf: $NO_2 \cdot Na$ mw: 69.00

PROP: Sltly yellowish or white crystals, sticks or powder; slt salty taste. Mp: 271°, bp: decomp @ 320°, d: 2.168. Deliquescent in air; sol in water, sltly sol in alc.

SYNS: ANTI-RUST ◇ DIAZOTIZING SALTS ◇ DUSITAN SODNY (CZECH) ◇ ERINITRIT ◇ FILMERINE ◇ NATRIUM NITRIT (GERMAN) ◇ NCI-C02084 ◇ NITRITE de SODIUM (FRENCH) ◇ NITROUS ACID, SODIUM SALT

USE IN FOOD:

Purpose: Antimicrobial agent, color fixative in meat and meat products, preservative.

Where Used: Bacon, ham (canned), meat (cured), poultry.

Regulations: FDA - 21CFR 172.175. Limitation of 10 ppm in smoked cured tunafish, 200 ppm in smoked cured sablefish, 200 ppm in smoked cooked salmon, 200 ppm in smoked cooked shad, 200 ppm in meat curing preparations for home use [500 ppm sodium nitrate]. 21CFR 181.34. USDA - 9CFR 318.79, 381.147. Limitation of 2 pounds per 100 gallons of pickle at 10 percent pump level; 1 ounce per 100 pounds meat, 0.25 ounces to 100 pounds chopped meat. The use of nitrites, nitrates, or combination shall not result in more than 200 ppm of nitrite, calculated as sodium nitrate in finished product.

EPA Genetic Toxicology Program.

DOT Classification: Oxidizer; Label: Oxidizer

SAFETY PROFILE: Human poison by ingestion. Experimental poison by ingestion, subcutaneous, intravenous, and intraperitoneal routes. An experimental neoplastigen, tumorigen, and teratogen. Human systemic effects by ingestion: motor activity changes, coma, decreased blood pressure with possible pulse rate increase without fall in blood pressure, arteriolar or venous dilation, nausea or vomiting, and blood methemoglobinemia-carboxhemoglobinemia. Experimental reproductive effects. Human mutagenic data. An eye irritant. They have been implicated in an increased incidence of cancer. They may react with organic amines in the body to form carcinogenic nitrosamines.

Flammable; a strong oxidizing agent. In contact with organic matter, will ignite by friction. When heated to decomposition it emits toxic fumes of NO_x and Na_2O.

TOXICITY DATA and CODEN

eye-rbt 500 mg/24H MLD 28ZPAK -,15,72
mmo-omi 50 mmol/L JGMIAN 128,1401,82
dns-hmn:hla 6 mmol/L FCTOD7 21,551,83
orl-rat TDLo:660 mg/kg (1-22D preg):TER AJHEAA 62,1045,72
orl-rat TDLo:11 g/kg (1-22D preg/21D post): REP TOXID9 4,89,84
orl-mus TDLo:280 mg/kg (1-14D preg):TER TJADAB 18,367,78
orl-rat TDLo:40 g/kg/56W-C:NEO ZKKOBW 90,87,77
orl-rat TD :63 g/kg/95W-C:ETA JJIND8 64,1435,80
orl-hmn TDLo:14 mg/kg:CNS,CVS,GIT DMWOAX 74,961,49
orl-man TDLo:1714 μg/kg/70M:CVS JCINAO 16,73,37
orl-hmn LDLo:71 mg/kg:CNS,GIT,BLD 34ZIAG -,543,69
orl-chd LDLo:22 mg/kg LANCAO 2,162,17
orl-rat LD50:85 mg/kg AIHAAP 30,470,69

SIZ025 CAS: 408-35-5
SODIUM PALMITATE
mf: $C_{16}H_{31}NaO_2$ mw: 278.47

PROP: White to yellow powder.

SYNS: SODIUM HEXADECANOATE ◇ SODIUM PENTADECANECARBOXYLATE ◇ SODIUM SALT OF HEXADECANOIC ACID

USE IN FOOD:

Purpose: Anticaking agent, binder, constituent of paperboard, emulsifier, stabilizer.

Where Used: Packaging materials, various foods.

Regulations: FDA - 21CFR 186.1771. GRAS when used in accordance with good manufacturing practice.

SAFETY PROFILE: When heated to decomposition it emits acrid smoke and irritating fumes.

TOXICITY DATA and CODEN

dni-gpg:kdy 300 μmol/L FCTXAV 14,431,76

SIZ050
SODIUM PANTOTHENATE

USE IN FOOD:

Purpose: Dietary supplement, nutrient.

Where Used: Various.

Regulations: FDA - 21CFR 182.5772. GRAS when used in accordance with good manufacturing practice.

SAFETY PROFILE: A nuisance dust.

SJA000 CAS: 131-52-2
SODIUM PENTACHLOROPHENATE
DOT: 2567
mf: $C_6Cl_5O \cdot Na$ mw: 288.30

PROP: Tan powder.

SYNS: DOW DORMANT FUNGICIDE ◇ DOWICIDE G-ST ◇ NAPCLOR-G ◇ PENTACHLOROPHENATE SODIUM ◇ PENTACHLOROPHENOL, SODIUM SALT ◇ PENTACHLOROPHENOXY SODIUM ◇ PENTAPHENATE ◇ SANTOBRITE ◇ SODIUM PCP ◇ SODIUM PENTACHLOROPHENATE (DOT) ◇ SODIUM PENTACHLOROPHENOL ◇ SODIUM PENTACHLOROPHENOLATE ◇ SODIUM PENTACHLOROPHENOXIDE ◇ WEEDBEADS

USE IN FOOD:

Purpose: Animal glue adjuvant.

Where Used: Packaging materials.

Regulations: FDA - 21CFR 178.3120, 178.3900. For use as a preservative only.

EPA Extremely Hazardous Substances List. Chlorophenol compounds are on the Community Right-To-Know List. EPA Genetic Toxicology Program.

DOT Classification: ORM-A; Label: None: Poison B; Label: Poison

SAFETY PROFILE: Poison by ingestion, inhalation, skin contact, intravenous, intraperitoneal, subcutaneous, and intratracheal routes. Experimental reproductive effects. Mutagenic data. When heated to decomposition it emits toxic fumes of Cl^- and Na_2O.

TOXICITY DATA and CODEN

mrc-bcs 5 ng/disc/24H MUREAV 40,19,76
orl-rat TDLo: 360 mg/kg (8-19D preg): REP CHYCDW 13,8,79
orl-rat LD50: 126 mg/kg CHYCDW 13,8,79
ihl-rat LD50: 11700 μg/kg BECTA6 15,463,76

SJE000 CAS: 7775-27-1
SODIUM PERSULFATE
DOT: 1505
mf: $O_8S_2 \cdot 2Na$ mw: 238.10

PROP: White, crystalline powder. Sol in water; decomp by alc.

SYNS: PERSULFATE de SODIUM (FRENCH) ◇ SODIUM PEROXYDISULFATE

USE IN FOOD:

Purpose: Denuding agent.

Where Used: Tripe.

Regulations: USDA - 9CFR 318.7. Sufficient for purpose.

ACGIH TLV: TWA 5 mg(S_2O_8)/m^3 DOT Classification: Oxidizer; Label: Oxidizer

SAFETY PROFILE: Poison by intraperitoneal and intravenous routes. A powerful oxidizer; can cause fires. When heated to decomposition it emits toxic fumes of SO_x and Na_2O.

TOXICITY DATA and CODEN

ipr-mus LD50: 226 mg/kg COREAF 257,791,63
ivn-rbt LDLo: 178 mg/kg MEIEDD 10,1239,83

SJK000 CAS: 9003-04-7
SODIUM POLYACRYLATE

USE IN FOOD:

Purpose: Boiler water additive.

Where Used: Various.

Regulations: FDA - 21CFR 173.310. Use at a level not in excess of the amount reasonably required to accomplish the intended effect.

SAFETY PROFILE: An eye irritant. When heated to decomposition it emits toxic fumes of Na_2O.

TOXICITY DATA and CODEN

eye-rbt 2 mg MOD PSTGAW 20,16,53

SJK385 CAS: 304-59-6
SODIUM POTASSIUM TARTRATE
mf: $C_4H_4KNaO_6 \cdot 4H_2O$ mw: 282.22

PROP: Colorless crystals or white crystalline powder; cooling, salty taste. Sol in water; insol in alc.

SYN: ROCHELLE SALT

USE IN FOOD:

Purpose: Buffer, emulsifier, pH control agent, sequestrant.

Where Used: Cheese, jams, jellies, margarine, oleomargarine.

Regulations: FDA - 21CFR 184.1804. GRAS when used in accordance with good manufacturing practice. USDA - 9CFR 318.7. Sufficient for purpose.

SAFETY PROFILE: When heated to decomposition it emits acrid smoke and irritating fumes.

SJL500 CAS: 137-40-6
SODIUM PROPIONATE
mf: $C_3H_5O_2 \cdot Na$ mw: 96.07

PROP: Transparent crystals or granules; nearly odorless. Very sol in water; sltly sol in alc.

SYNS: NATRIUMPROPIONAT (GERMAN) ◇ PROPANOIC ACID, SODIUM SALT

USE IN FOOD:

Purpose: Antimicrobial agent, flavoring agent, mold and rope inhibitor, preservative.

Where Used: Baked goods, beverages (nonalcoholic), candy (soft), cheese, confections, dough (fresh pie), fillings, frostings, gelatins, jams, jellies, meat products, pizza crust, puddings.

Regulations: FDA - 21CFR 181.23, 184.1784. Use at a level not in excess of the amount reasonably required to accomplish the intended effect. USDA - 9CFR 318.7. Limitation of 0.32 percent alone or in combination with calcium propionate based on the weight of the flour brace used. 9CFR 381.147. Limitation of 0.3 percent alone or in combination with calcium propionate based on the weight of the flour used.

SAFETY PROFILE: Moderately toxic by skin contact and subcutaneous routes. An allergen. When heated to decomposition it emits toxic fumes of Na_2O.

TOXICITY DATA and CODEN

scu-mus LD50:2100 mg/kg ZGEMAZ 113,536,44
skn-rbt LD50:1640 mg/kg JIHTAB 31,60,49

SJT750 CAS: 533-96-0
SODIUM SESQUICARBONATE
mf: $Na_2CO_3NaHCO_3 \cdot 2H_2O$ mw: 226.03

PROP: White crystals, flakes, or crystalline powder. Sol in water.

USE IN FOOD:

Purpose: Alkali, neutralizer in dairy products, pH control agent, poultry scald agent.

Where Used: Cream, poultry.

Regulations: FDA - 21CFR 184.1792. GRAS when used in accordance with good manufacturing practice. USDA - 9CFR 381.147. Sufficient for purpose.

SAFETY PROFILE: A nuisance dust.

SJU000 CAS: 6834-92-0
SODIUM SILICATE
mf: $O_3Si \cdot 2Na$ mw: 122.07

SYNS: B-W ◇ CRYSTAMET ◇ DISODIUM METASILICATE ◇ DISODIUM MONOSILICATE ◇ METSO 20 ◇ METSO BEADS 2048 ◇ METSO BEADS, DRYMET ◇ METSO PENTABEAD 20 ◇ ORTHOSIL ◇ SODIUM METASILICATE ◇ SODIUM METASILICATE, ANHYDROUS ◇ WATER GLASS

USE IN FOOD:

Purpose: Boiler water additive, denuding agent, hog scald agent, preservative.

Where Used: Eggs, hog carcasses, tripe.

Regulations: FDA - 21CFR 173.310. Use at a level not in excess of the amount reasonably required to accomplish the intended effect. USDA - 9CFR 318.7. Sufficient for purpose.

SAFETY PROFILE: Poison by ingestion and intraperitoneal routes. A caustic material which is a severe eye, skin and mucous membrane irritant. Experimental reproductive effects. Ingestion causes gastrointestinal tract upset. When heated to decomposition it emits toxic fumes of Na_2O.

TOXICITY DATA and CODEN

skn-hmn 250 mg/24H SEV TXAPA9 31,481,75
skn-rbt 250 mg/24H SEV TXAPA9 31,481,75
orl-rat TDLo:15 g/kg (14W male/14W pre-3W post):REP JANSAG 36,271,73
scu-rat TDLo:9766 μg/kg (1D male):REP JRPFA4 7,21,64
orl-rat LD50:1280 mg/kg 14CYAT 3(2B),3065,82

SJV000 CAS: 7757-81-5
SODIUM SORBATE
mf: $C_6H_7O_2 \cdot Na$ mw: 134.12

SYN: SORBIC ACID, SODIUM SALT

USE IN FOOD:

Purpose: Preservative.

Where Used: Baked goods, cheese, margarine, oleomargarine.

Regulations: FDA - 21CFR 182.3795. GRAS when used in accordance with good manufacturing practice. USDA - 9CFR 318.7. Limitation of 0.1 percent individually, of if used in combination with its salts or benzoic acid or its salts, 0.2 percent (expressed as the acids in the weight of the finished product. Not allowed in cooked sausage.

SAFETY PROFILE: Moderately toxic by intraperitoneal route. Mildly toxic by ingestion. Mutagenic data. When heated to decomposition it emits toxic fumes of Na_2O.

TOXICITY DATA and CODEN

cyt-ham:lng 400 mg/L FCTOD7 22,501,84
msc-ham:lng 1 g/L CNREA8 44,3270,84
orl-rat LD50:7160 mg/kg JIHTAB 30,63,48

SJV500 CAS: 822-16-2
SODIUM STEARATE

mf: $C_{18}H_{36}O_2 \cdot Na$ mw: 306.52

SYNS: OCTADECANOIC ACID, SODIUM SALT ◇ SODIUM OCTADECANOATE ◇ STEARIC ACID, SODIUM SALT

USE IN FOOD:

Purpose: Anticaking agent, binder, emulsifier, masticatory substance in chewing gum base, stabilizer.

Where Used: Chewing gum, packaging materials, various foods.

Regulations: FDA - 21CFR 172.615, 172863. Must conform to FDA specifications for salts of fats or fatty acids derived from edible oils. 21CFR 181.29.

SAFETY PROFILE: Poison by intravenous and possibly other routes. When heated to decomposition it emits toxic fumes of Na_2O.

TOXICITY DATA and CODEN

ivn-dog LDLo:10 mg/kg FCTXAV 17,357,79

SJV700 CAS: 25383-99-7
SODIUM STEAROYL LACTYLATE

PROP: Cream-colored powder; caramel-like odor. Sol in hot oil or fat, dispersible in warm water.

USE IN FOOD:

Purpose: Dough conditioner, dough strengthener, emulsifier, stabilizer, whipping agent.

Where Used: Baked goods, cheese imitations, cheese substitutes, fillings, gravies, icings, milk or cream substitutes for beverage coffee, pancakes, potatoes (dehydrated), puddings, sauces, snack dips, toppings, waffles.

Regulations: FDA - 21CFR 172.846. Limitation of 0.5 part for each 100 part of flour in baked products; 0.5 part for each 100 part of flour in pancakes; 0.5 part for each 100 part of flour in waffles; 0.2 percent in icings; 0.2 percent in fillings; 0.2 percent in puddings; 0.2 percent in toppings; 0.3 percent in milk or cream substitutes for beverage coffee; 0.5 percent in dehydrated potatoes; 0.2 percent in snack dips; 0.2 percent in cheese substitutes; 0.2 percent in cheese imitations; 0.25 percent in sauces; 0.25 percent in gravies. Must conform to FDA specifications for fats or fatty acids derived from edible oils.

SAFETY PROFILE: When heated to decomposition it emits acrid smoke and irritating fumes.

SJV710
SODIUM STEARYL FUMARATE

mf: $C_{22}H_{39}NaO_4$ mw: 390.54

PROP: Fine white powder. Sol in methanol; insol in water.

USE IN FOOD:

Purpose: Dough conditioner.

Where Used: Bakery products (non-yeast leavened), bakery products (yeast-leavened), cereals (processed for cooking), flour (starch-thickened), potatoes (dehydrated).

Regulations: FDA - 21CFR 172.826. Limitation of 0.5 percent of flour for yeast-leavened bakery products; 1 percent of flour for non-yeast leavened bakery products; 1 percent of dehydrated potatoes; 1 percent of dry processed cereals for cooking; 0.2 percent for starch-thickened flour.

SAFETY PROFILE: When heated to decomposition emits toxic fumes of Na_2O.

SJW200
SODIUM SULFACHLOROPYRAZINE MONOHYDRATE

USE IN FOOD:

Purpose: Animal drug.

Where Used: Chickens.

Regulations: FDA - 21CFR 556.625. Limitation of zero in uncooked edible tissues of chickens.

SAFETY PROFILE: When heated to decomposition it emits acrid smoke and irritating fumes.

SJW475 CAS: 127-58-2
SODIUM SULFAMERAZINE
mf: $C_{11}H_{12}N_4O_2S \cdot Na$ mw: 287.32

PROP: Crystals; bitter, caustic taste. Hygroscopic. On prolonged exposure to humid air, it absorbs CO_2 with the liberation of sulfamerazine and becomes incompletely sol in water. Its solns are alkaline to phenolphthalein (pH 10 or more). One gram dissolves in 3.6 mL water. Sltly sol in alc; insol in ether, chloroform.

SYNS: 4-AMINO-N-(4-METHYL-2-PYRIMIDINYL)-BENZENESULFONAMIDE MONOSODIUM SALT ◇ N^1-(4-METHYL-2-PYRIMIDINYL)SULFANILAMIDE SODIUM SALT ◇ SODIUM SULPHAMERAZINE ◇ SOLUBLE SULFAMERAZINE ◇ SOLUMEDINE ◇ SULFAMERAZINE SODIUM

USE IN FOOD:

Purpose: Animal drug, animal feed drug.

Where Used: Animal feed, beef, chicken, pork, trout, turkey.

Regulations: FDA - 21CFR 556.660. Limitation of zero in trout. 21CFR 558.582.

SAFETY PROFILE: Moderately toxic by ingestion, subcutaneous, intraperitoneal, and intravenous routes. When heated to decomposition it emits toxic fumes of SO_x, NO_x and Na_2O.

TOXICITY DATA and CODEN

orl-mus LD50:2800 mg/kg AIPTAK 94,338,53

SJY000 CAS: 7757-82-6
SODIUM SULFATE (2:1)
mf: $O_4S \cdot 2Na$ mw: 142.04

PROP: White crystals or powder; odorless. Mp: 888°, d: 2.671. Sol in water, glycerin; insol in alc.

SYNS: DISODIUM SULFATE ◇ NATRIUMSUFAT (GERMAN) ◇ SALT CAKE ◇ SODIUM SULFATE ANHYDROUS ◇ SODIUM SULPHATE ◇ SULFURIC ACID, DISODIUM SALT ◇ THENARDITE ◇ TRONA

USE IN FOOD:

Purpose: Boiler water additive, caramel production agent, hog scald agent, paper manufacturing aid, poultry scald agent.

Where Used: Beverages (alcoholic), caramel, chewing gum, food packaging, hog carcasses, poultry.

Regulations: FDA - 21CFR 172.615, 173.310. Use in amounts not to exceed those required to produce the intended physical or other technical effect. 21CFR 186.1797. GRAS as an indirect additive when used in accordance with good manufacturing practice. USDA - 9CFR 318.7, 381.147. Sufficient for purpose.

EPA Genetic Toxicology Program.

SAFETY PROFILE: Moderately toxic by intravenous route. Mildly toxic by ingestion. An experimental teratogen. Experimental reproductive effects. When heated to decomposition it emits toxic fumes of SO_x and Na_2O.

TOXICITY DATA and CODEN

par-mus TDLo:60 mg/kg (8D preg):TER JPMSAE 62,1626,73
orl-mus LD50:5989 mg/kg SKEZAP 4,15,63

SJZ000 CAS: 7757-83-7
SODIUM SULFITE (2:1)
mf: $O_3S \cdot 2Na$ mw: 126.04

PROP: Hexagonal prisms or white powder; odorless with salty, sulfurous taste. Bp: decomp, d: 2.633 @ 15.4°. Sol in water; sltly sol in alc.

SYNS: ANHYDROUS SODIUM SULFITE ◇ DISODIUM SULFITE ◇ EXSICATED SODIUM SULFITE ◇ NATRIUMSULFID (GERMAN) ◇ SODIUM SULFITE ANHYDROUS ◇ SODIUM SULPHITE ◇ SULFTECH ◇ SULFUROUS ACID, SODIUM SALT (1:2)

USE IN FOOD:

Purpose: Antioxidant, boiler water additive, preservative.

Where Used: Chewing gum, fruits (fresh), meat, vegetables (fresh).

Regulations: FDA - 21CFR 172.615, 173.310. 21CFR 182.3798. GRAS when used in accordance with good manufacturing practice except that it is not used in meats or in foods recognized as sources of vitamin B_1 or on fruits and vegetables intended to be sold or served raw or to be presented to consumers as fresh.

EPA Genetic Toxicology Program.

SAFETY PROFILE: Poison by intravenous and subcutaneous routes. Moderately toxic by ingestion and intraperitoneal routes. Human mutagenic data. When heated to decomposition it emits very toxic fumes of Na_2O and SO_x. A reducing agent.

TOXICITY DATA and CODEN

mmo-omi 600 ppm POASAD 34,114,53
cyt-dom:oth 250 mg/L ENVRAL 9,84,75
ivn-rat LD50:115 mg/kg JPETAB 101,101,51
orl-rbt LDLo:2825 mg/kg AHYGAJ 57,87,06

SJZ050
SODIUM SULFOACETATE derivatives of MONO and DIGLYCERIDES

SYN: MONO and DIGLYCERIDES, SODIUM SULFOACETATE DERIVATIVES

USE IN FOOD:

Purpose: Flavoring agent.

Where Used: Meat, poultry.

Regulations: USDA - 9CFR 318.7, 381.147. Limitation of 0.5 percent.

SAFETY PROFILE: When heated to decomposition it emits acrid smoke and irritating fumes.

SKI000 CAS: 7772-98-7
SODIUM THIOSULFATE

mf: $O_3S_2 \cdot 2Na$ mw: 158.10

PROP: Colorless crystals or crystalline powder. Sol in water; insol in alc.

SYNS: HYPO ◇ SODIUM HYPOSULFITE ◇ SODIUM THIOSULFATE ANHYDROUS

USE IN FOOD:

Purpose: Antioxidant, formulation aid, reducing agent, sequestrant.

Where Used: Alcoholic beverages, table salt.

Regulations: FDA - 21CFR 184.1807. GRAS with a limitation of 0.00005 percent in alcoholic beverages; 0.1 percent in table salt when used in accordance with good manufacturing practice.

SAFETY PROFILE: Moderately toxic by subcutaneous route. Incompatible with metal nitrates; sodium nitrite. When heated to decomposition it emits very toxic fumes of Na_2O and SO_x.

TOXICITY DATA and CODEN

scu-rbt LDLo:4000 mg/kg AIPTAK 5,161,1899

SKM500 CAS: 7785-84-4
SODIUM TRIMETAPHOSPHATE

mf: $O_9P_3 \cdot 3Na$ mw: 305.88

PROP: White crystals or white crystalline powder. Sol in water.

SYN: TRIMETAPHOSPHATE SODIUM

USE IN FOOD:

Purpose: Starch-modifying agent.

Where Used: Various.

Regulations: GRAS when used in accordance with good manufacturing practice.

SAFETY PROFILE: Poison by intravenous route. Moderately toxic by intraperitoneal route. When heated to decomposition it emits toxic fumes of PO_x and Na_2O.

TOXICITY DATA and CODEN

ipr-rat LD50:3650 mg/kg JPETAB 108,117,53
ivn-rbt LDLo:240 mg/kg AEPPAE 169,238,33

SKN000 CAS: 13573-18-7
SODIUM TRIPOLYPHOSPHATE

mf: $O_{10}P_3 \cdot 5Na$ mw: 367.86

PROP: White granules or powder. Sltly hygroscopic; sol in water.

SYNS: ARMOFOS ◇ NATRIUMTRIPOLYPHOSPHAT (GERMAN) ◇ PENTASODIUM TRIPHOSPHATE ◇ POLY ◇ POLYGON ◇ SODIUM TRIPHOSPHATE ◇ STPP ◇ TRIPHOSPHORIC ACID, SODIUM SALT ◇ TRIPOLY ◇ TRIPOLYPHOSPHATE

USE IN FOOD:

Purpose: Boiler water additive, cooked out juices retention agent, hog scald agent, poultry scald agent, texturizer.

Where Used: Angel food cake mix, beef (cooked), beef (fresh), beef for further cooking, beef patties, desserts, gelling juices, goat, ham (canned), hog carcasses, lamb, lima beans, meat loaf, meat products, meat toppings, meringues, mutton, peas (canned), pork, pork (cured), poultry, poultry food products, sausage products, veal.

Regulations: FDA - 21CFR 173.310, 182.6810. GRAS when used in accordance with good manufacturing practice. USDA - 9CFR 318.7. Sufficient for purpose in hog scald water. In meat food products, where allowed, limitation of 5 percent of phosphate in pickle at 10 percent pump level, 0.5 percent of phosphate in product (only clear solution may be injected into product. 9CFR 381.147. Limitation of of 0.5 percent in total poultry product.

SAFETY PROFILE: Poison by intravenous route. Moderately toxic by ingestion, subcutaneous and intraperitoneal routes. Ingestion of large doses of sodium phosphates causes catharsis. Sodium meta and pyrophosphates can cause hemorrhages from the intestine if taken internally in large doses. When heated to decomposition it emits toxic fumes of PO_x and Na_2O.

TOXICITY DATA and CODEN

orl-rat LD50:6500 mg/kg AIHAAP 30,470,69

SKU000 CAS: 110-44-1
SORBIC ACID
mf: $C_6H_8O_2$ mw: 112.14

PROP: Colorless needles or white powder; characteristic odor. Bp: 228° (decomp), mp: 134.5°, flash p: 260°F (COC), vap press: 0.01 mm @ 20°, vap d: 3.87. Sol in hot water; very sol in alc, ether.

SYNS: (2-BUTENYLIDENE)ACETIC ACID ◇ CROTYLIDENE ACETIC ACID ◇ HEXADIENIC ACID ◇ HEXADIENOIC ACID ◇ 2,4-HEXADIENOIC ACID ◇ trans-trans-2,4-HEXADIENOIC ACID ◇ 1,3-PENTADIENE-1-CARBOXYLIC ACID ◇ 2-PROPENYLACRYLIC ACID ◇ SORBISTAT

USE IN FOOD:

Purpose: Preservative.

Where Used: Baked goods, beverages (carbonated), beverages (still), bread, cake batters, cake fillings, cake topping, cheese, cottage cheese (creamed), fish (smoked or salted), fruit juices (fresh), fruits (dried), margarine, oleomargarine, pickled goods, pie crusts, pie fillings, salad dressings, salads (fresh), sausage (dry), sea food cocktail, syrups (chocolate dairy), wine.

Regulations: FDA - 21CFR 181.23, 182.3089. GRAS when used in accordance with good manufacturing practice. USDA - 9CFR 318.7. Limitation of 0.1 percent individually, of if used in combination with its salts or benzoic acid or its salts, 0.2 percent (expressed as the acids in the weight of the finished product. Not allowed in cooked sausage. BATF - 27CFR 240.1051. Limitation of 300 mg/1000 gallons of wine.

SAFETY PROFILE: Moderately toxic by intraperitoneal and subcutaneous routes. Mildly toxic by ingestion. An experimental tumorigen. Experimental reproductive effects. Mutagenic data. A severe human and experimental skin irritant. Combustible when exposed to heat or flame; can react with oxidizing materials. To fight fire, use water. When heated to decomposition it emits acrid smoke and irritating fumes.

TOXICITY DATA and CODEN

skn-man 150 mg/1H SEV JPPMAB 10,719,58
cyt-ham:lng 1050 mg/L FCTOD7 22,501,84
sce-ham:lng 1050 mg/L FCTOD7 22,501,84
orl-rat TDLo:4154 g/kg (2Y pre):REP FCTXAV 13,31,75
scu-rat TDLo:1040 mg/kg/65W-I:ETA BJCAAI 20,134,66
orl-rat LD50:7360 mg/kg JIHTAB 30,63,48

SKU700 CAS: 1338-41-6
SORBITAN C
mf: $C_{24}H_{46}O_6$ mw: 430.70

PROP: Cream to tan-colored waxy solid; bland odor and taste. Insol in cold water, mineral spirits, acetone; dispersible in warm water; sol above 50° in mineral oil, ethyl acetate.

SYNS: ANHYDRO-d-GLUCITOL MONOOCTADECANOATE ◇ ANHYDROSORBITOL STEARATE ◇ ARLACEL 60 ◇ ARMOTAN MS ◇ CRILL 3 ◇ CRILL K 3 ◇ DREWSORB 60 ◇ DURTAN 60 ◇ EMSORB 2505 ◇ GLYCOMUL S ◇ HODAG SMS ◇ IONET S 60 ◇ LIPOSORB S ◇ LIPOSORB S-20 ◇ MONTANE 60 ◇ MS 33 ◇ MS 33F ◇ NEWCOL 60 ◇ NIKKOL SS 30 ◇ NISSAN NONION SP 60 ◇ NONION SP 60 ◇ NONION SP 60R ◇ RIKEMAL S 250 ◇ SORBITAN MONOOCTADECANOATE ◇ SORBITAN MONOSTEARATE (FCC) ◇ SORBITAN STEARATE ◇ SORBON S 60 ◇ SORGEN 50 ◇ SPAN 55 ◇ SPAN 60

USE IN FOOD:

Purpose: Defoaming agent, emulsifier, rehydration aid, stabilizer.

Where Used: Cake fillings, cake icings, cake mixes, cakes, chocolate coatings, coffee whiteners, confectionery coatings, cream fillings, desserts (nonstandardized frozen), edible oil (whipped topping), fruit (raw), milk or cream substitutes for beverage coffee, vegetables (raw), yeast (active dry).

Regulations: FDA - 21CFR 172.515, 172.842. Limitation of 0.4 percent in whipped edible oil topping; 1 percent in nonstandardized confectionery coatings; 0.61 percent in cakes; 0.61 percent in cake mixes; 0.4 percent in milk or cream substitutes for beverage coffee; 0.7 percent in cake icings; 0.7 percent in cake fillings (additional limitations when used in combination with polysorbate 60, polysorbate 65, and/or sorbitan monostearate), 1 percent in active dry yeast. 21CFR 172.878, 172.886. Use at a level not in excess of the amount reasonably required to accomplish the intended effect. Must conform to FDA specifications for fats or fatty acids derived from edible oils. 21CFR 173.340.

EPA Genetic Toxicology Program.

SAFETY PROFILE: Very mildly toxic by ingestion. Experimental reproductive effects. When heated to decomposition it emits acrid smoke and irritating fumes.

TOXICITY DATA and CODEN

orl-rat TDLo:635 g/kg (MGN):REP JONUAI 60,489,56
orl-rat TDLo:1270 g/kg (84D pre-21D post): REP JONUAI 60,489,56
orl-rat LD50:31 g/kg FOREAE 21,348,56

SKV100 CAS: 1338-43-8
SORBITAN MONOOLEATE

USE IN FOOD:

Purpose: Emulsifier.

Where Used: Sugar liquor or juice.

Regulations: FDA - 21CFR 173.75. Limitation of 0.70 ppm in cane or beet sugar juice and 1.4 ppm in cane or beet sugar liquor.

SAFETY PROFILE: When heated to decomposition it emits acrid smoke and irritating fumes.

SKV400 CAS: 87-79-6
SORBOSE

mf: $C_6H_{12}O_6$ mw: 180.16

PROP: Orthorhombic, bisphenoidal crystalline solid.

USE IN FOOD:

Purpose: Constituent of cotton and cotton fabrics.

Where Used: Packaging materials.

Regulations: FDA - 21CFR 186.1839. GRAS when used in accordance with good manufacturing practice.

SAFETY PROFILE: When heated to decomposition it emits acrid smoke and irritating fumes.

SKW825
SOYBEAN OIL (UNHYDROGENATED)

PROP: From the seed of the legume *Glycine max*. Amber-colored oil.

USE IN FOOD:

Purpose: Coating agent, emulsifying agent, formulation aid, texturizer.

Where Used: Cooking oil, margarine, salad oil, shortening.

Regulations: GRAS when used in accordance with good manufacturing practice.

SAFETY PROFILE: When heated to decomposition it emits acrid smoke and irritating fumes.

SKW840
SOY PROTEIN, ISOLATED

USE IN FOOD:

Purpose: Binder, emulsifier, extender, stabilizer.

Where Used: Meatballs, poultry, sausage, snack foods, spaghetti (frozen), whipped toppings.

Regulations: GRAS when used in accordance with good manufacturing practice. USDA - 9CFR 318.7. Limitation of 2 percent in sausage as provided for in 9CFR 319. 9CFR 381.147. Sufficient for purpose.

SAFETY PROFILE: When heated to decomposition it emits acrid smoke and irritating fumes.

SKY000 CAS: 8008-79-5
SPEARMINT OIL

PROP: From steam distillation of the plant *Mentha spicata* L. (Common Spearmint), or of *Mentha cardiaca* Gerard ex Baker (Scotch Spearmint) (Fam. *Labiatae*). Contains principally carvone, phellandrene, limonene, and either dihydrocarveol acetate or dihydrocuminic acetate (FCTXAV 16, 637,78). Colorless or greenish-yellow liquid; odor and taste of spearmint.

SYN: OIL of SPEARMINT

USE IN FOOD:

Purpose: Flavoring agent.

Where Used: Various.

Regulations: FDA - 21CFR 182.10, 182.20. GRAS when used at a level not in excess of the amount reasonably required to accomplish the intended effect.

SAFETY PROFILE: Mildly toxic by ingestion. Mutagenic data. A skin irritant and an allergen. When heated to decomposition it emits acrid smoke and irritating fumes. Used as a flavoring agent.

TOXICITY DATA and CODEN

skn-rbt 500 mg/24H MOD FCTXAV 16,637,78
dnr-bcs 10 mg/disc TOFOD5 8,91,85
orl-rat LD50:5 g/kg FCTXAV 16,637,78

SLB500 CAS: 84837-04-7
SPIKE LAVENDER OIL

PROP: From steam distillation of the plant *Lavandula latifolia* Vill. (*Lavandula spica*, D.C.)(Fam. *Labiatae*). The main constituents are linalool and cineole. (FCTXAV 14,443,76). Yellow liquid; lavender odor. D: 0.893-0.909, refr index: 1.463 @ 20°. Sol in fixed oils, propylene glycol; sltly sol in glycerin, mineral oil.

SYNS: LAVENDER OIL, SPIKE ◇ OIL of SPIKE LAVENDER

USE IN FOOD:

Purpose: Flavoring agent.

Where Used: Various.

Regulations: FDA - 21CFR 182.20. GRAS when used at a level not in excess of the amount reasonably required to accomplish the intended effect.

SAFETY PROFILE: Moderately toxic by ingestion. A skin irritant. When heated to decomposition it emits acrid smoke and irritating fumes.

TOXICITY DATA and CODEN

skn-rbt 500 mg/24H MOD FCTXAV 14,443,76
orl-rat LD50:3800 mg/kg FCTXAV 14(5),443,76

SLI325 CAS: 21736-83-4
STANILO
mf: $C_{14}H_{24}N_2O_7 \cdot 2ClH$ mw: 405.32

SYNS: DECAHYDRO-4a,7,9-TRIHYDROXY-2-METHYL-6,8-BIS(METHYLAMINO)-4H-PYRANO(2,3-b)(1,4)BENZODIOXIN-4-ONE DIHYDROCHLORIDE, (2R-(2-α,4a-β,5a-β,6-β,7-β,8-β,9-α,9a-α,10a-β))- ◇ SPECTINOMYCIN DIHYDROCHLORIDE ◇ SPECTINOMYCIN HYDROCHLORIDE

USE IN FOOD:

Purpose: Animal drug.

Where Used: Chicken.

Regulations: FDA - 21CFR 556.600. Limitation of 0.1 ppm in chickens.

SAFETY PROFILE: Moderately toxic by intraperitoneal route. When heated to decomposition it emits toxic fumes of NO_x and HCl.

TOXICITY DATA and CODEN

ipr-rat LD50:2020 mg/kg NIIRDN 6,382,82
ipr-mus LD50:2350 mg/kg NIIRDN 6,382,82

SLI350
STANNOUS STEARATE

SYN: TIN STEARATE

USE IN FOOD:

Purpose: Stabilizer.

Where Used: Packaging materials.

Regulations: FDA - 21CFR 181.29. Limitation of 50 ppm of tin as a migrant in finished food.

SAFETY PROFILE: When heated to decomposition it emits acrid smoke and irritating fumes.

SLJ000 CAS: 7081-44-9
STAPHYBIOTIC
mf: $C_{19}H_{17}ClN_3O_5S \cdot Na \cdot H_2O$ mw: 475.91

SYNS: BACTOPEN ◇ BRL-1621 ◇ 6-(3-(o-CHLOROPHENYL)-5-METHYL-4-ISOXAZOLECARBOXAMIDEO)-3,3-DIMETHYL-7-OXO-4-THIA-1-AZABICYCLO(3.2.0)HEPTANE-2-CARBOXYLIC ACID, SODIUM SALT, MONOHYDRATE ◇ CLOXACILLIN SODIUM MONOHYDRATE ◇ CLOXAPEN ◇ CLOXYPEN ◇ EKVACILLIN ◇ GELSTAPH ◇ METHOCILLIN-S ◇ ORBENIN SODIUM HYDRATE ◇ P-25 ◇ PROSTAPHLIN-A ◇ SODIUM CLOXACILLIN MONOHYDRATE ◇ STAPHOBRISTOL-250 ◇ TEGOPEN ◇ TEPOGEN

USE IN FOOD:

Purpose: Animal drug.

Where Used: Beef, milk.

Regulations: FDA - 21CFR 556.165. Tolerance of 0.01 ppm in uncooked edible tissues of cattle and in milk.

SAFETY PROFILE: Moderately toxic by intraperitoneal, intramuscular, subcutaneous and intravenous routes. Mildly toxic by ingestion. When heated to decomposition it emits very toxic fumes of Cl^-, NO_x, Na_2O and SO_x.

TOXICITY DATA and CODEN

orl-rat LD50:5 g/kg NIIRDN 6,232,82

SLJ700
STARTER DISTILLATE

PROP: Steam distillate of culture of *Streptococcus lactis S. cremoris, S. lactis subsp. diacetylactis, Leuconostoc citrovorum,* and *L. dextranicum.*

SYN: BUTTER STARTER DISTILLATE

USE IN FOOD:

Purpose: Adjuvant, flavoring agent.

Where Used: Oleomargarine.

Regulations: FDA - 21CFR 184.1848. GRAS when used in accordance with good manufacturing practice. USDA - 9CFR 318.7. Sufficient for purpose.

SAFETY PROFILE: When heated to decomposition it emits acrid smoke and irritating fumes.

SLK000 CAS: 57-11-4
STEARIC ACID
mf: $C_{18}H_{36}O_2$ mw: 284.54

PROP: White, amorphous solid; slt odor and taste of tallow. Mp: 69.3°, bp: 383°, flash p: 385°F (CC), d: 0.847, autoign temp: 743°F, vap press: 1 mm @ 173.7°, vap d: 9.80. Sol in alc, ether, acetone, chloroform; insol in water.

SYNS: CENTURY 1240 ◇ DAR-CHEM 14 ◇ EMERSOL 120 ◇ GLYCON DP ◇ GLYCON S-70 ◇ GLYCON TP ◇ GROCO 54 ◇ 1-HEPTADECANECARBOXYLIC ACID ◇ HYDROFOL ACID 1655 ◇ HY-PHI 1199 ◇ HYSTRENE 80 ◇ INDUSTRENE 5016 ◇ KAM 1000 ◇ KAM 2000 ◇ KAM 3000 ◇ NEO-FAT 18-61 ◇ NEO-FAT 18-S ◇ OCTADECANOIC ACID ◇ PEARL STEARIC ◇ STEAREX BEADS ◇ STEAROPHANIC ACID ◇ TEGOSTEARIC 254

USE IN FOOD:

Purpose: Component in the manufacture of other food-grade additives, defoaming agent, flavoring agent, lubricant.

Where Used: Chewing gum.

Regulations: FDA - 21CFR 172.615. Must conform to 21CFR 172.860. specifications for fats or fatty acids derived from edible oils. 21CFR 184.1090. GRAS when used in accordance with good manufacturing practice.

EPA Genetic Toxicology Program.

SAFETY PROFILE: Poison by intravenous route. An experimental tumorigen. A human and experimental skin irritant. Combustible when exposed to heat or flame. Heats spontaneously. To fight fire, use CO_2, dry chemical. When heated to decomposition it emits acrid smoke and irritating fumes.

TOXICITY DATA and CODEN

skn-hmn 75 mg/3D-I MLD 85DKA8 -,127,77
skn-rbt 500 mg/24H MOD FCTXAV 17,357,79
imp-mus TDLo:400 mg/kg:ETA BJCAAI 17,127,63
ivn-rat LD50:21500 μg/kg FCTXAV 17,357,79

SLN100
STEARYL CITRATE

USE IN FOOD:

Purpose: Flavor preservative, plasticizer, sequestrant.

Where Used: Oleomargarine, packaging materials.

Regulations: FDA - 21CFR 181.27. Use in accordance with good manufacturing practice. 21CFR 182.6851. Limitation of 0.15 percent as a sequestrant.

SAFETY PROFILE: When heated to decomposition it emits acrid smoke and irritating fumes.

SLN175
STEARYL-2-LACTYLIC ACID

USE IN FOOD:

Purpose: Emulsifier.

Where Used: Shortening.

Regulations: USDA - 9CFR 318.7. Limitation of 3 percent in shortening to be used for cake icings and fillings.

SAFETY PROFILE: When heated to decomposition it emits acrid smoke and irritating fumes.

SLN200
STEARYL MONOGLYCERIDYL CITRATE

PROP: Soft tan to white, waxy solid; tasteless. Insol in water; sol in chloroform, ethylene glycol.

USE IN FOOD:

Purpose: Emulsion stabilizer.

Where Used: Margarine, oleomargarine, shortening.

Regulations: FDA - 21CFR 172.755. 21CFR 172.860. Must conform to specifications for fats or fatty acids derived from edible oils. USDA - 9CFR 318.7. Sufficient for purpose, 0.5 percent in oleomargarine or margarine.

SAFETY PROFILE: When heated to decomposition it emits acrid smoke and irritating fumes.

SLW500 CAS: 57-92-1
STREPTOMYCIN
mf: $C_{21}H_{39}N_7O_{12}$ mw: 581.67

PROP: An antibiotic. It is a base and readily forms salts with anions.

SYNS: AGRIMYCIN 17 ◇ CHEMFORM ◇ GEROX ◇ HOKKO-MYCIN ◇ NSC 14083 ◇ STREPCEN ◇ STREPTOMICINA (ITALIAN) ◇ STREPTOMYCIN A ◇ STREPTOMYCINE ◇ STREPTOMYCINUM ◇ STREPTOMYZIN (GERMAN)

USE IN FOOD:

Purpose: Animal drug.

Where Used: Chicken, eggs, pork, turkeys.

Regulations: FDA - 21CFR 556.610. Limitation of zero in chicken, turkeys, swine, and eggs. USDA CES Ranking: A-3 (1986).

EPA Genetic Toxicology Program.

SAFETY PROFILE: Poison by intravenous and subcutaneous routes. Moderately toxic by ingestion, and intraperitoneal routes. An experimental teratogen. Human systemic effects by ingestion and intraperitoneal routes: change in vestibular functions, blood pressure decrease, eosinophiis, respiratory depression and other pulmonary changes. Human reproductive and teratogenic effects by unspecified routes: developmental abnormalities of the eye and ear and effects on newborn including postnatal measures or effects. Toxic to kidneys and central nervous system. Has been implicated in aplastic anemia. Experimental reproductive effects. Human mutagenic data. When heated to decomposition it emits toxic fumes of NO_x.

TOXICITY DATA and CODEN

mmo-sat 10 g/L SCIEAS 172,1058,71
mrc-esc 200 μg/L IDZAAW 52,417,77
unr-wmn TDLo:620 mg/kg (23-39W preg): TER SJRDAH 50,61,69
scu-rat TDLo:22500 μg/kg (6-14D preg):REP DPHFAK 23,383,71
scu-rat TDLo:225 mg/kg (6-14D preg):TER DPHFAK 23,383,71
orl-hmn TDLo:400 mg/kg/28D-I:EAR RMSRA6 73,820,53
ipr-hmn TDLo:143 mg/kg:CVS,PUL BMJOAE 1,556,61
orl-rat LD50:9 g/kg 85ARAE 4,35,76/77

SMR000 CAS: 9003-55-8
STYRENE POLYMER with 1,3-BUTADIENE

SYNS: AFCOLAC B 101 ◇ ANDREZ ◇ BASE 661 ◇ 1,3-BUTADIENE-STYRENE COPOLYMER ◇ BUTADIENE-STYRENE POLYMER ◇ 1,3-BUTADIENE-STYRENE POLYMER ◇ BUTADIENE-STYRENE RESIN ◇ BUTADIENE-STYRENE RUBBER (FCC) ◇ BUTAKON 85-71 ◇ DIAREX 600 ◇ DIENOL S ◇ DOW 209 ◇ DOW LATEX 612 ◇ DST 50 ◇ DURANIT ◇ EDISTIR RB 268 ◇ ETHENYLBENZENE POLYMER with 1,3-BUTADIENE ◇ GOODRITE 1800X73 ◇ HISTYRENE S 6F ◇ HYCAR LX 407 ◇ K 55E ◇ KOPOLYMER BUTADIEN STYRENOVY (CZECH) ◇ KRO 1 ◇ LITEX CA ◇ LYTRON 5202 ◇ MARBON 9200 ◇ NIPOL 407 ◇ PHAROS 100.1 ◇ PLIOFLEX ◇ PLIOLITE S5 ◇ POLYBUTADIENE-POLYSTYRENE COPOLYMER ◇ POLYCO 2410 ◇ RICON 100 ◇ SBS ◇ SD 354 ◇ S6F HISTYRENE RESIN ◇ SKS 85 ◇ SOIL STABILIZER 661 ◇ SOLPRENE 300 ◇ STYRENE-BUTADIENE COPOLYMER ◇ STYRENE-1,3-BUTADIENE COPOLYMER ◇ STYRENE-BUTADIENE POLYMER ◇ SYNPOL 1500 ◇ THERMOPLASTIC 125 ◇ TR 201 ◇ UP 1E ◇ VESTYRON HI

USE IN FOOD:

Purpose: Manufacture of paper and paperboard, masticatory substance in chewing gum base.

Where Used: Chewing gum, packaging materials.

Regulations: FDA - 21CFR 172.615. Use in amounts not to exceed those required to produce the intended physical or other technical effect. 21CFR 181.30.

IARC Cancer Review: Human Inadequate Evidence IMEMDT 19,231,79.

SAFETY PROFILE: An eye irritant. When heated to decomposition it emits acrid smoke and irritating fumes.

TOXICITY DATA and CODEN

eye-rbt 500 mg/24H MLD 28ZPAK -,257,72

SMU100
STYRYLPYRIDINIUM CHLORIDE, DIETHYLCARBAMAZINE

USE IN FOOD:

Purpose: Animal feed drug.

Where Used: Animal feed.

Regulations: FDA - 21CFR 558.565.

SAFETY PROFILE: When heated to decomposition it emits acrid smoke and irritating fumes.

SMY000 CAS: 110-15-6
SUCCINIC ACID

mf: $C_4H_6O_4$ mw: 118.10

PROP: Colorless or white crystals; odorless with acid taste. Mp: 185°, bp: 235° (decomp), d: 1.564 @ 15°/4°. Sol in water; very sol in alc, ether, acetone, glycerin.

SYNS: AMBER ACID ◇ BERNSTEINSAURE (GERMAN) ◇ BUTANEDIOIC ACID ◇ 1,2-ETHANEDICARBOXYLIC ACID ◇ ETHYLENESUCCINIC ACID

USE IN FOOD:

Purpose: Flavor enhancer, miscellaneous and general-purpose food chemical, neutralizing agent, pH control agent.

Where Used: Beverages, condiments, meat products, relishes, sausages (hot).

Regulations: FDA - 21CFR 184.1091. GRAS with a limitation of 0.084 percent in condiments and relishes, 0.0061 percent in meat products when used in accordance with good manufacturing practice.

SAFETY PROFILE: Moderately toxic by subcutaneous route. A severe eye irritant. When heated to decomposition it emits acrid smoke and irritating fumes.

TOXICITY DATA and CODEN

eye-rbt 1179 µg SEV AJOPAA 29,1363,46
orl-rat LD50:2260 mg/kg KODAK* #82-0158

SNB000 CAS: 123-25-1
SUCCINIC ACID, DIETHYL ESTER

mf: $C_8H_{14}O_4$ mw: 174.22

PROP: Colorless, mobile liquid; pleasant odor. Flash p: 230°F. Sol in alc, ether, fixed oils, water.

SYNS: BUTANEDIOIC ACID, DIETHYL ESTER ◇ DIETHYL SUCCINATE (FCC) ◇ ETHYL SUCCINATE ◇ FEMA No. 2377

USE IN FOOD:

Purpose: Flavoring agent.

Where Used: Various.

Regulations: FDA - 21CFR 172.515. Use at a level not in excess of the amount reasonably required to accomplish the intended effect.

SAFETY PROFILE: Mildly toxic by ingestion. A skin and eye irritant. Combustible liquid. When heated to decomposition it emits acrid smoke and irritating fumes.

TOXICITY DATA and CODEN

skn-rat 500 mg/24H MLD FCTXAV 16,637,78
eye-rbt 500 mg/24H MLD AMIHBC 4,119,51
orl-rat LD50:8530 mg/kg AMIHBC 4,119,51

SNF700
SUCCINYLATED MONOGLYCERIDES

PROP: Off-white colored waxy solid; bland taste. Mp: 60°. Sol in warm methanol, ether, n-propanol.

USE IN FOOD:

Purpose: Dough conditioner, emulsifier.

Where Used: Bread dough, liquid shortening, plastic shortening.

Regulations: FDA - 21CFR 172.830. Limitation of 3 percent by weight of shortening; 0.5 percent by weight of flour. Must conform to FDA specifications for fats or fatty acids derived from edible oils.

SAFETY PROFILE: When heated to decomposition it emits acrid smoke and irritating fumes.

SNG600
SUCCISTEARIN

SYN: STEAROYL PROPLENE GLYCOL HYDROGEN SUCCINATE

USE IN FOOD:

Purpose: Emulsifier.

Where Used: Cake mixes, cakes, edible oils, fillings, icings, pastries, shortening, toppings.

Regulations: FDA - 21CFR 172.765. Use in accordance with good manufacturing practice.

SAFETY PROFILE: When heated to decomposition it emits acrid smoke and irritating fumes.

SNH000 CAS: 57-50-1
SUCROSE

mf: $C_{12}H_{22}O_{11}$ mw: 342.34

PROP: White crystals; sweet taste. D: 1.587 @ 25°/4°, mp: 170-186° (decomp). Sol in water, alc; insol in ether.

SYNS: BEET SUGAR ◇ CANE SUGAR ◇ CONFECTIONER'S SUGAR ◇ α-d-GLUCOPYRANOSYL β-d-FRUCTOFURANOSIDE ◇ (α-d-GLUCOSIDO)-β-d-FRUCTOFURANOSIDE ◇ GRANULATED SUGAR ◇ NCI-C56597 ◇ ROCK CANDY ◇ SACCHAROSE ◇ SACCHARUM ◇ SUGAR

USE IN FOOD:

Purpose: Flavoring agent, hog scald agent.

Where Used: Hog carcasses, meat, poultry.

Regulations: GRAS when used in accordance with good manufacturing practice. USDA - 9CFR 318.7, 381.147. Sufficient for purpose.

EPA Genetic Toxicology Program.

ACGIH TLV: TWA 10 mg/m^3 (total dust)

SAFETY PROFILE: Mildly toxic by ingestion. An experimental teratogen. Experimental reproductive effects. Mutagenic data. Vigorous reaction with nitric acid or sulfuric acid (forms carbon monoxide and carbon dioxide). When heated to decomposition it emits acrid smoke and irritating fumes.

TOXICITY DATA and CODEN

mma-sat 600 μg/plate PMRSDJ 1,343,81
dnr-smc 300 mg/L PMRSDJ 1,502,81
orl-rat TDLo: 1548 g/kg (21D pre/1-22D preg): TER IJMDAI 16,789,80
orl-mam TDLo: 54810 mg/kg (15-35D preg): TER TJADAB 30,203,84
orl-rat LD50: 29700 mg/kg TXAPA9 7,609,65

SNH100
SUCROSE FATTY ACID ESTERS

USE IN FOOD:

Purpose: Emulsifier, protective coating, texturizer.

Where Used: Apples, baked goods, bananas, dairy product analogs, frozen dairy desserts, pears, pineapples, whipped milk products.

Regulations: FDA - 21CFR 172.859. Use at a level not in excess of the amount reasonably required to accomplish the intended effect.

SAFETY PROFILE: When heated to decomposition it emits acrid smoke and irritating fumes.

SNH875 CAS: 116-45-0
SULFABROMOMETHAZINE SODIUM

mf: $C_{12}H_{13}BrN_4O_2S$ mw: 357.22

PROP: Crystals. Decomp 250-252°. Sol in alkaline solns.

SYNS: 4-AMINO-N-(5-BROMO-4,6-DIMETHYL-2-PYRIMIDINYL)BENZENESILFANILAMIDE ◇ N[1]-(5-BROMO-4,6-DIMETHYL-2-PYRIMIDINYL)BENZENESULFONAMIDE ◇ 5-BROMOSULFAMETHAZINE

USE IN FOOD:

Purpose: Animal drug.

Where Used: Beef, milk.

Regulations: FDA - 21CFR 556.620. Limitation of 0.1 ppm in uncooked edible tissues of cattle, 0.01 ppm in milk.

SAFETY PROFILE: When heated to decomposition it emits acrid smoke and irritating fumes.

SNH900 CAS: 80-32-0
SULFACHLORPYRIDAZINE

mf: $C_{10}H_9ClN_4O_2S$ mw: 284.74

SYNS: 4-AMINO-N-(6-CHLORO-3-PYRIDAZINYL)BENZENESULFONAMIDE ◇ 3-CHLORO-6-SULFANILAMIDOPYRIDAZINE ◇ CONSULFA ◇ COSULID ◇ NEFROSUL ◇ PRINZONE ◇ SONILYN ◇ VETSULID

USE IN FOOD:

Purpose: Animal drug.

Where Used: Beef, pork.

Regulations: FDA - 21CFR 556.630. Limitation of 0.1 ppm in uncooked edible tissues of calves and swine.

SAFETY PROFILE: When heated to decomposition it emits acrid smoke and irritating fumes.

SNJ100 CAS: 963-14-4
SULFAETHOXYPYRIDAZINE

mf: $C_{12}H_{14}N_4O_3S$ mw: 294.34

PROP: Crystals. Mp: 183-184°. Sol in water at 37°.

SYNS: 4-AMINO-N-(6-ETHOXY-3-PYRIDAZINYL)BENZENESULFONAMIDE ◇ N[1]-(6-ETHOXY-3-PYRIDAZINYL) SULFANILAMIDE ◇ 6-ETHOXY-3-SULFANILAMIDOPYRIDAZINE

USE IN FOOD:

Purpose: Animal drug.

Where Used: Beef, milk, pork.

Regulations: FDA - 21CFR 556.650. Limitation of zero in uncooked edible tissues of swine and in milk. Limitation of 0.1 ppm in uncooked edible tissues of cattle. 21CFR 558.579.

SAFETY PROFILE: When heated to decomposition it emits acrid smoke and irritating fumes.

SNJ500 CAS: 57-68-1
SULFAMETHAZINE

mf: $C_{12}H_{14}N_4O_2S$ mw: 278.36

PROP: Crystals; odorless. Mp: 176° (also a range reported of from 178-179°, 198-199°, and 205-207°). Sol in acetone, water, ether; sltly sol in alc.

SYNS: A-502 ◇ 2-(p-AMINOBENZENESULFONAMIDO)-4,6-DIMETHYLPYRIMIDINE ◇ 6-(4′-AMINOBENZOL-SULFONAMIDO)-2,4-DIMETHYLPYRIMIDIN (GERMAN) ◇ (p-AMINOBENZOLSULFONYL)-2-AMINO-4,6-DIMETHYLPYRIMIDIN (GERMAN) ◇ AZOLMETAZIN ◇ CREMOMETHAZINE ◇ DIAZYL ◇ N[1]-(4,6-DIMETHYL-2-PYRIMIDINYL)SULFANILAMIDE ◇ N-(4,6-DIMETHYL-2-PYRIMIDYL)SULFANILAMIDE ◇ 4,6-DIMETHYL-2-SULFANILAMI-

DOPYRIMIDINE ◇ DIMEZATHINE ◇ MERMETH ◇ METAZIN ◇ NCI-C56600 ◇ NEASINA ◇ PIRMAZIN ◇ PRIMAZIN ◇ SA 111 ◇ SEAZINA ◇ SPANBOLET ◇ SULFADIMERAZINE ◇ SULFADIMETHYLDIAZINE ◇ SULFADIMETHYLPYRIMIDINE ◇ SULFADIMETINE ◇ SULFADIMEZINE ◇ SULFADIMIDINE ◇ SULFADINE ◇ SULFADSIMESINE ◇ SULFA-ISODIMERAZINE ◇ SULFAISODIMIDINE ◇ SULFAMETHIAZINE ◇ SULFAMETHIN ◇ SULFAMEZATHINE ◇ 2-SULFANILAMIDO-4,6-DIMETHYLPYRIMIDINE ◇ SULFISOMIDIN ◇ SULFISOMIDINE ◇ SULFODIMESIN ◇ SULFODIMEZINE ◇ SULMET ◇ SULPHADIMETHYLPYRIMIDINE ◇ SULPHADIMIDINE ◇ SUPERSEPTIL ◇ VERTOLAN

USE IN FOOD:

Purpose: Animal drug.

Where Used: Beef, chicken, pork, turkey.

Regulations: FDA - 21CFR 556.670. Limitation of 0.1 ppm in chickens, turkeys, cattle, and swine. USDA CES Ranking: B-1 (1985).

SAFETY PROFILE: Moderately toxic by intravenous and intraperitoneal routes. Mildly toxic by ingestion. An experimental tumorigen and teratogen. Experimental reproductive effects. When heated to decomposition it emits very toxic fumes of SO_x and NO_x.

TOXICITY DATA and CODEN

orl-rat TDLo:6850 mg/kg (6-15D preg):TER NTIS** PB83-151035
orl-rbt TDLo:25200 mg/kg (6-19D preg):TER NTIS** PB83-105690
imp-rat TDLo:5000 mg/kg:ETA ACRAAX 37,258,52
orl-mus LD50:50 g/kg NIIRDN 6,392,82

SNK500 CAS: 5329-14-6
SULFAMIC ACID

mf: H_3NO_3S mw: 97.10
DOT: 2967

PROP: White crystals. Mp: 200° (decomp), bp: decomp, d: 203 @ 12°.

SYNS: AMIDOSULFONIC ACID ◇ AMIDOSULFURIC ACID ◇ AMINOSULFONIC ACID ◇ KYSELINA AMIDOSULFONOVA (CZECH) ◇ KYSELINA SULFAMINOVA (CZECH) ◇ SULFAMIDIC ACID ◇ SULPHAMIC ACID (DOT)

USE IN FOOD:

Purpose: Paper manufacturing aid.

Where Used: Food packaging.

Regulations: FDA - 21CFR 186.1093. GRAS as an indirect additive when used in accordance with good manufacturing practice.

DOT Classification: Corrosive Material; Label: Corrosive

SAFETY PROFILE: Poison by intraperitoneal route. Moderately toxic by ingestion. A human skin irritant. A corrosive irritant to skin, eyes, and mucous membranes. Violent or explosive reactions with chlorine; metal nitrates + heat; metal nitrites + heat; fuming HNO_3. When heated to decomposition it emits very toxic fumes of SO_x and NO_x.

TOXICITY DATA and CODEN

skn-hmn 4%/5D-I MLD JIHTAB 25,26,43
skn-rbt 500 mg/24H SEV 28ZPAK -,18,72
eye-rbt 250 μg/24H SEV 28ZPAK -,18,72
orl-rat LD50:3160 mg/kg 28ZPAK -,18,72

SNN300 CAS: 122-11-2
6-SULFANILAMIDO-2,4-DIMETHOXYPYRIMIDINE

mf: $C_{12}H_{14}N_4O_4S$ mw: 310.36

PROP: Crystals from dil alc. Mp: 201-203°. Sol in dil HCl and in aq solns of sodium carbonate. Solubility in water at 37° (mg/100 mL): 4.6 at pH 4.10; 29.5 at pH 6.7; 58.0 at pH 7.06.

SYNS: ABCID ◇ AGRIBON ◇ ALBON ◇ 4-AMINO-N-(2,6-DIMETHOXY-4-PYRIMIDINYL)BENZENESULFONAMIDE ◇ ARNOSULFAN ◇ BACTROVET ◇ DEPOSUL ◇ DIASULFA ◇ DIASULFYL ◇ DIMETAZINA ◇ 2,6-DIMETHOXY-4-(p-AMINOBENZENESULFONAMIDO)PYRIMIDINE ◇ N^1-(2,6-DIMETHOXY-4-PYRIMIDINYL)SULFANILAMIDE ◇ DIMETHOXYSULFADIAZINE ◇ 2,4-DIMETHOXY-6-SULFANILAMIDO-1,3-DIAZINE ◇ 2,6-DIMETHOXY-4-SULFANILAMIDOPYRIMIDINE ◇ DINOSOL ◇ DORISUL ◇ FUXAL ◇ MADRIBON ◇ MADRIGID ◇ MADRIQID ◇ MADROXIN ◇ MADROXINE ◇ MAXULVET ◇ MEMCOZINE ◇ METOXIDON ◇ NEOSTREPAL ◇ OMNIBON ◇ PERSULFEN ◇ RADONIN ◇ REDIFAL ◇ ROSCOSULF ◇ SCANDISIL ◇ SDM ◇ SDMO ◇ SUDINE ◇ SULDIXINE ◇ SULFADIMETHOXIN ◇ SULFADIMETHOXINE ◇ SULFADIMETHOXYDIAZINE ◇ SULFADIMETOSSINA (ITALIAN) ◇ SULFADIMETOXIN ◇ SULFASOL ◇ SULFASTOP ◇ SULFOPLAN ◇ SULPHADIMETHOXINE ◇ SULXIN ◇ SYMBIO ◇ THERACANZAN

USE IN FOOD:

Purpose: Animal drug, animal feed drug.

Where Used: Animal feed, beef, catfish, chicken, duck, salmonids, turkey.

Regulations: FDA - 21CFR 556.640. Limitation of 0.1 ppm in chickens, turkeys, cattle, ducks, salmonids, and catfish. 21CFR 558.575. (with ormetoprim).

SAFETY PROFILE: Moderately toxic by intraperitoneal, intravenous, and subcutaneous routes. An experimental teratogen. Experimental reproductive effects. When heated to decomposition it emits toxic fumes of SO_x and NO_x.

TOXICITY DATA and CODEN

orl-rat TDLo: 3 g/kg (9-14D preg): TER SEIJBO 13,17,73
orl-mus TDLo: 12 g/kg (7-12D preg): TER OYYAA2 7,1005,73
ipr-mus LD50: 866 mg/kg NIIRDN 6,386,82

SNQ600 CAS: 122-16-7
SULFANITRAN

mf: $C_{14}H_{13}N_3O_5S$ mw: 335.34

PROP: Crystals. Mp: 239-240°. Sol in acetone, ethanol, methanol, water, ether.

SYNS: 4-ACETAMINOBEZENESULFON ◇ ACETYL(P-NITROPHENYL)SULFANILAMIDE ◇ APNPS ◇ N-(p-NITROPHENYL)SULFANILAMIDE ◇ N-[4-[[(4-NITROPHENYL) AMINO]SULFONYL]ACETAMIDE

USE IN FOOD:

Purpose: Animal drug.

Where Used: Chicken.

Regulations: FDA - 21CFR 556.680. Limitation of zero in chickens.

SAFETY PROFILE: When heated to decomposition it emits acrid smoke and irritating fumes.

SNQ850 CAS: 59-40-5
SULFAQUINOXALINE

mf: $C_{14}H_{12}N_4O_2S$ mw: 300.36

SYNS: 2-p-AMINOBENZENESULFONAMIDOQUINOXALINE ◇ COMPOUND-3-120 ◇ N-(2-QUINOXALINYL) SULFANILAMIDE ◇ SULFABENZPYRAZINE ◇ SULFACOX ◇ SULFALINESULQUIN

USE IN FOOD:

Purpose: Animal feed drug.

Where Used: Animal feed.

Regulations: 21CFR 558.586. USDA CES Ranking: B-3 (1986).

SAFETY PROFILE: Moderately toxic by ingestion. When heated to decomposition it emits very toxic fumes of NO_x and SO_x.

TOXICITY DATA and CODEN

orl-rat TD50: 1370 mg/kg MahWM# 16NOV82

SNW600
SULFOMYXIN

SYN: N-SULFOMETHYL-POLYMYXIN B SODIUM SALT

USE IN FOOD:

Purpose: Animal drug.

Where Used: Chicken, turkey.

Regulations: FDA - 21CFR 556.700. Limitation of zero in uncooked edible tissues of chickens and turkeys.

SAFETY PROFILE: When heated to decomposition it emits acrid smoke and irritating fumes.

SOH500 CAS: 7446-09-5
SULFUR DIOXIDE

DOT: 1079
mf: O_2S mw: 64.06

PROP: Colorless gas or liquid under pressure; pungent odor. Mp: −75.5°, bp: −10.0°, d (liquid): 1.434 @ 0°, vap d: 2.264 @ 0°, vap press: 2538 mm @ 21.1°. Sol in water.

SYNS: BISULFITE ◇ FERMENICIDE LIQUID ◇ FERMENICIDE POWDER ◇ SCHWEFELDIOXYD (GERMAN) ◇ SIARKI DWUTLENEK (POLISH) ◇ SULFUROUS ACID ANHYDRIDE ◇ SULFUROUS ANHYDRIDE ◇ SULFUROUS OXIDE ◇ SULFUR OXIDE ◇ SULPHUR DIOXIDE, LIQUEFIED (DOT)

USE IN FOOD:

Purpose: Bleaching agent, preservative.

Where Used: Beverages, fruits (fresh), meat, vegetables (fresh), wine.

Regulations: FDA - 21CFR 182.3862 GRAS when used in accordance with good manufacturing practice except that it is not used in meats or in foods recognized as sources of vitamin B_1 or on fruits and vegetables intended to be sold or served raw or to be presented to consumers as fresh. BATF - 27CFR 240.1051. Limitation proscribed in 27CFR 4.22(b)(f).

EPA Extremely Hazardous Substances List. EPA Genetic Toxicology Program.

OSHA PEL: TWA 5 ppm; STEL 5 ppm ACGIH TLV: TWA 2 ppm; STEL 5 ppm DFG MAK: 2 ppm (5 mg/m^3) DOT Classification: Nonflammable Gas; Label: Nonflammable Gas; Poison A; Label: Poison Gas NIOSH REL: (Sulfur Dioxide) TWA 0.5 ppm

SAFETY PROFILE: A poison gas. Moderately toxic experimentally by inhalation. Mildly toxic to humans by inhalation. An experimental tumorigen and teratogen. Human systemic effects by inhalation: pulmonary vascular resistance, respiratory depression and other pulmonary changes. It chiefly affects the upper respiratory tract and the bronchi. It may cause edema of the lungs or glottis, and can produce respiratory paralysis. Experimental reproductive effects. Human mutagenic data. A corrosive irritant to eyes, skin, and mucous membranes. This material is so irritating that it provides its own warning of toxic concentration. Levels of 400-500 ppm are immediately dangerous to life and 50-100 ppm is considered to be the maximum permissible concentration for exposures of 30-60 minutes. Excessive exposures to high enough concentration of this material can be fatal.

A nonflammable Gas. Will react with water or steam to produce toxic and corrosive fumes. When heated to decomposition it emits toxic fumes of SO_x.

TOXICITY DATA and CODEN

eye-rbt 6 ppm/4H/32D MLD JPCAAC 10,17,60
oms-esc 2 mmol/L CBINA8 43,289,83
mmo-omi 10 mmol/L MUREAV 39,149,77
ihl-rat TCLo: 4 mg/m^3/24H (72D pre): REP HYSAAV 35(4-6),277,70
ihl-rbt TCLo: 70 ppm/7H (6-18D preg): TER NTIS** CONF-771017
ihl-mus TCLo: 500 ppm/5M/30W-I: ETA BJCAAI 21,606,67
ihl-hmn TCLo: 3 ppm/5D: PUL TXAPA9 22,319,72
ihl-man TCLo: 4 ppm/1M: PUL JAPYAA 17,252,62
ihl-hmn LCLo: 1000 ppm/10M: PUL CTOXAO 5,198,72
ihl-rat LC50: 2520 ppm/1H NTIS** AD-A148-952

SOI500 CAS: 7664-93-9
SULFURIC ACID
mf: H_2O_4S mw: 98.08
DOT: 1830/1832

PROP: Colorless oily liquid; odorless. Mp: 10.49°, d: 1.834, vap press: 1 mm @ 145.8°, bp: 290°, decomp @ 340°. Misc with water and alc (liberating great heat).

SYNS: ACIDE SULFURIQUE (FRENCH) ◇ ACIDO SOLFORICO (ITALIAN) ◇ BOV ◇ DIPPING ACID ◇ HYDROOT ◇ MATTING ACID (DOT) ◇ NORDHAUSEN ACID (DOT) ◇ OIL of VITRIOL (DOT) ◇ SCHWEFELSAEURELOESUNGEN (GERMAN) ◇ SPENT SULFURIC ACID (DOT) ◇ SULPHURIC ACID ◇ VITRIOL BROWN OIL ◇ VITRIOL, OIL of (DOT) ◇ ZWAVELZUUROPLOSSINGEN (DUTCH)

USE IN FOOD:

Purpose: Acid, pH control agent, processing aid.

Where Used: Beverages (alcoholic), cheese.

Regulations: FDA - 21CFR 184.1095. GRAS with a limitation of 0.014 percent in alcoholic beverages, 0.0003 percent in cheese when used in accordance with good manufacturing practice. BATF - 27CFR 240.1051a. See 27CFR 240.486.

OSHA PEL: TWA 1 mg/m^3 ACGIH TLV: TWA 1 mg/m^3; STEL 3 mg/m^3 DFG MAK: 1 mg/m^3 DOT Classification: Corrosive Material; Label: Corrosive NIOSH REL: (Sulfuric Acid) TWA 1 mg/m^3

SAFETY PROFILE: Human poison by unspecified route. Experimental poison by inhalation. Moderately toxic by ingestion. A severe eye irritant. Extremely irritating, corrosive, and toxic to tissue resulting in rapid destruction of tissue, causing severe burns. Repeated contact with dilute solutions can cause a dermatitis, and repeated or prolonged inhalation of a mist of sulfuric acid can cause inflammation of the upper respiratory tract leading to chronic bronchitis. Severe exposure may cause a chemical pneumonitis; erosion of the teeth due to exposure to strong acid fumes has been recognized in industry.

This is a very powerful, acidic oxidizer which can ignite or explode on contact with many materials. When heated it emits highly toxic fumes; will react with water or steam to produce heat; can react with oxidizing or reducing materials. When heated to decomposition it emits toxic fumes of SO_x.

TOXICITY DATA and CODEN

eye-rbt 100 mg rns SEV TXCYAC 23,281,82
ihl-hmn TCLo: 3 mg/m^3/24W BJIMAG 18,63,61
unr-man LDLo: 135 mg/kg 85DCAI 2,73,70
orl-rat LD50: 2140 mg/kg AIHAAP 30,470,69
ihl-rat LC50: 510 mg/m^3/2H 85GMAT -,107,82

SON500 CAS: 151-21-3
SULFURIC ACID, MONODODECYL ESTER, SODIUM SALT
mf: $C_{12}H_{26}O_4S{\cdot}Na$ mw: 289.43

PROP: White to cream-colored crystals, flakes or powder; slt odor. Sol in water.

SYNS: AQUAREX METHYL ◇ AVIROL 118 CONC ◇ CARSONOL SLS ◇ CONCO SULFATE WA ◇ CYCLORYL 21 ◇ DETERGENT 66 ◇ DODECYL ALCOHOL, HYDROGEN SULFATE, SODIUM SALT ◇ DODECYL SODIUM SULFATE ◇ DODECYL SULFATE, SODIUM SALT ◇ DREFT ◇ DUPONOL ◇ EMERSAL 6400 ◇ EMULSIFIER NO. 104 ◇ HEXAMOL SLS ◇ IRIUM ◇ LANETTE WAX-S ◇ LAURYL SODIUM SULFATE ◇ LAURYL SULFATE, SODIUM SALT ◇ MAPROFIX 563 ◇ MAPROFIX WAC-LA ◇ NCI-C50191 ◇ NEUTRAZYME ◇ ORVUS WA PASTE ◇ PRODUCT NO. 161 ◇ QUOLAC EX-UB ◇ REWOPOL NLS 30 ◇ RICHONOL C ◇ SIPEX OP ◇ SIPON WD ◇ SLS ◇ SODIUM DODECYL SULFATE ◇ SODIUM LAURYL SULFATE (FCC) ◇ SODIUM MONODODECYL SULFATE ◇ SOLSOL NEEDLES ◇ STANDAPOL 112 CONC ◇ STEPANOL WAQ ◇ STERLING WAQ-COSMETIC ◇ SULFOPON WA 1 ◇ SULFOTEX WALA ◇ TARAPON K 12 ◇ TEXAPON ZHC ◇ TREPENOL WA ◇ ULTRA SULFATE SL-1

USE IN FOOD:

Purpose: Emulsifier, hog scald agent, poultry scald agent, surface-active agent, surfactant, wetting agent, whipping agent.

Where Used: Angel food cake, beverage bases (dry, fumaric acid-acidulated), citrus fruit (fresh), egg white (frozen), egg white (liquid), egg white solids, fats (crude), fruit juice drink (fumaric acid-acidulated), hog carcasses, marshmallows, poultry, vegetable oils (crude).

Regulations: FDA - 21CFR 172.210. Must comply with 21CFR 172.822. 21CFR 172.822. Limitation of 1000 ppm in egg white solids, 125 ppm in frozen egg whites, 125 ppm in liquid egg whites, 0.5 percent in marshmallows, 25 ppm in fumaric acid-acidulated dry beverage base, 25 ppm in fumaric acid-acidulated fruit juice drink, 10 ppm in crude vegetable oils and fats. USDA - 9CFR 318.7, 381.147. Sufficient for purpose.

SAFETY PROFILE: Poison by intravenous and intraperitoneal routes. Moderately toxic by ingestion. An experimental teratogen. Experimental reproductive effects. A human skin irritant. An experimental eye and severe skin irritant. A mild allergen. Mutagenic data. When heated to decomposition it emits toxic fumes of SO_x and Na_2O.

TOXICITY DATA and CODEN

skn-hmn 250 mg/24H MLD TXAPA9 31,481,75
skn-mus 25 mg/24H MOD JSCCA5 23,371,72
eye-rbt 250 μg MLD AROPAW 34,99,45
mmo-omi 200 mg/L JDREAF 55,266,76
dni-gpg:kdy 60 μmol/L FCTXAV 14,431,76
skn-mus TDLo:480 mg/kg (6-13D preg):TER TRENAF 27(2),113,76
orl-rat LD50:1288 mg/kg FCTXAV 5,763,67

SOP000 CAS: 2312-35-8
SULFUROUS ACID, 2-(p-tert-BUTYLPHENOXY)CYCLOHEXYL-2-PROPYNYL ESTER

DOT: 2765
mf: $C_{19}H_{26}O_4S$ mw: 350.51
DOT: 2765

SYNS: BPPS ◇ 2-(p-tert-BUTYLPHENOXY)CYCLOHEXYL PROPARGYL SULFITE ◇ 2-(p-tert-BUTYLPHENOXY)CYCLOHEXYL 2-PROPYNYL SULFITE ◇ COMITE ◇ 2-(4-(1,1-DIMETHYLETHYL)PHENOXY)CYCLOHEXYL 2-PROPYNYL ESTER, SULFUROUS ACID ◇ 2-(4-(1,1-DIMETHYLETHYL)PHENOXY)CYCLOHEXYL 2-PROPYNYL SULFITE ◇ DO 14 ◇ ENT 27,226 ◇ NAUGATUCK D-014 ◇ OMAIT ◇ OMITE ◇ PROPARGITE (DOT) ◇ UNIROYAL D014 ◇ U.S. RUBBER D-014

USE IN FOOD:

Purpose: Insecticide.

Where Used: Animal feed, apple pomace (dried), citrus pulp (dried), figs (dried), grape pomace (dried), hops (dried), raisins, tea (dried).

Regulations: FDA - 21CFR 193.370. Insecticide residue tolerance of 9 ppm in dried figs, 30 ppm in dried hops, 25 ppm in raisins, 10 ppm in dried tea. 21CFR 561.330. Limitation of 80 ppm in dried apple pomace, 40 ppm in dried citrus pulp and dried grape pomace when used for animal feed.

DOT Classification: ORM-E; Label: None

SAFETY PROFILE: Poison by skin contact. Moderately toxic by ingestion and possibly other routes. When heated to decomposition it emits toxic fumes of SO_x.

TOXICITY DATA and CODEN

orl-rat LD50:1480 mg/kg TXAPA9 14,515,69
skn-rat LD50:250 mg/kg WRPCA2 9,119,70

SOU875
SUNFLOWER OIL (UNHYDROGENATED)

PROP: From the seed of *Helianthus annuus*. Amber-colored liquid.

USE IN FOOD:

Purpose: Coating agent, emulsifying agent, formulation aid, texturizer.

Where Used: Margarine, shortening.

Regulations: GRAS when used in accordance with good manufacturing practice.

SAFETY PROFILE: When heated to decomposition it emits acrid smoke and irritating fumes.

SPE600
SYNTHETIC PARAFFIN and SUCCINIC DERIVATIVES

USE IN FOOD:

Purpose: Protective coating.

Where Used: Grapefruit, lemons, limes, muskmelons, oranges, sweet potatoes, tangerines.

Regulations: FDA - 21CFR 172.275. Use at a level not in excess of the amount reasonably required to accomplish the intended effect.

SAFETY PROFILE: When heated to decomposition it emits acrid smoke and irritating fumes.

T

TAB260
TAGETES MEAL and EXTRACT

PROP: Extracted from the dried ground flower petals of the Aztec marigold *Tagetes erecta* L. Mp: 53.5-55.0°.

USE IN FOOD:

Purpose: Color additive.

Where Used: Chicken feed.

Regulations: FDA - 21CFR 73.295. Feed must be supplemented sufficiently with xanthophyll and associated carotenoids to accomplish the intended effect.

SAFETY PROFILE: When heated to decomposition it emits acrid smoke and irritating fumes.

TAB750 CAS: 14807-96-6
TALC (powder)
mf: $H_2O_3Si \cdot 3/4Mg$ mw: 96.33

PROP: White to grayish-white, fine powder; odorless and tasteless. Powdered native hydrous magnesium silicate. Insol in water, cold acids, or alkalies.

SYNS: AGALITE ◇ AGI TALC, BC 1615 ◇ ALPINE TALC USP, BC 127 ◇ ALPINE TALC USP, BC 141 ◇ ALPINE TALC USP, BC 662 ◇ ASBESTINE ◇ C.I. 77718 ◇ DESERTALC 57 ◇ EMTAL 596 ◇ FIBRENE C 400 ◇ LO MICRON TALC 1 ◇ LO MICRON TALC, BC 1621 ◇ LO MICRON TALC USP, BC 2755 ◇ METRO TALC 4604 ◇ METRO TALC 4608 ◇ METRO TALC 4609 ◇ MISTRON FROST P ◇ MISTRON RCS ◇ MISTRON 2SC ◇ MISTRON STAR ◇ MISTRON SUPER FROST ◇ MISTRON VAPOR ◇ MP 12-50 ◇ MP 25-38 ◇ MP 45-26 ◇ NCI-C06008 ◇ NO. 907 METRO TALC ◇ NYTAL ◇ OOS ◇ OXO ◇ PURTALC USP ◇ SIERRA C-400 ◇ SNOWGOOSE ◇ STEAWHITE ◇ SUPREME DENSE ◇ TALCUM

USE IN FOOD:

Purpose: Anticaking agent, coating agent, lubricant, release agent, surface-finishing agent, texturizing agent.

Where Used: Various.

Regulations: Not food grade if derived from deposits containing asbestiform minerals.

ACGIH TLV: TWA 2 mg/m^3, respirable dust (use asbestos TLV if asbestos fibers are present)

SAFETY PROFILE: The talc with less than 1 percent asbestos is regarded as a nuisance dust. An experimental tumorigen. A human skin irritant. Prolonged or repeated exposure can produce a form of pulmonary fibrosis (talc pneumoconiosis) which may be due to asbestos content.

TOXICITY DATA and CODEN

skn-hmn 300 μg/3D-I MLD 85DKA8 -,127,77
ihl-rat TCLo:11 mg/m^3/1Y-I:ETA 43GRAK -,389,79
imp-rat TDLo:200 mg/kg:ETA JJIND88 67,965,81

TAC000 CAS: 8002-26-4
TALL OIL

PROP: Composition: Rosin acids, oleic and linoleic acids. Dark brown liquid; acrid odor. D: 0.95, flash p: 360°F.

SYNS: LIQUID ROSIN ◇ TALLOL

USE IN FOOD:

Purpose: Drying oil.

Where Used: Food packaging.

Regulations: FDA - 21CFR 181.26. Use at a level not in excess of the amount reasonably required to accomplish the intended effect. 21CFR 186.1557. GRAS as an indirect additive when used in accordance with good manufacturing practice.

SAFETY PROFILE: A mild allergen. Combustible when exposed to heat or flame; can react with oxidizing materials. To fight fire, use dry chemical, CO_2. When heated to decomposition it emits acrid smoke and irritating fumes.

TAC100
TALLOW

PROP: Off-white fat from rendering of beef or mutton.

SYN: BLEACHED-DEODORIZED TALLOW

USE IN FOOD:

Where Used: Various.

Regulations: GRAS when used in accordance with good manufacturing practice.

Purpose: Coating agent, emulsifying agent, formulation aid, texturizer.

SAFETY PROFILE: When heated to decomposition it emits acrid smoke and irritating fumes.

TAD500 CAS: 8008-31-9
TANGERINE OIL

PROP: Expressed from the peels of Dancy and related varities of *Citrus reticulata Blanco*. The components include d-limonene, n-octylaldehyde, n-decylaldehyde, citral, linalool, citronella, cadinene, terpenes, aldehydes, alcohols, and esters (FCTXAV 16, 637,78). Red orange to brown orange liquid; orange-like odor. Sol in fixed oils, mineral oil; sltly sol in propylene glycol; insol in glycerin.

SYNS: TANGERINE OIL, COLDPRESSED (FCC) ◇ TANGERINE OIL, EXPRESSESED (FCC)

USE IN FOOD:

Purpose: Flavoring agent.

Where Used: Various.

Regulations: FDA - 21CFR 182.20. GRAS when used at a level not in excess of the amount reasonably required to accomplish the intended effect.

SAFETY PROFILE: A skin irritant. When heated to decomposition it emits acrid smoke and irritating fumes.

TOXICITY DATA and CODEN

skn-mus 500 mg open FCTXAV 16,637,78
skn-rbt 500 mg/24H FCTXAV 16,637,78

TAD750 CAS: 1401-55-4
TANNIC ACID
mf: $C_{76}H_{52}O_{46}$ mw: 1701.28

PROP: From the nutgalls of *Quercus infectoria Oliver* or seed pods of *Caesalpinia spinosa* or the nutgalls of various sumac species. Yellowish-white or brown, bulky powder or flakes; odorless with astringent taste. Mp: 200°, flash p: 390°F (OC), autoign temp: 980°F. Very sol in water, alc, acetone; almost insol in benzene, chloroform, ether, petr ether, carbon disulfide.

SYNS: d'ACIDE TANNIQUE (FRENCH) ◇ GALLOTANNIC ACID ◇ GALLOTANNIN ◇ GLYCERITE ◇ TANNIN

USE IN FOOD:

Purpose: Boiler water additive, clarifying agent, fat rendering aid, flavor enhancer, flavoring agent, pH control agent.

Where Used: Apple juice, baked goods, beer, candy (hard), candy (soft), cough drops, fats (rendered), fillings, frozen dairy desserts and mixes, gelatins, meat products, non-alcoholic beverages, puddings, wine.

Regulations: FDA - 21CFR 173.310, 184.1097. GRAS with a limitation of 0.01 percent in baked goods, 0.015 percent in alcoholic beverages, 0.005 percent in nonalcoholic beverages and for gelatins, puddings, and fillings, 0.04 percent in frozen dairy desserts and mixes, 0.04 percent in soft candy, 0.013 percent in hard candy and cough drops, 0.001 percent in meat products, when used in accordance with good manufacturing practice. USDA - 9CFR 318.7. Sufficient for purpose. BATF - 27CFR 240.1051. Limitation of 3.0 g/L calculated as gallic acid equivalents (GAE) in apple juice or wine. Limitation of 0.8 g/L in white wine and 0.3 g/L in rose wine.

IARC Cancer Review: Animal Sufficient Evidence IMEMDT 10,253,76. EPA Genetic Toxicology Program.

SAFETY PROFILE: Poison by ingestion, intramuscular, intravenous, and subcutaneous routes. Moderately toxic by parenteral route. An experimental carcinogen and tumorigen. Experimental reproductive effects. Mutagenic data. Combustible when exposed to heat or flame. To fight fire, use water. Incompatible with salts of heavy metals; oxidizing materials. When heated to decomposition it emits acrid smoke and irritating fumes.

TOXICITY DATA and CODEN

dns-rat-orl 25 g/kg JJIND8 74,1283,85
dni-mus-ipr 76 mg/kg IJEBA6 17,1141,79
orl-rat TDLo:112 g/kg (49D pre/1-21D preg): REP CRSBAW 172,470,78
orl-mus TDLo:1478 g/kg (11W male/11W pre):REP CPGPAY 4,393,73
scu-rat TDLo:4450 mg/kg/17W-I:CAR AMSHAR 3,353,53
scu-mus TDLo:750 mg/kg/12W-I:ETA BJCAAI 14,147,60
orl-rat LD50:2260 mg/kg 34ZIAG -,571,69

TAF750 CAS: 87-69-4
TARTARIC ACID
mf: $C_4H_6O_6$ mw: 150.10

PROP: Colorless to translucent crystals or white powder; odorless with an acid taste. Sol in water, alc.

SYNS: 2,3-DIHYDROSUCCINIC ACID ◇ 2,3-DIHYDROXY-BUTANEDIOC ACID

USE IN FOOD:

Purpose: Acid, firming agent, flavor enhancer, flavoring agent, humectant, pH control agent, sequestrant.

Where Used: Baking powder, beverages (grape and lime flavored), jellies (grape flavored), poultry, wine.

Regulations: FDA - 21CFR 184.1099. GRAS When used in accordance with good manufacturing practice. USDA - 9CFR 318.7, 9CFR 381.147. Sufficient for purpose. BATF - 27CFR 240.1051. Use as prescribed in 27CFR 240.364, 240.512.

SAFETY PROFILE: Moderately toxic by intravenous route. Mildly toxic by ingestion. When heated to decomposition it emits acrid smoke and irritating fumes.

TOXICITY DATA and CODEN

orl-dog LDLo:5000 mg/kg JAPMA8 39,275,50

TAV750 CAS: 126-92-1
TERGITOL 08

mf: $C_8H_{18}O_4S \cdot Na$ mw: 233.31

SYNS: EMERSAL 6465 ◇ 2-ETHYL-1-HEXANOL HYDROGEN SULFATE, SODIUM SALT ◇ 2-ETHYL-1-HEXANOL SULFATE SODIUM SALT ◇ 2-ETHYLHEXYL SODIUM SULFATE ◇ MONO(2-ETHYLHEXYL)SULFATE SODIUM SALT ◇ NCI-C50204 ◇ NIA PROOF 08 ◇ PROPASTE 6708 ◇ SIPEX BOS ◇ SODIUM ETASULFATE ◇ SODIUM ETHASULFATE ◇ SODIUM(2-ETHYLHEXYL)ALCOHOL SULFATE ◇ SODIUM 2-ETHYLHEXYL SULFATE ◇ SULFURIC ACID, MONO(2-ETHYLHEXYL)ESTER, SODIUM SALT (8CI) ◇ TERGEMIST ◇ TERGIMIST ◇ TERGITOL ANIONIC 08

USE IN FOOD:

Purpose: Lye peeling agent, poultry scald agent, washing water agent.

Where Used: Fruits, poultry, vegetables.

Regulations: FDA - 21CFR 173.315. Limitation of 0.2 percent in wash water. USDA - 9CFR 381.147. Sufficient for purpose.

SAFETY PROFILE: Poison by intraperitoneal route. Moderately toxic by ingestion and skin contact. A skin and eye irritant. When heated to decomposition it emits very toxic fumes of SO_x and Na_2O.

TOXICITY DATA and CODEN

skn-rbt 500 mg open MOD UCDS** 1/20/72
eye-rbt 250 μg MLD AROPAW 34,99,45
orl-mus LD50:1550 mg/kg GISAAA 43(7),70,78

TBC575
TERPENE RESIN, NATURAL

PROP: Extracted from wood. Mp: 155°.

USE IN FOOD:

Purpose: Color diluent, component of polypropylene film, masticatory substance in chewing gum base, moisture barrier.

Where Used: Ascorbic acid powder, chewing gum, fruits, gelatin capsules (soft), vegetables.

Regulations: FDA - 21CFR 73.1. 21CFR 172.280. Limitation of 0.07 percent of weight of capsule; 7 percent of ascorbic acid and salts. 21CFR 172.615, 178.3930.

SAFETY PROFILE: When heated to decomposition it emits acrid smoke and irritating fumes.

TBC580
TERPENE RESIN, SYNTHETIC

PROP: Polymers of α-pinene, β-pinene, and/or dipentine (FCC III). Mp: 112-118°. Sol in benzene.

USE IN FOOD:

Purpose: Color diluent, component of polypropylene film, masticatory substance in chewing gum base, moisture barrier.

Where Used: Ascorbic acid powder, chewing gum, fruits, gelatin capsules (soft), vegetables.

Regulations: FDA - 21CFR 73.1. 21CFR 172.280. Limitation of 0.07 percent of weight of capsule; 7 percent of ascorbic acid and salts. 21CFR 172.615, 178.3930.

SAFETY PROFILE: When heated to decomposition it emits acrid smoke and irritating fumes.

TBD500 CAS: 8006-39-1
TERPINEOL

mf: $C_{10}H_{18}O$ mw: 154.28

PROP: A mixture of α, β, and γ isomers (FCTXAV 12,807,74). Colorless, viscous liquid; lilac odor. D: 0.930-0.936, refr index: 1.482, flash p: 196°F. Sltly sol in water, glycerin.

SYNS: FEMA No. 3045 ◇ p-MENTH-1-EN-8-OL ◇ MIXTURE OF p-METHENOLS ◇ α-TERPINEOL (FCC) ◇ TERPINEOLS

USE IN FOOD:

Purpose: Flavoring agent.

Where Used: Various.

Regulations: FDA - 21CFR 172.515. Use at a level not in excess of the amount reasonably required to accomplish the intended effect.

SAFETY PROFILE: Mildly toxic by ingestion. A skin irritant. Combustible liquid. When heated to decomposition it emits acrid smoke and irritating fumes.

TOXICITY DATA and CODEN

skn-rbt 500 mg/24H MOD FCTXAV 12,807,74
orl-rat LD50:4300 mg/kg FCTXAV 12,807,74

TBE250 CAS: 80-26-2
TERPINYL ACETATE
mf: $C_{12}H_{20}O_2$ mw: 196.32

PROP: Colorless liquid; sweet, herbaceous odor. D: 0.966 @ 20/4°, refr index: 1.464, mp: <−50°, bp: 220° decomp, flash p: 212°F. Insol in water; sol in alc, fixed oils, mineral oil, propylene glycol.

SYNS: FEMA No. 3047 ◇ α-TERPINEOL ACETATE

USE IN FOOD:

Purpose: Flavoring agent.

Where Used: Various.

Regulations: FDA - 21CFR 172.515. Use at a level not in excess of the amount reasonably required to accomplish the intended effect.

SAFETY PROFILE: Mildly toxic by ingestion. Combustible liquid. When heated to decomposition it emits acrid smoke and irritating fumes.

TOXICITY DATA and CODEN

orl-rat LDLo:4160 mg/kg FCTXAV 2,327,64

TBE600
TERPINYL PROPIONATE
mf: $C_{13}H_{22}O_2$ mw: 210.32

PROP: Colorless to sltly yellow liquid; sweet, floral, lavender-like odor. D: 0.944, refr index: 1.461, flash p: +212°F. Sol in glycerin; misc in alc, chloroform, ether, fixed oils; sltly sol in propylene glycol; insol in water @ 240°.

SYNS: FEMA No. 3053 ◇ MENTHEN-1-YL-8 PROPIONATE

USE IN FOOD:

Purpose: Flavoring agent.

Where Used: Various.

Regulations: FDA - 21CFR 172.515. Use at a level not in excess of the amount reasonably required to accomplish the intended effect.

SAFETY PROFILE: Combustible liquid. When heated to decomposition it emits acrid smoke and irritating fumes.

TBG000 CAS: 57-85-2
TESTOSTERONE PROPIONATE
mf: $C_{22}H_{32}O_3$ mw: 344.54

SYNS: AGOVIRIN ◇ ANDROGEN ◇ ANDROSAN ◇ Δ^4-ANDROSTENE-17-β-PROPIONATE-3-ONE ◇ ANDROTESTON ◇ ANDROTEST P ◇ ANDRUSOL-P ◇ ANERTAN ◇ AQUAVIRON ◇ BIO-TESTICULINA ◇ ENARMON ◇ HOMANDREN (amps) ◇ HORMOTESTON ◇ MASENATE ◇ NASDOL ◇ NEO-HOMBREOL ◇ NSC 9166 ◇ OKASA-MASCUL ◇ ORCHIOL ◇ ORCHISTIN ◇ ORETON ◇ ORETON PROPIONATE ◇ 17-(1-OXOPROPOXY)-(17-β)-ANDROST-4-EN-3-ONE ◇ PANESTIN ◇ PERANDREN ◇ PROPIOKAN ◇ RECTHORMONE TESTOSTERONE ◇ STERANDRYL ◇ SYNANDROL ◇ SYNERONE ◇ TELIPEX ◇ TESTAFORM ◇ TESTEX ◇ TESTODET ◇ TESTODRIN ◇ TESTOGEN ◇ TESTONIQUE ◇ TESTORMOL ◇ TESTOSTERON PROPIONATE ◇ TESTOSTERONE-17-PROPIONATE ◇ TESTOSTERONE-17-β-PROPIONATE ◇ TESTOVIRON ◇ TESTOXYL ◇ TESTREX ◇ TOSTRIN ◇ TP ◇ UNITESTON ◇ VULVAN

USE IN FOOD:

Purpose: Animal drug.

Where Used: Beef.

Regulations: FDA - 21CFR 556.710. Limitation of no residue in excess of the following increments above the concentrations naturally present in untreated animals: 0.60 ppb in muscle, 2.6 ppb in fat, 1.9 ppb in kidney, and 1.3 ppb in liver of heifers.

IARC Cancer Review: Animal Sufficient Evidence IMEMDT 21,519,79. EPA Genetic Toxicology Program.

SAFETY PROFILE: Moderately toxic by ingestion and intraperitoneal routes. An experimental carcinogen, neoplastigen, tumorigen, and teratogen. Human male reproductive effects by intramuscular and parenteral routes: changes in spermatogenesis, testes, epididymis, and sperm duct. Human female reproductive effects by intramuscular and parenteral routes: menstrual cycle changes or disorders and effects on fertility. Experimental reproductive effects. Mutagenic data. When heated to decomposition it emits acrid smoke and irritating fumes.

TOXICITY DATA and CODEN

cyt-ofs-unr 500 μg BEXBBO 15,329,80
dnd-rat:lvr 300 μmol/L SinJF# 26OCT82
ims-man TDLo:10714 μg/kg (30D male):REP CCPTAY 5,295,72
ims-wmn TDLo:3 mg/kg (6D pre):REP PSEBAA 37,689,38
orl-rat TDLo:100 mg/kg (17-20D preg):TER ECJPAE 24,77,77
scu-rat TDLo:10 mg/kg:NEO JSONDX 5,396,84
imp-rat TDLo:432 mg/kg/48W-C:ETA CNREA8 37,1929,77
orl-rat LD50:1000 mg/kg NIIRDN 6,487,82

TBJ600
(1R,3S)3[(1′RS)(1′,2′,2′,2′-TETRABROMOETHYL)]-2,2-DIMETHYLCYCLOPROPANECARBOXYLIC ACID (S)-α-CYANO-3-PHENOXYBENZYL ESTER

USE IN FOOD:

Purpose: Insecticide.

Where Used: Cottonseed oil.

Regulations: FDA - 21CFR 193.418. Limitation of 0.20 ppm in cottonseed oil.

SAFETY PROFILE: When heated to decomposition emits toxic fumes of Br^-, CN^-.

TBQ250 CAS: 127-18-4
1,1,2,2-TETRACHLOROETHYLENE

DOT: 1897
mf: C_2Cl_4 mw: 165.82

PROP: Colorless liquid; chloroform-like odor. Mp: −23.35°, bp: 121.20°, d: 1.6311 @ 15°/4°, vap press: 15.8 mm @ 22°, vap d: 5.83.

SYNS: ANKILOSTIN ◇ ANTISOL 1 ◇ CARBON BICHLORIDE ◇ CARBON DICHLORIDE ◇ CZTEROCHLOROETYLEN (POLISH) ◇ DIDAKENE ◇ DOW-PER ◇ ENT 1,860 ◇ ETHYLENE TETRACHLORIDE ◇ FEDAL-UN ◇ NCI-C04580 ◇ NEMA ◇ PERAWIN ◇ PERCHLOORETHYLEEN, PER (DUTCH) ◇ PERCHLOR ◇ PERCHLORAETHYLEN, PER (GERMAN) ◇ PERCHLORETHYLENE ◇ PERCHLORETHYLENE, PER (FRENCH) ◇ PERCHLOROETHYLENE (ACGIH, DOT) ◇ PERCLENE ◇ PERCLOROETILENE (ITALIAN) ◇ PERCOSOLVE ◇ PERK ◇ PERKLONE ◇ PERSEC ◇ RCRA WASTE NUMBER U210 ◇ TETLEN ◇ TETRACAP ◇ TETRACHLOORETHEEN (DUTCH) ◇ TETRACHLORAETHEN (GERMAN) ◇ TETRACHLOROETHENE ◇ TETRACHLOROETHYLENE (DOT) ◇ TETRACLOROETENE (ITALIAN) ◇ TETRALENO ◇ TETRALEX ◇ TETRAVEC ◇ TETROGUER ◇ TETROPIL

USE IN FOOD:

Purpose: Adjuvant in foamed plastic, blowing agent.

Where Used: Packaging materials.

Regulations: FDA - 21CFR 178.3010. Limitation of 0.3 percent in finished foamed polyethylene.

IARC Cancer Review: Animal Limited Evidence IMEMDT 20,491,79. NCI Carcinogenesis Bioassay (gavage); Clear Evidence: mouse NCITR* NCI-CG-TR-13,77; (inhalation); Clear Evidence: mouse, rat NTPTR* NTP-TR-311,86; (gavage); Inadequate Studies: rat NCITR* NCI-CG-TR-13,77. EPA Genetic Toxicology Program. Community Right-To-Know List.

OSHA PEL: TWA 25 ppm ACGIH TLV: TWA 50 ppm (skin); STEL 200 ppm DFG MAK: 50 ppm (345 mg/m^3); BAT: blood 100 μg/dl DOT Classification: Poison B; Label: St. Andrews Cross; ORM-A; Label: None

NIOSH REL: (Tetrachloroethylene) Minimize workplace exposure.

SAFETY PROFILE: Experimental poison by intravenous route. Moderately toxic to humans by inhalation with the following effects: local anesthetic, conjunctiva irritation, general anesthesia, hallucinations, distorted perceptions, coma, and pulmonary changes. Moderately experimentally toxic by ingestion, inhalation, intraperitoneal and subcutaneous routes. An experimental carcinogen and teratogen. Experimental reproductive effects. Human mutagenic data. An eye and severe skin irritant. Can cause dermatitis, particularly after repeated or prolonged contact with the skin. Irritates the gastrointestinal tract upon ingestion. When heated to decomposition it emits highly toxic fumes of Cl^-.

TOXICITY DATA and CODEN

skn-rbt 810 mg/24H SEV JETOAS 9,171,76
eye-rbt 162 mg MLD JETOAS 9,171,76
mmo-sat 50 μL/plate NIOSH* 5AUG77
dns-hmn:lng 100 mg/L NTIS** PB82-185075
ihl-rat TCLo:1000 ppm/24H (14D pre/1-22D preg):TER APTOD9 19,A21,80
ihl-rat TCLo:900 ppm/7H (7-13D preg):REP TJADAB 19,41A,79
ihl-mus TCLo:300 ppm/7H (6-15D preg):TER TXAPA9 32,84,75
orl-mus TDLo:195 g/kg/50W-I:CAR NCITR* NCI-CG-TR-13,77

orl-mus TD :240 g/kg/62W-I:CAR NCITR* NCI-CG-TR-13,77
ihl-hmn TCLo:96 ppm/7H:PNS,EYE,CNS NTIS** PB257-185
ihl-man TCLo:280 ppm/2H:EYE,CNS AMIHBC 5,566,52
ihl-man LDLo:2857 mg/kg:CNS,PUL MLDCAS 5,152,72
orl-rat LD50:8850 mg/kg NPIRI* 1,96,74

TBQ750 CAS: 1897-45-6
TETRACHLOROISOPHTHALONITRILE
mf: $C_8Cl_4N_2$ mw: 265.90

SYNS: BRAVO ◇ BRAVO 6F ◇ BRAVO-W-75 ◇ CHLOROALONIL ◇ CHLOROTHALONIL ◇ CHLORTHALONIL (GERMAN) ◇ DAC 2797 ◇ DACONIL ◇ DACONIL 2787 FLOWABLE FUNGICIDE ◇ DACOSOIL ◇ 1,3-DICYANOTETRACHLOROBENZENE ◇ EXOTHERM ◇ EXOTHERM TERMIL ◇ FORTURF ◇ NCI-C00102 ◇ NOPCOCIDE ◇ SWEEP ◇ TCIN ◇ m-TCPN ◇ TERMIL ◇ 2,4,5,6-TETRACHLORO-3-CYANOBENZONITRILE ◇ m-TETRACHLOROPHTHALONITRILE ◇ TPN (pesticide)

USE IN FOOD:

Purpose: Fungicide.

Where Used: Broccoli, cabbage, cantaloupe, carrots, cauliflower, celery, citrus oil, cucumber, lettuce, onions, potatoes, tomatoes, watermelon.

Regulations: FDA - 21CFR 193.84. Fungicide residue tolerance of 10 ppm in citrus oil. (Expired 12/31/1983)

IARC Cancer Review: Animal Limited Evidence IMEMDT 30,319,83. NCI Carcinogenesis Bioassay (feed); Clear Evidence: rat NCITR* NCI-CG-TR-41,78. Cyanide and its compounds are on the Community Right-To-Know List. EPA Genetic Toxicology Program.

SAFETY PROFILE: Moderately toxic by intraperitoneal route. Mildly toxic by ingestion. An experimental carcinogen. When heated to decomposition it emits very toxic fumes of Cl^-, NO_x, and CN^-.

TOXICITY DATA and CODEN

orl-rat TDLo:142 g/kg/80W-C:CAR NCITR* NCI-CG-TR-41,78
orl-rat LD50:10 mg/kg 85ARAE 4,75,76

TBW100 CAS: 961-11-5
TETRACHLORVINPHOS
mf: $C_{10}H_9Cl_4O_4P$ mw: 365.96

SYNS: 2-CHLORO-1-(2,4,5-TRICHLOROPHENYL)VINYL DIMETHYL PHOSPHATE ◇ 2-CHLORO-1-(2,4,5-TRICHLOROPHENYL(VINYL PHOSPHORIC ACID DIMETHYL ESTER ◇ O,O-DIMETHYL-O-2-CHLOR-1-(2,4,5-TRICHLORPHENYL)-VINYL-PHOSPHAT (GERMAN) ◇ IPO 8 ◇ NCI C00168 ◇ PHOSPHORIC ACID, 2-CHLORO-1-(2,4,5-TRICHLOROPHENYL)ETHENYL DIMETHYL ESTER ◇ 2,4,5-TRICHLORO-α-(CHLOROMETHYLENE)BENZYL PHOSPHATE

USE IN FOOD:

Purpose: Feed additive, insecticide.

Where Used: Animal feed, beef, pork.

Regulations: FDA - 21CFR 561.91. Limitation of 0.00015 lb/100 lb of body weight of cattle and horses when used in animal feed. Limitation of 0.00011 lb/100 lb of body weight of swine when used in animal feed.

NCI Carcinogenesis Bioassay (feed); Results Positive: Mouse, Rat (NCITR* NCI-CG-TR-33,78). Community Right-To-Know List.

SAFETY PROFILE: Poison by ingestion. An experimental carcinogen, neoplastigen, and tumorigen. Experimental reproductive effects. When heated to decomposition it emits toxic fumes of Cl^- and PO_x.

TOXICITY DATA and CODEN

orl-mus TDLo:692 g/kg (2Y male):REP FAATDF 5,840,85
orl-rat TDLo:240 g/kg,80W-C:NEO NCITR* NCI-CG-TR-33,78
orl-mus TDLo:450 g/kg/67W-C:CAR NCITR* NCI-CG-TR-33,78
orl-mus TD :1384 g/kg/2Y-C:CAR FAATDF 5,840,85
orl-rat LD50:4 g/kg 85GYAZ -,34,71

TBX250 CAS: 64-75-5
TETRACYCLINE HYDROCHLORIDE
mf: $C_{22}H_{24}N_2O_8 \cdot ClH$ mw: 480.94

PROP: Very sol in water; sol in methanol, ethanol; insol in ether, hydrocarbon solvents.

SYNS: ACHROMYCIN ◇ ACHROMYCIN HYDROCHLORIDE ◇ AMBRACYN ◇ ARTOMYCIN ◇ BRISTACYCLINE ◇ CEFRACYCLINE TABLETS ◇ CHLORHYDRATE de TETRACYCLINE (FRENCH) ◇ CYCLOPAR ◇ DIACYCINE ◇ DUMOCYCIN ◇ MEDAMYCIN ◇ MEPHACYCLIN ◇ NCI-C55561 ◇ PALTET ◇ PANMYCIN HYDROCHLORIDE ◇ PARTREX ◇ PIRACAPS ◇ POLYCYCLINE HYDROCHLORIDE ◇ QIDTET ◇ QUADRACYCLINE ◇ QUATREX ◇ REMICYCLIN ◇ RICYCLINE ◇ RO-CYCLINE ◇ SK-TETRACYCLINE ◇ STECLIN HYDROCHLORIDE ◇ SUBAMYCIN ◇ SUPRAMYCIN ◇ T-250 CAPSULES ◇ TC HYDROCHLORIDE ◇ TEFILIN ◇ TELINE ◇ TELOTREX ◇ TETRABAKAT ◇ TETRABLET ◇ TETRACAPS ◇ TETRACICLINA CLORI-

DRATO (ITALIAN) ◇ TETRACOMPREN ◇ TETRACYCLINE CHLORIDE ◇ TETRA-D ◇ TETRALUTION ◇ TETRA-WEDEL ◇ TETROSOL ◇ TOPICYCLINE ◇ TOTOMYCIN ◇ TRIPHACYCLIN ◇ U-5965 ◇ UNICIN ◇ UNIMYCIN ◇ VETQUAMYCIN-324

USE IN FOOD:

Purpose: Animal drug.

Where Used: Beef, chicken, lamb, pork, turkey.

Regulations: FDA - 21CFR 556.720. Limitation of 0.25 ppm in calves, swine, sheep, chickens, and turkeys. USDA CES Ranking: B-3 (1986).

SAFETY PROFILE: Poison by intraperitoneal, intravenous, and possibly other routes. Moderately toxic by ingestion and subcutaneous routes. An experimental teratogen. Experimental reproductive effects. Mutagenic data. When heated to decomposition it emits very toxic fumes of HCl and NO_x.

TOXICITY DATA and CODEN

dnd-esc 50 μmol/L MUREAV 89,95,81
msc-mus:mmr 100 mg/L MUREAV 40,261,76
orl-rat TDLo:14 g/kg (3D male/3D pre-22D preg):TER TXAPA9 7,409,65
orl-rat TDLo:6 g/kg (9-14D preg):REP KSRNAM 4,516,70
ipr-mus TDLo:900 mg/kg (8-13D preg):TER IYKEDH 3,75,72
orl-rat LD50:6443 mg/kg TXAPA9 18,185,71
ipr-rat LD50:318 mg/kg GNRIDX 2,26,68

TBY250 CAS: 27196-00-5
TETRADECANOL, mixed isomers

mf: $C_{14}H_{30}O$ mw: 214.44

SYNS: MYRISTYL ALCOHOL (mixed isomers) ◇ TETRADECYL ALCOHOL

USE IN FOOD:

Purpose: Intermediate.

Where Used: Various.

Regulations: FDA - 21CFR 172.864. Use at a level not in excess of the amount reasonably required to accomplish the intended effect.

SAFETY PROFILE: Mildly toxic by ingestion and skin contact. Combustible when exposed to heat or flame; can react with oxidizing materials. To fight fire, use CO_2, dry chemical. When heated to decomposition it emits acrid smoke and irritating fumes.

TOXICITY DATA and CODEN

orl-rat LD50:33 g/kg AIHAAP 30,470,69
skn-rbt LD50:7130 mg/kg AIHAAP 30,470,69

TCA500 CAS: 139-08-2
TETRADECYL DIMETHYL BENZYLAMMONIUM CHLORIDE

mf: $C_{23}H_{42}N{\bullet}Cl$ mw: 368.11

SYNS: ARQUAD DM14B-90 ◇ N,N-DIMETHYL-N-TETRADECYLBENZENEMETHANAMINIUM, CHLORIDE (9CI) ◇ NISSAN CATION M2-100

USE IN FOOD:

Purpose: Antimicrobial agent.

Where Used: Beets, sugarcane, sugarcane juice (raw).

Regulations: FDA - 21CFR 172.165. Limitation of 3.0-12.0 ppm. 21CFR 173.320. Limitation of 0.60 ppm based on weight of raw sugarcane or raw beets.

SAFETY PROFILE: A skin and eye irritant. When heated to decomposition it emits very toxic fumes of NO_x, NH_3 and Cl^-.

TCE250 CAS: 112-60-7
TETRAETHYLENE GLYCOL

mf: $C_8H_{18}O_5$ mw: 194.26

PROP: Colorless to pale straw-colored liquid. Bp: 327.3°, fp: −6°, flash p: 360°F (OC), d: 1.1248 @ 20/20°, vap press: 1 mm @ 153.9°. Misc in water.

SYNS: HI-DRY ◇ 2,2′-(OXYBIS(ETHYLENEOXY))DIETHANOL

USE IN FOOD:

Purpose: Finishing agent (twine).

Where Used: Meat.

Regulations: FDA - 21CFR 178.3940. Limitation of 0.7 percent by weight of twine.

SAFETY PROFILE: Mildly toxic by ingestion. A skin and eye irritant. Combustible when exposed to heat or flame; can react with oxidizing materials. To fight fire, use alcohol foam, water, CO_2, dry chemical. When heated to decomposition it emits acrid smoke and irritating fumes.

TOXICITY DATA and CODEN

skn-rbt 550 mg open MLD UCDS** 3/3/69
eye-rbt 565 mg AJOPAA 29,1363,46
orl-rat LD50:29 g/kg UCDS** 3/3/69

TCR750 CAS: 109-99-9

TETRAHYDROFURAN

DOT: 2056

mf: C_4H_8O mw: 72.12

PROP: Colorless, mobile liquid; ether-like odor. Bp: 65.4°, flash p: 1.4°F (TCC), lel: 1.8%, uel: 11.8%, fp: −108.5°, d: 0.888 @ 20/4°, vap press: 114 mm @ 15°, vap d: 2.5, autoign temp: 610°F. Misc with water, alc, ketones, esters, ethers, and hydrocarbons.

SYNS: BUTYLENE OXIDE ◇ CYCLOTETRAMETHYLENE OXIDE ◇ DIETHYLENE OXIDE ◇ 1,4-EPOXYBUTANE ◇ FURANIDINE ◇ HYDROFURAN ◇ NCI-C60560 ◇ OXACYCLOPENTANE ◇ OXOLANE ◇ RCRA WASTE NUMBER U213 ◇ TETRAHYDROFURAAN (DUTCH) ◇ TETRAHYDROFURANNE (FRENCH) ◇ TETRAIDROFURANO (ITALIAN) ◇ TETRAMETHYLENE OXIDE ◇ THF

USE IN FOOD:

Purpose: Solvent.

Where Used: Packaging materials.

Regulations: FDA - 21CFR 178.3950. Limitation of 1.5 percent in film.

OSHA PEL: TWA 200 ppm; STEL 250 ppm ACGIH TLV: TWA 200 ppm; STEL 250 ppm DFG MAK: 200 ppm (590 mg/m^3) DOT Classification: Flammable Liquid; Label: Flammable Liquid

SAFETY PROFILE: Moderately toxic by ingestion and intraperitoneal routes. Mildly toxic by inhalation. Human systemic effects by inhalation: general anesthesia. Mutagenic data. Irritant to eyes and mucous membranes. Narcotic in high concentrations.

Flammable liquid. A very dangerous fire hazard when exposed to heat, flames, oxidizers. Explosive in the form of vapor when exposed to heat or flame. In common with other ethers, unstabilized tetrahydrofuran forms thermally explosive peroxides on exposure to air. Stored THF must always be tested for peroxide prior to distillation. Caustic alkalies deplete the inhibitor in THF and may subsequently cause an explosive reaction. Violent reaction with metal halides. Can react with oxidizing materials. To fight fire, use foam, dry chemical, CO_2. When heated to decomposition it emits acrid smoke and irritating fumes.

TOXICITY DATA and CODEN

mmo-esc 1 µmol/L GTPZAB 26(1),43,82
ihl-hmn TCLo:25000 ppm:CNS 34ZIAG -,580,69
orl-rat LDLo:3000 mg/kg TPKVAL 5,21,63
ihl-rat LC50:21000 ppm/3H SSEIBV 20,141,84

TCU600

TETRAHYDROLINALOOL

mf: $C_{10}H_{22}O_2$ mw: 158.29

PROP: Colorless liquid; floral odor. D: 0.923, refr index: 1.431, flash p: 183°F. Sol in alc, fixed oils; insol in water.

SYNS: 3,7-DIMETHYL-3-OCTANOL ◇ FEMA No. 3060

USE IN FOOD:

Purpose: Flavoring agent.

Where Used: Various.

Regulations: FDA - 21CFR 172.515. Use at a level not in excess of the amount reasonably required to accomplish the intended effect.

SAFETY PROFILE: Combustible liquid. When heated to decomposition it emits acrid smoke and irritating fumes.

TCW750 CAS: 33401-94-4

(E)-4,5,6-TETRAHYDRO-1-METHYL-2-(2-(2-THIENYL)ETHENYL)PYRIMIDINE

mf: $C_{11}H_{14}N_2S{\bullet}C_4H_6O_6$ mw: 356.43

SYNS: BANMINTH ◇ CP 10423-18 ◇ PYRANTEL TARTRATE ◇ PYREQUAN TARTRATE ◇ (E)-1,4,5,6-TETRAHYDRO-1-METHYL-2-(2-(2-THIENYL)VINYL)PYRIMIDINE TARTARATE (1:1)

USE IN FOOD:

Purpose: Animal drug.

Where Used: Pork.

Regulations: FDA - 21CFR 556.560. Limitation of 10 ppm in swine liver and kidney and 1 ppm in swine muscle. 21CFR 558.485.

SAFETY PROFILE: Poison by ingestion and intravenous routes. When heated to decomposition it emits very toxic fumes of NO_x and SO_x.

TOXICITY DATA and CODEN

orl-rat LD50:170 mg/kg AUVJA2 46,297,70

TDE500 CAS: 16423-68-0

TETRAIODOFLUORESCEIN SODIUM SALT

mf: $C_{20}H_6I_4O_5{\bullet}2Na$ mw: 879.84

PROP: Brown powder. Sol in water, conc sulfuric acid.

SYNS: AIZEN ERYTHROSINE ◇ CALCOCID ERYTHROSINE N ◇ CANACERT ERYTHROSINE BS ◇ 9-(o-CARBOXYPHENYL)-6-HYDROXY-2,4,5,7-TETRAIODO-3-ISOXANTHONE ◇ C.I. 45430 ◇ C.I. ACID RED 51 ◇ CILEFA PINK B ◇ D & C RED NO. 3 ◇ DOLKWAL ERYTHROSINE ◇ DYE F D & C RED NO. 3 ◇ E 127 ◇ EBS ◇ EDICOL SUPRA ERYTHROSINE A ◇ ERYTHROSIN ◇ ERYTHROSINE B-FO (BIOLOGICAL STAIN) ◇ FD & C RED NO. 3 (FCC) ◇ FOOD RED 14 ◇ HEXACERT RED NO. 3 ◇ HEXACOL ERYTHROSINE BS ◇ LB-ROT 1 ◇ MAPLE ERYTHROSINE ◇ NEW PINK BLUISH GEIGY ◇ 1427 RED ◇ 1671 RED ◇ 2',4',5',7'-TETRAIODOFLUORESCEIN, DISODIUM SALT ◇ USACERT RED NO. 3

USE IN FOOD:

Purpose: Color additive.

Where Used: Candy, cherries, confections.

Regulations: FDA - 21CFR 74.303, 81.1. Use in accordance with good manufacturing practice.

EPA Genetic Toxicology Program.

SAFETY PROFILE: Poison by intravenous route. Moderately toxic by ingestion and possibly other routes. Human mutagenic data. When heated to decomposition it emits very toxic fumes of Na_2O and I^-.

TOXICITY DATA and CODEN

dnr-bcs 2 mg/disc TRENAF 27,153,76
dni-hmn:leu 500 mg/L NEZAAQ 30,574,75
orl-rat LD50:1840 mg/kg SCPHA4 47,39,79

TDV725 CAS: 1124-11-4
TETRAMETHYLPYRAZINE
mf: $C_8H_{12}N_2$ mw: 136.22

PROP: White crystals or powder; fermented soybean odor. Mp: 85-90°, sol in alc, propylene glycol, fixed oils; sltly sol in water.

SYNS: FEMA No. 3237 ◇ 2,3,5,6-TETRAMETHYL PYRAZINE (FCC)

USE IN FOOD:

Purpose: Flavoring agent.

Where Used: Various.

Regulations: GRAS when used at a level not in excess of the amount reasonably required to accomplish the intended effect.

SAFETY PROFILE: Poison by intravenous, and intraperitoneal routes. Moderately toxic by ingestion. When heated to decomposition it emits toxic fumes of NO_x.

TOXICITY DATA and CODEN

orl-rat LD50:1910 mg/kg DCTODJ 3,249,80
ipr-mus LD50:800 μg/kg IJCREE 23,119,85

TDW500 CAS: 108-62-3
2,4,6,8-TETRAMETHYL-1,3,5,7-TETROXOCANE
DOT: 1332
mf: $C_8H_{16}O_4$ mw: 176.24

SYNS: ACETALDEHYDE, TETRAMER ◇ ANTIMILACE ◇ ARIOTOX ◇ CEKUMETA ◇ CORRY'S SLUG DEATH ◇ HALIZAN ◇ META ◇ METACETALDEHYDE ◇ METALDEHYD (GERMAN) ◇ METALDEHYDE (DOT) ◇ METALDEIDE (ITALIAN) ◇ METASON ◇ NAMEKIL ◇ SLUG-TOX

USE IN FOOD:

Purpose: Pesticide.

Where Used: Strawberries.

Regulations: FDA - 21CFR 193.280. Pesticide residue tolerance of zero in strawberries.

DOT Classification: Flammable Solid; Label: Flammable Solid

SAFETY PROFILE: Human poison by ingestion. Moderately toxic experimentally by ingestion and possibly other routes. Human systemic effects by ingestion: convulsions or effect on seizure threshold. Experimental reproductive effects. Mutagenic data. A flammable solid. When heated to decomposition it emits acrid smoke and irritating fumes.

TOXICITY DATA and CODEN

mmo-smc 5 ppm RSTUDV 6,161,76
orl-rat TDLo:61 g/kg (MGN):REP TXCYAC 4,97,75
orl-hmn LDLo:43 mg/kg 32ZWAA 8,159,74
orl-chd LDLo:100 mg/kg:CNS 85GYAZ -,158,71
orl-rat LD50:630 mg/kg PCOC** -,701,66

TEE500 CAS: 7722-88-5
TETRASODIUM PYROPHOSPHATE, ANHYDROUS
mf: $O_7P_2 \cdot 4Na$ mw: 265.90

PROP: White crystalline powder. Mp: 988°, d: 2.534. Sol in water; insol in alc.

SYNS: ANHYDROUS TETRASODIUM PYROPHOSPHATE ◇ NATRIUMPYROPHOSPHAT ◇ PHOSPHOTEX ◇ PYROPHOSPHATE ◇ SODIUM PYROPHOSPHATE (FCC) ◇ TETRANATRIUMPYROPHOSPHAT (GERMAN) ◇ TETRASODIUM DIPHOSPHATE ◇ TETRASODIUM PYROPHOSPHATE (ACGIH) ◇ TSPP ◇ VICTOR TSPP

USE IN FOOD:

Purpose: Boiler water additive, buffer, dietary supplement, emulsifier, nutrient, sequestrant, stabilizer.

Where Used: Cheese, chocolate drink powders, malted milk, tuna.

Regulations: FDA - 21CFR 173.310, 21CFR 181.29, 182.6789. GRAS when used in accordance with good manufacturing practice.

ACGIH TLV: TWA 5 mg/m^3

SAFETY PROFILE: Poison by ingestion, intraperitoneal, intravenous and subcutaneous routes. It is not a cholinesterase inhibitor. When heated to decomposition it emits toxic fumes of PO_x and Na_2O.

TOXICITY DATA and CODEN

orl-rat LD50:4000 mg/kg ARZNAD 7,172,57

TER500 CAS: 19525-20-3

THIABENDAZOLE HYDROCHLORIDE

mf: $C_{10}H_7N_3S{\bullet}ClH$ mw: 237.72

SYN: 2-(4-THIAZOLYL)-BENZIMIDAZOLE, HYDROCHLORIDE

USE IN FOOD:

Purpose: Animal drug, mold retardant.

Where Used: Animal feed, apple pomace (dried), apples, bananas, beef, citrus fruit, citrus molasses, citrus pulp (dried), goat, grape pomace (dried), grape pomace (wet), lamb, milk, pears, pheasants, pork, potato processing waste, potatoes, rice hulls, sugar beet pulp (dried), sugar beets, wheat milled fractions (except flour).

Regulations: FDA - 21CFR 556.730. Limitation of 0.1 ppm in cattle, goats, sheep, pheasant, and swine. Limitation of 0.05 ppm in milk. 21CFR 558.615. 21CFR 561.380. Limitation of 33 ppm in dried apple pomace, 3.5 ppm in dried sugar beet pulp, 20 ppm in citrus molasses, 35 ppm in dried citrus pulp, 150.0 ppm in dry or wet grape pomace, 30 ppm in potato processing waste, 8 ppm in rice hulls, 3.0 ppm in wheat milled fractions (except flour) when used for animal feed.

SAFETY PROFILE: Poison by intravenous route. Moderately toxic by ingestion and intraperitoneal routes. When heated to decomposition it emits very toxic fumes of HCl, SO_x and NO_x.

TOXICITY DATA and CODEN

orl-rat LD50:3600 mg/kg TXAPA9 7,53,65

TET000 CAS: 67-03-8

THIAMINE CHLORIDE HYDROCHLORIDE

mf: $C_{12}H_{17}N_4OS{\bullet}ClH{\bullet}Cl$ mw: 337.30

PROP: Small white hygroscopic crystals or crystalline powder; nut-like odor. Mp: 248° (decomp). Sol in water, glycerin; sltly sol in alc; insol in ether, benzene.

SYNS: THIAMINE DICHLORIDE ◇ THIAMINE HYDROCHLORIDE (FCC) ◇ THIAMIN HYDROCHLORIDE ◇ THIAMINIUM CHLORIDE HYDROCHLORIDE ◇ USAF CB-20 ◇ VITAMIN B^1 ◇ VITAMIN B HYDROCHLORIDE

USE IN FOOD:

Purpose: Dietary supplement, flavoring agent, nutrient.

Where Used: Wine.

Regulations: FDA - 21CFR 182.5875, 184.1875. GRAS when used at a level not in excess of the amount reasonably required to accomplish the intended effect. BATF - 27CFR 240.1051. Limitation of 0.005 pounds/1000 gallons of wine.

SAFETY PROFILE: Poison by intravenous and intraperitoneal routes. Mildly toxic by ingestion. When heated to decomposition it emits very toxic fumes of HCl, Cl^-, SO_x and NO_x.

TOXICITY DATA and CODEN

orl-mus LD50:8224 mg/kg IZVIAK 37,82,67

TET500 CAS: 532-43-4

THIAMINE NITRATE

mf: $C_{12}H_{17}N_4OS{\bullet}NO_3$ mw: 327.40

PROP: White crystals or crystalline powder; slt characteristic odor. Mp: 196-200° (decomp). Non-hygroscopic; sltly sol in water, alc and chloroform.

SYNS: 3-(4-AMINO-2-METHYLPYRIMIDYL-5-METHYL)-4-METHYL-5,β-HYDROXYETHYLTHIAZOLIUM NITRATE ◇ THIAMINE MONONITRATE (FCC) ◇ THIAMINE NITRATE (SALT) ◇ VITAMIN B1 MONONITRATE ◇ VITAMIN B1 NITRATE

USE IN FOOD:

Purpose: Dietary supplement, nutrient.

Where Used: Crackers, egg substitutes (frozen), enriched flour.

Regulations: FDA - 21CFR 182.5878, 184.1878. GRAS when used in accordance with good manufacturing practice.

SAFETY PROFILE: Poison by intravenous and intraperitoneal routes. A powerful oxidizer. When heated to decomposition it emits very toxic fumes of NO_x and SO_x.

TOXICITY DATA and CODEN

ivn-rbt LD50:113 mg/kg PSEBAA 68,153,48

TET800 CAS: 55297-95-5
THIAMUTILIN
mf: $C_{28}H_{47}NO_4S$ mw: 493.82

PROP: Crystals from acetone. Mp: 147-148° (after stirring in ethyl acetate and drying at 60° and 80° overnight).

SYNS: 14-DEOXY-14-((2-DIETHYLAMINOETHYL-THIO)-ACETOXY)MUTILINE ◇ 14-DESOSSI-14-((2-DIETILAMINO-ETIL)MERCAPTO-ACETOSSI)MUTILIN IDROGENO FUMARATO (ITALIAN) ◇ 14-DESOXY-14-((DIETHYLAMINO-ETHYL)-MERCAPTO ACETOXYL)-MUTILIN HYDROGEN FUMARATE ◇ DYNALIN INJECTABLE ◇ DYNAMUTILIN ◇ 81723 HFU ◇ SQ 14055 ◇ SQ 22947 ◇ TIAMULIN ◇ TIAMULINA (ITALIAN)

USE IN FOOD:

Purpose: Animal drug, animal feed drug.

Where Used: Animal feed, pork.

Regulations: FDA - 21CFR 556.736. Limitation of 0.4 ppm in liver of swine. 21CFR 558.600.

SAFETY PROFILE: Poison by intramuscular and intravenous routes. Moderately toxic by ingestion and subcutaneous routes. When heated to decomposition it emits toxic fumes of SO_x and NO_x.

TOXICITY DATA and CODEN

orl-rat LD50:2230 mg/kg RZOVBM 8,251,80

TEX000 CAS: 148-79-8
2-(THIAZOL-4-YL)BENZIMIDAZOLE
mf: $C_{10}H_7N_3S$ mw: 201.26

PROP: White-to-tan; odorless. Mp: 304°. Insol in water; sltly sol in alc, acetone; very sltly sol in ether, chloroform.

SYNS: APL-LUSTER ◇ ARBOTECT ◇ 4-(2-BENZIMIDAZO-LYL)THIAZOLE ◇ BOVIZOLE ◇ EPROFIL ◇ EQUIZOLE ◇ LOMBRISTOP ◇ MERTEC ◇ METASOL TK-100 ◇ MINTEZOL ◇ MINZOLUM ◇ MK 360 ◇ MYCOZOL ◇ NEMAPAN ◇ OMNIZOLE ◇ POLIVAL ◇ TBDZ ◇ TECTO ◇ THIABEN ◇ THIABENDAZOLE (USDA) ◇ THIABENZOLE ◇ 2-(4-THIAZOLYL)BENZIMIDAZOLE ◇ 2-(4′-THIAZOLYL)BENZIMIDAZOLE ◇ 2-(4-THIAZOLYL)-1H-BENZIMIDAZOLE ◇ THIBENZOLE ◇ TOP FORM WORMER

USE IN FOOD:

Purpose: Fungicide.

Where Used: Animal feed.

Regulations: USDA CES Ranking: B-2 (1987).

EPA Genetic Toxicology Program.

SAFETY PROFILE: Poison by ingestion. An experimental teratogen. Experimental reproductive effects. Mutagenic data. When heated to decomposition it emits toxic fumes of SO_x and NO_x.

TOXICITY DATA and CODEN

oms-asn 80 μmol/L BBACAQ 543,82,78
cyt-ham:lng 45 mg/L/48H GMCRDC 27,95,81
orl-rat TDLo:1250 mg/kg (6-15D preg):TER JPFCD2 14,563,79
orl-mus TDLo:250 mg/kg (9D preg):TER TRENAF 34,343,83
orl-rat LD50:3100 mg/kg TXAPA9 7,53,65

TEX250 CAS: 72-14-0
N[1]-2-THIAZOLYLSULFANILAMIDE
mf: $C_9H_9N_3O_2S_2$ mw: 255.33

SYNS: 2-(p-AMINOBENZENESULFONAMIDO)THIAZOLE ◇ 2-(p-AMINOBENZENESULPHONAMIDO)THIAZOLE ◇ 4-AMINO-N-2-THIAZOLYLBENZENESULFONAMIDE ◇ AZOSEPTALE ◇ CERAZOL (suspension) ◇ CHEMOSEPT ◇ DUATOK ◇ ELEUDRON ◇ FORMOSULFATHIAZOLE ◇ M+B 760 ◇ NEOSTREPSAN ◇ NORSULFASOL ◇ NORSULFAZOLE ◇ PLANOMIDE ◇ POLISEPTIL ◇ RP 2990 ◇ STREPTOSILTHIAZOLE ◇ SULFAMUL ◇ 2-SULFANILAMIDOTHIAZOLE ◇ 2-(SULFANILYLAMI-NO)THIAZOLE ◇ SULFATHIAZOL ◇ SULFATHIAZOLE (USDA) ◇ 2-SULFONAMIDOTHIAZOLE ◇ SULPHATHIA-ZOLE ◇ SULZOL ◇ THIACOCCINE ◇ THIAZAMIDE ◇ THIOZAMIDE ◇ USAF SN-9

USE IN FOOD:

Purpose: Animal drug.

Where Used: Pork.

Regulations: FDA - 21CFR 556.690. Limitation of 0.1 ppm in swine. USDA CES Ranking: B-1 (1987).

EPA Genetic Toxicology Program.

SAFETY PROFILE: Human poison by unspecified route. Experimental poison by intraperitoneal route. Moderately toxic by intravenous, subcutaneous and parenteral routes. Mildly toxic by ingestion. An experimental tumorigen. Human systemic effects by unspecified route: conjuctiva irritation, tubule changes and allergic skin dermatitis. Mutagenic data. When heated to decomposition it emits very toxic fumes of NO_x and SO_x.

TOXICITY DATA and CODEN

dnd-esc 50 µmol/L MUREAV 89,95,81
pic-omi 5 mg/L JGMIAN 8,116,53
par-rat TDLo:500 mg/kg:ETA ACRAAX 37,258,52
par-mus TDLo:500 mg/kg:ETA ACRAAX 37,258,52
unr-man LDLo:250 mg/kg/23D-I: EYE,KID,SKN JPBAA7 59,501,47
ipr-rat LDLo:1250 mg/kg HBTXAC 5,164,59
orl-mus LD50:4500 mg/kg ARZNAD 21,571,71

TEX600 CAS: 51707-55-2
THIDIAZURON

mf: $C_9H_8N_4OS$ mw: 220.27

SYNS: DEFOLIT ◇ DROPP ◇ N-PHENYL-N′-1,2,3-THIADIAZOL-5-YL-UREA ◇ SN 49537 ◇ (N-1,2,3-THIADIAZOLYL-5)-N′-PHENYLUREA

USE IN FOOD:

Purpose: Defoliant.

Where Used: Animal feed, cottonseed hulls.

Regulations: FDA - 21CFR 561.385. Limitation of 0.8 ppm in cottonseed hulls when used for animal feed.

SAFETY PROFILE: Moderately toxic by ingestion. Experimental reproductive effects. When heated to decomposition it emits toxic fumes of SO_x and NO_x.

TOXICITY DATA and CODEN

orl-rat TDLo:813 mg/kg (76D male):REP GISAAA 49(1),72,84
orl-rat LD50:5350 mg/kg GISAAA 49(1),72,84

TFD500 CAS: 123-28-4
THIOBIS(DODECYL PROPIONATE)

mf: $C_{30}H_{58}O_4S$ mw: 514.94

PROP: White crystalline flakes; characteristic sweetish odor. Sol in organic solvents; insol in water.

SYNS: BIS(DODECYLOXYCARBONYLETHYL) SULFIDE ◇ DIDODECYL-3,3′-THIODIPROPIONATE ◇ DILAURYLESTER KYSELINY β′,β′-THIODIPROPIONOVE (CZECH) ◇ DILAURYL THIODIPROPIONATE ◇ DILAURYL-β-THIODIPROPIONATE ◇ DILAURYL-β′,β′-THIODIPROPIONATE ◇ DILAURYL-3,3′-THIODIPROPIONATE

USE IN FOOD:

Purpose: Antioxidant.

Where Used: Fats, oils, packaging materials.

Regulations: FDA - 21CFR 181.24. Limitation of 0.005 percent migrating from food packages.

SAFETY PROFILE: An eye irritant. When heated to decomposition it emits toxic fumes of SO_x.

TOXICITY DATA and CODEN

eye-rbt 500 mg/24H MOD 28ZPAK -,174,72

TFU750 CAS: 72-19-5
l-THREONINE

mf: $C_4H_9NO_3$ mw: 119.14

PROP: An essential amino acid. Colorless crystals or white crystalline powder; slt sweet taste. Mp: 255-257° with decomp. Sol in water; very sol in hot water; insol in alc, chloroform, ether.

SYNS: l-2-AMINO-3-HYDROXYBUTYRIC ACID ◇ THREONINE

USE IN FOOD:

Purpose: Dietary supplement, nutrient.

Where Used: Various.

Regulations: FDA - 21CFR 172.310. Limitation 5.0 percent by weight.

SAFETY PROFILE: Moderately toxic by intraperitoneal route. When heated to decomposition it emits toxic fumes of NO_x.

TOXICITY DATA and CODEN

ipr-rat LD50:3098 mg/kg ABBIA4 58,253,55

TFX500 CAS: 8007-46-3
THYME OIL

PROP: From distillation of flowering plant *Thymus vulgaris* L. (Fam. *Labiatae*). Colorless to reddish-brown liquid; pleasant odor, sharp taste. D: 0.930 @ 25/25°, refr index: 1.495 @ 20°.

SYNS: OIL of THYME ◇ THYMIAN OEL (GERMAN) ◇ THYM OIL

USE IN FOOD:

Purpose: Flavoring agent.

Where Used: Appetizers, cheese, sauces, soups.

Regulations: FDA - 21CFR 182.10, 182.20. GRAS when used at a level not in excess of the amount reasonably required to accomplish the intended effect.

SAFETY PROFILE: Moderately toxic by ingestion. Mutagenic data. An allergen and an irritant. Combustible when exposed to heat or flame. When heated to decomposition it emits acrid smoke and irritating fumes.

TOXICITY DATA and CODEN

dnr-bcs 2 mg/disc TOFOD5 8,91,85
orl-rat LD50:2840 mg/kg PHARAT 14,435,59

TFX750 CAS: 8007-46-3
THYME OIL RED

PROP: Main constituents are thymol, carvacrol. Found in plants *Thymus vulgaris L.* and *Thymus zygis L.* (FCTXAV 12,807,74).

SYN: SPANISH THYME OIL

USE IN FOOD:

Purpose: Flavoring agent.

Where Used: Various.

Regulations: FDA - 21CFR 182.10, 182.20. GRAS when used at a level not in excess of the amount reasonably required to accomplish the intended effect.

SAFETY PROFILE: Mildly toxic by ingestion. A severe skin irritant. When heated to decomposition it emits acrid smoke and irritating fumes.

TOXICITY DATA and CODEN

skn-mus 500 mg SEV FCTXAV 12,807,74
orl-rat LD50:4700 mg/kg FCTXAV 12,807,74

TGC000 CAS: 7772-99-8
TIN(II) CHLORIDE (1:2)
mf: Cl_2Sn mw: 189.59
DOT: 1759

PROP: Colorless crystals. D: 2.71, mp: 37-38°. Sol in less than its own weight of water; very sol in hydrochloric acid (dilute or conc); sol in alc, ethyl acetate, glacial acetic acid, sodium hydroxide solution.

SYNS: C.I. 77864 ◇ NCI-C02722 ◇ STANNOUS CHLORIDE (FCC) ◇ STANNOUS CHLORIDE, SOLID (DOT) ◇ TIN DICHLORIDE ◇ TIN PROTOCHLORIDE

USE IN FOOD:

Purpose: Antioxidant, reducing agent.

Where Used: Asparagus, beverages (carbonated).

Regulations: FDA - 21CFR 172.180. Limitation of 20 ppm in asparagus packed in glass. 21CFR 184.1845. GRAS with a limitation of 0.0015 percent calculated as tin when used in accordance with good manufacturing practice.

NTP Carcinogenesis Bioassay (feed); No Evidence: mouse, rat NTPTR* NTP-TR-231,82. EPA Genetic Toxicology Program.

OSHA PEL: TWA 2 $mg(Sn)/m^3$ ACGIH TLV: TWA 2 $mg(Sn)/m^3$ DOT Classification: ORM-B; Label: None

SAFETY PROFILE: Poison by ingestion, intraperitoneal, intravenous and subcutaneous routes. Experimental reproductive effects. Human mutagenic data. Potentially explosive reaction with metal nitrates. Violent reactions with hydrogen peroxide; ethylene oxide; hydrazine hydrate; nitrates; K; Na. When heated to decomposition it emits toxic fumes of Cl^-.

TOXICITY DATA and CODEN

sln-smc 6 mmol/L MUTAEX 1,21,86
dnd-hmn:leu 10 μmol/L CBINA8 46,189,83
itt-rat TDLo:15167 μg/kg (1D male):REP JRPFA4 7,21,64
orl-rat LD50:700 mg/kg FOREAE 7,313,42

TGH000 CAS: 13463-67-7
TITANIUM OXIDE
mf: O_2Ti mw: 79.90

PROP: White amorphous powder. Mp: 1860° (decomp), d: 4.26. Insol in water, hydrochloric acid, dil sulfuric acid, alc.

SYNS: 1700 WHITE ◇ A-FIL CREAM ◇ ATLAS WHITE TITANIUM DIOXIDE ◇ AUSTIOX ◇ BAYERITIAN ◇ BAYERTITAN ◇ BAYTITAN ◇ CALCOTONE WHITE T ◇ C.I. 77891 ◇ C.I. PIGMENT WHITE 6 ◇ COSMETIC WHITE C47-5175 ◇ C-WEISS 7 (GERMAN) ◇ FLAMENCO ◇ HOMBITAN ◇ HORSE HEAD A-410 ◇ KH 360 ◇ KRONOS TITANIUM DIOXIDE ◇ LEVANOX WHITE RKB ◇ NCI-C04240 ◇ RAYOX ◇ RUNA RH20 ◇ RUTILE ◇ TIOFINE ◇ TIOXIDE ◇ TITANDIOXID (SWEDEN) ◇ TITANIUM DIOXIDE (ACGIH, FCC) ◇ TRIOXIDE(S) ◇ TRONOX ◇ UNITANE O-110 ◇ ZOPAQUE

USE IN FOOD:

Purpose: Color additive.

Where Used: Candy, creamed-type canned products, ham salad spread (canned), icings, poultry salads, poultry spreads, sugar syrup.

Regulations: FDA - 21CFR 73.575. Limitation of 1 percent by weight of food. USDA - 9CFR 318.7. Limitation of 0.5 percent in canned ham salad spread, creamed-type canned products. 9CFR 381.147. Limitation of 0.5 percent in poultry salads and poultry spreads.

NCI Carcinogenesis Bioassay (feed); No Evidence: mouse, rat NCITR* NCI-CG-TR-97,79. EPA Genetic Toxicology Program. Community Right-To-Know List.

OSHA PEL: TWA 10 mg/m^3 Total Dust; 5 mg/m^3 Respirable Dust ACGIH TLV: TWA 10 mg/m^3 of total dust when toxic impurities are not present, e.g., quartz < 1%

SAFETY PROFILE: An experimental carcinogen, neoplastigen, and tumorigen. A human skin irritant. A nuisance dust. Violent or incandescent reaction with metals (e.g., aluminum; calcium; magnesium; potassium; sodium; zinc; lithium).

TOXICITY DATA and CODEN

skn-hmn 300 μg/3D-I MLD 85DKA8 -,127,77
ihl-rat TCLo: 250 mg/m^3/6H/2Y-I: CAR TXAPA9 79,179,85
ims-rat TDLo: 360 mg/kg/2Y-I: NEO NCIUS* PH 43-64-886,JUL,68
ims-rat TD : 260 mg/kg/84W-I: ETA NCIUS* PH 43-64-886,AUG,69

TGJ050 CAS: 58-95-7
d-α-TOCOPHERYL ACETATE
mf: $C_{31}H_{52}O_3$ mw: 472.75

PROP: From vacuum steam distillation and acetylation of edible vegetable oil products. (FCC III) Colorless to yellow oil; odorless. Mp: 25°. insol in water, sol in alc; misc with acetone, chloroform, ether, vegetable oil.

USE IN FOOD:

Purpose: Dietary supplement, nutrient.

Where Used: Various.

Regulations: FDA - 21CFR 182.5892, 182.8892. GRAS when used in accordance with good manufacturing practice.

SAFETY PROFILE: When heated to decomposition it emits acrid smoke and irritating fumes.

TGJ055
dl-α-TOCOPHERYL ACETATE
mf: $C_{31}H_{52}O_3$ mw: 472.75

PROP: From vacuum steam distillation and acetylation of edible vegetable oil products. (FCC III) Colorless to yellow oil; odorless. Insol in water; sol in alc; misc with acetone, chloroform, ether, vegetable oil.

USE IN FOOD:

Purpose: Dietary supplement, nutrient.

Where Used: Various.

Regulations: FDA - 21CFR 182.5892, 182.8892. GRAS when used in accordance with good manufacturing practice.

SAFETY PROFILE: When heated to decomposition it emits acrid smoke and irritating fumes.

TGJ060 CAS: 4345-03-3
dl-α-TOCOPHERYL ACID SUCCINATE
mf: $C_{33}H_{54}O_5$ mw: 530.79

PROP: From vacuum steam distillation and succinylation of edible vegetable oil products. (FCC III) Colorless to white crystalline powder; odorless and tasteless. Mp: 75°. Insol in water; very sol in chloroform, sol in acetone, alc, ether, vegetable oil.

USE IN FOOD:

Purpose: Dietary supplement, nutrient.

Where Used: Various.

Regulations: FDA - 21CFR 182.5890, 182.8890. GRAS when used in accordance with good manufacturing practice.

SAFETY PROFILE: When heated to decomposition it emits acrid smoke and irritating fumes.

TGK750 CAS: 108-88-3
TOLUENE
mf: C_7H_8 mw: 92.15
DOT: 1294

PROP: Colorless liquid; benzol-like odor. Mp: −95 to −94.5°, bp: 110.4°, flash p: 40°F (CC), ULC: 75-80, lel: 1.27%, uel: 7%, d: 0.866 @ 20°/4°, autoign temp: 996°F, vap press: 36.7 mm @ 30°, vap d: 3.14. Insol in water; sol in acetone; misc in absolute alc, ether, chloroform.

SYNS: ANTISAL 1a ◇ METHACIDE ◇ METHYLBENZENE ◇ METHYLBENZOL ◇ NCI-C07272 ◇ PHENYLMETHANE ◇ RCRA WASTE NUMBER U220 ◇ TOLUEEN (DUTCH) ◇ TOLUEN (CZECH) ◇ TOLUOL ◇ TOLUOL (DOT) ◇ TOLUOLO (ITALIAN) ◇ TOLU-SOL

USE IN FOOD:

Purpose: Adjuvant in foamed plastic, blowing agent.

Where Used: Packaging materials.

Regulations: FDA - 21CFR 178.3010. Limitation of 0.35 percent in finished foamed polyethylene.

Community Right-To-Know List. EPA Genetic Toxicology Program.

OSHA PEL: TWA 100 ppm; STEL 150 ppm ACGIH TLV: TWA 100 ppm; STEL 150 ppm; BEI: toluene in venous blood end of shift 1 mg/L DFG MAK: 100 ppm (380 mg/m^3); BAT: blood end of shift 340 μg/dl DOT Classification: Flammable Liquid; Label: Flammable Liquid NIOSH REL: (Toluene) TWA 100 ppm; CL 200 ppm/10M

SAFETY PROFILE: Moderately toxic by intravenous, subcutaneous, and possibly other routes. Mildly toxic by inhalation. An experimental teratogen. Human systemic effects by inhalation: CNS recording changes, hallucinations or distorted perceptions, motor activity changes, antipsychotic, psychophysiological test changes and bone marrow changes. Experimental reproductive effects. Mutagenic data. A human eye irritant. An experimental skin and severe eye irritant.

Flammable liquid. A very dangerous fire hazard when exposed to heat, flame or oxidizers. Explosive in the form of vapor when exposed to heat or flame. Can react vigorously with oxidizing materials. To fight fire, use foam, CO_2, dry chemical. When heated to decomposition it emits acrid smoke and irritating fumes.

TOXICITY DATA and CODEN

eye-hmn 300 ppm JIHTAB 25,282,43
skn-rbt 500 MOD FCTOD7 20,563,82
eye-rbt 2 mg/24H SEV 28ZPAK -,23,72
oms-grh-ihl 562 mg/L MUREAV 113,467,83
dns-rat:lvr 30 μmol/L SinJF# 26OCT82
ihl-rat TCLo:1500 mg/m^3/24H (1-8D preg): TER TXCYAC 11,55,78
orl-mus TDLo:9 g/kg (6-15D preg):TER TJADAB 19,41A,79
ihl-hmn TCLo:200 ppm:BRN,CNS,BLD JAMAAP 123,1106,43
ihl-man TCLo:100 ppm:CNS WEHRBJ 9,131,72
orl-rat LD50:5000 mg/kg AMIHAB 19,403,59
ihl-rat LCLo:4000 ppm/4H AIHAAP 30,470,69

TGW250 CAS: 131-79-3
1-(o-TOLYLAZO)-2-NAPHTHYLAMINE

mf: $C_{17}H_{15}N_3$ mw: 261.35

SYNS: A.F. YELLOW NO. 3 ◇ CERISOL YELLOW TB ◇ C.I. 11390 ◇ C.I. FOOD YELLOW 11 ◇ DOLKWAL YELLOW OB ◇ EXT. D & C YELLOW NO. 10 ◇ FD & C YELLOW NO. 4 ◇ JAUNE OB ◇ 1-(2-METHYLPHENYL)AZO-2-NAPHTHALENAMINE ◇ 1-((2-METHYLPHENYL)AZO)-2-NAPHTHALENAMINE ◇ 1-(2-METHYLPHENYL)AZO-2-NAPHTHYLAMINE ◇ OIL YELLOW OB ◇ o-TOLUENE-1-AZO-2-NAPHTHYLAMINE ◇ YELLOW OB

USE IN FOOD:

Purpose: Color additive.

Where Used: Various.

Regulations: FDA - 21CFR 81.10. Provisional listing terminated.

IARC Cancer Review: Animal Sufficient Evidence IMEMDT 8,287,75. EPA Genetic Toxicology Program.

SAFETY PROFILE: Moderately toxic by ingestion, intraperitoneal, and subcutaneous routes. An experimental carcinogen, and tumorigen. Mutagenic data. When heated to decomposition it emits toxic fumes of NO_x.

TOXICITY DATA and CODEN

mma-sat 50 μg/plate CANCAR 49,1970,82
scu-rat TDLo:700 mg/kg/2Y-I:ETA FEPRA7 16,367,57
orl-rbt LDLo:1000 mg/kg JBCHA3 27,403,16

THA250 CAS: 103-93-5
p-TOLYL ISOBUTYRATE

mf: $C_{11}H_{14}O_2$ mw: 178.25

PROP: Colorless liquid; characteristic odor. D: 0.990-0.996, refr index: 1.485, flash p: +212°F. Sol in alc; insol in water.

SYNS: p-CRESYL ISOBUTYRATE ◇ FEMA No. 3075 ◇ ISOBUTYRIC ACID, p-TOLYL ESTER ◇ PARACRESYL ISOBUTYRATE

USE IN FOOD:

Purpose: Flavoring agent.

Where Used: Various.

Regulations: FDA - 21CFR 172.515. Use at a level not in excess of the amount reasonably required to accomplish the intended effect.

SAFETY PROFILE: Moderately toxic by ingestion and skin contact. Combustible liquid. When heated to decomposition it emits acrid smoke and irritating fumes.

TOXICITY DATA and CODEN

orl-rat LD50:4000 mg/kg FCTXAV 13,773,75
skn-rbt LD50:3970 mg/kg FCTXAV 13,773,75

THH750 CAS: 8001-35-2

TOXAPHENE

DOT: 2761

mf: $C_{10}H_{10}Cl_8$ mw: 413.80

PROP: Yellow, waxy solid; pleasant piney odor. Mp: 65-90°. Almost insol in water; very sol in aromatic hydrocarbons.

SYNS: ALLTEX ◇ ALLTOX ◇ ATTAC 6 ◇ CAMPHECHLOR ◇ CAMPHOCHLOR ◇ CAMPHOCLOR ◇ CAMPHOFENE HUILEUX ◇ CHEM-PHENE ◇ CHLORINATED CAMPHENE 60% (ACGIH) ◇ CHLOROCAMPHENE ◇ CRESTOXO ◇ CRISTOXO 90 ◇ ESTONOX ◇ FASCO-TERPENE ◇ GENIPHENE ◇ GY-PHENE ◇ HERCULES TOXAPHENE ◇ KAMFOCHLOR ◇ MELIPAX ◇ MOTOX ◇ NCI-C00259 ◇ OCTACHLOROCAMPHENE ◇ PCC ◇ PENPHENE ◇ PHENACIDE ◇ PHENATOX ◇ POLYCHLORCAMPHENE ◇ POLYCHLORINATED CAMPHENES ◇ POLYCHLOROCAMPHENE ◇ RCRA WASTE NUMBER P123 ◇ STROBANE-T-90 ◇ TOXADUST ◇ TOXAFEEN (DUTCH) ◇ TOXAKIL ◇ TOXAPHEN (GERMAN) ◇ TOXON 63 ◇ TOXYPHEN ◇ VERTAC TOXAPHENE 90

USE IN FOOD:

Purpose: Insecticide.

Where Used: Soybean oil.

Regulations: FDA - 21CFR 193.450. Insecticide residue tolerance of 6 ppm in soybean oil. USDA CES Ranking: A-2 (1985).

IARC Cancer Review: Human Limited Evidence IMEMDT 20,327,79; Animal Sufficient Evidence IMEMDT 20,327,79. NCI Carcinogenesis Bioassay (feed); Clear Evidence: mouse, rat NCITR* NCI-CG-TR-37,79.

OSHA PEL: TWA 0.5 mg/m^3; STEL 1 mg/m^3 (skin) ACGIH TLV: TWA 0.5 mg/m^3; STEL 1 mg/m^3 (skin) DOT Classification: ORM-A; Label: None

SAFETY PROFILE: Human poison by ingestion and possibly other routes. Experimental poison by ingestion, intraperitoneal and possibly other routes. Moderately toxic experimentally by inhalation and skin contact. An experimental carcinogen, tumorigen and teratogen. May be a human carcinogen. Human systemic effects by ingestion and skin contact: somnolence, convulsions or effect on seizure threshold, coma and allergic skin dermatitis. A skin irritant; absorbed through the skin. Human mutagenic data. Liver injury has been reported. Lethal amounts of toxaphene can enter the body through the mouth, lungs, and skin. When heated to decomposition it emits toxic fumes of Cl^-.

TOXICITY DATA and CODEN

skn-mam 500 mg MOD JAMAAP 149,1135,52
mmo-sat 100 μg/plate ENMUDM 8(Suppl 7),1,86
sce-hmn:lym 10 μmol/L ARTODN 52,221,83
orl-rat TDLo:280 mg/kg (10W male):REP GISAAA 45(5),14,80
orl-rat TDLo:150 mg/kg (7-16D preg):TER BECTA6 15,660,76
orl-mus TDLo:100 mg/kg (8D preg):TER TCMUD8 5,3,85
orl-mus TDLo:6600 mg/kg/80W-C:CAR NCITR* NCI-CG-TR-37,79
orl-mus TD :13 g/kg/80W-C:CAR NCITR* NCI-CG-TR-37,79
skn-hmn TDLo:657 mg/kg:SKN CMEP** -,1,56
orl-hmn LDLo:28 mg/kg:CNS 34ZIAG -,598,69
orl-man LDLo:29 mg/kg CMEP** -,-,56
orl-rat LD50:55 mg/kg GISAAA 44(4),51,79
skn-rat LD50:600 mg/kg SPEADM 74-1,-,74

THJ250 CAS: 9000-65-1

TRAGACANTH GUM

PROP: from the shrub *Astragalus gummifier* Labillardiere. Powder is white, pieces are white to pale yellow, translucent, and horny; odorless with mucilaginous taste.

SYNS: GUM TRAGACANTH ◇ TRAGACANTH

USE IN FOOD:

Purpose: Emulsifier, preservative, thickener.

Where Used: Baked goods, citrus beverages, condiments, fats, fruit fillings, gravies, meat products, oils, relishes, salad dressings, sauces.

Regulations: FDA - 21CFR 184.1351. GRAS with a limitation of 0.2 percent in baked goods, 0.7 percent in condiments and relishes, 1.3 percent in fats and oils, 0.8 gravies and sauces, 0.2 percent in meat products, 0.1 percent in all other foods when used in accordance with good manufacturing practice.

SAFETY PROFILE: Mildly toxic by ingestion. A mild allergen. Combustible when exposed to heat or flame. When heated to decomposition it emits acrid smoke and irritating fumes.

TOXICITY DATA and CODEN

orl-rat LD50:16400 mg/kg 85AIAL -,45,73

THL600
TRENBOLONE
mf: $C_{18}H_{22}O_2$ mw: 270.38

PROP: Crystals. Mp: 186°.

SYNS: 4,9,11,-ESTRATRIEN-17β-OL-3-ONE ◇ 17β-HYDROXYESTRA-4,9,11-TRIEN-3-ONE ◇ TRIENBOLONE ◇ TRIENOLONE

USE IN FOOD:

Purpose: Animal drug.

Where Used: Beef.

Regulations: FDA - 21CFR 556.739. Limitation of 50 ppb in muscle, 100 ppb in liver, 300 ppb in kidney, and 400 ppb in fat cattle.

SAFETY PROFILE: When heated to decomposition it emits acrid smoke and irritating fumes.

THM500 CAS: 102-76-1
TRIACETYL GLYCERIN
mf: $C_9H_{14}O_6$ mw: 218.23

PROP: Colorless oily liquid; slt fatty odor and taste. Mp: −*78°, bp: 258°, flash p: 280°F (COC), d: 1.161, autoign temp: 812°F, vap d: 7.52. Sol in water; misc with alc, ether, chloroform.

SYNS: ENZACTIN ◇ FEMA No. 2007 ◇ FUNGACETIN ◇ GLYCERINE TRIACETATE ◇ GLYCEROL TRIACETATE ◇ GLYCERYL TRIACETATE ◇ GLYPED ◇ KESSCOFLEX TRA ◇ KODAFLEX TRIACETIN ◇ 1,2,3-PROPANETRIOL TRIACETATE ◇ TRIACETIN (FCC) ◇ VANAY

USE IN FOOD:

Purpose: Adjuvant, flavoring agent, formulation aid, humectant, plasticizer, solvent, vehicle.

Where Used: Baked goods, baking mixes, beverage bases, beverages (alcoholic), beverages (nonalcoholic), candy (hard), candy (soft), chewing gum, confections, fillings, frostings, frozen dairy desserts, frozen dairy dessert mixes, gelatins, puddings.

Regulations: FDA - 21CFR 181.27, 184.1903. GRAS when used in accordance with good manufacturing practice.

SAFETY PROFILE: Poison by ingestion. Moderately toxic by intraperitoneal, subcutaneous, and intravenous routes. An eye irritant. Combustible when exposed to heat, flame or powerful oxidizers. To fight fire use alcohol foam, water, CO_2, dry chemical. When heated to decomposition it emits acrid smoke and irritating fumes.

TOXICITY DATA and CODEN

eye-rbt 116 mg JPETAB 82,377,44
orl-rat LD50: 3000 mg/kg AMIHAB 21,28,60

TIG750 CAS: 60-01-5
TRIBUTYRIN
mf: $C_{15}H_{26}O_6$ mw: 302.41

PROP: Colorless, oily liquid; bitter taste. Mp: −75°, d: 1.0356 @ 20/20°, bp: 305-310°, flash p: +212°F. Insol in water; very sol in alc, ether, chloroform.

SYNS: BUTANOIC ACID, 1,2,3-PROPANETRIYL ESTER ◇ BUTYRIC ACID TRIESTER with GLYCERIN ◇ BUTYRYL TRIGLYCERIDE ◇ FEMA No. 2223 ◇ GLYCEROL TRIBUTYRATE ◇ KODAFLEX ◇ TRIBUTYROIN

USE IN FOOD:

Purpose: Flavoring agent, adjuvant.

Where Used: Baked goods, beverages (alcoholic), beverages (nonalcoholic), candy (soft), fats, oils, frozen dairy desserts, frozen dairy dessert mixes, gelatins, puddings.

Regulations: FDA - 21CFR 172.515. Use at a level not in excess of the amount reasonably required to accomplish the intended effect.

SAFETY PROFILE: Poison by intravenous route. Moderately toxic by ingestion. An experimental tumorigen. Combustible liquid. When heated to decomposition it emits acrid smoke and irritating fumes.

TOXICITY DATA and CODEN

orl-rat TDLo: 177 g/kg/3W-C:ETA JNCIAM 10,361,49
orl-rat LDLo: 3200 mg/kg KODAK* -,-,71

TIH600
TRICALCIUM SILICATE

USE IN FOOD:

Purpose: General-purpose food additive.

Where Used: Table salt.

Regulations: FDA - 21CFR 182.2906. Limitation of 2 percent in table salt.

SAFETY PROFILE: A nuisance dust.

TIO750 CAS: 79-01-6
TRICHLOROETHYLENE
DOT: 1710
mf: C_2HCl_3 mw: 131.38

PROP: Clear, colorless, mobile liquid; characteristic sweet odor of chloroform. D: 1.4649 @ 20°/4°, bp: 86.7°, flash p: 89.6°F (does not burn), lel: 12.5%, uel: 90% @ > 30°, mp: −73°, fp: −86.8°, autoign temp: 788°F, vap press: 100 mm @ 32°, vap d: 4.53, refr index: 1.477 @ 20°. Immiscible with water; misc with alc, ether, acetone, carbon tetrachloride.

SYNS: ACETYLENE TRICHLORIDE ◇ ALGYLEN ◇ ANAMENTH ◇ BENZINOL ◇ BLACOSOLV ◇ CECOLENE ◇ 1-CHLORO-2,2-DICHLOROETHYLENE ◇ CHLORYLEA ◇ CHORYLEN ◇ CIRCOSOLV ◇ CRAWHASPOL ◇ DENSINFLUAT ◇ 1,1-DICHLORO-2-CHLOROETHYLENE ◇ DOW-TRI ◇ DUKERON ◇ ETHINYL TRICHLORIDE ◇ ETHYLENE TRICHLORIDE ◇ FLECK-FLIP ◇ FLUATE ◇ GERMALGENE ◇ LANADIN ◇ LETHURIN ◇ NARCOGEN ◇ NARKOSOID ◇ NCI-CO4546 ◇ NIALK ◇ PERM-A-CHLOR ◇ PETZINOL ◇ RCRA WASTE NUMBER U228 ◇ THRETHYLENE ◇ TRIAD ◇ TRIASOL ◇ TRICHLOORETHEEN (DUTCH) ◇ TRICHLOORETHYLEEN, TRI (DUTCH) ◇ TRICHLORAETHEN (GERMAN) ◇ TRICHLORAETHYLEN, TRI (GERMAN) ◇ TRICHLORAN ◇ TRICHLORETHENE (FRENCH) ◇ TRICHLORETHYLENE, TRI (FRENCH) ◇ TRICHLOROETHENE ◇ 1,2,2-TRICHLOROETHYLENE ◇ TRI-CLENE ◇ TRICLORETENE (ITALIAN) ◇ TRICLOROETILENE (ITALIAN) ◇ TRIELINA (ITALIAN) ◇ TRILENE ◇ TRIMAR ◇ TRI-PLUS ◇ VESTROL ◇ VITRAN ◇ WESTROSOL

USE IN FOOD:

Purpose: Extraction solvent.

Where Used: Coffee (decaffeinated instant), coffee (decaffeinated), spice oleoresins.

Regulations: FDA - 21CFR 173.290. Limitation of 25 ppm in decaffeinated coffee, 10 ppm in decaffeinated instant coffee, 30 ppm in spice oleoresins or 30 ppm for all chlorinated solvents used.

IARC Cancer Review: Animal Limited Evidence IMEMDT 20,545,79; Human Inadequate Evidence IMEMDT 20,545,79; Animal Sufficient Evidence IMEMDT 11,263,76. NCI Carcinogenesis Bioassay (gavage); No Evidence: rat NCITR* NCI-CG-TR-2,76; (gavage); Clear Evidence: mouse NCITR* NCI-CG-TR-2,76. Community Right-To-Know List. EPA Genetic Toxicology Program.

OSHA PEL: TWA 50 ppm; STEL 200 ppm ACGIH TLV: TWA 50 ppm; STEL 200 ppm; BEI: trichloroethanol in urine end of shift 320 mg/g creatinine, trichloroethylene in end-exhaled air prior to shift and end of work week 0.5 ppm DFG MAK: 50 ppm (270 mg/m^3); BAT: blood end of work week and end of shift 500 μg/dl DOT Classification: ORM-A; Label: None; Poison B; Label: St. Andrews Cross NIOSH REL: (Trichloroethylene) TWA 250 ppm; (Waste Anesthetic Gases) CL 2 ppm/1H

SAFETY PROFILE: Experimental poison by intravenous and subcutaneous routes. Moderately toxic experimentally by ingestion and intraperitoneal routes. Mildly toxic to humans by ingestion and inhalation. Mildly toxic experimentally by inhalation. An experimental carcinogen, tumorigen, and teratogen. Human systemic effects by ingestion and inhalation: eye effects, somnolence, hallucinations or distorted perceptions, gastrointestinal changes and jaundice. Experimental reproductive effects. Human mutagenic data. An eye and severe skin irritant. Inhalation of high concentrations causes narcosis and anesthesia.

High concentrations of trichloroethylene vapor in high-temperature air can be made to burn mildly if plied with a strong flame. Though such a condition is difficult to produce, flames or arcs should not be used in closed equipment which contains any solvent residue or vapor. When heated to decomposition it emits toxic fumes of Cl^-.

TOXICITY DATA and CODEN

skn-rbt 500 mg/24H SEV 28ZPAK -,28,72
eye-rbt 20 mg/24H MOD 28ZPAK -,28,72
mmo-asn 2500 ppm MUREAV 155,105,85
sln-asn 17500 ppm MUREAV 155,105,85
orl-rat TDLo:2688 mg/kg (1-22D preg/21D post):REP TOXID9 4,179,84
ihl-rat TCLo:1800 ppm/24H (1-2D preg):TER APTOD9 19,A22,80
ihl-rat TCLo:100 ppm/4H (8-21D preg):TER BJANAD 54,337,82
ihl-rat TCLo:150 ppm/7H/2Y-I:CAR INHEAO 21,243,83
orl-mus TDLo:455 g/kg/78W-I:CAR NCITR* NCI-CG-TR-2,76
orl-man TDLo:2143 mg/kg:GIT 34ZIAG -,602,69
ihl-hmn TCLo:6900 mg/m^3/10M:CNS AHBAAM 116,131,36
ihl-hmn TCLo:160 ppm/83M:CNS AIHAAP 23,167,62

ihl-hmn TDLo: 812 mg/kg: CNS,GIT,LIV BMJOAE 2,689,45
ihl-man TCLo: 110 ppm/8H: EYE,CNS BJIMAG 28,293,71
ihl-man LCLo: 2900 ppm NZMJAX 50,119,51
orl-rat LC50: 3670 mg/kg 28ZPAK -,28,72
ihl-rat LCLo: 8000 ppm/4H AIHAAP 30,470,69

TIQ250 CAS: 52-68-6
((2,2,2-TRICHLORO-1-HYDROXYETHYL) DIMETHYLPHOSPHONATE)
mf: $C_4H_8Cl_3O_4P$ mw: 257.44

SYNS: AEROL 1 (PESTICIDE) ◇ AGROFOROTOX ◇ ANTHON ◇ BAY 15922 ◇ BILARCIL ◇ BOVINOX ◇ BRITON ◇ BRITTEN ◇ CEKUFON ◇ CHLORAK ◇ CHLOROFOS ◇ CHLOROFTALM ◇ CHLOROPHOS ◇ CHLOROPHTHALM ◇ CHLOROXYPHOS ◇ CICLOSOM ◇ CLOROFOS (RUSSIAN) ◇ COMBOT EQUINE ◇ DANEX ◇ DEP (PESTICIDE) ◇ DEPTHON ◇ DETF ◇ O,O-DI-METHYL-(1-HYDROXY-2,2,2-TRICHLORAETHYL)PHOS-PHONSAEURE ESTER (GERMAN) ◇ O,O-DIMETHYL-(1-HY-DROXY-2,2,2-TRICHLORATHYL)-PHOSPHAT (GERMAN) ◇ O,O-DIMETHYL-(1-HYDROXY-2,2,2-TRICHLOROETHYL) PHOSPHONATE ◇ O,O-DIMETHYL-(2,2,2-TRICHLOOR-1-HYDROXY-ETHYL)-FOSFONAAT (DUTCH) ◇ O,O-DIME-THYL-(2,2,2-TRICHLOR-1-HYDROXY-AETHYL)PHOSPHO-NAT (GERMAN) ◇ DIMETHYL-2,2,2-TRICHLORO-1-HY-DROXYETHYLPHOSPHONATE ◇ O,O-DIMETIL-(2,2,2-TRICLORO-1-IDROSSI-ETIL)-FOSFONATO (ITALIAN) ◇ DIMETOX ◇ DIPTERAX ◇ DIPTEREX ◇ DIPTEREX 50 ◇ DIPTEVUR ◇ DITRIFON ◇ DYLOX ◇ DYLOX-METASYS-TOX-R ◇ DYREX ◇ DYVON ◇ ENT 19,763 ◇ EQUINO-ACID ◇ FLIBOL E ◇ FLIEGENTELLER ◇ FOROTOX ◇ FOSCHLOR ◇ FOSCHLOREM (POLISH) ◇ 1-HYDROXY-2,2,2-TRICHLO-ROETHYLPHOSPHONIC ACID DIMETHYL ESTER ◇ HYPODERMACID ◇ LEIVASOM ◇ LOISOL ◇ MASOTEN ◇ MAZOTEN ◇ METHYL CHLOROPHOS ◇ METIFONATE ◇ METRIFONATE ◇ METRIPHONATE ◇ NCI-C54831 ◇ NEGUVON ◇ POLFOSCHLOR ◇ PROXOL ◇ RICIFON ◇ RITSIFON ◇ SATOX 20WSC ◇ SOLDEP ◇ SOTIPOX ◇ TRICHLOORFON (DUTCH) ◇ TRICHLORFON (USDA) ◇ 2,2,2-TRICHLORO-1-HYDROXYETHYL-PHOSPHONATE, DIMETHYL ESTER ◇ TRICHLOROPHON ◇ TRICHLOR-PHENE ◇ TRICHLORPHON ◇ TRINEX ◇ TUGON ◇ TUGON FLY BAIT ◇ VERMICIDE BAYER 2349 ◇ VOLFARTOL ◇ VOTEXIT ◇ WOTEXIT

USE IN FOOD:

Purpose: Pesticide.

Where Used: Animal feed, citrus pulp (dried).

Regulations: FDA - 21CFR 561.190. Limitation of 2.5 ppm in dried citrus pulp when used for animal feed. USDA CES Ranking: B-3 (1985).

IARC Cancer Review: Animal Inadequate Evidence IMEMDT 30,207,83. Community Right-To-Know List. EPA Extremely Hazardous Substances List. EPA Genetic Toxicology Program.

DOT Classification: ORM-A; Label: None

SAFETY PROFILE: Poison by ingestion, inhalation, intraperitoneal, subcutaneous, intravenous, and intramuscular routes. Moderately toxic by skin contact and possibly other routes. An experimental carcinogen and teratogen. Experimental reproductive effects. Human mutagenic data. An eye irritant. When heated to decomposition it emits very toxic fumes of Cl^- and PO_x.

TOXICITY DATA and CODEN

eye-rbt 120 mg/6D-I MLD BUMMAB 9,7,55
mmo-ssp 20 mmol/L MUREAV 117,139,83
dns-hmn: oth 4 mol/L PSSCBG 15,439,84
orl-rat TDLo: 78400 μg/kg (16-22D preg): REP NUPOBT 17,135,79
orl-rat TDLo: 80 mg/kg (9D preg): TER EVHPAZ 13,121,76
orl-mus TDLo: 1500 mg/kg (10-14D preg): TER EVHPAZ 30,105,79
orl-rat TDLo: 186 mg/kg/6W-I: CAR ARGEAR 41,311,73
ims-rat TDLo: 183 mg/kg/6W-I: CAR ARGEAR 41,311,73
orl-rat LD50: 150 mg/kg FMCHA2 -,C242,83
ihl-rat LC50: 1300 μg/m^3 ARGEAR 48,112,78
skn-rat LD50: 2000 mg/kg ARGEAR 48,112,78

TIR000 CAS: 299-84-3
TRICHLOROMETAFOS
mf: $C_8H_8Cl_3O_3PS$ mw: 321.54

PROP: White powder. Mp: 41°, vap press: 8 × 10^{-4} mm.

SYNS: DERMAFOSU (POLISH) ◇ DERMAPHOS ◇ DIMETHYL TRICHLOROPHENYL THIOPHOSPHATE ◇ O,O-DIMETHYL-O-(2,4,5-TRICHLOROPHENYL)THIO-PHOSPHATE ◇ O,O-DIMETHYL-O-(2,4,5-TRICHLORPHE-NYL)-THIONOPHOSPHAT(GERMAN) ◇ DOW ET 57 ◇ ECTORAL ◇ ENT 23,284 ◇ ETROLENE ◇ FENCHLOORFOS (DUTCH) ◇ FENCHLORFOS ◇ FENCHLORFOSU (POLISH) ◇ FENCHLOROPHOS ◇ FENCHLORPHOS ◇ KARLAN ◇ KORLAN ◇ KORLANE ◇ NANCHOR ◇ NANKER ◇ NANKOR ◇ RONNEL (ACGIH) ◇ THIOPHOSPHATE de O,O-DIMETHYLE et de O-(2,4,5-TRICHLOROPHENYLE) (FRENCH) ◇ O-(2,4,5-TRICHLOOR-FENYL)-O,O-DIMETHYL-MONOTHIOFOSFAAT (DUTCH) ◇ 2,4,5-TRICHLOROPHE-NOL, O-ESTER WITH O,O-DIMETHYL PHOSPHOROTHIOATE

◇ O-(2,4,5-TRICHLOR-PHENYL)-O,O-DIMETHYL-MONO-THIOPHOSPHAT (GERMAN) ◇ O-(2,4,5-TRICLORO-FENIL)-O,O-DIMETIL-MONOTIOFOSFATO (ITALIAN) ◇ TROLEN ◇ TROLENE ◇ VIOZENE

USE IN FOOD:

Purpose: Animal feed drug.

Where Used: Animal feed.

Regulations: FDA - 21CFR 558.525.

Chlorophenol compounds are on the Community Right-To-Know List.

OSHA PEL: TWA 10 mg/m^3 ACGIH: TWA 10 mg/m^3

SAFETY PROFILE: Poison by ingestion, intraperitoneal and possibly other routes. Moderately toxic by skin contact. A cholinesterase inhibitor. An experimental teratogen. Experimental reproductive effects. When heated to decomposition it emits very toxic fumes of Cl^-, PO_x, and SO_x.

TOXICITY DATA and CODEN

orl-rat TDLo:368 mg/kg (16-22D preg):REP NUPOBT 17,577,79
orl-rat TDLo:6 g/kg (6-15D preg):TER JTEHD6 10,111,82
orl-mam TDLo:2100 mg/kg (30-50D preg): REP AVSCA7 24,99,83
orl-rat LD50:625 mg/kg GISAAA 45(6),14,80
skn-rat LD50:2000 mg/kg WRPCA2 9,119,70

TIW500 CAS: 93-76-5
2,4,5-TRICHLOROPHENOXYACETIC ACID

DOT: 2765
mf: $C_8H_5Cl_3O_3$ mw: 255.48

PROP: Crystals; light tan solid. Mp: 151-153°. Usually has 2,3,7,8-TCDD as a minor component.

SYNS: ACIDE 2,4,5-TRICHLORO PHENOXYACETIQUE (FRENCH) ◇ ACIDO (2,4,5-TRICLORO-FENOSSI)-ACETICO (ITALIAN) ◇ BCF-BUSHKILLER ◇ BRUSHTOX ◇ DACAMINE ◇ DEBROUSSAILLANT CONCENTRE ◇ DECAMINE 4T ◇ DED-WEED BRUSH KILLER ◇ DINOXOL ◇ ESTERON BRUSH KILLER ◇ FARMCO FENCE RIDER ◇ FORRON ◇ FORTEX ◇ FRUITONE A ◇ INVERTON 245 ◇ LINE RIDER ◇ PHORTOX ◇ RCRA WASTE NUMBER U232 ◇ REDDON ◇ REDDOX ◇ SPONTOX ◇ SUPER D WEEDONE ◇ 2,4,5-T (ACGIH, DOT, USDA) ◇ TIPPON ◇ TORMONA ◇ TRANSAMINE ◇ TRIBUTON ◇ (2,4,5-TRICHLOOR-FENOXY)-AZIJNZUUR (DUTCH) ◇ (2,4,5-TRICHLOR-PHENOXY)-ESSIGSAEURE (GERMAN) ◇ TRINOXOL ◇ TRIOXON ◇ TRIOXONE ◇ VERTON 2T ◇ WEEDAR ◇ WEEDONE

USE IN FOOD:

Purpose: Herbicide.

Where Used: Various.

Regulations: USDA CES Ranking: A-3 (1985).

IARC Cancer Review: Animal Inadequate Evidence IMEMDT 15,273,77; Human Inadequate Evidence IMEMDT 15,273,77; Human Limited Evidence IMEMDT 41,357,86. EPA Genetic Toxicology Program.

OSHA PEL: TWA 10 mg/m^3 ACGIH TLV: TWA 10 mg/m^3 DFG MAK: 10 mg/m^3 DOT Classification: ORM-A; Label: None

SAFETY PROFILE: Poison by ingestion. An experimental neoplastigen, tumorigen, and teratogen. May be a human and experimental carcinogen. Experimental reproductive effects. Mutagenic data. Readily absorbed by inhalation, and ingestion routes, slowly by skin contact. Signs of intoxication include weakness, lethargy, anorexia, diarrhea, ventricular fibrillation and/or cardiac arrest and death. The teratogenicity is due in part to 2,3,7,8-TCDD, which is present as a contaminant (ARENAA 17,123,72). When heated to decomposition it emits toxic fumes of Cl^-.

TOXICITY DATA and CODEN

mmo-bcs 1 nmol/plate MSERDS 5,93,81
cyt-dmg-orl 250 ppm MUREAV 65,83,79
orl-rat TDLo:100 mg/kg (8D preg):REP PHBHA4 9,357,72
orl-rat TDLo:27600 μg/kg (10-15D preg):TER EVSRBT 2,708,73
orl-mus TDLo:450 mg/kg (6-15D preg):TER NSAPCC 272,243,72
orl-mus TDLo:3379 mg/kg/33W-C:NEO BJCAAI 33,626,76
scu-mus TDLo:215 mg/kg:ETA NTIS** PB223-159
orl-rat LD50:300 mg/kg RREVAH 10,97,65

TIX500 CAS: 93-72-1
α-(2,4,5-TRICHLOROPHENOXY)PROPIONIC ACID

mf: $C_9H_7Cl_3O_3$ mw: 269.51

PROP: Crystals. Mp: 182°. Sltly water-sol.

SYNS: ACIDE 2-(2,4,5-TRICHLORO-PHENOXY) PROPIONIQUE (FRENCH) ◇ ACIDO 2-(2,4,5-TRICLORO-FENOSSI)-PROPIONICO (ITALIAN) ◇ AMCHEM 2,4,5-TP ◇ AQUA-VEX ◇ COLOR-SET ◇ DED-WEED ◇ DOUBLE STRENGTH ◇ FENOPROP ◇ FENORMONE ◇ FRUITONE T ◇ HERBI-

CIDES, SILVEX ◇ KURAN ◇ KURON ◇ KUROSAL ◇ MILLER NU SET ◇ PROPON ◇ RCRA WASTE NUMBER U233 ◇ SILVEX (USDA) ◇ SILVI-RHAP ◇ STA-FAST ◇ 2,4,5-TC ◇ 2,4,5-TCPPA ◇ 2,4,5-TP ◇ 2-(2,4,5-TRICHLOOR-FENOXY)-PROPIONZUUR (DUTCH) ◇ 2-(2,4,5-TRICHLORO-PHENOXY)PROPIONIC ACID ◇ 2,4,5-TRICHLOROPHE-NOXY-α-PROPIONIC ACID ◇ 2-(2,4,5-TRICHLOR-PHE-NOXY)-PROPIONSAEURE (GERMAN) ◇ WEED-B-GON

USE IN FOOD:

Purpose: Herbicide.

Where Used: Various.

Regulations: USDA CES Ranking: A-3 (1986).

IARC Cancer Review: Human Limited Evidence IMEMDT 41,357,86.

SAFETY PROFILE: Moderately toxic by ingestion and possibly other routes. May be a human carcinogen. An experimental teratogen. Experimental reproductive effects. When heated to decomposition it emits toxic fumes of Cl^-.

TOXICITY DATA and CODEN

orl-mus TDLo: 1617 mg/kg (12-15D preg): TER AECTCV 6,33,77
scu-mus TDLo: 1617 mg/kg (12-15D preg): TER AECTCV 6,33,77
orl-rat LD50: 650 mg/kg RREVAH 10,97,65

TJH000 CAS: 13121-70-5
TRICYCLOHEXYLHYDROXYSTANNANE
mf: $C_{18}H_{34}OSn$ mw: 385.21

SYNS: CYHEXATIN ◇ DOWCO-213 ◇ ENT 27,395-X ◇ M 3180 ◇ PLICTRAN ◇ PLYCTRAN ◇ TCTH ◇ TRICYCLO-HEXYLHYDROXYTIN ◇ TRICYCLOHEXYLTIN HYDROXIDE ◇ TRICYCLOHEXYLZINNHYDROXID (GERMAN)

USE IN FOOD:

Purpose: Insecticide.

Where Used: Animal feed, apple pomace (dried), citrus pulp (dried), hops (dried), prunes (dried).

Regulations: FDA - 21CFR 193.430. Insecticide residue tolerance of 90 ppm in dried hops, 4 ppm in dried prunes. 21CFR 561.400. Limitation of 8 ppm in dried apple pomace and dried citrus pulp when used for animal feed.

OSHA PEL: TWA 5 mg/m^3 ACGIH TLV: TWA 5 mg/m^3 NIOSH REL: (Organotin Compounds) TWA 0.1 $mg(Sn)/m^3$.

SAFETY PROFILE: Poison by ingestion, inhalation and intraperitoneal routes. Moderately toxic by skin contact. Experimental reproductive effects. When heated to decomposition it emits acrid smoke and irritating fumes.

TOXICITY DATA and CODEN

orl-rat TDLo: 108 mg/kg (MGN): REP EQSFAP 4,80,75
orl-rat LD50: 180 mg/kg KSKZAN 16(2),59,78
ihl-rat LC50: 244 mg/m^3 GISAAA 49(2),74,84
skn-rat LD50: 1880 mg/kg GISAAA 47(7),80,82

TJJ400
2-TRIDECENAL
mf: $C_{13}H_{24}O$ mw: 196.33

PROP: White to yellow liquid; oily, citrus odor. D: 0.842-0.862, refr index: 1.457. Sol in alc, fixed oils; insol in water.

SYN: FEMA No. 3082

USE IN FOOD:

Purpose: Flavoring agent.

Where Used: Various.

Regulations: FDA - 21CFR 172.515. Use at a level not in excess of the amount reasonably required to accomplish the intended effect.

SAFETY PROFILE: When heated to decomposition it emits acrid smoke and irritating fumes.

TJK700
TRIETHANOLAMINE DODECYLBENZENE SULFONATE

USE IN FOOD:

Purpose: Hog scald agent.

Where Used: Hog carcasses.

Regulations: USDA - 9CFR 318.7. Sufficient for purpose.

SAFETY PROFILE: When heated to decomposition it emits toxic fumes of SO_x.

TJP750 CAS: 77-93-0
TRIETHYL CITRATE
mf: $C_{12}H_{20}O_7$ mw: 276.32

PROP: Colorless oily liquid; odorless. Bp: 294°, flash p: 303°F (COC), d: 1.136 @ 25°, vap press: 1 mm @ 107.0°. Sltly sol in water; misc in alc, ether.

SYNS: CITROFLEX 2 ◇ ETHYL CITRATE ◇ 2-HY-DROXY,1,2,3-PROPANETRICARBOXYLIC ACID, TRIETHYL ESTER ◇ TEC

USE IN FOOD:

Purpose: Plasticizer, sequestrant.

Where Used: Egg white (dried).

Regulations: FDA - 21CFR 181.27, 182.1911. Limitation of 0.25 percent in dried egg whites. GRAS when used in accordance with good manufacturing practice.

SAFETY PROFILE: Moderately toxic by intraperitoneal route. Mildly toxic by ingestion and inhalation. Combustible liquid when exposed to heat or flame. To fight fire, use dry chemical, CO_2. When heated to decomposition it emits acrid smoke and irritating fumes.

TOXICITY DATA and CODEN

orl-rat LD50:5900 mg/kg IYKEDH 16,214,85
ihl-rat LC50:1300 ppm/6H FCTXAV 17,357,79

TKB310 CAS: 1493-13-6
TRIFLUOROMETHANE SULFONIC ACID
mf: CHF_3O_3S mw: 150.08

SYN: TRIFLIC ACID

USE IN FOOD:

Purpose: Catalyst.

Where Used: Cocoa butter substitute from palm oil.

Regulations: FDA - 21CFR 173.395. Limitation of 0.2 ppm as fluoride in cocoa butter substitute from palm oil.

SAFETY PROFILE: A corrosive irritant to the skin, eyes, and mucous membranes. A strong acid. Violent reaction with acyl chlorides or aromatic hydrocarbons evolves toxic hydrogen chloride gas. When heated to decomposition it emits toxic fumes of F^- and SO_x.

TKL100 CAS: 26644-46-2
TRIFORINE
mf: $C_{10}H_{14}Cl_6N_4O_2$ mw: 434.95

PROP: White crystals. Sol in water.

SYN: N,N-[1,4-PIPERAZINEDIYL-BIS(2,2,2-TRICHLOROETHYLIDENE]BIS[FORMAMIDE]

USE IN FOOD:

Purpose: Fungicide.

Where Used: Animal feed, hops (dried), hops (spent).

Regulations: FDA - 21CFR 193.476. Limitation of 60 ppm in dried hops. 21CFR 561.442. Limitation of 60 ppm in spent hops when used for animal feed.

SAFETY PROFILE: When heated to decomposition emits toxic fumes of NO_x, Cl^-.

TKO250 CAS: 1421-63-2
2′,4′,5′-TRIHYDROXYBUTYROPHENONE
mf: $C_{10}H_{12}O_4$ mw: 196.22

PROP: Yellow-tan crystals. Mp: 149-153°, d: 6.0 lb/gal @ 20°. Very sltly sol in water; sol in alc, propylene glycol.

SYN: THBP ◇ 2,4,5-TRIHYDROXYBUTYROPHENONE ◇ USAF EK

USE IN FOOD:

Purpose: Antioxidant.

Where Used: Various.

Regulations: FDA - 21CFR 172.190. Limitation of 0.02 percent of the oil or fat content of the food including the volatile oil content of the food. 21CFR 181.24. Limitation of 0.005 percent migrating from food packages.

SAFETY PROFILE: Poison by intraperitoneal route. Mutagenic data. When heated to decomposition it emits acrid smoke and irritating fumes.

TOXICITY DATA and CODEN

mma-sat 167 μg/plate ENMUDM 8(Suppl 7),1,86
ipr-mus LD50:200 mg/kg NTIS** AD277-689

TKP500 CAS: 102-71-6
TRIHYDROXYTRIETHYLAMINE
mf: $C_6H_{15}NO_3$ mw: 149.22

PROP: Pale yellow viscous liquid. Mp: 21.2°, bp: 360°, flash p: 355°F (CC), d: 1.1258 @ 20/20°, vap press: 10 mm @ 205°, vap d: 5.14.

SYNS: DALTOGEN ◇ NITRILO-2,2′,2″-TRIETHANOL ◇ 2,2′,2″-NITRILOTRIETHANOL ◇ STEROLAMIDE ◇ THIOFACO T-35 ◇ TRIAETHANOLAMIN-NG ◇ TRIETHANOLAMIN ◇ TRIETHANOLAMINE ◇ TRIETHYLOLAMINE ◇ TRI(HYDROXYETHYL)AMINE ◇ 2,2′,2″-TRIHYDROXYTRIETHYLAMINE ◇ TRIS(2-HYDROXYETHYL)AMINE ◇ TROLAMINE

USE IN FOOD:

Purpose: Flume wash water additive.

Where Used: Sugar beets.

Regulations: FDA - 21CFR 173.315. Limitation of 2 ppm in wash water.

Cyanide and its compounds are on the Community Right-To-Know List. EPA Genetic Toxicology Program.

SAFETY PROFILE: Moderately toxic by intraperitoneal route. Mildly toxic by ingestion. An experimental carcinogen. Liver and kidney damage has been demonstrated in animals from chronic exposure. A human and experimental skin irritant. An eye irritant. Combustible liquid when exposed to heat or flame; can react vigorously with oxidizing materials. To fight fire, use alcohol foam, CO_2, dry chemical. When heated to decomposition it emits toxic fumes of NO_x and CN^-.

TOXICITY DATA and CODEN

skn-hmn 15 mg/3D-I MLD 85DKA8 -,127,77
skn-rbt 560 mg/24H MLD TXAPA9 19,276,71
eye-rbt 10 mg MLD TXAPA9 55,501,80
orl-mus TDLo:16 g/kg/64W-C:CAR CNREA8 38,3918,78
orl-mus TD :154 g/kg/61W-C:CAR CNREA8 38,3918,78
orl-rat LD50:8 g/kg NTIS** PB158-507

TLP500 CAS: 147-47-7
2,2,4-TRIMETHYL-1,2-DIHYDROQUINOLINE
mf: $C_{12}H_{15}N$ mw: 173.28

SYNS: ACETONE ANIL ◇ 1,2-DIHYDRO-2,2,4-TRIMETHYLQUINOLINE ◇ FLECTOL H ◇ NCI-C60902

USE IN FOOD:

Purpose: Packaging adhesives.

Where Used: Packaging materials.

Regulations: FDA - 21CFR 189.220. Prohibited from indirect addition to human food from food-contact surfaces.

SAFETY PROFILE: Moderately toxic by ingestion. When heated to decomposition it emits toxic fumes of NO_x.

TOXICITY DATA and CODEN

orl-rat LD50:2000 mg/kg HYSAAV 31,183,66

TLX800
2,4,5-TRIMETHYL Δ-3-OXAZOLINE
mf: $C_6H_{11}NO$ mw: 113.16

PROP: Yellow-orange liquid; powerful, musty, nut-like odor. D: 0.911-0.932, refr index: 1.414-1.435. Sol in alc, propylene glycol, water; insol in fixed oils.

SYN: FEMA No. 3525

USE IN FOOD:

Purpose: Flavoring agent.

Where Used: Various.

Regulations: GRAS when used at a level not in excess of the amount reasonably required to accomplish the intended effect.

SAFETY PROFILE: When heated to decomposition emits toxic fumes of NO_x.

TME270 CAS: 14667-55-1
2,3,5-TRIMETHYLPYRAZINE
mf: $C_7H_{10}N_2$ mw: 122.19

PROP: Colorless to slightly yellow liquid; baked potato, peanut odor. D: 0.960-0.990 @ 20°, refr index: 1.503, flash p: +153°F. Sol in water; organic solvents.

SYNS: FEMA No. 3244 ◇ TRIMETHYLPYRAZINE

USE IN FOOD:

Purpose: Flavoring agent.

Where Used: Various.

Regulations: GRAS when used at a level not in excess of the amount reasonably required to accomplish the intended effect.

SAFETY PROFILE: Moderately toxic by ingestion. Combustible liquid. When heated to decomposition emits toxic fumes of NO_x.

TOXICITY DATA and CODEN

orl-rat LD50:806 mg/kg DCTODJ 3,249,80

TNK750 CAS: 1934-21-0
TRISODIUM-3-CARBOXY-5-HYDROXY-1-p-SULFOPHENYL-4-p-SULFOPHENYLAZOPYRAZOLE
mf: $C_{16}H_9N_4O_9S_2 \cdot 3Na$ mw: 534.38

PROP: Yellow-orange powder. Sol in water, conc sulfuric acid.

SYNS: ACID LEATHER YELLOW T ◇ ACILAN YELLOW GG ◇ AIREDALE YELLOW T ◇ AIZEN TARTRAZINE ◇ ATUL TARTRAZINE ◇ BUCACID TARTRAZINE ◇ CALCOCID YELLOW XX ◇ CANACERT TARTRAZINE ◇ 3-CARBOXY-5-HYDROXY-1-p-SULFOPHENYL-4-o-SULFOPHENYLAZOPYRAZOLE TRISODIUM SALT ◇ C.I. 19140 ◇ C.I. FOOD YELLOW 4 ◇ CURON FAST YELLOW 5G ◇ D & C YELLOW NO. 5 ◇ DOLKWAL TARTRAZINE ◇ EDICOL SUPRA TARTRAZINE N ◇ EGG YELLOW A ◇ EUROCERT TARTRAZINE ◇ F D & C YELLOW NO. 5

(FCC) ◇ FENAZO YELLOW T ◇ FOOD YELLOW NO. 4 ◇ HEXACOL TARTRAZINE ◇ HYDRAZINE YELLOW ◇ KARO TARTRAZINE ◇ KITON YELLOW T ◇ LAKE YELLOW ◇ MAPLE TARTRAZOL YELLOW ◇ NAPHTHOCARD YELLOW O ◇ OXANAL YELLOW T ◇ SHULTZ NO. 737 ◇ SUGAI TARTRAZINE ◇ TARTAR YELLOW ◇ TARTRAZINE ◇ TARTRAZOL YELLOW ◇ TRISODIUM SALT of 3-CARBOXY-5-HYDROXY-1-SULFOPHENYLAZOPYRAZOLE ◇ UNITERTRACID YELLOW TE ◇ USACERT YELLOW NO. 5 ◇ VONDACID TARTRAZINE ◇ WOOL YELLOW ◇ XYLENE FAST YELLOW GT ◇ YELLOW LAKE 69

USE IN FOOD:

Purpose: Color additive.

Where Used: Butter, cheese, ice cream.

Regulations: FDA - 21CFR 74.705. Use in accordance with good manufacturing practice.

EPA Genetic Toxicology Program.

SAFETY PROFILE: Mildly toxic by ingestion. An experimental teratogen. Human systemic effects by ingestion: paresthesia and changes in teeth and supporting structures. Experimental reproductive effects. Human mutagenic data. When heated to decomposition it emits very toxic fumes of NO_x, SO_x and Na_2O.

TOXICITY DATA and CODEN

cyt-hmn:lym 100 mg/L SOGEBZ 11,528,75
cyt-ham:lng 2100 mg/L GMCRDC 27,95,81
orl-rat TLDo:18500 mg/kg (7-22D preg/21D post):TER JTEHD6 2,1211,77
orl-rat TDLo:1280 mg/kg (7-14D preg):REP OYYAA2 24,399,82
orl-rat TLDo:20 g/kg (7-14D preg):REP OYYAA2 24,399,82
orl-hmn TDLo:14 µg/kg:PNS,MSK ANAEA3 17,719,59
orl-mus LD50:12750 mg/kg FAONAU 38B,90,66

TNM250 CAS: 7601-54-9
TRISODIUM PHOSPHATE

mf: $O_4P\cdot 3Na$ mw: 163.94

PROP: White crystals or crystalline powder; odorless. Sol in water; insol in alc.

SYNS: DRI-TRI ◇ EMULSIPHOS 440/660 ◇ NUTRIFOS STP ◇ PHOSPHORIC ACID, TRISODIUM SALT ◇ SODIUM PHOSPHATE ◇ SODIUM PHOSPHATE, ANHYDROUS ◇ SODIUM PHOSPHATE, TRIBASIC (DOT, FCC) ◇ TRIBASIC SODIUM PHOSPHATE ◇ TRINATRIUMPHOSPHAT (GERMAN) ◇ TRISODIUM ORTHOPHOSPHATE ◇ TROMETE ◇ TSP

USE IN FOOD:

Purpose: Boiler water additive, buffer, denuding agent, dietary supplement, emulsifier, fat rendering aid, hog scald agent, nutrient, poultry scald agent.

Where Used: Cereals, cheese, evaporated milk, fats (rendered animal), hog carcasses, poultry, tripe.

Regulations: FDA - 21CFR 173.310, 182.5778, 182.8778. GRAS when used in accordance with good manufacturing practice. USDA - 9CFR 318.7, 381.147. Sufficient for purpose.

DOT Classification: ORM-E; Label: None

SAFETY PROFILE: Moderately toxic by intravenous route. Mutagenic data. A strong, caustic material. When heated to decomposition it emits toxic fumes of Na_2O and PO_x.

TOXICITY DATA and CODEN

sln-dmg-orl 11 pph DRISAA 20,87,46
ivn-rbt LDLo:1580 mg/kg HBAMAK 4,1289,35

TNP250 CAS: 786-19-6
TRITHION

mf: $C_{11}H_{16}ClO_2PS_3$ mw: 342.87

PROP: Amber liquid. Bp: 82° @ 0.1 mm, d: 1.29 @ 20°. Essentially insol in water; misc in common solvents.

SYNS: ACARITHION ◇ AKARITHION ◇ CARBOFENOTHION ◇ CARBOFENOTHION (DUTCH) ◇ S-((p-CHLOROPHENYLTHIO)METHYL)-O,O-DIETHYL PHOSPHORODITHIOATE ◇ DAGADIP ◇ O,O-DIAETHYL-S-((4-CHLORPHENYL-THIO)-METHYL)DITHIOPHOSPHAT (GERMAN) ◇ O,O-DIETHYL-S-(4-CHLOOR-FENYL-THIO)-METHYL)-DITHIOFOSFAAT (DUTCH) ◇ O,O-DIETHYL-S-p-CHLORFENYLTHIOMETHYLESTER KYSELINY DITHIOFOSFORECNE (CZECH) ◇ O,O-DIETHYL-S-p-CHLORLPHENYLTHIOMETHYL DITHIOPHOSPHATE ◇ O,O-DIETHYL-S-(p-CHLOROPHENYLTHIOMETHYL) PHOSPHORODITHIOATE ◇ O,O-DIETIL-S-((4-CLORO-FENIL-TIO)-METILE)-DITIOFOSFATO (ITALIAN) ◇ DITHIOPHOSPHATE de O,O-DIETHYLE et de (4-CHLORO-PHENYL) THIOMETHYLE (FRENCH) ◇ ENDYL ◇ ENT 23,708 ◇ GARRATHION ◇ LETHOX ◇ NEPHOCARP ◇ OLEOAKARITHION ◇ TRITHION MITICIDE

USE IN FOOD:

Purpose: Insecticide, miticide.

Where Used: Animal feed, citrus meal, citrus pulp (dehydrated), grapefruit, lemons, limes, oranges, tangelos, tangerines, tea (dried).

Regulations: FDA - 21CFR 193.50. Insecticide residue tolerance of 20 ppm on dried tea. 21CFR 561.51. Limitation of 10 ppm in dehydrated citrus pulp and citrus meal when used for cattle feed.

EPA Farm Worker Field Reentry FEREAC 39,16888,74. EPA Extremely Hazardous Substances List.

SAFETY PROFILE: Poison by ingestion, skin contact, and intraperitoneal routes. Moderately toxic by subcutaneous route. A cholinesterase inhibitor. When heated to decomposition it emits very toxic fumes of SO_x, PO_x, and Cl^-.

TOXICITY DATA and CODEN

orl-rat LD50:6800 μg/kg FMCHA2 -,C246,83
skn-rat LD50:27 mg/kg TXAPA9 2,88,60
skn-rbt LD50:1270 mg/kg PCOC** -,200,66

TNW500 CAS: 54-12-6
dl-TRYPTOPHAN
mf: $C_{11}H_{12}N_2O_2$ mw: 204.25

PROP: White crystals or crystalline powder; odorless. Sol in water, dil acids, alkalies; sltly sol in alc. Optically inactive.

USE IN FOOD:

Purpose: Dietary supplement, nutrient.

Where Used: Various.

Regulations: FDA - 21CFR 172.310. Limitation 1.6 percent by weight.

EPA Genetic Toxicology Program.

SAFETY PROFILE: An experimental carcinogen. When heated to decomposition it emits toxic fumes of NO_x.

TOXICITY DATA and CODEN

orl-rat TDLo:844 g/kg/92W-C:CAR CNREA8 39,1207,79

TNX000 CAS: 73-22-3
l-TRYPTOPHANE
mf: $C_{11}H_{12}N_2O_2$ mw: 204.25

PROP: An essential amino acid; occurs in isomeric forms. Mp: decomp 289°.The l and dl forms are: White crystals or crystalline powder; slt bitter taste; (dl) sltly sol in water; (l) Sol in water, hot alc, alkali hydroxides; insol in chloroform.

SYNS: l-α-AMINO-3-INDOLEPROPRIONIC ACID ◇ α'-AMINO-3-INDOLEPROPRIONIC ACID ◇ α-AMINO-INDOLE-3-PROPRIONIC ACID ◇ 2-AMINO-3-INDOL-3-YL-PROPRIONIC ACID ◇ EH 121 ◇ INDOLE-3-ALANINE ◇ 1-β-3-INDOLYLALANINE ◇ NCI-C01729 ◇ (−)-TRYPTOPHAN ◇ l-TRYPTOPHAN (FCC) ◇ TRYPTOPHANE

USE IN FOOD:

Purpose: Dietary supplement, nutrient.

Where Used: Various.

Regulations: FDA - 21CFR 172.310. Limitation 1.6 percent by weight.

NCI Carcinogenesis Bioassay (feed); No Evidence: mouse, rat NCITR* NCI-CG-TR-71,78.

SAFETY PROFILE: Moderately toxic by intraperitoneal route. An experimental tumorigen and teratogen. Experimental reproductive effects. Human mutagenic data. When heated to decomposition it emits toxic fumes of NO_x.

TOXICITY DATA and CODEN

mmo-omi 5 g/L MGGEAE 121,117,73
dni-rat:lvr 100 μmol/L CNREA8 45,337,85
ipr-rat TDLo:15 g/kg (30D pre):REP IYKEDH 11,646,80
ivn-rbt TDLo:30 g/kg (30D male):REP IYKEDH 11,635,80
scu-rat TDLo:9500 mg/kg/2Y-C:ETA,TER VOONAW 20(8),75,74
imp-mus TDLo:80 mg/kg:ETA CALEDQ 17,101,82

TOA510
TUNG NUT OIL

SYN: CHINAWOOD OIL

USE IN FOOD:

Purpose: Drying oil.

Where Used: Packaging materials.

Regulations: FDA - 21CFR 181.26.

SAFETY PROFILE: Toxic by ingestion. Contact causes dermatitis. Ingestion causes nausea, vomiting, cramps, diarrhea and tenesmus, thirst, dizziness, lethargy and disorientation. Large doses can cause fever, tachycardia and respiratory effects. Combustible when exposed to heat or flame. Can react with oxidizing materials.

TOD625
TURMERIC

PROP: From solvent extraction of dried ground rhizome of *Curcuma ionga L.* Bright yellow powder or yellow-orange to brown liquid; mustard taste. Misc in water.

SYN: OLEORESIN TUMERIC

USE IN FOOD:

Purpose: Color additive.

Where Used: Casings, fats (rendered), fish, meat, poultry, rice dishes.

Regulations: FDA - 21CFR 73.600. Use in accordance with good manufacturing practice. USDA - 9CFR 318.7. Sufficient for purpose.

SAFETY PROFILE: Human mutagenic data. When heated to decomposition it emits acrid smoke and irritating fumes.

TOXICITY DATA and CODEN

cyt-hmn:lym 6 mg/L FCTXAV 14,9,76
cyt-mam:oth 10 mg/L/4H FCTXAV 14,9,76

TOE600 CAS: 1401-69-0
TYLOSIN
mf: $C_{45}H_{77}NO_{17}$ mw: 904.23

PROP: Crystals from water. Mp: 128-132°. Sol in water at 25°: 5 mg/mL. Sol in lower alc, esters, and ketones, in chlorinated hydrocarbons, benzene, ether.

SYNS: TYLAN ◇ TYLON

USE IN FOOD:

Purpose: Animal drug.

Where Used: Beef, chicken, eggs, milk, pork, turkey.

Regulations: FDA - 21CFR 556.740. Limitation of 0.2 ppm in chickens, turkeys, cattle, swine, eggs. Limitation of 0.05 ppm in milk. 21CFR 558.625. USDA CES Ranking: Z-3 (1986).

SAFETY PROFILE: Poison by intravenous route. Moderately toxic by ingestion and intraperitoneal routes. When heated to decomposition it emits toxic fumes of NO_x.

TOXICITY DATA and CODEN

orl-mus LD50:10 g/kg 85GDA2 2,135,80

TOE750 CAS: 11032-12-5
TYLOSIN HYDROCHLORIDE
mf: $C_{45}H_{77}NO_{17}\cdot ClH$ mw: 940.69

PROP: Crystals. Mp: 141-145°.

USE IN FOOD:

Purpose: Animal feed drug.

Where Used: Animal feed.

Regulations: USDA CES Ranking: Z-3 (1986).

SAFETY PROFILE: Poison by intravenous route. When heated to decomposition it emits very toxic fumes of NO_x and HCl.

TOXICITY DATA and CODEN

ivn-mus LD50:582 mg/kg FCTXAV 4,1,66

TOE810
TYLOSIN and SULFAMETHAZINE

USE IN FOOD:

Purpose: Animal feed drug.

Where Used: Animal feed.

Regulations: FDA - 21CFR 558.630.

SAFETY PROFILE: When heated to decomposition it emits acrid smoke and irritating fumes.

TOG300 CAS: 60-18-4
l-TYROSINE
mf: $C_9H_{11}NO_3$ mw: 181.21

PROP: Colorless, silky needles or white crystalline powder. Sol in water, dil mineral acids, alkaline solutions; sltly sol in alc.

SYNS: l-β-(p-HYDROXYPHENYL)ALANINE ◇ TYROSINE ◇ l-p-TYROSINE ◇ p-TYROSINE

USE IN FOOD:

Purpose: Dietary supplement, nutrient.

Where Used: Various.

Regulations: FDA - 21CFR 172.310. Limitation 4.3 percent by weight.

SAFETY PROFILE: Experimental reproductive effects. When heated to decomposition it emits acrid smoke and irritating fumes.

TOXICITY DATA and CODEN

orl-rat LDLo:20750 mg/kg (125-19D preg):REP TOIZAG 30,518,83
scu-mus LDLo:150 mg/kg (3D pre):REP BIRSB5 20(Suppl 1), 133A,79
orl-rbt LDLo:3 g/kg (29-31D preg):REP BINEAA 12,282,68

U

UJA200 CAS: 57455-37-5
ULTRAMARINE BLUE
mf: $Na_7Al_6Si_6O_{24}S_3$ mw: 971.50

PROP: Calcined mixture of kaolin, sulfur, sodium carbonate, and carbon above 700°.

USE IN FOOD:

Purpose: Color additive.

Where Used: Packaging materials, salt for animal feed.

Regulations: FDA - 21CFR 73.50. Limitation of 0.5 percent of salt. 21CFR 178.3970. Use at a level not in excess of the amount reasonably required to accomplish the intended effect.

SAFETY PROFILE: A nuisance dust.

UJA800
γ-UNDECALACTONE
mf: $C_{11}H_{20}O_2$ mw: 184.28

PROP: Colorless to slightly yellow liquid; peach odor. D: 0.825, refr index: 1.430, flash p: 279°F. Sol in fixed oils, propylene glycol; insol in glycerine, water @ 223°.

SYNS: ALDEHYDE C-14 PURE ◇ FEMA No. 3091 ◇ PEACH ALDEHYDE

USE IN FOOD:

Purpose: Flavoring agent.

Where Used: Beverages, candy, gelatins, ice cream, puddings.

Regulations: FDA - 21CFR 172.515. Use at a level not in excess of the amount reasonably required to accomplish the intended effect.

SAFETY PROFILE: Combustible liquid. When heated to decomposition it emits acrid smoke and irritating fumes.

UJJ000 CAS: 112-44-7
1-UNDECANAL
mf: $C_{11}H_{22}O$ mw: 170.33

PROP: Colorless to sltly yellow liquid; sweet, fatty, floral odor. Mp: −4°, bp: 117° @ 18 mm, flash p: 235°F (COC), d: 0.830 @ 20/4°, refr index: 1.430, vap press: 0.04 mm @ 20°, vap d: 5.94. Sol in fixed oils, propylene glycol; glycerin, water @ 223°. Reported in lemon and mandarin oils (FCTXAV 11,477,73).

SYNS: ALDEHYDE-14 ◇ 1-DECYL ALDEHYDE ◇ FEMA No. 3092 ◇ HENDECANAL ◇ HENDECANALDEHYDE ◇ UNDECANAL ◇ n-UNDECANAL ◇ UNDECANALDEHYDE ◇ UNDECYL ALDEHYDE ◇ N-UNDECYL ALDEHYDE ◇ UNDECYLIC ALDEHYDE

USE IN FOOD:

Purpose: Flavoring agent.

Where Used: Various.

Regulations: FDA - 21CFR 172.515. Use at a level not in excess of the amount reasonably required to accomplish the intended effect.

SAFETY PROFILE: A skin irritant. Combustible liquid when exposed to heat or flame. To fight fire, use CO_2, dry chemical. When heated to decomposition it emits acrid smoke and irritating fumes.

TOXICITY DATA and CODEN

skn-rbt 500 mg MLD FCTXAV 11,1079,73

ULJ000 CAS: 112-45-8
10-UNDECENAL
mf: $C_{11}H_{20}O$ mw: 168.31

PROP: Colorless to light yellow liquid; rose odor. D: 0.840-0.850, refr index: 1.441-1.447, flash p: 212°F. Sol in fixed oils, propylene glycol; insol in water @ 235°, glycerin.

SYNS: ALDEHYDE C-11, UNDECYLENIC ◇ FEMA No. 3095 ◇ HENDECENAL ◇ 1-UNDECEN-10-AL ◇ UNDECYLENALDEHYDE ◇ 10-UNDECYLENEALDEHYDE ◇ UNDECYLENIC ALDEHYDE

USE IN FOOD:

Purpose: Flavoring agent.

Where Used: Various.

Regulations: FDA - 21CFR 172.515. Use at a level not in excess of the amount reasonably required to accomplish the intended effect.

SAFETY PROFILE: A skin irritant. Combustible liquid. When heated to decomposition it emits acrid smoke and irritating fumes.

TOXICITY DATA and CODEN

skn-rbt 500 mg MLD FCTXAV 11,1079,73

ULS875
2-UNDECENOL

mf: $C_{11}H_{22}O$ mw: 170.30

PROP: White to sltly yellow liquid; oily, sweet, floral odor. D: 0.847, refr index: 1.450 @ 22°. Insol in water.

USE IN FOOD:

Purpose: Flavoring agent.

Where Used: Various.

Regulations: FDA - 21CFR 172.515. Use at a level not in excess of the amount reasonably required to accomplish the intended effect.

SAFETY PROFILE: When heated to decomposition it emits acrid smoke and irritating fumes.

UNA000 CAS: 112-42-5
UNDECYL ALCOHOL

mf: $C_{11}H_{24}O$ mw: 172.35

PROP: Colorless liquid; fatty-floral odor. D: 0.820-0.840, refr index: 1.437-1.443, mp: 19°, bp: 131° @ 15 mm, flash p: 234°F. Sol in fixed oils; insol in water.

SYNS: ALCOHOL C-11 ◇ FEMA No. 3097 ◇ HENDECANOIC ALCOHOL ◇ 1-HENDECANOL ◇ HENDECYL ALCOHOL ◇ n-HENDECYLENIC ALCOHOL ◇ n-UNDECANOL

USE IN FOOD:

Purpose: Flavoring agent.

Where Used: Various.

Regulations: FDA - 21CFR 172.515. Use at a level not in excess of the amount reasonably required to accomplish the intended effect.

SAFETY PROFILE: Moderately toxic by ingestion. A skin irritant. Combustible liquid. When heated to decomposition it emits acrid smoke and irritating fumes.

TOXICITY DATA and CODEN

skn-rbt 500 mg/24H MOD FCTXAV 16,637,78
orl-rat LD50:3000 mg/kg JIHTAB 26,269,44

USS000 CAS: 57-13-6
UREA

mf: CH_4N_2O mw: 60.07

PROP: White crystals. Mp: 132.7°, bp: decomp, d: (solid) 1.335. Sol in water, alc; sltly sol in ether.

SYNS: CARBAMIDE ◇ CARBAMIDE RESIN ◇ CARBAMIMIDIC ACID ◇ CARBONYL DIAMIDE ◇ CARBONYLDIAMINE ◇ ISOUREA ◇ NCI-C02119 ◇ PRESPERSION, 75 UREA ◇ PSEUDOUREA ◇ SUPERCEL 3000 ◇ UREAPHIL ◇ UREOPHIL ◇ UREVERT ◇ VARIOFORM II

USE IN FOOD:

Purpose: Fermentation aid, formulation aid, yeast nutrient.

Where Used: Alcoholic beverages, gelatin products, wine, yeast-raised bakery products.

Regulations: FDA - 21CFR 184.1923. GRAS when used in accordance with good manufacturing practice. BATF - 27CFR 240.1051. Limitation of 2 pounds/1000 gallons of wine.

EPA Genetic Toxicology Program.

SAFETY PROFILE: Moderately toxic by ingestion, intravenous, and subcutaneous routes. An experimental carcinogen and neoplastigen. Human reproductive effects by intraplacental route: fertility effects. Experimental reproductive effects. Human mutagenic data. A human skin irritant. When heated to decomposition it emits toxic fumes of NO_x.

TOXICITY DATA and CODEN

skn-hmn 22 mg/3D-I MLD 85DKA8 -,127,77
dni-hmn:lym 600 mmol/L PNASA6 79,1171,82
cyt-hmn:leu 50 mmol/L CNREA8 25,980,65
ipc-wmn TDLo:1400 mg/kg (16W preg):REP OBGNAS 43,765,74
ipc-wmn TDLo:1600 mg/kg (16W preg):REP BJPCAL 26,24,72
orl-rat TDLo:821 g/kg/1Y-C:NEO JEPTDQ 3(5-6),149,80
orl-mus TDLo:394 g/kg/1Y-C:CAR JEPTDQ 3(5-6),149,80
orl-rat LD50:14300 mg/kg OYYAA2 13,749,77

V

VAD000 CAS: 54965-21-8
VALBAZEN
mf: $C_{12}H_{15}N_3O_2S$ mw: 265.36

PROP: Colorless crystals. Mp: 208-210°.

SYNS: ALBENDAZOLE (USDA) ◇ METHYL 5-(PROPYLTHIO)-2-BENZIMIDAZOLECARBAMATE ◇ ((PROPYLTHIO)-5-1H-BENZIMIDAZOLYL-2) CARBAMATE de METHYLE (FRENCH) ◇ (5-(PROPYLTHIO)-1H-BENZIMIDAZOL-2-YL) CARBAMIC ACID METHYL ESTER ◇ 5-(PROPYLTHIO)-2-CARBOMETHOXYAMINOBENZIMIDAZOLE ◇ SKF 62979 ◇ ZENTAL

USE IN FOOD:

Purpose: Animal drug.

Where Used: Various.

Regulations: USDA CES Ranking: A-2 (1987).

SAFETY PROFILE: Moderately toxic by ingestion. An experimental teratogen. Experimental reproductive effects. When heated to decomposition it emits toxic fumes of SO_x and NO_x.

TOXICITY DATA and CODEN

orl-rat TDLo:71 mg/kg (8-15D preg):TER ARCVBP 12,159,81
orl-rat TDLo:85 mg/kg (8-15D preg):REP ARCVBP 12,159,81
orl-dom TDLo:20 mg/kg (17D preg):REP AMSHAR 28,226,80
orl-rat LD50:2400 mg/kg APFRAD 40,55,82

VAQ000 CAS: 109-52-4
VALERIC ACID
DOT: 1760
mf: $C_5H_{10}O_2$ mw: 102.15

PROP: Colorless, mobile liquid; penetrating, rancid odor. D: 0.940 @ 20/4°, refr index: 1.405-1.14 @ 25°, mp: −34.5°, bp: 186.4°, flash p: 203°F. Sol in water; misc in alc, ether.

SYNS: BUTANECARBOXYLIC ACID ◇ 1-BUTANECARBOXYLIC ACID ◇ FEMA No. 3101 ◇ PENTANOIC ACID ◇ n-PENTANOIC ACID ◇ PROPYLACETIC ACID ◇ VALERIANIC ACID ◇ n-VALERIC ACID

USE IN FOOD:

Purpose: Flavoring agent.

Where Used: Various.

Regulations: FDA - 21CFR 172.515. Use at a level not in excess of the amount reasonably required to accomplish the intended effect.

DOT Classification: Corrosive Material; Label: Corrosive

SAFETY PROFILE: Moderately toxic by ingestion, intravenous, and subcutaneous routes. Mildly toxic by inhalation. A corrosive irritant to skin, eyes, and mucous membranes. Combustible liquid. When heated to decomposition it emits acrid smoke and irritating fumes. Used in perfumes.

TOXICITY DATA and CODEN

orl-mus LD50:600 mg/kg 85GMAT -,119,82
ihl-mus LC50:4100 mg/m^3/2H 85GMAT -,119,82

VAV000 CAS: 108-29-2
4-VALEROLACTONE
mf: $C_5H_8O_2$ mw: 100.13

PROP: Colorless, mobile liquid; sweet, herbaceous odor. Mp: −31°, bp: 205-206.5°, flash p: 205°F (COC), d: 1.047-1.054, refr index: 1.43, vap d: 3.45. Misc in alc, fixed oils, water.

SYNS: FEMA No. 3103 ◇ 4-HYDROXYPENTANOIC ACID LACTONE ◇ 4-HYDROXYVALERIC ACID LACTONE ◇ γ-METHYL-γ-BUTYROLACTONE ◇ 4-METHYL-γ-BUTYROLACTONE ◇ γ-PENTALACTONE ◇ 4-PENTANOLIDE ◇ γ-VALEROLACTONE (FCC)

USE IN FOOD:

Purpose: Flavoring agent.

Where Used: Various.

Regulations: FDA - 21CFR 172.515. Use at a level not in excess of the amount reasonably required to accomplish the intended effect.

SAFETY PROFILE: Moderately toxic by ingestion. A skin irritant. Combustible liquid when exposed to heat or flame; can react with oxidizing materials. To fight fire, use water, foam, CO_2, dry chemical. When heated to decomposition it emits acrid smoke and irritating fumes.

TOXICITY DATA and CODEN

skn-rbt 500 mg/24H MLD FCTOD7 20(Suppl),847,82
orl-rat LD50:8800 mg/kg JIHTAB 27,263,45

VBP000 CAS: 72-18-4
VALINE
mf: $C_5H_{11}NO_2$ mw: 117.17

PROP: White, crystalline solid; characteristic taste. Mp (dl): 298° (decomp), mp (l): 315°, d (l): 1.230. Sol in water; very sltly sol in alc; insol in ether. An essential amino acid.

SYNS: l-(+)-α-AMINOISOVALERIC ACID ◇ l-VALINE (FCC)

USE IN FOOD:

Purpose: Dietary supplement, nutrient.

Where Used: Various.

Regulations: FDA - 21CFR 172.310. Limitation 7.4 percent by weight.

SAFETY PROFILE: Mutagenic data. When heated to decomposition it emits toxic fumes of NO_x.

TOXICITY DATA and CODEN

oms-omi 10 mmol/L CBINA8 16,201,77
ipr-rat LD50:5390 mg/kg ABBIA4 58,253,55

VFK000 CAS: 121-33-5
VANILLIN
mf: $C_8H_8O_3$ mw: 152.16

PROP: White, crystalline needles; vanilla odor. D: 1.056, bp: 285°, mp: 80-81°. Sol in 125 parts water, 20 parts glycerin, 2 parts 95% alc, chloroform, ether.

SYNS: FEMA No. 3107 ◇ 4-HYDROXY-m-ANISALDEHYDE ◇ 4-HYDROXY-3-METHOXYBENZALDEHYDE ◇ LIOXIN ◇ 3-METHOXY-4-HYDROXYBENZALDEHYDE ◇ METHYLPROTOCATECHUALDEHYDE ◇ VANILLA ◇ VANILLALDEHYDE ◇ VANILLIC ALDEHYDE ◇ p-VANILLIN ◇ ZIMCO

USE IN FOOD:

Purpose: Flavoring agent.

Where Used: Various.

Regulations: FDA - 21CFR 182.60. Use at a level not in excess of the amount reasonably required to accomplish the intended effect.

SAFETY PROFILE: Moderately toxic by ingestion, intraperitoneal, subcutaneous, and intravenous routes. Experimental reproductive effects. Human mutagenic data. When heated to decomposition it emits acrid smoke and irritating fumes.

TOXICITY DATA and CODEN

sce-hmn:lym 750 μmol/L MUREAV 169,129,86
scu-rat TDLo:20 mg/kg (4D pre):REP JSICAZ 19,264,60
orl-rat LD50:1580 mg/kg FCTXAV 2,327,64

VRF000 CAS: 11006-76-1
VIRGINIAMYCIN

PROP: White powder. Decomp @ 138-140°. Sltly sol in water and dil acid; sol in methanol, ethanol, acetone, benzene; almost insol in ligroin.

SYNS: ANTIBIOTIC NO. 899 ◇ ESKALIN V ◇ MIKAMYCIN ◇ OSTREOGRYCIN ◇ PATRICIN ◇ PRISTINAMYCIN ◇ PYOSTACINE ◇ RP7293 ◇ SKF 7988 ◇ STAFAC ◇ STAPHYLOMYCIN ◇ STAPYOCINE ◇ STREPTOGRAMIN ◇ VERNAMYCIN ◇ VIRGIMYCIN

USE IN FOOD:

Purpose: Animal drug.

Where Used: Chicken, pork.

Regulations: FDA - 21CFR 556.750. Limitation in swine of 0.4 ppm in kidney, skin and fat, 0.3 ppm in liver, 0.1 ppm in muscle. Limitation in broiler chickens of 0.5 ppm in kidney, 0.3 ppm in liver, 0.2 ppm in skin and fat, 0.1 ppm in muscle. 21CFR 558.635.

SAFETY PROFILE: Moderately toxic by ingestion, intraperitoneal and subcutaneous routes.

TOXICITY DATA and CODEN

orl-mus LD50:2100 mg/kg 85ERAY 1,383,78
ipr-mus LD50:450 mg/kg MEIEDD 10,1432,83

VSZ000 CAS: 68-19-9
VITAMIN B_{12} COMPLEX
mf: $C_{63}H_{88}CoN_{14}O_{14}P$ mw: 1355.55

PROP: The anti-pernicious anemia vitamin. All vitamin B_{12} compounds contain the cobalt atom in its trivalent state. There are at least three active forms: cyanocobalamin, hydroxycobalamin, and nitrocobalamin. Dark red crystals or crystalline powder. Very hygroscopic; sltly sol in water; sol in alc; insol in acetone, chloroform, ether.

SYNS: ANACOBIN ◇ B-12 ◇ BERUBIGEN ◇ BETALIN 12 CRYSTALLINE ◇ BEVATINE-12 ◇ BEVIDOX ◇ BYLADOCE ◇ CABADON M ◇ COBADOCE FORTE ◇ COBALIN

◇ COBAMIN ◇ COBIONE ◇ CRYSTWEL ◇ CYANOCOBALAMIN ◇ CYCOLAMIN ◇ CYKOBEMINET ◇ CYREDIN ◇ CYTACON ◇ CYTAMEN ◇ CYTOBION ◇ DEPINAR ◇ 5,6-DIMETHYLBENZIMIDAZOLYCOBAMIDE CYANIDE ◇ DIMETHYLBENZIMIDAZOYLCOBAMIDE ◇ DOBETIN ◇ DOCEMINE ◇ DOCIBIN ◇ DOCIGRAM ◇ DODECABEE ◇ DODECAVITE ◇ DODEX ◇ DUCOBEE ◇ DUODECIBIN ◇ EMBIOL ◇ EMOCICLINA ◇ ERITRONE ◇ ERYCYTOL ◇ ERYTHROTIN ◇ EUHAEMON ◇ HEMOMIN ◇ HEPAGON ◇ HEPAVIS ◇ HEPCOVITE ◇ MACRABIN ◇ MEGABION ◇ MEGALOVEL ◇ MILBEDOCE ◇ NAGRAVON ◇ NORMOCYTIN ◇ PERNAEMON ◇ PERNAEVIT ◇ PERNIPURON ◇ PLECYAMIN ◇ POYAMIN ◇ REBRAMIN ◇ REDAMINA ◇ REDISOL ◇ RHODACRYST ◇ RUBESOL ◇ RUBRAMIN ◇ RUBRIPCA ◇ RUBROCITOL ◇ SYTOBEX ◇ VIBALT ◇ VIBISONE ◇ VIRUBRA ◇ VITAMIN B_{12} (FCC) ◇ VITARUBIN ◇ VITA-RUBRA ◇ VITRAL

USE IN FOOD:

Purpose: Dietary supplement, nutrient.

Where Used: Various.

Regulations: FDA - 21CFR 182.5945, 184.1945. GRAS when used in accordance with good manufacturing practice.

Cobalt and its compounds are on the Community Right-To-Know List. EPA Genetic Toxicology Program.

NIOSH REL: (Cobalt) TWA 0.1 mg/m^3

SAFETY PROFILE: Poison by subcutaneous route. Moderately toxic by intraperitoneal route. An experimental teratogen. Experimental reproductive effects. When heated to decomposition it emits very toxic fumes of PO_x and NO_x.

TOXICITY DATA and CODEN

orl-rat TDLo:115 mg/kg (1-22D preg):REP
NATUAS 242,263,73
ims-mus TDLo:168 mg/kg (10D preg):TER
JPMSAE 67,377,78
ims-mus TDLo:84 mg/kg (11D preg):TER
JPMSAE 67,377,78

VSZ450 CAS: 59-02-9
VITAMIN E

mf: $C_{29}H_{50}O_2$ mw: 430.79

PROP: dl-Form: Sltly viscous, pale yellow oil; d-form: red liquid; odorless. Natural α-tocopherol has been crystallized. Mp: 2.5-3.5°, d: (25°/4°) 0.950, bp: (0.1) 200-220°. Practically insol in water; freely sol in oils, fats, acetone, alc, chloroform, ether, other fat solvents. Gradually darkens on exposure to light.

SYNS: ALMEFROL ◇ ANTISTERILITY VITAMIN ◇ COVI-OX ◇ DENAMONE ◇ EMIPHEROL ◇ ENDO E ◇ EPHYNAL ◇ EPROLIN ◇ EPSILAN ◇ ESORB ◇ ETAMICAN ◇ ETAVIT ◇ EVION ◇ EVITAMINUM ◇ ILITIA ◇ PHYTOGERMINE ◇ PROFECUNDIN ◇ SPAVIT ◇ SYNTOPHEROL ◇ d-α-TOCOPHEROL (FCC) ◇ dl-α-TOCOPHEROL (FCC) ◇ (R,R,R)-α-TOCOPHEROL ◇ α-TOCOPHEROL ◇ (2R, 4'R,8'R)-α-TOCOPHEROL ◇ TOKOPHARM ◇ 5,7,8-TRIMETHYLTOCOL ◇ VASCUALS ◇ VERROL ◇ VITAPLEX E ◇ VITAYONON ◇ VITEOLIN

USE IN FOOD:

Purpose: Antioxidant, dietary supplement, nutrient, preservative.

Where Used: Bacon, fats (rendered animal), pork fat (rendered), poultry.

Regulations: FDA - 21CFR 182.3890, 184.1890. GRAS with a limitation to inhibition of nitrosamine formation and in pump-cured bacon when used in accordance with good manufacturing practice. USDA - 9CFR 318.7. Limitation of 0.03 percent in rendered animal fat. A 30 percent concentration in vegetable oils shall be used when added as an antioxidant to produce products designated as lard or rendered pork fat. Limitation of 500 ppm in pump cured bacon. 9CFR 381.147. Limitation of 0.03 percent in poultry based on fat content.

SAFETY PROFILE: Experimental reproductive effects. Mutagenic data. When heated to decomposition it emits acrid smoke and irritating fumes.

TOXICITY DATA and CODEN

dnd-rat-ivn 27 nmol/kg EXPEAM 31,1023,75
cyt-mus-ipr 2 g/kg/4W FOBLAN 26,94,80
orl-rat TDLo:7500 mg/kg (1-20D preg):REP
NYKZAU 69,293,73

W

WBL100 CAS: 8002-80-0
WHEAT GLUTEN

USE IN FOOD:

Purpose: Dough conditioner, formulation aid, nutrient supplement, processing aid, stabilizer, surface-finishing agent, texturizing agent, thickener.

Where Used: Various.

Regulations: FDA - 21CFR 184.1322. GRAS when used in accordance with good manufacturing practice.

SAFETY PROFILE: When heated to decomposition it emits acrid smoke and irritating fumes.

WBL150
WHEY, DRY

SYNS: DRY WHEY ◇ DRIED WHEY

USE IN FOOD:

Purpose: Binder, extender.

Where Used: Beef with barbecue sauce, bockwurst, chili con carne, loaves (nonspecific), pork with barbecue sauce, poultry, sausage, sausage (imitation), soups, stews.

Regulations: FDA - 21CFR 184.1979. GRAS when used in accordance with good manufacturing practice. USDA - 9CFR 318.7. Limitation of 3.5 percent individually or collectively with other binders in sausage and bockwurst. Sufficient for purpose in imitation sausage, soups, stews, nonspecific loaves. Limitation of 8 percent individually or collectively with other binders in chili con carne, pork or beef with barbecue sauce. 9CFR 381.147. Sufficient for purpose in poultry.

SAFETY PROFILE: When heated to decomposition it emits acrid smoke and irritating fumes.

WBL155
WHEY, PROTEIN CONCENTRATE

USE IN FOOD:

Purpose: Binder.

Where Used: Beef with barbecue sauce, bockwurst, chili con carne, loaves (nonspecific), pork with barbecue sauce, sausage, sausage (imitation), soups, stews.

Regulations: FDA - 21CFR 184.1979c. GRAS when used in accordance with good manufacturing practice. USDA - 9CFR 318.7. Limitation of 3.5 percent individually or collectively with other binders in sausage and bockwurst. Sufficient for purpose in imitation sausage, soups, stews, nonspecific loaves. Limitation of 8 percent individually or collectively with other binders in chili con carne, pork or beef with barbecue sauce.

SAFETY PROFILE: When heated to decomposition it emits acrid smoke and irritating fumes.

WBL160
WHEY, REDUCED LACTOSE

SYN: REDUCED LACTOSE WHEY

USE IN FOOD:

Purpose: Binder.

Where Used: Beef with barbecue sauce, bockwurst, chili con carne, loaves (nonspecific), pork with barbecue sauce, sausage, sausage (imitation), soups, stews.

Regulations: FDA - 21CFR 184.1979a. GRAS when used in accordance with good manufacturing practice. USDA - 9CFR 318.7. Limitation of 3.5 percent individually or collectively with other binders in sausage and bockwurst. Sufficient for purpose in imitation sausage, soups, stews, nonspecific loaves. Limitation of 8 percent individually or collectively with other binders in chili con carne, pork or beef with barbecue sauce.

SAFETY PROFILE: When heated to decomposition it emits acrid smoke and irritating fumes.

WBL165
WHEY, REDUCED MINERALS

SYN: REDUCED MINERALS WHEY

USE IN FOOD:

Purpose: Binder.

Where Used: Beef with barbecue sauce, bockwurst, chili con carne, loaves (nonspecific), pork with barbecue sauce, sausage, sausage (imitation), soups, stews.

Regulations: FDA - 21CFR 184.1979b. GRAS when used in accordance with good manufacturing practice. USDA - 9CFR 318.7. Limitation of 3.5 percent individually or collectively with other binders in sausage and bockwurst. Sufficient for purpose in imitation sausage, soups, stews, nonspecific loaves. Limitation of 8 percent individually or collectively with other binders in chili con carne, pork or beef with barbecue sauce.

SAFETY PROFILE: When heated to decomposition it emits acrid smoke and irritating fumes.

X

XAK800 CAS: 11138-66-2

XANTHAN GUM

PROP: Produced by fermentation of a carbohydrate with *Xanthomonas campestris*. Cream-colored powder. Sol in hot or cold water.

USE IN FOOD:

Purpose: Binder, bodying agent, emulsifier, extender, foam stabilizer, stabilizer, suspending agent, thickener.

Where Used: Baked goods, batter or breading mixes, beverages, chili (canned), chili with beans (canned), desserts, fish pates, gravies, jams, jellies, meat pates, milk products, pizza topping mixes, poultry, salad dressings, salads (meat), sauces, sauces (meat), stews (canned or frozen).

Regulations: FDA - 21CFR 172.695. Use at a level not in excess of the amount reasonably required to accomplish the intended effect. USDA - 9CFR 318.7. Limitation of 8 percent. 9CFR 381.147. Sufficient for purpose in poultry.

SAFETY PROFILE: When heated to decomposition it emits acrid smoke and irritating fumes.

XDJ100

3-(2-XENOLYL)-1,2-EPOXYPROPANE

USE IN FOOD:

Purpose: Plasticizer.

Where Used: Packaging materials.

Regulations: FDA - 21CFR 181.27. Use in accordance with good manufacturing practice.

SAFETY PROFILE: When heated to decomposition it emits acrid smoke and irritating fumes.

XPJ000 CAS: 87-99-0

XYLITOL

mf: $C_5H_{12}O_5$ mw: 152.17

PROP: White crystals or crystalline powder; sweet taste with cooling sensation. Mp: 92-96°. Sol in water; sltly sol in alc.

SYNS: KLINIT ◇ XYLITE (SUGAR)

USE IN FOOD:

Purpose: Nutritive sweetener.

Where Used: Various.

Regulations: FDA - 21CFR 172.395. Use at a level not in excess of the amount reasonably required to accomplish the intended effect

SAFETY PROFILE: Moderately toxic by intravenous route. Mildly toxic by ingestion. When heated to decomposition it emits acrid smoke and irritating fumes. A sugar.

TOXICITY DATA and CODEN

orl-mus LD50:22 g/kg TOLED5 18(Suppl 1),37,83
ivn-rbt LD50:4000 mg/kg FEPRA7 31,726,72

Z

ZAT100 CAS: 9010-66-6
ZEIN

PROP: Powder. Insol in water, alc; sol in glycols, glycol ethers.

USE IN FOOD:

Purpose: Glaze, surface-finishing agent.

Where Used: Confections, grain, nuts, panned goods.

Regulations: FDA - 21CFR 184.1984. GRAS when used in accordance with good manufacturing practice.

SAFETY PROFILE: When heated to decomposition it emits acrid smoke and irritating fumes.

ZBJ000 CAS: 7440-66-6
ZINC
DOT: 1383/1436
af: Zn aw: 65.37

PROP: Bluish-white, lustrous, metallic element. Mp: 419.8°, bp: 908°, d: 7.14 @ 25°, vap press: 1 mm @ 487°. Stable in dry air.

SYNS: BLUE POWDER ◇ C.I. 77945 ◇ C.I. PIGMENT BLACK 16 ◇ C.I. PIGMENT METAL 6 ◇ EMANAY ZINC DUST ◇ GRANULAR ZINC ◇ JASAD ◇ MERRILLITE ◇ PASCO ◇ ZINC DUST ◇ ZINC POWDER ◇ ZINC, POWDER OR DUST, NON-PYROPHORIC (DOT) ◇ ZINC, POWDER OR DUST, PYROPHORIC (DOT)

USE IN FOOD:

Purpose: Pesticide.

Where Used: Various.

Regulations: USDA CES Ranking: D-4 (1985).

Zinc and its compounds are on the Community Right-To-Know List. EPA Genetic Toxicology Program.

DOT Classification: Flammable Solid; Label: Dangerous When Wet, non-pyrophoric; Flammable Solid; Label: Spontaneously Combustible, pyrophoric

SAFETY PROFILE: Human systemic effects by ingestion: cough, dyspnea, and sweating. A human skin irritant. Pure zinc powder, dust, fume is relatively nontoxic to humans by inhalation. The difficulty arises from oxidation of zinc fumes immediately prior to inhalation or presence of impurities such as Cd, Sb, As, Pb. Inhalation may cause sweet taste, throat dryness, cough, weakness, generalized aches, chills, fever, nausea, vomiting.

Flammable in the form of dust when exposed to heat or flame. May ignite spontaneously in air when dry. Explosive in the form of dust when reacted with acids. To fight fire, use special mixtures of dry chemical. When heated to decomposition it emits toxic fumes of ZnO.

TOXICITY DATA and CODEN

skn-hmn 300 μg/3D-I:MLD 85DKA8 -,127,77
ihl-hmn TCLo:124 mg/m^3/50M:PUL,SKN AHYGAJ 72,358,10

ZIA750 CAS: 4468-02-4
ZINC GLUCONATE
mf: $C_{12}H_{22}O_{14}Zn$ mw: 455.68

PROP: White granular or crystalline powder. Sol in water; very sltly sol in alc.

USE IN FOOD:

Purpose: Dietary supplement, nutrient.

Where Used: Various.

Regulations: FDA - 21CFR 182.5988, 182.8988. GRAS when used in accordance with good manufacturing practice.

SAFETY PROFILE: When heated to decomposition it emits toxic fumes of ZnO.

ZJA100 CAS: 56329-42-1
ZINC METHIONINE SULFATE

USE IN FOOD:

Purpose: Dietary supplement, nutrient.

Where Used: Vitamin tablets.

Regulations: FDA - 21CFR 172.399. Use in accordance with good manufacturing practice.

SAFETY PROFILE: When heated to decomposition it emits very toxic fumes of SO_x.

ZJS400
ZINC ORTHOPHOSPHATE

USE IN FOOD:

Purpose: Stabilizer.

Where Used: Packaging materials.

Regulations: FDA - 21CFR 181.29. Limitation of 50 ppm zinc as a migrant in finished food.

SAFETY PROFILE: When heated to decomposition it emits acrid smoke and irritating fumes.

ZKA000 CAS: 1314-13-2
ZINC OXIDE

mf: OZn mw: 81.37

PROP: Odorless, white or yellowish powder. Mp: >1800°, d: 5.47. Insol in water, alc; sol in dil acetic or mineral acids, ammonia.

SYNS: AKRO-ZINC BAR 85 ◇ AMALOX ◇ AZO-33 ◇ AZODOX-55 ◇ CALAMINE (spray) ◇ CHINESE WHITE ◇ C.I. 77947 ◇ C.I. PIGMENT WHITE 4 ◇ CYNKU TLENEK (POLISH) ◇ EMANAY ZINC OXIDE ◇ EMAR ◇ FLOWERS OF ZINC ◇ KADOX-25 ◇ K-ZINC ◇ OZIDE ◇ OZLO ◇ PASCO ◇ PROTOX TYPE 166 ◇ SNOW WHITE ◇ WHITE SEAL-7 ◇ ZINCITE ◇ ZINCOID ◇ ZINC OXIDE FUME ◇ ZINC WHITE

USE IN FOOD:

Purpose: Dietary supplement, nutrient.

Where Used: Various.

Regulations: FDA - 21CFR 182.5991, 182.8991. GRAS when used in accordance with good manufacturing practice.

Zinc and its compounds are on the Community Right-To-Know List.

OSHA PEL: TWA 5 mg/m^3; STEL 10 mg/m^3 ACGIH TLV: TWA 5 mg/m^3; STEL 10 mg/m^3 (fume); 10 mg/m^3 (total dust) DFG MAK: 5 mg/m^3 NIOSH REL: TWA (Zinc Oxide) 5 mg/m^3; CL 15 mg/m^3/15M

SAFETY PROFILE: Poison by intraperitoneal route. An experimental teratogen. Human systemic effects by inhalation of freshly formed fumes: metal fume fever with chills, fever, tightness of chest, cough, dyspnea and other pulmonary changes. Mutagenic data. A skin and eye irritant. When heated to decomposition it emits toxic fumes of ZnO.

TOXICITY DATA and CODEN

skn-rbt 500 mg/24H MLD 28ZPAK -,10,72
eye-rbt 500 mg/24H MLD 28ZPAK -,10,72
dnd-esc 3000 ppm MUREAV 89,95,81
cyt-rat-ihl 100 μg/m^3 CYGEDX 12(3),46,78
orl-rat TDLo:6846 mg/kg (1-22D preg):TER JONUAI 98,303,69
ihl-hmn TCLo:600 mg/m^3:PUL JIDHAN 9,88,27
ipr-rat LD50:240 mg/kg ZDKAA8 38(9),18,78

ZMJ100
ZINC RESINATE

USE IN FOOD:

Purpose: Stabilizer.

Where Used: Packaging materials.

Regulations: FDA - 21CFR 181.29. Limitation of 50 ppm zinc as a migrant in finished food.

SAFETY PROFILE: When heated to decomposition it emits acrid smoke and irritating fumes.

ZNA000 CAS: 7733-02-0
ZINC SULFATE

DOT: 9161
mf: $O_4S{\cdot}Zn$ mw: 161.43

PROP: Rhombic, colorless crystals or crystalline powder. Mp: decomp @ 740°, d: 3.74 @ 15°. Sol in water; almost insol in alc.

SYNS: BONAZEN ◇ BUFOPTO ZINC SULFATE ◇ OP-THAL-ZIN ◇ SULFATE de ZINC (FRENCH) ◇ SULFURIC ACID, ZINC SALT (1:1) ◇ VERAZINC ◇ WHITE COPPERAS ◇ WHITE VITRIOL ◇ ZINC SULPHATE ◇ ZINC VITRIOL ◇ ZINKOSITE

USE IN FOOD:

Purpose: Dietary supplement, nutrient.

Where Used: Egg substitutes (frozen).

Regulations: FDA - 21CFR 182.5997, 182.8997. GRAS when used in accordance with good manufacturing practice.

Zinc and its compounds are on the Community Right-To-Know List. EPA Genetic Toxicology Program.

DOT Classification: ORM-E; Label: None

SAFETY PROFILE: Poison by intraperitoneal, subcutaneous, and intravenous routes. Moderately toxic by ingestion. An experimental tumorigen and teratogen. Human systemic effects by ingestion: increased pulse rate without blood pressure decrease, blood pressure decrease, acute pulmonary edema, normocytic anemia, hypermotility, diarrhea and other gastrointestinal changes. Experimental reproductive effects. Human mutagenic data. An eye irritant. When heated to decomposition it emits toxic fumes of SO_x and ZnO.

TOXICITY DATA and CODEN

eye-rbt 420 μg MOD JAPMA8 45,474,56
sln-dmg-orl 5 mmol/L MUREAV 90,91,81
mmo-smc 100 mmol/L MUREAV 117,149,83
orl-rat TDLo:333 mg/kg (1-18D preg):REP NURIBL 13,33,76
scu-ham TDLo:15 mg/kg (8D preg):TER ENVRAL 35,45,85
ivn-ham TDLo:2 mg/kg/(8D preg):TER ENVRAL 6,95,73
scu-rbt TDLo:3625 μg/kg/5D-C:ETA COREAF 236,1387,53
orl-hmn TDLo:45 mg/kg/7D-C: CVS,GIT,BLD BMJOAE 1,754,78
orl-hmn TDLo:106 mg/kg:CVS,PUL,GIT BMJOAE 1,1390,77
orl-rat LD50:2949 mg/kg TOERD9 1,371,78

ZNJ000 CAS: 7446-20-0
ZINC SULFATE HEPTAHYDRATE (1:1:7)

mf: $O_4SZn{\bullet}7H_2O$ mw: 287.57

PROP: Colorless crystals or crystalline powder; odorless . D: 1.97; mp: 100°. Decomp >500°. Insol in alc; glycerin.

SYNS: SULFURIC ACID, ZINC SALT (1:1), HEPTAHYDRATE ◇ WHITE VITRIOL ◇ ZINC SULFATE ◇ ZINC SULFATE (1:1) HEPTAHYDRATE ◇ ZINC VITRIOL

USE IN FOOD:

Purpose: Dietary supplement, nutrient.

Where Used: Various.

Regulations: FDA - 21CFR 182.5997, 182.8997. GRAS when used in accordance with good manufacturing practice.

Zinc and its compounds are on the Community Right-To-Know List.

SAFETY PROFILE: Human poison by an unspecified route. Poison experimentally by subcutaneous, intravenous, and intraperitoneal routes. Moderately toxic by ingestion. When heated to decomposition it emits toxic fumes of SO_x and ZnO.

TOXICITY DATA and CODEN

sln-dmg-orl 5 mmol/L MUREAV 90,91,81
dni-mus-ipr 20 g/kg ARGEAR 51,605,81
unr-man LDLo:221 mg/kg 85DCAI 2,73,70
orl-rat LDLo:2200 mg/kg HBAMAK 4,1419,35

Section III

Purpose Served in Foods Cross-Index
Food Type Cross-Index
CAS Number Cross-Index
Synonym Cross-Index
Coden Bibliographic Citations

Purpose Served in Foods Cross-Index

Food Type Cross-Index

CAS Number Cross-Index

Synonym Cross-Index

2-(p-AMINOBENZENESULFONAMIDO)THIAZOLE see TEX250
2-(p-AMINOBENZENESULPHONAMIDO)THIAZOLE see TEX250
o-AMINOBENZOIC ACID, ETHYL ESTER see EGM000
2-AMINOBENZOIC ACID METHYL ESTER see APJ250
o-AMINOBENZOIC ACID METHYL ESTER see APJ250
2-AMINOBENZOIC ACID-3-PHENYL-2-PROPENYL ESTER see API750
6-(4'-AMINOBENZOL-SULFONAMIDO)-2,4-DIMETHYLPYRIMIDIN (GERMAN) see SNJ500
(p-AMINOBENZOLSULFONYL)-2-AMINO-4,6-DIMETHYLPYRIMIDIN (GERMAN) see SNJ500
AMINOBENZYLPENICILLIN see AIV500
d-(-)-α-AMINOBENZYLPENICILLIN see AIV500
AMINOBENZYLPENICILLIN TRIHYDRATE see AOD125
α-AMINOBENZYLPENICILLIN TRIHYDRATE see AOD125
4-AMINO-N-(5-BROMO-4,6-DIMETHYL-2-PYRIMIDINYL)BENZENESILFANILAMIDE see SNH875
2-AMINOBUTANE see BPY000
4-AMINO-6-tert-BUTYL-3-(METHYLTHIO)-1,2,4-TRIAZIN-5-ON see AJB750
4-AMINO-6-tert-BUTYL-3-METHYLTHIO-as-TRIAZIN-5-ONE see AJB750
4-AMINO-6-tert-BUTYL-3-(METHYLTHIO)-1,2,4-TRIAZIN-5-ONE see AJB750
(4-((AMINOCARBONYL)AMINO)PHENYL)ARSONIC ACID see CBJ000
3-AMINO-N-(α-CARBOXYPHENETHYL)SUCCINAMIC ACID N-METHYL ESTER, stereoisomer see ARN825
4-AMINO-N-(6-CHLORO-3-PYRIDAZINYL)BENZENESULFONAMIDE see SNH900
AMINOCYCLOHEXANE see CPF500
O-4-AMINO-4-DEOXY-A-D-GLUCOPYRANOSYL-(1-8)-O-(8R)-2-AMINO-2,3,7-TRIDEOXY-7-(METHYLAMINO)-D-GLYCERO-A-D-ALLO-OCTODIALDO-1,5:8,4-DIPYRANOSYL-(1-4)-2-DEOXY-D-STREPAMINE see AQP885
4-AMINO-N-(2,6-DIMETHOXY-4-PYRIMIDINYL)BENZENESULFONAMIDE see SNN300
4-AMINO-6-(1,1-DIMETHYLETHYL)-3-(METHYLTHIO)-1,2,4-TRIAZIN-5(4H)-ONE see AJB750
2-AMINOETANOLO (ITALIAN) see MRH500
2-AMINOETHANOL see MRH500
4-AMINO-N-(6-ETHOXY-3-PYRIDAZINYL)BENZENESULFONAMIDE see SNJ100
β-AMINOETHYL ALCOHOL see MRH500
2-AMINOGLUTARAMIC ACID see GFO050
l-2-AMINOGLUTARAMIDIC ACID see GFO050
α-AMINOGLUTARIC ACID see GFO000
l-2-AMINOGLUTARIC ACID see GFO000
α-AMINOGLUTARIC ACID HYDROCHLORIDE see GFO025
l-2-AMINOGLUTARIC ACID HYDROCHLORIDE see GFO025
l-1-AMINO-4-GUANIDOVALERIC ACID see AQV980
AMINOHEXAHYDROBENZENE see CPF500
α-AMINOHYDROCINNAMIC ACID see PEC750
α-AMINO-p-HYDROXYBENZYLPENICILLIN TRIHYDRATE see AOA100
l-2-AMINO-3-HYDROXYBUTYRIC ACID see TFU750
(2S-(2-α,5-α,6-β(S*)))-6-((AMINO(4-HYDROXYPHENYL)ACETYL)AMINO)-3,3-DIMETHYL-7-OXO-4-THIA-1-AZABICYCLO(3.2.0)HEPTANE-2-CARBOXYLIC ACID TRIHYDRATE see AOA100
l-2-AMINO-3-HYDROXYPROPANOIC ACID see SCA355
l-N-(p-(((-2-AMINO-4-HYDROXY-6-PTERIDINYL)METHYL)-AMINO)BENZOYL)GLUTAMIC ACID see FMT000
l-α-AMINO-4(OR 5)-IMIDAZOLEPROPIONIC ACID see HGE700
l-α-AMINO-4(OR 5)-IMIDAZOLEPROPIONIC ACID MONOHYDROCHLORIDE see HGE800
α-AMINO-INDOLE-3-PROPRIONIC ACID see TNX000
l-α-AMINO-3-INDOLEPROPRIONIC ACID see TNX000
α'-AMINO-3-INDOLEPROPRIONIC ACID see TNX000
2-AMINO-3-INDOL-3-YL-PROPRIONIC ACID see TNX000
α-AMINOISOCAPROIC ACID see LES000
l-(+)-α-AMINOISOVALERIC ACID see VBP000
l-α-AMINO-γ-METHYLMERCAPTOBUTYRIC ACID see MDT750
2-AMINO-3-METHYLPENTANOIC ACID see IKX000
2-AMINO-4-METHYLPENTANOIC ACID see LES000
4-AMINO-N-(4-METHYL-2-PYRIMIDINYL)-BENZENESULFONAMIDE MONOSODIUM SALT see SJW475
3-(4-AMINO-2-METHYLPYRIMIDYL-5-METHYL)-4-METHYL-5,β-HYDROXYETHYLTHIAZOLIUM NITRATE see TET500
2-AMINO-4-(METHYLTHIO)BUTYRIC ACID see MDT750
l(−)-AMINO-γ-METHYLTHIOBUTYRIC ACID see MDT750
2-AMINO-4-METHYLVALERIC ACID see LES000
α-AMINO-β-METHYLVALERIC ACID see IKX000
α-AMINO-γ-METHYLVALERIC ACID see LES000
l-2-AMINO-4-METHYLVALERIC ACID see LES000
dl-2-AMINO-3-METHYLVALERIC ACID see IKX010
dl-2-AMINO-4-METHYLVALERIC ACID see LER000
d-(-)-α-AMINOPENICILLIN see AIV500
2-AMINOPENTANEDIOIC ACID see GFO000
2-AMINOPENTANEDIOIC ACID HYDROCHLORIDE see GFO025
6-(d(-)-α-AMINOPHENYLACETAMIDO)PENICILLANIC ACID see AIV500
AMINOPHENYLARSINE ACID see ARA250
p-AMINOPHENYLARSINE ACID see ARA250
p-AMINOPHENYLARSINIC ACID see ARA250
4-AMINOPHENYLARSONIC ACID see ARA250
p-AMINOPHENYLARSONIC ACID see ARA250
(4-AMINOPHENYL)ARSONIC ACID SODIUM SALT see ARA500
(AMINOPHENYLMETHYL)-PENICILLIN see AIV500
α-AMINO-β-PHENYLPROPIONIC ACID see PEC750
dl-α-AMINO-β-PHENYLPROPIONIC ACID see PEC500
1-AMINOPROPANE-1,3-DICARBOXYLIC ACID see GFO000
1-AMINOPROPANE-1,3-DICARBOXYLIC ACID HYDROCHLORIDE see GFO025
l-2-AMINOPROPANOIC ACID see AFH625
dl-2-AMINOPROPANOIC ACID see AFH600
1-[(4-AMINO-2-PROPYL-5-PYRIMIDINYL)METHYL]-2-METHYLPYRIDINIUM CHLORIDE see AOD175
1-(4-AMINO-2-n-PROPYL-5-PYRIMIDINYLMETHYL)-2-PICOLINIUM CHLORIDE see AOD175
l-α-AMINOSUCCINAMIC ACID see ARN810
l-AMINOSUCCINIC ACID see ARN850
dl-AMINOSUCCINIC ACID see ARN830
AMINOSULFONIC ACID see SNK500
4-AMINO-N-2-THIAZOLYLBENZENESULFONAMIDE see TEX250
4-AMINO-3,5,6-TRICHLOROPICOLINIC ACID see AMU250
4-AMINO-3,5,6-TRICHLORO-2-PICOLINIC ACID see AMU250
4-AMINO-3,5,6-TRICHLORPICOLINSAEURE (GERMAN) see AMU250
AMINOZIDE see DQD400
AMIPENIX S see AIV500
AMIRAL see CJO250
AMIXICOTYN see NCR000
AMMONIA AQUEOUS see ANK250
AMMONIA SOLUTION (DOT) see ANK250
AMMONIATED GLYCYRRHIZIN see AMY700
AMMONIUM ALGINATE see ANA300
AMMONIUM BICARBONATE (1:1) see ANB250
AMMONIUMCARBONAT (GERMAN) see ANE000
AMMONIUM CARBONATE see ANB250, ANE000
AMMONIUMCHLORID (GERMAN) see ANE500
AMMONIUM CHLORIDE see ANE500
AMMONIUM CITRATE see ANF800
AMMONIUM CITRATE, DIBASIC (DOT) see ANF800
AMMONIUMGLUTAMINAT (GERMAN) see MRF000
AMMONIUM HYDROGEN CARBONATE see ANB250
AMMONIUM HYDROXIDE see ANK250
AMMONIUM MURIATE see ANE500
AMMONIUM PHOSPHATE see ANR500

BENZYLACETALDEHYDE see HHP000
BENZYL ACETATE see BDX000
BENZYL ALCOHOL see BDX500
BENZYL ALCOHOL BENZOIC ESTER see BCM000
BENZYL ALCOHOL CINNAMIC ESTER see BEG750
BENZYL BENZENECARBOXYLATE see BCM000
BENZYL BENZOATE (FCC) see BCM000
BENZYL n-BUTANOATE see BED000
BENZYL n-BUTYRATE see BED000
BENZYL CARBINOL see PDD750
BENZYLCARBINOL ISOBUTYRATE see PDF750
BENZYLCARBINYL ACETATE see PFB250
BENZYLCARBINYL ISOBUTYRATE see PDF750
BENZYLCARBINYL-α-TOLUATE see PDI000
BENZYL CINNAMATE see BEG750
BENZYL DIMETHYL CARBINOL see DQQ200
BENZYLDIMETHYLDODECYLAMMONIUM CHLORIDE see BEM000
BENZYLDIMETHYLSTEARYLAMMONIUM CHLORIDE see DTC600
BENZYL ETHANOATE see BDX000
BENZYL ETHER see BEO250
BENZYLETS see BCM000
5-BENZYL-3-FURYL METHYL(±)-cis,trans-CHRYSANTHEMATE see BEP500
(5-BENZYL-3-FURYL) METHYL-2,2-DIMETHYL-3-(2-METHYLPROPENYL)-CYCLOPROPANECARBOXYLATE see BEP500
BENZYL-o-HYDROXYBENZOATE see BFJ750
BENZYLIDENEACETALDEHYDE see CMP969
BENZYLIDENE GLYCEROL see BBA000
BENZYL ISOBUTYRATE (FCC) see IJV000
BENZYL ISOVALERATE (FCC) see ISW000
BENZYL-3-METHYLBUTANOATE see ISW000
BENZYL-3-METHYL BUTYRATE see ISW000
BENZYL-2-METHYL PROPIONATE see IJV000
BENZYL OXIDE (CZECH) see BEO250
7-(BENZYLOXY)-6-N-BUTYL-1,4-DIHYDRO-4-OXO-3-QUINOLINECARBOXYLIC ACID METHYL ESTER see NCN600
7-(BENZYLOXY)-6-N-BUTYL-4-HYDROXY-3-QUINOLINECARBOXYLIC ACID METHYL see NCN600
BENZYL PHENYLACETATE see BFD400
BENZYL γ-PHENYLACRYLATE see BEG750
BENZYL PHENYLFORMATE see BCM000
BENZYL PROPIONATE see BFD800
BENZYL SALICYLATE see BFJ750
BENZYLSTEARYLDIMETHYLAMMONIUM CHLORIDE see DTC600
BENZYL VIOLET see BFN250
BENZYL VIOLET 3B see BFN250
BEOSIT see BCJ250
BEPANTHEN see PAG200
BEPANTHENE see PAG200
BEPANTOL see PAG200
BERELEX see GEM000
BERGAMIOL see DTD600
BERGAMOT OIL RECTIFIED see BFO000
BERGAMOTTE OEL (GERMAN) see BFO000
BERMAT see CJJ250
BERNSTEINSAEURE-2,2-DIMETHYLHYDRAZID (GERMAN) see DQD400
BERNSTEINSAURE (GERMAN) see SMY000
BEROL 478 see DJL000
BERTHOLITE see CDV750
BERUBIGEN see VSZ000
BETACIDE P see HNU500
BETALIN 12 CRYSTALLINE see VSZ000
BETULA OIL see MPI000
BEVATINE-12 see VSZ000
BEVIDOX see VSZ000
BEXOL see BBQ500
BFV see FMV000
BHA (FCC) see BQI000
BHC (USDA) see BBP750
BHC see BBQ500
γ-BHC see BBQ500
BH 2,4-D see DFY600
BH DALAPON see DGI400
B-HERBATOX see SFS000
BHT (FOOD GRADE) see BFW750
BI-58 see DSP400
BIACETYL see BOT500
BIBESOL see DRK200
BICAM ULV see DQM600
BICHLORURE d'ETHYLENE (FRENCH) see DFF900
BICORTONE see PLZ000
BICYCLO(2.2.1)HEPTENE-2-DICARBOXYLIC ACID, 2-ETHYLHEXYLIMIDE see OES000
BIFURON see NGG500
BIG DIPPER see DVX800
BILARCIL see TIQ250
BILOBRAN see DOL600
BINOTAL see AIV500
BIO 5,462 see BCJ250
BIOCETIN see CDP250
BIOCOLINA see CMF750
BIOFANAL see NOH500
BIOFUREA see NGE500
BIOMITSIN see CMA750
BIOMYCIN see CMA750
BIONIC see NDT000
BIOPHENICOL see CDP250
BIOPHYLL see CKN000
BIOQUIN see BLC250
BIOQUIN 1 see BLC250
BIO-SOFT D-40 see DXW200
BIOSOL VETERINARY see NCG000
BIOSTAT see HOH500
BIOSTEROL see REU000
BIO-TESTICULINA see TBG000
BIOTIN see BGD100
d-BIOTIN see BGD100
2-BIPHENYLOL, SODIUM SALT see BGJ750
BIRCH TAR OIL see BGO750
BIRCH TAR OIL, RECTIFIED (FCC) see BGO750
S-(1,2-BIS(AETHOXY-CARBONYL)-AETHYL)-O,O-DIMETHYL-DITHIOPHASPHAT (GERMAN) see CBP000
2,4-BIS(AETHYLAMINO)-6-CHLOR-1,3,5-TRIAZIN (GERMAN) see BJP000
BIS AMINE see MQY750
2,2-BIS(p-ANISYL)-1,1,1-TRICHLOROETHANE see DOB400
BIS(2-BENZOTHIAZOLYLTHIO)ZINC see BHA750
BIS(n-BUTYL)SEBACATE see DEH600
BIS(2-CARBOXYETHYL) SULFIDE see BHM000
1,3-BIS((p-CHLOROBENZYLIDENE)AMINO)GUANIDINE see RLK890
O,O-BIS(2-CHLOROETHYL)-O-(3-CHLORO-4-METHYL-7-COUMARINYL) PHOSPHATE see DFH600
1,6-BIS(5-(p-CHLOROPHENYL)BIGUANIDINO)HEXANE see BIM250
1,6-BIS(p-CHLOROPHENYLDIGUANIDO)HEXANE see BIM250
1,1-BIS(CHLOROPHENYL)-2,2,2-TRICHLOROETHANOL see BIO750
1,1-BIS(4-CHLOROPHENYL)-2,2,2-TRICHLOROETHANOL see BIO750
1,1-BIS(p-CHLOROPHENYL)-2,2,2-TRICHLOROETHANOL see BIO750
BIS(S-(DIETHOXYPHOSPHINOTHIOYL)MERCAPTO)-METHANE see EMS500
1,4-BIS(3,4-DIHYDROXYPHENYL)-2,3-DIMETHYLBUTANE see NBR000
BIS(DIMETHYLCARBAMODITHIOATO-S,S')ZINC see BJK500
BIS(DIMETHYLDITHIOCARBAMATE de ZINC) (FRENCH) see BJK500

BIS(DIMETHYLDITHIOCARBAMATO)ZINC see BJK500
2,6-BIS(1,1-DIMETHYLETHYL)-4-METHYLPHENOL see BFW750
BIS(N,N-DIMETIL-DITIOCARBAMMATO) DI ZINCO (ITALIAN) see BJK500
BIS(DITHIOPHOSPHATE de O,O-DIETHYLE) de S,S'-(1,4-DIOXANNE-2,3-DIYLE) (FRENCH) see DVQ709
BIS(DODECANOYLOXY) DI-n-BUTYLSTANNANE see DDV600
BIS(DODECYLOXYCARBONYLETHYL) SULFIDE see TFD500
S-(1,2-BIS(ETHOXY-CARBONYL)-ETHYL)-O,O-DIMETHYL-DITHIOFOSFAAT (DUTCH) see CBP000
S-1,2-BIS(ETHOXYCARBONYL)ETHYL-O,O-DIMETHYL THIOPHOSPHATE see CBP000
2,4-BIS(ETHYLAMINO)-6-CHLORO-s-TRIAZINE see BJP000
BIS(2-ETHYLHEXYL)-1,2-BENZENEDICARBOXYLATE see BJS000
BIS(ETHYLHEXYL) ESTER of SODIUM SULFOSUCCINIC ACID see DJL000
BIS(2-ETHYLHEXYL)PHTHALATE see BJS000
BIS(2-ETHYLHEXYL)SODIUM SULFOSUCCINATE see DJL000
BIS(2-ETHYLHEXYL)-S-SODIUM SULFOSUCCINATE see DJL000
1,4-BIS(2-ETHYLHEXYL) SODIUM SULFOSUCCINATE see DJL000
1,4-BIS(2-ETHYLHEXYL)SULFOBUTANEDIOIC ACID ESTER, SODIUM SALT see DJL000
S-(1,2-BIS(ETOSSI-CARBONIL)-ETIL)-O,O-DIMETIL-DITIOFOSFATO (ITALIAN) see CBP000
N,N-BIS(2-HYDROXYETHYL)DODECAN AMIDE see BKE500
BIS(2-HYDROXYETHYL)LAURAMIDE see BKE500
N,N-BIS(HYDROXYETHYL)LAURAMIDE see BKE500
N,N-BIS(2-HYDROXYETHYL)LAURAMIDE see BKE500
N,N-BIS(β-HYDROXYETHYL)LAURAMIDE see BKE500
BIS[HYDROXYETHYLPOLY(ETHYLENEOXY)ETHYLPROPYLENEGLYCOL see PJK151
BIS[HYDROXYETHYLPOLY(ETHYLENEOXY)ETHYLPROPYLENEGLYCOL see PJK150
2,4-BIS(ISOPROPYLAMINO)-6-CHLORO-s-TRIAZINE see PMN850
2,4-BIS(ISOPROPYLAMINO)-6-METHYLMERCAPTO-s-TRIAZINE see BKL250
4,6-BIS(ISOPROPYLAMINO)-2-METHYLMERCAPTO-s-TRIAZINE see BKL250
2,4-BIS(ISOPROPYLAMINO)-6-METHYLTHIO-s-TRIAZINE see BKL250
2,4-BIS(ISOPROPYLAMINO)-6-METHYLTHIO-1,3,5-TRIAZINE see BKL250
BIS(LAUROYLOXY)DIBUTYLSTANNANE see DDV600
BIS(LAUROYLOXY)DI(n-BUTYL)STANNANE see DDV600
BIS(MERCAPTOBENZOTHIAZOLATO)ZINC see BHA750
1,1-BIS(p-METHOXYPHENYL)-2,2,2-TRICHLOROETHANE see DOB400
2,2-BIS(p-METHOXYPHENYL)-1,1,1-TRICHLOROETHANE see DOB400
N,N'-BIS(1-METHYLETHYL)-6-METHYL-THIO-1,3,5-TRIAZINE-2,4-DIAMINE see BKL250
BIS(6-METHYLHEPTYL)ESTER OF PHTHALIC ACID see ILR100
N,N-BIS(4-NITRPHENYL)UREA, compd with 4,6-DIMETHYL-2(1H)-PYRIMIDINONE (1:1) see NCW100
BISODIUM TARTRATE see BLC000
BISOFLEX 81 see BJS000
BISOFLEX DOP see BJS000
BISOLVOMYCIN see HOI000
BIS(8-OXYQUINOLINE)COPPER see BLC250
2,4-BIS(PROPYLAMINO)-6-CHLOR-1,3,5-TRIAZIN (GERMAN) see PMN850
BIS(8-QUINOLINATO)COPPER see BLC250
BIS(8-QUINOLINOLATO)COPPER see BLC250
BIS(8-QUINOLINOLATO-N(^{1}),O(^{8}))-COPPER see BLC250
BIS(TRIS(β,β-DIMETHYLPHENETHYL)TIN)OXIDE see BLU000
BIS(TRIS(2-METHYL-2-PHENYLPROPYL)TIN)OXIDE see BLU000
BISULFITE see SOH500
BISULFITE de SODIUM (FRENCH) see SFE000
BITEMOL see BJP000
BITEMOL S 50 see BJP000
BITTER ALMOND OIL see BLV500
BITTER ALMOND OIL CAMPHOR see BCP250
BITTER FENNEL OIL see FAP000
BITTER ORANGE OIL see OGY010
BIXA ORELLANA see BLW000
B-K LIQUID see SHU500
γ-BL see BOV000
BLACK OXIDE of IRON see IHD000
BLACK PEPPER OIL see BLW250
BLACOSOLV see TIO750
BLADAN see EMS500
BLANDLUBE see MQV750
BLANOSE BWM see SFO500
BLATTERALKOHOL see HFE000
BLEACHED-DEODORIZED LARD see LBE300
BLEACHED-DEODORIZED TALLOW see TAC100
BLEACHED LARD see LBE300
BLENDED RED OXIDES of IRON see IHD000
BLEX see DIN800
BLO see BOV000
BLOAT GUARD see PJK150, PJK151
BLOC see FAK100
BLON see BOV000
BLOTIC see MKA000
1206 BLUE see FAE000
1311 BLUE see DXE400
12070 BLUE see DXE400
BLUE COPPER see CNP250
BLUE POWDER see ZBJ000
BLUE STONE see CNP250
BLUE VITRIOL see CNP250
B-NINE see DQD400
BNM see MHV500
BO-ANA see FAG759
BOIS d'INDE see BAT500
BOIS de ROSE OIL see BMA550
BOLETIC ACID see FOU000
BOLINAN see PKQ250
BOLSTAR see ENI000
BONAID see BOO632
BONAPICILLIN see AIV500
BONAZEN see ZNA000
BOND CH 18 see AAX250
BONOMOLD OE see HJL000
BONOMOLD OP see HNU500
BOOKSAVER see AAX250
BORDEN 2123 see AAX250
BORER SOL see DFF900
BORNYL ACETATE see BMD100
l-BORNYL ACETATE see BMD100
BOROLIN see AMU250
BOURBONAL see EQF000
BOV see SOI500
BOVILENE see FAQ500
BOVINE RENNET see RCZ100
BOVINOX see TIQ250
BOVIZOLE see TEX000
BPPS see SOP000
BRAVO see TBQ750
BRAVO 6F see TBQ750
BRAVO-W-75 see TBQ750
BRAZIL WAX see CCK640
BRELLIN see GEM000
BREVINYL see DRK200
BRILLIANT BLUE FCD NO. 1 see FAE000
BRILLIANT BLUE FCF see FAE000

CEKUDIFOL see BIO750
CEKUFON see TIQ250
CEKUGIB see GEM000
CEKUMETA see TDW500
CEKUSAN see BJP000, DRK200
CEKUTHOATE see DSP400
CEKUTROTHION see DSQ000
CEKUZINA-S see BJP000
CEKUZINA-T see PMC325
CELANEX see BBQ500
CELANOL DOS 75 see DJL000
CELERY SEED OIL see CCS660
CELINHOL -A see OAV000
CELLILASE and CARBOHYDRASE see CBS405
CELLOFAS see SFO500
CELLOGEL C see SFO500
CELLOIDIN see CCU250
CELLPRO see SFO500
CELLUFIX FF 100 see SFO500
CELLUGEL see SFO500
CELLULOSE GEL see CCU100
CELLULOSE GLYCOLIC ACID, SODIUM SALT see SFO500
CELLULOSE GUM see SFO500
CELLULOSE, MICROCRYSTALLINE see CCU100
CELLULOSE NITRATE see CCU250
CELLULOSE, POWDERED see CCU150
CELLULOSE SODIUM GLYCOLATE see SFO500
CELLULOSE TETRANITRATE see CCU250
CELLU-QUIN see BLC250
CELMIDE see EIY500
CELON E see EIV000
CELON H see EIV000
CELON IS see EIV000
CELPHIDE see AHE750
CELPHOS see AHE750
CELTHIGN see CBP000
CENOLATE see ARN125
CENTURY 1240 see SLK000
CENTURY CD FATTY ACID see OHU000
CEP see CDS125
2-CEPA see CDS125
CEPHA see CDS125
CEPHAPIRIN see CCX500
CEPHA 10LS see CDS125
CEPHROL see CMT250, RLU800
CERASYNT see HKJ000
CERASYNT 1000-D see OAV000
CERASYNT S see OAV000
CERAZOL (suspension) see TEX250
CERELOSE see GFG000
CERISOL YELLOW AB see PEJ750
CERISOL YELLOW TB see TGW250
CET see BJP000
CETAFFINE see HCP000
CETAL see HCP000
CETALOL CA see HCP000
CETONE V see AGI500
CETYL ALCOHOL see HCP000
CETYLIC ACID see PAE250
CETYLIC ALCOHOL see HCP000
CETYLOL see HCP000
CEVIAN A 678 see AAX250
CEVIAN HL see ADY500
CEVITAMIC ACID see ARN000
CEVITAMIN see ARN000
CEYLON ISINGLASS see AEX250
CGA 15324 see BNA750
CHA see CPF500
CHALK see CAO000
CHAMOMILE-GERMAN OIL see CDH500
CHAMOMILE OIL see CDH500
CHAMOMILE OIL (ROMAN) see CDH750
CHANNEL BLACK see CBT750
CHARCOAL, ACTIVATED (DOT) see AEC500
CHAVICOL METHYL ETHER see AFW750
CHEELOX BF see EIV000
CHEELOX BR-33 see EIV000
CHELADRATE see EIX500
CHELAFER see FBC100
CHELAPLEX III see EIX500
CHELATON III see EIX500
CHEL-IRON see FBC100
CHELON 100 see EIV000
CHEMAGRO 1,776 see BSH250
CHEMAGRO 2353 see DFV400
CHEMAGRO B-1776 see BSH250
CHEMANOX 11 see BFW750
CHEMATHION see CBP000
CHEM BAM see DXD200
CHEMCOCCIDE see RLK890
CHEMCOLOX 200 see EIV000
CHEMFORM see DOB400, SLW500
CHEMICETIN see CDP250
CHEMICETINA see CDP250
CHEMI-CHARL see SHM500
CHEMOCCIDE see RLK890
CHEMOFURAN see NGE500
CHEMOSEPT see TEX250
CHEM-PHENE see THH750
CHEM RICE see DGI000
CHEM-TOL see PAX250
CHEQUE see MQS225
CHEVRON RE 12,420 see DOP600
CHILE SALTPETER see SIO000
CHIMCOCCIDE see RLK890
CHINA CLAY see KBB600
CHINAWOOD OIL see TOA510
CHINESE ISINGLASS see AEX250
CHINESE SEASONING see MRL500
CHINESE WHITE see ZKA000
CHININDIHYDROCHLORID (GERMAN) see QIJ000
CHIPCO 26019 see GIA000
CHIPCO TURF HERBICIDE 'D' see DFY600
CHIPMAN 11974 see BDJ250
CHLOMIN see CDP250
CHLOMYCOL see CDP250
CHLOOR (DUTCH) see CDV750
CHLOORDAAN (DUTCH) see CDR750
CHLOOR-METHAAN (DUTCH) see CHX500
CHLOORWATERSTOF (DUTCH) see HHL000
CHLOPHEN see PJL750
CHLOR (GERMAN) see CDV750
CHLORACETIC ACID see CEA000
2-CHLORAETHYL-PHOSPHONSAEURE (GERMAN) see CDS125
CHLORAK see TIQ250
CHLORAMEX see CDP250
CHLORAMFICIN see CDP250
CHLORAMFILIN see CDP250
CHLORAMP (RUSSIAN) see AMU250
CHLORAMPHENICOL see CDP250
d-CHLORAMPHENICOL see CDP250
d-threo-CHLORAMPHENICOL see CDP250
CHLORAMPHENICOL MONOPALMITATE see CDP700
CHLORAMPHENICOL PALMITATE see CDP700
CHLORAMSAAR see CDP250
CHLORASOL see CDP250
CHLORA-TABS see CDP250
CHLORATE OF SODA (DOT) see SFS000
CHLORATE SALT of SODIUM see SFS000
CHLORAX see SFS000
4-CHLORBUTAN-1-OL (GERMAN) see CEU500
CHLORDAN see CDR750
γ-CHLORDAN see CDR750
CHLORDANE see CDR750
CHLORDANE, LIQUID (DOT) see CDR750
CHLORDIMEFORM see CJJ250
CHLORE (FRENCH) see CDV750

CHLORESENE see BBQ500
CHLORESSIGSAEURE-N-(METHOXYMETHYL)-2,6-DIAETHYLANILID (GERMAN) see CFX000
CHLORETHEPHON see CDS125
2-CHLORETHYLPHOSPHONIC ACID see CDS125
CHLOREXTOL see PJL750
CHLORFENAMIDINE see CJJ250
CHLORHEXIDIN (CZECH) see BIM250
CHLORHEXIDINE see BIM250
CHLORHYDRATE de TETRACYCLINE (FRENCH) see TBX250
CHLORICOL see CDP250
CHLORID AMONNY (CZECH) see ANE500
CHLORID DRASELNY (CZECH) see PLA500
CHLORIDE de CHOLINE (FRENCH) see CMF750
CHLORINATED BIPHENYL see PJL750
CHLORINATED CAMPHENE 60% (ACGIH) see THH750
CHLORINATED DIPHENYL see PJL750
CHLORINATED DIPHENYLENE see PJL750
CHLORINDAN see CDR750
CHLORINE see CDV750
CHLORINE MOL. see CDV750
CHLOR KIL see CDR750
CHLOR-METHAN (GERMAN) see CHX500
CHLORNITROMYCIN see CDP250
CHLOROACETIC ACID see CEA000
α-CHLOROACETIC ACID see CEA000
CHLOROACETIC ACID, LIQUID (DOT) see CEA000
CHLOROACETIC ACID, SOLID (DOT) see CEA000
CHLOROALONIL see TBQ750
CHLORO BIPHENYL see PJL750
CHLORO-1,1-BIPHENYL see PJL750
2-CHLORO-4,6-BIS(ETHYLAMINO)-s-TRIAZINE see BJP000
1-CHLORO, 3,5-BISETHYLAMINO-2,4,6-TRIAZINE see BJP000
2-CHLORO-4,6-BIS(ETHYLAMINO)-1,3,5-TRIAZINE see BJP000
4-CHLORO-1-BUTANE-OL see CEU500
4-CHLOROBUTANOL see CEU500
4-CHLORO-1-BUTANOL see CEU500
CHLOROCAMPHENE see THH750
CHLOROCAPS see CDP250
3-CHLOROCHLORDENE see HAR000
5-CHLORO-N-(2-CHLORO-4-NITROPHENYL)-2-HYDROXYBENZAMIDE see DFV400
5-CHLORO-2′-CHLORO-4′-NITROSALICYLANILIDE see DFV400
4-CHLORO-α-(4-CHLOROPHENYL)-α-(TRICHLOROMETHYL)BENZENEMETHANOL see BIO750
CHLOROCID see CDP250
CHLOROCOL see CDP250
CHLORODANE see CDR750
1-CHLORO-2,2-DICHLOROETHYLENE see TIO750
2-CHLORO-2′,6′-DIETHYL-N-(METHOXYMETHYL) ACETANILIDE see CFX000
2-CHLORO-N-(2,6-DIETHYL)PHENYL-N-METHOXYMETHYLACETAMIDE see CFX000
2-CHLORO-N-(2,6-DIETHYLPHENYL)-N-(METHOXYMETHYL)ACETAMIDE see CFX000
S-(2-CHLORO-1-(1,3-DIHYDRO-1,3-DIOXO-2H-ISOINDOL-2-YL)ETHYL)-O,O-DIETHYL PHOSPHORODITHIOATE see DBI099
7-CHLORO-4-(DIMETHYLAMINO)-1,4,4a,5,5a,6,11,12a-OCTAHYDRO-2-NAPHTHACENECARBOXAMIDE see CMA750
2-CHLOROETHANEPHOSPHONIC ACID see CDS125
CHLOROETHANOIC ACID see CEA000
2-CHLOROETHANOL HYDROGEN PHOSPHATE ESTER with 3-CHLORO-7-HYDROXY-4-METHYLCOUMARIN see DFH600
2-CHLOROETHANOL PHOSPHATE DIESTER ESTER with 3-CHLORO-7-HYDROXY-4-METHYLCOUMARIN see DFH600
2-CHLORO-1-(3-ETHOXY-4-NITROPHENOXY)-4-(TRIFLUOROMETHYL)BENZENE see OQU100
2-CHLORO-4-ETHYLAMINEISOPROPYLAMINE-s-TRIAZINE see PMC325
1-CHLORO-3-ETHYLAMINO-5-ISOPROPYLAMINO-s-TRIAZINE see PMC325
2-CHLORO-4-ETHYLAMINO-6-ISOPROPYLAMINO-s-TRIAZINE see PMC325
1-CHLORO-3-ETHYLAMINO-5-ISOPROPYLAMINO-2,4,6-TRIAZINE see PMC325
2-CHLORO-4-ETHYLAMINO-6-ISOPROPYLAMINO-1,3,5-TRIAZINE see PMC325
6-CHLORO-N-ETHYL-N′-(1-METHYLETHYL)-1,3,5-TRIAZINE-2,4-DIAMINE (9CI) see PMC325
CHLOROFOS see TIQ250
CHLOROFTALM see TIQ250
CHLOROHYDRIC ACID see HHL000
3-CHLORO-7-HYDROXY-4-METHYLCOUMARIN BIS(2-CHLOROETHYL)PHOSPHATE see DFH600
3-CHLORO-7-HYDROXY-4-METHYL-COUMARIN-O,O-DIETHYL PHOSPHOROTHIOATE see CNU750
3-CHLORO-7-HYDROXY-4-METHYL-COUMARIN-O-ESTER with O,O-DIETHYL PHOSPHOROTHIOATE see CNU750
CHLOROJECT L see CDP250
CHLOROMAX see CDP250
S-(6-CHLORO-3-(MERCAPTOMETHYL)-2-BENZOXAZOLINONE)-O,O-DIETHYL PHOSPHORODITHIOATE see BDJ250
CHLOROMETHANE see CHX500
4-CHLORO-5-(METHYLAMINO)-2-(α,α,α-TRIFLUORO-m-TOLYL)-3(2H)-PYRIDAZINONE see NNQ100
CHLOROMETHYLATED AMINATED STRYRNE-DIVINYLBENZENE RESIN see CIF775
3-CHLORO-4-METHYL-7-COUMARINYL DIETHYLPHOSPHATE see CIK750
3-CHLORO-4-METHYL-7-COUMARINYL DIETHYL PHOSPHOROTHIOATE see CNU750
O-3-CHLORO-4-METHYL-7-COUMARINYL-O,O-DIETHYL PHOSPHOROTHIOATE see CNU750
3-CHLORO-4-METHYL-7-HYDROXYCOUMARIN DIETHYL THIOPHOSPHORIC ACID ESTER see CNU750
N′-(4-CHLORO-2-METHYLPHENYL)-N,N-DIMETHYLMETHANIMIDAMIDE see CJJ250
3-CHLORO-4-METHYL-UMBELLIFERONE BIS(2-CHLOROETHYL)PHOSPHATE see DFH600
3-CHLORO-4-METHYLUMBELLIFERONE-O-ESTER WITH-O,O-DIETHYL PHOSPHOROTHIOATE see CNU750
CHLOROMYCETIN see CDP250
CHLORONITRIN see CDP250
2-CHLORO-4-NITROBENZAMIDE see AFH400
2-CHLORO-4-NITROPHENYLAMIDE-6-CHLOROSALICYLIC ACID see DFV400
N-(2-CHLORO-4-NITROPHENYL)-5-CHLOROSALICYLAMIDE see DFV400
3-(6-CHLORO-2-OXOBENZOXAZOLIN-3-YL)METHYL-O,O-DIETHYL PHOSPHOROTHIOLOTHIONATE see BDJ250
CHLOROPENTAFLUORETHANE (ACGIH, DOT) see CJI500
CHLOROPENTAFLUOROETHANE see CJI500
CHLOROPHEN see PAX250
CHLOROPHENAMADIN see CJJ250
CHLOROPHENAMIDINE see CJJ250
1-(4-CHLOROPHENOXY)-3,3-DIMETHYL-1-(1,2,4-TRIAZOL-1-YL)-2-BUTAN-2-ONE see CJO250
1-(4-CHLOROPHENOXY)-3,3-DIMETHYL-1-(1H-1,2,4-TRIAZOL-1-YL)-2-BUTANONE see CJO250
N-(((4-CHLOROPHENYL)AMINO)CARBONYL)-2,6-DIFLUOROBENZAMIDE see CJV250
(2-CHLOROPHENYL)-α-(4-CHLOROPHENYL)-5-PYRIMIDINEMETHANOL see FAK100
α-(2-CHLOROPHENYL)-α-(4-CHLOROPHENYL)-5-PYRIMIDINEMETHANOL see FAK100
1-(4-CHLOROPHENYL)-3-(2,6-DIFLUOROBENZOYL)UREA see CJV250
6-(3-(o-CHLOROPHENYL)-5-METHYL-4-ISOXAZOLECARBOXAMIDEO)-3,3-DIMETHYL-7-OXO-4-THIA-1-

AZABICYCLO(3.2.0)HEPTANE-2-CARBOXYLIC ACID, SODIUM SALT, MONOHYDRATE see SLJ000
S-((p-CHLOROPHENYLTHIO)METHYL)-O,O-DIETHYL PHOSPHORODITHIOATE see TNP250
4-CHLOROPHENYL-2,4,5-TRICHLOROPHENYL SULFONE see CKM000
p-CHLOROPHENYL-2,4,5-TRICHLOROPHENYL SULFONE see CKM000
p-CHLOROPHENYL-2,4,5-TRICHLOROPHENYL SULPHONE see CKM000
CHLOROPHOS see TIQ250
S-(2-CHLORO-1-PHTHALIMIDOETHYL)-O,O-DIETHYL PHOSPHORODITHIOATE see DBI099
CHLOROPHTHALM see TIQ250
CHLOROPHYL, GREEN see CKN000
CHLOROPHYLL see CKN000
CHLOROPOTASSURIL see PLA500
2-CHLORO-4-(2-PROPYLAMINO)-6-ETHYLAMINO-s-TRIAZINE see PMC325
CHLOROPTIC see CDP250
CHLOROS see SHU500
3-CHLORO-6-SULFANILAMIDOPYRIDAZINE see SNH900
7-CHLOROTETRACYCLINE see CMA750
CHLOROTHALONIL see TBQ750
N'-(4-CHLORO-o-TOLYL)-N,N-DIMETHYLFORMAMIDINE see CJJ250
2-CHLORO-1-(2,4,5-TRICHLOROPHENYL)VINYL DIMETHYL PHOSPHATE see TBW100
2-CHLORO-1-(2,4,5-TRICHLOROPHENYL(VINYL PHOSPHORIC ACID DIMETHYL ESTER see TBW100
CHLOROVULES see CDP250
CHLOROWODOR (POLISH) see HHL000
CHLOROX see SHU500
CHLOROXONE see DFY600
CHLOROXYPHOS see TIQ250
CHLORPHENAMIDINE see CJJ250
CHLORPYRIFOS (ACGIH, DOT, USDA) see DYE000
CHLORPYRIFOS-METHYL see DUJ200
CHLORSAURE (GERMAN) see SFS000
CHLORTETRACYCLINE see CMA750
CHLORTHALONIL (GERMAN) see TBQ750
CHLORTHIEPIN see BCJ250
N'-(4-CHLOR-o-TOLYL)-N,N-DIMETHYLFORMAMIDIN (GERMAN) see CJJ250
CHLORTOX see CDR750
CHLORURE d'ETHYLENE (FRENCH) see DFF900
CHLORURE de METHYLE (FRENCH) see CHX500
CHLORURE de METHYLENE (FRENCH) see MDR000
CHLORURE PERRIQUE (FRENCH) see FAU000
CHLORVINPHOS see DRK200
CHLORWASSERSTOFF (GERMAN) see HHL000
CHLORYLEA see TIO750
CHOCOLA A see REU000
CHOLAXINE see GEY000
CHOLECALCIFEROL see CMC750
CHOLEIC ACID see DAQ400
CHOLEREBIC see DAQ400
CHOLIC ACID, MONOSODIUM SALT see SFW000
CHOLIC ACID, SODIUM SALT see SFW000
CHOLINE BITARTRATE see CMF300
CHOLINE CHLORHYDRATE see CMF750
CHOLINE CHLORIDE (FCC) see CMF750
CHOLINE HYDROCHLORIDE see CMF750
CHOLINIUM CHLORIDE see CMF750
CHOLOREBIC see DAQ400
CHONDRUS see CCL250
CHONDRUS EXTRACT see CCL250
CHORYLEN see TIO750
2-CHROMANONE see HHR500
CHROME ALUM see PLB500
CHROME POTASH ALUM see PLB500
CHROMIC POTASSIUM SULFATE see PLB500
CHROMIC POTASSIUM SULPHATE see PLB500
CHROMIUM POTASSIUM SULFATE (1:1:2) see PLB500
CHROMIUM POTASSIUM SULPHATE see PLB500
CHRYSOMYKINE see CMA750
CHRYSON see BEP500
CHRYSRON see BEP500
C.I. 1956 see CKN000
C.I. 7581 see DXE400
C.I. 10355 see DVX800
C.I. 11380 see PEJ750
C.I. 11390 see TGW250
C.I. 12156 see DOK200
C.I. 15985 see FAG150
C.I. 16035 see FAG050
C.I. 19140 see TNK750
C.I. 42053 see FAG000
C.I. 42090 see FAE000
C.I. 42640 see BFN250
C.I. 45430 see TDE500
C.I. 73015 see DXE400
C.I. 75300 see COG000
C.I. 77180 see CAD000
C.I. 77400 see CNI000
C.I. 77491 see IHD000
C.I. 77575 see LCF000
C.I. 77713 see MAD500
C.I. 77718 see TAB750
C.I. 77775 see NCW500
C.I. 77864 see TGC000
C.I. 77891 see TGH000
C.I. 77945 see ZBJ000
C.I. 77947 see ZKA000
C.I. ACID BLUE 74 see DXE400
C.I. ACID BLUE 9, DISODIUM SALT see FAE000
C.I. ACID RED 51 see TDE500
CIBA 2059 see DUK800
CIBA 8514 see CJJ250
CICLOESANO (ITALIAN) see CPB000
CICLOSOM see TIQ250
CIDEX see GFQ000
CIDOCETINE see CDP250
C.I. FOOD BLUE 1 see DXE400
C.I. FOOD BLUE 2 see FAE000
C.I. FOOD GREEN 3 see FAG000
C.I. FOOD VIOLET 2 see BFN250
C.I. FOOD YELLOW 4 see TNK750
C.I. FOOD YELLOW 10 see PEJ750
C.I. FOOD YELLOW 11 see TGW250
CILEFA PINK B see TDE500
CIMEXAN see CBP000
1,8-CINEOL see CAL000
CINEOLE see CAL000
1,8-CINEOLE see CAL000
CINNAMAL see CMP969
CINNAMALDEHYDE see CMP969
CINNAMEIN see BEG750
CINNAMIC ACID see CMP975
trans-CINNAMIC ACID BENZYL ESTER see BEG750
CINNAMIC ACID, ISOBUTYL ESTER see IIQ000
CINNAMIC ALCOHOL see CMQ740
CINNAMON BARK OIL see CCO750
CINNAMON BARK OIL, CEYLON TYPE (FCC) see CCO750
CINNAMON LEAF OIL see CMQ500
CINNAMON LEAF OIL, Ceylon see CMQ500
CINNAMON LEAF OIL, Seychelles see CMQ500
CINNAMON OIL see CCO750
CINNAMYL ACETATE see CMQ730
CINNAMYL ALCOHOL see CMQ740
CINNAMYL ALCOHOL ANTHRANILATE see API750
CINNAMYL ALCOHOL, FORMATE see CMR500
CINNAMYL ALCOHOL, SYNTHETIC see CMQ740
CINNAMYL ALDEHYDE see CMP969
CINNAMYL-2-AMINOBENZOATE see API750
CINNAMYL-o-AMINOBENZOATE see API750
CINNAMYL ANTHRANILATE (FCC) see API750
CINNAMYL FORMATE see CMR500

CRISATRINA see PMC325
CRISAZINE see PMC325
CRISODIN see DOL600
CRISODRIN see DOL600
CRISTALLOSE see SAB500
CRISTOXO 90 see THH750
CRISULFAN see BCJ250
CRISURON see DGC400
CROCUS MARTIS ADSTRINGENS see IHD000
CRODACID see MSA250
CRODACOL-CAS see HCP000
CRODACOL-S see OAX000
CROP RIDER see DFY600
CROSPOVIDONE see PKQ150
CROTILIN see DFY600
CROTYLIDENE ACETIC ACID see SKU000
CRTRON see EDA000
CRYPTOGIL OL see PAX250
CRYSTAL CHROME ALUM see PLB500
CRYSTALETS see REZ000
CRYSTALLINA see EDA000
CRYSTALLOSE see SAB500
CRYSTAL O see CCP250
CRYSTAL PROPANIL-4 see DGI000
CRYSTAMET see SJU000
CRYSTHION 2L see ASH500
CRYSTHYON see ASH500
CRYSTOL CARBONATE see SFO000
CRYSTOSOL see MQV750
CRYSTWEL see VSZ000
CTC see CMA750
CUBEB OIL see COE175
CUBIC NITER see SIO000
CUCUMBER ALCOHOL see NMV780
CUCUMBER ALDEHYDE see NMV760
CUMAFOS (DUTCH) see CNU750
CUMALDEHYDE see COE500
CUMAN see BJK500
CUMAN L see BJK500
CUMENE ALDEHYDE see COF000
p-CUMIC ALDEHYDE see COE500
CUMINALDEHYDE see COE500
CUMINIC ALDEHYDE (FCC) see COE500
CUMIN OIL see COF325
CUMINYL ALDEHYDE see COE500
CUMMIN see COF325
CUNILATE see BLC250
CUNILATE 2472 see BLC250
CUPRIC-8-HYDROXYQUINOLATE see BLC250
CUPRIC-8-QUINOLINOLATE see BLC250
CUPRIC SULFATE see CNP250
CUPROUS IODIDE see COF680
CURACRON see BNA750
CURALIN M see MQY750
CURATERR see FPE000
CURCUMA OIL see COG000
CURCUMIN see COG000
CURCUMINE see COG000
CURENE 442 see MQY750
CURON FAST YELLOW 5G see TNK750
CUTICURA ACNE CREAM see BDS000
C-WEISS 7 (GERMAN) see TGH000
CYAMOPSIS GUM see GLU000
CYANAMIDE see COH500
CYANASET see MQY750
CYANOAMINE see COH500
CYANOCOBALAMIN see VSZ000
CYANO(4-FLUORO-3-PHENOXYPHENYL)METHYL-3-(2,2-DICHLOROETHENYL)-2,2-DIMETHYLCYCLOPROPANECARBOXYLATE see CON750
CYANOGENAMIDE see COH500
CYANOGEN NITRIDE see COH500
α-CYANO-3-PHENOXYBENZYL-2-(4-CHLOROPHENYL)ISOVALERATE PYDRIN see FAR100
α-CYANO-3-PHENOXYBENZYL-2-(4-CHLOROPHENYL)-3-METHYLBUTYRATE see FAR100
CYANO(3-PHENOXYPHENYL)METHYL 4-CHLORO-α-(1-METHYLETHYL)BENZENEACETATE see FAR100
(+)CYANO(3-PHENOXYPHENYL)METHYL(±)-1-(DIFLUOROMETHOXY)-α-(1-METHYLETHYL)BENZENEACETATE see COQ380
CYANOPHOS see DRK200
CYANURAMIDE see MCB000
CYANUROTRIAMIDE see MCB000
CYANUROTRIAMINE see MCB000
CYAZIN see PMC325
CYCLAL CETYL ALCOHOL see HCP000
CYCLALIA see CMS850
CYCLAMAL see COU500
CYCLAMATE see CPQ625, SGC000
CYCLAMATE CALCIUM see CAR000
CYCLAMATE, CALCIUM SALT see CAR000
CYCLAMATE SODIUM see SGC000
CYCLAMEN ALDEHYDE see COU500
CYCLAMIC ACID see CPQ625
CYCLAMIC ACID SODIUM SALT see SGC000
CYCLAN see CAR000
CYCLOCHEM GMS see OAV000
α-CYCLOCITRYLIDENEACETONE see IFW000
β-CYCLOCITRYLIDENEACETONE see IFX000
CYCLODAN see BCJ250
CYCLOGEST see PMH500
CYCLOHEXAAN (DUTCH) see CPB000
CYCLOHEXAN (GERMAN) see CPB000
CYCLOHEXANAMINE see CPF500
CYCLOHEXANE see CPB000
cis-1,2,3,5-trans-4,6-CYCLOHEXANEHEXOL see IDE300
CYCLOHEXANESULFAMIC ACID, CALCIUM SALT see CAR000
CYCLOHEXANESULFAMIC ACID, MONOSODIUM SALT see SGC000
CYCLOHEXANESULPHAMIC ACID see CPQ625
CYCLOHEXANESULPHAMIC ACID, MONOSODIUM SALT see SGC000
CYCLOHEXANOL ACETATE see CPF000
CYCLOHEXANOLAZETAT (GERMAN) see CPF000
CYCLOHEXANYL ACETATE see CPF000
CYCLOHEXYL ACETATE see CPF000
CYCLOHEXYLAMIDOSULPHURIC ACID see CPQ625
CYCLOHEXYLAMINE see CPF500
CYCLOHEXYLAMINESULPHONIC ACID see CPQ625
3-CYCLOHEXYL-6-(DIMETHYLAMINO)-1-METHYL-s-TRIAZINE-2,4(1H,3H)-DIONE see HFA300
3-CYCLOHEXYL-6-(DIMETHYLAMINO)-1-METHYL-1,3,5-TRIAZINE-2,4(1H,3H)-DIONE see HFA300
CYCLOHEXYLSULFAMIC ACID (9CI) see CPQ625
CYCLOHEXYL SULPHAMATE SODIUM see SGC000
CYCLOHEXYLSULPHAMIC ACID see CPQ625
N-CYCLOHEXYLSULPHAMIC ACID see CPQ625
CYCLOHEXYLSULPHAMIC ACID, CALCIUM SALT see CAR000
CYCLOOCTAFLUOROBUTANE see CPS000
CYCLOPAR see TBX250
2-CYCLOPENTENYL-4-HYDROXY-3-METHYL-2-CYCLOPENTEN-1-ONE CHRYSANTHEMATE see POO000
3-(2-CYCLOPENTEN-1-YL)-2-METHYL-4-OXO-2-CYCLOPENTEN-1-YL CHRYSANTHEMUMATE see POO000
3-(2-CYCLOPENTENYL)-2-METHYL-4-OXO-2-CYCLOPENTENYL CHRYSANTHEMUMMONOCARBOXYLATE see POO000
CYCLOPENTENYLRETHONYL CHRYSANTHEMATE see POO000
CYCLORYL 21 see SON500
CYCLOSIA see CMS850
CYCLOTEN see HMB500
CYCLOTETRAMETHYLENE OXIDE see TCR750
CYCOLAMIN see VSZ000
CYFEN see DSQ000

DECORTIN H see PMA000
DECORTISYL see PLZ000
DECYL ALCOHOL see DAI600
n-DECYL ALCOHOL see DAI600
1-DECYL ALDEHYDE see DAG000, UJJ000
DECYL BENZENE SODIUM SULFONATE see DAJ000
DECYLIC ACID see DAH400
n-DECYLIC ACID see DAH400
DECYLIC ALCOHOL see DAI600
DECYL OCTYL ALCOHOL see OAX000
6-DECYLOXY-7-ETHOXY-4-HYDROXY-3-QUINOLINECARBOXYLID ACID ETHYL ESTER see DAI495
DEDEVAP see DRK200
DED-WEED see DFY600, DGI400, TIX500
DED-WEED BRUSH KILLER see TIW500
DED-WEED LV-69 see DFY600
DEE-OSTEROL see EDA000
DEE-RON see EDA000
DEE-RONAL see EDA000
DEE-ROUAL see EDA000
DEF see BSH250
DEF DEFOLIANT see BSH250
DE-FEND see DSP400
DEFILIN see DJL000
DE-FOL-ATE see SFS000
DEFOLIT see TEX600
DEGALOL see DAQ400
DE-GREEN see BSH250
DEHP see BJS000
DEHYDRACETIC ACID see MFW500
DEHYDROACETIC ACID (FCC) see MFW500
DEHYDROACETIC ACID, SODIUM SALT see SGD000
7-DEHYDROCHOLESTROL, ACTIVATED see CMC750
Δ^1-DEHYDROCORTISOL see PMA000
1-DEHYDROCORTISONE see PLZ000
Δ-1-DEHYDROCORTISONE see PLZ000
1-DEHYDROHYDROCORTISONE see PMA000
Δ^1-DEHYDROHYDROCORTISONE see PMA000
6-DEHYDRO-16-METHYLENE-6-METHYL-17-ACETOXYPROGESTERONE see MCB380
DEHYQUART STC-25 see DTC600
DEIQUAT see DWX800
DEKORTIN see PLZ000
DELCORTOL see PMA000
DELICIA see AHE750
DELONIN AMIDE see NCR000
DELPHINIC ACID see ISU000
DELSTEROL see CMC750
DELTA-CORTEF see PMA000
DELTACORTELAN see PLZ000
DELTACORTENOL see PMA000
DELTACORTISONE see PLZ000
DELTACORTONE see PLZ000
DELTACORTRIL see PMA000
DELTA-DOME see PLZ000
DELTA F see PMA000
DELTALIN see EDA000
DELTAMYCIN A see CBT250
DELTA-STAB see PMA000
DELTISONE see PLZ000
DEMETON-S-METHYL-SULFOXID (GERMAN) see DAP000
DEMETON-O-METHYL SULFOXIDE see DAP000
DEMETON-S-METHYL SULFOXIDE see DAP000
DEMETON-METHYL SULPHOXIDE see DAP000
DEMETON-S see DAP200
DEMOS-L40 see DSP400
DENAMONE see VSZ450
DENDRITIS see SFT000
DENSINFLUAT see TIO750
DEODOPHYLL see CKN000
DEODORIZED WINTERIZED COTTONSEED OIL see CNU000
DEOXYCHOLATIC ACID see DAQ400
DEOXYCHOLIC ACID (FCC) see DAQ400
7-α-DEOXYCHOLIC ACID see DAQ400
14-DEOXY-14-((2-DIETHYLAMINOETHYL-THIO) ACETOXY)MUTILINE see TET800
9-DEOXY-12,13-EPOXY-9-OXOLEUCOMYCIN V 3-ACETATE 4^B-(3-METHYLBUTANOATE) see CBT250
4-DEOXYTETRONIC ACID see BOV000
DEPARAL see CMC750
DEPC see DIZ100
DEPINAR see VSZ000
DEPOSUL see SNN300
DEP (PESTICIDE) see TIQ250
DEPTHON see TIQ250
DERATOL see EDA000
DERIBAN see DRK200
DERMAFOSU (POLISH) see TIR000
DERMAGINE see OAV000
DERMAPHOS see TIR000
DERMOFURAL see NGE500
DERRIBANTE see DRK200
DESERTALC 57 see TAB750
DESOLET see SFS000
DESORMONE see DFY600
14-DESOSSI-14-((2-DIETILAMINOETIL)MERCAPTO-ACETOSSI)MUTILIN IDROGENO FUMARATO (ITALIAN) see TET800
DESOXYCHOLIC ACID see DAQ400
DESOXYCHOLSAEURE (GERMAN) see DAQ400
14-DESOXY-14-((DIETHYLAMINOETHYL)-MERCAPTO ACETOXYL)-MUTILIN HYDROGEN FUMARATE see TET800
DESPHEN see CDP250
DESTRUXOL BORER-SOL see DFF900
DETALUP see EDA000
DETERGENT 66 see SON500
DETERGENT HD-90 see DXW200
DETF see TIQ250
DETIA GAS EX-B see AHE750
DETMOL-EXTRAKT see BBQ500
DETMOL MA see CBP000
DETMOL MA 96% see CBP000
DETOX 25 see BBQ500
DETOXARGIN see AQW000
DETREOMYCINE see CDP250
DETREOPAL see CDP700
DEVIGON see DSP400
DEVIKOL see DRK200
DEVIPON see DGI400
DEVISULPHAN see BCJ250
DEVORAN see BBQ500
DEXA-CORTIDELT HOSTACORTIN H see PMA000
DEXPANTHENOL (FCC) see PAG200
DEXTRANS see DBD800
DEXTRINS see DBD800
DEXTRO CALCIUM PANTOTHENATE see CAU750
DEXTROMYCETIN see CDP250
DEXTRONE see DWX800
DEXTROPUR see GFG000
DEXTROSE (FCC) see GFG000
DEXTROSE, ANHYDROUS see GFG000
DEXTROSOL see GFG000
DFA see DVX800
DHA see MFW500
DHA-SODIUM see SGD000
DHMS see DME000
DHS see MFW500
DIACETIC ETHER see EFS000
DIACETYL (FCC) see BOT500
DIACETYLMETHANE see ABX750
DIACETYL TARTARIC ACID ESTERS of MONO- and DIGLYCERIDES see DBH700
DIACTOL see EDA000
DIACYCINE see TBX250
DI-ADRESON F see PMA000

O,O-DIAETHYL-S-(2-AETHYLTHIO-AETHYL)-DITHIOPHOSPHAT (GERMAN) see EKF000
O,O-DIAETHYL-S-(2-AETHYLTHIO-AETHYL)-MONOTHIOPHOSPHAT (GERMAN) see DAP200
O,O-DIAETHYL-S-(AETHYLTHIO-METHYL)-DITHIOPHOSPHAT (GERMAN) see PGS000
DIAETHYLAMINOAETHANOL (GERMAN) see DJH600
O,O-DIAETHYL-O-(3-CHLOR-4-METHYL-CUMARIN-7-YL)-MONOTHIOPHOSPHAT (GERMAN) see CNU750
O,O-DIAETHYL-S-(6-CHLOR-2-OXO-BEN(b)-1,3-OXALIN-3-YL)-METHYL-DIT HIOPHOSPHAT (GERMAN) see BDJ250
O,O-DIAETHYL-S-((4-CHLOR-PHENYL-THIO)-METHYL)DITHIOPHOSPHAT (GERMAN) see TNP250
O,O-DIAETHYL-O-(2-ISOPROPYL-4-METHYL-PYRIMIDIN-6-YL)-MONOTHIOPHOSPHAT (GERMAN) see DCL000
O,O-DIAETHYL-O-(2-ISOPROPYL-4-METHYL)-6-PYRIMIDYL-THIONOPHOSPHAT (GERMAN) see DCL000
O,O-DIAETHYL-S-(3-THIA-PENTYL)-DITHIOPHOSPHAT (GERMAN) see EKF000
DIAETHYLTHIOPHOSPHORSAEUREESTER des AETHYLTHIOGLYKOL (GERMAN) see DAP200
O,O-DIAETHYL-O-3,5,6-TRICHLOR-2-PYRIDYLMONOTHIOPHOSPHAT (GERMAN) see DYE000
DIAFURON see NGG500
DIAKARMON see GEY000
DIALIFOR see DBI099
DIALUMINUM SULPHATE see AHG750
DIALUMINUM TRISULFATE see AHG750
DIALUX see ADY500
DIAMIDE see HGS000
DIAMINE see HGS000
1,2-DIAMINOAETHAN (GERMAN) see EEA500
DI(-4-AMINO-3-CHLOROPHENYL)METHANE see MQY750
DI-(4-AMINO-3-CLOROFENIL)METANO (ITALIAN) see MQY750
4,4′-DIAMINO-3,3′-DICHLORODIPHENYLMETHANE see MQY750
1,2-DIAMINO-ETHAAN (DUTCH) see EEA500
1,2-DIAMINOETHANE see EEA500
1,2-DIAMINO-ETHANO (ITALIAN) see EEA500
2,6-DIAMINOHEXANOIC ACID HYDROCHLORIDE see LJO000
DIAMMONIUM CARBONATE see ANE000
DIAMMONIUM CITRATE see ANF800
DIAMMONIUM HYDROGEN PHOSPHATE see ANR500
DIAMMONIUM SULFATE see ANU750
DIANAT (RUSSIAN) see MEL500
DIANATE see MEL500
DIANISYLTRICHLORETHANE see DOB400
2,2-DI-p-ANISYL-1,1,1-TRICHLOROETHANE see DOB400
DIANON see DCL000
DIAREX 600 see SMR000
DIASTATIN see NOH500
DIASULFA see SNN300
DIASULFYL see SNN300
DIATER see DGC400
DIATERR-FOS see DCL000
DIATOMACEOUS EARTH see DCJ800
DIATOMACEOUS SILICA see DCJ800
DIATOMITE see DCJ800
DIAZAJET see DCL000
DIAZATOL see DCL000
DIAZIDE see DCL000
DIAZINON (ACGIH, DOT) see DCL000
DIAZINONE see DCL000
DIAZITOL see DCL000
DIAZOL see DCL000
DIAZOTIZING SALTS see SIQ500
DIAZYL see SNJ500
DIBA see DNH125
DIBAM see SGM500
DIBASIC AMMONIUM PHOSPHATE see ANR500
DIBASIC SODIUM PHOSPHATE see SIM500
DIBENZOYLPEROXID (GERMAN) see BDS000
DIBENZOYL PEROXIDE see BDS000
DIBENZOYLPEROXYDE (DUTCH) see BDS000
DIBENZYLETHER (CZECH) see BEO250
DIBROM see DRJ600
1,2-DIBROMAETHAN (GERMAN) see EIY500
O-(1,2-DIBROM-2,2-DICHLORAETHYL)-O,O-DIMETHYLPHOSPHAT (GERMAN) see DRJ600
DIBROMOCYANOACETAMIDE see DDM000
α,α-DIBROMO-α-CYANOACETAMIDE see DDM000
1,2-DIBROMO-2,2-DICHLOROETHYL DIMETHYL PHOSPHATE see DRJ600
O-(1,2-DIBROMO-2,2-DICLORO-ETIL)-O,O-DIMETIL-FOSTATO (ITALIAN) see DRJ600
1,2-DIBROMOETANO (ITALIAN) see EIY500
DIBROMOETHANE see EIY500
α,β-DIBROMOETHANE see EIY500
sym-DIBROMOETHANE see EIY500
2,2-DIBROMO-3-NITRILOPROPIONAMIDE see DDM000
DIBROMURE d'ETHYLENE (FRENCH) see EIY500
O-(1,2-DIBROOM-2,2-DICHLOOR-ETHYL)-O,O-DIMETHYL-FOSFAAT (DUTCH) see DRJ600
1,2-DIBROOMETHAAN (DUTCH) see EIY500
DIBUTYLATED HYDROXYTOLUENE see BFW750
DIBUTYLBIS(LAUROYLOXY)STANNANE see DDV600
DIBUTYLBIS(LAUROYLOXY)TIN see DDV600
DIBUTYL-o-(o-CARBOXYBENZOYL) GLYCOLATE see BQP750
DIBUTYL-o-CARBOXYBENZOYLOXYACETATE see BQP750
2,6-DI-tert-BUTYL-p-CRESOL (ACGIH) see BFW750
2,6-DI-tert-BUTYL-1-HYDROXY-4-METHYLBENZENE see BFW750
3,5-DI-tert-BUTYL-4-HYDROXYTOLUENE see BFW750
2,6-DI-terc. BUTYL-p-KRESOL (CZECH) see BFW750
2,6-DI-tert-BUTYL-4-METHYLPHENOL see BFW750
2,6-DI-tert-BUTYL-p-METHYLPHENOL see BFW750
DIBUTYL SEBACATE see DEH600
DI-n-BUTYL SEBACATE see DEH600
DI-n-BUTYLTIN DI(DODECANOATE) see DDV600
DIBUTYLTIN DILAURATE (USDA) see DDV600
DIBUTYLTIN LAURATE see DDV600
DIBUTYL-ZINN-DILAURAT (GERMAN) see DDV600
DIC 1468 see AJB750
DICALCIUM PHOSPHATE see CAW100
DICALITE see SCH000
DICAMBA (DOT) see MEL500
DICARBETHOXYMETHANE see EMA500
DICARBONIC ACID DIETHYL ESTER see DIZ100
DICARZOL see DSO200
1,2-DICHLOORETHAAN (DUTCH) see DFF900
(2,4-DICHLOOR-FENOXY)-AZIJNZUUR (DUTCH) see DFY600
3-(3,4-DICHLOOR-FENYL)-1,1-DIMETHYLUREUM (DUTCH) see DGC400
3,6-DICHLOOR-2-METHOXY-BENZOEIZUUR (DUTCH) see MEL500
(2,2-DICHLOOR-VINYL)-DIMETHYL-FOSFAAT (DUTCH) see DRK200
DICHLOORVO (DUTCH) see DRK200
1,2-DICHLOR-AETHAN (GERMAN) see DFF900
3,3′-DICHLOR-4,4′-DIAMINODIPHENYLMETHAN (GERMAN) see MQY750
DICHLOREMULSION see DFF900
DICHLORFENIDIM see DGC400
DICHLORFOS (POLISH) see DRK200
3,6-DICHLOR-3-METHOXY-BENZOESAEURE (GERMAN) see MEL500
DI-CHLOR-MULSION see DFF900
2′,5-DICHLOR-4′-NITRO-SALIZYLSAEUREANILID (GERMAN) see DFV400
d-(-)-threo-2-DICHLOROACETAMIDO-1-p-NITROPHENYL-1,3-PROPANEDIOL see CDP250
3,6-DICHLORO-o-ANISIC ACID see MEL500
DICHLOROCHLORDENE see CDR750

1,1-DICHLORO-2-CHLOROETHYLENE see TIO750
3,3′-DICHLORO-4,4′-DIAMINODIPHENYLMETHANE see MQY750
DICHLORODIFLUOROMETHANE see DFA600
3,5-DICHLORO-2,6-DIMETHYL-4-PYRIDINOL see MII500
1,2-DICHLOROETHANE see DFF900
α,β-DICHLOROETHANE see DFF900
sym-DICHLOROETHANE see DFF900
DICHLORO-1,2-ETHANE (FRENCH) see DFF900
2,2-DICHLOROETHENOL DIMETHYL PHOSPHATE see DRK200
2,2-DICHLOROETHENYL DIMETHYL PHOSPHATE see DRK200
2,2-DICHLOROETHENYL PHOSPHORIC ACID DIMETHYL ESTER see DRK200
DI-(2-CHLOROETHYL)-3-CHLORO-4-METHYL-7-COUMARINYL PHOSPHATE see DFH600
DI-(2-CHLOROETHYL)-3-CHLORO-4-METHYLCOUMARIN-7-YL PHOSPHATE see DFH600
O,O-DI(2-CHLOROETHYL)-7-(3-CHLORO-4-METHYLCOUMARINYL)PHOSPHATE see CIK750
O,O-DI(2-CHLOROETHYL)-O-(3-CHLORO-4-METHYLCOUMARIN-7-YL) PHOSPHATE see DFH600
DICHLOROETHYLENE see DFF900
d-(-)-threo-2,2-DICHLORO-N-(β-HYDROXY-α-(HYDROXYMETHYL))-p-NITROPHENETHYLACETAMIDE see CDP250
DICHLOROKELTHANE see BIO750
DICHLOROMETHANE (DOT) see MDR000
2,5-DICHLORO-6-METHOXYBENZOIC ACID see MEL500
3,6-DICHLORO-2-METHOXYBENZOIC ACID see MEL500
2′,5-DICHLORO-4′-NITROSALICYLANILIDE see DFV400
3-(3,4-DICHLOROPHENOL)-1,1-DIMETHYLUREA see DGC400
DICHLOROPHENOXYACETIC ACID see DFY600
2,4-DICHLOROPHENOXYACETIC ACID (DOT) see DFY600
1,6-DI(4′-CHLOROPHENYLDIGUANIDO)HEXANE see BIM250
3-(3,4-DICHLOROPHENYL)-1,1-DIMETHYLUREA see DGC400
N′-(3,4-DICHLOROPHENYL)-N,N-DIMETHYLUREA see DGC400
1-(3,4-DICHLOROPHENYL)-3,3-DIMETHYLUREE (FRENCH) see DGC400
3-(3,5-DICHLOROPHENYL)-5-ETHENYL-5-METHYL-2,4-OXAZOLIDINEDIONE see RMA000
3-(3,5-DICHLOROPHENYL)-N-(1-METHYLETHYL)-2,4-DIOXO-1-IMIDAZOLIDINECARBOXAMIDE see GIA000
3-(3,5-DICHLOROPHENYL)-5-METHYL-5-VINYL-2,4-OXAZOLIDINEDIONE see RMA000
N-(3,4-DICHLOROPHENYL)PROPANAMIDE see DGI000
1-(2-(2,4-DICHLOROPHENYL)-2-(2-PROPENYLOXY)ETHYL)-1H-IMIDAZOLE see FPB875
N-(3,4-DICHLOROPHENYL)PROPIONAMIDE see DGI000
DI-(p-CHLOROPHENYL)TRICHLOROMETHYLCARBINOL see BIO750
DICHLOROPHOS (ACGIH, USDA) see DRK200
DICHLOROPROPIONANILIDE see DGI000
3,4-DICHLOROPROPIONANILIDE see DGI000
3′,4′-DICHLOROPROPIONANILIDE see DGI000
α-DICHLOROPROPIONIC ACID see DGI400
2,2-DICHLOROPROPIONIC ACID see DGI400
α,α-DICHLOROPROPIONIC ACID see DGI400
2,2-DICHLOROPROPIONIC ACID, SODIUM SALT see DGI600
α,α-DICHLOROPROPIONIC ACID SODIUM SALT see DGI600
3,6-DICHLORO-2-PYRIDINECARBOXYLIC ACID see CMX896
4,4′-DICHLORO-α-(TRICHLOROMETHYL)BENZHYDROL see BIO750
DICHLOROVAS see DRK200
(2,2-DICHLORO-VINIL)DIMETILFOSFATO (ITALIAN) see DRK200
2,2-DICHLOROVINYL ALCOHOL, DIMETHYL PHOSPHATE see DRK200
2,2-DICHLOROVINYL DIMETHYL PHOSPHATE see DRK200
2,2-DICHLOROVINYL DIMETHYL PHOSPHORIC ACID ESTER see DRK200
DICHLOROVOS see DRK200
2,4-DICHLORPHENOXYACETIC ACID see DFY600
(2,4-DICHLOR-PHENOXY)-ESSIGSAEURE (GERMAN) see DFY600
3-(3,4-DICHLOR-PHENYL)-1,1-DIMETHYL-HARNSTOFF (GERMAN) see DGC400
1-(2-(2,4-DICHLORPHENYL)-2-PROPENYLOXY)AETHYL)-1H-IMIDAZOLE see FPB875
DICHLORPHOS see DRK200
(2,2-DICHLOR-VINYL)-DIMETHYL-PHOSPHAT (GERMAN) see DRK200
O-(2,2-DICHLORVINYL)-O,O-DIMETHYLPHOSPHAT (GERMAN) see DRK200
DICHLORVOS (ACGIH, DOT) see DRK200
DICHLOSALE see DFV400
3,3′-DICLORO-4,4′-DIAMINODIFENILMETANO (ITALIAN) see MQY750
1,2-DICLOROETANO (ITALIAN) see DFF900
3-(3,4-DICLORO-FENYL)-1,1-DIMETIL-UREA (ITALIAN) see DGC400
DICOFOL see BIO750
DICOPUR see DFY600
DICORTOL see PMA000
DICOTOX see DFY600
1,3-DICYANOTETRACHLOROBENZENE see TBQ750
DICYSTEINE see CQK325
DIDAKENE see TBQ250
DIDODECYL-3,3′-THIODIPROPIONATE see TFD500
DIENOL S see SMR000
DIETHANOLLAURAMIDE see BKE500
N,N-DIETHANOLLAURAMIDE see BKE500
N,N-DIETHANOLLAURIC ACID AMIDE see BKE500
DIETHION see EMS500
S-(1,2-DI(ETHOXYCARBONYL)ETHYL DIMETHYL PHOSPHOROTHIOLOTHIONATE see CBP000
DIETHYL see BOR500
DIETHYL ACETALDEHYDE see DHI000
DIETHYLACETIC ACID see DHI400
DIETHYLAMINOETHANOL see DJH600
2-(DIETHYLAMINO)ETHANOL see DJH600
N-DIETHYLAMINOETHANOL see DJH600
β-DIETHYLAMINOETHANOL see DJH600
2-N-DIETHYLAMINOETHANOL see DJH600
DIETHYLAMINOETHANOL (DOT) see DJH600
2-DIETHYLAMINOETHANOL (ACGIH) see DJH600
β-DIETHYLAMINOETHYL ALCOHOL see DJH600
2-DIETHYLAMINO-6-METHYLPYRIMIDIN-4-YL DIMETHYL PHOSPHOROTHIONATE see DIN800
O-(2-DIETHYLAMINO-6-METHYLPYRIMIDIN-4-YL)-O,O-DIMETHYL PHOSPHOROTHIOATE see DIN800
O-(2-(DIETHYLAMINO)-6-METHYL-4-PYRIMIDINYL)-O,O-DIMETHYL PHOSPHOROTHIOATE see DIN800
O,O-DIETHYL-S-(4-CHLOOR-FENYL-THIO)-METHYL)-DITHIOFOSFAAT (DUTCH) see TNP250
O,O-DIETHYL-O-(3-CHLOOR-4-METHYL-CUMARIN-7-YL)MONOTHIOFOSFAAT (DUTCH) see CNU750
O,O-DIETHYL-S-((6-CHLOOR-2-OXO-BENZOXAZOLIN-3-YL)-METHYL)-DITHIO FOSFAAT (DUTCH) see BDJ250
O,O-DIETHYL-S-p-CHLORFENYLTHIOMETHYLESTER KYSELINY DITHIOFOSFORECNE (CZECH) see TNP250
O,O-DIETHYL-S-p-CHLORLPHENYLTHIOMETHYL DITHIOPHOSPHATE see TNP250
O,O-DIETHYL-S-(6-CHLOROBENZOXAZOLINYL-3-METHYL)DITHIOPHOSPHATE see BDJ250
DIETHYL-3-CHLORO-4-METHYL-7-COUMARINYL PHOSPHATE see CIK750
O,O-DIETHYL-O-(3-CHLORO-4-METHYLCOUMARIN-7-YL)PHOSPHATE see CIK750

2,2-DI-(p-METHOXYPHENYL)-1,1,1-TRICHLOROETHANE see DOB400
DI(p-METHOXYPHENYL)-TRICHLOROMETHYL METHANE see DOB400
3-(DIMETHOXYPHOSPHINYLOXY)N-METHYL-cis-CROTONAMIDE see DOL600
1,2-DIMETHOXY-4-PROPENYLBENZENE see IKR000
N[1]-(2,6-DIMETHOXY-4-PYRIMIDINYL)SULFANILAMIDE see SNN300
DIMETHOXYSULFADIAZINE see SNN300
2,4-DIMETHOXY-6-SULFANILAMIDO-1,3-DIAZINE see SNN300
2,6-DIMETHOXY-4-SULFANILAMIDOPYRIMIDINE see SNN300
DIMETHYLACETIC ACID see IJU000
2,5-DIMETHYL-3-ACETYLFURAN see ACI400
O,S-DIMETHYLACETYLPHOSPHOROAMIDOTHIOATE see DOP600
O,O-DIMETHYL-S-(2-AETHYLSULFINYL-AETHYL)-THIOLPHOSPHAT (GERMAN) see DAP000
DIMETHYLAMINE-EPICHLOROHYDRIN COPOLYMER see DOR500
N-DIMETHYL AMINO-β-CARBAMYL PROPIONIC ACID see DQD400
2-(DIMETHYLAMINO)-N-(((METHYLAMINO)CARBONYL) OXY)-2-OXOETHANIMIDOTHIOIC ACID METHYL ESTER see DSP600
m-(((DIMETHYLAMINO)METHYLENE)AMINO)PHENYLMETHYL CARBAMATE,HYDROCHLORIDE see DSO200
3-DIMETHYLAMINOMETHYLENEIMINOPHENYL-N-METHYLCARBAMATE, HYDROCHLORIDE see DSO200
2-DIMETHYLAMINO-1-(METHYLTHIO)GLYOXAL-o-METHYLCARBAMOYLMONOXIME see DSP600
DIMETHYLAMINOSUCCINAMIC ACID see DQD400
N-(DIMETHYLAMINO)SUCCINAMIC ACID see DQD400
N-DIMETHYLAMINO-SUCCINAMIDSAEURE (GERMAN) see DQD400
O-(4-((DIMETHYLAMINO)SULFONYL)PHENYL) O,O-DIMETHYL PHOSPHOROTHIOATE see FAG759
2-(2,6-DIMETHYLANILINO)-5,6-DIHYDRO-4H-1,3-THIAZINE see DMW000
DIMETHYL ANTHRANILATE (FCC) see MGQ250
O,O-DIMETHYL-S-(BENZAZIMINOMETHYL) DITHIOPHOSPHATE see ASH500
2,2-DIMETHYL-1,3-BENZDIOXOL-4-YL-N-METHYLCARBAMATE see DQM600
5,6-DIMETHYLBENZIMIDAZOLYCOBAMIDE CYANIDE see VSZ000
DIMETHYLBENZIMIDAZOYLCOBAMIDE see VSZ000
2,2-DIMETHYL-1,3-BENZODIOXOL-4-OL METHYLCARBAMATE see DQM600
2,2-DIMETHYLBENZO-1,3-DIOXOL-4-YL METHYLCARBAMATE see DQM600
O,O-DIMETHYL-S-(1,2,3-BENZOTRIAZINYL-4-KETO) METHYL PHOSPHORODITHIOATE see ASH500
DIMETHYL BENZYL CARBINOL see DQQ200
DIMETHYL BENZYL CARBINYL ACETATE see DQQ375
DIMETHYL BENZYL CARBINYL BUTYRATE see DQQ380
DIMETHYLBENZYLOCTADECYLAMMONIUM CHLORIDE see DTC600
β,γ-DIMETHYL-α,Δ-BIS(3,4-DIHYDROXYPHENYL)BUTANE see NBR000
DIMETHYLCARBAMODITHIOIC ACID, ZINC COMPLEX see BJK500
DIMETHYLCARBAMODITHIOIC ACID, ZINC SALT see BJK500
DIMETHYLCARBINOL see INJ000
O,O-DIMETHYL-O-2-CHLOR-1-(2,4,5-TRICHLORPHENYL)-VINYL-PHOSPHAT (GERMAN) see TBW100
2,2-DIMETHYL-7-COUMARANYL N-METHYLCARBAMATE see FPE000
DIMETHYL DIALKYL AMMONIUM CHLORIDE see DRI500
2,3-DIMETHYL-1,4-DIAZINE see DTU400
2,5-DIMETHYL-1,4-DIAZINE see DTU600
DIMETHYL 1,2-DIBROMO-2,2-DICHLORETHYL PHOSPHATE see DRJ600
O,O-DIMETHYL-O-(1,2-DIBROMO-2,2-DICHLOROETHYL) PHOSPHATE see DRJ600
DIMETHYL DICARBONATE see DRJ850
O,O-DIMETHYL-O-2,2-DICHLORO-1,2-DIBROMOETHYL PHOSPHATE see DRJ600
DIMETHYL-2,2-DICHLOROETHENYL PHOSPHATE see DRK200
1,1-DIMETHYL-3-(3,4-DICHLOROPHENYL)UREA see DGC400
DIMETHYL DICHLOROVINYL PHOSPHATE see DRK200
DIMETHYL-2,2-DICHLOROVINYL PHOSPHATE see DRK200
O,O-DIMETHYL DICHLOROVINYL PHOSPHATE see DRK200
O,O-DIMETHYL-O-2,2-DICHLOROVINYL PHOSPHATE see DRK200
O,O-DIMETHYL-O-(2,2-DICHLOR-VINYL)-PHOSPHAT (GERMAN) see DRK200
2,2-DIMETHYL-2,3-DIHYDROBENZOFURAN-7-YL ESTER, METHYLCARBAMIC ACID see FPE000
2,2-DIMETHYL-2,3-DIHYDRO-7-BENZOFURANYL-N-METHYLCARBAMATE see FPE000
O,O-DIMETHYL-S-(3,4-DIHYDRO-4-KETO-1,2,3-BENZOTRIAZINYL-3-METHYL) DITHIOPHOSPHATE see ASH500
O,O-DIMETHYL-S-1,2-DIKARBETOXYLETHYLDITIOFOSFAT (CZECH) see CBP000
DIMETHYL DIKETONE see BOT500
O,O-DIMETHYL-O-(p-(N,N-DIMETHYLSULFAMOYL) PHENYL)PHOSPHOROTHIOATE see FAG759
DIMETHYLDITHIOCARBAMATE ZINC SALT see BJK500
DIMETHYLDITHIOCARBAMIC ACID, SODIUM SALT see SGM500
DIMETHYLDITHIOCARBAMIC ACID, ZINC SALT see BJK500
O,O-DIMETHYLDITHIOPHOSPHATE DIETHYLMERCAPTOSUCCINATE see CBP000
DIMETHYLDITHIOPHOSPHORIC-ACID N-METHYLBENZAZIMIDE ESTER see ASH500
O,O-DIMETHYLDITHIOPHOSPHORYLACETIC ACID-N-MONOMETHYLAMIDE SALT see DSP400
O,O-DIMETHYL-DITHIOPHOSPHORYLESSIGSAEURE MONOMETHYLAMID (GERMAN) see DSP400
DIMETHYLENEDIAMINE see EEA500
DIMETHYLENE OXIDE see EJN500
O,O-DIMETHYL-S-(2-ETHTHIONYLETHYL) PHOSPHOROTHIOATE see DAP000
DIMETHYL-S-(2-ETHTHIONYLETHYL) THIOPHOSPHATE see DAP000
2-(4-(1,1-DIMETHYLETHYL)PHENOXY)CYCLOHEXYL 2-PROPYNYL ESTER, SULFUROUS ACID see SOP000
2-(4-(1,1-DIMETHYLETHYL)PHENOXY)CYCLOHEXYL 2-PROPYNYL SULFITE see SOP000
O,O-DIMETHYL-S-(2-ETHYLSULFINYL-ETHYL)-MONOTHIOFOSFAAT (DUTCH) see DAP000
O,O-DIMETHYL-S-(2-(ETHYLSULFINYL)ETHYL) PHOSPHOROTHIOATE see DAP000
O,O-DIMETHYL-S-(2-ETHYLSULFINYL)ETHYL THIOPHOSPHATE see DAP000
O,O-DIMETHYL-S-ETHYLSULPHINYLETHYL PHOSPHOROTHIOLATE see DAP000
DIMETHYLFORMALDEHYDE see ABC750
DIMETHYLGLYOXAL see BOT500
2,6-DIMETHYL-5-HEPTENAL see DSD775
3,7-DIMETHYL-7-HYDROXYOCTANAL see CMS850
O,O-DIMETHYL-(1-HYDROXY-2,2,2-TRICHLORAETHYL) PHOSPHONSAEURE ESTER (GERMAN) see TIQ250
O,O-DIMETHYL-(1-HYDROXY-2,2,2-TRICHLORATHYL)-PHOSPHAT (GERMAN) see TIQ250
O,O-DIMETHYL-(1-HYDROXY-2,2,2-TRICHLOROETHYL) PHOSPHONATE see TIQ250
DIMETHYLKETAL see ABC750
DIMETHYLKETOL see ABB500
DIMETHYL KETONE see ABC750

1,4-DIOSSAN-2,3-DIYL-BIS(O,O-DIETIL-DITIOFOSFATO) (ITALIAN) see DVQ709
DIOSUCCIN see DJL000
DIOTHENE see PJS750
DIOTILAN see DJL000
DIOVAC see DJL000
1,4-DIOXAAN-2,3-DIYL-BIS(O,O-DIETHYL-DITHIOFOS-FAAT) (DUTCH) see DVQ709
2,3-p-DIOXANDITHIOL S,S-BIS(O,O-DIETHYL PHOSPHORO-DITHIOATE) see DVQ709
1,4-DIOXAN-2,3-DIYL-BIS(O,O-DIAETHYL-DITHIOPHOS-PHAT) (GERMAN) see DVQ709
1,4-DIOXAN-2,3-DIYL-BIS(O,O-DIETHYLPHOSPHOROTHIOLOTHIONATE) see DVQ709
1,4-DIOXAN-2,3-DIYL-O,O,O′,O′-TETRAETHYL DI(PHOS-PHOROMITHIOATE) see DVQ709
2,3-p-DIOXANE-S,S-BIS(O,O-DIETHYLPHOSPHOROI-THIOATE) see DVQ709
p-DIOXANE-2,3-DITHIOL-S,S-DIESTER WITH O,O-DIETHYL PHOSPHORODITHIOATE see DVQ709
p-DIOXANE-2,3-DIYL ETHYL PHOSPHORODITHIOATE see DVQ709
DIOXATHION see DVQ709
DIOXYMETHYLENE-PROTOCATECHUIC ALDEHYDE see PIW250
DIPEGYL see NCR000
DIPEPTIDE SWEETENER see ARN825
DIPHENYLAMINE see DVX800
N,N-DIPHENYLAMINE see DVX800
DIPHENYL-2-ETHYLHEXYL PHOSPHATE see DWB800
DIPHENYLGLYOXAL PEROXIDE see BDS000
DIPHENYL KETONE see BCS250
DIPHENYLMETHANONE see BCS250
DIPHOSPHORIC ACID, DISODIUM SALT see DXF800
DIPIGYL see NCR000
DIPOFENE see DCL000
DIPOLYOXYETHYLATEDPOLYPROPYLENEGLYCOL ETHER see PJK150, PJK151
DIPOTASSIUM DICHLORIDE see PLA500
DIPOTASSIUM MONOPHOSPHATE see PLQ400
DIPOTASSIUM PERSULFATE see DWQ000
DIPOTASSIUM PHOSPHATE see PLQ400
DIPPING ACID see SOI500
DIPRAM see DGI000
4-(DI-N-PROPYLAMINO)-3,5-DINITRO-1-TRIFLUORO-METHYLBENZENE see DUV600
N,N-DI-N-PROPYL-2,6-DINITRO-4-TRIFLUOROMETHYLA-NILINE see DUV600
N,N-DIPROPYL-4-TRIFLUOROMETHYL-2,6-DINITROANI-LINE see DUV600
DIPTERAX see TIQ250
DIPTEREX see TIQ250
DIPTEREX 50 see TIQ250
DIPTEVUR see TIQ250
DIQUAT (ACGIH,DOT) see DWX800
DIQUAT DIBROMIDE see DWX800
DIREKTAN see NDT000
DIREX 4L see DGC400
DIRIMAL see OJY100
DISATABS TABS see REU000
DISODIUM CARBONATE see SFO000
DISODIUM CITRATE see DXC400
DISODIUM DIACID ETHYLENEDIAMINETETRAACETATE see EIX500
DISODIUM DIHYDROGEN ETHYLENEDIAMINETETRAACE-TATE see EIX500
DISODIUM DIHYDROGEN(ETHYLENEDINITRILO)TETRAACETATE see EIX500
DISODIUM DIHYDROGEN PYROPHOSPHATE see DXF800
DISODIUM DIPHOSPHATE see DXF800
DISODIUM EDATHAMIL see EIX500
DISODIUM EDETATE see EIX500
DISODIUM EDTA (FCC) see EIX500
DISODIUM ETHYLENEBIS(DITHIOCARBAMATE) see DXD200
DISODIUM ETHYLENE-1,2-BISDITHIOCARBAMATE see DXD200
DISODIUM ETHYLENEDIAMINETETRAACETATE see EIX500
DISODIUM ETHYLENEDIAMINETETRAACETIC ACID see EIX500
DISODIUM (ETHYLENEDINITRILO)TETRAACETATE see EIX500
DISODIUM (ETHYLENEDINITRILO)TETRAACETIC ACID see EIX500
DISODIUM GMP see GLS800
DISODIUM-5′-GMP see GLS800
DISODIUM GUANYLATE (FCC) see GLS800
DISODIUM-5′-GUANYLATE see GLS800
DISODIUM HYDROGEN CITRATE see DXC400
DISODIUM HYDROGEN PHOSPHATE see SIM500
DISODIUM IMP see DXE500
DISODIUM INDIGO-5,5-DISULFONATE see DXE400
DISODIUM INOSINATE see DXE500
DISODIUM-5′-INOSINATE see DXE500
DISODIUM INOSINE-5′-MONOPHOSPHATE see DXE500
DISODIUM INOSINE-5′-PHOSPHATE see DXE500
DISODIUM METASILICATE see SJU000
DISODIUM MONOHYDROGEN PHOSPHATE see SIM500
DISODIUM MONOSILICATE see SJU000
DISODIUM ORTHOPHOSPHATE see SIM500
DISODIUM PHOSPHATE see SIM500
DISODIUM PHOSPHORIC ACID see SIM500
DISODIUM PYROPHOSPHATE see DXF800
DISODIUM PYROSULFITE see SII000
DISODIUM SALT of EDTA see EIX500
DISODIUM SALT of 1-INDIGOTIN-S,S′-DISULPHONIC ACID see DXE400
DISODIUM SEQUESTRENE see EIX500
DISODIUM SULFATE see SJY000
DISODIUM SULFITE see SJZ000
DISODIUM TARTRATE see BLC000
DISODIUM l-(+)-TARTRATE see BLC000
DISODIUM TETRACEMATE see EIX500
DISODIUM VERSENATE see EIX500
DISODIUM VERSENE see EIX500
DISPERSED BLUE 12195 see FAE000
DISPERSED VIOLET 12197 see BFN250
DISTEARIN see OAV000
DISTEARYL THIODIPROPIONATE see DXG650
DISTILLED LIME OIL see OGO000
DISTOL 8 see EIV000
DISULFATON see EKF000
DISULFOTON (ACGIH, DOT) see EKF000
DI-SYSTON see EKF000
DISYSTOX see EKF000
DITHANE D-14 see DXD200
DITHANE A-40 see DXD200
DITHIOCARBONIC ANHYDRIDE see CBV500
DITHIODEMETON see EKF000
β,β′-DITHIODIALANINE see CQK325
DITHIOPHOSPHATE de O,O-DIETHYLE et de (4-CHLORO-PHENYL) THIOMETHYLE (FRENCH) see TNP250
DITHIOPHOSPHATE de O,O-DIETHYLE et de S-(2-ETHYL-THIO-ETHYLE) (FRENCH) see EKF000
DITHIOPHOSPHATE de O,O-DIETHYLE et d'ETHYLTHIO-METHYLE (FRENCH) see PGS000
DITHIOPHOSPHATE de O,O-DIMETHYLE et de S-(1,2-DICAR-BOETHOXYETHYLE) (FRENCH) see CBP000
DITHIOPHOSPHATE de O,O-DIMETHYLE et de S(-N-METHYL-CARBAMOYL-METHYLE) (FRENCH) see DSP400
DITHIOSYSTOX see EKF000
DI(TRI-(2,2-DIMETHYL-2-PHENYLETHYL)TIN)OXIDE see BLU000
DITRIFON see TIQ250
DIUREX see DGC400
DIUROL see DGC400

FA see FMV000
Fa 100 see EQR500
FAC 5273 see PIX250
FACTITIOUS AIR see NGU000
FACTOR PP see NCR000
FALITHION see DSQ000
FALL see SFS000
FAMFOS see FAG759
FAMOPHOS see FAG759
FAMOPHOS WARBEX see FAG759
FAMPHOS see FAG759
FAMPHUR see FAG759
FANFOS see FAG759
FANNOFORM see FMV000
FARMCO see DFY600
FARMCO ATRAZINE see PMC325
FARMCO DIURON see DGC400
FARMCO FENCE RIDER see TIW500
FARMCO PROPANIL see DGI000
FARMICETINA see CDP250
FARNESOL see FAG875
FARNESYL ALCOHOL see FAG875
FASCIOLIN see CBY000
FASCO-TERPENE see THH750
FAST GREEN FCF see FAG000
FATTY ACIDS see FAG900
FB/2 see DWX800
FC 12 see DFA600
FC-C 318 see CPS000
F D &C BLUE NO. 1 see FAE000
F D & C BLUE No. 1 see FAE000
F D & C BLUE No. 2 (FCC) see DXE400
F D & C GREEN NO. 3 see FAG000
F D & C RED NO. 3 (FCC) see TDE500
F D & C RED No. 40 see FAG050
F D & C VIOLET NO. 1 see BFN250
F D & C YELLOW NO. 3 see PEJ750
F D & C YELLOW NO. 4 see TGW250
F D & C YELLOW NO. 5 (FCC) see TNK750
F D & C YELLOW No. 6 see FAG150
FECAMA see DRK200
FEDACIN see NGE500
FEDAL-UN see TBQ250
FEGLOX see DWX800
α-FELLANDRENE see MCC000
FEMA No. 2003 see AAG250
FEMA No. 2005 see MDW750
FEMA No. 2006 see AAT250
FEMA No. 2007 see THM500
FEMA No. 2008 see ABB500
FEMA No. 2009 see ABH000
FEMA No. 2011 see AEN250
FEMA No. 2026 see AGC500
FEMA No. 2031 see AGH250
FEMA No. 2032 see AGA500
FEMA No. 2033 see AGI500
FEMA No. 2034 see AGJ250
FEMA No. 2037 see AGM500
FEMA No. 2045 see ISV000
FEMA No. 2055 see ILV000
FEMA No. 2058 see MHV250
FEMA No. 2060 see IHP400
FEMA No. 2061 see AOG500
FEMA No. 2063 see AOG600
FEMA No. 2069 see IHS000
FEMA No. 2073 see AOJ900
FEMA No. 2075 see IHU100
FEMA No. 2082 see AON350
FEMA No. 2084 see IME000
FEMA No. 2085 see ITB000
FEMA No. 2086 see PMQ750
FEMA No. 2097 see AOX750
FEMA No. 2098 see AOY400
FEMA No. 2099 see MED500
FEMA No. 2109 see ARN000
FEMA No. 2127 see BAY500
FEMA No. 2134 see BCS250
FEMA No. 2135 see BDX000
FEMA No. 2137 see BDX500
FEMA No. 2138 see BCM000
FEMA No. 2140 see BED000
FEMA No. 2141 see IJV000
FEMA No. 2142 see BEG750
FEMA No. 2149 see BFD400
FEMA No. 2150 see BFD800
FEMA No. 2151 see BFJ750
FEMA No. 2152 see ISW000
FEMA No. 2159 see BMD100
FEMA No. 2160 see IHX600
FEMA No. 2170 see BOV250
FEMA No. 2174 see BPU750
FEMA No. 2175 see IIJ000
FEMA No. 2178 see BPW500
FEMA No. 2179 see IIL000
FEMA No. 2183 see BQI000
FEMA No. 2184 see BFW750
FEMA No. 2186 see BQM500
FEMA No. 2187 see BSW500
FEMA No. 2188 see BRQ350
FEMA No. 2190 see BQP000
FEMA No. 2193 see IIQ000
FEMA No. 2203 see DTC800
FEMA No. 2209 see BBA000
FEMA No. 2210 see IJF400
FEMA No. 2213 see IJN000
FEMA No. 2218 see ISX000
FEMA No. 2219 see BSU250
FEMA No. 2220 see IJS000
FEMA No. 2221 see BSW000
FEMA No. 2222 see IJU000
FEMA No. 2223 see TIG750
FEMA No. 2224 see CAK500
FEMA No. 2229 see CBA500
FEMA No. 2245 see CCM000
FEMA No. 2249 see MCD379, MCD500
FEMA No. 2252 see CCN000
FEMA No. 2286 see CMP969
FEMA No. 2288 see CMP975
FEMA No. 2293 see CMQ730
FEMA No. 2294 see CMQ740
FEMA No. 2295 see API750
FEMA No. 2299 see CMR500
FEMA No. 2301 see CMR850
FEMA No. 2302 see CMR800
FEMA No. 2306 see CMS750
FEMA No. 2307 see CMS845
FEMA No. 2309 see CMT250
FEMA No. 2311 see AAU000
FEMA No. 2312 see CMT600
FEMA No. 2313 see CMT900
FEMA No. 2314 see CMT750
FEMA No. 2316 see CMU100
FEMA No. 2341 see COE500
FEMA No. 2356 see CQI000
FEMA No. 2361 see DAF200
FEMA No. 2362 see DAG000, DAG200
FEMA No. 2365 see DAI600
FEMA No. 2366 see DAI350
FEMA No. 2370 see BOT500
FEMA No. 2371 see BEO250
FEMA No. 2375 see EMA500
FEMA No. 2376 see DJY600
FEMA No. 2377 see SNB000
FEMA No. 2379 see DKV150
FEMA No. 2381 see HHR500
FEMA No. 2391 see DTE600
FEMA No. 2392 see DQQ375
FEMA No. 2393 see DQQ200
FEMA No. 2394 see DQQ380
FEMA No. 2401 see DXS700

FEMA No. 3060 see TCU600
FEMA No. 3073 see MNR250
FEMA No. 3075 see THA250
FEMA No. 3082 see TJJ400
FEMA No. 3091 see UJA800
FEMA No. 3092 see UJJ000
FEMA No. 3095 see ULJ000
FEMA No. 3097 see UNA000
FEMA No. 3101 see VAQ000
FEMA No. 3102 see ISU000
FEMA No. 3103 see VAV000
FEMA No. 3107 see VFK000
FEMA No. 3126 see ADA350
FEMA No. 3135 see DAE450
FEMA No. 3149 see EIL000
FEMA No. 3155 see ENF200
FEMA No. 3164 see HAV450
FEMA No. 3183 see MEX350
FEMA No. 3202 see ADA375
FEMA No. 3212 see NMV775
FEMA No. 3213 see NNA300
FEMA No. 3215 see ODQ800
FEMA No. 3237 see TDV725
FEMA No. 3244 see TME270
FEMA No. 3264 see DAI360
FEMA No. 3271 see DTU400
FEMA No. 3272 see DTU600
FEMA No. 3273 see DTU800
FEMA No. 3289 see HBI800
FEMA No. 3291 see BOV000
FEMA No. 3302 see MFN285
FEMA No. 3309 see MOW750
FEMA No. 3317 see NMV760
FEMA No. 3326 see ABC750
FEMA No. 3332 see BOV700
FEMA No. 3354 see HFM600
FEMA No. 3355 see HKC600
FEMA No. 3356 see NMV790
FEMA No. 3379 see NNA532
FEMA No. 3386 see PPS250
FEMA No. 3391 see ACI400
FEMA No. 3406 see MLA300
FEMA No. 3432 see IIN300
FEMA No. 3465 see NNA530
FEMA No. 3467 see ODW025
FEMA No. 3491 see EKF575
FEMA No. 3497 see HFE550
FEMA No. 3498 see ISZ000
FEMA No. 3499 see HFR200
FEMA No. 3500 see HFQ600
FEMA No. 3525 see TLX800
FEMA No. 3558 see MLA250
FEMA No. 3559 see MCB750
FEMA No. 3565 see DKV175
FEMA No. 3581 see OCY100
FEMA No. 3583 see OEG100
FEMA No. 3587 see ODW030
FEMA No. 3612 see ODW040
FEMA No. 3632 see PDF790
FEMA No. 7071 see DTV300
FEMESTRONE see EDP000
FENAMIN see PMC325
FENAMINE see PMC325
FENAMIPHOS see FAK000
FENARIMOL see FAK100
FENARSONE see CBJ000
FENASAL see DFV400
FENATROL see PMC325
FENAZO BLUE XI see FAE000
FENAZO YELLOW T see TNK750
FENBENDAZOL see FAL100
FENBENDAZOLE see FAL100
FENBUTATIN OXIDE see BLU000
FENCHEL OEL (GERMAN) see FAP000
FENCHLOORFOS (DUTCH) see TIR000
FENCHLORFOS see TIR000
FENCHLORFOSU (POLISH) see TIR000
FENCHLOROPHOS see TIR000
FENCHLORPHOS see TIR000
FENCLOR see PJL750
FENICOL see CDP250
FENITOX see DSQ000
FENITROTHION see DSQ000
FENITROTION (HUNGARIAN) see DSQ000
FENNEL OIL see FAP000
FENOLOVO see HON000
FENOPROP see TIX500
FENORMONE see TIX500
FENPROSTALENE see FAQ500
FENTHION see FAQ999
FENTIN HYDROXIDE see HON000
FENVALERATE see FAR100
FEOSOL see FBO000, IHM000
FEOSPAN see IHM000
FEOSTAT see FBJ100
FERGON see FBK000
FERGON PREPARATIONS see FBK000
FER-IN-SOL see FBO000, IHM000
FERKETHION see DSP400
FERLUCON see FBK000
FERMENICIDE LIQUID see SOH500
FERMENICIDE POWDER see SOH500
FERMENTATION ALCOHOL see EFU000
FERMENTATION BUTYL ALCOHOL see IIL000
FERNESTA see DFY600
FERNIMINE see DFY600
FERNISOLONE see PMA000
FERNOXONE see DFY600
FERO-GRADUMET see FBO000, IHM000
FEROTON see FBJ100
FERRALYN see IHM000
FERRIC AMMONIUM CITRATE see FAS700
FERRIC AMMONIUM CITRATE, GREEN see FAS700
FERRIC CHLORIDE see FAU000
FERRIC CHLORIDE, SOLID (DOT) see FAU000
FERRIC CHLORIDE, SOLID, ANHYDROUS (DOT) see FAU000
FERRIC CHOLINE CITRATE see FBC100
FERRIC CITRATE see FAW100
FERRIC ORTHOPHOSPHATE see FAZ500
FERRIC OXIDE see IHD000
FERRIC PHOSPHATE see FAZ500
FERRIC PYROPHOSPHATE see FAZ525
FERRIC SODIUM PYROPHOSPHATE see SHE300
FERRIC SULFATE see FBA000
FERROCHOLINATE see FBC100
FERROFUME see FBJ100
FERRO-GRADUMET see IHM000
FERROLIP see FBC100
FERRONAT see FBJ100
FERRONE see FBJ100
FERRONICUM see FBK000
FERROSULFAT (GERMAN) see IHM000
FERROSULFATE see IHM000
FERROTEMP see FBJ100
FERRO-THERON see IHM000
FERROUS ASCORBATE see FBH050
FERROUS CARBONATE see FBH100
FERROUS CITRATE see FBJ075
FERROUS FUMARATE see FBJ100
FERROUS GLUCONATE see FBK000
FERROUS LACTATE see LAL000
FERROUS SULFATE (FCC) see FBO000
FERROUS SULFATE (DOT, FCC) see IHM000
FERROUS SULFATE HEPTAHYDRATE see FBO000
FERRUGO see IHD000

FERRUM see FBJ100
FERSAMAL see FBJ100
FERSOLATE see IHM000
FESOFOR see FBO000
FESOTYME see FBO000
FIBRENE C 400 see TAB750
FICAM see DQM600
FICIN see FBS000
FICUS PROTEASE see FBS000
FICUS PROTEINASE see FBS000
FILMERINE see SIQ500
FINE GUM HES see SFO500
FINTINE HYDROXYDE (FRENCH) see HON000
FINTIN HYDROXID (GERMAN) see HON000
FINTIN HYDROXYDE (DUTCH) see HON000
FINTIN IDROSSIDO (ITALIAN) see HON000
FIR NEEDLE OIL, CANADIAN TYPE see FBU850
FIR NEEDLE OIL, SIBERIAN see FBV000
FIRON see FBJ100
FIXOL see CMS850
FLAMENCO see TGH000
FLAMYCIN see CMA750
FLANOGEN ELA see CCL250
FLAVAXIN see RIK000
FLAVAZONE see NGE500
FLAVOMYCIN see MRA250
FLAVOPHOSPHOLIPOL see MRA250
FLEBOCORTID see HHR000
FLECK-FLIP see TIO750
FLECTOL H see TLP500
FLEXIMEL see BJS000
FLEXOL DOP see BJS000
FLEXOL PLASTICIZER DIP see ILR100
FLEXOL PLASTICIZER DOP see BJS000
FLIBOL E see TIQ250
FLIEGENTELLER see TIQ250
FLIT 406 see CBG000
FLO-GARD see SCH000
FLO-MOR see PAI000
FLO PRO V SEED PROTECTANT see CCC500
FLORALTONE see GEM000
FLORDIMEX see CDS125
FLOREL see CDS125
FLORES MARTIS see FAU000
FLOWERS OF ZINC see ZKA000
FLUATE see TIO750
FLUAZIFOP-BUTYL see FDA885
FLUCYTHRINATE see COQ380
FLUKOIDS see CBY000
FLUOMETURON see DUK800
FLUOPERIDOL see FLU000
1-(3-(4-FLUOROBENZOYL)PROPYL)-4-(2-PYRIDYL)PIPERAZINE see FLU000
FLUOROCARBON-12 see DFA600
FLUOROCARBON-115 see CJI500
1-(4-FLUOROPHENYL)-4-(4-(2-PYRIDINYL)-1-PIPERAZINYL)-1-BUTANONE see FLU000
4'-FLUORO-4-(4-(2-PYRIDYL)-1-PIPERAZINYL)BUTYROPHENONE see FLU000
FLY-DIE see DRK200
FLY FIGHTER see DRK200
FMC-1240 see EMS500
FMC 5273 see PIX250
FMC 5462 see BCJ250
FMC 5488 see CKM000
FMC 10242 see FPE000
FMC 17370 see BEP500
FMC 33297 see AHJ750
FMC 41655 see AHJ750
FOLACIN see FMT000
FOLATE see FMT000
FOLCYSTEINE see FMT000
FOLETHION see DSQ000
FOLIC ACID see FMT000
FOLIONE see MND275
FOLLICORMON see EDP000
FOLLIDRIN see EDP000
FOMREZ SUL-4 see DDV600
FONOLINE see MQV750
FOOD BLUE 2 see FAE000
FOOD BLUE DYE NO. 1 see FAE000
FOOD RED 14 see TDE500
FOOD STARCH, MODIFIED see FMU100
FOOD YELLOW NO. 4 see TNK750
FORAAT (DUTCH) see PGS000
FOREDEX 75 see DFY600
FORLIN see BBQ500
FORMAGENE see PAI000
FORMAL see CBP000
FORMALDEHYD (CZECH, POLISH) see FMV000
FORMALDEHYDE see FMV000
FORMALDEHYDE, solution (DOT) see FMV000
FORMALIN see FMV000
FORMALIN 40 see FMV000
FORMALIN (DOT) see FMV000
FORMALINA (ITALIAN) see FMV000
FORMALINE (GERMAN) see FMV000
FORMALIN-LOESUNGEN (GERMAN) see FMV000
FORMALITH see FMV000
FORMETANATE HYDROCHLORIDE see DSO200
FORMIATE de METHYLE (FRENCH) see MKG750
FORMIC ACID see FNA000
FORMIC ACID, CINNAMYL ESTER see CMR500
FORMIC ACID, CITRONELLYL ESTER see CMT750
FORMIC ACID-3,7-DIMETHYL-6-OCTEN-1-YL ESTER see CMT750
FORMIC ACID, ETHYL ESTER see EKL000
FORMIC ACID, GERANIOL ESTER see GCY000
FORMIC ACID, HEPTYL ESTER see HBO500
FORMIC ACID, ISOBUTYL ESTER see IIR000
FORMIC ACID, ISOPENTYL ESTER see IHS000
FORMIC ALDEHYDE see FMV000
FORMIC ETHER see EKL000
FORMOL see FMV000
FORMOLA 40 see DFY600
FORMOSULFATHIAZOLE see TEX250
FORMVAR 1285 see AAX250
α-FORMYLETHYLBENZENE see COF000
FORMYLIC ACID see FNA000
2-FORMYLQUINOXALINE-1,4-DIOXIDE CARBOMETHOXYHYDRAZONE see FOI000
FORMYL VIOLET S4BN see BFN250
FOROTOX see TIQ250
FORRON see TIW500
FOR-SYN see BEP500
FORTEX see TIW500
FORTHION see CBP000
FORTIGRO see FOI000
FORTION NM see DSP400
FORTODYL see EDA000
FORTRACIN see BAC250
FORTURF see TBQ750
FOSCHLOR see TIQ250
FOSCHLOREM (POLISH) see TIQ250
FOS-FALL "A" see BSH250
FOSFAMID see DSP400
FOSFONO 50 see EMS500
FOSFORZUUROPLOSSINGEN (DUTCH) see PHB250
FOSFOTHION see CBP000
FOSFOTION see CBP000
FOSFOTOX see DSP400
FOSFURI di ALLUMINIO (ITALIAN) see AHE750
FOSFURI di MAGNESIO (ITALIAN) see MAI000
FOSSIL FLOUR see SCH000
FOSTEX see BDS000
FOSTION MM see DSP400
FOZALON see BDJ250

LINCOMYCINE (FRENCH) see LGD000
LINDAGRAIN see BBQ500
LINDAN see DRK200
LINDANE (ACGIH, DOT, USDA) see BBQ500
LINE RIDER see TIW500
LINGUSORBS see PMH500
LINOLEAMIDE see LGF900
LINOLEIC ACID see LGG000
9,12-LINOLEIC ACID see LGG000
LINOLEIC ACID AMIDE see LGF900
LINSEED OIL see LGK000
LINTOX see BBQ500
LIOXIN see VFK000
LIPO GMS 410 see OAV000
LIPO-LUTIN see PMH500
LIPOSORB S see SKU700
LIPOSORB O-20 see PKG250
LIPOSORB S-20 see PKG750, SKU700
LIPOTRIL see CMF750
LIQUAMYCIN INJECTABLE see HOI000
LIQUIDOW see CAO750
LIQUID ROSIN see TAC000
LIQUIMETH see MDT750
LIROPON see DGI400
LIROPREM see PAX250
LISACORT see PLZ000
LITAC see ADY500
LITEX CA see SMR000
LITHOGRAPHIC STONE see CAO000
LOBAMINE see ADG375
LOCUST BEAN GUM see LIA000
LOISOL see TIQ250
LOMBRISTOP see TEX000
LO MICRON TALC 1 see TAB750
LO MICRON TALC, BC 1621 see TAB750
LO MICRON TALC USP, BC 2755 see TAB750
LOROL see DXV600
LOROL 20 see OEI000
LOROL 22 see DAI600
LOROL 24 see HCP000
LOROL 28 see OAX000
LOROMISIN see CDP250
LOROXIDE see BDS000
LORSBAN see DYE000
LOVAGE OIL see LII000
LO-VEL see SCH000
LOVOSA see SFO500
LOW ERUCIC ACID RAPESEED OIL see RBK200
LOXANOL K see HCP000
LOXON see DFH600
LUCEL (polysaccharide) see SFO500
LUCIDOL see BDS000
LUCORTEUM SOL see PMH500
LUDOX see SCH000
LUPERCO see BDS000
LUPEROX FL see BDS000
LURAN see ADY500
LURGO see DSP400
LUSTRAN see ADY500
LUTALYSE see POC750
LUTEAL HORMONE see PMH500
LUTEOHORMONE see PMH500
LUTEOSAN see PMH500
LUTEX see PMH500
LUTOCYCLIN see PMH500
LUTOSOL see INJ000
LUTROL see PJT000
LUTROMONE see PMH500
LUVISKOL see PKQ250
LUXON see DFH600
LXON see DFH600
LYCOID DR see GLU000
LYE see PLJ500
LYE (DOT) see SHS000
LYE SOLUTION see SHS500
LYGOMME CDS see CCL250
l-LYSINE HYDROCHLORIDE see LJO000
LYSINE MONOHYDROCHLORIDE see LJO000
l-LYSINE MONOHYDROCHLORIDE see LJO000
LYSOFORM see FMV000
LYTRON 5202 see SMR000

M-74 see EKF000
M 140 see CDR750
M 410 see CDR750
M 3180 see TJH000
M-4209 see CBT250
MA-1214 see DXV600
MAA see AFI850
MACE OIL (FCC) see OGW000
MACRABIN see VSZ000
MACROGOL 1000 see PJT250
MACROGOL 4000 see PJT750
MACRONDRAY see DFY600
MADAGASCAR LEMONGRASS OIL see LEH000
MADHURIN see SAB500
MADRIBON see SNN300
MADRIGID see SNN300
MADRIQID see SNN300
MADROXIN see SNN300
MADROXINE see SNN300
MAFU see DRK200
MAG see MRF000
MAGBOND see BAV750
MAGMASTER see MAD500
MAGNAMYCIN see CBT250
MAGNAMYCIN A see CBT250
MAGNESIA see MAH500
MAGNESIA ALBA see MAD500
MAGNESIA MAGMA see MAG750
MAGNESIA USTA see MAH500
MAGNESITE see MAD500
MAGNESIUM CARBONATE see MAD500
MAGNESIUM(II) CARBONATE (1:1) see MAD500
MAGNESIUM CARBONATE, PRECIPITATED see MAD500
MAGNESIUM CHLORIDE see MAE250
MAGNESIUMFOSFIDE (DUTCH) see MAI000
MAGNESIUM GLYCEROPHOSPHATE see MAG100
MAGNESIUM HYDRATE see MAG750
MAGNESIUM HYDROGEN PHOSPHATE see MAG550
MAGNESIUM HYDROXIDE see MAG750
MAGNESIUM OXIDE see MAH500
MAGNESIUM PHOSPHATE, DIBASIC see MAH775
MAGNESIUM PHOSPHATE, TRIBASIC see MAH780
MAGNESIUM PHOSPHIDE see MAI000
MAGNESIUM SILICATE HYDRATE see MAJ000
MAGNESIUM STEARATE see MAJ030
MAGNESIUM SULFATE (1:1) see MAJ250
MAGNESIUM SULPHATE see MAJ250
MAGNEZU TLENEK (POLISH) see MAH500
MAJOL PLX see SFO500
MALACIDE see CBP000
MALAFOR see CBP000
MALAGRAN see CBP000
MALAKILL see CBP000
MALAMAR see CBP000
MALAPHELE see CBP000
MALAPHOS see CBP000
MALASOL see CBP000
MALASPRAY see CBP000
MALATHION see CBP000
MALATHION (ACGIH,DOT) see CBP000
MALATHIOZOO see CBP000
MALATHON see CBP000
MALATION (POLISH) see CBP000
MALATOL see CBP000
MALATOX see CBP000
MALDISON see CBP000

(3,4-METHYLENEDIOXY-6-PROPYLBENZYL) (BUTYL) DIETHYLENE GLICOL ETHER see PIX250
3,4-METHYLENEDIOXY-6-PROPYLBENZYL n-BUTYL DIETHYLENEGLYCOL ETHER see PIX250
METHYLENE GLYCOL see FMV000
3-METHYLENE-7-METHYL-1,6-OCTADIENE see MRZ150
METHYLENE OXIDE see FMV000
S,S′-METHYLENE O,O,O′,O′-TETRAETHYL PHOSPHORODITHIOATE see EMS500
8-METHYLENE-4,11,11-(TRIMETHYL)BICYCLO(7.2.0)-UNDEC-4-ENE see CCN000
METHYL-18-EPIRESERPATE METHYL ETHER HYDROCHLORIDE see MQR200
METHYL ESTER of p-HYDROXYBENZOIC ACID see HJL500
METHYL ESTER OF WOOD ROSIN see MFT500
METHYL ESTER OF WOOD ROSIN, PARTIALLY HYDROGENATED (FCC) see MFT500
METHYL ETHOXOL see EJH500
4-(1-METHYLETHYL)-BENZALDEHYDE (9CI) see COE500
3-(1-METHYLETHYL)-1H-2,1,3-BENZOTHIAZAIN-4(3H)-ONE-2,2-DIOXIDE see MJY500
METHYL ETHYL CELLULOSE see MJY550
METHYLETHYLENE GLYCOL see PML000
METHYL ETHYLENE OXIDE see ECF500
(E)-1-METHYLETHYL-3-(((ETHYLAMINO)METHOXYPHOSPHINOTHIOYL)OXY-2-BUTENOATE see MKA000
1-(METHYLETHYL)-ETHYL 3-METHYL-4-(METHYLTHIO) PHENYL PHOSPHORAMIDATE see FAK000
METHYL ETHYL KETONE (ACGIH) see BOV250
METHYLETHYLMETHANE see BOR500
METHYL EUGENOL (FCC) see AGE250
METHYL FORMATE see MKG750
METHYL FORMATE (DOT) see MKG750
METHYLFORMIAAT (DUTCH) see MKG750
METHYLFORMIAT (GERMAN) see MKG750
METHYL GLYCOL see EJH500, PML000
METHYLGLYKOL (GERMAN) see EJH500
METHYL GUTHION see ASH500
METHYL HEPTENONE see MKK000
6-METHYL-5-HEPTEN-2-ONE see MKK000
METHYL HEPTINE CARBONATE see MND275
METHYL HEXYL KETONE (FCC) see ODG000
Δ^1-6-α-METHYLHYDROCORTISONE see MOR500
METHYL HYDROXIDE see MDS250
METHYL-o-HYDROXYBENZOATE see MPI000
METHYL p-HYDROXYBENZOATE see HJL500
2-METHYL-3-HYDROXY-4,5-BIS(HYDROXYMETHYL)PYRIDINE HYDROCHLORIDE see PPK500
2-METHYL-3-HYDROXY-4-PYRONE see MNG750
METHYL-ISOBUTYL-CETONE (FRENCH) see HFG500
METHYLISOBUTYLKETON (DUTCH, GERMAN) see HFG500
METHYL ISOBUTYL KETONE (ACGIH,DOT) see HFG500
METHYL ISOEUGENOL (FCC) see IKR000
1-METHYL-4-ISOPROPENYLCYCLOHEXAN-3-OL see MCE750
1-1-METHYL-4-ISOPROPENYL-6-CYCLOHEXEN-2-ONE see MCD500
d-1-METHYL-4-ISOPROPENYL-6-CYCLOHEXEN-2-ONE see MCD379
p-METHYLISOPROPYL BENZENE see CQI000
1-METHYL-4-ISOPROPYLBENZENE see CQI000
1-METHYL-4-ISOPROPYL-1,3-CYCLOHEXADIENE see MLA250
1-METHYL-4-ISOPROPYLCYCLOHEXADIENE-1,3 see MLA250
1-METHYL-4-ISOPROPYLCYCLOHEXADIENE-1,4 see MCB750
2-METHYL-5-ISOPROPYL-1,3-CYCLOHEXADIENE see MCC000
6-METHYL-3-ISOPROPYLCYCLOHEXANOL see DKV150
5-METHYL-2-ISOPROPYL-2-HEXENAL see MLA300
α-METHYL-p-ISOPROPYLHYDROCINNAMALDEHYDE see COU500
2-METHYL-5-ISOPROPYLPHENOL see CCM000
2-METHYL-3-(p-ISOPROPYLPHENYL)PROPIONALDEHYDE see COU500
METHYL KETONE see ABC750
2-METHYLMERCAPTO-4,6-BIS(ISOPROPYLAMINO)-s-TRIAZINE see BKL250
4-METHYLMERCAPTO-3-METHYLPHENYL DIMETHYL THIOPHOSPHATE see FAQ999
METHYL METHANOATE see MKG750
4-METHYL-1-METHOXYBENZENE see MGP000
METHYL METHYLAMINOBENZOATE see MGQ250
METHYL N-((METHYLAMINO)CARBONYL)OXY)-ETHANIMIDO) THIOATE see MDU600
METHYL-N-METHYL ANTHRANILATE see MGQ250
METHYL 2-METHYLBUTANOATE see MLL600
METHYL 2-METHYLBUTYRATE see MLL600
METHYL-N-((METHYLCARBAMOYL)OXY)-THIOACETIMIDATE see MDU600
cis-1-METHYL-2-METHYL CARBAMOYL VINYL PHOSPHATE see DOL600
S-METHYL N-[(METHYLCARBAMOYLOOXY]THIOACETIMIDATE see MDU600
d-2-METHYL-5-(1-METHYLENENYL)-CYCLOHEXANONE see DKV175
7-METHYL-3-METHYLENE-1,6-OCTADIENE see MRZ150
1-METHYL-4-(1-METHYLETHENYL)-(S)-CYCLOHEXENE see MCC500
(R)-1-METHYL-4-(1-METHYLETHENYL)-CYCLOHEXENE see LFU000
(S)-2-METHYL-5-(1-METHYLETHENYL)-2-CYCLOHEXEN-1-ONE see MCD379
(R)-2-METHYL-5-(1-METHYLETHENYL)-2-CYCLOHEXEN-1-ONE (9CI) see MCD500
5-METHYL-2-(1-METHYLETHYL)CYCLOHEXANOL see MCF750
5-METHYL-2-(1-METHYLETHYL)-CYCLOHEXANOL (1-α,2-β,5-α) see MCG000
(1R-(1-α,2-β,5-α))-5-METHYL-2-(1-METHYLETHYL)CYCLOHEXANOL see MCG250
(R-(1α,2β,5α))-5-METHYL-2-(1-METHYLETHYL)-CYCLOHEXANOL ACETATE (9CI) see MCG750
trans-5-METHYL-2-(1-METHYLETHYL)-CYCLOHEXANONE see MCG275
1-METHYL-2-(1-METHYLETHYL)-5-NITRO-1H-IMIDAZOLE see IGH000
2-METHYL-2-(METHYLTHIO)PROPANAL-O-((METHYLAMINO)CARBONYL)OXIME see CBM500
2-METHYL-2-(METHYLTHIO)PROPIONALDEHYDE O-(METHYLCARBAMOYL)OXIME see CBM500
2-METHYL-2-(METHYLTHIO)PROPIONALDEHYDE OXIME see CBM500
2-METHYL-2-METHYLTHIO-PROPIONALDEHYD-O-(N-METHYL-CARBAMOYL)-OXIM (GERMAN) see CBM500
N-METHYL-1-NAFTYL-CARBAMAAT (DUTCH) see CBM750
METHYL NAMATE see SGM500
N-METHYL-1-NAPHTHYL-CARBAMAT (GERMAN) see CBM750
N-METHYL-1-NAPHTHYL CARBAMATE see CBM750
N-METHYL-α-NAPHTHYLCARBAMATE see CBM750
METHYL-2-NAPHTHYL KETONE see ABC500
β-METHYL NAPHTHYL KETONE see ABC500
METHYL-β-NAPHTHYL KETONE (FCC) see ABC500
N-METHYL-α-NAPHTHYLURETHAN see CBM750
METHYLNITROPHOS see DSQ000
METHYL n-NONYL ACETALDEHYDE see MQI550
4-METHYLNORVALINE see LES000
METHYL 2-OCTINATE see MND275
METHYL 2-OCTYNOATE see MND275
METHYLOL see MDS250
METHYLOLPROPANE see BPW500
METHYL OXIRANE see ECF500
METHYL OXITOL see EJH500
METHYL p-OXYBENZOATE see HJL500

2-METHYL-3-OXY-γ-PYRONE see MNG750
METHYLPARABEN (FCC) see HJL500
METHYL PARAHYDROXYBENZOATE see HJL500
METHYL PARASEPT see HJL500
4-METHYL-2-PENTANON (CZECH) see HFG500
4-METHYL-PENTAN-2-ON (DUTCH, GERMAN) see HFG500
2-METHYL-4-PENTANONE see HFG500
4-METHYL-2-PENTANONE (FCC) see HFG500
METHYL PENTYL KETONE see HBG000
4-METHYLPHENOL METHYL ETHER see MGP000
α-METHYL PHENYLACETALDEHYDE see COF000
METHYL PHENYLACETATE (FCC) see MHA500
4-METHYLPHENYL ACETATE see MNR250
p-METHYLPHENYL ACETATE see MNR250
1-((2-METHYLPHENYL)AZO)-2-NAPHTHALENAMINE see TGW250
1-(2-METHYLPHENYL)AZO-2-NAPHTHALENAMINE see TGW250
1-(2-METHYLPHENYL)AZO-2-NAPHTHYLAMINE see TGW250
METHYLPHENYLCARBINOL see PDE000
METHYL PHENYLCARBINYL ACETATE see MNT075
METHYL PHENYL ETHER see AOX750
3-METHYL-3-PHENYLGLYCIDIC ACID ETHYL ESTER see ENC000
METHYL PHENYL KETONE see ABH000
2-METHYL-3-PHENYL-2-PROPENAL see MIO000
METHYL-3-PHENYLPROPENOATE see MIO500
METHYL PIRIMIPHOS see DIN800
METHYLPREDNISOLONE see MOR500
6-α-METHYLPREDNISOLONE see MOR500
2-METHYLPROPANAL see IJS000
2-METHYL-1-PROPANAL see IJS000
2-METHYLPROPANOIC ACID see IJU000
2-METHYL PROPANOL see IIL000
2-METHYLPROPAN-1-OL see IIL000
2-METHYL-1-PROPANOL see IIL000
2-METHYLPROPIONALDEHYDE see IJS000
2-METHYLPROPIONIC ACID see IJU000
α-METHYLPROPIONIC ACID see IJU000
2-METHYLPROPIONIC ACID, ETHYL ESTER see ELS000
2-METHYLPROPYL ACETATE see IIJ000
2-METHYL-1-PROPYL ACETATE see IIJ000
2-METHYLPROPYL ALCOHOL see IIL000
1-METHYLPROPYLAMINE see BPY000
2-METHYLPROPYL BUTYRATE see BSW500
METHYL-PROPYL-CETONE (FRENCH) see PBN250
β-METHYLPROPYL ETHANOATE see IIJ000
METHYL-n-PROPYL KETONE see PBN250
METHYL PROPYL KETONE (ACGIH, DOT) see PBN250
METHYL 5-(PROPYLTHIO)-2-BENZIMIDAZOLECARBAMATE see VAD000
METHYLPROTOCATECHUALDEHYDE see VFK000
2-METHYLPYRAZINE see MOW750
N^1-(4-METHYL-2-PYRIMIDINYL)SULFANILAMIDE SODIUM SALT see SJW475
2-METHYL PYROMECONIC ACID see MNG750
METHYL 2-PYRROLYL KETONE see ADA375
METHYL SALICYLATE see MPI000
4-METHYLSALINOMYCIN see NBO600
METHYLTHEOBROMIDE see CAK500
1-METHYLTHEOBROMINE see CAK500
7-METHYLTHEOPHYLLINE see CAK500
2-METHYLTHIO-ACETALDEHYD-O-(METHYLCARBAMOYL)-OXIM (GERMAN) see MDU600
l-γ-METHYLTHIO-α-AMINOBUTYRIC ACID see MDT750
2-METHYLTHIO-4,6-BIS(ISOPROPYLAMINO)-s-TRIAZINE see BKL250
2-METHYLTHIO-PROPIONALDEHYD-O-(METHYLCARBAMOYL)-OXIM (GERMAN) see MDU600
METHYL-α-TOLUATE see MHA500
α-METHYL-α-TOLUIC ALDEHYDE see COF000
METHYL-p-TOLYL ETHER see MGP000
METHYL-p-TOLYL KETONE see MFW250
METHYLTRIMETHYLENE GLYCOL see BOS500
2-METHYLUNDECANAL see MQI550
METHYL VIOLOGEN (2+) see PAI990
METHYL ZIMATE see BJK500
METHYL ZINEB see BJK500
METHYL ZIRAM see BJK500
METHYPHENYLMETHANOL see PDE000
METICORTELONE see PMA000
METI-DERM see PMA000
METIFONATE see TIQ250
METIL CELLOSOLVE (ITALIAN) see EJH500
METILCLORPINDOL see MII500
4,4-METILENE-BIS-o-CLOROANILINA (ITALIAN) see MQY750
METILETILCHETONE (ITALIAN) see BOV250
METIL (FORMIATO di) (ITALIAN) see MKG750
METILISOBUTILCHETONE (ITALIAN) see HFG500
METILMERCAPTOFOSOKSID see DAP000
N-METIL-1-NAFTIL-CARBAMMATO (ITALIAN) see CBM750
4-METILPENTAN-2-ONE (ITALIAN) see HFG500
2-METIL-2-TIOMETIL-PROPIONALDEID-O-(N-METIL-CARBAMOIL)-OSSIMA (ITALIAN) see CBM500
METILTRIAZOTION see ASH500
METIONE see ADG375
2-METOKSY-4-ALLILOFENOL (POLISH) see EQR500
METOKSYCHLOR (POLISH) see DOB400
METOKSYETYLOWY ALKOHOL (POLISH) see EJH500
METOMIL (ITALIAN) see MDU600
METOSERPATE HYDROCHLORIDE see MQR200
2-METOSSIETANOLO (ITALIAN) see EJH500
METOX see DOB400
METOXIDON see SNN300
METOXYDE see HJL500
METRIBUZIN see AJB750
METRIFONATE see TIQ250
METRIPHONATE see TIQ250
METRISONE see MOR500
METRO TALC 4604 see TAB750
METRO TALC 4608 see TAB750
METRO TALC 4609 see TAB750
METSO 20 see SJU000
METSO BEADS 2048 see SJU000
METSO BEADS, DRYMET see SJU000
METSO PENTABEAD 20 see SJU000
METYLENU CHLOREK (POLISH) see MDR000
METYLESTER KYSELINY SALICYLOVE (CZECH) see MPI000
METYLOETYLOKETON (POLISH) see BOV250
METYLOIZOBUTYLOKETON (POLISH) see HFG500
METYLOPROPYLOKETON (POLISH) see PBN250
METYLOWY ALKOHOL (POLISH) see MDS250
METYLU BROMEK (POLISH) see BNM500
METYLU CHLOREK (POLISH) see CHX500
MEXENE see BJK500
MEYPRALGIN R/LV see SEH000
MEZENE see BJK500
MGA see MCB380
MGA 100 (STEROID) see MCB380
MGK-264 see OES000
MIBK see HFG500
MIBOLERON see MQS225
MIBOLERONE see MQS225
MICALEX see AHD650
MICOCHLORINE see CDP250
MICROCETINA see CDP250
MICRO-CHECK 12 see CBG000
MICROCRYSTALLINE WAX see PCT600
MICROTHENE see PJS750
MIERENZUUR (DUTCH) see FNA000
MIK see HFG500
MIKAMYCIN see VRF000

MILBAM see BJK500
MILBAN see BJK500
MILBEDOCE see VSZ000
MILBOL see BIO750
MILBOL 49 see BBQ500
MILCHSAURE (GERMAN) see LAG000
MIL-DU-RID see BGJ750
MILK ACID see LAG000
MILK-CLOTTING ENZYME from BACILLUS CEREUS see MQU075
MILK-CLOTTING ENZYME from ENDOTHIA PARASITICA see MQU100
MILK-CLOTTING ENZYME from MUCOR MIEHEI see MQU120
MILK-CLOTTING ENZYME from MUCOR PUSILLUS see MQU125
MILK OF MAGNESIA see MAG750
MILLER NU SET see TIX500
MILMER see BLC250
MILOGARD see PMN850
MILTON see SHU500
MINERAL OIL see MQV750
MINERAL OIL, WHITE (FCC) see MQV750
MINOPHAGEN A see AQW000
MINTEZOL see TEX000
MINUS see SEH000
MINZOLUM see TEX000
MIRACLE see DFY600
MISTRON 2SC see TAB750
MISTRON FROST P see TAB750
MISTRON RCS see TAB750
MISTRON STAR see TAB750
MISTRON SUPER FROST see TAB750
MISTRON VAPOR see TAB750
MITIGAN see BIO750
MITION see CKM000
MIXTURE OF p-METHENOLS see TBD500
MK-188 see RBF100
MK 360 see TEX000
MK 933 see ITD875
MMA see MGQ250
MOCA see MQY750
MODANE SOFT see DJL000
MODIFIED HOP EXTRACT. see HGK750
MODIFIED POLYACRYLAMIDE RESINS see MRA075
MODOCOLL 1200 see SFO500
MOENOMYCIN see MRA250
MOENOMYCIN A see MRA250
MOLASSES ALCOHOL see EFU000
MOLATOC see DJL000
MOLCER see DJL000
MOLDEX see HJL500
MOLECULAR CHLORINE see CDV750
MOL-IRON see FBO000
MOLLAN O see BJS000
MOLOFAC see DJL000
MOLOL see MQV750
MOLTEN ADIPIC ACID see AEN250
MOLURAME see BJK500
MOMO-tert-BUTYLHYDROQUINONE see BRM500
MON 0573 see PHA500
MONAMID 150-LW see BKE500
MONASIRUP see PMQ750
MONATE see MRL750
MONAWET MD 70E see DJL000
MONELAN see MRE225
MONELGIN see OAV000
MONENSIC ACID see MRE225
MONENSIN A see MRE225
MONENSIN (USDA) see MRE225
MONITAN see PKG250
MONOAETHANOLAMIN (GERMAN) see MRH500
MONOAMMONIUM CARBONATE see ANB250
MONOAMMONIUM GLUTAMATE see MRF000
MONOAMMONIUM l-GLUTAMATE see MRF000
MONOAMMONIUM GLYCYRRHIZINATE see AMY700
MONOBROMOMETHANE see BNM500
MONOBUTYL GLYCOL ETHER see BQQ500
MONOCALCIUM PHOSPHATE see CAW110
MONOCHLOORAZIJNZUUR (DUTCH) see CEA000
MONOCHLORACETIC ACID see CEA000
MONOCHLORESSIGSAEURE (GERMAN) see CEA000
MONOCHLOROACETIC ACID see CEA000
MONOCHLOROETHANOIC ACID see CEA000
MONOCHLOROMETHANE see CHX500
MONOCHLOROPENTAFLUOROETHANE (DOT) see CJI500
MONOCIL 40 see DOL600
MONOCRON see DOL600
MONOCROTOPHOS (ACGIH) see DOL600
MONO- and DIGLYCERIDES see MRH215
MONO- and DIGLYCERIDES, MONOSODIUM PHOSPHATE derivatives see MRH218
MONO and DIGLYCERIDES, SODIUM SULFOACETATE derivatives see SJZ050
MONO-, DI-, and TRIPOTASSIUM CITRATE see MRH225
MONO-, DI-, and TRISODIUM CITRATE see MRH230
MONO-, DI-, and TRISTEARYL CITRATE see MRH235
MONOETHANOLAMINE see MRH500
MONO(2-ETHYLHEXYL)SULFATE SODIUM SALT see TAV750
MONOFURACIN see NGE500
MONOGLYCERIDE CITRATE see MRI300
MONOHYDROXYMETHANE see MDS250
MONOISOPROPYL CITRATE see MRI785
N-MONOMETHYLAMIDE OF O,O-DIMETHYLDITHIOPHOSPHORYLACETIC ACID see DSP400
MONOMETHYL ETHER of ETHYLENE GLYCOL see EJH500
MONOPLEX DBS see DEH600
MONOPOTASSIUM GLUTAMATE see MRK500
MONOPOTASSIUM l-GLUTAMATE (FCC) see MRK500
MONOPOTASSIUM PHOSPHATE see PLQ405
MONOPROPYLENE GLYCOL see PML000
MONOPYRROLE see PPS250
MONOSAN see DFY600
MONOSODIOGLUTAMMATO (ITALIAN) see MRL500
MONOSODIUM ACID METHANEARSONATE see MRL750
MONOSODIUM ACID METHARSONATE see MRL750
MONOSODIUM ASCORBATE see ARN125
MONOSODIUM DIHYDROGEN PHOSPHATE see SGL500
MONOSODIUM GLUCONATE see SHK800
MONOSODIUM GLUTAMATE see MRL500
MONOSODIUM-l-GLUTAMATE (FCC) see MRL500
α-MONOSODIUM GLUTAMATE see MRL500
MONOSODIUM METHANEARSONATE see MRL750
MONOSODIUM METHANEARSONIC ACID see MRL750
MONOSODIUM METHYLARSONATE see MRL750
MONOSODIUM NOVOBIOCIN see NOB000
MONOSODIUM PHOSPHATE see SGL500
MONOSODIUM PHOSPHATE DERIVATIVES of MONO- and DIGLYCERIDES see MRH218
MONOSORB XP-4 see SGL500
MONOSTEARIN see OAV000
MONSANTO CP 47114 see DSQ000
MONTANE 60 see SKU700
MONTANOX 80 see PKG250
MONTAR see PJL750
MONTEBAN see NBO600
MONTMORILLONITE see BAV750
MONTROSE PROPANIL see DGI000
MOON see GGA000
MOPARI see DRK200
MORANTEL TARTRATE see MRN260
MORANTREL TARTRATE see MRN260
MORBOCID see FMV000
MOREPEN see AOD125
MORONAL see NOH500

NCI-C00497 see DOB400
NCI-C00511 see DFF900
NCI-C00522 see EIY500
NCI-C00566 see BCJ250
NCI-C01729 see TNX000
NCI-C02073 see EFE000
NCI-C02084 see SIQ500
NCI-C02119 see USS000
NCI-C02722 see TGC000
NCI-C02733 see CAK500
NCI-C02799 see FMV000
NCI-C02813 see PIX250
NCI-C03134 see EFU000
NCI-C03372 see IAQ000
NCI-C03510 see API750
NCI-C03598 see BFW750
NCI-C03827 see DQD400
NCI-C04240 see TGH000
NCI-C04580 see TBQ250
NCI-C04591 see CBV500
NCI-C04897 see PLZ000
NCI-C06008 see TAB750
NCI-C06111 see BDX500
NCI-C06508 see BDX000
NCI-C07272 see TGK750
NCI-C08640 see CBM500
NCI-C08651 see FAQ999
NCI-C08662 see CNU750
NCI-C08673 see DCL000
NCI-C08695 see DUK800
NCI-C50011 see BCP250
NCI-C50088 see EJN500
NCI-C50099 see ECF500
NCI-C50102 see MDR000
NCI-C50168 see CNU000
NCI-C50191 see SON500
NCI-C50204 see TAV750
NCI-C50384 see EFT000
NCI-C50395 see GLU000
NCI-C50419 see LIA000
NCI-C50442 see BJK500
NCI-C50453 see EQR500
NCI-C50464 see AGJ250
NCI-C50475 see AEX250
NCI-C50715 see MCB000
NCI-C50748 see AQQ500
NCI-C52733 see BJS000
NCI-C54717 see ISV000
NCI-C54728 see DTD800
NCI-C54808 see ARN000
NCI-C54831 see TIQ250
NCI-C54933 see PAX250
NCI-C55163 see CCP250
NCI-C55323 see BKE500
NCI-C55425 see GFQ000
NCI-C55505 see SEM000
NCI-C55561 see TBX250
NCI-C55572 see LFU000
NCI-C55685 see PDE000
NCI-C55812 see MIP750
NCI-C55823 see GEM000
NCI-C55878 see BOV000
NCI-C55890 see HHR500
NCI-C56064 see NGE500
NCI-C56086 see AOD125
NCI-C56111 see CMP969
NCI-C56133 see BAY500
NCI-C56177 see FPG000
NCI-C56291 see BSU250
NCI-C56326 see AAG250
NCI-C56348 see DTC800
NCI-C56473 see HOH500
NCI-C56484 see OGW000
NCI-C56508 see HMY000
NCI-C56575 see CAL000
NCI-C56597 see SNH000
NCI-C56600 see SNJ500
NCI-C60015 see COG000
NCI-C60048 see DJX000
NCI-C60071 see MRL750
NCI-C60195 see PEC500
NCI-C60231 see CEA000
NCI-C60286 see PKG250
NCI-C60297 see CNV000
NCI-C60402 see EEA500
NCI-C60560 see TCR750
NCI-C60571 see HEN000
NCI C60582 see PKQ250
NCI-C60902 see TLP500
NCI-C60946 see AFW750
NCI-C60968 see IJS000
NCI-C60979 see IKQ000
NCI-C61018 see NMW500
NCI-C61029 see PMT750
NCI-C61143 see MAU250
NCI-C61176 see ARA500
NCI-CO0077 see CBG000
NCI-CO4546 see TIO750
NDGA see NBR000
NDRC-143 see AHJ750
NEANTINE see DJX000
NEASINA see SNJ500
NEAT OIL of SWEET ORANGE see OGY000
NECATORINA see CBY000
NECATORINE see CBY000
NECCANOL SW see DXW200
NEDCIDOL see DCL000
NEFCO see NGE500
NEFROSUL see SNH900
NEFTIN see NGG500
NEGUVON see TIQ250
NEKAL WT-27 see DJL000
NEMA see TBQ250
NEMACUR see FAK000
NEMAPAN see TEX000
NEMATOLYT see PAG500
NEMICIDE see NBU500
NENDRIN see EAT500
NEOBIOTIC see NCG000
NEOCIDOL see DCL000
NEOCOMPENSAN see PKQ250
NEO-CULTOL see MQV750
neo-FAT 8 see OCY000
NEO-FAT 10 see DAH400
NEO-FAT 12 see LBL000
NEO-FAT 18-61 see SLK000
NEO-FAT 90-04 see OHU000
NEO-FAT 92-04 see OHU000
NEO-FAT 18-S see SLK000
NEO-HOMBREOL see TBG000
NEOLOID see CCP250
NEO-MANTLE CREME see NCG000
NEOMIX see NCG000
NEOMYCINE SULFATE see NCG000
NEOMYCIN SULFATE see NCG000
NEOMYCIN SULPHATE see NCG000
NEO-SCABICIDOL see BBQ500
NEOSTREPAL see SNN300
NEOSTREPSAN see TEX250
NEPHIS see EIY500
NEPHOCARP see TNP250
NEQUINATE see NCN600
NERACID see CBG000
NERAL see DTC800
NERKOL see DRK200
NEROL (FCC) see DTD200
NEROLIDOL see NCN700
NERVANAID B LIQUID see EIV000

NSC 9169 see HOI000
NSC-9704 see PMH500
NSC 10023 see PLZ000
NSC 14083 see SLW500
NSC-19987 see MOR500
NSC 60380 see DUJ200
NSC-70731 see LGD000
NSC-82261 see GCS000
NSC-177023 see NBU500
NSC 190935 see CJJ250
NSC 195022 see BEP500
NSC 195106 see FAK000
NSC-528986 see AIV500
NTA see SIP500
NU-BAIT II see MDU600
NUCIDOL see DCL000
NUDRIN see MDU600
NUJOL see MQV750
NULLAPON B see EIV000
NULLAPON BFC CONC see EIV000
NULLAPON BF-78 see EIV000
NUOPLAZ DOP see BJS000
NUSYN-NOXFISH see PIX250
NUTMEG OIL see NOG500
NUTMEG OIL, EAST INDIAN see NOG500
NUTRASWEET see ARN825
NUTRIFOS STP see TNM250
NUTROSE see SFQ000
NUVA see DRK200
NUVACRON see DOL600
NUVANOL see DSQ000
NUVAPEN see AIV500
NYACOL see SCH000
NYACOL 830 see SCH000
NYACOL 1430 see SCH000
NYCOLINE see PJT200
NYMCEL S see SFO500
NYSTAN see NOH500
NYSTATIN see NOH500
NYSTATINE see NOH500
NYSTAVESCENT see NOH500
NYTAL see TAB750

OBSTON see DJL000
OCHRE see IHD000
OCIMUM BASILICUM OIL see BAR250
1,2,4,5,6,7,8,8-OCTACHLOOR-3a,4,7,7a-TETRAHYDRO-4,7-endo-METHANO-INDAAN (DUTCH) see CDR750
OCTACHLOR see CDR750
OCTACHLOROCAMPHENE see THH750
OCTACHLORODIHYDRODICYCLOPENTADIENE see CDR750
1,2,4,5,6,7,8,8-OCTACHLORO-2,3,3a,4,7,7a-HEXAHYDRO-4,7-METHANO-1H-INDENE see CDR750
1,2,4,5,6,7,8,8-OCTACHLORO-2,3,3a,4,7,7a-HEXAHYDRO-4,7-METHANOINDENE see CDR750
1,2,4,5,6,7,8,8-OCTACHLORO-3a,4,7,7a-HEXAHYDRO-4,7-METHYLENE INDANE see CDR750
OCTACHLORO-4,7-METHANOHYDROINDANE see CDR750
OCTACHLORO-4,7-METHANOTETRAHYDROINDANE see CDR750
1,2,4,5,6,7,8,8-OCTACHLORO-4,7-METHANO-3a,4,7,7a-TETRAHYDROINDANE see CDR750
1,2,4,5,6,7,8,8-OCTACHLORO-3a,4,7,7a-TETRAHYDRO-4,7-METHANOINDAN see CDR750
1,2,4,5,6,7,8,8-OCTACHLORO-3a,4,7,7a-TETRAHYDRO-4,7-METHANOINDANE see CDR750
1,2,4,5,6,7,10,10-OCTACHLORO-4,7,8,9-TETRAHYDRO-4,7-METHYLENEINDANE see CDR750
1,2,4,5,6,7,8,8-OCTACHLOR-3a,4,7,7a-TETRAHYDRO-4,7-endo-METHANO-INDAN (GERMAN) see CDR750
OCTACIDE 264 see OES000
cis,cis-9,12-OCTADECADIENOIC ACID see LGG000
9,12-OCTADECADIENOIC ACID see LGG000
cis-9,cis-12-OCTADECADIENOIC ACID see LGG000
OCTADECANOIC ACID see SLK000
OCTADECANOIC ACID, BUTYL ESTER see BSL600
OCTADECANOIC ACID, MONOESTER with 1,2,3-PROPANETRIOL see OAV000
OCTADECANOIC ACID, SODIUM SALT see SJV500
OCTADECANOL see OAX000
1-OCTADECANOL see OAX000
n-OCTADECANOL see OAX000
9,10-OCTADECENOIC ACID see OHU000
cis-9-OCTADECENOIC ACID see OHU000
cis-OCTADEC-9-ENOIC ACID see OHU000
cis-Δ^9-OCTADECENOIC ACID see OHU000
9-OCTADECENOIC ACID CALCIUM SALT see CAU300
OCTA DECYL ALCOHOL see OAX000
n-OCTADECYL ALCOHOL see OAX000
OCTADECYLAMINE see OBC000
N-OCTADECYLAMINE see OBC000
N-OCTADECYL-N-BENZYL-N,N-DIMETHYLAMMONIUM-CHLORIDE see DTC600
OCTADECYLDIMETHYLBENZYLAMMONIUM CHLORIDE see DTC600
OCTAFLUOROCYCLOBUTANE (DOT) see CPS000
OCTA-KLOR see CDR750
γ-OCTALACTONE see OCE000
OCTALENE see AFK250
1-OCTANAL see OCO000
OCTANALDEHYDE see OCO000
OCTAN n-BUTYLU (POLISH) see BPU750
OCTAN ETYLU (POLISH) see EFR000
OCTAN KOBALTNATY (CZECH) see CNA500
OCTANOIC ACID see OCY000
OCTANOIC ACID ALLYL ESTER see AGM500
OCTANOIC ACID, ETHYL ESTER see ENY000
OCTANOIC ACID-2-PROPENYL ESTER see AGM500
OCTANOL see OEI000
1-OCTANOL (FCC) see OEI000
3-OCTANOL see OCY100
OCTANOL-3 see OCY100
n-OCTANOL see OEI000
1-OCTANOL ACETATE see OEG000
OCTANOLIDE-1,4 see OCE000
2-OCTANONE see ODG000
n-OCTANYL ACETATE see OEG000
trans-2-OCTEN-1-AL see ODQ800
cis-3-OCTEN-1-OL see ODW025
1-OCTEN-3-YL ACETATE see ODW030
1-OCTEN-3-YL BUTYRATE see ODW040
OCTIC ACID see OCY000
OCTILIN see OEI000
n-OCTOIC ACID see OCY000
OCTOIL see BJS000
OCTOWY ALDEHYD (POLISH) see AAG250
OCTOWY KWAS (POLISH) see AAT250
OCTYL ACETATE see OEG000
1-OCTYL ACETATE see OEG000
3-OCTYL ACETATE see OEG100
n-OCTYL ACETATE see OEG000
OCTYL ALCOHOL see OEI000
OCTYL ALCOHOL ACETATE see OEG000
OCTYL ALCOHOL, NORMAL-PRIMARY see OEI000
n-OCTYL ALDEHYDE see OCO000
N-OCTYL BICYCLOHEPTENE DICARBOXIMIDE see OES000
N-OCTYLBICYCLO-(2.2.1)-5-HEPTENE-2,3-DICARBOXIMIDE see OES000
OCTYL CARBINOL see NNB500
OCTYL FORMATE see OEY100
OCTYL GALLATE see OFA000
n-OCTYLIC ACID see OCY000
OCTYL PHTHALATE see BJS000
ODA see SAB500
ODB see EDP000
ODORLESS LIGHT PETROLEUM HYDROCARBONS see OGI200

PENTANEN (DUTCH) see PBK250
PENTANI (ITALIAN) see PBK250
PENTANOIC ACID see VAQ000
n-PENTANOIC ACID see VAQ000
4-PENTANOLIDE see VAV000
2-PENTANONE see PBN250
PENTAPHENATE see SJA000
PENTAPOTASSIUM TRIPHOSPHATE see PLW400
PENTASODIUM TRIPHOSPHATE see SKN000
PENTASOL see PAX250
PENTIFORMIC ACID see HEU000
PENTREX see AIV500
PENTREXL see AIV500
PENTYLCARBINOL see HFJ500
α-PENTYLCINNAMALDEHYDE see AOG500
PENTYLFORMIC ACID see HEU000
PENTYL HEXANOATE see IHU100
PENWAR see PAX250
PEPPERMINT CAMPHOR see MCF750
PEPPERMINT OIL see PCB250
PERAGAL ST see PKQ250
PERANDREN see TBG000
PERATOX see PAX250
PERAWIN see TBQ250
PERCHLOORETHYLEEN, PER (DUTCH) see TBQ250
PERCHLOR see TBQ250
PERCHLORAETHYLEN, PER (GERMAN) see TBQ250
PERCHLORETHYLENE see TBQ250
PERCHLORETHYLENE, PER (FRENCH) see TBQ250
PERCHLOROETHYLENE (ACGIH, DOT) see TBQ250
PERCHLOROMETHANE see CBY000
PERCHLORURE de FER (FRENCH) see FAU000
PERCIPITATED CALCIUM PHOSPHATE see CAW120
PERCLENE see TBQ250
PERCLOROETILENE (ITALIAN) see TBQ250
PERCOLATE see PHX250
PERCOSOLVE see TBQ250
PERCUTACRINE see PMH500
PERFECTA see MQV750
PERFECTHION see DSP400
PERFLUOROCYCLOBUTANE see CPS000
PERGACID VIOLET 2B see BFN250
PER-GLYCERIN see LAM000
PERHYDROGERANIOL see DTE600
PERHYDROL see HIB000
PERISTON see PKQ250
PERK see TBQ250
PERKLONE see TBQ250
PERLITE see PCJ400
PERM-A-CHLOR see TIO750
PERMACIDE see PAX250
PERMAGARD see PAX250
PERMA KLEER 50 CRYSTALS see EIV000
PERMA KLEER 50 CRYSTALS DISODIUM SALT see EIX500
PERMA KLEER TETRA CP see EIV000
PERMASAN see PAX250
PERMASEAL see PJH500
PERMATOX DP-2 see PAX250
PERMETHRIN (USDA) see AHJ750
PERMETRINA (PORTUGUESE) see AHJ750
PERMETRIN (HUNGARIAN) see AHJ750
PERMITE see PAX250
PERNAEMON see VSZ000
PERNAEVIT see VSZ000
PERNIPURON see VSZ000
PERONE see HIB000
PEROSSIDO di BENZOILE(ITALIAN) see BDS000
PEROSSIDO di IDROGENO (ITALIAN) see HIB000
PEROXAN see HIB000
PEROXIDE see HIB000
PEROXYDE de BENZOYLE (FRENCH) see BDS000
PEROXYDE d'HYDROGENE (FRENCH) see HIB000
PEROXYDISULFURIC ACID DIPOTASSIUM SALT see DWQ000
PERSADOX see BDS000
PERSEC see TBQ250
PERSULFATE de SODIUM (FRENCH) see SJE000
PERSULFEN see SNN300
PERUSCABIN see BCM000
PERUVIAN BALSAM see BAE750
PESTMASTER see EIY500
PESTMASTER EDB-85 see EIY500
PESTMASTER (OBS.) see BNM500
PETERSILIENSAMEN OEL (GERMAN) see PAL750
PETITGRAIN OIL, PARAGUAY TYPE see PCR100
PETNAMYCETIN see CDP250
PETROGALAR see MQV750
PETROHOL see INJ000
PETROLATUM see PCR200
PETROLATUM, LIQUID see MQV750
PETROLEUM BENZIN see BCD500
PETROLEUM DISTILLATES (NAPHTHA) see BCD500
PETROLEUM ETHER (DOT) see BCD500
PETROLEUM NAPHTHA (DOT) see BCD500
PETROLEUM SPIRIT (DOT) see BCD500
PETROLEUM WAX see PCT600
PETROLEUM WAX, SYNTHETIC (FCC) see PCT600
PETZINOL see TIO750
PEVITON see NDT000
PFEFFERMINZ OEL (GERMAN) see PCB250
PFIKLOR see PLA500
PFIZERPEN A see AIV500
PG 12 see PML000
PGF2-α THAM see POC750
PGF2-α TRIS SALT see POC750
PGF2-α TROMETHAMINE see POC750
PH 60-40 see CJV250
PHANAMIPHOS see FAK000
PHAROS 100.1 see SMR000
PHASOLON see BDJ250
α-PHELLANDRENE (FCC) see MCC000
PHENACETALDEHYDE DIMETHYL ACETAL see PDX000
PHENACIDE see THH750
PHENASAL see DFV400
PHENATOX see THH750
PHENETHANOL see PDD750
2-PHENETHYL ACETATE see PFB250
β-PHENETHYL ACETATE see PFB250
PHENETHYL ALCOHOL see PDD750
2-PHENETHYL ALCOHOL see PDD750
α-PHENETHYL ALCOHOL see PDE000
β-PHENETHYL ALCOHOL see PDD750
PHENETHYLCARBAMID (GERMAN) see EFE000
PHENETHYL ESTER ISOVALERIC ACID see PDF775
PHENETHYL ISOBUTYRATE see PDF750
PHENETHYL ISOVALERATE see PDF775
2-PHENETHYL 2-METHYLBUTYRATE see PDF790
PHENETHYL PHENYLACETATE see PDI000
PHENETHYL SALICYLATE see PDK200
p-PHENETOLCARBAMID (GERMAN) see EFE000
p-PHENETOLCARBAMIDE see EFE000
p-PHENETOLECARBAMIDE see EFE000
p-PHENETYLUREA see EFE000
PHENITROTHION see DSQ000
PHENOCHLOR see PJL750
PHENOLCARBINOL see BDX500
PHENOX see DFY600
3-PHENOXYBENZYL (±)-3-(2,2-DICHLOROVINYL)-2,2-DIMETHYLCYCLOPROPANECARBOXYLATE see AHJ750
PHENOXYETHYL ISOBUTYRATE see PDS900
(3-PHENOXYPHENYL)METHYL-3-(2,2-DICHLORETHENYL)-2,2-DIMETHYLCYCLOPROPANECARBOXYLATE see AHJ750
PHENVALERATE see FAR100
PHENYLACETALDEHYDE (FCC) see BBL500
PHENYLACETALDEHYDE DIMETHYL ACETAL see PDX000
PHENYLACETIC ACID see PDY850
omega-PHENYLACETIC ACID see PDY850

RHODIA RP 11974 see BDJ250
RHODINOL (FCC) see DTF400
RHODINOL see CMT250, RLU800
RHODINOL ACETATE see RHA000
RHODINYL ACETATE see RHA000
RHODINYL FORMATE see RHA500
RHODOCIDE see EMS500
RHODOPAS M see AAX250
RHYUNO OIL see SAD000
RIBIPCA see RIK000
RIBODERM see RIK000
RIBOFLAVIN see RIK000
RIBOFLAVINE see RIK000
RIBOFLAVINEQUINONE see RIK000
RIBOFLAVIN 5′-PHOSPHATE ESTER MONOSODIUM SALT see RIF500
RIBOFLAVIN 5′-PHOSPHATE SODIUM see RIF500
RICE BRAN WAX see RJF800
RICHAMIDE 6310 see BKE500
RICHONATE 1850 see DXW200
RICHONOL C see SON500
RICIFON see TIQ250
RICINUS OIL see CCP250
RICIRUS OIL see CCP250
RICKETON see CMC750
RICON 100 see SMR000
RICYCLINE see TBX250
RIKEMAL S 250 see SKU700
RIMIDIN see FAK100
RIOMITSIN see HOH500
RIPERCOL-L see NBU500
RISELECT see DGI000
RISTAT see HMY000
RITSIFON see TIQ250
RIVOMYCIN see CDP250
RL-50 see MRL500
RO 7-1554 see IGH000
RO-AMPEN see AIV500, AOD125
ROBENIDINE see RLK890
ROBIMYCIN see EDH500
ROCHELLE SALT see SJK385
ROCK CANDY see SNH000
ROCK SALT see SFT000
RO-CYCLINE see TBX250
RODANIN S-62 (CZECH) see IAQ000
RODINOL see CMT250, RLU800
RODOCID see EMS500
ROGODIAL see DSP400
ROGOR see DSP400
ROGUE see DGI000
ROLAMID CD see BKE500
ROLL-FRUCT see CDS125
ROMAN VITRIOL see CNP250
ROMPHENIL see CDP250
ROMPUN see DMW000
ROMULGIN O see PKG250
RONILAN see RMA000
RONNEL (ACGIH) see TIR000
ROP 500 F see GIA000
ROPTAZOL see NGG500
ROSANIL see DGI000
ROSCOSULF see SNN300
ROSE GERANIUM OIL ALGERIAN see GDA000
ROSEMARIE OIL see RMU000
ROSEMARY OIL see RMU000
ROSEN OEL (GERMAN) see RNA000
ROSE OIL see RNA000
ROSMARIN OIL (GERMAN) see RMU000
ROTATE see DQM600
ROTERSEPT see BIM250
ROTOX see BNM500
ROUGE see IHD000
ROVRAL see GIA000
ROXARSONE (USDA) see HMY000
ROXION U.A. see DSP400
ROYALTAC see DAI600
ROZTOZOL see CKM000
RP 2990 see TEX250
RP7293 see VRF000
RP 8167 see EMS500
RP 26019 see GIA000
RS 141 see CJJ250
RUBBER HYDROCHLORIDE see PJH500
RUBBER HYDROCHLORIDE POLYMER see PJH500
RUBESOL see VSZ000
RUBIGAN see FAK100
RUBIGO see IHD000
RUBITOX see BDJ250
RUBRAMIN see VSZ000
RUBRIPCA see VSZ000
RUBROCITOL see VSZ000
RUE OIL and HERB see RQU750
RUM ETHER see EOB050
RUNA RH20 see TGH000
RUTILE see TGH000
RYOMYCIN see HOH500
RYZELAN see OJY100

S115 see NDT000
S 276 see EKF000
S 75M see SFO500
S 112A see DSQ000
S 1752 see FAQ999
S-3151 see AHJ750
S 5602 see FAR100
S 5660 see DSQ000
S 10165 see DGI000
SA see SAI000
SA 111 see SNJ500
SACARINA see BCE500
SACCAHARIMIDE see BCE500
SACCHARIN see SAB500
SACCHARINA see BCE500
SACCHARIN ACID see BCE500
SACCHARINE see BCE500
SACCHARINE SOLUBLE see SAB500
SACCHARINNATRIUM see SAB500
SACCHARINOL see BCE500
SACCHARINOSE see BCE500
SACCHARIN, SODIUM see SAB500
SACCHARIN, SODIUM SALT see SAB500
SACCHARIN SOLUBLE see SAB500
SACCHAROIDUM NATRICUM see SAB500
SACCHAROL see BCE500
SACCHAROSE see SNH000
SACCHARUM see SNH000
SACHSISCHBLAU see DXE400
SADH see DQD400
SADOFOS see CBP000
SADOPHOS see CBP000
SAFFLOWER OIL see SAC000
SAFFLOWER OIL (UNHYDROGENATED) (FCC) see SAC000
SAFFRON see SAC100
SAFROL see SAD000
SAFROLE see SAD000
SAFROLE MF see SAD000
SAFROTIN see MKA000
SAGE OIL see SAE500
SAGE OIL, DALMATIAN TYPE see SAE500
SAGE OIL, SPANISH TYPE see SAE000
SALACHLOR see SHJ000
SAL AMMONIA see ANE500
SAL AMMONIAC see ANE500
SALBEI OEL (GERMAN) see SAE000, SAE500
SALICYLALDEHYDE see EOB050
SALICYLIC ACID see SAI000
SALICYLIC ACID, ISOBUTYL ESTER see IJN000
SALICYLIC ACID, ISOPENTYL ESTER see IME000

TOPAZONE see NGG500
TOPEX see BDS000
TOP FLAKE see SFT000
TOP FORM WORMER see TEX000
TOPICHLOR 20 see CDR750
TOPICLOR see CDR750
TOPICLOR 20 see CDR750
TOPICYCLINE see TBX250
TOPITRACIN see BAC250
TORAK see DBI099
TORDON see AMU250
TORDON 10K see AMU250
TORDON 22K see AMU250
TORDON 101 MIXTURE see AMU250
TORMONA see TIW500
TORQUE see BLU000
TOSTRIN see TBG000
TOTACILLIN see AIV500
TOTALCICLINA see AIV500
TOTAPEN see AIV500
TOTOMYCIN see TBX250
TOXADUST see THH750
TOXAFEEN (DUTCH) see THH750
TOXAKIL see THH750
TOXAPHEN (GERMAN) see THH750
TOXAPHENE see THH750
TOXICHLOR see CDR750
TOXON 63 see THH750
TOXYPHEN see THH750
TP see TBG000
2,4,5-TP see TIX500
TPN (pesticide) see TBQ750
TPTH see HON000
TR 201 see SMR000
TRAFARBIOT see AOD125
TRAGACANTH see THJ250
TRAGACANTH GUM see THJ250
TRAGAYA see SEH000
TRAMISOL see NBU500
TRAMISOLE see NBU500
TRANSAMINE see DFY600, TIW500
TRAUMANASE see BMN775
TRAVEX see SFS000
TREFANOCIDE see DUV600
TREFICON see DUV600
TREFLAM see DUV600
TREFLAN see DUV600
TREFLANOCIDE ELANCOLAN see DUV600
TRENBOLONE see THL600
TREOMICETINA see CDP250
TREPENOL WA see SON500
TRIACETIN (FCC) see THM500
TRIACETYL GLYCERIN see THM500
TRIAD see TIO750
TRIADIMEFON see CJO250
TRIAETHANOLAMIN-NG see TKP500
2,4,6-TRIAMINO-s-TRIAZINE see MCB000
TRIASOL see TIO750
TRIATOMIC OXYGEN see ORW000
TRIAZINE A 384 see BJP000
TRIAZINE A 1294 see PMC325
TRIBASIC SODIUM PHOSPHATE see TNM250
TRIBUTON see DFY600, TIW500
S,S,S-TRIBUTYL PHOSPHOROTRITHIOATE see BSH250
S,S,S-TRIBUTYL TRITHIOPHOSPHATE see BSH250
TRIBUTYRIN see TIG750
TRIBUTYROIN see TIG750
TRICALCIUM PHOSPHATE see CAW120
TRICALCIUM SILICATE see TIH600
TRICARBALLYLIC ACID-β-ACETOXYTRIBUTYL ESTER see ADD750
TRICARBAMIX Z see BJK500
2,2,2-TRICHLOOR-1,1-BIS(4-CHLOOR FENYL)-ETHANOL (DUTCH) see BIO750
TRICHLOORETHEEN (DUTCH) see TIO750
TRICHLOORETHYLEEN, TRI (DUTCH) see TIO750
(2,4,5-TRICHLOOR-FENOXY)-AZIJNZUUR (DUTCH) see TIW500
2-(2,4,5-TRICHLOOR-FENOXY)-PROPIONZUUR (DUTCH) see TIX500
O-(2,4,5-TRICHLOOR-FENYL)-O,O-DIMETHYL-MONOTHIOFOSFAAT (DUTCH) see TIR000
TRICHLOORFON (DUTCH) see TIQ250
TRICHLORAETHEN (GERMAN) see TIO750
TRICHLORAETHYLEN, TRI (GERMAN) see TIO750
TRICHLORAN see TIO750
1,1,1-TRICHLOR-2,2-BIS(4-CHLORPHENYL)-AETHANOL (GERMAN) see BIO750
2,2,2-TRICHLOR-1,1-BIS(4-CHLOR-PHENYL)-AETHANOL (GERMAN) see BIO750
1,1,1-TRICHLOR-2,2-BIS(4-METHOXY-PHENYL)-AETHAN (GERMAN) see DOB400
TRICHLORETHENE (FRENCH) see TIO750
TRICHLORETHYLENE, TRI (FRENCH) see TIO750
TRICHLORFON (USDA) see TIQ250
N-(TRICHLOR-METHYLTHIO)-PHTHALIMID (GERMAN) see CBG000
3,5,6-TRICHLORO-4-AMINOPICOLINIC ACID see AMU250
1,1,1-TRICHLORO-2,2-BIS(p-ANISYL)ETHANE see DOB400
2,2,2-TRICHLORO-1,1-BIS(4-CHLOROPHENYL)-ETHANOL (FRENCH) see BIO750
2,2,2-TRICHLORO-1,1-BIS(4-CLORO-FENIL)-ETANOLO (ITALIAN) see BIO750
1,1,1-TRICHLORO-2,2-BIS(p-METHOXYPHENOL)ETHANOL see DOB400
1,1,1-TRICHLORO-2,2-BIS(p-METHOXYPHENYL)ETHANE see DOB400
2,4,5-TRICHLORO-α-(CHLOROMETHYLENE)BENZYL PHOSPHATE see TBW100
1,2,4-TRICHLORO-5-((4-CHLOROPHENYL)SULFONYL)-BENZENE see CKM000
2,2,2-TRICHLORO-1,1-DI-(4-CHLOROPHENYL)ETHANOL see BIO750
1,1,1-TRICHLORO-2,2-DI(4-METHOXYPHENYL)ETHANE see DOB400
TRICHLOROETHENE see TIO750
TRICHLOROETHYLENE see TIO750
1,2,2-TRICHLOROETHYLENE see TIO750
1,1′-(2,2,2-TRICHLOROETHYLIDENE)BIS(4-METHOXYBENZENE) see DOB400
((2,2,2-TRICHLORO-1-HYDROXYETHYL) DIMETHYLPHOSPHONATE) see TIQ250
2,2,2-TRICHLORO-1-HYDROXYETHYL-PHOSPHONATE, DIMETHYL ESTER see TIQ250
TRICHLOROMETAFOS see TIR000
N-TRICHLOROMETHYLMERCAPTO-4-CYCLOHEXENE-1,2-DICARBOXIMIDE see CBG000
N-(TRICHLOROMETHYLMERCAPTO)-Δ^4-TETRAHYDROPHTHALIMIDE see CBG000
N-TRICHLOROMETHYLTHIOCYCLOHEX-4-ENE-1,2-DICARBOXIMIDE see CBG000
N-((TRICHLOROMETHYL)THIO)-4-CYCLOHEXENE-1,2-DICARBOXIMIDE see CBG000
N-TRICHLOROMETHYLTHIO-cis-Δ^4-CYCLOHEXENE-1,2-DICARBOXIMIDE see CBG000
TRICHLOROMETHYLTHIO-1,2,5,6-TETRAHYDROPHTHALAMIDE see CBG000
N-((TRICHLOROMETHYL)THIO)TETRAHYDROPHTHALIMIDE see CBG000
N-TRICHLOROMETHYLTHIO-3A,4,7,7A-TETRAHYDROPHTHALIMIDE see CBG000
2,4,5-TRICHLOROPHENOL, O-ESTER WITH O,O-DIMETHYL PHOSPHOROTHIOATE see TIR000
2,4,5-TRICHLOROPHENOXYACETIC ACID see TIW500
2-(2,4,5-TRICHLOROPHENOXY)PROPIONIC ACID see TIX500
2,4,5-TRICHLOROPHENOXY-α-PROPIONIC ACID see TIX500

VITAMIN A1 see REU000
VITAMIN A ACETATE see REZ000
trans-VITAMIN A ACETATE see REZ000
VITAMIN A1 ALCOHOL see REU000
all-trans-VITAMIN A ALCOHOL see REU000
VITAMIN A ALCOHOL ACETATE see REZ000
VITAMIN Bc see FMT000
VITAMIN B^1 see TET000
VITAMIN B2 see RIK000
VITAMIN B3 see NCR000
VITAMIN B-5 see CAU750
VITAMIN B_{12} (FCC) see VSZ000
VITAMIN B_{12} COMPLEX see VSZ000
VITAMIN B HYDROCHLORIDE see TET000
VITAMIN B6-HYDROCHLORIDE see PPK500
VITAMIN B1 MONONITRATE see TET500
VITAMIN B1 NITRATE see TET500
VITAMIN C see ARN000, ARN125
VITAMIN C SODIUM see ARN125
VITAMIN D2 (FCC) see EDA000
VITAMIN D3 see CMC750
VITAMIN E see VSZ450
VITAMIN G see RIK000
VITAMIN M see FMT000
VITAMIN PP see NCR000
VITAMISIN see ARN000
VITAPLEX E see VSZ450
VITAPLEX N see NDT000
VITARUBIN see VSZ000
VITA-RUBRA see VSZ000
VITASCORBOL see ARN000
VITAVAX see CCC500
VITAVEL-A see REU000
VITAVEL-D see EDA000
VITAYONON see VSZ450
VITEOLIN see VSZ450
VITINC DAN-DEE-3 see CMC750
VITON see BBQ500
VITPEX see REU000
VITRAL see VSZ000
VITRAN see TIO750
VITRIOL BROWN OIL see SOI500
VITRIOL, OIL OF (DOT) see SOI500
VITRIOL RED see IHD000
VM & P NAPHTHA see BCD500
VOGAN see REU000
VOGAN-NEU see REU000
VOGEL'S IRON RED see IHD000
VOLATILE OIL OF MUSTARD see AGJ250
VOLCLAY see BAV750
VOLCLAY BENTONITE BC see BAV750
VOLFARTOL see TIQ250
VONDACID TARTRAZINE see TNK750
VONDCAPTAN see CBG000
VONDURON see DGC400
VOPCOLENE 27 see OHU000
VOTEXIT see TIQ250
O-V STATIN see NOH500
VULCACURE see BJK500
VULKACITE L see BJK500
VULKACIT NPV/C2 see IAQ000
VULKACIT ZM see BHA750
VULKASIL see SCH000
VULNOPOL NM see SGM500
VULVAN see TBG000
VYDATE see DSP600
VYDATE L INSECTICIDE/NEMATICIDE see DSP600
VYDATE L OXAMYL INSECTICIDE/NEMATOCIDE see DSP600

W 6658 see BJP000
WACHOLDERBEER OEL (GERMAN) see JEA000
WAMPOCAP see NDT000
WAPNIOWY TLENEK (POLISH) see CAU500
WARBEX see FAG759
WARECURE C see IAQ000
WARKEELATE PS-43 see EIV000
WASSERSTOFFPEROXID (GERMAN) see HIB000
WATER GLASS see SJU000
WATERSTOFPEROXYDE (DUTCH) see HIB000
WATTLE GUM see AQQ500
WAXSOL see DJL000
WECOLINE 1295 see LBL000
WECOLINE OO see OHU000
WEED 108 see MRL750
WEED-AG-BAR see DFY600
WEEDAR see TIW500
WEEDAR-64 see DFY600
WEEDBEADS see SJA000
WEED-B-GON see DFY600, TIX500
WEED-E-RAD see MRL750
WEEDEX A see PMC325
WEEDEZ WONDER BAR see DFY600
WEED-HOE see MRL750
WEEDONE see PAX250, TIW500
WEEDONE LV4 see DFY600
WEED TOX see DFY600
WEEDTRINE-D see DWX800
WEEDTROL see DFY600
WEEVILTOX see CBV500
WEGLA DWUSIARCZEK (POLISH) see CBV500
WEST INDIAN LEMONGRASS OIL see LEH000
WESTROSOL see TIO750
WETAID SR see DJL000
WH 7286 see DMW000
WHEAT GLUTEN see WBL100
WHEY, DRY see WBL150
WHEY, PROTEIN CONCENTRATE see WBL155
WHEY, REDUCED LACTOSE see WBL160
WHEY, REDUCED MINERALS see WBL165
1700 WHITE see TGH000
WHITE CAUSTIC see SHS000
WHITE CAUSTIC SOLUTION see SHS500
WHITE CEDAR OIL see CCQ500
WHITE COPPERAS see ZNA000
WHITE CRYSTAL see SFT000
WHITE MINERAL OIL see MQV750
WHITE PETROLATUM see PCR200
WHITE SEAL-7 see ZKA000
WHITE SHELLAC see SCC700
WHITE VITRIOL see ZNA000, ZNJ000
WILKINITE see BAV750
WILLOSETTEN see SAB500
WINACET D see AAX250
WINE ETHER see ENW000
WINE YEAST OIL see CNG825
WING STOP B see SGM500
WINTERGREEN OIL (FCC) see MPI000
WINTERGREEN OIL, SYNTHETIC see MPI000
WITAMINA PP see NCR000
WITCIZER 312 see BJS000
WITCONOL MS see OAV000
WL 18236 see MDU600
WL 43479 see AHJ750
WL 43775 see FAR100
WOCHEM NO. 320 see OHU000
WOJTAB see PLZ000
WONUK see PMC325
WOOD ALCOHOL (DOT) see MDS250
WOOD NAPHTHA see MDS250
WOOD SPIRIT see MDS250
WOOL FAT see LAU550
WOOL VIOLET see BFN250
WOOL YELLOW see TNK750
WORM-CHEK see NBU500
WOTEXIT see TIQ250
WY-5103 see AIV500
WYACORT see MOR500

CODEN Bibliographic Citations

AACHAX Antimicrobial Agents and Chemotherapy. (Ann Arbor, MI) 1961-70. For publisher information, see AMACCQ

AACRAT Anesthesia and Analgesia; Current Research. (International Anesthesia Research Society, 3645 Warrensville Center Rd., Cleveland, OH 44122) V.36- 1957-

AAGAAW Antimicrobial Agents Annual. (New York, NY) 1960-60. For Publisher information, see AMACCQ

AAOPAF AMA Archives of Ophthalmology. (Chicago, IL) V.44, No.4-63, 1950-60. For publisher information, see AROPAW

ABANAE Antibiotics Annual. (New York, NY) 1953-60. For publisher information, see AMACCQ

ABBIA4 Archives of Biochemistry and Biophysics. (Academic Press, 111 5th Ave., New York, NY 10003) V.31- 1951-

ABCHA6 Agricultural and Biological Chemistry. (Maruzen Co. Ltd., P.O.Box 5050 Tokyo International, Tokyo 100-31, Japan) V.25- 1961-

ABHYAE Abstracts on Hygiene. (Bureau of Hygiene and Tropical Diseases, Keppel St., London WC1E 7HT, England) V.1- 1926-

ABMGAJ Acta Biologica et Medica Germanica. (Berlin, Ger. Dem. Rep.) V.1-41, 1958-82. For publisher information, see BBIADT

ABMPAC Advances in Biological and Medical Physics. (Academic Press, 111 5th Ave., New York, NY 10003) V.1- 1948-

ACATA5 Acta Anatomica. (S. Karger AG, Arnold-Boecklin-St 25, CH-4000 Basel 11, Switzerland) V.1- 1945-

ACEDAB Acta Endocrinologica, Supplementum. (Periodica, Skolegade 12E, 2500 Copenhagen-Valby, Denmark) No.1- 1948-

ACENA7 Acta Endocrinologica. (Periodica, Skolegade 12E, 2500 Copenhagen-Valby, Denmark) V.1- 1948-

ACHAAH Acta Haematologica. (S. Karger AG, Arnold-Boecklin-St 25, CH-4000 Basel 11, Switzerland) V.1- 1948-

ACHTA6 Antibiotica et Chemotherapia (Basel, 1954-70). (Basel, Switzerland) V.1-16, 1954-70. For publisher information, see ANBCB3

ACPADQ Acta Pathologica, Microbiologica et Immunologica Scandinavica, Section A: Pathology. (Munksgaard, 35 Noerre Soegade, DK-1370, Copenhagen K, Denmark) V.90- 1982-

ACRAAX Acta Radiologica. (Stockholm, Sweden) V.1-58, 1921-62. For publisher information, see ACRDA8

ACRDA8 Acta Radiologica: Diagnosis. (Box 7449, S-10391 Stockholm, Sweden) NS.V.1- 1963-

ACYTAN Acta Cytologica. (Williams & Wilkins Co., 428 E. Preston St., Baltimore, MD 21202) V.1- 1957-

ADCHAK Archives of Disease in Childhood. (British Medical Journal, 1172 Commonwealth Avenue, Boston, MA 02134) V.1- 1926-

ADTEAS Advances in Teratology. (Academic Press, 111 5th Ave., New York, NY 10003) V.1-5, 1966-72, Discontinued

ADVEA4 Acta Dermato-Venereologica. (Almqvist & Wiksell International, Box 62, S-101 20 Stockholm, Sweden) V.1- 1920-

AECTCV Archives of Environmental Contamination and Toxicology. (Springer-Verlag New York, Inc., Service Center, 44 Hartz Way, Secaucus, NJ 07094) V.1- 1973-

AEHA** U. S. Army, Environmental Hygiene Agency Reports. (Edgewood Arsenal, MD 21010)

AEHLAU Archives of Environmental Health. (Heldreff Publications, 4000 Albemarle St., N.W., Washington, D.C. 20016) V.1- 1960-

AEMBAP Advances in Experimental Medicine and Biology. (Plenum Publishing Corp., 233 Spring St., New York, NY 10013) V.1- 1967-

AEMIDF Applied and Environmental Microbiology. (American Society for Microbiology, 1913 I St., N.W., Washington, DC 20006) V.31- 1976-

AEPPAE Naunyn-Schmiedeberg's Archiv fuer Experimentelle Pathologie und Pharmakologie. (Berlin, Germany) V.110-253, 1925-66. For publisher information, see NSAPCC

AEXPBL Archiv fuer Experimentelle Pathologie und Pharmakologie. (Leipzig, Germany) V.1-109, 1873-1925. For publisher information, see NSAPCC

AFDOAQ Association of Food and Drug Officials of the United States, Quarterly Bulletin. (Editorial Committee of the Association, P.O. Box 20306, Denver, CO 80220) V.1- 1937-

AFREAW Advances in Food Research. (Academic Press, 111 5th Ave., New York, NY 10003) V.1- 1948-

AFSPA2 Archivo di Farmacologia Sperimentale e Scienze Affini. (Rome, Italy) V.1-82, 1902-1954. Discontinued.

AGGHAR Archiv fuer Gewerbepathologie und Gewerbehygiene. (Berlin, Germany) V.1-18, 1930-61. For publisher information, see IAEHDW

AHBAAM Archiv fuer Hygiene und Bakteriologie. (Munich, Germany) V.101-154, 1929-71. For publisher information, see ZHPMAT

AHYGAJ Archiv fuer Hygiene. (Munich, Germany) V.1-100, 1883-1928. For publisher information, see ZHPMAT

AICCA6 Acta Unio Internationalis Contra Cancrum. (Louvain, Belgium) V.1-20, 1936-64. For publisher information, see IJCNAW

AIDZAC Aichi Ika Daigaku Igakkai Zasshi. Journal of

the Aichi Medical Univ. Assoc. (Aichi Ika Daigaku, Yazako, Nagakute-machi, Aichi-gun, Aichi- Ken 480-11, Japan) V.1- 1973-

AIHAAP American Industrial Hygiene Association Journal. (AIHA, 475 Wolf Ledges Pkwy., Akron, OH 44311) V.19- 1958-

AIHOAX Archives of Industrial Hygiene and Occupational Medicine. (Chicago, IL) V.1-2, No.3, 1950. For publisher information, see AEHLAU

AIHQA5 American Industrial Hygiene Association Quarterly. (Baltimore, MD) V.7-18, 1946-57. For publisher information, see AIHAAP

AIMDAP Archives of Internal Medicine. (American Medical Association, 535 N. Dearborn St., Chicago, IL 60610) V.1- 1908-

AIMEAS Annals of Internal Medicine. (American College of Physicians, 4200 Pine St., Philadelphia, PA 19104) V.1- 1927-

AIPAAV Annales de l'Institut Pasteur. (Paris, France) V.1-123, 1887-1972. For publisher information, see ANMBCM

AIPTAK Archives Internationales de Pharmacodynamie et de Therapie. (Editeurs, Institut Heymans de Pharmacologie, De Pintelaan 135, B-9000 Ghent, Belgium) V.4- 1898-

AISSAW Annali dell'Istituto Superiore di Sanita. (Istituto Poligrafico dello Stato, Libreria dello Stato, Piazza Verdi, 10 Rome, Italy) V.1- 1965-

AITEAT Archivum Immunologiae et Therapiae Experimentalis. (Ars Polona-RUCH, P.O. Box 1001, P-00 068 Warsaw, 1, Poland) V.10- 1962-

AJANA2 American Journal of Anatomy. (Alan R. Liss Inc., 150 5th Ave., New York, NY 10011) V.1- 1901-

AJCAA7 American Journal of Cancer. (New York, NY) V.15-40, 1931-40. For publisher information, see CNREA8

AJCNAC American Journal of Clinical Nutrition. (American Society for Clinical Nutrition, Inc., 9650 Rockville Pike, Bethesda, MD 20814) V.2- 1954-

AJCPAI American Journal of Clinical Pathology. (J.B. Lippincott Co., (Keystone Industrial Park, Scanton, PA 18512) V.1- 1931-

AJDCAI American Journal of Diseases of Children. (American Medical Assoc., 535 N. Dearborn St., Chicago, IL 60610) V.1-80(3), 1911-50; V.100- 1960-

AJDDAL American Journal of Digestive Diseases. (Plenum Publishing Corp., 233 Spring St., New York, NY 10013) V.5-22, 1938-55., V.1-23, 1956-78

AJEBAK Australian Journal of Experimental Biology and Medical Science. (Univ. of Adelaide Registrar, Adelaide, S.A. 5000, Australia) V.1- 1924-

AJEPAS American Journal of Epidemiology. (Johns Hopkins Univ., 550 N. Broadway, Suite 201, Baltimore, MD 21205) V.81- 1965-

AJHEAA American Journal of Public Health. (American Public Health Assoc., Inc., 1015 15th St., N.W., Washington, D.C. 20005) V.2-17, 1912-27; V.61- 1971-

AJHYA2 American Journal of Hygiene. (Baltimore, MD) V.1-80, 1921-64. For publisher information, see AJEPAS

AJINO* Ajinomoto Co., Inc. (9 W. 57th St., Suite 4625, New York, NY 10019)

AJMSA9 American Journal of the Medical Sciences. (Charles B. Slack, Inc., 6900 Grove Rd., Thorofare, NJ 08086) V.1- 1841-

AJOGAH American Journal of Obstetrics and Gynecology. (C.V. Mosby Co., 11830 Westline Industrial Dr., St. Louis, MO 63141) V.1- 1920-

AJOPAA American Journal of Ophthalmology. (Ophthalmic Publishing Co., 435 N. Michigan Ave., Chicago, Il 60611) V.1- 1918-

AJPAA4 American Journal of Pathology. (Harper & Row, Medical Dept., 2350 Virginia Ave., Hagerstown, MD 21740) V.1- 1925-

AJPHAP American Journal of Physiology. (American Physiological Society, 9650 Rockville Pike, Bethesda, MD 20814) V.1- 1898-

AJVRAH American Journal of Veterinary Research. (American Veterinary Medical Association, 930 N. Meacham Road, Schaumburg, IL 60196) V.1- 1940-

AKGIAO Akushcherstvo i Ginekologiya. (V/O Mezhdunarodnaya Kniga, Kuznetskii Most 18, Moscow G-200, USSR) No. 1- 1936-

AMACCQ Antimicrobial Agents & Chemotherapy. (American Society for Microbiology, 1913 I St., N.W., Washington, DC 20006) V.1- 1972-

AMAHA5 Acta Microbiologica Academiae Scientiarum Hungaricae. (Akademiai Kiado, P.O. Box 24, Budapest 502, Hungary) V.1- 1954-

AMASA4 Acta Medica Academiae Scientiarum Hungaricae. (Kultura, POB 149 H-1389, Budapest, Hungary) V.1- 1950-

AMBIEH Antibiotiki i Meditsinskaya Bioteckhnologiya. Antibiotics abd Medical Biotechnology. (V/O Mezhdunarodnaya Kinga, 113095 Moscow, USSR), V.30- 1985-

AMBOCX Ambio. A Journal of the Human Environment, Research & Management. (Pergamon Press, Inc., Maxwell House, Fairview Park, Elmsford, NY 10523) V.1- 1972-

AMBPBZ Acta Pathologica et Microbiologica Scandinavica, Section A: Pathology. (Copenhagen K, Denmark) V.78-89, 1970-81. For publisher information, see ACPADQ

AMCYC* Toxicological Information on Cyanamid Insecticides. (American Cyanamid Co., Agricultural Division, Princeton, NJ 08540)

AMDCA5 AMA Journal of Diseases of Children. (Chicago, IL) V.91-99, 1956-60. For publisher information, see AJDCAI

AMICCW Archives of Microbiology. (Springer-Verlag New York, Inc., Service Center, 44 Hartz Way, Secaucus, NJ 07094) V.95- 1974-

AMIHAB AMA Archives of Industrial Health. (Chicago, IL) V.11-21, 1955-60. For publisher information, see AEHLAU

AMIHBC AMA Archives of Industrial Hygiene & Occupational Medicine. (Chicago, IL) V.2-10, 1950-54. For publisher information, see AEHLAU

AMLTAS Annales de Medecine Legale et de Criminologie.

(Paris, France) V. 31-8, 1951-8. For publisher information, see MLDCAS

AMNTA4 American Naturalist. (University of Chicago Press, 5801 S. Ellis Ave., Chicago, IL 60637) V.1- 1867-

AMOKAG Acta Medicia Okayama. (Okayama University Medical School, 2-5-1 Shikata-cho, Okayama 700, Japan) V.8- 1952-

AMONDS Applied Methods in Oncology. (Elsevier North Holland, Inc., 52 Vanderbilt Ave., New York, NY 10017) V.1- 1978-

AMPLAO AMA Archives of Pathology. (American Medical Association, 535 N. Dearborn St., Chicago, IL 60610) V.50, No.4-V.69, 1950-60. For publisher information, see APLMAS

AMPMAR Archives des Maladies Professionnelles de Medecine du Travail et de Securite Sociale. (Masson et Cie, Editeurs, 120 Blvd. Saint-Germain, P-75280, Paris 06, France) V.7- 1946-

AMRL** Aerospace Medical Research Laboratory Report. (Aerospace Technical Div., Air Force Systems Command, Wright-Patterson Air Force Base, OH 45433)

AMSHAR Acta Morphologica Academiae Scientiarum Hungaricae. (Akademiai Kiado, P.O. Box 24, H-1389 Budapest 502, Hungary) V.1- 1951-

AMTUA3 Acta Medica Turcica. (Dr. Ayhan Okcuoglu, Cocuk Hastalikari Klinig i, c/o Ankara University Tip Facultesi, PK 48, Cebeci, Ankara, Turkey) V.1- 1964-

ANAEA3 Annals of Allergy. (American College of Allergists, Box 20671, Bloomington, MN 55420) V.1- 1943-

ANESAV Anesthesiology. (J.B. Lippincott Co., Keystone Industrial Park, Scranton, PA 18512) V.1- 1940-

ANMBCM Annales de Microbiologie (Paris). (Masson et Cie, Editeurs, 120 Blvd. Saint-Germain, P-75280, Paris 06, France) V.124- 1973-

ANTBAL Antibiotiki. (Moscow, USSR) V.1-29, 1956-84. For publisher information, see AMBIEH

ANTCAO Antibiotics & Chemotherapy. (Washington, DC) V.1-12, 1951-62. For publisher information, see CLMEA3

ANYAA9 Annals of the New York Academy of Sciences. (The Academy, Exec. Director, 2 E. 63rd St., New York, NY 10021) V.1- 1877-

AOGLAR Acta Obstetrica et Gynaecologica Japonica, English Edition. (Tokyo, Japan) V.16-23, 1969-76. For publisher information, See NISFAY

AOHYA3 Annals of Occupational Hygiene. (Pergamon Press, Headington Hill Hall, Oxford OX3 OBW, England) V.1- 1958-

APFRAD Annales Pharmaceutiques Francaises. (Masson et Cie, Editeurs, 120 Blvd. Saint-Germain, P-75280, Paris 06, France) V.1- 1943-

APJAAG Acta Pathologica Japonica. (Nippon Byori Gakkai, 7-3-1, Hongo, Bunkyo-Ku, Tokyo 113, Japan) V.1- 1951-

APJUA8 Acta Pharmaceutica Jugoslavica. (Jugoslovenska Knjiga, P.O. Box 36, Terazije 27, YU-11001 Belgrade, Yugoslavia) V.1- 1951-

APLMAS Archives of Pathology and Laboratory Medicine. (American Medical Association, 535 N. Dearbon St., Chicago, Il. 60610) V.1-5, No. 2, 1926-28, V.100- 1976-

APMBAY Applied Microbiology. (Washington, DC) V.1- 30, 1953-75. For publisher information, see AEMIDF

APMIAL Acta Pathologica et Microbiologica Scandinavica. (Copenhagen, Denmark) V.1-77, 1924-69. For publisher information, see AMBPBZ

APMUAN Acta Pathologica et Microbiologica Scandinavica, Supplementum. (Munksgaard, 35 Noerre Soegade, DK-1370 Copenhagen K, Denmark) No.1- 1926-

APPHAX Acta Poloniae Pharmaceutica. (Ars Polona-RUCH, P.O. Box 1001, P-00 068 Warsaw, 1, Poland) V.1- 1937-

APSCAX Acta Physiologica Scandinavica. (Karolinska Institutet, S-10401 Stockholm, Sweden) V.1- 1940-

APSVAM Acta Paediatrica Scandinavica. (Almqvist & Wiksell, P.O. Box 62, 26 Gamla Brogatan, S-101, 20 Stockholm, Sweden) V.54- 1965-

APTOA6 Acta Pharmacologica et Toxicologica. (Munksgaard, 35 Noerre Soegade, DK-1370, Copenhagen K, Denmark) V.1- 1945-

APTOD9 Abstracts of Papers, Society of Toxicology. Annual Meetings. (Academic Press, 111 5th Ave., New York, NY 10003)

APYPAY Acta Physiologica Polonica. (Panstwowy Zaklad Wydawnictw Lekarskich, ul. Dluga 38-40, P-00 238 Warsaw, Poland) V.1- 1950-

ARANDR Archives of Andrology. (Elsevier North Holland, Inc., 52 Vanderbilt Ave., New York, NY 10017) V.1- 1978-

ARCVBP Annales Recherches Veterinaires. (Institut National de la Recherche Agronomique, Service des Publ., route de Saint-Cyr, 78000 Versailles, France) V.1- 1970-

ARDSBL American Review of Respiratory Disease. (American Lung Association, 1740 Broadway, New York, NY 10019) V.80- 1959-

ARGEAR Archiv fuer Geschwulstforschung. (VEB Verlag Volk und Gesundheit Neue Gruenstr. 18, DDR-102 Berlin, German Democratic Republic) V.1- 1949-

ARMCAH Annual Review of Medicine. (Annual Reviews, Inc., 4139 El Camino Way, Palo Alto, CA 94306) V.1- 1950-

ARMKA7 Archiv fuer Mikrobiologie. (Springer (Berlin)) V.1-13, 1930-43; V.14-94, 1948-73. For publisher information, see AMICCW

AROPAW Archives of Ophthalmology. (American Medical Assn., 535 N. Dearborn St., Chicago, IL 60610) V.1-44, No. 3, 1929-50; V.64- 1960-

ARPAAQ Archives of Pathology. (American Medical Assn., 535 N. Dearborn St., Chicago, IL 60610) V.5, no.3-V.50, no.3, 1928-50; V.70-99, 1960-75. For publisher information, see APLMAS

ARSIM* Agricultural Research Service, USDA Information Memorandum. (Beltsville, MD 20705)

ARSUAX Archives of Surgery. (American Medical Association, 535 N. Dearborn St., Chicago, IL 60610) V.1- 61, 1920-50; V.81- 1960-

ARTODN Archives of Toxicology. (Springer-Verlag, Heidelberger, Pl. 3, D-1 Berlin 33, Germany) V.32- 1974-

ARZNAD Arzneimittel-Forschung. (Edition Cantor Verlag fur Medizin und Naturwissenschaften KG, D-7960 Aulendorf, Germany) V.1- 1951-

ASBIAL Archivio di Science Biologiche. (Cappelli Editore, Via Marsili 9, I-40124 Bologna, Italy) V.1- 1919-

ASPHAK Archives des Sciences Physiologique. (Paris, France) V.1-28, 1947-74. Discontinued.

ASTTA8 ASTM Special Technical Publication. (American Society for Testing Materials, 1916 Race St., Philadelphia. PA 19103) No.1- 1911-

ATENBP Atmospheric Environment. Air Pollution, Industrial Aerodynamics, Micrometerology, Aerosols. (Pergamon Press, Headington Hill Hall, Oxford OX3 OBW, England) V.1- 1967-

ATMPA2 Annals of Tropical Medicine & Parasitology. (Academic Press, 24-28 Oval Rd., London NWl 7DX, England) V.1- 1907-

ATSUDG Archives of Toxicology, Supplement. (Springer-Verlag, Heidelberger Pl. 3, D-1000 Berlin 33, Fed. Rep. Ger.) No. 1- 1978-

ATXKA8 Archiv fuer Toxikologie. (Berlin, Germany) V.15-31, 1954-74. For publisher information, see ARTODN

AUPJB7 Australian Paediatric Journal. (Royal Childrens Hospital, Parkville, Victoria 3052, Australia) V.1- 1965-

AUVJA2 Australian Veterinary Journal. (Australian Veterinary Association, Executive Director, 134-136 Hampden Road, Artarmon, N.S.W. 2064, Australia) V.2- 1927-

AVERAG American Veterinary Review. (Chicago, IL) V.1-47, 1877-1915. For publisher information, see JAVMA4

AVPCAQ Advances in Pharmacology and Chemotherapy. (Academic Press, 111 5th Ave., New York, NY 10003) V.7- 1969-

AVSCA7 Acta Veterinaria Scandinavica. (Danske Dyrlargeforening, Alhambravej 15, DK-1826, Copenhagen V, Denmark) V.1- 1959-

AVSUAR Acta Dermato-Venereologica, Supplementum. (Almqvist & Wiksell Periodical Co., P.O. Box 62, 26 Gamla Brogatan, S-101 20 Stockholm, Sweden) No.1- 1929-

AXVMAW Archiv fuer Experimentella Veterinaermedizin. (S. Hirzel Verlag, Postfach 506, DDR-701 Leipzig, German Democratic Republic) V.6- 1952-

BAFEAG Bulletin de l'Association Francaise pour l'Etude du Cancer. (Paris, France) V.1-52, 1908-65. For publisher information, see BUCABS

BANRDU Banbury Report. (Cold Spring Harbor Laboratory, POB 100, Cold Spring Harbor, NY 11724) V.1- 1979-

BBACAQ Biochimica et Biophysica Acta. (Elsevier Publishing Co., POB 211, Amsterdam C, Netherlands) V.1- 1947-

BBIADT Biomedica Biochimica Acta. (Akademie-Verlag GmbH, Postfach 1233, DDR-1086 Berlin, Ger. Dem. Rep.) V.42- 1983-

BBRCA9 Biochemical and Biophysical Research Communications. (Academic Press Inc., 111 5th Ave., New York, NY 10003) V.1- 1959-

BCFAAI Bollettino Chimico Farmaceutico. (Societa Editoriale Farmaceutica, Via Ausonio 12, 20123 Milan, Italy) V.33- 1894-

BCPCA6 Biochemical Pharmacology. (Pergamon Press, Headington Hill Hall, Oxford OX3 OBW, England) V.1- 1958-

BCTKAG Bromatologia i Chemia Toksykologiczna. (Ars Polona-RUCH, P.O. Box 1001, P-00 068 Warsaw, 1, Poland) V.4- 1971-

BECTA6 Bulletin of Environmental Contamination & Toxicology. (Springer- Verlag New York, Inc., Service Center, 44 Hartz Way, Secaucus, NJ 07094) V.1- 1966-

BESAAT Bulletin of the Entomological Society of America. (The Society, 4603 Calvert Rd., College Park, MD 20740) V.1- 1955-

BEXBAN Bulletin of Experimental Biology & Medicine. Translation of BEBMAE. (Plenum Publishing Corp., 233 Spring St., New York, NY 10013) V.41- 1956-

BEXBBO Biochemistry and Experimental Biology. (Piccin Medical Books, Via Brunacci, 12, 35100 Padua, Italy) V.10- 1971/72-

BIATDR Bulletin of the International Association of Forensic Toxicologists.

BIJOAK Biochemical Journal. (Biochemical Society, P.O. Box 32, Commerce Way, Whitehall Rd., Industrial Estate, Colchester CO2 8HP, Essex, England) V.1- 1906-

BIMADU Biomaterials. (Quadrant Subscription Services Ltd., Oakfield House, Perrymount Rd., Haywards Heath, W. Sussex, RH16 3DH, U.K.) V.1- 1980-

BINEAA Biologia Neonatorum. (Basel, Switzerland) V.1-14, No.5/6, 1956- 1969, For publisher information, see BNEOBV

BIOFX* BIOFAX Industrial Bio-Test Laboratories, Inc., Data Sheets. (1810 Frontage Rd., Northbrook, IL 60062)

BIOGAL Biologico. (Instituto Biologica, Av. Rodriques Alves, 1252, C.P. 4185, Sao Paulo, Brazil) V.1- 1935-

BIOJAU Biophysical Journal. (Rockefeller Univ. Press, 1230 York Ave., New York, NY 10021) V.1- 1960-

BIORAK Biochemistry. Translation of BIOHAO. (Plenum Publishing Corp., 233 Spring St., New York, NY 10013) V.21- 1956-

BIREBV Biology of Reproduction. (Society for the Study of Reproduction, 309 West Clark Street, Champaign, IL 61820) V.1- 1969-

BIRSB5 Biology of Reproduction, Supplement. (Champaign, Ill.) For publisher information, see BIREBV

BISNAS BioScience. (American Instutite of Biological Sciences, 1401 Wilson Blvd., Arlington, VA 22209) V.14- 1964-

BIZEA2 Biochemische Zeitschrift. (Berlin, Germany) V.1-346, 1906-67. For publisher information, see EJBCAI

BJANAD British Journal of Anesthesia. (Scientific & Medical Division, Macmillan Publishers Ltd., Houndmills, Basingstoke, Hampshire RG21 2XS, England) V.1- 1923-

BJCAAI British Journal of Cancer. (H.K. Lewis & Co., 136 Gower St., London WC1E 6BS, England) V.1- 1947-

BJEPA5 British Journal of Experimental Pathology. (H.K. Lewis & Co., 136 Gower St., London WC1E 6BS, England) V.1- 1920-

BJIMAG British Journal of Industrial Medicine. (British Medical Journal, 1172 Commonwealth Ave., Boston, MA 02134) V.1- 1944-

BJOGAS British Journal of Obstetrics and Gynaecology. (British Journal of Obstetrics and Gynaecology, 27 Sussex Place, Regent's Park, London NW1 4RG, England) V.82- 1975-

BJPCAL British Journal of Pharmacology & Chemotherapy. (London, England) V.1-33, 1946-68. For publisher information, see BJPCBM

BJPCBM British Journal of Pharmacology. (MacMillan Journals Ltd., Houndmills Estate, Basingstroke, Hampshire RG21 2XS, England) V.34- 1968-

BJSUAM British Journal of Surgery. (John Wright & Sons Ltd., 42-44 Triangle West, Bristol BS8 1EX, England) V.1- 1913-

BJURAN British Journal of Urology. (Williams & Wilkens Co., 428 E. Preston St., Baltimore, MD 21202) V.1- 1929-

BLLIAX Bratislavske Lekarske Listy. (PNS-Ustredna Expedicia Tlace, Gottwaldovo Namestie 48/7, CS-884 19 Bratislava, Czechoslovakia) V.1- 1921-

BMJOAE British Medical Journal. (British Medical Association, BMA House, Travistock Square, London WC1H 9JR, England) V.1- 1857-

BMRII* "U. S. Bureau of Mines Report of Investigation No. 2979," Patty, F.A., and W.P. Yant, 1929

BNEOBV Biology of the Neonate. (S. Karger AG, Postfach, CH-4009 Basel, Switzerland) V.15- 1970-

BPOSA4 British Poultry Science. (Longman Group Ltd., Journals Division, 43-45 Annandale St., Edinburgh EH47 4AT, Scotland) V.1- 1960-

BRMEAY Bruxelles-Medical. Bruxelles-Medical, Boite 1, 1180 Brussels, Belgium) V.1- 1921-

BSFDA3 Bulletin de la Societe Francaise de Dermatologie et de Syphiligraphie. (Paris, France) V.1-83, 1890-1976. Discontinued.

BSIBAC Bolletino della Societe Italiana di Biologia Sperimentale. (Casa Editrice Libraria V. Idelson, Via Alcide De Gasperi, 55, 80133 Naples, Italy) V.2- 1927-

BSRSA6 Bulletin de la Society Royale des Sciences de Liege. (Societe Royale des Sciences de Liege, Universite de Liege, 15, Ave des Tilleuls, B-4000 Liege, Belgium) V.1- 1932-

BUCABS Bulletin du Cancer. (Masson et Cie, Editeurs, 120 Blvd. Saint- Germain, P-75280, Paris 06, France) V.53- 1966-

BUMMAB Bulletin of the University of Miami School of Medicine and Jackson Memorial Hospital. (Coral Gables, FL) V.1-20, 1939-66. Discontinued.

BUYRAI Bulletin of Parenteral Drug Association. (The Association, Western Saving Fund Bldg., Broad & Chestnut Sts., Philadelphia, PA 19107) V.1- 1946-

BWHOA6 Bulletin of the World Health Organization. (WHO, 1211 Geneva 27, Switzerland) V.1- 1947-

CAANAT Comptes Rendus de l'Association des Anatomistes. (Paris, France) V.1- 1916(?)-

CALEDQ Cancer Letters. (Elsevier Publishing, P.O. Box 211, Amsterdam C, Netherlands) V.1- 1975-

CANCAR Cancer. (J. B. Lippincott Co., E. Washington Sq., Philadelphia, PA 19105) V.1- 1948-

CBCCT* "Summary Tables of Biological Tests," National Research Council Chemical-Biological Coordination Center. (National Academy of Science Library, 2101 Constitution Ave., N.W., Washington, DC 20418)

CBINA8 Chemico-Biological Interactions. (Elsevier Publishing, P.O. Box 211, Amsterdam C, Netherlands) V.1- 1969-

CBTIAE Contributions from Boyce Thompson Institute. (Yonkers, NY) V.1-24, 1925-71. Discontinued.

CCPTAY Contraception. (Geron-X, Publishers, P.O. Box 1108, Los Altos, CA 94022) V.1- 1970-

CCROBU Cancer Chemotherapy Reports, Part 1. (Washington, DC) V.52, no.6-V.59, 1968-75. For publisher information, see CTRRDO

CECED9 Commission of the European Communities, Report EUR. (Office for Official Publications, POB 1003, Luxembourg 1, Luxembourg) 1967-

CEHYAN Ceskoslovenska Hygiena. Czechoslovak Hygiene. (Ve Smeckach 30, Prague 1, Czech). V.1- 1956-

CELLB5 Cell (Cambridge, Mass.). (Massachusetts Institute of Technology Press, 28 Carleton St., Cambridge, MA 02142) V.1- 1974-

CENEAR Chemical & Engineering News. (American Chemical Society, 1155 16th St., N.W., Washington, DC 20036) V.20- 1942-

CGCGBR Cytogenetics and Cell Genetics. (S. Karger AG, Arnold-Boecklin Str. 25, CH-4011 Basel, Switzerland) V.12- 1973-

CHABA8 Chemical Abstracts. (Chemical Abstracts Service, Box 3012, Columbus, OH 43210) V.1- 1907-

CHDDAT Comptes Rendus Hebdomadaires des Seances de l'Academie des Sciences, Serie D. (Centrale des Revues Dunod-Gauthier-Villars, 24-26 Blvd. de l'Hopital, 75005 Paris, France) V.262- 1966-

CHHTAT Zhonghua Yixue Zazhi. Chinese Medical Journal. (Guozi Shudian, Beijing, Peop. Rep. China) V.1- 1915-

CHINAG Chemistry & Industry. (Society of Chemical Industry, 14 Belgrave Sq., London SW1X 8PS, England) V.1-21, 1923-43; No.1- 1944-

CHMTBL Chemical Technology. (American Chemical Society, 1155 16th St., N.W., Washington, DC 20036) V.1- 1917-

CHROAU Chromosoma. (Springer-Verlag, Heidelberger Platz 3, D-1000 Berlin 33, Federal Republic of Germany) V.1- 1939-

CHRTBC Chromosomes Today. (Elsevier Scientific Publishing Co., P.O. Box 211 Amsterdam, The Netherlands) V.1- 1966-

CHTHBK Chemotherapy. (S. Karger AG, Arnold-Boecklin-St 25, CH-4000, Basel 11, Switzerland) V.13- 1968-

CHTPBA Chimica Therapeutica. (Editions DIMEO, Arcueil, France) V.1-8, 1965-73.

CHYCDW Zhonghua Yufangyixue Zazhi. Chinese Journal of Preventive Medicine. (42 Tung Szu Hsi Ta Chieh, Beijing, Peop. Rep. China) Beginning history not known.

CIGET* Ciba-Geigy Toxicology Data/Indexes, 1977. (Ciba-Geigy Corp., Ardsley, NY 10502)

CIIT** Chemical Industry Institute of Toxicology, Docket Reports. (POB 12137, Research Triangle Park, NC 27709)

CIRUAL Circulation Research. (American Heart Association, Publishing Director, 7320 Greenville Ave., Dallas, TX 75231) V.1- 1953-

CISCB7 CIS, Chromosome Information Service. (Maruzen Co. Ltd., POB 5050, Tokyo International, Tokyo 100-31, Japan) No.1- 1961-

CIWYAO Carnegie Institute of Washington, Year Book. (Carnegie Institution of Washington, 1530 P St., N.W. Washington, DC 20005) V.1- 1902-

CJCMAV Canadian Journal of Comparative Medicine. (360 Bronson Ave., Ottawa K1R 6J3, Ontario, Canada) V.1-3, 1937-39; V.32- 1968-

CJPPA3 Canadian Journal of Physiology & Pharmacology. (National Research Council of Canada, Ottawa K1A OR6, Ontario, Canada) V.42- 1964-

CLBIAS Clinical Biochemistry. (Canadian Society of Clinical Chemists, 151 Slater St., Ottawa K1P 5H3, Ontario, Canada) V.1- 1967-

CLCEAL Casopis Lekaru Ceskych. Journal of Czech Physicians. (ARTIA, Ve Smeckach 30, Prague 1, Czechoslovakia) V.1- 1862-

CLDND* Compilation of LD50 Values of New Drugs. (J.R. MacDougal, Dept. of National Health & Welfare, Food & Drug Divisions, 35 John St., Ottawa, Ontario, Canada)

CLMEA3 Clinical Medicine. (Clinical Medicine Publications, 444 Frontage Rd., Northfield, IL 60093) V.69- 1962-

CLPTAT Clinical Pharmacology & Therapeutics. (C.V. Mosby Co., 11830 Westline Industrial Dr., St. Louis, MO 63141) V.1- 1960-

CMAJAX Canadian Medical Association Journal. (CMA House, Box 8650, Ottawa K1G OG8, Ontario, Canada) V.1- 1911-

CMBID4 Cellular and Molecular Biology. (Pergamon Press Ltd., Headington Hill Hall, Oxford OX3 0BW, England) v.22- 1977-

CMEP** "Clinical Memoranda on Economic Poisons," U. S. Dept. HEW, Public Health Service, Communicable Disease Center, Atlanta, GA, 1956

CMJODS Chinese Medical Journal (Beijing, English Edition). New Series. (Guozi Shudian, Beijing, Peop. Rep. China) V.1- 1975- (Adopted vol. no. 92 in 1979)

CMMUAO Chemical Mutagens. Principles and Methods for Their Detection (Plenum Publishing Corp., 233 Spring St., New York, NY 10013) V.1- 1971-

CMSHAF Chemosphere. (Pergamon Press, Headington Hill Hall, Oxford OX3 OEW, England) V.1- 1971-

CNJGA8 Canadian Journal of Genetics and Cytology. (Genetics Society of Canada, 151 Slater St., Suite 907, Ottawa, Ont. K1P 5H4, Canada) V.1- 1959-

CNJMAQ Canadian Journal of Comparative Medicine and Veterinary Science. (Gardenvale, Quebec, Canada) V.4-32, 1940-68. For publisher information, see CJCMAV

CNREA8 Cancer Research. (Waverly Press, Inc., 428 E. Preston St. Baltimore, MD 21202) V.1- 1941-

COREAF Comptes Rendus Hebdomadaires des Seances, Academie des Sciences. (Paris, France) V.1-261, 1835-1965. For publisher information, see CHDDAT

CORTBR Clinical Orthopaedics and Related Research. (J. B. Lippincott Co., E. Washington Sq., Philadelphia, PA 19105) No.26- 1963-

CPBTAL Chemical & Pharmaceutical Bulletin. (Pharmaceutical Society of Japan, 12-15-501, Shibuya 2-chome, Shibuya-ku, Tokyo, 150, Japan) V.6- 1958-

CPEDAM Clinical Pediatrics. (J. B. Lippincott Co., E. Washington Sq., Philadelphia, PA 19105) V.1- 1962-

CPGPAY Comparative and General Pharmacology. (New York, NY) V.1-5, 1970-74. For publisher information, see GEPHDP

CRNGDP Carcinogenesis. (Information Retrieval, 1911 Jefferson Davis Highway, Arlington, VA 22202) V.1- 1980-

CRSBAW Comptes Rendus des Seances de la Societe de Biologie et de Ses Filiales. (Masson et Cie, Editeurs, 120 Blvd. Saint-Germain, P-75280, Paris 06, France) V.1- 1849-

CSLNX* U. S. Army Armament Research & Development Command, Chemical Systems Laboratory, NIOSH Exchange Chemicals. (Aberdeen Proving Ground, MD 21010)

CTOXAO Clinical Toxicology. (New York, NY) V.1-18, 1968-81. For publisher information, see JTCTDW

CTRRDO Cancer Treatment Reports. (U. S. Government Printing Office, Supt. of Doc., Washington, DC 20402) V.60- 1976-

CYGEDX Cytology and Genetics. English Translation of Tsitologiya i Genetika. (Allerton Press, Inc., 150 Fifth Ave., New York, NY 10011) V.8- 1974-

CYLPDN Zhongguo Yaoli Xuebao. Acta Pharmacologica Sinica. (Shanghai K'o Hsueh Chi Shu Ch'u Pan She, 450 Shui Chin Erh Lu, Shanghai 200020, Peop Rep. China) V.1- 1980-

CYTOAN Cytologia. (Maruzen Co. Ltd., P.O. Box 5050, Tokyo International, Tokyo 100-31, Japan) V.1- 1929-

DABBBA Dissertation Abstracts International, B: The Sciences and Engineering. (University Microfilms, A Xerox Co., 300 N. Zeeb Rd., Ann Arbor, MI 48106) V.30- 1969-

DADEDV Drug and Alcohol Dependence. (Elsevier Sequoia SA, POB 851, CH-1001 Lausanne, Switzerland) V.1- 1975-

DANKAS Doklady Akademii Nauk S.S.S.R. (v/o "Mezhdunarodnaya Kniga," Kuznetskii Most 18, Moscow G-200, U.S.S.R.) V.1- 1933-

DBANAD Doklady Bolgarskoi Akademii Nauk. (Hemus, Blvd. Russki 6, Sofia, Bulgaria) V.1- 1948-

DBTEAD Diabete. (Le Raincy, France) V.1-22, 1953-1974

DCTODJ Drug and Chemical Toxicology. (Marcel Dekker, POB 11305, Church St. Station, New York, NY 10249) V.1- 1977/78-

DHEFDK HEW Publication (FDA. United States). (Washington, DC) 19??-1979(?). For publisher information, see HPFSDS

DIAEAZ Diabetes. (American Diabetes Assoc., 600 5th Ave., New York, NY 10020) V.1- 1952-

DICPBB Drug Intelligence and Clinical Pharmacy. (Drug Intelligence and Clinical Pharmacy, Inc., University of Cincinnati, Cincinnati, OH 45267) V.3- 1969-

DIGEBW Digestion. (S. Karger AG, Arnold-Boecklin Street 25, CH-4011 Basel, Switzerland) V.1- 1968-

DKBSAS Doklady Biological Sciences (English Translation). (Plenum Publishing Corp., 233 Spring St., New York, NY 10013) V.112- 1957-

DMWOAX Deutsche Medizinische Wochenschrift. (Georg Thieme Verlag, Herdweg 63, Postfach 732, 7000 Stuttgart 1, Federal Republic of Germany) V.1- 1875-

DOEAAH Down to Earth. A Review of Agricultural Chemical Progress. (Dow Chemical U.S.A., 1703 S. Saginaw Rd., Midland, MI 48640) V.1- 1945-

DOWCC* Dow Chemical Company Reports. (Dow Chemical U.S.A., Health and Environment Research, Toxicology Research Lab., Midland, MI 48640)

DPHFAK Dissertationes Pharmaceuticae et Pharmacologicae. (Warsaw, Poland) V.18-24, 1966-72. For publisher information, see PJPPAA

DRFUD4 Drugs of the Future. (J.R. Prous, S.A. International Publishers, Apartado de Correos 1641, Barcelona, Spain) V.1- 1975/76-

DRISAA Drosophila Information Service (Cold Spring Harbor Laboratory, POB 100, Cold Spring Harbor, NY 11724) No.1- 1934-

DRSTAT Drug Standards. (Washington, DC) V.19-28, 1951-60. For publisher information, see JPMSAE

DRUGAY Drugs. International Journal of Current Therapeutics and Applied Pharmacology Reviews. (ADIS Press Ltd., 18/F., Tung Sun Commercial Centre, 194-200 Lockhart Road, Wanchai, Hong Kong)

DTLVS* "Documentation of Threshold Limit Values for Substances in Workroom Air." For publisher information, see 85INA8

DTLWS* "Documentation of the Threshold Limit Values for Substances in Workroom Air," Supplements. For publisher information, see 85INA8

DTTIAF Deutsche Tieraerztliche Wochenschrift. (M. Verlag und H. Schaper, Postfach 260669, 3 Hanover 26, Germany) V.1- 1893-

EAPHA6 Eastern Pharmacist. (Eastern Pharmacist, 507, Ashok Bhawan, 93, Neru Place, New Delhi 110019, India) V.1- 1958-

EbeAG# Personal Communication to NIOSH from A.G. Ebert, International Glutamate Technical Committee, 85 Walnut St., Watertown, MA 02172

ECBUDQ Ecological Bulletins. (Swedish National Science Research Council, Stockholm) Number 19- 1975-

ECEBDI Experimental Cell Biology. (Phiebig Inc., POB 352, White Plains, NY 10602) V.44- 1976-

ECJPAE Endocrinologia Japonica. (Japan Publications Trading Co., 1255 Howard St., San Francisco, CA 94103) V.1- 1954-

EDRCAM Endocrine Research Communcations. (Marcel Dekker, POB 11305, Church St. Station, New York, NY 10249) V.1- 1974-

EESADV Ecotoxicology and Environmental Safety. (Academic Press, 111 5th Ave., New York, NY 10003) V.1- 1977-

EGESAQ Egeszsegtudomany. (Kultura, POB 149, H-1389 Budapest, Hungary) V.1- 1957-

EJBCAI European Journal of Biochemistry. (Springer-Verlag, Heidelberger, Pl. 3, D-1 Berlin 33, Germany) V.1- 1967-

EJCAAH European Journal of Cancer. (Pergamon Press, Headington Hill Hall, Oxford OX3 OEW, England) V.1- 1965-

EJCODS European Journal of Cancer and Clinical Oncology. (Pergamon Press Ltd., Headington Hill Hall, Oxford OX3 0BW, England) V.17, No.7- 1981-

EJTXAZ European Journal of Toxicology and Environmental Hygiene. (Paris, France) v.7-9, 1974-76. For publisher information, see TOERD9

EKFMA7 Eksperimental'naya i Klinicheskaya Farmakoterapiya. (Akademiya Nauk Latviiskoi SSR, Inst., Organicheskogo Sinteza, ul. Aizkraukles 21, Riga, U.S.S.R.) No.1- 1970-

EMSUA8 Experimental Medicine & Surgery. (Brooklyn Medical Press, 600 Lafayette Ave., Brooklyn, NY 11216) V.1- 1943-

ENDOAO Endocrinology. (Williams & Wilkins Co., Dept. 260, P.O. Box 1496, Baltimore, MD 21203) V.1- 1917-

ENMUDM Environmental Mutagenesis. (Alan. R. Liss, Inc., 150 Fifth Ave., New York, NY 10011) V.1- 1979-

ENPBBC Environmental Physiology and Biochemistry. (Copenhagen, Denmark) V.2-5, No. 6, 1972-5, Discontinued.

ENVRAL Environmental Research. (Academic Press, 111 5th Ave., New York, NY 10003) V.1- 1967-

EQSFAP Environmental Quality & Safety. (Academic Press, 111 5th Ave., New York, NY 10003) V.1- 1972-

EQSSDX Environmental Quality & Safety, Supplement. (Academic Press, 111 5th Ave., New York, NY 10003) V.1- 1975-

ESKHA5 Eisei Shikenjo Hokoku. Bulletin of the National Hygiene Sciences. (Kokuritsu Eisei Shikenjo, 18-1 Kamiyoga 1 chome, Setagaya-ku, Tokyo, Japan) V.1- 1886-

ETOCDK Environmental Toxicology and Chemistry. (Pergamon Press Inc., Maxwell House, Fairview Park, Elmsford, NY 10523) V.1- 1982-

EVHPAZ EHP, Environmental Health Perspectives. Subseries of DHEW Publications. (U.S. Government Printing Office, Superintendent of Documents, Washington, DC 20402) No.1- 1972-

EVSRBT Environmental Science Research. (Plenum Publishing Corp., 233 Spring St., New York, NY 10013) V.1- 1972-

EXPEAM Experientia. (Birkhaeuser Verlag, P.O. Box 34, Elisabethenst 19, CH-4010, Basel, Switzerland) V.1- 1945-

FAATDF Fundamental and Applied Toxicology (Official Journal of the Society of Toxicology, 475 Wolf Ledges Parkway, Akron, OH 44311) V.1- 1981-

FAONAU Food and Agriculture Organization of United Nations, Report Series. (FAO-United Nations, Room 101, 1776 F Street, NW, Washington, DC 20437)

FATOAO Farmakologiya i Toksikologiya (Moscow). (v/o

"Mezhdunarodnaya Kniga," Kuznetskii Most 18, Moscow G-200, U.S.S.R.) V.2- 1939-

FAVUAI Fiziologicheski Aktivnye Veshchestva. Physiologically Active Substances. (Akademiya Nauk Ukrainskoi S.S.R., Kiev, U.S.S.R.) No.1- 1966-

FAZMAE Fortschritte der Arzneimittelforschung. (Birkhauser Verlag, P.O. Box 34, Elisabethenst 19, CH-4010, Basel, Switzerland) V.1- 1959-

FCTOD7 Food and Chemical Toxicology. (Pergamon Press, Headington Hill Hall, Oxford OX3 OBW, England) V.20- 1982-

FCTXAV Food and Cosmetics Toxicology. (Pergamon Press Ltd., Maxwell House, Fairview Park Elmsford, NY 10523) V.1-19, 1963-81. For Publisher information, see FCTOD7

FDRLI* Food & Drug Research Labs., Papers. (Waverly, NY 14892)

FDWU** Uber die Wirkung Verschiedener Gifte Auf Vogel, Ludwig Forchheimer Dissertation. (Pharmakologischen Institut der Universitat Wurzburg, Germany, 1931)

FEPRA7 Federation Proceedings, Federation of American Societies for Experimental Biology. (9650 Rockville Pike, Bethesda, MD 20014) V.1- 1942-

FEREAC Federal Register. (U. S. Government Printing Office, Sup. of Doc., Washington, DC 20402) V.1- 1936-

FESTAS Fertility & Sterility. (American Fertility Society, 1608 13th Ave. S. Birmingham, AL 35205) V.1- 1950-

FKIZA4 Fukuoka Igaku Zasshi. (Fukuoka Igakkai, c/o Kyushu Daigaku Igakubu, Tatekasu, Fukuoka-shi, Fukuoka, Japan) V.33- 1940-

FLCRAP Fluorine Chemistry Reviews. (Marcel Dekker Inc., 305 E. 45th St., New York, NY 10017) V.1- 1967-

FMCHA2 Farm Chemicals Handbook. (Meister Publishing, 37841 Euclid Ave., Willoughy, OH 44094)

FMDZAR Fortschritte der Medizin. (Dr. Schwappach und Co., Wessobrunner Str 4, 8035 Gauting vor Muenchen, Germany) V.1- 1883-

FMLED7 FEMS Microbiology Letters. (Elsevier Scientific Publishing Co., POB 211, Amsterdam, Neth.) V.1- 1977-

FNSCA6 Forensic Science. (Elsevier Sequoia SA, P.O. Box 851, CH 1001, Lausanne 1, Switzerland) V.1- 1972-

FOBLAN Folia Biologica (Prague). (Academic Press, 111 Fifth Ave., New York, NY 10003) V.1- 1955-

FOMAAB Food Manufacture. (Bayard, S. Co., 20 Vesey St., New York, NY 10007) V.1- 1927-

FOMIAZ Folia Microbiologica (Prague). (Academia, Vodickova 40, CS-112 29 Prague 1, (Czechoslavakia) V.4- 1959-

FOMOAJ Folia Morphologica (Warsaw). (Panstwowy Zaklad Wydawnictw Lekarskich, ul. Druga 38-40, P-00 238 Warsaw, Poland) V.1- 1929-

FOREAE Food Research. (Champaign, IL) V.1-25, 1936-60. For publisher information, see JFDSAZ

FOTEAO Food Technology. (Institute of Food Technolgists, 221 N. La Salbe St., Chicago, IL 60601) V.1- 1947-

FRPPAO Farmaco, Edizione Pratica. (Casella Postale 114, 27100 Pavia, Italy) V.8- 1953-

FRPSAX Farmaco, Edizione Scientifica. (Casella Postale 114, 27100 Pavia, Italy) V.8- 1953-

FRXXBL French Demande Patent Document. (Commissioner of Patents and Trademarks, Washington, DC 20231)

FRZKAP Farmatsevtichnii Zhurnal (Kiev). (v/o "Mezhdunarodnaya Kniga, Kuznetskii Most 18, Moscow G-200, USSR) V.3- 1930-

FSASAX Fette, Seifen, Anstrichmittel. (Industrieverlag von Hernhaussen KG, Roedingsmarkt 24, 2 Hamburg 11, Fed. Rep. Ger.) V.55- 1953-

GANNA2 Gann. Japanese Journal of Cancer Research. (Tokyo, Japan) V.1-75, 1907-84. For publisher information, see JJCREP

GASTAB Gastroenterology. (Elsevier North Holland, Inc., 52 Vanderbilt Avenue, New York, NY 10017) V.1- 1943-

GENRA8 Genetical Research. (Cambridge University Press, P.O. Box 92, Bentley House, 200 Euston Rd., London NW1 2DB, England) V.1- 1960-

GEPHDP General Pharmacology. (Pergamon Press Inc., Maxwell House Fairview Park, Elmsford, NY 10523) V.1- 1970-

GISAAA Gigiena i Sanitariya. (English Translation is HYSAAV). (v/o "Mezhdunarodnaya Kniga," Kuznetskii Most 18, Moscow G-200, U.S.S.R.) V.1- 1936-

GMCRDC Gann Monograph on Cancer Research. (Japan Scientific Societies Press, Hongo 6-2-10, Bunkyo-ku, Tokyo 113, Japan) No. 11- 1971-

GNAMAP Gigiena Naselennykh Mest. Hygiene in Populated Places. (Kievskii Nauchno - Issledovatel'skii Institut Obshchei i Kommunol'noi Gigieny, Kiev, U.S.S.R.) V.7- 1967-

GNRIDX Gendai no Rinsho. (Tokyo, Japan) V.1-10, 1967-76(?)

GSAMAQ Geological Society of America, Memoir. (The Society, Colorado Bldg., P.O. Box 1719, Boulder, CO 80302) No.1- 1934-

GSLNAG Gaslini(Genoa). Rivista de Pediatria e di Specialita Pediatriche. (Istituto "Giannina Gaslini", Via Cinque Maggio, 39, 16148 Genoa, Italy) V.1- 1969-?

GTPZAB Gigiena Truda i Professional'nye Zabolevaniia. Labor Hygiene and Occupational Diseases. (v/o "Mezhdunarodnaya Kniga," Kuznetskii Most 18, Moscow G-200, U.S.S.R.) V.1- 1957-

GUCHAZ "Guide to the Chemicals Used in Crop Protection," Information Canada, 171 Slater St., Ottawa, Ontario, Canada (Note that although the Registry cites data from the 1968 edition, this reference has been superseded by the 1973 edition.)

GWXXBX German Offenlegungsschrift Patent Document. (U.S. Patent Office, Science Library, 2021 Jefferson Davis Highway, Arlington, VA 22202)

HAONDL Hematological Oncology. (John Wiley & Sons Ltd., Baffins Lane, Chichester, Sussex, PO19 1UD, U.K.) V.1- 1983-

HBAMAK "Abdernalden's Handbuch der Biologischen Arbeitsmethoden." (Leipzig, Germany)

HBTXAC "Handbook of Toxicology, Volumes I-V," Philadelphia, W.B. Saunders, 1956-1959

HDTU** Pharmakologische Prufung von Analgetika, Gunter Herrlen Dissertation. (Pharmakologischen Institut der Universitat Tubingen, Germany, 1933)

HDWU** Beitrage zur Toxikologie des Athylenoxyds und der Glykole, Arnold Hofbauer Dissertation. (Pharmakologischen Institut der Universitat Wurzburg, Germany, 1933)

HEREAY Hereditas. (J.L. Toernqvist Book Dealers, S-26122 Landskrona, Sweden) V.1- 1947-

HKXUDL Huanjing Kexue Xuebao. Environmental Sciences Journal. (Guoji Shudian, POB 399, Beijing, Peop. Rep. China) V.1- 1981-

HPFSDS HHS Publication (FDA. United States). (U.S. Government Printing Office, Superintendent of Documents, Washington, DC 20402) 1980-

HSZPAZ Hoppe-Seyler's Zeitschrift fuer Physiologische Chemie. (Walter de Gruyter & Co., Genthiner Street 13, D-1000, Berlin 30, Federal Republic of Germany) V.21- 1895/96-

HUGEDQ Human Genetics. (Springer-Verlag, Neuenheimer Landst 28-30, D-6900 Heidelberg 1, Germany) V.31- 1976-

HUMAA7 Humangenetik. (Heidelberg, Germany) V.1-30, 1964-1975. For publisher information, see HUGEDQ

HYSAAV Hygiene & Sanitation: English Translation of Gigiena Sanitariya. (Springfield, VA) V.29-36, 1964-71. Discontinued.

IAEC** Interagency Collaborative Group on Environmental Carcinogenesis, National Cancer Institute, Memorandum, June 17, 1974

IAEHDW International Archives of Occupational and Environmental Health. Springer-Verlag, Heidelberger, Platz 3, D-1000 Berlin 33, Federal Republic of Germany) V.35- 1975-

IAPUDO IARC Publications. (World Health Organization, CH-1211 Geneva 27, Switzerland) No.27- 1979-

IAPWAR International Journal of Air and Water Pollution. (London, England) V.4, No.1-4, 1961. For publisher information, see ATENBP

IDZAAW Idengaku Zasshi. Japanese Journal of Genetics. (Genetics Society of Japan, Nippon Iden Gakkai, Tanida 111, Mishima-shi, Shizuoka 411, Japan) V.1- 1921-

IECHAD Industrial & Engineering Chemistry. (Washington, DC) V.15-62, 1923-70. For publisher information, see CHMTBL

IGIBA5 Igiena. (Editura Medicala, St 13 Decembrie 14, Bucharest, Romania) V.5- 1956-

IIZAAX Iwate Igaku Zasshi. Journal of the Iwate Medical Association. (Iwate Igakkai, c/o Iwate Ika Daigaku, Uchimaru, Morioka, Japan) V.1- 1947-

IJBBBQ Indian Journal of Biochemistry and Biophysics. (Council of Scientific and Industrial Research, Publication & Information Director, Hillside Rd., New Delhi 110012, India) V.8- 1971-

IJCAAR Indian Journal of Cancer. (Dr. D.J. Jussawalla, Hospital Ave, Parel, Bombay 12, India) V.1- 1963-

IJCNAW International Journal of Cancer. (International Union Against Cancer, 3 rue du Conseil-General, 1205 Geneva, Switzerland) V.1- 1966-

IJCREE International Journal of Crude Drug Research. (Swets & Zeitlinger B. V., POB 825, 2160 SZ Lisse, Netherlands) V.20- 1982-

IJDEBB International Journal of Dermatology. (J. B. Lippincott Co., E. Washington Sq., Philadelphia, PA 19105) V.9- 1970-

IJEBA6 Indian Journal of Experimental Biology. V.1- 1963- For publisher information, see IJBBBQ

IJLAA4 Indian Journal Animal Science. (Indian Council of Agricultural Research, Krishi Bhavan, New Delhi 110 001, India) V.39- 1969-

IJMDAI Israel Journal of Medical Sciences. (P.O. Box 2296, Jerusalem, Israel) V.1- 1965-

IJMRAQ Indian Journal of Medical Research. (Indian Council of Medical Research, P.O. Box 4508, New Delhi 110016, India) V.1- 1913-

IJNMCI International Journal of Nuclear Medicine and Biology. (Pergamon Press Inc., Maxwell House, Fairview Park, Elmsford, NY 10523) V.1- 1973-

IJVNAP International Journal for Vitamin & Nutrition Research. (Verlag Hans Huber, Laenggasstr. 76, CH-3000 Bern 9, Switzerland) V.41- 1971-

IMEMDT IARC Monographs on the Evaluation of Carcinogenic Risk of Chemicals to Man. (World Health Organization, Internation Agency for Research on Cancer, Lyon, France) (Single copies can be ordered from WHO Publications Centre U.S.A., 49 Sheridan Avenue, Albany, NY 12210)

IMMUAM Immunology. (Blackwell Scientific Publications, Osney Mead, Oxford OX2 OEL, England) V.1- 1958-

IMSUAI Industrial Medicine & Surgery. (Chicago, IL/Miami, FL) V.18-42, 1949-73. For publisher information see IOHSA5

INHEAO Industrial Health. (2051 Kizukisumiyoshi-cho, Nakahara-ku, Kawasaki, Japan) V.1- 1963-

INJFA3 International Journal of Fertility. (Allen Press, 1041 New Hampshire, St., Lawrence, KS66044) V.1- 1955-

INJPD2 Indian Journal of Pharmacology. (Indian Pharmacological Society, Jawaharlal Institute of Postgraduate Medical Education and Research, Pharmacology Dept., Pondicherry 605006, India) V.1- 1968(?)-

INMEAF Industrial Medicine. (Chicago, IL) V.1-18, 1932-49. For publisher information, see IOHSA5

IOHSA5 International Journal of Occupational Health & Safety. (Medical Publications, Inc., 3625 Woodhead Dr., Northbrook, IL 60093) V.43- 1974-

ITCSAF In Vitro. (Tissue Culture Association, 12111 Parklawn Dr., Rockville, MD 20852) V.1- 1965-

IVEJAC Indian Veterinary Journal. (Indian Veterinary Journal, Dr. R. Krishnamurti, G.M.V.C., 10 Avenue Rd., Madras 600 034, India) V.1- 1924-

IYKEDH Iyakuhin Kenkyu. (Nippon Koteisho Kyokai, 12-15, 2-chome, Shibuya-Ku, Tokyo 150, Japan) V.1- 1970-

IZVIAK Internationale Zeitschrift fuer Vitaminforschung. International Review of Vitamin Research. (Bern, Switzerland) V.19-40, 1947-70. For publisher information, see IJVNAP

JACIBY Journal of Allergy & Clinical Immunology. (C.V. Mosby Co., 11830 Westline Industrial Dr., St. Louis, MO 63141) V.48- 1971-

JACTDZ Journal of the American College of Toxicology. (Mary Ann Liebert, Inc., 500 East 85th St., New York, NY 10028) V.1- 1982-

JAFCAU Journal of Agricultural & Food Chemistry. (American Chemical Society Publications, 1155 16th St., N.W., Washington, DC 20036) V.1- 1953-

JAINAA Journal of the Anatomical Society of India. (Dept. of Anatomy, Medical College, Aurangabad MS, India) V.1- 1952-

JAJBAD Journal of Antibiotics, Series B. (Tokyo, Japan) V.6-20, 1953-67. For publisher information, see JJANAX

JAMAAP JAMA, Journal of the American Medical Association. (American Medical Association, 535 N. Dearborn St., Chicago, IL 60610) V.1- 1883-

JANCA2 Journal of the Association of Official Analytical Chemists. (The Assoc., Box 540, Benjamin Franklin Station, Washington, DC 20044) V.49- 1966-

JANSAG Journal of Animal Science. (American Society of Animal Science, 309 West Clark Street, Champaign, IL 61820) V.1- 1942-

JANTAJ Journal of Antibiotics. (Japan Antibiotics Research Association, 2-20-8 Kamiosaki, Shinagawa-ku, Tokyo, Japan) V.2-5, 1948-52; V.21- 1968-

JAOCA7 Journal of the American Oil Chemists' Society. (American Oil Chemists' Society, 508 South 6th St., Champaign, IL 61820) V.24-56, 1947-79.

JAPMA8 Journal of the American Pharmaceutical Assoc., Scientific Edition. (Washington, DC) V.29-49, 1940-60. For publisher information, see JPMSAE

JAPRAN Journal of Small Animal Practice. (Blackwell Scientific Publications Ltd., POB 88, Oxford, UK) V.1- 1960-

JAPYAA Journal of Applied Physiology. (American Physiological Society, 9650 Rockville Pike, Bethesda, MD 20014) V.1- 1948-

JASIAB Journal of Agricultural Science. (Cambridge University Press, London or Cambridge University Press, New York) V.1- 1905-

JAVMA4 Journal of the American Veterinary Medical Association. (American Veterinary Medical Assoc., 600 S. Michigan Ave, Chicago, IL 60605) V.48- 1915-

JBCHA3 Journal of Biological Chemistry. (American Society of Biological Chemists, Inc., 428 E. Preston St., Baltimore, MD 21202) V.1- 1905-

JCEMAZ Journal of Clinical Endocrinology and Metabolism. (Williams & Wilkins Co., 428 East Preston St., Baltimore, MD 21202) V.12- 1952-

JCENA4 Journal of Clinical Endocrinology. (Springfield, IL) V.1-11, 1941-51. For publisher information, see JCEMAZ

JCINAO Journal of Clinical Investigation. (Rockefeller University Press, 1230 York Avenue, New York, NY 10021) V.1- 1924-

JCLBA3 Journal of Cell Biology. (Rockefeller University Press, 1230 York Avenue, New York, NY 10021) V.12- 1962-

JCLLAX Journal of Cellular Physiology. (Alan R. Liss, Inc., 150 Fifth Avenue, New York, NY 10011) V.67- 1966-

JCPCBR Journal of Clinical Pharmacology. (Hall Associates, P.O. Box 482, Stamford, CN 06904) V.13- 1973-

JCPHB8 Journal of Clinical Pharmacology and Journal of New Drugs. (Albany, NY) V.7-10, 1967-70. For publisher information, see JCPCBR

JCPRB4 Journal of the Chemical Society, Perkin Transactions 1. (Chemical Society, Publications Sales Office, Burlington House, London W1V 0BN, England) No.1- 1972-

JCROD7 Journal of Cancer Research and Clinical Oncology. (Springer-Verlag, Heidelberger Pl. 3, D-1 Berlin 33, Germany) V.93- 1979-

JCTODH Journal of Combustion Toxicology. (Technomic Publishing Co., 265 Post Rd. W., Westport, CT 06880) V.3, no.1- 1976-

JDGRAX Journal of Drug Research. (Drug Research & Control Center, 6, Abou Hazem St., Pyramids Ave., POB 29, Cairo, Egypt) V.2- 1969-

JDREAF Journal of Dental Research. (American Association for Dental Research, 734 Fifteenth St., NW, Suite 809, Washington, D.C. 20005) V.1- 1919-

JEENAI Journal of Economic Entomology. (Entomological Society of America, 4603 Calvert Rd., College Park, MD 21201) V.1- 1908-

JEPTDQ Journal of Environmental Pathology and Toxicology. (Park Forest South, IL) V.1-5, 1977-81.

JETOAS Journal Europeen de Toxicologie. (Paris, France) V.1-6, 1968-72. For publisher information, see TOERD9

JFDSAZ Journal of Food Science. (Institute of Food Technologists, Subscrip. Dept., Suite 2120, 221 N. La Salle St., Chicago, IL 60601) V.26- 1961-

JFIBA9 Journal of Fish Biology. (Academic Press, 24-28 Oval Rd., London NW1 7DX, England) V.1- 1969-

JFMAAQ Journal of the Florida Medical Association. (The Assoc., POB 2411, Jacksonville, FL 32203) V.1- 1914-

JGHUAY Journal de Genetique Humaine. (Editeurs Medecine et Hygiene, 78, ave de la Rosaraie, 1211 Geneva 4, Switzerland) V.1- 1952-

JGMIAN Journal of General Microbiology (P.O. Box 32, Commerce Way, Colchester CO2 8HP, UK) V.1- 1947-

JHEMA2 Journal of Hygiene, Epidemiology, Microbiology and Immunology. (Avicenum, Zdravotnicke Nakladatelstvi, Malostranske namesti 28, Prague 1, Czechoslovakia) V.1- 1957-

JIDEAE Journal of Investigative Dermatology. (Williams & Wilkins Co., 428 E. Preston St., Baltimore, MD 21202) V.1- 1938-

JIDHAN Journal of Industrial Hygiene. (Baltimore, MD/New York, NY) V.1-17, 1919-35. For publisher information, see AEHLAU

JIHTAB Journal of Industrial Hygiene and Toxicology. (Baltimore, MD/New York, NY) V.18-31, 1936-49. For publisher information, see AEHLAU

JIMSAX Journal of the Irish Medical Association. (The Association, I.M.A. House, 10 Fitzwilliam Pl., Dublin, Ireland) V.28- 1951-

JISMAB Journal of the Iowa State Medical Society. (Iowa Medical Society, 1001 Grand Ave., West Des Moines, IA 50265) V.1- 1911-

JJANAX Japanese Journal of Antibiotics. (Japan Antibiotics Research Assoc., 2-20-8 Kamiosaki, Shinagawa-ku, Tokyo, Japan) V.21- 1968-

JJATDK JAT, Journal of Applied Toxicology. (Heyden and Son, Inc., 247 S. 41st St., Philadelphia, PA 19104) V.1- 1981-

JJCREP Japanese Journal of Cancer Research (Gann). (Elsevier Science Publishers B.V., POB 211, 1000 AE Amsterdam, Netherlands) V.76- 1985-

JJIND8 JNCI, Journal of the National Cancer Institute. (U.S. Government Printing Office, Superintendent of Documents, Washington, DC 20402) V.61- 1978-

JJPAAZ Japanese Journal of Pharmacology. (Nippon Yakuri Gakkai, c/o Kyoto Daigaku Igakubu Yakurigaku Kyoshitu Sakyo-ku, Kyoto 606, Japan) V.1- 1951-

JLCMAK Journal of Laboratory and Clinical Medicine. (C.V. Mosby Co., 11830 Westline Industrial Dr., St. Louis, MO 63141) V.1- 1915-

JMCMAR Journal of Medicinal Chemistry. (American Chemical Society Pub., 1155 16th St., N.W., Washington, DC 20036) V.6- 1963-

JMOBAK Journal of Molecular Biology. (Academic Press, 111 Fifth Ave., New York, NY 10003) V.1- 1959-

JNCIAM Journal of the National Cancer Institute. (Washington, DC) V.1-60, No.6, 1940-78. For publisher information, see JJIND8

JNDRAK Journal of New Drugs. (Albany, NY) V.1-6, 1961-66. For publisher information, see JCPCBR

JNMDBO Journal of Medicine. (PJD Publications, POB 966, Westbury, NY 11590) V.1- 1970-

JOALAS Journal of Allergy, Including Allergy Abstracts. (St. Louis, MO) V.1-47, 1929-71. For publisher information, see JACIBY

JOANAY Journal of Anatomy. (Cambridge University Press, The Pitt Building, Trumpington Street, Cambridge CB2 1RP, UK) V.51- 1916-

JOBAAY Journal of Bacteriology. (American Society for Microbiology, 1913 I St., N.W., Washington, DC 20006) V.1- 1916-

JOCMA7 Journal of Occupational Medicine. (American Occupational Medical Association, 150 N. Wacker Dr., Chicago, IL 60606) V.1- 1959-

JOENAK Journal of Endocrinology. (Biochemical Society Publications, P.O. Box 32, Commerce Way, Whitehall Industrial Estate, Colchester CO2 8HP, Essex, England) V.1- 1939-

JOGBAS Journal of Obstetrics and Gynaecology of the British Commonwealth. (London, England) V.68-81, 1961-74. For publisher information, see BJOGAS

JOHEA8 Journal of Heredity. (American Genetic Association, 818 Eighteenth St., N.W. Washington, D.C. 20006) V.5- 1914-

JOHYAY Journal of Hygiene. (Cambridge University Press, P.O. Box 92, Bentley House, 200 Euston Rd., London NW1 2DB, England) V.1- 1901-

JOIMA3 Journal of Immunology. (Williams & Wilkins Co., 428 E. Preston St., Baltimore, MD 21202) V.1- 1916-

JONUAI Journal of Nutrition. (Journal of Nutrition, Subscription Dept., 9650 Rockville Pike, Bethesda, MD 20014) V.1- 1928-

JOPDAB Journal of Pediatrics. (C.V. Mosby Co., 11830 Westline Industrial Dr., St. Louis, MO 63141) V.1- 1932-

JOPHDQ Journal of Pharmacobio-Dynamics. (Pharmaceutical Society of Japan, 12-15-501, Shibuya 2-chome, Shibuya-Ku, Tokyo 150, Japan) V.1- 1978-

JOURAA Journal of Urology. (Williams & Wilkins Co., 428 E. Preston St., Baltimore, MD 21202) V.1- 1917-

JOVIAM Journal of Virology. (American Society for Microbiology, 1913 I St., N.W., Washington, DC 20006) V.1- 1967-

JPBAA7 Journal of Pathology & Bacteriology. (London, England) V.1-96, 1892-1968. For publisher information, see JPTLAS

JPCAAC Journal of the Air Pollution Control Association. (The Association, 4400 5th Ave., Pittsburgh, PA 15213) V.5- 1955-

JPCEAO Journal fuer Praktische Chemie. (Johann Ambrosius Barth Verlag, Postfach 109, DDR-701, Leipzig, Ger. Dem. Rep.) V.1- 1834- (Several new series, but continuous vol. nos. also used)

JPETAB Journal of Pharmacology & Experimental Therapeutics. (Williams & Wilkins Co., 428 E. Preston St., Baltimore, MD 21202) V.1- 1909/10-

JPFCD2 Journal of Environmental Science and Health, Part B: Pesticides, Food Contaminants, and Agricultural Wastes. (Marcel Dekker, POB 11305, Church St. Station, New York, NY 10249) V.B11- 1976-

JPHAA3 Journal of the American Pharmaceutical Association. (American Pharmaceutical Assoc., 2215 Constitution Ave, N.W., Washington, DC 20037) V.1-28, 1912-39; New Series V.1-17, 1961-77.

JPMRAB Japanese Journal of Medical Sciences, Part IV: Pharmacology. (Tokyo, Japan) V.1-16, 1926-43. For publisher information, see JJPAAZ

JPMSAE Journal of Pharmaceutical Sciences. (American Pharmaceutical Assoc., 2215 Constitution Ave., N.W., Washington, DC 20037) V.50- 1961-

JPPMAB Journal of Pharmacy & Pharmacology. (Pharmaceutical Society of Great Britain, 1 Lambeth High Street, London SEI 5JN, England) V.1- 1949-

JPTLAS Journal of Pathology. (Longman Group Ltd., Subscriptions (Journals Department, Fourth Avenue, Harlow, Essex, CM19 5AA, UK) V.97- 1969-

JRPFA4 Journal of Reproduction and Fertility. (Journal of Reproduction and Fertility Ltd., 22 New Market Rd., Cambridge CB5 8D7, England) V.1- 1960-

JSCCA5 Journal of the Society of Cosmetic Chemists. (Society of Cosmetic Chemists, 50 E. 41st St., New York, NY 10017) V.1- 1947-

JSICAZ Journal of Scientific & Industrial Research, Section C: Biological Sciences. (New Delhi, India) V.14-21, 1955-62. For publisher information, see IJEBA6

JSONDX Journal of Soviet Oncology. (New York) V. 1-5, 1980-84.

JSOOAX Journal of the Chemical Society, Section C: Organic. (London, England) No.1-24, 1966-71. For publisher information, see JCPRB4

JTCTDW Journal of Toxicology, Clinical Toxicology. (Marcel Dekker, POB 11305, Church St. Station, New York, NY 10249) V.19- 1982-

JTEHD6 Journal of Toxicology and Environmental Health. (Hemisphere Publ., 1025 Vermont Ave., N.W., Washington, DC 20005) V.1- 1975/76-

JTSCDR Journal of Toxicological Sciences. (Editorial Office, Higashi Nippon Gakuen Univ., 7F Fuji Bldg., Kita 3, Nishi 3, Sapporo 060, Japan) V.1- 1976-

JZKEDZ Jitchuken, Zenrinsho Kenkyuho. Central Institute for Experimental Animals, Preclinical Reports. (The Institute, 1433 Nogawa, Takatsu- Ku, Kawasaki 211, Japan) V.1- 1974/75-

KAIZAN Kaibogaku Zasshi. Journal of Anatomy. (Nihon Kaibo Gakkai, c/o Tokyo Daigaku Igakubu Kaibogaku Kyoshitsu, 7-3-1, Hongo, Bunkyo-ku, Tokyo, Japan) V.1- 1928-

KCRZAE Kauchuk i Rezina. (v/o "Mezhdunarodnaya Kniga," Kuznetskii Most 18, Moscow G-200, USSR) V.11- 1937-

KEKHB8 Kanagawa-ken Eisei Kenkyusho Kenkyu Hokoku. Bulletin of Kanagawa Prefectural Public Health Laboratories. (Kanagawa Prefectural Public Health Laboratories, 52-2, Nakao-cho, Asahi-ku, Yokohama 221, Japan) No.1- 1971-

KHFZAN Khimiko-Farmatsevticheskii Zhurnal. Chemical Pharmaceutical Journal. (v/o "Mezhdunarodnaya Kniga," Kuznetskii Most 18, Moscow G-200, U.S.S.R.) V.1- 1967-

KHZDAN Khigiena i Zdraveopazvane. (Hemus, blvd Russki 6, Sofia, Bulgaria) V.9- 1966-

KJMSAH Kyushu Journal of Medical Science. (Fukuoka, Japan) V.6-15, 1955-64. Discontinued.

KLWOAZ Klinische Wochenscrift. (Springer-Verlag, Heidelberger Pl. 3, D-1 Berlin 33, Germany) V.1- 1922-

KODAK* Kodak Company Reports. (343 State St., Rochester, NY 14650)

KRKRDT Kriobiologiya i Kriomeditsina. (Izdatel'stvo Naukova Dumka, ul Repina 3, Kiev, USSR) No.1- 1975-

KRMJAC Kurume Medical Journal. (Kurume Igakkai, c/o Kurume Daigaku Igakubu, 67, Asahi-machi, Kurume, Japan) V.1- 1954-

KSKZAN Khimiya v Sel'skom Khozyaistve. Chemistry in Agriculture. (V/O "Mezhdunarodnaya Kniga", Kuznetskii Most 18, Moscow G-200, U.S.S.R.) No.1- 1963-

KSRNAM Kiso to Rinsho. Clinical Report. (Yubunsha Co., Ltd., 1-5, Kanda Suda-Cho, Chiyoda-ku, KS Bldg., Tokyo 101, Japan) V.1- 1960-

KUMJAX Kumamoto Medical Journal. (Kumamoto Daigaku Igakubu, Library, Kumamoto, Japan) V.1- 1938-

LANCAO Lancet. (7 Adam St., London WC2N 6AD, England) V.1- 1823-

LBANAX Laboratory Animals. (Biochemical Society Book Depot, POB 32, Commerce Way, Colchester Essex CO2 8HP, England) V.1- 1967-

LIFSAK Life Sciences. (Pergamon Press, Maxwell House, Fairview Park, Elmsford, NY 10523) V.1-8, 1962-69; V.14- 1974-

LONZA# Personal Communication from LONZA Ltd., CH-4002, Basel, Switzerland, to NIOSH, Cincinnati, OH 45226

LPDSAP Lipids. (American Oil Chemists' Society, 508 South Sixth St., Champaign, IL 61820) V.1- 1966-

MACPAJ Mayo Clinic Proceedings. (Mayo Foundation, Plummer Bldg., Mayo Clinic, Rochester, MN 55901) V.39- 1964-

MAGDA3 Mechanisms of Ageing and Development. (Elsevier Sequoia SA, POB 851, Ch 1001, Lausanne 1, Switzerland) V.1- 1972-

MahWM# Personal Communication from W.M. Mahlburg, Hopkins Agricultural Chemical Co., P.O. Box 7532, Madison, WI 53707 to NIOSH, Cincinnati, OH 45226, November 16, 1982

MarJV# Personal Communication from Josef V. Marhold, VUOS, 539-18, Pardubice, Czechoslavakia, to the Editor of RTECS, Cincinnati, OH, March 29, 1977

MccSB# Personal Communication from Susan B. McCollister, Dow Chemical U.S.A., Midland, MI 48640, to NIOSH, Cincinnati, OH 45226, June 15, 1984

MDMIAZ Medycyna Doswiadczalna i Mikrobiologia. (Ars Polona-RUCH, POB 1001, 1, P-00068 Warsaw, 1, Poland) V.1- 1949-

MDWTAG Medycyna Weterynaryjna. Veterinary Medicine. (Ars Polona-RUCH, POB 1001, P-00068 Warsaw, Poland) V.1- 1945-

MDZEAK Medizin und Ernaehrung. (Stuttgart, Germany) V.1-13, 1959-72. Discontinued.

MECHAN Medicinal Chemistry, A Series of Reviews. (New York, NY) V.1-6, 1951-63. Discontinued.

MEHYDY Medical Hypotheses. (Churchill Livingstone Inc., 19 W. 44 St., New York, NY 10036) V.1- 1975-

MEIEDD Merck Index. (Merck & Co., Inc., Rahway, NJ 07065) 10th ed.- 1983- Previous eds. had individual CODENS

MEKLA7 Medizinische Klinik. (Urban & Schwarzenberg, Pettenkoferst 18, D-8000 Munich 15, Germany) V.1- 1904-

MELAAD Medicina del Lavoro. Industrial Medicine. (Via S. Barnaba, 8 Milan, Italy) V.16- 1925-

MEMOAQ Medizinische Monatsschrift. (Wissenschaftliche Verlagsgesellschaft mbH, Postfach 40, 7000 Stuttgart 1, Germany) V.1- 1947-

MEPAAX Medycyna Pracy. Industrial Medicine. (ARs-Polona-RUSH, POB 1001, 00-068 Warsaw 1, Poland) V.1- 1950-

MEXPAG Medicina Experimentalis. (Basel, Switzerland) V.1-11, 1959-64; V.18-19, 1968-69. For publisher information, see JNMDBO

MGGEAE Molecular & General Genetics. (Springer-Verlag, Heidelberger Pl. 3, D-1 Berlin 33, Germany) V.99- 1967-

MILEDM Microbios Letters. (Faculty Press, 88 Regent St., Cambridge, England) V.1- 1976-

MJAUAJ Medical Journal of Australia. (P.O. Box 116, Glebe 2037, NSW 2037, Australia) V.1- 1914-

MLDCAS Medecine Legale et Dommage Corporel. (Paris, France) V.1-7, 1968-74. Discontinued.

MMAPAP Mycopathologia et Mycologia Applicata. (The Hague, Netherlands) V.5-54, No.4, 1950-74, For publisher information, see MYCPAH

MMWOAU Muenchener Medizinische Wochenscrift. (Munich, Fed. Rep. Ger.) V.33-115, 1886-1973.

MPPBAB Meditsinskaya Parazitologiya i Parazitarnye Bolezni. Medical Parasitology and Parasitic Diseases. (v/o "Mezhdunarodnaya Kniga," Kuznetskii Most 18, Moscow G-200, U.S.S.R.) V.1- 1932-

MRLR** U. S. Army, Chemical Corps Medical Laboratories Research Reports. (Army Chemical Center, Edgewood Arsenal, MD)

MSERDS Microbiology Series. (Marcel Dekker, Inc., POB 11305, Church St. Station, New York, NY 10249) V.1- 1973-

MUREAV Mutation Research. (Elsevier/North Holland Biomedical Press, P.O. Box 211, 1000 AE Amsterdam, Netherlands) V.1- 1964-

MUTAEX Mutagenesis. (IRL Press Ltd. 1911 Jefferson Davis Highway, Suite 907, Arlington, VA 22202) V.1- 1986-

MVCRB3 "Fluorescent Whitening Agents, Proceedings of A Symposium." MVC-Report, Miljoevardscentrum, Stockholm Center for Environmental Sciences, No. 2, 1973

MYCPAH Mycopathologia. (Dr. W. Junk bv Publishers, POB 13713, 2501 ES The Hague, Netherlands) V.1- 1938-

MZUZA8 Meditsinskii Zhurnal Uzbekistana (v/o "Mezhdunarodnaya Kniga," Kuznetskii Most 18, Moscow G-200, U.S.S.R.) No.1- 1957-

NAIZAM Nara Igaku Zasshi. Journal of the Nara Medical Association. (Nara Kenritsu Ika Daigaku, Kashihara, Nara, Japan) V.1- 1950-

NAREA4 Nutrition Abstracts and Reviews. (Central Sales Branch, Commonwealth Agricultrual Bureaux, Farnham Royal, Slough SL2 3BN, England) V.1- 1931-

NARHAD Nucleic Acids Research. (Information Retrieval Inc., 1911 Jefferson Davis Highway, Arlington, VA 22202) V.1- 1974-

NATUAS Nature. (Macmillan Journals Ltd., Brunel Rd., Basingstoke RG21 2XS, UK) V.1- 1869-

NATWAY Naturwissenschaften. (Springer-Verlag, Heidelberger Platz 3, D-1000 Berlin 33, Federal Republic of Germany) V.1- 1913-

NCIAL* Progress Report Submitted to the National Cancer Institute by Arthur D. Little, Inc. (15 Acorn Park, Cambridge, MA 02140)

NCIBR* Progress Report for Contract No. NIH-NCI-E-68-1311, Submitted to the National Cancer Institute by Bio-Research Consultants, Inc. (9 Commercial Ave., Cambridge, MA 02141)

NCIHL* Progress Report Submitted to the National Cancer Institute by Hazelton Laboratories, Inc.

NCILB* Progress Report for Contract No. NIH-NCI-E-C-72-3252, Submitted to the National Cancer Institute by Litton Bionetics, Inc. (Bethesda, MD)

NCIMR* Progress Report Submitted to the National Cancer Institute by Mason Research Institute. (Worcester, MA)

NCISP* National Cancer Institute Screening Program Data Summary, Developmental Therapeutics Program, Bethesda, MD 20205

NCITR* National Cancer Institute Carcinogenesis Technical Report Series. (Bethesda, MD 20014) No. 0-205. For publisher information, see NTPTR*

NCIUS* Progress Report for Contract NO. PH-43-64-886, Submitted to the National Cancer Institute by the Institute of Chemical Biology, University of San Francisco. (San Francisco, CA 94117)

NCNSA6 National Academy of Sciences, National Research Council, Chemical- Biological Coordination Center, Review. (Washington, DC)

NDRC** National Defense Research Committee, Office of Scientific Research and Development, Progress Report.

NEACA9 News Edition, American Chemical Society, (Easton, PA) V.18-19, 1940-41. For publisher information, see CENEAR

NEJMAG New England Journal of Medicine. (Massachusetts Medical Society, 10 Shattuck St., Boston, MA 02115) V.198- 1928-

NEOLA4 Neoplasma. (Karger-Libri AG, Scientific Booksellers, Arnold-Boecklin-Strasse 25, CH-4000 Basel 11, Switzerland) V.4- 1957-

NETOD7 Neurobehavioral Toxicology. (ANKHO International, Inc., P.O. Box 426, Fayetteville, NY 13066) V.1-2, 1979-80, For publisher information, See NTOTDY

NEURAI Neurology. (Modern Medicine Publications, Inc., 757 Third Avenue, New York, NY 10017) V.1- 1951-

NEZAAQ Nippon Eiseigaku Zasshi. Japanese Journal of Hygiene. (Nippon Eisei Gakkai, c/o Kyoto Daigaku Igakubu Koshu Eisergaku Kyoshits, Yoshida Konoe-cho, Sakyo-ku, Kyoto, Japan) V.1- 1946-

NHOZAX Nippon Hoigaku Zasshi. Japanese Journal of Legal Medicine. (Nippon Hoi Gakkai, c/o Tokyo Daigaku Igakubu Hoigaku Kyoshitsu, 7-3-1, Hongo, Bunkyo-ku, Tokyo, Japan) V.1- 1944-

NIHBAZ National Institutes of Health, Bulletin. (Bethesda, MD)

NIIRDN "Drugs in Japan. Ethical Drugs, 6th Edition 1982," Edited by Japan Pharmaceutical Information Center. (Yakugyo Jiho Co., Ltd., Tokyo, Japan)

NIOSH* National Institute for Occupational Safety and Health, U. S. Dept. of Health, Education, and Welfare, Reports and Memoranda.

NISFAY Nippon Sanka Fujinka Gakkai Zasshi. Journal of Japanese Obstetrics and Gynecology. (Nippon Sanka Fujinka Gakkai, c/o Hoken Kaikan Building., 1-1 Sadohara-cho, Ichigaya, Shinjuku-ku, Tokyo 162, Japan) V.1- 1949-

NKEZA4 Nippon Koshu Eisei Zasshi. Japanese Journal

of Public Health. (Nippon Koshu Eisei Gakkai, 1-29-8 Shinjuku, Shinjuku-ku, Tokyo 160, Japan) V.1- 1954-

NKRZAZ Chemotherapy (Tokyo). (Nippon Kagaku Ryoho Gakkai, 2-20-8 Kamiosaki, Shinagawa-Ku, Tokyo, 141, Japan) V.1- 1953-

NNGADV Nippon Noyaku Gakkaishi. (Pesticide Science Society of Japan, 43-11, 1-chome, Komagome, Toshima-ku, Tokyo 170, Japan) V.1- 1976-

NPIRI* Raw Material Data Handbook, V.1 Organic Solvents, 1974. (National Association of Printing Ink Research Institute, Francis McDonald Sinclair Memorial Laboratory, Lehigh University, Bethlehem, PA 18015)

NRTXDN Neurotoxicology. (Pathotox Publishers, Inc., 2405 Bond St., Park Forest South, IL 60464) V.1- 1979-

NSAPCC Naunyn-Schmiedeberg's Archives of Pharmacology. (Springer Verlag, Heidelberger, Pl. 3, D-1 Berlin 33, Germany) V.272- 1972-

NTIS** National Technical Information Service. (Springfield, VA 22161) (Formerly U. S. Clearinghouse for Scientific & Technical Information)

NTOTDY Neurobehavioral Toxicology and Teratology. (ANKHO International Inc., P.O. Box 426, Fayetteville, NY 13066) V.3- 1981-

NTPTB* NTP Technical Bulletin. (National Toxicology Program, Landow Bldg. 3A-06, 7910 Woodmont Ave., Bethesda, Maryland 20205)

NTPTR* National Toxicology Program Technical Report Series. (Research Triangle Park, NC 27709) No.206-

NUCADQ Nutrition and Cancer. (Franklin Institute Press, POB 2266, Phildelphia, PA 19103) V.1- 1978-

NULSAK Nucleus (Calcutta). (Dr. A.K. Sharma, c/o Cytogenetics Laboratory, Department of Botany, University of Calcutta, 35 Ballygunge Circular Rd., Calcutta 700 019, India) V.1- 1958-

NUPOBT Neuropatologia Polska. (Ars-Polona-RUCH, POB 1001, 00-068 Warsaw 1, Poland) V.1- 1963-

NURIBL Nutrition Reports International. (Geron-X, Inc., POB 1108. Los Altos, CA 94022) V.1- 1970-

NYKZAU Nippon Yakurigaku Zasshi. Japanese Journal of Pharmacology. (Nippon Yakuri Gakkai, 2-4-16, Yayoi, Bunkyo-Ku, Tokyo 113, Japan) V.40- 1944-

NZMJAX New Zealand Medical Journal. (Otago Daily Times & Witness Newspapers, P.O. Box 181, Dunedin C1, New Zealand) V.1- 1900-

OBGNAS Obstetrics and Gynecology. (Elsevier/North Holland, Inc., 52 Vanderbilt Avenue, New York, NY 10017) V.1- 1953-

OEKSDJ Osaka-furitsu Koshu Eisei Kenkyusho Kenkyu Hokoku, Shokuhin Eisei Hen. (Osaka-furitsu Koshu Eisei Kenkyusho, 1-3-69 Nakamichi, Higashinari-ku, Osaka 537, Japan) No.1- 1970-

OIZAAV Osaka Igaku Zasshi. (Osaka, Japan)

ONCOBS Oncology. (S. Karger AG, Postfach CH-4009 Basel, Switzerland) V.21- 1967-

ONCODU Revista de Chirurgui, Oncologie, Radiologie, ORL, Oftalmologie, Stomatologie, Seria: Oncologia. (Rompresfilatelia, ILEXIM, POB 136-137, Bucharest, Romania) V.13, No.4- 1974-

OSDIAF Osaka Shiritsu Daigaku Igaku Zasshi. Journal of the Osaka City Medical Center. (Osaka-Shiritsu Daigaku Igakubu, 1-4-54, Asahi-cho, Abeno-Ku, Osaka, Japan) V.4-23, 1955-74.

OYYAA2 Oyo Yakuri. Pharmacometrics. (Oyo Yakuri Kenkyukai, Tohoku Daigaku, Kitayobancho, Sendai 980, Japan) V.1- 1967-

PAACA3 Proceedings of the American Association for Cancer Research. (Waverly Press, 428 E. Preston St., Baltimore, MD 21202) V.1- 1954-

PAREAQ Pharmacological Reviews. (Williams & Wilkins, 428 E. Preston St., Baltimore, MD 21202) V.1- 1949-

PATHAB Pathologica. (Via Alessandro Volta, 8 Casella Postale 894, 16128 Genoa, Italy) V.1- 1908-

PBBHAU Pharmacology, Biochemistry and Behavior. (ANKHO International Inc., P.O. Box 426, Fayetteville, NY 13066) V.1- 1973-

PBPHAW Progress in Biochemical Pharmacology. (S. Karger AG, Postfach CH-4009 Basel, Switzerland) V.1- 1965-

PCBPBS Pesticide Biochemistry and Physiology. (Academic Press, 111 5th Ave., New York, NY 10003) V.1- 1971-

PCBRD2 Progress in Clinical and Biological Research. (Allan R. Liss, Inc., 150 5th Ave., New York, NY 10011) V.1- 1975-

PCJOAU Pharmaceutical Chemistry Journal. English Translation of KHFZAN. (Plenum Publishing Corp., 233 Spring St., New York, NY 10013) No.1- 1967-

PCOC** Pesticide Chemicals Official Compendium, Association of the American Pesticide Control Officials, Inc. (Topeka, Kansas, 1966)

PEDIAU Pediatrics. (P.O. Box 1034, Evanston, IL 60204) V.1- 1948-

PEREBL Pediatric Research. (Williams & Wilkins Co., 428 E. Preston St., Baltimore, MD 21202) V.1- 1967-

PESTC* Pesticide & Toxic Chemical News. (Food Chemical News, Inc., 400 Wyatt Bldg., 777 14th St., N.W. Washington, DC 20005) V.1- 1972-

PESTD5 Proceedings of the European Society of Toxicology. (Amsterdam, Netherlands) V.16-18, 1975-77, Discontinued.

PEXTAR Progress in Experimental Tumor Research. (S. Karger AG, Postfach CH-4009 Basel, Switzerland) V.1- 1960-

PHABDI Proceedings of the Hungarian Annual Meeting for Biochemistry. (Magyar Kemikusok Egyesulete, Anker Koz 1, 1061 Budapest, Hungary) V.1- 1961-

PHARAT Pharmazie. (VEB Verlag Volk und Gesundheit, Neue Gruenstr 18, 102 Berlin, E. Germany) V.1- 1946-

PHBHA4 Physiology and Behavior. (Pergamon Press Inc., Maxwell House, Fairview Park, Elmsford, NY 10523) V.1- 1966-

PHJOAV Pharmaceutical Journal. (Pharmaceutical Press, 17 Bloomsbury Sq., London WC1A 2NN, England) V.131- 1933-

PHMCAA Pharmacologist. (American Society for Pharmacology and Experimental Therapeutics, 9650 Rockville Pike, Bethesda, MD 20014) V.1- 1959-

PHMGBN Pharmacology: International Journal of Experimental and Clinical Pharmacology. (S. Karger AG, Postfach CH-4009 Basel, Switzerland) V.1- 1968-

PHRPA6 Public Health Reports. (U.S. Government Printing Office, Supt. of Doc., Washington, DC 20402) V.1- 1878-

PHTHDT Pharmacology & Therapeutics. (Pergamon Press Ltd., Headington Hall, Oxford OX3 0BW, England) V.4- 1979-

PHYTAJ Phytopathology. (Phytopathological Society, 3340 Pilot Knob Rd., St. Paul, MN 55121) V.1- 1911-

PJPPAA Polish Journal of Pharmacology and Pharmacy. (ARS-Polona-Rush, POB 1001, 00-068 Warsaw 1, Poland) V.25- 1973-

PMDCAY Progress in Medical Chemistry. (American Elsevier Publishing Co., 52 Vanderbilt Ave., New York, NY 10017) V.1- 1961-

PMJMAQ Proceedings of the Annual Meeting of the New Jersey Mosquito Extermination Association. (New Brunswick, NJ) V.1-61, 1914-74.

PMRSDJ Progress in Mutation Research. (Elsevier North Holland, Inc., 52 Vanderbilt Ave., New York, NY 10017) V.1- 1981-

PNASA6 Proceedings of the National Academy of Sciences of the United States of America. (The Academy, Printing & Publishing Office, 2101 Constitution Ave., Washington, DC 20418) V.1- 1915-

POASAD Proceedings of the Oklahoma Academy of Science. (Oklahoma Academy of Science, c/o James F. Lowell, Executive Secretary-Treasurer, Southwestern Oklahoma State University, Weatherford, Oklahoma 73096) V.1- 1910/1920-

POMDAS Postgraduate Medicine. (McGraw-Hill, Inc., Distribution Center, Princeton Rd., Hightstown, NJ 08520) V.1- 1947-

POSCAL Poultry Science. (Poultry Science Assoc., Texas A & M University, College Station, TX 77843) V.1- 1921-

PPGDS* PPG Industries, Inc., Material Safety Data Sheet. (PPG Industries, Inc., Chemicals Group, One Gateway Center, Pittsburgh, PA 15222)

PREBA3 Proceedings of the Royal Society of Edinburgh, Section B. (Royal Society of Edinburgh, 22 George St., Edinburgh, Scotland) V.61- 1943-

PRGLBA Prostaglandins. (Geron-X, Inc., P.O. Box 1108, Los Altos, California 94022) V.1- 1972-

PRKHDK Problemi na Khigienata. Problems in Hygiene. (Durzhavno Izdatel'stvo Meditsina i Zizkultura, Pl. Slaveikov 11, Sofia, Bulgaria) V.1- 1975-

PRLBA4 Proceedings of the Royal Society of London, Series B, Biological Sciences. (The Society, 6 Carlton House Terrace, London SW1Y 5AG, England) V.76- 1905-

PROTA* ''Problemes de Toxicologie Alimentaire,'' Truhaut, R., Paris, France, L'evolution Pharmaceutique, (1955?)

PSDTAP Proceedings of the European Society for the Study of Drug Toxicity. (Princeton, NJ 08540) V.1-15, 1963-74. For publisher information, see PESTD5

PSEBAA Proceedings of the Society for Experimental Biology and Medicine. (Academic Press, 111 5th Ave., New York, NY 10003) V.1- 1903/04-

PSSCBG Pesticide Science. (Blackwell Scientific Publications Ltd., Osney Mead, Oxford OX2 OEL, England) V.1- 1970-

PSTGAW Proceedings of the Scientific Section of the Toilet Goods Association. (The Toilet Goods Assoc., Inc., 1625 I St., N.W., Washington, DC 20006) No.1-48, 1944-67. Discontinued.

PWPSA8 Proceedings of the Western Pharmacology Society. (Univ. of California Dept. of Pharmacology, Los Angeles, CA 94122) V.1- 1958-

PYTCAS Phytochemistry. An International Journal of Plant Biochemistry. (Pergamon Press Ltd., Headington Hill Hall, Oxford OX3 OEW, England) V.1- 1961-

QJPPAL Quarterly Journal of Pharmacy & Pharmacology. (London, England) V.2-21, 1929-48. For publisher information, see JPPMAB

RABIDH Revista de Igiena, Bacteriologie, Virusologie, Parazitologie, Epidemiologie, Pneumoftiziologie, Seria: Igiena. (Editura Medicala, Str.13 Decembrie 14, Bucharest, Romania) V.23- 1974-

RCBIAS Revue Canadienne de Biologie. (Les Presses de l'Universite de Montreal, P.O. Box 6128, 101 Montreal 3, Quebec, Canada) V.1- 1942-

RCCRDT Revista do Centro de Ciencias Rurais (Universidade Federal Santa Maria). Review of the Center for Rural Sciences. (Universidade Federal de Santa Maria, Centro de Ciencias Rurais, 97.100 Santa Maria, Brazil) V.1- 1971-

RCPBDC Research Communications in Psychology, Psychiatry and Behavior. (PJD Publications, P.O. Box 966, Westbury, NY 11590) V.1- 1976-

RCRVAB Russian Chemical Reviews. (Chemical Society, Publications Sales Office, Burlington House, London W1V 0BN, England) V.29- 1960-

RCTEA4 Rubber Chemistry and Technology. (Div. of Rubber Chemistry, American Chemical Society, University of Akron, Akron, OH 44325) V.1- 1928-

REPMBN Revue d'Epidemiologie, Medecine Sociale et Sante Publique. (Masson et Cie, Editeurs, 120 Blvd. Saint-Germain, P-75280, Paris 06, France) V.1- 1953-

REXMAS Research in Experimental Medicine. (Springer-Verlag, Heidelberger, Pl. 3, D-1 Berlin 33, Germany) V.157- 1972-

RMSRA6 Revue Medicale de la Suisse Romande. (Societe Medicale de La Suisse Romande, 2 Bellefontaine, 1000 Lausanne, Switzerland) V.1- 1881-

RMVEAG Recueid de Medecine Veterinaire. (Editions Vigot Freres, 23 rue de l'Ecole-de-Medecine, Paris 6, France) V.1- 1824-

RPOBAR Research Progress in Organic Biological and Medicinal Chemistry. (New York) V.1-3, 1964-72. Discontinued.

RPTOAN Russian Pharmacology and Toxicology. Translation of FATOAO. (Euromed Publications, 97 Moore Park Rd., London SW6 2DA, England) V.30- 1967-

RPZHAW Roczniki Panstwowego Zakladu Higieny. (Ars

Polona-RUSH, POB 1001, 00-068 Warsaw 1, Poland) V.1- 1950-

RREVAH Residue Reviews. (Springer Verlag New York, Inc., Service Center, 44 Hartz Way, Secaucus, NJ 07094) V.1- 1962-

RSTUDV Rivista di Scienza e Technologia degli Alimenti e di Nutrizione Umana. Review of Science and Technology of Food and Human Nutrition. (Bologna, Italy) V.5-6, 1975-76. Discontinued

RZOVBM Rivista di Zootecnia e Veterinaria. Zootechnical and Veterinary Review. (Rivista di Zootecnia e Veterinaria, Via Viotti 3/5, 20133 Milan, Italy) Number 1- 1973-

SAIGBL Sangyo Igaku. Japanese Journal of Industrial Health. (Japan Association of Industrial Health, c/o Public Health Building, 78Shinjuku 1-29-8, Shinjuku-Ku, Tokyo, Japan) V.1- 1959-

SAKNAH Soobshcheniia Akademii Nauk Gruzinskoi S.S.R. (v/o "Mezhdunarodnaya Kniga," Kuznetskii Most 18, Moscow G-200, U.S.S.R.) V.2- 1941-

SAMJAF South African Medical Journal. (Medical Association of South Africa, Secy., P.O. Box 643, Cape Town, S. Africa) V.6- 1932-

SCCUR* Shell Chemical Company. Unpublished Report. (2401 Crow Canyon Rd., San Romon, CA 94583)

SCIEAS Science. (American Assoc. for the Advancement of Science, 1515 Massachusetts Ave., NW, Washington, DC 20005) V.1- 1895-

SCPHA4 Scientia Pharmaceutica. (Oesterreichische Apotheker- Verlagsgesellschaft MBH, Spitalgasse 31, 1094 Vienna 9, Austria) V.1- 1930-

SEIJBO Senten Ijo. Congenital Anomalies. (Nihon Senten Ijo Gakkai, Kyoto 606, Japan) V.1- 1960-

SFCRAO Symposium on Fundamental Cancer Research (Williams and Wilkins Co., 428 E. Preston St., Baltimore, Md. 21202) V.1- 1947-

SHELL* Shell Chemical Company, Technical Data Bulletin. (Agricultural Div., Shell Chemical Co., 2401 Crow Canyon Rd., San Ramon, CA 94583)

SinJF# Personal Communication from J.F. Sina, Merck Institute for Therapeutic Research, West Point, PA 19486, to the Editor of RTECS, Cincinnati, OH, on October 26, 1982

SJRDAH Scandinavian Journal of Respiratory Diseases. (Munksgaard, 6 Norregade, DK-1165 Copenhagen K, Denmark) V.47- 1966-

SKEZAP Shokuhin Eiseigaku Zasshi. Journal of the Food Hygiene Society of Japan. (Nippon Shokuhin Eisei Gakkai, c/o Kokuritsu Eisei Shikenjo, 18-1, Kamiyoga 1-chome, Setagaya-Ku, Tokyo, Japan) V.1- 1960-

SMDJAK Scottish Medical Journal. (Longman Group Journal Division, 4th Ave., Harlow, Essex CM20 2JE, England) V.1- 1956-

SMSJAR Scottish Medical and Surgical Journal. (Edinburgh, Scotland) For publisher information, see SMDJAK

SMWOAS Schweizerische Medizinische Wochenschrift. (Schwabe & Co., Steintorst 13, 4000 Basel 10, Switzerland) V.50- 1920-

SOGEBZ Soviet Genetics. Translation of GNKAA5. (Plenum Publishing Corp., 233 Spring St., New York, NY 10013). V.2- 1966-

SOMEAU Sovetskaia Meditsina. (v/o "Mezhdunarodnaya Kniga," Kuznetskii Most 18, Moscow G-200, U.S.S.R.) V.1- 1937-

SOVEA7 Southwestern Veterinarian. (College of Veterinary Medicine, Texas A & M University, College Station, TX 77843) V.1- 1948-

SPEADM Special Publication of the Entomological Society of America. (4603 Calvert Rd., College Park, MD 20740) (Note that although the Registry cites data from the 1974 edition, this reference has been superseded by the 1978 edition.)

SRTCDF Scientific Report. (Dr. C.R. Krishna Murti, Director, Industrial Toxicology Research Centre, Luchnow-226001, India) History unknown

SSCMBX Scripta Scientifica Medica. (Durzhavno Izdatelstvo Meditsina i Fizkultura, Pl. Slaveikov II, Sofia, Bulgaria) V.4(2)- 1965-

SSEIBV Sumitomo Sangyo Eisie. Sumitomo Industrial Health. (Sumitomo Byoin Sangyo Eisei Kenkyushitsu, 5-2-2, Nakanoshima, Kita-ku, Osaka, 530, Japan) No.1- 1965-

SSHZAS Seishin Shinkeigaku Zasshi. (Nippon Seishin Shinkei Gakkai, c/o Toyo Bunko, 2-28-21, Honkomagome, Bunkyo-Ku, Tokyo 113, Japan) V.39- 1935-

STBIBN Studia Biophysica. (Akademie-Verlag GmbH, Liepziger Str. 3-4, DDR-108 Berlin, German Democratic Republic) V.1- 1966-

STEDAM Steroids. (Holden-Day Inc., 500 Sansome St., San Francisco, CA 94111) V.1- 1963-

SWEHDO Scandinavian Journal of Work, Environment and Health. (Haartmaninkatu 1, FIN-00290 Helsinki 29, Finland) V.1 1975-

SYSWAE Shiyan Shengwa Xuebao. Journal of Experimental Biology. (Guozi Shudian, China Publications Center, P.O. Box 399, Peking, Peop. Rep. China) V.1- 1953- (Suspended 1966-77)

TABIA2 Tabulae Biologicae. (The Hague, Netherlands) V.1-22, 1925-63. Discontinued.

TAKHAA Takeda Kenkyusho Ho. Journal of the Takeda Research Laboratories. (Takeda Yakuhin Kogyo K. K., 4-54 Juso-nishino-cho, Higashi Yodogawa-Ku, Osaka 532, Japan) V.29- 1970-

TCMUD8 Teratogenesis, Carcinogenesis, and Mutagenesis. (Alan R. Liss, Inc., 150 Fifth Ave., New York, NY 10011) V.1- 1980-

TFAKA4 Trudy Instituta Fiziologii, Akademiya Nauk Kazakhshoi SSR. Transactions of the Institute of Physiology, Academy of Sciences of the Kazakh. SSR. (The Academy, Alma-Ata, USSR) V.1- 1955-

TGANAK Tsitologiya i Genetika. (v/o "Mezhdunarodnaya Kniga", Kuznetskii Most 18, Moscow G-200, USSR) V.1- 1967-

TGNCDL "Handbook of Organic Industrial Solvents," 2nd ed., Chicago, IL, National Association of Mutual Casualty Companies, 1961

THAGA6 Theoretical & Applied Genetics. (Spinger Verlag New York, Inc., Service Center, 44 Hartz Way, Secaucus, NJ 07094) V.38- 1968-

THERAP Therapie. (Doin, Editeurs, 8 Place de l'Odeon, Paris 6, France) V.1- 1946-

THMOAM Therapeutische Monatshefte. (Berlin, Germany) V.1-33, 1887-1919. For publisher information, see KLWOAZ

TIDZAH Tokyo Ika Daigaku Zasshi. Journal of Tokyo Medical College. (Tokyo Ika Daigaku Igakkai, 1-412, Higashi Okubo, Shinjuku-ku, Tokyo, Japan) 1918-

TIEUA7 Tieraerztliche Umschau. (Terra-Verlag, Neuhauser. Str. 21, 7750 Constance, Germany) V.1- 1946-

TIUSAD Tin & Its Uses. (Tin Research Institute, 483 W. 6th Ave., Columbus, OH 43201) No.1- 1939-

TJADAB Teratology, A Journal of Abnormal Development. (Wistar Institute Press, 3631 Spruce St., Philadelphia, PA 19104) V.1- 1968-

TJIZAF Tokyo Joshi Ika Daigaku Zasshi. Journal of Tokyo Women's Medical College. (Society of Tokyo Women's Medical College, c/o Tokyo Joshi Ika Daigaku Toshokan, 10, Kawada-cho, Shinjuku-ku, Tokyo 162, Japan.) V.1- 1931-

TMPRAD Tropenmedizin und Parasitologie. (Georg Thieme Verlag, Postfach 732, Herdweg 63, 7000 Stuttgart, Germany) V.25- 1974-

TOANDB Toxicology Annual. (Marcel Dekker, POB 11305, Church St. Station, New York, NY 10249) 1974/75-

TobJS# Personal Communication to the Editor, Toxic Substances List from Dr. J.S. Tobin, Director, Health and Safety, FMC Corporation, New York, NY 14103, November 9, 1973

TOERD9 Toxicological European Research. (Editions Ouranos, 12 bis, rue Jean-Jaures, 92807 Puteaux, France) V.1- 1978-

TOFOD5 Tokishikoroji Foramu. Toxicology Forum. (Saiensu Foramu, c/o Kida Bldg., 1-2-13 Yushima, Bunkyo-ku, Tokyo, 113, Japan) V.6- 1983-

TOIZAG Toho Igakkai Zasshi. Journal of Medical Society of Toho University. (Toho Daigaku Igakubu Igakkai, 5-21-16, Omori, Otasku, Tokyo, Japan) V.1- 1954-

TOLED5 Toxicology Letters. (Elseveir Scientific Publishing Co., P.O. Box 211, Amsterdam, Netherlands) V.1- 1977-

TOPADD Toxicologic Pathology. (E.I. du Pont de Nemours Co., Inc., Elkton Rd., Newark, DE 19711) V.6(3/4)- 1978-

TOXID9 Toxicologist. (Society of Toxicology, Inc., 475 Wolf Ledge Parkway, Akron, OH 44311) V.1- 1981-

TPKVAL Toksikologiya Novykh Promyshlennykh Khimicheskikh Veshchestv. Toxicology of New Industrial Chemical Sciences. (Akademiya Meditsinskikh Nauk S.S.R., Moscow, U.S.S.R.) No.1- 1961-

TRBMAV Texas Reports on Biology and Medicine. (Texas Reports, University of Texas Medical Branch, Galveston, TX 77550) V.1- 1943-

TRENAF Kenkyu Nenpo - Tokyo-toritsu Eisei Kenkyusho. Annual Report of Tokyo Metropolitan Research Laboratory of Public Health. (24-1, 3 Chome, Hyakunin-cho, Shin-Juku-Ku, Tokyo, Japan) V.1- 1949/50-

TRIPA7 Tin Research Institute, Publication. (The Institute, Fraser Rd., Greenford UB6 7AQ, Middlesex, England) No.1- 1934-

TRSTAZ Transactions of the Royal Society of Tropical Medicine and Hygiene. (The Society, Manson House, 26 Portland Pl., London W1N 4EY, England) V.14- 1920-

TSITAQ Tsitologiya. Cytology. (v/o "Mezhdunarodnaya Kniga", Kuznetskii Most 18, Moscow G-200, USSR) V.1 1959-

TUMOAB Tumori. (Casa Editrice Ambrosiana, Via G. Frua 6, 20146 Milan, Italy) V.1- 1911-

TXAPA9 Toxicology and Applied Pharmacology. (Academic Press, 111 5th Ave., New York, NY 10003) V.1- 1959-

TXCYAC Toxicology. (Elsevier/North-Holland Scientific Publishers Ltd., 52 Vanderbilt Ave., New York, NY 10017) V.1- 1973-

TXMDAX Texas Medicine. (Texas Medical Assoc., 1905 N. Lamar Blvd., Austin, TX 78705) V.60- 1964-

TYKNAQ Tohoku Yakka Daigaku Kenkyu Nempo. Annual Report of the Tohoku College of Pharmacy. (77, Odawara Minamikotaku, Hara-machi, Sendai, Japan) No.10- 1963-

UCDS** Union Carbide Data Sheet. (Industrial Medicine & Toxicology Dept., Union Carbide Corp., 270 Park Ave., New York, NY 10017)

UCPHAQ University of California, Publications in Pharmacology. (Berkeley, CA) V.1-3, 1938-57. Discontinued.

UICMAI UICC Monograph Series. (Springer Verlag New York, Inc., Service Center, 44 Hartz Way, Secaucus, NJ 07094)

UPJOH* "Compounds Available for Fundamental Research, Volume II-6, Antibiotics, A Program of Upjohn Company Research Laboratory." (Kalamazoo, MI 49001)

VCTDC* Vanderbilt, R.T., Co., Technical Data Sheet. (230 Park Ave., New York, NY 10017)

VDGIA2 Verhandlungen der Deutschen Gesellschaft fuer Innere Medizin. (J.F. Bergmann Verlag, Trogerstrasse 56, 8000 Munich 80, Federal Republic of Germany) V.33-52, 1921-40; V.54- 1948-

VETNAL Veterinariya. (v/o "Mezhdunarodnaya Kniga," Kuznetskii Most 18, Moscow G-200, U.S.S.R.) V.1- 1924-

VHAGAS Verhandlungen der Anatomischen Gesellschaft. (VEB Gustav Fischer, Postfach 176, DDR-69 Jena, E. Germany) V.1- 1887-

VHTODE Veterinary and Human Toxicology. (American College of Veterinary Toxicologists, Office of the Secretary-Treasurer, Comparative Toxicology Laboratory, Kansas State University, Manhatten, Kansas 66506) V.19- 1977-

VIHOAQ Vitamins and Hormones. (Academic Press, 111 5th Ave., New York, NY 10003) V.1- 1943-

VMDNAV Veterinarno-Meditsinski Nauki. (Hemus, Blvd. Russki 6, Sofia, Bulgaria) V.1- 1964-

VOONAW Voprosy Onkologii. Problems of Onkology. (v/o "Mezhdunarodnaya Kniga," Kuznetskii Most 18, Moscow G-200, U.S.S.R.) V.1- 1955-

VPITAR Voprosy Pitaniya. Problems of Nutrition. (v/o "Mezhdunarodnaya Kniga," Kuznetskii Most 18, Moscow G-200, U.S.S.R.) V.1- 1932-

VRDEA5 Vrachebnoe Delo. Medical Profession. (v/o "Mezhdunarodnaya Kniga," Kuznetskii Most 18, Moscow G-200, U.S.S.R.) No.1- 1918-

WEHRBJ Work, Environment, Health. (Helsinki, Finland) V.1-11, 1962-74. For publisher information, see SWEHDO

WILEAR Wiadomosci Lekarskie. (Ars Polona-RUSH, POB 1001, P-00068 Warsaw 1, Poland) V.1- 1948-

WKWOAO Wiener Klinische Wochenschrift. (Springer-Verlag, Postfach 367, A-1011 Vienna, Austria) V.1- 1888-

WMHMAP Wissenschaftliche Zeitschrift der Martin-Luther Universitaet, Halle- Wittenberg, Mathematisch-Naturwissenschaftliche Reihe. (Deutsche Buch-Export und-Import GmbH, Leninstr. 16, 701 Leipzig, E. Germany) V.1- 1951-

WRPCA2 World Review of Pest Control. (London, England) V.1-10, 1962-71. Discontinued.

XAWPA2 U.S. Army Chemical Warfare Laboratories. (Army Chemical Center, MD)

XENOBH Xenobiotica. (Taylor & Francis Ltd., 4 John St., London WC1N 2ET, England) V.1- 1971-

XEURAQ U. S. Atomic Energy Commission, University of Rochester, Research and Development Reports. (Rochester, NY)

XPHBAO U.S. Public Health Service, Public Health Bulletin. (Washington, DC)

XPHPAW U.S. Public Health Service Publication. (U.S. Government Printing Office, Supt. of Doc., Washington, DC 20402) No.1- 1950-

YACHDS Yakuri to Chiryo. Pharmacology and Therapeutics. (Raifu Saiensu Pub. Co., 5-3-3 Yaesu, Chuo-ku, Tokyo 104, Japan) V.1- 1973-

YJBMAU Yale Journal of Biology and Medicine. (The Yale Journal of Biology and Medicine, Inc., 333 Cedar Street, New Haven, Connecticut 06510) V.1- 1928-

YKKZAJ Yakugaku Zasshi. Journal of Pharmacy. (Nippon Yakugakkai, 12-15-501, Shibuya 2-chome, Shibuya-ku, Tokyo 150, Japan) No.1- 1881-

YKYUA6 Yakkyoku. Pharmacy. (Nanzando, 4-1-11, Yushima, Bunkyo-ku, Tokyo, Japan) V.1- 1950-

ZANZA9 Zeitschrift Angewandte Zoologie. (Duncker und Humblot, Postfach 410329, D-1000 Berlin 41, Fed. Rep. Germany) V.41- 1954-

ZAPPAN Zentralblatt fuer Allgemeine Pathologie und Pathologische Anatomie. (VEB Gustav Fischer Verlag, Postfach 176, Villengang 2, 69 Jena, E. Germany) V.1- 1890-

ZBPIA9 Zentralblatt fuer Bakteriologie, Parasitenkunde, Infektionskrankheiten und Hygiene, Abteilung II. Naturwissenschattliche: Allgemeine, Landwirtschattliche und Technische Miknobiologie. (VEB Gustav Fischer Verlag, Postfach 176, DDR-69 Jena, Ger. Dem. Rep.) V.107-132, 1952-77.

ZDKAA8 Zdravookhranenie Kazakhstana. Public Health of Kazakhstan. (V/O "Mezhdunarodnaya Kniga", Kuznetskii Most 18, Moscow G-200, U.S.S.R.) V.1- 1941-

ZDVKAP Zdravookhranenie. Public Health. (v/o "Mezhdunarodnaya Kniga", Kuznetskii Most 18, Moscow G-200, USSR) V.1- 1958-

ZEKBAI Zeitschrift fuer Krebsforschung. (Berlin, Germany) V.1-75, 1903-71. For publisher information, see JCROD7

ZERNAL Zeitschrift fuer Ernaehrungswissenschaft. (Steinkopff Verlag, Postfach 1008, 6100 Darmstadt, Germany) V.1- 1960-

ZGEMAZ Zeitschrift fuer die Gesamte Experimentelle Medizin. (Berlin, Germany) V.1-139, 1913-65. For publisher information, see REXMAS

ZGSHAM Zeitschrift fuer Gesundheitstechnik und Staedtehygiene. (Berlin, Germany) V.21-27, 1929-35. For publisher information, see ZANZA9

ZHPMAT Zentralblatt fuer Bakteriologie, Parasitenkunde, Infektionskrankran- heiten und Hygiene, Abteilung 1: Originale, Reihe B: Hygiene, Praeventive Medizin. (Gustav Fischer Verlag, Postfach 72-01-43, D-7000 Stuttgart 70, Fed. Rep. Germany) V.155- 1971-

ZHYGAM Zeitschrift fuer die Gesamte Hygiene und Ihre Grenzgebiete. (VEB Georg Thieme, Hainst 17/19, Postfach 946, 701 Leipzig, E. Germany) V.1- 1955-

ZIEKBA Zeitschrift fuer Immunitaetsforschung, Experimentelle und Klinische Immunologie. (Fischer, Gustav, Verlag, Postfach 53, Wollgrasweg 49, 7000 Stuttgart-Hohenheim, Germany) V.141- 1970-

ZIETA2 Zeitschrift fuer Immunitaetsforschung und Experimentelle Therapie. (Stuttgart, Germany) 1924-V.124, 1962. For publisher information, see ZIEKBA

ZKKOBW Zeitschrift fuer Krebsforschung und Klinische Onkologie. (Berlin, Germany) V.76-92, 1971-78. For publisher information, see JCROD7

ZKMAAX Zhurnal Eksperimental'noi i Klinicheskoi Meditsiny. (V/O Mezhdunarodnaya Kniga, 121200 Moscow, USSR) V.2- 1962-

ZLUFAR Zeitschrift fuer Lebensmittel-Untersuchung und-Forschung. (Springer- Verlag, Heidelberger, Pl. 3, D-1 Berlin 33, Germany) V.86- 1943-

ZMMPAO Zentralblatt fuer Bakteriologie, Parasitenkunde, Infektionskrankh eiten und Hygiene, Abteilung 1: Originale, Reihe A: Medizinische Mikrobiologie und Parasitologie. (Gustav Fischer Verlag, Postfach 72-01-43, D-7000 Stuttgart 70, Fed. Rep. Germany) V.217- 1971-

ZMOAAN Zeitschrift fuer Morphologie und Anthropologie. (Schweizerbart Verlagsbuchhandlung, Johannesst 3A, D-7000, Stuttgart 1, Germany) V.1- 1899-

ZTMPA5 Zeitschrift fuer Tropenmedizin und Parasitologie. (Stuttgart, Germany) V.1-24, 1949-73. For publisher information, see TMPRAD

13BYAH "Morphological Precursors of Cancer," ed. Lucio Severi, Proceedings of an International Conference, Universita Degli Studi, Perugia, Italy, June 26-30, 1961, Perugia, Univ. of Perugia, 1962

13ZGAF "Gigiena i Toksikologiya Novykh Pestitsidov i Klinika Otravlenii, Doklady Vsesoyuznoi Konferentsii," Komiteta Po Izucheniiu i Reglamenta-tsii Indokhimikatov Glavnoi Gosudarstvennoi Sanitarnoi Inspektsii U.S.S.R., 1962

14CYAT "Industrial Hygiene and Toxicology," 2nd rev. ed., Patty, F.A., ed., New York, Interscience Publishers, 1958-63

14KTAK "Boron, Metallo-Boron Compounds and Boranes," R.M. Adams, New York, Wiley, 1964
17QLAD "Ratsionl'noe Pitanie," (Rational Nutrition) P.D. Leshchenko, ed., USSR, 1967
26UZAB "Pesticides Symposia," collection of papers presented at the Sixth & Seventh Inter-American Conferences on Toxicology and Occupational Medicine, Miami, Florida, Univ. of Miami, School of Medicine, 1968-70
27ZIAQ "Drug Dosages in Laboratory Animals - A Handbook," C.D. Barnes and L.G. Eltherington, Berkeley, Univ. of California Press, 1965, 1973
27ZQAG "Psychotropic Drugs and Related Compounds," E. Usdin, and D.H. Efron, 2nd Ed., Washington, DC, 1972
27ZTAP "Clinical Toxicology of Commercial Products - Acute Poisoning," Gleason, et al., 3rd Ed., Baltimore, Williams, and Wilkins, 1969. (Note that although the Registry cites data from the 1969 edition, this reference has been superseded by the 4th edition, 1976.)
27ZWAY "Heffter's Handbuch der Experimentelle Pharmakologie."
27ZXA3 "Handbook of Poisoning: Diagnosis and Treatment," R.H. Dreisbach, Los Altos, California, Lange Medical Publications
28ZEAL "Pesticide Index," E.H. Frear, ed., State College, PA, College Science Publications, 1969 (Note that although the Registry cites data from the 1969 edition, this reference has been superseded by the 5th edition, 1976.)
28ZOAH "Chemicals in War," A.M. Prentiss, 1937
28ZPAK "Sbornik Vysledku Toxixologickeho Vysetreni Latek A Pripravku," J.V. Marhold, Institut Pro Vychovu Vedoucicn Pracovniku Chemickeho Prumyclu Praha, Czechoslovakia, 1972
29ZVAB "Handbook of Analytical Toxicology," Irvings Sunshine, ed., Cleveland, Ohio, Chemical Rubber Company, 1969
29ZWAE "Practical Toxicology of Plastics," R. Lefaux, Cleveland, Ohio, Chemical Rubber Company, 1968
30ZDA9 "Chemistry of Pesticides." N.N. Melnikov, New York, Springer-Verlag, 1971
31ZOAD "Pesticide Manual," Worcesteshire, England, British Crop Protection Council, 1968
31ZTAS "Alcohols: Their Chemistry, Properties and Manufacture," J.A. Monick, New York, Reinhold Book, 1968
32ZDAL "Aldrin Dieldrin Endrin and Telodrin: An Epidemiological and Toxicological Study of Long-Term Occupational Exposure," K.W. Jager, New York, Elsevier, 1970
32ZWAA "Handbook of Poisoning: Diagnosis and Treatment," R.H. Dreishach, 8th Ed., Los Altos, California, Lange Medical Publications, 1974
32ZXAD "Transactions of the 36th Annual Meeting of the American Conference of Governmental Industrial Hygienists," Miami Beach, Fla., May 12-17, 1974
34LXAP "Insecticide Biochemistry and Physiology," C.F. Wilkinson, ed., New York, Plenum, 1976
34ZIAG "Toxicology of Drugs and Chemicals," W.B. Deichmann, New York, Academic Press, 1969
36PYAS "Pharmacology of Steroid Contraceptive Drugs," Garattini, S. and H.W. Berendes, eds., New York, Raven Press, 1977. (Monograph Mario Negri Inst. Pharm. Res.)
36SBA8 "Antifungal Compounds," M.R. Siegel and H.D. Sisler, eds., 2 vols., New York, Marcel Dekker, 1977
37ASAA "Kirk-Othmer Encyclopedia of Chemical Technology, 3rd Edition," Grayson, M. and D. Eckroth, eds., New York, Wiley, 1978
40YJAX "Evaluation of Embryotoxicity, Mutagenicity and Carcinogenicity Risks In New Drugs," Proceedings of the Symposium on Toxicological Testing for Safety of New Drugs, 3rd, Prague, Apr. 6-8, 1976, Benesova, O., et al., eds., Prague, Czechoslovakia, Univerzita Karlova, 1979
41HTAH "Aktual'nye Problemy Gigieny Truda. Current Problems of Labor Hygiene," Tarasenko, N.Y., ed., Moscow, Pervyi Moskovskii Meditsinskii Inst., 1978
43GRAK "Dusts and Disease," Proceedings of the Conference on Occupational Exposures to Fibrous and Particulate Dust and Their Extension into the Environment, 1977, Lemen, R., and J.M. Dement, eds., Park Forest South, IL, Pathotox Publishers, 1979
46GFA5 "Safety Problems Related to Chloramphenicol and Thiamphenicol Therapy," Najean, Y., et al., eds., New York, Raven Press, 1981
47YKAF "Sporulation and Germination," Proceedings of the Eighth International Spore Conference, 1980, Washington, DC, American Society of Microbiology, 1981
50EXAK "Formaldehyde Toxicity," Conference, 1980, Gibson, J.E., ed., Washington, DC, Hemisphere Publishing Corp., 1983
85AIAL "Toxicity of Pure Foods," E.M. Boyd, Cleveland, Ohio, CRC Press, 1973
85ALAU "Carbamate Insecticides: Chemistry, Biochemistry, and Toxicology," R.J. Kuhr, and H.W. Dorough, Cleveland, Ohio, CRC Press, 1976
85ARAE "Agricultural Chemicals, Books I, II, III, and IV," W.T. Thomson, Fresno, CA, Thomson Publications, 1976/77 revision
85CYAB "Chemistry of Industrial Toxicology," H.B. Elkins, 2nd Ed., New York, J. Wiley, 1959
85DAAC "Bladder Cancer, A Symposium," K.F. Lampe et al., eds., Fifth Inter- American Conference on Toxicology and Occupational Medicine, University of Miami, School of Medicine, Coral Gables, Florida, Aesculapius Pub., 1966
85DCAI "Poisoning; Toxicology, Symptoms, Treatments," J.M. Arena, 2nd Ed., Springfield, Illinois, C. C. Thomas, 1970
85DJA5 "Malformations Congenitales des Mammiferes," H. Tuchmann-Duplessis, Paris, Masson et Cie, 1971
85DKA8 "Cutaneous Toxicity," V.A. Drill and P. Lazar, eds., New York, Academic Press, 1977
85DPAN "Wirksubstanzen der Pflanzenschutz und Schadlingsbekampfungsmittel," Werner Perkow, Berlin, Verlag Paul Parey, 1971-1976
85ERAY "Antibiotics: Origin, Nature, and Properties," T. Korzyoski, Z. Kowszyk-Gindifer, and W. Kurylowicz, eds., Washington DC, American Society for Microbiology, 1978

85FZAT ''Index of Antibiotics from Actinomycetes,'' Umezawa, H. et al., eds., Tokyo, University of Tokyo Press, 1967
85GDA2 ''CRC Handbook of Antibiotic Compounds, Volumes 1-9,'' Berdy, Janos. Boca Raton, Florida, CRC Press, 1980
85GMAT ''Toxicometric Parameters of Industrial Toxic Chemicals Under Single Exposure,'' Izmerov, N.F., et al. Moscow, Centre of International Projects, GKNT, 1982
85GYAZ 'Pflanzenschutz-und Schaedlingsbekaempfungsmittel: Abriss einer Toikologie und Therapie von Vergiftungen,' 2nd ed., Klimmer, O. R., Hattingen, Fed. Rep. Ger., Hyndt-Verlag, 1971
85GYAZ ''Pflanzenschutz-und Schaedlingsbekaempfungsmittel: Abriss einer Toxikologie und Therapie von Vergiftungen,'' 2nd ed., Klimmer, O.R., Hattingen, Fed. Rep. Ger., Hundt-Verlag, 1971
85INA8 ''Documentation of the Threshold Limit Values and Biological Exposure Indices,'' 5th ed., Cincinnati, Ohio, American Conference of Governmental Industrial Hygienists, Inc., 1986

A Guide to Using This Book

Entry Number — Entries are indexed in order by this alphanumeric code.
See Introduction: paragraph 1, *p. xi*

Entry Name — A complete entry name and synonym cross-index is located in Section III.
See Introduction: paragraph 2, *p. xi*

mf: — the molecular formula
mw: — the molecular weight
See Introduction: paragraphs 5 *and* 6, *pp. xi-xii*

Purpose: — The reason the substance is added to foods. A complete purpose cross-index is located in Section III.
See Introduction: paragraph 9, *p. xii*

Where Used: — The foods in which the substance is used. A complete use cross-index is located in Section III.
See Introduction: paragraph 10, *p. xii*

Regulations: — U.S. Federal regulations which apply to the substance when used as a food additive.
See Introduction: paragraph 11, *p. xii*

Safety Profiles: — These are text summaries of the toxicity, fire, reactivity, incompatibilities and other dangerous properties of the entry.
See Introduction: paragraph 14, *p. xv*